SMALL DIESEL ENGINE SERVICE MANUAL

(3rd Edition)

CONTENTS

Diesel Engine Fundamentals 3
Trouble-Shooting 9
Continental 10
Deutz 20
Farymann 28
Isuzu 38
Kirloskar 59
Kubota 64
Lister-Petter 113
Lombardini 185
Mitsubishi 231
MWM 242
Onan 257
Perkins 287
Peugeot 314
Slanzi 328
Volkswagen 351
Westerbeke 360
Wisconsin 391
Yanmar 422

Published by
INTERTEC PUBLISHING CORPORATION
P. O. BOX 12901, OVERLAND PARK, KS 66212

Copyright ©1991 by Intertec Publishing Corp. Printed in the United States of America.
Library of Congress Catalog Card Number 91-055369.
Cover photograph courtesy of Lister-Petter.

All rights reserved. Reproduction or use, without express permission, of editorial or pictorial content, in any manner, is prohibited. No patent liability is assumed with respect to the use of the information contained herein. While every precaution has been taken in the preparation of this book, the publisher assumes no responsibility for errors or omissions. Neither is any liability assumed for damages resulting from use of the information contained herein. Publication of the servicing information in this manual does not imply approval of the manufacturers of the products covered.

All instructions and diagrams have been checked for accuracy and ease of application; however, success and safety in working with tools depend to a great extent upon individual accuracy, skill and caution. For this reason the publishers are not able to guarantee the result of any procedure contained herein. Nor can they assume responsibility for any damage to property or injury persons occasioned from the procedures. Persons engaging in the procedures do so entirely at their own risk.

Dual Dimension

This service manual provides specifications in both the U.S. Customary and Metric (SI) systems of measurement. The first specification is given in the measuring system perceived by us to be the preferred system when servicing a particular component, while the second specification (given in parenthesis) is the converted measurement. For instance, a specification of "0.011 inch (0.28 mm)" would indicate that we feel the preferred measurement, in this instance, is the U.S. system of measurement and the metric equivalent of 0.011 inch is 0.28 mm.

DIESEL ENGINE FUNDAMENTALS

PRINCIPLE OF OPERATION

The diesel engines covered in this manual are all technically known as "Internal Combustion Reciprocating Engines." The source of power is heat formed by the burning of a combustible mixture of petroleum products and air. This burning takes place in a closed cylinder containing a piston. Expansion resulting from the heat of combustion applies downward pressure on the piston to produce reciprocating (up-and-down) motion of the piston. The reciprocating motion is transmitted to the crankshaft and converted to rotary motion by means of a connecting rod and the offset "crank" on the crankshaft.

Diesel engines may be classified as "compression ignition." Their operation depends upon the air temperature in the engine cylinder being increased above the combustion point of the fuel to ignite the fuel-air mixture, rather than using an electrical spark to ignite the fuel-air mixture as in "spark ignition" engines. In diesel engines, there is no premixing of the fuel and air outside the cylinder; air only is taken into the cylinder through the intake manifold. The pressure created by the piston moving upward on the compression stroke causes a rapid temperature increase of the air in the cylinder. When the piston nears the top of the cylinder, finely atomized fuel is injected into the cylinder where it mixes with the air and is ignited by the heat of the compressed air. Expansion resulting from the heat of combustion then forces the piston downward on the power stroke and turns the crankshaft.

As with spark ignition engines, diesel engines are produced in both 2-stroke cycle and 4-stroke cycle design. However, since all of the engines covered in this manual are of 4-stroke cycle design, only this type will be discussed.

With the 4-stroke cycle design, two revolutions of the engine crankshaft (four piston strokes) are required to provide one power stroke of the piston. See Fig. FN1. The four-stroke cycle begins with the engine intake valve open and the piston being pulled to the bottom of the cylinder by crankshaft rotation (intake stroke). Atmospheric pressure (see also TURBOCHARGING paragraph) pushes air into the cylinder past the open intake valve. As the piston reaches the bottom of the stroke, the intake valve closes and crankshaft rotation forces the piston upward compressing the air trapped in the cylinder (compression stroke). As the piston nears the end of the compression stroke, the compressed air in the cylinder becomes very hot. At this point, a timed injection of fuel enters the cylinder and is ignited by the temperature of the air. The combustion process begins and as the piston reaches top dead center (TDC) and starts downward, the burning fuel-air mixture expands rapidly and forces the piston down (power stroke), imparting a turning motion to the engine crankshaft. As the piston reaches the bottom of the cylinder, the exhaust valve opens. The crankshaft continues to turn from inertia of the rotating engine flywheel, moving the piston up and expelling the burned fuel-air mixture out through the open exhaust valve (exhaust stroke). The exhaust valve closes and the intake valve opens as piston reaches the top of cylinder. The piston then is pulled back down (intake stroke) by the crankshaft and connecting rod, beginning another 4-stroke engine operating cycle.

The above is a simplified explanation of how a four-stroke cycle diesel engine operates for a single cylinder engine. Engines with two, three, four or five cylinders will be found in the service section of this manual. In these multi-cylinder engines, the crankshaft operates each piston in a sequence referred to as the "firing order" of the engine. This provides a balance between the piston power strokes of each cylinder. For example, the single cylinder engine has one power stroke in two revolutions of the engine crankshaft. A two-cylinder engine will have a power stroke on every revolution of the crankshaft and a four-cylinder engine will have two power strokes for each crankshaft revolution.

DIESEL FUEL SYSTEM

The basic diesel fuel system consists of a fuel supply tank (1—Fig. FN2), an injection pump (3) capable of creating a high pressure flow of fuel, and a fuel injector (4) for each engine cylinder which receives the fuel from the pump and sprays it into the engine cylinder. The diesel fuel system may include one or more fuel filters (2) to protect the system from contamination, and a low pressure pump that feeds or lifts fuel from the supply tank to the fuel injection pump. On some diesel engines, the fuel feed pump is mounted on the injection pump and is actuated by a cam lobe on the fuel injection pump camshaft, or it may be an integral part of the fuel injection pump.

In order for the diesel engine to operate properly, the fuel injection system must supply the correct quantity of fuel, time the fuel delivery, atomize the fuel and distribute the fuel evenly through the cylinder.

FUEL INJECTION SYSTEM TYPES

There are two basic types of diesel fuel injection systems: direct injection and indirect injection. The cylinder head configuration is different with each type of fuel injection system.

DIRECT FUEL INJECTION. Direct fuel injection utilizes an injector with the nozzle tip extending directly into the engine combustion chamber; see Fig. FN3. Usually, the combustion cham-

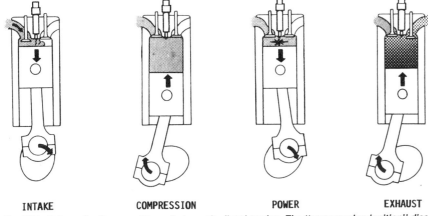

INTAKE   COMPRESSION   POWER   EXHAUST

Fig. FN1—Schematic diagram of four-stroke cycle diesel engine. The "compression ignition" diesel engine depends on the heat produced from compressing the air in the cylinder to ignite the fuel as it is sprayed into the cylinder.

Diesel Engine Fundamentals

ber is located in the piston crown and the cylinder head surface is flat. The injector nozzle protrudes from the head surface directly into the combustion chamber. The injector nozzle tip has three to five orifices (holes) of approximately 0.25-0.30 mm (0.010-0.012 inch) diameter. The spray holes are spaced equidistantly around the nozzle tip. The spray from each orifice should form a cone and the spray from all orifices should be alike. The fuel spray is finely atomized so that it will ignite readily from the heat of the compressed air and will burn steadily through the engine power stroke.

INDIRECT FUEL INJECTION. The injector nozzle for indirect injection systems does not extend into the engine combustion chamber. The indirect injection nozzle sprays a single spray cone into a precombustion chamber (3—Fig. FN4), or to an air cell (3—Fig. FN5) directly across the cylinder head combustion chamber from the injector nozzle. The fuel is concentrated in the center of the spray cone, but the outer circumference of the cone is finely atomized so as to provide instant combustion. Some indirect system injectors have two spray orifices, one for the major volume of fuel that enters the air cell and another small orifice that atomizes the fuel that directly enters the combustion chamber.

With the precombustion chamber design, fuel is sprayed into the precombustion chamber and ignites there, but cannot completely burn due to lack of air. What combustion does occur heats the fuel and forces it into the cylinder where complete combustion occurs on the power stroke of the piston.

With the air cell design, the fine atomization surrounding the core of the fuel spray ignites in the engine combustion chamber, increasing the temperature and pressure in the combustion chamber and air cell. The major portion of the fuel enters the air cell, where partial combustion of the fuel occurs. This partial combustion heats the remainder of the fuel and forces it out of the air cell into the cylinder where complete combustion occurs on the power stroke.

DIESEL FUEL SYSTEM COMPONENTS

FUEL INJECTION PUMPS. There are a number of different fuel injection pump designs, but basically all incorporate a cam actuated plunger operating in a cylinder with extremely close tolerances and capable of pumping fuel under high pressure to the fuel injector. A clean supply of fuel free of water is very important to the precision parts of the diesel injection system.

A single cylinder diesel engine requires a fuel injection pump with only one pumping element. See Fig FN6. Some multi-cylinder engines are fitted with a separate pump, comparable to the single cylinder engine pump, for each cylinder of the engine. Many multi-cylinder engines utilize a common pump housing containing a pumping element

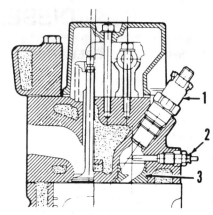

Fig. FN4—Cross-sectional view of cylinder head, injector (1) and precombustion chamber (3) typical of indirect injection engines. A glow plug (2) may be used as a starting aid on indirect injection engines.

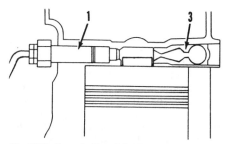

Fig. FN5—Some indirect injection engines may be equipped with an air cell (3) rather than a precombustion chamber. Injector (1) sprays fuel across the combustion chamber into the air cell where partial combustion of the fuel occurs.

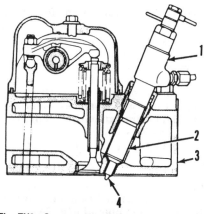

Fig. FN3—Cross-sectional view of cylinder head and injector typical of direct injection engines. Note that nozzle tip extends into the combustion chamber.
1. Injector
2. Nozzle sleeve
3. Cylinder head
4. Nozzle tip

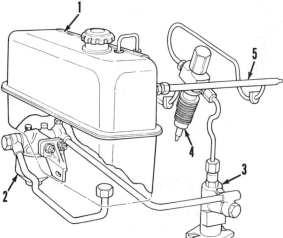

Fig. FN2—Drawing of basic fuel injection system components.
1. Fuel tank
2. Fuel filter
3. Fuel injection pump
4. Fuel injector
5. Fuel return line

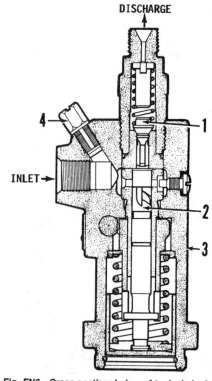

Fig. FN6—Cross-sectional view of typical single plunger injection pump used on some engines.
1. Delivery valve
2. Pump plunger
3. Pump body
4. Air bleed screw

SMALL DIESEL

SERVICE MANUAL

Diesel Engine Fundamentals

(in-line type) for each cylinder of the engine. A different arrangement for multicylinder engines is a pump with only one pumping element that uses a distributor type arrangement within the pump to direct a timed pressure pulse to the injector of each cylinder in the firing order of the engine. See Fig. FN9.

SINGLE OR MULTIPLE ELEMENT PUMPS. On engines with a separate pump for each cylinder or with an in-line type pump having a pumping element for each cylinder, a camshaft is used to actuate the pump plunger(s). When the engine is running, there is a constant supply of fuel provided by a lift pump or gravity available at the inlet port and spill port of the pumping element (Fig. FN7). With the plunger at its lowest point (A) in the cylinder barrel, fuel is free to flow into the barrel through the inlet port. As the plunger is forced upward (B) by the camshaft, the top of the plunger closes the inlet port and spill port, trapping the fuel in the barrel. Further movement of the plunger causes the pressure of the trapped fuel to increase until it overcomes the delivery valve spring pressure and flows out the top of the barrel to the injector. The injector nozzle valve is forced open by the pressurized fuel and sprays a quantity of fuel into the cylinder corresponding to the amount of fuel delivered by the pump plunger. The injection of fuel into the cylinder continues until pump plunger has risen to position (C), where a helical land on the plunger (which is connected to the top of the plunger by a vertical slot) uncovers the spill port in the barrel and allows pressurized fuel in the barrel to escape back to inlet side of the pumping element. This causes an immediate drop in pressure, resulting in the delivery valve and injector valve closing under spring pressure.

The total length of the plunger stroke is always constant, but the effective pumping stroke can be controlled by altering the position of the bottom of the plunger helix in relation to the spill port. The position of the helix is changed by rotating the plunger in the barrel, and the point of fuel cut-off can then be made to occur earlier (less fuel delivery) or later (more fuel delivery) in the stroke.

With the engine running under load, the plunger is rotated to the left (Fig. FN8) providing a long effective pumping stroke before the helix uncovers the spill port. At idle or very light load, plunger is rotated to the right so that the helix uncovers the spill port early in the stroke, reducing the amount of fuel pumped to the injector. To stop the engine, plunger is moved further to the right so that vertical slot in plunger is aligned with the spill port during the complete plunger stroke, preventing any build-up of pressure in the barrel.

The rotation of the helix is accomplished by a toothed quadrant on the pumping element mating with a control rack, which is controlled by the engine governor. On a multi-cylinder engine, all pumps are controlled by a common control rack.

DISTRIBUTOR TYPE PUMP. On distributor type injection pumps, a drive shaft rotates a fuel transfer pump, distributor rotor and pump element in a hydraulic head to deliver fuel to each injector. See Fig. FN9. The operating principles for typical distributor type pump are outlined below:

Fuel entering the injection pump is pressurized by the positive displacement transfer pump (2). The transfer pump pressure is controlled by pump speed and a pressure regulating valve in the pump end plate (1). The pressurized fuel flows through the pump housing, providing cooling and lubrication of injection pump cam, rollers, governor and other parts. Part of the excess fuel not required for injection is bypassed back to the inlet side of the pump and the remainder to the fuel tank.

The pressurized fuel for charging the injection pumping element flows through passages in the hydraulic head to a metering valve (8), which is controlled by governor action. The position of the metering valve and transfer pump pressure determines the quantity of fuel which is available to charge the pumping element. As the rotor turns, its charging channel comes into register with the inlet port to the pumping elements (Fig. FN10). As fuel enters the pumping cylinder, the two plungers are forced outward a distance proportionate to the quantity of fuel to be injected. The quantity of fuel entering the pumping cylinder is controlled by fuel pressure at the charging passage, by the amount of time available for charging, and by the maximum outward travel of the plungers which is limited by the roller shoes contacting an adjustable stop.

As the rotor continues to turn, the inlet port is closed and the single distributor port (Fig. FN10) in the rotor aligns with an outlet port in the hydraulic head. An internal cam ring (7—Fig. FN9) mounted in the pump housing, which has as many lobes as there are engine cylinders, operates the opposed pump plungers through cam rollers carried in shoes sliding in the rotor body. The rollers (4) simultaneously contact the diametrically opposed cam lobes and force the plungers (6) toward each other, discharging the fuel through the hydraulic head discharge port, into the fuel injector line (3) and opening the injector nozzle valve. The cam lobes are contoured to allow a very slight outward movement of the plungers at the end of the injection cycle. This gives a sharp cut-off of fuel to the injector and prevents "dribble" at injector nozzle tip.

INJECTOR. The injector must atomize the fuel to provide good combustion and must spray the fuel into the cylinder in such a manner that it mixes thoroughly with the air. Most injector nozzles used on modern diesel engines

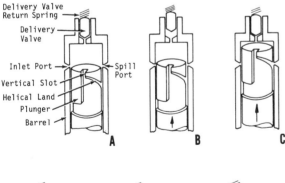

Fig. FN7—Injection pump plunger is forced upward by a camshaft. Beginning of injection occurs when top of plunger closes inlet and spill ports in pump barrel. End of injection occurs when helical land on plunger uncovers the spill port.

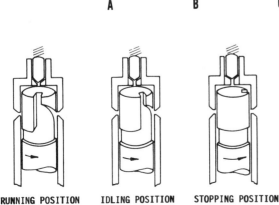

Fig. FN8—Injection pump plunger is rotated to alter the position of the plunger helix in relation to the spill port, which varies the effective length of the pumping stroke and controls the amount of fuel delivery.

Diesel Engine Fundamentals

SMALL DIESEL

are closed, inward opening type. See Figs. FN11 and FN12. They use a spring (S) to hold the injector nozzle valve (V) on its seat, and are operated hydraulically when fuel pressure acting on one side of the nozzle valve causes the valve to lift off its seat against spring pressure. The fuel pressure required to open the nozzle valve is determined by the spring pressure against the top end of the valve and is normally adjustable.

Hole type nozzles (Fig. FN11) have more than one spray hole in the nozzle tip. The size and angle of the holes can be varied by the manufacturer to achieve the spray pattern within the cylinder needed for proper combustion. This type nozzle is normally used in engines with open combustion chambers and direct injection.

Pintle type nozzles (Fig. FN12) have a single orifice hole that sprays fuel in a direct line from the injector. This type injector is normally used on indirect injection engines using a precombustion chamber or air cell.

GOVERNOR

Engine speed and power output are determined by the amount of fuel injected into the cylinder. To protect the engine from overspeeding or exceeding its designed power output limit, a means of controlling the amount of fuel delivered to the engine must be used. Depending upon the type of fuel injection pump, various methods are used to meter the amount of fuel that is delivered per each stroke of the pumping element, and a governor is then used to control the fuel metering system. A flyweight type governor is most commonly used, and may be contained in the injection pump housing or may be part of the engine.

The governor maintains a nearly constant engine speed by varying the amount of fuel supplied to the engine to satisfy varying load demands placed on the engine. In operation, governor main spring tension (which increases or decreases relative to the position of throttle control lever on variable speed governors) tends to move the fuel metering linkage to increase the amount of fuel being delivered to the engine, while centrifugal force created by the spinning governor flyweights tends to move the fuel metering linkage to decrease the amount of fuel delivery. A constant engine speed is maintained when the centrifugal force created by the spinning flyweights counterbalances the tension of governor main spring.

An increase in the load being applied to the engine causes a decrease in engine speed which results in a corresponding decrease in flyweight centrifugal force, allowing the flyweights to

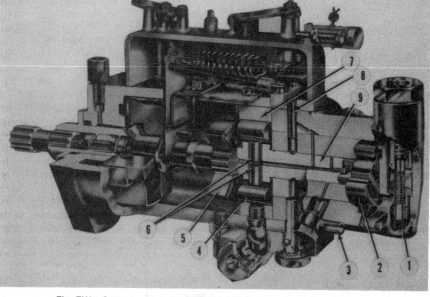

Fig. FN9—Cutaway view of a rotary distributor type fuel injection pump.
1. End plate & pressure regulating valve
2. Fuel transfer pump
3. Injector line
4. Cam roller
5. Rotor
6. Pump plungers
7. Cam ring
8. Metering valve
9. Pump rotor

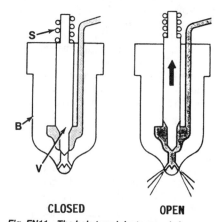

Fig. FN11—The hole type injector nozzle has more than one spray hole in nozzle tip and is most often used in direct injection engines.

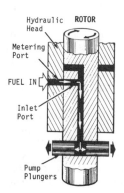 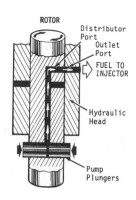

Fig. FN10—On the charging cycle, inlet port in pump rotor is in register with fuel metering port in hydraulic head. On pumping cycle, distributor port in rotor is in register with one of the outlet ports in hydraulic head.

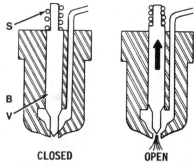

Fig. FN12—Pintle type injector nozzle has a single hole in nozzle tip and is most often used in indirect injection engines.

move inward. Movement of the flyweights is transmitted through the governor linkage to move the injection pump fuel metering control to obtain increased fuel delivery to the engine, resulting in increased power and speed. The engine speed increases until the centrifugal force of the flyweights and governor spring tension are again balanced.

When the load on the engine is reduced, engine speed increases and the opposite reactions take place in the governor. As the governor flyweight centrifugal force exceeds the governor spring force, the governor linkage moves the fuel metering controls to reduce the fuel delivery rate to the engine and maintain desired speed setting.

TURBOCHARGING

Power obtained from an internal combustion engine is limited by the amount of air, thus oxygen, in the cylinder for combustion of the fuel. Engines with normal air intake through air cleaner and intake ports, called "naturally aspirated" engines, are limited by atmospheric pressure to the amount of air that can be taken in on the intake stroke. This limit is reduced at higher elevations because of decrease in atmospheric pressure. Increasing the amount of fuel above that which will be burned causes excessive exhaust smoke and loss of engine power. If the fuel delivery system is adjusted for proper air/fuel ratio at sea level or lower elevations, fuel delivery must be decreased for the engine to operate efficiently at higher elevations.

The turbocharger utilizes a turbine blade (rotor) powered by the engine exhaust to drive a turbine blade in the engine air intake to compress the air entering the engine cylinder, thus increasing the volume of air going into the engine cylinder on the intake stroke. As the volume of air in the cylinder is increased, more fuel can be injected and burned to provide more power per unit of engine displacement. The turbocharger is especially beneficial for high altitude engine operation, but will increase power output at any elevation.

COLD WEATHER STARTING

As diesel engines rely solely on the temperature of air compressed in the cylinder to ignite the fuel when it is injected into the cylinder, cold temperatures have an adverse effect on engine starting. To improve diesel engine starting characteristics in cold ambient temperatures, engine starting aids are used. Several different methods used to aid in cold weather starting are:

a. Increasing cranking speed with more powerful starting motor and battery. An optional, more powerful electric starter and higher capacity battery are available on some engines. The higher cranking speeds increases compression temperature for easier starting.

b. Heating the air entering the engine through the intake manifold by an electric heater. If the air at beginning of compression stroke is warmer, the temperature at end of compression stroke will be relatively higher and result in easier ignition of fuel injected into the cylinder. This method is usually utilized in direct injection engines.

c. Use of electric "glow plugs" in the engine cylinder to cause ignition of the fuel injected into the cylinder. This method is most generally used in indirect injection engines.

d. Adding a small amount of engine oil in the engine intake to increase sealing of the piston rings and increase compression pressure, thereby raising air temperature to aid ignition of fuel injected in cylinder.

e. Using ether sprayed into the air intake to provide a more volatile and easily ignited fuel-air mixture. This method is not recommended on glow plug equipped engines as extreme engine damage can result.

ENGINE COOLING

To remove heat transferred from the combustion process to the engine cylinder and cylinder head, a cooling system must be provided. Two different systems are used which are air cooling and liquid cooling.

AIR COOLING. Engines may be cooled either by air flow around cooling fins cast as a part of the engine cylinder, and on some engines, cooling fins on the cylinder head. The air flow can be created by fan blades attached to the engine flywheel, or by a belt driven cooling fan. The air cooled engine is fitted with a cover, or shroud, to concentrate air flow from the cooling fan to the engine parts requiring cooling.

On air cooled engines, maintenance of the cooling system is extremely important for engines operating in dirty or trashy conditions. In time, the air intake screen or air passage between the engine cooling fins and cover or shrouding may be blocked with trash or dirt. If this happens, the air cooled engine will overheat and severe engine damage will occur.

LIQUID COOLING. Some engines are cooled by a liquid flowing through passages within the cylinder block and cylinder head. A system for removing heat from the cooling liquid must be used and several different methods are utilized.

The simplest method of removing heat from engine coolant is fitting a large jacket or hopper to the engine cylinder; heat is transferred from the coolant liquid through the water hopper material to surrounding air.

On many engines used for industrial purposes, the coolant, after picking up heat from the engine, is passed through a radiator. Many small tubes attached to two radiator manifold tanks carry liquid through the cooling fin area. The radiator has a fan pulling or pushing air through the cooling fins, and heat from coolant passing through the radiator tubes is transmitted to the air. On most engines, a pump is used to circulate the cooling liquid through the engine, to the radiator where heat is removed, and back to the engine. Some engines utilize "thermosiphon" cooling where circulation through the radiator and engine is created by percolation of the heated coolant.

Some engines for marine use are cooled from water pumped up through an inlet in the bottom of the marine vessel directly through the engine liquid cooling system and then the water is discharged back to source, normally some or all with the engine exhaust. On other marine liquid cooled engines, the water pumped through the bottom of the vessel passes through a heat exchanger where heat is removed from the engine's fresh water cooling liquid. The heat exchanger method is very well suited to uses where the marine vessel is operated in salt, or sea water. Water pumps used for this purpose have a rubber impeller which should be inspected and renewed, if necessary, on a periodic basis.

Where salt or sea water is used for engine cooling, the salt will tend to corrode metal parts quite rapidly. To reduce the corrosive effect, zinc anodes are installed in the sea water flow so that the chemical corrosion will be concentrated on the anodes rather than other engine or heat exchanger parts. The anodes should be inspected periodically and renewed whenever they are reduced to about 25 percent of their original size.

ENGINE TEMPERATURE CONTROL. Air cooled engines do not normally need any method of engine temperature control except by circulating fan speed changing with engine speed.

In liquid cooled engines, the engine coolant must be kept within a desired

Diesel Engine Fundamentals

temperature range for efficient operation. Therefore, except those having a water hopper or thermosiphon system, most systems have a thermostat valve that stops or reduces flow of coolant through the engine until engine temperature reaches a level predetermined by design of the thermostat. If the thermostat fails, flow of engine coolant may either be blocked causing the engine to overheat, or flow may not be controlled at all causing the engine to operate at coolant temperature too low. In either case, engine will be damaged if operated with a defective thermostat. Some engine manufacturers recommend periodic replacement of the cooling system thermostat.

DIESEL FUEL

There are two universal diesel fuel classifications: API Number 1 and Number 2. The engine manufacturer will specify which fuel should be used in their engines. Ambient temperature in which the engine is operating may affect choice of fuel type. Number 1 diesel fuel will flow at a much lower temperature than Number 2 diesel fuel and will be recommended for some engines operating in very low ambient temperatures. However, in higher ambient and engine operating temperatures, Number 2 diesel fuel is needed for its lubricating quality and to reduce leakage past the fuel injection pump and injector elements. Distillate, heating oils or kerosene should never be substituted for diesel fuel.

Extreme care must be taken in the handling and storage of diesel fuel. Diesel fuel should never be stored in a container previously used to store gasoline. Diesel fuel storage tanks should be of a design to inhibit condensation within the tank; water is very damaging to the diesel engine fuel system. Sediment will remain in suspension in diesel fuel for a number of hours and may cause early clogging of the fuel filter element. The diesel engine fuel tank should be refilled at end of operation period and before storage so that there is no room for air to enter the tank and cause condensation.

TROUBLE-SHOOTING

The following are problems which may occur during operation of the engine and their possible causes.

1. Engine hard to start or will not start.
   a. Low cranking speed due to faulty starting system components.
   b. Incorrect oil viscosity (too heavy) for ambient temperature.
   c. No fuel in tank or incorrect grade of fuel.
   d. Fuel filter plugged with water or dirt.
   e. Air in fuel system.
   f. Injection pump timing incorrect.
   g. Injection pump control rack stuck or governor linkage broken.
   h. Faulty injector nozzle operation.
   i. Faulty injection pump or pump drive.
   j. Low engine compression.

2. Engine misses.
   a. Water in fuel system.
   b. Air in fuel system.
   c. Faulty injector nozzle.
   d. Faulty injection pump.

3. Engine lacks power.
   a. Fuel filter plugged or other restriction in fuel lines.
   b. Dirty air cleaner or other restriction in intake system.
   c. Restriction in exhaust system.
   d. Wrong grade of fuel.
   e. Injection pump timing retarded.
   f. Faulty injector pump or injector.
   g. Incorrect valve tappet clearance.
   h. High idle no-load speed too low.
   i. Low compression.

4. Engine knocks.
   a. Faulty injection timing.
   b. Foreign material in cylinder.
   c. Valve sticking and contacting piston.
   d. Worn engine bearings.
   e. Excessive crankshaft end play.
   f. Excessive clearance between piston and cylinder.

5. Blue exhaust smoke.
   a. Piston rings worn, stuck or broken.
   b. Cylinder bore worn.

6. White exhaust smoke.
   a. Water in fuel.
   b. Low cetane fuel.
   c. Incomplete combustion due to low compression or incorrect injection timing.

7. Black exhaust smoke.
   a. Engine overloaded.
   b. Air filter plugged.
   c. Faulty injection pump delivering excessive amount of fuel.
   d. Faulty injector nozzle.

8. Engine uses too much oil.
   a. Incorrect oil viscosity (too light) for ambient temperature.
   b. Restricted air intake.
   c. Restricted oil return passage in cylinder head.
   d. Worn pistons and rings.
   e. Worn valve guides or valve stem seals.

9. Engine overheating — water cooled.
   a. Faulty radiator cap.
   b. Radiator fins plugged.
   c. Faulty thermostat.
   d. Coolant level too low.
   e. Engine lubricating oil level too low.
   f. Water pump belt slipping.
   g. Engine overloaded.
   h. Incorrect injection timing.

10. Engine overheating — air cooled.
    a. Cooling air being recirculated.
    b. Fins of cylinder head or cylinder plugged.
    c. Cooling air inlet or outlet blocked.
    d. Lubricating oil level too low.
    e. Engine overloaded.
    f. Incorrect injection timing.

11. Lubricating oil pressure too low.
    a. Incorrect oil viscosity (too light) for ambient temperature.
    b. Low oil level.
    c. Oil diluted with fuel.
    d. Engine overheating.
    e. Engine oil cooler (if used) plugged.
    f. Insufficient output from oil pump or leaking connections.
    g. Relief valve not seating.
    h. Worn engine bearings.

CONTINENTAL

TELEDYNE TOTAL POWER
3409 Democrat Road
P.O. Box 181160
Memphis, Tennessee 38181-1160

Model	No. Cyls.	Bore	Stroke	Displ.
TMD13	2	91 mm (3.58 in.)	103.2 mm (4.06 in.)	1350 cc (82 cu. in.)
TMD20	3	91 mm (3.58 in.)	103.2 mm (4.06 in.)	2000 cc (123 cu. in.)
TMD27	4	91 mm (3.58 in.)	103.2 mm (4.06 in.)	2680 cc (164 cu. in.)
TMDT27	4	91 mm (3.58 in.)	103.2 mm (4.06 in.)	2680 cc (164 cu. in.)

The TMD Series of engines are water cooled, four cylinder, four stroke cycle, indirect injection diesel engines. Model TMDT27 is turbocharged. Crankshaft rotation is counterclockwise as viewed from flywheel (PTO) end. Number 1 cylinder is at timing gear end of engine. Fuel injection system pump and filter may be either CAV or Stanadyne RoosaMaster.

MAINTENANCE

LUBRICATION

For engine lubrication, API classification CD motor oils are recommended. Oil sump capacity, including filter, is 4.73 liters (5 quarts) on 2-cylinder engines, 5.68 liters (6 quarts) on 3-cylinder engines and 6.65 liters (7 quarts) on 4-cylinder engines. Under normal use, oil and filter should be changed after every 100 hours of use in clean environment, every 50 hours in dirty environment. Under light duty use (25 percent of maximum continuous rating), oil and filter change interval can be extended to 200 hours. Oil and filter must always be changed after first 50 hours use of new or rebuilt engine.

For intermittent use, oil viscosity should be based on ambient temperature. Oil viscosity for engines in continuous use should be selected based on sump oil temperature.

INTERMITTENT USE. If single grade oils are used, use SAE 10W in temperatures of −24° to −7° C (−10° to +20° F), SAE 20W/20 oil in temperatures of −15° to +10° C (5° to 52° F), SAE 30 in temperatures of 0° to 33° C (32° to 91° F) or SAE 40 oil in temperatures above 22° C (72° F).

Multigrade SAE 20W-40 or SAE 20W-50 oils can be used in temperatures of −15° to +32° C (5° to 90° F), and SAE 15W-40 or 15W-50 oil in temperatures of −18° to 32° C (0° to 90° F), 10W-40 or 10W-50 oils in temperatures of −24° to 27° C (−10° to 80° F), or SAE 5W-20 oil in temperatures below −10° C (14° F).

CONTINUOUS USE. Single grade SAE 20 oil should be used if oil sump temperature is 55°-71° C (130°-160° F), SAE 30 if oil sump temperature is 71°-99° C (160°-210° F), or SAE 40 if oil sump temperature is 99°-121° C (210°-250° F). Multigrade oils should not be used if sump temperature is above 99° C (210° F).

ENGINE SPEED ADJUSTMENT

Slow idle speed should be 800-1000 rpm and is adjusted by turning idle speed screw (I—Fig. C1-1 or C1-2). Maximum governed engine speed is determined by engine application, and is adjusted by turning high speed screw (H—Fig. C1-1 or C1-2).

On RoosaMaster pumps used on welder or generator set engine, pump is equipped with a speed droop control as shown in Fig. C1-3. This control may be adjusted externally to increase or decrease governor control spring (S) tension to provide sharpest governor regulation possible without surge. Control may be adjusted with engine running. Turning the speed droop screw (D) clockwise broadens governor regulation and reduces instability. After adjusting speed droop setting, high idle speed stop screw must be readjusted.

The RoosaMaster pump may also be equipped with a maximum fuel delivery adjusting screw (torque screw or

Fig. C1-1—View of typical CAV-DPA fuel injection pump used on some engines. Loosen vent screws (V) to bleed air. Turn screw (I) to adjust low idle speed and screw (H) to adjust high rpm.

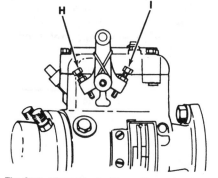

Fig. C1-2—View of typical RoosaMaster fuel injection pump used on some engines. Turn screw (I) to adjust low idle speed and screw (H) to adjust high rpm.

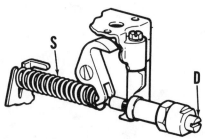

Fig. C1-3—View showing location of speed droop governor control on RoosaMaster fuel injection pump.

SERVICE MANUAL

Continental

smoke stop) as shown in Fig. C1-4. Screw (T) may be adjusted to compensate for high altitude operation. If there is excessive exhaust smoke under load, turn screw in (clockwise) until exhaust smoke is reduced to acceptable level. For operation at altitudes under 1000 meters (3300 feet), back screw out to obtain maximum power to point where exhaust smoke occurs, then turn screw back in to obtain acceptable smoke level.

FUEL SYSTEM

FUEL FILTERS. An externally mounted fuel filter is used on all engines. Under normal conditions, the filter should be renewed after each 400 hours of engine operation; renew filter more often if dirty fuel is encountered.

On CAV-DPA filter and water trap type, remove filter bowl and discard the dirty element. Clean inside of bowl and center tube. Fit new upper and lower sealing washers and install new "O" ring on center stud. Place new filter in position, install the bowl and tighten center stud. Bleed air from the fuel system.

On Stanadyne fuel filter (not a water trap), release the two spring clamps and remove the old filter element. Be sure mounting face is clean, position new element in metal canister and secure with the spring clamps.

BLEED FUEL SYSTEM. Procedure to bleed fuel system will vary depending on whether engine is equipped with a CAV-DPA or Stanadyne fuel system. Refer to the appropriate following paragraph:

CAV-DPA SYSTEM. To bleed fuel system, remove bleed screw (A—Fig. C1-5) on filter body or loosen supply line fitting (B) at injection pump, whichever point is highest.

NOTE: On filters with four fuel line connection bosses, the plug must be removed from the unused boss, regardless if at lower point than pump inlet.

Operate priming lever on fuel transfer pump or activate electric fuel pump until fuel flows free of air, then install plug or tighten pump inlet connection. Loosen injection pump bleed screws (C and D) and operate fuel transfer pump until fuel free of air flows from bleed screws, then tighten screws. Air should be purged from low pressure fuel system at this point and engine should start. If engine fails to start, bleed high pressure system as follows: Loosen fittings for high pressure injection lines at injectors. Crank engine with engine stop control in run position and speed control in full speed position until fuel flows from injector lines, then tighten fittings and start engine.

STANADYNE SYSTEM. To bleed system, turn ignition switch to "on" position and loosen filter outlet (A—Fig. C1-6) or fuel injection pump inlet connection (B), whichever is higher, and allow fuel to flow until free of air. Low pressure fuel system should be free of air at this point and engine should start. If engine fails to start, bleed high pressure fuel system as follows: Tighten the fitting and loosen injector high pressure line fittings at the injectors. Set speed control to full speed position and with ignition switch "on", crank engine until fuel flows from injector lines. Tighten the fittings and start engine.

INJECTION PUMP TIMING. Procedure to time the fuel injection will vary depending on whether engine is equipped with a CAV-DPA or Stanadyne fuel system. Refer to the appropriate following paragraph:

CAV-DPA FUEL SYSTEM. The injection pump should be properly timed when scribe marks on timing gear cover, pump adapter plate and pump flange are aligned. Loosen adapter plate and injection pump mounting nuts, then rotate pump or plate as required to align marks.

If timing marks are not present, or believed to be false, turn engine so number 1 piston is at TDC on compression

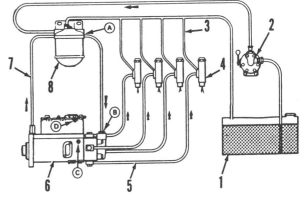

Fig. C1-5—Schematic diagram of CAV-DPA fuel system and bleed screw locations. Refer to text for air bleeding procedure. Stanadyne Roosa-Master fuel system is shown in Fig. C1-6.

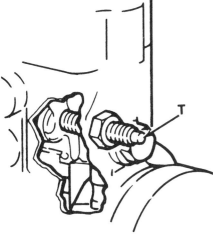

Fig. C1-4—View showing torque screw (also called delivery limit or smoke stop screw) on RoosaMaster fuel injection pump.

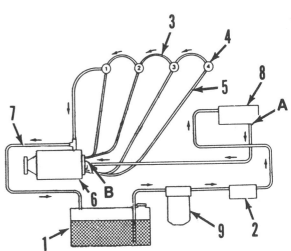

Fig. C1-6—Schematic diagram for Stanadyne Roosa-Master fuel system and air bleed locations. Refer to text for air bleeding procedure. Schematic diagram for CAV-DPA system is shown in Fig. C1-5.

Illustrations Courtesy Teledyne Total Power

Continental

stroke. Remove inspection plate on side of injection pump. Back engine up about 50 degrees BTDC, then turn engine slowly in normal direction of rotation until 13-15 degrees BTDC timing mark is aligned with mark on crankshaft pulley or pointer in bell housing. The line "A" on pump drive shaft should be aligned with flat end of snap ring as shown in Fig. C1-7. If not, loosen pump mounting bolts and turn pump so that "A" mark and snap ring are aligned. Tighten mounting bolts and install pump cover plate. Restamp pump to adapter plate scribe marks.

Engine manufacturer recommends using diesel timing light to check injection timing with engine running. Injection timing should occur at 13-15 degrees BTDC with engine running at 1600 rpm no load.

If injection timing cannot be adjusted as outlined, refer to INJECTION PUMP section as injection pump gear may not be correctly timed.

STANADYNE FUEL SYSTEM. To check injection pump timing, turn engine so number 1 piston is at TDC of compression stroke and remove the cover plate from side of RoosaMaster fuel injection pump. The two timing marks should be aligned in timing window (Fig. C1-8) and the scribed marks on pump mounting flange and adapter should be aligned. If timing marks in window are not aligned, refer to INJECTION PUMP section as injection pump gear may not be correctly timed. If scribe marks on pump mounting flange and adapter are not aligned, loosen pump mounting bolts and rotate pump as necessary, then tighten bolts.

Engine manufacturer recommends using a "AVL Model 876" diesel timing light with transducer placed near pump on number 1 fuel injection line to check injection timing with engine running. A different timing light may give different reading. For injection pumps with variable speed governor and with engine running at 600-950 rpm no-load, injection timing should be 3-4 degrees BTDC on Models TMD13 and TMD20, and 4-5 degrees BTDC on Model TMD27. For pumps with fixed speed governors (welders and generator sets) with engine running at 1850 rpm no-load, injection timing should be 5-6 degrees BTDC for Models TMD13 and TMD20, and 8-9 degrees BTDC for Model TMD27. If necessary to reset timing, loosen pump mounting bolts, turn pump to obtain proper timing and tighten bolts. Scribe new lines across pump flange and adapter.

COOLING SYSTEM

All models are water cooled and use a pump mounted on front of engine to circulate coolant. A mixture of ethylene glycol antifreeze and clean water should be used in engine coolant.

A thermostat is located under the coolant outlet. Thermostat should start to open when placed in water and heated to 81°-93° C (180°-200° F). Most engines are equipped with radiator caps having a pressure release at either 31 kPa (4.5 psi) or 103 kPa (15 psi). At these pressures, water temperature would be 107° C (224° F) or 120° C (248° F).

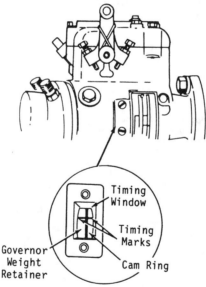

Fig. C1-8—View of RoosaMaster fuel injection pump with side cover removed showing timing marks.

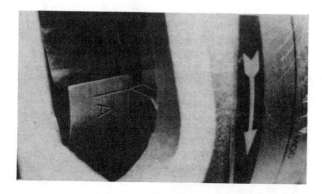

Fig. C1-7—Injection timing for CAV-DPA pump is correct when scribed line "A" aligns with end of snap ring and number 1 piston is at TDC on compression stroke.

SMALL DIESEL

BELT TENSION

Fan and alternator belt tension measured midway between water pump and crankshaft pulleys should be approximately 13 mm (1/2 inch) using thumb pressure. Belt tension is adjusted by repositioning alternator.

REPAIRS

TIGHTENING TORQUES

Refer to the following table for special tightening torques. All fasteners are metric.

Camshaft nut	88-95 N·m (65-70 ft.-lbs.)
Connecting rod	61-68 N·m (45-50 ft.-lbs.)
Crankshaft pulley	163-176 N·m (120-130 ft.-lbs.)
Cylinder head	See text
Flywheel	68-75 N·m (50-55 ft.-lbs.)
Flywheel housing	61-68 N·m (45-50 ft.-lbs.)
Glow plug	31-38 N·m (23-28 ft.-lbs.)
Injection pump gear	27-34 N·m (20-25 ft.-lbs.)
Injector	68-75 N·m (50-55 ft.-lbs.)
Main bearing cap	150-162 N·m (110-120 ft.-lbs.)
Manifolds:	
M8	20-24 N·m (15-18 ft.-lbs.)
M10	34-40 N·m (25-30 ft.-lbs.)
Oil pan, sheet steel	14-19 N·m (10-14 ft.-lbs.)
Oil pump	20-24 N·m (15-18 ft.-lbs.)
Rocker arm cover	7-8 N·m (5-8 ft.-lbs.)
Rocker arm supports	23-27 N·m (17-20 ft.-lbs.)
Timing gear cover	34-40 N·m (25-30 ft.-lbs.)
Water pump	34-40 N·m (25-30 ft.-lbs.)

WATER PUMP

R&R AND OVERHAUL. To remove water pump, drain coolant, disconnect hoses and remove drive belt and fan. Remove radiator if necessary. Unbolt and remove water pump.

Refer to Fig. C1-9 and disassemble pump as follows: Remove cover (1) from rear side of pump. Using suitable puller, remove pulley (10) from shaft (9). Remove snap ring (6), then press shaft and

Illustrations Courtesy Teledyne Total Power

SMALL DIESEL

bearing (9) forward out of impeller (3) and pump housing (7). Note that housing will be damaged if shaft is forced out rear of pump. Remove seal assembly (4) and "O" ring (5).

Reverse disassembly procedure for reassembly. Lubricate seal and shaft with soap suds to prevent damage to seal when installing pump shaft. A thin film of lubricant applied to face of seal will facilitate seating and sealing. Support rear end of shaft when installing pulley and front end of shaft when installing impeller.

VALVE ADJUSTMENT

Valve clearance should be adjusted with engine running at idle speed and at normal operating temperature. Remove rocker arm cover and adjust intake valve clearance to 0.36 mm (0.014 inch) and exhaust valve clearance to 0.46 mm (0.018 inch).

CYLINDER HEAD

REMOVE AND REINSTALL. Drain cooling system and remove radiator hoses. Remove intake and exhaust manifolds. Disconnect glow plug wiring. Remove fuel injector lines and bleed-off lines; immediately cap or plug all fuel openings. Remove fuel injectors with modified socket (see INJECTORS paragraph). Unscrew and remove glow plug located below each injector. Remove rocker arm cover, rocker arm assembly and push rods. Snap push rods to one side to break oil seal in the lifter ball socket so lifters will not be pulled up with push rods. Remove cylinder head bolts, then lift cylinder head from engine. Refer to VALVE SYSTEM paragraph.

With head removed, precombustion chambers (20—Fig. C1-12) can be removed by carefully inserting a long thin drift through injector bores and tapping chambers from head. Renew precombustion chambers if cracked or burned.

Before reinstalling head, be sure that gasket surfaces of head and block are clean. Inspect head for cracks and renew if necessary. Check flatness with straightedge and feeler gage. Maximum permissible distortion lengthwise is 0.10 mm (0.004 inch). Crosswise distortion limit is 0.076 mm (0.003 inch). If these limits are exceeded, resurface or renew cylinder head. If precombustion chamber inserts have been removed, be sure they are installed and fully seated; inserts should be flush to 0.076 mm (0.003 inch) above head surface.

Be sure threads of cylinder head cap screws have been properly cleaned. If tap is used to clean holes in block, use M10 × 1.5 and M12 × 1.75 class 6G taps.

Install cylinder head by reversing removal procedure. Insert a long headless M12 × 1.75 guide stud at each end of cylinder block and place head gasket on the guides. Lubricate cylinder head cap screws with light engine oil and install finger tight. Remove the two guide studs and install remaining cap screws. Refer to Fig. C1-13 and tighten cap screws in sequence shown in three equal steps to a final cold torque of 68-75 N·m (50-55 ft.-lbs.) for M10 bolts and 122-129 N·m (90-95 ft.-lbs.) for M12 bolts. After engine is completely assembled, run engine until normal operating temperature is reached, remove rocker arm cover and re-tighten the head bolts in proper sequence to a torque of 61-68 N·m (45-50 ft.-lbs.) for M10 bolts and 109-115 N·m (80-85 ft.-lbs.) for M12 bolts.

Continental

VALVE SYSTEM

When disassembling cylinder head, keep all valve components in order so they can be reinstalled in original positions if reused.

Intake valve face and seat angle is 30 degrees and exhaust valve face and seat angle is 45 degrees. Valve seat width should be 1.6-2.4 mm (1/16 to 3/32 inch). If seat width is over 2.4 mm (3/32 inch), seat may be reconditioned using 15, 30 and 60 or 75 degree stones. Seat contact area should be centered on valve face.

Valve guide inner diameter is 8.692-8.717 mm (0.342-0.343 inch) and maximum allowable wear diameter is 8.775 mm (0.345 inch). Valve guides are renewable; when properly installed, bottom end of guide should be 43 mm (1.690 inches) from gasket surface of cylinder head. Guide length is 60.4 mm (2-3/8 inches) and outside diameter is 16.675-16.70 mm (0.6565-0.6575 inch).

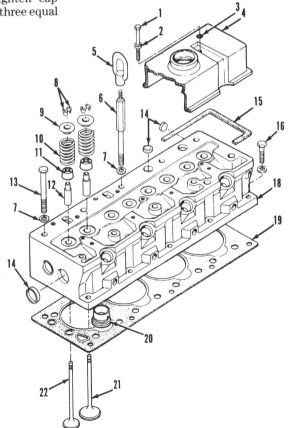

Fig. C1-12—Exploded view of 4-cylinder head. Other models are similar. M10 bolts (16) are used along fuel injector side of head and longer M12 bolts (13) are used in other head bolt locations. Special bolt (6) is used at center of manifold side of head as adapter for engine lifting eye (5). All core hole cup plugs (14) should be renewed whenever engine is overhauled.

1. Rocker cover bolts
2. Washers
3. Sealing rings
4. Rocker arm cover
5. Lifting eye
6. Adapter bolt
7. M12 washers
8. Split valve locks
9. Valve spring retainers
10. Valve springs
11. Valve stem seals
12. Valve guides
13. M12 head bolts
14. Cup plugs
15. Cover gasket
16. M10 bolts
17. M10 washers
18. Cylinder head
19. Head gasket
20. Precombustion chambers
21. Intake valve
22. Exhaust valve

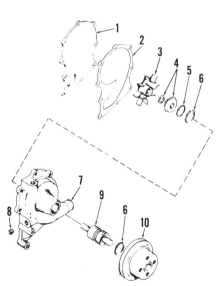

Fig. C1-9—Exploded view of water pump assembly.

1. Cover
2. Gasket
3. Impeller
4. Seal assy.
5. "O" ring
6. Snap ring
7. Pump housing
8. Plug
9. Shaft assy.
10. Pulley

Illustrations Courtesy Teledyne Total Power

Continental

Intake valve stem diameter is 8.642-8.660 mm (0.3402-0.3409 inch) with wear limit of 8.592 mm (0.3383 inch). Specified intake valve stem to guide clearance is 0.032-0.075 mm (0.0013-0.0030 inch) with desired clearance of 0.053 mm (0.002 inch) and wear limit of 0.125 mm (0.005 inch).

Exhaust valve stem diameter is 8.622-8.640 mm (0.3394-0.3402 inch) with wear limit of 8.575 mm (0.3376 inch). Specified exhaust valve stem to guide clearance is 0.052-0.095 mm (0.002-0.0037 inch) with desired clearance of 0.073 mm (0.003 inch) and wear limit of 0.142 mm (0.0056 inch).

Valve spring pressure should be 235 N (53 pounds) at valve closed height of 42 mm (1.65 inches) and 466 N (105 pounds) at valve open height of 32.88 mm (1.29 inches).

Maximum allowable rocker arm to rocker arm shaft clearance is 0.13 mm (0.005 inch). When reassembling, be sure dowel pin at end of shaft fits in notch of end rocker shaft support.

TURBOCHARGER

Model TMDT-27 engines are equipped with a Garrett/AiResearch turbocharger mounted near flywheel end on manifold side of engine. When starting engine with turbocharger, do not run engine above slow idle speed until normal engine oil pressure is obtained. Operation of turbocharger without sufficient oil supply for even a few seconds can cause turbocharger bearing failure.

If necessary to remove turbocharger, proceed as follows: Disconnect oil drain line at engine block and allow oil to drain. Disconnect oil pressure (feed) line at engine block. Remove air inlet ducting between air cleaner and turbocharger. Remove compressor outlet hose at intake manifold. Disconnect exhaust pipe from turbine outlet. Support turbocharger, disconnect turbine inlet pipe from exhaust manifold and remove the turbocharger from engine. Cover, plug or cap all turbocharger openings.

Unless shop is equipped with special tools, specifications and trained personnel necessary, disassembly of turbocharger should not be attempted.

To install turbocharger, first remove all protective coverings from turbocharger openings and do not allow any dirt or foreign material to enter openings during installation. Mount turbine inlet pipe on turbine housing flange using new stainless steel gasket and grade 8 mounting bolts; tighten bolts to a torque of 54 N·m (40 ft.-lbs.). Mount turbocharger assembly on exhaust outlet flange using new stainless steel gasket and grade 8 bolts; tighten bolts to a torque of 54 N·m (40 ft.-lbs.). Connect hose between compressor outlet and intake manifold and tighten hose clamps securely. Connect oil pressure (feed) line from engine to oil inlet port of turbocharger housing.

CAUTION: Do not use any thread sealers on line fittings.

Without allowing engine to start, crank engine until steady stream of oil is flowing from turbocharger oil drain line port. If removed, use new gasket and attach oil drain line flange to turbocharger. Tighten bolts to a torque of 34 N·m (25 ft.-lbs.). Install oil drain line, making sure it does not have any sharp bends or kinks and tighten clamps securely. Using new stainless steel gasket, connect exhaust pipe to turbocharger. Connect ducting between air cleaner and turbocharger. Start engine, operate at idle speed until normal oil pressure is reached and check all connections for leaks.

INJECTORS

Refer to Fig. C-15 for cross-sectional view of cylinder head showing location of injector (1), glow plug (2) and precombustion chamber (3).

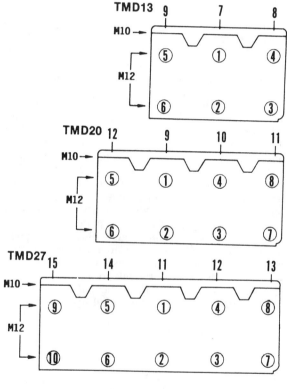

Fig. C1-13—Drawings showing cylinder head bolt tightening sequence for two, three and four cylinder engines.

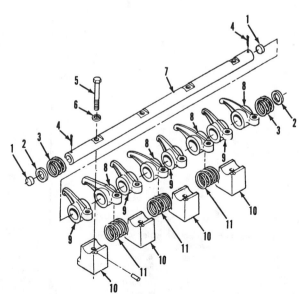

Fig. C1-14—Exploded view of rocker arm assembly for 4-cylinder engine; others are similar.

1. End plugs
2. End washers
3. End springs
4. Cotter pins
5. Rocker shaft bolts
6. Washers
7. Rocker arm shaft
8. Intake rocker arms
9. Exhaust rocker arms
10. Rocker shaft supports
11. Rocker shaft springs

SMALL DIESEL

Illustrations Courtesy Teledyne Total Power

SERVICE MANUAL

Continental

REMOVE AND REINSTALL. First, thoroughly clean injector, lines and surrounding area using suitable solvent and compressed air. Remove high pressure injector lines and fuel leak-off lines from injectors. Immediately cap or plug all openings to prevent dirt entry. Use die grinder to cut reliefs at each side inside of deep well socket (Fig. C1-16) so socket will slide down over injector body to reach nozzle nut without interfering with leak-off nipples on top of body. Unscrew and remove injectors from cylinder head. If not removed with injector, pull heat shield (9—Fig. C1-16A) from bottom of injector bore.

Before installing injector, be sure injector bore and seating surface in cylinder head are clean and free of carbon. Install a new heat shield (9). Install injector and tighten to a torque of 68-75 N·m (50-55 ft.-lbs.). Reconnect fuel lines, but do not tighten high pressure line fittings at the injectors. With fuel control in "Run" position, crank engine with starter until fuel flows from loosened connections, then tighten fuel line fittings.

TESTING. A faulty injector may be located on the engine by loosening high pressure fuel line fitting on each injector, in turn, with engine running. This will allow fuel to escape from the loosened fitting rather than enter the cylinder. The injector that least affects engine operation when its line is loosened is the faulty injector. If a faulty injector is found and considerable time has elapsed since injectors have been serviced, it is recommended that all injectors be removed and serviced, or that new or reconditioned units be installed.

A complete job of testing and adjusting injectors requires use of special test equipment. Only clean, approved testing oil should be used to test injectors. Injector nozzles should be tested for opening pressure, seat leakage and spray pattern.

WARNING: Fuel emerges from injector with sufficient force to penetrate the skin. When testing injector, keep yourself clear of nozzle spray.

Before conducting test, operate tester lever until oil flows from open line, then attach injector. Close valve to tester gage and operate tester lever a few quick strokes to be sure nozzle valve is not stuck. When operating properly, injector nozzle will emit a buzzing sound and cut off quickly with no fluid leakage at seat.

OPENING PRESSURE. Open valve to tester gage and operate tester lever slowly while observing gage reading. Opening pressure should be 14140-14700 kPa (2050-2130 psi). If not within limits, remove nozzle nut (8—Fig. C1-16A) and vary shim (2) thickness as necessary. All injectors in an engine should be set as close as possible to same opening pressure.

SEAT LEAKAGE. Injector nozzle should not leak when held at a pressure slightly lower than observed opening pressure for a period of 10 seconds. If drop appears on nozzle tip or test oil dribbles from nozzle, nozzle valve is not seating and injector should be overhauled or renewed. Slight wetting of nozzle tip is acceptable on a used injector.

SPRAY PATTERN. Spray from nozzle should be in a straight small conical pattern and be well atomized. If pattern is wet, ragged, intermittent or is to one side, nozzle must be overhauled or renewed.

OVERHAUL. Hard or sharp tools, emery cloth, grinding compound or other than approved solvents or lapping compounds must never be used. An approved nozzle cleaning kit is available through a number of specialized sources.

Wipe all dirt and loose carbon from exterior of injector assembly. Refer to Fig. C1-16A for exploded view of injector and proceed as follows:

Secure injector body (1) in a soft jawed vise or holding fixture and remove nozzle nut (8). Place all parts in clean calibrating oil or diesel fuel as they are removed. Use particular care not to mix components if more than one injector is disassembled at a time.

Clean exterior surfaces with a brass wire brush. Soak parts in an approved carbon solvent, if necessary, to loosen hard carbon deposits. Rinse parts immediately in clean calibrating oil or diesel fuel after cleaning to neutralize solvent and prevent etching of polished surfaces. Assemble injector while parts are still wet under oil or fuel. Tighten nozzle nut (8) to a torque of 68-75 N·m (50-55 ft.-lbs.). Retest injector before installing in engine.

GLOW PLUGS

Each cylinder is equipped with a glow plug (2—Fig. C1-15). Tip of glow plug extends into precombustion chamber (3). Glow plugs are connected in parallel with each glow plug grounded through mounting threads. Before suspecting a glow plug malfunction, be sure that current is reaching glow plugs. Check each glow plug with an ohmmeter; plug not showing continuity is burned out. When installing, tighten glow plugs to a torque of 31-38 N·m (23-28 ft.-lbs.).

INJECTION PUMP

REMOVE AND REINSTALL. To remove injection pump, first thoroughly clean pump, lines, injectors and sur-

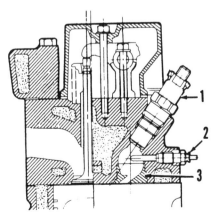

Fig. C1-15—Cross-sectional view of cylinder head showing fuel injector (1), glow plug (2) and precombustion chamber (3).

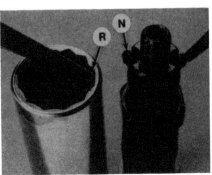

Fig. C1-16—When removing fuel injectors, cut a relief (R) inside socket to provided clearance between socket and leak-off nipples (N).

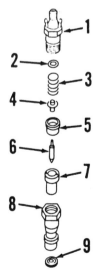

Fig. C1-16A—Exploded view of fuel injector.
1. Body
2. Shims
3. Spring
4. Spring seat
5. Spacer
6. Valve
7. Nozzle
8. Nozzle nut
9. Heat shield

Illustrations Courtesy Teledyne Total Power

Continental   SMALL DIESEL

rounding area. Disconnect throttle linkage and engine stop cable. Disconnect fuel supply line and return line and remove injector lines. Immediately cap or plug all openings to prevent entry of dirt. Remove pump adapter plate retaining nuts and remove injection pump, adapter plate and gear assembly. Remove cap screws retaining drive gear to pump and remove the gear and adapter plate.

The injection pump should be tested and overhauled by a shop qualified in diesel injection equipment repair.

Two different makes of fuel injection pumps, CAV and RoosaMaster, have been used. Refer to appropriate following paragraph for information on installing each type pump. After pump is installed and engine is running, verify timing with diesel engine timing light.

NOTE: If installing new injection pump, refer to INJECTION PUMP TIMING paragraph after installing pump. Set pump timing and restamp scribe marks on pump flange and adapter.

CAV-DPA INJECTION PUMP. To install pump, proceed as follows: Place adapter plate on pump with new gasket and loosely install retaining nuts. Install gear on pump shaft so that pin in gear hub engages slot (S—Fig. C1-17) in shaft. Tighten gear retaining cap screws to a torque of 27-34 N·m (20-25 ft.-lbs.). Turn engine so that number 1 piston is at TDC on compression stroke; the beveled tooth on camshaft injection pump drive gear should be centered in injection pump mounting opening. Remove bolt (5—Fig. C1-20) from front of timing cover and insert timing pin in hole. Install new "O" ring on injection pump adapter plate. Turn injection pump gear and shaft so that pump can be installed with timing pin entering hole in pump drive gear. Install pump and rotate adapter plate so that scribe marks on adapter plate and timing gear housing are aligned, then install and tighten adapter plate bolts. Turn pump so that scribe marks on pump flange and adapter plate are aligned, then tighten pump mounting bolts. Reconnect pump lines and linkage. Bleed system and start engine.

ROOSAMASTER INJECTION PUMP. To install pump, proceed as follows: Place adapter plate on pump with new gasket and loosely install retaining nuts. Install gear on pump shaft so that pin in gear hub engages slot (S—Fig. C1-17) in shaft and tighten gear retaining cap screws to a torque of 27-34 N·m (20-25 ft.-lbs.). Install new "O" ring on pump adapter plate. Turn engine so number 1 piston is at TDC on compression stroke; the beveled tooth on camshaft injection pump drive gear should be centered in injection pump mounting opening. Remove cover plate on side of pump and turn pump shaft so that internal timing marks (Fig. C1-8) are aligned. Install pump and adapter plate, being sure that pump internal timing marks stay aligned, and tighten adapter plate bolts. Turn pump to align scribed marks on pump flange and adapter plate and tighten pump mounting bolts. Install pump side cover, throttle linkage, stop control and fuel lines. Bleed fuel system and start engine.

OIL PAN

The oil pan is sealed by RTV form-in-place gasket. Apply a 2 mm (3/32 inch) bead of gasket material as shown in Fig. C1-18. Gasket surfaces of oil pan and crankcase must be clean and dry before applying gasket material. Tighten sheet steel oil pan bolts to a torque of 14-19 N·m (10-14 ft.-lbs.).

PISTON AND ROD UNITS

Piston and connecting rod are removed as a unit after removing cylinder head and oil pan. If not present, stamp cylinder numbers on camshaft side of rod and cap for reassembly. Remove ring ridge, if present, from top of cylinder. Remove rod cap and push piston and rod up out of top of block.

When assembling piston and rod, piston pin installation is easier if piston is heated prior to inserting piston pin. Piston pin retaining rings (3—Fig. C1-19) should be renewed after removal. Install piston and rod so that notch (N) or arrow on top of piston is toward front (timing gear) end of engine. Tighten connecting rod cap screws to a torque of 61-68 N·m (45-50 ft.-lbs.).

NOTE: Do not exceed the maximum torque figure of 68 N·m (50 ft.-lbs.).

PISTON, PIN AND RINGS

All pistons are equipped with two compression rings (1—Fig. C1-19) and one oil control ring (2). The piston pin (4) is fully floating in piston and rod and is retained by a snap ring (3) at each end of pin bore in piston.

Piston skirt to cylinder bore clearance is measured using a 13 mm (1/2 inch) wide, 0.08 mm (0.003 inch) thick feeler gage and a pull scale of about 7 kg (15 pounds) capacity. Pistons should be

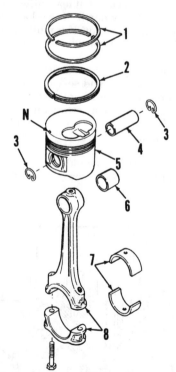

Fig. C1-19—Exploded view of piston and rod unit. Notch (N) or arrow on top of piston must be toward timing gear end of engine.

1. Compression rings
2. Oil control ring
3. Retaining rings
4. Piston pin
5. Piston
6. Piston pin bushing
7. Crankpin bearing inserts
8. Connecting rod

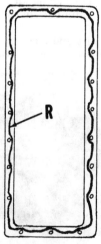

Fig. C1-18—Apply RTV form-in-place gasket in a 2 mm (3/32 inch) bead (R) to oil pan flange as shown.

Fig. C1-17—When installing fuel injection pump, slot (S) must engage pin in pump gear.

Illustrations Courtesy Teledyne Total Power

SERVICE MANUAL

Continental

fitted at room temperature of 20°-21° C (68°-70° F). Invert piston, keeping pin bore parallel to crankshaft, and insert piston with feeler gage 90 degrees from pin bore into cylinder until piston skirt is about 50 mm (2 inches) from top of cylinder. Slowly withdraw gage with pull scale and observe scale reading. Piston fit is correct when a pull of 2.3-4.5 kg (5-10 pounds) is required to withdraw feeler gage.

Piston ring end gap should be 0.40-0.65 mm (0.016-0.025 inch) for compression rings and 0.3-0.6 mm (0.012-0.024 inch) for oil control ring. Piston ring side clearance should be 0.050-0.082 mm (0.002-0.003 inch) for all rings.

NOTE: Some rings are taper faced. These rings will be marked "TOP" or "PIP" on side to be facing up when assembled on piston. Failure to install taper face rings correctly will result in high oil consumption.

Piston pin diameter is 28.571-28.575 mm (1.1249-1.250 inch) with wear limit of 28.562 mm (1.1245 inch). Pin bore in piston should be 28.578-28.583 mm (1.1251-1.1253 inch).

Pistons and rings are available in oversizes of 0.50. 0.75 and 1.00 mm (0.020, 0.030 and 0.040 inch) as well as standard size. Piston pins are available in 0.08 or 0.13 mm (0.003 or 0.005 inch) oversize as well as standard size. Pin bores in piston and rod may be honed to fit oversize pin as alternative to renewing piston and connecting rod pin bushing.

CYLINDERS

The pistons and rings ride directly in unsleeved cylinder bores. Standard cylinder inside diameter is 91.000-91.039 mm (3.5827-3.5842 inches). Cylinder should be rebored and honed to fit next oversize piston and ring set if cylinder bore wear is 0.20 mm (0.008 inch) or more. Check cylinder bore diameter for wear at a point about 6 mm (1/4 inch) below ring travel shoulder (ring ridge) at intervals of about 45 degrees. Difference between maximum measurement and standard or oversize diameter is amount of cylinder wear. If wear is less than 0.20 mm (0.008 inch), all traces of ring travel shoulder (ring ridge) should be removed and cylinder deglazed before re-ringing engine.

CONNECTING ROD AND BEARINGS

The connecting rod is fitted with a renewable piston pin bushing (6—Fig. C1-19) and a precision fit crankpin bearing insert (7). Cap is retained to rod by cap screws.

Crankpin bearing oil clearance should be 0.016-0.080 mm (0.0006-0.0031 inch) with wear limit of 0.091 mm (0.0036 inch). Standard crankpin diameter is 49.187-49.212 mm (1.9365-1.9375 inches) with wear limit of 49.162 mm (1.9355 inches). Rod side clearance on crankpin should be 0.15-0.28 mm (0.006-0.011 inch).

Connecting rod bearings are available in undersizes of 0.25, 0.50, 0.75 and 1.00 mm (0.010, 0.020, 0.030 and 0.040 inch) as well as standard size.

TIMING COVER

To remove timing gear cover, first remove alternator, belt, crankshaft pulley and the three front oil pan bolts. Then, unbolt and remove cover from front of engine. Remove crankshaft oil seal (3—Fig. C1-20) from timing cover.

To reassemble, be sure gasket surfaces are clean and free of old gasket material. Install new crankshaft oil seal with lip of seal to inside. Lubricate lip of seal with clean engine oil. Apply 2 mm (3/32 inch) bead of RTV gasket material to front end of oil pan. Stick new gasket to timing gear cover, then install cover. Tighten oil pan bolts to a torque of 14-19 N·m (10-14 ft.-lbs.) and tighten timing cover retaining bolts securely. Install crankshaft pulley and tighten nut to a torque of 163-176 N·m (120-130 ft.-lbs.). Install alternator and drive belt.

CAMSHAFT AND CAM FOLLOWERS

The camshaft rides directly in unbushed bores in cylinder block. Camshaft gear (2—Fig. C1-21) and injection pump drive gear (3) are retained on front end of shaft with a hex nut.

To remove camshaft, proceed as follows: Remove timing cover and cylinder head as previously outlined. If equipped with mechanical diaphragm type fuel feed pump, disconnect fuel lines from pump, then unbolt and remove pump from side of cylinder block. Remove cam followers (7) and keep them in order for possible reinstallation. Camshaft end play is controlled by retainer/thrust plate (4) and should be checked prior to removing camshaft. Specified end play is 0.038-0.178 mm (0.0015-0.007 inch). Unscrew camshaft nut, then use suitable pullers to remove camshaft gear and injection pump gear off front end of camshaft. Remove camshaft retainer plate and camshaft.

Camshaft front bearing journal diameter is 47.486-47.511 mm (1.8695-1.8705 inches), center journal diameter is 44.311-44.336 mm (1.7445-1.7455 inches) and rear journal diameter is 42.723-42.749 mm (1.682-1.683 inches). Journal

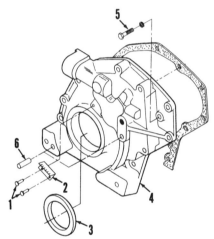

Fig. C1-20—Remove bolt (5) from timing cover to insert timing pin.

1. Rivets
2. Timing indicator
3. Crankshaft oil seal
4. Timing cover
5. Timing hole plug bolt
6. Dowel

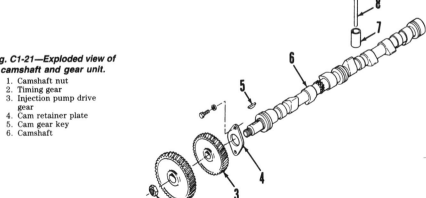

Fig. C1-21—Exploded view of camshaft and gear unit.

1. Camshaft nut
2. Timing gear
3. Injection pump drive gear
4. Cam retainer plate
5. Cam gear key
6. Camshaft

Continental

wear limit is 0.025 mm (0.001 inch) under new shaft diameter. Camshaft journal to block bore oil clearance should be 0.089-0.139 mm (0.0035-0.0055 inch). Camshaft end play should be 0.038-0.178 mm (0.0015-0.007 inch); renew thrust plate if scored, or if end play was excessive on disassembly.

Cam follower (tappet) outside diameter is 25.311-25.324 mm (0.9965-0.9970 inch) and block bore diameter is 25.344-25.364 mm (0.9978-0.9986 inch), with normal operating clearance of 0.020-0.053 mm (0.0008-0.0021 inch). Wear limit is 0.13 mm (0.005 inch).

To reinstall camshaft, reverse removal procedure while noting the following: If renewing camshaft, always install new cam followers (tappets). If reusing camshaft and cam followers, be sure to reinstall cam followers in same position from which they were removed. Lubricate camshaft and cam followers prior to installation. Install camshaft thrust plate and tighten retaining screws. Inspect gears and renew if worn or pitted. Install fuel injection pump drive gear (3) with beveled tooth to rear. Align timing marks on crankshaft and camshaft gears as shown in Fig. C1-22. Insert a suitable pry bar through fuel transfer pump opening in cylinder block to hold camshaft forward while bumping injection pump drive gear and camshaft gear onto front end of shaft. Use care to prevent camshaft from bumping expansion plug loose in rear of cylinder block.

NOTE: Do not attempt to push gears on with camshaft nut as this may break off threaded portion of cast iron camshaft.

Tighten camshaft nut to a torque of 88-95 N·m (65-70 ft.-lbs.). Check camshaft end play and correct if not within limits.

OIL PUMP

The oil pump is mounted on the front end of the cylinder block. Inner impeller (4—Fig. C1-23) of pump fits over crankshaft and is driven by a hardened key in crankshaft.

To remove oil pump, first remove timing gear cover, camshaft timing gear, crankshaft gear and oil pan. Unbolt and remove oil pickup, then unbolt and remove pump and relief valve assembly from front of block. Be careful not to lose the hardened oil pump drive key from crankshaft.

The oil pump and relief valve may be disassembled to check for wear, scoring or other damage. Check pump drive key and key slot in inner impeller for wear. If any part of the pump or relief valve is damaged, the complete assembly must be renewed.

Early pickup screen is sealed to pump inlet by adapter plate and "O" ring; late pickup screen is sealed to pump inlet with two "O" rings. Both the early and late units are serviced as complete assemblies, although the flat screen of early models may be removed for cleaning.

Reverse removal procedure to reinstall pump. Be sure hardened key is inserted in crankshaft and that pump is seated squarely in counterbore in front of block. Tighten retaining cap screws to a torque of 20-24 N·m (15-18 ft.-lbs.).

FLYWHEEL AND FLYWHEEL HOUSING

The flywheel housing is secured to the cylinder block by special cap screws. The cap screws used in the upper holes have sealing bands; oil leakage will result if these cap screws are not used. Flywheel runout should not exceed 0.20 mm (0.008 inch). Tighten flywheel housing cap screws to a torque of 61-68 N·m (45-50 ft.-lbs.) and flywheel bolts to a torque of 68-75 N·m (50-55 ft.-lbs.).

REAR CRANKSHAFT SEAL

The rear crankshaft oil seal may be renewed without removing crankshaft. To remove seal, remove flywheel, flywheel housing and oil pan. Remove rear main bearing cap with puller and remove seal halves from cap and block. Take care not to damage rear main bearing insert or the cap aligning dowel pins.

To install seal, first check for and remove any burrs at edge of surface (C—Fig. C1-24) on both block and cap. If crankshaft has been removed, apply small amount of sealing compound to seal contacting edge (E) of block and cap. Apply graphite grease to oil seal lip. If crankshaft was not removed, use light coat of grease instead of sealing compound in seal groove (E) to assist in sliding seal into place with lip to inside of engine. Install lower seal half in cap with lip to inside of engine. Apply small amount of RTV gasket material to block contact surface (B) of bearing cap, being sure to seal portion of cap between edge of seal and outer edge of cap. Main bearing caps of some engines may have a groove running from edge of seal to outside edge of cap; fill groove of these

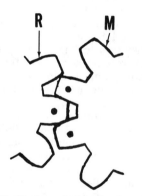

Fig. C1-22—Single marked tooth of crankshaft gear (R) must mesh between two marked teeth of camshaft gear (M).

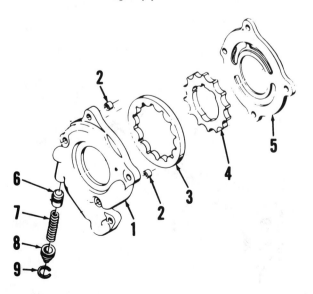

Fig. C1-23—Exploded view of typical oil pump and relief valve assembly. Some models may have threaded plug instead of plug (8) and snap ring (9).

1. Pump body
2. Hollow dowel pins
3. Outer rotor
4. Inner rotor
5. Cover
6. Pressure relief valve
7. Spring
8. Plug
9. Snap ring

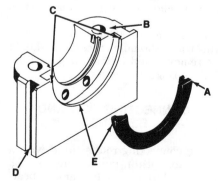

Fig. C1-24—Refer to text for installation of crankshaft seal (A) and rear main bearing cap (B).

SERVICE MANUAL

Continental

caps with RTV sealing compound. Apply light coat of cement (National Oil Seal or EC-847) to butting ends of oil seal halves and allow cement to become tacky before installing cap. Install cap and tighten retaining cap screws to a torque of 150-162 N·m (110-120 ft.-lbs.). With cap installed, force RTV into side seal slots (see Fig. C1-25) until it flows out corner chamfers. Dip curing inserts (pipe cleaners) in water and insert in RTV compound in side seal slots until about 16 mm (5/8 inch) of inserts protrude from slot. Cut off ends of inserts flush with pan rail. Fill dowel holes in main bearing cap with RTV.

Complete reassembly by reversing oil pan, flywheel housing and flywheel removal procedure.

CRANKSHAFT AND BEARINGS

To remove crankshaft, remove oil pump, piston and rod units, flywheel and flywheel housing. The crankshaft can then be removed after removing main bearing caps.

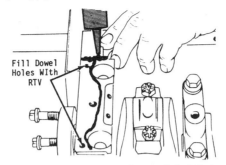

Fig. C1-25—After main bearing cap is installed, fill side slots (D—Fig. C1-24) and dowel holes with RTV compound.

Standard main bearing journal diameter is 72.944-72.974 mm (2.8718-2.8730 inches) with wear limit of 72.918 mm (2.8708 inches). Crankshaft journal to main bearing oil clearance should be 0.058-0.133 mm (0.0023-0.0052 inch) with desired clearance being 0.096 mm (0.0038 inch). Crankshaft end play, controlled by flange on center main bearing, should be 0.04-0.17 mm (0.0015-0.0067 inch).

Refer to REAR CRANKSHAFT SEAL paragraph for proper installation of rear main bearing cap. Tighten main bearing cap screws to a torque of 150-162 N·m (110-120 ft.-lbs.)

Illustrations Courtesy Teledyne Total Power

DEUTZ

7585 Ponce de Leon Circle
Atlanta, Georgia 30340

Model	No. Cyls.	Bore	Stroke	Displ.
F1L 208 D	1	80 mm (3.150 in.)	82 mm (3.228 in.)	413 cc (25.2 cu. in.)
F1L 210 D	1	95 mm (3.740 in.)	95 mm (3.740 in.)	673 cc (41.0 cu. in.)

These engines are air-cooled, four-stroke, direct injection diesels. Crankshaft rotation is counterclockwise when facing flywheel.

MAINTENANCE

LUBRICATION

Crankcase oil must meet or exceed API service classification CC or CD. The oil viscosity grade is determined by the ambient temperature expected between oil changes. For operation in temperatures below −5° (23°F), SAE 10W oil is recommended. For temperatures ranging between −10°C (14°F) to 10°C (50°F), SAE 20W-20 should be used. When temperatures are between 5°C (40°F) to 30°C (85°F), SAE 30 oil is recommended. For temperature above 25°C (77°F), SAE 40 oil should be used. Multiviscosity oil, such as 20W-40, is approved for all temperature operation, however, the oil must meet MIL-L46152 specification.

Manufacturer recommends governing oil change intervals on oil grade used (CC or CD), operating conditions (normal or severe) and sulfur content of fuel (above or below 0.5%). On all new or overhauled engines, first oil change should be after the first 20-30 hours of operation for severe duty applications or after the first 40-60 hours of operation for normal duty applications. Then, the second oil change is performed at the normal interval for all engines as follows: For normal duty operation and using fuel with less than 0.5% sulfur content, oil change interval is every 200 hours for "CC" classification oil or every 300 hours for "CD" oil. If fuel sulfur content exceeds 0.5% interval is reduced to every 100 hours for "CC" oil or every 200 hours for "CD" oil. For engines used in severe duty applications and using fuel with less than 0.5% sulfur content, oil change interval is every 100 hours of operation when using "CC" oil or every 200 hours if using "CD" oil. If fuel sulfur content exceeds 0.5%, use of "CC" oil is not recommended for severe duty applications. Oil change interval using "CD" oil is reduced to 100 hours of operation. In all applications regardless of hours of operation, oil should be changed at least once every 12 months.

Engine oil should be drained while engine is warm. Crankcase capacity is approximately 2.1 liters (2.2 quarts) for F1L 208 D and 2.4 liters (2.5 quarts) for F1L 210 D. Engine filter should be renewed at the same time oil is changed.

AIR CLEANER

The oil bath type air cleaner should be service after every 10 to 60 hours of operation depending on dust conditions. Engine should be stopped at least one hour before removing filter bowl to allow oil to drain from filter element into bowl. Clean bowl and refill with clean oil to the oil level mark. If filter element is heavily contaminated, it should be removed and cleaned in diesel fuel. Allow diesel fuel to drain out of filter before reinstalling.

FUEL SYSTEM

FUEL FILTER AND BLEEDING. On engines equipped with a fuel tank mounted on the engine, a renewable filter element is mounted internally in bottom of tank. It is recommended that filter element be renewed after every 200 hours of operation. Fuel should be drained from tank before removing filter.

On engines using a remote mounted fuel tank, an externally mounted fuel filter is used. Manufacturer recommends renewing filter after every 1200 hours of operation or sooner if loss of power is evident.

Whenever fuel filter is serviced or fuel lines are disconnected, air must be bled from fuel system in the following manner. If equipped with external type filter, loosen bleed screw (1 – Fig. D1-1) on filter mounting bracket. When air-free fuel flows from bleeder, retighten screw. On all engines, loosen bleed screw (2 – Fig. D1-2) on injection pump. When air-free fuel flows from bleeder, retighten screw. If engine fails to start at this point, loosen high pressure fuel line at injector. With pump controls in "run" position, crank engine until fuel is discharged at loosened connection. Retighten fuel line fitting and start engine.

ENGINE SPEED ADJUSTMENT. The engine must be at normal operating

Fig. D1-1—View of air bleed screw (1) on filter housing (if so equipped).

Fig. D1-2—View of injection pump air bleed screw (2).

SERVICE MANUAL

Deutz

Fig. D1-3—Idle speed is adjusted using screw (1) and maximum speed is adjusted using screw (2). Be sure governor rod (3) moves freely.

temperature and an accurate tachometer used when adjusting engine speed setting. The engine speed settings depend on the engine application. Recommended engine speed should be stamped on the engine nameplate.

Low idle speed is normally 800-1000 rpm. Idle speed is adjusted using adjusting screw (1–Fig. D1-3). Note that some engines are not equipped with a low idle adjusting screw.

Maximum speed is adjusted using adjusting screw (2). Maximum rated speed must not exceed 3600 rpm on F1L 208 D engine or 3000 rpm on F1L 210 D engine. Be sure to install a new seal wire on maximum speed screw to secure adjustment.

INJECTION PUMP TIMING. To check injection pump static timing, first turn crankshaft in normal direction until piston is on compression stroke. Disconnect fuel inlet pipe and injector high pressure pipe from injection pump. Connect a nozzle tester pump to fuel inlet of injection pump and attach a drip tube to high pressure outlet of pump. Move speed control lever to full-speed position. Do not actuate excess fuel starting button. Actuate tester pump and note that fuel should flow from drip tube. Slowly rotate crankshaft until fuel flow from injection pump outlet just stops. AT this point beginning of injection occurs, and timing marks on crankshaft pulley and front cover should be aligned as shown in Fig. D1-4.

Injection timing is adjusted by changing thickness of pump mounting shims. Reducing shim thickness will advance timing and increasing shim thickness will retard timing.

REPAIRS

TIGHTENING TORQUES

Refer to the following table for special tightening torques. Metric fasteners are used throughout the engine.

Connecting rod:
  F1L 208 D .... 35 N·m
                    (26 ft.-lbs.)
  F1L 210 D .... 45 N·m
                      (33 ft.-lbs.)

Cylinder head:
  F1L 208 D .... 40 N·m
                      (30 ft.-lbs.)
  F1L 210 D .... 55 N·m
                      (40 ft.-lbs.)

Crankshaft pulley .... 40 N·m
                      (30 ft.-lbs.)

Flywheel:
  F1L 208 D .... 160 N·m
                      (118 ft.-lbs.)
  F1L 210 D .... 180 N·m
                      (133 ft.-lbs.)

Front cover .... 23 N·m
                      (17 ft.-lbs.)

Injection pump .... 23 N·m
                      (17 ft.-lbs.)

Injector .... 35 N·m
                      (26 ft.-lbs.)

Main bearing flange .... 35 N·m
                      (26 ft.-lbs.)

Oil sump .... 23 N·m
                      (17 ft.-lbs.)

Rocker arm bracket .... 30 N·m
                      (22 ft.-lbs.)

Rocker cover .... 23 N·m
                      (17 ft.-lbs.)

COMPRESSION PRESSURE

Compression pressure may be checked to establish relative condition of engine before proceeding with engine disassembly. Engine should be run briefly to ensure there is a normal fill of oil on rings and cylinder before checking pressure. Remove injector and install Deutz special adapter 100 080 with pressure gage. Cranking speed must be at least 150 rpm. Compression pressure should be 1900-2100 kPa (275-305 psi).

VALVE ADJUSTMENT

Valve clearance should be adjusted with engine cold. Recommended clearance is 0.15 mm (0.006 inch) for intake and 0.20 mm (0.008 inch) for exhaust.

To adjust clearance, remove rocker cover and rotate crankshaft until piston is at TDC on compression stroke. Loosen rocker arm adjusting screw locknut. Turn adjusting screw as required until appropriate size feeler gage can be inserted between valve and rocker arm with a slight drag. Tighten locknut and recheck clearance.

On hand-start engines, a decompression device is fitted to cylinder head to partially open the intake valve during engine starting. With intake valve closed and decompression device in disengaged position, measure clearance between stop pin of rocker arm and decompressor cam with a feeler gage. Clearance should be 0.7 mm (0.027 inch). If necessary, adjust stop pin to obtain desired clearance. To check decompressor operation, measure distance from rocker arm to top of rocker arm housing with decompressor engaged and disengaged. The difference between the two readings is distance valve is being opened which should be 0.4-0.6 mm (0.016-0.023 inch).

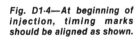

Fig. D1-4—At beginning of injection, timing marks should be aligned as shown.

Deutz

SMALL DIESEL

CYLINDER HEAD AND VALVE SYSTEM

To remove cylinder head, first remove fuel tank (if equipped), air cleaner and exhaust pipe. Remove injector pipe and injector. Remove rocker cover, rocker arm bracket and push rods. Remove stud nuts and lift off cylinder head. Remove valves from cylinder head and inspect all parts for wear or other damage. Cylinder head is equipped with renewable valve seat inserts and valve guides. Refer to Fig. D1-5.

Valve face and seat angles are 45° for intake and exhaust. Renew valve if head margin is less than 0.7 mm (0.027 inch). Refer to the following table for valve and valve guide dimensions.

F1L 208 D

Intake:
  Valve stem OD......6.953-6.975 mm
    (0.2737-0.2746 inch)
  Valve guide ID.......7.00-7.015 mm
    (0.2756-0.2762 inch)
  Clearance-desired...0.025-0.062 mm
    (0.0010-0.0024 inch)
  Wear limit..............0.12 mm
    (0.005 inch)

Exhaust:
  Valve stem OD......6.938-6.960 mm
    (0.2732-0.2740 inch)
  Valve guide ID.......7.00-7.015 mm
    (0.2756-0.2762 inch)
  Clearance-desired...0.040-0.077 mm
    (0.0016-0.0030 inch)
  Wear limit..............0.15 mm
    (0.006 inch)

F1L 210 D

Intake:
  Valve stem OD......7.953-7.975 mm
    (0.3131-0.3140 inch)
  Valve guide ID.......8.00-8.015 mm
    (0.3150-0.3155 inch)
  Clearance-desired...0.025-0.062 mm
    (0.0010-0.0024 inch)
  Wear limit..............0.12 mm
    (0.005 inch)

Exhaust:
  Valve stem OD......7.938-7.960 mm
    (0.3125-0.3134 inch)
  Valve guide ID.......8.00-8.015 mm
    (0.3150-0.3155 inch)
  Clearance-desired...0.040-0.077 mm
    (0.0016-0.0030 inch)
  Wear limit..............0.15 mm
    (0.006 inch)

Valve guides should have an interference fit in cylinder head bores of 0.027-0.055 mm (0.001-0.002 inch). Guides with 0.02 mm (0.0008 inch) oversize outside diameter are available for F1L 210 D engines. Use suitable tools to remove and install valve guides. Cylinder head should first be heated to 240°-260°C (465°-500°F). Press new guides in until snap ring contacts cylinder head surface. Be sure guides are installed with chamfered end facing outward. After installation, guide bore must be reamed to provide recommended clearance for valve stem.

NOTE: When renewing valve guides and valve seat inserts, both operations should be performed together so cylinder head is heated only once.

When renewing valve seat inserts, suitable tools must be used to avoid damage to cylinder head. Grind valve seats to obtain recommended valve seating width of 0.8-1.0 mm (0.031-0.039 inch). Maximum allowable seat width is 1.5 mm (0.059 inch).

Install valves and measure distance top of valve head is recessed below surface of cylinder head. Distance should be 0.8-1.0 mm (0.031-0.039 inch) with a maximum limit of 2.0 mm (0.079 inch). If recession is excessive, renew valve seat. If recession is less than specified, grind valve seat as necessary.

Fig. D1-6—Tighten cylinder head stud nuts in steps following sequence shown.

Check valve springs for distortion, sign of overheating and other damage. Spring free length should be 44.1-44.5 mm (1.736 inches) for F1L 208 D engine and minimum allowable length is 42 mm (1.653 inches). On F1L 210 D engine, spring free length should be 51.3-51.7 mm (2.020-2.035 inches) and minimum allowable length is 49 mm (1.929 inches).

To reinstall cylinder head, reverse the removal procedure. Be sure push rod cover tubes are properly seated. Tighten cylinder head stud nuts in steps following sequence shown in Fig. D1-6. Final tightening torque is 40 N·m (30 ft.-lbs.) for F1L 208 D or 55 N·m (40 ft.-lbs.) for F1L 210 D.

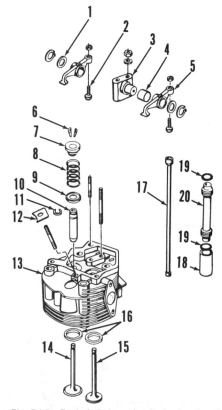

Fig. D1-5—Exploded view of cylinder head and valve system

1. Washer
2. Adjusting screw
3. Rocker arm bracket
4. Bushing
5. Rocker arm
6. Valve keepers
7. Spring retainer
8. Valve spring
9. Spring seat
10. Valve guide
11. Snap ring
12. Injector retainer plate
13. Cylinder head
14. Intake valve
15. Exhaust valve
16. Valve seat inserts
17. Push rod
18. Cam follower
19. Gasket
20. Push rod tube

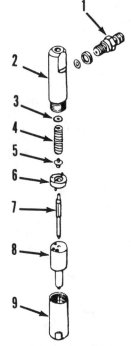

Fig. D1-7—Exploded view of injector assembly.

1. High pressure inlet
2. Injector body
3. Shim
4. Spring
5. Spring seat
6. Spacer
7. Nozzle needle
8. Nozzle body
9. Nozzle nut

SERVICE MANUAL

Deutz

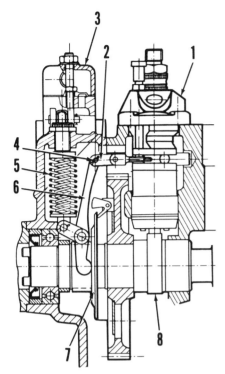

Fig. D1-8—Cross-sectional view of injection pump and governor assembly.

1. Injection pump
2. Fork
3. Governor cover
4. Pin
5. Governor springs
6. Governor lever
7. Governor plate
8. Camshaft

INJECTOR

R&R AND OVERHAUL. Before removing injector, thoroughly clean injector and surrounding area. Immediately plug all openings as fuel lines are disconnected. Be sure to remove seal washer from cylinder head bore after injector is removed.

Be sure to renew copper seal washer when reinstalling injector. Tighten injector clamp nut to 35 N·m (26 ft.-lbs.).

To disassemble injector, secure body (2–Fig. D1-7) in a vise. Remove nozzle nut (9), then withdraw nozzle valve, spacer, spring and shims.

Clean exterior surfaces with a brass wire brush. It may be necessary to soak parts in an approved carbon solvent to loosen hard carbon deposits. Rinse parts in clean diesel fuel after cleaning with solvent. Be sure nozzle needle (7) moves smoothly in nozzle body (8). When needle is pulled about halfway out of body and then released, it should slide down to its seat from its own weight. If needle sticks, reclean or renew nozzle valve assembly.

When reassembling injector, all parts should be wet with diesel fuel. Check injector operation as outlined in TESTING paragraph.

TESTING. A complete job of testing and adjusting injector requires use of special test equipment. Injector should be checked for opening pressure, seat leakage and spray pattern.

WARNING: Fuel emerges from injector with sufficient force to penetrate the skin. When testing injector, keep yourself clear of nozzle spray.

Connect injector to test pump, then operate tester lever several quick strokes to purge air from the injector and to make sure nozzle valve is not stuck. Then operate tester lever slowly while observing tester gage. Nozzle opening pressure should be 19000-22000 kPa (2755-3190 psi). Opening pressure is adjusted by increasing or decreasing thickness of shims (3–Fig. D1-7).

To check nozzle for leakage, operate tester lever slowly to maintain pressure at about 2750 kPa (400 psi) below opening pressure and observe nozzle tip for leakage. If a drop forms on nozzle tip within a 10 second period, nozzle valve is not seating and must be overhauled or renewed. A slight wetness at tip is allowable if a drop does not form.

Operate tester lever several quick strokes (about one stroke per second) and check spray pattern. Spray should be uniform, well atomized and form a cone shape. Nozzle should produce a buzzing sound when operating properly.

INJECTION PUMP

A CIPA injection pump is used on all engines. It is recommended that injection pump be tested and serviced only by a shop qualified in diesel fuel injection repair.

To remove injection pump, remove fuel lines from pump and immediately plug all openings. Remove governor housing cover and lift out governor inner and outer springs (5–Fig. D1-8). Disconnect governor arm (6) from fork (2), then screw fork out and move pump control rack to center position. Remove pump retaining nuts and lift out pump. Retain pump mounting shims for use in reassembly.

To reinstall injection pump, proceed as follows: If original pump shim pack is not available or if a new pump is being installed, place mounting gasket on crankcase and measure distance from surface of gasket to base circle of pump cam. Compare measurement with dimension scribed on pump body and assemble shim pack to obtain the required dimension. The distance with gasket and shims in place cannot be smaller than 82.6 mm (3.252 inches). Install pump and tighten nuts to 23 N·m (17 ft.-lbs.). Turn in fork (2–Fig. D1-9) and connect governor arm (6). Press governor plate (7) forward and check governor lever play (P) at pin (4). Adjust fork in or out to provide 0.2-0.4 mm (0.008-0.016 inch) free play at pin. Reinstall springs and governor cover. Bleed air from fuel system and check injection pump static timing. Check engine speed settings and adjust if necessary.

TIMING GEARS AND FRONT COVER

To remove engine front cover, remove governor cover (3–Fig. D1-8) and lift out governor springs (5). Disconnect governor lever (6) from fork (2). Remove governor pivot shaft and lift out governor lever. Remove front pulley. Remove camshaft cover and balance weight cover with shims. Remove front cover mounting screws, then use a suitable puller to withdraw cover with bearings. Heat cover before tapping out bearings.

Inspect gears for excessive wear and other damage. To remove balance weight and bearing, use a suitable slide hammer puller to remove balancer assembly (2–Fig. D1-10) from crankcase. Remove crankshaft gear retaining ring, then use a suitable puller to pull gear (3) off crankshaft. Refer to

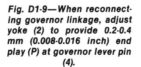

Fig. D1-9—When reconnecting governor linkage, adjust yoke (2) to provide 0.2-0.4 mm (0.008-0.016 inch) end play (P) at governor lever pin (4).

Deutz   SMALL DIESEL

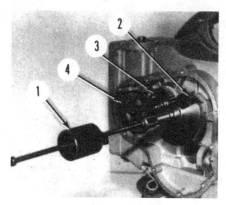

Fig. D1-10—Use a slide hammer puller to remove balance weight assembly.
1. Slide hammer puller
2. Balance weight assy.
3. Crankshaft gear
4. Oil pump gear

Fig. D1-12—To adjust backlash between oil pump drive gear (1) and crankshaft gear (2), loosen pump mounting screws and rotate pump in crankcase. Note position of small head pump mounting screw (S).

CAMSHAFT section for removal of camshaft gear. Remove oil pump gear (4) retaining nut and withdraw gear.

To reinstall gears, proceed as follows: Heat crankshaft gear to about 150°C (300°F) before reinstalling. Heat bearings before installing onto balance weight shaft and camshaft. Make certain timing marks (Fig. D1-11) on crankshaft gear (1), balance weight gear (2) and camshaft gear (3) are aligned as shown. Install oil pump drive gear (1–Fig. D1-12) and check for specified backlash of 0.10-0.15 mm (0.004-0.006 inch) between crankshaft gear and oil pump gear. To adjust backlash, loosen pump mounting cap screws and rotate pump housing in crankcase as necessary. Tighten pump mounting screws to 10 N·m (8 ft.-lbs.). Tighten oil pump drive gear nut to 23 N·m (17 ft.-lbs.).

Heat front cover slightly before reinstalling. Tighten mounting screws to 23 N·m (17 ft.-lbs.). Check and adjust balance weight and camshaft end play as follows: Measure distance (A–Fig. D1-13) from surface of front cover to race of ball bearing. Measure distance (B–Fig. D1-15) of bearing retainer with gasket in place as shown. Distance "A" must be 0.15-0.20 mm (0.006-0.008 inch) larger than distance "B". If the distance is too large, install shims between bearing retainer and bearing. If the distance is too small, install thicker gasket between bearing retainer and front cover. Tighten bearing retainer cap screws to 23 N·m (17 ft.-lbs.). Camshaft end play is determined in the same manner by measuring from surface of front cover to race of bearing (A–Fig. D1-14) and subtracting bearing retainer dimension (B–Fig. D1-15). The difference must be within range of 0.15-0.20 mm (0.006-0.008 inch). Adjustment procedure is the same as that for balance weight shaft. Tighten camshaft cover mounting screws to 23 N·m (17 ft.-lbs.). Complete installation by reversing the removal procedure.

CAMSHAFT

To remove camshaft, first remove cylinder head, push rods, cam follower retainer plate and cam followers. Remove injection pump and engine front cover as previously outlined. Remove camshaft assembly from crankcase.

Inspect camshaft (9–Fig. D1-16) and bearings for excessive wear. Cam height for F1L 208 D engine should be 6.85-6.95 mm (0.2697-0.2736 inch) for intake and exhaust. Cam height for F1L 210 D engine should be 6.55-6.65 mm (0.2579-0.2618 inch). Check governor weights (6), plate (5) and disc (4) for wear and damage and renew if necessary.

To reinstall camshaft, reverse the removal procedure. Be sure to align timing gear marks as shown in Fig. D1-11. Check and adjust camshaft end play as outlined in TIMING GEARS AND FRONT COVER section.

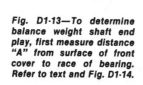

Fig. D1-13—To determine balance weight shaft end play, first measure distance "A" from surface of front cover to race of bearing. Refer to text and Fig. D1-14.

Fig. D1-11—Make certain timing marks are aligned as shown when reassembling engine.
1. Crankshaft gear
2. Balance weight gear
3. Camshaft gear

Fig. D1-14—To determine camshaft end play, measure distance "A" from surface of front cover to race of bearing. Refer to text.

Deutz

SMALL DIESEL

Fig. D1-15—Measure bearing retainer flange distance "B" with gasket in place. Refer to text.

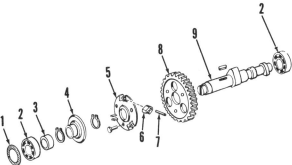

Fig. D1-16—Exploded view of camshaft assembly.
1. Shim
2. Bearings
3. Spacer
4. Governor disc
5. Flyweight plate
6. Flyweight
7. Pivot pin
8. Camshaft gear
9. Camshaft

Ring Side Clearance

F1L 208 D:
  Top ring.............0.09-0.11 mm
    (0.0035-0.0043 inch)
  Maximum...............0.15 mm
    (0.006 inch)
  Second ring..........0.05-0.07 mm
    (0.0020-0.0027 inch)
  Maximum...............0.15 mm
    (0.006 inch)
  Slotted oil ring.......0.03-0.05 mm
    (0.0012-0.0020 inch)
  Maximum...............0.15 mm
    (0.006 inch)

F1L 210 D:
  Top ring.............0.11-0.13 mm
    (0.0044-0.0051 inch)
  Maximum...............0.20 mm
    (0.008 inch)
  Second ring..........0.07-0.09 mm
    (0.0028-0.0035 inch)
  Maximum...............0.15 mm
    (0.006 inch)
  Slotted oil ring.......0.03-0.05 mm
    (0.0012-0.0020 inch)
  Maximum...............0.15 mm
    (0.006 inch)

Piston ring end gap should be checked using a feeler gage with ring inserted squarely into cylinder bore. On F1L 208 D engine, ring end gap should be 0.25-0.50 mm (0.010-0.020 inch) and maximum allowable gap is 1.0 mm (0.040 inch) for all rings. On F1L 210 D engine, ring end gap should be 0.25-0.55 mm (0.010-0.021 inch) and maximum allowable gap is 1.0 mm (0.040 inch) for all rings.

The piston pin is a transition fit in piston bore. Heating piston to about 80°C (175°F) will make removal and installation easier. Piston pin outer diameter is 27.994-28.000 mm (1.1021-1.1023 inches) and piston bore diameter is 28.000-28.006 mm (1.1023-1.1026 inches).

When reassembling, note that combustion cavity is offset in piston crown.

PISTON, PISTON RING, RINGS AND CYLINDER

To disassemble, remove cylinder head and air shroud. Withdraw cylinder from piston and crankcase. Retain shims used between bottom of cylinder and crankcase for use in reassembly. Remove piston pin retaining rings. Press out piston pin and remove piston from connecting rod.

Cylinder standard bore diameter is 80.00-80.015 mm (3.1496-3.1502 inches) for F1L 208 D and 95.00-95.015 mm (3.7402-3.7407 inches) for F1L 210 D. Cylinder should be renewed if wear exceeds 0.15 mm (0.006 inch). Piston skirt standard diameter should be 79.805-79.825 mm (3.1419-3.1427 inches) for F1L 208 D and 94.780-94.800 mm (3.7315-3.7322 inches) for F1L 210 D. Pistons and cylinders are available in oversizes of 0.50 mm (0.020 inch) and 1.0 mm (0.040 inch).

With rings installed on piston, measure side clearance between rings and ring grooves using a feeler gage. Compare measurements with the following dimensions and renew if necessary.

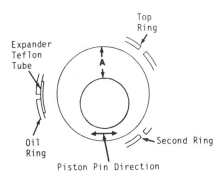

Fig. D1-17—Stagger ring end gaps around piston as shown. Note that combustion cavity in piston crown is offset and wider part of crown (A) should be towards injection pump side of engine.

Fig. D1-18—To check piston crown clearance, secure cylinder using spacers (1) and stud nuts. Measure distance from top of cylinder to piston crown using a depth gage (2).

Deutz

SMALL DIESEL

Fig. D1-19—Make certain identification numbers (1) on rod and cap match and are on the same side.

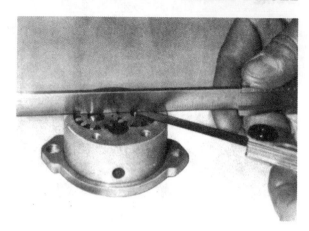

Fig. D1-20—Use a straightedge and feeler gage to measure oil pump end clearance.

Install piston so wider part of piston crown (A – Fig. D1-17) is towards injection pump side of engine. If connecting rod is removed, make certain side of rod and cap stamped with identification numbers (Fig. D1-19) is opposite from wider side of piston crown. Install rings onto piston making sure second compression ring is installed with side marked "TOP" facing upward and chrome plated ring is installed in top ring groove. Stagger ring end gaps around piston as shown in Fig. D1-17. Position cylinder so side with flattened fins is towards camshaft side of crankcase.

If piston or cylinder was renewed, piston crown clearance should be checked and adjusted as follows: Assemble three shims onto bottom of cylinder, then install cylinder onto piston. Using suitable spacer tubes (1 – Fig. D1-18) on cylinder studs, tighten stud nuts securely to clamp cylinder against crankcase. Rotate crankshaft until piston is at top dead center, then measure distance between piston crown and top of cylinder. Recommmended distance is 1.0-1.2 mm (0.040-0.047 inch). If measured distance is too large, lift cylinder slightly and cut one of the shims with side cutters. Remove the cut shim and recheck piston crown dimension. When specified dimension is obtained, reinstall cylinder head.

CONNECTING ROD

The connecting rod is equipped with a renewable, precision insert type bearing in the big end. Bearings are available in undersizes of 0.25, 0.50, 0.75 and 1.00 mm (0.010, 0.020, 0.030 and 0.040 inch) as well as standard size. Bearing standard bore diameter is 40.025-40.052 mm (1.5758-1.5768 inches) for F1L 208 D and 51.955-51.994 mm (2.0455-2.0470 inches) for F1L 210 D. Recommended clearance between bearing and crankpin is 0.025-0.068 mm (0.0010-0.0026 inch) for F1L 208 D and 0.010-0.069 mm (0.0004-0.0027 inch) for F1L 210 D. Maximum allowable operating clearance for both engines is 0.12 mm (0.0047 inch).

Inside diameter of small end bore is 28.005-28.015 mm (1.1026-1.1029 inches) for F1L 208 D and outside diameter of piston pin is 27.994-28.000 mm (1.1021-1.1023 inches). Desired clearance of pin in rod bore is 0.005-0.021 mm (0.0002-0.0008 inch) with a wear limit of 0.05 mm (0.0020 inch). Inside diameter of small end bore for F1L 210 D engine is 30.005-30.015 mm (1.1813-1.1817 inches) and outside diameter of piston pin is 29.994-30.000 mm (1.1809-1.1811 inches). Desired clearance of piston pin in rod bore is 0.005-0.021 mm (0.0002-0.0008 inch) with a wear limit of 0.05 mm (0.0020 inch).

Connecting rod side clearance on crankshaft should be 0.20-0.40 mm (0.008-0.016 inch) for all engines. Renew connecting rod if side clearance exceeds 0.60 mm (0.023 inch).

When reinstalling connecting rod, be sure identification numbers (Fig. D1-19) on big end of rod and on cap are on the same side and face away from injection pump side of engine.

NOTE: If connecting rod bearing is renewed, manufacturer recommends renewing connecting rod cap screws.

Tighten connecting rod cap screws to 35 N·m (26 ft.-lbs.) on F1L 208 D engine or 45 N·m (33 ft.-lbs.) on F1L 210 D engine.

CRANKSHAFT AND MAIN BEARINGS

To remove crankshaft, remove cylinder head, oil sump, connecting rod and piston and engine front cover. Remove crankshaft gear and flywheel using suitable pullers. Use two jackscrews in threaded holes of crankshaft rear bearing retainer to pull retainer housing from crankcase. Withdraw crankshaft from crankcase.

Crankpin standard diameter is 39.984-40.000 mm (1.5742-1.5748 inches) for F1L 208 D engine or 51.925-51.945 mm (2.0443-2.0450 inches) for F1L 210 D engine. Journal out-of-round limit is 0.05 mm (0.002 inch). Recommended clearance between connecting rod bearing and crankpin is 0.025-0.068 mm (0.0010-0.0026 inch) for F1L 208 D engine or 0.010-0.069 mm (0.0004-0.0027 inch) for F1L 210 D engine. Maximum allowable clearance for both engines is 0.12 mm (0.0047 inch). Undersize bearings are available.

Main journal standard diameter is 46.995-47.011 mm (1.8502-1.8508 inches) for F1L 208 D engine or 54.993-55.006 mm (2.1651-2.1656 inches) for F1L 210 D engine. Journal out-of-round limit is 0.05 mm (0.002 inch). Standard inside diameter of main bearings is 47.030-47.039 mm (1.8516-1.8519 inches) for F1L 208 D engine or 55.030-55.079 mm (2.1665-2.1684 inches) for F1L 210 D engine. Recommended clearance between main journals and bearings is 0.019-0.084 mm (0.0008-0.0033 inch) for F1L 208 D engine or 0.024-0.086 mm (0.0009-0.0034 inch) for F1L 210 D engine. Maximum allowable main bearing clearance is 0.12 mm (0.0047 inch) for both engines. Undersize bearings are available.

When renewing main bearings, crankcase and bearing retainer should be heated slightly prior to installing new bearings. Press front bearing in until flush with outer surface of crankcase bore and install rear bearing flush with inner surface of retainer housing.

Crankshaft end thrust is taken by thrust washers mounted in crankcase at front and in bearing retainer housing at rear. Recommended end play is 0.15-0.20 mm (0.006-0.008 inch) for both engines. Maximum allowable end play is 0.40 mm (0.016 inch). Standard thickness of thrust washers is 1.95-2.00 mm (0.077-0.079 inch). If end play is excessive, insert a shim behind thrust washers. If end play is less than recom-

SERVICE MANUAL

Fig. D1-21—Measure clearance between oil pump body and gears with a feeler gage. Refer to text.

mended, assemble thicker mounting gasket for rear bearing retainer.

When reinstalling crankshaft, lubricate bearings and journals with oil prior to assembly. Be sure side of thrust washers with oil grooves is towards crankshaft. Tighten bearing retainer nuts to 35 N·m (26 ft.-lbs.). Be sure timing gear marks are aligned as shown in Fig. D1-11. Complete installation by reversing the removal procedure.

OIL PUMP

To remove oil pump, drain the engine oil and remove oil pan and engine front cover. Remove oil pickup tube. Remove pump mounting screws and withdraw pump assembly from crankcase.

Use a straightedge to check pump rear cover for distortion or wear. Renew cover and pickup assembly if warped or excessively worn. Use a straightedge and feeler gage to measure gear end clearance (Fig. D1-20). Specified end clearance is 0.07-0.15 mm (0.003-0.006 inch). Renew components as required if end clearance exceeds 0.15 mm (0.006 inch). Use a feeler gage to measure clearance between gears and housing as shown in Fig. D1-21. Specified clearance is 0.03-0.10 mm (0.001-0.004 inch) and maximum allowable clearance is 0.15 mm (0.006 inch). Backlash between pump gears should be 0.06-0.12 mm (0.0023-0.0047 inch) with a wear limit of 0.20 mm (0.008 inch).

To reinstall pump, reverse the removal procedure. Tighten pump cover mounting screws to 8.5 N·m (75 in.-lbs.). Be sure pump mounting screw with smaller head (Fig. D1-12) is installed in outer hole. Rotate pump in crankcase to obtain recommended backlash of 0.10-0.15 mm (0.004-0.006 inch) between pump drive gear and crankshaft gear. Tighten pump mounting screws to 10 N·m (88 in.-lbs.).

Engine oil pressure with oil at 80°C (175°F) and engine running at 3200 rpm should be 300-500 kPa (44-72 psi).

FARYMANN

PARTS R PARTS, INC.
Farymann Diesel Parts & Engines
1017 Spray Avenue
Beachwood, New Jersey 08722

Model	No. Cyls.	Bore	Stroke	Displ.
15A (K54)	1	75 mm (2.953 in.)	55 mm (2.165 in.)	243 cc (14.8 cu. in.)
18A/C (K64)	1	82 mm (3.228 in.)	55 mm (2.165 in.)	290 cc (17.7 cu. in.)

Engines covered in this section are single cylinder, air-cooled, diesel engines. Crankshaft rotation is counter-clockwise at flywheel end of engine.

Metric fasteners are used throughout engine.

MAINTENANCE

LUBRICATION

Recommended engine oil is SAE 10W for temperatures below 0°C (32°F), SAE 20W-20 for temperatures between 0°C (32°F) and 30°C (86°F), and SAE 30 for temperatures above 30°C (86°F). API classification for oil should be CD. Fill oil sump to full level on dipstick. Manufacturer recommends renewing oil after first 20 hours of operation and after every 100 hours thereafter.

Both models are equipped with an oil pump to provide pressurized oil for lubrication. The oil pickup (2–Fig. F1-1) may be removed from the gear cover for periodic cleaning and inspection.

ENGINE SPEED ADJUSTMENT

Low idle speed is determined by detents in the speed control lever and is not adjustable. Maximum governed speed is determined by engine application.

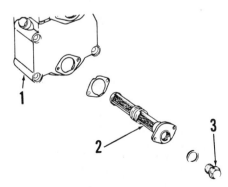

Fig. F1-1—Oil pickup and screen (2) is removable from front cover (1). Oil drain plug (3) screws into pickup mounting flange.

FUEL SYSTEM

FUEL FILTER. A renewable fuel filter is located in the fuel tank. Renew filter after every 50 hours of operation or sooner if required.

BLEED FUEL SYSTEM. Due to gravity feed type of fuel system, air should be bled from system as fuel injection pump operates. However, bleeding time may be shortened by loosening then retightening fuel line fittings, starting first at fuel tank and working to fuel injection pump. Retighten fitting when air-free fuel flows.

INJECTION PUMP TIMING

Injection pump timing is adjusted using shim gaskets (G–Fig. F1-2) between pump body and mounting surface on crankcase. To check injection timing proceed as follows: Disconnect high pressure line from injection pump, then unscrew delivery valve holder (1) and remove spring (6), shim (5), spring guide (4) and delivery valve (7); do not remove valve seat. Reinstall delivery valve holder (1) and connect a suitable spill pipe to valve holder. Aim spill pipe at a receptacle to catch discharged fuel. Move throttle control to full open position. Rotate engine flywheel slowly in counterclockwise direction until fuel just stops flowing from spill pipe. Measure distance from reference mark on flywheel housing to top dead center mark on flywheel. Measured distance should be 18 mm (0.708 inch) if maximum governed rpm is 2000 rpm or less, 21 mm (0.827 inch) if maximum governed rpm is between 2000 rpm and 3300 rpm, or 27 mm (1.063 inches) if maximum governed rpm is 3300 rpm or more. Measured distance should be within plus or minus 2 mm (0.079 inch) of desired distance.

If injection timing is incorrect, remove injection pump and remove or install shims (G–Fig. F1-2) as required. Adding shims will retard injection timing while removing shims advances injection timing. Be sure to reinstall removed pump parts after checking timing.

Tighten delivery valve holder to 34-39 N·m (25-29 ft.-lbs.)

REPAIRS

TIGHTENING TORQUES

Refer to the following table for special tightening torques.

Connecting rod..............30 N·m
(22 ft.-lbs.)
Cylinder head................30 N·m
(22 ft.lbs.)
Injection pump..............30 N·m
(22 ft.-lbs.)
Injector retainer plate.........6 N·m
(50 in.-lbs.)
Main bearing support.........30 N·m
(22 ft.-lbs.)
Rocker arm cover.............9 N·m
(80 in.-lbs.)

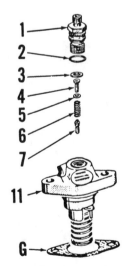

Fig. F1-2—Partial exploded view of fuel injection pump. Injection timing is adjusted using shim gaskets (G).

1. Delivery valve holder
2. "O" ring
3. Gasket
4. Spring guide
5. Shim
6. Spring
7. Delivery valve
11. Fuel injection pump

SERVICE MANUAL

Farymann

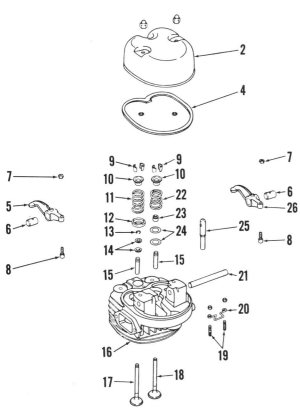

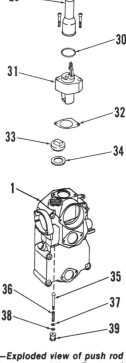

Fig. F1-3—Exploded view of cylinder head.

2. Rocker arm cover
4. Gasket
5. Exhaust rocker arm
6. Bushing
7. Locknut
8. Adjuster
9. Valve keys
10. Retainer
11. Exhaust valve spring
12. Valve rotator
13. Retaining ring
14. Washers
15. Valve guide
16. Cylinder head
17. Exhaust valve
18. Intake valve
19. Studs
20. Push rod tube retainer
21. Rocker arm shaft
22. Intake valve spring
23. Valve seal
24. Washers
25. Breather
26. Intake rocker arm

Fig. F1-4—Exploded view of push rod tube and excess fuel assemblies. Excess fuel components (35 through 39) are not used on Model 15A430.

1. Front cover
27. "O" ring
28. Push rods
29. Push rod tube
30. "O" ring
31. Tappet housing & compression release
32. Gasket
33. Oil fill plug
34. Gasket
35. Excess fuel shaft
36. Spring
37. Washer
38. "O" ring
39. Knob

VALVE CLEARANCE

Valve clearance may be adjusted after removing rocker arm cover. Piston should be at top dead center on compression stroke and compression release must be in "OFF" position. Turn adjuster screw (8–Fig. F1-3) until clearance between rocker arm and valve stem end is 0.1 mm (0.004 inch) for both valves with engine cold.

COMPRESSION RELEASE

Compression release (31–Fig. F1-4) is manually operated to raise the intake valve to aid engine starting. Compression release is not adjustable. Install a new "O" ring (30) on push rod tube when reassembling.

CYLINDER HEAD AND VALVE SYSTEM

Valve face and valve seat angles are 45 degrees for intake and exhaust valves. Minimum allowable depth of valve in head measured from top of valve to cylinder head surface is 0.0-0.1 mm (0-0.004 inch). Valve guides (15–Fig. F1-3) are renewable and identical. Valve springs are dissimilar. Note that rotator (12) is located under the exhaust valve spring and valve stem seal (23) is used on the intake valve.

Install a new "O" ring (27–Fig. F1-4) around push rod tube before installing cylinder head. Model 18A430 is equipped with a head gasket while Model 15A430 is not so equipped. Tighten cylinder head retaining nuts in a crossing pattern to a torque of 30 N·m (22 ft.-lbs.). Install clip (20–Fig. F1-3) so clip bears against end of push rod tube. Push rod nearer cylinder is connected to intake valve rocker arm.

INJECTOR

REMOVE AND REINSTALL. To remove injector, first clean dirt from injector, injection line, return line and cylinder head. Disconnect return line and high pressure injection line and immediately cap or plug all openings. Unscrew retainer plate and remove injector and asbestos washer.

Reverse removal procedure to reinstall injector. Install a new asbestos washer (8–Fig. F1-5). Tighten injector retainer plate nuts to 6 N·m (50 in.-lbs.).

TESTING. WARNING: Fuel leaves the injection nozzle wiht sufficient force to penetrate the skin. When testing, keep yourself clear of nozzle spray.

If a suitable test stand is available, injector, operation may be checked. Only clean, approved testing oil should be used to test injector. When operating properly during test, injector nozzle will emit a buzzing sound and cut off quickly with no fluid leakage at seat.

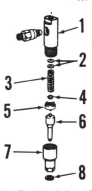

Fig. F1-5—Exploded view of injector.

1. Injector body
2. Shims
3. Spring
4. Push piece
5. Spacer
6. Nozzle & valve
7. Nozzle holder nut
8. Asbestos washer

Opening pressure should be 20270 kPa (2940 psi). Opening pressure is adjusted by varying thickness of shims (2–Fig. F1-5).

OVERHAUL. Clamp injector body (1–Fig. F1-5) in a vise with nozzle pointing upward, unscrew nozzle holder nut (7), then remove injector components shown in Fig. F1-5. Thoroughly clean all

Farymann   SMALL DIESEL

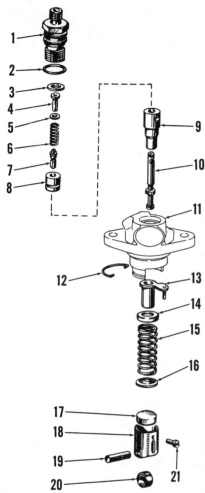

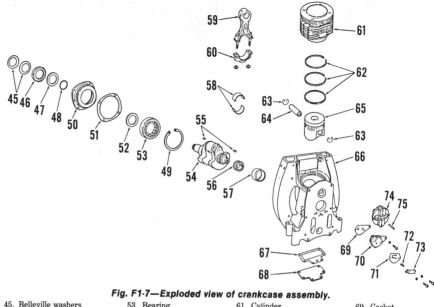

Fig. F1-7—Exploded view of crankcase assembly.

45. Belleville washers	53. Bearing	61. Cylinder	69. Gasket
46. Seal	54. Crankshaft	62. Piston rings	70. Oil pump
47. "O" ring retainer	55. Keys	63. Piston pin retainer	71. Relief valve plate
48. "O" ring	56. Gear	64. Piston pin	72. Relief valve spring
49. Snap ring	57. Bushing	65. Piston	73. Spring retainer
50. Main bearing support	58. Rod bearing	66. Crankcase	74. Governor flyweight assy.
51. Gasket	59. Connecting rod	67. Gasket	75. Pin
52. Washer	60. Rod cap	68. Cover	

Fig. F1-6—Exploded view of fuel injection pump.
1. Delivery valve holder
2. "O" ring
3. Gasket
4. Spring guide
5. Shim
6. Spring
7. Delivery valve
8. Delivery valve seat
9. Barrel
10. Plunger
11. Body
12. Clip
13. Control sleeve
14. Spring seat
15. Spring
16. Spring retainer
17. Spacer
18. Tappet
19. Pin
20. Roller
21. Pin

parts in a suitable solvent. Clean inside orifice of nozzle tip with a wooden cleaning stick. When reassembling injector, make certain all components are clean and wet with clean diesel fuel oil. Tighten nozzle holder nut (7) to 24-29 N·m (18-22 ft.-lbs.).

INJECTION PUMP

R&R AND OVERHAUL. To remove fuel injection pump, disconnect fuel lines and immediately cap or plug all openings. Unscrew pump retaining nuts and remove fuel injection pump. Do not lose shim gaskets (G–Fig. F1-2). Pump components (17 through 21–Fig. F1-6) will remain in crankcase, and timing gear cover must be removed for access.

Refer to Fig. F1-6 for an exploded view of fuel injection pump. The injection pump should be tested and overhauled by a shop qualified in diesel fuel injection pump repair.

Model 18A430 is equipped with an excess of fuel device (35 through 39–Fig. F1-4) to aid starting. Pushing up on starter knob (39) forces shaft (35) against governor arm which moves fuel injection pump control arm to intermediate position for additional fuel.

Reverse removal procedure to reinstall pump. Tighten pump mounting nuts to 30 N·m (22 ft.-lbs.) If pump is renewed or overhauled, or original shim gaskets are not used, refer to INJECTION PUMP TIMING section and adjust pump timing.

CYLINDER, PISTON, PIN AND RINGS

R&R AND OVERHAUL. The cylinder is removable after removing cylinder head. After cylinder is removed, cover crankcase opening to prevent entry of foreign material. Extract piston pin retainers (63–Fig. F1-7), remove piston pin, then separate piston from connecting rod.

Nominal standard cylinder bore diameter measured at top of cylinder is 74.86-74.88 mm (2.947-2.948 inches) on Model 15A430 governed at 1800 rpm or less, and 74.91-74.93 mm (2.949-2.950 inches) on Model 15A430 governed at engine speeds above 1800 rpm. Nominal standard cylinder bore diameter measured at top of cylinder is 81.88-81.90 mm (3.2236-3.2244 inches) on Model 18A430 governed at 1800 rpm or less, and 81.94-81.96 mm (3.226-3.227 inches) on Model 18A430 governed at engine speeds above 1800 rpm. Cylinder and piston are available in standard size only.

Piston ring end gap should be 0.2-0.8 mm (0.008-0.031 inch) with piston ring located 25 mm (1 inch) from top of cylinder.

Reverse removal procedure for installation. With cylinder installed and held against crankcase, measure height of piston from top edge of cylinder. Piston on Model 15A430 should be beneath top of cylinder a distance of

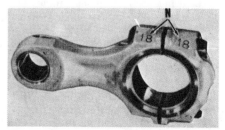

Fig. F1-8—View of connecting rod and cap showing location of stamped numbers (N) which must be on same side after assembly.

SERVICE MANUAL

Farymann

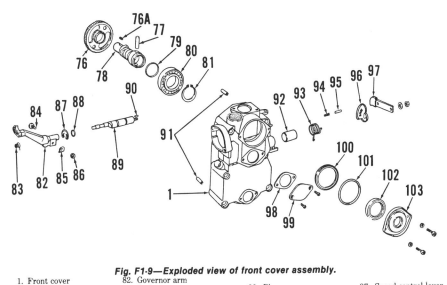

Fig. F1-9—Exploded view of front cover assembly.

1. Front cover
76. Gear
76A. Key
77. Starter pin
78. Camshaft
79. "O" ring
80. Bearing
81. Snap ring
82. Governor arm
83. Spring
84. Flat-head screw
85. Lockwasher
86. Locknut
87. Retaining ring
88. "O" ring
89. Shaft
90. Pin
91. Dowel pins
92. Bushing
93. Spring
94. Spring
95. Detent pin
96. Detent plate
97. Speed control lever
98. Gasket
99. Cover
100. "O" ring retainer
101. "O" ring
102. Seal
103. Starter guide

Fig. F1-12—Align flywheel mark (M) with threaded hole (H) when using alternate front cover installation method outlined in text.

0.475-0.800 mm (0.019-0.031 inch). Piston height above top edge of cylinder on Model 18A430 should be 0.650-0.975 mm (0.026-0.038 inch). Note that Model 15A430 is not equipped with a cylinder head gasket while Model 18A430 is so equipped.

CONNECTING ROD

REMOVE AND REINSTALL. To remove connecting rod, drain oil and remove cylinder. Detach cover (68–Fig. F1-7) on crankcase bottom. Reach through cover opening, unscrew rod cap retaining nuts and withdraw rod cap through cover opening. Remove piston and rod unit from top of engine. If required, separate piston from rod.

An insert type bearing is used in connecting rod big end while a non-renewable bushing is located in rod small end. Standard crankpin diameter is 40.020-40.030 mm (1.5756-1.5760 inches.) Big end bearing clearance of both models should be 0.03-0.05 mm (0.0012-0.0019 inch). Connecting rod big end bearing is available in standard size and undersizes of 0.25 and 0.50 mm.

Reverse removal procedure to install connecting rod while noting the following: The connecting rod side with a paint mark should be towards the fuel injection pump. Install rod cap so stamped numbers (N–Fig. F1-8) on rod and cap are on same side. Tighten connecting rod nuts to 30 N·m (22 ft.-lbs.).

FRONT COVER

REMOVE AND REINSTALL. To remove front cover (1–Fig. F1-9), drain oil, then remove rocker arm cover and extract push rods. Remove push rod tube hold down clip (20–Fig. F1-3) and slide push rod tube up into cylinder head. Remove fuel injection pump. Unscrew six front cover screws and remove front cover; it may be necessary to rotate engine crankshaft if governor mechanism snags.

When reinstalling front cover, camshaft and crankshaft gears must be correctly timed using the following procedure: Rotate crankshaft so top dead center mark on flywheel is aligned with reference mark on crankcase as shown in Fig. F1-10. Rotate camshaft gear so gear mark is aligned with mark in tappet housing bore of front cover as shown in Fig. F1-11. Install gasket and front cover on crankcase being careful not disturb camshaft gear (three long screws are located in top two holes and bottom left hole). After tightening front cover screws, recheck timing marks (M–Fig. F1-10 and T–Fig F1-11). Gear timing is correct if marks (M–Fig. F1-10) are within 6.35 mm (¼ inch) of

Fig. F1-10—Piston is at top dead center when marks (M) on flywheel and crankcase are aligned.

Fig. F1-11—View of timing marks (T) on camshaft gear and tappet housing opening in front cover.

Fig. F1-13—Position camshaft gear so timing marks (T) on gear and tappet housing opening are three teeth apart as shown. Refer to text.

Farymann   SMALL DIESEL

alignment when marks (T–Fig. F1-11) are aligned.

NOTE: Do not attempt to force front cover into place. Interference of governor mechanism may prevent front cover from mating with crankcase. Use procedure in following paragraph to install front cover.

If front cover cannot be installed using preceding procedure, use the following procedure: Rotate engine crankshaft so flywheel top dead center mark (M–Fig. F1-12) is aligned with threaded hole (H) in crankcase. Position camshaft gear so gear mark is three teeth to left of timing mark in tappet housing bore in cover as shown in Fig. F1-13. Install gasket and front cover on crankcase being careful not ot disturb camshaft gear (three long screws are located in top two holes and bottom left hole). After tightening front cover screws, recheck timing marks. Gear timing is correct if marks (M–Fig. F1-10) are within 6.35 mm (¼ inch) of alignment when marks (T–Fig. F1-11) are aligned.

Complete remainder of assembly by reversing disassembly procedure.

OIL PUMP

R&R AND OVERHAUL. Remove the front cover as previously outlined for access to the oil pump. Remove oil pump components (69 through 73–Fig. F1-7). Be careful when handling oil pressure relief valve plate (71) and spring (72) as oil pressure may be affected. Oil pump (70) is available only as a unit assembly. Install oil pump by reversing removal procedure. Do not overtighten oil pump screws

Oil pressure may be measured by removing plug in injection pump side of crankcase and connecting a suitable gage to plug hole. Oil pressure with engine at normal operating temperature should be at least 80 kPa (12 psi) at low idle speed and no more than 400 kPa (58 psi) at maximum engine speed. Oil pressure is not adjustable.

GOVERNOR

R&R AND OVERHAUL. Remove front cover as previously outlined for access to governor. Governor shaft assembly may be disassembled after detaching retaining ring (87–Fig. F1-9). Remove pin (75–Fig. F1-7) and unscrew governor flyweight assembly (74) from crankshaft; **note that stud has left-hand threads.**

Governor flyweight assembly (74) is available as a unit assembly. Flyweight springs are available, however, be sure correct governor flyweight assembly or springs are installed so engine will operate at rpm stamped on engine identification plate.

Reassemble governor as follows: Install flyweight assembly (74) on crankshaft and tighten to 34 N·m (25 ft.-lbs.); **note that stud has left-hand threads.** Install flathead screw (84–Fig. F1-9), lockwasher (85) and nut (86) so flat head is opposite governor arm. Place governor arm (82) in front cover. Install pin (90), spring (93), plate (96), lever (97) and "O" ring (88) on shaft plate (89). Insert shaft assembly into front cover and governor arm (82). Turn lever (97) so it points toward bottom of front cover (away from camshaft), then install spring (83) so small spring end engages slot in shaft end and long spring end is on same side as nut (86). Install retaining ring (87), then engage long end of spring (93) with boss on side of front cover. Engage short end of spring (93) in middle notch of plate (96) then install retaining ring (87) in shaft groove. Engage long end of spring (93) with pin on side of front cover. Note that spring (93) should return lever (97) from stop position to idle position and it may be necessary to relocate short spring end in another notch to increase spring tension. Install front cover as previously outlined.

With front cover installed, adjust position of flat-head screw (84) as follows: Remove cover (99) then rotate crankshaft so governor flyweights are positioned with mating surfaces vertical. Loosen locknut (86). Using a small screwdriver, spread flyweights apart as far as they will go. Turn screw (84) so it just contacts pin (75–Fig. F1-7), release flyweights, then turn screw (84–Fig. F1-9) in an additional ½ turn. Without disturbing screw, tighten locknut.

CAMSHAFT AND TAPPETS

R&R AND OVERHAUL. To remove camshaft, remove front cover as outlined in FRONT COVER section. Remove tappet and compression release housing (31–Fig. F1-4) and components (100 through 103–Fig. F1-9). Remove snap ring (81) and press camshaft (78) out of bearing (80). Then, press bearing (80) out of front cover (1).

Inspect cam lobes and bearing journals and renew camshaft if excessively worn or damaged. Inspect tappets and tappet housing bores. The manual starter handle is supported by guide (103) and starter end engages pin (77) in camshaft end. Inspect guide (103), pin (77) and camshaft end for damage.

Reverse removal procedure for installation. It is necessary to twist rather than drive guide (103) into position in front cover. Install front cover as outlined in FRONT COVER section to properly time camshaft and crankshaft gears.

CRANKSHAFT AND CRANKCASE

R&R AND OVERHAUL. To remove crankshaft, drain oil and remove flywheel, front cover and connecting rod. Remove governor flyweight assembly (74–Fig. F1-7) from crankshaft. Remove Belleville washers (45), washers (47 and 52) and "O" ring (48). Unscrew main bearing support nuts and remove main bearing support (50). If main bearing support will not remove easily, jack screws may be threaded into holes provided in support to push support free from crankcase. Remove crankshaft.

The governor end of crankshaft is supported by bushing (57). Standard crankshaft journal diameter at governor end should be 40.020-40.030 mm (1.5756-1.5760 inches). Clearance between crankshaft journal and bushing should be 0.04-0.06 mm (0.0016-0.0023 inch). Bushing (57) is available in standard size and undersizes of 0.25 and 0.50 mm.

Reverse removal procedure to install crankshaft while noting the following: Tighten main bearing support nuts to 30 N·m (22 ft.-lbs.). Install "O" ring retainer (47) with beveled face towards "O" ring. Install Belleville washers (45) with concave faces towards each other.

SERVICE MANUAL

FARYMANN

Model	No. Cyls.	Bore	Stroke	Displ.
25A (L14)	1	80 mm (3.150 in.)	82 mm (3.228 in.)	412 cc (25.1 cu. in.)
36A (A10)	1	95 mm (3.740 in.)	82 mm (3.228 in.)	581 cc (35.5 cu. in.)
36E (A20)	1	95 mm (3.740 in.)	82 mm (3.228 in.)	581 cc (35.5 cu. in.)
71A (R10)	2	95 mm (3.740 in.)	82 mm (3.228 in.)	1162 cc (45.7 cu. in.)
95A (S10)	2	105 mm (4.134 in.)	90 mm (3.543 in.)	1558 cc (95.1 cu. in.)

Engines covered in this section are four-stroke, air-cooled diesel engines. Crankshaft rotation is counterclockwise at flywheel end of engine. Number 1 cylinder on two-cylinder models is nearer flywheel.

Metric fasteners are used throughout engine.

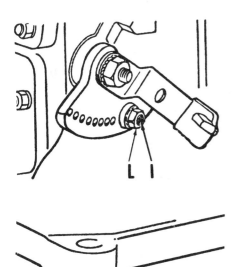

Fig. F2-1—Adjust low idle speed by loosening locknut (L), then turn screw (I) so detent can be relocated in plate.

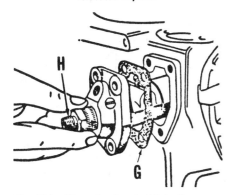

Fig. F2-2—View showing fuel injection pump timing shim gaskets (G) and delivery valve holder (H).

MAINTENANCE

LUBRICATION

Recommended engine oil is SAE 10W for temperatures below 0°C (32°F), SAE 20W-20 for temperatures between 0°C (32°F) and 30°C (86°F), and SAE 30 for temperatures above 30°C (86°F). API classification for oil should be CD. Fill oil sump to full level on dipstick. Manufacturer recommends renewing oil after first 20 hours of operation and after every 100 hours thereafter.

ENGINE SPEED ADJUSTMENT

Low idle speed may be adjusted by loosening locknut (L–Fig. F2-1) and turning screw (I) so detent can be relocated in plate. Maximum governed speed is determined by governor and engine application.

FUEL SYSTEM

FUEL FILTER. A renewable fuel filter is located in the fuel tank. Renew filter after every 500 hours of operation or sooner if required.

BLEED FUEL SYSTEM. To bleed fuel system, loosen fuel line fitting at fuel injection pump or bleed screw on pump, if so equipped. On gravity flow systems, open fuel valve until air-free fuel flows at pump then tighten fuel line fitting or bleed screw. On engines equipped with a fuel transfer pump, open fuel valve and operate pump primer lever until air-free fuel flows at pump then tighten fuel line fitting or bleed screw. On all engines, loosen high pressure injection line at injector then rotate engine crankshaft to operate fuel injection pump until fuel is discharged from injection line. Retighten injection line.

INJECTION PUMP TIMING

Injection pump timing is adjusted using shim gaskets (G–Fig. F2-2) between pump body and mounting surface on crankcase. To check injection timing proceed as follows: Disconnect high pressure line for number 1 cylinder from injection pump, then unscrew delivery valve holder (H) and remove spring and delivery valve; do not remove valve seat. Reinstall delivery valve holder (H) and connect a suitable spill pipe to valve

MODEL	MAXIMUM GOVERNED RPM					
	1500	1800	2000	2500	2800	3000
25A430	15mm (0.590in.)	20mm (0.787in.)	25mm (0.984in.)	30mm (1.181in.)	35mm (1.378in.)	40mm (1.575in.)
36A430	10mm (0.394in.)	15mm (0.590in.)	22mm (0.866in.)	30mm (1.181in.)	33mm (1.299in.)	37mm (1.457in.)
36E435	10mm (0.394in.)	14mm (0.551in.)	20mm (0.787in.)	25mm (0.984in.)	27mm (1.063in.)	30mm (1.181in.)
71A437	21mm (0.827in.)	21mm (0.827in.)	23mm (0.905in.)	27mm (1.063in.)	—	—
95A437	25mm (0.984in.)	29mm (1.142in.)	31mm (1.220in.)	38mm (1.496in.)	—	—

Fig. F2-3—Table listing distance between reference mark on flywheel housing and TDC mark on flywheel used when checking injection timing. Refer to text.

Farymann

SMALL DIESEL

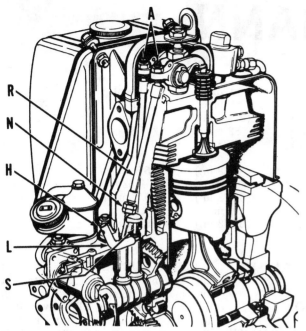

Fig. F2-4—View of valve train for single-cylinder models showing compression release adjustment point. Refer to Fig. F2-6 for twin cylinder models.

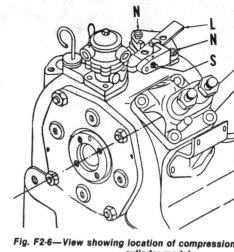

Fig. F2-6—View showing location of compression release on twin-cylinder models.

holder. Aim spill pipe at a receptacle to catch discharged fuel. Move throttle control to full open position. Rotate engine flywheel slowly in counterclockwise direction until fuel just stops flowing from spill pipe. Measure distance from reference mark on flywheel housing to top dead center mark on flywheel. Refer to Fig. F2-3 for desired measured distance for correct injection timing. Measured distance should be within plus or minus 2 mm (0.079 inch) of distance listed in Fig. F2-3.

If injection timing is incorrect, remove injection pump and remove or install shims (G – Fig. F2-2) as required. Adding shims will retard injection timing while removing shims advances injection timing. Be sure to reinstall delivery valve and spring into pump. Tighten delivery valve holder to 34-39 N·m (25-29 ft.-lbs.).

REPAIRS

TIGHTENING TORQUES

Refer to the following table for special tightening torques.

Connecting rod:
  95A437 . . . . 80 N·m
                        (60 ft.-lbs.)
  All other models . . . . 65 N·m
                        (48 ft.-lbs.)
Cylinder head:
  25A430 . . . . 45 N·m
                        (33 ft.-lbs.)
  36A430, 36E435 . . . . 54 N·m
                        (40 ft.-lbs.)
  71A437 . . . . 58 N·m
                        (43 ft.-lbs.)
  95A437 . . . . 65 N·m
                        (48 ft.-lbs.)
Injection pump:
  25A430, 71A437 . . . . 19 N·m
                        (14 ft.-lbs.)
  All other models . . . . 28 N·m
                        (20 ft.-lbs.)
Injector retainer plate . . . . 30 N·m
                        (22 ft.-lbs.)
Main bearing support . . . . 30 N·m
                        (22 ft.-lbs.)
Rocker arm bracket . . . . 80 N·m
                        (60 ft.-lbs.)
Rocker arm cover . . . . 10 N·m
                        (90 ft.-lbs.)

VALVE CLEARANCE

Valve clearance may be adjusted after removing rocker arm cover. Piston should be at top dead center on compression stroke and compression release must be in "OFF" position. Turn adjusting screws (A – Fig. F2-4) until clearance between rocker arm and valve stem end is 0.1 mm (0.004 inch) for both valves with engine cold.

COMPRESSION RELEASE

On single-cylinder models, the intake valve is held open by rotating compression release lever (L – Fig. F2-4) located in valve tappet housing (H). The compression release lever forces spindle (S) against nuts (N) on intake push rod (R) thereby opening the intake valve. The intake valve should open 1.0 mm (0.040 inch) when actuated by compression release lever. Remove rocker arm shaft, extract intake push rod and turn push rod nuts (N) to adjust intake valve opening.

When the compression release lever (L – Fig. F2-5 and F2-6) on twin-cylinder models is turned, tangs (T – Fig. F2-5) on spindle force cam followers (F) to open exhaust valves on Model 71A437 or intake valves on Model 95A437. To adjust compression release, loosen nuts (N – Fig. F2-5 and F2-6), then move lever (L) until resistance is felt indicating spindle tangs are contacting cam followers. Retighten nuts (N). Turn adjusting screw (S – Fig. F2-6) so exhaust valves on Model 71A437 or intake valves on Model 95A437 are opened 1.0 mm (0.040 inch) when compression release lever is operated.

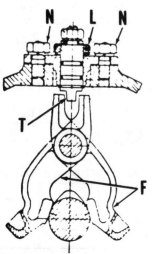

Fig. F2-5—Cross-sectional view of compression release on twin-cylinder models.

SERVICE MANUAL

Farymann

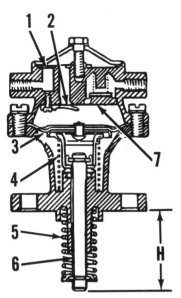

Fig. F2-7—Cross-sectional view of fuel transfer pump. Measure plunger height (H) as outlined in text.

1. Screen
2. Inlet valve
3. Diaphragm
4. Spring
5. Spring
6. Plunger
7. Exhaust valve

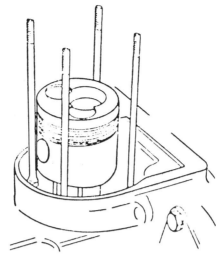

Fig. F2-8—Install piston so indentations in crown are positioned as shown.

Fig. F2-9—Assemble cap on rod so locating pin and hole (L) are properly mated.

CYLINDER HEAD AND VALVE SYSTEM

Cylinder heads and valve components are identical on twin-cylinder engines, however, do not interchange components when servicing engines.

Valve face and valve seat angles are 45° for intake and exhaust valves. Minimum allowable depth of valve in head, measured from top of valve to cylinder head surface, on Model 25A430 is 1.1-1.2 mm (0.043-0.047 inch) for intake valve and 0.4-0.5 mm (0.016-0.020 inch) for exhaust valve. Minimum allowable valve depth on Models 36A430 and 36E435 is 1.1 mm (0.043 inch) for intake valve and 1.4 mm (0.055 inch) for exhaust valve. Minimum allowable valve depth on Models 71A437 and 95A437 is 1.0-1.1 mm (0.040-0.043 inch) for all valves. Valve seats are renewable and must be installed with head heated to 80°-90°C (176°-194°F).

Install a new "O" ring around push rod tube before installing cylinder head. Tighten cylinder head retaining nuts in a crossing pattern starting on injector side of head. Tighten nuts to torque listed in TIGHTENING TORQUES section. Push rod nearer cylinder is connected to exhaust valve rocker arm on single-cylinder models or to intake valve rocker arm on twin-cylinder models.

FUEL TRANSFER PUMP

REMOVE AND REINSTALL. Clean area around fuel pump then disconnect fuel lines. Immediately cap or plug all openings to prevent contamination. Remove fuel pump being careful not to lose or damage shims between pump and crankcase. Refer to Fig. F2-7 for a cross-sectional view of pump.

Before installing fuel pump, measure plunger height (H–Fig. F2-7) and distance from pump mounting surface on crankcase to pump lobe base circle on camshaft. Install shims between pump and crankcase so plunger height is 1 mm (0.040 inch) greater than distance from crankcase surface to cam. For instance, if plunger height is 62.5 mm (2.460 inches) and distance from crankcase to cam is 61.0 mm (2.400 inches), then a 0.5 mm (0.020 inch) shim would be installed. Do not overtighten mounting screws when installing pump. Bleed fuel system as previously outlined after connecting fuel lines.

INJECTOR

R&R AND OVERHAUL. To remove injector, first clean dirt from injector, injection line, return line and cylinder head. Disconnect return line and injection line and immediately cap or plug all openings. Remove injector.

If a suitable test stand is available, injector may be checked.

WARNING: Fuel leaves the injection nozzle with sufficient force to penetrate the skin. When testing, keep yourself clear of nozzle spray.

Only clean, approved testing oil should be used to test injector. When operating properly during test, injector nozzle will emit a buzzing sound and cut off quickly with no fluid leakage at seat. Opening pressure should be (17720 kPa (2570 psi).

Thoroughly clean all injector components in a suitable solvent. Clean inside orifice end of nozzle tip with a wooden cleaning stick. When reassembling injector, make certain all components are clean and wet with clean diesel fuel oil.

INJECTION PUMP

R&R AND OVERHAUL. To remove injection pump, clean area around injection pump, disconnect all fuel lines from pump and immediately cap or plug all openings to prevent contamination. Remove crankcase breather, then reach through crankcase opening and disconnect governor link from pump. Unscrew pump retaining screws and remove pump being careful not to lose or damage shim gaskets (G–Fig. F2-2).

The injection pump should be tested and overhauled by a shop qualified in diesel fuel injection pump repair.

When installing pump, engage governor link with pump control rack. Tighten injection pump mounting screws to torque specified in TIGHTENING TORQUES section. If pump is renewed or overhauled, or if original shim gaskets are not use, refer to INJECTION PUMP TIMING section and adjust pump timing.

CYLINDER, PISTON, PIN AND RINGS

R&R AND OVERHAUL. The cylinder is removable after removing cylinder head. After cylinder is removed, cover crankcase opening to prevent entry of foreign material. Detach snap rings retaining piston pin, remove piston pin, then separate piston from connecting rod. Do not interchange com-

Farymann — SMALL DIESEL

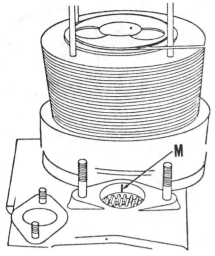

Fig. F2-10—Camshaft gear mark on single-cylinder models must align with mark (M) in tappet housing opening in crankcase when piston is at top dead center.

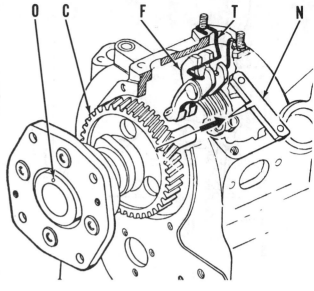

Fig. F2-11—Install tools 7480634 (T) and 7480676 (N) as outlined in text for twin-cylinder camshaft service.
- C. Camshaft gear
- F. Cam followers
- N. Tool 7480676
- O. Camshaft "O" mark
- T. Tool 7480634

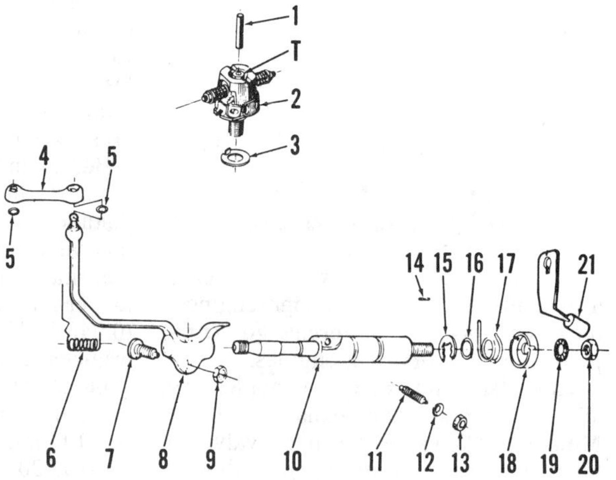

ponents between cylinders on twin-cylinder models.

Nominal standard cylinder bore diameter is 80.02-80.04 mm (3.1504-3.1512 inches) on Model 25A430, 105.00-105.02 mm (4.1338-4.1346 inches) on Model 95A437, and 95.04-95.06 mm (3.7417-3.7425 inches) on all other models. Piston ring end gap should be 0.3-1.0 mm (0.012-0.040 inch) on Model 25A430, 0.5-1.5 mm (0.020-0.060 inch) on Model 95A437, and 0.4-1.0 mm (0.016-0.040 inch) on all other models.

Position piston on rod so combustion chamber in piston crown is located as shown in Fig. F2-8 and install piston pin. Compress piston rings and slide cylinder down over piston and studs. With cylinder installed and held against crankcase, position piston at top dead center and measure distance from piston crown to top edge of cylinder. Piston crown height should be 0.7-1.0 mm (0.028-0.040 inch) on Model 25A430, 0.8-1.05 mm (0.031-0.041 inch) on Models 36A430 and 36E435, 0.775-1.05 mm (0.030-0.041 inch) on Model 71A437 and 0.775-1.075 mm (0.030-0.042 inch) on Model 95A437.

CONNECTING ROD

REMOVE AND REINSTALL. The connecting rod and crankshaft on Model 25A430 are a unit assembly; refer to crankshaft section for removal. To remove connecting rod on all other models, remove cylinder(s), then remove crankcase side cover. Reach through cover opening, unscrew rod cap retaining screws, remove rod cap through side cover opening and remove rod and piston.

Models 36A430 and 36E435 are

Fig. F2-12—Exploded view of single-cylinder model governor mechanism. Twin-cylinder models are similar.
1. Pin
2. Flyweight assy.
3. Tab washer
4. Link
5. Clips
6. Governor spring
7. Screw
8. Governor arm
9. Nut
10. Shaft
11. Stud
12. Lockwasher
13. Nut
14. Pin
15. "E" ring
16. "O" ring
17. Spring
18. Spring retainer
19. Lockwasher
20. Nut
21. Lever

equipped with a roller type rod bearing while Models 71A437 and 95A437 are equipped with an insert type bearing. Undersize rod bearings are available for Models 71A437 and 95A437. Standard crankpin journal diameter is 64.985-65.015 mm (2.5585-2.5596 inches) on Model 71A437 and 74.985-75.015 mm (2.9522-2.9533 inches) on Model 95A437 with a bearing clearance of 0.03-0.05 mm (0.0012-0.0020 inch) for both models.

Reverse removal procedure to install connecting rod. Be sure locating pin and hole (L–Fig. F2-9) are properly mated when installing rod cap on rod. Tighten rod screws to 80 N·m (60 ft.-lbs.) on Model 95A437 or to 65 N·m (48 ft.-lbs.) on all other models.

CAMSHAFT AND TAPPETS

Single-Cylinder Models

REMOVE AND REINSTALL. To remove camshaft, remove cylinder head and push rod tube. Unscrew retaining nuts and remove tappet housing from crankcase. Remove fuel injection pump. Unscrew camshaft bearing housing screws and remove camshaft assembly. If bearing housing will not remove easily, jackscrews may be threaded into holes provided in housing to push housing free.

Inspect cam lobes and renew camshaft if excessively worn or damaged. Remove snap rings to separate bearing from camshaft bearing housing. The gear end of the camshaft rides in a bushing in the crankcase and is accessible after removing crankshaft. Inspect tappets and tappet housing bores.

Reassembly is reverse of disassembly. When reinstalling camshaft assembly in crankcase, rotate crankshaft so piston is at top dead center, then insert camshaft so mark on camshaft gear is aligned with mark (M–Fig. F2-10) in tappet housing opening of crankcase. On Model 25A430, the camshaft mark will be slightly to left of crankcase mark (M).

SERVICE MANUAL

Farymann

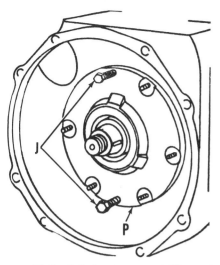

Fig. F2-13—Main bearing support (P) may be dislodged using jackscrews (J) in holes provided.

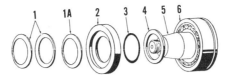

Fig. F2-15—Exploded view of thrust ring components. Belleville washer (1A) is only used on Model 95A437.

1. Belleville washers
1A. Belleville washer
2. Thrust ring
3. "O" ring
4. Thrust washer
5. Crankshaft
6. Bearing

Insert camshaft assembly and mate camshaft and crankshaft gears so mark on camshaft gear is aligned with pointer on tool (N). Remove special tools and complete assembly.

GOVERNOR

Single-Cylinder Models

REMOVE AND REINSTALL. To remove governor, remove breather on top of crankcase then reach through opening and disconnect governor link (4 – Fig. F2-12) from injection pump. Detach speed control lever (21), spring retainer (18) and spring (17) from shaft (10). Remove crankcase end cover adjacent to camshaft end. Unscrew and remove stud (11) from shaft. Detach "E" ring (15) and withdraw governor shaft (10) from side of crankcase while removing governor fork (8) assembly through end of crankcase. Remove pin (1), turn governor flyweight stud (T) clockwise (left-hand threads) and remove governor flyweight assembly.

Flyweight assembly is designed according to engine governed speed and is available as a unit assembly. Be sure correct flyweight assembly is installed.

Reverse disassembly procedure to install governor components. Long end of governor spring (6) must engage notch in crankcase while short end of spring is against back side of governor arm (8) so spring tension will force fork arm towards end of crankcase. With all governor components installed, adjust position of screw (7) as follows: Rotate crankshaft so flyweights (2) are positioned with mating surfaces vertical. Loosen locknut (9). Using a small screwdriver, spread flyweights apart as far as they will go. Turn screw (7) so it just contacts pin (1), release flyweights, then turn screw in an additional ½ turn. Without disturbing screw, tighten locknut.

Twin-Cylinder Models

REMOVE AND REINSTALL. Governor service is similar to single-cylinder governor outlined in preceding section. However, governor is mounted on oil pump and governor/oil pump assembly is removed as a unit through crankcase end cover opening. Separate governor from oil pump after removal. Carefully align oil pump mounting screw holes when installing governor/oil pump assembly. Reverse disassembly procedure to install governor. Refer to single-cylinder governor section and adjust position of screw (7 – Fig. F2-12) as outlined.

CRANKSHAFT AND CRANKCASE

R&R AND OVERHAUL. To remove crankshaft, remove camshaft, flywheel, governor and connecting rod (On Model 25A430 the connecting rod must remain with crankshaft). Unscrew main bearing support nuts and remove main bearing support (P – Fig. 2-13). If main bearing support will not remove easily, jackscrews (J) may be threaded into holes provided in support to push support free from crankcase. Remove crankshaft.

The governor end of crankshaft on Models 71A437 and 95A437 is supported by a bushing. Standard crankshaft journal diameter at governor end should be 64.985-65.015 mm (2.5585-2.5596 inches) on Model 71A437 or 79.985-80.015 mm (3.1490-3.1502 inches) on Model 95A437. Clearance between crankshaft journal and bushing should be 0.04-0.08 mm (0.0016-0.0032 inch). Bushing is available in standard size and in 0.5 and 1.0 mm undersizes.

Flywheel end main bearing on Models 71A437 and 95A437, and main bearings on all other models are antifriction type bearings. Inspect bearings and crankshaft on all models and renew if damaged or excessively worn. Inner bearing races should be heated to approximately 100°C (212°F) before installation on crankshaft. The crankshaft gear is removable and should be heated to approximately 180°C (360°F) before installation on crankshaft. Install crankshaft gear so when gear is viewed from crankshaft end, keyway (K – Fig. F2-14) is centered on two gear teeth (T).

Connecting rod, rod bearing and crankshaft are a unit assembly on Model 25A430 and individual components may not be serviced separately.

Reverse removal procedure for installation. Tighten main bearing support retaining nuts to 30 N·m (22 ft.-lbs.). Note assembly of thrust ring components at flywheel end as shown in Fig. F2-15. Outer circumference of thrust ring should be smooth as it contacts crankcase oil seal. Install Belleville washers (1) so concave sides face each other. Belleville washer (1A) is only used on Model 95A437 and must be installed with concave face towards thrust ring (2).

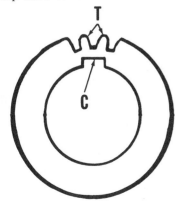

Fig. F2-14—When viewed from end of crankshaft, crankshaft gear keyway (C) should be centered between two gear teeth (T).

ISUZU

ISUZU MOTORS LIMITED
22-10 Minami-oi 6-chome
Shinagawa-ku, Tokyo, Japan

Model No.	Cyls.	Bore	Stroke	Displ.
2AA1	2	86 mm (3.386 in.)	84 mm (3.307 in.)	975 cc (59.49 cu. in.)
2AB1	2	86 mm (3.386 in.)	102 mm (4.016 in.)	1184 cc (72.24 cu. in.)
3AA1	3	86 mm (3.386 in.)	84 mm (3.307 in.)	1463 cc (89.27 cu. in.)
3AB1	3	86 mm (3.386 in.)	102 mm (4.016 in.)	1777 cc (108.43 cu. in.)

All models listed are four-stroke, indirect injection, water cooled, overhead valve in-line diesel engines. Firing order is 1-3-2 for 3-cylinder engines. Crankshaft rotation is clockwise as viewed from front (timing cover) end.

MAINTENANCE

LUBRICATION

ENGINE. Engine oil capacity is 3.6 liters (3.8 quarts) for 2-cylinder models and 6.2 liters (6.5 quarts) for 3-cylinder models. Recommended engine lubricant is good quality oil with API classification CC or CD. Below −30° C (−20° F), SAE 5W-20 motor oil is recommended. SAE 10W-30 or 10W-40 motor oil may be used at temperatures above −30° C (−20° F), and SAE 15W-40 or 15W-50 oil may be used at temperatures above −15° C (0° F).

FUEL INJECTION PUMP. Injection pump housing and governor lubricant should be changed after each 200 hours or once each month, whichever occurs first. To drain oil, remove drain plugs located on bottom of pump housing and governor housing.

NOTE: A small amount of diesel fuel will leak past the injection pump plungers to lubricate these parts and will tend to dilute the oil in the injection pump housing; if dilution of oil with diesel fuel is excessive, worn fuel injection pump plungers or fuel feed pump push rod should be suspected.

Refill both the injection pump cam and governor compartments at breather cap openings (Fig. IZ1) with high quality diesel engine oil. Maintain oil level at indicator line on dipsticks. Approximate capacities are 0.16 liter (0.17 quart) for governor housing and 0.11 liter (0.12 quart) for pump cam housing.

FUEL SYSTEM

The diesel fuel system consists of three basic units: fuel filter, injection pump and injection nozzles. All models use an in-line fuel injection pump with internal mechanical governor. Two-cylinder models are fitted with a Diesel Kiki modified Bosch type PES2K injection pump, and 3-cylinder models have a Diesel Kiki modified Bosch type PES-A pump. Both injection pumps are of similar construction except for number of pumping elements. A cam operated fuel feed pump with integral manual primer pump is mounted on the side of the fuel injection pump and delivers fuel from the fuel tank to the fuel filter assembly.

NOTE: Fuel injection pump is sealed at factory. Seals broken during warranty period will void warranty on fuel injection equipment.

FUEL FEED PUMP. To operate manual primer pump (Fig. IZ1), turn primer pump plunger counterclockwise until plunger is free, then a spring under the plunger will force it upward and fuel will enter the manual pump. Push plunger down to pump fuel and continue to actuate plunger until bleeding of system is accomplished. Then, push plunger down and turn clockwise to lock plunger into place.

FUEL FILTER. Fuel filter assembly (Fig. IZ2) is located on upper right side of engine. A strainer (2) is located in the fuel inlet bolt (3). An overflow valve (11) normally remains closed, but will open when flow from fuel feed pump becomes excessive, allowing fuel to bypass back to fuel tank. Overflow valve will also bypass fuel when fuel filter element becomes clogged, resulting in lack of fuel supply to engine.

Loosen plug (6) and drain water and sediment from fuel filter body after each 100 hours of operation. The filter element (9) should be renewed after each 600 hours of operation, or sooner if a loss of engine power is evident. If water enters fuel filter, the filter element material will swell, closing off fuel supply. In this instance, element must be renewed regardless of number of hours of operation.

To remove filter element, first thoroughly clean outside of filter assembly. Shut off fuel supply valve, then unscrew center pipe (7) and remove cover (8), filter body (4) and element (9) taking care not to lose spring (5) and the two flat washers located at each end of

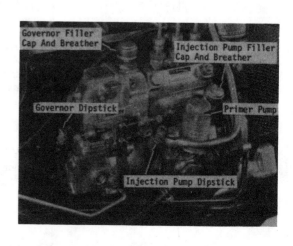

Fig. IZ1—Fuel injection pump camshaft and governor assembly is lubricated by engine oil.

SERVICE MANUAL

Isuzu

element. Remove inlet banjo bolt (3) and clean fuel strainer (2), then reinstall bolt and strainer. Install new element and bleed air from system as outlined in following paragraph.

BLEED FUEL SYSTEM. To bleed air from fuel system, open fuel supply valve and loosen bleed plug (12—Fig. IZ2) in top of fuel bypass valve (11). Operate manual primer pump (4—Fig. IZ3) until fuel flows freely from filter bleed plug. Close the plug and open air bleed screw (3—Fig. IZ3) on fuel injection pump. Operate primer pump until fuel flows free of air bubbles, then close air bleed screw on pump and lock manual primer pump plunger in place.

If engine fails to start or runs unevenly, loosen high pressure fuel lines at injectors. Open throttle and crank engine until fuel flowing from loosened injector lines is free of air, then tighten injector line fittings.

INJECTION PUMP TIMING

Injection pump should be correctly timed when installed with "Z" marks on pump drive gear and engine camshaft gear aligned and with notches (scribe marks) on pump housing and mounting flange aligned as shown in Fig. IZ4. Some engines may have an inspection hole in front of timing cover (Fig. IZ5) so that timing marks may be checked without removing timing cover.

If timing marks are suspected of being inaccurate or a new fuel injection pump is being installed, check pump flow timing as follows: Disconnect number one fuel injection line from pump and remove delivery valve holder, spring and delivery valve. Reinstall delivery valve holder. Turn engine crankshaft in normal direction to bring number one piston up on compression stroke while operating manual primer pump. When fuel stops flowing from number one outlet of pump (beginning of injection), stop turning crankshaft and check position of crankshaft pulley timing marks (Fig. IZ6). Beginning of injection should be 18 degrees BTDC.

NOTE: Some applications may require different injection timing; set timing according to engine or equipment operator's manual if different from 18 degrees BTDC.

If timing is not correct, loosen pump mounting bolts and turn pump housing on mounting flange to obtain correct timing. Scribe new alignment marks on pump housing and mounting flange after proper timing is obtained.

GOVERNOR

The flyweight type governor is located in the rear end of the fuel injection pump assembly. Check equipment operator's manual for recommended high and low idle speeds according to engine application.

If necessary to adjust governed speeds, proceed as follows: Start engine and allow to run until engine operating temperature is reached. Loosen locknut on low idle adjusting screw (1—Fig. IZ3), hold pump lever against screw and adjust screw to obtain desired low idle speed. Engine surging at low idle speed can be corrected by removing cap from rear of governor housing and turning idling sub spring adjusting screw (Fig. IZ7).

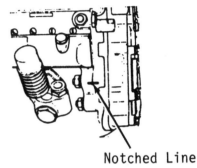

Fig. IZ4—When checking injection pump timing, be sure scribe marks on pump housing and mounting flange are aligned.

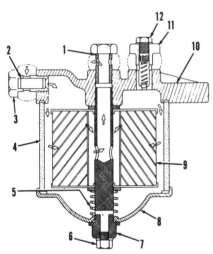

Fig. IZ2—Cross-sectional view of fuel filter assembly. A strainer (2) is located in inlet fuel line banjo fitting bolt (3).

1. Banjo bolt
2. Strainer
3. Banjo bolt
4. Filter body
5. Coil spring
6. Drain plug
7. Center pipe
8. Lower cover
9. Filter element
10. Upper cover
11. Overflow valve
12. Air bleed screw

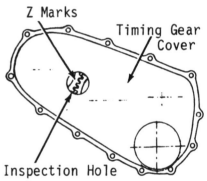

Fig. IZ5—With number one piston at TDC of compression stroke, "Z" marks on pump drive gear and camshaft gear must be aligned.

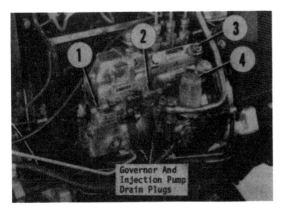

Fig. IZ3—To bleed air from injection pump, open air bleed screw (3) and actuate hand primer pump (4).
1. Low speed adjusting screw
2. High speed adjusting screw
3. Air bleed screw
4. Primer pump

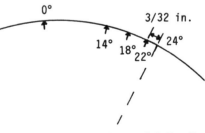

Fig. IZ6—Drawing showing crankshaft pulley marks. Farthest mark from TDC is 22 degrees BTDC. If engine or equipment operator's manual specifies mark not stamped on pulley, measure 2.4 mm (3/32 inch) on pulley circumference for each 2 degrees difference in specification from stamped mark.

Isuzu

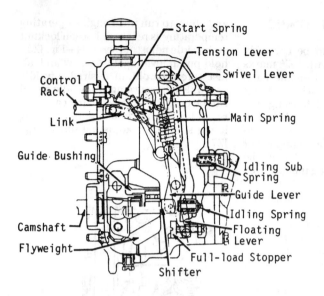

Fig. IZ7—Cross-sectional view of governor assembly in fuel injection pump.

To adjust high idle (no-load) speed, loosen locknut on adjusting screw (2—Fig. IZ3). Hold pump lever against screw and adjust screw to obtain desired high idle speed.

After adjustments are made, tighten adjusting screw locknuts and reseal adjusting screws with seal wires.

REPAIRS

TIGHTENING TORQUES

Refer to the following table for special tightening torques. All fasteners are metric.

Connecting rod bolts ...... 74-83 N·m
  (55-61 ft.-lbs.)
Main bearing cap bolts .. 157-176 N·m
  (116-130 ft.-lbs.)
Crankshaft pulley bolt ...... 147 N·m
  (108 ft.-lbs.)
Camshaft gear cap screw .... 108 N·m
  (80 ft.-lbs.)
Flywheel bolts .... 83 N·m
  (61 ft.-lbs.)
Cylinder head bolts .... 76-81 N·m
  (56-60 ft.-lbs.)
Rocker arm shaft support .... 25 N·m
  (18 ft.-lbs.)
Valve rocker arm cover ...... 13 N·m
  (10 ft.-lbs.)
Timing gear cover .... 19 N·m
  (14 ft.-lbs.)
Camshaft thrust plate .... 69 N·m
  (51 ft.-lbs.)
Injection pump delivery
  valve holder .... 30-34 N·m
  (22-25 ft.-lbs.)
Governor flyweight nut ... 49-58 N·m
  (36-43 ft.-lbs.)
Injector nozzle cap nut ...... 61 N·m
  (45 ft.-lbs.)
Injector nozzle retaining
  nut .... 73 N·m
  (54 ft.-lbs.)
Water pump mounting
  bolts .... 41 N·m
  (30 ft.-lbs.)

WATER PUMP

To remove water pump, drain coolant from engine block and disconnect radiator hose from pump. Loosen alternator adjusting bracket bolt and loosen fan belt. Remove radiator (if necessary), fan, spacer, belt and pulley. Unbolt and remove water pump.

To disassemble water pump, use suitable puller to remove fan pulley hub (9—Fig. IZ8). Remove bearing set screw (8). Remove rear cover (1); then, using drift smaller than water pump shaft diameter, press shaft and bearing assembly (6) out toward front (pulley) end of pump body.

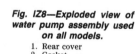

Fig. IZ8—Exploded view of water pump assembly used on all models.
1. Rear cover
2. Gasket
3. Impeller
4. Seal assy.
5. Water slinger
6. Bearing & shaft assy.
7. Pump housing
8. Set screw
9. Fan hub

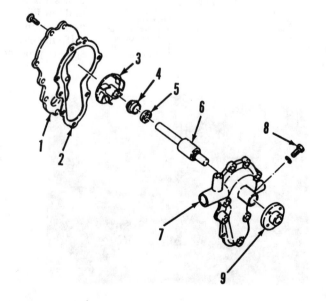

SMALL DIESEL

Water pump repair kit consists of new bearing (6), seal (4), impeller (3) and gaskets. Position slinger (5) on rear end of shaft and bearing assembly (6), then reassemble pump as follows: Align set screw hole in bearing with hole in pump body, then press bearing into place. Install set screw (8). Apply a thin coat of liquid gasket such as "Belco Bond No. 4" on outer diameter of seal unit, then install seal over bearing shaft and into rear of pump body. Support front end of bearing shaft and press impeller onto rear end of shaft until rear face of impeller is 0.4-0.6 mm (0.016-0.023 inch) below rear face of pump body. Support rear end of bearing shaft and press fan pulley hub onto front end of shaft so that distance from fan mounting surface to rear face of pump, including thickness of gasket and rear cover, is 91.7-92.3 mm (3.610-3.634 inches). Apply suitable liquid gasket sealer to both sides of gasket and install rear cover. Apply liquid gasket sealer to both sides of pump mounting gasket, then reinstall water pump by reversing removal procedure.

Adjust fan belt tension so that hand pressure midway between fan and alternator pulleys will deflect belt about 5 mm (3/16 inch).

THERMOSTAT

To remove thermostat, drain cooling system, then unbolt and remove water outlet elbow and withdraw thermostat. Thermostat should start to open at 75°-78° C (167°-172° F) and be fully open at 90° C (194° F).

VALVE ADJUSTMENT

Clearance between valve stem end and rocker arm should be 0.45 mm (0.018 inch) for both intake and exhaust valves with engine cold. To adjust clear-

SERVICE MANUAL

Isuzu

ance, remove valve rocker arm cover and bring No. 1 piston up on compression stroke until TDC (0 degrees) mark on crankshaft pulley is aligned with timing pointer. On three-cylinder models, adjust the four valves as indicated in Fig. IZ9A. Then, turn crankshaft one complete revolution until TDC mark is again aligned with timing pointer and adjust the two remaining valves as shown in Fig. IZ9B. For two-cylinder models, follow adjustment procedure shown for front and center cylinders of three-cylinder models.

CYLINDER HEAD

To remove cylinder head, drain coolant and disconnect radiator hose and bypass hose. Close fuel supply valve and remove fuel filter assembly. Disconnect wiring to glow plugs and remove the glow plugs and fuel injectors. Remove rocker shaft oil feed pipe. Remove intake and exhaust manifolds. Remove valve rocker arm cover, rocker arm shaft assembly and push rods. Loosen cylinder head bolts in reverse order of sequence shown in Fig. IZ10, then remove bolts and cylinder head. Remove head gasket and thoroughly clean head and block surfaces.

Inspect cylinder head for cracks, distortion or other damage. Cylinder head surface should be flat within 0.05 mm (0.002 inch). If cylinder head gasket surface is distorted in excess of 0.1 mm (0.004 inch), resurface or renew head.

NOTE: If new hot plugs (Fig. IZ11) are to be installed, perform this operation before surfacing head.

Intake and exhaust manifold joining faces should be flat within 0.05 mm (0.002 inch) and must be resurfaced or renewed if distortion is 0.2 mm (0.008 inch) or more. Inspect hot plugs for cracks or other damage. Drive hot plugs out using thin drift pin as shown in Fig. IZ11 if renewal is necessary. Be sure cylinder head bore for hot plug is clean and free of burrs. When installing new hot plug, be sure to align ball on hot plug with groove in cylinder head. Using protective steel plate, press hot plug into position. After installing hot plug, face of plug should be ground flush with cylinder head.

Reinstall cylinder head by reversing removal procedure. Lubricate bolt threads with engine oil, then install bolts and tighten to a final torque of 76-81 N·m (56-60 ft.-lbs.) in three steps in sequence shown in Fig. IZ10.

VALVE SYSTEM

Valve face angle for both intake and exhaust valves is 45 degrees. Valve stem diameter is 7.950-7.967 mm (0.313-0.3136 inch); wear limit is 7.88 mm (0.310 inch) for intake valves and 7.85 mm (0.309 inch) for exhaust valves.

Check the valve guides and renew if cracked or excessively worn. Desired valve stem to guide clearance is 0.05 mm (0.002 inch) for intake valves and 0.08 mm (0.003 inch) for exhaust valves. Wear limit for stem to guide clearance is 0.2 mm (0.008 inch) for intake valves and 0.25 mm (0.010 inch) for exhaust valves. Valve thickness (margin) of new valve is 1.3 mm (0.051 inch); replace valve if less than 1.0 mm (.039 inch) thick at outer edge.

To renew guides, use suitable piloted driver and drive old guides out toward top of head. Install new guides into top of head and drive in until top end of guide is 13 mm (0.512 inch) above top of head for intake guides and 14 mm (0.551 inch) above top of head for exhaust guides. Use suitable valve guide reamer to obtain desired valve stem to guide clearance.

Check valve seats and renew if cracked or excessively worn. Valve seating surface should be 3.1 mm (0.122 inch) below cylinder head surface; maximum allowable seating distance is 3.6 mm (0.142 inch). To remove old seats, use special valve seat puller or using arc welder, run a short bead along valve seating surface. Heat from the arc weld will shrink the seat and it can be removed easily.

If old valve seat was loose in cylinder head, or valve seat bore was damaged during removal of seat, it is possible to re-cut the cylinder head bore using proper sized cutter and install an oversized seat. Counterbore diameter should

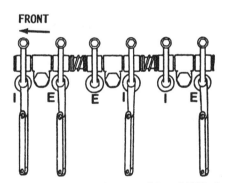

Fig. IZ9A—With number one piston at TDC of compression stroke, adjust the four valves indicated on three-cylinder models. On two-cylinder models, ignore the number three (rear) cylinder adjustment.

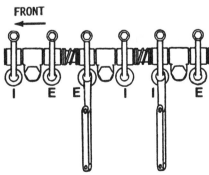

Fig. IZ9B—With number one piston at TDC of exhaust stroke, adjust the two valves as indicated on three-cylinder engines. On two-cylinder models, ignore the number three (rear) cylinder adjustment.

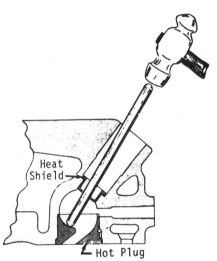

Fig. IZ11—If necessary to renew hot plugs, drive old plugs out with thin drift. Heat shield can be easily driven out with hot plug removed.

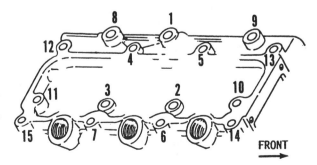

Fig. IZ10—Tighten cylinder head cap screws in progressive sequence starting with center bolt. A three-cylinder head is shown; use similar sequence for two-cylinder models. Reverse sequence when loosening cylinder head bolts.

41

Isuzu

be 0.06-0.16 mm (0.0025-0.006 inch) smaller than outside diameter of new valve seat. Be sure counterbore is clean and drive new seat into place using piloted valve seat driver tools. Chilling seat in dry ice prior to installation will shrink seat and ease installation. Valve seating surface must be reground after installing new seat.

Valve seating surface can be reconditioned using 30, 45 and 60 degree grinding stones or cutters. Valve seat width should be 1.2-1.5 mm (0.047-0.059 inch); maximum seat width should be less than 2 mm (0.079 inch).

INJECTORS

REMOVE AND REINSTALL. To remove injectors, first thoroughly clean injectors, fuel lines and surrounding area using suitable solvent and compressed air. Remove fuel leak-off line and high pressure lines from injectors. Cap or plug all openings immediately to prevent entry of dirt. Unscrew and remove injectors from cylinder head.

Prior to reinstalling injectors, remove old nozzle sealing washers from head (if not removed with injector). Be sure cylinder head bores are clean and check condition of heat shields (see Fig. IZ11). Heat shields can be removed from above if care is taken and suitable puller is used. Install new nozzle washers in head with corrugated side up. Install injector and tighten to a torque of 73.5 N·m (54 ft.-lbs.). When reinstalling fuel lines, leave fittings for high pressure lines loose at injectors and crank engine until fuel emerges from loose fittings, then tighten injector line fittings.

TESTING. A faulty injector may be located on the engine by loosening high pressure fuel line fitting on each injector, in turn, with engine running. This will allow fuel to escape from the loosened fitting rather than enter the cylinder. The injector that least affects engine operation when its line is loosened is the faulty injector. If a faulty injector is found and considerable time has elapsed since injectors have been serviced, it is recommended that all injectors be removed and serviced, or that new or reconditioned units be installed.

A complete job of testing and adjusting fuel injectors requires use of special test equipment. Only clean, approved testing oil should be used in tester tank. Injection nozzle should be tested for opening pressure, spray pattern, seat leakage and back leakage. Check injection nozzle as outlined in the following paragraphs.

WARNING: Fuel emerges from injector with sufficient force to penetrate the skin. When testing injector, keep yourself clear of nozzle spray.

OPENING PRESSURE. Before conducting test, operate tester lever until fuel flows from tester line, then attach injection nozzle to tester. Pump tester lever a few quick strokes to purge air and to be sure nozzle valve is not stuck and spray hole is open.

Operate tester lever slowly and observe pressure gage reading when nozzle opens. On all models, opening pressure should be 11,700 kPa (1706 psi).

To adjust opening pressure, remove cap nut (3—Fig. IZ12) and turn adjusting screw (5) in to increase pressure or out to decrease pressure.

SPRAY PATTERN. The injector nozzle tip used on all models has one orifice with zero degree spray angle. To check spray pattern, operate tester handle and observe spray. Spray pattern must be uniform and free of streaks or drops. If irregular pattern is observed, check for partially clogged or damaged spray hole or improperly seating nozzle valve.

SEAT LEAKAGE. Wipe nozzle tip dry, then operate tester lever slowly to maintain pressure at 1000 kPa (150 psi) below nozzle opening pressure and hold tester at this pressure for ten seconds. There should be no drops or noticeable accumulation of oil on tip of injector. If drop forms, disassemble and overhaul

Fig. IZ12—Exploded view of fuel injector assembly used on all models.

1. Banjo bolt
2. Gaskets
3. Cap nut
4. Gasket
5. Adjusting screw
6. Nut
7. Washer
8. Spring
9. Inlet screen
10. Inlet connector
11. Push rod
12. Nozzle holder body
13. Retaining bushing
14. Nozzle & valve assy.
15. Nozzle nut

SMALL DIESEL

injector. Slight wetting of nozzle tip is acceptable on a used injector.

BACK LEAKAGE. Operate tester to obtain pressure just below opening pressure, then close pump isolator valve. Pressure drop should not exceed 4900 kPa (700 psi) in five seconds. If back leakage exceeds this limit, renew or overhaul injector.

OVERHAUL. First, thoroughly clean outside of injector. Nozzle valve and body are individually fit and these parts must be kept together.

Clamp injector in a soft-jawed vise. Remove cap nut (3—Fig. IZ12), adjusting screw (5), spring (8) and push rod (11). Unscrew nozzle nut (15) and remove nozzle valve assembly (14) and injector retainer nut (13). Soak parts in diesel fuel or approved carbon solvent, then brush off outside of parts with brass wire brush.

Examine lapped pressure surfaces of nozzle valve body and valve for obvious damage. Check to see that nozzle valve will slide smoothly downward in valve body by its own weight. If valve binds or slides too fast, renew nozzle assembly. Check spring (8) for distortion, corrosion or other damage. Check for bent or worn push rod (11) and renew if damage is noted. Be sure nozzle spray orifice is clean. Nozzle orifice diameter is 1 mm (0.039 inch).

Thoroughly flush all parts in clean diesel fuel. Make sure all sealing surfaces are absolutely clean. Using new gaskets, reassemble injector by reversing disassembly procedure. Tighten nozzle nut to a torque of 61 N·m (45 ft.-lbs.) and tighten cap nut (3) finger tight until nozzle opening pressure is readjusted.

FUEL FEED PUMP

All models are equipped with a fuel feed pump mounted on the side of the fuel injection pump. The feed pump is actuated by a cam on the fuel injection pump camshaft and may also be operated manually by a hand primer pump (2—Fig. IZ13).

Refer to Fig. IZ13 for exploded view of the fuel feed pump. Pump can be tested by installing a fuel pressure gage in the drain plug opening of the fuel filter (see Fig. IZ2). With engine running at rated speed, pressure should be approximately 157 kPa (23 psi).

To remove fuel feed pump, first thoroughly clean outside of pump and surrounding area. Cap or plug all openings to prevent entry of dirt. Unbolt and remove fuel feed pump from fuel injection pump.

To disassemble pump, remove snap ring (22—Fig. IZ13) and remove tappet

SERVICE MANUAL

Isuzu

(20) taking care not to drop tappet guides (18), pin (19) and roller (21). Filter (15) is located inside fuel inlet banjo bolt (14) and can be removed by turning counterclockwise after removing bolt. Remainder of disassembly is obvious from inspection of unit and reference to exploded view.

Inspect all parts for evidence of sticking, scoring or excessive wear. Excessive wear of push rod (13) and pump body will allow leakage of diesel fuel into pump cam chamber and result in dilution of injection pump lubricating oil.

Reassemble and reinstall by reversing disassembly and removal procedures. Bleed system of air before attempting to start engine.

INJECTION PUMP

To remove injection pump, first thoroughly clean pump and fuel lines, then proceed as follows: Remove inspection plate and tachometer drive housing from engine timing gear cover. Turn crankshaft so that TDC mark (0 degree mark) on crankshaft pulley (see Fig. IZ6) and "Z" marks on camshaft gear and injection pump drive gear (see Fig. IZ5) are aligned. Close fuel supply valve and disconnect fuel lines from injection pump and fuel feed pump. Remove fuel injector high pressure lines. Disconnect throttle control rod from pump governor arm. Unbolt pump mounting flange from engine front plate and remove pump, mounting flange and drive gear as an assembly. Remove pump drive gear and mounting flange from pump.

The injection pump should be tested and overhauled by a shop qualified in diesel injection equipment repair.

To reinstall pump, attach mounting flange with scribe mark on pump and flange aligned as shown in Fig. IZ4, then install pump drive gear. Install the assembly on engine front plate with TDC mark (0 degree mark) on crankshaft pulley aligned with timing pointer and with "Z" marks (see Fig. IZ5) on pump drive gear and engine camshaft gear aligned. Bleed air from fuel system before tightening fuel injector high pressure lines.

TIMING GEAR COVER

To remove timing gear cover, first drain engine coolant and lubricating oil, then proceed as follows: Disconnect radiator hoses and remove radiator if necessary. Remove tachometer drive (if used), fan and fan belt. Remove crankshaft pulley bolt and remove pulley from crankshaft using suitable puller. Unbolt and remove timing gear cover.

Install new crankshaft oil seal with lip to inside and lubricate seal and crankshaft pulley. Reinstall timing cover by reversing removal procedure and tighten cover bolts to a torque of 19 N·m (14 ft.-lbs.). Tighten crankshaft pulley bolt to a torque of 147 N·m (108 ft.-lbs.).

TIMING GEARS

Timing gear train consists of crankshaft gear, idler gear, camshaft gear and fuel injection pump drive gear.

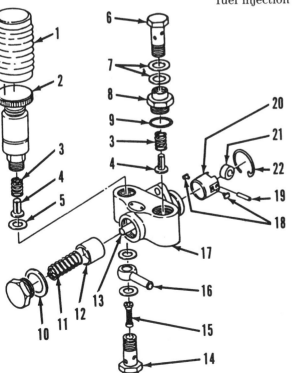

Fig. IZ13—Exploded view of fuel feed pump mounted on outer side of fuel injection pump. Manually actuated plunger (2) is used to prime fuel system.

1. Bellows
2. Primer pump
3. Spring
4. Check valve
5. Gasket
6. Banjo bolt
7. Gaskets
8. Adapter
9. "O" ring
10. Gasket
11. Spring
12. Piston
13. Push rod
14. Banjo bolt
15. Screen
16. Banjo fitting
17. Pump body
18. Tappet guides
19. Tappet pin
20. Tappet body
21. Cam roller
22. Snap ring

Before disassembly of timing gears, use a dial indicator to check gear backlash, end play of camshaft and idler gear, and idler gear to shaft clearance. Desired backlash is 0.1 mm (0.004 inch); maximum allowable backlash is 0.3 mm (0.012 inch). Camshaft and idler gear end play should be 0.05-0.115 mm (0.002-0.005 inch); maximum allowable end play is 0.2 mm (0.008 inch). Oil clearance between idler gear bushing and shaft should be 0.025-0.085 mm (0.001-0.003 inch) with maximum allowable oil clearance of 0.2 mm (0.008 inch).

To renew camshaft gear and/or thrust plate, remove gear retaining bolt from front end of camshaft and pull gear from shaft. Unbolt and remove thrust plate from front face of cylinder block. When reassembling, lubricate thrust surfaces and be sure timing marks on camshaft gear are facing outward. Tighten camshaft thrust plate to a torque of 69 N·m (51 ft.-lbs.) and cam gear bolt to a torque 108 N·m (80 ft.-lbs.).

To renew idler gear, bushings, idler gear shaft or plate, remove the two plate retaining bolts and gear from shaft, then remove shaft from block. Standard idler gear shaft diameter is 44.945-44.975 mm (1.769-1.770 inches), renew shaft if wear exceeds 0.1 mm (0.004 inch). Standard idler gear bushing inside diameter is 45-45.03 mm (1.772-1.773 inches); renew bushing or gear and bushing assembly if wear exceeds 0.1 mm (0.004 inch). Be sure oil hole in shaft is aligned with oil hole in block when installing idler gear shaft.

If necessary to renew crankshaft gear, use suitable puller to remove gear from shaft. Heat new gear prior to installation, be sure timing mark is facing outward and drive the gear into place using a hollow driver.

When installing camshaft and timing gears, be sure "X" marks on crankshaft gear and idler gear, "Y" marks on idler gear and camshaft gear and "Z" marks on camshaft gear and injection pump drive gear are aligned.

OIL PUMP AND RELIEF VALVE

An internally mounted gerotor type oil pump is driven by a gear on engine camshaft. The pressure relief valve located in oil filter base limits oil pressure to 412-461 kPa (60-67 psi). Refer to Fig. IZ14 for exploded view of oil pump and to Fig. IZ15 for exploded view of oil filter base and relief valve assembly.

To remove oil pump, remove oil pan and remove the oil pressure tube from pump to cylinder block. Remove retaining bolt and withdraw oil pump from block.

Isuzu

To disassemble pump, remove screen (9—Fig. IZ14) and unbolt strainer case (11). Remove bolt from pump cover (8) and remove cover, spacer (7) and outer rotor (6). Drive pin (2) out of pump shaft and remove pinion (1). Slide shaft (3) out toward rotor end of pump body. Remove the pin (10) and inner rotor (5) from shaft.

Inspect all parts for excessive wear, scoring or other damage and renew as necessary. Rotor shaft to pump body clearance should be 0.04 mm (0.0016 inch); maximum allowable clearance is 0.2 mm (0.008 inch). Outer rotor to pump body clearance should be 0.02 (0.0008 inch) with maximum allowable clearance of 0.27 mm (0.010 inch). Clearance between tips of inner and outer rotor should be 0.02 mm (0.0008 inch) and must not exceed 0.13 mm (0.005 inch).

To reassemble pump, reverse disassembly procedure. Be sure pump shaft turns smoothly after pump is assembled and prime pump with heavy oil before installing in engine.

GOVERNOR

The governor is an integral part of the fuel injection pump assembly. Refer to cross-sectional view of governor assembly in Fig. IZ7. Other than external speed adjustments, pump should be serviced at authorized diesel service center.

PISTON AND ROD UNITS

Connecting rod and piston assemblies are removed from above after removing cylinder head and oil pan.

Pistons have a mark on top indicating front side of piston. When reinstalling piston and rod units, be sure mark on top of piston is toward front of engine and side of rod with stamped cylinder number is toward camshaft side of engine. Tighten connecting rod cap bolts to a torque of 74-83 N·m (54-61 ft.-lbs.).

PISTONS, RINGS AND CYLINDERS

All engines are equipped with aluminum alloy pistons having three compression rings and one oil control ring. Cylinder bores are not equipped with sleeves. Standard cylinder bore is 86 mm (3.386 inch). Rebore cylinders and install oversize pistons and rings if cylinder bore wear exceeds 0.2 mm (0.008 inch).

Desired piston skirt to cylinder bore clearance is 0.104-0.124 mm (0.004-0.005 inch). Standard piston skirt diameter, measured at right angle to piston pin and 52 mm (2 inches) below top of piston is 85.854-85.883 mm (3.3801-3.3812 inches).

Recommended piston ring end gap is 0.2-0.4 mm (0.008-0.016 inch) for compression rings and 0.1-0.3 mm (0.004-0.012 inch) for oil control ring. Desired piston ring side clearance in piston groove is 0.045-0.075 mm (0.0018-0.0029 inch) for top and second compression rings, 0.030-0.060 mm (0.0012-0.0024 inch) for third compression ring and 0.020-0.054 mm (0.0008-0.0021 inch) for oil control ring. Maximum allowable ring side clearance is 0.3 mm (0.012 inch) for compression rings and 0.15 mm (0.006 inch) for oil control ring. Renew piston if side clearance is excessive.

When installing rings, oil control ring can be installed either side up; place expander in groove and install ring with end gap at 180 degrees away from expander ends. Install third compression ring with undercut down (Fig. IZ16). Install second compression ring with "N" mark up. Install top compression ring either side up. Be sure to space ring end gaps at 90 degrees apart.

Assemble piston to connecting rod so that "front" mark on top of piston will be toward front of engine and cylinder number stamped on side of rod will be toward camshaft side of engine.

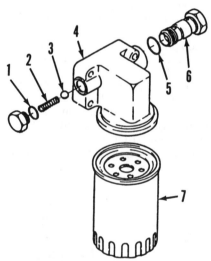

Fig. IZ15—Engine oil pressure relief valve (6) is located in oil filter base (4).

1. "O" ring
2. Bypass valve spring
3. Bypass valve ball
4. Filter base
5. "O" ring
6. Relief valve assy.
7. Oil filter

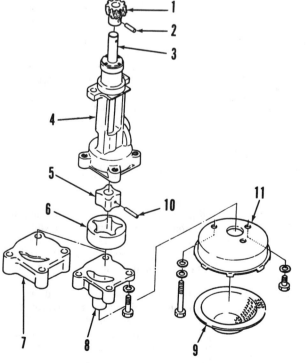

Fig. IZ14—Exploded view of gerotor type oil pump used on all models. Some engines may not have spacer (7).

1. Drive pinion
2. Pin
3. Drive shaft
4. Pump housing
5. Inner rotor
6. Outer rotor
7. Spacer
8. Pump cover
9. Strainer screen
10. Spring pin
11. Strainer case

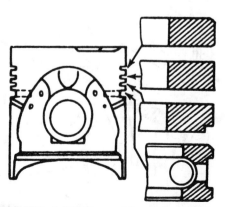

Fig. IZ16—Install rings on piston as shown above. Refer to text.

SERVICE MANUAL

PISTON PINS AND BUSHINGS

Piston pins are retained in pistons by snap rings and pins are available in standard size only. Standard pin diameter is 26.995-27.0 mm (1.0628-1.0630 inches) with wear limit of 26.97 mm (1.0618 inches). Pin fit in piston bore should be a slight interference fit. After heating piston to about 80° C (175° F), pin should be a thumb press fit in piston.

Connecting rod is fitted with renewable piston pin bushing. Install bushing with oil hole aligned with oil hole in rod and finish hone bushing to fit pin. Desired fit is clearance of 0.008-0.020 mm (0.0003-0.0008 inch). Maximum wear limit is 0.05 mm (0.002 inch).

CONNECTING RODS AND BEARINGS

Connecting rod bearings are renewable precision insert type. Connecting rods should be checked for distortion using a connecting rod alignment fixture. Distortion should not exceed 0.2 mm (0.008 inch) per 100 mm (4 inches) of length and misalignment should not exceed 0.15 mm (0.006 inch) per 100 mm (4 inches). Side clearance of rod on crankpin should be 0.18-0.30 mm (0.007-0.012 inch) with maximum allowable side clearance being 0.35 mm (0.014 inch).

Standard crankpin diameter is 52.918-52.93 mm (2.0834-2.0839 inches). Desired connecting rod bearing oil clearance is 0.029-0.082 mm (0.0011-0.0032 inch), with maximum allowable oil clearance being 0.12 mm (0.0047 inch).

CAMSHAFT AND CAM FOLLOWERS

Drain coolant and engine lubricating oil. Remove valve cover, rocker arm shaft assembly and push rods. Remove timing gear cover and oil pan. Remove oil tube and oil pump. Turn engine upside down to move cam followers away from camshaft. Remove camshaft thrust plate retaining bolts, then slide camshaft, gear and thrust plate assembly forward out of engine block. Remove cam followers from below, taking care to identify followers by position from which they were removed so they can be reinstalled in original locations.

Inspect cam followers for excessive wear, pitting or scoring of cam lobe contact surfaces and check follower barrels for excessive wear. Minimum allowable cam follower diameter is 12.95 mm (0.510 inch) and maximum clearance in block bore is 0.1 mm (0.004 inch). Cam followers should be renewed if a new camshaft is being installed.

Check camshaft journals, lobes and oil pump drive gear for excessive wear or other damage. Standard height of cam lobes is 40.6 mm (1.598 inches); camshaft should be renewed if any lobe measures 40.2 mm (1.583 inches) or less. Standard journal diameter is 48 mm (1.890 inches); wear limit is 47.6 mm (1.874 inches).

Camshaft end play can be measured with feeler gage placed between thrust plate and front journal. If end play clearance exceeds 0.2 mm (0.008 inch), renew thrust plate.

Thoroughly lubricate cam followers, cam and thrust plate and install by reversing removal procedure. Be sure timing gear marks are aligned as described in TIMING GEARS paragraph. On assembly, tighten timing gear cap screw to a torque of 108 N·m (80 ft.-lbs.).

CRANKSHAFT AND MAIN BEARINGS

Crankshaft of two-cylinder engines is supported in three main bearings and 3-cylinder engine crankshaft is supported in four main bearings. Main bearings are renewable precision insert type with crankshaft end play controlled by thrust inserts located in cylinder block on center bearing of two-cylinder engines and rear intermediate bearing of three-cylinder models. Bearing inserts are available in standard and undersizes.

Refer to CONNECTING RODS AND BEARINGS paragraph for crankpin information. Standard main journal diameter is 69.92-69.932 mm (2.7527-2.7532 inches) and desired main bearing oil clearance is 0.019-0.064 mm (0.0008-0.0025 inch). Oil clearance wear limit is 0.12 mm (0.0047 inch). Crankshaft end play should be 0.04-0.198 mm (0.0016-0.008 inch) with maximum allowable end play being 0.3 mm (0.012 inch). Check crankshaft runout at center main journal; runout should be less than 0.03 mm (0.0012 inch). Straighten or renew crankshaft if runout exceeds 0.06 mm (0.0024 inch).

When installing crankshaft, coat thrust bearings with grease and stick into position at center main bearing in block for two-cylinder engines, or at rear intermediate main bearing for three-cylinder engines. Be sure thrust bearings are installed with bearing metal surface (oil groove side) toward thrust face of crankshaft. Apply a light coat of liquid silicone gasket material to side surfaces of front and rear bearing caps and back side of oil pan packing (arch seals) as shown in Fig. IZ17. Apply a small amount of liquid silicone gasket material to outside diameter of crankshaft rear oil seal, apply grease to seal lip and place seal on rear end of crankshaft with lip to inside (forward). Place upper bearing inserts in cylinder block, lubricate bearings and crankshaft journals and lower crankshaft squarely into block. Take care not to dislodge thrust bearings and make sure that rear oil seal is in proper position. Place lower main bearing inserts in main bearing caps, install the caps and tighten evenly to a torque of 157-176 N·m (116-130 ft.-lbs.). Check to see that crankshaft turns freely.

CRANKSHAFT OIL SEALS

FRONT SEAL. To renew crankshaft front oil seal, remove timing gear cover and install new seal with lip inward (to rear) using a suitable seal driver. Lubricate lip of seal and crankshaft pulley seal contact surface prior to reassembly.

REAR OIL SEAL. To renew crankshaft rear oil seal, remove flywheel and engine rear plate. The oil seal can then be removed from rear of cylinder block and main bearing cap. Apply a thin coat of liquid silicone gasket material to outer circumference of new seal, lubricate seal lip and crankshaft oil seal surface, then install seal using suitable driver. Complete reassembly by reversing disassembly procedure.

ELECTRICAL SYSTEM

GLOW PLUGS

All models are equipped with glow plugs for easier cold starts. The tip of the

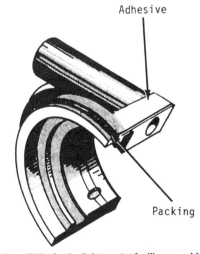

Fig. IZ17—Apply light coat of silicone rubber adhesive to side surfaces of front and rear main bearing caps and to back side of packing (pan arch gasket).

Isuzu

Small Diesel

glow plug extends into the hot plug chamber in the cylinder head. Fuel injected into the hot plug chamber will start burning on contact with the red hot glow plug.

Glow plugs are wired in parallel. Glow plug circuit will continue to operate even if one glow plug is burned out on 2-cylinder engine or two plugs are burned out on a 3-cylinder engine. However, if a glow plug burns out, it will take a longer time for the indicator light to glow red when glow plug circuit is actuated. A shorted glow plug will cause indicator to glow red in a short period of time. The glow plug indicator light is used as the fuse in the circuit; if indicator light is inoperative, glow plugs will not heat up.

To check glow plugs, disconnect glow plug connector strap and check for continuity between glow plug terminal and ground. Normal glow plug resistance is 1.8 ohms.

NOTE: Do not check glow plugs by applying full current to each individual plug as excessive current may burn out the wiring circuit. If tester is not available, test with a 10 ampere fuse connected to the lead wire of the glow plug. If fuse burns out, glow plug is shorted and must be renewed.

REGULATOR

A mechanical, adjustable regulator is used. Refer to Figs. IZ19 and IZ20 for adjustment settings.

Voltage setting
 @ 21° C (70° F)..........13.5 Volts
 @ 66° C (150° F).......14.5 Volts
Field relay:
 Yoke gap................0.9 mm
  (0.035 in.)
 Core gap............0.8-1.0 mm
  (0.31-0.39 in.)
 Point gap............0.4-0.6 mm
  (0.016-0.024 in.)
Voltage regulator:
 Yoke gap................0.9 mm
  (0.035 in.)
 Core gap............0.6-1.0 mm
  (0.024-0.039 in.)
 Point gap............0.3-0.4 mm
  (0.012-0.016 in.)

ALTERNATOR

To disassemble alternator, remove brush cover (12—Fig. IZ21) and brushes (10). Remove through-bolts and separate front cover (5) and rotor (7) from rear cover (14) and stator (9). Remove pulley (1) and fan (2), then separate rotor (7) from front frame (5). Remove rectifier retaining bolts, then withdraw stator (9) and diode holder (13) as an assembly.

Tag stator coil wires, then unsolder wires from diode terminals.

To test diodes, use ohmmeter to measure resistance between each diode terminal and holder in each direction by switching test connections. Diode is normal if resistance is nearly zero ohms in one direction and high resistance in other direction. If normal reading is not obtained on any diode, renew the rectifier assembly.

To test stator, check for continuity between each coil lead. If there is no continuity, renew stator. Standard resistance between stator terminals is 0.26 ohms. Check for continuity between coil leads and stator frame. If continuity is found, renew stator.

To test rotor, check continuity between slip rings. Standard resistance is 4.4 ohms. Check for continuity between slip rings and rotor shaft or frame. Renew rotor if continuity is found.

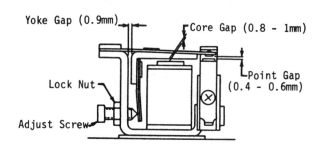

Fig. IZ19—Adjust field relay to recommended settings as shown.

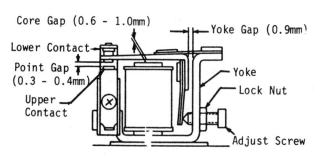

Fig. IZ20—Adjust voltage regulator to recommended settings as shown.

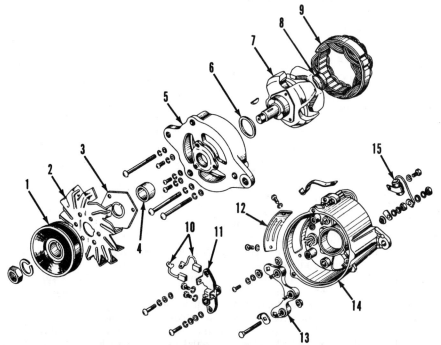

Fig. IZ21—Exploded view of alternator typical of that used on all models.

1. Pulley
2. Fan
3. Retainer plate
4. Bearing
5. Front cover
6. Spacer
7. Rotor
8. Slip rings
9. Stator
10. Brushes
11. Brush holder
12. Brush cover
13. Rectifier diode assy.
14. Rear cover
15. Condenser

SERVICE MANUAL

Check brushes for wear or other damage. Brush length when new is 14.5 mm (0.570 inch); minimum usable brush length is 9.5 mm (0.374 inch).

To reassemble alternator, reverse disassembly procedure. When soldering stator wires to diode terminals, be careful not to overheat diodes. Tighten rotor shaft nut to a torque of 35-39 N·m (26-29 ft.-lbs.).

STARTING MOTOR

Starter motor no-load test specifications are as follows:

Volts 12
Amperes .... 65 (max.)
Rpm .... 4500 (min.)

To disassemble starter, disconnect magnetic switch wires. Remove shift lever pin (4—Fig. IZ22). Unbolt and remove magnetic switch (1). Remove through-bolts and cover (20). Remove brush holder assembly (18) and gear case (6). Remove retaining clip, pinion stop (12) and pinion (7). Remove center housing (8) and armature (13).

Check armature for open circuits using ohmmeter as follows: Touch test leads to adjoining segments. Meter reading should be zero indicating continuity. Repeat test for all segments. If open circuit is found, renew armature. Check for grounded circuit by placing one test lead on armature shaft and touch each of the commutator segments with other lead. If ohmmeter indicates continuity, renew armature.

Place armature in "V" blocks and check runout of commutator with dial indicator. Renew armature or resurface commutator if runout is 0.4 mm (0.015 in.) or more. Standard outside diameter of commutator is 44 mm (1.732 inches); minimum allowable diameter is 41 mm (1.614 inches). Standard depth of undercut mica on commutator is 0.5-0.8 mm (0.020-0.031 inch) and minimum allowable depth is 0.2 mm (0.008 inch).

Check field windings for continuity between coil leads; renew field windings if open circuit exists. Renew windings if showing signs of being burned.

Standard brush length is 18 mm (0.710 inch); renew brushes if worn to 12 mm (0.470 inch) in length. Check brush spring tension which should be 7.8 N (28 ounces); renew spring if tension is low, or spring is distorted or damaged.

To reassemble starter motor, reverse disassembly procedure. Connect a 6 volt battery to magnetic switch to engage the switch and force drive pinion out (starter drive will not turn using 6 volt power source). Pinion clearance (C—Fig. IZ23) should be 0.2-1.5 mm (0.008-0.059 inch). If necessary, adjust magnetic switch plunger nut (1) to obtain desired pinion clearance.

Fig. IZ22—Exploded view of starter motor used on all models.

1. Magnetic switch
2. "O" ring
3. Shift lever
4. Pin
5. Bushing
6. Gear case
7. Pinion gear & clutch
8. Center housing
9. Bushing
10. "O" ring
11. Frame
12. Pinion stop
13. Armature
14. Field coil
15. Washers
16. Brush spring
17. Brush
18. Brush holder
19. Bushing
20. Rear cover

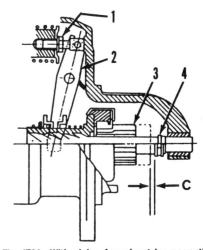

Fig. IZ23—With pinion forced out by magnetic switch, clearance "C" should be within range of 0.2-1.5 mm (0.008-0.059 inch). Refer to text.

1. Plunger nut
2. Shift lever
3. Pinion
4. Pinion stop

ISUZU

Model	No. Cyls.	Bore	Stroke	Displ.
2KA1	2	70 mm (2.756 in.)	70 mm (2.756 in.)	538 cc (32.8 cu. in.)
2KB1	2	70 mm (2.756 in.)	76 mm (2.992 in.)	584 cc (35.6 cu. in.)
2KC1	2	74 mm (2.913 in.)	76 mm (2.992 in.)	653 cc (39.8 cu. in.)
3KA1	3	70 mm (2.913 in.)	70 mm (2.992 in.)	808 cc (49.3 cu. in.)
3KB1	3	70 mm (2.913 in.)	76 mm (2.992 in.)	877 cc (53.5 cu. in.)
3KC1	3	74 mm (2.913 in.)	76 mm (2.992 in.)	980 cc (59.78 cu. in.)

All models are four-stroke, in-line, overhead camshaft, water cooled, indirect injection diesel engines. Firing order is 1-3-2 on three-cylinder models.

MAINTENANCE

LUBRICATION

Recommended engine lubricant is good quality oil with API classification CC or CD. Below −30° C (−20° F), SAE 5W-20 motor oil is recommended. SAE 10W-30 or 10W-40 motor oil may be used at temperatures above −30° C (−20° F), and SAE 15W-40 or 15W-50 oil may be used at temperatures above −15° C (0° F).

FUEL SYSTEM

FUEL FILTER. Refer to Fig. IZ30 for exploded view of fuel filter assembly with cartridge type filter. Filter cartridge should be renewed after each 600 hours of operation, or earlier if engine is misfiring or not running evenly.

BLEED FUEL SYSTEM. To bleed air from fuel system, remove bleed screw (1—Fig. IZ30) on top of filter and open fuel supply valve. When fuel is flowing freely without air bubbles, install and tighten filter bleed screw and open bleed screw (9—Fig. IZ31) on fuel injection pump. When fuel is flowing without air bubbles at injection pump bleed screw, tighten the screw. If engine will not start at this time, loosen the injection lines at the fuel injectors. Move throttle control to ''Run'' position and crank engine until fuel is flowing from all open lines. Tighten the fuel lines and start engine.

INJECTION PUMP TIMING

Injection should occur at 16 degrees BTDC on all models. To check static timing, proceed as follows: Thoroughly clean injection pump and lines. Remove the No. 1 fuel injection line. Remove the No. 1 fuel delivery valve holder (1—Fig. IZ31), remove the delivery valve spring (4), then replace the delivery valve holder. Open the fuel supply valve; fuel should flow from the open delivery valve holder. Slowly turn engine crankshaft in normal direction until fuel stops flowing. Check to see that 16 degrees BTDC timing mark on engine crankshaft pulley is aligned with timing pointer.

Timing is adjusted by changing thickness of the shim (S—Fig. IZ32) under fuel injection pump mounting flange. Timing will be changed about 1 degree with each 0.1 mm change in shim thickness. Increasing shim thickness will retard timing, decreasing shim thickness will advance timing. Shims are available in 11 thicknesses ranging from 0.2 mm to 1.2 mm in steps of 0.1 mm, and have a mark to indicate thickness.

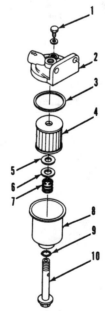

Fig. IZ30—Exploded view of fuel filter assembly used on all models.

1. Air bleed screw
2. Filter body
3. Seal ring
4. Filter element
5. Seal ring
6. Washer
7. Spring
8. Filter cover
9. Seal ring
10. Center bolt

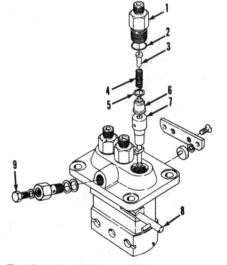

Fig. IZ31—View of three-cylinder fuel injection pump showing air bleed screw (9) and delivery valve spring (4). Pump for two-cylinder engine is similar.

1. Delivery valve holder
2. ''O'' ring
3. Delivery valve stop
4. Delivery valve spring
5. Gasket
6. Delivery valve assy.
7. Plunger assy.
8. Pump control rack
9. Air bleed screw

SERVICE MANUAL

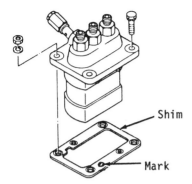

Fig. IZ32—Injection pump timing is varied by changing thickness of shim under the injection pump flange. A mark on the shim indicates thickness.

After correct timing is obtained, reinstall delivery valve spring and tighten delivery valve holder to a torque of 45 N·m (33 ft.-lbs.).

GOVERNOR

The flyweight type governor is mounted on the front face of the fuel injection pump drive gear. Engine idle and no-load governed speeds will vary with engine application; therefore, it will be necessary to check engine or equipment operator's manual for engine speed specifications and adjustment.

REPAIRS

TIGHTENING TORQUES

Refer to the following special tightening torques when reassembling engine. All fasteners are metric.

Connecting rod .... 39 N·m
(29 ft.-lbs.)
Main bearing cap bolts (1) .... 88 N·m
(65 ft.-lbs.)
Cylinder head bolts (2),
 new bolts .... 98 N·m
(72 ft.-lbs.)
When reusing cylinder
 head bolts .... 113 N·m
(83.3 ft.-lbs.)
Cylinder head to timing
 cover bolt .... 20 N·m
(15 ft.-lbs.)
Camshaft cap bolts (3) .... 20 N·m
(15 ft.-lbs.)
Intake & exhaust manifolds .. 20 N·m
(15 ft.-lbs.)
Valve cover .... 8-14 N·m
(6-10 ft.-lbs.)
Camshaft timing sprocket
 bolt .... 103 N·m
(76 ft.-lbs.)
Governor flyweight bolts .... 7.8 N·m
(69 in.-lbs)

Timing gear idler .... 25 N·m
(19 ft.-lbs.)
Injection pump drive
 gear .... 69-88 N·m
(50-65 ft.-lbs.)
Injection pump stop
 lever plate .... 0.8 N·m
(7 in.-lbs.)
Injection nozzle .... 49 N·m
(36 ft.-lbs.)
Glow plugs .... 15 N·m
(10 ft.-lbs.)
Timing cover to block bolts ... 20 N·m
(15 ft.-lbs.)
Engine oil pump mounting
 bolts .... 20 N·m
(15 ft.-lbs.)
Crankshaft pulley .... 176 N·m
(130 ft.-lbs.)
Rear oil seal plate .... 17 N·m
(13 ft.-lbs.)
Flywheel housing to block ... 40 N·m
(30 ft.-lbs.)
Flywheel bolts .... 98 N·m
(72 ft.-lbs.)
Starting motor mounting ..... 39 N·m
(29 ft.-lbs.)

(1) Tighten in sequence shown in Fig. IZ69.
(2) Tighten in sequence shown in Fig. IZ44.
(3) Tighten a small amount at a time in sequence shown in Fig. IZ38.

WATER PUMP

To remove water pump, drain engine coolant and remove radiator if necessary. Loosen fan belt adjustment and remove fan belt and fan. Unbolt and remove water pump assembly. Water pump body casting may be either aluminum or cast iron.

To disassemble water pump, use suitable pullers to remove fan pulley hub (1—Fig. IZ33). Remove bearing set screw (3) and rear cover (9). If water pump housing is aluminum, heat water pump in hot water to a temperature of 80-90° C (176-194° F). On all pumps, use a drift smaller than water pump shaft diameter and press shaft and bearing assembly (2) out front (pulley) end of pump body.

Water pump repair kit consists of new bearing (2), seal (7), impeller (8) and gaskets. Reassemble pump using repair kit as follows: Align set screw hole in bearing with hole in pump body and press bearing into place, then install set screw (3). Apply a thin coat of liquid gasket such as "Belco Bond No. 4" on outer diameter of seal unit, then install seal over bearing shaft and into rear of pump body. Support front end of bearing shaft and press impeller onto rear end of shaft until rear face of impeller is 0.4-0.6 mm (0.016-0.023 inch) below rear face of pump body. Support rear end of bearing shaft and press fan pulley hub onto front end of shaft so that distance from fan mounting surface to rear face of pump, including thickness of gasket and rear cover, is 91.7-92.3 mm (3.610-3.634 inches). Apply suitable liquid gasket sealer to both sides of gasket (6) and install rear cover. Apply liquid gasket sealer to both sides of pump mounting gasket, then reinstall water pump by reversing removal procedure.

THERMOSTAT

To remove thermostat, drain engine coolant, then unbolt and remove water outlet housing from cylinder block. The thermostat should start to open at 75-78° C (167-172° F) and be fully open at 90° C (194° F).

VALVE ADJUSTMENT

To check valve adjustment, remove valve cover and turn crankshaft so "TDC" mark on crankshaft pulley is aligned with pointer as shown in Fig. IZ34. If timing mark on rear end of camshaft (see Fig. IZ35) is aligned with joining line of bearing cap at left side of engine (as viewed from rear), number one

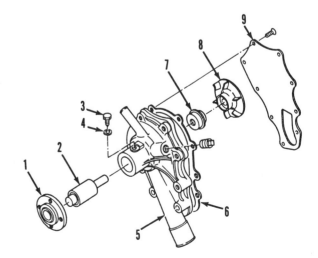

Fig. IZ33—Exploded view of water pump assembly used on all models.
1. Fan pulley hub
2. Shaft & bearing assy.
3. Bearing set screw
4. Lockwasher
5. Pump housing
6. Gasket
7. Seal assy.
8. Impeller
9. Rear cover

Isuzu — SMALL DIESEL

piston is on TDC on compression stroke. If mark on rear end of camshaft is opposite the position shown in Fig. IZ35, number one piston is on exhaust stroke.

With number one piston on compression stroke, measure clearance between both intake and exhaust cam lobes and tappets for No. 1 (front) cylinder. Also measure No. 2 cylinder intake valve clearance on all engines, and on 3-cylinder engines, measure No. 3 cylinder exhaust valve clearance.

Rotate crankshaft one complete revolution so number one piston is on exhaust stroke (mark on rear end of camshaft will be opposite position shown in Fig. IZ35). Check exhaust valve clearance on No. 2 cylinder on all engines, and on 3-cylinder engines, check intake valve clearance on No. 3 cylinder. Desired valve clearance cold is 0.25 mm (0.010 inch) for both intake and exhaust valves.

Record clearance of any valve if clearance is not within specifications. If valve clearance is less than 0.1 mm (0.004 inch) or more than 0.4 mm (0.016 inch), clearance must be adjusted by changing shim (9—Fig. IZ36).

To adjust clearance, first remove camshaft (7) and tappets (8) as outlined in following paragraph. Remove shims (9) from tappets for valves not within clearance specifications, taking care not to mix parts from different valves. Using a micrometer, measure thickness of shim removed; then, add measured valve clearance (in millimeters) and subtract 0.25 mm to determine thickness of shim to install for proper clearance. Shims are available in 49 different thicknesses with minimum thickness of 1.0 mm and in graduations of 0.025 mm.

Reinstall tappets and camshaft, then recheck clearance after reinstalling camshaft sprocket. With clearance within specifications, complete reassembly of engine.

CAMSHAFT AND TAPPETS

To remove camshaft, remove valve cover and proceed as follows: Turn engine so that "TDC" mark on crankshaft pulley is aligned with pointer as shown in Fig. IZ34 and timing mark on rear end of camshaft is in position shown in Fig. IZ35. Refer to Fig. IZ37 and remove the chain tensioner plug from front of timing cover. Using a screwdriver, move lock lever to the right and tilt tension lever to the left to create slack in timing chain. Remove camshaft sprocket from camshaft, leaving sprocket in chain and chain guide. Loosen cam bearing cap bolts a small amount at a time and in reverse order of tightening sequence shown in Fig. IZ38, then remove camshaft. Remove valve tappets taking care not to lose or exchange adjusting shims and place tappets and shims in container marked for tappet position.

Inspect cam bearing caps and bearing surfaces in cylinder head for scoring, excessive wear or other damage. Inspect camshaft for excessive wear or obvious damage and check camshaft against the following specifications: Nominal cam lobe height (maximum diameter) is 45.75 mm (1.801 inches) for intake lobes and 46.68 mm (1.838 inches) for exhaust lobes. Renew camshaft if intake lobe height measures 45.25 mm (1.7815 inches) or less, or if exhaust lobe height

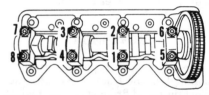

Fig. IZ38—When installing camshaft, tighten bolts a small amount at a time in sequence shown. When removing camshaft, loosen bolts in reverse order of tightening sequence.

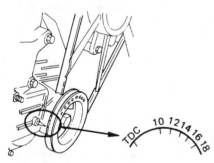

Fig. IZ34—Timing marks are located on crankshaft pulley as shown; timing pointer (circle) is located in timing cover.

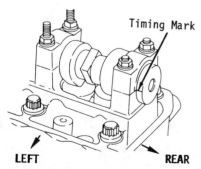

Fig. IZ35—When timing mark on rear end of camshaft is positioned as shown, number one piston should be on TDC of compression stroke. Turn crankshaft one complete revolution to bring piston to TDC of exhaust stroke.

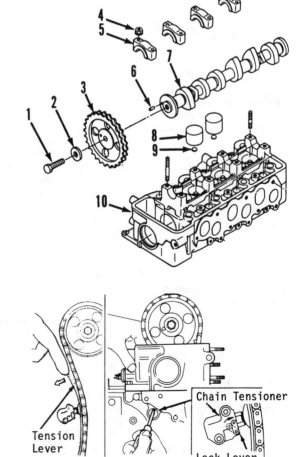

Fig. IZ36—Valve clearance is adjusted by removing camshaft and changing thickness of shim (9) under tappet (8).

1. Camshaft bolt
2. Washer
3. Camshaft sprocket
4. Nuts
5. Cam bearing caps
6. Dowel pin
7. Camshaft
8. Valve tappet
9. Adjusting shim
10. Cylinder head

Fig. IZ37—Chain tensioner can be locked in released position by working through hole in timing gear cover. Refer to text.

SERVICE MANUAL

Isuzu

measures 46.18 mm (1.818 inches) or less. Nominal camshaft journal diameter is 26 mm (1.0236 inches). Camshaft bearing oil clearance should be 0.040-0.082 mm (0.0016-0.0032 inch), with wear limit being 0.12 mm (0.0047 inch). Maximum allowable runout of center cam bearing journal is 0.1 mm (0.004 inch).

Reinstall valve tappets with shim thickness as removed unless changing shim thickness to correct tappet to cam clearance. Check to see that TDC mark on crankshaft pulley is aligned with pointer and place camshaft on cylinder head with timing mark on rear end of cam in position shown in Fig. IZ35. Align locating pin (6—Fig. IZ36) on camshaft with hole in timing sprocket (3). Install and partially tighten cam sprocket bolt. Position the camshaft caps (5) in proper order and install and tighten the cap bolts a small amount at a time in sequence shown in Fig. IZ38 until caps are drawn against cylinder head, then tighten the bolts to a torque of 20 N·m (15 ft.-lbs.). Release the tensioner and install chain tensioner plug. Tighten the cam sprocket bolt to a torque of 103 N·m (76 ft.-lbs.). Recheck valve clearance and if correct, reinstall valve cover.

CYLINDER HEAD

To remove cylinder head, first drain engine coolant and proceed as follows: Thoroughly clean fuel injection pump and injector lines using suitable solvent and compressed air. Disconnect fuel injector lines and immediately cap or plug all fuel openings. Unscrew fuel injectors and glow plugs from head. Remove camshaft as outlined in preceding paragraph. Remove the bolts retaining cylinder head to timing cover, then loosen cylinder head bolts a little bit at a time following sequence shown in Fig. IZ39. When bolts are completely loose, remove bolts and cylinder head.

Thoroughly clean gasket material from head and check cylinder head for warpage (distortion) using straightedge and feeler gage (Fig. IZ41). Distortion of head surface should be less than 0.075 mm (0.003 inch). Renew or resurface head if distortion exceeds 0.15 mm (0.006 inch); maximum amount that can be removed from combination of head and block surfaces is 0.3 mm (0.012 inch). Refer to VALVE SYSTEM paragraphs and check distance (R—Fig. IZ45) that face of valve head is below head surface after resurfacing head. If head is resurfaced, use oversize thickness gasket when reinstalling head.

Set camshaft in cylinder head to check alignment of cam bearing surfaces; camshaft journals should contact all bearing surfaces. Renew head or align bore cam bearings if cam bearings are out of line.

With cylinder head removed, hot plugs can be removed as shown in Fig. IZ42. When hot plugs are removed, heat shields can be tapped from fuel injector bores as shown in Fig. IZ43. To install new hot plug, be sure bore in head is clean and free of burrs, then align lock ball in hot plug with groove in head. Tap hot plug into head using plastic faced hammer, then use bench press and protective metal pad to press hot plug into proper position. After installation, face of hot plug must be surface ground flush with cylinder head surface. Tap new heat shield into place using brass drift.

To install cylinder head, reverse removal procedure, using an oversize thickness gasket if cylinder head and/or block mating surfaces have been reground, or if piston protrusion above block at TDC of stroke exceeds 0.94 mm (0.037 inch). Lubricate bolt threads and install all bolts finger tight, including the two timing cover bolts (9 and 10—Fig. IZ44). Tighten cylinder head bolts (except bolts to timing cover) to a torque of 49 N·m (36 ft.-lbs.) in sequence shown in Fig. IZ44, then to final torque of 98 N·m (72 ft.-lbs.) if using new bolts or 113 N·m (83 ft.-lbs.) if reusing head bolts. Then, tighten the two cylinder head to timing cover bolts to a torque of 20 N·m (15 ft.-lbs.). Complete reassembly by reversing removal procedures.

VALVE SYSTEM

Valve face and seat angle is 45 degrees for both intake and exhaust valves. Valves seat in renewable seat inserts in cylinder head. Recondition seats using 45, 60 and 30 degree stones or cutters. Valve seat width should be 1.75 mm (0.070 inch); maximum allowable seat width is 2.5 mm (0.098 inch).

After reconditioning seats, check distance (R—Fig. IZ45) valve head is recessed below cylinder head surface. Standard valve head recess is 0.7 mm (0.027 inch). If valve head recess is 1.4 mm (0.055 inch) or more, install new valve and/or valve seat.

To remove old valve seat, use arc welder to run a bead around portions of the seat as shown in Fig. IZ46. Seat can be easily removed after cooling. Using suitable piloted valve seat driver, install new seat with chamfered outer edge down. Then, regrind seating surface for full valve contact and proper seat width.

Standard valve head thickness (margin) is 1.0 mm (0.039 inch). Renew valve if thickness is 0.7 mm (0.027 inch) or less after reconditioning. Nominal valve stem diameter is 7.0 mm (0.275 inch). Renew intake valve if stem is worn to 6.85 mm (0.270 inch) and exhaust valve if stem is worn to 6.8 mm (0.268 inch). Desired valve stem to guide clearance is 0.023-0.056 mm (0.0009-0.002 inch) for

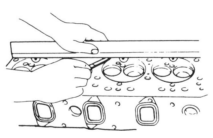

Fig. IZ41—Cylinder head surface should be checked for distortion using a straightedge and feeler gage as shown.

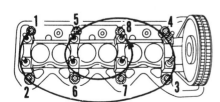

Fig. IZ39—When removing cylinder head, loosen bolts a little at a time in sequence shown to prevent distortion of head.

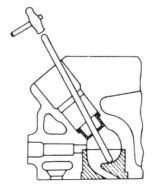

Fig. IZ42—Hot plugs may be driven out of cylinder head using drift pin inserted through injector bore and heat shield as shown.

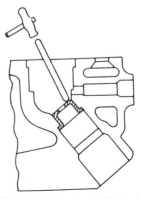

Fig. IZ43—Carefully tap heat shield out of injector bore as shown after removing hot plug.

Isuzu

intake valves, and maximum allowable clearance is 0.15 mm (0.006 inch). Exhaust valve stem to guide clearance should be 0.03-0.063 mm (0.0012-0.0025 inch), and maximum allowable exhaust stem clearance is 0.2 mm (0.008 inch).

Intake and exhaust valve spring free length new is 47.4 mm (1.866 inches). Renew spring if free length is 46 mm (1.811 inches) or less. Be sure to install springs with closer spaced coils next to cylinder head.

INJECTORS

REMOVE AND REINSTALL. To remove fuel injectors, first thoroughly clean injectors, lines and surrounding area using suitable solvent and compressed air. Disconnect fuel return line from fittings on side of injectors, and remove high pressure lines from pump to injectors. Immediately cap or plug all openings to prevent entry of dirt. Remove injectors from cylinder head by unscrewing the injector retaining nuts (9—Fig. IZ47).

Clean injector bores in cylinder head before reinstalling injectors, taking care not to damage the heat shields (see Fig. IZ43). Install injectors and tighten retaining nuts to a torque of 49 N·m (36 ft.-lbs.). If necessary to align return line fittings to line, hold injector body (3—Fig. IZ47) and loosen nut (1). Align fittings with return line and install line, then tighten nut (1). Leave high pressure lines loose, crank engine until air-free fuel flows from lines, then tighten the line fittings.

TESTING. A faulty injector may be located on the engine by loosening high pressure fuel line fitting on each injector, in turn, with engine running. This will allow fuel to escape from the loosened fitting rather than enter the cylinder. The injector that least affects engine operation when its line is loosened is the faulty injector. If a faulty injector is found and considerable time has elapsed since injectors have been serviced, it is recommended that all injectors be removed and serviced, or that new or reconditioned units be installed.

A complete job of testing and adjusting an injection nozzle requires use of special test equipment. Use only a clean approved testing oil in injector tester tank. Injection nozzle should be checked for opening pressure, spray pattern, seat leakage and back leakage. Test injection nozzle as outlined in following paragraphs:

WARNING: Fuel emerges from injector with sufficient force to penetrate the skin. When testing injector, keep yourself clear of nozzle spray.

OPENING PRESSURE. Before conducting test, operate tester lever until fuel flows from tester line, then attach injection nozzle to line. Pump lever a few quick strokes to purge air and be sure that nozzle valve is not stuck and spray hole is open.

Operate tester lever slowly and observe gage reading when nozzle opens. On all models, opening pressure should be 11700 kPa (1700 psi). If necessary to adjust opening pressure, remove nozzle retaining nut (9—Fig. IZ47), nozzle (8), spacer (7), spring seat (6), spring (5) and shims (4). Add shims to increase opening pressure or decrease shim thickness to decrease pressure. Adjusting shims are available in 38 different thicknesses ranging from 0.5 to 1.24 mm; they are graduated in steps of 0.02 mm. Injector opening pressure varies approximately

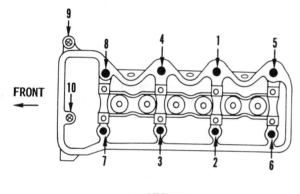

Fig. IZ44—Cylinder head bolt tightening sequences for two-cylinder and three-cylinder models are shown. Do not tighten bolts (9 and 10) to timing cover until head bolts are properly torqued.

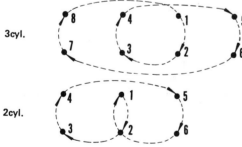

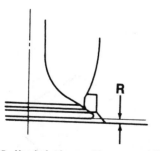

Fig. IZ45—Head of valve must be recessed (R) at least 0.7 mm (0.028 in.) below flush with cylinder head; maximum allowable recess is 1.4 mm (0.055 in.).

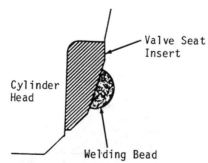

Fig. IZ46—To remove valve seat, run arc-weld bead around seat. Seat will shrink when cool and can be removed easily.

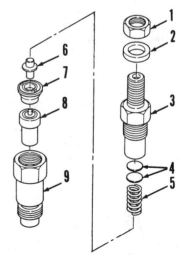

Fig. IZ47—Exploded view of fuel injection nozzle used on all models.

1. Inlet fitting nut
2. Washer
3. Injector body
4. Adjusting shims
5. Spring
6. Spring seat
7. Spacer
8. Nozzle & valve assy.
9. Retaining nut

SERVICE MANUAL

Isuzu

470 kPa (68 psi) with each 0.02 mm change in shim thickness. Change shim thickness as necessary, reassemble nozzle and recheck opening pressure.

SPRAY PATTERN. The throttling type nozzle should emit a tight conical spray pattern with no branches, splits or dribbles. Refer to Fig. IZ48 for drawing showing correct spray pattern and possible incorrect patterns. If incorrect spray pattern is observed, check for partially clogged or damaged spray hole or improperly seating or sticking nozzle valve.

SEAT LEAKAGE. Wipe nozzle tip dry, then operate tester lever slowly to maintain a gage pressure approximately 1000 kPa (150 psi) below nozzle opening pressure for a period of 10 seconds. If a drop forms or undue wetness appears on nozzle tip, disassemble and clean injector. Slight wetting of nozzle tip is acceptable on a used injector.

OVERHAUL. First, thoroughly clean outside of injector. Hold injector body (3—Fig. IZ47) and remove return line nut (1), washer (2) and fitting. Unscrew nozzle retaining nut (9) and disassemble nozzle. Place all parts in clean calibrating oil or diesel fuel and keep all parts of each injector separate from others.

Clean exterior surfaces with a brass wire brush. Soak parts in an approved carbon solvent if necessary to loosen hard carbon deposits. Rinse parts in clean diesel fuel or calibrating oil immediately after cleaning to neutralize the solvent.

Clean nozzle spray hole from inside using a hardwood scraper. Scrape carbon from nozzle chamber using a hooked brass scraper. Clean valve seat using a brass scraper, then polish seat using wood polishing stick and mutton tallow.

Back flush nozzle using a reverse flush adapter on injector tester. Reclean all parts by rinsing thoroughly in clean diesel fuel or calibrating oil, and reassemble while parts are immersed in the fluid. Be sure to assemble with same shims as were removed. Tighten nozzle retaining nut to a torque of 49 N·m (36 ft.-lbs.) and retest injector.

INJECTION PUMP

To remove injection pump, first thoroughly clean pump, injection lines and surrounding area. Close fuel supply valve. Disconnect and remove fuel injector lines and fuel supply and return lines from pump. Immediately cap or plug all openings to prevent entry of dirt. Disconnect engine stop control from lever on plate below pump (see Fig. IZ49), then unbolt and remove plate from engine. Working through opening, remove the link plate and set spring as shown in Fig. IZ50. Unbolt pump from cylinder block and lift pump upward, taking care not to lose or damage shim between pump flange and cylinder block. Be sure to place pump in clean area.

The injection pump should be tested and overhauled by a shop qualified in diesel injection equipment repair.

To reinstall pump, reverse removal procedure and install pump with same thickness shim as when removed unless change of timing is indicated. Shim may be reused up to three times. When installing throttle control lever plate, tighten the front upper cap screw as indicated in Fig. IZ49 after tightening the other three cap screws. Bleed the fuel system as outlined in MAINTENANCE section.

TIMING GEAR COVER

First, thoroughly clean timing gear cover and surrounding area. Remove fan, fan belt and crankshaft pulley. Remove front oil pan bolts that are threaded into bottom of timing gear cover. Disconnect fuel shut-off control from lever on plate at side of engine behind timing gear cover, remove the plate and remove the link plate and spring as shown in Fig. IZ50. Remove the lower nut from accessory drive plate (see Fig. IZ51). Unbolt and remove timing gear cover from front of engine.

To reinstall timing gear cover, proceed as follows: If oil pan has not been removed, remove all old gasket material from front edge of pan and cut portion from new pan gasket to fit. Lower ends of timing cover gasket must be trimmed so that gasket will be aligned with bolt holes. Apply gasket sealer to both sides of gasket and a small bead of sealer between joint of pan and crankcase. Set timing gear cover in place being sure link plate and set spring enter pump connecting hole. Refer to Fig. IZ52. Install timing gear cover bolts and tighten to a torque of 20 N·m (15 ft.-lbs.) in sequence shown in Fig. IZ53. Complete reassembly by reversing disassembly procedure.

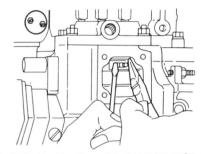

Fig. IZ50—Working through fuel stop lever plate opening, remove link plate and set spring.

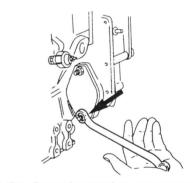

Fig. IZ51—Remove lower nut from accessory cover prior to removing engine timing gear cover.

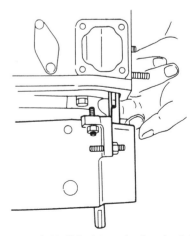

Fig. IZ52—Guide link plate and set spring into pump connecting hole when reinstalling timing gear cover assembly.

Fig. IZ48—Drawing showing correct and incorrect injector nozzle spray patterns.

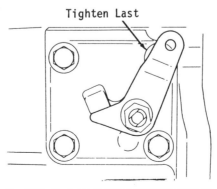

Fig. IZ49—Fuel stop lever is mounted in plate below fuel injection pump. Upper rear bolt should be tightened last when installing plate.

Isuzu

SMALL DIESEL

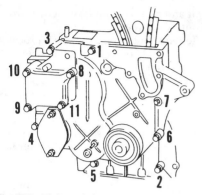

Fig. IZ53—Tighten timing gear cover bolts in sequence shown.

GOVERNOR

The flyweight type governor is mounted on the front of the fuel injection pump drive gear as shown in Fig. IZ54 and can be serviced after removing engine timing gear cover. Governor linkage is mounted in the timing gear cover. Check flyweight assembly and linkage for excessive wear and binding, and renew as necessary. Refer to MAINTENANCE section for governor adjustment.

INJECTION PUMP CAMSHAFT

To remove injection pump camshaft (8—Fig. IZ54), first remove fuel injection pump, engine oil pump and timing gear cover, then proceed as follows: Unbolt and remove governor flyweight assembly (5) from front of drive gear (6), then remove the hex nut, washer and gear from front of camshaft (8). Remove the bearing retaining bolt (at top side of bearing) from front face of crankcase then bump camshaft and bearing forward out of block. If necessary to renew camshaft bushing, use suitable camshaft bearing tools to remove old bushing. Install new bushing with notch in end up and toward rear of engine as shown in Fig. IZ55.

Reinstall fuel injection pump camshaft, ball bearing and bearing retaining bolt. Be sure camshaft turns freely before proceeding further. Complete reassembly by reversing removal procedure, making sure to observe all special tightening torques.

TIMING GEARS

The crankshaft gear (18—Fig. IZ54), idler gear (21) and fuel injection pump drive gear (6) are accessible after removing engine timing gear cover. Check gear backlash, idler gear bearing clearance and idler gear end play before removing gears. Backlash in gears should be 0.06 mm (0.0024 inch) with maximum allowable backlash being 0.2 mm (0.008 inch). Idler gear to shaft oil clearance should be 0.025-0.085 mm (0.001-0.003 inch), with maximum allowable clearance being 0.2 mm (0.008 inch).

Remove the two bolts from idler gear, then remove retaining plate (22), gear (21) and shaft (19) from front of cylinder block. Use suitable gear puller to remove crankshaft gear (18). Remove governor flyweight assembly (5), gear retaining bolt and injection pump drive gear (6) from front of fuel injection pump camshaft.

Install idler gear shaft (19) in block with oil hole in shaft aligned with oil hole in block (to left as viewed from front of engine). When installing gears, be sure "X" marks on crankshaft gear and idler gear, and "Y" marks on idler gear and fuel injection pump drive gear align as shown in Fig. IZ56. Complete reassembly by reversing disassembly procedure, making sure all fasteners are tightened to specified torque.

CAMSHAFT DRIVE CHAIN AND SPROCKETS

With crankshaft gear and idler gear assembly removed, timing chain (9—Fig. IZ54), camshaft sprocket (10) and crank-

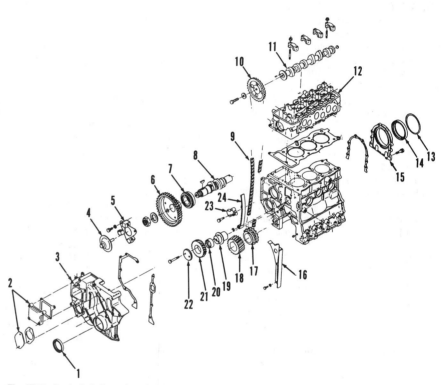

Fig. IZ54—Exploded view of engine timing gears. Governor flyweight assembly (5) is mounted on injection pump camshaft drive gear (6).

1. Crankshaft front oil seal
2. Inspection hole covers
3. Timing gear cover
4. Sleeve
5. Governor flyweight assy.
6. Injection pump timing gear
7. Ball bearing
8. Injection pump camshaft
9. Timing chain
10. Camshaft sprocket
11. Valve camshaft
12. Cylinder head
13. Spacer
14. Crankshaft rear oil seal
15. Seal retainer
16. Timing chain guide
17. Crankshaft sprocket
18. Crankshaft timing gear
19. Idler gear shaft
20. Bushing
21. Idler gear
22. Retainer plate
23. Chain tensioner
24. Tension lever

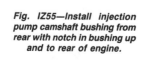

Fig. IZ55—Install injection pump camshaft bushing from rear with notch in bushing up and to rear of engine.

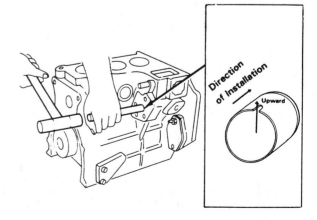

54

SERVICE MANUAL

Isuzu

shaft sprocket (17) can be removed. The chain tensioner (23), tension lever (24) and chain guide (16) can now be removed. Renew both sprockets and chain if either shows excessive wear. Check chain contact surfaces of guide and tension lever and renew if excessive wear is noted. If tensioner pivot pin is excessively worn, install new pin in block with identification punch mark up and so pin protrudes 37.5 mm (1.476 inch) as shown in Fig. IZ58. Be sure plunger in chain tensioner is working smoothly.

Install chain guide, tensioner and tension lever; lock tensioner in retracted position. With keyway in crankshaft up, install crankshaft sprocket with chamfered inside diameter in as shown in Fig. IZ59. Install camshaft sprocket and timing chain with marked links aligned with marks on crankshaft sprocket and camshaft sprocket as shown in Fig. IZ60, then release chain tensioner. Complete reassembly by reversing disassembly procedure, being sure to tighten all fasteners to specified torque.

OIL PUMP AND RELIEF VALVE

The externally mounted gerotor type oil pump, filter base and relief valve assembly is attached to engine crankcase and the pump is driven from the rear end of the fuel injection pump camshaft. To remove pump, remove oil filter canister and remove the three bolts (see Fig. IZ61) holding pump to engine crankcase.

Refer to Fig. IZ62 for exploded view of pump and disassemble by removing remaining two cap screws (13 and 14) from pump.

Clearance between lobes of inner and outer rotor (Fig. IZ63) should be less than 0.2 mm (0.008 inch). Limit for clearance between outer rotor and pump body (Fig. IZ64) is 0.4 mm (0.016 inch), and end clearance of pump rotors to cover (Fig. IZ65) should be less than 0.15 mm (0.006 inch).

Remove relief valve cap (8—Fig. IZ62); spring (10) and valve (11). Valve should slide freely in bore and be free of score marks. With engine at normal operating temperature, oil pressure should be 147 kPa (21 psi) at idle speed and 295 kPa (42 psi) at medium speed.

Clean all parts thoroughly and reassemble by reversing disassembly procedure. Fill pump with clean heavy weight oil and turn pump shaft back and forth to be sure pump is primed. Tighten oil pump mounting bolts to a torque of 20 N·m (15 ft.-lbs.).

PISTON AND ROD UNITS

Piston and connecting rod units are removed from above after removing cyl-

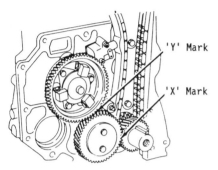

Fig. IZ56—View showing timing gear timing mark location.

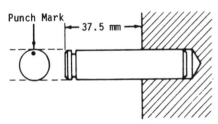

Fig. IZ58—Install new tension lever pin with punched mark side up.

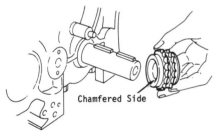

Fig. IZ59—Install crankshaft timing sprocket with chamfered side in.

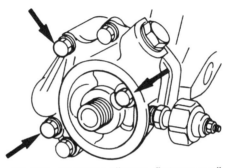

Fig. IZ61—To remove oil pump, first remove oil filter and remove the three bolts indicated by arrows.

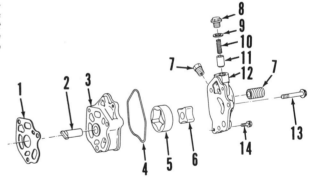

Fig. IZ62—Exploded view of Model 3KC1 oil pump assembly; pumps for other models are similarly constructed. Flat on shaft (2) engages slot in rear end of injection pump camshaft.

1. Gasket
2. Drive shaft
3. Pump body
4. Sealing ring
5. Outer rotor
6. Inner rotor
7. Plugs
8. Relief valve cap
9. Gasket
10. Spring
11. Relief valve
12. Pump cover
13. Cover bolt
14. Cover bolt

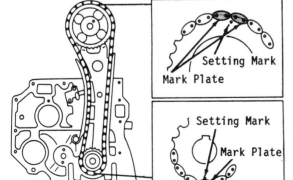

Fig. IZ60—Marked plates on timing chain and marks on timing sprockets are aligned as shown to correctly time engine camshaft.

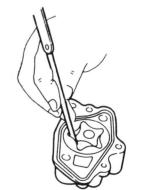

Fig. IZ63—Check clearance between inner and outer rotor lobes using a feeler gage.

55

Isuzu

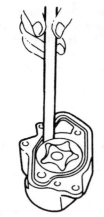

Fig. IZ64—Use feeler gage to check clearance between outer rotor and pump body.

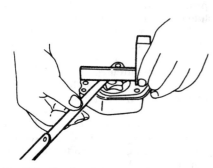

Fig. IZ65—Check end clearance of inner and outer rotors using straightedge and feeler gage.

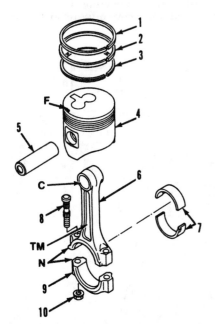

Fig. IZ66—Drawing showing piston and connecting rod unit. Trade mark (ISUZU) on rod and front mark (F) on piston are to be installed toward front of engine. Also note chamfered (C) end of pin bore in rod and cylinder number (N) location on rod.

C. Large chamfer
F. Piston front mark
N. Cylinder number
TM. Trade mark
1. Top compression ring
2. Second compression ring
3. Oil control ring
4. Piston
5. Piston pin
6. Connecting rod
7. Bearing inserts
8. Connecting rod bolt
9. Connecting rod cap
10. Nut

inder head and oil pan. Piston pin is retained by press fit in connecting rod. The rod and piston should not be disassembled unless necessary to renew piston or connecting rod.

NOTE: Disassembly or reassembly should not be attempted without proper piston pin inserting tools.

To assemble rod and piston, refer to Fig. IZ66. Heat connecting rod to a temperature of 180°-220°C (335°-425°F). Pin should be pressed in from front side of piston (a notch [F] on piston crown indicates front side of piston) and large chamfered side (C) of connecting rod.

PISTONS, PINS, RINGS AND CYLINDERS

Refer to previous paragraph on piston and connecting rod unit. Check pistons for cracking, scoring or excessive wear in ring grooves or on piston skirt. To determine piston-to-cylinder clearance, measure piston skirt diameter at a point 55.2 mm (2.173 inches) from top face of piston and at right angle to pin bore, and measure cylinder bore diameter at several locations. Piston skirt to cylinder bore clearance should be 0.102-0.132 mm (0.0040-0.0052 inch). Pistons and rings are available in standard size and oversizes of 0.25 mm (0.010 inch) and 0.50 mm (0.020 inch).

Piston pin bore to pin clearance should be 0.016-0.020 mm (0.0006-0.0008 inch) with wear limit of 0.075 mm (0.0029 inch). Pin is a press fit in connecting rod. Piston pin nominal diameter is 21 mm (0.8268 inch). Standard inside diameter of connecting rod small end is 20.964-20.977 mm (0.8254-0.8258 inch).

Piston ring side clearance in groove should be 0.045-0.085 mm (0.0018-0.0033 inch) for top ring, 0.030-0.070 (0.0012-0.0027 inch) for second ring, and 0.02-0.06 mm (0.0008-0.0024 inch) for the oil ring. Maximum allowable side clearance in groove is 0.3 mm (0.012 inch) for compression rings and 0.15 mm (0.006 inch) for the oil ring. Renew piston if ring side clearance is excessive.

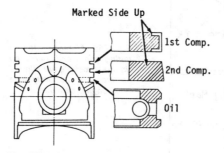

Fig. IZ67—Install piston rings in grooves as shown.

SMALL DIESEL

Ring end gap should be 0.2-0.4 mm (0.008-0.016 inch) for all rings; maximum allowable ring end gap is 2.0 mm (0.079 inch) for compression rings and 1.0 mm (0.039 inch) for the oil ring. Refer to Fig. IZ67 for placement of rings on piston. Install oil ring expander with ends butted together, then install oil ring with end gap positioned 180 degrees away from expander ends. Install compression rings with marked face up. Space ring end gaps equally around piston, but not over pin bore.

Cylinder bores are not sleeved and can be rebored to oversize if worn excessively or scored. All cylinders should be rebored to the same oversize. Nominal cylinder diameter is 74 mm (2.913 inches) for Models 2KC1 and 3KC1, and 70 mm (2.756 inches) for all other models. Rebore cylinders to appropriate oversize if cylinder bore wear exceeds 0.4 mm (0.015 inch).

CONNECTING RODS AND BEARINGS

Connecting rod crankpin bearings are precision insert type and can be renewed after removing engine oil pan and connecting rod caps. Standard crankpin diameter is 42.925-42.94 mm (1.6899-1.6905 inches), with wear limit of 42.87 mm (1.6878 inches). Bearing to crankpin oil clearance should be 0.035-0.073 mm (0.0014-0.0029 inch); maximum allowable clearance is 0.12 mm (0.0047 inch).

Be sure bearing surfaces are clean and well lubricated before reinstalling rod to crank. Check to see that cylinder number on rod and cap are aligned. Side of rod with ISUZU mark should face front of engine. Tighten connecting rod bolts to a torque of 39 N·m (29 ft.-lbs.).

CRANKSHAFT AND MAIN BEARINGS

The crankshaft is supported in three main bearings in two-cylinder models and four main bearings (see Fig. IZ68) in three-cylinder models. Crankshaft end play is controlled by thrust rings (2) inserted between block and second (from front) crankshaft main journal.

To remove crankshaft, first remove cylinder head, oil pan, rod and piston units, flywheel, rear oil seal retainer, timing gear cover and camshaft drive chain. Check to be sure bearing caps are identified (cylinder number except on thrust bearing cap should be identified by punch marks), then unbolt and remove bearing caps and lift crankshaft from block.

Inspect crankshaft for scoring or other obvious damage and if crankshaft appears normal, check crankshaft against

SERVICE MANUAL

Isuzu

the following specifications: Main bearing journal standard diameter is 51.920-51.935 mm (2.044-2.0447 inches). Wear limit for journals is 51.86 mm (2.042 inches). Standard oil clearance for main bearings should be 0.029-0.072 mm (0.001-0.0028 inch); wear limit is clearance of 0.12 mm (0.0047 inch). Crankshaft end play should be 0.06-0.26 mm (0.0024-0.010 inch), maximum allowable end play is 0.3 mm (0.012 inch).

Refer to CONNECTING RODS AND BEARINGS paragraph for crankpin information.

To reinstall crankshaft, make sure block, bearing caps and crankshaft are clean and free of any nicks or burrs. Place bearing inserts (1) with oil hole and groove in cylinder block. Lubricate bearings and crank journals, then set crankshaft into bearings. Insert thrust rings (2) between block and bearing shoulder at second main journal from front; be sure bearing metal side (side with oil grooves) of each thrust ring is toward crankshaft journal shoulder. Install plain bearing inserts (6) in main bearing caps. Lubricate bearings and install caps in proper order with arrow on caps pointed toward front of engine. Tighten main bearing cap bolts to a torque of 88 N·m (65 ft.-lbs.) in sequence shown in Fig. IZ69. Be sure crankshaft turns freely, then complete reassembly by reversing disassembly procedure.

CRANKSHAFT REAR OIL SEAL

The crankshaft rear oil seal and seal spacer are retained in a seal plate bolted to rear end of cylinder block. Seal can be renewed after removing flywheel and seal retainer plate. If crankshaft oil seal surface is worn or grooved, position of seal should be changed by installing seal in retainer plate without the spacer. Remove all old gasket material from cylinder block and seal plate. Install new seal in plate using a suitable driver with lip of seal inward (toward front of engine). Lubricate seal lip and crankshaft, then install seal retainer plate using new gasket. Tighten the plate retaining bolts to a torque of 17 N·m (13 ft.-lbs.) in sequence shown in Fig. IZ70. Complete reassembly by reversing removal procedure.

CRANKSHAFT FRONT OIL SEAL

The crankshaft front oil seal is located in timing gear cover and, if care is taken not to damage timing gear cover, can be renewed after removing crankshaft pulley. Install new seal with lip inward (to rear of engine). Check crankshaft pulley for wear at seal contact surface and renew or recondition pulley as necessary. Lubricate seal lip and pulley seal contact surface, install pulley and tighten pulley bolt to a torque of 176 N·m (130 ft.-lbs.).

ELECTRICAL SYSTEM

GLOW PLUGS

Glow plugs are parallel connected with each individual glow plug grounding through its mounting threads. Indicator light should glow about 30 seconds after control is actuated if units are operating satisfactorily. If indicator light fails to glow, check for open circuit at switch, indicator lamp or glow plug connections.

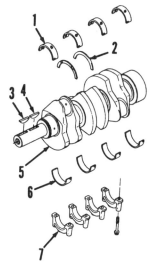

Fig. IZ68—View showing crankshaft and main bearings for three-cylinder engine. Two-cylinder models have only one intermediate bearing. Thrust rings (2) are located at second main journal on all models.

1. Main bearing insert, upper
2. Thrust ring
3. Key
4. Key
5. Crankshaft
6. Main bearing insert, lower
7. Main bearing cap

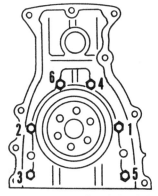

Fig. IZ70—Tighten rear oil seal plate bolts in sequence shown.

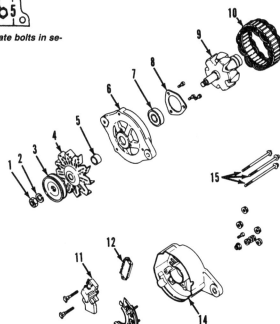

Fig. IZ71—Exploded view of Hitachi alternator typical of that used on all models.

1. Hex nut
2. Lockwasher
3. Belt pulley
4. Fan
5. Spacer
6. Front cover
7. Bearing
8. Retainer
9. Rotor
10. Stator
11. Brush holder
12. Cover
13. Rectifier assy.
14. Rear cover assy.

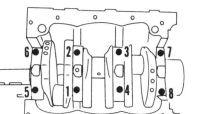

Fig. IZ69—Main bolt tightening sequence for three-cylinder models. Sequence for two-cylinder models is for bolts 1 to 6 only.

Isuzu

ALTERNATOR AND REGULATOR

Refer to Fig. IZ71 for exploded view of Hitachi LR120-23 alternator with built in voltage regulator used on all models. Regulator has no provisions for adjustment. Specified alternator output is 20 amperes at 5000 alternator rpm. Regulated voltage should be 14.2-14.8 volts.

New brush length is 16 mm (0.630 inch). Renew brushes if worn to 9 mm (0.354 inch); a wear limit line is etched on the brushes. Standard slip ring diameter is 31.6 mm (1.244 inch); wear limit is 30.6 mm (1.205 inch).

To check stator assembly (10), use ohmmeter and check for continuity between each stator coil terminal. Renew stator if no continuity exists. Check for continuity between terminals and stator core. If continuity exists, unit is shorted out and must be renewed.

To check rotor (9), use ohmmeter to check for continuity between slip rings. If no continuity is found, there is an open circuit in winding and rotor must be renewed. Check for continuity between slip rings and rotor core or shaft. Unit is shorted out and must be renewed if continuity exists. Rotor coil resistance should be 5.2 ohms at 20° C (68° F). Slip ring standard diameter is 31.6 mm (1.244 inches) and wear limit is 30.6 mm (1.205 inches).

Using ohmmeter, check diodes in rectifier assembly (13). Ohmmeter should show infinite reading in one direction and continuity when leads are reversed. Renew rectifier assembly if continuity exists in both directions, or if infinite resistance reading is noted in both directions.

SMALL DIESEL

STARTING MOTOR

Either a Hitachi S114-385 or S114-387 starting motor is used. Refer to Fig. IZ72 for exploded view of starting motor. Starter service specifications are as follows:

No-load test
  Volts 12
  Current draw,
    maximum . . . . 60 amperes
  Rpm, minimum
    SR114-385 6000
    SR114-387 7000
Brush length, standard . . . . 16 mm
                              (0.630 in.)
Wear limit . . . . 12 mm
                              (0.472 in.)
Pinion gap . . . . 0.3-2.5 mm
                        (0.012-0.098 in.)

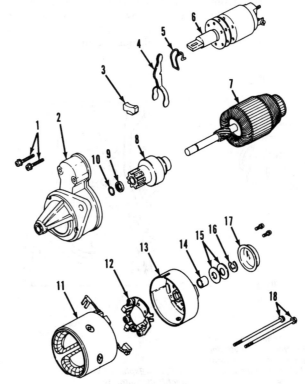

Fig. IZ72—Exploded view of Hitachi starter motor typical of that used on all models.

1. Bolts
2. Gear case
3. Dust cover
4. Shift lever
5. Torsion spring
6. Magnetic switch
7. Armature assy.
8. Pinion
9. Pinion stop
10. Clip
11. Field coil assy.
12. Brush holder
13. Cover
14. Bushing
15. Thrust washers
16. "E" ring
17. Dust cover

SERVICE MANUAL

KIRLOSKAR

1401 Cherry Hill Road
Baltimore, Maryland 21225

Model	No. Cyls.	Bore	Stroke	Displ.
KA-27	1	68 mm (2.677 in.)	76 mm (2.992 in.)	276 cc (16.8 cu. in.)

Model KA-27 is an air-cooled, single-cylinder diesel engine. Crankshaft rotation may be clockwise or counterclockwise depending on application.

MAINTENANCE

LUBRICATION

Recommended crankcase oil is SAE 10W for temperatures below 5°C (41°F), SAE 30 for temperatures between 5°C (41°F) and 35°C (95°F) or SAE 40 for temperatures above 35°C (95°F). Crankcase oil should pass specifications for MIL-L-2104A. Crankcase oil capacity is 1.8 liters (1.9 quarts).

Oil is pressure fed to engine components by a gear type pump. Oil pressure should be 245-294 kPa (35-43 psi) with engine warm and running at operating speed. An oil pressure gage may be connected after removing plug inside of oil filter housing. To adjust oil pressure, remove oil filter housing and turn adjusting screw in back of filter housing.

The renewable oil filter element should be renewed after every 500 hours of operation.

ENGINE SPEED ADJUSTMENT

To adjust idle speed, remove cover (C–Fig. KR1-1), loosen locknut, then turn idle speed screw (I). Idle speed should be 1200-1500 rpm. Do not turn out screw (I) too far as it must remain in contact with lug on governor arm so engine can be stopped. Maximum speed on variable speed models is adjusted by turning high speed adjusting screw (H–Fig. KR1-2). Maximum governed speed under load should be 1500, 2000, 2500, 2800, 3000 or 3600 rpm depending on engine application.

FUEL SYSTEM

FUEL FILTER. A renewable fuel filter is located below fuel tank on side of engine. Manufacturer recommends renewing fuel filter element after every 500 hours of operation.

BLEED FUEL SYSTEM. Due to gravity feed type of fuel system, air should be bled from system as fuel injection pump operates. However, bleeding time may be shortened by loosening then retightening fuel line fittings, starting first at fuel tank and working to fuel injection pump. Retighten fittings when air-free fuel flows.

INJECTION PUMP TIMING. Injection pump timing adjusted using shims (G–Fig. KR1-3). To check injection pump timing, proceed as follows: Disconnect high pressure line from injection pump, then unscrew delivery valve holder (H) and remove spring and delivery valve assembly. Reinstall

Fig. KR1-1—Remove cover (C) for access to idle speed screw (I) on governor. Refer to text.

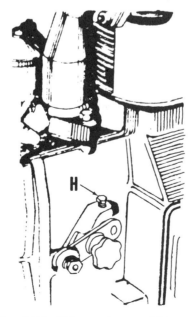

Fig. KR1-2—High speed on variable speed engines is adjusted by turning adjusting screw (H).

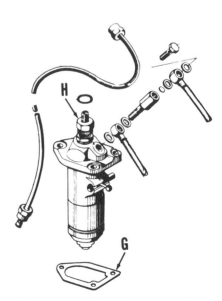

Fig. KR1-3—Injection timing is adjusted by varying thickness of shim gasket (G). Refer to text for timing procedure.

Kirloskar

delivery valve holder (H) and connect a suitable spill pipe to valve holder. Aim spill pipe at a receptacle to catch discharged fuel. Move throttle control to full speed position. Rotate engine flywheel slowly in running direction (engines are designed for clockwise or counterclockwise rotation) until fuel just stops flowing from spill pipe. Note rated rpm of engine and refer to following table for desired injection timing:

Rated Rpm	Injection Timing
1500	24°BTDC
2000	24°BTDC
2500	28°BTDC
2800	28°BTDC
3000	31°BTDC
3600	34°BTDC

If injection timing is incorrect, remove injection pump and remove or install shims (G) as required. Adding shims will retard injection timing while removing shims advances injection timing. Be sure to reassemble delivery valve in pump after testing.

AIR FILTERS

Manufacturer recommends cleaning air filter after every 50 hours of operation and renewing filter after 500 hours.

REPAIRS

TIGHTENING TORQUES

Refer to the following table for special tightening torques.

Connecting rod.............40 N·m
(30 ft.-lbs.)
Cylinder head..............40 N·m
(30 ft.-lbs.)
Flywheel.................176 N·m
(130 ft.-lbs.)
Injection pump............25 N·m
(18 ft.-lbs.)
Injector retainer plate.....25 N·m
(18 ft.-lbs.)
Main bearing support......25 N·m
(18 ft.-lbs.)
Rocker arm stand..........78 N·m
(58 ft.-lbs.)

VALVE CLEARANCE

Valve clearance may be adjusted after removing rocker arm cover. Rotate crankshaft until piston is at top dead center on compression stroke, then turn rocker arm adjusting screws until appropriate thickness feeler gage can be inserted between rocker arm and end of valve stem. Recommended clearance is 0.10 mm (0.004 inch) for intake and 0.15 mm (0.006 inch) for exhaust with engine cold.

CYLINDER HEAD AND VALVE SYSTEM

Valve face and valve seat angles are 45° for intake and exhaust. Valve seat width should be 2.05-2.33 mm (0.080-0.090 inch). Valve seats are renewable. Intake valve stem diameter is 8.956-8.968 mm (0.3526-0.3530 inch) and exhaust valve stem diameter is 8.918-8.930 mm (0.3511-0.3515 inch). Intake and exhaust valve guide inside diameter is 9.013-9.035 mm (0.3548-0.3557 inch). Intake valve stem clearance is 0.045-0.079 mm (0.0018-0.0031 inch) and exhaust valve stem clearance is 0.083-0.117 mm (0.0033-0.0046 inch). Maximum valve stem clearance for both valves is 0.15 mm (0.006 inch). Valve guides are renewable. Valve guide outer diameter is 12.050-12.062 mm (0.4744-0.4749 inch) and cylinder head bore is 12.000-12.027 mm (0.4724-0.4735 inch).

Clearance between rocker shaft and rocker arm bushing should be 0.065-0.101 mm (0.0025-0.0040 inch). Maximum allowable clearance is 0.15 mm (0.006 inch).

SMALL DIESEL

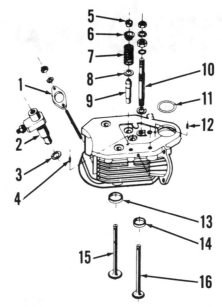

Fig. KR1-4—Exploded view of cylinder head. Rocker arm assembly is not shown.

1. Injector retainer
2. Injector
3. Copper washer
4. Dowel
5. Keys
6. Spring retainer
7. Valve spring
8. Washer
9. Valve guide
10. Rocker stand stud
11. Push pin rod tube seal
12. Dowel
13. Intake valve seat
14. Exhaust valve seat
15. Intake valve
16. Exhaust valve

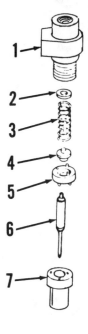

Fig. KR1-5—Exploded view of injector.

1. Body
2. Shim
3. Spring
4. Push pin
5. Spacer
6. Valve
7. Nozzle
8. Nozzle holder

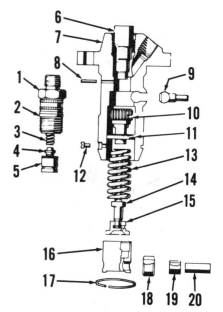

Fig. KR1-6—View of injection pump.

1. Delivery valve holder
2. "O" ring
3. Spring
4. Delivery valve
5. Delivery valve seat
6. Barrel
7. Pump body
8. Pin
9. Control rack
10. Pinion
11. Spring seat
12. Pin
13. Spring
14. Plunger
15. Spring retainer
16. Tappet
17. Retaining ring
18. Outer roller
19. Inner roller
20. Pin

Tighten cylinder head nuts in a crossing pattern to 40 N·m (30 ft.-lbs.). Measure clearance between piston at

SERVICE MANUAL

Kirloskar

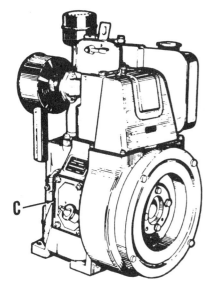

Fig. KR1-7—Remove cover (C) for access to connecting rod.

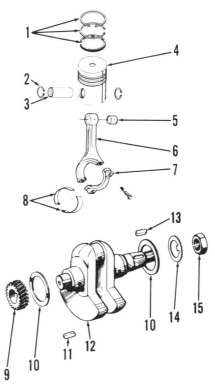

Fig. KR1-8—Exploded view of crankshaft assembly.

1. Piston rings
2. Snap ring
3. Piston pin
4. Piston
5. Bushing
6. Connecting rod
7. Rod cap
8. Rod bearing
9. Gear
10. Thrust washers
11. Key
12. Crankshaft
13. Key
14. Tab washer
15. Nut

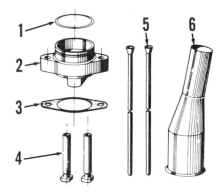

Fig. KR1-9—Exploded view of tappets and push rods.

1. Push rod tube seal
2. Valve tappet guide
3. Gasket
4. Valve tappets
5. Push rods
6. Push rod tube

top dead center and cylinder head by inserting a lead wire (solder) or other suitable tool through injector opening in head. Clearance between head and piston should be 0.7-0.8 mm (0.028-0.031 inch). Head gaskets are available in varying thicknesses to adjust clearance. Note that push rod nearer cylinder operates exhaust valve while outer push rod operates intake valve. Tighten rocker arm stand nut to 78 N·m (58 ft.-lbs.).

INJECTOR

REMOVE AND REINSTALL. To remove injector, first clean dirt from injector, injection line, return line and cylinder head. Disconnect return line and injection line and immediately cap or plug all openings. Remove injector retainer plate and carefully withdraw injector from head.

Use a new copper washer (3–Fig. KR1-4) when installing injector. Tighten nuts securing injector retainer plate to 25 N·m (18 ft.-lbs.).

TESTING. WARNING: Fuel leaves the injection nozzle with sufficient force to penetrate the skin. When testing, keep yourself clear of nozzle spray.

If a suitable test stand is available, injector operation may be checked. Only clean, approved test oil should be used to test injector. When operating properly during test, injector nozzle will emit a buzzing sound and cut off quickly with no fluid leakage at seat.

Opening pressure with a new spring (3–Fig. KR1-5) should be 14220-14710 kPa (2060-2135 psi) while opening pressure with a used spring should be approximately 13240 kPa (1920 psi). Opening pressure is adjusted by varying number and thickness of shims (2). Valve should not show leakage at orifice spray holes for 10 seconds at 12260 kPa (1780 psi).

OVERHAUL. Clamp injector body (1–Fig. KR1-5) in a vise with nozzle tip pointing upward. Remove nozzle holder nut (8). Remove nozzle (7) with valve (6) and spacer (5). Invert injector body (1) and remove push pin (4), spring (3) and shims (2). Thoroughly clean all parts in a suitable solvent. Clean inside orifice end of nozzle with a wooden cleaning stick. Orifice spray holes may be cleaned by inserting a cleaning wire slightly smaller than spray holes.

When reassembling injector, make certain all components are clean and wet with clean diesel fuel oil. Tighten nozzle nut (8) to 59 N·m (44 ft.-lbs.).

INJECTION PUMP

REMOVE AND REINSTALL. Remove breather adjacent to pump, then reach through breather opening in crankcase and disconnect governor link (4–Fig. KR1-13) from pump. Disconnect fuel lines from pump and immediately cap all openings to prevent contamination. Remove pump being careful not to lose timing shims.

Injection pump should be serviced by a shop experienced in fuel injection pump repair.

Refer to INJECTION PUMP TIMING section after pump installation if original timing shims were not installed or timing is believed incorrect.

CYLINDER, PISTON, PIN AND RINGS

R&R AND OVERHAUL. The cylinder is removable after removing cylinder head as previously outlined. After cylinder is removed, cover crankcase opening to prevent entry of foreign material. Detach snap rings (2–Fig. KR1-8), then use a suitable puller to withdraw piston pin (3). Remove piston from connecting rod.

Piston ring end gap should be 0.25-0.40 mm (0.010-0.015 inch) for compression rings and 0.20-0.35 mm (0.008-0.013 inch) for oil ring. Maximum piston ring end gap is 0.6 mm (0.023 inch) for compression rings and 0.5 mm (0.019 inch) for oil ring. Piston ring side clearance should be 0.11-0.14 mm (0.004-0.005 inch) for top compression ring, 0.09-0.12 mm (0.004-0.005 inch) for second compression ring and 0.05-0.08 mm (0.002-0.003 inch) for oil ring. Maximum ring side clearance is 0.25 mm (0.010 inch) for top compression ring, 0.20 mm (0.008 inch) for second compression ring and 0.15 mm (0.006 inch) for oil ring.

Piston pin bore diameter in piston is 23.996-24.009 mm (0.9447-0.9452 inch) and piston pin outer diameter is 23.996-24.000 mm (0.9447-0.9449 inch). Clearance between piston pin and connecting rod should be 0.030-0.054 mm (0.0012-0.0021 inch); maximum

Kirloskar

SMALL DIESEL

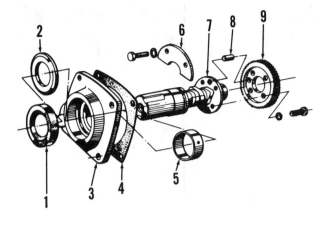

Fig. KR1-10—Exploded view of bushing-supported camshaft used on some engines.
1. Seal
2. Oil deflector
3. Bearing housing
4. Gasket
5. Bushing
6. Thrust plate
7. Camshaft
8. Dowel
9. Gear

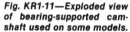

Fig. KR1-11—Exploded view of bearing-supported camshaft used on some models.
1. Seal
2. Snap ring
3. Bearing housing
4. Gasket
5. Bearing
6. Oil deflector
7. Camshaft
8. Starter pin
9. Gear
10. Snap ring
11. Key
12. Sleeve

allowable clearance is 0.15 mm (0.006 inch). Refer to CONNECTING ROD section.

Cylinder standard bore diameter is 68.06-68.08 mm (2.6795-2.6803 inches). Standard piston diameter is 67.97 mm (2.676 inches). Piston clearance should be 0.09-0.11 mm (0.0035-0.0043 inch).

Assembly is reverse of disassembly. Refer to CYLINDER HEAD section for cylinder head installation.

CONNECTING ROD

R&R AND OVERHAUL. To remove connecting rod, remove cylinder as outlined in previous section, then remove side cover (C–Fig. KR1-7) for access to connecting rod big end. Unscrew rod screws and remove rod cap through side cover opening, then remove rod out top of engine.

Connecting rod big end clearance between rod bearing and crankpin should be 0.031-0.083 mm (0.0012-0.0032 inch). Undersize rod bearings are available. The connecting rod small end bushing is renewable. Inner diameter of small end bushing is 24.03-24.05 mm (0.9461-0.9468 inch) and clearance between piston pin and rod bushing should be 0.03-0.054 mm (0.0012-0.0021 inch).

Install piston and rod so numbered side of rod is away from side cover opening. Install rod cap through side cover opening then install and tighten rod cap screws to 40 N·m (30 ft.-lbs.) Attach side cover to crankcase and refer to previous sections for installation of cylinder and cylinder head.

CAMSHAFT, TAPPETS AND PUSH RODS

R&R AND OVERHAUL. To remove push rods and tappets, remove cylinder head and detach push rod tube (6–Fig. KR1-9). Unscrew retaining nuts and remove tappet guide (2) and tappets from crankcase. Remove fuel injection pump. Unscrew camshaft bearing housing (3–Fig. KR1-10 or KR1-11) screws and remove camshaft assembly. If bearing housing will not remove easily, jackscrews may be threaded into holes provided in housing to push housing free.

Two different camshafts are used and are not interchangeable. The camshaft shown in Fig. KR1-10 is supported at outer end by bushing (5) while the camshaft shown in Fig. KR1-11 is supported at outer end by ball bearing (5). To disassemble bushing supported camshaft assembly, remove thrust plate (6–Fig. KR1-10) and slide camshaft out of housing (3). Detach gear (9) from camshaft. Bushing (5) may be renewed if necessary. To disassemble bearing supported camshaft, unscrew Allen screw and remove steel sleeve (12–Fig. KR1-11). Pull gear (9) off shaft, detach snap ring (10) and slide shaft out of bearing (5). Remove snap rings (2) and remove bearing (5) from housing (3).

Inspect components and renew if required. Inner end of camshaft (7–Fig. KR1-10) rides in crankcase bushing while steel sleeve (12–Fig. KR1-11) attached to camshaft (7) rides in crankcase bushing. Clearance between crankcase bushing and camshaft (7–Fig. KR1-10) journal or camshaft sleeve (12–Fig. KR1-11) should not exceed 0.20 mm (0.008 inch). Renew tappet guide (2–Fig. KR1-9) if tappet bores are out-of-round more than 0.05 mm (0.002 inch).

Reassembly is reverse of disassembly. When installing camshaft assembly in crankcase, rotate crankshaft so piston is at top dead center then insert camshaft so "I" mark on gear is aligned with mark in tappet guide opening of crankcase as shown in Fig. KR1-12.

GOVERNOR

R&R AND OVERHAUL. To remove governor, remove breather on top of crankcase then reach through opening and disconnect governor link (4–Fig. KR1-13) from injection pump. Detach speed control lever (21), spring retainer (18) and spring (17) from shaft (10). Remove end cover (E–Fig. KR1-14). Unscrew and remove stud (11–Fig. KR1-13) from shaft. Detach "E" ring (15) and withdraw governor shaft (10) from side of crankcase while removing governor fork (8) assembly through end of crankcase. The governor flyweight assembly (2) may be removed by turning stud (T) after removing pin (1).

Governor flyweight assembly (2) is designed according to engine governed speed and direction of crankshaft rotation. Governed speed may be 1500, 2000, 2500, 2800, 3000 or 3600 rpm. Flyweight assembly is available only as a

Fig. KR1-12—Align "I" mark on camshaft gear with mark (M) on crankcase. Piston must be at TDC. Refer to text.

SERVICE MANUAL

Kirloskar

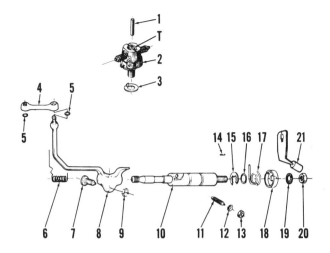

Fig. KR1-13—Exploded view of governor mechanism.
1. Pin
2. Flyweight assy.
3. Tab washer
4. Link
5. Clips
6. Governor spring
7. Screw
8. Governor arm
9. Nut
10. Shaft
11. Idle speed screw
12. Lockwasher
13. Nut
14. Pin
15. "E" ring
16. "O" ring
17. Spring
18. Spring retainer
19. Lockwasher
20. Nut
21. Lever

Fig. KR1-14—Remove end cover (E) for access to governor and oil pump.

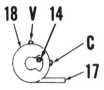

Fig. KR1-15—Position short end of spring (17) in slot (V) of spring retainer on variable speed engines or in slot (C) on constant speed engines.

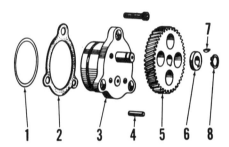

Fig. KR1-16—View of oil pump. Dowel pins (4) are inserted through pump end plates and body.
1. "O" ring
2. Gasket
3. Oil pump
4. Dowel pins (2)
5. Gear
6. Spacer
7. Key
8. Snap ring

unit assembly. When installing a new governor, be sure correct flyweight assembly is installed.

Reverse disassembly procedure to install governor components. Long end of governor spring (6) must engage notch in crankcase while short end of spring is against back side of governor arm (8) so spring tension will force fork arm towards end of crankcase. Refer to Fig. KR1-15 when installing spring (17–Fig. KR1-13) and spring retainer (18). Short end of spring (17) on constant speed engines should engage slot (C–Fig. KR1-15) of spring retainer (18) while short spring end on variable speed engines should engage slot (V). Long end of spring on all engines must engage lug on side of crankcase so lever (21–Fig. KR1-13) is tensioned upwards. After governor installation, refer to ENGINE SPEED ADJUSTMENT section.

OIL PUMP

R&R AND OVERHAUL. To remove oil pump, remove end cover (E–Fig. KR1-14), then unscrew three Allen screws securing oil pump and remove oil pump. If oil pump is difficult to remove due to binding or a close fit, it will be necessary to remove crankshaft so pump may be tapped loose from inside crankcase.

To disassemble oil pump, detach snap ring (8–Fig. KR1-16) and remove gear (5). Unscrew oil pump housing screws, then tap gently on shaft to dislodge oil pump components. Oil pump is available only as a unit assembly and should be renewed if components are damaged or excessively worn. Gear backlash should not exceed 0.12 mm (0.005 inch). Two different oil pumps are used according to crankshaft rotation. Be sure new pump matches crankshaft rotation.

Reverse disassembly procedure for assembly.

CRANKSHAFT

R&R AND OVERHAUL. To remove crankshaft, unscrew flywheel nut and use a suitable puller to remove flywheel.

NOTE: On engines with clockwise crankshaft rotation, the flywheel nut has left-hand threads. Remove connecting rod and governor as outlined in previous sections. At flywheel end, unscrew bearing support retaining nuts and install jackscrews in holes provided in bearing support. Loosen, then remove bearing support. Withdraw crankshaft from crankcase.

Crankshafts are designed for clockwise or counterclockwise rotation. Be sure correct crankshaft is installed if renewal is required.

Use a suitable puller to remove crankshaft gear. Heat gear prior to assembly to ease installation.

Main bearing clearance is 0.053-0.093 mm (0.0020-0.0036 inch). Standard crankpin journal diameter is 39.975-39.991 mm (1.5738-1.5744 inches) and rod bearing clearance is 0.031-0.083 mm (0.0012-0.0032 inch). Undersize main and rod bearings are available.

Crankshaft end play should be 0.15-0.30 mm (0.006-0.012 inch) and is controlled by thrust washers (10–Fig. KR1-7). Thrust washer thickness is 2.31-2.36 mm (0.091-0.093 inch).

Install crankshaft by reversing removal procedure. Tighten main bearing support nuts to 25 N·m (18 ft.-lbs.).

KUBOTA

550 West Artesia Blvd.
P.O. Box 7020
Compton, California 90224

Model	No. Cyls.	Bore	Stroke	Displ.
EA400-N	1	78 mm (3.071 in.)	84 mm (3.307 in.)	401 cc (24.5 cu. in.)
EA400-NB	1	78 mm (3.071 in.)	84 mm (3.307 in.)	401 cc (24.5 cu. in.)
EA450-N	1	84 mm (3.307 in.)	84 mm (3.307 in.)	465 cc (28.4 cu. in.)
EA450-NB	1	84 mm (3.307 in.)	84 mm (3.307 in.)	465 cc (28.4 cu. in.)
EA500-N	1	86 mm (3.386 in.)	90 mm (3.543 in.)	522 cc (31.9 cu. in.)
EA500-NB	1	86 mm (3.386 in.)	90 mm (3.543 in.)	522 cc (31.9 cu. in.)
EA600-N	1	92 mm (3.622 in.)	90 mm (3.543 in.)	598 cc (36.5 cu. in.)
EA600-NB	1	92 mm (3.622 in.)	90 mm (3.543 in.)	598 cc (36.5 cu. in.)

K1-1—View showing location of oil drain plug (D) and oil pressure relief valve (R).

All engines in this section are liquid-cooled, single-cylinder, four-stroke diesel engines. Crankshaft rotation is counterclockwise from flywheel end.

MAINTENANCE

LUBRICATION

Depending on ambient temperature, recommended crankcase oil is SAE 20W, SAE 30 or SAE 10W-30 with API classification CC. Crankcase capacity is 1.9 liters (2 quarts) for Models EA400-N, EA400-NB, EA450-N and EA450-NB or 2.3 liters (2.4 quarts) for all other models.

All models are equipped with a pressure lubrication system. The oil drain plug (D–Fig. K1-1) also serves as the oil pickup and a filter screen is attached to end of drain plug. The drain plug and filter should be removed and filter cleaned and oil renewed after every 100 hours of operation.

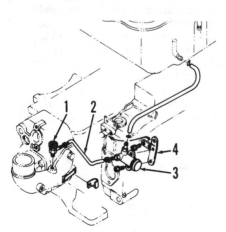

K1-2—View of fuel primer circuit.
1. Fuel jet valve
2. Fuel line
3. Primer pump
4. Bracket

Fig. K1-3—Exploded view of fuel system. Refer to Fig. K1-2 for fuel primer circuit.
1. Injector
2. High pressure fuel line
3. Inlet fuel line
4. Injection pump
5. Shim gasket
6. Fuel valve
7. "O" ring
8. "O" ring
9. Filter
10. Filter canister
11. Nut
12. Fuel return line

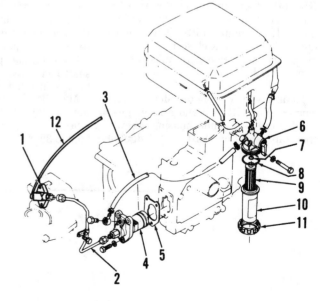

Service Manual

Kubota

FUEL SYSTEM

FUEL PRIMER. All models are equipped with a fuel primer to aid starting. To operate fuel primer, open fuel valve (1–Fig. K1-2), push primer pump (3) button, then close fuel valve. The primer pump button should be depressed 5 to 6 times for smaller engines and 8 to 9 times for larger engines. Use of primer pump will depend on ambient temperature.

FUEL FILTER. A renewable fuel filter is located below fuel valve as shown in Fig. K1-3. Unscrew nut (11), detach canister (10) and remove filter (9). Manufacturer recommends renewing fuel filter after every 100 hours of operation.

BLEED FUEL SYSTEM. Refer to Fig. K1-3 for view of fuel system. To bleed air from system, first loosen fuel line (3) fitting at injection pump. Open fuel shut-off valve, then retighten fitting when fuel flows from fuel inlet line. Loosen high pressure fuel line (2) to injector (1). Rotate crankshaft to operate injection pump until fuel is discharged from loosened connector. Retighten fitting and start engine.

INJECTION PUMP TIMING. Injection pump timing is adjusted using pump mounting shims (5–Fig. K1-3). To check injection pump timing proceed as follows: Disconnect high pressure line from injection pump, then unscrew delivery valve holder (1–Fig. K1-4) and remove valve spring (3) and delivery valve assembly (5). Reinstall delivery valve holder (1) and connect a suitable spill pipe to valve holder. Aim spill pipe at a receptacle to catch discharged fuel. Move throttle control to full speed position. Rotate engine flywheel slowly in counterclockwise direction until fuel just stops flowing from spill pipe. Mark (M–Fig. K1-5) on fan cover should be aligned with "F" mark on flywheel which should provide injection timing of 19-21

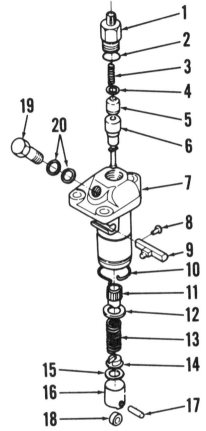

Fig. K1-4—Exploded view of injection pump.
1. Delivery valve holder
2. "O" ring
3. Spring
4. Gasket
5. Delivery valve
6. Plunger
7. Pump body
8. Pin
9. Control rack
10. Clip
11. Control sleeve
12. Washer
13. Spring
14. Spring retainer
15. Shim
16. Tappet
17. Pin
18. Roller
19. Banjo bolt
20. Gaskets

Fig. K1-5—View showing location of timing mark (M) on fan cover.

Fig. K1-6—Exploded view of governor mechanism. Spring (10) connects to pin (7—K1-6A).
1. Pin
2. Washer
3. Flyweight
4. Weight carrier
5. Pin
6. Push rod
7. Spring
8. Pin
9. Lever
10. Spring
11. Shaft
12. Ball
13. Throttle shaft
14. Key
15. "O" ring
16. Idle speed screw
17. Spring
18. Throttle lever

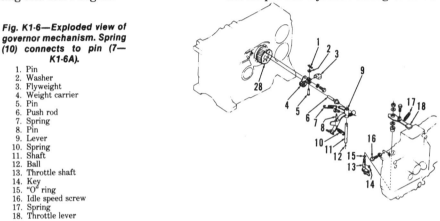

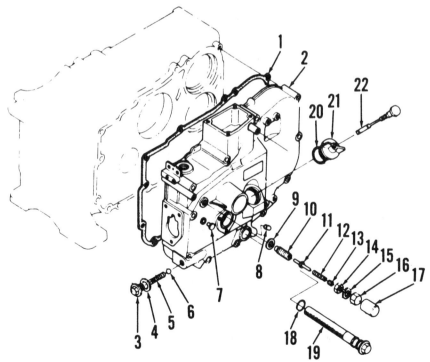

Fig. K1-6A—Exploded view of gearcase components.
1. Gasket
2. Gearcase
3. Plug
4. Gasket
5. Oil pressure relief spring
6. Oil pressure relief ball
7. Pin
8. Plug
9. Gasket
10. Fuel limiter spring housing
11. Pin
12. Spring
13. Screw
14. Nut
15. Gasket
16. Cap nut
17. Cap
18. "O" ring
19. Filter & drain plug
20. "O" ring
21. Fill plug
22. Dipstick

Illustrations Courtesy Kubota

Kubota

SMALL DIESEL

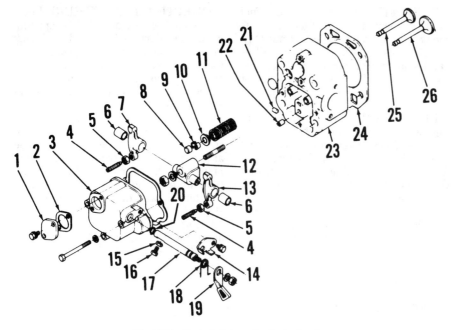

Fig. K1-7—Exploded view of cylinder head.

1. Cover
2. Gasket
3. Rocker arm
4. Adjusting screw
5. Locknut
6. Bushing
7. Rocker arm, ex.
8. Cap
9. Retainer keys
10. Spring retainer
11. Valve spring
12. Rocker stand
13. Rocker arm, int.
14. Bracket
15. Locknut
16. Screw
17. Shaft
18. Spring
19. Compression release lever
20. "O" ring
21. Plug
22. Pin
23. Head
24. Gasket
25. Exhaust valve
26. Intake valve

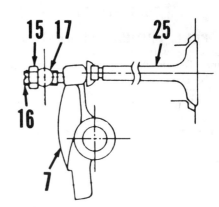

Fig. K1-8—Loosen locknut (15) and turn adjusting screw (16) to adjust compression release. Refer to text.

degrees BTDC. If injection timing is incorrect, remove injection pump and remove or install shims (5–Fig. K1-3) as required. Adding shims will retard injection timing while removing shims advances injection timing. Each shim alters injection timing approximately 1½ degrees. Tighten injection pump mounting screws to 24-27 N·m (18-20 ft.-lbs.). Reinstall delivery valve and spring and tighten delivery valve holder to 45 N·m (33 ft.-lbs.).

GOVERNOR

All models are equipped with a flyweight type governor attached to the crankshaft gear. Refer to K1-6 for an exploded view of governor mechanism.

Maximum no-load governed speed should be 2540-2580 rpm while low idle speed should be less than 1000 rpm.

Fuel limiting device (components 9 through 17–Fig. K1-6A) should be adjusted by loosening nut (14) then turning spring housing (10) so excessive smoke is not produced when engine is slightly overloaded.

COMPRESSION RELEASE

All models are equipped with a compression release which holds the exhaust valve open slightly when compression release lever (19–Fig. K1-7) is rotated. To adjust compression release, rotate engine crankshaft so piston is at top dead center on compression. Remove cover (1), loosen locknut (15–Fig. K1-8) and back off adjusting screw (16). Rotate compression release lever to engaged position. Turn adjusting screw (16) in until it contacts exhaust rocker arm (7), then turn screw 1½ additional turns. Tighten locknut and check operation of compression release.

COOLING SYSTEM

All models are equipped with a liquid type cooling system. A radiator is mounted above the engine and an engine driven fan circulates air through the radiator. Coolant is circulated by thermo-siphon. On "NB" models, alternator coils are mounted behind the fan and the fan also functions as the alternator rotor.

A coolant drain valve is located on underside of the cylinder head. Cooling system capacity is 1.6 liters (1.7 quarts) for Models EA400-N, EA400-NB, EA450-N and EA450-NB or 2.1 liters (2.2 quarts) for all other models. Recommended pressure rating for radiator cap is 88.25 kPa (13 psi).

Fan belt tension should be adjusted so belt deflects 5-10 mm (3/16-3/8 inch) when using finger pressure against belt at midpoint between fan pulley and tension pulley. Relocate tension pulley to adjust belt tension.

AIR FILTER

All models are equipped with a dry type renewable air filter. Manufacturer recommends blowing out filter after every 100 to 200 hours of operation and renewing filter after six cleanings or one year.

REPAIRS

TIGHTENING TORQUES

Refer to the following table for tightening torques.

Connecting rod:
EA400-N, EA400-NB,
EA450-N, EA450-NB....30-34 N·m
(22-25 ft.-lbs.)
All other models.........49-54 N·m
(36-40 ft.-lbs.)
Crankcase cover...........9-10 N·m
(7-8 ft.-lbs.)
Cylinder head:
EA400-N, EA400-NB
EA450-N, EA450-NB...98-118 N·m
(73-87 ft.-lbs.)
All other models.......137-157 N·m
(101-115 ft.-lbs.)
Flywheel................294-392 N·m
(217-289 ft.-lbs.)
Gearcase cover..........24-27 N·m
(18-20 ft.-lbs.)
Injection pump..........24-27 N·m
(18-20 ft.-lbs.)
Injector................24-27 N·m
(18-20 ft.-lbs.)
Main bearing carrier....24-27 N·m
(18-20 ft.-lbs.)
Rocker arm stand........40-45 N·m
(30-33 ft.-lbs.)

COMPRESSION PRESSURE

Compression pressure at cranking speed (100-200 rpm) should be at least 2530 kPa (377 psi) for Models EA400-N and EA400-NB, 2432 kPa (353 psi) for Models EA450-N and EA450-NB or 2235 kPa (324 psi) for all other models.

SERVICE MANUAL

Kubota

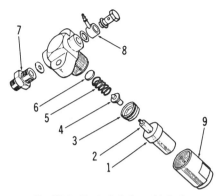

Fig. K1-9—Exploded view of injector.
1. Nozzle body
2. Nozzle valve
3. Spacer
4. Pressure pin
5. Spring
6. Shim
7. Pressure fitting
8. Return fitting
9. Nozzle nut

CYLINDER HEAD

R&R AND OVERHAUL. Drain coolant, remove muffler and disconnect fuel return line from injector. Disconnect both ends of high pressure fuel line. Remove air cleaner and intake pipe. Remove rocker cover (3–Fig. K1-7), then unscrew rocker stand retaining nut and remove rocker arm assembly. Unscrew cylinder head retaining nuts and remove cylinder head.

Check flatness of head surface using a straightedge placed along sides and across mating surface. If a feeler gage of 0.05 mm (0.002 inch) thickness will pass under straightedge, then head must be resurfaced. If head is resurfaced, check valve-to-piston clearance as outlined in following paragraph.

To check valve-to-piston clearance, install head and gasket with valves and springs installed but with injector removed. Tighten head retaining nuts to specified torque of 98-118 N·m (73-87 ft.-lbs.) for Models EA400-N, EA400-NB, EA450-N and EA450-NB or 137-157 N·m (101-115 ft.-lbs.) for all other models. Using a suitable measuring gage such as soft solder or Plastigage, insert gage between each valve and piston crown while rotating crankshaft through top dead center. Minimum allowable valve-to-piston clearance is 0.65 mm (0.026 inch). Note that crush thickness of head gasket should be 1.25-1.45 mm (0.049-0.057 inch) for Models EA400N, EA400-NB, EA450-N and EA450-NB or 1.35-1.55 mm (0.053-0.061 inch) for all other models.

Reverse removal procedure to reinstall cylinder head. Tighten cylinder head nuts to specified torque using a crossing pattern. Tighten rocker arm stand nut to specified torque. Adjust valve clearance as outlined in following section and compression release as outlined in a previous section. Operate engine for 30 minutes, then retorque cylinder heat nuts.

VALVE CLEARANCE

Valve clearance should be adjusted with engine cold and piston at top dead center on compression. Remove rocker cover and turn adjusting screws in rocker arms until appropriate size feeler gage can be inserted between rocker arm and valve stem end. Clearance for both valves is 0.16-0.20 mm (0.006-0.008 inch) for Models EA400-N, EA400-NB, EA450-N and EA450-NB or 0.20-0.23 mm (0.008-0.009 inch) for all other models.

VALVE SYSTEM

Both valves ride directly in cylinder head. Maximum allowable clearance between valve stem and valve guide is 0.1 mm (0.004 inch). Valve guide diameter should be 7.010-7.025 mm (0.2760-0.2765 inch) on EA400-N, EA400-NB, EA450-N and EA450-NB or 8.015-8.030 mm (0.3156-0.3161 inch) for all other models. Valve stem diameter should be 6.960-6.975 mm (0.2740-0.2746 inch) on EA400-N, EA400-NB, EA450-N and EA450-NB or 7.960-7.975 mm (0.3134-0.3140 inch) for all other models.

Valves seat directly in head. Valve seat and face angles are 45 degrees. Valve seat width should be 1.4 mm (0.055 inch). When depth of valve head from cylinder head surface exceeds 1.5 mm (0.059 inch), then head surface should be machined.

Valve springs are interchangeable. Valve spring free length should be 38.5 mm (1.515 inches) while installed height is 33 mm (1.299 inches). Valve spring pressure at installed height should be 66.7 N (15 pounds) with a minimum allowable pressure of 56.9 N (12.8 pounds).

Rocker arm bushings (6–Fig. K1-7) are renewable. Bushing ID should be 14.002-14.050 mm (0.551-0.553 inch). Diameter of rocker stand (12) shafts should be 13.973-13.984 mm (0.5501-0.5505 inch). Maximum clearance between shaft and bushing is 0.15 mm (0.006 inch).

INJECTOR

WARNING: Fuel emerges from injector with sufficient force to penetrate the skin. When testing injector, keep yourself clear of nozzle spray.

REMOVE & REINSTALL. Before removing an injector or loosening injector lines, thoroughly clean injector, lines and surrounding area using compressed air and a suitable solvent.

To remove injector unit, first remove high pressure line leading from pump to injector. Disconnect bleed line by removing banjo bolt or by pulling line from banjo nipple fitting (8–Fig. K1-9). Remove two stud nuts securing ears of injector body to left side of cylinder head and withdraw injector unit.

When installing injector, make sure machined seating surface in cylinder head bore is completely clean and free from carbon buildup. Use a new copper washer underneath injector nozzle. Turn retaining stud nuts both finger tight, then tighten alternately and evenly one-sixth turn at a time to a torque of 24-27 N·m (18-20 ft.-lbs.). Start and run engine, listening for pressure leaks around nozzle seating washer. Correct pressure leaks by checking to be sure stud nuts are tightened evenly and injector unit is not cocked.

TESTING. A complete job of testing and adjusting the injector requires use of special test equipment. Only clean, approved testing oil should be used in tester tank. Nozzle should be tested for opening pressure, seat leakage and spray pattern. When tested, nozzle should open with a high-pitched buzzing sound, and cut off quickly at end of injection with a minimum of seat leakage and a controlled amount of back leakage.

Before conducting test, operate tester lever until fuel flows, then attach injector. Close valve to tester gage and pump tester lever a few quick strokes to be sure nozzle valve is not stuck, and that possibilities are good that injector can be returned to service without disassembly.

OPENING PRESSURE. Open valve to tester gage and operate tester lever slowly while observing gage reading. Opening pressure should be 11770-12260 kPa (1705-1775 psi).

Opening pressure is adjusted by adding or removing shims in shim pack (6–Fig. K1-9). Adding or removing one 0.1 mm (0.004 inch) thickness shim will change opening pressure approximately 980 kPa (140 psi).

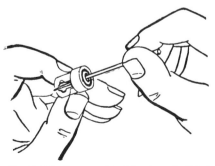

Fig. K1-10—Use a pointed hardwood stick to clean spray hole as shown.

Illustrations Courtesy Kubota

Kubota — SMALL DIESEL

SEAT LEAKAGE. Operate tester lever slowly to maintain a pressure of 1000 kPa (145 psi) below nozzle opening pressure. Observe nozzle tip for leakage. If a drop forms within a period of 10 seconds, nozzle valve is not seating and must be overhauled or renewed.

SPRAY PATTERN. Spray pattern should be well atomized and slightly conical, emerging in a straight axis from nozzle tip. If pattern is wet, ragged or intermittent, nozzle must be overhauled or renewed.

OVERHAUL. Hard or sharp tools, emery cloth, grinding compound or other than approved solvents or lapping compounds must never be used. An approved nozzle cleaning kit is available through a number of specialized sources.

Wipe all dirt and loose carbon from exterior of nozzle and holder assembly. Refer to Fig. K1-9 for exploded view and proceed as follows:

Secure nozzle in a soft-jawed vise or holding fixture and remove cap nut (9). Place all parts in clean calibrating oil or diesel fuel as they are removed.

Clean exterior surfaces with a brass wire brush. Soak injector in an approved carbon solvent, if necessary, to loosen hard carbon deposits. Rinse parts in clean diesel fuel or calibrating oil immediately after cleaning to neutralize the solvent and prevent etching of polished surfaces.

Clean nozzle spray hole from inside using a pointed hardwood stick or wood splinter as shown in Fig. K1-10. Scrape carbon from pressure chamber using hooked scraper as shown in Fig. K1-11. Clean valve seat using brass scraper as shown in Fig. K1-12, then polish seat using wood polishing stick and mutton tallow as in Fig. K1-13.

Reclean all parts by rinsing thoroughly in clean diesel fuel or calibrating oil and assemble while parts are immersed in clean fluid. Make sure adjusting shim pack is intact. Tighten nozzle retaining nut (9 – Fig. K1-9) to a torque of 79-98 N·m (58-72 ft.-lbs.). Do not overtighten, distortion may cause valve to stick and no amount of overtightening can stop a leak caused by scratches or dirt. Retest assembled injector as previously outlined.

INJECTION PUMP

All models are equipped with the injection pump shown in Fig. K1-4. The injection pump should be tested and overhauled by a shop qualified for diesel injection pump repair.

If rack (9) is removed and must be reinstalled, marks (M – Fig. K1-14) on pump body and rack must be aligned and master tooth (T – Fig. K1-15) must align with mark (M) on rack.

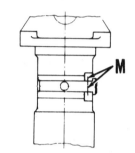

Fig. K1-14—Marks (M) on injection pump body and rack should align.

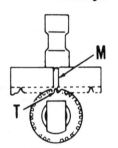

Fig. K1-15—Mark (M) on rack should align with control sleeve master tooth (T).

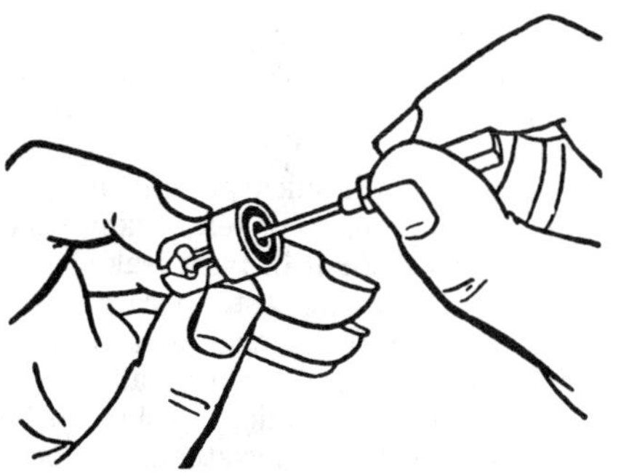

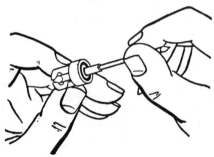

Fig. K1-11—Use a hooked scraper to clean carbon from pressure chamber.

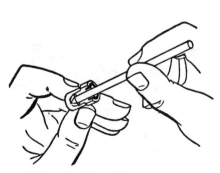

Fig. K1-12—Clean seat using brass scraper as shown.

Fig. K1-13—Polish seat using polishing stick and mutton tallow.

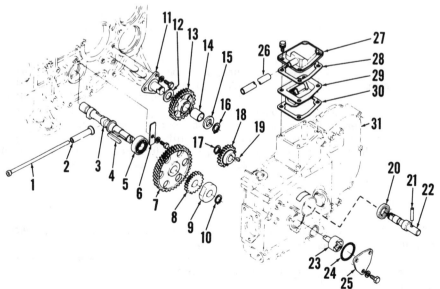

Fig. K1-16—Exploded view of camshaft, idler and oil pump assemblies.

1. Push rod
2. Tappet
3. Camshaft
4. Key
5. Bearing
6. Bearing retainer
7. Gear
8. Gear
9. Injection pump cam
10. Snap ring
11. Idler shaft
12. Washer
13. Idler gear
14. Bushing
15. Slotted washer
16. Snap ring
17. Snap ring
18. Starter gear
19. Key
20. Bearing
21. Pin
22. Starter shaft
23. Oil pump
24. "O" ring
25. Pump cover
26. Hose
27. Breather cover
28. Gasket
29. Breather assy.
30. Gasket
31. Gearcase

SERVICE MANUAL

Kubota

The injection pump tappet is actuated by cam (9 – Fig. K1-16) which is attached to camshaft (3). Inspect cam each time injection pump is removed.

Tighten injection pump screws to 24-27 N·m (18-20 ft.-lbs.) and refer to INJECTION PUMP TIMING section.

OIL PUMP

R&R AND OVERHAUL. Oil pump (23 – Fig. K1-16) is housed in gearcase (31) and driven by a slot in the end of camshaft (3). To remove pump, unscrew cover (25) and extract pump assembly from cover.

The pump is available as a unit assembly only. Note the following measurements to detect pump wear. Inner rotor to outer rotor clearance (Fig. K1-17) should not exceed 0.2 mm (0.008 inch). Outer rotor to gear cover clearance (Fig. K1-18) should not exceed 0.24 mm (0.009 inch). With a straightedge placed across oil pump as shown in Fig. K1-19, gap between either rotor and gear cover should not exceed 0.25 mm (0.010 inch).

The oil pressure relief valve ball (6 – Fig. K1-6A) and spring (5) are located in gear cover. Oil pressure is not adjustable. Minimum allowable oil pressure is 196 kPa (28 psi).

GOVERNOR

All models are equipped with a flyweight type governor as shown in Fig. K1-6. The flyweight assembly is attached to crankshaft gear (28) and actuates governor lever (9) which controls the position of fuel injection pump rack (9 – Fig. K1-4).

Governor components are accessible after removing gearcase. Inspect components and renew any which are excessively worn or damaged.

GEAR TRAIN

Should the camshaft gear, crankshaft gear, idler gear or either balancer gear be removed, refer to Fig. K1-20 and align timing marks on gears shown. Note that timing dots are found on camshaft and crankshaft gears while numbers are used on balancer gears. Timing dots and a number are stamped on the idler gear.

BALANCER SHAFTS

R&R AND OVERHAUL. The balancer shafts are accessible after removing gearcase. Remove idler gear (13 – Fig. K1-16), then unscrew bearing

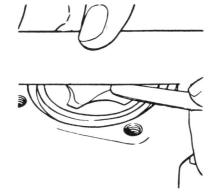

Fig. K1-19—Place a straightedge on oil pump housing and measure between rotor faces and housing. Gap must not exceed 0.25 mm (0.010 inch).

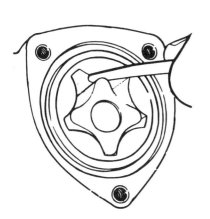

Fig. K1-17—Inner rotor to outer rotor clearance should not exceed 0.2 mm (0.008 inch). Measure as shown.

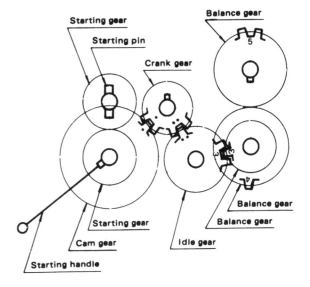

Fig. K1-20—Diagram of gear train showing timing marks.

Fig. K1-18—Outer rotor to oil pump housing clearance should not exceed 0.24 mm (0.009 inch). Measure as shown.

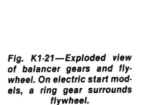

Fig. K1-21—Exploded view of balancer gears and flywheel. On electric start models, a ring gear surrounds flywheel.

1. Cover
2. Pin
3. Rope pulley
4. Nut
5. Lockwasher
6. Flywheel
7. Key
8. Bearing
9. Balancer
10. Bearing
11. Gear
12. Snap ring
13. Gear
14. Bearing retainer
15. Lockwasher
16. Screw

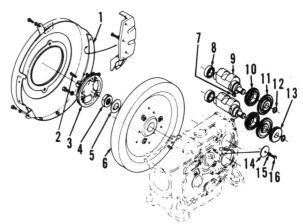

Kubota SMALL DIESEL

1. Seal housing
2. "O" ring
3. Bearing
4. "O" rings
5. Oil sleeve
6. Snap ring
7. "O" ring
8. Hollow dowel
9. Bearing carrier
10. "O" ring
11. "O" ring
12. Snap ring
13. Washer
14. Bearing
15. Screw
16. Crankshaft
17. Weights
18. Bearing
19. Bushing
20. Retainer
21. Piston pin
22. Piston rings
23. Piston
24. Connecting rod
25. Plug
26. Key
27. Bearing
28. Gear
29. Sleeve
30. "O" rings
31. "O" ring
32. Hollow dowel
33. Plug
34. Hollow dowel
35. "O" ring
36. Coupler
37. Oil pressure light
38. Bearing retainer
39. Cylinder block
40. Gasket
41. Cover

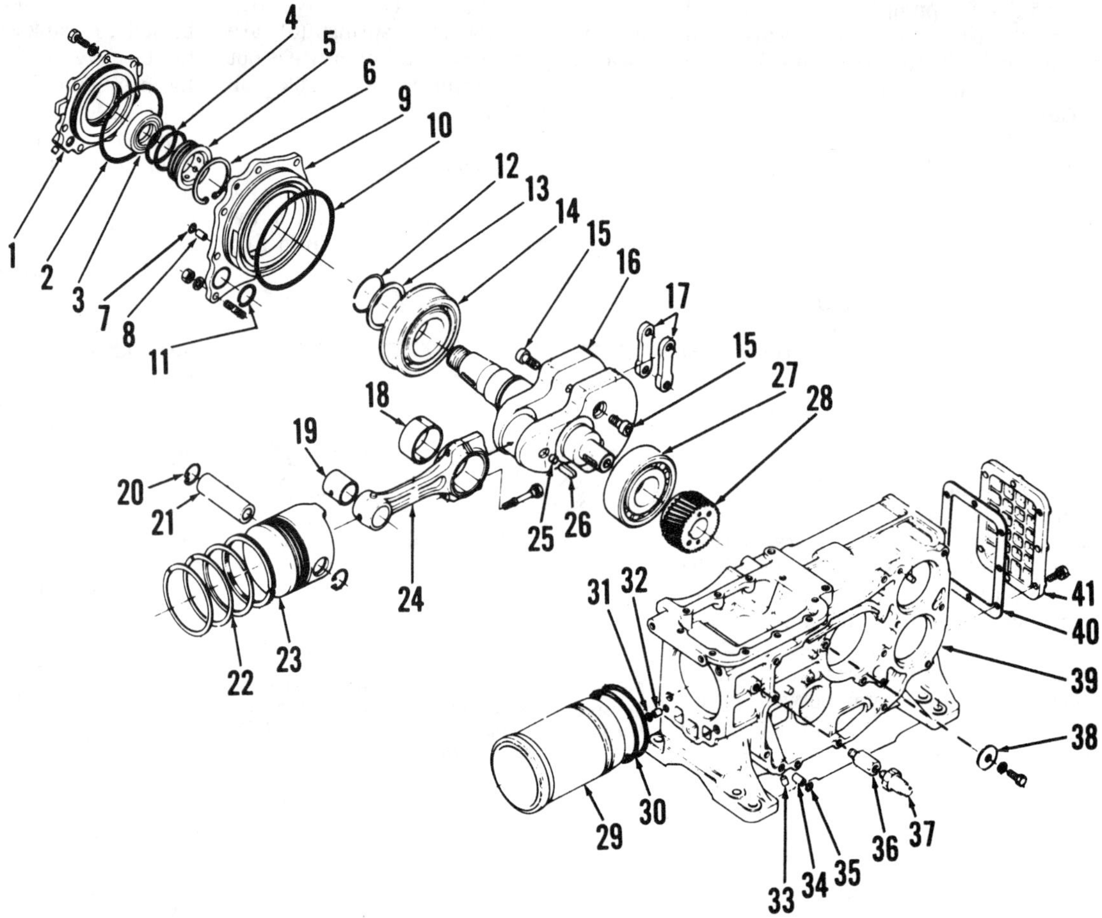

Fig. K1-22—Exploded view of crankshaft, piston, rod and block assemblies.

retainer (14–Fig. K1-21) and extract balancer shafts (9). Inspect shafts, gears and bearings and renew if damaged. When reinstalling balancer assemblies, refer to GEAR TRAIN section for proper timing of balancer gears.

PISTON AND ROD UNIT

REMOVE AND REINSTALL. Piston and connecting rod are removed as a unit after removing cylinder head, crankcase cover (41–Fig. K1-22) and balancer shafts (9–Fig. K1-21). Unscrew rod cap retaining screws, detach rod cap and extract piston and rod from head end of engine.

When assembling piston and rod, arrow on original equipment piston crown should be on same side as rod and cap alignment marks as shown in Fig. K1-23. Install piston and rod in engine so piston arrow (A) points towards engine top. Replacement piston does not have arrow on crown and may be installed in either direction. Install rod cap so marks (M) on rod and cap are aligned and tighten rod screws to 30-34 N·m (22-25 ft.-lbs.) on EA400-N, EA400-NB, EA450-N and EA450-NB or to 49-54 N·m (36-40 ft.-lbs.) on all other models.

PISTON, RINGS AND SLEEVE

All models are equipped with an aluminum piston which uses three compression rings and an oil control ring. Piston and rings are available in standard size only.

Piston pin boss inner diameter should be 25.00-25.013 mm (0.9843-0.9847 inch) on Models EA400-N, EA400-NB, EA450-N and EA450-NB. Renew piston if diameter exceeds 25.04 mm (0.986 inch). On all other models, pin boss inner diameter should be 27.000-27.021 inch (1.0630-1.0638 inches) and maximum allowable diameter is 27.04 mm (1.0646 inches).

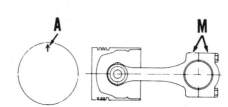

Fig. K1-23—Assemble rod, cap and piston so marks (M) on rod and cap are on same side as piston arrow (A). Install in engine so marks (M) and arrow (A) are up.

Desired piston ring end gap is 0.2-0.4 mm (0.008-0.016 inch) for all rings. Side clearance of compression rings in piston grooves should be 0.04-0.07 mm (0.0016-0.0027 inch) and oil ring side clearance should be 0.02-0.05 mm (0.0008-0.0020 inch).

Cylinder sleeve inside diameter should be 78.000-78.019 mm (3.0709-3.0716 inches) on Models EA400-N and EA400-NB, 84.000-84.022 mm (3.3071-3.3080 inches) on Models EA450-N and EA450-NB, 86.000-86.022 mm (3.3858-3.3867 inches) on Models EA500-N and EA500-

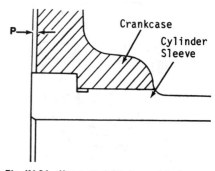

Fig. K1-24—Use a straightedge and feeler gage to measure cylinder sleeve protrusion beyond crankcase surface. Refer to text.

SERVICE MANUAL

Kubota

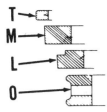

Fig. K1-25—View of ring configuration for top compression ring (T), middle compression ring (M), lower compression ring (L) and oil control ring (O).

NB or 92.000-92.022 mm (3.6220-3.6229 inches) on Models EA600-N and EA600-NB. Renew cylinder sleeve if wear exceeds 0.20 mm (0.008 inch). With sleeve installed in crankcase, measure protrusion (P – Fig. K1-24) of sleeve beyond crankcase surface using a straightedge and feeler gage. Desired sleeve protrusion is 0.08-0.18 mm (0.003-0.007 inch). Renew sleeve if protrusion is not within recommended limits.

When installing rings onto piston, note shape and location of rings as shown in Fig. K1-25. Stagger ring end gaps evenly around piston.

PISTON PIN

A full floating piston pin is used on all models. Piston pin outside diameter is 25.002-25.011 mm (0.9843-0.9847 inch) on Models EA400-N, EA400-NB, EA450-N and EA450-NB. On all other models, pin diameter is 27.002-27.011 mm (1.0630-1.0634 inches). Clearance between pin and connecting rod bushing should be 0.014-0.038 mm (0.0006-0.0015 inch) with a maximum allowable clearance of 0.20 mm (0.008 inch).

CONNECTING ROD AND BEARINGS

The connecting rod is equipped with a renewable bushing in the small end and insert type bearings in the big end. Inner diameter of standard size big end bearing should be 44.010-44.056 mm (1.7327-1.7345 inches) for Models EA400-N, EA400-NB, EA450-N and EA450-NB, or 48.010-48.056 mm (1.8902-1.8920 inches for all other models. Clearance between bearing and crankpin should be 0.035-0.097 mm (0.0014-0.0038 inch) for all models. Bearings are available in 0.25 and 0.50 mm undersizes.

Inner diameter of small end bushing is 25.025-25.040 mm (0.9852-0.9858 inch) for Models EA400-N, EA400-NB, EA450-N and EA450-NB, or 27.025-27.040 mm (1.0640-1.0646 inches) for all other models. Clearance between bushing and piston pin should be 0.014-0.038 mm (0.0006-0.0015 inch) for all models.

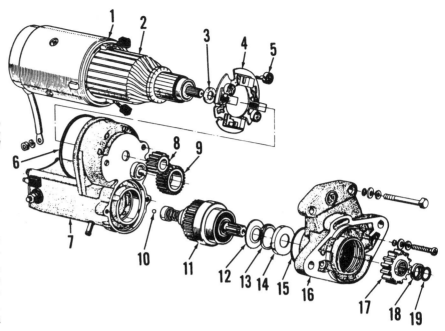

Fig. K1-26—Exploded view of electric starter.

1. Frame
2. Armature
3. Felt washer
4. Brush assy.
5. Brush spring
6. Gasket
7. Switch housing assy.
8. Gear
9. Gear
10. Ball
11. Drive assy.
12. Thrust washer
13. Spring
14. Spring holder
15. "O" ring
16. End frame
17. Gear
18. Collar
19. Retainer

CAMSHAFT

The camshaft is supported by a ball bearing at the gear end while the opposite end rides directly in the cylinder block.

R&R AND OVERHAUL. To remove camshaft, remove cylinder head and push rods. Remove gearcase (31 – Fig. K1-16), then unscrew bearing retainer (6). Rotate camshaft one turn to force tappets away from lobes and withdraw camshaft while being careful not to dislodge tappets.

Camshaft lobe height for intake and exhaust lobes is 27 mm (1.0630 inches) with a wear limit of 26.5 mm (1.0433 inches) for Models EA400-N, EA400-NB, EA450-N and EA450-NB, or 33.5 mm (1.3189 inches) with a wear limit of 33 mm (1.2992 inches) for all other models.

Diameter of plain bearing end of camshaft is 21.967-21.980 mm (0.8648-0.8654 inch) while bearing bore in crankcase is 22.000-22.021 mm (0.8661-0.8670 inch). Desired clearance is 0.02-0.054 mm (0.0008-0.0021 inch) while maximum allowable clearance is 0.25 mm (0.010 inch).

Refer to GEAR TRAIN section when installing camshaft for proper alignment of gear timing marks. Check backlash between camshaft and crankshaft gears.

Backlash should be 0.048-0.14 mm (0.0019-0.0055 inch) for all models. Gear must be renewed if backlash exceeds 0.3 mm (0.012 inch).

CRANKSHAFT AND BEARINGS

R&R AND OVERHAUL. To remove crankshaft, remove flywheel, balancer shafts and piston with rod as previously outlined. Remove seal housing (1 – Fig. K1-22) and unscrew bearing carrier (9) retaining nuts, then withdraw crankshaft and bearing assembly from crankcase. Disassemble as required.

When inspecting components, be sure oil sleeve (5) is clean and serviceable as pressurized oil is directed through sleeve from oil passage in seal housing (1) to crankshaft for rod bearing lubrication. Inner diameter of sleeve should be 48.000-48.025 mm (1.8898-1.8907 inches) while corresponding diameter of crankshaft should be 47.959-47.975 mm (1.8881-1.8888 inches). Maximum allowable clearance between sleeve and crankshaft is 0.15 mm (0.006 inch). Be sure "O" rings (4) seal effectively or low pressure may result.

Note balancer weights (17) attached to crankshaft on Models EA400-N, EA400-NB, EA450-N and EA450-NB.

Crankshaft crankpin diameter is 43.959-43.975 mm (1.7307-1.7313 inch-

Kubota

SMALL DIESEL

es) for Models EA400-N, EA400-NB, EA450-N and EA450-NB, or 47.959-47.975 mm (1.8881-1.8888 inches) for all other models. Maximum allowable clearance between crankpin and bearing is 0.25 mm (0.010 inch). Rod bearings are available in standard size and 0.25 and 0.50 mm undersizes.

Crankshaft end play should be 0.05-0.46 mm (0.002-0.018 inch). Renew main bearing retainer washer (13) if end play is not within specified limits.

Refer to GEAR TRAIN section for proper timing of crankshaft gear.

ALTERNATOR

Models EA400-NB, EA450-NB, EA500-NB and EA600-NB are equipped with an alternator mounted behind the fan. The fan serves as the alternator rotor. Alternator output should be 8 volts with 3.1 amperes current at alternator speed of 6800 rpm.

ELECTRIC STARTER

Models EA400-NB, EA450-NB, EA500-NB and EA600-NB are equipped with an electric starter as shown in Fig. K1-26. Minimum brush length is 10 mm (0.394 inch) while wear limit of commutator is 29 mm (1.141 inches) diameter. With no load imposed on starter and using an 11.5 volt source, the starter shaft should rotate at 3500 rpm or more while drawing less than 90 amperes current.

KUBOTA

Model	No. Cyls.	Bore	Stroke	Displ.
D650-B	3	64 mm (2.520 in.)	70 mm (2.756 in.)	675 cc (41.2 cu. in.)
D750-B	3	68 mm (2.677 in.)	70 mm (2.756 in.)	762 cc (46.5 cu. in.)
D850-B	3	72 mm (2.835 in.)	70 mm (2.756 in.)	855 cc (52.2 cu. in.)
DH850-B	3	72 mm (2.835 in.)	70 mm (2.756 in.)	855 cc (52.2 cu. in.)
D950-B	3	75 mm (2.953 in.)	70 mm (2.756 in.)	927 cc (56.5 cu. in.)
Z500-B	2	68 mm (2.677 in.)	70 mm (2.756 in.)	508 cc (31.0 cu. in.)
Z600-B	2	72 mm (2.835 in.)	70 mm (2.756 in.)	570 cc (34.8 cu. in.)
ZB500C-B	2	68 mm (2.677 in.)	70 mm (2.756 in.)	508 cc (31.0 cu. in.)
ZB600C-B	2	72 mm (2.835 in.)	70 mm (2.756 in.)	570 cc (34.8 cu. in.)

All engines in this section are liquid-cooled four-stroke diesel engines. Crankshaft rotation is counterclockwise viewed from flywheel end. Firing order is 1-2-3 for three-cylinder models.

MAINTENANCE

LUBRICATION

Recommended crankcase oil is SAE 10W for ambient temperatures of 0°C (32°F) and below, SAE 20 for temperatures of 0°-25°C (32°-77°F) and SAE 30 for temperatures above 25°C (77°F). SAE 10W-30 oil may be substituted for recommended straight weight oils for use in all temperatures. Use oil with API classification CC/CD. Crankcase capacity is 4.6 liters (4.8 quarts) for Models D650-B, D750-B, D850-B and D950-B, 3.7 liters (3.9 quarts) for Models DH850-B, 2.6 liters (2.7 quarts) for Models Z500-B and Z600-B or 3.1 liters (3.3 quarts) for Models ZB500C-B and ZB600C-B. Manufacturer recommends renewing engine oil and oil filter after every 150 hours of operation.

All models are equipped with a pressure lubrication system. Oil is directed from the oil pump to the oil pressure relief valve and oil filter. Oil is routed through an oil passage in the cylinder block to the main bearings, then through crankshaft oil passages to the crankpins. The cylinder block oil passage also routes oil to the cylinder head to lubricate the valve train.

Normal oil pressure is 295-440 kPa (43-64 psi) with engine at operating speed and temperature. Minimum allowable pressure is 196 kPa (28 psi). Refer to OIL PUMP section for pump service.

FUEL SYSTEM

FUEL FILTER. A combination of fuel shut-off valve and filter assembly (2 – Fig. K2-1) is attached to engine. Fuel filter element is renewable and manufacturer recommends renewing filter after every 400 hours of operation. However, if engine indicates signs of fuel starvation (loss of power or surging), renew filter regardless of hours of operation. After renewing filter, air must be bled from system as outlined in the following paragraph.

BLEED FUEL SYSTEM. To bleed air from fuel system, first loosen injection pump bleed screw (4 – Fig. K2-1) and open fuel shut-off valve. With engine stop lever in STOP position, rotate crankshaft with starter or manually to operate fuel transfer pump (3). When air-free fuel flows from bleed screw, retighten bleed screw.

If engine fails to start at this point, loosen high pressure line (6) fittings at

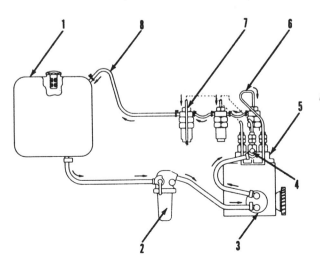

Fig. K2-1—Drawing of typical fuel system.
1. Fuel tank
2. Filter assy.
3. Fuel pump
4. Bleed screw
5. Injection pump
6. High pressure line
7. Injector
8. Fuel return line

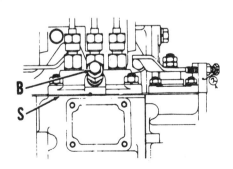

Fig. K2-2—Injection timing is adjusted by adding or deleting shims (S) between pump and engine block. Air may be bled from injection pump by loosening bleed screw (B).

Kubota

injectors. With engine stop lever in RUN position, crank engine to operate injection pump. Retighten injector line fittings when fuel is discharged from injector lines, then start engine.

INJECTION PUMP TIMING. Injection pump timing is adjusted using shims (S – Fig. K2-2). To check injection pump timing proceed as follows: Disconnect high pressure line for number 1 cylinder from injection pump, then unscrew delivery valve holder (1 – Fig. K2-3) and remove valve spring (3) and delivery valve assembly (5). Reinstall delivery valve holder (1) and connect a suitable spill pipe to valve holder. Aim spill pipe at a receptacle to catch discharged fuel. Move throttle control to full open position. Rotate engine flywheel slowly in counterclockwise direction until fuel just stops flowing from spill pipe. Beginning of injection occurs at this point. If timing is correct, "F1" mark (Fig. K2-4) on flywheel should be aligned with timing mark (M) on cylinder block rear plate.

To adjust injection timing, remove injection pump and add shim gaskets to retard timing or remove shims to advance timing. Each shim alters timing approximately 1½ crankshaft degrees.

When reinstalling injection pump, be sure pump control rack pin properly engages governor operating fork. Tighten pump mounting nuts or screws to 24-27 N·m (18-20 ft.-lbs.). Reinstall delivery valve and spring and tighten delivery valve holder to 40-49 N·m (30-36 ft.-lbs.). Bleed air from system as previously outlined.

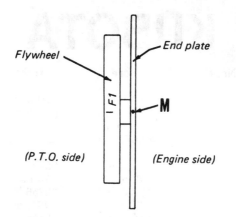

Fig. K2-4—View of "F1" injection timing mark on flywheel and timing mark (M) on end plate of models so equipped. Refer to text.

SMALL DIESEL

GOVERNOR

All models are equipped with a flyball type governor mounted on front end of fuel injection pump camshaft. Refer to Fig. K2-5 for an exploded view of governor linkage.

Slow idle speed for all models is 800 rpm and is adjusted by turning speed screw (37—Fig. K2-5). High idle speed and maximum fuel limiting stop screws are sealed and should be adjusted by qualified personnel only. High idle (no-load) speed should be 3780 rpm for Model DH850-B, 3500 rpm for Models Z600-B, ZB600C-B or 3200 rpm for all other models. Turn high idle speed screw (40) for adjustment.

Maximum fuel limiting stop should be set to prevent excessive smoke level at slight overload. To make adjustment, remove seal cap, loosen jam nut and turn adjuster (19) in to lower smoke level or out to raise smoke level.

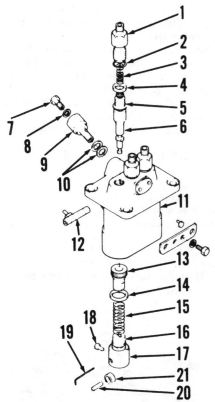

Fig. K2-3—Exploded view of fuel injection pump.

1. Delivery valve holder
2. "O" ring
3. Spring
4. Shim
5. Delivery valve assy.
6. Plunger
7. Air bleed screw
8. Gasket
9. Adapter
10. Gaskets
11. Pump body
12. Control rack
13. Control sleeve
14. Washer
15. Spring
16. Spring seat
17. Tappet
18. Guide pin
19. Pin
20. Pin
21. Roller

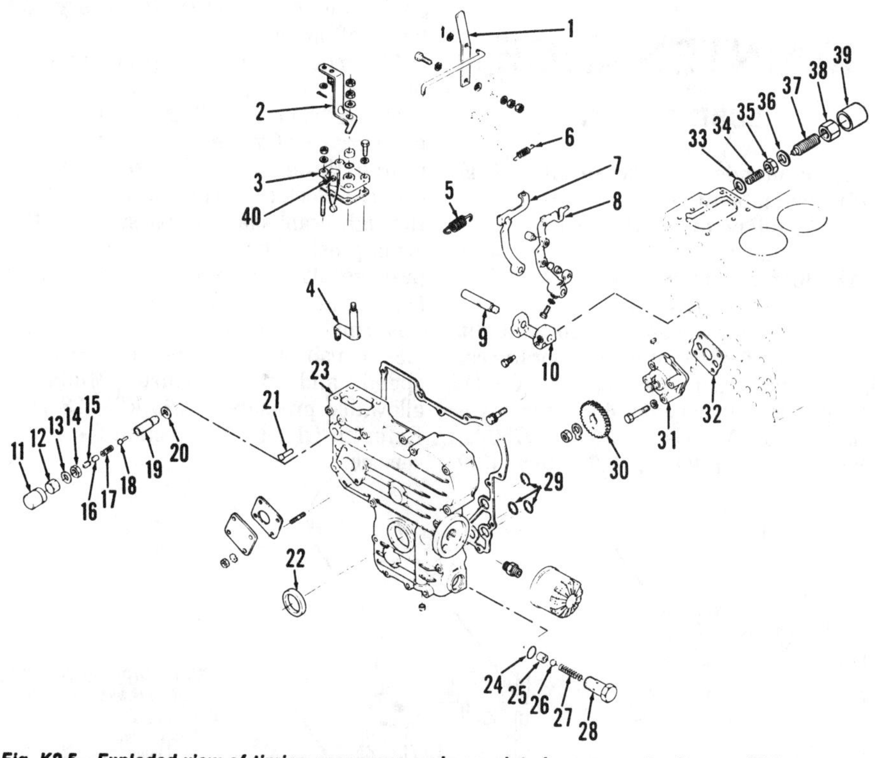

Fig. K2-5—Exploded view of timing gear cover and associated components. Be sure "O" rings (29) are in place when installing cover (23).

1. Speed control
2. Control lever
3. Plate
4. Control arm
5. Governor spring
6. Start spring
7. Governor arm
8. Pump control arm
9. Pivot pin
10. Pivot block
11. Cap
12. Cap nut
13. Gasket
14. Locknut
15. Maximum fuel limiting screw
16. Spacer
17. Spring
18. Pin
19. Maximum fuel limiting body
20. Gasket
21. Start spring pin
22. Seal
23. Timing gear cover
24. "O" ring
25. Oil pressure relief valve seat
26. Relief valve ball
27. Spring
28. Valve body
29. "O" rings
30. Pump gear
31. Oil pump
32. Gasket
33. Gasket
34. Spring
35. Locknut
36. Gasket
37. Low idle speed screw
38. Cap nut
39. Cap
40. High idle speed screw

SERVICE MANUAL

Kubota

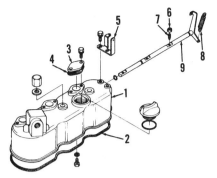

K2-6—View of valve cover and compression release components.

1. Valve cover
2. Gasket
3. Cover
4. Gasket
5. Bracket
6. Locknut
7. Compression release screw
8. Spring
9. Compression release shaft

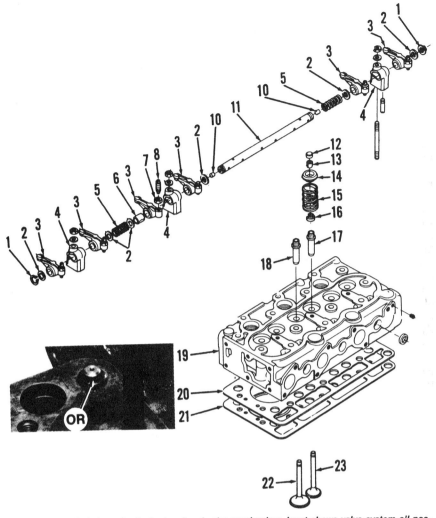

Fig. K2-7—Exploded view of cylinder head and valve mechanism. Inset shows valve system oil passage pipe and "O" ring (OR) at left corner of cylinder block.

1. Snap ring
2. Washer
3. Rocker arm
4. Shaft stand
5. Spring
6. Bushing
7. Locknut
8. Adjusting screw
9. Set screw
10. (blank)
11. Rocker shaft
12. Valve cap
13. Retainer keys
14. Retainer
15. Spring
16. Seal
17. Exhaust valve guide
18. Intake valve guide
19. Cylinder head
20. Shim
21. Gasket
22. Intake valve
23. Exhaust valve

COMPRESSION RELEASE

All models are equipped with a compression release which holds exhaust valves open slightly when compression release lever (9–Fig. K2-6) is rotated. To adjust compression release, rotate engine crankshaft so piston in cylinder being adjusted is at top dead center on compression. Remove cover (3), loosen locknut (6) and back out adjusting screw (7). Rotate compression release lever to engaged position. Turn adjusting screw (7) in until it contacts exhaust rocker arm, then turn screw an additional 1½ turns. Tighten locknut and adjust remaining cylinders. Check operation of compression release being sure exhaust valves do not contact pistons.

AIR FILTER

All models are equipped with a dry type renewable air filter. Manufacturer recommends blowing out filter after every 100 to 200 hours of operation and renewing filter after six cleanings or one year.

COOLING SYSTEM

All engines are liquid-cooled. Some early models used a thermo-siphon system (without water pump) to circulate coolant, while late models are equipped with forced coolant circulation provided by a centrifugal water pump. On all models, radiator pressure rating is 88 kPa (13 psi).

Cooling system should be drained and refilled after every 500 hours of operation or once a year. It is recommended that a mixture of water and ethylene glycol antifreeze be used even if temperatures below freezing are not expected. The antifreeze solution contains additives which help prevent corrosion of cooling system components.

Fan belt tension is correct if a pressure of 59-68 N (13-15 pounds) applied against belt at midpoint between crankshaft pulley and alternator pulley deflects belt approximately 8 mm (5/16 inch).

REPAIRS

TIGHTENING TORQUES

Refer to the following table for special tightening torques.

Connecting rod............27-30 N·m
(20-22 ft.-lbs.)
Crankshaft pulley nut....138-157 N·m
(102-115 ft.-lbs.)
Crankshaft seal carrier.....10-12 N·m
(8-9 ft.-lbs.)
Cylinder head.............See Text
Flywheel.................54-59 N·m
(40-43 ft.-lbs.)
Injection pump............24-27 N·m
(18-20 ft.-lbs.)
Injector nozzle nut.........30-49 N·m
(22-36 ft.-lbs.)
Main bearing retainer......30-34 N·m
(22-25 ft.-lbs.)
Rocker arm brackets.......17-20 N·m
(13-15 ft.-lbs.)

COMPRESSION PRESSURE

Compression pressure should be checked with all injectors removed while cranking engine with starter at 200 to 300 rpm. Compression pressure should be 2667-3138 kPa (387-455 psi) for Models D650-B, D750-B, Z500-B and ZB500C-B; 2746-3236 kPa (398-469 psi) for Models D850-B, DH850-B, Z600-B and ZB600C-B; or 2550-2942 kPa (370-427 psi) for Model D950-B. Difference in compression pressure among cylinders should not exceed 10 percent.

Illustrations Courtesy Kubota

Kubota SMALL DIESEL

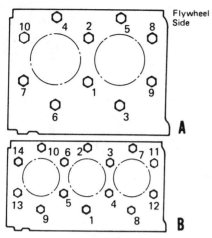

Fig. K2-8—Follow sequence in diagram A when tightening cylinder head fasteners on two-cylinder models or diagram B on three-cylinder models.

VALVE CLEARANCE

Valve clearance should be adjusted with engine cold and piston at top dead center on compression. Remove rocker cover, then turn rocker arm adjusting screw until a feeler gage of appropriate thickness can be inserted between rocker arm and end of valve. Recommended clearance is 0.145-0.185 mm (0.006-0.007 inch) for both intake and exhaust.

Repeat procedure for remaining cylinders. On two-cylinder engines, rotate crankshaft one complete revolution (360°) to position second piston at TDC on compression. On three-cylinder engines, rotating crankshaft 2/3 turn (240°) will place the next piston in firing order (1-2-3) at TDC on compression.

CYLINDER HEAD

To remove cylinder head, proceed as follows: Drain cooling system and remove upper radiator hose. Unbolt and remove thermostat housing, intake and exhaust manifolds. Remove fuel injection lines, injector fuel return lines and glow plugs. Remove rocker arm cover, rocker arm shaft assembly and the push rods. Unbolt and remove cylinder head assembly from block. Take care not to damage the valve mechanism oil pipe (see Fig. K2-7) extending from left corner of cylinder block. Remove the cylinder head gasket, cylinder head shim (20), if fitted, and oil pipe sealing "O" ring (OR).

Before reinstalling cylinder head, completely clean all traces of old cylinder head gasket and carbon from both head and block surfaces. Be careful not to damage valve system oil tube protruding from corner of cylinder block. Clean carbon from top of pistons, taking care to clean all loose carbon from cylinders. Closely check cylinder head for cracks. Use a straightedge and feeler gage to check cylinder head for any warped or twisted condition. Do not place straightedge over precombustion chamber inserts. Maximum allowable deviation from flatness is 0.05 mm (0.002 inch) per 100 mm (3.94 inches). Surface grind head if not within limits. Refer to VALVE SYSTEM paragraphs for valve head recess.

If engine has been overhauled with new pistons and/or connecting rods, piston to cylinder head clearance should be checked. Piston to cylinder head clearance should be 0.60-0.80 mm (0.024-0.031 inch). Engine manufacturer recommends checking piston to cylinder head clearance after installing cylinder head and before fuel injectors are reinstalled. Turn crankshaft so that piston is at bottom of compression stroke and insert a soft lead wire into cylinder through injector opening. Turn the crankshaft by hand so that piston goes past top dead center, then withdraw the lead wire and measure flattened thickness. This thickness is equal to piston to cylinder head clearance. If clearance is less than specified, remove cylinder head and reinstall with a 0.2 mm (0.008 inch) shim between head gasket and cylinder head.

To reinstall cylinder head, reverse removal procedure and note the following: Do not apply any sealing compound to head gasket surfaces. Place a new "O" ring (OR–Fig. K2-7) on valve system oil tube. Place new head gasket on block and place new 0.002 mm (0.008 inch) shim, if needed, on top of gasket. Carefully lower the cylinder head on the two stud bolts. Oil the threads of the cylinder head cap screws and nuts and install all finger tight. Refer to the appropriate tightening sequence in Fig. K2-8 and tighten the head bolts and nuts in several steps until final torque of 39-44 N·m (29-32 ft.-lbs.) is reached on Z500-B, D650-B and D750-B engines or 65-69 N·m (48-51 ft.-lbs.) on all other engines. Loosen locknuts and back rocker arm adjusting screws out. Install push rods and rocker arms and tighten rocker arm nuts to a torque of 10-11 N·m (7-8 ft.-lbs.). Adjust valve clearance as previously described. Complete reassembly by reversing disassembly procedure. Start and run engine until normal operating temperature is reached, then retorque cylinder head bolts and nuts to specified torque.

VALVE SYSTEM

Intake and exhaust valves seat directly in cylinder head. Valve face and seat angles are 45 degrees. Desired valve seat width is 2.1 mm (0.083 inch). After refacing valves and seats, check distance valve head is recessed from cylinder head surface. Valve should be recessed 0.9-1.1 mm (0.035-0.043 inch). If recession exceeds 1.3 mm (0.051 inch), a shim washer of appropriate thickness should be added between valve springs and cylinder head to maintain correct valve spring tension and cylinder head surface should be machined to obtain recommended recession.

Valve stem diameter should be 6.960-6.975 mm (0.2740-0.2765 inch) for intake and exhaust. Inside diameter of valve guides should be 7.010-7.025 mm (0.2759-0.2765 inch). Recommended clearance between valve stems and guides is 0.035-0.065 mm (0.0014-0.0026 inch) and maximum allowable clearance is 0.1 mm (0.004 inch). When renewing guides, note that intake and exhaust guides are not interchangeable. After installation, check guide inside diameter and ream to provide recommended clearance.

Intake and exhaust valve springs are identical. Valve spring free length should be 35.1-35.6 mm (1.382-1.402 inches) while installed height is 31 mm (1.220 inches). Spring pressure at installed height should be 74 N (16.6 pounds) with a minimum allowable pressure of 63 N (14.1 pounds). Renew spring if distorted or otherwise damaged.

Rocker arm bushings (6 – Fig. K2-7) are renewable. Bushing ID should be 10.997-11.038 mm (0.4330-0.4346 inch). Rocker shaft diameter should be 10.973-10.984 mm (0.4320-0.4324 inch). Desired clearance between rocker arm shaft and bushings is 0.013-0.065 mm (0.0005-0.0026 inch) while maximum allowable clearance is 0.15 mm (0.006 inch). When renewing rocker arm bushings, make certain the oil hole in bushing is aligned with hole in rocker arm.

INJECTOR

WARNING: Fuel emerges from injector with sufficient force to penetrate the skin. When testing injector, keep yourself clear of nozzle spray.

REMOVE AND REINSTALL. Before removing an injector, or loosening injector lines, thoroughly clean injector, lines and surrounding area using compressed air and a suitable solvent.

To remove injector unit, first remove high pressure line leading from injection pump to injector. Disconnect bleed line by removing nut and banjo fitting, or by pulling line(s) from banjo nipple fitting (2 – Fig. K2-9). With pressure and bleedback lines removed, unscrew injector

SERVICE MANUAL

Kubota

from its mounting position on cylinder head. Note that a special nozzle holder socket wrench, Code No. 07916-30841, is available from manufacturer for use in removing and installing injectors.

When installing injector, make sure that machined seating surface in cylinder head is completely clean and free from carbon buildup. Use a new copper washer underneath injector nozzle and tighten injector carefully to 29-49 N·m (22-36 ft.-lbs.).

TESTING. A complete job of testing and adjusting the injector requires use of special test equipment. Only clean, approved testing oil should be used in tester tank. Nozzle should be tested for opening pressure, seat leakage and spray pattern. When tested, nozzle should open with high-pitched buzzing sound, and cut-off quickly at end of injection with a minimum of seat leakage.

Before conducting test, operate tester lever until fuel flows, then attach injector. Close valve to tester gage and pump tester lever a few quick strokes to be sure nozzle valve is not stuck, and that possibilities are good that injector can be returned to service without disassembly.

OPENING PRESSURE. Open valve to tester gage, then operate tester lever slowly while observing gage reading. Opening pressure should be 13720-14710 kPa (1990-2133 psi).

Opening pressure is adjusted by adding or removing shims (5 – Fig. K2-9). Adding or removing one 0.1 mm (0.004 inch) shim will change opening pressure approximately 980 kPa (140 psi).

SEAT LEAKAGE. Operate tester lever slowly to maintain pressure at 1000 kPa (150 psi) below nozzle opening pressure and observe nozzle tip for leakage. If a drop forms within a period of 10 seconds, nozzle valve is not seating and must be overhauled or renewed.

SPRAY PATTERN. Spray pattern should be well atomized and slightly conical, emerging in a straight axis from nozzle tip. If pattern is wet, ragged or intermittent, nozzle must be overhauled or renewed.

OVERHAUL. Hard or sharp tools, emery cloth, grinding compound or other than approved solvents or lapping compounds must never be used. An approved nozzle cleaning kit is available through a number of specialized sources.

Wipe all dirt and loose carbon from exterior of nozzle and holder assembly. Refer to Fig. K2-9 for exploded view and proceed as follows:

Secure pressure fitting (4) in a soft-jawed vise or holding fixture and remove nozzle nut (10). Place all parts in clean calibrating oil or diesel fuel as they are removed, using a compartmented pan and using extra care to keep parts from each injector together and separate from other units.

Clean exterior surfaces with a brass wire brush. Soak parts in an approved carbon solvent, if necessary, to loosen hard carbon deposits. Rinse parts in clean diesel fuel or calibrating oil immediately after cleaning to neutralize the solvent and prevent etching of polished surfaces.

Clean nozzle spray hole from inside using a pointed hardwood stick or wood splinter as shown in Fig. K2-10. Scrape carbon from pressure chamber using hooked scraper as shown in Fig. K2-11. Clean valve seat using brass scraper as shown in K2-12, then polish seat using wood polishing stick and mutton tallow as in Fig. K2-13.

Back flush nozzle using reverse flusher adapter. Reclean all parts by rinsing thoroughly in clean diesel fuel or calibrating oil and assemble while parts are immersed in clean fluid. Make sure adjusting shim pack is intact. Tighten nozzle retaining nut (10 – Fig. K2-9) to a torque of 59-78 N·m (44-57 ft.-lbs.). Do not overtighten as distortion may cause

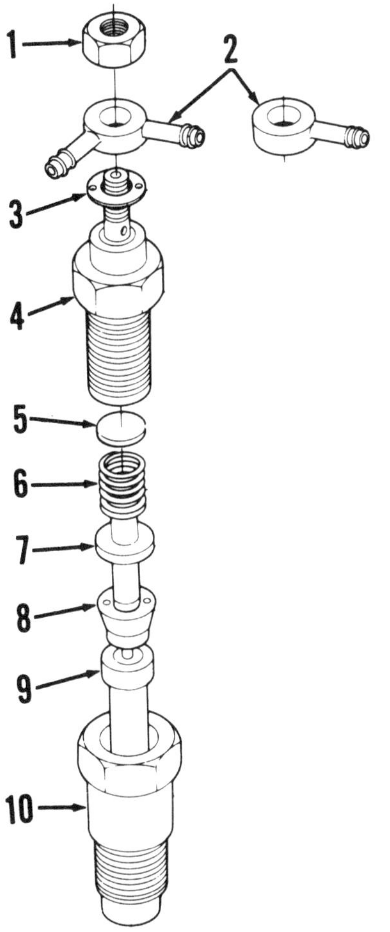

Fig. K2-9—Exploded view of injector.

1. Nut
2. Bypass fitting
3. Washer
4. Pressure fitting
5. Shim
6. Spring
7. Pressure pin
8. Spacer
9. Nozzle & valve
10. Nozzle nut

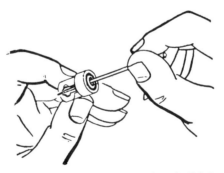

Fig. K2-10—Use a pointed hardwood stick to clean spray hole as shown.

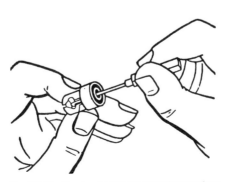

Fig. K2-11—Use hooked scraper to clean carbon from pressure chamber.

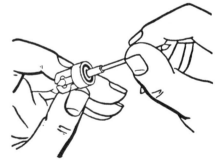

Fig. K2-12—Clean valve seat using brass scraper as shown.

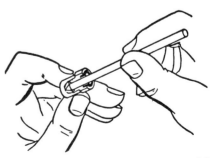

Fig. K2-13—Polish seat using polishing stick and mutton tallow.

Illustrations Courtesy Kubota

Kubota

valve to stick, and no amount of overtightening can stop a leak caused by scratches or dirt. Retest assembled injector as previously outlined.

GLOW PLUGS

Glow plugs are parallel connected with each individual glow plug grounding through its mounting threads. Indicator light should glow after about 30 seconds if units are operating satisfactorily.

If indicator light fails to glow, check for an open circuit at switch, indicator lamp or glow plug connections. Using an ohmmeter, check resistance of each glow plug in turn. Resistance between glow plug terminal and body should be about 1.6 ohms. If resistance is zero, glow plug is shorted. If resistance is infinite, glow plug element is broken.

INJECTION PUMP

All models are equipped with an injection pump similar to type shown in Fig. K2-3. The injection pump should be tested and overhauled by a shop qualified in diesel injection pump repair.

The injection pump tappets are actuated by lobes on injection pump camshaft (12 – Fig. K2-17). Inspect camshaft each time injection pump is removed.

Tighten pump retaining nuts to 24-27 N·m (18-20 ft.-lbs.) and refer to INJECTION PUMP TIMING section.

TIMING GEARS AND COVER

To remove timing gear cover, first remove cover just below injection pump and detach governor spring (5 – Fig. K2-5) from governor arm. Remove speed control plate (3) and governor spring. Disconnect start spring (6) from timing gearcase. Drain coolant and disconnect radiator hoses, then relocate radiator to provide clearance for crankshaft pulley removal. Remove fan belt and fan. Unscrew crankshaft pulley nut, then use a suitable puller to remove pulley. Unbolt and remove timing gear cover.

Backlash between any two gears should be 0.04-0.11 mm (0.002-0.004 inch). If backlash exceeds 0.2 mm (0.008 inch), renew gears as needed. Idler gear bushings (3 – Fig. K2-19) and shaft (6) are renewable. When installing new idler gear bushings, press bushings in from each side of gear until flush with outer surfaces of gear. Tighten idler shaft mounting cap screws to 10-12 N·m (8-9 ft.-lbs.). If crankshaft gear is being renewed, heat new gear to about 100°C (212°F) before driving gear onto crankshaft.

Refer to Fig. K2-14 for proper alignment of timing marks on crankshaft, idler, camshaft and injection pump camshaft gears. Renew crankshaft oil seal (22 – Fig. K2-5) and lubricate seal lip with oil prior to reassembly. Be sure crankshaft oil slinger (13 – Fig. K2-20), "O" ring (12) and sleeve (11) are properly located on crankshaft. Make certain the three "O" rings (29 – Fig. K2-5) are in place. Tighten timing gear cover cap screws to 10-11 N·m (7-8 ft.-lbs.). Lubricate threads of crankshaft pulley nut with engine oil, then tighten nut to 138-157 N·m (102-115 ft.-lbs.).

OIL PUMP

R&R AND OVERHAUL. To remove oil pump, first remove timing gear cover as previously outlined. Use a suitable puller to remove pump drive gear.

Inspect pump rotors and body for wear or other damage. Clearance between inner and outer rotor, measured as shown in Fig. K2-15, should be 0.11-0.15 mm (0.004-0.006 inch) with an allowable limit of 0.2 mm (0.008 inch). Clearance between outer rotor and pump body, measured as shown in Fig. K2-16, should be 0.07-0.15 mm (0.003-0.006 inch) with an allowable limit of 0.25 mm (0.010 inch). Individual pump components are not available; pump must be serviced as a unit assembly.

Oil pressure relief ball (26 – Fig. K2-5) and spring (27) are located in timing gear cover. Oil pressure should be 196-441 kPa (28-64 psi) and is not adjustable.

Fig. K2-15—Measure clearance between inner and outer oil pump rotors as shown. Desired clearance is 0.11-0.15 mm (0.004-0.006 inch).

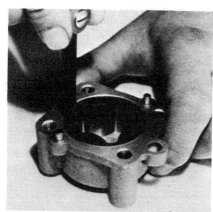

Fig. K2-16—Measure clearance between outer rotor and oil pump body as shown. Desired clearance is 0.07-0.15 mm (0.003-0.006 inch).

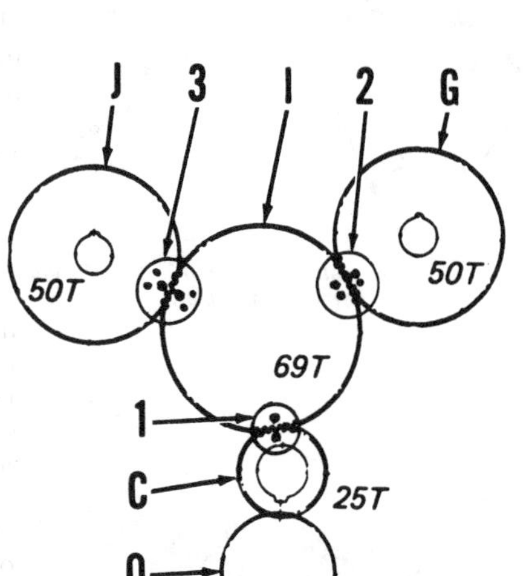

Fig. K2-14—Diagram of drive gears showing proper alignment of timing marks. Note three marks (3) on injection pump gear (J) and idler gear (I); two marks (2) on camshaft gear (G) and idler gear (I); single marks on crankshaft gear (C) and idler gear (I). No marks are used on oil pump gear (O).

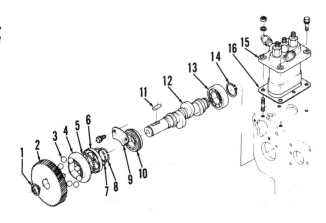

Fig. K2-17—Exploded view of injection pump camshaft and governor components.

1. Snap ring
2. Gear
3. Large governor balls (8)
4. Sleeve
5. Small governor balls (30)
6. Governor case
7. Retainer ring
8. Retainer ring
9. Bearing retainer
10. Bearing
11. Key
12. Injection pump camshaft
13. Bearing
14. Snap ring
15. Injection pump
16. Shim gasket

Service Manual

INJECTION PUMP CAMSHAFT

R&R AND OVERHAUL. To remove camshaft, first remove fuel injection pump and timing gear cover as outlined in previous paragraphs. Remove fuel transfer pump from side of crankcase. Remove cap screws securing governor arm pivot block (10–Fig. K2-5) and withdraw governor control arm assembly. Remove cap screws from camshaft retainer plate (9–Fig. K2-17), then withdraw camshaft (12), bearings and governor as an assembly. Remove snap ring (1), then separate gear (2) and governor assembly from camshaft being careful not to lose governor balls (3 and 5).

Inspect camshaft and bearings for excessive wear and other damage and renew if necessary.

To reinstall injection pump camshaft, reverse the removal procedure. Refer to TIMING GEARS AND COVER section to properly align timing marks during installation.

GOVERNOR

All models are equipped with a flyball type governor as shown in Fig. K2-17. Ball movement against governor sleeve (4) actuates governor lever (8–Fig. K2-5) which is connected to fuel injection control rack. Flyball movement is balanced by governor spring (5).

Governor components are accessible after removing timing gear cover. Inspect components and renew any which are excessively worn or damaged.

NOTE: There are 30 balls (5) 3.97 mm (0.156 inch) in diameter and eight balls (3) 13.44 mm (0.530 inch) in diameter contained in governor ball case.

PISTON AND ROD UNITS

REMOVE AND REINSTALL. Piston and connecting rod are removed as a unit after removing oil pan, oil pickup and cylinder head. Be sure to remove carbon deposits and ring ridge (if present) from top of cylinder before attempting to remove piston. Unscrew rod cap retaining screws, detach rod cap and push piston and rod out top of crankcase. Identify piston and rod units and the cylinder from which they were removed so units can be reinstalled in their original locations.

Note that numbers are stamped on sides of rod and cap and should be on same side when assembled. Install piston and rod units so numbers on rod and cap are toward fuel injection pump side of engine. Lubricate threads of connecting rod cap screws with engine oil, then tighten to 27-30 N·m (20-22 ft.-lbs.).

PISTON, PIN AND RINGS

All models are equipped with two compression rings and an oil control ring. Pistons are aluminum and cam-ground. Piston and rings are available in 0.5 mm (0.020 inch) oversize as well as standard size.

Insert piston rings squarely into cylinder bore, then measure end gap with a feeler gage. On all models except D950-B, recommended end gap is 0.25-0.40 mm (0.010-0.016 inch) for compression rings and 0.20-0.40 mm (0.008-0.016 inch) for oil control ring. On Model D950-B, end gap for compression rings should be 0.30-0.45 mm (0.012-0.018 inch) and 0.25-0.40 mm (0.010-0.016 inch) for oil control ring.

Use a feeler gage to check side clearance of second compression ring and oil control ring in piston ring grooves. Top compression ring is a keystone type ring and side clearance is not measured. Recommended side clearance is 0.085-0.112 mm (0.0033-0.0044 inch) for second compression ring and 0.020-0.055 mm (0.001-0.002 inch) for oil control ring.

A full floating piston pin is used on all models. Piston pin boss inner diameter should be 20.000-20.013 mm (0.7874-0.7879 inch) with a wear limit of 20.03 mm (0.7886 inch). Outside diameter of piston pin should be 20.002-20.011 mm (0.7875-0.7878 inch).

When reinstalling rings, note location and shape of rings as shown in Fig. K2-18. Stagger ring end gaps around piston 180 degrees from each other.

CYLINDER LINER

All models are equipped with dry type cylinder liners. Use suitable removal and installation tools to renew defective liners. When renewing liner, be sure to thoroughly clean outer surface of liner and cylinder bore, then apply a light coat of oil to liner and bore surfaces. Press liner into crankcase so top of liner is flush with top of crankcase within plus or minus 0.025 mm (0.001 inch). After installation, liner must be honed to final size.

Standard inner diameter of cylinder liner is 64.000-64.019 mm (2.5197-2.5204 inches) for Model D650-B; 68.00-68.019 mm (2.6772-2.6779 inches) for Models D750-B, Z500-B and ZB500C-B; 72.000-72.019 mm (2.8346-2.8353 inches) for Models D850-B, DH850-B, Z600-B and ZB600C-B or 75.000-75.019 mm (2.9528-2.9535 inches) on Model D950-B. Maximum allowable liner wear is 0.15 mm (0.006 inch). Standard cylinder liner may be bored for 0.5 mm (0.020 inch) oversize piston installation.

CONNECTING ROD AND BEARINGS

Connecting rods are equipped with a renewable bushing in the small end and a precision, insert type bearing in the big end. Inner diameter of big end bearing should be 37.004-37.050 mm (1.4569-1.4587 inches). Clearance between crankpin and bearing should be 0.029-0.091 mm (0.0011-0.0036 inch) while maximum allowable clearance is 0.20 mm (0.008 inch). Bearings are available in 0.20 and 0.40 mm (0.008 and 0.016 inch) undersizes.

Small end bushing inner diameter should be 20.025-20.040 mm (0.7884-0.7890 inch). Recommended clearance between piston pin and bushing is 0.014-0.038 mm (0.0006-0.0015 inch) while maximum allowable clearance is 0.15 mm (0.006 inch). When renewing bushing, make certain oil hole in bushing is aligned with hole in end of rod. Bushing must be honed to provide desired clearance for piston pin.

CAMSHAFT

R&R OVERHAUL. To remove camshaft, first remove cylinder head and timing gear cover as previously outlined. Remove cam followers (tappets) from their bores. Unscrew camshaft retainer plate (9–Fig. K2-19) cap screws, then withdraw camshaft and gear assembly. If necessary, press camshaft gear (8) off camshaft.

Camshaft lobe height should be 26.88 mm (1.0583 inches) with a wear limit of 26.83 mm (1.0536 inches). Camshaft bearing journal diameter should be 32.934-32.950 mm (1.2966-1.2972 inch-

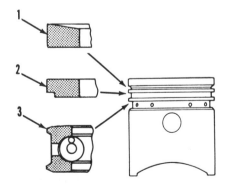

Fig. K2-18—Drawing showing correct installation of piston rings. Top ring (1) is keystone type and can be installed correctly only one way. Second compression ring (2) must be installed with notched edge downward. Oil control ring (3) may be installed either side up.

Kubota

SMALL DIESEL

es) while inner diameter of camshaft bearing bores is 33.000-33.025 mm (1.2992-1.3002 inches). Maximum allowable clearance between camshaft journal and bearing bore is 0.15 mm (0.006 inch). With camshaft supported in V-blocks at outer bearing journals, maximum allowable runout measured at either center bearing journal is 0.02 mm (0.0008 inch).

When installing camshaft gear on camshaft, first install retainer (9—Fig. K2-19). Heat camshaft gear to approximately 80°C (175°F), then push gear onto camshaft. Refer to TIMING GEARS AND COVER section to align timing gear marks during camshaft installation.

CRANKSHAFT AND BEARINGS

The crankshaft is carried in a sleeve type bearing (15—Fig. K2-20) pressed into front bearing bore of crankcase and in insert type bearings (28 and 30) in bearing carriers (25, 26 and 27) for intermediate (center) and rear main journals. Two-cylinder models have one center bearing, and three-cylinder models have two center bearings. Crankshaft end play is controlled by thrust washers (29) at front and rear of rear main bearing carrier (10 and 27). Before removing crankshaft, check crankshaft end play.

To remove crankshaft, remove timing gear cover, crankshaft timing gear, piston and rod units and flywheel. Refer to Fig. K2-22 and remove the rear oil seal retainer (3). Remove the cap screw (31—Fig. K2-21) from bottom of each center bearing web, then remove crankshaft and bearing carriers as an assembly from rear end of crankcase taking care not to damage front main bearing remaining in crankcase. Note that the main bearing carriers are a tight fit in cylinder block to prevent oil loss between oil passages in block and carriers.

Identify main bearing carriers as to their position in crankcase to ensure correct reassembly; bearing carrier halves are not interchangeable. Unbolt and remove the bearing carriers from crankshaft. Note that crankshaft bearing thrust washers are located at each side of rear main bearing carrier.

Crankshaft main bearing journal diameter is 43.934-43.950 mm (1.7297-1.7303 inches). Desired oil clearance of journals to main bearings is 0.034-0.106 mm (0.0013-0.0042 inch). Maximum allowable oil clearance is 0.2 mm (0.008 inch). Crankshaft end play should be 0.15-0.31 mm (0.006-0.012 inch) with maximum allowable end play being 0.5 mm (0.020 inch). Crankpin diameter is 36.959-36.975 mm (1.4551-1.4557 inches) with desired bearing oil clearance of 0.029-0.091 mm (0.0011-0.0036 inch) and maximum allowable clearance of 0.2 mm (0.008 inch).

Main bearings and crankpin bearings are available in undersizes of 0.2 and 0.4 mm (0.008 and 0.016 inch) and thrust washers are available in oversize thickness of 0.2 and 0.4 mm (0.008 and 0.016 inch). Crankshaft may be reground undersize and with overwidth thrust surfaces at rear main journal to accommodate undersize bearings and oversize thrust washers.

To renew front main bearing, use suitable driver to remove old bearing from crankcase and install new bearing with oil hole in bearing aligned with oil passage in cylinder block. Bearing should be 4.2-4.5 mm (0.165-0.177 inch)

Fig. K2-21—View showing location of locating screw (31) which secures bearing carrier to block.

Fig. K2-19—Exploded view of camshaft and idler gear assemblies.
1. Snap ring
2. Slotted washer
3. Bushing
4. Idler gear
5. Washer
6. Idler shaft
7. Snap ring
8. Gear
9. Retainer
10. Plug
11. Key
12. Push rod
13. Cam follower
14. Camshaft

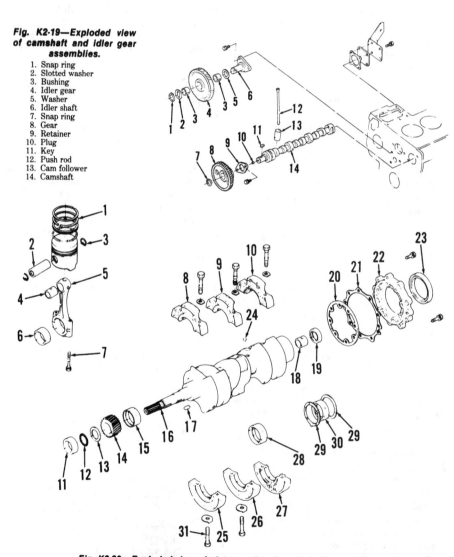

Fig. K2-20—Exploded view of piston, rod and crankshaft assembly.
1. Piston rings
2. Piston pin
3. Retainer
4. Bushing
5. Connecting rod
6. Bearing
7. Screw
8. Upper bearing carrier half, front
9. Upper bearing carrier half, middle
10. Upper bearing carrier half, rear
11. Spacer
12. "O" ring
13. Oil slinger
14. Gear
15. Bearing
16. Crankshaft
17. Key
18. Bushing
19. Seal
20. Gasket
21. Gasket
22. Seal carrier
23. Seal
24. Plug
25. Lower bearing carrier half, front
26. Lower bearing carrier half, middle
27. Lower bearing carrier half, rear
28. Main bearing
29. Thrust washers
30. Rear main bearing
31. Locator screw

SERVICE MANUAL

Kubota

below flush with front face of crankcase and positioned with seam toward camshaft side of engine.

Install bearing inserts in bearing carriers and stick thrust washers in recesses of rear bearing carrier with light grease. Be sure that grooved side of thrust washers face outward. Lubricate crankshaft and bearings and install carriers in correct position on crankshaft with face of carriers having bolt holes toward rear end of crankshaft. Tighten the bearing carrier bolts to a torque of 20-24 N·m (15-17 ft.-lbs.). Lubricate front main bearing, then insert crankshaft and bearing supports into crankcase from rear taking care not to damage front main bearing.

Install new rear oil seal with lip to inside in rear seal retainer plate. Lubricate seal and carefully install retainer over rear of crankshaft with arrow mark up as shown in Fig. K2-22. Be sure to remove the two jackscrews from cover. Tighten retainer plate cap screws to a torque of 10-12 N·m (7-9 ft.-lbs.). Install the center bearing support retaining cap screw(s) in cylinder block webs and tighten to a torque of 30-34 N·m (22-25 ft.-lbs.). Tighten flywheel retaining cap screws to a torque of 54-59 N·m (40-43 ft.-lbs.). Complete reassembly of engine by reversing disassembly procedure.

ELECTRICAL SYSTEM

The alternator should produce 14 volts with a minimum charging current of 8.5 amperes.

Refer to Fig. K2-23 for an exploded view of electric starter motor used on all models. With no load imposed on starter, the starter shaft should rotate at 5000 rpm or more while drawing less than 50 amperes current and 11 volts.

Minimum brush length is 10.5 mm (0.413 inch). Pinion engagement depth is adjusted by turning hook (H–Fig. K2-23). With starter pinion in engaged position, distance between collar (8) and pinion should be 0.1-0.4 mm (0.004-0.016 inch). Turn hook (H) so pinion engagement depth is correct.

Fig. K2-22—Be sure mark (1) is positioned at top when reinstalling crankshaft rear seal retainer plate (3). Cap screws (2) are used in jackscrew holes to remove retainer plate.

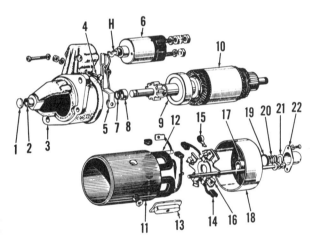

Fig. K2-23—Exploded view of electric starter.
1. Cap
2. Bushing
3. Drive housing
4. Pivot bolt
5. Fork
6. Solenoid
7. Snap ring
8. Collar
9. Starter drive
10. Armature
11. Frame
12. Field coils
13. Field magnets
14. Brush
15. Brush spring
16. Brush plate
17. Bushing
18. End frame
19. Packing
20. Spring
21. Spring retainer
22. Cap

KUBOTA

Model	No. Cyls.	Bore	Stroke	Displ.
DH1101-B	3	76 mm (2.99 in.)	82 mm (3.23 in.)	1115 cc (68 cu. in.)
D1102-B, D1102-BC	3	76 mm (2.99 in.)	82 mm (3.23 in.)	1115 cc (68 cu. in.)
D1301-B, D1302-B	3	82 mm (3.23 in.)	82 mm (3.23 in.)	1299 cc (79.3 cu. in.)
D1402-B	3	85 mm (3.35 in.)	82 mm (3.23 in.)	1395 cc (85.1 cu. in.)
S2200-B	6	76 mm (2.99 in.)	82 mm (3.23 in.)	2231 cc (136.1 cu. in.)
S2600-B	6	82 mm (3.23 in.)	82 mm (3.23 in.)	2598 cc (158.5 cu. in.)
V1501-B	4	76 mm (2.99 in.)	82 mm (3.23 in.)	1487 cc (90.7 cu. in.)
V1502-B, V1502-BC	4	76 mm (2.99 in.)	82 mm (3.23 in.)	1487 cc (90.7 cu. in.)
V1701-B, V1702-B	4	82 mm (3.23 in.)	82 mm (3.23 in.)	1732 cc (105.7 cu. in.)
V1902-B	4	85 mm (3.35 in.)	82 mm (3.23 in.)	1861 cc (113.6 cu. in.)
VT1502-B	4	76 mm (2.99 in.)	82 mm (3.23 in.)	1487 cc (90.7 cu. in.)
Z751-B	2	76 mm (2.99 in.)	82 mm (3.23 in.)	743 cc (45.3 cu. in.)
Z851-B	2	82 mm (3.23 in.)	82 mm (3.23 in.)	866 cc (52.8 cu. in.)

All engines in this section are liquid-cooled, four-stroke diesel engines having two, three, four or six-cylinders.

MAINTENANCE

LUBRICATION

Recommended crankcase oil is SAE 10W or SAE 10W-30 when ambient temperature is below 0°C (32°F), SAE 20 when temperature is between 0°-25°C (32°-77°F) or SAE 30 when temperature is above 25°C (77°F). Manufacturer recommends using oil with API classification CD.

All models are equipped with a pressure lubrication system. Oil is directed from the oil pump to the oil pressure relief valve and oil filter. Oil is routed through a passage in cylinder block to the crankshaft and to the cylinder head to lubricate the valve train. Normal oil pressure at rated speed

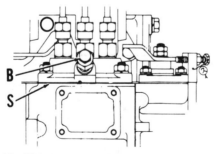

Fig. K3-1—Injection timing is adjusted by adding or deleting shims (S) between pump and engine block.

is 295-441 kPa (43-64 psi) on Models S2200-B, S2660-B, Z751-B and Z851-B or 295-392 kPa (43-57 psi) on all other models.

FUEL SYSTEM

FUEL FILTER. A renewable fuel filter is mounted on side of engine. Manufacturer recommends renewing filter element after every 400 hours of operation. However, if engine indicates signs of fuel starvation (loss of power or surging), renew filter regardless of hours of operation. After renewing filter, air must be bled from system as outlined in the following paragraph.

BLEED FUEL SYSTEM. To bleed air from fuel system, first remove bleed screw on fuel filter mounting bracket. Open fuel valve, then reinstall bleed screw when air-free fuel flows. Loosen injection pump bleed screw (B–Fig. K3-1). On models equipped with a fuel transfer pump, rotate crankshaft with starter to operate fuel transfer pump. On all models, tighten bleed screw after fuel flows from bleeder.

If engine fails to start at this point, loosen high pressure lines at injectors. With throttle control lever in full speed position, crank engine to operate injection pump. Retighten injector line fittings when fuel is discharged from injector lines, then start engine.

SERVICE MANUAL

Kubota

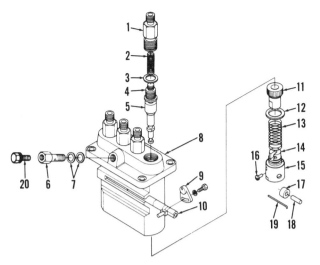

Fig. K3-2—Exploded view of injection pump used on four-cylinder engines. Other models are similar. A jetstart cock is installed in place of bleed screw (20) on some models.

1. Delivery valve holder
2. Spring
3. Gasket
4. Delivery valve assembly
5. Pumping element
6. Inlet fuel fitting
7. Gaskets
8. Pump body
9. Plate
10. Control rack
11. Control sleeve
12. Washer
13. Spring
14. Spring seat
15. Tappet
16. Guide pin
17. Roller
18. Pin
19. Retainer wire
20. Bleed screw

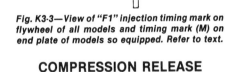

Fig. K3-3—View of "F1" injection timing mark on flywheel of all models and timing mark (M) on end plate of models so equipped. Refer to text.

INJECTION PUMP TIMING. Injection pump timing is adjusted using shims (S—Fig. K3-1). To check injection pump timing, proceed as follows: Disconnect high pressure line for number 1 cylinder from injection pump, then unscrew delivery valve holder (1—Fig. K3-2) and valve assembly (4). Remove valve spring (2) and delivery valve. Reinstall delivery valve holder (1) and connect a suitable spill pipe to valve holder. Aim spill pipe at a receptacle to catch discharged fuel. Move throttle control to full speed position. Rotate engine flywheel slowly in counterclockwise direction until fuel just stops flowing from spill pipe. On models equipped with a flywheel housing, remove timing cover. Flywheel mark "F1" should be aligned with timing mark (M—Fig. K3-3) on cylinder block rear plate or timing pointer on flywheel housing.

To adjust injection timing, remove injection pump and add or delete shims (S—Fig. K3-1). Each shim alters timing approximately 1½ crankshaft degrees. Add shims to retard timing or delete shims to advance timing. Be sure pump control rack pin properly engages operating fork when installing pump. Tighten pump retaining nuts or screws to 24-27 N·m (18-20 ft.-lbs.). Reinstall delivery valve and spring and tighten delivery valve holder to 40-49 N·m (30-36 ft.-lbs.).

GOVERNOR

All models are equipped with a flyball type governor mounted on front end of fuel injection pump camshaft. Refer to Fig. K3-4 for an exploded view of governor linkage.

Slow idle speed for all models is 800 rpm and is adjusted by turning idle speed screw (30—Fig. K3-4). High idle speed and maximum fuel limiting stop screws are sealed and should be adjusted by qualified personnel only. High idle speed should be 2800 rpm for Models S2200-B and S2600-B or 3000 rpm for all other models. Turn high idle speed screw (3) for adjustment.

Maximum fuel limiting stop should be set to prevent excessive smoke level at slight overload. To make adjustment, remove seal cap, loosen jam nut and turn adjusting screw (21) in to lower smoke level or out to raise smoke level.

Some models are equipped with a "torque rise" adjustment screw (12). Position of screw is set at factory and manufacturer does not recommend further adjustment.

COMPRESSION RELEASE

All models are equipped with a compression release which holds exhaust valves open slightly when compression release lever (1—Fig. K3-5) is rotated. To adjust compression release, rotate engine crankshaft so piston in cylinder being adjusted is at top dead center on compression. Remove cover (8), loosen locknut (3) and back out adjusting screw (2). Rotate compression release lever to engaged position. Turn adjusting screw (2) in until it contacts exhaust rocker arm, then turn screw an additional 1½ turns. Tighten locknut. Repeat procedure to adjust remaining cylinders.

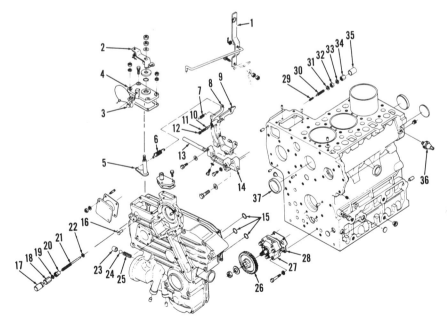

Fig. K3-4—Exploded view of timing gear cover and associated components.

1. Speed control
2. Control lever
3. High idle adjusting screw
4. Plate
5. Control arm
6. Governor spring
7. Start spring
8. Governor arm
9. Governor lever
10. Plunger
11. Spring
12. Set screw
13. Pivot pin
14. Pivot block
15. "O" rings
16. Pin
17. Cap
18. Cap nut
19. Gasket
20. Locknut
21. Maximum fuel limiting screw
22. Gasket
23. Valve seat
24. Relief valve ball
25. Spring
26. Gear
27. Oil pump
28. Gasket
29. Spring
30. Low idle adjusting screw
31. Gasket
32. Locknut
33. Gasket
34. Cap nut
35. Cap
36. Oil pressure sender

Illustrations Courtesy Kubota

Kubota

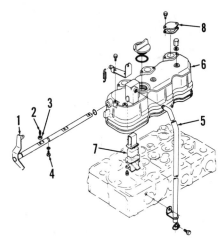

Fig. K3-5—Exploded view of valve cover and compression release components.

1. Compression release shaft
2. Compression release screw
3. Locknut
4. Screw
5. Vent tube
6. Valve cover
7. Breather
8. Cover

Check operation of compression release being sure exhaust valves do not contact pistons.

COOLING SYSTEM

All engines are liquid-cooled with Models Z751-B and Z851-B using thermosiphon circulation while all other models use forced circulation with a centrifugal water pump.

Recommended pressure rating for radiator cap is 88.25 kPa (13 psi). On all models except Z751-B and Z851-B, thermostat opening temperature is about 82°C (180°F). Fan belt tension is correct if finger pressure applied to belt at midpoint between alternator pulley and crankshaft pulley deflects belt about 8 mm (5/16 inch).

AIR FILTER

All models are equipped with a dry type renewable air filter. Manufacturer recommends blowing out filter after every 100 to 200 hours of operation and renewing filter after six cleanings or one year.

REPAIRS

TIGHTENING TORQUES

Connecting rod . . . . 37-41 N·m
(27-30 ft.-lbs.)
Crankshaft pulley nut:
S2200-B, S2600-B . . . . . . 196-215 N·m
(145-158 ft.-lbs.)
All other models . . . . 137-156 N·m
(101-115 ft.-lbs.)
Crankshaft seal carrier . . . . . 24-27 N·m
(18-20 ft.-lbs.)

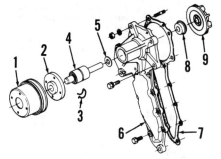

Fig. K3-6—Exploded view of water pump used on all engines except Models Z751-B and Z851-B.

1. Pulley
2. Flange
3. Retainer clip
4. Shaft & bearing assy.
5. Slinger
6. Pump body
7. Gasket
8. Seal
9. Impeller

Cylinder head:
S2200-B, S2600-B . . . . 74-83 N·m
(55-61 ft.-lbs.)
All other models . . . . 74-78 N·m
(55-58 ft.-lbs.)
Flywheel . . . . 98-108 N·m
(73-80 ft.-lbs.)
Injection pump . . . . 24-27 N·m
(18-20 ft.-lbs.)
Injector . . . . 30-49 N·m
(22-36 ft.-lbs)
Main bearing retainer
screw . . . . 64-68 N·m
(48-50 ft.-lbs.)
Main bearing case screws . . . 30-34 N·m
(22-25 ft.-lbs.)
Oil pump . . . . 10-12 N·m
(8-9 ft.-lbs.)

SMALL DIESEL

Rocker arm stand . . . . 18-20 N·m
(13-15 ft.-lbs)
Rocker arm cover . . . . 7-9 N·m
(5-7 ft.-lbs.)
Timing gear cover . . . . 24-27 N·m
(18-20 ft.-lbs.)

WATER PUMP

All Models Except Z751-B and Z851-B

R&R AND OVERHAUL. To remove water pump, drain coolant and if necessary, relocate radiator for access to pump. Remove fan belt, fan, spacer and pulley then unbolt and remove pump from timing gear cover.

Using a suitable puller, pull flange (2–Fig. K3-6) off shaft. Disengage retainer clip (3), then push shaft and bearing assembly (4) out front of housing. Disassemble as required.

Shaft and bearings (4) are available only as a unit assembly.

COMPRESSION PRESSURE

Compression pressure at cranking speed (200-300 rpm) should be 2746-3236 kPa (398-469 psi) for Models Z751-B, DH1101-B, V1501-B and S2200-B; 2844-3334 kPa (412-483) psi for Models Z851-B, D1301-B, V1701-B and S2600-B; or 2944-3234 kPa (427-469

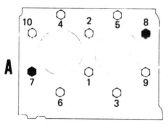

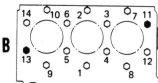

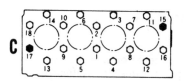

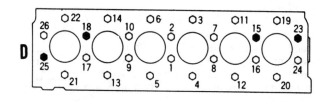

Fig. K3-7—When tightening cylinder head fasteners, follow sequence in drawing (A) for two-cylinder engines; drawing (B) for three-cylinder engines; drawing (C) for four-cylinder engines; drawing (D) for six-cylinder engines. Studs are located at blackened locations.

SERVICE MANUAL

Kubota

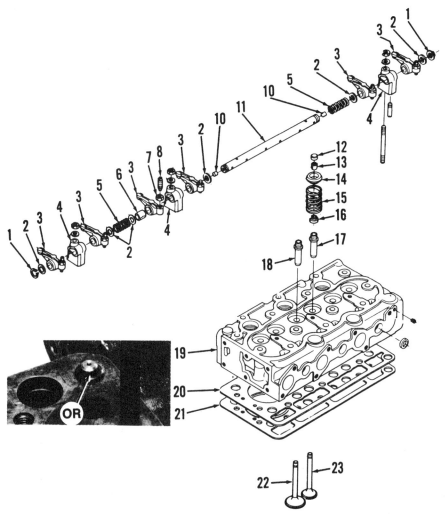

Fig. K3-8—Exploded view of cylinder head and valve mechanism. Inset shows valve system oil passage pipe and "O" ring at front end of cylinder block.

1. Snap ring
2. Washer
3. Rocker arm
4. Shaft stand
5. Spring
6. Bushing
7. Locknut
8. Adjusting screw
9. Set screw
10. Set screw
11. Rocker shaft
12. Valve cap
13. Retainer keys
14. Retainer
15. Spring
16. Seal
17. Exhaust valve guide
18. Intake valve guide
19. Cylinder head
20. Shim
21. Gasket
22. Intake valve
23. Exhaust valve

psi) on all other models. Compression pressure readings among cylinders should not vary more than 10 percent.

CYLINDER HEAD

R&R AND OVERHAUL. To remove cylinder head, proceed as follows: Drain cooling system and remove upper radiator hose. Unbolt and remove thermostat housing, intake and exhaust manifolds. Remove fuel injection lines, injector fuel return lines and glow plugs. Remove rocker arm cover, rocker arm shaft assembly and the push rods. Unbolt and remove cylinder head assembly from block. Take care not to damage the valve mechanism oil pipe (see Fig. K3-8) extending from left corner of cylinder block. Remove the cylinder head gasket, cylinder head shim (20), if fitted, and oil pipe sealing "O" ring (OR).

Before reinstalling cylinder head, completely clean all traces of old cylinder head gasket and carbon from both head and block surfaces. Be careful not to damage valve system oil tube protruding from corner of cylinder block. Clean carbon from top of pistons, taking care to clean all loose carbon from cylinders. Closely check cylinder head for cracks. Use a straightedge and feeler gage to check cylinder head for any warped or twisted condition. Do not place straightedge over precombustion chamber inserts. Maximum allowable deviation from flatness is 0.05 mm (0.002 inch) per 100 mm (3.94 inches). Surface grind head if not within limits. Refer to VALVE SYSTEM paragraphs for valve head recess.

If engine has been overhauled with new pistons and/or connecting rods, piston to cylinder head clearance should be checked. Piston to cylinder head clearance should be 0.7-0.9 mm (0.028-0.035 inch). Engine manufacturer recommends checking piston to cylinder head clearance after installing cylinder head and before fuel injectors are reinstalled. Turn crankshaft so that piston is at bottom of compression stroke and insert a soft lead wire into cylinder through injector opening. Turn the crankshaft by hand so that piston goes past top dead center, then withdraw the lead wire and measure flattened thickness. This thickness is equal to piston to cylinder head clearance. If clearance is less than specified, remove cylinder head and reinstall with a 0.2 mm (0.008 inch) shim between head gasket and cylinder head.

To reinstall cylinder head, reverse removal procedure and note the following: Do not apply any sealing compound to head gasket surfaces. Place a new "O" ring on valve system oil tube. Place new head gasket on block and place new 0.2 mm (0.008 inch) shim, if needed, on top of gasket. Carefully lower the cylinder head on the two stud bolts. Oil the threads of the cylinder head cap screws and nuts and install all finger tight. Refer to the appropriate tightening sequence in Fig. K3-7 and tighten the head bolts and nuts in several steps until final torque of 74-83 N·m (55-61 ft.-lbs.) is reached on S2200 and S2600 engines or 74-78 N·m (55-58 ft.-lbs.) on all other engines. Loosen locknuts and back rocker arm adjusting screws out. Install push rods and rocker arms and tighten rocker arm nuts to a torque of 24-27 N·m (18-20 ft.-lbs.). Adjust valve clearance as described below. Complete reassembly by reversing disassembly procedure. Start and run engine until normal operating temperature is reached, then retorque cylinder head bolts and nuts to specified torque.

VALVE CLEARANCE

Valve clearance should be adjusted with engine cold and piston at top dead center on compression. Remove rocker cover and rotate crankshaft so number 1 piston is at TDC on compression. Turn rocker arm adjusting screws until clearance between rocker arm and valve stem end cap is 0.18-0.22 mm (0.007-0.009 inch) for both valves. Rotate crankshaft until next piston in firing order (1-2-3 for three-cylinder engines, 1-3-4-2 for four-cylinder engines or 1-5-3-6-2-4 for six-cylinder engines) is at TDC on compression. Repeat adjustment procedure for each of the remaining cylinders.

Kubota — SMALL DIESEL

VALVE SYSTEM

Intake and exhaust valves seat directly in cylinder head. Valve face and seat angles are 45 degrees. Desired valve seat width is 2.1 mm (0.083 inch). After refacing valves and seats, check distance valve head is recessed from cylinder head surface. Valve should be recessed 1.1-1.3 mm (0.043-0.051 inch). If recession exceeds 1.6 mm (0.063 inch), a shim washer of appropriate thickness should be added between valve spring and cylinder head to maintain correct valve spring tension and cylinder head surface should be machined to obtain recommended recession.

Valve stem diameter should be 7.960-7.975 mm (0.3134-0.3140 inch) for intake and exhaust. Inside diameter of valve guides should be 8.015-8.030 mm (0.3156-0.3161 inch). Recommended clearance between valve stems and guides is 0.04-0.07 mm (0.0016-0.0028 inch) and maximum allowable clearance is 0.1 mm (0.004 inch). When renewing guides, note that intake and exhaust guides are not interchangeable. After installation, check guide inside diameter and ream to provide recommended clearance for valves.

Intake and exhaust valve springs are identical. Valve spring free length should be 41.7-42.2 mm (1.642-1.661 inches) while installed height is 35.15 mm (1.384 inches). Load required to compress spring to installed height should be 100-117.7 N (22.5-26.5 pounds). Renew spring if it fails to meet test load requirements or if it is distorted or otherwise damaged.

Rocker arm bushings (6–Fig. K3-8) are renewable. Inside diameter of bushing should be 14.002-14.043 mm (0.5513-0.5529 inch). Rocker shaft diameter should be 13.973-13.984 mm (0.5501-0.5506 inch). Recommended clearance between rocker arm shaft and bushings is 0.01-0.07 mm (0.0004-0.0028 inch) while maximum allowable clearance is 0.15 mm (0.006 inch). When renewing rocker arm bushings, make certain oil holes in bushing and rocker arm are aligned.

INJECTOR

WARNING: Fuel emerges from injector with sufficient force to penetrate the skin. When testing injector, keep yourself clear of nozzle spray.

REMOVE AND REINSTALL. Before removing an injector, or loosening injector lines, thoroughly clean injector, lines and surrounding area using compressed air and a suitable solvent.

To remove injector unit, first remove high pressure line leading from injection pump to injector. Disconnect bleed line by removing nut and banjo fitting, or by pulling line(s) from banjo nipple fitting (2–Fig. K3-9). With pressure and bleed-back lines removed, unscrew injector from its mounting position on cylinder head. A special nozzle holder socket wrench, Code No. 07916-30841, is available from manufacturer for use in removing and installing injectors.

When installing injector, make sure that machined seating surface in cylinder head is completely clean and free from carbon buildup. Use a new copper washer underneath injector nozzle and tighten injector carefully to 30-49 N·m (22-36 ft.-lbs.).

TESTING. A complete job of testing and adjusting the injector requires use of special test equipment. Only clean, approved testing oil should be used in tester tank. Nozzle should be tested for opening pressure, seat leakage and spray pattern. When tested, nozzle should open with a high-pitched buzzing sound, and cut off quickly at end of injection with a minimum of seat leakage.

Before conducting test, operate tester lever until fuel flows, then attach injector. Close valve to tester gage and pump tester lever a few quick strokes to be sure nozzle valve is not stuck, and that possibilities are good that injector can be returned to service without disassembly.

OPENING PRESSURE. Open valve to tester gage, then operate tester lever slowly while observing gage reading. Opening pressure should be 13730-14710 kPa (1990-2130 psi).

Opening pressure is adjusted by adding or removing shims (5-Fig. K3-9). Adding or removing one 0.1 mm (0.004 inch) thickness shim will change opening pressure approximately 980 kPa (140 psi).

SEAT LEAKAGE. Operate tester lever slowly to maintain pressure at 1000 kPa (150 psi) below nozzle opening pressure and observe nozzle tip for leakage. If a drop forms within a period of 10 seconds, nozzle valve is not seating and must be overhauled or renewed.

SPRAY PATTERN. Spray pattern should be well atomized and slightly conical, emerging in a straight axis from nozzle tip. If pattern is wet, ragged or intermittent, nozzle must be overhauled or renewed.

OVERHAUL. Hard or sharp tools, emery cloth, grinding compound or other than approved solvents or lapping compounds must never be used. An approved nozzle cleaning kit is available through a number of specialized sources.

Wipe all dirt and loose carbon from exterior of nozzle and holder assembly. Refer to Fig. K3-9 for exploded view and proceed as follows:

Secure nozzle in soft jawed vise or holding fixture and remove cap nut (10). Place all parts in clean calibrating oil or diesel fuel as they are removed. Use compartmented pan and use extra care to keep parts from each injector together and separate from other units which are disassembled at the same time.

Clean exterior surfaces with a brass wire brush. Soak parts in an approved carbon solvent, if necessary, to loosen hard carbon deposits. Rinse parts in clean diesel fuel or calibrating oil immediately after cleaning to neutralize the solvent and prevent etching of polished surfaces.

Clean nozzle spray hole from inside using a pointed hardwood stick or wood splinter as shown in Fig. K3-10. Scrape

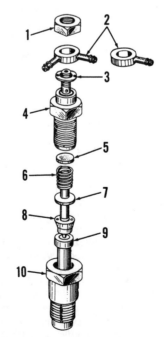

Fig. K3-9—Exploded view of injector.

1. Nut
2. Bypass fitting
3. Washer
4. Pressure fitting
5. Shim
6. Spring
7. Pressure pin
8. Spacer
9. Nozzle & valve
10. Nozzle nut

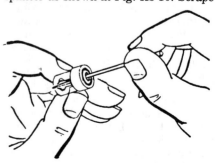

Fig. K3-10—Use a pointed hardwood stick to clean spray hole as shown.

Service Manual

Kubota

carbon from pressure chamber using hooked scraper as shown in Fig. K3-11. Clean valve seat using brass scraper as shown in Fig. K3-12, then polish seat using wood polishing stick and mutton tallow as in Fig. K3-13.

NOTE: Nozzle valve needle must slide freely in nozzle valve body. If needle sticks, reclean or renew nozzle valve assembly (9—Fig. K3-9).

Back flush nozzle using reverse flusher adapter. Reclean all parts by rinsing thoroughly in clean diesel fuel or calibrating oil and assemble while parts are immersed in clean fluid. Make sure adjusting shim pack is intact. Tighten nozzle retaining nut (10–Fig. K3-9) to a torque of 59-78 N·m (44-58 ft.-lbs.). Do not overtighten as distortion may cause valve to stick, and no amount of overtightening can stop a leak caused by scratches or dirt. Retest assembled injector as previously outlined.

GLOW PLUGS

Glow plugs are parallel connected with each individual glow plug grounding through its mounting threads. The glow plug indicator light will glow after about 30 seconds if glow plugs are operating satisfactorily, and will fail to glow if circuit is open.

If indicator light fails to glow, check for loose connections at switch, indicator lamp and glow plugs. Using an ohmmeter, check resistance of each glow plug. Resistance between glow terminal and body should be 1.5-1.6 ohms. If resistance is zero, glow plug is shorted. If resistance is infinite, glow plug element is broken.

INJECTION PUMP

All models are equipped with an injection pump similar to type shown in Fig. K3-2. Before removing pump, thoroughly clean pump, fuel lines and surrounding area. After disconnecting fuel lines, immediately plug all openings to prevent entrance of dirt.

The injection pump should be tested and overhauled by a shop qualified in diesel injection pump repair. The injection pump is actuated by lobes on the injection pump camshaft (14 – Fig. K3-18). Inspect camshaft whenever injection pump is removed.

To reinstall injection pump, reverse the removal procedure. Be sure pump control rack pin engages slot in governor linkage. Tighten retaining nuts and cap screws to 24-27 N·m (18-20 ft.-lbs.). Bleed air from system and check pump timing as outlined in previous section.

TIMING GEARS AND COVER

REMOVE AND REINSTALL. To remove timing gear cover, first drain coolant and relocate radiator. Remove injection pump cover from side of crankcase and detach governor spring (6 – Fig. K3-4) from governor arm (8). Detach throttle control linkage, then remover speed control plate (4) and governor spring. Remove start spring (7) from timing gear cover. Remove fan belt and fan. Unscrew crankshaft pulley nut, then remove pulley using a suitable puller. Remove retaining cap screws, then withdraw timing gear cover.

Backlash between any two gears should be 0.04-0.11 mm (0.0017-0.0045 inch). Maximum allowable backlash is 0.15 mm (0.006 inch). Idler gear bushings (3 – Fig. K3-14) and stub shaft (6) are renewable. Refer to appropriate sections for service procedures covering remainder of gears.

When reinstalling gears, make certain timing marks on crankshaft gear, camshaft gear, injection pump camshaft gear and idler gear are aligned as shown in Fig. K3-15. Be sure to renew crankshaft "O" ring (14 – Fig. K3-20) and oil seal (12). Lubricate oil seal lip and "O" ring with oil prior to reassembly. Be sure the three "O" rings (15 – Fig. K3-4) are in place in front cover, then reinstall cover and tighten cap screws to 24-27 N·m (18-20 ft.-lbs.). Tighten crankshaft pulley nut to 196-215 N·m (145-158 ft.-lbs.) on Models S2200-B and S2600-B or 137-156 N·m (101-115 ft.-lbs.) on all other models. Complete installation by reversing removal procedure.

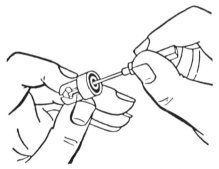

Fig. K3-11—Use a hooked scraper to clean carbon from pressure chamber.

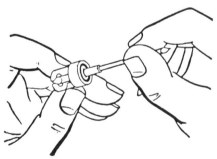

Fig. K3-12—Clean valve seat using brass scraper as shown.

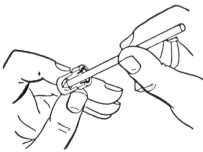

Fig. K3-13—Polish nozzle seat using polishing stick and mutton tallow.

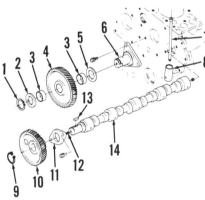

Fig. K3-14—Exploded view of camshaft and idler gear assemblies.

1. Snap ring
2. Slotted washer
3. Bushing
4. Idler gear
5. Washer
6. Idler shaft
7. Push rod
8. Tappet
9. Snap ring
10. Gear
11. Retainer
12. Plug
13. Key
14. Camshaft

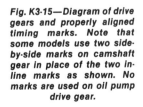

Fig. K3-15—Diagram of drive gears and properly aligned timing marks. Note that some models use two side-by-side marks on camshaft gear in place of the two in-line marks as shown. No marks are used on oil pump drive gear.

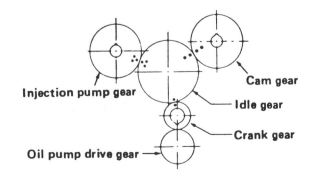

Illustrations Courtesy Kubota

Kubota · SMALL DIESEL

OIL PUMP

To remove oil pump, first remove timing gear cover as previously outlined. Unbolt and remove oil pump assembly.

Clearance between inner and outer rotor, measured as shown in Fig. K3-16, should be 0.04-0.13 mm (0.0016-0.0051 inch) on Models V1502-B, V1702-B and V1902-B or 0.10-0.16 mm (0.0039-0.0063 inch) on all other models. Maximum allowable clearance is 0.20 mm (0.008 inch) for all models. Clearance between outer rotor and pump body, measured as shown in Fig. K3-17, should be 0.11-0.19 mm (0.0043-0.0075 inch) for all models. Maximum allowable clearance is 0.25 mm (0.010 inch). Individual pump components are not available; pump must be serviced as a unit assembly.

An oil pressure relief valve is located in the oil filter mounting pad.

GOVERNOR AND FUEL CAMSHAFT

All models are equipped with a flyball type governor as shown in Fig. K3-18. Ball movement against governor sleeve (6) actuates governor lever (9–Fig. K3-4) which is connected to injection pump control rack. Flyball movement is balanced by governor spring (6).

Governor components are accessible after removing timing gear cover and injection pump camshaft gear (3–Fig. K3-18). Note that 39 loose balls (7) are contained in ball case (8).

Examine sleeve (6), balls (5) and ball travel surface on back of gear for furrowing or pitting and renew if damaged. Be sure governor linkage moves freely. Inspect fuel camshaft (14) and bearings for wear or other damage and renew if necessary.

PISTON AND ROD UNITS

Connecting rod and piston units may be removed from above after removing cylinder head and oil pan. Be sure to remove ring ridge (if present) from top of cylinder before attempting to push piston out. Be sure to identify piston and rod and the cylinder from which it was removed so units can be reinstalled in their original positions. Before removing piston from connecting rod, mark top of piston in relation to alignment marks on connecting rod (Fig. K3-19) and identify piston and connecting rod as a pair to ensure correct assembly.

When reinstalling piston and rod units, be sure numbers stamped on sides of rod and cap are on the same side and aligned with mark made on top of piston. Install piston and rod unit so numbers on rod and cap face away from camshaft side of engine. Lubricate threads of connecting rod cap screws with engine oil, then tighten evenly to 37-41 N·m (27-30 ft.-lbs.).

PISTON AND RINGS

All models are equipped with two compression rings and an oil control ring surrounding an aluminum, cam-ground piston. Piston and rings are available in standard size and 0.5 mm (0.020 inch) oversize.

Recommended piston ring end gap is 0.03-0.45 mm (0.012-0.018 inch) for top and second compression rings and 0.25-0.45 mm for oil control ring. Maximum allowable ring end gap is 1.25 mm (0.049 inch) for all rings.

To check ring groove wear, measure side clearance of second compression ring and oil control ring in piston ring grooves. Top compression ring is a keystone type ring and side clearance is not measured. Recommended ring groove side clearance for second compression ring is 0.09-0.12 mm (0.004-0.005 inch) and 0.02-0.05 mm (0.001-0.002 inch) for oil control ring. Renew piston if side clearance is excessive.

When installing rings onto piston, be sure manufacturer's name or "TOP" mark on ring is towards top of piston. Make sure notched outer edge on second compression ring (Fig. K3-19) is facing down. Position joint of coil expander opposite of oil control ring end gap. Stagger rings on piston so end gaps are 90 degrees apart with no gap in line with piston pin.

PISTON PIN

A full floating piston pin is used on all models. Pin is retained in piston by a snap ring at each end of pin.

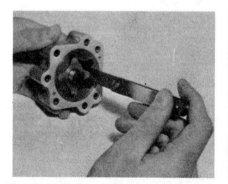

Fig. K3-16—Measure clearance between inner and outer pump rotors as shown. Refer to text.

Fig. K3-17—Measure clearance between outer rotor and oil pump body as shown. Refer to text.

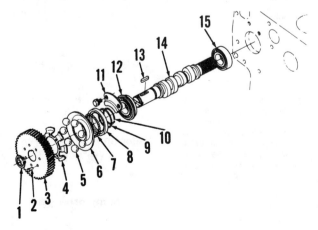

Fig. K3-18—Exploded view of injection pump camshaft and governor components.

1. Snap ring
2. Pin
3. Gear
4. Ball guide
5. Large governor balls
6. Sleeve
7. Small governor balls
8. Ball case
9. Retainer ring
10. Retainer ring
11. Bearing retainer
12. Bearing
13. Key
14. Injection pump camshaft
15. Bearing

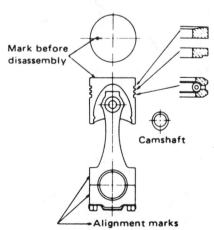

Fig. K3-19—Before removing piston, mark piston crown in relation to alignment marks on connecting rod. Marks must face away from camshaft side of engine.

Service Manual

Piston pin outside diameter should be 23.002-23.011 mm (0.9056-0.9059 inch). Inside diameter of piston pin bosses should be 23.002-23.013 mm (0.9055-0.9060 inch) while the wear limit is 23.053 mm (0.9076 inch). Desired piston pin fit in piston bosses is 0.011 mm (0.0004 inch) interference to 0.011 mm (0.0004 inch) clearance.

It is recommended that piston be heated prior to reassembly to ease installation of pin.

CYLINDER LINER

All models are equipped with dry type cylinder liners. Use suitable removal and installation tools to renew defective liners. When renewing liner, be sure to thoroughly clean outer surface of liner and cylinder bore, then apply a light coat of oil to liner and bore surfaces. Press new liner into crankcase so top of liner is flush with top of crankcase within plus or minus 0.025 mm (0.001 inch). After installation, liner must be honed to final size.

Standard inner diameter of cylinder liner is 76.0-76.019 mm (2.9921-2.9929 inches) for Models Z751-B, DH1101-B, D1102-B, V1501-B, V1502-B and S2200-B; 82.0-82.022 mm (3.2283-3.2292 inches) for Models Z851-B, D1301-B, D1302-B, V1701-B, V1702-B and S2600-B; 85.0-85.022 mm (3.3465-3.3473 inches) for Models D1402-B and V1902-B. Maximum allowable liner wear is 0.15 mm (0.006 inch). Standard diameter cylinder liner may be bored for 0.5 mm (0.020 inch) oversize piston installation.

CONNECTING ROD AND BEARINGS

Connecting rods are equipped with a renewable bushing in the piston pin end and a precision, insert type bearing in the crankpin end.

Inner diameter of crankpin end bearing should be 44.010-44.052 mm (1.7327-1.7343 inches). Clearance between crankpin and bearing should be 0.035-0.093 mm (0.0014-0.0037 inch) while maximum allowable clearance is 0.2 mm (0.008 inch).

Small end bushing inner diameter should be 23.025-23.040 mm (0.9065-0.9071 inch). Recommended clearance between piston pin and bushing is 0.014-0.038 mm (0.0006-0.0015 inch) while maximum allowable clearance is 0.15 mm (0.006 inch). When renewing bushing, make certain oil hole in bushing is aligned with hole in end of rod. After installation, bushing must be honed to provide desired piston pin clearance.

CAMSHAFT

R&R AND OVERHAUL. To remove camshaft, remove cylinder head and timing gear cover as previously outlined then remove tappets. Unscrew camshaft retainer plate (11—Fig. K3-14) cap screws and withdraw camshaft from cylinder block. If necessary, press camshaft gear (10) off camshaft.

Camshaft lobe height should be 33.36 mm (1.3134 inches) for intake and exhaust. Cam lobe wear limit is 33.31 mm (1.3114 inches). Camshaft bearing journal diameter should be 39.934-39.950 mm (1.5722-1.5728 inches) while inner diameter of bearing bores should be 40.000-40.025 mm (15.748-1.5758 inches). Renew camshaft if clearance between camshaft journal and bearing exceeds 0.15 mm (0.006 inch).

When installing camshaft gear on camshaft, first install retainer (11–Fig. K3-14). Heat gear to approximately 80°C (175°F), then push gear onto camshaft.

To reinstall camshaft, reverse the removal procedure. Refer to TIMING GEARS AND COVER section to properly align timing marks during installation.

CRANKSHAFT AND BEARINGS

The crankshaft is carried in a sleeve type bearing (17—Fig. K3-20) pressed into front bearing bore of crankcase and in insert type bearings (25 and 26) in bearing carriers (28 and 31) for intermediate (center) and rear main journals. Two-cylinder models have one center bearing, three-cylinder models have two center bearings, four-cylinder models have three center bearings and six-cylinder models have five center bearings. Crankshaft end play is controlled by thrust washers (27) at front and rear of rear main bearing carrier (31). Before removing crankshaft, check crankshaft end play.

To remove crankshaft, remove timing gear cover, crankshaft timing gear, piston and rod units and flywheel. Remove the rear oil seal retainer (23—Fig. K3-20). Remove the cap screw (30—Fig. K3-21) from bottom of each center bearing web and, on Models S2200 and S2600, from side of block at each center bearing. Then carefully remove crankshaft from rear end of crankcase taking care not to damage front main bearing remaining in crankcase. Note that the center and rear main bearing carriers are

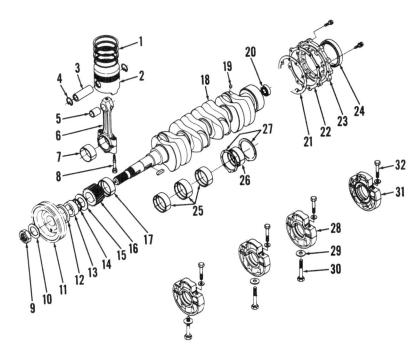

Fig. K3-20—Exploded view of piston, connecting rod and crankshaft assembly. On Models S2200-B and S2600-B, main bearing carriers (28) are retained by side screws as well as locating screws (30).

1. Piston rings
2. Piston
3. Piston pin
4. Retainer
5. Bushing
6. Connecting rod
7. Bearing
8. Screw
9. Nut
10. Washer
11. Pulley
12. Oil seal
13. Spacer
14. "O" ring
15. Oil slinger
16. Gear
17. Bearing
18. Crankshaft
19. Plug
20. Ball bearing
21. Gasket
22. Gasket
23. Seal carrier
24. Oil seal
25. Bearings
26. Bearing
27. Thrust washers
28. Bearing carriers
29. Washer
30. Screw
31. Rear bearing carrier
32. Screw

Kubota

a tight fit in cylinder block to prevent oil loss between oil passages in block and carriers.

Identify main bearing carriers as to their position in crankcase to ensure correct reassembly; bearing carrier halves are not interchangeable. Unbolt and remove the bearing carriers from crankshaft. Note that crankshaft bearing thrust washers are located at rear main bearing carrier.

Crankshaft main bearing journal diameter is 51.921-51.940 mm (2.0441-2.0449 inches). Desired oil clearance of journals to main bearings is 0.040-0.104 mm (0.0016-0.0041 inch). Maximum allowable oil clearance is 0.2 mm (0.008 inch). Crankshaft end play should be 0.15-0.31 mm (0.006-0.012 inch) with maximum allowable end play being 0.5 mm (0.020 inch). Crankpin diameter is 33.959-33.975 mm (1.3370-1.3376 inches) with desired bearing oil clearance of 0.019-0.081 mm (0.00075-0.003 inch) and maximum allowable clearance of 0.2 mm (0.008 inch).

Main bearings and crankpin bearings are available in undersizes of 0.2 and 0.4 mm (0.008 and 0.016 inch) and thrust bearings are available in oversize thickness of 0.2 and 0.4 mm (0.008 and 0.016 inch). Crankshaft may be reground undersize and with overwidth thrust surfaces at rear main journal to accommodate undersize bearings and oversize thrust washers.

To renew front main bearing, use suitable driver to remove old bearing from crankcase and install new bearing with oil hole in bearing aligned with oil passage in cylinder block. Bearing should be 4.2-4.5 mm (0.165-0.177 inch) below flush with front face of crankcase and positioned with seam toward camshaft side of engine.

Install bearing inserts in bearing carriers and stick thrust washers to rear bearing carrier with light grease. Be sure that grooved side of thrust washers face outward. Lubricate crankshaft and bearings and install carriers in correct position on crankshaft with side having bolt holes toward rear end of crankshaft. Tighten the main bearing cap bolts to a torque of 30-34 N·m (22-25 ft.-lbs.). Lubricate front main bearing and insert crankshaft and bearing supports into crankcase from rear taking care not to damage front main bearing.

Install new rear oil seal with lip to inside in rear seal retainer plate. Lubricate seal and carefully install retainer plate over rear of crankshaft with arrow mark on rear face of plate upward. Be sure to remove the two jackscrews from cover. Tighten retainer cap screws to a torque of 10-12 N·m (7-9 ft.-lbs.). Install the center bearing support retaining cap screw(s) in cylinder block and tighten to a torque of 64-68 N·m (47-50 ft.-lbs.). Tighten flywheel cap screws to a torque of 98-108 N·m (72-80 ft.-lbs.). Complete reassembly of engine by reversing disassembly procedure.

SMALL DIESEL

ELECTRICAL SYSTEM

ALTERNATOR AND REGULATOR

ALTERNATOR. An alternator rated at 20 amperes output is used on Models Z751-B, Z851-B and DH1101-B while all other models are equipped with a 25-ampere output alternator. Refer to Fig. K3-22 for an exploded view of typical alternator used on all models.

New brush length is 15.5 mm (0.610 inch); renew brush if worn to 10 mm (0.394 inch) or less. To check stator assembly, use an ohmmeter and check continuity between "N" terminal and each stator lead. If continuity does not exist, renew stator. Check continuity between "N" terminal and stator frame. If continuity exists, renew stator.

Using an ohmmeter, check diodes in rectifier assembly (14). Ohmmeter should read infinity during one test then continuity in other test with ohmmeter leads reversed. Individual diodes are not available and rectifier must be renewed as an assembly.

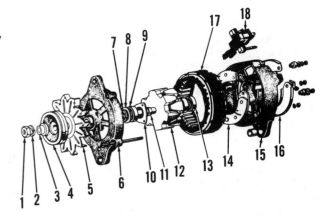

Fig. K3-22—Exploded view of typical alternator.
1. Nut
2. Lockwasher
3. Spacer
4. Pulley
5. Fan
6. End frame
7. Oil felt
8. Cover
9. Bearing
10. Bearing retainer
11. Spacer
12. Rotor
13. Bearing
14. Rectifier
15. Housing
16. Cover
17. Stator
18. Brush assy.

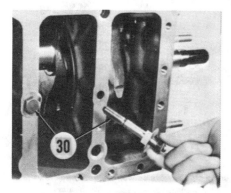

Fig. K3-21—View showing locating screws (30) which secure main bearing carriers to crankcase. Models S2200-B and S2600-B are equipped with screws in side of block that secure bearing carriers.

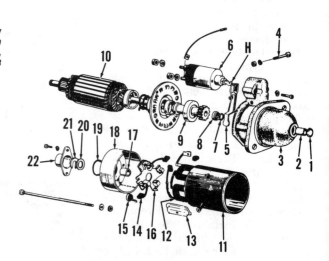

Fig. K3-23—Exploded view of electric starter used on Models D1102-B, D1102-BC, DH1101-B, D1302-B, Z751-B and Z851-B.
1. Cap
2. Bushing
3. Drive housing
4. Pivot bolt
5. Fork
6. Solenoid
7. Snap ring
8. Collar
9. Starter drive
10. Armature
11. Frame
12. Field coils
13. Field magnets
14. Brush
15. Brush spring
16. Brush plate
17. Bushing
18. End frame
19. Packing
20. Spring
21. Spring retainer
22. Cap

SERVICE MANUAL

Kubota

REGULATOR. Voltage regulator is available only as an assembly and adjustment is not normally required. Output voltage is controlled at 13.6-14.6 volts, with a rated output of 10 amperes.

With wiring disconnected or regulator removed, check regulator using an ohmmeter as follows:

Touch ohmmeter leads to IG and F terminals of regulator. Ohmmeter should read zero. If cover is removed and upper voltage control points manually opened, 11 ohms resistance should exist across resistor.

Touch ohmmeter leads to L and E terminals of regulator. Ohmmeter should read zero. If cover is removed and light relay points opened, 100 ohms resistance should exist across voltage regulator coil.

Touch ohmmeter leads to N an E terminals of regulator. Reading should be approximately 32 ohms.

Infinite resistance should exist between B terminal and any other terminal unless regulator cover is removed and light relay armature pushed down to connect lower set of points. With armature depressed, zero resistance should exist between L and B terminals and 100 ohms resistance should exist between E and B terminals.

STARTER. Refer to Fig. K3-23 or K3-24 for an exploded view of electric starter.

MODEL Z751-B. Minimum brush length is 10.7 mm (0.421 inch) while minimum commutator diameter is 32.5 mm (1.279 inch). With no load imposed on starter and using an 11 volt source, the starter shaft should rotate at 5000 rpm or more while drawing less than 50 amperes current.

Pinion engagement depth is adjusted by turning hook (H—Fig. K3-23). With starter pinion in engaged position, distance between collar (8) and pinion should be 0.1-0.4 mm (0.004-0.016 inch). Turn hook (H) so pinion engagement depth is correct.

MODELS Z851-B, DH1101-B, D1102-B, D1102-BC and D1302-B. Minimum brush length is 12.7 mm (0.5 inch) while minimum commutator diameter is 32.5 mm (1.279 inch). With no load imposed on starter, the starter shaft should rotate at 6000 rpm or more while drawing less than 45 amperes current at a voltage of 11.0.

Pinion engagement depth is adjusted by turning hook (H—Fig. K3-23). With starter pinion in engaged postion, distance between collar (8) and pinion should be 0.1-0.4 mm (0.004-0.016 inch). Turn hook (H) so pinion engagement depth is correct.

ALL OTHER MODELS. Minimum brush length is 12.7 mm (0.5 inch) while minimum commutator diameter is 29 mm (1.142 inch). With no load imposed on starter and using an 11.5 volt source, the starter shaft should rotate at 3500 rpm or more while drawing less than 90 amperes current.

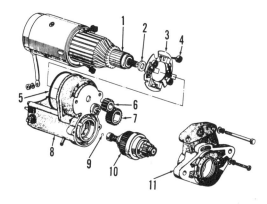

Fig. K3-24—Exploded view of electric starter used on all models except D1102-B, D1102-BC, DH1101-B, D1302-B, Z751-B and Z851-B.

1. Armature
2. Gasket
3. Brush plate
4. Brush spring
5. Gasket
6. Drive gear
7. Driven gear
8. Switch assy.
9. Ball
10. Clutch assy.
11. Gear housing

KUBOTA

Model	No. Cyls.	Bore	Stroke	Displ.
D600-B	3	64 mm (2.520 in.)	62.2 mm (2.450 in.)	600 cc (36.61 cu. in.)
D640-B	3	66 mm (2.600 in.)	62.2 mm (2.450 in.)	638 cc (38.93 cu. in.)
Z400-B	2	64 mm (2.520 in.)	62.2 mm (2.450 in.)	400 cc (24.41 cu. in.)
Z430-B	2	66 mm (2.600 in.)	62.2 mm (2.450 in.)	425 cc (25.93 cu. in.)

All engines in this section are liquid cooled, four-stroke, indirect injection diesel engines. Crankshaft rotation is counterclockwise as viewed from flywheel end. Firing order is 1-2-3 for three-cylinder models.

MAINTENANCE

LUBRICATION

Recommended engine lubricating oil is API classification CC or CD. For ambient temperatures below 0° C (32° F), use SAE 10W oil. In temperatures between 0° and 25° C (32°-77° F), SAE 20 oil is recommended. For temperatures above 25° C (77° F), use SAE 30 oil. Multi-grade SAE 10W-30 motor oil may be used in all temperatures.

Crankcase capacity is 1.87 liters (2 quarts) for 2-cylinder models and 2.91 liters (3 quarts) for 3-cylinder models. Engine oil and filter should be changed after each 100 hours of operation.

FUEL SYSTEM

FUEL FILTER. A combination fuel shut-off valve and filter assembly (see Fig. K4-1) is attached to engine. Filter element (9) should be removed and cleaned after each 100 hours of operation and be renewed after each 400 hours. To clean filter element, first thoroughly clean outside of filter assembly. Close fuel shut-off valve (7). Unscrew bowl retaining ring (12) and remove bowl (11) and filter element (9). Be careful not to lose the spring (10) located between filter element and bottom of bowl. Wash bowl and rinse element in kerosene, then reassemble filter and bleed air from system.

BLEED FUEL SYSTEM. To bleed air from fuel system, be sure tank is full and turn fuel shut-off valve lever (7—Fig. K4-1) to open position. Open fuel bleed screw (3) and allow fuel to flow until free of air. Tighten fuel filter bleed screw and open bleed screw (S—Fig. K4-2) on fuel injection pump until fuel flows freely. If equipped with fuel feed pump, crank engine with engine stop lever at "STOP" position until fuel flows from open bleed screw. Tighten the injection pump bleed screw and start engine.

If the engine will not start, loosen fuel injection lines at injectors and crank engine with engine stop control in "RUN" position until fuel spurts from loosened fittings. Tighten fuel injection line fittings and start engine.

INJECTION PUMP TIMING. Injection pump timing is adjusted by varying thickness of shims (2—Fig. K4-2) between fuel injection pump and crankcase. To check injection pump timing, remove the fuel injection lines. With speed control lever at maximum speed position, turn flywheel in normal direction of rotation until fuel fills the open fuel delivery valve holders (1). Continue turning flywheel slowly until fuel begins to flow over open delivery valve holder, stop turning engine and check flywheel timing mark. The "FI" mark on flywheel for that cylinder should then be aligned with punch mark on engine rear plate as shown in Fig. K4-3. If not, remove fuel injection pump and change shim thickness; decreasing shim thick-

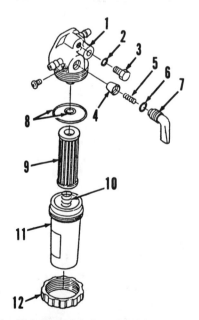

Fig. K4-1—Exploded view of fuel filter and fuel shut-off valve assembly.

1. Filter base
2. "O" ring
3. Air bleed screw
4. Shut-off valve
5. Spring
6. "O" ring
7. Valve lever
8. Filter seal rings
9. Filter element
10. Spring
11. Filter bowl
12. Bowl retainer

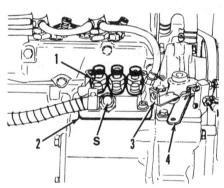

Fig. K4-2—Injection timing is adjusted by means of shims (2) located between injection pump flange and crankcase.

S. Air bleed screw
1. Delivery valve holders
2. Timing shims
3. Stop control lever
4. Throttle lever
5. High speed adjusting screw

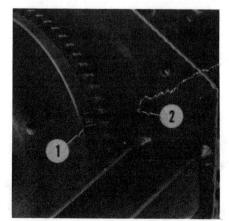

Fig. K4-3—View showing fuel injection (FI) timing mark on flywheel for number 1 cylinder aligned with punch mark (M) on engine rear plate.

SERVICE MANUAL

Kubota

ness will advance timing and increasing shim thickness will retard timing. Changing shim thickness by 0.15 mm (0.006 inch) will change timing about 1.5 to 2 degrees. Refer to preceding paragraph and bleed fuel system.

GOVERNOR

All models are equipped with a flyball type governor mounted at rear side of the fuel injection pump drive (timing) gear. Refer to Fig. K4-4 for exploded view of governor and to Fig. K4-5 for cross-sectional view.

Slow idle speed for all models is 800 rpm and is adjusted by turning adjusting screw (3). The idle speed adjusting screw is located in engine crankcase at rear side of fuel injection pump, and spring (6) on end of screw contacts pump control rack. When the engine stop lever is moved to stop position, an arm on lower end of lever shaft (13) pushes the injection pump control rack to the rear compressing the idle speed adjusting screw spring to stop pump fuel delivery.

Maximum no-load governed speed is 3800 rpm for all models. Rated speed under load is 3600 rpm. The throttle lever stop screw (5—Fig. K4-2) is adjusted and sealed with a wire and lead disc at the factory. The stop screw is located in the throttle lever cover at front of fuel injection pump. Strength of governor spring (16—Fig. K4-4 or K4-5) should be suspected if engine will not attain maximum no-load speed.

The maximum fuel limit stop screw (32) is adjusted at the factory and may need to be reset if engine is operating at high altitudes, or if engine smokes excessively under load. Remove seal cap (28), loosen jam nut (29) and turn screw in to lower smoke level. The fuel limit stop screw is located in engine timing gear cover.

AIR FILTER

All models are equipped with a dry type renewable element air filter. The filter should be removed and cleaned after each 100 hours operation, or more often in very dusty operating conditions. The dust can be cleaned from filter element by an air jet directed to inside surface of element. Air pressure must not exceed 690 kPa (100 psi).

The filter element can be washed in a solution of Kubota Filter Cleaner or Donaldson ND-1500 dissolved in water. Allow the filter to soak in the solution for 15 minutes, then rinse by agitating in clear water. Allow filter to dry before reusing.

The filter element should be renewed at least once each year or at each six cleaning period.

REPAIRS

TIGHTENING TORQUES

Refer to the following table for special tightening torques.

Connecting rod ....15-18 N·m
  (11-13 ft.-lbs.)
Crankshaft pulley cap
  screw ....98-108 N·m
  (72-79 ft.-lbs.)
Crankshaft seal carrier ....10-11 N·m
  (7-8 ft.-lbs.)
Cylinder head ....39-44 N·m
  (29-32 ft.-lbs.)
Flywheel ....54-59 N·m
  (40-43 ft.-lbs.)
Injection pump..........10-11 N·m
  (7-8 ft.-lbs.)
Injector nozzle nut ......49-68 N·m
  (36-50 ft.-lbs.)
Main bearing retainer .....20-23 N·m
  (15-17 ft.-lbs.)
Main bearing caps, center .12-15 N·m
  (9-11 ft.-lbs.)
Main bearing cap, rear ....20-23 N·m
  (15-17 ft.-lbs.)
Rocker arm brackets ......10-11 N·m
  (7-8 ft.-lbs.)

WATER PUMP

To remove water pump, drain coolant and, if necessary, relocate radiator for access to water pump. Remove fan belt, fan and pulley, then unbolt and remove pump from timing gear cover.

Using suitable pullers, remove pulley flange (1—Fig. K4-6) from shaft. Press shaft and bearing assembly (2) out toward rear of housing. Remove the impeller (5) and seal (4) from shaft.

Reassemble using new shaft and bearing assembly, seal and impeller. Press shaft and bearing assembly into housing. Position new seal over rear end of shaft and press into place using hollow driver that contacts only outer rim of

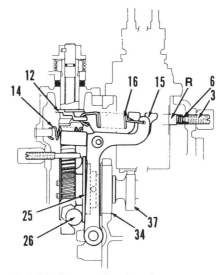

Fig. K4-5—Cross section view of governor components in idling position with pump rack (R) contacting idle speed spring (6). Refer to Fig. K4-4 for exploded view and parts identification.

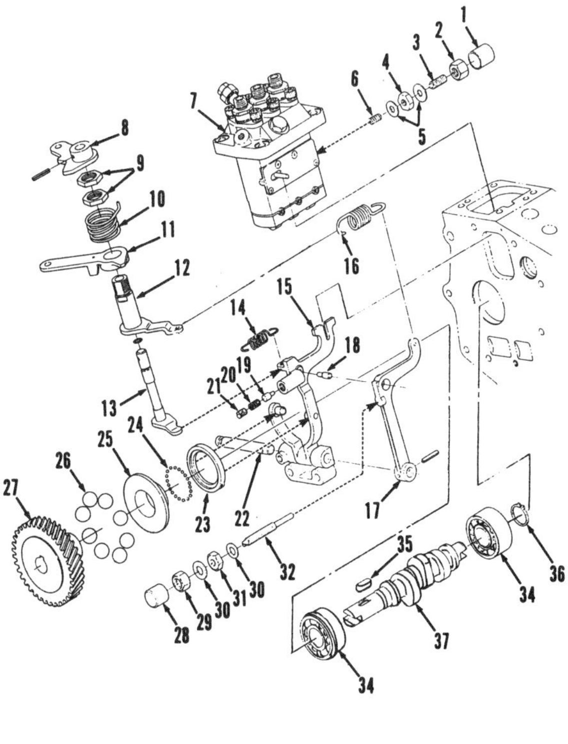

Fig. K4-4—Exploded view of governor components. Refer also to Fig. K4-5 for cross-sectional view.

1. Seal cap
2. Locknut
3. Idle speed screw
4. Locknut
5. Washers
6. Spring
7. Fuel injection pump
8. Stop control lever
9. Locknuts
10. Spring
11. Throttle lever
12. Throttle arm
13. Stop control arm
14. Start spring
15. Governor fork lever
16. Governor spring
17. Governor spring lever
18. Pin
19. Plunger
20. Spring
21. Screw
22. Lever pivot pin
23. Thrust bearing
24. Thrust bearing balls
25. Sliding sleeve
26. Governor balls
27. Injection pump gear
28. Seal cap
29. Locknut
30. Washers
31. Locknut
32. Smoke stop screw
34. Ball bearings
35. Key
36. Snap ring
37. Injection pump camshaft

Kubota

seal. Support front end of shaft and press impeller onto shaft. Support rear end of shaft and press flange onto shaft. Reinstall pump with new gasket.

COMPRESSION PRESSURE

Compression pressure should be checked with all injectors removed, battery in fully charged condition and with valve clearance properly adjusted. Move stop control lever to stop position. Turn engine with starter and check compression pressure of all cylinders. Starter should turn engine at 200-300 rpm. Compression of engine in good condition should be approximately 3100 kPa (450 psi) and variation between cylinders should not exceed 10 percent. Low limit for compression pressure is 2320 kPa (337 psi).

If compression pressure is below low limit, pour a small amount of oil in each cylinder through injector opening and recheck the compression of each cylinder (wet test). If compression is increased considerably, wear of pistons, rings and cylinders should be suspected. If compression readings do not increase, leaking valves should be suspected.

VALVE CLEARANCE

Valve clearance can be checked after removing rocker arm cover. Valve clearance with engine cold should be 0.15-0.18 mm (0.006-0.007 inch) for both intake and exhaust valves. Turn crankshaft to position number 1 piston at top dead center (1/TC mark on flywheel aligned with punch mark on engine rear plate as shown in Fig. K4-7A) of compression stroke, then check clearance with feeler gage between rocker arm and tip of valve stem. Loosen adjusting screw locknut, turn adjusting screw to obtain correct clearance, tighten nut and recheck clearance. Readjust if necessary.

Repeat procedure for remaining cylinders. On two-cylinder engines, turning crankshaft one complete revolution will place number two piston at TDC on compression stroke. On three-cylinder engines, turning crankshaft 2/3 turn (240 degrees) will place next piston in firing order (1-2-3) at TDC on compression stroke.

CYLINDER HEAD

To remove cylinder head, proceed as follows: Drain cooling system and remove upper radiator hose. Unbolt and remove thermostat housing, intake and exhaust manifolds. Remove fuel injection lines, injector fuel return lines and glow plugs. Remove rocker arm cover, rocker arm shaft assembly and the push rods. Take care not to lose the lash caps from upper end of valve stems. Unbolt and remove cylinder head assembly from block. Take care not to damage the valve mechanism oil pipe (see Fig. K4-7) extending from front left corner of cylinder block. Remove the cylinder head gasket and oil pipe sealing "O" ring (OR).

Before reinstalling cylinder head, completely clean all traces of old cylinder head gasket and carbon from both

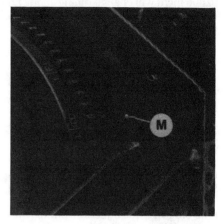

Fig. K4-7A—View showing top dead center timing mark (TC) for number one piston aligned with punch mark (M) on engine rear plate.

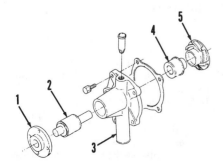

Fig. K4-6—Exploded view of water pump assembly.

1. Pulley flange
2. Shaft & bearing assy.
3. Pump housing
4. Seal assy.
5. Impeller

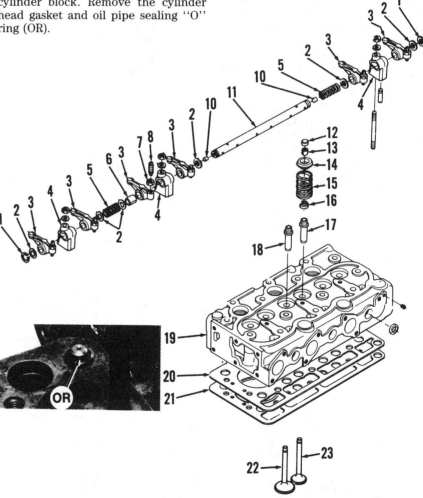

Fig. K4-7—Exploded view of cylinder head assembly. "O" ring (OR) seals oil delivery pipe between cylinder block and head.

1. Snap ring
2. Washer
3. Rocker arm
4. Shaft support
5. Spring
6. Spring
7. Locknut
8. Adjusting screw
10. Set screw
11. Rocker shaft
12. Valve cap
13. Retainer keys
14. Retainer
15. Spring
16. Seal
17. Exhaust valve guide
18. Intake valve guide
19. Cylinder head
20. Shim
21. Gasket
22. Intake valve
23. Exhaust valve

SERVICE MANUAL

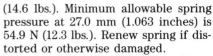

head and block surfaces. Be careful not to damage valve system oil tube protruding from corner of cylinder block. Clean carbon from top of pistons, taking care to clean all loose carbon from cylinders. Closely check cylinder head for cracks. Use straightedge and feeler gage to check cylinder head for any warped or twisted condition. Do not place straightedge over precombustion chamber inserts. Maximum allowable deviation from flatness is 0.05 mm (0.002 inch) per 100 mm (3.94 inch). Surface grind head if not within limits. Refer to VALVE SYSTEM paragraphs for valve head recess.

Piston to cylinder head clearance should be 0.60-0.80 mm (0.024-0.031 inch). Engine manufacturer recommends checking piston to cylinder head clearance after installing cylinder head and before fuel injectors are reinstalled. Turn crankshaft so that piston is at bottom of compression stroke and insert a soft lead wire into cylinder through injector opening as shown in Fig. K4-8. Turn the crankshaft by hand so that piston goes past top dead center, then withdraw the lead wire and measure flattened thickness. This thickness is equal to piston to cylinder head clearance. If clearance is less than specified, remove cylinder head and reinstall with a 0.2 mm (0.008 inch) shim between head gasket and cylinder head.

To reinstall cylinder head, reverse removal procedure and note the following: Do not apply any sealing compound to head gasket surfaces. Place a new "O" ring (OR—Fig. K4-7) on valve system oil tube. Place new gasket on block and place new 0.2 mm (0.008 inch) shim, if needed, on top of gasket. Carefully lower the cylinder head on the two stud bolts. Oil the threads of the cylinder head cap screws and nuts and install all finger tight. Refer to the appropriate tightening sequence in Fig. K4-9 and tighten the head bolts and nuts in several steps until final torque of 39-44 N·m (29-32 ft.-lbs.) is reached. Loosen locknuts and back rocker arm adjusting screws out. Be sure that lash caps are on valve stems. Install push rods and rocker arms and tighten rocker arm nuts to a torque of 10-11 N·m (7-8 ft.-lbs.). Adjust valve clearance as previously described. Complete reassembly by reversing disassembly procedure. Start engine and run until normal operating temperature is reached, then retorque cylinder head bolts and nuts to specified torque.

VALVE SYSTEM

Intake and exhaust valves seat directly in cylinder head. Lash caps are fitted to upper end of valve stems. Valve guides are renewable. Intake and exhaust valve springs are alike. Rocker arm bushings are renewable.

Valve stem diameter is 5.969-5.980 mm (0.2350-0.2354 inch) and guide inside diameter is 6.010-6.025 mm (0.2366-0.2372 inch), providing a clearance of 0.030-0.056 mm (0.0012-0.0022 inch). Maximum allowable stem to guide clearance is 0.1 mm (0.004 inch). If necessary to renew guides, use suitable piloted valve guide driver and drive old guides out top of head. Install new guides from top side of head so that upper ends are flush with valve cover gasket surface of head. Ream guides after installation for desired stem to guide clearance. Grind valve seats after installing new guides.

Valve face angle is 44.5 degrees and seat angle is 45 degrees. Desired valve seat width is 1.4 mm (0.055 inch). Seat width may be narrowed by reducing outside diameter of seat using a 15 degree stone.

After grinding valves and seats, install valves in head and check distance valve heads are recessed below cylinder head surface. Distance should be 0.75-0.95 mm (0.030-0.037 inch); limit for valve head recess is 1.2 mm (0.047 inch). If distance is excessive with new valve, manufacturer recommends replacement of cylinder head.

Valve spring free length should be 31.6 mm (1.244 inches). Renew springs if free length is 28.4 mm (1.118 inches) or less. Spring pressure at compressed height of 27.0 mm (1.063 inches) should be 64.7 N (14.6 lbs.). Minimum allowable spring pressure at 27.0 mm (1.063 inches) is 54.9 N (12.3 lbs.). Renew spring if distorted or otherwise damaged.

Rocker arm bushing ID should be 10.50-10.54 mm (0.413-0.415 inch). Rocker shaft diameter should be 10.47-10.48 mm (0.412-0.413 inch). Desired clearance between rocker arm shaft and bushings is 0.016-0.068 mm (0.0006-0.0027 inch); maximum allowable clearance is 0.15 mm (0.006 inch). When renewing rocker arm bushings, be sure oil hole in bushing is aligned with oil hole in rocker arm.

INJECTORS

REMOVE AND REINSTALL. To remove injectors, first thoroughly clean injectors, fuel lines and surrounding area using suitable solvent and compressed air. Remove fuel leak-off line and high pressure lines from injectors. Cap or plug all openings immediately to prevent entry of dirt. Unscrew and remove injectors from cylinder head.

Prior to reinstalling injectors, remove old nozzle sealing washers from head (if not removed with injector) and be sure cylinder head bores are clean. Install new nozzle washers in head with corrugated side up. Install injector and tighten to a torque of 49-68 N·m (36-50 ft.-lbs.). When reinstalling fuel lines, leave fittings for high pressure lines loose at injectors and crank engine until fuel emerges from loose fittings, then tighten injector line fittings.

TESTING. A faulty injector may be located on the engine by loosening high pressure fuel line fitting on each injector, in turn, with engine running. This

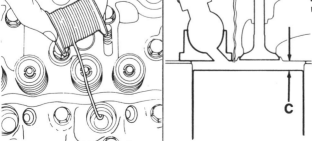

Fig. K4-8—View showing engine manufacturer's recommended method for checking piston to cylinder head clearance (C) by inserting soft lead wire through injector opening. Refer to text.

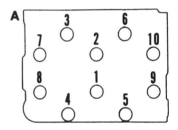

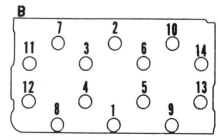

Fig. K4-9—Cylinder head bolt tightening sequence for two-cylinder (A) and three-cylinder (B) engines.

Kubota SMALL DIESEL

will allow fuel to escape from the loosened fitting rather than enter the cylinder. The injector that least affects engine operation when its line is loosened is the faulty injector. If a faulty injector is found and considerable time has elapsed since injectors have been serviced, it is recommended that all injectors be removed and serviced, or that new or reconditioned units be installed.

A complete job of testing and adjusting fuel injectors requires use of special test equipment. Only clean, approved testing oil should be used in tester tank. Injection nozzle should be tested for opening pressure, spray pattern and seat leakage. Check injection nozzle as outlined in the following paragraphs.

WARNING: Fuel emerges from injector with sufficient force to penetrate the skin. When testing injector, keep yourself clear of nozzle spray.

OPENING PRESSURE. Before conducting test, operate tester lever until fuel flows from tester line, then attach injection nozzle to tester. Pump tester lever a few quick strokes to purge air and to be sure nozzle valve is not stuck and spray hole is open.

Operate tester lever slowly and observe pressure gage reading when nozzle opens. On all models, opening pressure should be 13730-14710 kPa (1990-2133 psi).

Opening pressure is adjusted by changing thickness of adjusting washer (5—Fig. K4-10). Washers are available in 43 different thicknesses of 0.90-1.95 mm (0.0354-0.0768 inch) in steps of 0.025 mm (0.001 inch). Installing a thicker washer will increase opening pressure, a thinner washer will decrease pressure.

SPRAY PATTERN. The injector nozzle tip used on all models has one orifice. To check spray pattern, operate tester handle and observe spray. Spray pattern must be uniform, well atomized and slightly conical. If irregular pattern is observed, check for partially clogged or damaged spray hole or improperly seating nozzle valve.

SEAT LEAKAGE. Wipe nozzle tip dry, then operate tester lever slowly to maintain pressure of 12750 kPa (1850 psi) for ten seconds. There should be no drops or noticeable accumulation of oil on tip of injector. If drop forms, disassemble and overhaul injector. Slight wetting of tip is acceptable on used injector.

OVERHAUL. First, thoroughly clean outside of injector. Clamp injector body (4—Fig. K4-10) in a soft-jawed vise. Remove nozzle nut (10) and separate components from nozzle body. Place all parts in clean calibrating oil or diesel fuel as they are removed, keeping parts from each injector together and separate from other units. Nozzle valve (9) and body are individually fit and these parts must not be interchanged.

Clean exterior surfaces with a brass wire brush. Soak parts in an approved carbon solvent if necessary to loosen hard carbon deposits. Rinse parts in clean diesel fuel or calibrating oil to neutralize solvent and prevent etching of polished surfaces.

Clean nozzle spray hole from inside using a pointed hardwood stick or wood splinter as shown in Fig. K4-11. Scrape carbon from pressure chamber using hooked scraper as shown in Fig. K4-12. Clean valve seat using brass scraper as shown in Fig. K4-13, then polish seat using wood polishing stick and mutton tallow as shown in Fig. K4-14.

Examine lapped pressure surfaces of nozzle and valve (9—Fig. K4-10) for obvious damage. Check to see that nozzle valve will slide smoothly downward in valve body by its own weight. If valve binds or slides too fast, renew nozzle assembly. Check spring (6) for distortion, corrosion or other damage. Check for worn pressure pin (7) and renew if damage is noted. Be sure nozzle spray orifice is clean.

Back flush nozzle using reverse flush adapter on injector tester. Thoroughly rinse all parts in clean diesel fuel. Make sure all sealing surfaces are absolutely clean. Reassemble injector by reversing disassembly procedure. Tighten nozzle nut (10) to a torque of 49-68 N·m (36-50 ft.-lbs.). Retest assembled injector as previously outlined.

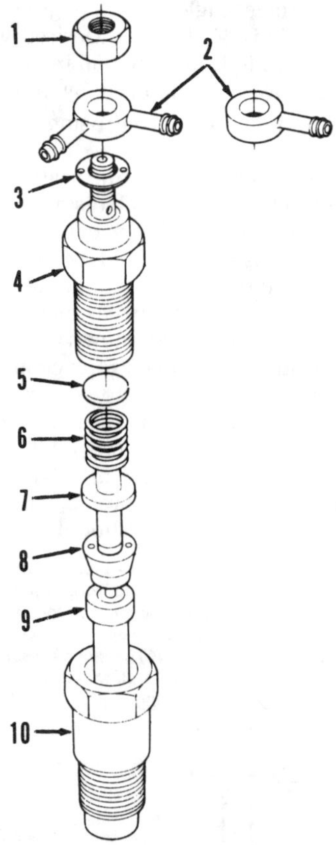

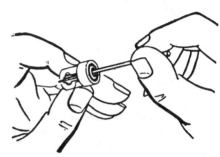

Fig. K4-11—Use a pointed hardwood stick to clean spray hole.

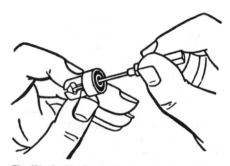

Fig. K4-12—Use hooked scraper to clean carbon from pressure chamber.

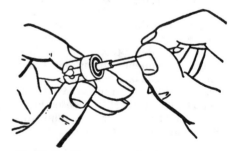

Fig. K4-13—Clean valve seat with brass scraper.

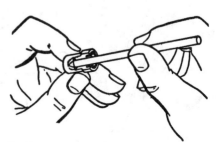

Fig. K4-14—Polish valve seat using polishing stick and mutton tallow.

Fig. K4-10—Exploded view of fuel injector.
1. Nut
2. Return line fitting
3. Sealing washer
4. Injector body
5. Shim washer
6. Spring
7. Push rod
8. Spacer
9. Nozzle & valve
10. Nozzle nut

Illustrations Courtesy Kubota

SERVICE MANUAL

GLOW PLUGS

Glow plugs (Fig. K4-15) are parallel connected with each glow plug grounding through its mounting threads. Indicator light should glow about 30 seconds after turning glow plug switch to "ON" if units are operating satisfactorily.

If indicator light fails to glow, check for burned out indicator lamp, or for an open circuit at switch, indicator lamp or glow plug connections. Resistance between glow plug terminal and body should be about 1.6 ohms when glow plug is cold. If resistance is zero, glow plug is shorted. If resistance is infinite, glow plug element is burned out.

INJECTION PUMP

To remove injection pump, first thoroughly clean pump, injection lines and surrounding area. Close fuel supply valve. Disconnect and remove fuel injection lines and fuel supply line from pump. Immediately cap or plug all openings to prevent entry of dirt. Disconnect stop and throttle control linkage from levers on plate in front of pump. Unbolt pump from cylinder block and lift pump upward, taking care not to lose or damage shim between pump flange and cylinder block. Be sure to place pump in clean area.

The injection pump should be tested and overhauled by a shop qualified in diesel injection equipment repair.

To reinstall pump, use same thickness shim as when removed unless change of timing is indicated. Move pump rack pin (3—Fig. K4-16) forward so that rear end of rack (2) is flush with pump housing. Move control levers so that slot of governor lever (4) is aligned with notch in crankcase. Insert pump in crankcase opening so that rack pin engages slot in governor fork lever and idle spring (1) contacts rear end of pump rack. Release control levers and be sure pump is in place. Install retaining nuts and tighten to a torque of 10-11 N·m (7-8 ft.-lbs.). Install fuel lines and governor linkage, leaving injection line fittings at injectors loose to bleed the fuel system as outlined in MAINTENANCE section.

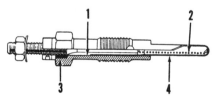

Fig. K4-15—Cutaway view of glow plug used for cold weather starting.
1. Insulating powder
2. Heating coil
3. Housing
4. Metal tube

TIMING GEAR COVER

To remove timing gear cover, proceed as follows: Drain cooling system and remove radiator if necessary to provide clearance for crankshaft pulley removal. Remove fan belt, fan and alternator. Disconnect linkage from throttle and stop control arms on plate (2—Fig. K4-17) at front of fuel injection pump. Unbolt the plate and carefully lift the plate and levers assembly, then disconnect the governor spring (1) from throttle lever arm on bottom of plate. Disconnect the start spring (2—Fig. K4-18) from pin in front of timing cover. Lock engine from turning, flatten the lock tab washer and remove the crankshaft pulley cap screw. Note alignment marks (3) on end of crankshaft and pulley. Use suitable pullers to remove crankshaft pulley. Remove cap screw (1—Fig. K4-19) inside throttle/stop lever plate opening and

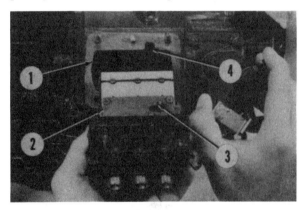

Fig. K4-16—View showing installation of fuel injection pump. Refer to text.
1. Idle speed spring
2. Pump rack
3. Pump rack pin
4. Notch in governor fork

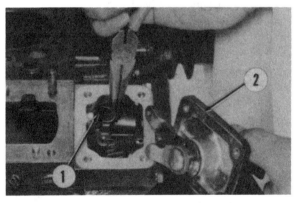

Fig. K4-17—Detaching governor spring (1) from throttle arm at bottom of control lever plate (2).

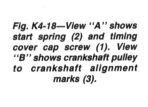

Fig. K4-18—View "A" shows start spring (2) and timing cover cap screw (1). View "B" shows crankshaft pulley to crankshaft alignment marks (3).

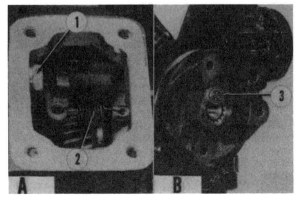

one cap screw (2) at lower left side of fan pulley (as viewed from front of engine). Remove the remaining timing gear cover bolts and remove cover from engine.

With cover removed, clean all traces of old gasket from cover and front of cylinder block. Inspect seal sleeve on crankshaft and renew if necessary. Install new crankshaft oil seal in timing gear cover with lip to inside.

To install cover, use new gasket, apply grease to inside of crankshaft seal and carefully position cover over crankshaft to avoid damage to crankshaft seal. Install the retaining bolts and tighten to a torque of 10-11 N·m (7-8 ft.-lbs.). Install crankshaft pulley with punch marks (3—Fig. K4-18) on pulley and crankshaft aligned. Oil pulley retaining cap screw threads and install cap screw with new lock tab washer. Tighten cap screw to a torque of 98-108 N·m (72-79

Kubota — SMALL DIESEL

ft.-lbs.) and bend washer against one flat of cap screw head. Reconnect start spring (2—Fig. K4-18). Place new gasket on throttle/stop lever plate, hold plate in position and reconnect the governor spring (1—Fig. K4-17). Install cover retaining cap screws with copper washers. Complete remainder of assembly by reversing disassembly procedure.

TIMING GEARS

The timing gears are accessible for inspection after removing timing gear cover. Remove the fuel injection pump and engine rocker arm shaft assembly to relieve pressure on timing gears. Check timing gear backlash with dial indicator. Desired timing gear backlash is 0.04-0.12 mm (0.0016-0.0047 inch); maximum allowable backlash between any two gears is 0.15 mm (0.006 inch). Desired end play of idler gear on shaft is 0.2-0.5 mm (0.008-0.020 inch); maximum allowable end play is 0.6 mm (0.024 inch). Oil clearance of idler gear on shaft should be 0.016-0.045 mm (0.0006-0.0018 inch) with maximum oil clearance being 0.05 mm (0.002 inch). Idler gear bushings, thrust collar and shaft are renewable.

To remove idler gear, remove snap ring (1—Fig. K4-20) and pull gear (4) and thrust collar (2) from shaft. If renewing bushings, press each bushing in until flush with outside surfaces of gear. Unbolt and remove idler shaft (3) from front of crankcase if necessary to renew shaft. Reinstall gear with timing marks aligned as shown in Fig. K4-21.

Crankshaft timing gear, washer, "O" ring and crankshaft oil seal sleeve can be removed using suitable pullers engaging back side of gear. When installing gear, install idler gear with timing marks aligned as shown in Fig. K4-21. Then install washer, "O" ring and oil seal sleeve with chamfered inside diameter of sleeve toward "O" ring.

To remove and install camshaft gear and fuel injection pump gear, refer to CAMSHAFT and INJECTION PUMP CAMSHAFT paragraphs.

OIL PUMP, RELIEF VALVE AND FILTER

The oil pump is located on front of engine cylinder block inside the timing gear cover. Oil pressure relief valve is located under the oil filter and is retained by the filter adapter nut and washer as shown in Fig. K4-22. Engine oil pressure should be 196-441 kPa (28-64 psi) at rated engine speed with minimum allowable pressure being 69 kPa (10 psi). To check pressure, remove the oil pressure switch and install a master oil pressure gage as shown in Fig. K4-23.

To remove oil pump, first remove timing cover as previously outlined, then unbolt and remove pump and gear assembly from front of crankcase. Remove screw retaining port plate to back of oil pump and remove the plate. Measure clearance between outer rotor and pump housing; desired clearance is 0.15-0.21 mm (0.006-0.008 inch) with maximum allowable clearance of 0.3 mm (0.012 inch). Measure clearance between inner and outer rotors with feeler gage; maximum allowable clearance is 0.25 mm (0.010 inch). Rotor end clearance should be 0.08-0.13 mm (0.003-0.005 inch). To measure rotor end clearance, place strip of Plastigage between rotors and port plate and install pump on crankcase. Remove the pump and check clearance with Plastigage indicator strip.

If any clearance is not within specification limits, renew the complete pump assembly.

CAMSHAFT

To remove engine camshaft, first remove cylinder head and timing gear cover as previously outlined. Remove cam followers (tappets) from their bores and identify each by position from which they were removed. Unbolt camshaft retainer plate and withdraw camshaft, plate and camshaft gear from front of cylinder block. If necessary, press camshaft out of gear and retainer plate.

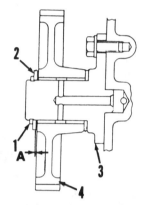

Fig. K4-20—Cross-sectional view of idler gear and mounting shaft assembly.

1. Snap ring
2. Thrust collar
3. Idler gear shaft
4. Idler gear

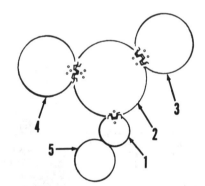

Fig. K4-21—Drawing showing timing marks for crankshaft gear (1), idler gear (2), camshaft gear (3) and fuel injection pump gear (4). Timing of oil pump gear (5) is not required.

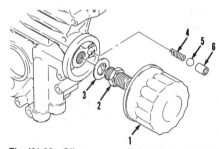

Fig. K4-22—Oil pressure relief valve (4, 5 and 6) can be removed after removing oil filter (1), adapter (2) and washer (3).

Fig. K4-19—Cap screw (1) inside timing cover and water pump cap screw (2) must be removed to remove timing gear cover.

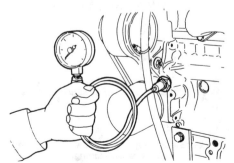

Fig. K4-23—Remove oil pressure switch and install master pressure gage to determine oil pressure.

SERVICE MANUAL

Camshaft lobe height should be 26.88 mm (1.0583 inches); renew camshaft if lobe height is less than 26.83 mm (1.0563 inches). Camshaft bearing journal diameter should be 32.934-32.950 mm (1.2966-1.2972 inches). Desired oil clearance of camshaft journals in bearing bore is 0.05-0.09 mm (0.002-0.003 inch) with maximum allowable clearance being 0.15 mm (0.006 inch). With camshaft supported in V-blocks at outer bearing journals, camshaft runout at center journals should be 0.01 mm (0.0004 inch); maximum allowable runout is 0.08 mm (0.003 inch).

Check bottom of cam followers (tappets) for abnormal wear and if noted, carefully inspect camshaft lobes for scoring or abnormal wear pattern. Renew cam followers if renewing camshaft.

When installing camshaft gear on camshaft, first install retainer plate. Then heat gear to approximately 80° C (175° F) and push gear onto shaft, being sure that timing marks on gear are forward and that keyway aligns with key on camshaft. Install camshaft, retainer plate and gear with timing marks of gear and idler gear aligned as shown in Fig. K4-21. Complete reassembly of engine by reversing disassembly procedure.

INJECTION PUMP CAMSHAFT

To remove injection pump camshaft, first remove fuel injection pump and timing gear cover as previously outlined. Remove fuel feed pump if so equipped from below fuel injection pump opening. Remove cap screws from retainer plate (1—Fig. K4-24) and governor fork lever bracket (5), then withdraw injection pump camshaft, gear and governor fork assembly from front of crankcase.

Inspect camshaft lobes, gear and bearings for excessive wear or other damage and renew parts as necessary. Cam lobe diameter, new, is 30 mm (1.181 inches).

When installing, first hook the governor spring to fork lever (7—Fig. K4-25) as shown. Leave governor fork lever bracket bolts loose until fuel injection pump camshaft and injection pump are installed. Be sure injection pump camshaft gear and idler gear timing marks are aligned as shown in Fig. K4-21. Adjust governor fork lever assembly so that pins on forks are centered on governor sleeve, and fork lever (6—Fig. K4-25) clears injection pump rack plate by a minimum of 1.5 mm (0.059 inch). Tighten the bolts and complete reassembly of engine by reversing disassembly procedure.

GOVERNOR

All models are equipped with a flyball type governor at back side of fuel injection pump gear as shown in Fig. K4-26. Refer to FUEL INJECTION PUMP CAMSHAFT paragraph for removal and installation procedure. Governor balls (26) are carried in bores of rear face of fuel injection pump drive gear (27). Ball movement with variation in engine speed against governor sleeve (25) actuates governor lever (15) which is connected to fuel injection pump control rack. Force of flyball movement through governor sleeve is balanced by governor spring (16).

Governor components are accessible after removing fuel injection pump camshaft as previously outlined. Inspect governor parts and renew any that are excessively worn or damaged. Be careful not to lose any of the thrust bearing balls (24) or governor flyballs (26) when disassembling fuel injection pump camshaft and governor unit.

PISTON AND ROD UNITS

Piston and connecting rods are removed from above as a unit after removing cylinder head, oil pan and oil pickup. Identify piston and rod units so they can be reinstalled in original positions if found to be serviceable. Original installed pistons are marked with cylinder number (see Fig. K4-27).

Note that numbers stamped on rod and cap must be toward same side when cap is installed on connecting rod. Numbers on connecting rod and cap should be toward camshaft side of engine. Lubricate threads on connecting rod cap screws and tighten to a torque of 15-18 N·m (11-14 ft.-lbs.).

CONNECTING RODS AND BEARINGS

Connecting rods (see Fig. K4-28) are fitted with a renewable bushing (1) at piston pin end and a precision shell type insert bearing (3) at crankpin end. Piston pin diameter is 18.001-18.009 mm (0.7087-0.7090 inch) and crankpin diameter is 33.959-33.975 mm (1.3370-1.3376 inches).

Oil clearance between bushing and piston pin should be 0.016-0.039 mm (0.0006-0.0015 inch) with maximum allowable clearance being 0.15 mm (0.006 inch). Oil clearance between crankpin bearing and crankpin should be 0.019-0.081 mm (0.00075-0.0032 inch) with maximum allowable clearance being 0.2 mm (0.008 inch).

When checking connecting rod for twist or misalignment, maximum allowable twist or bend is 0.05 mm (0.002 inch) per 100 mm (3.94 inch).

If installing new piston pin bushing, be sure oil hole in bushing is aligned with hole in rod, then hone bushing to inside diameter of 18.025-18.040 mm (0.7097-0.7102 inch) to obtain desired bushing to piston pin clearance.

PISTON, PIN AND RINGS

All models are equipped with cam ground aluminum pistons fitted with two compression rings and one oil control ring. Pistons and rings are available in oversize of 0.50 mm (0.020 inch) as well as standard size. Piston pins are available only in standard size of 18.001-18.009 mm (0.7087-0.7090 inch). Piston pin bore is 18.000-18.011 mm (0.7087-0.7091 inch) with wear limit of 18.050

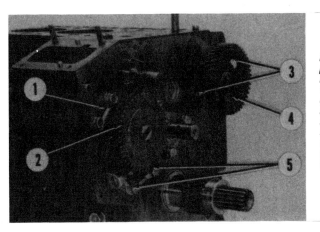

Fig. K4-24—Remove retainer plate (1) and governor lever mounting cap screws (5) to withdraw fuel injection pump camshaft, gear and governor assembly (2). Engine camshaft retainer plate cap screws (3) may be removed through holes in timing gear (4) to remove camshaft, retainer plate and gear assembly.

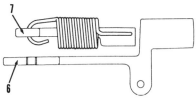

Fig. K4-25—Be sure governor spring is hooked in lever (7) as shown before installing camshaft, gear and governor assembly. Governor fork lever (6) must clear injection pump rack plate by 1.5 mm (0.059 inch).

Kubota

SMALL DIESEL

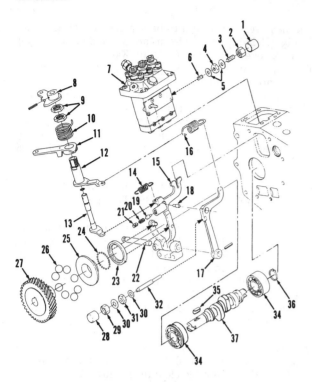

Fig. K4-26—Exploded view of governor components.

1. Seal cap
2. Locknut
3. Idle speed screw
4. Locknut
5. Washers
6. Spring
7. Fuel injection pump
8. Stop control lever
9. Locknuts
10. Spring
11. Throttle lever
12. Throttle arm
13. Stop control arm
14. Start spring
15. Governor fork lever
16. Governor spring
17. Governor spring lever
18. Pin
19. Plunger
20. Spring
21. Screw
22. Lever pivot pin
23. Thrust bearing
24. Thrust bearing balls
25. Sliding sleeve
26. Governor balls
27. Injection pump gear
28. Seal cap
29. Locknut
30. Washers
31. Locknut
32. Smoke stop screw
34. Ball bearings
35. Key
36. Snap ring
37. Injection pump camshaft

CYLINDERS

Some engines do not have cylinder liners (sleeves) and others are equipped with thin wall dry type cylinder liners (sleeves). Standard cylinder inside diameter is 64.000-64.019 mm (2.5197-2.5204 inches) for Models Z-400 and Z-430, and 66.000-66.019 mm (2.5984-2.5992 inches) for Models D600 and D640. Renew cylinder sleeves or rebore cylinders to 0.50 mm (0.020 inch) oversize if scored or wear exceeds 0.15 mm (0.006 inch) more than standard diameter.

Use suitable sleeve removal and installation tools to renew worn or damaged sleeves on engines so equipped. Thoroughly clean cylinder bore and outside of sleeve, then press sleeve into crankcase so top of sleeve is flush with top of crankcase.

CRANKSHAFT AND BEARINGS

The crankshaft is carried in a sleeve type bearing (1—Fig. K4-30) pressed into front bearing bore of crankcase and in insert type bearings (4 and 6) in bearing carriers for intermediate (center) and rear main journals. Two-cylinder models have one center bearing and three-cylinder models have two center bearings. Crankshaft end play is controlled by thrust washers (5) at front and rear of rear main bearing carrier. An oil seal sleeve (3) is pressed onto rear end of crankshaft. Before removing crankshaft, check crankshaft end play.

To remove crankshaft, remove timing gear cover, crankshaft timing gear, piston and rod units and flywheel. Refer to Fig. K4-31 and remove the rear oil seal retainer (1). Remove the cap screw from bottom of each center bearing web, then carefully remove crankshaft from rear end of crankcase taking care not to damage front main bearing remaining in

Fig. K4-27—Factory installed pistons have cylinder number (A) on top of piston and on cylinder block. Assemble connecting rod and piston so that mark (B) on side of rod is opposite number on piston.

mm (0.7106 inch). When inserting pin, first heat piston to a temperature of approximately 80° C (175° F). Install retaining snap rings in piston pin bore grooves with sharp edge of snap rings out.

Piston ring side clearance in groove should be 0.085-0.112 mm (0.003-0.004 inch) for second compression ring and 0.02-0.06 mm (0.0008-0.0024 inch) for the oil control ring. Top compression ring is keystone type. Maximum allowable side clearance in groove is 0.15 mm (0.006 inch) for all rings. Renew piston if ring side clearance is excessive. Ring end gap should be 0.25-0.40 mm (0.010-0.016 inch) for all rings with maximum allowable end gap being 1.25 mm (0.049 inch). Refer to Fig. K4-29 for proper installation of piston rings. Place oil ring spring expander in groove with ends butted together, then install oil ring on top of expander with ends 180 degrees away from expander ends. Position ring end gaps equally spaced around piston.

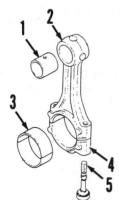

Fig. K4-28—Exploded view of connecting rod and bearings.

1. Piston pin bushing
2. Connecting rod
3. Crankpin bearing insert
4. Connecting rod cap
5. Rod cap screws

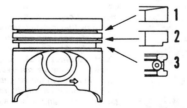

Fig. K4-29—View showing correct installation of piston rings.

1. Keystone ring
2. Compression ring
3. Oil ring & expander

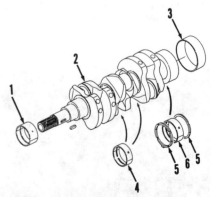

Fig. K4-30—Crankshaft front main bearing (1) is pressed in front of engine cylinder block. Center (4) and rear (6) inserts are supported in bearing carriers (see Fig. K4-32). Thrust washers (5) are fitted in rear bearing carrier. Oil seal sleeve (3) on rear end of crankshaft (2) is renewable.

SERVICE MANUAL

Kubota

crankcase. Note that the center and rear main bearing carriers are a tight fit in cylinder block to prevent oil loss between oil passages in block and carriers.

Carefully identify main bearing carriers as to their position on crankshaft to ensure correct reassembly. Bearing carrier halves are not interchangeable. Unbolt bearing carriers and remove the carriers and bearings from crankshaft. Note that crankshaft bearing thrust washers are located at rear main bearing carrier.

Crankshaft front and center main bearing journal diameter is 39.934-39.950 mm (1.5722-1.5728 inches); rear bearing journal diameter is 43.934-43.950 mm (1.7297-1.7303 inches). Desired oil clearance of main journals to front main bearing is 0.034-0.106 mm (0.0013-0.0042 inch) and to center and rear main bearings is 0.034-0.092 mm (0.0013-0.0036 inch). Maximum allowable oil clearance for all main bearings is 0.2 mm (0.008 inch). Crankshaft end play should be 0.15-0.31 mm (0.006-0.012 inch) with maximum allowable end play being 0.5 mm (0.020 inch). Crankpin diameter is 33.959-33.975 mm (1.3370-1.3376 inches) with desired bearing oil clearance of 0.019-0.081 mm (0.00075-0.003 inch). Check crankshaft main bearing alignment by supporting shaft in V-blocks at front and rear journals and checking runout at a center journal with a dial indicator. Runout should be 0.02 mm (0.0008 inch) or less with maximum allowable runout being 0.08 mm (0.003 inch). Inspect oil seal wear sleeve at rear end of crankshaft and renew sleeve if it is grooved from contact with oil seal or otherwise damaged.

Main bearings and crankpin bearings are available in undersizes of 0.2 and 0.4 mm (0.008 and 0.016 inch) and thrust bearings are available in oversize thickness of 0.2 and 0.4 mm (0.008 and 0.016 inch). Crankshaft may be reground undersize and with overwidth thrust surfaces at rear main journal to accommodate undersize bearings and oversize thrust washers.

To renew front main bearing, use suitable driver to remove old bearing from crankcase and install new bearing with oil hole in bearing aligned with oil passage in cylinder block. Bearing should be 4.2-4.5 mm (0.165-0.177 inch) below flush with front face of crankcase and positioned with seam toward camshaft side of engine.

Install bearing inserts in bearing carriers and stick thrust washers (5—Fig. K4-32) to rear bearing carrier (7) with light grease. Lubricate crankshaft and bearings and install carriers in correct position on crankshaft with face of carriers having Japanese character identification or side having bolt holes toward rear end of crankshaft. Tighten the center bearing cap bolts to a torque of 12-15 N·m (9-11 ft.-lbs.) and rear support cap bolts to a torque of 20-23 N·m (15-17 ft.-lbs.). Lubricate front main bearing and insert crankshaft and bearing supports into crankcase from rear taking care not to damage front main bearing.

Install new rear oil seal with lip to inside in rear seal retainer plate. Lubricate seal and carefully install retainer plate over rear of crankshaft with arrow mark up as shown in Fig. K4-31. Be sure to remove the two jackscrews from retainer. Tighten retainer plate cap screws to a torque of 10-11 N·m (7-8 ft.-lbs.). Install the center bearing support retaining cap screw(s) in cylinder block webs and tighten to a torque of 20-23 N·m (15-17 ft.-lbs.). Complete reassembly of engine by reversing disassembly procedure.

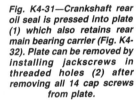

Fig. K4-31—Crankshaft rear oil seal is pressed into plate (1) which also retains rear main bearing carrier (Fig. K4-32). Plate can be removed by installing jackscrews in threaded holes (2) after removing all 14 cap screws from plate.

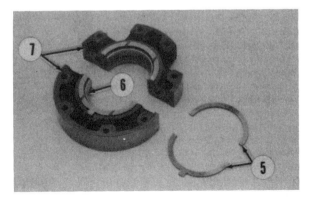

Fig. K4-32—View showing thrust washers (5) installed in rear face of bearing carrier (7). Second set of thrust washers are installed in front face of carrier.

ELECTRICAL SYSTEM

ALTERNATOR. To check alternator no-load output, disconnect the lead wires from the alternator. Place a voltmeter between the two alternator leads; the alternator should develop an AC voltage of 20 volts or more at 5200 rpm. If AC voltage is less than 20 volts, renew the alternator. With the charging circuit fully connected and with a fully charged 12-volt battery, regulated voltage should be 14 to 15 volts. If not within specified voltage, renew the alternator.

STARTING MOTOR. No load rpm should be 7000 rpm or higher at a current draw of less than 53 amperes with supply voltage of 11.5 volts. Brush length, new, is 17 mm (0.67 inch); minimum brush length is 11.5 mm (0.453 inch). Pinion clearance with starter pinion in engaged position is 0.5-2.0 mm (0.020-0.079 inch). Adjust pinion clearance by adding or removing washers between solenoid switch and front end frame.

Illustrations Courtesy Kubota

KUBOTA

Model	No. Cyls.	Bore	Stroke	Displ.
D1302-DI	3	82 mm (3.228 in.)	82 mm (3.228 in.)	1299 cc (79.3 cu. in.)
D1402-DI	3	85 mm (3.346 in.)	82 mm (3.228 in.)	1395 cc (85.1 cu. in.)
V1702-DI	4	82 mm (3.228 in.)	82 mm (3.228 in.)	1732 cc (105.7 cu. in.)
F2302-DI-B	5	85 mm (3.346 in.)	82 mm (3.228 in.)	2326 cc (141.9 cu. in.)

All models are liquid cooled, four stroke, direct injection diesel engines. Firing order is 1-2-3 on 3-cylinder engines, 1-3-4-2 on 4-cylinder engine and 1-3-5-4-2 on 5-cylinder engine. Cylinder number 1 is at timing gear end of engine.

MAINTENANCE

LUBRICATION

Recommended engine lubricant is API classification CC or CD motor oil. Below 0° C (32° F), use SAE 10W motor oil. In temperatures between 0° and 25° C (32 to 77° F), SAE 20 oil is recommended. In temperatures above 25° C (77° F), use SAE 30 oil. SAE 10W-30 oil may be used for all ambient operating temperatures.

Engine oil and filter should be changed after each 200 hours of use on 3 and 4-cylinder engines and after each 75 hours use on 5-cylinder engine. Engine lubricant capacity is 5.7 liters (6 quarts) for 3-cylinder, 6.5 liters (6.8 quarts) for 4-cylinder and 10.8 liters (11.5 quarts) for 5-cylinder engine.

FUEL SYSTEM

Refer to Fig. K5-1 for drawing showing fuel system for three and four-cylinder models; system for five-cylinder engine is similar except that it does not have air bleed line (A) running from filter base to fuel tank.

FUEL FILTER. A replaceable element type fuel filter (Fig. K5-2) with fuel shut-off valve is used on 3 and 4-cylinder models; element should be renewed after each 300 hours of use. Filter cartridge (Fig. K5-3) on 5-cylinder should be renewed after each 400 hours. However, if engine shows signs of fuel starvation (loss of power or surging), renew filter regardless of hours of operation. After renewing filter element, bleed the system as described in following paragraph.

BLEED FUEL SYSTEM. To bleed fuel system, first be sure that tank is full of fuel and open the fuel supply valve. On 5-cylinder engine, loosen the air vent plug (3—Fig. K5-3) on top of fuel filter and tighten plug when fuel flows freely without air bubbles. On all models, loosen the air vent valve (B—Fig. K5-1). Crank the engine for 10 seconds at 30 second intervals until all air has bled from system, then close the pump air vent valve. If the injector lines have been disconnected, loosen fuel line fittings at injectors and crank engine until fuel spurts from loosened fittings. Tighten the fittings and crank engine.

INJECTION PUMP TIMING. To check injection timing, remove the fuel injection lines and move throttle lever to maximum fuel position. Turn engine clockwise (facing crankshaft pulley end) until fuel fills up the hole in one of the open delivery valves (1—Fig. K5-4). Continue to turn the engine slowly until fuel flows over top of fuel delivery valve, then stop turning engine and check flywheel timing mark. The fuel injection mark (FI) on flywheel for that cylinder should then align with mark of timing window in flywheel housing. If not, remove fuel injection pump and add shims (2) to retard timing or decrease shim thickness to advance timing. Adding or removing one 0.15 mm (0.006 in.) shim will change timing by 1.5 degrees. On 3

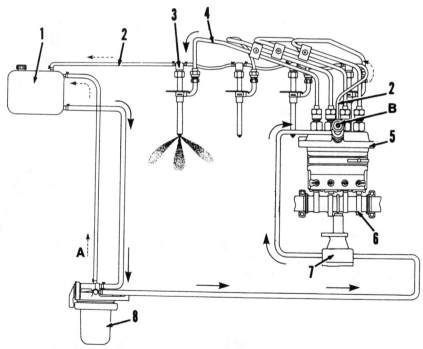

Fig. K5-1—Drawing showing fuel system of four-cylinder engine; three-cylinder models are similar. Five-cylinder engines do not have air bleed line (A) running from fuel filter back to tank.

A. Air bleed line
B. Air bleed screw
1. Fuel tank
2. Fuel return line
3. Fuel injector
4. Injection line
5. Injection pump
6. Pump camshaft
7. Fuel feed pump
8. Fuel filter

SERVICE MANUAL

Kubota

and 4-cylinder engines, approximately 3.6 mm (0.142 inch) on flywheel circumference equals 1.5 degrees of crankshaft rotation. On 5-cylinder engine, 4.4 mm (0.173 inch) equals 1.5 degrees of crankshaft rotation. After timing is correct, reinstall fuel injection lines and bleed lines if necessary.

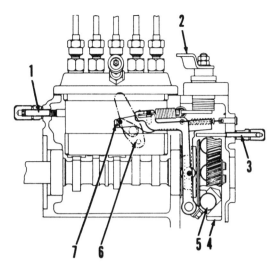

Fig. K5-5—Cutaway view showing engine speed control mechanism. Refer also to Fig. K5-6.
1. Idle speed adjusting screw
2. Throttle lever
3. Maximum fuel delivery screw
4. Injection pump gear
5. Governor flyballs
6. Stop lever
7. Governor lever

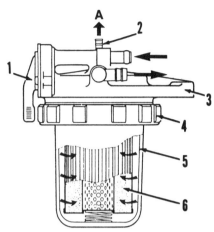

Fig. K5-2—Cutaway view of renewable element fuel filter assembly used on 3 and 4-cylinder models.

A. Air bleed line
1. Fuel shut-off valve
2. Air bleed nipple
3. Filter base
4. Bowl retainer
5. Filter bowl
6. Filter element

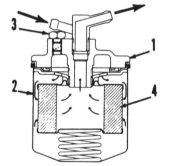

Fig. K5-3—Cross-sectional view of cartridge type fuel filter used on 5-cylinder engine.
1. Filter base
2. Filter cartridge
3. Air bleed screw
4. Filter element

GOVERNOR

All models are equipped with a flyball type governor mounted on the fuel injection pump camshaft behind the pump drive gear. Refer to cross-sectional view of governor components in Fig. K5-5. Engine idle speed is controlled by an idle compensator (see Fig. K5-6) which provides an idle speed of approximately 1400 rpm when engine is cold and 850-900 rpm when engine oil temperature is above 27° C (81° F).

Idle speed should be adjusted only after engine is at operating temperature. If hot idle speed is not within 850 to 900 rpm, loosen the locknut (2—Fig. K5-6) and turn cam (1) to provide desired hot idle speed and tighten locknut. Cold idle speed is not affected by turning compensator cam. If both cold and hot idle speeds are too low or too high, remove cap seal and loosen locknut on idle speed adjusting screw (1—Fig. K5-5) in crankcase at rear of fuel injection pump. Turn screw in or out to obtain desired idle speed with engine cold, then recheck hot idle speed and adjust cam (1—Fig. K5-6) if necessary to correct hot idle speed.

High no-load speed is adjusted by loosening locknut (2—Fig. K5-7) and turning adjusting screw (3) with throttle in wide open position. High speed adjusting screw is secured by lead seal (1) at factory. Install new wire and seal after making adjustment.

Maximum fuel delivery is adjusted by removing seal cap, loosening locknut and turning adjusting screw (3—Fig. K5-5). If engine is being operated at higher

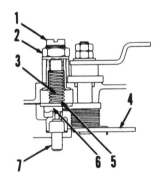

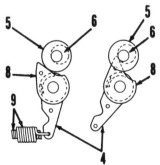

Fig. K5-6—Drawing showing cutaway view of throttle control lever and idle speed compensator. Lower left view shows compensating lever (8) stopped against cold idle cam; view at right shows hot idle compensating lever position.

1. Idle speed cam
2. Locknut
3. Thermostat spring
4. Throttle lever arm
5. Hot idle cam
6. Cold idle cam
7. Thermostat
8. Compensating lever
9. Governor springs

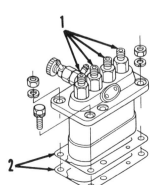
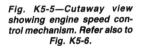
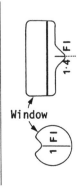

Fig. K5-4—Thickness of shims (2) is varied to adjust injection timing. Timing window in flywheel housing for 3-cylinder and 4-cylinder models is shown at right. A "TC" (top dead center) mark and a "FI" (fuel injection) mark appear on flywheel for each cylinder. Top window shows "FI" mark for cylinders 1 and 4 for 4-cylinder engine; bottom window is "FI" mark for cylinder 1 on 3-cylinder engine.

Illustrations Courtesy Kubota

103

Kubota

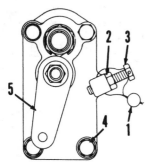

Fig. K5-7—High speed no-load adjusting screw (3) is sealed (1) at factory.
1. Lead wire seal
2. Locknut
3. Adjusting screw
4. Throttle lever/idle compensator plate
5. Throttle lever

elevations or smokes excessively under full load, adjusting screw should be turned in to reduce maximum fuel delivery which will increase engine performance.

AIR FILTER

All models are equipped with a dry type renewable element air filter. The filter should be removed and cleaned after each 100 hours operation, or more often in very dusty operating conditions. The dust can be cleaned from filter element by an air jet directed to inside surface of element. Air pressure must not exceed 690 kPa (100 psi).

The filter element can be washed in a solution of Donaldson ND-1500 dissolved in water. Allow the filter to soak in the solution for 15 minutes, then rinse by agitating in clear water. Allow filter to dry completely before reinstalling.

The filter element should be renewed at least once each year or at each sixth cleaning period.

REPAIRS

TIGHTENING TORQUES

Refer to the following table for special tightening torques.

Connecting rod............37-41 N·m
 (27-30 ft.-lbs.)
Crankshaft pulley:
 Model F2302.........196-215 N·m
  (144-159 ft.-lbs.)
 All other models......137-156 N·m
  (101-115 ft.-lbs.)
Crankshaft seal carrier....24-27 N·m
 (18-20 ft.-lbs.)
Cylinder head:
 Model F2302..........88-92 N·m
  (65-68 ft.-lbs.)
 All other models........79-83 N·m
  (58-61 ft.-lbs.)

Flywheel...............98-108 N·m
 (72-79 ft.-lbs.)
Injection pump..........10-11 N·m
 (7-8 ft.-lbs.)
Injector nozzle bracket....25-29 N·m
 (18-21 ft.-lbs.)
Main bearing retainer
 bolt..................64-68 N·m
  (47-50 ft.-lbs.)
Main bearing caps........30-34 N·m
 (22-25 ft.-lbs.)
Rocker arm bracket......24-27 N·m
 (18-20 ft.-lbs.)
Water pump shaft nut.....69-78 N·m
 (51-57 ft.-lbs.)

WATER PUMP

To remove water pump, first drain coolant and remove radiator if necessary. Remove fan belt and fan, then unbolt and remove pump from timing gear cover. Refer to appropriate following paragraph for disassembly and reassembly.

Model F2302-DI-B

Refer to Fig. K5-8 and clamp pulley (8) in vise at diameter indicated by arrows (B). Unscrew hex nut (1), then remove lockwasher and pulley from pump shaft; take care not to lose key (9). Remove internal snap ring (7) and press shaft (2) and bearings out of impeller (5) and housing (4) toward front end of housing. Remove bearings from shaft and remove seal (6) from rear end of housing.

To reassemble pump, install bearings on pump shaft. Install shaft and bearings in housing and secure with snap ring with sharp edge of snap ring to outside. Install new seal over shaft and into housing. Press impeller onto rear end of shaft so that clearance (C) between impeller and housing is 0.4-0.5 mm (0.016-0.020 inch). Insert key in shaft and install pulley, lockwasher and hex nut. Tighten the nut to a torque of 69-78 N·m (51-57 ft.-lbs.).

All Other Models

Using a suitable puller, remove flange (1—Fig. K5-9) from shaft (3). Disengage retainer clip (2) and press shaft and bearing assembly (3) forward out of housing (5) and impeller (7). Remove seal (6) from rear of housing.

Reassemble using new shaft and bearing assembly, seal and impeller.

COMPRESSION PRESSURE

Compression pressure should be checked with all injectors removed, battery in fully charged condition and with valve clearance properly adjusted. Move stop control lever to stop position. Turn engine with starter and check compression pressure of all cylinders. Starter should turn engine at 200-300 rpm. Compression of engine in good condition should be 2750-3040 kPa (400-440 psi), and variation between cylinders should not exceed 10 percent. Low limit for compression pressure is 2160 kPa (313 psi).

If compression pressure is below low limit, pour a small amount of oil in each cylinder through injector opening and recheck the compression of each cylinder (wet test). If compression is increased considerably, wear of pistons, rings and cylinders should be suspected. If compression readings do not increase, leaking valves should be suspected.

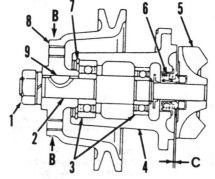

Fig. K5-8—Cross-sectional view of water pump assembly for 5-cylinder engine. Arrows (B) show where pulley can be clamped in vise for removal of nut (1).
1. Hex nut
2. Pump shaft
3. Ball bearings
4. Pump housing
5. Impeller
6. Seal assy.
7. Snap ring
8. Pulley
9. Pulley key

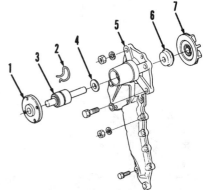

Fig. K5-9—Exploded view of water pump used on 3 and 4-cylinder engines.
1. Pulley flange
2. Wire retaining ring
3. Shaft & bearing assy.
4. Water slinger
5. Pump housing
6. Seal assy.
7. Impeller

SERVICE MANUAL

VALVE CLEARANCE

Valve clearance can be checked with a feeler gage after removing rocker arm cover. Valve clearance with engine cold should be 0.18-0.22 mm (0.007-0.009 inch) for both intake and exhaust valves. Turn flywheel so that number 1 piston is at top dead center of compression stroke. On 3 and 5-cylinder models, the "1 TC" mark on flywheel will be aligned with housing window mark; on 4-cylinder engines, use the "1.4 TC" mark. Refer to Fig. K5-10, K5-11 or K5-12 for three, four or five-cylinder engine and adjust valves as indicated by "X". Then, turn the flywheel one complete turn until timing marks on flywheel and housing are again aligned and adjust remaining valves.

CYLINDER HEAD

To remove cylinder head, proceed as follows: Drain cooling system and remove upper radiator hose. Unbolt and remove thermostat housing, intake and exhaust manifolds. Remove fuel injection lines, injector fuel return lines and glow plugs. Remove rocker arm cover, rocker arm shaft assembly and the push rods. Take care not to loose the lash caps from upper end of valve stems. Unbolt and remove cylinder head assembly from block. Take care not to damage the valve mechanism oil pipe (see Fig. K5-13) extending from front left corner of cylinder block. Remove the cylinder head gasket and oil pipe sealing "O" ring (OR).

Before reinstalling cylinder head, completely clean all traces of old cylinder head gasket and carbon from both head and block surfaces. Be careful not to damage valve system oil tube protruding from cylinder block. Clean carbon from top of pistons, taking care to clean all loose carbon from cylinders. Check cylinder head for cracks. Use a straightedge and feeler gage to check cylinder head for any warped or twisted condition. Do not place straightedge over precombustion chamber inserts.

Maximum allowable deviation from flatness is 0.05 mm (0.002 inch) per 100 mm (3.94 inches). Surface grind head if not within limits. Refer to VALVE SYSTEM paragraphs for valve head recess specifications.

Piston to cylinder head clearance should be 0.60-0.70 mm (0.024-0.028 inch) on 5-cylinder engines and 0.60-0.80 mm (0.024-0.031 inch) on 3 and 4-cylinder models. Engine manufacturer recommends checking piston to cylinder head clearance as follows: Turn crankshaft so that piston is at top of cylinder and position four strips of lead wire on top of piston as shown in Fig. K5-15. Turn crankshaft so that piston is down and install cylinder head. Turn crankshaft until piston passes top dead center, then remove the cylinder head. Measure flattened thickness of lead wires. This is equal to piston to cylinder head clearance. If clearance is less than specified, install cylinder head with a 0.2 mm (0.008 inch) shim between head gasket and cylinder head.

To reinstall cylinder head, reverse removal procedure and note the following: Do not apply any sealing compound to head gasket surfaces. Place a new "O" ring (OR—Fig. K5-13) on valve system oil tube at front left corner of cylinder block. Place new head gasket on block and if necessary, position new 0.2 mm (0.008 inch) shim on top of gasket. Carefully lower the cylinder head on the two stud bolts. Oil the threads of the cylinder head cap screws and nuts and install all finger tight. Refer to the appropriate tightening sequence in Fig. K5-14 and tighten the head bolts and nuts in several steps until final torque of 88-92 N·m (65-68 ft.-lbs.) on Model F2302 or 79-83 N·m (58-61 ft.-lbs.) on other models is reached. Loosen locknuts and back rocker arm adjusting screws out. Be sure that lash caps are on valve stems. Install push rods and rocker arms and tighten rocker arm nuts to a torque of 24-27 N·m (18-20 ft.-lbs.). Adjust valve clearance as previously described. Start engine and run until normal operating temperature is reached, then retighten cylinder head bolts and nuts to specified torque.

VALVE SYSTEM

Intake and exhaust valves seat on renewable seat inserts in cylinder head. Lash caps are fitted to upper end of valve stems. Valve guides are renewable. Intake and exhaust valve springs are alike. Rocker arm bushings are renewable.

Valve stem diameter, new, is 7.960-7.975 mm (0.3134-0.3140 inch) and guide inside diameter is 8.0015-8.030 mm (0.3156-0.3161 inch), providing a clearance of 0.040-0.070 mm (0.0016-0.0028 inch). Maximum allowable stem to guide clearance is 0.1 mm (0.004 inch). If necessary to renew guides, use suitable

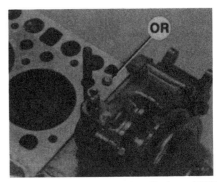

Fig. K5-13—View showing valve oil tube and "O" ring (OR) at the left front corner of cylinder block.

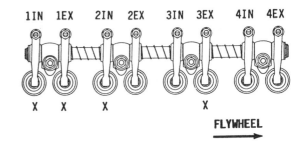

Fig. K5-11—Adjust 4-cylinder engine valves indicated by "X" with number 1 piston at TDC of compression stroke. Adjust remaining valves after turning crankshaft one revolution to place number 1 piston at TDC of exhaust stroke.

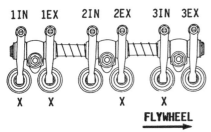

Fig. K5-10—With number 1 piston at TDC of compression stroke, adjust 3-cylinder engine valves indicated by "X" above. Turn crankshaft one revolution to TDC of exhaust stroke and adjust remaining valves.

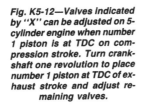

Fig. K5-12—Valves indicated by "X" can be adjusted on 5-cylinder engine when number 1 piston is at TDC on compression stroke. Turn crankshaft one revolution to place number 1 piston at TDC of exhaust stroke and adjust remaining valves.

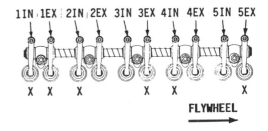

Kubota

piloted valve guide driver and drive old guides out top side of head. Install new guides from top side of head so that upper ends are flush with valve cover gasket surface of head. Ream guides after installation for desired stem to guide clearance.

Valve face angle is 44.5 degrees and seat angle is 45 degrees. Desired valve seat width is 1.4 mm (0.055 inch). Grind valve seats after installing new guides, then narrow seat width to 1.4 mm (0.055 inch) by reducing outside diameter of seat using a 15 degree stone.

After grinding valves and seats, install valves in head and check distance valve heads are recessed below cylinder head surface. Distance should be 0.65-0.85 mm (0.026-0.034 inch); limit for valve head recess is 1.15 mm (0.045 inch). If distance is excessive with new valve, valve seats should be renewed.

To remove old seats, use special valve seat puller or using arc welder, run a short bead along valve seating surface. Heat from the arc weld will shrink the seat and it can be removed easily. Be sure counterbore in head is clean and free of burrs. Chill new insert with dry ice and heat cylinder head to approximately 80° C (175° F), then drive seat into place. If correctly installed, intake seat should be 2.85-2.95 mm (0.112-0.116 inch) below head surface and exhaust seat should be 2.65-2.75 mm (0.104-0.108 inch) below head surface. After head is cool, grind valve seat to obtain correct valve recess as outlined in preceding paragraph.

Valve spring free length should be 41.7-42.2 mm (1.642-1.661 inch). Renew springs if free length is 41.2 mm (1.622 inches) or less. Spring pressure at compressed height of 35.5 mm (1.384 inches) should be 117 N (26.5 pounds). Minimum allowable spring pressure is 100 N (22.5 pounds). Renew spring if distorted or otherwise damaged.

Rocker arm bushing ID should be 14.0-14.04 mm (0.551-0.553 inch). Rocker shaft diameter should be 13.97-13.98 mm (0.5501-0.5505 inch). Desired clearance between rocker arm shaft and bushings is 0.018-0.070 mm (0.0007-0.0028 inch); maximum allowable clearance is 0.15 mm (0.006 inch). When renewing rocker arm bushings, be sure oil hole in bushing is aligned with oil hole in rocker arm.

INJECTORS

REMOVE AND REINSTALL. Prior to removing injectors, thoroughly clean injectors, fuel lines and surrounding area using solvent and compressed air. Remove the fuel return lines and the fuel injection lines from pump and injectors

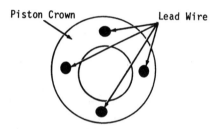

Fig. K5-15—Place soft lead wire strips at positions indicated on piston crown to check piston to cylinder head clearance. Refer to text.

and immediately cap or plug all openings. Remove cap screws (1—Fig. K5-16) from injector locating clamps and remove the clamps. Pull injectors from cylinder head.

NOTE: Unless the carbon stop seal has failed causing injector to stick, the injectors can be easily removed by hand. If injectors cannot be removed by hand, use Kubota injector puller 07916-32721 or equivalent. DO NOT attempt to pry injector from cylinder head or damage to injector could result.

When installing injector, be sure injector bore and compression seal (3) seat in cylinder head are clean and free of carbon or other foreign material. Install new compression seal and carbon seal (4) on injector; installation of carbon seal requires special Kubota tool 07916-32741 or equivalent. Insert injector into bore with a twisting motion. Install and align clamp spacer (5), locating clamp (2) and cap screw (1). Tighten cap screw to a torque of 25-29 N·m (18-21 ft.-lbs.). Reconnect fuel return line and install injection line, leaving fitting at injector loose until after bleeding air from line.

TESTING. A faulty injector may be located on the engine by loosening high pressure fuel line fitting on each injector, in turn, with engine running. This will allow fuel to escape from the loosened fitting rather than enter the cylinder. The injector that least affects engine operation when its line is loosened is the faulty injector. If a faulty injector is found and considerable time has elapsed since injectors have been serviced, it is recommended that all injectors be removed and serviced, or that new or reconditioned units be installed.

A complete job of nozzle testing and adjusting requires use of an approved nozzle tester and testing oil. The nozzle should be tested for spray pattern, opening pressure and seat leakage. Injector should produce a distinct audible chatter when being tested and cut off

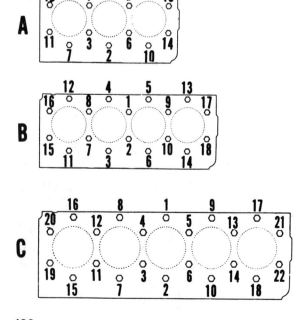

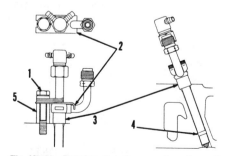

Fig. K5-14—Cylinder head bolt tightening sequence for 3-cylinder (A), 4-cylinder (B) and 5-cylinder (C) engines. When removing cylinder head, loosen head bolts in reverse order of tightening sequence shown.

Fig. K5-16—Drawing showing mounting of pencil type direct injection fuel injector.

1. Clamp cap screw
2. Locating clamp
3. Compression seal
4. Carbon dam seal
5. Spacer

SERVICE MANUAL

Kubota

quickly at end of injection with minimum seat leakage.

SPRAY PATTERN. Attach injector to tester, close valve to tester gage and operate tester lever at approximately 30 downward strokes per minute and observe spray pattern. A finely atomized spray (A—Fig. K5-17) should emerge at each of the four nozzle spray holes and a distinct chatter should be observed as tester is operated. If spray is not symmetrical (B or C—Fig. K5-17) and is streaked, or if injector does not chatter, refer to Fig. K5-18 and clean nozzle spray orifices. First, use a 0.19 mm (0.008 inch) wire protruding from pin vise approximately 1 mm (0.040 inch). Insert the cleaning wire in each orifice and rotate pin vise until wire fits into orifices freely. Repeat the process with a 0.22 mm (0.009 inch) cleaning wire. Finally, clean the orifices with a 0.24 mm (0.010 inch) wire and wipe off and inspect nozzle tip. If tip appears to be in good condition, repeat spray pattern test. Renew nozzle if it cannot be cleaned satisfactorily. If spray pattern is correct, proceed with nozzle testing.

OPENING PRESSURE. The correct opening pressure is 22410-23440 kPa (3250-3400 psi). Open valve to tester gage, operate tester lever slowly and observe tester gage. If opening pressure is not within specifications, adjust opening pressure as follows: Secure injector in holding fixture and remove leak-off cap (4—Fig. K5-19). Remove lift adjusting screw locknut (2) and loosen pressure adjusting screw locknut (6). Back lift adjusting screw (3) out two to three turns to prevent interference while adjusting opening pressure. Operate tester lever slowly and turn pressure adjusting screw (5) to obtain correct opening pressure. Temporarily tighten pressure adjusting screw locknut to hold pressure setting. Gently turn valve lift adjusting screw in until it bottoms, then operate injector tester to raise gage pressure to 24130-26200 kPa (3500-3800 psi). Fuel should not dribble from nozzle. Back lift adjusting screw out 1/2 turn and recheck opening pressure. If correct, hold pressure adjusting screw (5) and tighten locknut (6) to a torque of 8-9 N·m (70-80 in.-lbs.). Hold lift adjusting screw (3) and tighten locknut (2) to a torque of 4-5 N·m (35-45 in.-lbs.).

SEAT LEAKAGE. Attach injector to tester with injector nozzle at a 45 degree angle above horizontal. Operate tester to raise gage pressure to 19000-20000 kPa (2750-2900 psi) for a period of five seconds. If fuel dribbles from tip within that time, renew the nozzle. Slight wetting of tip is acceptable on a used injector.

GLOW PLUGS

Glow plugs (Fig. K5-20) are parallel connected with each glow plug grounding through its mounting threads. Indicator light should glow about 30 seconds after turning glow plug switch on if units are operating satisfactorily.

If indicator light fails to glow, check for burned out indicator lamp, or for an open circuit at switch, indicator lamp or glow plug connections. Resistance between glow plug terminal and body should be about 1.6 ohms when glow plug is cold. If resistance is zero, glow plug is shorted. If resistance is infinite, glow plug element is burned out.

INJECTION PUMP

To remove injection pump, first thoroughly clean pump, injection lines and surrounding area. Close fuel supply valve. Disconnect and remove fuel injection lines and fuel supply line from pump. Immediately cap or plug all openings to prevent entry of dirt. Disconnect stop control linkage from lever on plate (3—Fig. K5-21) below injection pump and remove the plate with stop control lever (2). Unbolt pump from cylinder block and lift pump upward, taking care not to lose or damage shim between pump flange and cylinder block. Be sure to place pump in clean area.

The injection pump should be tested and overhauled by a shop qualified in diesel injection equipment repair.

To reinstall pump, use same thickness shim (6—Fig. K5-21) as when removed unless change of timing is indicated. Apply gasket sealer to both sides of shim. Move pump rack forward so that rear end of rack is flush with pump housing. Move control levers so that slot of governor lever is aligned with notch (5) in pump mounting face of crankcase. Insert pump in crankcase opening so that rack pin (4) engages slot in governor fork lever and idle spring contacts rear

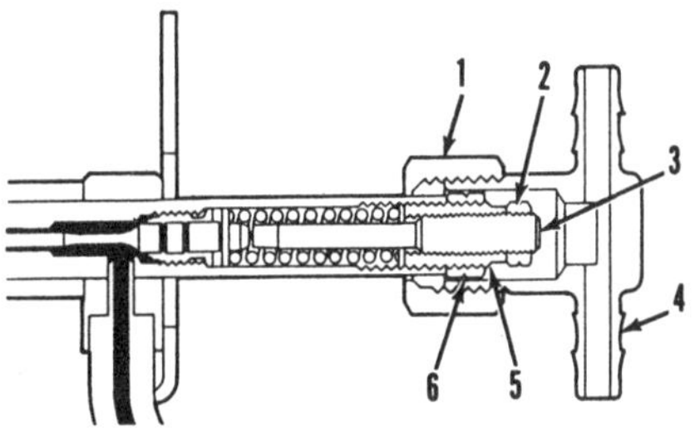

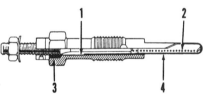

Fig. K5-19—Drawing showing cross section of pencil type fuel injector.
1. Leak-off cap nut
2. Lift screw locknut
3. Lift adjusting screw
4. Leak-off cap
5. Pressure adjusting screw
6. Adjusting screw locknut

Fig. K5-20—Cutaway view of glow plug used for cold weather starting.
1. Insulating powder
2. Heating coil
3. Housing
4. Metal tube

Fig. K5-17—Correct nozzle spray pattern is shown at "A." "B" and "C" are examples of faulty spray pattern.

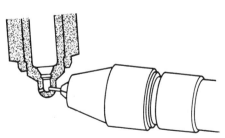

Fig. K5-18—Cross section of injector nozzle showing cleaning the nozzle orifice with cleaning wire in a pin vise. Refer to text.

Fig. K5-21—Removing stop control lever plate.
1. Governor fork
2. Stop lever
3. Stop lever plate
4. Control rack pin
5. Groove
6. Shim

Kubota

SMALL DIESEL

end of pump rack. Release control levers and be sure pump is in place. Install retaining nuts and tighten to a torque of 10-11 N·m (7-8 ft.-lbs.). Install stop control plate and lever with arm (2—Fig. K5-21) at right side of governor fork lever (1). Install fuel lines leaving injection line fittings at injectors loose to bleed air from fuel system as outlined in MAINTENANCE section.

TIMING GEAR COVER

To remove timing gear cover, proceed as follows: Drain cooling system and remove radiator, fan and alternator. Disconnect linkage from stop control lever on plate (3—Fig. K5-21) below fuel injection pump and remove the plate stop lever assembly. Working through plate opening, unhook both the inner (1—Fig. K5-22) and outer (2) governor springs from governor lever (3). Disconnect throttle linkage from lever on plate at front of fuel injection pump. Unbolt the plate and carefully lift the plate (2—Fig. K5-23) with idle compensator (1), throttle lever and governor spring (5) from timing gear cover. Remove thermostat housing (4) and idle compensator thermostat (3) from cover. Disconnect the start spring (6—Fig. K5-23) from pin in front of timing cover. Lock engine from turning and remove the crankshaft pulley cap screw or nut. On Model F2302, note alignment marks on end of crankshaft and pulley. Use suitable pullers to remove crankshaft pulley. Remove oil filter cartridge. Remove timing gear cover bolts and remove cover from engine.

With cover removed, clean all traces of old gasket from cover and front of cylinder block. Inspect seal sleeve on crankshaft and renew if necessary. Install new crankshaft oil seal in timing gear cover with lip to inside.

To install cover, use gasket sealer on both sides of new gasket and use grease to stick the three "O" rings (1—Fig. K5-24) to timing cover. Lubricate inside of crankshaft seal and carefully position cover over crankshaft to avoid damage to crankshaft seal. Install the retaining bolts and tighten evenly. Install crankshaft pulley; on Model F2302, align punch marks (M—Fig. K5-25) on pulley and crankshaft. Oil pulley retaining cap screw threads and install cap screw or nut with new washer. Tighten Model F2302 cap screw or nut to a torque of 196-215 N·m (144-159 ft.-lbs.); tighten nut on other models to a torque of 137-156 N·m (101-115 ft.-lbs.). Bend washer against one flat of cap screw head or nut. Reconnect start spring (6—Fig. K5-23). Place new gasket on throttle lever plate and install plate with governor spring connected to inside throttle arm. Install stop lever cover with arm to right of governor fork lever. Complete remainder of assembly by reversing disassembly procedure.

TIMING GEARS

The timing gears are accessible for inspection after removing timing gear cover. Remove the fuel injection pump and engine rocker arm shaft assembly to relieve pressure on timing gears. Check timing gear backlash with dial indicator. Desired timing gear backlash is 0.04-0.11 mm (0.0016-0.0043 inch); maximum allowable backlash between any two gears is 0.15 mm (0.006 inch). Desired end play of idler gear on shaft is 0.21-0.51 mm (0.008-0.020 inch); maximum allowable end play is 0.9 mm (0.035 inch). Oil clearance of idler gear on shaft should be 0.020-0.054 mm (0.0008-0.0021 inch) with maximum oil clearance being 0.1 mm (0.004 inch). Idler gear bushings, thrust collar and shaft are renewable.

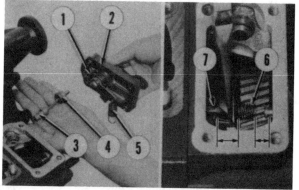

Fig. K5-23—Removing throttle and idle compensator plate. Disconnect start control spring (6) from governor lever (7) or pin in timing cover prior to removing cover.

1. Idle compensator cams
2. Plate
3. Thermostat
4. Thermostat housing
5. Governor springs
6. Start control spring
7. Fork lever

Fig. K5-24—Stick the three "O" rings (1) to timing cover with light grease to prevent dropping them out of place when installing cover.

Fig. K5-25—On F2302 engine, align punch marks (M) on crankshaft and pulley.

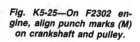

Fig. K5-22—Removing inner (1) and outer (2) governor springs from governor fork lever (2).

Illustrations Courtesy Kubota

SERVICE MANUAL

To remove idler gear (2—Fig. K5-26), remove snap ring and pull gear with thrust collar from shaft. If renewing bushings, press each bushing in until flush with outside surfaces of gear. Unbolt and remove idler shaft from front of crankcase if necessary to renew shaft. Reinstall gear with timing marks aligned as shown in Fig. K5-27.

Crankshaft timing gear (3—Fig. K5-26) can be removed using suitable pullers engaging back side of gear. When installing gear, remove idler gear and heat crankshaft gear to a temperature of approximately 80° C (175° F). Install heated gear on crankshaft with timing mark out. Install idler gear with timing marks aligned as shown in Fig. K5-27. Then refer to Fig. K5-26 and install collar (9) on models using crankshaft pulley retaining nut, oil slinger (6), "O" ring (7) and oil seal sleeve (8) with chamfered inside diameter of sleeve toward "O" ring.

To remove and install camshaft gear and fuel injection pump gear, refer to CAMSHAFT and INJECTION PUMP CAMSHAFT paragraphs.

OIL PUMP, RELIEF VALVE AND FILTER

The oil pump (10—Fig. K5-26) is located on front of engine cylinder block inside the timing gear cover. The oil pressure relief valve is located in the oil filter adapter and is retained by hex cap nut on Model F2302 and by a cap screw and plate on other models. Engine oil pressure should be 294-441 kPa (43-64 psi) at rated engine speed with minimum allowable pressure being 245 kPa (36 psi). To check pressure, remove the oil pressure switch and install a master oil pressure gage. Pressure relief valve spring free length should be 35 mm (1.38 inches); renew spring if free length is less than 30 mm (1.18 inches).

To remove oil pump, first remove timing cover as previously outlined. Flatten tab on washer, unscrew gear retaining nut and remove drive gear (4) with puller. Then unbolt and remove pump from front of crankcase and remove pump cover. Measure clearance between outer rotor and pump housing (see Fig. K5-28). Desired clearance is 0.11-0.19 mm (0.0043-0.0075 inch) with maximum allowable clearance of 0.25 mm (0.010 inch). Measure clearance between inner and outer rotors with feeler gage as shown in Fig. K5-29. Desired clearance is 0.10-0.16 mm (0.004-0.006 inch) with maximum allowable clearance of 0.20 mm (0.008 inch). Rotor end clearance should be 0.105-0.150 mm (0.004-0.006 inch). To measure rotor end clearance, place strip of Plastigage on front face of inner rotor, assemble pump, then disassemble and check with Plastigage indicator strip as shown in Fig. K5-30.

If any clearance is not within specification limits, renew the complete pump assembly.

CAMSHAFT

To remove engine camshaft, first remove cylinder head and timing gear cover as previously outlined. Remove cam followers (tappets) from their bores and identify each by position from which they were removed. Unbolt camshaft retainer plate and withdraw camshaft, plate and camshaft gear from front of cylinder block. If necessary, press camshaft out of gear and retainer plate.

Camshaft exhaust lobe height on all models should be 33.46-33.48 mm (1.317-1.318 inches); wear limit is 33.42 mm (1.316 inches). On Model F2302, intake lobe height is same as exhaust. On all other models, intake lobe height should be 33.26-33.28 mm (1.309-1.310 inches) with wear limit of 33.22 mm (1.308 inches). Camshaft bearing journal diameter should be 39.934-39.950 mm (1.572-1.573 inches) with wear limit of 39.88 mm (1.570 inches). Desired oil clearance of camshaft journals in bearing bore is 0.05-0.09 mm (0.002-0.003 inch) with maximum allowable clearance being 0.15 mm (0.006 inch). With camshaft supported in V-blocks at outer bearing journals, camshaft runout at center journals should be 0.01 mm (0.0004 inch) or less. Maximum allowable runout is 0.08 mm (0.003 inch).

Check bottom of cam followers (tappets) for abnormal wear and if noted, carefully inspect camshaft lobes for

Fig. K5-26—View showing crankshaft oil slinger, "O" ring and seal sleeve installation. Inset shows spacer collar used only on engines with crankshaft pulley retaining nut.

1. Injection pump gear
2. Idler gear
3. Crankshaft gear
4. Oil pump gear
6. Oil slinger
7. "O" ring
8. Oil seal sleeve
9. Spacer collar
10. Oil pump assy.

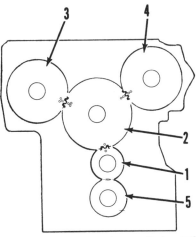

Fig. K5-27—Drawing showing gear timing marks. Oil pump drive gear (5) is not timed.

1. Crankshaft gear
2. Idler gear
3. Injection pump gear
4. Camshaft gear
5. Oil pump gear

Fig. K5-28—Checking oil pump outer rotor to housing clearance with feeler gage. Refer to text for specifications.

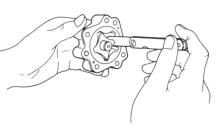

Fig. K5-29—Using feeler gage to check inner to outer rotor clearance. Refer to text for specifications.

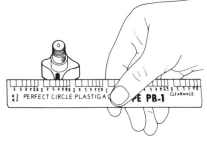

Fig. K5-30—Checking rotor end clearance using Plastigage. Refer to text for specifications.

Illustrations Courtesy Kubota

Kubota

SMALL DIESEL

scoring or abnormal wear pattern. Renew cam followers if renewing camshaft.

When installing camshaft gear on camshaft, first install retainer plate. Then heat gear to approximately 80° C (175° F) and push gear onto shaft, being sure that timing marks on gear are forward and that keyway aligns with key on camshaft. Install camshaft, retainer plate and gear with timing marks of gear and idler gear aligned as shown in Fig. K5-27. Complete reassembly of engine by reversing disassembly procedure.

INJECTION PUMP CAMSHAFT

To remove injection pump camshaft, first remove fuel injection pump and timing gear cover as previously outlined. Remove fuel feed pump from below fuel injection pump opening. Remove cap screws from camshaft bearing retainer plate and remove the plate. Remove cap screws from governor fork lever bracket (31—Fig. K5-31), then withdraw injection pump camshaft (30), gear (20), governor assembly (21 through 24) and governor fork (28 and 29) assembly from front of crankcase. Remove snap ring (19) and press camshaft out of gear and governor flyball assembly (21 through 24). Be careful not to lose any of the flyballs (21) or thrust bearing balls (23).

Inspect camshaft lobes, gear and bearings for excessive wear or other damage and renew parts as necessary. When reassembling camshaft, governor flyball unit and gear, stick thrust bearing balls in ball case (24) and to rear side of sleeve (22) with grease and place unit on camshaft. Stick flyballs (21) in bores on rear side of governor gear and press gear onto camshaft, aligning keyway in gear with key in camshaft. Install retaining snap ring (19).

When installing camshaft, position governor fork assembly on camshaft and install in crankcase. Leave governor fork lever bracket bolts loose until fuel injection pump camshaft and injection pump are installed. Be sure injection pump camshaft gear and idler gear timing marks are aligned as shown in Fig. K5-27. Adjust governor fork lever assembly so that pins on fork (29—Fig. K5-31) are centered on governor sleeve (22) and fork lever (29) does not bind against injection pump. Tighten the bolts and complete reassembly of engine by reversing disassembly procedure.

GOVERNOR

All models are equipped with a flyball type governor at back side of fuel injection pump gear as shown in Fig. K5-31. Governor balls (21) are carried in bores of rear face of fuel injection pump drive gear (20). As engine speed varies, the balls move against governor sleeve (22). Movement of governor sleeve actuates governor fork lever (29) which engages fuel injection pump control rack pin. Force of flyball movement through governor sleeve is balanced by governor springs (25 and 26).

Governor components are accessible after removing fuel injection pump camshaft as previously outlined. Inspect governor parts and renew any that are excessively worn or damaged. Be careful not to lose any of the thrust bearing balls (23) or governor flyballs (21) when disassembling fuel injection pump camshaft and governor unit.

PISTON AND ROD UNITS

Piston and connecting rods are removed from above as a unit after removing cylinder head, oil pan and oil pickup. Identify piston and rod units so they can be reinstalled in original positions if found to be serviceable.

Note that numbers stamped on rod and cap must be toward same side when cap is installed on connecting rod. "FW" mark on piston crown must be toward flywheel end of engine and numbers on rod and cap must be toward fuel injection pump side of engine. Lubricate threads on connecting rod cap screws and tighten to a torque of 15-18 N·m (11-14 ft.-lbs.).

CONNECTING RODS AND BEARINGS

Connecting rods are fitted with a renewable bushing (1—Fig. K5-32) at piston pin end and a precision shell type insert bearing (2) at crankpin end. Piston pin diameter is 25.002-25.011 mm (0.98433-0.98469 inch) and crankpin diameter is 46.959-46.975 mm (1.8488-1.8494 inches). Connecting rod side play on crankpin should be 0.4-0.6 mm (0.016-0.024 inch) with maximum allowable side play being 0.8 mm (0.031 inch).

Oil clearance between bushing and piston pin should be 0.015-0.039 mm (0.0006-0.0026 inch) with maximum al-

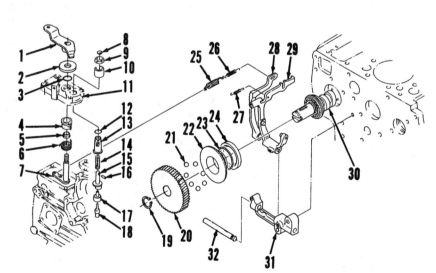

Fig. K5-31—Exploded view of governor components.

1. Throttle lever
2. Washer
3. "O" ring
4. Compensator lever
5. Spacer
6. Spring
7. Throttle arm
8. Cap
9. Locknut
10. Sleeve nut
11. Plate
12. "O" ring
13. Sleeve
14. Spring
15. Compensator cam
16. Pin
17. Thermostat housing
18. Thermostat
19. Snap ring
20. Fuel injection pump gear
21. Governor flyballs
22. Governor sleeve
23. Thrust bearing balls
24. Thrust collar
25. Governor outer spring
26. Governor inner spring
27. Start spring
28. Governor lever
29. Governor fork lever
30. Fuel injection pump cam
31. Governor lever bracket
32. Governor lever pin

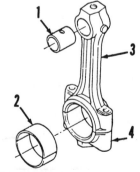

Fig. K5-32—Connecting rod is fitted with renewable piston pin bushing (1) and crankpin bearing insert (2).

1. Piston pin bushing
2. Crankpin bearing
3. Connecting rod
4. Rod cap

SERVICE MANUAL

Kubota

lowable clearance being 0.15 mm (0.006 inch). Oil clearance between crankpin bearing and crankpin should be 0.025-0.087 mm (0.0010-0.0034 inch) with maximum allowable clearance being 0.2 mm (0.008 inch).

When checking connecting rod for twist or misalignment, maximum allowable twist or bend is 0.05 mm (0.002 inch) per 100 mm (3.94 inches).

If installing new piston pin bushing, be sure oil hole in bushing is aligned with hole in rod, then hone bushing to inside diameter of 25.025-25.040 mm (0.98524-0.98583 inch) to obtain desired bushing to piston pin clearance.

PISTON, PIN AND RINGS

All models are equipped with cam ground aluminum pistons fitted with two compression rings and one oil control ring. Top compression ring is a keystone type ring. Pistons and rings are available in oversize of 0.50 mm (0.020 inch) as well as standard size. Piston pins are available only in standard size of 25.002-25.011 mm (0.9843-0.9847 inch). Piston pin bore is 25.000-25.013 mm (0.9843-0.9848 inch). When inserting pin, first heat piston to a temperature of approximately 80° C (175° F). Install retaining snap rings in piston pin bore grooves with sharp edge of snap rings out.

Piston ring side clearance in groove should be 0.093-0.120 mm (0.0036-0.0047 inch) for second compression ring and 0.020-0.052 mm (0.0008-0.0021 inch) for the oil control ring. Top compression ring is keystone type. Maximum allowable side clearance in groove is 0.2 mm (0.008 inch) for second compression ring and 0.15 mm (0.006 inch) for oil ring. Renew piston if ring side clearance is excessive.

Ring end gap should be 0.30-0.45 mm (0.012-0.018 inch) for compression rings and 0.25-0.45 mm (0.010-0.018 inch) for oil rings. Maximum allowable end gap is 1.25 mm (0.049 inch) for all rings. Place oil ring spring expander in groove with ends butted together, then install oil ring on top of expander with ends 180 degrees away from expander ends. Refer to Fig. K5-35 for installation of piston rings. Position ring end gaps equally spaced around piston.

CYLINDER LINER

All engines are equipped with dry type cylinder liners (sleeves). Standard cylinder inside diameter is 85.000-85.022 mm (3.3465-3.3473 inches) for Models D1402 and F2302, and 82.000-82.022 mm (3.2284-3.2292 inches) for Models D1302 and V1702. Renew or rebore cylinders to 0.50 mm (0.020 inch) oversize if scored or if wear exceeds 0.15 mm (0.006 inch) more than standard lower limit diameter.

Use suitable sleeve removal and installation tools to renew worn or damaged sleeves on engines so equipped. Thoroughly clean cylinder bore and outside of sleeve, then press sleeve into crankcase so top of sleeve is flush with top of crankcase. Hone sleeve after installation to final desired size.

CRANKSHAFT AND BEARINGS

The crankshaft is carried in a sleeve type bearing (1—Fig. K5-34) pressed into front bearing bore of crankcase and in insert type bearings (3) in bearing carriers (4, 5, 6 and 9) for intermediate (center) and rear main journals. Three-cylinder models have two center bearings, four-cylinder models have three center bearings and the 5-cylinder engine has four center bearings. Crankshaft end play is controlled by thrust washers (7) at front and rear of rear main bearing carrier (9). Before removing crankshaft, check crankshaft end play.

To remove crankshaft, remove timing gear cover, crankshaft timing gear, pis-

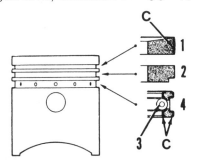

Fig. K5-33—Install piston rings after reference to ring cross-sectional view. Chrome ring surfaces are indicated by (C).

C. Chrome face
1. Keystone compression ring
2. Second compression ring
3. Oil ring expander spring
4. Oil ring

ton and rod units and flywheel. It is not necessary to remove flywheel housing. Refer to Fig. K5-35 and remove the rear oil seal retainer plate (1). Remove the cap screw from bottom of each center bearing web, then carefully remove crankshaft from rear of crankcase taking care not to damage front main bearing remaining in crankcase. Note that the center and rear main bearing carriers are a tight fit in cylinder block to prevent oil loss between oil passages in block and carriers.

Carefully identify main bearing carriers as to their position on the crankshaft to ensure correct reassembly. Main bearing carrier halves are not interchangeable. Unbolt and remove the bearing carriers from crankshaft. Note that crankshaft bearing thrust washers are located at rear main bearing carrier.

Crankshaft journal diameter is 51.92-51.94 mm (2.044-2.045 inches). Desired oil clearance of journals to main bearings is 0.040-0.118 mm (0.0016-0.0046 inch) for front journal and 0.040-0.104 mm (0.0016-0.0041 inch) for center and rear journals. Maximum allowable oil clearance is 0.2 mm (0.008 inch) for all journals. Crankshaft end play should be 0.15-0.31 mm (0.006-0.012 inch) with maximum allowable end play being 0.5 mm (0.020 inch). Crankpin diameter is 46.959-46.975 mm (1.8488-1.8494 inches) with desired bearing oil clearance of 0.025-0.087 mm (0.0010-0.0034 inch). Check crankshaft main bearing alignment by supporting shaft in V-blocks at front and rear journals and checking runout at a center journal with a dial indicator. Maximum allowable runout is 0.08 mm (0.003 inch).

Main bearings and crankpin bearings are available in undersizes of 0.2 and 0.4 mm (0.008 and 0.016 inch) and thrust bearings are available in oversize thickness of 0.2 and 0.4 mm (0.008 and 0.016 inch). Crankshaft may be reground undersize and with overwidth thrust sur-

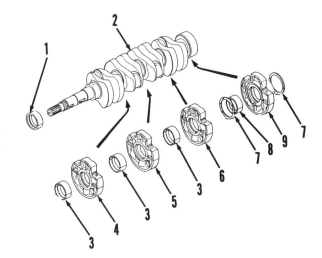

Fig. K5-34—Crankshaft front main bearing (1) is press fit in front face of cylinder block. Center and rear main bearings (3 and 8) are supported in bearing carriers (4, 5, 6, and 9) in 4-cylinder engines. Three-cylinder engines have two center main bearings and five-cylinder engine has four center main bearings.

1. Front main bearing
2. Crankshaft, 4-cylinder
3. Center main bearings
4. Bearing carrier
5. Bearing carrier
6. Bearing carrier
7. Thrust washers
8. Rear main bearing
9. Rear bearing carrier

Illustrations Courtesy Kubota

faces at rear main journal to accommodate undersize bearings and oversize thrust washers.

To renew front main bearing, use suitable driver to remove old bearing from crankcase and install new bearing with oil hole in bearing aligned with oil passage in cylinder block. Bearing should be 4.2-4.5 mm (0.165-0.177 inch) below flush with front face of crankcase and positioned with seam toward camshaft side of engine.

Install bearing inserts in bearing carriers and position thrust washers (7—Fig. K5-34) on rear bearing carrier (9), using light grease to hold them in place. Lubricate crankshaft and bearings and install carriers in correct position on crankshaft with face of carriers having bolt holes toward rear (flywheel) end of crankshaft. Tighten the center bearing cap bolts to a torque of 30-34 N·m (22-25 ft.-lbs.). Lubricate front main bearing and insert crankshaft and bearing supports into crankcase from rear taking care not to damage front main bearing.

Install new rear oil seal with lip to inside in rear seal retainer plate. Lubricate seal and carefully install retainer over rear of crankshaft with arrow mark up as shown in Fig. K5-35. Be sure to remove the two jackscrews from cover. Tighten cover cap screws to a torque of 10-11 N·m (7-8 ft.-lbs.). Install the center bearing support retaining cap screw(s) in cylinder block webs and tighten to a torque of 64-68 N·m (47-50 ft.-lbs.). Complete reassembly of engine by reversing disassembly procedure.

ELECTRICAL SYSTEM

ALTERNATOR. To check alternator no-load voltage, disconnect the lead wires from the alternator. Connect jumper lead from "F" terminal (Fig. K5-36) to "B" terminal, and from "E" terminal to ground "G." Place a voltmeter test leads "V" between "B" and "G" alternator terminals. Start engine and set alternator speed to 1050-1350 rpm. Disconnect battery ground cable and measure voltage. The alternator should develop 14 volts at 1050-1350 alternator rpm. Rated output cold is 25 amperes at 14 volts at alternator speed of 4000 rpm or less. Brush length new is 15.5 mm (0.61 inch); wear limit is 10 mm (0.394 inch).

VOLTAGE REGULATOR. Cut-in voltage is 4.5-5.8 volts. No-load regulated voltage is 13.8-14.8 volts. Point gap should be 0.30-0.45 mm (0.012-0.018 inch). With contacts open, resistance between terminals "IG" and "F", "L" and "E", and "B" and "L" should be zero ohms, and between "B" and "E" should be infinite ohms. With contacts closed, resistance between terminals "IG" and "F" should be approximately 11 ohms; between terminals "L" and "E" should be approximately 100 ohms; between terminals "B" and "L" should be zero ohms; and resistance between terminals "F" and "E" should be approximately 6 ohms. The resistance between terminals "N" and "E" should be 23 ohms.

STARTING MOTOR. Brush length, new, for Model F2302 is 19 mm (0.750 inch) with minimum brush length is 12.7 mm (0.500 inch). Brush length, new, for all other models is 15.5 mm (0.610 inch) with minimum wear length of 10 mm (0.394 inch).

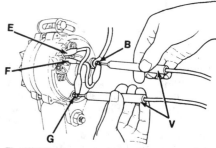

Fig. K5-36—Connect voltmeter test leads (V) between "B" terminal and ground "G" to check alternator no-load voltage. Refer to text.

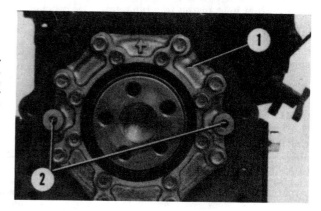

Fig. K5-35—Crankshaft rear oil seal is pressed into retainer plate (1) which also retains rear main bearing carrier (9—Fig. K5-34). Plate can be removed by installing jackscrews in threaded holes (2) after removing all 14 cap screws from plate.

SERVICE MANUAL

LISTER-PETTER

Lister-Petter

LISTER-PETTER
555 East 56 Highway
Olathe, Kansas 66061

Model	No. Cyls.	Bore	Stroke	Displ.
LT1	1	82.55 mm (3.25 in.)	76.2 mm (3.0 in.)	408 cc (25 cu. in.)
LV1	1	85.73 mm (3.37 in.)	82.55 mm (3.25 in.)	477 cc (29 cu. in.)
LV2	2	85.73 mm (3.37 in.)	82.55 mm (3.25 in.)	954 cc (58 cu. in.)

Lister diesel engine build type can be identified by the last two digits of engine number found on plate (Fig. L1) attached to air shield or fan shroud.

All models are four-stroke, vertical cylinder, air-cooled, direct injection diesel engines. Standard crankshaft rotation is counterclockwise from flywheel end. Metric fasteners are used throughout engine.

MAINTENANCE

STARTING PROCEDURE

Engine may be equipped with either a manual or electric starter. To start engine, refer to Fig. L2 on Models LT1 and LV1, rotate stop lever (A) fully clockwise to "excess fuel" start position. On Model LV2, an automatic excess fuel device is used. Excess fuel device is activated whenever stop lever is moved counterclockwise to "stop" position and then released. On models equipped with a variable speed control lever, move lever towards "fast" position. On all models, move compression release lever (B) towards the flywheel. On manual start engines, lightly oil end of starting shaft. Insert handle and crank engine. If equipped with electric start, crank engine with starter motor. When sufficient cranking speed is obtained, move compression release lever towards fuel tank and continue cranking until engine fires. Remove manual crank handle from the shaft.

CAUTION: Do not allow handle to continue to rotate on running shaft.

Turn stop lever back to "run" position (Models LT1 and LV1). Adjust speed as required if equipped with variable speed control.

LUBRICATION

Engines must use heavy-duty diesel lubricating oil that equals or exceeds API classification CC. If engine is operated under heavy load or high temperature, or if sulfur content in fuel exceeds 0.6 percent, API service classification CD oil is recommended. However, manufacturer does not recommend using CD classification oil in new or reconditioned engines until after the first oil change or 250 hours of operation. Multigrade oil may be used which meets specifications M1L-L46152-B or M1L-L2104C.

Recommended oil change interval is every 250 hours of operation with filter (if so equipped) renewed at same time. However, if engine is operated under heavy load and high temperature or over 3000 rpm, oil should be renewed after every 125 hours of operation. Crankcase capacity is 1.3 liters (1.4 quarts) for Model LT1 or 3.6 liters (3.8 quarts) for Models LV1 and LV2. Do not overfill.

If operating temperatures fall below −15°C (5°F), SAE 5W oil should be used. For temperatures between −15°C (5°F) and 4°C (40°F), use SAE 10W. If temperatures are between 4°C (40°F) and 30°C (85°F), SAE 20W-20 is recommended. For temperatures above 30°C (85°F), SAE 30 should be used.

Refer to Fig. L3 for a cut-away view of typical lubricating system.

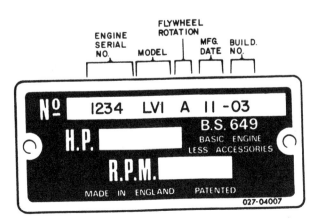

Fig. L1—View of typical engine identification plate. To ensure use of proper parts and specifications, use complete number.

Fig. L2—View of stop lever (A) and compression release lever (B).

Illustrations Courtesy Lister-Petter

Lister-Petter SMALL DIESEL

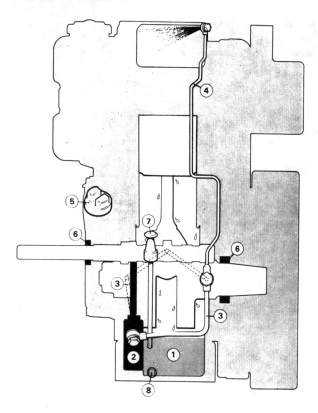

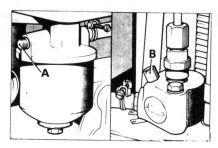

Fig. L3—Cut-away view of lubricating system.
1. Sump
2. Plunger type pump
3. Main bearing oil supply
4. Rocker arm oil supply
5. Filler plug
6. Oil seals
7. Dipstick
8. Drain plug

Fig. L4—Air bleed valves are located on fuel filter mounting bracket (A) and fuel pump (B). A self-bleeding system is used on LV1 engines and some LV2 engines.

Fig. L5—To time fuel pump, set control lever (A) to "START" position. Disconnect injector pipe and remove delivery valve holder (B), delivery valve (C) and spring (D). Refer to text for details.

FUEL SYSTEM

Number 2 diesel fuel is recommended for most conditions. Number 1 diesel fuel should be used for light duty or cold weather operation.

FILTER AND BLEEDING. Fuel filter element should be renewed after every 500 hours of operation. However, if engine shows signs of fuel starvation (surging or loss of power), renew filter regardless of hours of operation.

Fuel system is equipped with two air bleeder valves (Fig. L4). One valve is located on fuel filter mounting bracket and the other valve is on top side of fuel pump. A self-bleeding system is used in place of pump bleed valve on LV1 engines and some LV2 engines.

INJECTION TIMING. Bryce Berger fuel pump is mounted on side of engine between push rods. Two pumps are used on LV2 engines. Split shims at base of pump are used to adjust injection timing.

To check and adjust pump timing, set control lever to "Start" position. Turn flywheel until timing mark on flywheel is aligned with mark on fan shroud. Be sure both valves are closed on cylinder being timed. Disconnect fuel injector pipe at pump and injector, then remove delivery valve holder (B – Fig. L5), delivery valve (C) and spring (D). If fuel flows from pump, turn crankshaft until flow ceases. Reinstall delivery valve holder. Turn flywheel clockwise about ¼ turn, then slowly turn flywheel counterclockwise (normal direction of rotation) until fuel just stops flowing from delivery valve holder. At this point, beginning of injection occurs and timing marks on flywheel and fan shroud should be aligned. If not, shims below pump body must be added or removed to adjust timing. Add shims to retard timing, subtract shims to advance timing. When timing is correct, tighten pump mounting bolts to 9 N·m (80 in.-lbs.).

Reinstall delivery valve and spring and tighten holder to 54 N·m (40 ft.-lbs.). On LV2 engines, repeat procedure to check timing of second pump. Refer to the following chart for specified injection timing.

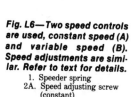

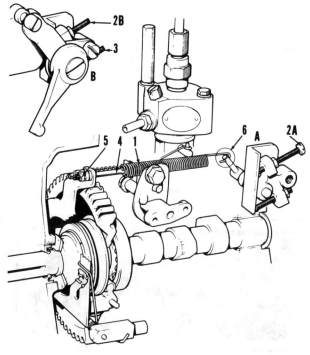

Fig. L6—Two speed controls are used, constant speed (A) and variable speed (B). Speed adjustments are similar. Refer to text for details.
1. Speeder spring
2A. Speed adjusting screw (constant)
2B. Idle speed screw (variable)
3. Maximum speed screw (variable)
4. Governor link
5. Link locknuts
6. Speed control lever pin

SERVICE MANUAL

Lister-Petter

LT1 Build No.	Speed Ranges Rpm	Degree BTDC
1,2,4	Up to 2500	24
3,5,6,7,8	2501 and over	27
10,27	Up to 1000	22
11,20	1001 to 1500	26
9,12,13,14, 16,17,19,21, 22,23,25, 26,28,	1501 to 3000	28
15,18,24	3001 to 3600	30

LV1 Build No.	Speed Ranges Rpm	Degree BTDC
9	Up to 1499	22
6	1500 to 1800	24
1,2,3,7,8	1801 to 3000	26
4,5	3001 to 3600	30

LV2 Build No.	Speed Ranges Rpm	Degree BTDC
6	1500 to 1800	24
1,2,3,4,7,8	1801 to 3000	26
5	3001 to 3600	30

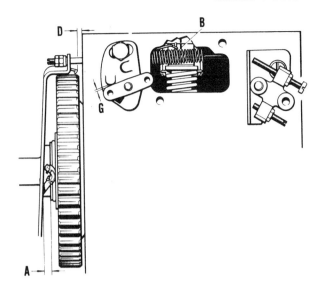

Fig. L7—To check or adjust governor, remove fuel pump inspection plate and timing gear end cover. Set distance (A) to 5.0 mm (0.197 inch) and align calibration mark (B). Governor lever stop to crankcase clearance (D) is 6.8-8.0 mm (0.236-0.315 inch). Refer to text for details.

GOVERNOR

To check or adjust governor settings, first remove fuel pump inspection cover and engine end cover. Place engine controls in "Start" position and tap camshaft gently towards flywheel to remove any end play.

Clearance (A–Fig. L7) between thrust sleeve and step on camshaft should be 5.0 mm (0.197 inch) with fuel pump calibration mark (B) aligned with mark on center of fuel pump body. Adjust by altering the length of governor link (C–Fig. L8). On LV2 engines, governor link must be adjusted so calibration marks (B–Fig. L7) are aligned in both pumps.

Set engine control lever to "Run" position. To be sure control lever on LV2 engines is in "Run" position, first turn lever to "Stop" and release. Then, move governor lever away from camshaft until excess fuel device clicks into "Run" position. With fuel pump calibration marks (B) aligned, loosen control lever plate retaining screw and move plate to obtain desired clearance (G–Fig. L7) as listed in Table 1 for load requirement and engine build number.

Excess fuel setting is adjusted by bending wire stop to obtain specified gap (D–Fig. L7) between governor lever stop plate and crankcase. With fuel pump calibration marks (B) aligned, gap (D) should be 6-8 mm (0.236-0.315 inch).

Engines may be equipped with either a constant speed or variable speed control. To adjust idle speed on variable speed control models, turn adjusting screw (2B–Fig. L6) counterclockwise

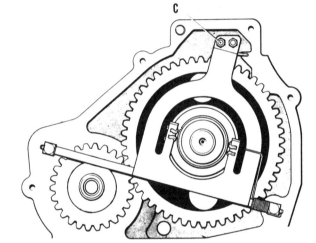

Fig. L8—To align injection pump calibration mark, adjust governor link at (C).

Table 1-Governor settings.

APPLICATION	LT1 Build No.	LV1 Build No.	LV2 Build No.	Clearance "G"
100% Load	—	9	—	0
	8	—	—	0.4 mm (0.016 in.)
	1,3,5	—	—	0.5 mm (0.020 in.)
	11,20	—	6	0.6 mm (0.024 in.)
	2,4	—	—	0.7 mm (0.028 in.)
	10,15,27	1,6,7	5	1.0 mm (0.040 in.)
	9,12,16 23,26,28	—	3,4	1.2 mm (0.047 in.)
90% Load	6,7,18,24	—	—	0.3 mm (0.012 in.)
	13,14,17, 19,21,22, 25	—	—	0.5 mm (0.020 in.)
	—	2,3,8	—	0.6 mm (0.024 in.)
	—	4,5	—	0.8 mm (0.031 in.)

Illustrations Courtesy Lister-Petter

Lister-Petter

until it no longer affects engine speed; governor will likely cause engine to "hunt". Turn screw clockwise until "hunting" stops and desired idle speed is obtained, then tighten locknut. Turning upper screw (2A) clockwise increases speed setting on constant speed engines.

The lower screw (3B) is used to limit the maximum speed on variable speed engines. Turning the screw clockwise will reduce maximum speed. Lower screw has no function on constant speed engines.

Governor weights and springs are serviced separately. For proper parts identification and adjustment, contact an authorized service department.

REPAIRS

TIGHTENING TORQUES

Refer to the following table for special tightening torques. Metric fasteners are used throughout engine.

```
Connecting rod:
 LT1.....................21 N·m
                       (15 ft.-lbs.)
 LV1, LV2...............24 N·m
                       (18 ft.-lbs.)
Cylinder head............41 N·m
                       (30 ft.-lbs.)
Flywheel................196 N·m
                      (145 ft.-lbs.)
Injection pipe nuts......28 N·m
                       (21 ft.-lbs.)
Injection pump............9 N·m
                       (80 in.-lbs.)
Injector clamp...........21 N·m
                       (15 ft.-lbs.)
Main bearing housing.....27 N·m
                       (20 ft.-lbs.)
Manifold upper nuts......21 N·m
                       (15 ft.-lbs.)
Manifold lower nuts.......9 N·m
                       (80 in.-lbs.)
```

VALVE ADJUSTMENT

To adjust valve clearance, turn crankshaft until piston is at TDC on compression stroke (both valves closed). Loosen locknut and turn rocker arm adjusting screw (23–Fig. L10) until appropriate size feeler gage can be inserted between rocker arm and valve stem end. Recommended clearance with engine cold is 0.13-0.18 mm (0.005-0.007 inch) for exhaust on all engines and 0.13-0.18 mm (0.005-0.007 inch) for intake on engines with operating speeds up to 3000 rpm or 0.05-0.10 mm (0.002-0.004 inch) on engines with operating speeds of 3001 to 3600 rpm. Repeat procedure for the other cylinder on LV2 engines.

SMALL DIESEL

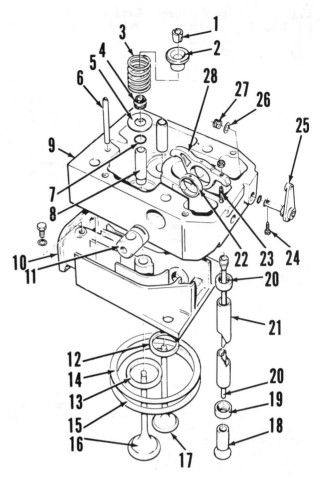

Fig. L10—Exploded view of cylinder head assembly used on LV1 and LV2 engines. LT1 engine is similar.

1. Valve keepers
2. Spring retainer
3. Valve spring
4. Valve seal (intake)
5. Retainer plate
6. Breather tube
7. "O" ring
8. Valve guide
9. Cylinder head (top)
10. Cylinder head (bottom)
11. Rocker arm stub shaft
12. Valve seat insert (exh.)
13. Valve seat insert (int.)
14. Shim
15. Head gasket
16. Intake valve
17. Exhaust valve
18. Tappet
19. Seals
20. Push rod
21. Push rod tube
22. Bushing
23. Adjusting screw
24. Adjusting screw
25. Compression release lever
26. Shim
27. Spring
28. Rocker arm

To adjust compression release lever, piston must be at TDC with both valves closed. With decompressor lever actuated, turn adjusting screw (24) until rocker arm just begins to depress valve, then turn screw clockwise one additional turn.

CYLINDER HEAD

A two-piece cylinder head is used. Top half is cast iron and holds the rocker arm assembly, valve springs, compression release lever and injector. Lower half is aluminum alloy and contains valve seats and a tapped hole in intake port for an oil priming cup used for cold starting. Valve guides are a press fit and hold head together. Refer to Fig. L10.

To remove cylinder head, first remove fuel tank, manifolds, air shield, cylinder head cover, injector clamp and injector. Remove cylinder head retaining nuts, then lift off cylinder head. Be sure to retain head gasket and shims.

To check piston-to-cylinder head clearance, place two pieces of lead wire approximately 1.2 mm (0.047 inch) thick on cylinder head as near as possible in line with piston pin. Hold wires in place with grease, then reinstall cylinder head and gasket. Tighten cylinder head nuts evenly to 41 N·m (30 ft.-lbs.). Turn piston past top dead center twice, then remove head and measure thickness of wires. Wire thickness (clearance) should be 0.71-0.79 mm (0.028-0.031 inch) for LT1 engines or 0.84-0.91 mm (0.033-0.036 inch) for LV1 and LV2 engines. Clearance is adjusted by installing shims (14–Fig. L10) between cylinder head and head gasket (15).

To reinstall cylinder head, reverse the removal procedure.

NOTE: No sealer is to be used on head gasket; however, a light coating of high melting point grease on the shims and cylinder head side of head gasket is recommended. Do not put grease on cylinder side of gasket. Thread sealer is recommended for cylinder head studs and nuts.

VALVE SYSTEM

The exhaust valve guide is a press fit into both halves of cylinder head. Intake guide is a press fit into the lower half. An "O" ring and retaining plate are located around top of intake guide. When renewing intake guide, "O" ring must be inserted into the top plate recess before pressing guide into head.

Illustrations Courtesy Lister-Petter

SERVICE MANUAL

Note that guides are marked "IN TOP" and "EX TOP" and must be installed accordingly with lettering on guide at the top. Using suitable tools, press guides in until they extend 12.35-12.60 mm (0.486-0.496 inch) above cylinder head surface on LT1 engine. On LV1 and LV2 engines, top of exhaust guide should be 13.15-13.40 mm (0.518-0.528 inch) above head surface and top of intake guide should project 10.15-10.40 mm (0.400-0.409 inch). After installation, check valve stem clearance in guide and ream guide if necessary. Desired operating clearance is 0.026-0.060 mm (0.001-0.002 inch) for intake and 0.043-0.076 mm (0.002-0.003 inch) for exhaust. Wear limit is 0.07 mm (0.003 inch) for intake and 0.09 mm (0.0035 inch) for exhaust.

Valves seat directly in cylinder head on LT1 engines while LV1 and LV2 engines are equipped with renewable valve seat inserts for both intake and exhaust. When renewing inserts, manufacturer recommends heating cylinder head to 150°C (300°F) and chilling inserts in dry ice before pressing insert into cylinder head counterbore.

Valve face and seat angles are 45° for intake and exhaust. Desired seat width is 1.65-2.29 mm (0.065-0.090 inch) for intake and 1.65-2.01 mm (0.065-0.079 inch) for exhaust. On LT1 and early LV1 and LV2 engines, intake valve should be recessed 0.84-1.15 mm (0.033-0.045 inch) below head surface and exhaust valve should be recessed 1.10-1.41 mm (0.043-0.055 inch). On late LV1 and LV2 engines, valve recession should be 1.00-1.31 mm (0.039-0.052 inch) for intake and 1.10-1.41 mm (0.043-0.055 inch) for exhaust. When servicing early LV1 and LV2 engines, manufacturer recommends grinding valve seats to obtain later engine valve recession values.

Valve spring free length should be 43.7-45.5 mm (1.720-1.791 inches). Renew spring if free length is 42.5 mm (1.673 inches) or less.

INJECTOR

To remove injector, first thoroughly clean injector and surrounding area. Remove cylinder head cover. Disconnect and remove high pressure fuel pipe and return pipe. Remove injector retainer clamp, then remove injector and seal washer.

When reinstalling injector, be sure to renew copper seal washer. Tighten retainer clamp screw to 21 N·m (15 ft.-lbs.).

WARNING: Fuel leaves the injector nozzle with sufficient force to penetrate the skin. When testing, keep yourself clear of nozzle spray.

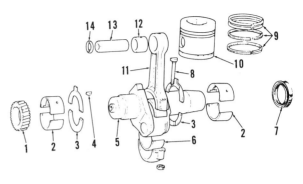

Fig. L11—Exploded view of crankshaft, connecting rod and piston used on LT1 engines.

1. Crankshaft
2. Main bearings
3. Thrust washers
4. Key
5. Crankshaft
6. Connecting rod bearings
7. Oil seal
8. Rod bolt
9. Piston rings (4)
10. Piston
11. Connecting rod
12. Piston pin bushing
13. Piston pin
14. Retaining ring

To check injector nozzle, connect to a hand test pump and slowly operate pump while observing pressure gage. Nozzle opening pressure for early LT1 engines should be 16210 kPa (2350 psi) for a new nozzle and 15200 kPa (2200 psi) for a used nozzle. On all other engines opening pressure should be 20265 kPa (2940 psi) for new nozzle and 19250 kPa (2790 psi) for a used nozzle. To check nozzle seat for leakage, operate tester to apply a pressure of about 12000 kPa (1750 psi). No leakage should appear at nozzle tip.

To check injector spray pattern, connect injector externally to injection pump using injector high pressure pipe. Set engine control to "Run" position and crank engine at about 60 rpm.

CAUTION: Make certain injector spray is directed away from your person.

If spray seems to be denser in one direction, the nozzle or pintle is dirty. Careful brushing of nozzle tip with a special brass brush will sometimes remove deposits causing the problem. But, if uneven spray continues, complete disassembly and cleaning will be necessary. Such repair should be done by an authorized service department.

INJECTION PUMP

All engines are equipped with a Bryce Berger fuel injection pump. On Model LV2, two pumps (one for each cylinder) are used. The injection pump should be tested and repaired by a shop qualified for diesel injection pump service.

To remove injection pump, first drain fuel from tank. Disconnect and remove fuel lines from pump. Remove fuel pump inspection covers from side of crankcase. Set pump operating lever in center position, then disconnect speeder spring and governor link from pump. Remove pump mounting bolts, then lift out pump. Retain adjusting shims (located under pump) for use in reassembly.

To reinstall pump, reverse the removal procedure. Be sure to reinstall correct shims. Tighten mounting bolts to 9 N·m (80 in.-lbs.). Check governor setting as outlined in GOVERNOR section and refer to INJECTION TIMING section to check and adjust pump timing.

PISTON, PISTION PIN, RINGS AND CYLINDER

A low expansion aluminum alloy piston with recessed combustion chamber is used in all engines. Piston is equipped with two compression rings and two oil control rings.

To remove piston without removing oil sump, proceed as follows: Remove fuel tank and cylinder head as previously outlined. Remove push rod guide and oil feed pipe. Identify air deflector shields so they can be properly reinstalled. Mark position of cylinder barrel to crankcase for use in reassembly, then withdraw cylinder barrel and air deflector shields. Remove piston pin retaining ring, push out pin and remove piston from connecting rod.

Thoroughly clean piston and cylinder barrel and check for scoring, wear or other damage. Piston diameter measured at bottom of skirt should be 82.398-82.423 mm (3.244-3.245 inches) for LT1 and 85.542-85.567 mm (3.3678-3.3668 inches) for LV1 and LV2 engines. Cylinder bore for engines with operating speeds up to 2500 rpm is 82.550-82.575 mm (3.250-3.251 inches) for LT1 engines and 85.750-85.775 mm (3.376-3.377 inches) for LV1 and LV2 engines. Piston clearance in cylinder bore should be 0.127-0.177 mm (0.005-0.007 inch) with maximum allowable wear clearance of 0.22 mm (0.0087 inch).

NOTE: Engines with operating speeds above 2500 rpm are equipped with cylinder barrels which provide an additional 0.025 mm (0.001 inch) piston-to-cylinder clearance over basic engine specifications. These cylinders are identified by a V-shaped notch in top cooling fin. The maximum piston-to-cylinder clearances given for basic engines should be increased by 0.05 mm (0.002 inch) and maximum ring end gaps increased by 0.16 mm (0.006 inch) for these engines.

Illustrations Courtesy Lister-Petter

Lister-Petter

SMALL DIESEL

Piston is fitted with four rings; a barrel faced, keystone type, chrome top ring; a taper faced second compression ring; a spring expander type oil scraper ring mounted above piston pin; and a slotted type oil scraper ring mounted below the piston pin. Piston ring end gap for basic engine (2500 rpm and below) should be 0.30-0.575 mm (0.012-0.022 inch) for top compression ring with a maximum allowable gap of 0.72 mm (0.028 inch). For all other rings, end gap should be 0.20-0.475 mm (0.008-0.018 inch) with a maximum wear limit of 0.67 mm (0.026 inch). Top piston ring groove is keystone type and a special gage is required to measure groove wear. All other ring grooves are rectangular and may be checked for wear by installing rings into grooves and measuring side clearance. Ring side clearance in groove for all rings should be 0.051-0.101 mm (0.002-0.004 inch) with a maximum wear clearance of 0.14 mm (0.0055 inch).

Piston pin diameter should be 28.5674-28.5725 mm (1.1247-1.1249 inches). Operating clearance in rod small end bushing is 0.0385-0.0556 mm (0.0015-0.0022 inch) while maximum allowable clearance is 0.064 inch (0.0025 inch). Pin is a push fit in piston and is retained by two snap rings.

When reassembling make certain piston is installed on connecting rod with the words "MANIFOLD SIDE" correctly positioned. Reinstall rings onto piston making certain second compression ring is installed with side marked "TOP" facing upward. Other rings may be installed either side up. Stagger ring end gaps 90° from each other. Complete installation by reversing the removal procedure.

CONNECTING ROD

A forged steel connecting rod is used with a precision, insert type bearing in the big end. Standard crankpin diameter is 53.977-53.990 mm (2.1251-2.1256 inches). Recommended rod bearing operating clearance is 0.023-0.067 mm (0.001-0.0026 inch) and maximum allowable clearance is 0.09 mm (0.0035 inch). Tighten connecting rod nuts to 21 N·m (15 ft.-lbs.) on Model LT1 and 24 N·m (18 ft.-lbs.) on Models LV1 and LV2.

Small end of rod is fitted with a copper faced bushing. Initial clearance for piston pin in bushing is 0.0385-0.0556 mm (0.0015-0.0022 inch) while maximum allowable clearance is 0.064 mm (0.0025 inch).

END COVER

The end cover (2 – Fig. L12) is bolted to gear end of engine and located by two dowels. The cover is fitted with an oil seal for the camshaft extension or crankshaft extension depending on engine type.

To remove end cover, first remove any driven equipment. Remove starter dog (2 – Fig. L13) on engines so equipped or spring pin (2 – Fig. L14) and starter han-

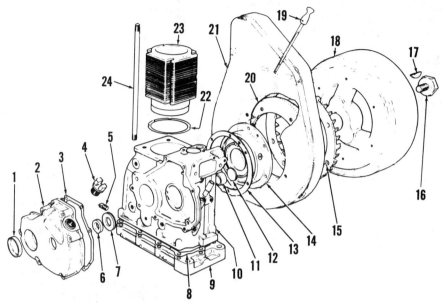

Fig. L12—Exploded view of cylinder block assembly for LT1 engines. Other models are similar.
1. Plug
2. End cover assy.
3. End cover gasket
4. Oil filler plug
5. Drain plug
6. Oil seal
7. Thrust washer
8. Sump gasket
9. Oil sump
10. Crankcase assy.
11. Camshaft bushing
12. Expansion plug
13. Shim
14. Main bearing housing
15. Fan
16. Flywheel screw
17. Key
18. Flywheel
19. Dipstick
20. Air seal
21. Fan shroud
22. Cylinder barrel gasket
23. Cylinder barrel
24. Stud

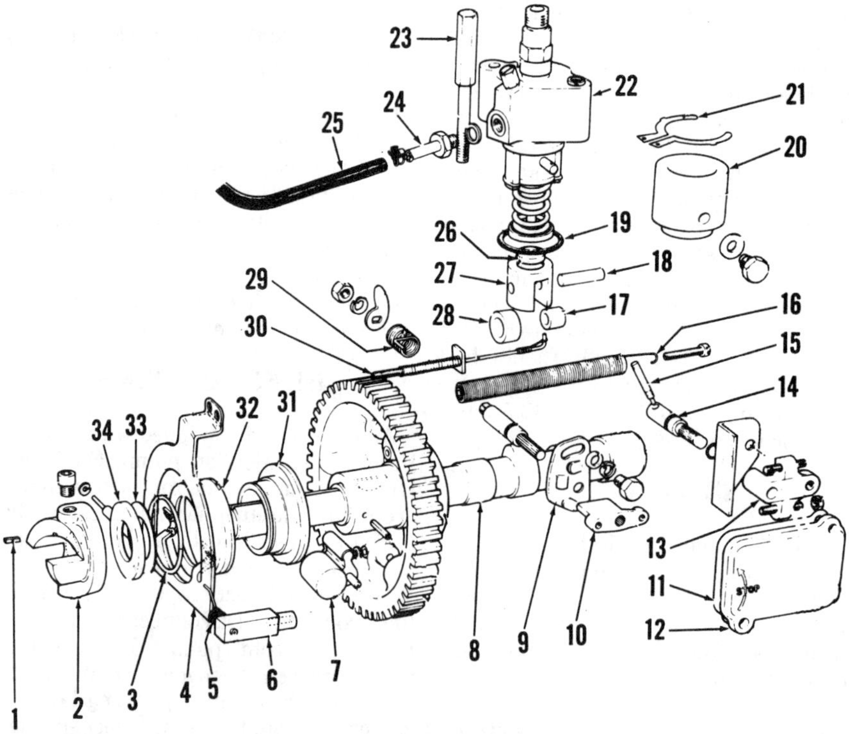

Fig. L13—Exploded view of camshaft, fuel pump and governor assemblies for LT1 engines.
1. Key
2. Starting dog collar
3. Spring ring
4. Governor lever assy.
5. Return spring
6. Pivot block
7. Governor weight
8. Camshaft assy.
9. Control plate
10. Control knob
11. Inspection door gasket
12. Fuel pump inspection door
13. Speed control lever
14. Speed control spindle
15. Speed control lever pin
16. Speeder spring
17. Roller bushing
18. Roller pin
19. Fuel pump "O" ring
20. Fuel pump tappet guide
21. Shims
22. Fuel pump
23. Fuel pump bolt
24. Inlet connector
25. Fuel pipe
26. Insert
27. Fuel pump tappet
28. Roller
29. Control lever spring
30. Governor link assy.
31. Governor sleeve
32. Thrust washer & pad
33. Shim
34. Thrust washer

SERVICE MANUAL

dle catch pin (3) from camshaft. Unbolt and remove end cover.

Inspect oil seal and renew if necessary. Be sure to protect oil seal lip when reinstalling cover. If cover must be renewed, the bores of the cover must be aligned with camshaft and crankshaft. Be sure shim (33–Fig. L13) and thrust washer (34) are properly positioned.

CAMSHAFT AND GOVERNOR

The camshaft is supported by a ball bearing at the gear end and a bushing sealed by an expansion plug at the flywheel end. On LV 2 engines, the camshaft is supported by a center bushing also. The governor weights are mounted on pins in the camshaft gear.

To remove camshaft, first remove fuel tank, cylinder head and end cover as previously outlined. Drain oil and remove sump, oil pump and push rod. Disconnect speeder spring (16–Fig. L13) and remove injection pump (22). Remove locating screw for fuel pump tappet guide (20) and remove tappet (27) and guide. Disconnect rocker arm oil supply pipe from crankcase. Remove retaining clips from push rod guides, then lift push rods and guides out. Lift off governor lever, link and speeder spring. Slide thrust sleeve assembly (31 and 32) off end of shaft. Lift and support tappets. Carefully withdraw camshaft, then remove tappets.

Inspect camshaft and tappets for excessive wear and other damage. Camshaft bearing journal diameter at flywheel end is 25.349-25.362 mm (0.9980-0.9985 inch). Recommended clearance in bushing is 0.038-0.051 mm (0.0015-0.002 inch) while maximum allowable clearance is 0.09 mm (0.0035 inch). On Model LV2, camshaft center bearing journal diameter is 36.942-36.995 mm (1.4544-1.4565 inches). Clearance in bushing should be 0.058-0.071 mm (0.0023-0.0028 inch) and maximum allowable clearance is 0.11 mm (0.004 inch). New bushings should be immersed in engine oil for four hours prior to installation. Special tools are available from manufacturer to properly remove and install bushings. Install new bushing with thinnest part of bearing wall (marked "O") towards the top. Apply sealer to recess in crankcase, then install new expansion plug.

To renew camshaft ball bearing, drive camshaft gear retaining pin into camshaft until it bottoms. Press gear off camshaft and remove Woodruff key, then press bearing off shaft. Drive spring pin out of camshaft. Press a new bearing onto camshaft. Install Woodruff key, then press gear onto camshaft until it bottoms against shoulder of shaft. Be

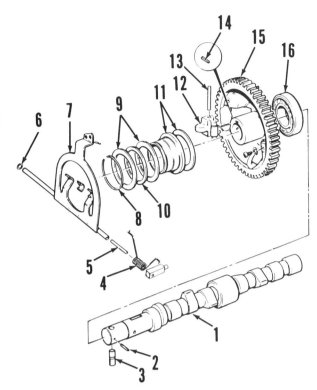

Fig. L14—Exploded view of camshaft and governor assembly for LV2 engines. LV1 engines are similar.

1. Camshaft
2. Pin
3. Starter catch pin
4. Spring
5. Pivot pin
6. Shim
7. Governor lever
8. Retaining ring
9. Thrust washers
10. Thrust pad
11. Sleeve assy.
12. Governor weight
13. Pivot pin
14. Spring pin
15. Camshaft gear
16. Ball bearing

sure gear retaining pin is driven in below surface of gear hub and any burrs on hub are removed, otherwise, operation of governor thrust sleeve will be affected.

To reinstall camshaft, first insert cam followers and support them in a raised position. Insert camshaft making sure "O" marks (Fig. L15) on camshaft gear and crankshaft gear are aligned. Install thrust sleeve assembly and governor lever assembly. Install shims (6–Fig. L14) as required to provide governor lever end play of 0.12-0.25 mm (0.005-0.010 inch). Protect oil seal in end cover, then reinstall end cover. Use a dial indicator to check camshaft end play which should be 0.08-0.20 mm (0.003-0.008 inch). Shims are available in different thicknesses to adjust end play, however, only one shim (33–Fig. L13) of appropriate thickness may be used on LT1 engines or two shims may be used on LV1 and LV2 engines. Complete installation by reversing the removal procedure.

CRANKSHAFT AND MAIN BEARINGS

The crankshaft is supported by two main bearings on LT1 and LV1 engines. An additional center main bearing is used on LV2 engines.

To remove crankshaft, remove fuel tank, oil sump, cylinder head, camshaft, flywheel and fan shroud. Remove cap screws securing main bearing housing, then use a suitable tool to push crankshaft out of crankshaft gear.

Remove main bearing housing. Retain shims for use in reassembly. On Model LV2, mark center main bearing housing and crankcase bore to make alignment easier during reassembly. Remove plug adjacent to dipstick and pull center bearing housing locating dowel from crankcase. On all models, carefully withdraw crankshaft from crankcase.

Inspect main bearings and crankshaft journals for excessive wear or other damage. Special tool is available from manufacturer for use in properly removing and installing main bearings. Crankshaft main bearing journal diameter when new is 53.955-53.967 mm (2.1242-1.1247 inches). Recommended operating clearance for main bearings is 0.034-0.089 mm (0.0014-0.0035 inch). Maximum allowable clearance is 0.12

Fig. L15—When installing camshaft or crankshaft gear, align "O" timing marks as shown.

Lister-Petter

SMALL DIESEL

mm (0.005 inch). Main bearings are two-piece inserts and are not interchangeable. Top half of bearing has an oil groove and hole which must be properly aligned. Undersize bearings are available.

Crankpin diameter should be 53.977-53.990 mm (2.1251-2.1256 inches). Desired rod bearing operating clearance is 0.023-0.067 mm (0.0009-0.0026 inch) while maximum allowable clearance is 0.09 mm (0.0035 inch).

Crankshaft end thrust is taken by steel-backed, copper-based, split thrust washers. Thickness of new thrust washer is 2.310-2.360 mm (0.091-0.093 inch). Renew thrust washer if thickness is less than 2.20 mm (0.087 inch).

When reassembling, note that center bearing housing (LV2) is marked "FLYWHEEL END" and must be installed accordingly. Align match marks on bearing housing and crankcase made during disassembly and install alignment dowel. On all models, be sure split thrust washers are installed with copper face towards crankshaft. Tighten main bearing housing mounting screws to 27 N·m (20 ft.-lbs.), then check crankshaft end play using a dial indicator. Desired end play is 0.13-0.23 mm (0.005-0.009 inch) for LT1, 0.178-0.254 mm (0.007-0.010 inch) for LV1 or 0.229-0.305 mm (0.009-0.012 inch) for LV2.

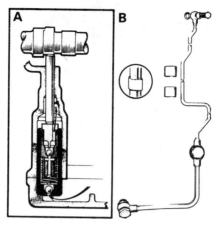

Fig. L16—A self-regulating plunger-type oil pump is used. An external pipe (B) carries oil to flywheel end main bearing and cylinder head. Gears, governor camshaft and underside of piston are splash lubricated.

End play is adjusted by installing shims (13–Fig. L12) between main bearing housing (14) and crankcase. Paper and metal shims are available. Note that paper shims must be installed between all metal surfaces for sealing purposes. Make certain the oil drain hole in main bearing housing is located at the bottom and protect oil seal lip when installing bearing housing. Heat crankshaft gear before reinstalling on crankshaft. Complete installation by reversing the removal procedure.

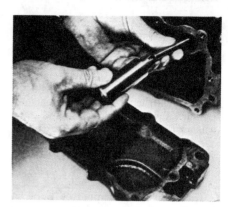

Fig. L17—To service oil pump, drain and carefully remove sump.

OIL PUMP

A self-regulating plunger type pump, operated by a push rod from the camshaft, is located in a drilling in the crankcase. Refer to Fig. L16. The pump is retained in the crankcase by the oil sump. The pump should deliver a minimum pressure of 70 kPa (10 psi) at 1000 rpm.

To service oil pump, drain oil and remove sump (Fig. L17). Withdraw oil pump assembly and renew if necessary.

SERVICE MANUAL

LISTER-PETTER

Lister-Petter

Model	No. Cyls.	Bore	Stroke	Displ.
ST1	1	95.25 mm (3.75 in.)	88.9 mm (3.50 in.)	0.63 liters (38.7 cu. in.)
ST2	2	95.25 mm (3.75 in.)	88.9 mm (3.50 in.)	1.27 liters (77.3 cu. in.)
ST3	3	95.25 mm (3.75 in.)	88.9 mm (3.50 in.)	1.90 liters (116 cu. in.)

Engines in this section are four-stroke, air-cooled type with vertical cylinders and direct fuel injection. Engine type can be indentified by the last two digits of engine number found on plate (Fig. L24) attached to fuel pump housing cover. Crankshaft rotation may be either clockwise or counterclockwise as viewed from flywheel end depending on engine build number.

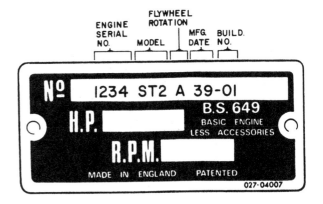

Fig. L24—View of typical engine identification plate. To ensure use of correct parts and specifications, use complete number.

MAINTENANCE

STARTING PROCEDURE

Engines may be equipped with either manual or electric start. In cases where final drive is from gear end, manual starting can be from flywheel end through a geared-up starter.

To start engine, pull control lever out and allow it to rotate to "Start" position. Move compression release levers toward flywheel. On engines equipped with variable speed control, move lever to "Fast" position. On manual start engines, lightly oil end of crank handle, then insert handle to crank engine. On all engines, crank engine slowly 3 to 20 turns to prime cylinders and lubrication system. After priming, crank engine until maximum speed is obtained and release compression release levers.

CAUTION: Do not allow crank handle to continue to rotate on running shaft.

When engine reaches normal speed, turn control lever to "Run" position and adjust speed as required.

NOTE: If engine is equipped with electric starter, do not crank engine more than 20 seconds at a time.

LUBRICATION

Under normal operating conditions, recommended engine oil is heavy-duty diesel lubricating oil that equals or exceeds API service classification CC. If engine is operated under heavy load and high temperature, or if sulfur content in fuel exceeds 0.6%, API service classification CD oil is recommended. However, manufacturer does not recommend using CD oil in new or reconditioned engines until after the first oil change or 250 hours of operation as lubricating characteristics of CD oil may inhibit running-in on engines operating under light load conditions.

Recommended oil change interval is every 250 hours of operation with filter change at the same time.

If operating temperatures fall below minus 15°C (5°F), SAE 5W oil should be used. For temperatures between minus 15°C (5°F) to 4°C (40°F) SAE 10W oil is recommended. From 4°C (40°F) to 30°C (85°F), use SAE 20W-20. Above 30°C (85°F), use SAE 30.

FUEL SYSTEM

Number 2 diesel fuel is recommended for most conditions. Number 1 diesel fuel should be used for light duty or cold weather operation.

FILTER AND BLEEDING. Manufacturer recommends renewing fuel filter after every 500 hours of operation. However, if engine indicates signs of fuel starvation (surging or loss of power), filter should be changed regardless of hours of operation. After servicing fuel filter, air must be bled from system in the following manner:

Make sure fuel tank is full. Loosen each bleeder valve (Fig. L26) one at a time on filter body and outlet banjo fitting. Retighten valves when air-free fuel flow is obtained. Loosen bleeder valve on each fuel pump (start nearest to fuel tank); tighten when air-free flow is obtained.

If engine fails to start at this point, loosen high pressure fuel line connections at injectors. With engine control lever in "Start" position, crank engine until fuel is discharged at loosened fittings. Retighten fuel line fittings and start engine.

INJECTION TIMING. A separate Bryce Berger fuel pump is used for each cylinder. It is recommended that all service of fuel pump(s), other than renewal, be carried out by an approved diesel injection service center.

To check injection pump timing, turn flywheel so "Z" mark (for cylinder/pump being checked) on flywheel is aligned with arrow on fan shroud. Refer to Fig. L27. Remove fuel pump delivery valve holder, delivery valve and spring. Reinstall delivery valve holder and set control lever in "Start" position. Turn crankshaft in opposite direction of normal rotation until fuel flows from delivery valve holder, then turn slowly in normal direction until fuel flow just stops. Beginning of injection occurs at this point and firing mark on flywheel

Illustrations Courtesy Lister-Petter

Lister-Petter

SMALL DIESEL

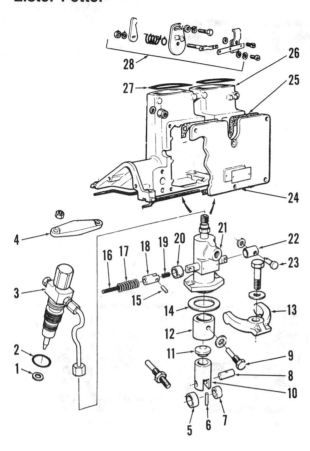

Fig. L25—Exploded view of typical fuel pump and housing assembly.

1. Injector gasket
2. Injector seal
3. Injector
4. Injector clamp
5. Roller
6. Retaining pin
7. Roller bushing
8. Roller pin
9. Tappet guide locating pin
10. Tappet
11. Tappet cap
12. Tappet guide
13. Fuel pump clamp
14. Shim
15. Pin
16. Fuel pump rod
17. Spring
18. Shackle
19. Spring
20. Spring clip
21. Fuel pump
22. Stop sleeve
23. Pin
24. Fuel pump housing cover
25. Gasket
26. Fuel pump housing
27. Gasket
28. Engine control assy.

should be aligned with mark on fan shroud if timing is correct.

If timing marks are not aligned, change thickness of shims (14—Fig. L25) between pump body and crankcase as necessary. To advance timing, remove shims; to retard timing, add shims. After correct timing is obtained, reinstall delivery valve and spring. Tighten delivery valve holder to 54 N·m (40 ft.-lbs.) and pump mounting screws to 9 N·m (80 in.-lbs.).

On multiple-cylinder engines, turn flywheel to firing position for next cylinder and repeat procedure for each pump.

FUEL INJECTOR. All engines are equipped with injectors using four spray hole nozzles. Diameter of spray holes is 0.25 mm (0.010 inch). To check injector spray pattern, remove injector and reconnect externally to fuel pump.

WARNING: Fuel emerges from injector with sufficient force to penetrate the skin. When checking injector, keep yourself clear of nozzle spray.

Crank engine about 60 rpm. Spray should be a very fine mist. All sprays should have the same appearance and length. If one hole is blocked or if fuel dribbles from nozzle tip, injector must be renewed or cleaned and rebuilt by an authorized service department.

Injector opening pressure setting is 19250-20270 kPa (2790-2940 psi). The leak down rate, measured with a hand test pump and gage, must drop from 15200 to 10135 kPa (2205 to 1470 psi) within 10 to 55 seconds.

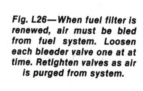

Fig. L26—When fuel filter is renewed, air must be bled from fuel system. Loosen each bleeder valve one at at time. Retighten valves as air is purged from system.

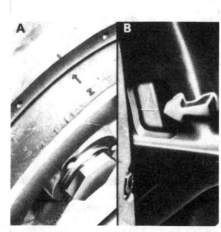

Fig. L27—When correctly timed, "Z" mark on flywheel should be aligned with mark on fan shroud at beginning of injection. View "A" is ST1 engine and View "B" is ST2 or ST3 engine.

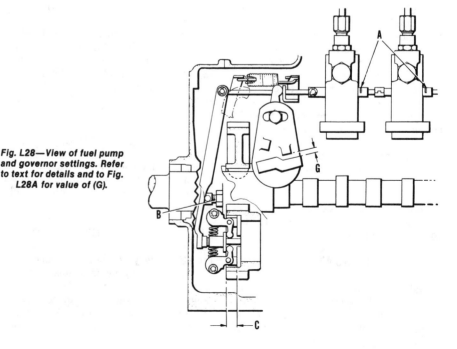

Fig. L28—View of fuel pump and governor settings. Refer to text for details and to Fig. L28A for value of (G).

SERVICE MANUAL

Lister-Petter

Fig. L29—Governor lever is carried in a fulcrum bearing in the crankcase.

Fig. L31—View of constant speed governor weight assembly.

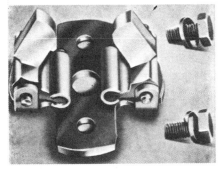

Fig. L32—Variable speed governor weight assembly used on ST engines.

GOVERNOR

Two types of flyweight governors are used: variable and constant speed. Governor assemblies are contained within the gear case. Governor weights and springs are serviced separately. To install correct weights and springs and adjust properly, contact an authorized service department.

Governor lever (Fig. L29) operating the fuel pump(s) is carried in a fulcrum bearing in the crankcase. The bearing extends approximately 19 mm (0.750 inch) from face of crankcase and is adjusted as follows:

Set engine control to "RUN" position and adjust linkage until calibration marks (A–Fig. L28) line up with sides of fuel pump bodies within 0.13 mm (0.005 inch). Pump racks must move freely after this adjustment. Adjust governor lever fulcrum (B) so that when calibration marks are in position, distance (C) is 12.7 mm (½-inch).

To set clearance (G–Fig. L28), maintain correct clearance and rotate locating plate until calibration marks (A) are in position. The full width of each mark must be visible. After making adjustments, pump racks and linkage must move freely.

The correct values of (G) are listed in Table 1.

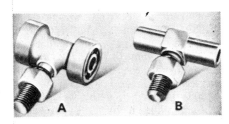

Fig. L30—Fulcrum bearing is fitted so bearing center line is approximately 19 mm (0.750 inch) from crankcase surface. Bearing (A) is used on three-cylinder engines; (B) is used on one- and two-cylinder engines.

REPAIRS

TIGHTENING TORQUES

Refer to the following table for special tightening torques. All fasteners are metric.

Connecting rod . . . . 43 N·m
  (32 ft.-lbs.)
Cylinder head . . . . 68 N·m
  (50 ft.-lbs.)
Crankshaft balance
  weights. . . . . 43 N·m
  (32 ft.-lbs.)
Flywheel. . . . . 405 N·m
  (300 ft.-lbs.)
Injector cap nut. . . . . 88 N·m
  (65 ft.-lbs.)
Injector clamp. . . . . 20 N·m
  (15 ft.-lbs.)
Rocker adjusting screw. . . . . . . 20 N·m
  (15 ft.-lbs.)

VALVE ADJUSTMENT

To adjust valve clearance, turn crankshaft until piston for cylinder being adjusted is at top dead center on compression stroke (both rocker arms loose). Loosen locknut and turn rocker arm adjusting screw (11–Fig. L33) until appropriate size feeler gage can be inserted between rocker arm and valve stem end. Recommended clearance with engine cold is 0.15-0.20 mm (0.006-0.008 inch) for intake and exhaust. On multiple cylinder engines, repeat procedure for each cylinder. Manufacturer recommends checking valve clearance after every 500 hours of operation.

To adjust compression release mechanism, proceed as follows: On engines with an oil filler hole in cylinder head cover, compression release is adjusted through this opening. Turn crankshaft until piston is at top dead center on compression stroke. Move lever to actuate compression release. Loosen locknut and turn adjusting screw down until it touches the rocker arm, then turn screw down an additional ¾ turn and tighten locknut. Repeat procedure for each cylinder.

When no oil filler hole is provided, adjust release lever screw (with cover removed) until it just touches rocker arm when cover is reinstalled and release lever is operated. Then, remove

Table 1—Governor settings.

Engine Speed Rpm	Clearance "G"	Movement of Rack corresponding to clearance
1200 - 2199	0.35 - 0.41 mm (0.014 - 0.016 in.)	1.14 - 1.29 mm (0.045 - 0.051 in.)
2200 - 2699	0.66 - 0.71 mm (0.026 - 0.028 in.)	2.11 - 2.26 mm (0.083 - 0.089 in.)
2700 - 3000	0.89 - 0.94 mm (0.035 - 0.037 in.)	2.84 - 3.0 mm (0.112 - 0.089 in.)
For engines driving fans or centrifugal pumps and marine engines, set as follows:		
1200 - 2199	*0.08 - 0.13 mm (0.003 - 0.005 in.)	0.23 - 0.38 mm (0.009 - 0.015 in.)
2200 - 2699	0.15 - 0.20 mm (0.006 - 0.008 in.)	0.51 - 0.66 mm (0.020 - 0.026 in.)
2700 - 3000	0.38 - 0.43 mm (0.015 - 0.017 in.)	1.22 - 1.37 mm (0.048 - 0.054 in.)
* All moisture Extraction Units should be set to this clearance.		

Lister-Petter — SMALL DIESEL

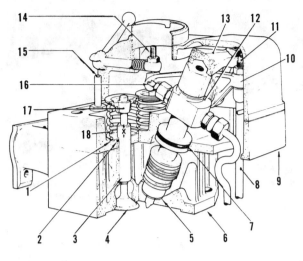

Fig. L33—Cross-sectional view of cylinder head assembly. When renewing valve guides, press in until top of guide extends 12.60-13.10 mm (0.495-0.515 inch) above surface of head (X).

1. Retaining plate
2. Valve guide seal
3. Intake valve guide
4. Intake valve
5. Injector gasket
6. Cylinder head
7. Injector seal
8. Push rod
9. Top plate
10. Rocker arm assy.
11. Rocker arm adjusting screw
12. Injector
13. Injector clamp
14. Compression release lever assy.
15. Breather tube
16. Leak-off pipe
17. Valve spring keepers
18. Valve spring

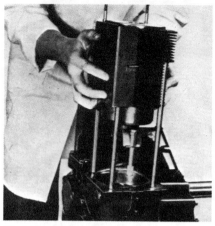

Fig. L34—When reinstalling piston, connecting rod and cylinder barrel, flat sided fins must face flywheel.

cover and turn screw down an additional ¾ turn.

After adjustment is completed, turn crankshaft by hand to make sure no interference exists between valve and piston when release lever is operated.

CYLINDER HEAD

A two-piece cylinder head is used. Top half (top plate) is cast iron and contains the valve gear and breather tube. Lower half is aluminum alloy and contains the valve seats. Valve guides are a press fit and hold the cylinder head halves together. Refer to Fig. L33.

Before removing cylinder heads, be sure to identify cylinder head position so heads can be reinstalled in their original location. As cylinder heads are removed, place shims and gasket from each cylinder with respective cylinder head for use in reassembly.

To check clearance between piston and cylinder head, place two pieces of soft lead wire 1.2 mm (0.048 inch) thick on head, clear of valves and as near as possible in line with piston pin. Hold wires in place with grease, then reinstall cylinder head and gasket. Tighten stud nuts evenly to specified torque, then crank piston past top dead center twice. Remove cylinder head and measure thickness of wire. Desired clearance is 0.89-0.97 mm (0.035-0.038 inch). To adjust clearance, install shims between cylinder head and head gasket.

During final assembly, a high melting point grease should be used to hold shims and head gasket in place. Apply grease sparingly and do not put grease on cylinder side of gasket. Manufacturer recommends that a thread sealer be used on cylinder head nuts, studs and area of top plate that contacts nuts. Before tightening cylinder head nuts on multiple cylinder engines, install the manifold or use a straightedge to ensure alignment of intake and exhaust ports. Then, tighten nuts evenly to 68 N·m (50 ft.-lbs.). Reinstall injectors after cylinder head nuts have been tightened. Complete installation by reversing the removal procedure.

VALVE SYSTEM

The valve guides are a press fit into the cylinder head. An "O" ring (2–Fig. L33) and retaining plate (1) are located around intake guide (3). Note that guides are not interchangeable. When installing new guides, use suitable tools and guide in from the top until top of guides extend 12.60-13.10 mm (0.495-0.515 inch) above surface of cylinder head casting. After guides are installed, check valve stem clearance in guide and ream guide if necessary. Desired operating clearance is 0.07-0.10 mm (0.003-0.004 inch). Wear limit is 0.13 mm (0.005 inch).

Valve face and seat angles are 45° for intake and exhaust. Desired seat width is 1.65-2.29 mm (0.065-0.090 inch) for intake and 1.35-1.78 mm (0.053-0.070 inch) for exhaust. With valve installed, top of intake valve should be 1.02-1.27 mm (0.040-0.050 inch) below combustion surface of cylinder head and exhaust valve should be 0.89-1.14 mm (0.035-0.045 inch) below head surface.

Rocker arm bushing bore should be 25.45-25.46 mm (1.0020-1.0025 inches) and rocker arm shaft diameter should be 25.36-25.37 mm (0.9985-0.9988 inch). Desired operating clearance is 0.08-0.10 mm (0.003-0.004 inch) with a wear limit of 0.20 mm (0.008 inch).

Intake and exhaust valve springs are identical. Spring free length should be 57.00-59.03 mm (2.244-2.324 inches). Renew valve springs if distorted, pitted or rusted.

PISTON, PIN, RINGS AND CYLINDER

The piston is equipped with five rings: one barrel faced, chrome, keystone type top ring; two tapered face compression rings; a spring expander type oil control ring above the piston pin and a slotted oil scraper ring below the piston pin. The full floating piston pin is retained in piston by two snap rings. The removable, cast iron cylinder is reborable for installation of oversize piston and rings.

When removing pistons and cylinders on multiple cylinder engines, mark cylinder number on piston, connecting rod and cylinder barrel. Components should be reinstalled in their original location if not renewed. To disassemble, remove cylinder head, air shields, crankcase side cover and lubricating oil pipes. Remove connecting rod cap, then rotate piston to TDC. Withdraw cylinder, piston and connecting rod as an assembly from crankcase. Retain all shims and gaskets with their respective cylinder for use in reassembly.

Thoroughly clean piston and cylinder, then check for scoring, wear and other damage. Piston standard diameter measured at bottom of skirt is 95.039-95.065 mm (3.7417-3.7427 inches) and standard cylinder bore is 95.27-95.30 mm (3.751-3.752 inches). Recommended clearance between piston skirt and cylinder is 0.211-0.262 mm (0.0083-0.0103 inch) with a wear limit of 0.35 mm (0.014 inch).

With piston ring positioned squarely in cylinder bore, measure ring end gap with feeler gage. Refer to the following table for recommended end gap dimensions.

Illustrations Courtesy Lister-Petter

SERVICE MANUAL

Lister-Petter

Fig. L35—To service camshaft, remove fuel pump(s), fuel pump tappet(s), tappet guide locating pin(s) and guide(s).

Fig. L37—Align timing marks (M) as shown when reinstalling timing gears.

CAMSHAFT

To remove camshaft, first remove cylinder head, fuel pump housing cover, crankcase side cover and end cover. If equipped with engine mounted fuel tank, disconnect fuel lines and remove fuel tank assembly. Disconnect speeder spring. Remove fuel pump(s), fuel pump tappet(s), tappet guide locating pin and guide (Fig. L35). Set oil pump to bottom of its travel and depress pump return spring (Fig. L36) until pump push rod is below level of camshaft bearing. Hold cam followers up with suitable tools, then carefully withdraw camshaft assembly.

Inspect camshaft and bushings for excessive wear and other damage. Camshaft journal diameter is 44.34-44.35 mm (1.7455-1.746 inches). Bushing inside diameter at flywheel end is 44.49-44.51 mm (1.7515-1.7525 inches). Desired operating clearance is 0.14-0.15 mm (0.0055-0.006 inch) and maximum allowable clearance is 0.23 mm (0.009 inch). Inner diameter of all other bushings is 44.53-44.56 mm (1.753-1.7545 inches). Desired clearance is 0.18-0.23 mm (0.007-0.009 inch) and maximum allowable clearance is 0.28 mm (0.011 inch).

When renewing camshaft bushings, soak bushings in engine oil for four hours, prior to installation. Using suitable tools, install bushings with thinnest part of bearing wall (marked with an "O") towards the top.

When reinstalling camshaft, reverse the removal procedure. Be sure to align "O" timing marks (M – Fig. L37) on camshaft gear and crankshaft gear.

CRANKSHAFT AND BEARINGS

The crankshaft is supported at gear end by split main bearing located in the crankcase and at flywheel end by a split bearing located in main bearing housing. One intermediate bearing is used on ST2 engines and two intermediate bearings are used on ST3 engines. The in-

Ring Gap	Desired	Maximum
Top	0.43-0.61 mm (0.017-0.024 in.)	0.89 mm (0.035 in.)
Compression	0.28-0.46 mm (0.011-0.018 in.)	0.76 mm (0.030 in.)
Scrapers	0.38-0.56 mm (0.015-0.022 in.)	0.76 mm (0.030 in.)

Piston pin diameter is 33.3324-33.3375 mm (1.3123-1.3125 inches). Recommended operating clearance in connecting rod bushing is 0.0435-0.0606 mm (0.0071-0.0024 inch) and maximum allowable clearance is 0.085 mm (0.0033 inch).

When reassembling, be sure piston is installed on connecting rod with the wording "CAMSHAFT SIDE" on top of piston on the same side as matching numbers stamped on connecting rod. One side of center compression rings is marked "TOP" and must be installed on piston accordingly. All other rings may be installed either side up. Lubricate piston and cylinder bore with clean oil, then install piston into cylinder. Assemble piston and cylinder to crankcase making certain cylinder is positioned with flat sided fins facing front and rear of engine and "CAMSHAFT SIDE" marking on piston is facing camshaft side of engine. Tighten connecting rod cap nuts to 43 N·m (32 ft.-lbs.).

CONNECTING ROD

The connecting rod is fitted with a bushing in the small end and a precision, insert type bearing in the big end.

Standard crankpin diameter is 63.487-63.500 mm (2.4995-2.500 inches). Recommended rod bearing clearance is 0.04-0.08 mm (0.0016-0.0032 inch) and maximum allowable clearance is 0.14 mm (0.0055 inch).

Piston pin diameter is 33.3324-33.3375 mm (1.3123-1.3125 inches) and bushing inner diameter is 33.881-33.393 mm (1.3142-1.3147 inches). Desired operating clearance is 0.0435-0.0606 mm (0.0017-0.0024 inch) and maximum allowable clearance is 0.085 mm (0.0033 inch).

Connecting rod and cap are marked with matching numbers. Make certain these numbers are on the same side and face camshaft side of engine. Tighten connecting rod cap nuts to 43 N·m (32 ft.-lbs.).

Fig. L36—To remove camshaft, set oil pump to bottom of its travel and depress pump return spring until pump push rod is below level of camshaft bushings.

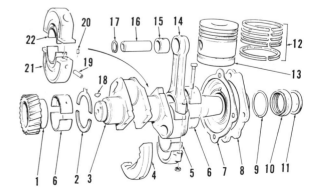

Fig. L38—Exploded view of typical crankshaft, connecting rod and piston assembly.

1. Crankshaft gear
2. Split thrust washer
3. Crankshaft assy.
4. Balance weight
5. Rod bearings
6. Main bearings
7. Shim
8. Main bearing housing
9. Oil slinger
10. Oil seal
11. Felt
12. Piston rings (5)
13. Piston
14. Connecting rod
15. Piston pin bushing
16. Piston pin
17. Retainer
18. Key
19. Locating dowel
20. Dowel
21. Center main bearing housing
22. Center main bearing

Illustrations Courtesy Lister-Petter

125

Lister-Petter SMALL DIESEL

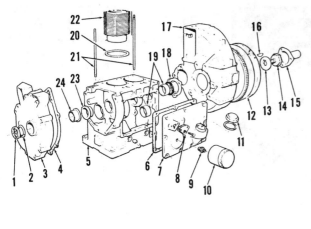

Fig. L39—Exploded view of typical crankcase assembly.
1. Oil seal
2. Oil slinger
3. End cover
4. End cover gasket
5. Crankcase assy.
6. Crankcase cover gasket
7. Crankcase cover
8. Plate
9. Union
10. Oil filter
11. Oil filler cap & seal
12. Flywheel assy.
13. Lockwasher
14. Flywheel retaining screw
15. Crankshaft extension
16. Key
17. Fan shroud
18. Camshaft end cover
19. Camshaft bushings
20. Cylinder barrel gasket
21. Cylinder studs
22. Cylinder barrel
23. Camshaft bushing
24. End cover bushing

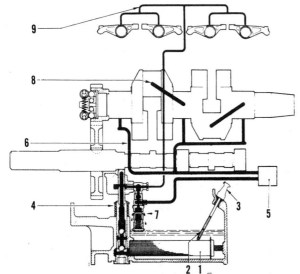

Fig. L40—Basic lubrication system on early model engines.
1. Screen
2. Oil level
3. Dipstick assy.
4. Oil pump assy.
5. Oil filter
6. Oil pipe to main bearings
7. Pressure relief valve
8. Oil passage to rod bearings
9. Oil pipe to rocker arms

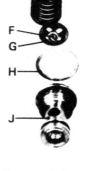
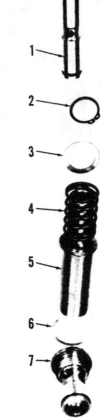

Fig. L41—A nonregulated oil pump (right) was used on early engines. The self-regulating pump (left) was introduced May 1978.

A. Push rod
B. Plunger cap
C. Ball valve
D. Plunger
E. Spring
F. Ball
G. Retaining plate
H. Washer
J. Plug

1. Push rod
2. Snap ring
3. Washer
4. Spring
5. Oil plunger assy.
6. Washer
7. Suction valve

termediate bearings are split type contained in a housing (21 – Fig. L38) which is located in crankcase by a hollow dowel (19). One end of dowel is threaded ¼ inch UNF so a bolt can be threaded into dowel for removal purposes.

End thrust is taken by split thrust washers at each end of crankshaft. Thickness of new thrust washers is 2.31-2.36 mm (0.091-0.093 inch). Main journal standard diameter is 63.47-63.50 mm (2.4988-2.500 inches). Desired main bearing clearance is 0.04-0.09 mm (0.0016-0.0035 inch) and wear limit is 0.16 mm (0.006 inch). Crankpin diameter is 63.487-63.500 mm (2.4995-2.500 inches). Desired rod bearing clearance is 0.04-0.08 mm (0.0016-0.0032 inch) and maximum allowable clearance is 0.14 mm (0.0055 inch).

When reassembling, be sure main bearings are installed with oil groove to the top and oil holes are aligned. Copper face of thrust washers must face crankshaft. Use grease to hold washers in place during assembly. On multiple cylinder engines, intermediate bearing housings are marked on one side "FLYWHEEL END" and must be installed accordingly. Align locating dowel hole in bearing housing with corresponding hole in side of crankcase. Install dowel with threaded end outward. Install main bearing housing then check crankshaft end play. Recommended end play is 0.13-0.25 mm (0.005-0.010 inch) for ST1 and ST2 engines or 0.18-0.30 mm (0.007-0.012 inch) for ST3 engines and is adjusted using shims (7 – Fig. L38) between bearing housing and crankcase. Apply sealing compound to bearing housing and shims during final assembly and make certain oil drain hole in housing is to the bottom. If crankshaft balance weights were removed, reinstall weights using new cap screws and tighten to 43 N·m (32 ft.-lbs.). Heat crankshaft gear before installing. Be sure to align timing marks on crankshaft gear and camshaft gear (Fig. L37). Tighten flywheel mounting screw to 405 N·m (300 ft.-lbs.).

OIL PUMP

Two types of plunger oil pumps are used. Early model pump requires a separate pressure regulator (Fig. L40). The self-regulating pump (A through J – Fig. L41) was introduced in ST1 engines in May 1978. By end of 1978, all multi-cylinder engines were equipped with self-regulating pumps.

To service oil pump, drain crankcase oil. Remove crankcase door and turn engine until oil pump push rod is at its highest point. On early models, compress return spring and remove snap ring. Loosen and remove suction valve (early type) or pump plug (late type) from bottom of crankcase.

NOTE: Pump assembly is under spring tension and care should be taken to prevent loss or damage to pump parts.

Clean and inspect all parts and renew any showing excessive wear or other damage. Reassemble by reversing disassembly procedure.

SERVICE MANUAL

LISTER-PETTER

Lister-Petter

Model	No. Cyls.	Bore	Stroke	Displ.
HR2	2	108 mm (4.25 in.)	114 mm (4.5 in.)	2089 cc (127.5 cu. in.)

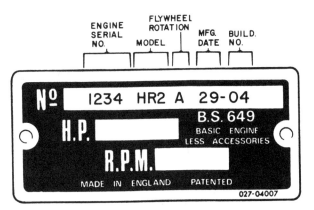

Fig. L70—View of typical engine identification plate used. To ensure use of proper parts and specifications, use complete number.

The HR2 is a four-stroke, air-cooled, direct injection diesel engine with vertical cylinders. Engine type can be identified by the last two digits of engine number found on plate attached to fuel pump; housing door (Fig. L70). Standard crankshaft rotation is counterclockwise as viewed from flywheel end.

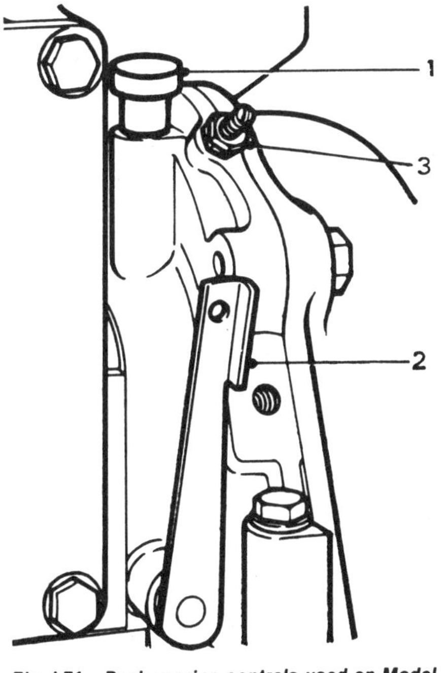

MAINTENANCE

STARTING PROCEDURES

HR engines have a manual crank start as standard equipment. Electric starters may be found on some engines.

Before starting, check fuel and oil levels. If engine is equipped with fuel primer pump, prime fuel filter by using lever on pump. Move compression release levers away from flywheel and crank engine slowly 3-10 times according to temperature and period of "down time" in order to prime oil system and combustion chambers. In cold weather, lift overload stop to allow pumps to deliver extra fuel (Fig. L71). After priming, crank engine until maximum speed is obtained and release compression release levers.

LUBRICATION

Under normal operating conditions, recommended engine oil is heavy-duty diesel lubricating oil that equals or exceeds API service classification CC. If engine is operated under continual heavy load and high ambient temperature, or if sulfur content in fuel exceeds 0.6%, API service classification CD oil is recommended. However, manufacturer does not recommend using CD oil in a new or reconditioned engine until after the first oil change or 250 hours of operation as lubricating properties of CD oil may inhibit running-in if engine is operating under light load conditions. Multigrade oil that meets MIL-L-2104B or MIL-L-2104C specifications may be used but is not approved for use in heavy-duty applications.

If operating temperatures fall below minus 15°C (5°F) SAE 5W oil should be used. From minus 15°C (5°F) to 4°C (40°F), use SAE 10W. For temperatures between 4°C (40°F) and 30°C (85°F), SAE 20W-20 is recommended. Above 30°C (85°F) use SAE 30.

Recommended oil change interval is every 250 hours of operation with filter renewal at the same time. Crankcase capacity is approximately 10.2 liters (10.8 quarts). Check oil level with dipstick; do not overfill.

Refer to Fig. L72 for a schematic view of lubrication system.

Fig. L71—Basic engine controls used on Model HR engines.
1. Overload trip
2. Control lever
3. Governor adjusting screw

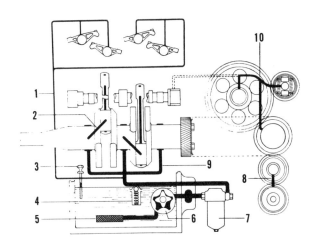

Fig. L72 – Diagram of typical lubrication system.
1. Oil pipe to rocker arm
2. Oil passage for rod bearings
3. Dipstick
4. Pressure relief valve
5. Pickup screen
6. Oil pump
7. Oil filter
8. Oil feed to idler gear
9. Oil pipe to main bearings
10. Oil pipe to gear train

Illustrations Courtesy Lister-Petter

Lister-Petter

SMALL DIESEL

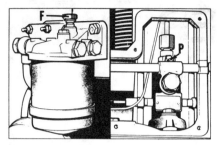

Fig. L73—To bleed air from fuel system, loosen bleed screw (F) on filter base and bleed screws (P) on fuel pumps.

Fig. L74—To check fuel pump timing, set control lever to "RUN" position and align timing marks on flywheel and fan shroud as shown.

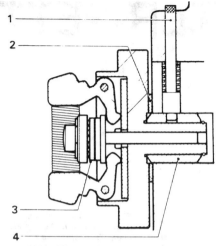

Fig. L76—Governor gear is driven directly by the camshaft and runs in a renewable bearing.
1. Spring-loaded dowel pin
2. Thrust washer
3. Thrust bearing & shaft assy.
4. Governor bearing

To check injection timing, turn the flywheel to firing position for cylinder/pump being checked (Fig. L74). Remove fuel injector pipe from pump. Remove delivery valve holder, delivery valve and spring, then reinstall delivery valve holder. Place control lever in "run" position. Note that fuel should flow from delivery valve holder at this time. Turn flywheel in opposite direction of rotation approximately ¼ turn slowly in normal direction until fuel just stops flowing from the delivery valve. At this point beginning of injection occurs and firing mark, FP on early models or Z on late models, on flywheel should be opposite arrow on fan shroud (Fig. L74) if timing is correct.

If timing marks are not aligned, change thickness of shims between pump body and crankcase as necessary. To advance timing, remove shims; to retard timing, add shims. Changing shim thickness 0.13 mm (0.005 inch) will change timing approximately 1½ crankshaft degrees. When timing is correct, reinstall delivery valve and tighten holder to 54 N·m (40 ft.-lbs.).

Turn flywheel to firing position for second cylinder and repeat timing procedure for the other pump.

FUEL SYSTEM

Number 2 diesel fuel is recommended for most operating conditions. Number 1 diesel fuel may be used for light duty or cold weather operation.

FILTER AND BLEEDING. Manufacturer recommends renewing fuel filter after every 1500 hours of operation or sooner if engine indicates signs of fuel starvation (surging or loss of power). After servicing fuel filter, air must be bled from system.

Whenever fuel filter is changed or fuel line opened, air must be bled from the system. To bleed system, fill fuel tank and remove fuel injection pump door. Loosen bleed screw (F–Fig. L73) on top of fuel filter body; tighten when air-free flow is obtained. Start with fuel injection pump nearest the filter and loosen bleed screws (P) one at a time. Tighten screw when air-free flow is obtained.

CAUTION: Care should be taken to prevent large flow of fuel into crankcase.

GOVERNOR

Two types of gear driven flyweight governors are used and are located in top of gear case. A constant or variable speed (Fig. L75) governor assembly rotates in a renewable bearing (4–Fig. L76) and is driven by the camshaft. The governor lever operates the fuel pumps. Governor weights and springs may be serviced separately. To ensure use of correct parts and adjustments, contact an authorized service department.

INJECTION TIMING. A separate Bryce Berger fuel injection pump is used for each cylinder. Fuel pumps are mounted on side of crankcase and are actuated by lobes on the engine camshaft. Injection timing is adjusted by changing thickness of pump mounting shim gaskets (18–Fig. L78).

Fig. L75—Two types of gear driven flyweight governors are used. A constant speed "A" or variable speed "B" governor is located at the top of the gear case.

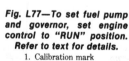

Fig. L77—To set fuel pump and governor, set engine control to "RUN" position. Refer to text for details.
1. Calibration mark
2. Shackle pin
3. Idling spring
4. Shackle
5. Adjusting sleeve
6. Connecting rod
7. Overload stop
8. Governor flyweight assy.
9. Rod
10. Overload trip
11. Speeder spring
12. Adjusting screw

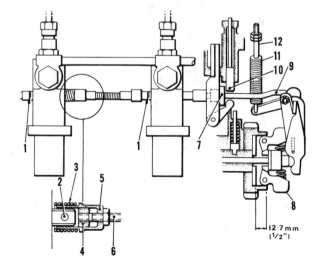

SERVICE MANUAL

Lister-Petter

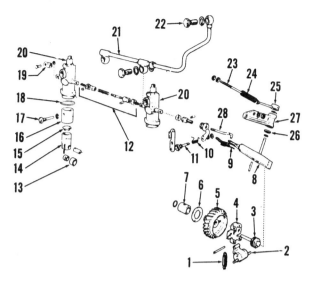

Fig. L78—Exploded view of fuel pump and governor assemblies.
1. Governor weight spring
2. Governor weight
3. Thrust bearing & shaft assy.
4. Weight carrier
5. Governor gear
6. Thrust washer
7. Bearing
8. Governor lever
9. Spring
10. Return spring
11. Stopping lever assy.
12. Rack assy.
13. Roller & bushing
14. Tappet
15. Tappet cap
16. Tappet guide
17. Locating pin
18. Shim
19. Stop sleeve
20. Fuel pump
21. Fuel line
22. Swivel union plug
23. Adjusting screw
24. Speeder spring
25. Governor lever
26. Ball bearing
27. Bracket

To calibrate fuel injection pump linkage and set governor, set engine control to "RUN" position. Adjust fuel injection pump linkage so calibration marks (1–Fig. L77) coincide with sides of pump bodies within 0.13 mm (0.005 inch). Rack must move freely after adjustment. Adjust overload stop (7–Fig. L77) so when it touches overload trip (10), calibration marks are aligned with edge of pump bodies.

To set governor, lift overload trip. Adjust rod (9) (remove cotter pin and screw in or out), with calibration marks aligned, so the distance between governor weight and bottom edge of thrust bearing is 12.7 mm (½ inch).

Engines equipped with variable speed governors use an adjustable idling spring (insert–Fig. L77). The spring rides against fuel injection pump and applies pressure on pump rack to eliminate "hunting" or surging. To adjust idling spring (3), main speeder spring (11) must be completely loosened. Rotate adjusting sleeve (5) until idle speed is about one-third rated engine speed. Main speeder spring should be adjusted so it just begins to increase engine speed.

REPAIRS

TIGHTENING TORQUES

Refer to the following table for special tightening torques. All fasteners are metric.

Connecting rod............92 N·m (68 ft.-lbs.)
Cylinder head............108 N·m (80 ft.-lbs.)
Flywheel............540 N·m (400 ft.-lbs.)
Fuel pump delivery valve holder............54 N·m (40 ft.-lbs.)
Injector cap nut............88 N·m (65 ft.-lbs.)
Injector clamp............20 N·m (15 ft.-lbs.)
Valve rocker cover............20 N·m (15 ft.-lbs.)

VALVE ADJUSTMENT

To adjust valve clearance, first remove cylinder head cover. Turn crankshaft until piston for cylinder being adjusted is at top dead center on compression stroke. Loosen locknut and turn rocker arm adjusting screw (17–Fig. L80) until appropriate size feeler gage can be inserted between rocker arm and valve stem end. Recommended clearance with engine cold is 0.38-0.43 mm (0.015-0.017 inch) for intake and exhaust. Repeat procedure for remaining cylinder. Manufacturer recommends adjusting valve clearance after the first 25 hours, 250 hours and every 500 hours thereafter until clearance remains constant. Then, adjustment interval can be extended to 1500 hours.

To adjust compression release mechanism (9–Fig. L79), proceed as follows: On engines with an oil filler hole in cylinder head covers, compression release is adjusted through these openings. Turn crankshaft until piston is at TDC on compression stroke. Move lever to actuate compression release. Loosen locknut and turn adjusting screw down until it touches the rocker arm, then turn screw an additional ¾ turn and tighten locknut. Repeat procedure for second cylinder.

When cylinder head cover does not have a filler opening, adjust compression release screw (with cover removed) until it just touches rocker arm when cover is reinstalled and release lever is operated. Then, remove cover and turn screw down an additional ¾ turn. Tighten locknut and reinstall cover. Repeat procedure for second cylinder. Tighten cylinder head cover nuts to 20 N·m (15 ft.-lbs.).

CYLINDER HEAD

Two-piece cylinder heads are used. Top half (top plate) is cast iron and contains the valve gear and breather tube. Lower half is aluminum alloy and contains the valve seat inserts. Valve guides are a press fit and hold the cylinder head valves together.

Be sure to identify cylinder heads before removal so they can be reinstalled in their original location. When heads are removed, place shims and gasket from each cylinder with respective cylinder head for use in reassembly. When servicing cylinder head, the two

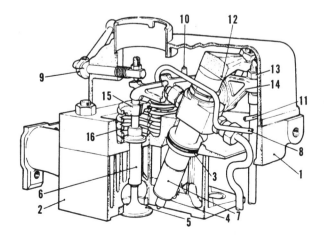

Fig. L79—Cross-sectional view of cylinder head assembly.
1. Top plate
2. Cylinder head
3. Injector seal
4. Injector gasket
5. Exhaust valve
6. Valve guide
7. Injector
8. Leak-off pipe
9. Compression release lever
10. Breather tube
11. Oil pipe to rocker arms
12. Injector clamp
13. Rocker arm adjusting screw
14. Rocker arm
15. Valve keepers
16. Valve spring

Illustrations Courtesy Lister-Petter

Lister-Petter

SMALL DIESEL

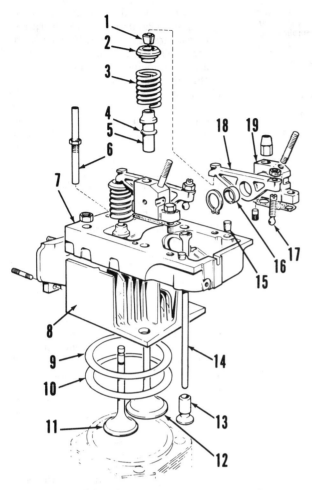

Fig. L80—Exploded view of cylinder head assembly.
1. Valve keeper
2. Retainer
3. Valve spring
4. Seal (intake only)
5. Valve guide
6. Breather tube
7. Top plate
8. Cylinder head
9. Shim
10. Head gasket
11. Exhaust valve
12. Intake valve
13. Cam follower
14. Push rod
15. Rubber plug
16. Bushing
17. Adjusting screw
18. Rocker arm
19. Rocker shaft bracket

rubber plugs (15—Fig. L80) in top face of head should not be removed.

Clearance between piston and cylinder head is adjusted using shims (9). To check clearance, place two pieces of soft lead wire 1.5 mm (0.062 inch) thick on head. Hold wires in place with a small amount of grease making sure they are clear of valves and in line with piston pin as near as possible. Reinstall head with gasket and tighten stud nuts to 108 N·m (80 ft.-lbs.). Crank piston past TDC twice, then remove cylinder head and measure thickness of wire. Desired clearance is 1.07-1.14 mm (0.042-0.045 inch). To adjust clearance, install shims (9) between cylinder head and the gasket (10).

During final assembly, a high melting point grease should be used to hold shims and head gasket in place. Use grease sparingly and do not apply grease on cylinder side of head gasket. Manufacturer recommends that a thread sealer be used on cylinder head nuts, studs and area of top plate in contact with nuts. Install nuts with grade symbol facing upward. Before tightening cylinder head nuts, install manifold or use a straightedge to ensure alignment of intake and exhaust ports. Then, tighten nuts evenly to 108 N·m (80 ft.-lbs.). Reinstall injectors after cylinder head nuts have been tightened. Complete installation by reversing the removal procedure.

VALVE SYSTEM

The valve guides are a press fit into the cylinder head valves. A rubber seal ring (4–Fig. L80) is used on the intake guide. The guides are not interchangeable. When renewing guides, apply sealing compound to outside of guide and use suitable installing tool to prevent damage to guide. Valve guide inner diameter should be 91.5-9.53 mm (0.3745-0.3753 inch). Valve stem diameter is 9.44-9.46 mm (0.3718-0.3723 inch). Desired clearance for valves in guides is 0.05-0.09 mm (0.0022-0.0037 inch) with a wear limit of 0.13 mm (0.005 inch).

Desired valve seat width is 2.78-3.20 mm (0.107-0.127 inch) for intake and 2.54-2.89 mm (0.100-0.114 inch) for exhaust. With valves installed, top of intake valve should be 0.96-1.27 mm (0.038-0.050 inch) below combustion surface cylinder head and exhaust valve should be 0.84-1.14 mm (0.033-0.045 inch) below head surface.

Rocker arm bushing bore should be 25.40-25.41 mm (1.000-1.0005 inch) and rocker shaft diameter should be 25.36-25.37 mm (0.9985-0.9988 inch). Desired operating clearance is 0.02-0.05 mm (0.001-0.002 inch) with a wear limit of 0.08 mm (0.003 inch).

Intake and exhaust valve springs are identical. Spring free length should be 56.95-59.13 mm (2.242-2.328 inches). Renew valve springs if distorted, pitted or rusted. Manufacturer recommends renewing valve springs after 6000 hours of operation.

INJECTOR

To remove injector, remove cylinder head cover and disconnect fuel lines. Plug all openings to prevent entry of dirt. Remove injector clamp and withdraw injector with seal ring and washer.

When reinstalling injector, renew copper washer and seal ring. Tighten injector clamp to 20 N·m (15 ft.-lbs.), then tighten injector high pressure line.

WARNING: Fuel leaves the injector nozzle with sufficient force to penetrate the skin. When testing, keep yourself clear of nozzle spray.

To check injector, connect to a hand test pump and operate tester lever a few quick strokes to purge air from injector and to make sure nozzle valve is not stuck. Then, operate tester lever slowly while observing tester gage. Opening pressure should be 17225-18235 kPa (2500-2645 psi). A plug in end of nozzle cap nut can be removed for access to pressure adjusting screw.

The injector nozzle has four spray holes each 0.27 mm (0.010 inch) in diameter. Spray pattern should be symmetrical and well atomized. If spray holes are blocked, injector should be overhauled or renewed.

If injector is disassembled, do not use wire brush, emery cloth or hard metal tools to clean injector. If more than one injector is disassembled at the same time, keep parts from each injector separated as they should not be interchanged. Nozzle spray holes may be cleaned by inserting a cleaning wire slightly smaller in diameter than the orifice hole diameter. Be sure nozzle needle slides freely in nozzle valve body. If needle sticks, reclean or renew nozzle valve assembly.

When reassembling injector, rinse all parts in clean diesel fuel and assemble while wet. Tighten nozzle holder nut to 88 N·m (65 ft.-lbs.). Retest injector as previously outlined.

INJECTION PUMPS

The Bryce Berger fuel injection pumps are located in a housing which is mount-

SERVICE MANUAL

Lister-Petter

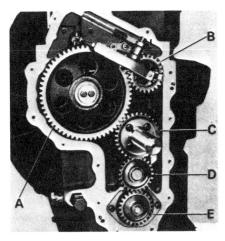

Fig. L81—Basic gear train used on HR engines. Initial backlash between gears is 0.02-0.23 mm (0.001-0.009 inch) with maximum backlash of 0.30 mm (0.0123 inch).

A. Camshaft gear
B. Governor gear
C. Crankshaft gear
D. Idler gear
E. Oil pump gear

ed to top of crankcase. To remove pumps, first shut off fuel supply or drain fuel tank. Remove pump housing cover. Disconnect fuel supply pipes and injector high pressure pipes. Disconnect linkage between pumps. Remove pump retaining clamps and lift out injection pumps. Keep adjusting shims with pumps as they are removed for use in reassembly.

It is recommended that fuel pumps be tested and serviced only by a shop qualified in diesel fuel injection repair.

To reinstall pump, reverse the removal procedure using original mounting shims. Be sure pump control racks move freely. Check pump timing as outlined in INJECTION TIMING paragraph. Adjust governor linkage as outlined in GOVERNOR section.

PISTON, PIN, RINGS AND CYLINDER

The piston is equipped with a barrel faced, chrome, keystone type top ring; two tapered face, rectangular center compression rings; a spring expander type oil control ring above the piston pin and a slotted oil scraper ring below the piston pin. The full floating piston pin is retained in piston by two snap rings.

NOTE: Mark cylinder number on piston, connecting rod and cylinder barrel prior to removal. All components should be reinstalled in their original location if not renewed.

To disassemble, remove oil sump and cylinder heads. Remove cooling air shrouds. Remove crankcase side cover with fuel transfer pump (if so equipped) and lubricating oil pipes. Remove connecting rod caps. Withdraw cylinder, piston and connecting rod as an assembly. Retain all shims and gaskets with their respective cylinder for use in reassembly.

Thoroughly clean piston and cylinder, then check for scoring, wear and other damage. Piston diameter at bottom of skirt should be 107.76-107.78 mm (4.2425-4.2435 inches) and the cylinder bore should be 108.03-108.05 mm (4.253-4.254 inches).

Piston top ring groove is a keystone type and a special gage is required to check groove wear. The groove width for center compression rings is 3.28-3.30 mm (0.129-0.130 inch). Recommended side clearance between compression rings is 3.28-3.30 mm (0.129-0.130 inch). Recommended side clearance between compression rings and piston grooves is 0.10-0.15 mm (0004-0.006 inch) with a wear limit of 0.20 mm (0.008 inch). Piston groove width for both oil scraper rings is 4.81-4.84 mm (0.1895-0.1905 inch). Side clearance for both oil scraper rings should be 0.05-0.10 mm (0.002-0.004 inch) with a wear limit of 0.15 mm (0.006 inch).

With ring positioned squarely in cylinder bore, measure ring end gap with a feeler gage. Refer to following table for recommended end gap dimensions.

Ring Gap	Desired
Top	0.66-0.91 mm (0.026-0.036 in.)
Maximum	1.14 mm (0.045 in.)
Compression	0.48-0.66 mm (0.019-0.026 in.)
Maximum	0.96 mm (0.038 in.)
Top scraper	0.66-0.91 mm (0.026-0.036 in.)
Maximum	1.22 mm (0.048 in.)
Bottom scraper	0.53-0.71 mm (0.21-0.028 in.)
Maximum	0.96 mm (0.038 in.)

Piston pin diameter is 39.684-39.689 mm (1.56235-1.56255 inch). Desired operating clearance in connecting rod bushing is 0.021-0.047 mm (0.0009-0.0018 inch) while maximum allowable clearance is 0.076 mm (0.003 inch).

When reassembling, be sure piston is installed on connecting rod with the wording "CAMSHAFT SIDE" on top of piston on the same side as "double" cylinder numbers stamped on connecting rod. One side of center compression rings is marked "TOP" and must be installed accordingly. Lubricate piston and cylinder bore with clean oil, then install piston into cylinder. Assemble piston and cylinder assembly onto crankcase making certain V-notch on top fin of cylinder barrel is towards flywheel end of engine and "CAMSHAFT SIDE" marking on piston and "double" cylinder numbers on connecting rod are all facing camshaft side of engine. Tighten connecting rod cap nuts to 92 N·m (68 ft.-lbs.).

CONNECTING ROD

A forged steel connecting rod is used with a precision, insert type bearing in the big end and a bushing in the small end.

Standard crankpin diameter is 69.82-69.83 mm (2.7488-2.7492 inches). Recommended rod bearing clearance is 0.05-0.09 mm (0.0020-0.0035 inch) and maximum allowable clearance is 0.14 mm (0.0055 inch).

Piston pin diameter is 39.684-39.689 mm (1.5624-1.5625 inches) and small end bushing inner diameter should be 39.71-39.73 mm (1.5634-1.5642 inches). Recommended operating clearance is 0.021-0.047 mm (0.0009-0.0018 inch) with a wear limit of 0.076 mm (0.003 inch).

Connecting rod and cap are marked with double cylinder numbers (11 and 22). Make certain these numbers are on the same side and face camshaft side of engine. Tighten connecting rod cap nuts to 90 N·m (68 ft.-lbs.).

TIMING GEARS AND END COVER

Various styles of end covers are used depending on engine application. The oil seal for the crankshaft extension is located in the end cover. Be sure to lubricate the lip of a new oil seal and protect seal when reinstalling cover.

The governor assembly and drive gear (B–Fig. L81) can be removed as a unit by first removing governor lever, then lift the spring loaded dowel pin (1–Fig. L76) above the bearing and withdraw complete assembly from crankcase. When reinstalling, make sure large hole in bearing (4) is towards the top and thrust washer (2) is positioned between the drive gear and crankcase.

The camshaft gear (A–Fig. L81) is retained by two screw cap screws, a locating plate and Woodruff key. The crankshaft gear (C) is retained by four cap screws and located on crankshaft by

Lister-Petter SMALL DIESEL

Fig. L82—Align timing marks (M) as shown when reinstalling timing gears.

Fig. L83—To service camshaft, remove fuel pumps, fuel pump tappets, tappet guide locating pins and guides.

two dowels. Be sure to align timing marks (M–Fig. L82) when reinstalling.

The oil pump idler gear (D–Fig. L81) is retained by a snap ring and thrust washer. The idler stub shaft is a press fit in crankcase and is also retained by a split pin inside the crankcase. The oil pump and gear (E) may be removed as an assembly after removing two mounting cap screws. Oil pump gear is retained by a split pin on oil pump shaft.

CAMSHAFT

To remove camshaft, remove cylinder heads, fuel pump housing cover, crankcase side cover and gearcase end cover. Release speeder spring, disconnect and remove governor link assembly. Disconnect fuel lines and remove fuel pumps and injectors. Remove fuel pump tappet guide locating pins and lift out tappet assemblies (Fig. L83). Remove oil supply pipe plug from camshaft locating bushing. Lift cam followers and carefully withdraw camshaft assembly and locating bushing.

The camshaft is carried in bronze bushings at the flywheel end and in crankcase dividers. A cast iron bushing (22–Fig. L84) is used at the gear end and is secured by oil supply line plug. All bushings, except locator bushing, are a press fit. Bronze bushing bores should be 44.40-44.42 mm (1.7483-1.7487 inches). Diameter of camshaft journals is 44.34-44.35 mm (1.7455-1.7460 inches). Desired operating clearance is 0.06-0.08 mm (0.0023-0.0032 inch) with a wear limit of 0.13 mm (0.005 inch). Inside diameter of locating bushing should be 34.92-34.95 mm (1.3750-1.3762 inches) and camshaft journal diameter is 34.86-34.87 mm (1.3725-1.3730 inches). Desired operating clearance is 0.05-0.09 mm (0.002-0.0037 inch) while maximum allowable clearance is 0.13 mm (0.005 inch).

When renewing bushings, soak bronze bushings in engine oil for four hours before installing. Install bushings with thinnest part of bearing wall (marked with a dot) towards the top. Install cast iron locating bushing with locating hole to the top. Camshaft end play should be 0.38 mm (0.015 inch).

Timing marks on camshaft and crankshaft gears should be aligned as shown in Fig. L82. Fuel pump timing should be checked as described in FUEL SYSTEM section.

CRANKSHAFT AND BEARINGS

The crankshaft is supported in split bushings at both ends of shaft. A center bearing is mounted in a split bearing carrier (1 and 2–Fig. L84) and is located in crankcase by hollow dowel pin.

To remove crankshaft, drain oil and remove oil sump. Remove cylinder heads, pistons and cylinders, timing gear cover (21–Fig. L85), crankshaft extension (22) and crankshaft gear, flywheel and fan shroud (10). Disconnect and remove main bearing lubricating oil pipes. Remove bearing housing (3) with shims and thrust washer. Thread a cap screw into the end of center bearing carrier locating dowel and withdraw dowel. Remove crankshaft assembly from crankcase.

Inspect crankshaft and bearings for excessive wear or other damage. If a bearing failure has occurred, be sure drilled oil passages in crankshaft are clean before reassembling.

Crankshaft main journal diameter when new is 82.54-82.55 mm (3.2496-3.250 inches). Main bearing bore should be 82.59-82.63 mm (3.2516-3.2531 inches). Recommended main bearing clearance is 0.04-0.09 mm (0.0016-0.0032 inch) and maximum allowable clearance is 0.15 mm (0.006 inch).

Crankpin diameter when new is 69.82-69.83 mm (2.7488-2.7492 inches). Recommended rod bearing clearance is 0.05-0.09 mm (0.0020-0.0035 inch) with a wear limit of 0.14 mm (0.0055 inch).

Crankshaft end thrust is taken by copper-faced split thrust washers. Thickness of new thrust washer is 3.10-3.17 mm (0.122-0.125 inch).

When renewing main bearings, make certain bearing shell with groove is to the top and oil holes are aligned.

To reassemble, proceed as follows: Be sure copper face of split thrust washer is towards crankshaft and tab is towards

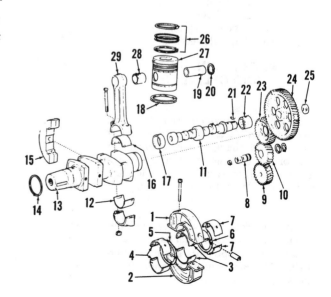

Fig. L84—Exploded view of crankshaft and camshaft assemblies.

1. Upper center main bearing housing
2. Lower center main bearing housing
3. Center main bearing
4. Main bearing
5. Split thrust washer
6. Thrust washer
7. Main bearing
8. Idler gear shaft & plug
9. Oil pump gear
10. Idler gear
11. Camshaft
12. Rod bearing
13. Crankshaft
14. Oil slinger
15. Balance weight
16. Rod bearing
17. Camshaft bushing
18. Piston ring
19. Piston pin
20. Retaining ring
21. Key
22. Camshaft bearing
23. Crankshaft gear
24. Camshaft gear
25. Locating plate
26. Piston rings
27. Piston
28. Rod bushing
29. Connecting rod

SERVICE MANUAL

Lister-Petter

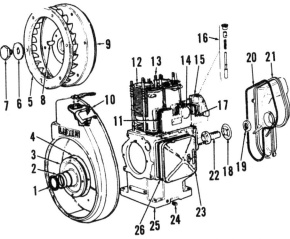

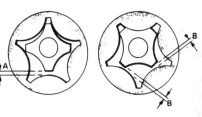

Fig. 85 – Exploded view of basic crankcase assembly.

7. Flywheel retaining screw
8. Key
9. Flywheel
10. Fan shroud
11. Gasket
12. Cylinder barrel
13. Stud
14. Fuel pump housing
15. End plate
16. Overload trip assy.
17. End plate gasket
18. Oil slinger
19. Oil seal
20. Gear cover gasket
21. Gear cover
22. Crankshaft extension
23. Crankcase side cover
24. Drain plug
25. Crankcase assy.
26. Gasket

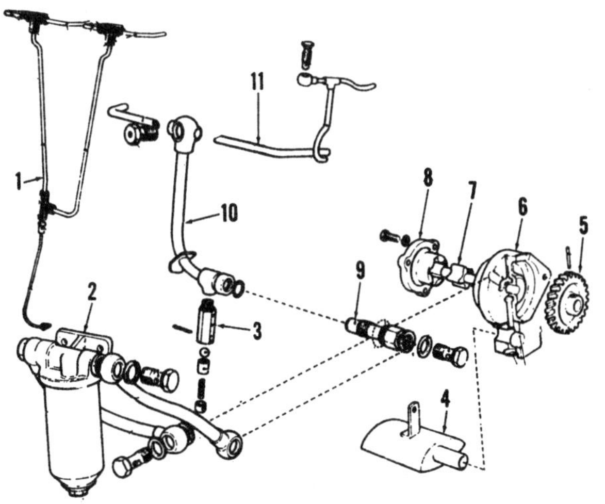

1. Felt washer
2. Oil retaining ring
3. Main bearing housing
4. Bearing housing gasket
5. Fan
6. Lockwasher

Fig. L86 – Exploded view of lubrication system used.

1. Oil pipe between cylinders
2. Oil filter & mounting bracket
3. Oil pressure relief valve assy.
4. Pick-up screen
5. Drive gear
6. Pump housing
7. Spindle & impeller
8. Eccentric ring
9. Connector
10. Main oil supply pipe
11. Oil pipe to main bearing, camshaft & governor

Fig. L87 – The gerotor type pump wear limits are as follows: A – 0.20 mm (0.008 inch); B – 0.18 mm (0.007 inch); End clearance – 0.15 mm (0.006 inch).

the top. Align locating hole in center bearing carrier with hole in side of crankcase and install dowel. Install main bearing housing, then check crankshaft end play. Recommended end play is 0.13-0.25 mm (0.005-0.010 inch) and is adjusted using shims (4 – Fig. L85) as required between bearing housing and crankcase. Apply sealing compound to shims during final assembly. Align timing marks (M – Fig. L82) on crankshaft gear and camshaft gear. Tighten flywheel mounting screw to 542 N·m (400 ft.-lbs.). Complete installation by reversing the removal procedure.

OIL PUMP

The gerotor type oil pump is driven by an idler gear from the crankshaft gear. Normal oil pressure is 205-315 kPa (30-45 psi). A pressure relief valve (3 – Fig. L86) is located in the oil gallery pipe.

To remove oil pump, drain the oil and remove the timing gear cover. Unbolt and remove pump assembly. Renew oil pump if clearances indicated in Fig. L87 exceed the following values.

Point A .... 0.20 mm
(0.008 in.)
Point B .... 0.18 mm
(0.007 in.)
End clearance .... 0.15 mm
(0.006 in.)

Illustrations Courtesy Lister-Petter

LISTER-PETTER

Model	No. Cyls.	Bore	Stroke	Displ.
TL2	2	98.42 mm (3.875 in.)	101.6 mm (4.000 in.)	1550 cc (94.5 cu. in.)
TL3	3	98.42 mm (3.875 in.)	101.6 mm (4.000 in.)	2320 cc (141.5 cu. in.)
TS1	1	95.25 mm (3.75 in.)	88.9 mm (3.50 in.)	633 cc (38.66 cu. in.)
TS2	2	95.25 mm (3.75 in.)	88.9 mm (3.50 in.)	1266 cc (77.31 cu. in.)
TS3	3	95.25 mm (3.75 in.)	88.9 mm (3.50 in.)	1899 cc (115.95 cu. in.)
TR1	1	98.42 mm (3.875 in.)	101.6 mm (4.00 in.)	773 cc (47.17 cu. in.)
TR2	2	98.42 mm (3.875 in.)	101.6 mm (4.00 in.)	1546 cc (94.35 cu. in.)
TR3	3	98.42 mm (3.875 in.)	101.6 mm (4.00 in.)	2319 cc (141.52 cu. in.)
TX2	2	100 mm (3.937 in.)	101.6 mm (4.00 in.)	1596 cc (97.77 cu. in.)
TX3	3	100 mm (3.937 in.)	101.6 mm (4.00 in.)	2394 cc (146.65 cu. in.)

Engines in this section are four-stroke, air-cooled type with vertical cylinders and direct fuel injection. "TL" and "TX" models use a belt driven axial flow fan to provide cooling air; the flywheel serves as cooling fan on "TS" and "TR" engines. Engine type can be identified by the last two digits (build number) found on plate (Fig. L3-1) attached to manifold air cowl. Standard crankshaft rotation is counterclockwise as viewed from flywheel end.

MAINTENANCE

STARTING PROCEDURE

Before starting, check fuel and oil levels. Move engine control to "START" and set variable speed control, if so equipped, to "FAST" position. On engines with compression release, move compression release lever to relieve compression pressure. On manual start engines, lightly oil end of crank handle, then insert handle to crank engine. With compression release models, turn engine as fast as possible and release compression lever. Remove manual start crank immediately when engine starts. On electric start engines, crank engine with starter motor until engine starts; do not use starting motor for more than 20 seconds at a time. On variable speed engines, adjust engine speed as required.

LUBRICATION

Under normal operating conditions, recommended engine oil is API service classification CC. If engine is operated under continual heavy load and high temperatures, or if sulfur content of fuel exceeds 0.6 percent, API classification CD oil is recommended.

Use SAE 5W oil in temperatures below −15° C (5° F). From −15° to 4° C (5° to 40° F), use SAE 10W motor oil. For temperatures of 4° to 30° C (40° to 85° F), SAE 20W oil is recommended. If temperature is above 30° C (85° F), use SAE 30 grade oil.

Recommended oil change interval is after each 250 hours of operation; renew filter at each oil change.

FUEL SYSTEM

Number 2 diesel fuel is recommended for most operating conditions. Number 1 diesel fuel may be used for light duty or cold weather operation.

FILTER AND BLEEDING. Some engines are equipped with a fuel filter attached to bottom of fuel tank; others have a cartridge type filter assembly mounted on flywheel end of engine. Fuel filter should be renewed after each 500 hours of operation, or sooner if engine shows signs of fuel starvation (surging or loss of power). After filter service, bleed air from system as follows:

Fill fuel tank. Loosen bleed screw (F—Fig. L3-3) on filter housing. Actuate primer lever on fuel feed pump if so equipped. Tighten screw when fuel flows from bleeder. Loosen bleed screw (P) on fuel pumps one at a time, starting with pump closest to filter. Tighten screw on pump when fuel flows and repeat with next pump.

If engine will not start at this point, loosen high pressure line fittings at fuel injectors, and with control lever in "START" position, crank engine until fuel flows from loosened fittings. Tighten fuel line fittings and start engine.

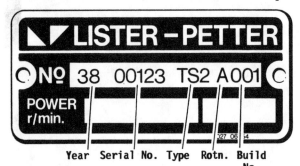

Fig. L3-1—View of typical engine identification plate. To ensure use of proper parts and specifications, use complete number.

SERVICE MANUAL

Lister-Petter

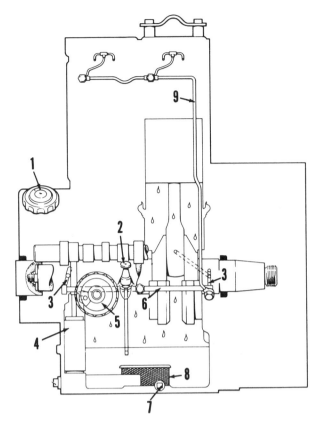

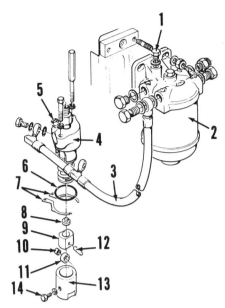

Fig. L3-2—Basic lubrication system typical of all models. An external pipe (9) supplies oil to valve train.
1. Filler cap
2. Dipstick
3. Oil passage drilling
4. Oil pump
5. Oil filter
6. Oil gallery
7. Drain plug
8. Pickup screen
9. Oil pipe to valve cover

Fig. L3-4—Partially exploded view of fuel injection pump and typical filter; some filters have only two ports instead of four shown.
1. Bleed screw
2. Fuel filter assy.
3. Fuel inlet hose
4. Injection pump
5. Bleed screw
6. "O" ring
7. Split shims
8. Tappet insert
9. Fuel pump tappet
10. Bushing
11. Roller
12. Pin
13. Tappet guide
14. Locating screw

Fig. L3-3—View at left shows tank mounted fuel filter and bracket mounted filter is at right. Refer to text for bleeding procedure.

INJECTION TIMING. A separate Bryce Berger fuel injection pump is used for each cylinder. Fuel injection pumps are located on push rod side of cylinder head and are actuated by lobes on engine camshaft. Split type pump mounting shims (7—Fig. L3-4) are used to adjust injection timing.

To check injection timing, turn flywheel to firing position for the cylinder being checked and align timing mark on flywheel with arrow (Fig. L3-6) on flywheel housing window. On cylinder being checked, remove fuel injector line, delivery valve holder (1—Fig. L3-5), delivery valve (3) and spring (2). Reinstall delivery valve holder, place control lever in "RUN" position and turn crankshaft opposite from normal direction of rotation until fuel flows from open delivery valve holder. Then, turn crankshaft slowly in normal direction of rotation until fuel flow stops; timing mark on flywheel should be aligned with arrow if injection timing is correct. Repeat procedure several times to be sure observation of timing is correct.

If timing marks are not aligned, loosen fuel injection pump mounting bolts and add shims to retard timing or remove shims to advance timing as necessary. After proper timing is obtained, reinstall delivery valve and spring and tighten delivery valve holder to a torque of 54 N·m (40 ft.-lbs.). Repeat procedure for each cylinder of engine.

GOVERNOR

To check or adjust governor settings, first remove injection pump inspection covers and timing gear end cover. Engine control lever should be in "START"

Fig. L3-5—Pump delivery valve (3) and spring (2) must be removed when checking injection timing. Delivery valve holder (1) is reinstalled without valve and spring during timing operation; refer to text.

Illustrations Courtesy Lister-Petter

135

position and camshaft bumped gently toward flywheel to remove any end play.

Check governor lever end play (E—Fig. L3-7) and install shims, if necessary, to obtain end play of 0.12-0.25 mm (0.005-0.010 inch).

Clearance (A) between thrust sleeve and step on camshaft should be 5.0 mm (0.197 inch) with fuel injection pump calibration mark (B) aligned with mark on center of pump for number 1 cylinder. Adjust by changing length of governor link (C) by turning locknuts. Then, adjust length of governor link between each pump until calibration marks (B) on all pumps are aligned.

Set engine control to "RUN" position. Loosen control lever plate (P) retaining screw and insert feeler gage between control lever and stop; thickness of feeler gage should be equal to recommended clearance (G) as listed in following table:

ENGINE RPM CLEARANCE (G)

"TL" Engines
1500, 1800 . . . . 0.6 mm
 (0.024 in.)
3000 . . . . 1.25 mm
 (0.050 in.)

TS1 Engines,
 Builds 01, 02, 03,
 04, 16, 60, & 61
2500 . . . . 0.25 mm
 (0.010 in.)
3000 . . . . 0.50 mm
 (0.020 in.)

TS2 and TS3 Engines,
 Builds 01, 02, 03,
 04, 60, & 61
2500 . . . . 0.35 mm
 (0.014 in.)
3000 . . . . 0.50 mm
 (0.020 in.)

"TS" Engines,
 All Other Builds
1500, 1800 . . . . 0.60 mm
 (0.024 in.)
2500 . . . . 0.80 mm
 (0.032 in.)
3000 . . . . 1.00 mm
 (0.039 in.)

TR1 Engines,
 Builds 01, 02, 03,
 04 & 16
2500 . . . . 0.25 mm
 (0.010 in.)
3000 . . . . 0.50 mm
 (0.020 in.)

TR2 and TR3 Engines,
 Builds 01, 02, 03 & 04
2500 . . . . 0.40 mm
 (0.016 in.)
3000 . . . . 0.30 mm
 (0.012 in.)

"TR" Engines,
 All Other Builds
1500, 1800 . . . . 0.60 mm
 (0.024 in.)
2500 . . . . 0.80 mm
 (0.032 in.)
3000 . . . . 1.00 mm
 (0.039 in.)

"TX" Engines,
 Builds 02, 04, 05,
 10 & 11
1500, 1800 . . . . 1.40 mm
 (0.056 in.)
2500 . . . . 1.75 mm
 (0.069 in.)
3000 . . . . 1.78 mm
 (0.070 in.)

"TX" Engines,
 Builds 20 & 21
2500, 2800 . . . . 1.77 mm
 (0.070 in.)

"TX" Engines,
 All Other Builds
2500 . . . . 1.04 mm
 (0.041 in.)
3000 . . . . 1.07 mm
 (0.042 in.)

While maintaining clearance (G), rotate control lever plate until fuel injection pump calibration marks (B) are aligned. If engine speed is between speeds shown above, use value for nearest engine rpm shown.

Engines may be equipped with either a constant speed or variable speed control. Turning adjusting screw (5C—Fig. L3-8) clockwise will increase speed on a constant speed governor control en-

Fig. L3-6—To check injection pump timing, turn crankshaft until flywheel timing mark for cylinder being checked is aligned with arrow on flywheel housing. Beginning of injection should occur at this point.

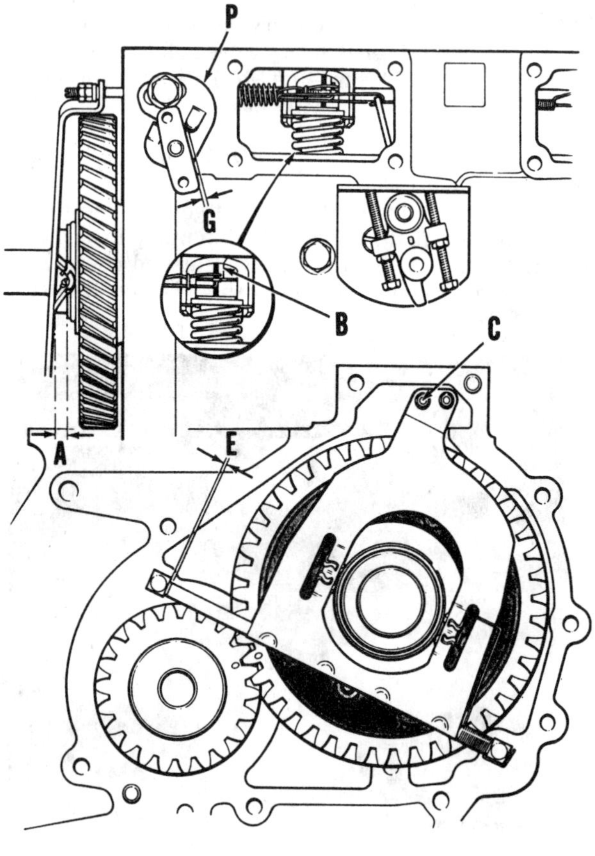

Fig. L3-7—To check and adjust fuel pump and governor linkage, remove fuel pump inspection covers and timing gear end cover. Refer to text for procedure.

SERVICE MANUAL

Lister-Petter

gine. The other screw has no function on constant speed engines.

Screw (5V) is used to adjust low idle speed on variable speed control engines. To adjust, turn screw counterclockwise until it no longer affects engine speed and governor causes engine to "hunt." Then, turn screw clockwise until "hunting" is eliminated and desired low idle speed is obtained and tighten adjusting screw locknut.

The lower screw (4) is used to limit maximum speed on variable speed control engines. Turning screw in (clockwise) reduces maximum speed and turning screw out (counterclockwise) will increase maximum speed.

Governor weights and springs are serviced separately. For proper parts identification and adjustment, contact an authorized service department.

REPAIRS

TIGHTENING TORQUES

Refer to the following table for special tightening torques.

Cylinder head:
"TX" engines . . . . 61 N·m
(45 ft.-lbs.)
All Others . . . . 68 N·m
(50 ft.-lbs.)
Connecting rod cap . . . . 43 N·m
(32 ft.-lbs.)
Balance weight bolts . . . . 57 N·m
(42 ft.-lbs.)
Flywheel:
"TL" engines . . . . 610 N·m
(450 ft.-lbs.)
"TS, TR & TX" engines . . . 475 N·m
(350 ft.-lbs.)
Injection pump . . . . 9 N·m
(80 in.-lbs.)
Injection pump delivery
valve . . . . 54 N·m
(40 ft.-lbs.)
Injector clamp bolts . . . . 21 N·m
(15 ft.-lbs.)
Injector line fittings . . . . 28 N·m
(21 ft.-lbs.)
Main bearing housing . . . . 27 N·m
(20 ft.-lbs.)
Main bearing dowel
locating plug . . . . 20 N·m
(15 ft.-lbs.)
Manifold nuts, lower . . . . 9 N·m
(80 in.-lbs.)
Manifold nuts, upper . . . . 21 N·m
(15 ft.-lbs.)
Oil sump . . . . 27 N·m
(20 ft.-lbs.)
"TX" crankshaft pulley . . . . . 196 N·m
(145 ft.-lbs.)
TS1 flywheel fan . . . . 4 N·m
(36 in.-lbs.)
TS2, TS3 flywheel fan . . . . 8 N·m
(72 in.-lbs.)

VALVE ADJUSTMENT

Valve clearance can be adjusted after removing cylinder head cover. Clearance for both intake and exhaust valves with engine cold should be 0.010-0.015 mm (0.004-0.006 inch) on "TX" models and 0.15-0.20 mm (0.006-0.008 inch) for all other models. Turn crankshaft to position number one piston at TDC on compression stroke (both valves closed) and adjust clearance, if necessary, on both valves. Repeat procedure for remaining cylinders. Firing order for 3-cylinder engines is 1-2-3.

To adjust compression release mechanism on models so equipped, turn crankshaft until piston is at TDC on compression stroke. Loosen adjuster locknut and move lever to actuate compression release. Turn adjusting screw (24—Fig. L3-10) until rocker arm just begins to depress valve, then turn screw one additional turn and tighten locknut. Repeat procedure for remaining cylinders.

CYLINDER HEAD

Two different cylinder head designs are used. Refer to the following paragraphs.

Models TX2 and TX3

A one-piece aluminum alloy cylinder head is used. Refer to Fig. L3-9 for exploded view of cylinder head assembly and cylinder barrel. Cylinder heads are retained by long cap screws (4). Early models have a plain washer (11) under the head bolts; Model TX2 after serial number 38 00138 and Model TX3 after serial number 38 00206 were fitted with two spring disc washers under each cylinder head bolt.

To remove cylinder heads, proceed as follows: Remove air cleaner, exhaust muffler, manifolds and cowling. Disconnect and remove fuel injection lines and immediately cap all openings. Remove fuel injectors and extract the injector sealing washer from bottom of each injector bore. Remove lubricating oil feed line from cylinder block to cylinder heads. Remove the valve rocker arm covers, rocker arm assemblies and push rods. Mark each head so that it may be reinstalled in original position. Unbolt and remove the cylinder heads and air baffle(s) from between the cylinder barrels. Remove the rocker arm push rod tubes and place with corresponding cylinder head. If cylinders are not to be removed, place suitable length tubes over cylinder head bolts, one for each cylinder, and clamp cylinders to block by inserting the bolt and tube.

To check piston to cylinder head clearance, place two pieces of soft lead wire of 1.2 mm (0.050 inch) thick on head so wires will clear valves and be in line with piston pin. Hold wires in place with grease. Install cylinder head and tighten head retaining bolts to a torque of 61 N·m (45 ft.-lbs.). Turn engine crankshaft so that piston goes past TDC twice, then remove cylinder head and check thickness of wire. If cylinder head

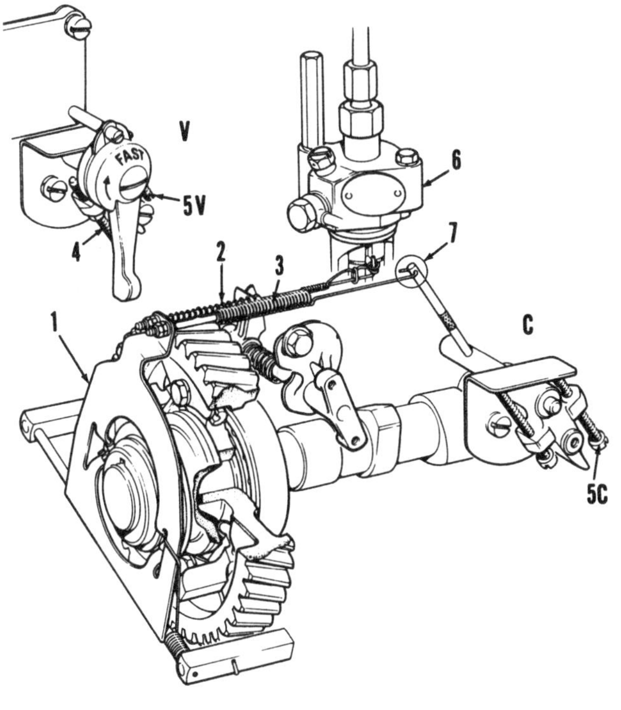

Fig. L3-8—A constant speed (C) or variable speed control (V) may be used. Refer to text for speed adjustment procedure. Be sure speeder (governor) spring (3) is connected properly through hole in spindle as shown at (7).

1. Governor lever
2. Governor link
3. Speeder (governor) spring
4. Maximum speed screw (variable)
5C. Speed adjusting screw (constant)
5V. Idle speed screw (variable)
6. Fuel injection pump
7. Speeder (governor) spring connection

Illustrations Courtesy Lister-Petter

Lister-Petter

SMALL DIESEL

clearance is correct, flattened wire should measure 0.81-0.95 mm (0.032-0.037 inch). To adjust clearance, it is necessary to change the shims between cylinder and crankcase. Cylinder shims (24—Fig. L3-9) are available in thicknesses of 0.07, 0.13 and 0.38 mm (0.003, 0.005 and 0.015 inch).

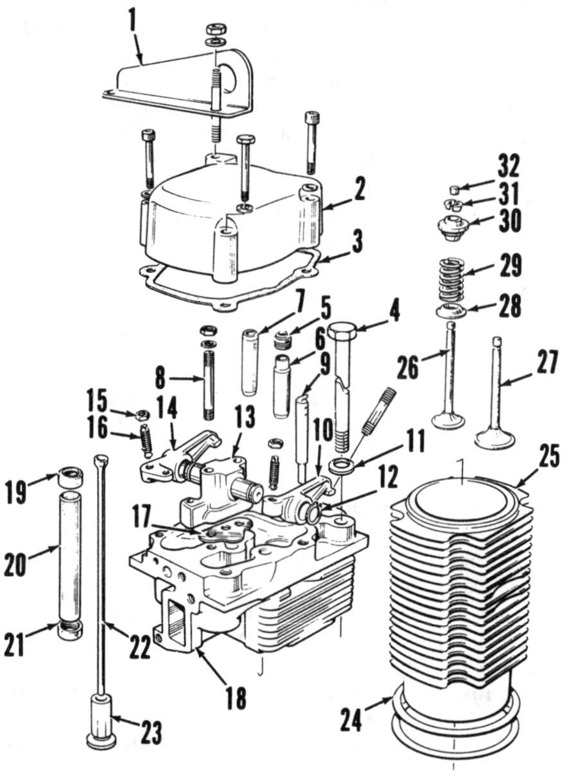

To reinstall heads, reverse removal procedure. Cylinder head bolts should be renewed whenever cylinder heads are removed; they must be renewed after fourth cylinder head removal. Install new rubber push rod seals in cylinder block and heads. Clean cylinder head and cylinder mating surfaces and remove all rocker arm support and valve cover gasket material. Lightly grease both ends of push rod tubes and insert them in original positions in cylinder block and in heads as heads are positioned on cylinders. Install head bolts finger tight, then align manifold surfaces with straightedge and install a manifold before tightening head bolts. Tighten cylinder head bolts at diagonal positions evenly to a torque of 61 N·m (45 ft.-lbs.). Install push rods and rocker arm assembly using new sealing plate and adjust valve clearance. Complete remainder of reassembly by reversing disassembly procedure.

Models "TL", "TS" and "TR"

A two-piece cylinder head is used. Refer to Fig. L3-10 for exploded view of cylinder head assembly. Top half (top plate) is cast iron and contains the valve gear and breather tube. Lower half is aluminum alloy and contains valve seat inserts. Valve guides are a press fit and hold the cylinder head halves together. Cylinder heads are retained by studs and nuts.

To remove cylinder heads, first remove air cleaner, exhaust muffler, manifolds and air cowling and air baffles. Remove injector lines from pumps and injectors, remove the injectors and immediately cap all openings. Remove the valve cover and valve rocker arm oil feed pipe from heads and block.

On 2 and 3-cylinder engines, be sure to identify cylinder head position so they can be reinstalled in their original positions. Note position of air baffles for correct reassembly. As cylinder heads are removed, place shims and gasket from each cylinder with corresponding cylinder head for use in reassembly. Remove the push rods and push rod tubes, keeping these parts with cylinder head in position from which they were removed. If cylinders are not to be removed, place suitable tube over one cylinder head stud for each cylinder and install cylinder head nut to hold cylinders from moving.

To check piston to cylinder head clearance, place 2 pieces of soft lead wire of 1.2 mm (0.050 inch) thick on head so wires will clear valves and be in line with piston pin. Hold wires in place with grease. Install cylinder head with

Fig. L3-9—Exploded view of Model "TX" cylinder head assembly.

1. Lift eye
2. Valve cover
3. Valve cover gasket
4. Cylinder head bolts
5. Intake valve seal
6. Intake valve guide
7. Exhaust valve guide
8. Rocker support stud
9. Breather tube
10. Intake rocker arm
11. Head bolt washers
12. Snap ring
13. Rocker arm support
14. Exhaust rocker arm
15. Locknut
16. Adjusting screws
17. Sealing plate
18. Cylinder head
19. Upper tube seal
20. Push rod tube
21. Lower tube seal
22. Push rod
23. Cam follower
24. Cylinder shims
25. Cylinder
26. Exhaust valve
27. Intake valve
28. Spring seat
29. Valve spring
30. Spring retainer
31. Split valve locks
32. Lash cap

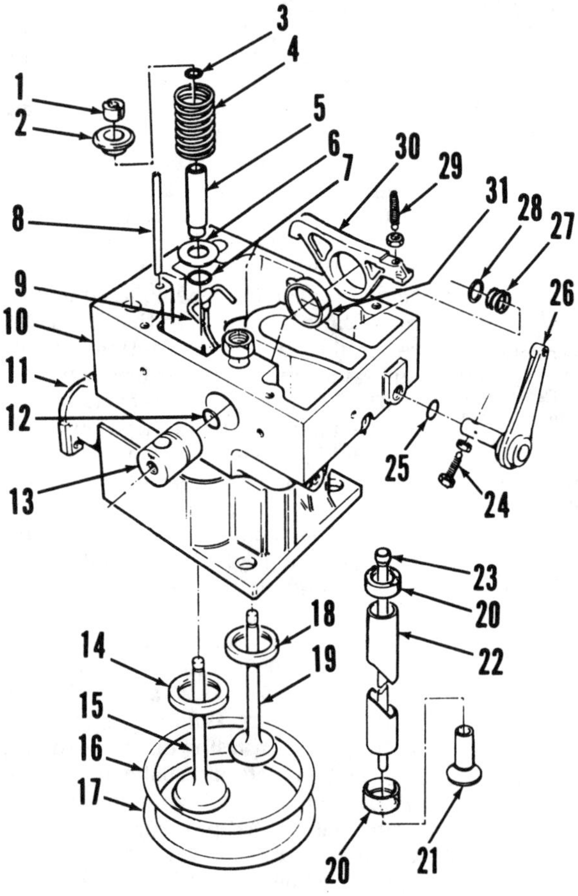

Fig. L3-10—Exploded view of cylinder head typical of all models except Model "TX." Compression release assembly (24 through 28) is not used on some models.

1. Split valve locks
2. Valve spring retainer
3. Seal ring (intake)
4. Valve spring
5. Valve guide
6. Spring seat washer
7. "O" ring (intake)
8. Breather tube
9. Oil pipe
10. Cylinder head upper half
11. Cylinder head lower half
12. "O" ring
13. Rocker stub shaft
14. Valve seat
15. Intake valve
16. Shim
17. Head gasket
18. Valve seat
19. Exhaust valve
20. Push rod tube seals
21. Cam follower
22. Push rod tube
23. Push rod
24. Adjusting screw
25. Seal ring
26. Compression release lever
27. Spring
28. Washer
29. Adjusting screw
30. Rocker arm
31. Bushing

Illustrations Courtesy Lister-Petter

SERVICE MANUAL

Lister-Petter

gasket and tighten head retaining nuts to a torque of 68 N·m (50 ft.-lbs.). Turn engine crankshaft so that piston goes past TDC twice, then remove cylinder head and check thickness of wire. If cylinder head clearance is correct, flattened wire should measure 0.889-0.965 mm (0.035-0.038 inch) on "TL" and "TR" engines and 0.813-0.889 mm (0.032-0.035 inch) on "TS" engines. To adjust clearance, it is necessary to change the shims (16—Fig. L3-10) between cylinder and cylinder head. Cylinder head shims are available in thicknesses of 0.076 or 0.254 mm (0.003 or 0.010 inch).

To install cylinder heads, proceed as follows: Insert new push rod tube seals in cylinder block and head. Apply light coat of grease to both ends of push rod tubes and insert them in cylinder block. Insert push rods in tubes. Apply a light coat of high melting point grease to shims and gasket to hold them in place in recess of cylinder head. Do not apply grease to cylinder side of gasket. Slide head downward onto studs making sure that push rod tubes and push rods are properly positioned. Install cylinder head nuts using thread sealer and with grade symbol upward. On 2 and 3-cylinder models, install nuts finger tight, then align manifold surfaces of heads with straightedge or install manifold. Then tighten nuts evenly to a torque of 68 N·m (50 ft.-lbs.). Adjust valve clearance and reinstall injectors after cylinder head nuts have been tightened. Complete installation by reversing removal procedure.

VALVE SYSTEM

Models TX2 and TX3

The valve guides are a press fit in cylinder head. If difficulty is encountered in pressing guide from head, heat head to 150° C (300° F), then press guide out toward bottom of head. Intake guide has small diameter at upper end to accommodate a valve stem seal. To install guides, heat head to 150° C (300° F) and press guides downward through head. Top end of guides must protrude 12.45-12.95 mm (0.489-0.509 inch) from machined face of head. Ream guides as necessary for proper valve stem clearance. Clearance should be 0.073-0.119 mm (0.003-0.005 inch) with maximum allowable clearance of 0.15 mm (0.006 inch).

Valve rocker arms are retained on support shaft (13—Fig. L3-9) by snap rings (12). Rocker arms can be removed without removing support. Remove retaining snap ring, loosen locknut (15), back adjusting screw (16) out fully and turn crankshaft to position piston at TDC on compression stroke. Rocker arm can then be removed from support. Rocker arm shaft diameter is 15.96-15.98 mm (0.6272-0.6280 inch). Desired rocker arm to shaft clearance is 0.020-0.067 mm (0.0008-0.0026 inch) with maximum allowable clearance (wear limit) of 0.1 mm (0.004 inch).

After refacing and reseating valves, valve heads should be 1.02-1.55 mm (0.040-0.061 inch) below combustion surface of cylinder head.

Intake and exhaust valve springs are identical. Spring free length should be approximately 48 mm (1.89 inches) and minimum free length is 46.6 mm (1.835 inches). Renew spring if distorted or otherwise damaged.

Models "TL", "TS" and "TR"

The valve guides are a press fit into the cylinder head upper and lower halves. An "O" ring (7—Fig. L3-10) and retaining plate (6) are located around the top of the intake valve guide. When renewing intake guide, new "O" must be inserted into the top plate recess before pressing guide into head. Note that guides are marked "IN TOP" and "EX TOP" and must be installed accordingly with lettering at top. Apply sealer to outside of guides, then use suitable tools to press guides in until top of exhaust guide extends 17.4-17.9 mm (0.685-0.704 inch) from top of plate surface and top of intake guide extends 12.4-12.9 mm (0.488-0.507 inch). After valve guides are installed, check valve stem to guide clearance and ream guide if necessary. Desired stem to guide clearance is 0.07-0.10 mm (0.003-0.004 inch) with wear limit of 0.13 mm (0.005 inch).

Valve face and seat angles are 45 degrees for both intake and exhaust. Desired seat width is 1.65-2.29 mm (0.065-0.090 inch) for intake and 1.35-1.78 mm (0.053-0.070 inch) for exhaust. Head of intake valve should be recessed 1.02-1.27 mm (0.040-0.050 inch) below combustion surface of cylinder head and head of exhaust valve should be recessed 0.89-1.14 mm (0.035-0.045 inch). If valve recession is not within specified limits, renew valve and/or valve seat.

The upper end of each intake valve stem and exhaust valve stem on 2 and 3-cylinder models are fitted with a sealing "O" ring (3). The "O" ring is installed on valve stem with spring retainer (2) and spring (4) compressed, then install valve keepers (1) and release valve spring tension.

The end of rocker stub shaft (13) is drilled and tapped M8 × 1.25. Shaft may be withdrawn using suitable slide hammer. Outer diameter of shaft is 25.436-25.444 mm (1.0014-1.0017 inches); inner diameter of rocker arm bushing should be 25.462-25.475 (1.0025-1.0029 inches). Clearance when new is 0.018-0.039 mm (0.0007-0.0015 inch) and wear limit is 0.07 mm (0.003 inch). When installing shaft, install new "O" ring (12) in recess on inside face of shaft and coat sealing groove nearest to bolt hole with "Wellseal." Hold rocker arm in position and drive stub shaft into place with mark "TOP" on outer end of shaft toward top of cylinder head.

Intake and exhaust valve springs are identical. Spring free length should be approximately 52 mm (2.047 inches) and minimum free length is 50.44 mm (1.986 inches). Renew spring if distorted or otherwise damaged.

INJECTORS

REMOVE AND REINSTALL. Procedure varies among models. First, thoroughly clean fuel lines, injector and surrounding area. Then, refer to the appropriate following paragraph.

MODELS "TL", "TS" AND "TR." Refer to Fig. L3-11. Remove cylinder head cover. Disconnect and remove fuel lines. Plug or cap all fuel openings to prevent entrance of dirt. Remove injector clamp bolt, then withdraw injector and copper sealing washer.

When installing injector, be sure injector bore in cylinder head is clean and stick new copper sealing washer to injector nozzle using small amount of high melting point grease. Install injector and

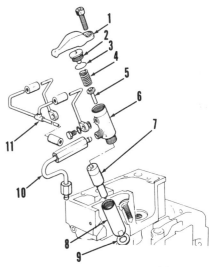

Fig. L3-11—Exploded view of fuel injector and lines typical of all models except "TX" models.
1. Injector clamp
2. Top plug
3. Seal ring
4. Spring
5. Push rod
6. Body
7. Nozzle & valve assy.
8. Nozzle nut
9. Seal washer
10. Fuel pressure pipe
11. Fuel return pipe

Lister-Petter

SMALL DIESEL

clamp bar, leaving clamp Allen screw finger tight. Install injection line, but do not tighten line fittings. Install new copper washers on each side of leak-off swivel union, but do not tighten. Install leak-off line to fuel tank. Tighten injector clamp screw to a torque of 21 N·m (15 ft.-lbs.). Tighten injection line fittings to a torque of 28 N·m (21 ft.-lbs.) and leak-off union to a torque of 4 N·m (36 in.-lbs.). Replace cylinder head cover. After initial run, retorque injector clamp screw.

MODELS TX2 AND TX3. To remove injector, pull fuel leak-off pipe from injector. Remove injector pipe clip from rocker arm cover, then remove fuel injector line (1—Fig. L3-12).

NOTE: Hold fuel delivery valve from turning while unscrewing line fitting at injection pump.

Remove the two injector clamp nuts and lift clamp (4) from studs and injector. Remove injector (7) and copper sealing washer (9) from cylinder head.

To install injector, first be sure injector bore in cylinder head is clean. Stick new copper sealing washer to injector nozzle with high melting point grease. Install injector and retaining clamp. Tighten the clamp nuts evenly to a torque of 8 N·m (72 in.-lbs.). Install injector line and tighten fittings to a torque of 28 N·m (21 ft.-lbs.).

TESTING. Injector nozzle should be tested for opening pressure, spray pattern, seat leakage and back leakage. Before connecting injector nozzle to tester, operate tester lever until fuel flows from open tester line. Then, connect injector, close valve to tester gage and operate tester lever a few quick strokes to purge air from nozzle and be sure nozzle valve is not stuck.

WARNING: Fuel emerges from injector with sufficient force to penetrate the skin. When testing injector, keep yourself clear of nozzle spray.

OPENING PRESSURE. Open valve to tester gage and operate tester lever slowly while observing gage reading. Nozzle opening pressure should be 19250-20265 kPa (2790-2940 psi).

SPRAY PATTERN. The injector nozzle has four spray holes. Spray patterns from all four holes should be alike. If pattern is uneven with one or more sprays different than the others, spray holes are partially blocked and should be cleaned. If one or more spray hole is partially or completely blocked, or if fuel dribbles from nozzle tip indicating seat leakage, injector should be overhauled or renewed.

BACK LEAKAGE. Operate tester lever to bring gage pressure to 16675 kPa (2495 psi), then time interval for gage pressure to drop to 13925 kPa (2060 psi). With calibration fluid temperature at 15.5° C (60° F), time interval should be between 6 and 27 seconds. If back leakage time is not within the time limits stated, injector should be overhauled or renewed.

NOTE: If temperature of calibrating oil is higher than 15.5° C (60° F), back leakage time will be reduced. If calibrating oil is colder than specified, time will be increased.

OVERHAUL. Refer to Fig. L3-13 for exploded view of injector used in "TX" Models and to Fig. L3-11 for all other models. Do not use wire brush, emery cloth or hard metal tools to clean injector. If more than one injector is disassembled at a time, keep parts from each injector separated as they are not interchangeable.

Injector nozzle spray hole diameter is 0.28 mm (0.011 inch) on "TL" engines, and 0.25 mm (0.010 inch) on all other models. Nozzle spray holes may be cleaned by inserting a cleaning wire slightly smaller in diameter than the orifice hole being cleaned. Be sure nozzle valve slides smoothly in nozzle valve body. If valve sticks, reclean or renew nozzle valve assembly.

When reassembling injector, rinse all parts in clean diesel fuel or calibrating oil and assemble while wet. Tighten injector top plug to a torque of 27 N·m (20 ft.-lbs.). Retest injector as previously outlined.

INJECTION PUMP

Refer to Fig. L3-4 for a partially exploded view of fuel injection pump. To remove fuel injection pump, first close fuel supply valve or drain fuel tank. Remove fuel supply pipe and fuel injector pipe from pump and immediately cap or plug openings. Remove injection pump inspection hole cover, then disconnect speeder spring and governor link. Remove pump mounting screws. Lift out fuel injection pump being sure that timing adjusting shims (7) are retained for reassembly.

It is recommended that the injection pump be tested and serviced by a shop qualified in diesel fuel injection pump repair.

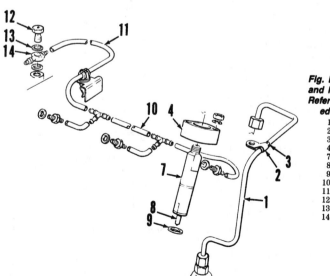

Fig. L3-12—View of injector and lines for "TX" models. Refer to Fig. L3-13 for exploded view of injector (7).

1. Injection line
2. Line clamp
3. Rubber sleeve
4. Injector clamp
7. Injector
8. Injector nozzle
9. Copper washer
10. Leak-off line
11. Return line to tank
12. Banjo bolt
13. Copper washer
14. Banjo fitting

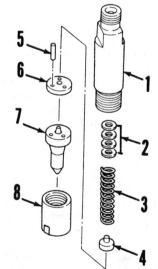

Fig. L3-13—Exploded view of injector typical of that used on "TX" models.

1. Injector body
2. Adjusting shims
3. Spring
4. Spring seat
5. Dowel pin
6. Spacer plate
7. Nozzle & valve assy.
8. Nozzle nut

SERVICE MANUAL

When reinstalling pump, be sure to renew "O" ring (6) and install original shim pack (7). Pump mounting screw with larger head should be installed closest to cylinder. Tighten pump mounting screws to a torque of 9 N·m (80 in.-lbs.). Complete installation by reversing removal procedure. Check injection timing and bleed air from system as outlined in MAINTENANCE section.

PISTONS, PINS, RINGS AND CYLINDERS

An aluminum alloy piston with a recessed combustion chamber in the crown is used on all engines. On "TL" engines, pistons are fitted with four rings; two compression rings and a spring expander type oil ring above piston pin and a slotted scraper type oil ring below the piston pin. The "TS" series pistons are equipped with five rings; three compression rings and a spring expander type ring above the piston pin and a slotted scraper type oil ring below the piston pin. Pistons for "TR" and "TX" engines have two compression rings and a slotted scraper type oil ring, all above the piston pin.

To remove pistons and cylinders, refer to CYLINDER HEAD paragraph and remove cylinder heads. Remove the crankcase oil sump. Mark each cylinder so that it may be reinstalled in same position. Remove the nuts and tubes or bolts and tubes if used to secure cylinder to crankcase. Remove connecting rod cap and remove cylinder, piston and rod assembly from top of block. On "TX" engines, be sure to keep the cylinder shims with corresponding cylinder. Remove piston and rod from cylinder. Remove snap rings from piston pin bore and push pin out of piston and rod.

A keystone type ring (tapered on both sides) is used in the top ring groove of all pistons. All other rings are rectangular and ring side clearance may be checked using new ring installed in piston groove. Hold outside edge of ring flush with piston surface and check side clearance with feeler gage. Piston ring end gap may be checked by inserting ring squarely into cylinder bore, then use a feeler gage to measure gap.

Refer to the following piston ring specifications:

Top ring end gap:
"TL" engines,
 desired . . . . 0.30-0.57 mm
 (0.012-0.022 in.)
 Max. allowable . . . . 0.85 mm
 (0.033 in.)
"TS" and "TR" engines,
 desired . . . . 0.38-0.66 mm
 (0.015-0.026 in.)

Max. allowable . . . . 0.98 mm
 (0.038 in.)
"TX" engines, desired . . 0.40-0.73 mm
 (0.016-0.029 in.)
Max. allowable . . . . 1.05 mm
 (0.041 in.)

Second ring end gap:
"TL" engines,
 desired . . . . 0.20-0.47 mm
 (0.008-0.018 in.)
 Max. allowable . . . . 0.75 mm
 (0.029 in.)
"TS", "TR" engines,
 desired . . . . 0.28-0.56 mm
 (0.011-0.022 in.)
 Max. allowable . . . . 0.90 mm
 (0.035 in.)
"TX" engines,
 desired . . . . 0.40-0.73 mm
 (0.016-0.029 in.)
 Max. allowable . . . . 1.02 mm
 (0.040 in.)

Top ring side clearance:
All models, desired . . . . 0.00-0.06 mm
 (0.0-0.0024 in.)
 Max. allowable . . . . 0.10 mm
 (0.004 in.)

Second ring side clearance:
"TL" engines,
 desired . . . . 0.08-0.13 mm
 (0.003-0.005 in.)
 Max. allowable . . . . 0.18 mm
 (0.007 in.)
"TS" engines,
 desired . . . . 0.122-0.172 mm
 (0.0048-0.0068 in.)
 Max. allowable . . . . 0.23 mm
 (0.009 in.)
"TR" engines,
 desired . . . . 0.076-0.127 mm
 (0.003-0.005 in.)
 Max. allowable . . . . 0.23 mm
 (0.009 in.)
"TX" engines,
 desired . . . . 0.070-0.102 mm
 (0.0027-0.004 in.)
 Max. allowable . . . . 0.18 mm
 (0.007 in.)

Check piston diameter at bottom of skirt and at right angle to piston pin. Standard piston diameter is 98.219-98.244 mm (3.8669-3.8679 inches) for "TL" engines, 95.039-95.065 mm (3.7417-3.7427 inches) for "TS" engine builds 01 to 43, 95.172-95.212 mm (3.7403-3.7418 inches) for "TS" engine builds 60-65, 98.283-98.323 mm (3.8625-3.8641 inches) for "TR" engines and 99.861-99.879 mm (3.9245-3.9252 inches) for "TX" engines.

Piston skirt to cylinder wall clearance specifications are as follows:
"TL" engines,
 desired . . . . 0.181-0.231 mm
 (0.007-0.009 in.)

Lister-Petter

Max. allowable . . . . 0.32 mm
 (0.013 in.)
"TS" builds 01-43,
 desired . . . . 0.210-0.261 mm
 (0.0083-0.010 in.)
 Max. allowable . . . . 0.35 mm
 (0.014 in.)
"TS" builds 60-65,
 desired . . . . 0.063-0.128 mm
 (0.0025-0.005 in.)
 Max. allowable . . . . 0.22 mm
 (0.008 in.)
"TX" engines,
 desired . . . . 0.121-0.164 mm
 (0.0048-0.0064 in.)
 Max. allowable . . . . 0.25 mm
 (0.010 in.)

Piston pin should be a hand push to slight interference fit in piston. It may be necessary to heat piston in boiling water to remove or install piston pin without excessive force. Be sure to securely install pin retaining snap ring at each end of piston pin bore.

Piston and rings are available in oversizes and the cylinder can be rebored and honed to fit oversize parts. Refer to desired piston skirt to cylinder clearance and desired piston ring end gap to correctly fit oversized parts. Piston pin is available only with new piston.

When reassembling, be sure piston is installed with the wording "CAMSHAFT SIDE" on top of piston toward the raised dots on side of connecting rod. Install compression rings with "TOP" marked side toward top of piston; oil rings may be installed either side up. Stagger piston ring end gaps at equal spacing around piston. Lubricate piston, piston rings and cylinder bore and install piston in cylinder with mark "CAMSHAFT SIDE" on top of piston to camshaft side of cylinder. On "TX" engines, install cylinder shims on bottom of cylinder using high melting point grease to hold shims in place. Lubricate crankpin and rod bearing and install piston and cylinder assembly. Tighten connecting rod nuts to a torque of 43 N·m (32 ft.-lbs.). Complete installation by reversing removal procedure.

CONNECTING RODS AND BEARINGS

A forged steel connecting rod is used with a precision insert type bearing at crankpin end and a renewable bushing in piston pin end.

Standard crankpin diameter is 63.487-63.500 mm (2.4995-2.500 inches). Recommended oil clearance is 0.04-0.08 mm (0.0016-0.0031 inch) with maximum allowable clearance of 0.14 mm (0.0055 inch).

Desired piston pin to bushing clearance is 0.0435-0.0606 mm (0.0017-0.0024 inch) with maximum allowable pin to bushing clearance of 0.09 mm (0.0035 inch). If renewing bushing, install in rod with oil hole in bushing and rod aligned. Ream or hone pin bushing to desired oil clearance. Connecting rod cap nut tightening torque is 43 N·m (32 ft.-lbs.).

END COVER

The end cover (5—Fig. L3-14 or L3-15) is located on gear end of crankcase with two dowels (1) and is retained by seven cap screws with copper sealing washers.

To remove end cover on engines with axial flow cooling fan, first remove cooling van drive belt and crankshaft pulley. Remove cap screws securing end cover to crankcase and axial flow fan to fan shroud, then remove end cover and fan as an assembly. Unbolt and remove fan housing from end cover if necessary.

To remove end cover on engines with flywheel mounted cooling fan, remove plastic dust cap (3—Fig. L3-14) from end cover. On early manual start engines, rotate camshaft until spring pin (18—Fig. L3-17) is visible through hole in cover. Drive the spring pin out, then turn camshaft 90 degrees and tap crank handle catch pin (17) out of shaft. On late model engines, remove the set screw from end of camshaft, turn camshaft until crank handle catch pin is visible and drive pin out. Unbolt and withdraw end cover.

Camshaft end play is controlled by a thrust washer (8—Fig. L3-14 or L3-15) and shim (9). Excessive camshaft end play affects governor operation and can cause uneven firing. On engines where camshaft does not extend through end cover, end play can be measured using a dial indicator inserted through bolt hole (B—Fig. L3-15) in end cover. Camshaft end play should be 0.08-0.20 mm (0.003-0.008 inch) on all models. Adjusting shims are available in thicknesses of 0.13, 0.25, 0.38, 0.50 and 0.63 mm (0.005, 0.010, 0.015, 0.020 and 0.025 inch); use only one shim.

Install new crankshaft oil seal (7—Fig. L3-13 or L3-14) with lip to inside. Lubricate seal lip with grease. Install end cover by reversing removal procedure.

GOVERNOR

The governor assembly is mounted on the front of camshaft drive gear. To service governor, first remove end cover as previously outlined. Remove fuel pump inspection covers from side of crankcase, then disconnect speeder (governor) spring and governor links at the fuel pumps. Remove governor lever

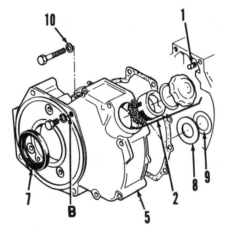

Fig. L3-15—Exploded view of end cover used on "TL" and "TX" engines. Refer to Fig. L3-14 for component identification.

pivot pins (6—Fig. L3-17) and shims (8). Remove governor lever (9) assembly. Remove governor sleeve (14) with thrust washers (12) and pad (13). Remove governor weight retainers (5), pivot pins (3) and weights (4).

Inspect all parts for excessive wear or other damage and renew as necessary.

To reinstall governor, reverse removal procedure while noting the following: Install governor sleeve (14) with thrust washer assembly facing outward. Install lever pivot pins (6) with drilled end to outside. Install shims (8) to provide governor lever end play of 0.12-0.25 mm (0.005-0.010 inch). Reconnect governor speeder spring and linkage as shown in Fig. L3-8. Refer to MAINTENANCE section for governor adjustment.

CAMSHAFT, FOLLOWERS AND BEARINGS

The camshaft is supported by a ball bearing at gear end and by bushings in crankcase at flywheel end and, on 2 and 3-cylinder models, one or two intermediate bushings in crankcase. The camshaft operates the valves, lubricating oil pump, fuel injection pumps and, if so equipped, the fuel transfer pump.

To remove camshaft, first remove the rocker arm assembly or cylinder head, push rods, end cover, governor linkage and the fuel injection pumps. Remove oil sump and withdraw lubricating oil pump and push rod. Remove tappet guide locating screws (2—Fig. L3-16), then withdraw tappet guides (1). Move cam followers away from camshaft and hold in that position with magnetic tools or clips, or invert engine so that followers move away from camshaft. Carefully withdraw camshaft and remove the

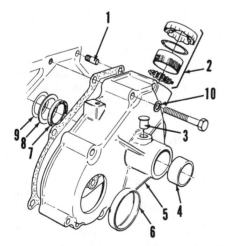

Fig. L3-14—Exploded view of end cover assembly used on "TR" and "TS" engines. Bushing (4) is used to support manual starter crank handle. Thrust washer (8) controls camshaft end play.

1. Dowel pin
2. Oil filler assy.
3. Plastic plug
4. Bushing
5. End cover
6. Cup plug
7. Oil seal
8. Thrust washer
9. Shim
10. Copper washer

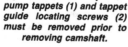

Fig. L3-16—Fuel injection pump tappets (1) and tappet guide locating screws (2) must be removed prior to removing camshaft.

SERVICE MANUAL

Lister-Petter

cam followers. Identify followers so they can be reinstalled in their original positions.

To remove camshaft gear and ball bearing, drive spring pin (15—Fig. L3-17) inward until it clears inside diameter of gear. Press shaft out of gear and bearing. Remove Woodruff key if necessary. Drive spring pin out of camshaft and retain for reassembly.

Inspect camshaft, cam followers and cam bushings for scoring, wear or other damage. Refer to the following camshaft journal and bearing oil clearance specifications:

Flywheel end, all engines:
Journal
  diameter .... 39.939-39.952 mm
              (1.5696-1.5701 in.)
Bushing ID ...... 39.975-40.060 mm
              (1.5710-1.5744 in.)
Desired oil
  clearance .... 0.023-0.121 mm
              (0.0009-0.0048 in.)
Wear limit,
  oil clearance .... 0.20 mm
              (0.008 in.)

Center bearings, "TX":
Journal
  diameter .... 46.740-46.755 mm
              (1.8369-1.8374 in.)
Bearing ID ..... 46.772-46.878 mm
              (1.8381-1.8423 in.)
Desired oil
  clearance .... 0.017-0.138 mm
              (0.0007-0.0054 in.)
Wear limit,
  oil clearance .... 0.22 mm
              (0.0086 in.)

Center bearings, except "TX":
Journal
  diameter ..... 44.3357-44.3484 mm
              (1.7424-1.7429 in.)
Bearing ID ...... 44.407-44.506 mm
              (1.7320-1.7491 in.)
Desired oil
  clearance .... 0.0586-0.1703 mm
              (0.0023-0.0067 in.)
Wear limit,
  oil clearance .... 0.22 mm
              (0.0086 in.)

To renew camshaft bushings, the flywheel, flywheel housing and main bearing housing must first be removed. Camshaft bushings should be replaced as a set. Soak bushings in clean engine oil for four hours prior to installation. Top of bushings are marked with a small dot on the outside of the bushing.

When installing camshaft drive gear, install ball bearing part way. Install key in shaft, align keyway of gear to key and then press gear and bearing onto shaft until gear bottoms on shoulder. Apply force to hub of gear only. Drive spring pin in gear hub and shaft until outer end of pin is about 1 mm (0.040 inch) below surface. Be sure gear hub is free of nicks or burrs that would affect sliding action of governor thrust sleeve.

To install camshaft, first insert cam followers in block and either retain them in raised position or turn engine over or onto flywheel end so they will stay in raised position. Lubricate camshaft lobes and journals and insert camshaft being sure that timing marks on camshaft gear and crankshaft gear are aligned as shown in Fig. L3-18.

Camshaft end play is controlled by a thrust washer (8—Fig. L3-14 or L3-15) and shim (9). Excessive camshaft end play affects governor operation and can cause uneven firing. On engines where camshaft does not extend through end

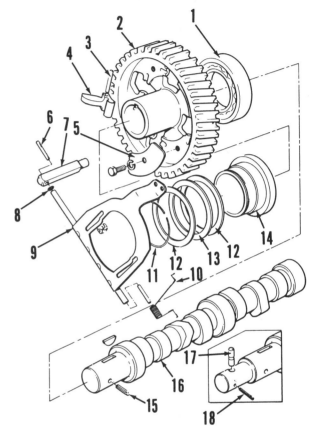

Fig. L3-17—Exploded view of typical camshaft and governor assembly. Catch pin (17) and spring pin (18) are used on some manual start models.

1. Ball bearing
2. Camshaft gear
3. Pivot pin
4. Governor weight
5. Weight retainer
6. Pivot pin
7. Pivot pin bracket
8. Shims
9. Governor lever
10. Spring
11. Retaining ring
12. Thrust washers
13. Thrust pad
14. Governor sleeve
15. Spring pin
16. Camshaft
17. Starter catch pin
18. Spring pin

Fig. L3-18—When reinstalling camshaft, make sure that timing marks (circled) on camshaft gear (2) and crankshaft gear (1) are aligned as shown.

Illustrations Courtesy Lister-Petter

Lister-Petter — SMALL DIESEL

cover, end play can be measured using a dial indicator inserted through bolt hole (B—Fig. L3-15) in end cover. Camshaft end play should be 0.08-0.20 mm (0.003-0.008 inch) on all models. Adjusting shims are available in thicknesses of 0.13, 0.25, 0.38, 0.50 and 0.63 mm (0.005, 0.010, 0.015, 0.020 and 0.025 inch); use only one shim.

Reinstall governor, thrust sleeve, shim, thrust washer and end cover. Check camshaft end play, and if within specifications, complete camshaft installation by reversing removal procedure.

CRANKSHAFT AND MAIN BEARINGS

The crankshaft is supported by main bearings located in crankcase at gear end and in the main bearing housing at flywheel end of crankcase. On 2-cylinder models, an intermediate bearing is retained in bearing support (18 and 18A—Fig. L3-19); 3-cylinder models have two intermediate bearings. All main bearings are two-piece precision insert type bearings; front and rear bearings are pressed into crankcase and bearing housing. Crankshaft end play is controlled by split type thrust washers at each end of crankshaft.

To remove crankshaft, remove cylinder heads, oil sump, pistons and rods, gear end cover, camshaft, flywheel and flywheel housing.

On 2 and 3-cylinder models, remove the intermediate bearing housing locating dowel (17—Fig. L3-19) at each bearing. Remove plug (15) and copper washer (16) at each intermediate bearing location. Thread a cap screw into the end of each dowel and pull dowel(s) from crankcase. Leave cap screw in dowels to be sure they will be reinstalled correct end first.

On all engines, use suitable puller to remove crankshaft timing gear and remove Woodruff key from shaft. Unbolt and remove crankshaft balance weights (10). Remove cap screws (1—Fig. L3-20) retaining rear main bearing housing (4). Remove short cap screws (2) and copper washers and thread two M6 bolts into threaded jackscrew holes in housing. Turn the M6 bolts evenly to remove housing. Retain shims (7) located between bearing housing and crankcase for reassembly. Remove split thrust washers from housing or crankshaft. Carefully remove crankshaft from rear end of cylinder block and remove split thrust washer (13—Fig. L3-19) at front end of shaft. On 2 and 3-cylinder models, unbolt and remove intermediate bearing supports from crankcase.

Inspect crankshaft, bearings and thrust washers for excessive wear or other damage. Crankshaft main journal diameter is 63.475-63.4870 mm (2.4946-2.4950 inches). Recommended main bearing oil clearance is 0.0386-0.0934 mm (0.0015-0.0037 inch) and wear limit is 0.16 mm (0.0063 inch). Crankshaft thrust washer standard thickness is 2.31-2.36 mm (0.091-0.093 inch). Crankshaft end play should be 0.229-0.305 mm (0.009-0.012 inch), with wear limit of 0.43 mm (0.017 inch).

To remove front main bearing inserts, use suitable bushing driver to press front bearing forward out of crankcase. Remove rear main bearing seal from housing and press rear main inserts out to rear side of housing. Special bearing removal and installation tools are available from engine manufacturer.

When installing bearings, note that bearings are different for each position and that top and bottom halves are not alike. Rear main bearing upper insert has tab so that bearing must be pressed into housing with bearing tab engaging slot in housing. Front upper bearing half has oil groove and oil hole; oil hole must be aligned with oil passage in block. The grooved main bearing half with oil hole must be installed in top half of center bearing housing with oil holes aligned. Do not install rear oil seal in main bearing housing until after housing is installed.

When installing crankshaft, be sure that all oil holes and surfaces are clean. Lubricate crankshaft journals and bearings. Assemble intermediate bearing supports with side marked "FLYWHEEL END" to rear of shaft and tighten cap screws to a torque of 27 N·m (20 ft.-lbs.). Apply light coat of high melting point grease to front thrust washers and insert them in front end of block with tab on upper half engaging slot in block, and with bearing surface toward rear. Carefully insert crankshaft from rear of block through thrust washer and front main bearing, taking care not to damage front bearing. On 2 and 3-cylinder models, dowel pin hole(s) in intermediate bearing(s) must be aligned with dowel pin hole(s) in block. Insert

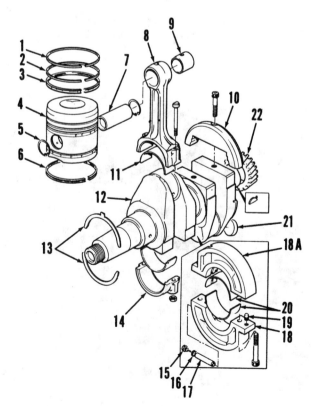

Fig. L3-19—Exploded view of typical crankshaft, connecting rod and piston assembly. Some models use threaded plugs instead of expansion plugs (21) to plug crankshaft oil passage drilling.

1. Top compression ring
2. Intermediate compression ring
3. Oil ring
4. Piston
5. Snap rings
6. Oil ring
7. Piston pin
8. Connecting rod
9. Rod pin bushing
10. Balance weight
11. Crankpin bearing insert
12. Crankshaft
13. Thrust washer
14. Rod cap
15. Dowel pin plug
16. Copper washer
17. Intermediate bearing dowel
18. Intermediate bearing lower housing
18A. Intermediate bearing upper housing
19. Dowel pin
20. Main bearing insert
21. Expansion plug
22. Crankshaft gear

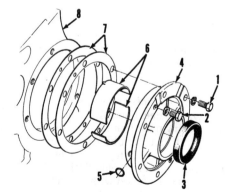

Fig. L3-20—Exploded view of rear main bearing housing and related parts. Shims (7 and 8) are available in thicknesses of 0.127 mm (0.005 inch) and 0.254 mm (0.010 inch).

1. Housing cap screws
2. Jackscrew hole plug bolts
3. Rear crankshaft seal
4. Main bearing housing
5. "O" ring
6. Main bearing insert
7. Shim
8. Crankcase

SERVICE MANUAL

Lister-Petter

each dowel pin through crankcase wall making sure that dowel pin fully enters intermediate bearing support. Stick thrust washers in front side of rear bearing housing with high melting point grease; be sure tab on upper washer engages slot in housing. Remove jackscrews from bearing housing and reinstall short cap screws and copper sealing washers. Install rear bearing housing with same thickness of shims removed on disassembly. Install two retaining cap screws and tighten to a torque of 27 N·m (20 ft.-lbs.). Check crankshaft end play and change shim thickness if necessary to bring end play within specified limits of 0.229-0.305 mm (0.009-0.012 inch). Turn crankshaft to be sure it rotates smoothly in bearings.

Install crankshaft balance weights (10—Fig. L3-19) with the part number sides facing each other. Tighten retaining bolts to a torque of 57 N·m (42 ft.-lbs.). Early models had one weight retaining bolt; later models have two bolts for each weight. Lock balance weight bolts by peening them with a center punch. Install remaining rear main bearing housing bolts and the two short cap screws with copper washers in the housing jackscrew holes. Install cap screw plugs with copper washers in intermediate bearing dowel holes in side of block and tighten to a torque of 20 N·m (15 ft.-lbs.).

Install Woodruff key for crankshaft gear in front end of shaft. Heat crankshaft gear and quickly install on crankshaft with keyway aligned with key and with timing mark out. Install camshaft with timing marks on camshaft gear aligned with timing mark on crankshaft gear as shown in Fig. L3-18. Complete the installation by reversing removal procedure.

OIL PUMP

A self-regulating plunger type oil pump is used on all models. The pump is located in a bore in the crankcase and is operated by a push rod from the camshaft. Refer to Fig. L3-21 for exploded view of the oil pump assembly. Normal oil pressure is 200 kPa (29 psi) at 1500 rpm and minimum oil pressure is 70 kPa (10 psi) with engine up to operating temperature and running at 1000 rpm.

To service oil pump, drain oil and remove sump. Withdraw oil pump, disassemble and inspect all parts. Reassemble using new parts as necessary.

AXIAL FLOW FAN

"TL" and "TX" Engines

The axial flow fan is belt driven from the crankshaft pulley. To remove fan, remove fan cowl, belt guard and belt. Remove fan housing mounting cap screws, then remove fan from engine.

To disassemble fan, remove fan pulley (17—Fig. L3-22). Press or tap fan shaft (5) out of fan (16) and fan body (14). Remove and inspect bearings and renew if necessary.

When reassembling fan, pack bearings with grease. Tighten pulley nut to 92 N·m (68 ft.-lbs.). Make sure that there is clearance of 0.25 mm (0.010 inch) between fan blades and fan body.

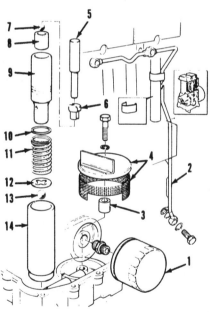

Fig. L3-21—Exploded view of typical engine oil pump and lubrication system components.

1. Oil filter
2. Oil line to valve system
3. Spacer
4. Oil pickup screen
5. Push rod
6. Plunger cap
7. Check valve ball
8. Valve seat
9. Pump plunger
10. Washer
11. Spring
12. Retaining plate
13. Check ball
14. Pump body

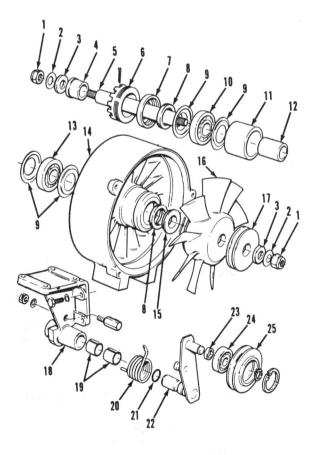

Fig. L3-22—Exploded view of axial flow cooling fan and belt tensioner used on "TL" and "TX" engines.

1. Nut
2. Spring washer
3. Flat washer
4. Seal bushing
5. Fan shaft
6. Nut
7. Seal carrier
8. Felt seal
9. Grease shields
10. Ball bearing
11. Outer spacer
12. Inner spacer
13. Roller bearing
14. Fan housing
15. Seal bushing
16. Fan
17. Pulley
18. Mounting bracket
19. Bushings
20. Tension spring
21. "O" ring
22. Idler arm
23. Spacer
24. Bearing
25. Idler pulley

/ Lister-Petter   SMALL DIESEL

LISTER-PETTER

Model	No. Cyls.	Bore	Stroke	Displ.
AA1	1	69.85 mm (2.75 in.)	57.15 mm (2.25 in.)	219 cc (13.4 cu. in.)
AB1	1	76.2 mm (3.00 in.)	57.15 mm (2.25 in.)	261 cc (15.9 cu. in.)
AC1	1	76.2 mm (3.00 in.)	66.68 mm (2.625 in.)	304 cc (18.6 cu. in.)
AC2	2	76.2 mm (3.00 in.)	66.68 mm (2.625 in.)	608 cc (37.2 cu. in.)
AC1Z, AC1ZS	1	76.2 mm (3.00 in.)	66.68 mm (2.625 in.)	304 cc (18.5 cu. in.)
AB1W	1	76.2 mm (3.00 in.)	57.15 mm (2.25 in.)	261 cc (15.9 cu. in.)
AC1W	1	76.2 mm (3.00 in.)	66.68 mm (2.625 in.)	304 cc (18.5 cu. in.)
AC2W	2	76.2 mm (3.00 in.)	66.68 mm (2.625 in.)	608 cc (37.0 cu. in.)

This section covers only the indirect injection versions of the engine models listed above. All models included in this section are four-stroke, vertical cylinder, indirect injection diesel engines. Models with a "W" suffix are water cooled; all others are air cooled. Standard crankshaft rotation is noted on the engine plate (see Fig. L5-1) with "A" indicating anticlockwise (counterclockwise) and "C" indicating clockwise as viewed from flywheel end. Both metric and U.S. Customary fasteners are used.

Lister-Petter diesel engine build numbers (engine types) can be identified by the last three digits on the plate (see Fig. L5-1) attached to the side of the crankcase below the oil filter.

MAINTENANCE

LUBRICATION

Under normal operating conditions, recommended engine oil is to conform with minimum performance level of API CC, MIL-L-46152A/B, MIL-L-2104B or DEF2101D specifications. Oils classified as API CD, MIL-L-2104C/D or Series 3 oils can inhibit the running in process of new or reconditioned engines, but are suitable for engines running under heavy loads, particularly at high ambient temperatures, after the first oil change.

Recommended oil grades are 5W or 5W/20 oils for all engines operating in temperatures below −15° C (5° F). In ambient temperature range of −15° to 4° C (5° to 39° F), SAE 5W or 5W/20 oil should be used in Model AA1 and AB1 engines and SAE 10W or 10W/30 oil should be used in Model AC engines. Between operating temperatures of 4° to 30° C (39° to 86° F), SAE 10W or 10W/30 can be used in all models; also, SAE 15W/40 can be used in AC engines. Above 30° C (86° F), SAE 10W/30, 15W/40 or 20W/40 can be used in all models; also, above this ambient temperature, single grade SAE 20 can be used in Model AA1 and SAE 30 in AC engines.

Oil and filter should be changed after each 250 hours of normal operation. Oil and filter should be changed in reconditioned engines after first 25 hours operation. Oil sump capacity is 1.9 liters (4 pints) for AA1 and AB1 models, 2.7 liters (5.7 pints) for AC1 models and 3.7 liters (6.5 pints for AC2 models.

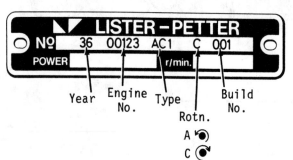

Fig. L5-1—View of typical engine identification plate. To ensure use of proper parts and specifications, use complete number.

FUEL SYSTEM

Number 2 diesel fuel is recommended for most conditions. Number 1 diesel fuel should be used for light duty or cold weather operation.

FUEL FILTER. Fuel filter element should be renewed after every 500 hours operation; however, filter may need to be renewed more frequently if fuel is not exceptionally clean. If engine shows signs of fuel starvation (surging or loss of power), renew fuel filter regardless of hours of operation. The fuel filter is mounted in the bottom of fuel tank on models with tank attached to engine.

BLEED FUEL SYSTEM. With a separately mounted fuel tank having a fuel lift pump, the pump priming lever must be operated to bleed air from fuel system. On engines fitted with a separately mounted fuel filter, loosen the two vent screws (S—Fig. L5-2) on top of filter. If engine is not fitted with a self-bleed fuel system, be sure fuel tank is full, move engine control lever to "RUN" position and vent fuel at pump through bleed screw (1—Fig. L5-3) until flow of fuel is free of air. If engine will not start, loosen fuel injector line fitting at injector. Crank engine until fuel flows from loosened fitting, then tighten fitting and crank engine.

INJECTION PUMP TIMING

Either a Bryce or I.E.S.A. fuel injection pump is used. Injection timing is varied by changing shim (2—Fig. L5-4) thickness between pump (1) mounting flange and crankcase. Setting of

Illustrations Courtesy Lister-Petter

SERVICE MANUAL

Lister-Petter

STOP/RUN lever to check injection pump timing will vary with engine model.

On Models AA1, AB1 and AB1W with fixed or 2-speed governor controls and running at speeds above 3000 rpm, move STOP/RUN lever toward STOP (vertical) position and secure lever at 10 degrees before vertical position. If running at speeds of 3000 rpm or below, move STOP/RUN lever to RUN (horizontal) position.

On Models AC1, AC1Z, AC1ZS and AC1W, move STOP/RUN lever to RUN position. On Models AC2 and AC2W with fixed or 2-speed governor controls, move STOP/RUN lever to midway between STOP and RUN positions.

On all models with variable speed governor controls, move control lever to full speed position. On all models, do not operate the overload stop lever.

To check timing, proceed as follows: Close fuel supply valve and remove pump-to-injector fuel line. Unscrew and remove delivery valve body (1—Fig. L5-5) and "O" ring (2). Lift out spring guide (4) (except early "B" series Bryce pump) and spring (5). Then, carefully remove delivery valve (6) without disturbing seat (7) or seal (3). Place the removed parts in clean diesel fuel. Place "O" ring (2) on pump delivery valve body and reinstall body in pump. Install a spill pipe (see Fig. L5-6) on the delivery valve body and turn crankshaft in normal direction of rotation so piston is 90 degrees BTDC on compression stroke.

Open fuel supply valve and if necessary, bleed fuel filter. Fuel should flow from spill pipe; if a fuel lift (feed) pump is installed, it is necessary to operate primer lever to obtain fuel flow. Turn crankshaft slowly until fuel stops flowing from spill pipe and check timing mark through flywheel housing timing hole (Fig. L5-7). Repeat check several times to be sure reading of timing is accurate. If timing is too slow, remove shims (2—Fig. L5-4). Increase shim thickness if timing is too far advanced. Changing shim thickness by 0.1 mm (0.004 inch) will change timing about 1 degree.

When timing is correct, reinstall delivery valve and spring using new "O" ring and tighten delivery valve body to a torque of 41 N·m (30 ft.-lbs.). Reinstall fuel injector line and bleed system before tightening line fitting at injector.

Refer to the following chart for injection pump timing specifications:

FUEL INJECTION TIMING

Model	Speed Range, RPM	Degrees BTDC
AA1	Up to 2200	24
AA1	2201-3300	27
AA1	3301-3600	33
AA1	Variable governor	33
AB1	Up to 2200	23
AB1	2201-2700	26
AB1	2701-3600	30

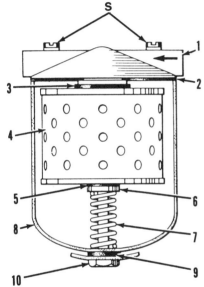

Fig. L5-2—Cross-sectional view of typical fuel filter used with remote mounted fuel tank. On models with fuel tank mounted on engine, filter is located in bottom of tank.

S. Bleed screws
1. Filter base
2. Bowl sealing ring
3. Upper filter seal
4. Filter element
5. Lower filter seal
6. Washer
7. Spring
8. Filter bowl
9. Bowl sealing ring
10. Center bolt

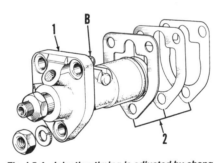

Fig. L5-4—Injection timing is adjusted by changing shim stack (2) thickness between pump flange (1) and engine crankcase. Pump rack ball (B) fits into fork in end of governor lever.

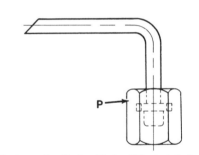

Fig. L5-6—A spill pipe (P) should be installed on delivery valve holder when checking injection timing.

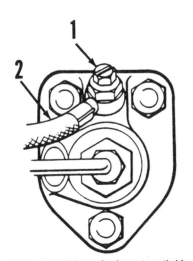

Fig. L5-3—On models not having automatic bleed, an air bleed screw (1) is located in fuel supply line (2) banjo bolt.

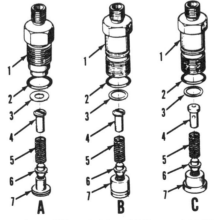

Fig. L5-5—Different styles of delivery valves are used in Bryce "A" and "B" series pumps (A), Bryce "E" series pump (B) and I.E.S.A. pump (C).

1. Delivery valve holder
2. "O" ring
3. Gasket
4. Spring guide
5. Spring
6. Delivery valve
7. Valve seat

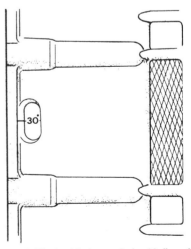

Fig. L5-7—Flywheel timing mark should align with mark in housing opening at beginning of injection.

Illustrations Courtesy Lister-Petter

Lister-Petter SMALL DIESEL

AB1	Variable governor	30
AC1	Up to 2200	26
AC1	2201-2800	28
AC1	2801-3300	30
AC1	3301-3600	33
AC1	Variable governor	30
AC1Z	All to Ser. #14649	32
AC1Z	All after #14649	30
AC1ZS	1000-1200	20
AC1ZS	1251-1800	25
AC2	2201-2800	30
AC2	2801-3200	32
AC2	3201-3600	34
AC2	Variable governor	32
AB1W	Up to 3300	29
AB1W	3301-3600	34
AB1W	Variable governor	29
AC1W	Up to 2200	26
AC1W	2201-2800	28
AC1W	2801-3300	30
AC1W	3301-3600	33
AC1W	Variable governor	30
AC2W	2201-2800	30
AC2W	2801-3200	32
AC2W	3201-3600	34
AC2W	Variable governor	32

GOVERNOR

The governor is a steel ball flyweight type located inside the engine crankcase and is gear driven from engine camshaft gear. Governor controls may be fixed single speed, fixed 2-speed or variable speed type, all with a Run/Stop control. An overload mechanism may be incorporated with governor controls; refer to following "OVERLOAD STOP" paragraph. Governor adjustments should be made only when engine is at normal operating temperature. Refer to the following paragraphs for adjustment of the different governor control units shown in Figs. L5-8 to L5-16.

FIXED SPEED CONTROL. Refer to Fig. L5-8. Loosen locknut (1) and turn adjuster (2) to obtain desired engine speed. Engine no-load speed should be 4 percent above rated speed shown on engine nameplate. Tighten locknut when desired speed is obtained.

VARIABLE SPEED (Type 1). To adjust Type 1 variable speed control, refer to Fig. L5-9 and proceed as follows: Set speed control to idle position. Loosen locknut and turn adjuster (1) to obtain idle speed of approximately 1200 rpm, then tighten locknut.

Set speed control to full speed position, loosen locknut and adjust screw (2) to obtain engine no-load speed that is 8 percent higher than rated speed shown on engine nameplate. Tighten locknut and install new locking wire and lead seal.

To adjust control cable, pull control rod (4) firmly in direction shown by arrow. There should then be a small amount of slack in control cable. If not, loosen locknut and turn adjuster (5) to obtain small amount of slack and tighten locknut. The speed control lever should then be moved slightly before cable starts to move speed control rod.

VARIABLE SPEED (Type 2). Refer to Fig. L5-10 and proceed as follows: Set speed control in idle position. Loosen locknut (1) and turn adjuster (2) to obtain idle speed of approximately 1200 rpm and tighten locknut.

Set speed control to full speed position, loosen locknut (5) and turn adjusting screw (4) to obtain no-load speed that is 8 percent higher than rated speed shown on engine nameplate. Tighten locknut and install new locking wire and lead seal. Recheck idle speed and readjust if necessary.

There should be a small amount of slack in control cable with speed control in idle position. If not, screw adjuster (6) in until control lever can be moved slightly before cable begins to move plunger.

TWO-SPEED CONTROL. To adjust 2-speed control shown in Fig. L5-11, proceed as follows: Set speed control (9) in idle speed position. Loosen locknut (1) and turn adjuster (2) to obtain idle speed of approximately 1200 rpm and tighten locknut.

Loosen locknut (5) and unscrew adjuster screw (4) so there can be full

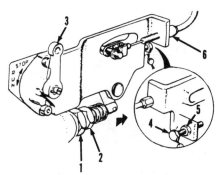

Fig. L5-10—Type 2 variable speed control.
1. Locknut
2. Idle speed adjusting screw
3. Control lever
4. High speed adjusting screw
5. Locknut
6. Cable adjuster

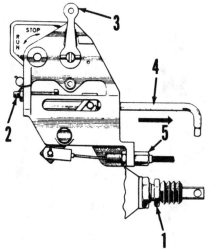

Fig. L5-9—View showing Type 1 variable speed control. Throttle control rod (4) engages hole in end of governor spring plunger.
1. Idle speed adjusting screw
2. High speed adjusting screw
3. Control lever
4. Throttle control rod
5. Cable adjuster

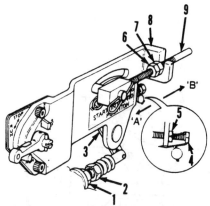

Fig. L5-11—Two-speed control in idle speed position. Lifting control arm (9) out of notch in bracket (8) and moving it so that adjusting nut (6) is against bracket with control lever back in notch places throttle in high speed position.
1. Locknut
2. Idle speed adjusting screw
3. Control rod
4. Control lever stop screw
5. Locknut
6. High speed adjusting nut
7. Locknut
8. Bracket
9. Operating arm

Fig. L5-8—View showing fixed speed control adjusting screw (2) and locknut (1).

SERVICE MANUAL

Lister-Petter

movement of speed control (9). Lift control (9) up and re-engage in bracket slot with nut (6) against outside of bracket (8). Loosen locknut (7) and turn adjuster nut (6) to obtain a no-load speed 8 percent above rated speed shown on engine nameplate and tighten locknut. Turn adjuster screw (4) in until it just touches arm (3), then back screw out slightly and tighten locknut (5). Install lock wire and lead seal.

VARIABLE/TWO-SPEED CONTROL. The control shown in Fig. L5-12 is available with a remote cable (9) and variable speed arm (1), or a two-speed control knob (2). To adjust, proceed as follows:

Set variable speed remote lever or 2-speed control knob in idle speed position. Loosen set screw (5) on idle stop (4) and move control rod (3) in direction of arrow to obtain 1200 rpm idle speed, move stop against bracket (8) and tighten set screw.

Set variable speed control to full speed position or move 2-speed control knob to RUN position. Loosen locknuts (7) and turn adjusting screw (6) to obtain no-load speed that is 8 percent higher than rated speed on engine nameplate. Tighten locknuts and install new lock wire and lead seal.

To adjust variable speed control cable (9), move control to idle speed position and turn cable adjuster so that there is a small movement of control lever before control rod (3) starts to move.

IDLE CONTROL. To adjust, loosen locknut (1—Fig. L5-13) and turn locknut and adjusting nut (2) to outer end of rod (3). Turn idle control lever (4) up to idle position. Loosen locknut (5) and turn adjuster (6) to obtain idle speed of approximately 1200 rpm and tighten locknut.

Loosen locknut (7), hold control rod (1) so that adjusting nut (8) is tight against bracket (9) and adjust nut (8) to obtain no-load speed that is 8 percent above rated speed shown on engine nameplate. Tighten locknut (7) and move lever (4) to position shown in Fig. L5-13. Turn nut (2) in until adjusting nut (8) is lightly touching bracket (9) and tighten locknut (1).

VARIABLE SPEED CONTROL. Refer to Fig. L5-14 for cable control and to Fig. L5-15 for adjusting (knurled) nut control of variable speed unit. Move locknuts (1) and knurled nut (2) to outer end of control rod (3). Loosen locknut (8) and turn adjuster (7) to obtain idle speed of 1200 rpm and tighten locknut. Tighten locknuts (1) at outer end of rod (3) and, on cable control only, tighten knurled nut (2) against locknuts. Knurled nut is used to vary engine speed on models without cable control.

Loosen locknut (6), pull rod (3) in direction of arrow until adjusting nut (5) is against bracket (4) and continue force on rod while turning nut (5) to obtain no-load speed of 8 percent above rated speed on engine nameplate. Tighten locknut (6); hold bracket (9) in proper position on cable control unit while tightening locknut.

The adjusting nut control unit may also be used as an adjustable fixed speed control by setting knurled nut to desired speed, then locking nuts (1) in place against knurled nut (2).

VARIABLE SPEED CONTROL (Spring Assisted). To adjust control shown in Fig. L5-16, proceed as follows: Move locknuts (4) and knurled nut (5) to outer end of control rod (6) and lock in this position. Set control lever to idle speed position. Loosen locknut (2) and set adjuster (3) so that control cable does not limit travel of control rod (6). Loosen locknut (8) and turn adjuster (9) to obtain idle speed of 1200 rpm, then tighten locknut.

Loosen locknut (11) and move control rod in direction of arrow so that adjusting nut (10) is tight against bracket (7). While holding control rod in this position, turn adjusting nut (10) to obtain no-load speed that is 8 percent above rated speed on engine nameplate. Position the plate (12) against adjusting nut

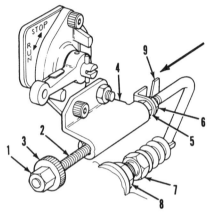

Fig. L5-14—Variable speed control adjusted for cable operation. Same unit may be set up for knurled nut (2) operation. Refer to Fig. L5-15.

1. Locknut
2. Control rod
3. Knurled nut
4. Bracket
5. High speed adjusting nut
6. Locknut
7. Idle speed adjuster
8. Locknut
9. Cable plate

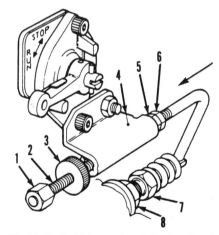

Fig. L5-15—Variable speed control set up for adjustment of operating speed by turning knurled nut (3). Same unit may be adjusted for fixed speed by turning knurled nut to obtain speed required, then tightening locknut (1) against knurled nut. Refer to Fig. L5-14 for same unit adjusted for cable operation.

1. Locknuts
2. Control rod
3. Knurled nut
4. Bracket
5. High speed adjusting nut
6. Locknut
7. Idle speed adjuster
8. Locknut

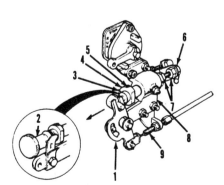

Fig. L5-12—Unit shown provides variable speed control when equipped with cable and lever (1). With control knob (2), control unit provides only two speeds.

1. Cable lever
2. Two-speed knob
3. Control rod
4. Idle speed stop
5. Set screw
6. High speed adjusting screw
7. Locknuts
8. Bracket
9. Cable adjuster

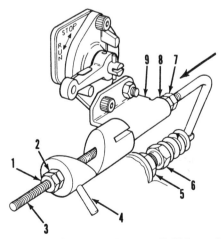

Fig. L5-13—Two speed idler control in high speed operation position.

1. Locknut
2. Adjusting nut
3. Control rod
4. Control lever
5. Locknut
6. Idle speed adjuster
7. Locknut
8. High speed adjusting nut
9. Bracket

Illustrations Courtesy Lister-Petter

Lister-Petter

(10) and tighten locknut (11). Turn cable adjuster (3) so engine control lever is capable of moving speed control from idle to maximum no-load position, then tighten nuts (2).

OVERLOAD STOP

Some engines are equipped with an overload stop (smoke stop or fuel injection limiter). On models so equipped, the exhaust should be a slight haze when engine at operating temperature is accelerated from idle to full speed position. The overload stop is factory adjusted and should not be disturbed unless wear or damage occurs to the mechanism.

Early model overload stops can only be adjusted when engine is disassembled; refer to OVERLOAD STOP paragraph in OVERHAUL section. On later models, refer to Fig. L5-17 or Fig. L5-18. Run engine under load until normal operating temperature is reached, then remove load from engine. Adjust governor speed control if necessary to give no-load speed of 8 percent above rated speed on engine nameplate. Move STOP/RUN control to decrease engine speed on fixed governor control, or move speed control lever to idle speed position. Allow engine to stabilize at idle speed, then quickly move STOP/RUN lever to RUN position or move variable speed control to full speed position. Turn overload stop spindle (2—Fig. L5-17 or L5-18) counterclockwise in small amounts until engine exhaust is black on acceleration. Repeat procedure as necessary. When black smoke is obtained on acceleration, turn spindle clockwise a small amount at a time and check exhaust on acceleration. Repeat procedure until exhaust no longer has black smoke and has a barely visible trace of smoke.

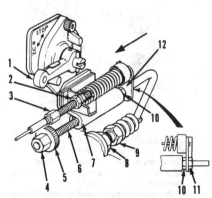

Fig. L5-16—Cable operated, spring assisted variable speed control.

1. Control lever
2. Locknut
3. Cable adjuster
4. Locknut
5. Knurled nut
6. Control rod
7. Bracket
8. Locknut
9. Idle speed adjuster
10. High speed adjusting nut
11. Locknut
12. Plate

REPAIRS

TIGHTENING TORQUES

When making engine repairs, observe the following special tightening torque values:

Cylinder head bolts:
 Air cooled . . . . .27 N·m
  (20 ft.-lbs.)
 Water cooled . . . . .30 N·m
  (22 ft.-lbs.)
Rocker support nut,
 water cooled . . . . .32 N·m
  (24 ft.-lbs.)
Timing cover bolts . . . . .12 N·m
  (9 ft.-lbs.)
Sump (oil pan) bolts . . . . .12 N·m
  (9 ft.-lbs.)
Connecting rod bolts . . . . .34 N·m
  (25 ft.-lbs.)
Intermediate main
 bearing housing . . . . .41 N·m
  (30 ft.-lbs.)
Camshaft gear bolt . . . . .36 N·m
  (27 ft.-lbs.)
Camshaft extension
 shaft bolt . . . . .19 N·m
  (14 ft.-lbs.)
Crankshaft gear bolt . . . . .36 N·m
  (27 ft.-lbs.)
Crankshaft extension
 shaft bolt . . . . .19 N·m
  (14 ft.-lbs.)

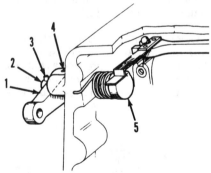

Fig. L5-17—View showing overload stop mechanism for fixed speed governor controls.

1. Overload stop lever
2. Spindle
3. Pin
4. Hole in lever
5. Cam

Fig. L5-19—Exploded view of flange mounted water pump used on some water cooled models. Refer to Fig. L5-20 for pedestal mounted pump assembly.

1. Adapter plate
2. Shims
3. Pump shaft washer
4. Water slinger
5. Pump body
6. Pump seal
7. Impeller
8. Gasket
9. Cover

SMALL DIESEL

Flywheel nut . . . . .210 N·m
  (155 ft.-lbs.)
Flywheel extension bolt . . . . .36 N·m
  (27 ft.-lbs.)
Fuel injection pump
 mounting bolts . . . . .18 N·m
  (13 ft.-lbs.)
Fuel injection pump
 delivery valve . . . . .41 N·m
  (30 ft.-lbs.)
Fuel injector stud nuts . . . . . . .18 N·m
  (13 ft.-lbs.)
Air cell plug . . . . .95 N·m
  (70 ft.-lbs.)

WATER PUMP AND COOLING SYSTEM

Models AB1W, AC1W and AC2W

Engine coolant for Models AB1W, AC1W and AC2W is circulated by a flange mounted pump (Fig. L5-19) or a pedestal mounted water pump (Fig. L5-20). Usual cooling installation is one or two high volume coolant tanks. A thermostat may be installed in the piping from engine to cooling tank. Temperature of engine outlet coolant should be maintained between 75° and 95° C (167° and 203° F).

FLANGE MOUNTED PUMP. To remove water pump, close off coolant lines, drain engine coolant and disconnect water lines from pump. Remove nuts from water pump body flange and slide pump from studs and drive shaft.

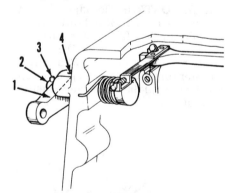

Fig. L5-18—View showing overload stop mechanism for variable speed governor controls. Refer to Fig. L5-17 for part identification.

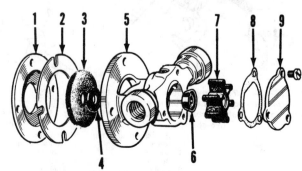

SERVICE MANUAL

Lister-Petter

Remove the pump adapter plate (1—Fig. L5-19), shims (2), washer (3) and water slinger (4). Remove the screws holding front cover (9) and remove cover and gasket (8). Remove impeller (7) and pump seal (6).

Check impeller, seal and water slinger for cracking or hardening of material. Check water pump drive shaft for wear or damage.

Using new parts as necessary, reassemble and reinstall pump as follows: Lubricate seal (6) with water pump grease and install in housing with lip to outside (toward impeller). Place water pump adapter plate (1) and shims (2) on mounting studs. Place washer (3) on pump drive shaft, then place water slinger (4) on shaft leaving a gap between washer and water slinger. Slide water pump body on drive shaft with impeller cam up, then install and tighten the mounting stud nuts. Using a straightedge, check to be sure pump drive shaft does not protrude beyond outer face of water pump; remove pump and add shims (2) if necessary. Lubricate impeller with water pump grease and install on shaft. Be sure gasket (8) is positioned correctly to cover impeller cam, then install gasket and cover (9).

PEDESTAL MOUNTED PUMP. To remove water pump, close off coolant lines, drain engine coolant and disconnect water lines from pump. Remove drive belt and pulley. Unbolt and remove pump assembly from mounting plate.

Remove cover (1—Fig. L5-20), gasket (2) and impeller (3); it may be necessary to use blunt tools to pry impeller from housing. Remove screw (6) and cam (5) from top of housing. Remove wear plate (4), seal (14) and "O" ring (13). Remove outer seal (10) and snap ring (9) from pump housing. Heat housing in hot water and press shaft and bearing assembly (12) from housing by pressing on impeller end of shaft. Remove inner seal (8) and water slinger (11) from housing.

Check all parts for wear or other damage and reassemble using necessary new parts as follows: Lubricate inner seal (8) with water pump grease and press seal into housing with outer lip of seal toward impeller cavity. Place water slinger (11) in drain area of pump housing and install slotted end of shaft and bearing assembly (12) through inner seal and water slinger. Heat pump housing in hot water, then press shaft and bearing assembly into housing. Install snap ring (9) with flat side toward bearing. Lubricate outer seal (10) with water pump grease and install with outer lip toward bearing so that outer face of seal is flush with housing. Lubricate "O" ring (13) and install in place. Lubricate seal (14) with water pump grease and install with lip toward impeller cavity. Install wear plate (4) with dimple toward impeller cavity and in line with cam retaining screw hole. Lightly coat top side and inner edge (groove side) of cam (5) with non-hardening gasket compound and install cam with groove over dimple on wear plate. Cam retaining screw should also be coated with non-hardening gasket compound. Lubricate impeller cavity and install impeller (3) with twisting motion until impeller drive screw engages slot in shaft. Be sure gasket (2) is positioned correctly to cover impeller cam, then install gasket and cover (1).

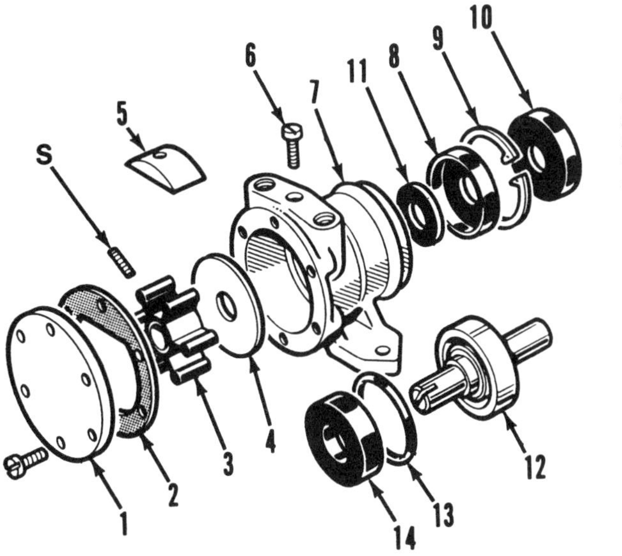

SALT WATER ANODES

Models AB1W, AC1W and AC2W

Where direct sea water cooling is used, zinc anodes must be installed in cylinder block and, on early models, cylinder head. These anodes must be checked at regular intervals and renewed if anode material is eroded by 75 percent of original size. Do not use any type of insulating material on threads of anode holders or adapters as this will reduce efficiency of the anode.

COLD START PRIMING PUMP

Refer to Fig. L5-21 for exploded view of optional cold start priming pump and oil reservoir unit. The reservoir also functions as rocker arm cover. To remove unit, proceed as follows: Disconnect external oil feed pipe (8). Remove the reservoir retaining bolts (1) and lift off cover (2), gasket (3) and anti-surge shim (4). Then remove the reservoir (5) and pump plunger (6) assembly. The plunger can now be removed from the reservoir.

Renew all "O" rings and gaskets and assemble as follows: Place outlet port sealing "O" ring (not shown) in cylin-

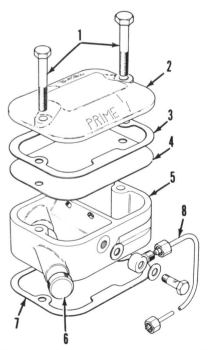

Fig. L5-21—Exploded view of optional cold start priming pump and reservoir. Priming plunger (12—Fig. L5-22) is standard equipment.

1. Mounting bolts
2. Cover
3. Gasket
4. Anti-surge shim
5. Reservoir
6. Pump plunger
7. Gasket
8. Oil supply line

Fig. L5-20—Exploded view of pedestal mounted water pump used on some water cooled engines; refer to Fig. L5-19 for flange mounted water pump.

S. Impeller drive screw
1. Cover
2. Gasket
3. Impeller
4. Wear plate
5. Impeller cam
6. Cam retaining screw
7. Pump body
8. Inner bearing seal
9. Snap ring
10. Outer bearing seal
11. Water slinger
12. Shaft & bearing assy.
13. "O" ring
14. Pump seal

der head. Fit the reservoir and plunger assembly on cylinder head with new gasket (7) and connect oil feed line to reservoir and cylinder head. Place the shim (4), gasket (3) and cover on reservoir and position plunger so that retaining bolt can be installed. The bolt fits in wide groove on plunger and keeps plunger from being fully withdrawn when bolt is in place. Tighten retaining bolts (1).

VALVE ADJUSTMENT

Valve rocker arm to valve tip clearance should be 0.1 mm (0.004 inch) for all models and adjustment may be checked after removing rocker arm cover(s) or cold start priming pump. To adjust valves, turn crankshaft until piston is at TDC of compression stroke and check gap between rocker arm and tip of valve stem. It may be necessary to bend a 0.1 mm (0.004 inch) gage strip in a right angle to reach valve gap. Loosen rocker arm adjusting screw locknut and turn screw to obtain correct clearance. Tighten locknut and recheck clearance. Repeat procedure for No. 2 cylinder on Models AC2 and AC2W.

CYLINDER HEAD

To remove cylinder head(s), proceed as follows: On models with engine mounted fuel tank, remove the tank. On Model AC2 only, remove the fuel tank bracket. On water cooled engines, close off coolant lines, drain coolant from engine and disconnect coolant pipes. On all engines, remove muffler, air cleaner and cylinder cowling. Remove pump to injector line(s), injector fuel return line and fuel injector(s); immediately cap or plug all fuel openings. Remove the manifolds and rocker arm cover(s) or cold start priming pump and reservoir (see Fig. L5-21) if so equipped. Remove cylinder head oiling line and standard equipment cold weather starting aid priming plungers (12—Fig. L5-22). Gradually loosen cylinder head stud nuts in sequence shown in Fig. L5-23, then remove nuts and washers. Remove rocker arm support assembly, "O" ring and the push rods. Lift cylinder head from engine and remove the push rod tubes on air cooled engines.

Piston to cylinder head (bumping) clearance should be checked before final installation of cylinder head. Turn crankshaft so that piston is about 6.35 mm (1/4 inch) before top dead center. Place three pieces of soft lead wire or solder at equal spacing on top of piston so they are not aligned with valve heads or combustion chamber. Install cylinder head with gasket and turn crankshaft so that piston moves past top dead center. Remove the cylinder head and measure thickness of flattened wires. The average of the three thicknesses is piston to head clearance. Clearance should be 0.56-0.66 mm (0.022-0.026 inch). If necessary, adjust clearance by adding or removing shims (33—Fig. L5-22).

When installing cylinder head, use new push rod tube seals on air cooled engines, new cylinder head gasket and rocker arm support "O" ring. Be sure gasket surfaces are cleaned thoroughly and place new head gasket on the stud bolts. Install cylinder head and push rod tubes with new tube seals. Lubricate push rods and place in tubes. Place rocker arm support on retaining stud with new sealing "O" ring and be sure support is aligned on dowel pin. Install the self-locking nut on rocker arm support stud. Install the remaining stud nuts and washers and turn down finger tight. Tighten the self-locking nut down just enough to be sure rocker arm support is level and is in contact with cylinder head. On two cylinder models, be sure the manifold bolting surfaces are aligned. Then, tighten each of the five cylinder head nuts down 1/4 turn at a time, following sequence shown in Fig. L5-23 until final torque value is reached. On air cooled models, all five cylinder head nuts, including the rocker arm support nut, should be tightened to a torque of 27 N·m (20 ft.-lbs.). Water cooled engine rocker support nut should be tightened to a torque of 32 N·m (24 ft.-lbs.) and the remaining nuts to a torque of 30 N·m (22 ft.-lbs.). Complete reassembly by reversing removal procedure.

NOTE: On water cooled models, two different brands of head gaskets (Reinz and Goetze) are available as shown in Fig. L5-24. The Goetze gasket is thicker than the Reinz gasket; therefore, Goetze gasket cannot be installed on an engine that was equipped with a Reinz gasket unless top of cylinder block is milled 0.51 mm (0.020 inch)

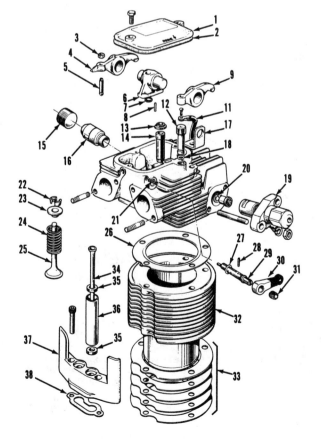

Fig. L5-22—Exploded view of cylinder head and related parts. Chain (11) keeps priming plunger (12) from being lost when plunger is removed to pour oil in priming cup. Optional priming pump is shown in Fig. L5-21.

1. Valve cover
2. Gasket
3. Locknut
4. Exhaust rocker arm
5. Adjusting screw
6. Rocker pedestal & shaft
7. Pedestal "O" ring
8. Pedestal dowel pin
9. Intake rocker arm
11. Plunger safety chain
12. Priming plunger
13. Oil seal (intake only)
14. Valve guide
15. Plug
16. Air cell
17. Lifting eye
18. Plunger "O" ring
19. Fuel injector
20. Injector seal washer
21. Lever stop pin
22. Split valve locks
23. Valve spring retainer
24. Valve spring
25. Valve
26. Head gasket
27. "O" ring
28. Pin
29. Compression relief shaft
30. Compression relief lever
31. Retaining sleeve
32. Cylinder
33. Cylinder shims
34. Push rod
35. Push rod tube seals
36. Push rod tube
37. Push rod tube seat
38. Gasket

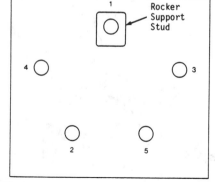

Fig. L5-23—View showing cylinder head nut tightening sequence for all models.

SMALL DIESEL

Lister-Petter

to maintain same compression as with Reinz gasket. Clearance between cylinder head and top of piston at TDC should be 0.66-0.076 mm (0.026-0.030 inch) with Goetze gasket and 0.56-0.66 mm (0.022-0.026 inch) with Reinz gasket.

After 20 running hours from cylinder head installation, cylinder head nuts should be retorqued. Loosen each nut 1/4 turn in sequence shown in Fig. L5-23, then tighten to specified torque in sequence.

VALVE SYSTEM

Valves seat in renewable seat inserts in the cylinder head. Valve face and seat angle is 45 degrees on all models. Seat width should be 0.99-1.447 mm (0.039-0.057 inch) on Model AA1 prior to serial #14927 and 0.635-1.066 mm (0.025-0.042 inch) on engine serial #14927 and up. Seat width on all other models should be 1.295-1.727 mm (0.051-0.068 inch). If necessary to renew valve seats, old seats must be removed and the cylinder head machined to accept service seats. Heat cylinder head to 150° C (300° F) prior to pressing new seats into head.

Distance from head of valve to cylinder head gasket surface (valve head clearance) should be 0.99-1.45 mm (0.039-0.057 inch) on Model AA1 prior to engine serial #14927 and 0.63-1.06 mm (0.025-0.042 inch) on Model AA1 engine serial #14927 and up. On all other models, valve head clearance should be 1.29-1.72 mm (0.051-0.068 inch). Maximum allowable valve head clearance is 1.65 mm (0.065 inch) for Model AA1 prior to serial #14927 and 1.27 mm (0.050 inch) for Model AA1 serial #14927 and up. For all other models, maximum allowable valve head clearance is 1.93 mm (0.076 inch).

If necessary to renew valve guides, first measure height of guide above valve spring surface of head. Then heat cylinder head in boiling water for at least two minutes. Support top side of head on wood blocks and using suitable driver, remove old guides toward top of head. To install new guides, be sure valve guide bores in head are clean, heat head in boiling water and drive new guides in from top of head so that height is same as old guide. New guides are prefinished and should not be reamed.

DECOMPRESSOR LEVER

To adjust decompressor lever, remove valve cover and turn crankshaft so that exhaust valve is fully closed. With decompressor lever in vertical position, pin on inner end of decompressor shaft (27—Fig. L5-22) should just lift exhaust rocker arm (4) sufficiently to take up valve clearance. If adjustment is necessary, hold decompressor lever (30) in vertical position and with screwdriver inserted in slot in outer end of shaft (29), turn shaft to obtain proper adjustment.

INJECTOR(S)

The pintle type nozzle emits a fine spray pattern which is directed to the hole in the air cell (16—Fig. L5-25) located directly across the cylinder head combustion chamber. A properly performing fuel injector is very important in these engines.

REMOVE AND REINSTALL. First, remove air cleaner assembly, engine cowling and if so equipped, engine mounted fuel tank. Thoroughly clean fuel injector lines and surrounding area. Disconnect and remove the injection pump to injector line and fuel return line from injector. Remove the two injector retaining nuts and pull injector from head. Remove the fuel injector sealing washers; refer to Figs. L5-26 and L5-27.

Be sure injector bore in cylinder head is clean and install new copper washer and heat shield as shown in Fig. L5-26, or new steel shield washer as shown in Fig. L5-27. Install injector and retaining nuts. Install pump to injector line and tighten fittings finger tight plus 1/3 turn with wrench. Tighten injector retaining nuts to a torque of 18 N·m (13 ft.-lbs.). Loosen injector line fitting at injector and bleed air from line, then tighten fitting.

TESTING. A complete job of testing and adjusting the injector requires use of special test equipment, although the injector spray pattern may be checked by attaching injector to line outside the engine and cranking engine while observing spray. Refer to SPRAY PATTERN paragraph. Recommended tester fluid is Shell Calibration Fluid B or C. Nozzle

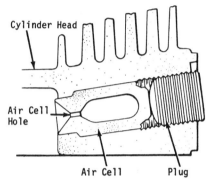

Fig. L5-25—Cross-sectional view of cylinder head and air cell. Refer to text for air cell cleaning procedure.

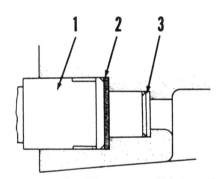

Fig. L5-26—Two different types of injector sealing have been used; refer also to Fig. L5-27. Note direction of heat shield at inner end of injector.
1. Injector
2. Copper washer
3. Heat shield

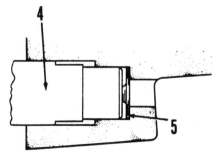

Fig. L5-27—Two different types of injector sealing have been used; refer also to Fig. L5-26. Note dimpled side of steel shield washer is toward injector.
4. Injector
5. Steel shield washer

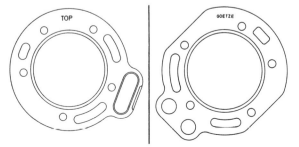

Fig. L5-24—"Reinz" (left) and "Goetze" (right) head gaskets used on water cooled models should not be interchanged without piston clearance modification because of different gasket thickness. Refer to text.

Illustrations Courtesy Lister-Petter

Lister-Petter — SMALL DIESEL

should be tested for opening pressure, seat leakage and spray pattern. When tested, nozzle should open with high pitch buzzing sound (chatter) and cut off quickly at end of injection with minimum seat leakage.

WARNING: Fuel emerges from injector with sufficient force to penetrate the skin. When testing injector, keep yourself clear of nozzle spray.

Before conducting test, operate tester lever until fuel flows from open line, then attach injector. Close valve to tester gage and pump tester lever a few quick strokes to be sure nozzle valve is not stuck, and that injector is suitable for further testing.

OPENING PRESSURE. Open valve to tester gage, then operate tester lever slowly while observing gage reading. Opening pressure should be 16200-18275 kPa (2350-2650 psi). Adjust opening pressure by removing cap nut (1—Fig. L5-28) and turning adjusting screw (3).

SPRAY PATTERN. Spray pattern should be a straight cone from nozzle tip, well atomized and free of split or drops. If pattern is not correct, injector must be disassembled and the nozzle cleaned or renewed. A poor spray pattern will result in fuel spray missing the hole in the air cell, causing loss of power and excessive exhaust smoke.

SEAT LEAKAGE. Operate tester lever slowly to a pressure 1000 kPa (150 psi) below observed opening pressure and hold pressure at this reading for 10 seconds. If drop of fuel forms at nozzle sufficient to cause continuous film on finger tip, clean or renew nozzle.

BACK LEAKAGE. Operate tester lever slowly to bring gage reading to 15175 kPa (2200 psi), then record time interval until gage reading drops to 10135 kPa (1470 psi). This time should be more than 6 seconds. If back leakage time is less than 6 seconds, nozzle should be renewed.

OVERHAUL. Wipe all dirt and loose carbon from exterior of injector assembly and secure injector body (7—Fig. L5-28) in fixture or soft jawed vise. Remove cap nut (1), adjusting screw (3) and sealing washer (2). Lift out spring seat (4), spring (5) and push rod (6). Unscrew nozzle nut (10) and remove nozzle body (9) and valve (8). It may be necessary to push nozzle out of nut using copper or brass hollow drift. Do not drive out by striking nozzle end face.

Clean exterior surfaces with a brass wire brush. Soak parts in an approved carbon solvent, if necessary, to loosen hard carbon deposits. Rinse parts in clean diesel fuel or calibrating oil immediately after cleaning to neutralize solvent and prevent etching of polished surfaces.

Refer to Fig. L5-29. Scrape carbon from pressure chamber (5) using a hooked scraper (4). Use suitable size wire or twist drill (2) to clean fuel passage (2) to pressure chamber. Clean nozzle spray hole using a spray hole cleaner (6) or hardwood stick.

Back flush nozzle using reverse flush adapter on tester. Reclean all parts by rinsing thoroughly in clean diesel fuel or calibrating oil and assemble while immersed in clean fluid. Set opening pressure before installing cap nut (1—Fig. L5-28).

FUEL AIR CELL

The fuel emitted from the injector nozzle is directed into an air cell directly across the cylinder head combustion chamber from the injector nozzle; refer to Fig. L5-25. The air cell hole can be cleaned of carbon with a soft wire after removing cylinder head. Do not remove the air cell unless damage to tip is visible. To renew air cell, remove the plug (15—Fig. L5-22) and use a drift to tap air cell (16) from head. Be sure bore is clean, install new cell and tighten retaining plug to a torque of 95 N·m (70 ft.-lbs.).

INJECTION PUMP

The fuel injection pump (see Fig. L5-31) is mounted on the right side of crankcase near timing gear cover and is operated from lobe(s) on engine camshaft. To remove injection pump, first

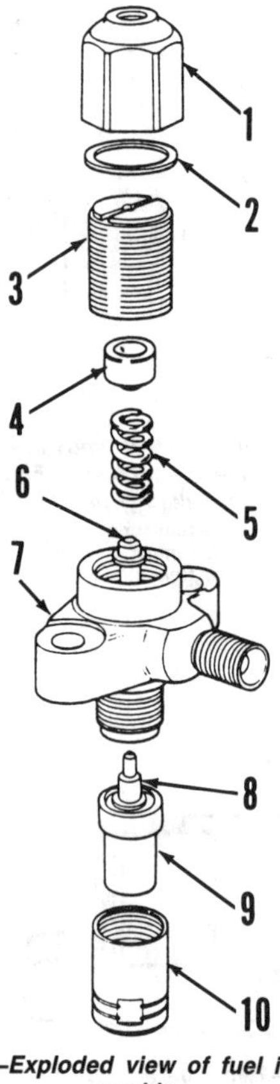

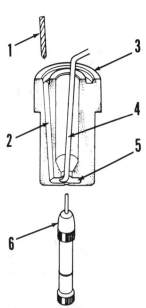

Fig. L5-28—Exploded view of fuel injector assembly.

1. Cap nut
2. Gasket washer
3. Adjusting screw
4. Spring seat
5. Spring
6. Push rod
7. Injector body
8. Nozzle valve
9. Nozzle body
10. Nozzle nut

Fig. L5-29—Clean nozzle fuel feed hole (2) with twist drill (1). Scrape carbon deposits from fuel chamber (5) with brass scraper (4). Clean spray hole with special spray hole cleaner (6).

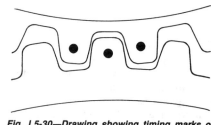

Fig. L5-30—Drawing showing timing marks on camshaft gear (two dots) and crankshaft gear (one dot).

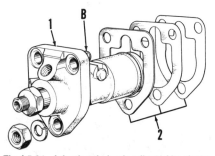

Fig. L5-31—Injection timing is adjusted by changing shim stack (2) thickness between pump flange (1) and engine crankcase. Pump rack ball (B) fits into fork in end of governor lever.

SERVICE MANUAL

Lister-Petter

thoroughly clean outside of engine and proceed as follows:

Close fuel supply valve or drain fuel tank and remove engine mounted fuel tank if so equipped. Remove air cowling on air cooled engines. Disconnect and remove fuel lines and immediately cap or plug all openings. Remove rocker arm oil supply line. Turn engine to TDC on exhaust stroke of piston (No. 1 cylinder on Models AC2 and AC2W). Remove injection pump mounting stud nuts. Move STOP/RUN lever (A—Fig. L5-32) to about 10 degrees before vertical position to align fuel pump rack ball with cut-away in crankcase, then remove pump and timing shims. Be careful not to lose or damage any of the shims.

Disassembly of pump is not recommended outside of an authorized diesel fuel system repair facility. A new or reconditioned injection pump should be installed if pump is faulty.

To install pump, be sure piston is at TDC on exhaust stroke (No. 1 piston on Models AC2 and AC2W). Place STOP/RUN lever at about 10 degrees before vertical position to align governor lever with cutaway in engine crankcase.

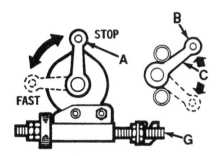

Fig. L5-32—View showing typical STOP/RUN lever (A), overload lever (B) and governor controls (G). Pushing overload lever (A) down to dotted position provides excess fuel delivery for aid in starting engine.

Place the timing shims (2—Fig. L5-31) on pump mounting studs. Install fuel injection pump, making sure ball (B) on pump rack engages governor lever fork. Install the retaining nuts and tighten evenly to a torque of 18 N·m (13 ft.-lbs.). Complete remainder of installation by reversing removal procedure. Bleed fuel system as outlined in MAINTENANCE section.

TIMING GEAR COVER

To remove timing gear cover (15—Fig. L5-33), proceed as follows: Drain oil from engine, drain fuel tank (on models equipped with engine mounted tank) and remove tank from engine. Remove manual starting gear (if used) from cover. Remove the eight socket head screws holding cover to crankcase, then carefully remove cover from crankcase and aligning dowel pins (12).

To reinstall timing gear cover, reverse removal procedure using a new timing cover gasket (17). Tighten cover retaining screws to a torque of 12 N·m (9 ft.-lbs.).

TIMING GEARS

Timing gears are accessible after removing the timing cover. The camshaft and crankshaft timing gears are retained by a flat washer and cap screw. The oil pump drive gear and governor drive gear are retained on respective shafts by a hex nut.

The crankshaft gear can be removed after removing retaining bolt and washer. Then install bolt without washer into end of crankshaft and thread puller bolts into crankshaft gear. When installing gear on crankshaft, first heat gear in hot oil, then be sure keyway is aligned and drive gear onto shaft using hollow driver. The single marked tooth on crankshaft gear must mesh between the two marked teeth on camshaft gear as shown in Fig. L5-30. Install gear retaining washer and bolt and tighten bolt to a torque of 36 N·m (27 ft.-lbs.).

Remove camshaft gear retaining bolt and washer, then install bolt back in end of camshaft to protect camshaft threads. Using suitable gear puller, remove camshaft gear. Take care not to lose key from camshaft. To install gear, remove bolt from camshaft and thread suitable threaded gear pusher tool and push gear onto camshaft with nut and thrust washer. The two marked teeth of camshaft gear must mesh between the single marked tooth of crankshaft gear as shown in Fig. L5-30. Reinstall gear retaining washer and bolt and tighten bolt to a torque of 36 N·m (27 ft.-lbs.).

NOTE: To remove camshaft gear on "Model Build" engines (AA1 after serial #160000, AB1 after serial #55000, AC1 after engine serial #60000, AC1Z after serial #20000, AC1ZS after serial #10000 and AC1W after serial #6000), it is first necessary to remove crankshaft gear. Other models have relief cut in crankshaft gear collar to allow clearance for camshaft gear.

To remove oil pump gear or governor gear, remove retaining nut and washer and pull gear from shaft. Be careful not to lose key from shaft. Install key, gear and washer on shaft and tighten the self-locking nut.

OIL PUMP AND RELIEF VALVE

The gerotor type oil pump is mounted in engine crankcase and oil pump gear (4—Fig. L5-34) is driven by engine cam-

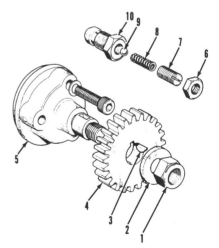

Fig. L5-34—Oil pump (5) is mounted in front face of crankcase and pump gear (4) is driven by camshaft timing gear. Oil pressure relief valve (6 through 10) is located at left side of oil pump.

1. Hex nut
2. Flat washer
3. Key
4. Pump gear
5. Oil pump
6. Locknut
7. Adjusting screw
8. Spring
9. Valve ball
10. Valve body

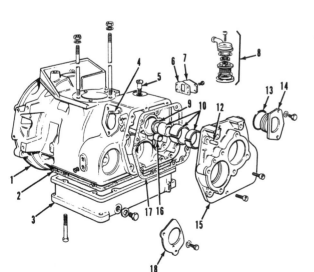

Fig. L5-33—Exploded view of crankcase (1) and timing gear cover (15) components.

1. Crankcase
2. Oil sump gasket
3. Oil sump
4. Injection pump stud
5. Plug
6. Gasket
7. Inspection cover
8. Breather assy.
9. Camshaft bore plug
10. Camshaft bearings
12. Dowel pin
13. "O" ring
14. Cover
15. Timing gear cover
16. Governor shaft bushing
17. Timing cover gasket
18. Cover plate

Illustrations Courtesy Lister-Petter

Lister-Petter

SMALL DIESEL

shaft gear. The oil pressure relief valve (6 through 10) is located in crankcase at side of engine oil pump. To remove the engine oil pump, it is first necessary to remove the camshaft gear. Remove oil pump gear retaining nut (1) and washer and pull gear from pump shaft if pump is to be disassembled for inspection. Unbolt and remove the pump (5) from engine crankcase. Install pump by reversing removal procedure.

The oil pump is serviced as a complete assembly only. Pump can be disassembled for inspection. Rotor end clearance should be 0.025-0.064 mm (0.001-0.0025 inch) with wear limit of 0.127 mm (0.005 inch). Clearance of inner rotor lobe to outer rotor lobe should be 0.051-0.127 mm (0.002-0.005 inch) and must not exceed 0.203 mm (0.008 inch). Shaft clearance in bearing bore should be 0.038-0.076 mm (0.0015-0.003 inch) and should not exceed 0.127 mm (0.005 inch). Shaft diameter new is 15.032-15.044 mm (0.5918-0.5923 inch).

GOVERNOR AND LINKAGE

The governor unit (see Fig. L5-36) can be removed after removing timing gear cover from engine crankcase. If governor is to be disassembled for inspection, remove the governor gear retaining nut (1), then pull gear (2) from governor shaft. Be careful not to lose the gear drive key (4). Unbolt and remove the governor housing (3) and flyweight assembly (5 through 12) from cylinder block.

Carefully inspect governor unit for wear or damage to bearing (6), drive plate (7), flyweight balls (8) and sliding cone (9). If any part is worn or damaged beyond further use, the complete governor assembly must be renewed. Inspect governor shaft bushing (16—Fig. L5-33) in engine crankcase and renew bushing if necessary. Install governor by reversing removal procedure.

With timing cover removed, inspect governor linkage. To service governor linkage, remove camshaft gear and governor assembly. Unbolt and remove governor controls (see Figs. L5-8 through L5-16) and the STOP/RUN control assembly. Loosen locknut (3—Fig. L5-37) and unscrew governor spring unit (2 through 7). Remove crankcase breather (7—Fig. L5-35) or plug if breather is not fitted. Loosen clamp bolt and remove governor arm (8) from bottom of lever shaft (6). Pull lever shaft up and place a wood block in crankcase under lower end of shaft. Using a sleeve that will fit over upper end of shaft, drive lever (5) downward until free of shaft. It may be necessary to use a thicker block than would originally fit under

shaft to completely free the lever from shaft.

Clean and inspect all linkage parts and renew as necessary. To reassemble, position lever (5) in crankcase and tap shaft (6) down through lever until it comes to stop. Install governor arm (8) on bottom of shaft and tighten clamp bolt enough to be sure arm will not fall off. Install governor spring unit (2

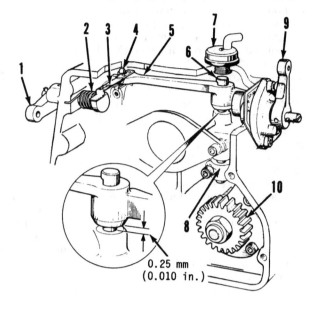

Fig. L5-35—Cutaway view of crankcase with timing cover removed showing typical later production governor components. Overload mechanism (1, 2 and 3) is not used on all models. End play of governor lever shaft should be 0.25 mm (0.010 inch).
1. Overload lever
2. Overload return spring
3. STOP/RUN leaf spring
4. Pump rack pin
5. Governor lever
6. Governor lever shaft
7. Breather or plug
8. Governor arm
9. STOP/RUN lever
10. Governor gear

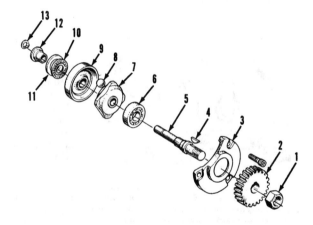

Fig. L5-36—Exploded view of typical governor assembly.
1. Hex nut
2. Governor gear
3. Bearing support
4. Key
5. Shaft
6. Ball bearing
7. Drive plate
8. Steel balls
9. Sliding cone
10. Thrust bearing
11. Thrust plate
12. Bushing
13. Snap ring

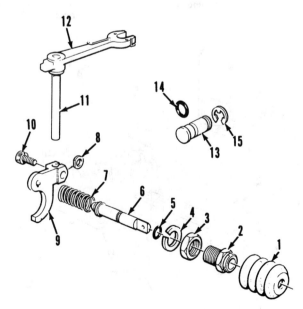

Fig. L5-37—Exploded view of typical governor linkage. Plug (13) is used to fill hole in crankcase on engines not equipped with overload stop mechanism.
1. Rubber boot
2. Idle adjuster
3. Locknut
4. Lockwasher
5. "O" ring
6. Spring plunger
7. Spring
8. Washer
9. Governor arm
10. Clamp screw
11. Lever shaft
12. Governor lever
13. Plug
14. "O" ring
15. Snap ring

SERVICE MANUAL

Lister-Petter

through 7—Fig. L5-37). To complete assembly and adjustment of linkage, the governor unit and fuel injection pump must be installed. Loosen clamp bolt in governor arm (8—Fig. L5-35) and turn governor spring adjuster (2) in until arm is against governor, pushing the sliding cone (9—Fig. L5-36) and rotating housing (7) together. Operate overload lever (1—Fig. L5-35), if so equipped, and push lever (5) and fuel injection pump rack pin (4) to maximum fuel position. With arm (8) and lever (5) in this position, tighten clamp bolt. Note that arm (8) should be positioned on shaft (6) to allow an up/down movement in shaft of 0.25 mm (0.010 inch). To set early type overload stop, refer to OVERLOAD STOP paragraph. Complete reassembly by reversing disassembly procedure.

OVERLOAD STOP

Early models may be equipped with an overload stop as shown in Fig. L5-38. This unit may be adjusted only with timing gear cover removed from engine. Late model overload stops (Figs. L5-17 and L5-18) are adjustable externally.

EARLY OVERLOAD STOP. To check overload stop adjustment, move STOP/RUN lever (L—Fig. L5-38) to STOP position and measure distance "X" between end of fuel pump rack and straightedge placed across gasket face of crankcase. Move STOP/RUN lever to RUN position and move governor lever and pump rack so that distance between end of pump rack and gasket face of crankcase is equal to dimension "X" plus "Y." Dimension "Y" is 11.2 mm (0.441 inch) on Models AA1, AB1 and AB1W, or 12.7 mm (0.500 inch) on Models AC1 and AC1W. The step on the overload cam (C) should then just touch the step in the governor leaf spring. If not, loosen cap screw (S) and rotate cam to proper position, then tighten cap screw.

LATE OVERLOAD STOP. To initially set late style overload stop, turn spindle (2—Fig. L5-17 or L5-18) to align pin (3) with hole in overload stop lever (4). Make final adjustment as outlined in OVERLOAD STOP paragraph in MAINTENANCE section.

PISTON AND ROD UNIT(S)

The piston and rod unit(s) can be removed after removing cylinder head(s) and oil sump (pan). Remove connecting rod cap and remove cylinder with piston and rod unit. Shims are installed between cylinder and crankcase to adjust piston to cylinder head clearance. Take care not to lose or damage shims as they are removed from cylinder studs and retain them for reassembly of engine.

Piston pin location in piston is not offset and the flat-top piston may be installed either way on connecting rod. To install piston and rod unit, be sure the piston ring end gaps are positioned equally around the piston and pin retaining snap rings are securely installed. Lubricate piston, rings and cylinder bore with clean engine oil and using ring compressor, install piston in cylinder with numbers on connecting rod and cap toward camshaft side of engine. Place rod bearing inserts in connecting rod and cap. Place cylinder shims on the cylinder studs, lubricate crankpin and install the cylinder, rod and piston unit. Install rod cap with assembly number on cap aligned with number on rod and tighten rod bolts to a torque of 34 N·m (25 ft.-lbs.).

Complete remainder of reassembly by reversing disassembly procedure.

PISTON, PIN, AND RINGS

Piston is fitted with two compression rings and one oil control ring. Piston for some Model AA1 engines may have a fourth ring groove below the piston pin, but this ring groove is not used.

With piston and rod unit removed as outlined in preceding paragraph, remove piston rings and the piston pin retaining snap rings. Check piston skirt for cracks, excessive wear or scoring. Check side clearance of new ring in top ring groove using a feeler gage. Renew piston if side clearance is more than 0.25 mm (0.010 inch). Piston pin should not show any visible wear and fit piston pin bore with a slight interference fit. Pistons and rings are available in oversizes of 0.5 mm (0.020 inch) and 1.0 mm (0.040 inch) as well as standard. Cylinder must be rebored and honed to install oversize piston and rings.

If new piston rings are being installed without reboring the cylinder, cylinder should be deglazed with medium grade carborundum cloth using hand motion to produce a matte surface or by use of a rotary brush cylinder hone. Wash cylinder in kerosene after deglazing to remove all traces of abrasive.

Piston ring end gap, new, should be 0.28-0.41 mm (0.011-0.016 inch) for Model AA1 and 0.30-0.43 mm (0.012-0.017 inch) for all other models. Maximum allowable ring end gap is 1.14 mm (0.045 inch) for all models. Install compression rings with step or bevel on inside of ring up, and with step on outside of ring down. Oil control rings and compression rings without step or bevel should be installed with identification mark up; rings without step, bevel or mark may be installed either side up. See Fig. L5-39.

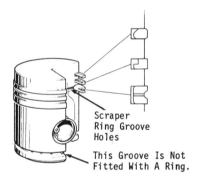

Fig. L5-39—Cross-sectional view of pistons and rings. Note ring shape and position for correct installation.

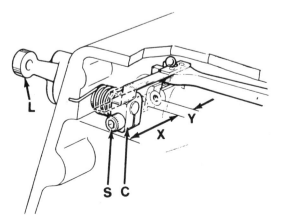

Fig. L5-38—Internal adjustment for early style overload lever (L) mechanism is shown. Refer to text for measurement of dimension "X" and adjustment of cam (C). Dimension "Y" is 11.2 mm (0.441 in.) for Models AA1 and AB1 and 12.7 mm (0.500 in.) for Model AC1. Adjustment must be made and cap screw (S) tightened securely before engine is assembled.

Lister-Petter — SMALL DIESEL

Place piston on rod and insert pin through bores in piston and rod. Be sure piston pin snap rings are securely installed in the grooves of piston pin bore. Sharp edge of snap ring should face away (out) from piston pin.

CYLINDER

Standard cylinder bore is 69.850-69.875 mm (2.750-2.751 inches) for Model AA1 and 76.20-76.23 mm (3.000-3.001 inches) for all other models. If cylinder wear is 0.25 mm (0.010 inch) or more, cylinder must be rebored to next oversize of 0.5 mm (0.020 inch) or 1.0 mm (0.040 inch), or renewed if wear exceeds rebore limit for next oversize.

CONNECTING RODS AND BEARINGS

Connecting rod is fitted with a pressed-in renewable bushing (5—Fig. L5-41) at piston pin end and a slip-in precision fit bearing insert (7) at crankpin end. Piston pin should be a hand push sliding fit in bushing. If necessary to renew piston pin bushing, install new bushing with oil hole aligned with hole in connecting rod. Ream or hone bushing after installation to inside diameter of 22.233-22.243 mm (0.8753-0.8757 inch).

Crankpin bearing oil clearance should be 0.025-0.090 mm (0.001-0.0035 inch). Crankpin bearings are available in undersizes of 0.25 mm (0.010 inch) and 0.5 mm (0.020 inch) as well as standard size.

CAMSHAFT AND BEARINGS

To remove camshaft, first remove rocker arm support and push rods and fuel injection pump. Remove timing gear cover. Remove camshaft gear retaining bolt (8—Fig. L5-40) and washer (7). To remove camshaft gear on "Model Build" engines (AA1 after serial #160000, AB1 after serial #55000, AC1 after engine serial #60000, AC1Z after serial #20000, AC1ZS after serial #10000 and AC1W after serial #6000), it is first necessary to remove crankshaft gear (12—Fig. L5-41). Hold cam followers (1—Fig. L5-40) up with magnet tools or invert engine so that followers fall away from camshaft. Working through holes in camshaft gear (6), remove cam thrust plate retaining screws (4) (some engines do not have a thrust plate), pull cam and gear assembly from crankcase and remove camshaft gear.

With camshaft removed, inspect camshaft bearings (10—Fig. L5-33) and renew if necessary. Camshaft bearings are available in standard size only and are a press fit in crankcase. Be sure when installing new bearings that oil holes in bearings are aligned with oil passages in cylinder block.

To install gear on camshaft, heat gear in hot oil, then press gear onto shaft. Reinstall camshaft and gear assembly making sure timing marks on camshaft and crankshaft gears are mated as shown in Fig. L5-30. Tighten thrust plate bolts if so equipped. Install gear retaining bolt and washer and tighten bolt to a torque of 36 N·m (27 ft.-lbs.). On models equipped with camshaft thrust plate (3—Fig. L5-40), camshaft end play should be 0.08-0.25 mm (0.003-0.010 inch).

CRANKSHAFT AND MAIN BEARINGS

On single cylinder models, crankshaft is supported in two main bearings (17 and 18—Fig. L5-41). On two-cylinder models, a third intermediate bearing is utilized. On all models, crankshaft end play is controlled by thrust washers (9) at each end of the crankshaft.

Crankshaft can be removed after removing crankshaft timing gear, rod and piston unit(s), flywheel and, on 2-cylinder models, the oil pickup pipe and screen and intermediate main bearing. When removing intermediate main bearing retaining bolts, note position of the two halves and remove them from crankshaft. Unbolt and remove the rear main bearing housing. Remove crankshaft from rear end of cylinder block and remove the two thrust washers from crankshaft.

Crankshaft main and rod journal standard diameter on all models is 41.262-41.275 mm (1.6245-1.6250 inches). Inspect the crankshaft closely and check bearing journals with micrometer for wear. Main bearings are available in undersizes of 0.25 mm (0.010 inch) and 0.50 mm (0.020 inch) as well as standard size. Crankshaft bearing oil clearance should be as follows:

Front bearing .... 0.05-0.088 mm
  (0.0021-0.0036 in.)
Intermediate
  bearing .... 0.093-0.137 mm
  (0.0037-0.0054 in.)
Rear bearing .... 0.04-0.078 mm
  (0.0016-0.0031 in.)
Crankpin bearing ..... 0.025-0.09 mm
  (0.0010-0.0035 in.)

To remove crankshaft front or rear main bearing, the crankcase or rear main bearing retainer should be heat-

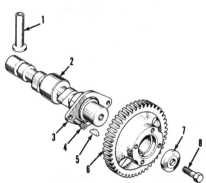

Fig. L5-40—Exploded view of camshaft assembly for single cylinder engine. Two cylinder models are similar except for number of cam lobes. Retainer plate (3) is not used on some engines.

1. Cam follower
2. Camshaft
3. Retainer plate
4. Cap screws
5. Woodruff key
6. Camshaft gear
7. Washer
8. Cap screw

Fig. L5-41—Exploded view of typical single cylinder crankshaft, piston and rod assemblies. Two cylinder models have intermediate (center) main bearing with split support. Bolts holding support halves together also retain support in crankcase.

1. Piston ring set
2. Snap ring
3. Piston
4. Connecting rod
5. Pin bushing
6. Piston pin
7. Crankpin bearing
8. Rod cap bolts
9. Thrust washers
10. Crankshaft
11. Woodruff key
12. Crankshaft gear
13. "O" ring
14. Washer
15. Crankshaft oil seal
16. Cap screw
17. Front main bearing
18. Rear main bearing

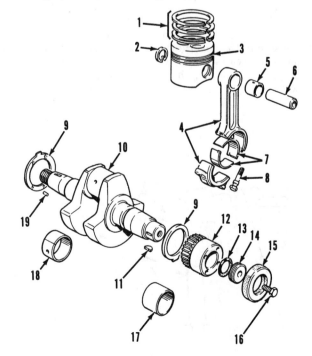

SERVICE MANUAL

Lister-Petter

ed in boiling water, then press bearing from crankcase or housing. To install new bearing, heat crankcase or bearing housing in boiling water and chill bearing in freezer. Press new bearing into crankcase or bearing housing so that oil hole in bearing is aligned with oil feed passage and with split side of bearing up. New bearings are prefinished and fit should be correct as installed.

To install crankshaft, first be sure all bearing journal surfaces and oil holes are clean. Place front crankshaft thrust washer on locating dowel pins inside front end of crankcase and with tab on thrust washer engaging slot in crankcase. Bearing material side of thrust washer must face crankshaft shoulder. Lubricate crankshaft journals and main bearings and insert front end of crank through thrust washer and front main bearing. Take care not to damage the bearing surfaces. On 2-cylinder engines, place top half of intermediate bearing housing with bearing into position in crankcase. Line up matching numbers on both halves of intermediate bearing housing and place bottom half in position. Install intermediate bearing bolts and tighten to a torque of 41 N·m (30 ft.-lbs.). Refer to CRANKSHAFT REAR OIL SEAL paragraph and install rear main bearing retainer. Be sure crankshaft turns smoothly and check crankshaft end play. End play should be 0.076-0.609 mm (0.003-0.024 inch). Complete reassembly of engine by reversing disassembly procedure.

FLYWHEEL

The flywheel is mounted on rear end of crankshaft on a taper fit and retained by a hex nut and lock-tab washer. To remove flywheel retaining nut, crankshaft must be held from turning. On Models AC1Z and AC1ZS, remove engine oil pan (sump) and place a block of wood between crankshaft and cylinder block. On all other models, insert a steel rod through timing hole in bell housing into hole in outer diameter of flywheel. Bend tab back on washer and remove flywheel retaining nut. Remove wood block or steel rod and using gear puller bolts threaded into puller holes at center of flywheel, remove flywheel from crankshaft. On models with flywheel alternator (see Fig. L5-42), be careful not to damage the plastic cooling fan or alternator stator windings. Remove flywheel key from crankshaft.

To install flywheel, insert key in slot of crankshaft and push flywheel onto shaft as far as possible. Use wood block or steel rod as in flywheel removal to keep crankshaft from turning. Place a new lock-tab washer on crankshaft, install nut and tighten to a torque of 210 N·m (155 ft.-lbs.).

CRANKSHAFT REAR OIL SEAL

The crankshaft rear oil seal is mounted in the rear main bearing housing. Remove flywheel and rear bearing housing to renew seal. Be careful not to lose or damage rear crankshaft thrust washer. Check condition of rear main bearing and bearing and oil seal surface of crankshaft. Remove oil seal from housing and press new seal in squarely and flush with outside of housing.

Lubricate seal liberally with Lithium No. 2 grease or soak seal in engine oil for 24 hours before installation. Place rear crankshaft thrust washer on locating dowel pins in bearing housing with bearing side away from housing; tab on thrust washer must engage slot in bearing housing. Using new gasket, install gasket and bearing housing over rear end of crankshaft taking care not to damage seal on keyway edges on crankshaft. Tighten bearing housing retaining nuts in diagonal pattern. Check crankshaft end play which should be 0.076-0.609 mm (0.003-0.024 inch). Complete reassembly by reversing disassembly procedure.

ELECTRICAL SYSTEM

FLYWHEEL ALTERNATOR

Refer to Fig. L5-42. Alternator rotor (3) is bolted to front side of engine flywheel and stator (5) and adapter (6) are retained on rear main bearing support (8) by long bolts (4). The voltage regulator (10) is mounted on bracket (9) outside of flywheel housing.

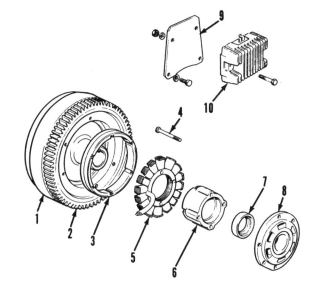

Fig. L5-42—Exploded view of flywheel and flywheel alternator used on some engines.
1. Flywheel
2. Starter ring gear
3. Alternator rotor
4. Stator mounting bolts
5. Alternator stator
6. Stator adapter
7. Crankshaft rear oil seal
8. Main bearing support
9. Mounting plate
10. Regulator

LISTER-PETTER

SMALL DIESEL

Lister-Petter

Model	No. Cyls.	Bore	Stroke	Displ.
LPA2	2	76 mm (2.99 in.)	80 mm (3.15 in.)	726 cc (44.3 cu. in.)
LPA3	3	76 mm (2.99 in.)	80 mm (3.15 in.)	1089 cc (66.5 cu. in.)
LPA4	4	76 mm (2.99 in.)	80 mm (3.15 in.)	1452 cc (88.6 cu. in.)
LPW2, LPWS2	2	86 mm (3.38 in.)	80 mm (3.15 in.)	930 cc (56.75 cu. in.)
LPW3, LPWS3	3	86 mm (3.38 in.)	80 mm (3.15 in.)	1395 cc (85.13 cu. in.)
LPW4, LPWS4	4	86 mm (3.38 in.)	80 mm (3.15 in.)	1860 cc (113.5 cu. in.)

Series "LPA" engines are four-stroke, air cooled type with vertical cylinders and direct fuel injection. Series "LPW" engines are similar except for being water cooled. Series "LPWS" engines are water cooled with indirect fuel injection. Engine type (build number) can be found on plate (see Fig. L6-1) attached to engine crankcase. Standard crankshaft rotation is (A) anticlockwise (counterclockwise) as viewed from flywheel end.

MAINTENANCE

LUBRICATION

Recommended engine lubricant is API classification CC motor oil for engines operating under normal conditions. For engines operating under high loads, in high ambient temperatures, or if sulfur content of fuel exceeds 0.5 percent, API classification CD oil should be used after the first oil change.

Ambient operating temperature should be used as guide to weight of motor oil. Below −15° C (4° F), use SAE 5W or SAE 5W-20 motor oil. In temperatures between −15° and 4° C (4°-39° F), use SAE 10W or SAE 10W-30 motor oil. Between 4° and 30° C (39°-86° F), SAE 20/20W or SAE 15W-50 motor oil should be used. Above 30° C (86° F), SAE 30, SAE 15W-40 or SAE 20W-40 oil should be used.

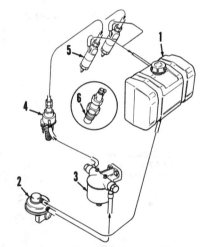

Fig. L6-2—Schematic drawing of typical fuel system.
1. Fuel tank
2. Fuel feed pump
3. Fuel filter
4. Injection pump (one per cylinder)
5. Direct injection nozzle
6. Indirect injection nozzle

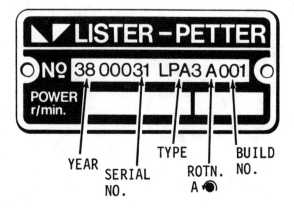

Fig. L6-1—View showing typical Lister nameplate which shows year, serial number, model number (type), engine rotation and build number (type).

FUEL SYSTEM

Refer to Fig. L6-2 for schematic view of the fuel system. The fuel system consists of a fuel pump and fuel injector for each cylinder, a fuel lift pump, filter and optional engine mounted fuel tank. Leak-off pipes for all injectors are joined and fuel is returned to tank through single line.

FUEL FILTER. Fuel filter element (4—Fig. L6-3) should be renewed after ever 500 hours of operation, or more often if a loss of engine power is evident. To change filter, close fuel supply valve or drain fuel tank. Unscrew center bolt (10) and remove filter bowl (8), element (4), sealing "O" rings (2, 3, and 5), wash-

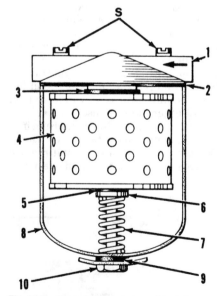

Fig. L6-3—Cross-sectional view of fuel filter assembly.
S. Bleed screws
1. Filter body
2. Bowl seal ring
3. Upper element seal
4. Filter element
5. Lower element seal
6. Washer
7. Spring
8. Filter bowl
9. Bowl bolt seal
10. Bowl retaining bolt

SERVICE MANUAL

Lister-Petter

er (6) and spring (7). Remove seal washer (9) from center bolt. A kit containing all new seals is available separately from filter. Discard element, clean filter bowl and assemble with a new element and sealing rings as shown in cross-sectional view.

BLEED FUEL SYSTEM. First, make sure fuel tank contains sufficient fuel. Open the two bleed screws (S—Fig. L6-4) on top of filter, then tighten the bleed screws when air-free fuel flows from bleed screws. Push a length of hose onto bleed nipple (N) at end of fuel distribution pipe and place other end of hose in suitable container. Move engine control to "RUN" position, open bleed nipple and start engine. When engine is running smoothly and no air bubbles are expelled with the fuel from the bleed nipple, close the nipple and remove hose.

If engine will not run smoothly with air bled from fuel distribution pipe, the fuel injection lines must be bled for each pump in turn with engine running. Refer to Fig. L6-5, loosen fuel injection line nut (A) at injection pump and loosen delivery valve holder (B) 1/4 turn. When no further air bubbles appear, tighten delivery valve holder to a torque of 47.5 N·m (35 ft.-lbs.) and tighten injector line fitting to a torque of 28.5 N·m (21 ft.-lbs.).

NOTE: Be sure injection pump does not turn when loosening and tightening line nut and delivery valve holder. If it does turn, refer to following INJECTION PUMP TIMING paragraph.

INJECTION PUMP TIMING

A fuel injection pump for each cylinder is mounted in the camshaft side of cylinder block between the valve push rods. The injection pumps are actuated by lobes on the engine camshaft. The fuel pumps are timed individually by a shim pack between pump flange and engine crankcase. The shim pack is provided by the injection pump manufacturer and should not be changed.

Pump timing can be maintained if original shim stack for that pump and cylinder are retained for installation. If original shim pack for injection pump has been lost or mixed with shims for other cylinders, each fuel pump must be timed as follows:

The distance (X—Fig. L6-6) between top face of crankcase and top of fuel injection pump tappet (cam follower) must be accurately measured at a time when tappet is in a predetermined position. To set fuel injection pump tappet to proper position for measuring distance "X", the piston for that cylinder should be positioned a specified distance BTDC on compression stroke. The following injection timing chart lists the specified distance BTDC on compression stroke for each model:

INJECTION TIMING CHART
Engine Type   Distance BTDC
Models LPA2,
 3 and 4; LPW2,
 3 and 4;
1500-3000 Rpm
 Fixed Speed . . . . .3.097 mm
                                 (0.122 in.)
3600 Rpm
 Fixed Speed . . . . .4.428 mm
                                 (0.174 in.)

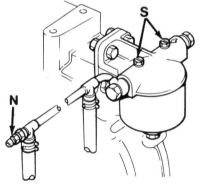

Fig. L6-4—View showing filter bleed screws (S) and line bleed nipple (N).

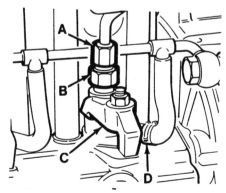

Fig. L6-5—Fuel injection pump for each cylinder is mounted between the valve push rod tubes.
A. Fuel line fitting
B. Delivery valve holder
C. Injection pump clamp
D. Return line clamp

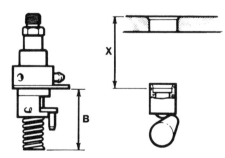

Fig. L6-6—If original pump shim pack is not available for determining shim pack thickness, refer to text for measuring dimension "X" which is subtracted from 51.2 mm (2.012 inch) (dimension "B").

Up to 3000 Rpm
 Variable Speed . . . . .3.097 mm
                                 (0.122 in.)

Model LPWS2:
All Speeds,
 Fixed & Variable . . . . .4.428 mm
                                 (0.174 in.)

Model LPWS3:
1500-3000 Rpm
 Fixed Speed . . . . .3.735 mm
                                 (0.147 in.)
3600 Rpm
 Fixed Speed . . . . .4.428 mm
                                 (0.174 in.)
Up to 3000 Rpm
 Variable speed . . . . .3.735 mm
                                 (0.147 in.)

Model LPWS4:
All Speeds,
 Fixed & Variable . . . . .3.375 mm
                                 (0.147 in.)

Remove injector from the cylinder corresponding to the injection pump being adjusted. Use a dial indicator and suitable probe on top of piston to determine piston TDC on compression stroke. Turn flywheel clockwise to lower piston beyond the distance specified in above chart, then turn flywheel counterclockwise until piston is positioned at the specified distance BTDC.

With piston set at proper distance before TDC, refer to Fig. L6-6 and measure distance "X" from top face of crankcase to injection pump tappet cap. Subtract distance "X" from 51.2 mm (2.012 inch) (which is dimension "B") to find correct thickness of shim pack to be installed between fuel pump plate and the crankcase. Shims are available in thicknesses of 0.075 mm (green shim), 0.125 mm (slate blue shim), and 0.250 mm (black shim).

Refer to INJECTION PUMP paragraphs in REPAIRS section for installation of fuel injection pump once correct shim pack thickness is determined for that pump. Repeat the procedure for each pump as necessary.

GOVERNOR

The governor is a flyweight type attached to front face of engine camshaft gear. Governor controls may be fixed single speed or variable speed type, all with a "Stop/Run" control. Refer to the following paragraphs for adjustment of the governor controls. Adjust governor speed control screws only after the Stop/Run control has been properly set.

STOP/RUN LEVER. Turn engine Stop/Run lever counterclockwise to the

Lister-Petter

"Stop" position. Loosen locknut (C—Fig. L6-8) and turn screw (A) until it just touches radius part of Stop/Run lever (B), then tighten locknut. Loosen locknut (D—Fig. L6-9) and back screw (C) away from lever. Turn the Stop/Run lever clockwise and place a special "G" setting gage, or a block machined to "G" setting thickness between radius part of lever (B) and head of adjusting screw (A). Turn screw (C) until the lever is holding the gage or block against screw (A), then tighten locknut (D). Refer to the following table for specified "G" setting.

ENGINE	"G" SETTING
LPA2	
Build 01	20 mm (0.787 in.)
Build 02, 05, 09, 10, 42, 44	22 mm (0.866 in.)
Build 08	21 mm (0.827 in.)
LPA3-LPA4	
Build 01	20 mm (0.787 in.)
Build 02, 05, 09, 42	22 mm (0.866 in.)
Build 07, 08	21 mm (0.827 in.)
LPW2-LPW3-LPW4	
Build 01	22.5 mm (0.886 in.)
Build 07, 08	25 mm (0.984 in.)
Build 02, 05, 09, 10	24.5 mm (0.965 in.)
LPWS2-LPWS3-LPWS4	
Build 01	24 mm (0.945 in.)
Build 02, 05, 09, 10	26.5 mm (1.043 in.)

FIXED SPEED CONTROL. Refer to Fig. L6-10. Loosen locknuts and turn adjuster screws (A and B) to obtain desired engine speed with both screws contacting the speed control arm. Engine no-load speed should be 4 percent above rated speed shown on engine nameplate. Tighten locknuts when desired speed is obtained.

VARIABLE SPEED. To adjust variable speed control, refer to Fig. L6-10 and proceed as follows: Set speed control to idle position. Loosen locknut on idle speed adjusting screw (A) and turn screw to obtain 900-1000 rpm, then tighten locknut.

Set speed control to full speed position. Loosen locknut and adjust screw (B) to obtain engine no-load maximum speed desired when screw contacts speed control lever, then tighten locknut. Turning the screw counterclockwise increases engine speed. Cable or rod remote control movement should operate speed control lever to move lever from idle to maximum no-load stops.

REPAIRS

TIGHTENING TORQUES

When making engine repairs, observe the following special tightening torque values:

Cylinder head bolts:
 Air cooled . . . . 47 N·m
  (35 ft.-lbs.)

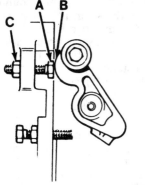

Fig. L6-8—Stop/Run control adjustment for control in stop position. Refer to Fig. L6-9 for adjustment of run position stop screw.
 A. Adjusting screw
 B. Stop/Run lever
 C. Adjusting screw lock nut

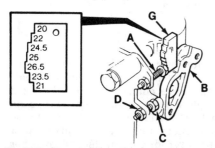

Fig. L6-9—View showing adjustment of run position adjusting screw (C) using special gage (G).
 A. Stop position screw
 B. Stop/Run lever
 C. Run position screw
 D. Locknut
 G. Gage block

SMALL DIESEL

 Water cooled . . . . 88 N·m
  (65 ft.-lbs.)
Cylinder head covers . . . . 9 N·m
  (78 in.-lbs.)
Rocker arm nut . . . . 34 N·m
  (25 ft.-lbs.)
Timing (end) cover bolts . . . . . 9 N·m
  (78 in.-lbs.)
Crankcase cover (door) bolts . . . 9 N·m
  (78 in.-lbs.)
Connecting rod bolts . . . . 24 N·m
  (18 ft.-lbs.)
Intermediate main bearing
 cap screws . . . . 21 N·m
  (15 ft.-lbs.)
Rear main bearing
 housing . . . . 27 N·m
  (20 ft.-lbs.)
Crankshaft pulley or
 pulley bolt . . . . 300 N·m
  (221 ft.-lbs.)
Crankshaft pulley stud . . . . 7 N·m
  (60 in.-lbs.)
Flywheel bolts . . . . 68 N·m
  (50 ft.-lbs.)
Flywheel housing bolts . . . . . . 78 N·m
  (58 ft.-lbs.)
Fuel injection pump
 clamp bolt . . . . 34 N·m
  (25 ft.-lbs.)
Fuel injection pump
 delivery valve . . . . 47 N·m
  (35 ft.-lbs.)
Fuel injector clamp nut . . . . . . 21 N·m
  (15 ft.-lbs.)
Fuel injector (LPWS) . . . . 68 N·m
  (50 ft.-lbs.)
Injector nozzle nut:
 Direct injection . . . . 47 N·m
  (35 ft.-lbs.)
 LPWS . . . . 81 N·m
  (60 ft.-lbs.)
Injector line fittings . . . . 28 N·m
  (21 ft.-lbs.)
Fuel lift pump bolts . . . . 21 N·m
  (15 ft.-lbs.)
Glow plug (LPWS) . . . . 27 N·m
  (20 ft.-lbs.)
Manifold bolts . . . . 9 N·m
  (78 in.-lbs.)
Water pump bolts . . . . 21 N·m
  (15 ft.-lbs.)

WATER PUMP AND COOLING SYSTEM

Models LPW-LPWS

LPW and LPWS engines are equipped with a closed cooling system consisting of radiator, fan, water pump and thermostat assembly. Refer to Fig. L6-11 for water pump installation. The fan is mounted on the water pump pulley and is retained by a left-hand thread nut. Either a pusher or puller fan assembly is used. Except for thermostat (3), water outlet (1) and gaskets, water pump and pulley are serviced as a complete assem-

Fig. L6-10—View showing idle speed adjusting screw (A) and high speed adjusting screw (B).

SERVICE MANUAL

bly only. Renew water pump assembly if seals are leaking, bearing is loose or noisy, or if other damage is noted.

To remove water pump, drain cooling system and remove radiator. Loosen alternator belt adjustment, then unbolt and remove water pump; there are five cap screws and two stud nuts. Install pump with new gasket and tighten retaining bolts and nuts to a torque of 21 N·m (15 ft.-lbs.). Tighten fan retaining nut to a torque of 30 N·m (22 ft.-lbs.).

THERMOSTAT

Models LPW-LPWS

Thermostat can be removed after draining cooling system at thermostat body drain plug (8—Fig. L6-11), disconnecting top radiator hose and removing water outlet (1) from top of water pump. Thermostat should start to open at 71° C (160° F) and be fully open at 85° C (185° F).

VALVE ADJUSTMENT

All engines are equipped with hydraulic valve lifters and no adjustment is necessary or possible. Removal of any part of valve gear will allow valve lifter to extend to full stroke. The engine should not be cranked for at least 45 minutes after reassembly to allow valve spring pressure to depress the lifters. When engine is reassembled with new lifters, the engine should be cranked for at least 15 seconds before attempting to start it.

CYLINDER HEAD

Air cooled engines have a separate cylinder head (Fig. L6-12) for each cylinder with all heads being alike. Water cooled engines have a one-piece (monobloc) cylinder head (Fig. L6-13).

To remove cylinder head(s), proceed as follows: Remove muffler, air cleaner and Series LPA cylinder cowling. On Series LPW and LPWS, drain cooling system and remove water pump assembly. On all engines, remove pump to injector line(s), injector fuel return line and fuel injector(s). Immediately cap or plug all fuel openings. Remove the manifolds and rocker arm cover(s). Remove the valve rocker arm nuts, rocker arms and push rods. Gradually loosen cylinder head bolts, then remove bolts and lift cylinder head(s) from engine and remove the push rod tubes.

Piston to cylinder head clearance should be 0.7-0.9 mm (0.027-0.035 inch), and is controlled by thickness of head gasket. The same thickness head gasket as was removed should be reinstalled unless piston or connecting rod have been renewed. On air cooled models, it is important that all cylinder head gaskets be of the same thickness. Head gasket thickness can be identified by the number of holes in corner of gasket as shown in Fig. L6-14. Refer to the following head gasket thickness chart:

Air cooled engines:
1 hole . . . . 0.38 mm
                                        (0.015 in.)
2 holes . . . . 0.25 mm
                                        (0.010 in.)
3 holes . . . . 0.50 mm
                                        (0.020 in.)
Water cooled engines:
1 hole . . . . 1.41 mm
                                        (0.056 in.)
2 holes . . . . 1.53 mm
                                        (0.060 in.)
3 holes . . . . 1.66 mm
                                        (0.065 in.)

Piston to cylinder head clearance should be checked before final installation of cylinder head if pistons or connecting rods have been renewed. Turn crankshaft so that piston is about 6.35 mm (1/4 inch) before top dead center. Using grease, stick two pieces of soft lead wire or solder 1.2 mm (0.048 inch) in diameter at spacing as wide as possible to cylinder head combustion surface in line with piston pin. Install cylinder head with gasket and turn engine so that piston moves past top dead center. Remove the cylinder head and measure thickness of flattened wires. The average of the two thicknesses is piston to head clearance. Clearance should be 0.7-0.9 mm (0.027-0.035 inch).

When installing cylinder head, use new push rod tube seals (20—Fig. L6-12 or L6-13) and cylinder head gasket(s) (15). Be sure gasket surfaces are cleaned thoroughly and place new head gasket on the block with gasket and block holes aligned. Lubricate inside of new push rod tube seals (see Fig. L6-15) with rubber lubricant. Install cylinder head and push rod tubes. Install the cylinder head bolts and turn down finger tight. Refer to Fig. L6-16 for head bolt location by bolt type; water cooled head is shown but bolt locations are same for air cooled head(s). On air cooled models, use straightedge to be sure the manifold flanges are aligned.

Refer to Fig. L6-17 or L6-18 for head bolt tightening sequence. On air cooled models, tighten bolts in sequence shown for each head, first to 8 N·m (6 ft.-lbs.), then final tighten all bolts to 47 N·m (35 ft.-lbs.). On water cooled engines, tighten bolts in sequence shown first to 8 N·m (6 ft.-lbs.), then to 47 N·m (35 ft.-

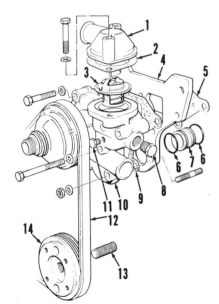

Fig. L6-11—View showing water pump and thermostat installation.

1. Water outlet
2. Gasket
3. Thermostat
4. Adapter plate
5. Gasket
6. "O" rings
7. Inlet sleeve
8. Thermostat housing drain plug
9. Gasket
10. Water pump assy.
11. Gasket
12. Fan belt
13. Pulley stud
14. Water pump pulley

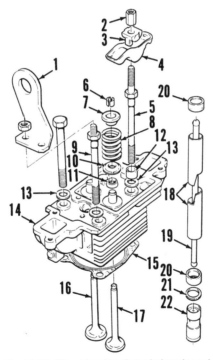

Fig. L6-12—View showing air cooled engine cylinder head. A separate head is used for each cylinder.

1. Lifting eye
2. Rocker arm nut
3. Rocker arm pivot
4. Rocker arm
5. Rocker arm stud/head bolt
6. Split valve locks
7. Valve spring retainer
8. Valve spring
9. Cylinder head/lift eye bolt
10. Valve spring seat
11. Valve stem shield
12. Spacer
13. Head bolt washers
14. Cylinder head
15. Head gasket
16. Intake valve
17. Exhaust valve
18. Push rod tube
19. Push rod
20. Push rod tube seals
21. Seal washer
22. Hydraulic lifter (tappet)

Lister-Petter

lbs.) and finally tighten bolts to 88 N·m (65 ft.-lbs.).

Lubricate rocker arms and push rods, install and tighten rocker arm nuts to a torque of 34 N·m (25 ft.-lbs.). Refer to VALVE ADJUSTMENT paragraph for information on hydraulic lifters. Complete reassembly by reversing removal procedure.

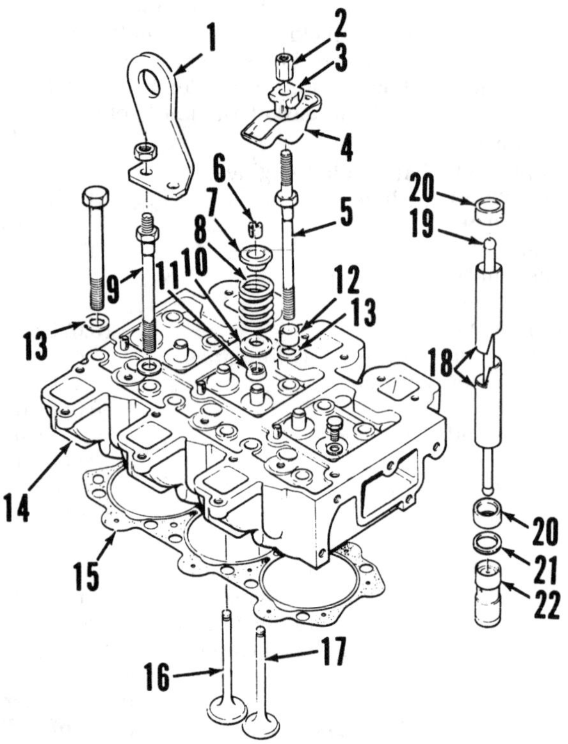

SMALL DIESEL

VALVE SYSTEM

The valve guides are a press fit in cylinder head. To remove guides, use suitable piloted valve guide driver and drive guides down from top of head. To install new guides, drive guides downward through head until top end of guides protrude 11.75-12.25 mm (0.462-0.481 inch) from machined face of head. Ream guides only as necessary for proper valve stem clearance. Clearance should be 0.025-0.095 mm (0.0010-0.0037 inch) with maximum allowable clearance of 0.165 mm (0.0065 inch).

Valve rocker arms are retained on top end of special cylinder head bolts by hex nuts. Rocker arm nuts are tightened to a torque of 34 N·m (25 ft.-lbs.) which will seat pivot (4—Fig. L6-11 or L6-12) against hex shoulder of bolt (5). Hydraulic valve lifters (22) are used and no rocker arm adjustment is necessary or provided for. Inspect rocker arm pivot, valve stem end and push rod contact points for scoring or wear.

Valve face and seat angles are 45 degrees for both intake and exhaust. If installing new guides, valve seats should not be reworked until after guides are installed. After refacing and reseating valves, intake valve heads should be 0.95-1.26 mm (0.037-0.049 inch) below combustion surface of cylinder head and exhaust valve heads should be 1.33-1.64 mm (0.053-0.064 inch) below combustion surface.

Intake and exhaust valve springs are alike. Spring free length should be 43.7-

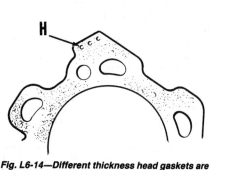

Fig. L6-13—Exploded view of monobloc water cooled 3-cylinder head; two and four cylinder heads are of similar design.

1. Lifting eye
2. Rocker arm nut
3. Rocker arm pivot
4. Rocker arm
5. Rocker arm stud/head bolt
6. Split valve locks
7. Valve spring retainer
8. Valve spring
9. Cylinder head/lift eye bolt
10. Valve spring seat
11. Valve stem shield
12. Spacer
13. Head bolt washers
14. Cylinder head
15. Head gasket
16. Intake valve
17. Exhaust valve
18. Push rod tube
19. Push rod
20. Push rod tube seals
21. Seal washer
22. Hydraulic lifter (tappet)

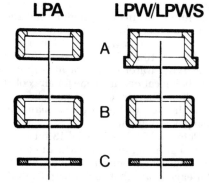

Fig. L6-14—Different thickness head gaskets are used to control piston to cylinder head clearance. Gasket thickness is indicated by number of holes (H) in corner of gasket.

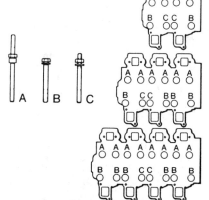

Fig. L6-16—Diagram showing location of different type cylinder head bolts. Water cooled heads are shown, but locations also apply to air cooled engines.

A. Rocker stud bolts
B. Cap screws
C. Lift eye bolts

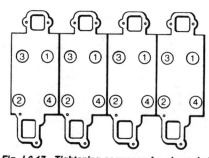

Fig. L6-15—View showing cross section of push rod tube seals.

A. Upper seal
B. Lower seal
C. Crankcase washer

Fig. L6-17—Tightening sequence for air cooled cylinder head bolts.

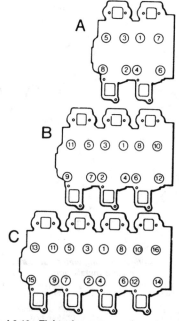

Fig. L6-18—Tightening sequence for water cooled cylinder head bolts.

SERVICE MANUAL

Lister-Petter

45.5 mm (1.720-1.791 inches) and minimum free length is 42.5 mm (1.673 inches). Renew spring if distorted or otherwise damaged.

NOTE: Engine manufacturer recommends that all valves and springs be renewed at major engine overhaul.

INJECTORS

REMOVE AND REINSTALL. Prior to removing injectors, thoroughly clean injectors, injector lines, fuel injection pump and surrounding area using suitable solvent and compressed air. Remove the fuel return line from top of injectors. Remove the fuel injection lines from injection pump and injectors.

On models with indirect injection nozzles (B—Fig. L6-19), unscrew (counterclockwise) fuel injector from cylinder head. It will be necessary to use special socket or a deepwell socket with points at opposite sides ground out with die grinder. The socket must clear the injector leak-off line nipples and engage the hex part of nozzle cap nut (8—Fig. L6-22). On models with direct injection nozzles (A—Fig. L6-21), remove the fuel injector retaining clamp and pull injector from head. Immediately cap or plug all fuel openings. Remove copper washers or heat shield from bottom of injector bores in cylinder head.

When installing fuel injectors, first be sure that injector bores in cylinder head are clean and that all old copper washers are removed from direct injection models and injector heat shield is removed for series "LPWS" indirect injection engines. Use small amount of high melting point grease to stick new copper washer or shield to injector nozzle. On indirect injection models, install the nozzle with new seal (see Fig. L6-20) and tighten to a torque of 68 N·m (50 ft.-lbs.).

On direct injection models, insert injector with copper washer in cylinder head bore and tighten clamp nuts finger tight. Install injection line leaving fittings finger tight. Install new "O" rings in cylinder head cover, replace the injection line clip and tighten clip nut to 9 N·m (78 in.-lbs.). Tighten the injector clamp nut to a torque of 21 N·m (15 ft.-lbs.), then tighten fuel injection line fittings to a torque of 28 N·m (21 ft.-lbs.). Bleed system as outlined in MAINTENANCE section.

TESTING. A complete job of testing, cleaning and adjusting the fuel injector requires use of special test equipment. Use only clean approved testing oil in tester tank. Injector should be tested for opening pressure, seat leakage, back leakage and spray pattern. Before connecting injector to test stand, operate tester lever until oil flows, then attach injector to tester line.

WARNING: Fuel emerges from injector with sufficient force to penetrate the skin. When testing injector, keep yourself clear of nozzle spray.

Close valve to tester gage and operate tester lever a few quick strokes to be sure nozzle valve is not stuck and injector is suitable for further tests.

OPENING PRESSURE. Open the valve to tester gage and operate tester lever slowly while observing gage reading. Opening pressure should be 20500-21500 kPa (2973-3118 psi) for new direct injection nozzles and 20000 kPa (2900 psi) for used direct injection nozzles. For indirect injection nozzles, opening pressure should be 12310-14000 kPa (1785-2030 psi) for new nozzles and 12100-13100 kPa (1755-1900 psi) for used nozzles. Opening pressure is adjusted by changing thickness of shim stack (2—Fig. L6-21 or L6-22). Adding shim thickness will increase pressure. After setting injection pressure, tighten nozzle nut to a torque of 47 N·m (35 ft.-lbs.).

SPRAY PATTERN. On indirect injection nozzles, the spray pattern should be conical, well atomized, and emerging in a straight axis from nozzle tip. If spray is drippy, ragged or to one side, nozzle must be cleaned or renewed.

On direct injection nozzles, spray pattern should be equal from all four nozzle tip holes and should be a fine mist. Place a sheet of blotting paper 30 cm (about 12 inches) below injector and operate tester one stroke. The spray pattern should be a perfect circle.

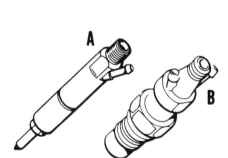

Fig. L6-19—Direct injection models use spray type nozzle (A); indirect injection Model LPWS uses pintle type nozzle (B).

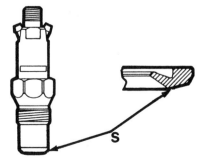

Fig. L6-20—Cross-sectional view of injector seal (heat shield) for Model LPWS indirect injection engines.

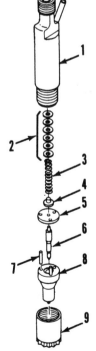

Fig. L6-21—Exploded view of direct injection spray type injector assembly.

1. Injector body
2. Adjusting shims
3. Spring
4. Spring seat
5. Spacer
6. Nozzle valve
7. Dowel pin
8. Nozzle body
9. Nozzle nut

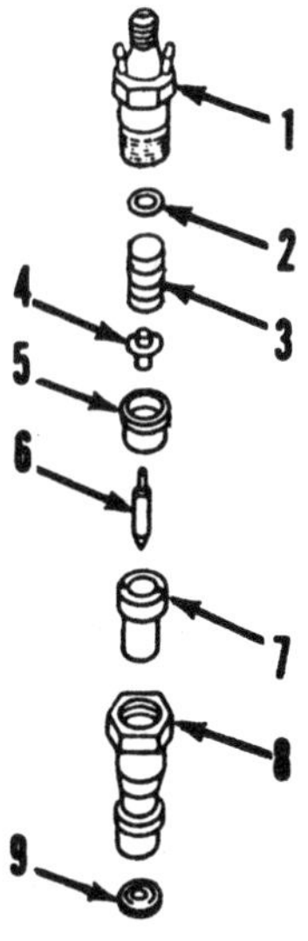

Fig. L6-22—Exploded view of pintle type nozzle used in Model LPWS engines.

1. Injector body
2. Adjusting shims
3. Spring
4. Spring seat
5. Spacer
6. Nozzle valve
7. Nozzle body
8. Nozzle nut
9. Heat shield

Illustrations Courtesy Lister-Petter

Lister-Petter — SMALL DIESEL

SEAT LEAKAGE. Wipe nozzle tip dry with clean blotting paper, then hold gage pressure to approximately 1965 kPa (285 psi) below opening pressure for 10 seconds. If drop of oil appears at nozzle tip, or oil drips from nozzle, clean or renew nozzle.

BACK LEAKAGE. Slowly bring tester gage pressure to 15175 kPa (2200 psi), then observe time interval for gage pressure to fall to 10100 kPa (1465 psi). With type "C" calibration fluid at a temperature of 15.5° C (60° F) in tester, leak-off time should be 10 to 40 seconds.

OVERHAUL. Hard or sharp tools, emery cloth, grinding compounds or other than approved tools, solvents and lapping compounds, must never be used. Wipe all dirt and loose carbon from injector. Refer to Fig. L6-21 for exploded view of direct injection nozzle and to Fig. L6-22 for indirect injection nozzle and proceed as follows:

Secure injector nozzle in holding fixture or soft jawed vise and loosen nozzle nut. Unscrew nut and remove internal parts from injector, noting placement for each part removed. Place all parts in clean diesel fuel or calibrating oil as they are removed, and take care not to mix parts with those from other injector assemblies.

Clean exterior surfaces with brass wire brush to loosen carbon deposits; soak parts in approved carbon solvent if necessary. Rinse parts in clean diesel fuel or calibrating oil after cleaning to neutralize carbon solvent.

On indirect injection nozzles, clean nozzle spray hole from inside using pointed hardwood stick. On direct injection nozzles, use a 0.2 mm (0.008 inch) cleaning wire in a pin vise to clean the four spray holes. Scrape carbon from nozzle pressure chamber using hooked scraper. Clean valve seat using brass scraper, then polish seat using wood polishing stick and mutton tallow.

Reclean all parts by rinsing thoroughly in clean diesel fuel or calibrating oil and assemble injector while immersed in fuel or oil. Reassemble all parts in position from which they were removed and tighten nozzle nut to a torque of 47 N·m (35 ft.-lbs.). Retest assembled injector; disassemble and re-clean or renew injector if test indicates malfunction.

INJECTION PUMP

A single plunger fuel injection pump is used for each cylinder. Pump timing is controlled by a shim pack thickness determined by the manufacturer and it is very important that this shim pack be kept with the pump for which it was intended. Also, the pump rack to governor linkage adjustment is made by turning the pump with clamp bolt loosened. Be sure to observe the following instructions.

REMOVE AND REINSTALL. If only one fuel injection pump is removed at a time (or if at least one pump is not disturbed), pump may be removed and reinstalled as follows:

Clean fuel injection pump, injection line and injector thoroughly. Shut off fuel supply, then disconnect fuel inlet line from injection pump. Remove the injection line clip from top of cylinder head cover. Hold injection pump delivery valve holder from turning, then unscrew injection line fittings and remove line. Remove the clamp holding injection pump in crankcase and lift pump out, taking care not to lose or damage shim pack located between pump flange and crankcase. Fuel pump tappet can be lifted from pump bore by using long nosed pliers expanded into inside of tappet, then lifting tappet from bore.

NOTE: Do not remove the fuel pump tappet guide stud; refer to Fig. L6-23.

Pump can be reinstalled as follows if at least one pump is left in position and retained by clamp. Refer to GOVERNOR paragraph in MAINTENANCE section to be sure speed control lever stops are properly adjusted. Lubricate injection pump tappet (cam follower) and insert in bore with roller end down and slot aligned with fuel pump tappet guide bolt (B—Fig. L6-23). Slowly turn engine until tappet is at lowest position. Install pump in bore with same shim pack thickness as removed and with the pump rack lever engaging slot in governor rack. Hold Stop/Run lever in Stop position. Turn pump carefully in counterclockwise direction until pump rack can be felt against stop. Hold pump in this position and install and tighten the clamp nut to a torque of 34 N·m (25 ft.-lbs.).

NOTE: If pump is not turned counterclockwise, or moved when clamp nut is being tightened, it is possible engine will not stop when Stop/Run lever is moved to Stop position.

SHIM PACK THICKNESS. If shim pack has been lost or shims mixed with pump from other cylinders, proceed as follows to determine proper shim pack thickness for each pump.

The distance (X—Fig. L6-24) between top face of crankcase and top of fuel injection pump tappet (cam follower) must be accurately measured at a time when tappet is in a predetermined position. To set fuel injection pump tappet to proper position for measuring distance "X", the piston for that cylinder should be positioned a specified distance BTDC on compression stroke. The following injection timing chart lists the specified distance BTDC on compression stroke for each model:

INJECTION TIMING CHART

Engine Type	Distance BTDC
Models LPA2, 3 and 4; LPW2, 3 and 4:	
1500-3000 Rpm Fixed Speed	3.097 mm (0.122 in.)
3600 Rpm Fixed Speed	4.428 mm (0.174 in.)
Up to 3000 Rpm Variable Speed	3.097 mm (0.122 in.)
Model LPWS2: All Speeds, Fixed & Variable	4.428 mm (0.174 in.)
Model LPWS3: 1500-3000 Rpm Fixed Speed	3.735 mm (0.147 in.)
3600 Rpm Fixed Speed	4.428 mm (0.174 in.)

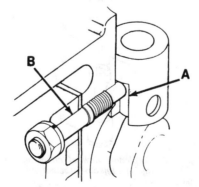

Fig. L6-23—Guide stud (B) fits in slot (A) of injection pump cam follower to keep follower from turning. Stud should not be removed.

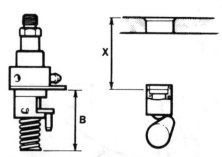

Fig. L6-24—If original pump shim pack is not available for determining shim pack thickness, refer to text for measuring dimension "X" which is subtracted from 51.2 mm (2.012 inch) (dimension "B").

SERVICE MANUAL

Lister-Petter

Up to 3000 Rpm
Variable speed............3.735 mm
(0.147 in.)

Model LPWS4:
All Speeds,
Fixed & Variable.........3.375 mm
(0.147 in.)

Remove injector from the cylinder corresponding to the injection pump being adjusted. Use a dial indicator and suitable probe on top of piston to determine piston TDC on compression stroke. Turn flywheel clockwise to lower piston beyond the distance specified in above chart, then turn flywheel counterclockwise until piston is positioned at the specified distance BTDC.

With piston set at proper distance before TDC, refer to Fig. L6-24 and measure distance "X" from top face of crankcase to injection pump tappet cap. Subtract distance "X" from 51.2 mm (2.012 inch) (which is dimension "B") to find correct thickness of shim pack to be installed between fuel pump plate and the crankcase. Shims are available in thicknesses of 0.075 mm (green shim), 0.125 mm (slate blue shim), and 0.250 mm (black shim).

Refer to previous paragraphs for installation of fuel injection pump once correct shim pack thickness is determined for that pump. Repeat the procedure for each pump as necessary.

R&R ALL PUMPS. If governor has been disassembled or if all fuel injection pumps have been removed at one time, it will be necessary to remove timing cover and set the governor rack in proper position. Front face of governor rack bushing (F—Fig. L6-25) should be exactly 55.5 mm (2.185 inches) from front end of crankcase (G). The special tool shown in Fig. L6-25 is provided by engine manufacturer or a tool can be fabricated for holding governor rack in this position. Set injection pump in block with correct shim thickness, making sure that fuel pump rack lever engages slot in governor rack. Carefully turn pump counterclockwise until pump rack is holding governor rack against stop. Hold pump in this position and tighten fuel pump clamp nut to a torque of 34 N·m (25 ft.-lbs.). Be sure pump does not turn from set position. Install other pumps in similar manner.

TIMING GEAR (END) COVER

Remove the fan drive belt. Lock flywheel from turning with special tool No. 317-50057 inserted through hole in flywheel housing to engage starter ring gear teeth, or wedge crankshaft from turning with suitable piece of wood. On early model engines, unscrew crankshaft pulley (left-hand thread); refer to Fig. L6-27 for special adapter. On later models, unscrew the crankshaft pulley bolt (left-hand threads) and remove the pulley. Remove flywheel locking tool.

Remove the seven stud nuts, lockwashers, cup and rubber washers or the bolts holding cover to crankcase, then carefully remove cover from crankcase and the two dowel pins. Remove crankshaft seal from cover.

To reinstall timing gear cover, install new crankshaft seal with lip to inside and use a new timing cover gasket. Install the rubber washers, cup washers, lockwashers and nuts or the bolts and tighten them finger tight. Tighten cover cap screws in a cross-over pattern to a torque of 9 N·m (78 in.-lbs.). Install flywheel locking tool. On early engines, install crankshaft pulley and tighten to a torque of 300 N·m (221 ft.-lbs.), then install pulley stud and tighten to a torque of 7 N·m (60 in.-lbs.). On later engines, install the crankshaft pulley and tighten bolt to a torque of 300 N·m (221 ft.-lbs.). Install fan drive belt.

GOVERNOR

The governor assembly is mounted on the front of camshaft drive gear; refer to Figs. L6-28, L6-29 and L6-30. To remove governor, first remove timing gear cover (end cover), cylinder head(s), push rods, hydraulic lifters and fuel injection pumps as previously outlined. Remove fuel lift pump and pump push rod. Unhook governor spring (9) from governor lever (14) and throttle lever (8). Remove

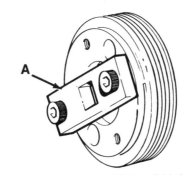

Fig. L6-27—Special adapter (A) is available for unscrewing crankshaft pulley on early models. Late model pulley is retained by cap screw.

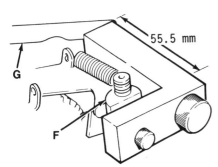

Fig. L6-25—After reassembling governor or when all injection pumps have been removed at one time, front end of governor bar (F) must be positioned at 55.5 mm (2.185 inch) from front face of cylinder block (G).

Fig. L6-28—Exploded view of governor linkage and controls. Refer to Fig. L6-30 for governor weights and sleeve.

1. Speed control lever
2. "O" ring
3. Bushing
4. Base plate
5. Locknuts
6. Idle speed screw
7. High speed screw
8. Governor spring lever
9. Governor spring
10. Shims
11. Spring pin
12. Governor rack spring
13. Bushing
14. Governor lever
15. Thrust collar
16. Rack toe
17. Fuel pump rack
18. Spring pin
19. Bearing block
20. Stop adjusting screw
21. Locknuts
22. Run adjusting screw
23. Spring clip
24. Trip lever
25. Pin
26. Stop/Run knob
27. Stop/Run lever
28. "O" ring
29. Bushing
30. Washers
31. Stop/Run internal lever
32. Spring
33. Pivot pin clip
34. Pivot pin support
35. Retainer
36. Pivot pin

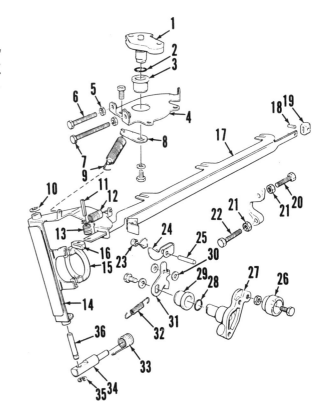

Illustrations Courtesy Lister-Petter

Lister-Petter

SMALL DIESEL

the governor lever upper pivot pin, taking care not to lose any shims at upper end of lever. Remove spring clip (33) and lower pivot pin (36) and remove the lever and thrust collar (15). Remove thrust washer (37—Fig. L6-30) and governor sleeve (38). Turn crankshaft so that weight pins (41) on camshaft gear (43) are in horizontal position, then unbolt weight plates (40) from camshaft gear and remove pins and weights (42).

Inspect all parts for excessive wear or other damage and renew as necessary.

To reinstall governor, reverse removal procedure while noting the following: Tighten weight plate retaining bolts (39—Fig. L6-30) to a torque of 9 N·m (78 in.-lbs.) and check to see that weights are free to move. Install governor sleeve (38) with thrust washer (37). Be sure governor lever is free on pivots; add or remove 0.25 mm shims (10—Fig. L6-28) if necessary to provide governor lever end play of 0.1-0.3 mm (0.004-0.012 inch) between top end of lever and upper lever pin bracket. Reconnect governor springs and linkage. If all fuel injection pumps have not been removed, loosen all pump retaining clamps. Position governor linkage and fuel injection pumps as described in R&R ALL PUMPS paragraph in FUEL INJECTION PUMP section. Complete the reassembly of engine by reversing disassembly procedure.

TIMING GEARS

The timing gears can be inspected after removing timing gear cover (end cover). The camshaft gear is not serviced separately from camshaft; refer to CAMSHAFT AND BEARINGS paragraph to remove and reinstall camshaft. The crankshaft gear can be removed with a suitable gear puller. To install crankshaft gear, first heat gear to a straw yellow color, then quickly install gear making sure that timing mark faces outward. Install camshaft with timing marks aligned as shown in Fig. L6-32. Refer to OIL PUMP AND RELIEF VALVE paragraph for renewing oil pump and gear.

CAMSHAFT AND BEARINGS

To remove camshaft, first remove cylinder head(s) and push rods, fuel lift pump, lift pump push rod and the fuel injection pumps. Remove the hydraulic lifters and fuel injection pump cam followers with magnetic tool. Turn engine control as far clockwise as it will go. Unhook the spring (32—Fig. L6-31) connecting engine control assembly (31) to camshaft thrust plate (44). Remove the governor lever assembly and the governor weights as outlined in preceding paragraph. Check camshaft end play with a dial indicator; camshaft end play should be 0.10-0.28 mm (0.003-0.011 inch). Excessive camshaft end play affects governor operation and can cause uneven firing. Working through holes in camshaft gear, remove the two cam thrust plate retaining screws and carefully pull cam and gear assembly from crankcase.

Camshaft, thrust plate and gear are available as a complete assembly only. If camshaft end play is excessive, gear teeth are worn or chipped, or if cam lobes or bearing journals are worn or scored, renew the complete assembly. With camshaft removed, inspect camshaft bearings and renew if necessary. Camshaft bearings are available in standard size only and are a press fit in crankcase. Be sure when installing new bearings that split in bearings are to top of bearing bore.

Refer to the following camshaft journal and bearing oil clearance specifications:

Journal diameter,
  all . . . . 34.965-34.980 mm
  (1.3765-1.3771 in.)
Front bushing ID . . 34.990-35.085 mm
  (1.3775-1.3812 in.)
Desired oil
  clearance . . . . 0.010-0.120 mm
  (0.0004-0.0047 in.)
Wear limit, oil clearance . . . . 0.17 mm
  (0.0067 in.)
Center & rear
  bushing ID . . . . 35.030-35.070 mm
  (1.3791-1.3807 in.)
Desired oil
  clearance . . . . 0.050-0.105 mm
  (0.002-0.004 in.)
Wear limit, oil
  clearance . . . . 0.17 mm
  (0.0067 in.)

Reinstall camshaft and gear assembly making sure timing marks on camshaft and crankshaft gears are aligned as shown in Fig. L6-32. Tighten thrust plate bolts to a torque of 9 N·m (78 in.-lbs.).

OIL PUMP AND RELIEF VALVE

The oil pump is mounted in engine crankcase and oil pump gear is driven by engine camshaft gear. The oil pressure relief valve (3—Fig. L6-33) is installed in rear of engine oil pump body (1).

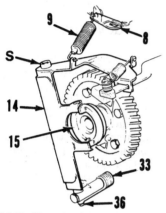

Fig. L6-29—View showing governor assembled; upper lever pivot pin and pin support are not shown. Shims (S) at top end of governor lever (14) control lever up and down play.

S. Shims
8. Governor spring lever
9. Governor spring
14. Governor lever
15. Thrust collar
33. Pivot pin clip
36. Pivot pin

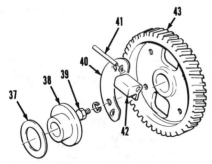

Fig. L6-30—Governor weights plates (40) are bolted to engine camshaft gear (43).

37. Thrust washer
38. Governor sliding collar
39. Weight plate bolts
40. Governor weight plates
41. Weight pivot pin
42. Governor weight
43. Camshaft gear

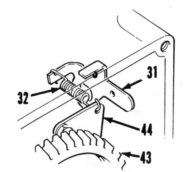

Fig. L6-31—To remove camshaft assembly, first unhook spring (32) from camshaft retainer plate (44).

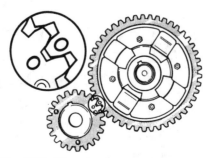

Fig. L6-32—View showing camshaft gear and crankshaft gear timing marks.

SERVICE MANUAL

Lister-Petter

To remove the engine oil pump, it is first necessary to remove the camshaft and crankcase side cover (door). Working through side cover opening, remove the oil pickup screen (2) and the pressure relief valve (3) from pump. Unbolt and remove the pump from engine crankcase. On early models, remove the copper sealing washer from pump inlet port when oil pickup screen is removed.

The oil pump is serviced as a complete assembly only; renew pump if it does not operate satisfactorily. Lubricating oil pressure with engine at normal operating temperature and running at 3000 rpm should be 248 kPa (36 psi); at slow idle speed, pressure should be 100 kPa (14.5 psi). Oil pressure relief valve parts are not serviced; renew valve as an assembly.

To install pump, insert pump in front of crankcase with cut-out section (S—Fig. L6-33) of pump body behind the pump gear toward top of crankcase and tighten the two pump mounting bolts to a torque of 9 N·m (78 in.-lbs.). On early engines, install new copper washer in pump inlet port, then install oil pickup and screen. Tighten pickup screen retaining nut with screen parallel to crankcase base. Install the relief valve assembly and tighten retaining nut. Install camshaft and crankcase side cover.

PISTON AND ROD UNITS

The piston and rod unit(s) can be removed after removing cylinder head(s) and crankcase side cover (door). Remove connecting rod cap and push piston and rod unit upward out of cylinder block as shown in Fig. L6-34.

Piston crown is marked "CAMSHAFT SIDE" and this side of piston must be to same side with identification marks (see Fig. L6-35) on connecting rod. To install piston and rod unit, be sure the piston ring end gaps are positioned equally around the piston and pin retaining snap rings are securely installed. Lubricate piston, rings and cylinder bore with clean engine oil and using ring compressor, install piston and rod unit in cylinder with "CAMSHAFT SIDE" mark and identification marks on connecting rod and cap toward camshaft side of engine. Place rod bearing inserts in connecting rod and cap. Install rod cap with assembly number on cap aligned with number on rod and tighten rod bolts to a torque of 24 N·m (18 ft.-lbs.). Complete installation by reversing removal procedure.

PISTONS, PINS, RINGS AND CYLINDERS

Aluminum alloy pistons (A—Fig. L6-36) with recessed combustion chamber in the piston crown are used on direct injection engines. Indirect injection "LPWS" models use piston (B) with valve reliefs in piston crown. Pistons have two compression rings and an expander type oil ring, all above the piston pin.

To remove piston assembly, remove cylinder head(s) and the crankcase side cover. Remove connecting rod cap and remove piston and rod assembly from top of block. Refer to Fig. L6-34. Keep rod cap with corresponding connecting rod. Remove snap rings from piston pin bore and push pin out of piston and rod.

A barrel faced chrome compression ring is used in the top ring groove of all pistons. Second ring is a tapered face ring. The oil ring has a coil spring type expander. Check ring side clearance using new ring installed in groove. Refer to the following piston ring specifications:

Piston ring end gap, all rings:
  Air cooled, desired ... 0.23-0.56 mm
    (0.009-0.022 in.)
  Wear limit .... 1.25 mm
    (0.049 in.)
  Water cooled, desired . 0.26-0.59 mm
    (0.010-0.023 in.)
  Wear limit .... 1.40 mm
    (0.055 in.)

Top ring side clearance:
  Air cooled, desired . 0.102-0.152 mm
    (0.004-0.006 in.)
  Wear limit .... 0.20 mm
    (0.008 in.)
  Water cooled, desired . 0.06-0.11 mm
    (0.0023-0.0043 in.)
  Wear limit .... 0.16 mm
    (0.0063 in.)

Second ring side clearance:
  All models, desired ... 0.05-0.10 mm
    (0.003-0.005 in.)
  Wear limit .... 0.15 mm
    (0.006 in.)

Oil ring side clearance:
  All models, desired ... 0.04-0.09 mm
    (0.0015-0.0035 in.)
  Wear limit .... 0.15 mm
    (0.006 in.)

Check piston diameter at bottom of skirt and at right angle to piston pin. Standard piston diameter is 75.85-75.86 mm (2.9862-2.9866 inches) for air cooled engines and 85.89-85.90 mm (3.3815-3.3818 inches) for water cooled engines. Piston skirt to cylinder wall clearance should be 0.14-0.175 mm (0.0055-0.0069 inch) for air cooled engines and 0.099-0.134 mm (0.0039-0.0053 inch) for water cooled engines. Wear limit is 0.4 mm (0.016 inch) for both air and water cooled engines.

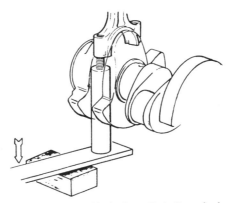

Fig. L6-34—Place block of wood in bottom of cylinder block and use prybar and pipe to push connecting rod upward.

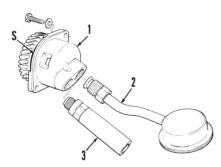

Fig. L6-33—Engine oil pump (1) is available as a complete assembly only. Oil pickup tube and screen (2) and system relief valve (3) are available separately from pump.

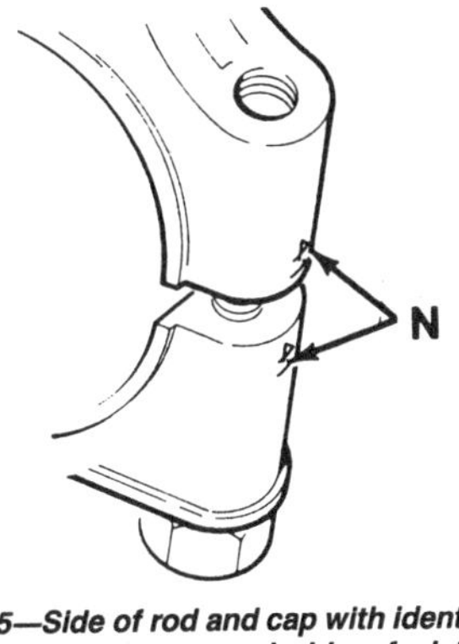

Fig. L6-35—Side of rod and cap with identification marks (N) must be toward side of piston with wording "CAMSHAFT SIDE" on piston crown.

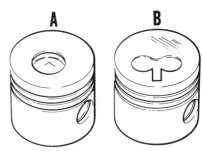

Fig. L6-36—Direct injection models have piston (A) with combustion chamber in crown. Piston (B) for Model LPWS has valve relief and fire notch only.

Illustrations Courtesy Lister-Petter

Lister-Petter

Piston pin should be a hand push fit in piston. Piston pin diameter is 24.9925-24.9975 mm (0.98396-0.98415 inch) for all models. Piston pin is available only with new piston. Be sure to securely install pin retaining snap ring (3—Fig. L6-38) at each end of piston pin bore.

Piston and rings are available in oversizes of 0.25 and 0.50 mm (0.010 and 0.020 inch) as well as standard size. The cylinder block should be rebored and honed to desired piston skirt to cylinder clearance and piston ring end gap.

When reassembling, be sure piston is installed with the wording "CAMSHAFT SIDE" on top of piston to the side of connecting rod having identification marks (see Fig. L6-35).

NOTE: Engine manufacturer states that rod bolts should be renewed on every occasion they are removed and that they must be renewed every fourth time.

Install compression rings with "TOP" marked side toward top of piston; oil rings may be installed either side up. See Fig. L6-37. Stagger piston ring end gaps at equal spacing around piston. Lubricate piston, piston rings and cylinder bore and install piston in cylinder with mark "CAMSHAFT SIDE" on top of piston to camshaft side of cylinder. Lubricate crankpin and rod bearing and install piston and rod assembly. Tighten connecting rod bolts to a torque of 24 N·m (18 ft.-lbs.). Complete installation by reversing removal procedure.

CONNECTING RODS AND BEARINGS

A forged steel connecting rod (6—Fig. L6-38) is used with a precision insert type bearing (7) at crankpin end and a renewable bushing (5) in piston pin end. Standard crankpin diameter is 49.985-50.000 mm (1.9679-1.9685 inches). Recommended oil clearance is 0.025-0.08 mm (0.001-0.003 inch) with maximum allowable clearance of 0.12 mm (0.0047 inch).

Desired piston pin to bushing clearance is 0.0075-0.0245 mm (0.0003-0.001 inch) with maximum allowable pin to bushing clearance of 0.05 mm (0.002 inch). If renewing bushing, install in rod with oil hole in bushing and rod aligned. Ream or hone pin bushing to desired oil clearance.

Engine manufacturer recommends renewing connecting rod bolts each time rod is disassembled and states that they must be renewed at fourth time rod is disassembled for any reason. Connecting rod bolt tightening torque is 24 N·m (18 ft.-lbs.).

CRANKSHAFT AND MAIN BEARINGS

The crankshaft is supported by main bearings located in crankcase at gear end and in the main bearing housing at flywheel end of crankcase, and by one or more intermediate bearings retained in bearing supports. Two-cylinder models have one intermediate bearing, 3-cylinder models have two intermediate bearings and 4-cylinder models have three intermediate bearings. All main bearings are two-piece precision insert type bearings; front and rear bearings are pressed into crankcase and rear bearing housing. Crankshaft end play is controlled by split type thrust washers at each end of crankshaft.

Fig. L6-38—Exploded view of crankshaft, bearings and the connecting rod and piston unit. All main bearings are split type inserts, although front main bearing is pressed into bore at front end of block and flywheel end bearing (21) is pressed into rear bearing support (23).

1. Piston ring set
2. Piston
3. Snap rings
4. Piston pin
5. Bushing
6. Connecting rod
7. Crankpin bearing insert
8. Crankshaft gear
9. Crankshaft gear key
10. Thrust washer upper half
11. Thrust washer lower half
12. Crankshaft
13. Support clamp bolts
14. Support upper half
15. Support dowel pins
16. Main bearing insert
17. Support lower half
18. Support dowel pin
19. Thrust washer lower half
20. Thrust washer upper half
21. Rear main bearing insert
22. Shim
23. Rear bearing support
24. Crankshaft rear oil seal

SMALL DIESEL

To remove crankshaft, remove cylinder heads, oil sump, pistons and rods, gear end cover, oil pump, camshaft, flywheel and flywheel housing. Then, proceed as follows:

Remove plug from outside of block at each intermediate bearing location. Thread a cap screw into the end of each dowel (18—Fig. L6-38) and pull dowel(s) from crankcase. Leave cap screws in dowels to aid in reinstalling dowels.

Use suitable puller to remove crankshaft timing gear and remove Woodruff key from shaft. Remove cap screws retaining rear main bearing housing (23) and pry it from crankcase with screwdriver using the two relief slots provided in the housing. Retain the metal shim (22) located between bearing housing and crankcase for reassembly. Remove split thrust washers (19 and 20) located between housing and crankshaft. Carefully remove crankshaft from rear end of cylinder block; remove split thrust washer at front end of shaft. Unbolt and remove intermediate bearing supports from crankshaft.

Inspect crankshaft, bearings and thrust washers for excessive wear or other damage. Crankshaft rear main journal diameter is 69.985-70.000 mm (2.7553-2.7559 inches). Front and intermediate main journal diameter is 54.985-55.000 mm (2.1647-2.1654 inches). Recommended rear main bearing oil clearance is 0.04-0.10 mm (0.0016-0.0039 inch) and wear limit is 0.14 mm

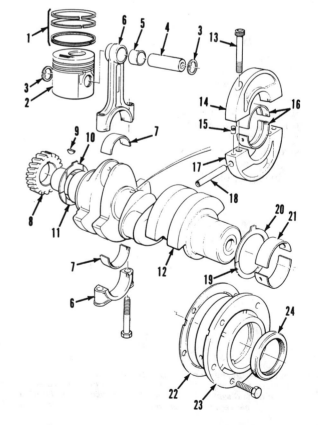

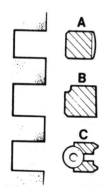

Fig. L6-37—Drawing showing correct installation of piston rings.

SERVICE MANUAL

Lister-Petter

(0.0055 inch). Front and intermediate bearing desired clearance is 0.035-0.095 mm (0.0014-0.0037 inch) and wear limit is 0.135 mm (0.0053 inch). Crankshaft thrust washer standard thickness is 2.31-2.36 mm (0.091-0.093 inch) and wear limit is 2.20 mm (0.086 inch). Crankshaft end play should be 0.18-0.38 mm (0.007-0.015 inch), and is maintained by condition of thrust surfaces, and by installation of either a 0.38 or 0.50 mm (0.015 or 0.020 inch) shim between rear main bearing housing flange and crankcase.

To remove front main bearing inserts, use suitable bushing driver to press front bearing forward out of crankcase. Remove rear main bearing seal from housing and press rear main inserts out to rear side of housing. Special bearing removal and installation tools are available from engine manufacturer.

When installing bearings, note that bearings for rear main journal, front journal and intermediate journals are different (all intermediate bearings are alike). Rear main bearing inserts have oil holes and inserts must be pressed into housing with oil holes in inserts aligned with oil holes in housing. Front upper bearing half has oil hole that must be aligned with oil passage in block. The grooved intermediate main bearing half with oil hole must be installed in top half of bearing support(s) with oil holes aligned.

NOTE: Do not install rear oil seal in main bearing housing until after housing is installed.

When installing crankshaft, be sure that all oil holes and surfaces are clean. Lubricate crankshaft journals and bearings. Assemble intermediate bearing supports with side marked "FLYWHEEL END" to rear of shaft and tighten cap screws to a torque of 21 N·m (15 ft.-lbs.). Apply light coat of high melting point grease to front thrust washer halves and insert them in front end of block with tab on upper half engaging slot in block, and with copper-lead bearing surface toward crankshaft shoulder. Carefully insert crankshaft from rear of block through thrust washer and front main bearing, taking care not to damage front bearing. Do not install bearing dowel pins until after crankshaft timing gear is installed.

Stick thrust washer halves in front side of rear bearing housing with high melting point grease; be sure tab on upper washer engages slot in housing. Use a new rear bearing housing shim of same thickness as removed on disassembly. Coat both sides of the new shim with Loctite 272 and fit on housing with flat side toward crankcase. Tighten retaining cap screws to a torque of 27 N·m (20 ft.-lbs.). Check crankshaft end play and change shim thickness if necessary to bring end play within specifications. Turn crankshaft to be sure it rotates smoothly in bearings. Install new rear main bearing seal with lip to inside of engine.

Install Woodruff key for crankshaft gear in front end of shaft. Heat crankshaft gear to a straw yellow color and quickly install on crankshaft with keyway aligned with key and with timing mark out. Install camshaft with timing marks on camshaft gear aligned with timing mark on crankshaft gear as shown in Fig. L6-32.

Dowel pin hole(s) in intermediate bearing(s) must be aligned with dowel pin hole(s) in block. Insert each dowel pin through crankcase wall making sure that dowel pin fully enters dowel pin hole in intermediate bearing support and not into the support cap screw head recess. Install cap screw plugs in intermediate bearing dowel holes in side of block and tighten to a torque of 20 N·m (15 ft.-lbs.). Complete the installation by reversing removal procedure.

LISTER-PETTER

Model	No. Cyls.	Bore	Stroke	Displ.
PH1	1	87.3 mm (3.437 in.)	100 mm (3.937 in.)	659 cc (40.2 cu.in.)
PH2	2	87.3 mm (3.427 in.)	100 mm (3.937 in.)	1318 cc (80.4 cu. in.)
PJ1, PJ1Z	1	96.8 mm (3.812 in.)	110 mm (4.33 in.)	810 cc (49.4 cu. in.)
PJ2, PJ2Z	2	96.8 mm (3.812 in.)	110 mm (4.33 in.)	1620 cc (98.9 cu. in.)
PJ3	3	96.8 mm (3.812 in.)	110 mm (4.33 in.)	2430 cc (148.2 cu. in.)

All engines are four-stroke, air-cooled direct injection diesels. Standard crankshaft rotation is clockwise at flywheel end. Firing order is 1-3-2 for PJ3 engine.

MAINTENANCE

LUBRICATION

Recommended engine oil is SAE 10W for temperatures below 5°C (41°F), SAE 20W-20 for temperatures between 5°C (41°F) and 30°C (86°F) and SAE30 for temperatures above 30°C (86°F). SAE 10W-30, 10W-40, 15W-40 and 20W-40 multigrade oils are also approved for use in all temperatures. All oil must be heavy duty diesel oil which meets MIL-L-46152-B specification.

NOTE: Use of oil meeting API specification CD or MIL-L-2104C specification is not recommended in new or overhauled engines until after the running-in period and first oil change, or in engines running at light load in low ambient temperatures. Oil meeting CD and MIL-L-2104C specifications should be used only in engines operating at high loads, speeds and ambient temperatures.

Manufacturer recommends renewing oil and oil filter (if so equipped) after the first 20 hours of operation on a new or overhauled engine. Thereafter, oil and oil filter should be renewed at intervals of 250 operating hours. Refer to the following table for approximate crankcase refill capacities.

Crankcase Capacity
PH1 .... 2.84 liters
    (5 pints)
  With filter .... 3.41 liters
    (6 pints)
PH2 .... 5.68 liters
    (10 pints)
  With filter .... 6.53 liters
    (11.5 pints)
PJ1 .... 3.4 liters
    (6 pints)
PJ1Z .... 2.85 liters
    (5 pints)
PJ2 .... 6.53 liters
    (11.5 pints)
PJ2Z .... 5.7 liters
    (10 pints)
PJ3 .... 7.67 liters
    (13.5 pints)

FUEL SYSTEM

FUEL FILTER. Manufacturer recommends renewing fuel filter element after every 500 hours of operation or sooner if loss of power is evident. Do not attempt to clean filter element. After renewing fuel filter, air must be bled from fuel system as outlined in the following paragraph.

BLEED FUEL SYSTEM. To bleed fuel system, first loosen the two bleed screws (1–Fig. PE-30) on top of fuel filter. When air-free fuel flows from the vents, retighten the screws. Loosen bleed screw (2) on fuel pump and actuate fuel feed pump lever (3), if so equipped, until fuel flows from vent, then retighten screw. Loosen the high pressure fuel line connection at the injector. With pump control in "RUN" position, crank engine until fuel is discharged at the loosened connection. Retighten injector line fitting and start engine.

ENGINE SPEED ADJUSTMENT

FIXED SPEED CONTROL. On fixed speed engines, the action of the centrifugal governor is balanced by the speeder spring (4–Fig. PE-31). The

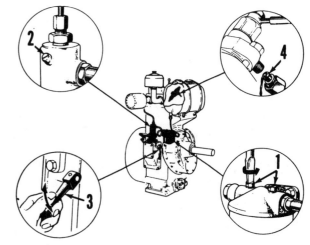

Fig. PE-30—Fuel system air venting points. Refer to text for procedure.

SERVICE MANUAL

Lister-Petter

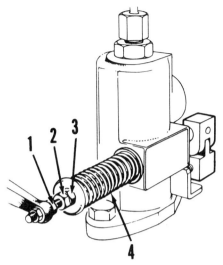

Fig. PE-31—Speeder spring (4) balances governor forces to regulate engine speed. Refer to text for adjustment.

1. Adjusting screw
2. Locknut
3. Nut
4. Speeder spring

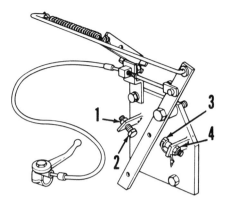

Fig. PE-33—View of variable speed control linkage adjustment points. Refer to text.

1. Locknut
2. Idle adjusting screw
3. Maximum speed adjusting screw
4. Locknut

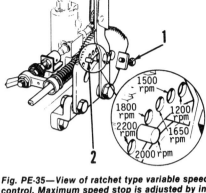

Fig. PE-35—View of ratchet type variable speed control. Maximum speed stop is adjusted by inserting dowel (2) in appropriate hole as shown in inset.

1. Idle speed adjusting screw
2. Dowel

spring is adjustable to allow small variation in a set speed range. If a speed outside this range is desired, a different spring and governor may be required.

To adjust fixed speed control, loosen locknut (2) and adjust nut (3) against the spring (4) to increase speed or away from spring to decrease speed. Maximum speed with no load should be set at 4.5 percent above rate full-load speed indicated on engine name plate.

VARIABLE FIXED SPEED (EARLY TYPE). To adjust engine speed, loosen locknut (1–Fig. PE-32). To decrease speed setting turn nuts (2 and 3) towards the flywheel until desired speed is obtained. To increase speed, turn nuts away from flywheel. Tighten nuts against bracket (4) to secure speed adjustment.

VARIABLE SPEED CONTROL. To adjust low idle speed, loosen locknuts (1–Fig. PE-33) and turn adjusting screw (2) clockwise to decrease speed or counterclockwise to increase speed. Recommended idle speed is 950 to 1000 rpm.

To adjust maximum no-load speed, loosen locknut (4) and move control lever to full speed position. Turn adjusting screw (3) clockwise to increase speed or counterclockwise to decrease speed. Maximum speed with no load should be set at 8 percent higher than specified rated full-load speed indicated on engine name plate.

VARIABLE SPEED CONTROL (EARLY TYPE). To adjust low idle speed, loosen locknut (5–Fig. PE-34) and turn adjusting screw (6) clockwise to increase speed or counterclockwise to decrease speed. Recommended idle speed is 500 to 600 rpm.

To adjust maximum no-load speed, loosen locknut (4) and move control lever to full speed position. Turn adjusting screw (3) clockwise to decrease speed or counterclockwise to increase speed setting. Maximum no-load speed should be set 8 percent above specified full-load speed indicated on engine name plate.

Be sure control cable allows full travel of control arm so it contacts both adjusting screws. If necessary, turn cable adjuster (1) until there is a small amount of slack in inner cable. Control lever should be able to move slightly before inner cable begins to move the control.

VARIABLE SPEED (RACHET TYPE). To adjust idle speed, turn adjusting screw (1–Fig. PE-35) clockwise to increase speed or counterclockwise to decrease speed. Recommended idle speed is 500 to 600 rpm.

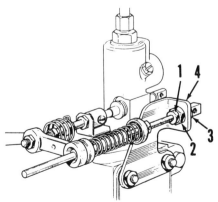

Fig. PE-32—View of early type adjustable fixed speed control.

1. Locknut
2. Nut
3. Nut
4. Bracket

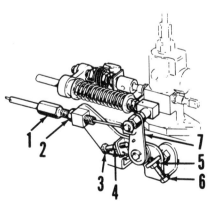

Fig. PE-34—View of early type variable speed control adjustment points. Refer to text.

1. Cable adjuster
2. Locknut
3. Maximum speed adjusting screw
4. Locknut
5. Locknut
6. Idle speed adjusting screw
7. Control arm

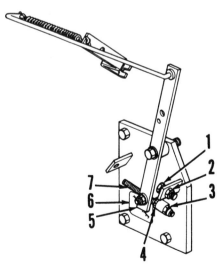

Fig. PE-36—Adjustable fixed speed control is used on some engines. Refer to text for adjustment procedure.

1. Maximum speed stop screw
2. Locknut
3. Locknut
4. Locknut
5. Control lever
6. Speed scale
7. Adjusting screw

Lister-Petter
SMALL DIESEL

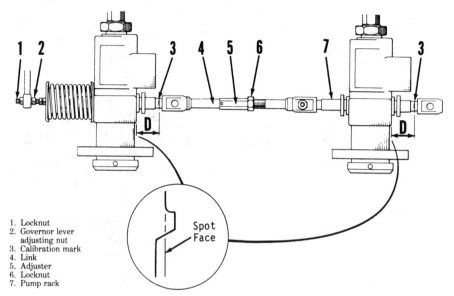

1. Locknut
2. Governor lever adjusting nut
3. Calibration mark
4. Link
5. Adjuster
6. Locknut
7. Pump rack

Fig. PE-37—On multiple cylinder engines, a separate injection pump is used for each cylinder. Pump timing and pump rack setting for each pump must be sychronized for proper engine operation. Refer to text.

To set maximum full-speed stop, move dowel (2) to appropriate hole in rachet bracket as indicated in inset.

ADJUSTABLE FIXED SPEED. To set the maximum no-load speed, loosen locknuts (2, 3 and 4 – Fig. PE-36). Turn the adjusting screw (7) in conjunction with the stop screw (1) until speed setting is 8 percent higher than the rated full-load speed indicated on the engine name plate. At this point, the bottom end of the lever (5) should point to the appropriate rated rpm graduation on the speed scale (6). If not, it will be necessary to adjust speeder spring tension and the adjusting rod until desired speed setting and graduation alignment on scale is obtained.

INJECTION PUMP TIMING

To time fuel injection pump, the following procedure must be followed.

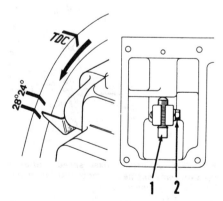

Fig. PE-38—Injection pump timing is adjusted by turning rocker adjuster screw (1). Refer to text for specified injection timing.

On multiple cylinder engines, the injection pump for each cylinder must be set one at a time. Remove the fuel pump cover. Position injection pump rack so calibration mark (3 – Fig. PE-37) is 20.6 mm (13/16 inch), setting distance "D", from spot face on pump body. Remove injector high pressure pipe from pump being timed. Remove the discharge fitting (1 – Fig. PE-39), spring and delivery valve (4) from pump, then reinstall discharge fitting. Do not remove the delivery valve seat (5). Attach a spill pipe to pump discharge fitting. Note that fuel will flow from the spill pipe and a suitable container should be used to catch the fuel. Turn the flywheel until appropriate piston is on compression stroke and specified timing mark on flywheel is aligned with pointer. Refer to pump injection timing table. Loosen pump rocker clamp bolt (2 – Fig. PE-38) and unscrew adjusting screw (1) until pump is at bottom of its stroke. With fuel flowing from pump discharge fitting (fuel feed pump, if so equipped, must be operated), turn adjusting screw until fuel flow just stops. Beginning of injection occurs at this point. Tighten clamp bolt being careful not to disturb adjusting screw setting. Remove spill pipe and discharge fitting. Reinstall delivery valve and spring and tighten discharge fitting to 54 N·m (40 ft.-lbs.). Repeat adjustment procedure for the other injection pump(s) on multiple cylinder engines.

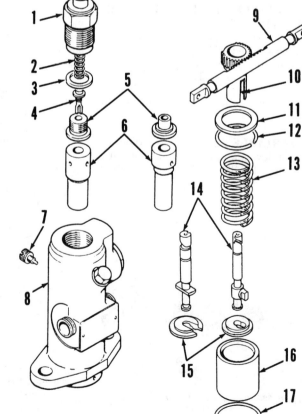

Fig. PE-39—Exploded view of injection pump assembly. A Bryce or CAV pump may be used. Both pumps are similar.

1. Discharge fitting
2. Spring
3. Gasket
4. Delivery valve
5. Delivery valve seat
6. Pumping element
7. Locating screw
8. Pump body
9. Rack
10. Pinion
11. Spring plate
12. Retaining ring (Bryce)
13. Spring
14. Plunger
15. Spring retainer
16. Tappet
17. Retaining ring

SERVICE MANUAL

Lister-Petter

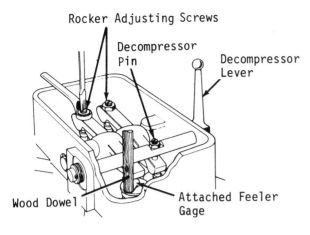

Fig. PE-40—Valve rocker clearance should be adjusted to 0.10 mm (0.004 inch) for both intake and exhaust. A special tool, fabricated from a wood dowel with a feeler gage attached as shown, will make measuring clearance easier on PH1 and PH2 engines.

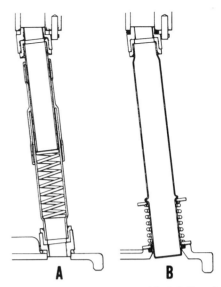

Fig. PE-42—View showing correct installation of early style (A) push rod tube assembly and late style (B) push rod tube assembly.

Injection timing (BTDC)
PH1 and PH2 engines
  Up to 1650 rpm .... 24°
  1650 to 2200 rpm .... 28°
PJ Engines
  Fixed speed up to
    1650 rpm .... 23°
  1651 to 2000 rpm .... 26°
  Variable and
    two speed .... 23°
PJZ Engines
  All speeds .... 23°

REPAIRS

TIGHTENING TORQUES

Refer to the following table for special tightening torques.

Connecting rod .... 77 N·m
                             (57 ft.-lbs.)
Crankshaft gear .... 62 N·m
                             (46 ft.-lbs)
Cylinder crankcase
  studs .... 34-41 N·m
                             (25-30 ft.-lbs.)
Cylinder head .... 75-88 N·m
                             (55-65 ft.-lbs.)
Fuel injector .... 20 N·m
                             (15 ft.-lbs.)
Inspection cover .... 22 N·m
                             (16 ft.-lbs.)

Fig. PE-41—Follow sequence shown when loosening or tightening cylinder head stud nuts.

VALVE ADJUSTMENT

Valve rocker clearance with engine cold should be 0.10 mm (0.004 inch) for intake and exhaust valves. Manufacturer recommends checking valve clearance and adjusting, if necessary, after every 250 hours of operation.

To adjust valve clearance, rotate crankshaft until piston is at top dead center on compression stroke (both valves closed). Loosen rocker adjusting screw locknut (Fig. PE-40) and turn adjusting screw until a 0.10 mm (0.004 inch) feeler gage can be inserted between valve stem end and rocker arm.

NOTE: Due to close confines of rocker box of PH1 and PH2 engines, an offset feeler gage or a fabricated tool as shown in Fig. PE-40 should be used to obtain accurate measurements.

Repeat adjustment procedure for each cylinder on multiple cylinder engines.

COMPRESSION RELEASE

To adjust compression release, first rotate crankshaft so piston is at TDC on compression stroke. Move decompressor lever (Fig. PE-40) to vertical position. Adjust decompressor pin until it contacts the exhaust rocker, then turn pin clockwise an additional ½ turn on PH1 and PH2 engines or an additional ⅔ turn on PH series engines. Make certain the valve is not depressed more than 0.63 mm (0.025 inch) or valve may hit the piston causing serious damage.

CYLINDER HEAD

A separate cylinder head is used for each cylinder on multiple cylinder engines. To remove a cylinder head proceed as follows: Remove cylinder air cowling. Remove intake and exhaust manifolds. Remove injector fuel lines and the oil supply pipe to the rockers. Plug all openings to prevent entry of dirt. Remove the fuel injector. Remove the rocker box and rocker assembly. Remove push rods, noting their position for reassembly, and push rod tubes. Gradually loosen cylinder head nuts in sequence shown in Fig. PE-41, then lift off cylinder head assembly.

When reinstalling cylinder head, always renew cylinder head gasket. On multiple cylinder engine, use a straightedge or attach manifold to heads to align manifold mounting surfaces. Tighten cylinder head nuts a quarter of a turn at a time following sequence shown in Fig. PE-41 until specified torque of 75-88 N·m (55-65 ft.-lbs.) is obtained. Install push rod tubes and adapters as shown in Fig. PE-42. Be sure push rods are reinstalled in their original positions. Push rods should not be interchanged. Complete installation by reversing the removal procedure. After engine has operated for about 20 hours, allow engine to cool, then retorque cylinder head nuts to specified torque following sequence shown in Fig. PE-41.

VALVE SYSTEM

As valve components are removed, identify all parts so they can be reinstalled in their original positions. Clean all parts and inspect for excessive wear or other damage.

Valve face and seat angles are 45° for intake and exhaust. Desired valve seating width is 1.33-1.47 mm (0.052-0.057 inch). With valves installed in cylinder head, distance (R – Fig. PE-43) from surface of cylinder head to valve head should be 0.81-1.17 mm (0.032-0.046 inch) on PH engines or 0.68-1.09 mm (0.027-0.043 inch) on PJ engines. If valves are recessed excessively, cylinder

Lister-Petter SMALL DIESEL

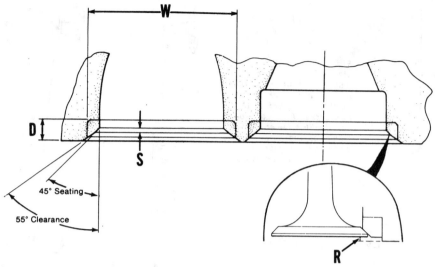

Fig. PE-43—Cylinder head ports must be machined as shown to install valve seat inserts. Refer to text for required dimensions.

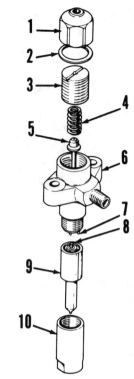

Fig. PE-44—Exploded view of fuel injector assembly.

1. Cap nut
2. Gasket
3. Adjusting screw
4. Spring
5. Spring seat & spindle
6. Body
7. Dowels
8. Nozzle needle
9. Nozzle body
10. Nozzle nut

head should be machined for installation of valve seat inserts. Refer to Fig. PE-43 and the following table for cylinder head dimensions required for installation of seat inserts.

PH1-PH2 Engines

Depth (D) .... 6.019-6.045 mm
    (0.237-0.238 in.)
Width (W) .... 35.687-35.712 mm
    (1.405-1.406 in.)
Valve seat (S) .... 1.33-1.47 mm
    (0.052-0.057 in.)

PJ1-PJ1Z-PJ2-PJ2Z-PJ3 Engines

Depth (D) .... 6.019-6.045 mm
    (0.237-0.238 in.)
Width (W) .... 42.16-42.18 mm
    (1.661-1.662 in.)
Valve seat (S) .... 1.33-1.47 mm
    (0.052-0.057 in.)

INJECTOR

REMOVE AND REINSTALL. Thoroughly clean injector, fuel lines and surrounding area prior to removing injector. Disconnect the pressure pipe and fuel leak-off pipe from the injector and immediately plug all openings to prevent entry of dirt. Remove retaining nuts and withdraw injector.

To reinstall injector, reverse the removal procedure. Tighten high pressure pipe union nuts finger tight, then tighten an additional third of a turn. Tighten injector retaining nuts to 20 N·m (15 ft.-lbs.).

TESTING. A suitable test stand is required to check injector operation. Only clean, approved testing oil should be used to test injector. Injector should be checked for proper opening pressure, seat leakage and spray pattern.

WARNING: Fuel leaves the injector nozzle with sufficient force to penetrate the skin. When testing, keep yourself clear of nozzle spray.

Attach injector to test pump and operate tester lever several strokes to purge air from nozzle and to make sure nozzle valve is not stuck. Operate tester lever slowly and observe gage pressure reading. On PH1 and PH2 engines that operate below 1100 rpm, opening pressure should be 13750-15200 kPa (1995-2205 psi). On all other engines, opening pressure should be 19700-21700 kPa (2850-3150 psi). Nozzle opening pressure is adjusted by removing the cap nut (1—Fig. PE-44) and turning adjusting screw (3).

To check nozzle seat leakage, wipe nozzle tip dry. Operate tester lever to maintain pressure at about 1000 kPa (150 psi) below opening pressure for a period of 10 seconds. If a drop forms at nozzle tip during this period, nozzle seat is not sealing and injector should be overhauled or renewed.

To check spray pattern, operate tester lever about six strokes in 10 seconds. The nozzle should open with a sharp chattering sound. The spray from each nozzle hole should be well atomized and free from ragged edges or solid squirts of fuel.

OVERHAUL. To disassemble injector, secure injector body (6—Fig. PE-44) in a vise or holding fixture. Remove nozzle nut (10) and withdraw nozzle valve assembly (8 and 9). Remove cap nut (1), adjusting screw (3), spring (4) and spindle (5). If disassembling more than one injector, be sure to keep parts from each injector separated in a compartmented pan. Parts must not be interchanged.

Thoroughly clean all parts in a suitable solvent. Use a brass wire brush to remove dirt and carbon from outer surfaces of injector. Do not use steel wire brush, emery cloth or sharp metal tools to clean injector components. Protect all polished surfaces from scratches or other damage. The sealing surfaces between nozzle valve (9) and body (6) must be clean and smooth. Clean the fuel feed hole (2—Fig. PE-45) by inserting a suitable wire or twist drill (1). Use a special

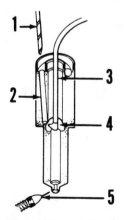

Fig. PE-45—Cross-sectional view of nozzle valve body showing areas to be cleaned during injector overhaul.

1. Twist drill
2. Fuel feed passage
3. Special scraper
4. Fuel chamber
5. Spray hole cleaner

SERVICE MANUAL

Lister-Petter

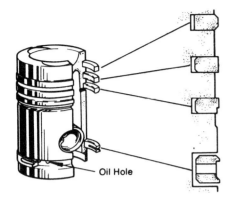

Fig. PE-47—Cross-sectional view showing correct installation of piston rings for PH1 and PH2 engines.

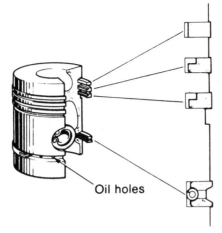

Fig. PE-48—Cross-sectional view showing correct installation of piston rings for PJ1, PJ1Z, PJ2, PJ2Z and PJ3 engines.

scraper (3) to clean deposits from nozzle fuel chamber (4). Clean the nozzle spray holes by inserting a cleaning wire (5) of slightly smaller diameter than the spray hole. Inspect nozzle needle for dull or very bright patches or any discoloration indicating overheating. When wet with diesel fuel, the needle must slide smoothly in bore of nozzle body. If needle sticks, reclean or renew nozzle valve assembly.

When reassembling injector, be sure all parts are clean and wet with diesel fuel as they are assembled. Recheck injector operation as outlined in TESTING section.

INJECTION PUMP

All engines are equipped with a Bryce or CAV fuel injection pump. Refer to Fig. PE-39 for an exploded view of injection pump. On multiple cylinder engines, a separate injection pump is used for each cylinder. The injection pump(s) is driven by a rocker arm which is actuated by the camshaft. It is recommended that injection pump be tested and repaired only by a shop qualified in diesel fuel injection repair.

REMOVE AND REINSTALL. Shut off fuel supply, then disconnect fuel inlet pipe and injector high pressure pipe from the pump. Plug all openings to prevent entry of dirt. On variable speed engines, remove the spring between governor fulcrum arm and pump rack extension of pump at flywheel end of engine. Disconnect pump linkage on multiple cylinder engines. Remove pump mounting cap screws and lift pump from the engine.

To reinstall pump, first rotate crankshaft to position pump rocker arm at the bottom of its stroke. Assemble pump to the mounting bracket and note that special screw with the extended plain shank is installed in right-hand position. Complete installation by reversing the removal procedure. Bleed air from fuel system. On multiple cylinder engines, adjust connecting linkage as outlined in the following paragraph. Check and adjust pump timing as previously outlined.

LINKAGE ADJUSTMENT. On multiple cylinder engines, adjust linkage beginning at the flywheel end pump as follows: Push pump rack towards the flywheel and hold against governor fulcrum arm adjusting screw (maximum fuel position). If equipped with overload stop, the hinged lower part of stop must be raised to allow full movement of rack Measure distance from calibration mark (3–Fig. PE-37) to spot face on side of pump (not the rack bushing). Distance (D) should be 12.7 mm (0.5 inch). If necessary, loosen locknut (1) and adjust governor lever screw (2) to obtain specified dimension. A special setting gage (number 392645) is available from the manufacturer for use in making this adjustment. The calibration mark on remaining pump(s) must also be 12.7 mm (0.5 inch) from spot face on pump body. To adjust, loosen locknut (6) and turn adjuster (5) to obtain correct rack setting.

PISTON, PIN, RINGS AND CYLINDER

To remove cylinder and piston with connecting rod, remove cylinder head and crankcase inspection cover. If equipped with straight connecting rod bearing cap (Fig. PE-49), remove injection pump(s) together with pump mounting bracket. Remove connecting rod bearing cap. Mark the cylinder and crankcase so cylinder can be reinstalled in it original position. Lift cylinder, piston and connecting rod as an assembly from crankcase. Retain shims used between cylinder and crankcase for use in reassembly. Withdraw piston from cylinder. If piston pin is difficult to remove, heat piston in hot water for a few minutes until pin can be pushed out.

Check cylinder bore for out-of-round, excessive wear or other damage. Cylinder standard inner diameter is 87.465-87.491 mm (3.4435-3.4445 inches) for PH1 and PH2 engines. On all other engines, standard cylinder bore is 96.965-96.990 mm (3.8175-3.8185 inches). If bore wear exceeds 0.25 mm (0.010 inch), cylinder may be rebored and honed for installation of oversize piston and rings. Pistons are available in oversizes of 0.508 mm (0.020 inch) and 1.016 mm (0.040 inch).

Piston ring end gap should be measured with piston ring positioned squarely in cylinder. On PH1 and PH2 engines, desired ring end gap is 0.74-0.94 mm (0.029-0.037 inch). On all other engines, desired ring end gap is 0.69-0.89 mm (0.027-0.035 inch) and maximum allowable end gap is 1.52 mm (0.060 inch).

To check piston ring groove wear, insert new rings into piston grooves and measure clearance between top of ring and top of groove with a feeler gage. Desired side clearance for compression rings is 0.08-0.13 mm (0.0033-0.0053 inch). Side clearance for oil scraper ring should be 0.06-0.11 mm (0.0025-0.0045 inch). Maximum side clearance for all rings is 0.25 mm (0.010 inch).

If piston is being reused, make certain oil drain holes in bottom ring groove are clean. When assembling rings onto piston, note shape of rings and refer to appropriate Fig. PE-47 or PE-48 for correct installation of rings. Stagger ring end gaps evenly around piston making sure they are not in line with piston pin bore.

To make piston pin installation easier, piston should be heated in hot water for a few minutes before assembly. If piston has valve recesses in the piston crown, be sure they are toward camshaft side of engine. Install original shim pack between bottom of cylinder and crankcase. Be sure assembly numbers on connecting rod and rod bearing cap match and are on the same side. If equipped with an offset rod bearing cap, be sure cap retaining bolt heads face away from injection pump side of engine. Tighten connecting rod cap screws to 77 N·m (57 ft.-lbs.).

If the cylinder, piston or connecting rod was renewed, or if original shim pack is not being used, check and adjust piston-to-cylinder head clearance as follows: Place three pieces of soft wire or solder on top of piston making sure they will not be aligned with valve heads. Install cylinder head and tighten stud nuts to specified torque. Rotate crankshaft by hand so piston goes past

Lister-Petter

SMALL DIESEL

Fig. PE-49—Engines may be equipped with connecting rod using a straight bearing cap or an offset cap.

top dead center. Remove cylinder head and measure thickness of the flattened wires. Average thickness should be 0.91-1.017 mm (0.036-0.042 inch). If necessary to adjust, add or remove shims between bottom of cylinder and crankcase.

Complete installation by reversing the removal procedure. After renewing piston rings, manufacturer recommends the following procedure to properly seat the rings. Operate engine for 2 minutes with no-load, then run for 10 minutes at half-load, and finally run for a minimum of 8 hours continuous at full-load.

CONNECTING ROD

Engines may be equipped with connecting rod having an offset bearing cap or a straight type bearing cap as shown in Fig. PE-49. The small end of connecting rod is fitted with a renewable bushing while the big end uses precision, thin wall, insert type bearing shells.

Inner diameter of small end bushing should be 30.035-30.048 mm (1.1825-1.1830 inches) for all engines. When renewing bushing, be sure hole in bushing is aligned with hole in the connecting rod.

Recommended connecting rod large end bearing operating clearance is 0.051-0.089 mm (0.002-0.0035 inch). No attempt should be made to modify bearing or connecting rod to obtain recommended clearance. Undersize bearings are available for service.

When reinstalling connecting rod, heat piston in hot water until piston pin will slide in freely. Be sure piston is positioned so valve recesses (if used) in piston crown are toward camshaft side of engine. If offset type connecting rod cap is used, cap retaining bolt heads should face away from camshaft side of engine. Make certain identification numbers on rod and cap match and are on the same side. Tighten rod cap bolts to 77 N·m (57 ft.-lbs.).

CAMSHAFT

To remove the camshaft, remove the gear cover and fuel pump bracket assembly. Remove rocker box, push rods and push rod tubes. Note position of push rods as they should not be interchanged when reassembling. Use rubber bands or other suitable devices to hold cam followers up away from camshaft and to prevent them from falling when camshaft is removed. If equipped with an SAE 5 bellhousing, the crankshaft gear and oil thrower must be removed. On PH1, PH2, PJ1Z and PJ2Z engines, a camshaft thrust plate (4–Fig. PE-50 is located between camshaft gear and crankcase. Thrust plate retaining screws (5) are accessible through holes in camshaft gear. Loosen screws and thrust plate will drop to end of slotted holes so camshaft can be removed. On all engines, withdraw camshaft and governor assembly from crankcase. If necessary, cam followers may be removed after camshaft is removed.

Inspect camshaft and gear for excessive wear or other damage. Check crankcase bearing bores for wear. A renewable bushing is used at gear end of crankcase. The camshaft journal at flywheel end operates directly in the crankcase bore. Camshaft and gear must be renewed as an assembly as gear is a shrink fit on the shaft and cannot be removed without damaging the gear or the camshaft. If equipped with a camshaft extension shaft, inspect support bearing and oil seal and renew if necessary. Heating extension shaft bearing in boiling water for about five minutes will make removal and installation of bearing easier. Be sure oil hole in housing is clean and is aligned with hole in bearing when pressing in a new bearing. Use suitable installing tool to avoid damaging bearing.

When reinstalling camshaft, lubricate bearing surfaces before inserting into crankcase. On PJ1, PJ2 and PJ3 engines, be sure thrust washers (3 and 6–Fig. PE-51) are held in proper position by pins (4). Make certain grooved side of inner washer is against thrust face of camshaft and the tang is located in plate recess. On PH1, PH2, PJ1Z and PJ2Z engines, install thrust plate (4–Fig. PE-50), if removed, and tighten retaining screws finger tight. On all engines, insert camshaft assembly aligning timing marks on camshaft gear and crankshaft gear. On PH1, PH2, PJ1Z and PJ2Z engines, raise thrust plate so upper edge engages groove in camshaft, then tighten retaining screws. Check camshaft end play by inserting a feeler gage between gear and thrust plate. Desired end play is 0.10-0.38 mm (0.004-0.015 inch). Renew thrust plate if

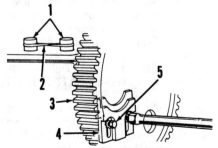

Fig. PE-50—A thrust plate (4) is located between camshaft gear and crankcase on PH1, PH2, PJ1Z and PJ2Z engines. A rubber band (2) may be used to support cam followers (1) when camshaft is removed.

1. Cam followers
2. Rubber band
3. Camshaft gear
4. Thrust plate
5. Retaining screws

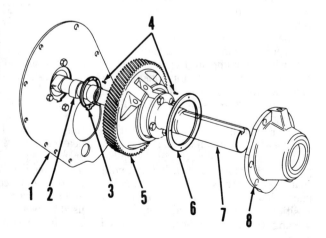

Fig. PE-51—Thrust washers (3 and 6) control end thrust of camshaft on PJ1, PJ2 and PJ3 engines. Some engines are equipped with camshaft extension shaft (7).

1. Plate
2. Camshaft
3. Thrust washer
4. Locating pins
5. Camshaft gear
6. Thrust washer
7. Extension shaft
8. Housing

SERVICE MANUAL
Lister-Petter

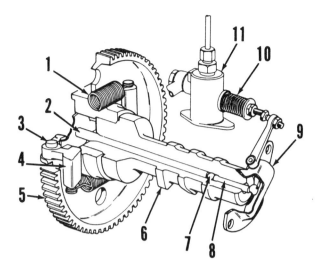

Fig. PE-52—Governor assembly is mounted on the camshaft gear.
1. Governor weight spring
2. Push rod
3. Pin
4. Governor weight
5. Camshaft gear
6. Camshaft
7. Steel ball
8. Push rod
9. Fulcrum arm
10. Speeder spring
11. Injection pump

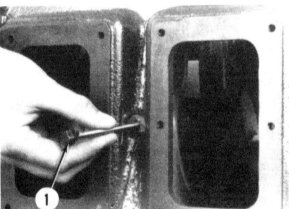

Fig. PE-53—On multiple cylinder engines, a locating screw (1) is used to position intermediate bearing housing in crankcase.

end play is excessive. On PJ1, PJ2 and PJ3 engines, install gear cover and extension shaft housing (if so equipped). Mount dial indicator on fuel pump side of crankcase with pointer against the side of one of the cam lobes. Desired end play is 0.13-0.71 mm (0.005-0.028 inch). Renew thrust washers if end play is excessive.

GOVERNOR

The governor assembly is mounted on the camshaft gear. Refer to CAMSHAFT section for removal procedure.

All components must move freely for proper governor operation. Remove governor push rods (2 and 8–Fig. PE-52) and steel ball (7). Be sure push rod hole through the camshaft is clean. Inspect for excessive wear and renew parts if necessary.

Refer to MAINTENANCE section for engine speed adjustment procedure.

CRANKSHAFT AND MAIN BEARINGS

To remove crankshaft, first remove cylinder head, cylinder, piston and connecting rod. Remove flywheel and gear end cover. Remove crankshaft gear using a suitable puller. On multiple cylinder engines, remove intermediate main bearing housing locating screw (1–Fig. PE-53). On all engines, mark crankshaft balance weights to identify assembly positions, then remove weights. Remove main bearing housing from flywheel end. Withdraw crankshaft through opening at flywheel end.

Inspect crankshaft and bearings for excessive wear or other damage. Oil pump drive gear is a shrink fit on crankshaft and should not be removed unless renewal is necessary. Heat the new gear prior to installation. Be sure thrust face of gear faces outward.

On PH1, PJ2, PJ1Z and PJ2Z engines, standard crankpin or intermediate main journal (if applicable) diameter is 60.312-60.325 mm (2.3745-2.3750 inches). Standard main journal diameter at gear end or flywheel end is 60.274-60.287 mm (2.3730-32.3735 inches). Desired rod bearing clearance is 0.051-0.089 mm (0.002-0.0035 inch). Main bearing clearance should be 0.051-0.114 mm (0.002-0.0045 inch).

On PJ1, PJ2 and PJ3 engines, standard diameter of crankpin or intermediate main journal (if applicable) is 60.312-60.325 mm (2.3745-2.3750 inches). Standard diameter of main journal at gear end is 60.274-60.287 mm (2.3730-2.3735 inches) while journal diameter at flywheel end is 104.661-104.673 mm (4.1205-4.1210 inches). Desired rod bearing clearance is 0.051-0.089 mm (0.002-0.0035 inch). Main bearing diametral clearance should be 0.051-0.115 mm (0.002-0.0045 inch) for gear end bearing, 0.080-0.135 mm (0.0028-0.0052 inch) for flywheel end bearing and 0.064-0.102 mm (0.0025-0.0040 inch) for intermediate bearing (if applicable).

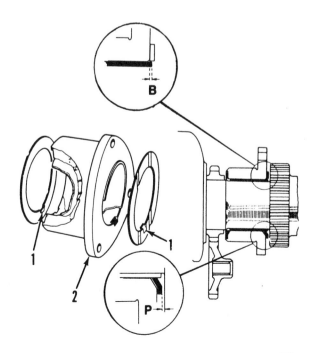

Fig. PE-54—Main bearing at gear end must protrude (B) beyond front surface of bearing housing (2) 1.27-1.52 mm (0.050-0.060 inch). Thrust washers (1) control crankshaft end play (P).

Illustrations Courtesy Lister-Petter

Lister-Petter   SMALL DIESEL

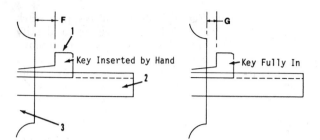

Fig. PE-55—On PH1, PH2, PJ1Z and PJ2Z engines, flywheel key (1) must be installed to dimensions shown. Refer to text.
F. 23.8 mm (15/16 in.)
G. 11.1 mm (7/16 in.)
1. Key
2. Crankshaft
3. Flywheel

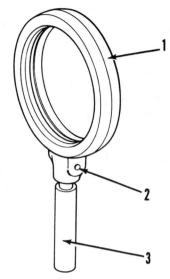

Fig. PE-57—Whenever oil pump is being serviced, plunger pin (2) connecting oil plunger (3) to crankshaft strap (1) should be renewed.

The main bearings are precision, thin wall, sleeve type bearings. Bearings are presized for proper fit and do not require further machining. Use suitable tools to remove and install bearings. Be sure bearing oil holes are aligned. The main bearing at gear end of crankcase must protrude between 1.27-1.52 mm (0.05-0.006 inch) beyond front surface of bearing housing as shown in (B–Fig. PE-54). Main and rod bearings are available in undersizes of 0.254, 0.508, 0.762 and 1.016 mm (0.010, 0.020, 0.030 and 0.040 inch) as well as standard size.

To reinstall crankshaft, proceed as follows: Install gear end bearing housing making sure oil hole is aligned with oil hole in crankcase. Install thrust washer (1–Fig. PE-54) making sure locating tangs are in bearing housing (2) notches and grooved side of washers face away from bearing housing. Use grease to hold washers in position. On two-cylinder engines, assemble intermediate main bearing housing with side marked TOP towards flywheel end. Tighten bearing housing nuts to 34 N·m (25 ft.-lbs.). On three-cylinder engines, be sure intermediate bearing housings are correctly installed with oil hole and locating screw hole in proper positions. Tighten bearing housing cap screws to 54 N·m (40 ft.-lbs.). On all engines, be sure all bearing surfaces are well lubricated, then carefully insert crankshaft into crankcase. Install a new oil seal in flywheel end bearing housing making sure seal face is flush with rear face of bearing housing. Lubricate seal lip, then reinstall bearing housing. Tighten retaining cap screws evenly. Install intermediate bearing housing locating screw (if applicable). Assemble balance weights in their original positions. On PH2 engine, tighten balance weight retaining bolts to 267 N·m (197 ft.-lbs.). On all other engines, tighten balance weight retaining bolts to 160 N·m (118 ft.-lbs.). Heat crankshaft gear before reinstalling. Brace opposite end of crankshaft with a wood block while driving gear onto crankshaft. Do not attempt to insert a wedge between crankshaft and crankcase during this operation as serious damage to crankcase or crankshaft could result. Be sure timing marks on teeth of crankshaft gear and camshaft gear are aligned. Make certain thrust washers are still in proper position after gear is installed. Tighten gear retaining cap screw to 62 N·m (46 ft.-lbs.). Crankshaft end play should be 0.2-0.5 mm (0.008-0.020 inch) and maximum allowable end play is 0.63 mm (0.025 inch). Complete installation by reversing the removal procedure.

FLYWHEEL

PH1-PH2-PJ1Z-PJ2Z Engines

REMOVE AND REINSTALL. To remove flywheel, first remove key using a tapered key drift or special key extractor tool available from manufacturer. Clean exposed portion of crankshaft, then withdraw flywheel.

To reinstall flywheel, lubricate crankshaft surface with small amount of oil. Push flywheel onto shaft making sure flywheel is bottomed against shoulder of crankshaft. A new key must be used when reinstalling flywheel and should be installed as follows: File a chamfer on all four corners of key to prevent key from binding in keyway. Insert key by hand as far as possible. At this point, distance (F-Fig. PE-55) between flywheel boss and head of key should be 23.8 mm (15/16 inch). If distance is greater, withdraw key and remove metal from bottom of key to obtain recommended measurement.

NOTE: Do not remove metal from tapered surface of key.

Install key and drive into flywheel until measurement (G) between flywheel and head of key is 11.1 mm (7/16 inch).

PJ1-PJ2-PJ3 Engines

REMOVE AND REINSTALL. The flywheel is located on crankshaft with dowels and retained by cap screws. Two studs should be used in place of two diagonally opposite retaining screws to act as guides and support weight of flywheel as it is being removed and reinstalled. Flywheel retaining cap screws should be tightened gradually in diagonal pattern to 43 N·m (32 ft.-lbs.) on PJ1 and PJ2 engines or 64 N·m (47 ft.-lbs.) on PJ3 engine.

OIL PUMP AND RELIEF VALVE

A plunger type oil pump was used on early production PH1 engines. The pump is mounted at the gear end of the engine and actuated by an eccentric on the crankshaft. All other engines are equipped with a rotary type oil pump. The pump is mounted at gear end of

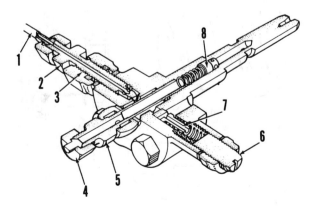

Fig. PE-56—Cross-sectional view of plunger type oil pump assembly used on early PH1 engines.
1. Oil pipe to valve rockers
2. Oil restrictor valve
3. Locknut
4. Union nut
5. Fulcrum pin
6. Pressure adjusting screw
7. Relief valve poppet
8. Check valve

SERVICE MANUAL

Lister-Petter

Fig. PE-58—Use two cap screws in threaded holes provided in pump flange to pull pump assembly (1) from crankcase.

Fig. PE-59—Exploded view of rotary type engine oil pump assembly.
1. Pump body
2. Dowels
3. Outer rotor
4. Inner rotor
5. Cover
6. Drive gear
7. Retaining pin

crankcase and driven by the crankshaft gear via an idler gear.

Plunger Pump

R&R AND OVERHAUL. To remove oil pump assembly, first drain the oil. Remove crankcase inspection cover. Remove the oil strainer. Disconnect oil pipes from pump and flywheel end main bearing. Remove crankshaft as outlined in CRANKSHAFT section. Withdraw pump assembly by pulling towards flywheel end of crankcase.

Disassemble pump and check components for wear. Count the number of turns required to remove relief valve adjusting screw (6–Fig. PE-56) so it can be reinstalled at approximately the same pressure setting. It is important that the pin (2–Fig. PE-57) securing the pump plunger (3) to crankshaft strap (1) be renewed as oil pump will not function if pin failure occurs.

Be sure all parts are clean and lubricated with oil before reassembling. Install relief valve adjusting screw same number of turns as required to remove screw. Reinstall pump by reversing the removal procedure.

Adjust the oil feed rate to valve rockers as follows: Loosen oil pipe union nut and locknut (3–Fig. PE-56). Turn oil restrictor valve (2) completely in, then turn valve back out 1/6 of a turn. Tighten locknut and oil pipe union nut.

Check and adjust engine oil pressure as follows: Remove union nut (4) and connect a pressure gage at this point. With engine at normal operating temperature and running at rated speed, oil pressure should be 240 kPa (35 psi). If necessary, turn adjusting screw (6) clockwise to increase pressure or counterclockwise to decrease pressure.

Rotary Oil Pump

R&R AND OVERHAUL. To remove oil pump, first drain oil from sump. Remove pump mounting cap screws, then screw two bolts into threaded holes provided in pump flange and pull pump from crankcase (Fig. PE-58). The pump drive idler gear is mounted on a shaft in the crankcase and retained by a washer and snap ring. Idler gear is accessible through crankcase inspection cover opening.

To disassemble pump, remove gear retaining pin (7–Fig. PE-59) and drive gear (6). Remove cover mounting screws and separate cover (5) from pump body. All parts should be lubricated with oil during assembly.

To reinstall pump, reverse the removal procedure.

Relief Valve

The oil pressure relief valve is located in the oil filter mounting bracket. When removing valve, count the number of turns required to remove the adjusting screw so it can be reinstalled at approximately the same pressure setting.

Engine oil pressure should be checked with engine at normal operating temperature and running at rated speed. Recommended oil pressure is 270-310 kPa (40-45 psi). If necessary, turn adjusting screw clockwise to increase pressure or counterclockwise to reduce pressure.

Lister-Petter SMALL DIESEL

SERVICING LISTER-PETTER ACCESSORIES

FUEL SYSTEM

LIFT PUMP. A fuel lift pump, when required, is fitted on the crankcase door and operated by a push rod (14–Fig. L90) or (B–Fig. L91) riding against the camshaft. Inlet side of pump is connected to main fuel supply and outlet feeds fuel filter. Complete pump repair kits are available. After completing service, check push rod clearance.

To check clearance, insert push rod and turn engine over until the rod is at its maximum outward travel. On ST and TS engines, install pump gasket and measure clearance. Pump clearance (D–Fig. L91) is 1/32 to 3/64 inch (0.8-1.2 mm). Add pump gaskets as necessary to obtain correct clearance.

On HR engines, with push rod at its maximum outward travel, measure

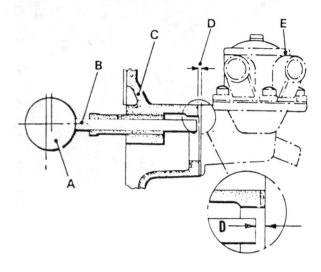

Fig. L91 – Cross-sectional view of lift pump used on ST engines. Push rod clearance (D) is 1/32 to 3/64 inch (0.8-1.2 mm).

A. Camshaft
B. Push rod
C. Crankcase door
D. Gasket
E. Lift pump

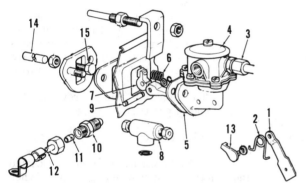

Fig. L90 – Exploded view of fuel lift pump.

1. Primer arm
2. Return spring
3. 5/16-inch union fitting
4. Lift pump
5. Gasket
6. Spring
7. Arm
8. "T" fitting & gasket
9. Pin
10. Union
11. Ferrel
12. Union nut
13. Lever
14. Push rod
15. Push rod adjusting screw

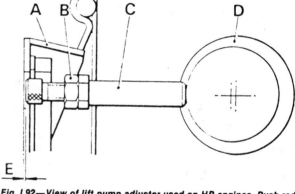

Fig. L92 – View of lift pump adjuster used on HR engines. Push rod clearance (E) is 0.018-0.020 inch (0.46-0.50 mm).

A. Crankcase door
B. Locknut
C. Push rod
D. Camshaft

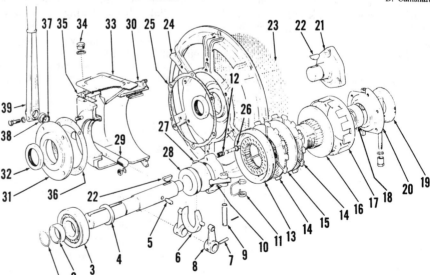

1. Retaining ring
2. Oil seal bushing
3. Ball bearing
4. Clutch shaft assy.
5. Locating pin
6. Sliding yoke
7. Pin
8. Operating lever
9. Pin
10. Clutch engaging arms
11. Spring
12. Locating pin spring
13. Pressure plate
14. Clutch plates
15. Driven disc
16. Clutch hub
17. Clutch body
18. Bearing
19. Oil slinger
20. Clutch drive member
21. Crankshaft extension
22. Key
23. Guard
24. Adapter
25. Gasket
26. Plunger
27. Adjusting ring
28. Engaging cone
29. Operating shaft
30. Clutch housing
31. End cover
32. Oil seal
33. Inspection cover
34. Plug
35. Gasket
36. End cover gasket
37. Seal
38. Seal retainer
39. Operating lever

Fig. L93 – Exploded view of multiplate clutch used on ST engines.

SERVICE MANUAL

Lister-Petter

distance from push rod to face of crankcase door (E – Fig. L92). Distance should be 0.018-0.020 inch (0.46-0.5 mm). To adjust, remove rod and lengthen or shorten adjuster as required. Reinstall push rod and pump.

CLUTCH. Some ST models are equipped with a multiplate set type clutch (Fig. L93). The lever operated clutch is self-locking. The oil capacity is approximately 5/8 pint of SAE 10 weight oil.

Some HR models are equipped with a single plate clutch for direct drive (Fig. L94) or with a reduction gear (Fig. L95).

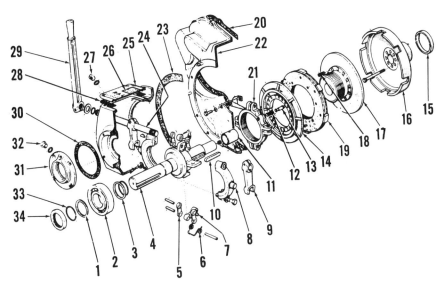

Fig. L94 — Exploded view of direct drive clutch used on HR engines.

1. Retaining washer
2. Ball bearing
3. Retaining ring & washer
4. Clutch shaft assy.
5. Link
6. Spring
7. Toggle lever & link
8. Yoke
9. Operating lever
10. Key
11. Bearing
12. Spring
13. Plunger
14. Pressure plate
15. Spigot ring
16. Clutch drive member
17. Clutch hub
18. Spring
19. Clutch plate
20. Gasket
21. Adjusting ring
22. Adapter
23. Gasket
24. Operating shaft assy.
25. Clutch housing
26. Inspection cover
27. Plug
28. Gasket
29. Operating lever
30. Gasket
31. End cover
32. Oil plug
33. Retaining ring
34. Oil seal

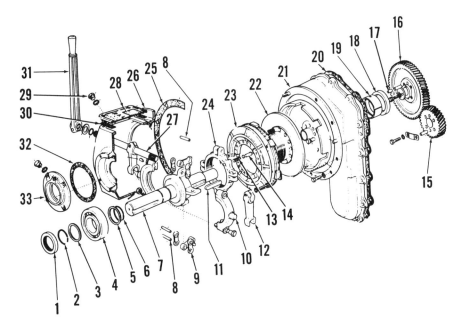

Fig. L95 — Exploded view of direct drive clutch with 2:1 reduction gear used on HR engines.

1. Oil seal
2. Retaining ring
3. Washer
4. Ball bearing
5. Washer
6. Retaining ring
7. Clutch shaft assy.
8. Fulcrum
9. Toggle lever & link
10. Operating yoke
11. Key
12. Operating lever
13. Spring
14. Plunger
15. Crankshaft gear assy.
16. Camshaft gear
17. Dowel
18. Spigot ring
19. Split bearing
20. Gasket
21. Adapter
22. Clutch hub
23. Clutch plate assy.
24. Adjusting ring
25. Gasket
26. Clutch housing
27. Operating shaft
28. Inspection cover
29. Plug
30. Gasket
31. Operating lever
32. Gasket
33. End cover assy.

Illustrations Courtesy Lister-Petter

Lister-Petter — SMALL DIESEL

Clutch lubrication is supplied by engine lubrication through a restrictor plug fitted in the end of crankshaft.

To adjust clutch, stop engine and remove inspection cover. Pull plunger (26 – Fig. L93 or 13 – Fig. L94) and rotate adjusting ring (27 – Fig. L93 or 21 – Fig. L94) clockwise one to three holes. Check "feel" of clutch lever. Do not adjust tighter than is necessary to transmit full power without slip. Clutch output shaft must turn freely in "neutral" position.

REDUCTION GEAR. A 2:1 or 3:1 reduction gear may be attached to flywheel end on Models ST2 and ST3 (Fig. L96). The gear assembly is secured to fan shroud and a splined drive engages with a shaft extension bolted to the flywheel. A multipurpose gear lubricant is used in gearbox.

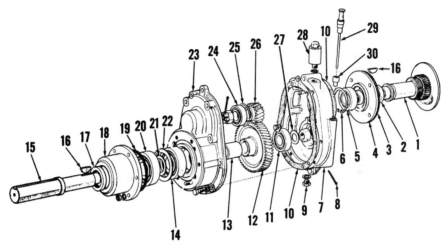

Fig. L96 — Exploded view of 2:1 or 3:1 reduction gear used on some ST engines.

1. Clutch shaft assy.
2. Oil seal
3. Gasket
4. Adapter
5. Bearing
6. Snap ring
7. Gear case
8. Cotter pin
9. Drain plug
10. Gasket
11. Bearing
12. Gear (2:1 or 3:1)
13. Spacer
14. Bearing
15. Secondary shaft
16. Key
17. Oil seal
18. Bearing housing
19. Oil ring
20. Bearing
21. Inner spacing washer
22. Outer spacing washer
23. Gear case end cover
24. Roller bearing
25. Retaining ring
26. Pinion (2:1 or 3:1)
27. Packing shim
28. Oil breather
29. Dipstick
30. Adapter

SERVICE MANUAL

LOMBARDINI

**BRIGGS & STRATTON
LOMBARDINI DIESEL, INC.
3402 Oakcliff Road B2
Doraville, Georgia 30340**

Model	No. Cyls.	Bore	Stroke	Displ.
3LD450, LDA450	1	85 mm (3.35 in.)	80 mm (3.15 in.)	454 cc (27.7 cu. in.)
3LD510, LDA510	1	85 mm (3.35 in.)	90 mm (3.54 in.)	501 cc (30.6 cu. in.)
3LD510/L, L8	1	85 mm (3.35 in.)	90 mm (3.54 in.)	510 cc (30.6 cu. in.)
4LD640, LDA96	1	95 mm (3.74 in.)	90 mm (3.54 in.)	638 cc (38.9 cu. in.)
4LD640/L, L10	1	95 mm (3.74 in.)	90 mm (3.54 in.)	638 cc (38.9 cu. in.)
FLD705, LDA100	1	100 mm (3.94 in.)	90 mm (3.54 in.)	707 cc (43.1 cu. in.)
4LD820, LDA820	1	102 mm (4.01 in.)	100 mm (3.94 in.)	817 cc (49.8 cu. in.)
4LD820/L, L14	1	102 mm (4.01 in.)	100 mm (3.94 in.)	817 cc (49.8 cu. in.)

All models are four-stroke, single-cylinder, air-cooled diesel engines. Crankshaft rotation is counterclockwise at pto end.

Metric fasteners are used throughout engine.

MAINTENANCE

LUBRICATION

Recommended engine oil is SAE 10W for temperatures below 0° (32°F), SAE 20W for temperatures between 0°C (32°F) and 20°C (68°F) and SAE 40 for temperatures above 20°C (68°F). API classification for oil should be CD. Oil sump capacity is 1.75 liters (1.85 quarts) on Models 3LD450, LDA450, 3LD510, LDA510, 3LD510/L and L8 or 2.6 liters (2.75 quarts) on all other models. Manufacturer recommends changing oil after every 300 hours of operation.

A renewable oil filter is mounted on side of engine crankcase. Manufacturer recommends renewing filter after every 300 hours of operation.

All models are equipped with a pressurized oil system. Minimum allowable oil pressure is 100 kPa (15 psi).

ENGINE SPEED ADJUSTMENT

Low idle speed is adjusted by turning idle speed screw (I–Fig. L1-1). Idle speed should be 1000-1100 rpm on all models. Maximum governed speed is adjusted by turning high speed screw (H). Set high idle (no-load) speed 180 rpm above specified governed full-load speed to compensate for governor droop. Normally, full-load speed should be 2200 rpm for Models 3LD510/L, L8, 4LD640/L, L10, 4LD820/L and L14; 2600 rpm for Models 4LD705, LDA100, 4LD820 and LDA820; 3000 rpm for Models 3LD450, LDA450, 3LD510, LDA510, 4LD640 and LDA96. Speeds may vary depending upon application.

Maximum fuel delivery is adjusted by loosening screws (S) and moving adjusting plate (41). Set plate so satisfactory engine acceleration and power are obtained without excessive smoke. Moving plate to the left increases fuel delivery.

FUEL SYSTEM

FUEL FILTER. The fuel filter is located in bottom of fuel tank as shown in Fig. L1-2. Manufacturer recommends renewing fuel filter after every 300 hours of operation. However, fuel filter renewal interval is more dependent upon fuel quality and cleanliness than length of service or operating condi-

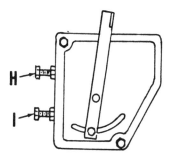

Fig. L1-1—Engine speed adjustment points. Refer to text for adjustment procedure.

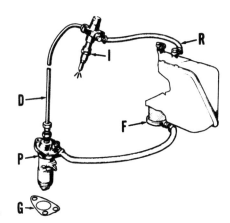

Fig. L1-2—Diagram of fuel system. Some models are also equipped with a fuel transfer pump (not shown), located between fuel tank and injection pump, which ensures constant fuel delivery to injection pump.

D. High pressure delivery line
F. Fuel filter
G. Shim gasket
I. Injector
P. Injection pump
R. Return line

Lombardini — SMALL DIESEL

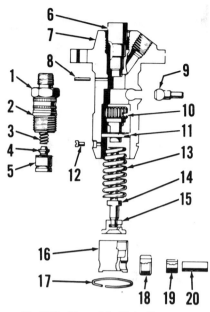

Fig. L1-3—View of fuel injection pump.
1. Delivery valve holder
2. "O" ring
3. Spring
4. Delivery valve
5. Delivery valve seat
6. Barrel
7. Pump body
8. Pin
9. Control rack
10. Pinion
11. Spring seat
12. Pin
13. Spring
14. Plunger
15. Spring retainer
16. Tappet
17. Snap ring
18. Outer roller
19. Inner roller
20. Pin

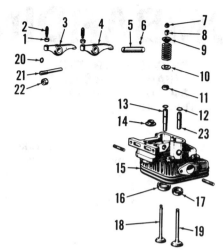

Fig. L1-5—Exploded view of cylinder head assembly.
1. Locknut
2. Adjuster
3. Intake rocker arm
4. Exhaust rocker arm
5. Rocker arm shaft
6. Set screw
7. Cap
8. Keys
9. Spring retainer
10. Washer
11. Oil seal
12. Snap ring
13. Intake valve guide
14. Spring seat
15. Cylinder head
16. Intake valve seat
17. Exhaust valve seat
18. Intake valve
19. Exhaust valve
20. "O" ring
21. Rocker shaft locating pin
22. Lock nut
23. Exhaust valve guide

tions. If engine indicates signs of fuel starvation (loss of power or surging), renew filter regardless of hours of operation. After renewing filter, air must be bled from system as outlined in following paragraph.

BLEED FUEL SYSTEM. To bleed air from fuel system, loosen bleed screw or fuel supply line fitting on injection pump. On gravity feed system, allow fuel to flow from fitting until free of air. On models equipped with fuel transfer pump, manually operate hand primer lever to purge air from fuel supply line. Retighten or bleed screw.

NOTE: If cam which drives transfer pump is at full lift, priming lever cannot be operated. Turn crankshaft one revolution to reposition camshaft.

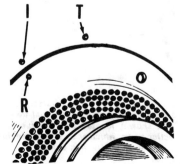

Fig. L1-4—View of timing marks located on air shroud. Refer to text for injection timing.

If engine fails to start, loosen high pressure fuel line at injector. Rotate engine crankshaft until fuel is discharged from injector line. Retighten injection line fitting and start engine.

INJECTION PUMP TIMING

Injection pump timing is adjusted using shim gaskets (G–Fig. L1-2) between pump body and mounting surface on crankcase. To check injection pump timing, unscrew high pressure delivery line (D) fitting from delivery valve holder (1–Fig. L1-3). Unscrew delivery valve holder and remove spring (3) and delivery valve (4), then screw delivery valve holder (1) into pump body. Move throttle control to full speed position. Rotate crankshaft in normal direction (counterclockwise at pto) so piston is on compression stroke. Note fuel in delivery valve holder will spill out. Stop crankshaft rotation at moment fuel ceases to flow out. At this point, beginning of injection occurs and timing dot (R–Fig. L1-4) on fan plate should be aligned with timing dot (I) on fan shroud. Injection timing should be 24°-26° BTDC on Models 3LD450, LDA 450, 3LD510, LDA510, 3LD510/L, 4LD640 and LDA96. On all other models, injection timing should be 22°-24° BTDC.

Reduce pump mounting shim gasket thickness to advance timing or increase gasket thickness to retard timing. Reinstall delivery valve and spring and tighten holder to a torque of 35 N·m (26 ft.-lbs.). Tighten injection pump retaining screws to 25 N·m (18 ft.-lbs.).

REPAIRS

TIGHTENING TORQUES

Refer to the following table for tightening torques.
Connecting rod:
  3LD450, LDA450, 3LD510,
  LDA510, 3LD510/L, L8 ..... 30 N·m
    (22 ft.-lbs.)
  All other models .... 45 N·m
    (33 ft.-lbs.)
Cylinder head:
  3LD450, LDA450, 3LD510,
  LDA510, 3LD510/L, L8 ..... 50 N·m
    (37 ft.-lbs.)
  All other models .... 78 N·m
    (58 ft.-lbs.)
Flywheel:
  3LD450, LDA450, 3LD510,
  LDA510, 3LD510/L, L8 .... 170 N·m
    (125 ft.-lbs.)
  All other models .... 345 N·m
    (255 ft.-lbs.)
Injection pump .... 25 N·m
    (18 ft.-lbs.)
Injector .... 25 N·m
    (18 ft.-lbs.)
Main bearing support:
  Flywheel side –
    3LD450, LDA450, 3LD510,
    LDA510, 3LD510/L, L8 ... 30 N·m
      (22 ft.-lbs.)
    All other models .... 35 N·m
      (26 ft.-lbs.)
  Gear train side –
    3LD450, LDA450, 3LD510,
    LDA510, 3LD510/L, L8 ... 25 N·m
      (18 ft.-lbs.)
    All other models .... 40 N·m
      (30 ft.-lbs.)
Oil pan .... 25 N·m
    (18 ft.-lbs.)
Oil pump .... 40 N·m
    (30 ft.-lbs.)
Oil pump gear .... 20 N·m
    (15 ft.-lbs.)
Rocker arm cover .... 20 N·m
    (15 ft.-lbs.)
Timing gear cover:
  3LD450, LDA450, 3LD510,
  LDA510, 3LD510/L, L8 ..... 25 N·m
    (18 ft.-lbs.)
  All other models .... 40 N·m
    (30 ft.-lbs.)

VALVE CLEARANCE ADJUSTMENT

Valve clearance should be 0.20 mm (0.008 inch) for intake and exhaust valve with engine cold. To adjust clearance, remove rocker arm cover and rotate crankshaft until piston is at TDC on compression stroke. Loosen rocker arm

Illustrations Courtesy Lombardini U.S.A., Inc.

SERVICE MANUAL

Lombardini

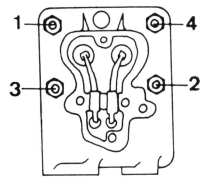

Fig. L1-6—Tighten cylinder head stud nuts in sequence shown.

adjusting screw jam nut, then turn adjusting screw until proper thickness feeler gage can be inserted between valve stem end and rocker arm. Hold adjusting screw and tighten jam nut.

COMPRESSION RELEASE

A manual compression release is located in the rocker arm cover so the exhaust valve can be held open to aid starting. Exhaust valve should be lowered approximately 1 mm (0.040 inch) from valve seat when compression release is operated.

CYLINDER HEAD AND VALVE SYSTEM

Do not remove cylinder head when hot as head may deform. To remove rocker arms, unscrew rocker shaft locating pin (21 – Fig. L1-5) on Models 3LD450, LDA450, 3LD510, LDA510, 3LD510/l and L8 or shaft locating set screw (6) on all other models. Withdraw rocker shaft using a suitable puller. Maximum allowable rocker arm to shaft clearance is 0.1 mm (0.004 inch).

Valve face angle is 45° for intake and exhaust. If valve head margin is less than 0.4 mm (0.016 inch), renew valve. Valve seat angle is 45° and recommended seat width is 1.4-1.6 mm (0.055-0.065 inch). Renewable valve seat inserts are used on intake and exhaust. When installing new inserts, cylinder head should first be heated in an oven to 160°-180°C (320°-355°F).

With valves installed, measure distance valve head is recessed from cylinder head mounting surface. Intake valve should be recessed 0.60-0.80 mm (0.024-0.031 inch) on Models 3LD450, LDA450, 3LD510, LDA510, 3LD510/L and L8 or 0.65-0.85 mm (0.026-0.033 inch) on all other models. Exhaust valve (except models with decompression) should be recessed 0.55-1.05 mm (0.022-0.041 inch) on Models 3LD450, LDA450, 3LD510, LDA510, 3LD510/L and L8 or 0.45-0.95 mm (0.018-0.037 inch) on all other models. On engines equipped with compression release, exhaust valve recession should be 0.55-0.95 mm (0.022-0.037 inch). If recession is less than specified, regrind valve seats. If valve recession exceeds specified limits, renew valve seat insert.

Valve stem diameter is 6.98-7.00 mm (0.2748-0.2756 inch) on Models 3LD450, LDA450, 3LD510, LDA510, 3LD510/L and L8 or 7.98-8.00 mm (0.3142-0.3149 inch) on all other models. Desired valve stem clearance in guides is 0.03-0.08 mm (0.001-0.003 inch) on all models. Valve guides are renewable and oversize guide must be machined to provide a 0.05-0.06 mm (0.0020-0.0023 inch) interference fit in cylinder head bore. The cylinder head should be heated to 160°-180°C (320°-355°F) prior to installing guides. Note locating ring (12 – Fig. L1-5) around top of each guide. After installation, inside diameter of guides must be reamed or honed to provide desired valve stem clearance.

Intake and exhaust valve springs are identical. Spring free length should be 45.6 mm (1.795 inches) on Models 3LD450, LDA450, 3LD510, LDA510, 3LD510/L and L8 or 52 mm (2.047 inches) on all other models. Spring pressure should be 160-170 N (36-38 pounds) at a length of 34.2 mm (1.346 inches).

Cylinder head mounting surface must not be deformed more than 0.3 mm (0.012 inch). Cylinder head and block mating surfaces may be lapped to improve fit.

When reinstalling cylinder head, be sure to examine push rod tube seals for damage and renew if needed. Tighten cylinder head nuts evenly using tightening sequence shown in Fig. L1-6. It is recommended that nuts be tightened in steps to a final torque of 50 N·m (37 ft.-lbs.) on Models 3LD450, LDA450, 3LD510, LDA510, 3LD510/L and L8 or 78 N·m (58 ft.-lbs.) on all other models.

INJECTOR

REMOVE AND REINSTALL. To remove injector, first clean dirt from injector, injection line, return line and cylinder head. Disconnect return and injection lines from injector and immediately cap or plug all openings. Unscrew injector retaining nuts and carefully remove injector from head being careful not to lose shims between injector and head.

Tighten injector retainer nuts to a torque of 25 N·m (18 ft.-lbs.). If accessible, measure protrusion of nozzle into combustion chamber. Nozzle tip should extend 2.5-3.0 mm (0.100-0.118 inch) on Models 3LD450 and LDA450; 3.0-3.5 mm (0.118-0.138 inch) on Models 3LD510, LDA510, 3LD510/L and L8;

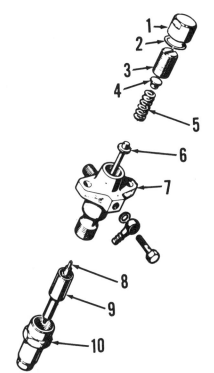

Fig. L1-7—Exploded view of fuel injector assembly.

1. Nut
2. Gasket
3. Adjuster
4. Spring seat
5. Spring
6. Push rod
7. Body
8. Needle
9. Nozzle valve
10. Nozzle nut

3.5-4.0 mm (0.138-0.158 inch) on all other models. Adjust position of nozzle by installing 0.5 mm (0.020 inch) shims between injector and cylinder head.

TESTING. A suitable test stand is required to check injector operation. Only clean, approved testing oil should be used to test injector. When operating properly, injector nozzle will emit a buzzing sound and should cut off quickly with no fluid leakage at nozzle valve seat.

WARNING: Fuel leaves the injector nozzle with sufficient force to penetrate the skin. When testing, keep yourself clear of nozzle spray.

Nozzle opening pressure should be 18635-19615 kPa (2705-2845 psi). Opening pressure is adjusted by turning adjuster screw (3 – Fig. L1-7). Nozzle valve should not show leakage at orifice spray holes for 10 seconds with pressure maintained at 2100 kPa (300 psi) below opening pressure. Spray pattern should be even and well atomized. If pattern is wet, ragged or intermittent, nozzle must be overhauled or renewed.

OVERHAUL. Refer to exploded view of Fig. L1-7. When disassembling injector, protect polished parts from damage. Thoroughly clean parts in a suitable solvent. Clean inside orifice end of nozzle valve with wooden cleaning stick. The

Lombardini SMALL DIESEL

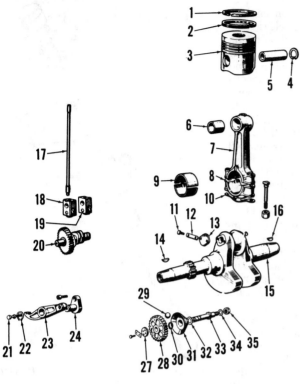

Fig. L1-8—Exploded view of crankshaft, governor and camshaft assemblies. Some models use flyweights instead of flyballs as shown.

1. Compression piston rings (3)
2. Oil control ring
3. Piston
4. Snap rings
5. Piston pin
6. Bushing
7. Connecting rod
8. Rod cap
9. Rod bearing
10. Lockplate
11. Screw
12. Lockplate
13. Plate
14. Key
15. Crankshaft
16. Key
17. Push rods
18. Exhaust cam follower
19. Intake cam follower
20. Camshaft
21. Screw
22. Snap ring
23. Rocker arm
24. Stud
27. Plate
28. Gear
29. Governor balls (6)
30. Snap ring
31. Cup
32. Washer
33. Governor shaft
34. Washer
35. Nut

orifice spray holes may be cleaned by inserting a cleaning wire slightly smaller that the 0.28 mm (0.011 inch) diameter holes.

When reassembling injector, make certain all parts are clean and wet with diesel fuel. Nozzle valve needle (8) must slide freely in valve body (9). If needle valve sticks, reclean or renew nozzle valve assembly.

INJECTION PUMP

Refer to Fig. L1-3 for an exploded view of injection pump. To remove pump, disconnect fuel lines. Plug all openings to prevent entry of dirt. Remove retaining screws and lift out pump. Be sure to retain pump mounting gaskets for use in reassembly. Pump timing is adjusted by changing gasket thickness.

The injection pump should be tested and serviced only by a shop qualified in diesel fuel injection repair.

When reinstalling pump, assemble correct thickness of mounting shim gaskets and be sure to engage control rack pin with governor arm. Tighten retaining screws to a torque of 25 N·m (18 ft.-lbs.). Refer to MAINTENANCE section to check injection timing.

PISTON AND ROD UNIT

REMOVE AND REINSTALL. Piston and connecting rod may be removed after removing cylinder head and oil pan.

When reinstalling piston and rod, note that depression in piston crown is closer to one side of piston. Install piston so depression side of piston is nearer injector side of engine. Some pistons also have an arrow embossed in top of piston so arrow is pointing towards intake side of engine.

Install cap on connecting rod so bearing tang grooves are on the same side. Tighten rod bolts to a torque of 30 N·m (22 ft.-lbs.) on Models 3LD450, LDA450, 3LD510, LDA510, 3LD510/L and L8 or 45 N·m (33 ft.-lbs.) on all other models.

PISTON, PIN AND RINGS

The piston is equipped with three compression rings and one oil control ring. The top compression ring is chrome plated.

Ring end gap should be measured with ring positioned squarely in cylinder. On Models 3LD450, LDA450, 3LD510, LDA510, 3LD510/L, L8, 4LD640, LDA96, 4LD640/L and L10, compression ring end gap should be 0.30-0.50 mm (0.011-0.020 inch). On all other models, compression ring end gap should be 0.40-0.65 mm (0.016-0.025 inch). Oil control ring end gap should be 0.25-0.50 mm (0.010-0.020 inch) on Models 3LD450, LDA450, 3LD510, LDA510, 3LD510/L and L8 or 0.30-0.60 mm (0.011-0.023 inch) on all other models.

To check piston ring groove wear, install new rings onto piston and check clearance between top of ring and top of groove. Renew piston if ring side clearance exceeds specified values as follows: 0.11-0.15 mm (0.004-0.006 inch) for top ring, 0.06-0.10 mm (0.002-0.004 inch) for second and third rings and 0.05-0.10 mm (0.002-0.004 inch) for oil control ring.

Refer to table in Fig. L1-9 for standard piston diameter and desired piston clearance in cylinder bore. Piston diameter is measured at bottom of skirt perpendicular to piston pin bore. Renew piston if wear exceeds 0.05 mm (0.002 inch). Pistons and rings are available in standard size and oversizes of 0.5 mm (0.020 inch) and 1.0 mm (0.040 inch).

Piston diameter is 22.995-23.000 mm (0.9053-0.9055 inch) on Models 3LD450, LDA450, 3LD510, LDA510, 3LD510/L and L8. On all other models, pin diameter is 27.995-28.000 mm (1.1022-1.1023 inches). Piston pin clearance in rod bushing should be

Fig. L1-9—Refer to chart for standard piston diameter and desired piston clearance in cylinder bore.

MODEL	PISTON O.D.	PISTON CLEARANCE
3LD450, LDA450	84.87 - 84.90 mm (3.341 - 3.342 in.)	0.10 - 0.15 mm (0.004 - 0.006 in.)
3LD510, LDA510	84.87 - 84.90 mm (3.341 - 3.342 in.)	0.10 - 0.15 mm (0.004 - 0.006 in.)
3LD510/L, L8	84.87 - 84.90 mm (3.341 - 3.342 in.)	0.10 - 0.015 mm (0.004 - 0.006 in.)
4LD640, LDA96	94.85 - 94.87 mm (3.734 - 3.735 in.)	0.13 - 0.17 mm (0.005 - 0.007 in.)
4LD640/L, L10	94.85 - 94.87 mm (3.734 - 3.735 in.)	0.13 - 0.17 mm (0.005 - 0.007 in.)
4LD705, LDA100	99.82 - 99.83 mm (3.9299 - 3.9303 in.)	0.17 - 0.20 mm (0.007 - 0.008 in.)
4LD820, LDA820	101.84 - 101.86 mm (4.009 - 4.010 in.)	0.14 - 0.18 mm (0.0055 - 0.007 in.)
4LD820/L, L14	101.84 - 101.86 mm (4.009 - 4.010 in.)	0.14 - 0.18 mm (0.0055 - 0.007 in.)

Illustrations Courtesy Lombardini U.S.A., Inc.

SERVICE MANUAL

Lombardini

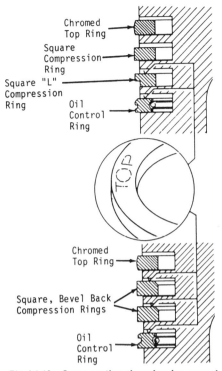

Fig. L1-10—Cross section view showing correct installation of different styles of piston rings.

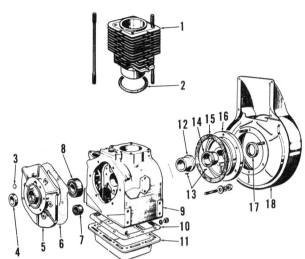

Fig. L1-11—Exploded view of typical crankcase.

1. Cylinder
2. Shim gasket
3. Plug
4. Seal
5. Pto bearing support
6. Gasket
7. Bearing
8. Bearing
9. Crankcase
10. Gasket
11. Oil pan
12. Main bearing
13. Dowel pins
14. Main bearing
15. Shim gasket
16. Main bearing support
17. Seal
18. Air shroud

0.020-0.035 mm (0.0008-0.0014 inch). Maximum allowable clearance is 0.07 mm (0.003 inch). Pin is a thumb push fit in piston.

When installing rings onto piston, stagger ring end gaps 180° apart. Be sure no end gaps are in line with piston pin bore. Refer to Fig. L1-10 for correct installation of rings on piston.

CONNECTING ROD

The connecting rod small end is fitted with a renewable bushing. Clearance between piston pin and connecting rod bushing should be 0.020-0.035 mm (0.0008-0.0014 inch). Maximum allowable clearance is 0.07 mm (0.003 inch).

A precision, insert type bearing is used in big end of connecting rod. Desired rod bearing clearance is 0.030-0.065 mm (0.0012-0.0025 inch) for Models 3LD450, LDA450, 3LD510, LDA510, 3LD510/L and L8 or 0.050-0.060 mm (0.0020-0.0023 inch) for all other models. Bearings are available in standard and undersizes.

CYLINDER

All models are equipped with a removable cylinder (1—Fig. L1-11). Standard cylinder diameter is 85.00-85.02 mm (3.346-3.347 inches) on Models 3LD450, LDA450, 3LD510, LDA510, 3LD510/L and L8; 95.00-95.02 mm (3.740-3.741 inches) on Models 4LD640, LDA96, 4LD640/L and L10; 100.00-100.02 mm (3.937-3.938 inches) on Models 4LD705 and LDA100; 102.00-102.02 mm (4.0157-4.0165 inches) on Models 4LD820, LDA820, 4LD820/L and L14. If taper or wear exceeds 0.1 mm (0.004 inch), rebore cylinder to an appropriate oversize.

With piston at top dead center, top of piston should be 0.75-0.90 mm (0.030-0.035 inch) below top surface of cylinder on Models 3LD450, LDA450, 3LD510, LDA510, 3LD510/L and L8 or 0.85-1.0 mm (0.034-0.039 inch) on all other models. Cylinder height is adjusted by varying cylinder mounting shim gaskets (2—Fig. L1-11).

TIMING GEARS

Gears are accessible after removing pto bearing support (5—Fig. L1-11). Crankshaft and camshaft gears are embossed with marks (M—Fig. L1-12) which should be aligned as shown. If crankshaft and camshaft gears are not marked, proceed as follows: If not previously removed, remove cylinder head, push rod tube and push rods. Position crankshaft so piston is at top dead center. Intake valve cam follower (nearer cylinder) should be opening (rising) and exhaust valve cam follower should be closing (going down). Both cam followers should be same height above crankcase when piston is at top dead center. If not, refer to CAMSHAFT section and remove camshaft, then reinstall camshaft so it is correctly timed with crankshaft. Mark crankshaft and camshaft gears for future reference.

CAMSHAFT, CAM FOLLOWERS AND PUSH RODS

To remove camshaft, first remove cylinder head, push rod tubes, push rods, cam followers and fuel injection pump. Remove pto bearing support (5—Fig. L1-11) and withdraw camshaft.

Inspect camshaft for excessive wear or other damage. Diameter of camshaft bearing journals is 17.96-17.98 mm (0.707-0.708 inch). Height of injection pump lobe should be 31.51 mm (1.240 inches). Height of intake lobe should be 33.95-34.05 mm (1.337-1.340 inches) on early models while intake lobe on late models is 33.68-33.78 mm (1.326-1.330 inches). Exhaust lobe height should be 33.45-33.55 mm (1.317-1.320 inches) on all models. If wear exceeds 0.1 mm (0.004 inch) on camshaft journals or lobes, renew camshaft.

Reassembly is the reverse of disassembly. Be certain timing marks (Fig. L1-12) on camshaft gear are aligned with mark on crankshaft gear. Camshaft end play should be 0.20-0.40 mm (0.008-0.016 inch). End play is adjusted by changing thickness of pto bearing support mounting gasket (6—Fig. L1-11). Note that cam follower rollers are offset to one side of holder and notched sliding surface is on opposite side of holder. Install cam followers so notched surfaces are together and intake cam follower is closer to cylinder barrel.

Fig. L1-12—View showing location of timing marks (M) on crankshaft and camshaft gears.

Illustrations Courtesy Lombardini U.S.A., Inc.

Lombardini SMALL DIESEL

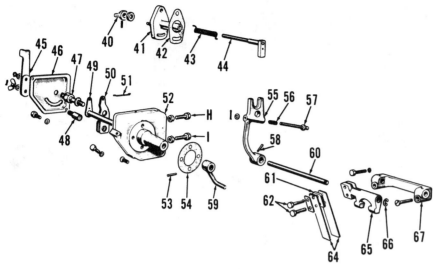
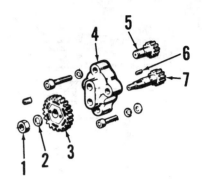

Fig. L1-13—Exploded view of governor and control linkage. Governor fork (65) contacts governor cup (31—Fig. L1-8).

40. Stop knob	48. Pivot screw	55. Arm
41. Plate	49. Arm	56. Spring
42. Gasket	50. Lever	57. Torque control rod
43. Spring	51. Link	58. Pin
44. Stop arm	52. Housing	59. Arm
45. Throttle lever	53. Pin	60. Shaft
46. Cover	54. Gasket	61. Spacer
47. Stud		

62. Screws
64. Spring plates
65. Fork
66. "E" ring
67. Bracket
I. Low idle speed screw
H. High speed screw

Fig. L1-16—Exploded view of oil pump.
1. Nut
2. Lockwasher
3. Gear
4. Pump body
5. Driven gear
6. Key
7. Drive gear

GOVERNOR

Most models are equipped with flyball type governor while some models may be equipped with flyweights. The governor shaft is shown in Fig. L1-8 while governor linkage is shown in Fig. L1-13. The crankshaft rotates flyball assembly (G–Fig. L1-14) which bears against fork (65–Fig. L1-13). As the flyballs move, the shaft attached to the fork is rotated thereby moving governor arm (55). Arm (55) mates with fuel injection control rack pin to regulate fuel flow. Throttle lever (45) operates through governor spring plates (64) to control engine speed.

To stop engine, stop knob (40) is turned counterclockwise which forces governor arm to move fuel injection pump control rack to no-fuel position. All models except 3LD510/L, L8, 4LD640/L, L10, 4LD820/L and L14 are equipped with a torque control rod (57) and spring (56) which allows the governor arm (59) additional movement for additional fuel usage under high torque load. By pulling stop knob (40) away from engine, stop arm (44) will slide off tip of torque control rod (57) and allow governor arm to move forward so maximum fuel is delivered during starting.

Governor mechanism is accessible after removing pto bearing support (5–Fig. L1-11), however, the oil pan must be removed for access to nut (35–Fig. L1-8) so governor shaft can be withdrawn from crankcase. Inspect governor components and renew any which are damaged or excessively worn. Mechanism must move freely for proper governor operation. Tighten governor shaft nut to a torque of 40 N·m (30 ft.-lbs.).

To adjust governor, pto bearing support (5–Fig. L1-11) and gasket must be removed. Move throttle lever (45–Fig. L1-13) to full throttle position. Loosen spring plate screws (62), then move governor arm (55) towards crankcase opening and measure distance (G–Fig. L1-15) from pto bearing support mating surface of crankcase to upper part of governor arm. Distance between crankcase surface and governor arm should be 22 mm (0.866 inch) on Models 3LD450, LDA450, 3LD510, LDA510, 3LD510/L and L8 or 28 mm (1.102 inches) on all other models. Retighten spring plate screws (62–Fig. L1-13).

OIL PUMP AND RELIEF VALVE

R&R AND OVERHAUL. To remove oil pump, remove pto bearing support (5–Fig. L1-11) and using a suitable puller remove pump gear (3–Fig. L1-16). Unscrew pump mounting screws and remove pump from crankcase bulkhead. Maximum clearance between gears and pump body should not exceed 0.15 mm (0.006 inch). Maximum clearance between ends of gears and mounting surface of pump body is 0.15 mm (0.006 inch).

Apply a thin coating of sealer to mounting surface of pump body. Install pump and tighten mounting screws to 40 N·m (30 ft.-lbs.). Tighten oil pump

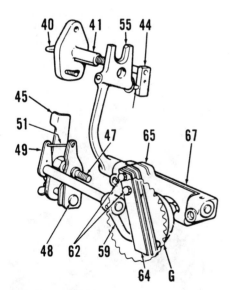

Fig. L1-14—Diagram of governor mechanism. Refer to text and Fig. L1-13 for parts identification.

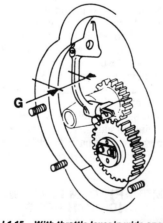

Fig. L1-15—With throttle lever in wide open position, distance (G) from pto bearing support mounting surface to upper part of governor arm should be 22 mm (0.866 inch) on Models 3LD450, LDA450, 3LD510, LDA510, 3LD510/L and L8 or 28 mm (1.102 inches) on all other models. Refer to text.

SERVICE MANUAL

Lombardini

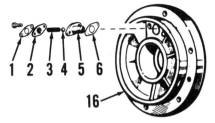

Fig. L1-17 — Exploded view of typical oil pressure relief valve. Some models use a poppet type valve instead of ball (4) and body (5).
1. Lockplate
2. Cover
3. Spring
4. Ball
5. Body
6. Gasket
16. Main bearing gasket

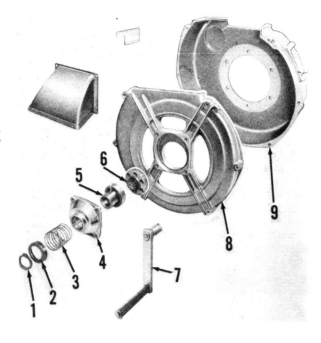

Fig. L1-18 — Exploded view of manual crank starter used on some models.
1. Snap ring
2. Cap
3. Spring
4. Flange
5. Ring gear
6. Pinion
7. Crank
8. Cover
9. Case

gear nut to 20 N·m (15 ft.-lbs.). Install timing gear cover.

The oil pressure relief valve is located on inner face of main bearing support (16 – Fig. L1-17). To remove main bearing support, remove crankshaft pulley or crank starter, flywheel and shroud. Unscrew retaining nuts and remove main bearing support. Inspect pressure relief valve components and renew if damaged or excessively worn. Reinstall relief valve by reversing disassembly procedure. With warm oil, minimum oil pressure should be 75 kPa (11 psi) at idle and 200 kPa (29 psi) at full throttle.

CRANKSHAFT AND BEARINGS

To remove crankshaft, remove crankshaft pulley or crank starter, then remove flywheel. Remove cylinder head, oil pan, piston and connecting rod as previously outlined. Remove pto bearing support (5 – Fig. L1-11) and air shroud (18). Remove main bearing support (16) and withdraw crankshaft from crankcase.

Thoroughly clean crankshaft making sure oil galleries are clear. Inspect for excessive wear and other damage.

Crankshaft main bearing journal standard diameter on Models 3LD450, LDA450, 3LD510, LDA510, 3LD510/L and L8 is 41.99-42.00 mm (1.6531-1.6535 inches) on pto end and 39.99-40.00 mm (1.5744-1.5748 inches) on flywheel end. Main bearings (12 and 14) are available in standard and 1.0 mm (0.040 inch) undersize. New bearings should be installed using a suitable installing tool such as part 7271.3595.047. After installation, check bearing ID and ream bearing to obtain recommended clearance of 0.04-0.06 mm (0.0016-0.0023 inch). Maximum allowable bearing clearance is 0.10 mm (0.004 inch).

On Models 4LD640, LDA96, 4LD640/L, L10, 4LD705, LDA100, 4LD820, 4LD820/L and L14, crankshaft main bearing journal standard diameter

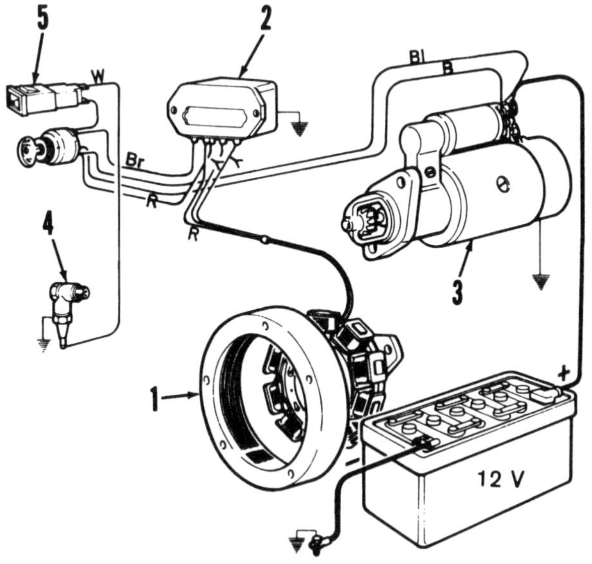

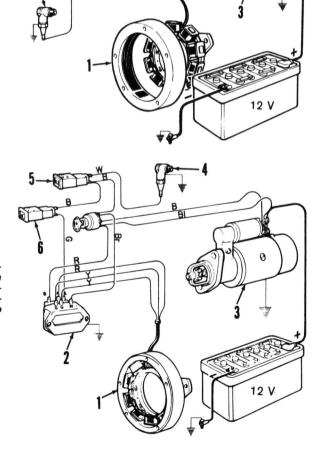

Fig. L1-19 — Wiring schematic for models equipped with internal alternator and electric starter.
1. Alternator
2. Regulator
3. Starter motor
4. Oil pressure switch
5. Oil pressure light

Fig. L1-20 — Wiring schematic for models equipped with internal alternator and electric starter using an alternator warning light (6). Refer to L1-19 legend.

Illustrations Courtesy Lombardini U.S.A., Inc.

191

Lombardini

is 44.99-45.00 mm (1.7713-1.7716 inches) for both main bearings. Main bearings are available in standard as well as 0.5 mm (0.020 inch) and 1.0 mm (0.040 inch) undersizes which should not require reaming. Recommended main bearing clearance is 0.06-0.08 mm (0.0024-0.0031 inch) on Models LDA100, LDA820, L10 and L14. On all other models, recommended main bearing clearance is 0.04-0.06 mm (0.0016-0.0023 inch).

To reinstall crankshaft, reverse the removal procedure. A special installing tool, part 7271.3595.047, is available from manufacturer to install bearing in retainer (16 – Fig. L1-11). Make certain lubrication oil holds in bearing and retainer are aligned. Renew oil seals (4 and 17) and lubricate lip of seals before reinstalling. Crankshaft end play should be 0.10-0.30 mm (0.004-0.012 inch) and is adjusted by varying thickness of bearing support gasket (15). On Models 3LD450, LDA450, 3LD510, LDA510, 3LD510/L and L8, tighten nuts on flywheel side bearing support to 30 N·m (22 ft.-lbs.) and gear train side bearing support to 25 N·m (18 ft.-lbs.) and gear train side bearing support to 40 N·m (30 ft.-lbs.).

MANUAL CRANK STARTER

Some engines are equipped with a crank type manual starter as shown in Fig. L1-18. Starter repair is evident after inspection of unit.

ALTERNATOR AND REGULATOR

Refer to Fig. L1-19 or L1-20 for wiring schematic typical of models equipped with internal alternator and electric starter motor. Note that circuit illustrated in Fig. L1-20 is for models equipped with an alternator warning light and Fig. L1-19 is for models not equipped with a warning light. Voltage regulator is different for the two circuits.

The internal alternator is contained in the flywheel. Rated output is 13.5 amperes. Stator and rotor are available only as a unit assembly.

SERVICE MANUAL

Lombardini

LOMBARDINI

Model	No. Cyls.	Bore	Stroke	Displ.
5LD675-2, LDA672	2	95 mm (3.740 in.)	95 mm (3.740 in.)	1346 cc (82.1 cu. in.)
5LD825-2, 832	2	100 mm (3.937 in.)	105 mm (4.134 in.)	1648 cc (100.5 cu. in.)
5LD825-2/L, L27	2	100 mm (3.937 in.)	105 mm (4.134 in.)	1648 cc (100.5 cu. in.)
5LD675-3, LDA673	3	95 mm (3.740 in.)	95 mm (3.740 in.)	2019 cc (123.2 cu. in.)
5LD825-3, 833	3	100 mm (3.937 in.)	105 mm (4.134 in.)	2472 cc (150.8 cu. in.)
5LD825-3/L L40	3	100 mm (3.037 in.)	105 mm (4.134 in.)	2472 cc (150.8 cu. in.)
5LD930-3	3	106 mm (4.173 in.)	105 mm (4.134 in.)	2778 cc (169.5 cu. in.)

Engines covered in this section are four-stroke, air-cooled diesel engines. Crankcase and cylinders are cast iron. Cylinder heads are aluminum. Crankshaft rotation is counterclockwise at pto end. Number 1 cylinder is nearest flywheel. Firing order is 1-3-2 on three cylinder models.

MAINTENANCE

LUBRICATION

Recommended engine oil is SAE 10W for temperatures below 0°C (32°F), SAE 20W for temperatures between 0°C (32°F) and 20°C (68°F) and SAE 40 for temperatures above 20°C (68°F). API classification for oil should be CD. Manufacturer recommends renewing oil after every 300 hours of operation. Oil sump capacities are as follows:

Model	Capacity
5LD825-2, 832, 5LD825-2/L, L27	3.5 L (3.7 qts.)
5LD675-2	4.0 L (4.2 qts.)
LDA672	4.5 L (4.7 qts.)
5LD825-3, 833, 5LD25-3/L, L40	5.5 L (5.8 qts.)
5LD675-3	6 L (6.3 qts.)
5LD930-3	6.3 L (6.6 qts.)
LDA673	6.5 L (6.8 qts.)

A renewable oil filter is mounted on side of crankcase. Manufacturer recommends renewing oil filter after every 300 hours of operation (at the same time oil is changed).

All models are equipped with a pressurized oil system. Refer to Fig. L2-1 for a diagram of the lubrication circuit.

ENGINE SPEED ADJUSTMENT

Low idle speed is adjusted by turning idle speed screw (2 – Fig. L2-2) located on top of injection pump control linkage housing. Low idle speed should be 900-1000 rpm on two-cylinder models and 700-800 rpm on three-cylinder models.

Maximum governed speed is adjusted by turning high speed screw (3). High idle (no-load) speed should be adjusted to 150 rpm over normal full-load speed to compensate for governor droop during operation. Maximum rated speed under load is normally 2200 rpm on Models 5LD825-2/L, L27, 5LD825-3/L and L40; 2600 rpm on Model 5LD930-3 and 3000 rpm on all other models. Rated speed under load may vary depending upon application.

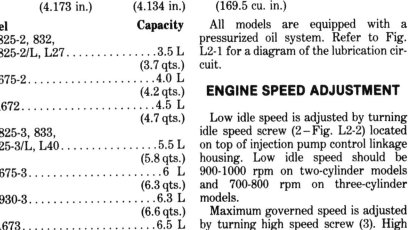

Fig. L2-1 – Drawing of lubrication system.

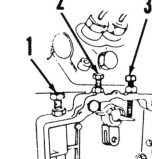

Fig. L2-2 – Drawing showing location of low idle adjusting screw (2), high idle adjusting screw (3) and torque control screw (1).

Illustrations Courtesy Lombardini U.S.A., Inc.

Lombardini

SMALL DIESEL

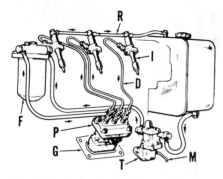

Fig. L2-3—Fuel circuit diagram for three-cylinder models. Other models are similar.

- D. High pressure delivery line
- F. Fuel filter
- G. Shim gasket
- I. Injector
- M. Pump primer lever
- P. Fuel injection pump
- R. Fuel return line
- T. Fuel transfer pump

FUEL SYSTEM

FUEL FILTER. All models are equipped with a single, renewable fuel filter. Manufacturer recommends renewing fuel filter after every 300 hours of operation. However, fuel filter renewal period is more dependent upon fuel quality and cleanliness than length of service or operating conditions. If engine indicates signs of fuel starvation (loss of power or surging), renew filter regardless of hours of operation. After renewing filter, air must be bled from system as outlined in the following paragraph.

BLEED FUEL SYSTEM. To bleed air from fuel system, loosen bleed screw on top of fuel filter mounting housing. Manually actuate fuel transfer pump primer lever (M–Fig. L2-3) until fuel flows from bleeder screw, then tighten screw. Loosen bleeder screw on injection pump. Actuate fuel pump primer lever and retighten bleed screw when a steady flow of fuel is obtained.

If engine fails to start, loosen high pressure line connections at injectors. Rotate crankshaft until fuel emerges from loosened connections. Retighten fuel line connections, then start engine.

INJECTION PUMP TIMING

Beginning of injection should occur at 27°-29° BTDC on Models 832 and 833. On all other models, beginning injection should be at 23°-25° BTDC.

To check injection timing, remove number 1 injection line from injection pump delivery valve holder (25–Fig. L2-4). Unscrew delivery valve holder and remove delivery valve (22) and spring (23). Do not remove valve seat (20). Reinstall delivery valve holder into pump. Place throttle at full speed position. While manually operating fuel transfer pump, turn flywheel in normal direction of rotation so number 1 piston is on compression stroke. Note that fuel should be flowing out of number 1 delivery valve holder. Stop turning flywheel at the exact point fuel ceases to flow from delivery valve. Beginning of injection occurs at this point and timing mark (I–Fig. L2-5 or L2-6) on flywheel bell housing or crankshaft front pulley should be aligned with reference mark (M).

To advance injection timing, remove injection pump and reduce the thickness of pump mounting shim gaskets (G–Fig. L2-3). Increase shim gasket thickness to retard injection timing. Reinstall delivery valve and spring. Tighten injection pump mounting cap screws to a torque of 25 N·m (18 ft.-lbs.). Bleed air from fuel system as previously outlined.

REPAIRS

Refer to the following table for tightening torques.

Camshaft gear.............196 N·m
    (145 ft.-lbs.)
Camshaft retainer..........20 N·m
    (15 ft.-lbs.)
Connecting rod..............49 N·m
    (36 ft.-lbs.)
Crankshaft drive gear.......490 N·m
    (360 ft.-lbs.)
Crankshaft pulley..........440 N·m
    (325 ft.-lbs.)
Cylinder head...............88 N·m
    (65 ft.-lbs.)
Exhaust manifold............20 N·m
    (15 ft.-lbs.)
Flywheel..................345 N·m
    (255 ft.-lbs.)
Idler gear.................137 N·m
    (100 ft.-lbs.)
Injection pump..............25 N·m
    (18 ft.-lbs.)

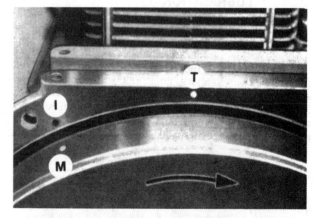

Fig. L2-5—View of timing marks located on flywheel of some models.
- I. Injection
- M. Reference mark
- T. Top dead center

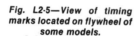

Fig. L2-4—Exploded view of injection pump delivery valve assembly.
- 14. Pump body
- 18. Packing
- 19. Barrel
- 20. Delivery valve seat
- 21. Gasket
- 22. Delivery valve
- 23. Spring
- 24. "O" ring
- 25. Delivery valve holder

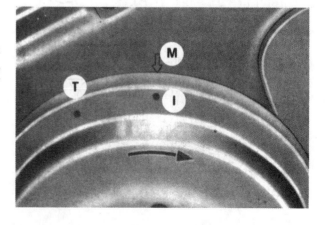

Fig. L2-6—View of timing marks on crankshaft pulley used on some models.
- I. Injection
- M. Reference mark
- T. Top dead center

SERVICE MANUAL

Lombardini

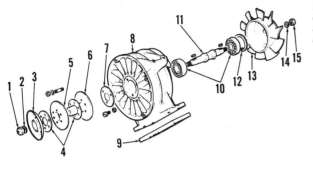

Fig. L2-7—Exploded view of cooling fan. Some models are equipped with two drive belts while others use a single belt.
1. Nut
2. Washer
3. Pulley half
4. Shims
5. Center pulley half
6. Drive pulley half
7. Washer
8. Fan housing
9. Spacer
10. Bearings
11. Shaft
12. Snap ring
13. Fan
14. Washer
15. Nut

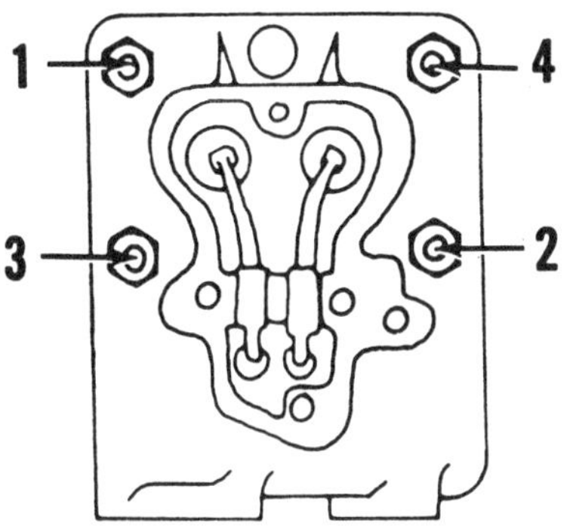

Fig. L2-9—Tighten cylinder head nuts evenly in steps using a crossing pattern. Final torque should be 88 N·m (65 ft.-lbs).

Injector.................20 N·m
   (15 ft.-lbs.)
Intake manifold...........20 N·m
   (15 ft.-lbs.)
Main bearing support:
   Center................49 N·m
      (36 ft.-lbs.)
   Flywheel end..........34 N·m
      (25 ft.-lbs.)
Oil pan..................30 N·m
   (22 ft.-lbs.)
Oil sump.................40 N·m
   (30 ft.-lbs.)
Oil pump gear............60 N·m
   (45 ft.-lbs.)
Rocker arm cover..........20 N·m
   (15 ft.-lbs.)
Timing gear cover........34 N·m
   (25 ft.-lbs.)

COOLING FAN

All models are equipped with a belt-driven axial cooling fan (Fig. L2-7) to force air around cylinder fins and through the oil cooler (if so equipped). The fan is attached to the fan shroud. The fan housing (8) is mounted on timing gear cover. The internal alternator (on models so equipped) is also mounted within fan housing. Overhaul is evident after inspection of the unit and referral to Fig. L2-7.

VALVE CLEARANCE ADJUSTMENT

Valve clearance should be 0.25-0.30 mm (0.010-0.012 inch) for intake and exhaust valves on all models. Clearance should be adjusted with engine cold. To adjust clearance, remove rocker arm covers and rotate crankshaft until number 1 piston is at TDC on compression stroke. Both valves should be closed (rocker arms loose). Loosen rocker arm adjusting screw locknut, then turn adjusting screw until proper thickness feeler gage can be inserted between rocker arm and end of valve. Hold adjusting screw and tighten locknut. Repeat procedure for remaining cylinders as follows: On two-cylinder engines, turn crankshaft one revolution (360°) to position number 2 piston at TDC on compression stroke. On three-cylinder engines, rotating crankshaft ⅔ turn (240°) will position next piston in firing order (1-3-2) at TDC on compression stroke.

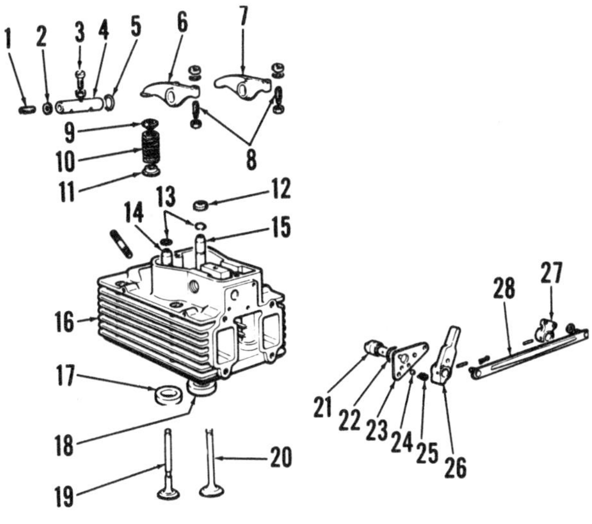

Fig. L2-8—Exploded view of cylinder head assembly.
1. Plug
2. Washer
3. Locating screw
4. Rocker arm shaft
5. Washer
6. Exhaust rocker arm
7. Intake rocker arm
8. Adjuster screw
9. Spring retainer
10. Valve spring
11. Spring seat
12. Seal
13. Snap ring
14. Exhaust valve guide
15. Intake valve guide
16. Cylinder head
17. Exhaust valve seat
18. Intake valve seat
19. Exhaust valve
20. Intake valve
21. Compression release
22. Washer
23. Plate
24. Ball
25. Spring
26. Lever
27. Arm
28. Link

CYLINDER HEAD AND VALVE SYSTEM

Cylinder head should not be removed when engine is hot as head may warp as a result. To remove rocker arms, remove shaft locating set screw (3 – Fig. L2-8). Pull rocker shaft (4) from head using a suitalbe puller. A special puller, part 7276.3595.049, is available. Clearance between rocker arm shaft and rocker arms should not exceed 0.1 mm (0.004 inch).

Valve face angle is 45° for intake and exhaust. If valve head margin is less than 0.40 mm (0.016 inch), renew valve. Valve seat angle is 45° and recommended seat width is 1.4-1.6 mm (0.055-0.065 inch). Renewable valve seat inserts are used on intake and exhaust. When installing new inserts, cylinder head should first be heated in an oven to 160°-180°C (320°-355°F).

With valves installed, measure distance valve head is recessed from cylinder head mounting surface. Intake valve should be recessed 0.65-0.85 mm (0.026-0.033 inch) and exhaust valve recession (models not equipped with compression release) should be 0.75-1.25 mm (0.030-0.049 inch). On models equipped with compression release, exhaust valve recession should be 0.55-0.95 mm (0.022-0.037 inch). If valve recession exceeds specified limits, renew seat insert. If recession is less than specified, grind valve seat as necessary.

Valve stem diameter is 8.98-9.00 mm (0.3535-0.3543 inch). Desired valve stem clearance in guide is 0.03-0.08 mm (0.001-0.003 inch). Valve guides are renewable and oversize guides are available. Outer diameter of oversized guides must be machined to provide a 0.05-0.06 mm (0.0020-0.0023 inch) interference fit in cylinder head bores. Prior to installing guides, heat cylinder head to 160°-180°C (320°-355°F). A locating ring around top of guide deter-

Illustrations Courtesy Lombardini U.S.A., Inc.

Lombardini SMALL DIESEL

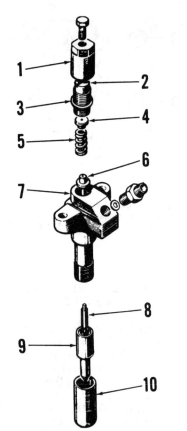

Fig. L2-10—Exploded view of injector.
1. Cap nut
2. Adjusting screw
3. Gasket
4. Spring seat
5. Spring
6. Push rod
7. Body
8. Valve
9. Nozzle
10. Nozzle nut

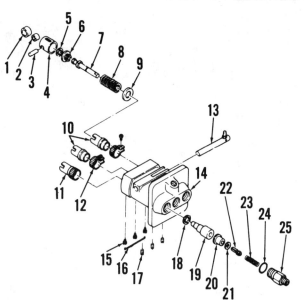

Fig. L2-11—Exploded view of typical fuel injection pump.
1. Outer roller
2. Inner roller
3. Pin
4. Guide
5. Shim
6. Spring retainer
7. Plunger
8. Spring
9. Spring seat
10. Control sleeve B
11. Control sleeve A
12. Sleeve B pinion
13. Control rack
14. Body
15. Guide pins
16. Retaining wire
17. Pins
18. Packing
19. Barrel
20. Delivery valve seat
21. Gasket
22. Delivery valve
23. Spring
24. "O" ring
25. Delivery valve holder

mines installed height of guides. After installation valve guide inside diameter must be reamed or honed to provide desired valve stem clearance.

Intake and exhaust valve springs are identical. Spring free length should be 55.4 mm (2.18 inches). Spring pressure should be 405-428 N (91-96 pounds) at a length of 26.4 mm (1.039 inches).

Before tightening cylinder head nuts, install manifolds so heads are properly aligned with manifolds. Then, tighten cylinder head retaining nuts in 20 N·m (15 ft.-lbs.) steps using a crossing pattern as shown in Fig. L2-9. Final torque should be 88 N·m (65 ft.-lbs.).

INJECTOR

REMOVE AND REINSTALL. To remove injector, first clean dirt from injector, injection line and cylinder head. Disconnect fuel return line and high pressure line and immediately cap or plug all openings. Unscrew injector retaining nuts, then remove injector being careful not to lose shims between injector and cylinder head.

If cylinder head is removed, measure protrusion of nozzle tip beyond combustion chamber surface. Protrusion should be 4.5-5.0 mm (0.177-0.197 inch) and is adjusted by installing shims between injector and cylinder head.

To reinstall injector, reverse the removal procedure. Tighten injector mounting cap screws to a torque of 20 N·m (15 ft.-lbs.).

TESTING. A suitable test stand is required to check injector operation. Only clean, approved testing oil should be used to test injectors.

WARNING: Fuel leaves the injector nozzle with sufficient force to penetrate the skin. When testing, keep yourself clear of nozzle spray.

Connect injector to tester and operate tester lever to purge air from nozzle and make sure nozzle valve is not stuck. When operating, injector nozzle will emit a buzzing sound and should cut off quickly with no leakage at the tip.

Opening pressure should be 20600-21575 kPa (2990-3130 psi). Opening pressure is adjusted by turning adjusting screw (2 – Fig. L2-10). Operate tester lever slowly to maintain pressure at 2900 kPa (400 psi) below opening pressure and check for leakage at nozzle tip. If fuel drips from nozzle tip within a 10 second period, nozzle valve must be overhauled or renewed. Operate tester handle to produce 4 to 6 nozzle opening cycles per second and observe spray pattern. Pattern should be evenly distributed and well atomized. If pattern is wet, ragged or intermittent, nozzle must be overhauled or renewed.

OVERHAUL. To disassemble injector, secure nozzle body (7 – Fig. L2-10) in a vise with nozzle tip pointing upward. Unscrew nozzle holder nut (10), then remove nozzle valve seat (9) and needle (8). Invert nozzle body and remove cap nut (1), adjusting screw (2), spring seat (4), spring (5) and push rod (6).

Thoroughly clean all parts in a suitable solvent. Protect polished surfaces from scratches or other damage. Clean inside the orifice end of nozzle with wooden cleaning stick. The orifice spray holes may be cleaned by inserting a cleaning wire through the holes. Begin cleaning with a small diameter wire and work up to a wire slightly smaller than the orifice hole inner diameter which is 0.28 mm (0.011 inch).

After cleaning, make certain nozzle needle slides freely in valve body. If needle valve sticks, reclean or renew nozzle assembly. When reassembling injector, make certain all parts are clean and wet with diesel fuel. Adjust opening pressure as outlined in TESTING paragraph.

INJECTION PUMP

Refer to Fig. L2-11 for an exploded view of fuel injection pump. To remove pump, disconnect fuel lines, remove retaining cap screws and lift pump from engine. Retain pump mounting shim gaskets for reassembly as thickness of gaskets determines injection timing.

The injection pump should be tested and overhauled by a shop qualified in diesel fuel injection pump repair.

When reinstalling pump, be sure to engage pin on pump control rack with governor fork. Tighten pump mounting cap screws to a torque of 30 N·m (22 ft.-lbs.). If pump is renewed or overhauled or if original shim gaskets are not used, refer to INJECTION PUMP TIMING paragraph and adjust pump timing.

SERVICE MANUAL

Lombardini

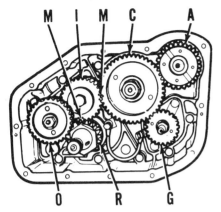

Fig. L2-12—View of timing gears. When correctly timed, marks (M) on idler gear (I) must be aligned with marks on crankshaft gear (R) and camshaft gear (C). Other gears do not require timing.

A. Auxiliary drive gear
C. Camshaft gear
G. Governor gear
I. Idler gear
M. Timing marks
O. Oil pump gear
R. Crankshaft gear

TIMING GEARS

To remove timing gear cover, remove fan belt guard and fan belt. Remove crankshaft pulley using a suitable puller. Remove timing gear cover mounting cap screws and withdraw cover.

Timing gears may be removed after removing gear retaining nuts. Use suitable pullers when removing gears to avoid damage to gears or shafts.

When installing gears, heat crankshaft gear to 180°C (360°F). Drive gear onto crankshaft, then tighten retaining nut to a torque of 490 N·m (360 ft.-lbs.). Install camshaft gear and idler gear making sure timing marks are aligned as shown in Fig. L2-12. Remainder of gears do not require timing.

If gears do not have timing marks, proceed as follows: If not previously removed, remove cylinder head and push rod tubes on number 1 cylinder. Rotate crankshaft so number 1 piston is at top dead center. Rotate camshaft so number 1 cylinder intake valve cam follower is rising and exhaust valve cam follower is going down, then stop rotation when cam followers are at same height above

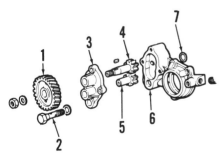

Fig. L2-13—Exploded view of oil pump.
1. Drive gear
2. Special screw
3. Cover
4. Gear & shaft
5. Gear
6. Pump body
7. "O" ring

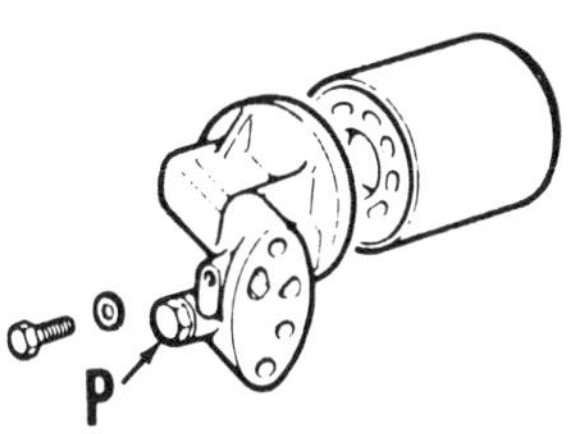

Fig. L2-14—Remove plug (P) for access to oil pressure relief valve in filter adapter.

crankcase surface. Install idler gear and mark contact teeth on crankshaft, idler and camshaft gears for future reference.

Tighten camshaft gear retaining nut to a torque of 195 N·m (145 ft.-lbs.), idler gear nut to 137 N·m (101 ft.-lbs.) and oil pump gear nut to 60 N·m (45 ft.-lbs.).

Inspect front oil seal located in timing gear cover and renew if necessary. Lubricate seal lip before reinstalling cover. Tighten cover retaining screws to a torque of 20 N·m (15 ft.-lbs.). Tighten front pulley retaining nut to 440 N·m (325 ft.-lbs.).

OIL PUMP

R&R OVERHAUL. The oil pump is mounted on the front of the engine and is accessible after removing the timing gear cover. Unscrew pump gear retaining nut, then using a suitable puller, remove pump gear (1–Fig. L2-13). Remove pump cover (3) and gears. Pump housing (6) surrounds the crankshaft and the crankshaft gear must be removed before pump housing can be removed. Note that screw (2) is drilled to allow oil flow through screw.

Oil clearance between oil pump housing and crankshaft should be 0.04-0.08 mm (0.0016-0.0031 inch). Renew pump housing if clearance exceeds 0.13 mm (0.005 inch). Maximum allowable backlash between gears is 0.15 mm (0.006 inch). Maximum allowable clearance between outer diameter of gears and pump housing bore is 0.15 mm (0.006 inch).

Assembly is reverse of disassembly procedure. Tighten pump mounting cap screws to a torque of 40 N·m (30 ft.-lbs.) and pump drive gear retaining nut to 60 N·m (45 ft.-lbs.).

Oil pressure relief valve is located in oil filter housing (Fig. L2-14). Normal oil pressure with oil hot and engine running at 2600 rpm is 285-335 kPa (44-51 psi). Oil pressure may be adjusted by removing plug (P) from filter base and turning adjusting screw.

PISTON AND ROD UNITS

REMOVE AND REINSTALL. Piston and connecting rod may be

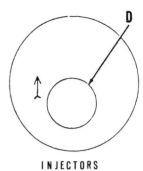

Fig. L2-15—Arrow on piston crown should point towards injection pump and depression (D) should be near injectors.

removed after removing cylinder head and oil pan. Be sure to identify piston and rod assemblies as they are removed so units can be reinstalled in their original locations.

When installing piston and rod, note that depression (D–Fig. L2-15) in piston crown is closer to one side of piston. Install piston so depression will be aligned with injector. Some pistons also have an arrow embossed on piston crown. Properly installed, arrow will point towards the injection pump.

Tighten connecting rod screws to 49 N·m (36 ft.-lbs.). Refer to CYLINDER section and measure piston height in cylinder.

PISTON, PIN AND RINGS

The piston is equipped with three compression rings and one oil control ring. The top compression ring is chrome plated.

Recommended ring end gap is 0.35-0.55 mm (0.014-0.021 inch) for top compression ring, 0.30-0.45 mm (0.012-0.017 inch) for second and third compression rings and 0.25-0.40 mm (0.010-0.015 inch) for oil control ring. Be sure ring is positioned squarely in cylinder bore when checking end gap.

Piston ring side clearance in piston ring groove should be 0.11-0.15 mm (0.004-0.006 inch) for top compression ring, 0.06-0.10 mm (0.002-0.004 inch) for intermediate compression rings and 0.05-0.10 mm (0.002-0.004 inch) for oil control ring. Renew piston if side clearance exceeds recommended values.

Standard piston diameter, measured at bottom of skirt perpendicular to piston pin bore, is 94.82-94.89 mm (3.733-3.736 inches) on Models 5LD675-2, LDA672, 5LD675-3 and LDA673; 105.76-105.78 mm (4.1638-4.1645 inches) on Model 5LD903-3; and 99.82-99.83 mm (3.9299-3.9303 inches) on all other models. Renew piston if wear exceeds

Illustrations Courtesy Lombardini U.S.A., Inc.

Lombardini

SMALL DIESEL

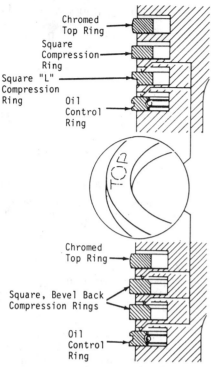

Fig. L2-16—Cross-sectional view of different types of piston rings showing correct installation in piston ring grooves.

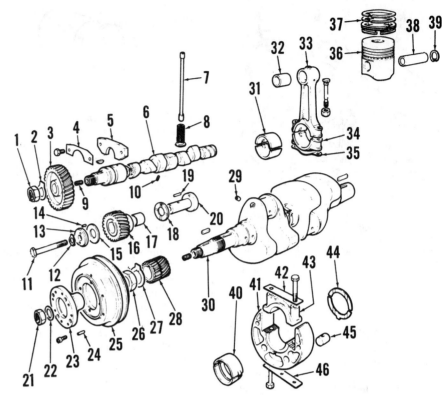

Fig. L2-17—Exploded view of crankshaft, camshaft and idler assemblies.

1. Nut
2. Tab washer
3. Camshaft gear
4. Lockplate
5. Retainer
6. Camshaft
7. Push rod
8. Cam follower
9. Plug
10. Plug
11. Screw
12. Tab washer
13. Spacer
14. Pin
15. Washer
16. Idler gear
17. Bushing
18. Thrust washer
19. Pin
20. Idler shaft
21. Nut
22. Washer
23. Hub
24. Pin
25. Pulley
26. Nut
27. Tab washer
28. Gear
29. Plug
30. Crankshaft
31. Bearing
32. Bushing
33. Connecting rod
34. Rod cap
35. Lockplate
36. Piston
37. Piston rings
38. Piston pin
39. Snap ring
40. Bearing
41. Main bearing support
42. Lockplate
43. Main bearing cap
44. Thrust washer
45. Nut
46. Lockplate

0.05 mm (0.002 inch). Pistons are available in standard size as well as oversizes. Weight of each piston must not vary more than 6 grams (0.2 ounce).

Standard piston pin diameter is 31.995-32.000 mm (1.2596-1.2598 inches). Clearance between piston pin and connecting rod bushing should be 0.02-0.03 mm (0.0008-0.0012 inch) and maximum allowable clearance is 0.07 mm (0.0027 inch). Pin should be a thumb push fit in piston.

When installing rings onto piston, stagger ring end gaps 180° apart. Be sure no end gaps are in line with piston pin bore. Refer to appropriate ring installation diagram shown in Fig. L2-16 for proper installation of rings.

CONNECTING ROD

The connecting rod small end is fitted with a renewable bushing. Clearance between piston pin and rod bushing should be 0.02-0.03 mm (0.0008-0.0012 inch) with a maximum allowable clearance of 0.07 mm (0.003 inch).

And insert type bearing is used in connecting rod big end. Desired rod bearing clearance is 0.04-0.07 mm (0.0016-0.0027 inch) with a maximum allowable clearance of 0.10 mm (0.004 inch). Big end bearings are available in standard and undersizes.

Weight of each connecting rod must not vary more than 10 grams (0.35 ounce).

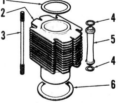

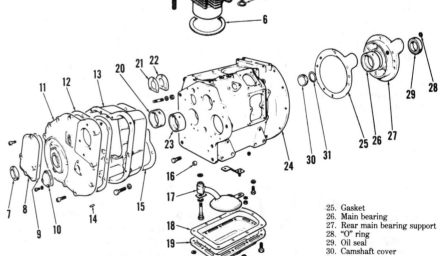

Fig. L2-18—Exploded view of typical crankcase assembly.

1. Head gasket
2. Cylinder
3. Stud
4. Seal
5. Push rod tube
6. Shim gasket
7. Seal
8. Cover
9. Gasket
10. Cover
11. Timing gear cover
12. Gasket
13. Gear housing
14. Pin
15. Gasket
16. Dowel
17. Oil pickup
18. Gasket
19. Oil pan
20. Camshaft bearing
21. Gasket
22. Cover
23. Main bearing
24. Crankcase
25. Gasket
26. Main bearing
27. Rear main bearing support
28. "O" ring
29. Oil seal
30. Camshaft cover
31. "O" ring

SERVICE MANUAL

Lombardini

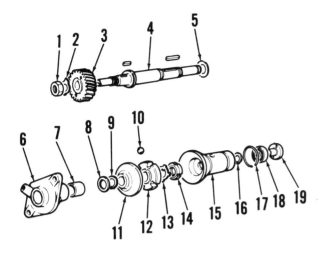

Fig. L2-19—Exploded view of governor shaft.

1. Nut
2. Tab washer
3. Governor shaft
4. Shaft
5. Washer
6. Retainer
7. Bushing
8. Washer
9. Snap ring
10. Balls (4)
11. Ball retainer
12. Ball carrier
13. Tab washer
14. Nut
15. Governor sleeve
16. Snap ring
17. Snap ring
18. Bearing
19. Spindle

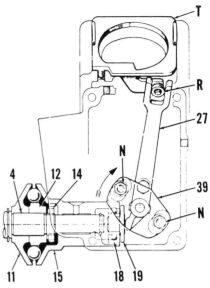

Fig. L2-21—View of governor linkage. Refer to text and Fig. L2-20.

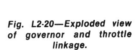

Fig. L2-20—Exploded view of governor and throttle linkage.

C. Torque control screw
H. High idle speed screw
I. Low idle speed screw
25. Start spring
26. Lever
27. Arm
28. Spindle
29. Pin
30. Lever
31. Governor spring
32. Gasket
33. Control housing
34. Dowel
35. Shaft
36. Spring
37. Pin
38. Arm
39. Eccentric
40. Pin
41. Lever
42. Shaft
43. Lever
44. Spring
45. Cover
46. Pin
47. Throttle lever
48. Stop lever

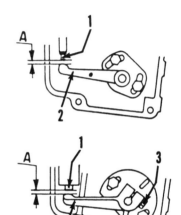

Fig. L2-22—Adjust torque control device to obtain recommended gap (A) between tip of torque control (1) and control lever (2). Refer to text.

1. Torque screw tip
2. Control lever
3. Clamp bolt

CYLINDERS

All models are equipped with removable cylinders. Cylinders may be bored oversize if worn or damaged.

Standard cylinder diameter is 95.00-95.02 mm (3.7401-3.7409 inches) on Models 5LD675-2, LDA672, 5LD675-3 and LDA673; 106.00-106.02 mm (4.1732-4.1740 inches) on Model 5LD930-3; and 100.00-100.02 mm (3.9370-3.9378 inches) on all other models. Cylinder should be rebored to appropriate oversize if taper or out-of-round exceeds 0.1 mm (0.004 inch).

After reinstalling pistons, check piston height in cylinder. With piston at TDC, top of piston should be 0.0-0.1 mm (0.0-0.004 inch) below top surface of cylinder. Piston height is adjusted by installing shim gaskets (6–Fig. L2-18) between cylinder and crankcase.

GOVERNOR

Refer to Figs. L2-19 and L2-20 for exploded view of flyball type governor and control linkage. Governor sleeve (15–Fig. L2-19) slides on governor shaft (4) according to flyball movement and forces spindle (19) to contact governor arm (27–Fig. L2-20). The control rack pin of the fuel injection pump engages the fork in governor arm. The throttle lever (47) operates through governor spring (31) to control engine speed.

The governor shaft assembly (Fig. L2-19) may be removed after removing timing gear cover. Unscrew retainer (6) screws and withdraw governor shaft. Inspect components for excessive wear or damage. Components must move easily without binding. When reinstalling governor shaft, tighten retainer screws to 20 N·m (15 ft.-lbs.).

To synchronize governor linkage with fuel injection pump, remove cover (45–Fig. L2-20) and fuel injection pump. Loosen nuts (N–Fig. L2-21) securing eccentric (39). Install special tool (T), part 7276.2003.004 on two-cylinder models or part 7276.2003.005 on three-cylinder models, in place of injection pump. Be sure tool roller (R-Fig. L2-21) engages fork on governor arm (27). Rotate eccentric (39) until all play is removed from governor but tool roller is still free in fork. Tighten nuts (N), remove tool and reinstall injection pump.

The torque control screw (C-Fig. L2-20) limits movement of governor lever (41) to control maximum fuel delivery under full load. On some models, torque control screw has a spring-loaded tip which allows additional movement of control lever and additional fuel delivery under high torque load. Some early models were equipped with a hydraulically actuated torque control device. When overhauling these

199

Lombardini

units, manufacturer recommends replacing hydraulic system with the mechanical system by renewing throttle housing assembly. The spring-loaded torque screw must be serviced as a unit assembly. The torque control screw must be matched to the engine setting.

To adjust torque control screw, remove housing cover (45 – Fig. L2-20). While operating engine at high idle (no-load) speed, turn torque control screw as needed to obtain specified gap (A – Fig. L2-22) between tip of screw (1) and control lever (2). On models equipped with hydraulically actuated device, loosen lever clamp bolt (3) and rotate lever to obtain recommended gap (A). Refer to the following table for recommended gap settings.

Model	Gap (A)
5LD675-2, LDA672	2.20-2.25 mm (0.087-0.088 in.)
5LD675-3, LDA673	2.30-2.40 mm (0.090-0.094 in.)
5LD825-2, 832	2.50-2.70 mm (0.098-0.106 in.)
5LD825-2, 832	3.40-3.60 mm (0.134-0.142 in.)
5LD930-3	3.40-3.60 mm (0.134-0.142 in.)

CAMSHAFT

To remove camshaft, first remove cylinder head, timing gear cover and fuel injection pump as previously outlined. Remove camshaft gear using a suitable puller. Remove push rod tubes and push rods, then use suitable tools to hold cam followers in a raised position. Remove fuel transfer pump. Remove camshaft retainer plate (5 – Fig. L2-17), then withdraw camshaft from crankcase. Remove oil pan for access to cam followers. Be sure to identify cam followers so they can be reinstalled in their original positions.

The camshaft front bearing journal operates in a renewable, sleeve bearing (20 – Fig. L2-18) which is pressed into front of crankcase. The camshaft center journal (three-cylinder models) and rear journal (all models) operate directly in unbushed bores in the crankcase.

Camshaft standard bearing journal diameter is 47.94-47.96 mm (1.8874-1.8882 inches) for front journal, 42.92-42.94 mm (1.6898-1.6905 inches) for middle journal (three-cylinder only) and 29.91-29.93 mm (1.1776-1.1783 inches) for rear journal. If clearance between front journal and bushing exceeds 0.2 mm (0.008 inch), renew bushing. Bushing is available in standard as well as undersizes of 0.25, 0.5 and 1.0 mm. After pressing new bushing into crankcase, check inside diameter and ream, if necessary, to provide operating clearance of 0.10-0.14 mm (0.004-0.005 inch).

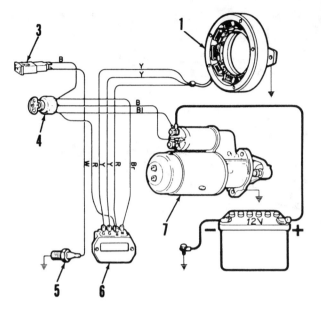

Fig. L2-23—Typical wiring schematic for models equipped with an internal alternator.
1. Alternator
3. Oil pressure light
4. Starting switch
5. Oil pressure switch
6. Regulator
7. Starter

Camshaft lobe height should be 34.3-34.4 mm (1.3504-1.3543 inches) for both intake and exhaust. Height of injection cam lobes should be 38 mm (1.496 inches).

Camshaft end play is controlled by retainer plate (5 – Fig. L2-17). Desired end play is 0.40-0.70 mm (0.016-0.027 inch). Renew retainer plate if end play exceeds 1.0 mm (0.039 inch).

Inspect face of cam followers for excessive wear, pitting or other damage. Outside diameter of cam followers is 19.96-19.98 mm (0.7858-0.7866 inch). Clearance between cam followers and crankcase bores should be 0.02-0.06 mm (0.0008-0.0023 inch).

When reinstalling camshaft, be sure to lubricate cams and bearing journals. Make certain retaining plate is installed with lubricating grooves facing the crankcase. Tighten retainer plate cap screws to a torque of 20 N·m (15 ft.-lbs.) and camshaft gear retaining nut to 196 N·m (145 ft.-lbs.). Be sure timing marks are aligned as outlined in TIMING GEAR paragraph. If camshaft is renewed, check injection pump timing as outlined in MAINTENANCE section.

CRANKSHAFT AND BEARINGS

To remove crankshaft, first remove cylinder head, oil pan, pistons and rods and timing gear cover. Remove idler gear, oil pump gear and crankshaft gear using suitable pullers. Remove oil pump. Remove nuts securing outer bearing support (27 – Fig. L2-18) and cap screws securing center bearing support (41 – Fig. L2-17). Carefully withdraw crankshaft and center bearing support assembly from crankcase. Separate main bearing cap (43) from bearing support (41).

Crankshaft main bearing journal diameter is 59.94-59.95 mm (2.3598-2.3602 inches) on Models LDA672 and LDA673. On all other models, standard main bearing journal diameter is 64.96-64.98 mm (2.5575-2.5582 inches). Desired main bearing clearance is 0.05-0.08 mm (0.002-0.003" inch) with a maximum allowable clearance of 0.10 mm (0.004 inch). Standard and undersize main bearings are available. Special tool 7271.3595.047 is available from manufacturer to install bearing (26 – Fig. L2-18) in bearing support (27). Be sure lubricating oil holes are aligned.

Standard crankpin journal diameter is 55.34-55.35 mm (2.1787-2.1791 inches). Recommended rod bearing clearance is 0.04-0.07 mm (0.001-0.0027 inch) with a maximum allowable clearance of 0.10 mm (0.004 inch). Standard and undersize bearings are available.

Crankshaft end play is controlled by thrust washer halves (44 – Fig. L2-17) mounted on center support (41) on two-cylinder models, or on support nearest timing gear end of engine on three-cylinder models. Crankshaft end play should be 0.05-0.40 mm (0.002-0.016 inch). Install new thrust washers if end play exceeds 0.5 mm (0.020 inch).

Main bearing cap (43) has a serrated parting face. Install cap in support so reference numbers on cap and support

SERVICE MANUAL

Lombardini

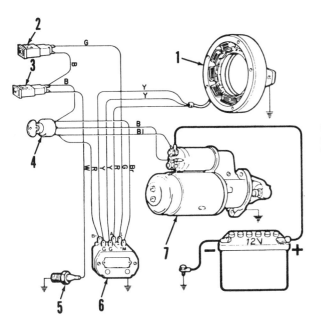

Fig. L2-24—Wiring schematic for models equipped with an alternator warning light (2). Refer to Fig. L2-23.

are on same side. Tighten bearing cap retaining screws to a torque of 49 N·m (36 ft.-lbs.). Reinstall crankshaft, then tighten center support cap screws to 49 N·m (36 ft.-lbs.). Lubricate lip of rear oil seal (29), then reinstall rear bearing plate (27) and tighten nuts to 34 N·m (25 ft.-lbs.). Refer to TIMING GEARS paragraph for correct installation of crankshaft gear. Complete reassembly by reversing the disassembly procedure.

ALTERNATOR AND REGULATOR

Refer to Fig. L2-23 or L2-24 for wiring schematic. Circuit illustrated in Fig. L2-24 uses and alternator warning light and voltage regulator is different than the regulator used in circuit shown in Fig. L2-23 without alternator warning light. Internal alternator (1) is located in blower fan housing.

LOMBARDINI

Model	No. Cyls.	Bore	Stroke	Displ.
6LD260, 500	1	70 mm (2.755 in.)	68 mm (2.677 in.)	262 cc (16 cu. in.)
6LD260/C	1	70 mm (2.755 in.)	68 mm (2.677 in.)	262 cc (16 cu. in.)
6LD325, 520	1	78 mm (3.070 in.)	68 mm (2.677 in.)	325 cc (19.8 cu. in.)
6LD325/C	1	78 mm (3.070 in.)	68 mm (2.677 in.)	325 cc (19.8 cu. in.)
6LD360, 530	1	82 mm (3.228 in.)	68 mm (2.677 in.)	359 cc (21.9 cu. in.)
6LD360/C, 6LD360/V	1	82 mm (3.228 in.)	68 mm (2.677 in.)	359 cc (21.9 cu. in.)
6LD500	1	86 mm (3.385 in.)	68 mm (2.677 in.)	395 cc (24.1 cu. in.)

Engines covered in this section are four-stroke air-cooled diesel engines. "C" series engines have pto drive on camshaft. Model 6LD360/V has crankshaft mounted vertically. On all engines, the cylinder head and cylinder block are aluminum while the cylinder is cast iron.

MAINTENANCE

LUBRICATION

Recommended engine oil is SAE 10W for temperatures below 0°C (32°F), SAE 20W for temperatures between 0°C (32°F) and 20°C (68°F) and SAE 40 for temperatures above 20°C (68°F). API classification for oil should be CD. Oil sump capacity is 0.7 liters (1.5 pints) on Model 6LD360/C, 1.2 liter (2.5 pints) on Model 6LD400 and 1.0 liter (2.1 pints) on all other models.

A renewable oil filter is located on side of engine block (Fig. L3-1). Manufacturer recommends renewing engine oil and filter after every 300 hours of operation.

ENGINE SPEED ADJUSTMENT

Idle speed is adjusted by turning idle speed screw (I–Fig. L3-2). Idle speed should be 1100-1400 rpm on Models 6LD260, 500, 6LD325, 520, 6LD360, 530, and 6LD360/V; 550-700 rpm on Models 6LD260/C, 6LD325/C and 6LD360/C; 900-1000 rpm on Model 6LD400.

Maximum governed speed is adjusted by turning high speed screw (H). Set maximum (no-load) speed 180 rpm above engine rated load speed to compensate for governor droop during operation. Maximum speed under load is normally 3600 rpm for all models except 6LD260/C, 6LD325/C and 6LD360/C which are normally rated at 1800 rpm. Governed speed may vary depending upon engine application.

FUEL SYSTEM

FUEL FILTER. The fuel filter may be located inside the fuel tank as shown in Fig. L3-3, or a cartridge type filter as shown in Fig. L3-4 may be used. Fuel

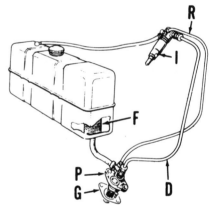

Fig. L3-3—Diagram of fuel system with filter located in tank.

D. High pressure line
F. Fuel filter
G. Shim gasket
I. Injector
P. Injection pump
R. Return line

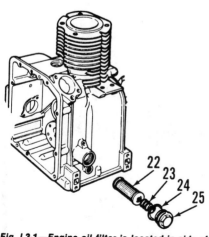

Fig. L3-1—Engine oil filter is located in side of crankcase.

22. Oil filter
23. Spring
24. "O" ring
25. Plug

Fig. L3-2—Turn screw (I) to adjust low idle speed and screw (H) to adjust high idle speeed. Refer to text for adjustment of torque control screw (C).

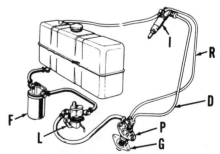

Fig. L3-4—Diagram of fuel system equipped with fuel filter cartridge and fuel transfer pump used on some models.

D. High pressure line
F. Fuel filter
G. Shim gasket
I. Injector
L. Fuel pump
P. Injection pump
R. Return line

Illustrations Courtesy Lombardini U.S.A., Inc.

SERVICE MANUAL

Lombardini

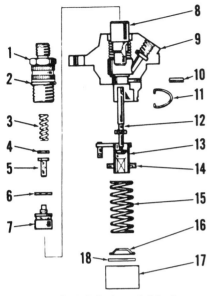

Fig. L3-5—Exploded view of injection pump used on all models. Pump is actuated by engine camshaft.

1. Delivery union
2. "O" ring
3. Spring
4. Washer
5. Delivery valve
6. Gasket
7. Delivery valve seat
8. Barrel
9. Pump body
10. Pin
11. Clip
12. Plunger
13. Control sleeve
14. Spring seat
15. Spring
16. Spring retainer
17. Tappet
18. Spacer

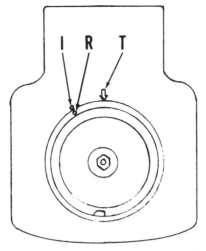

Fig. L3-6—Injection should occur when timing dot (R) of crankshaft pulley is aligned with timing dot (I) on fan shroud. Piston is at TDC when timing dot (R) and mark (T) are aligned.

filter should be renewed after every 300 hours of operation or sooner if required. After renewing filter, air must be bled from system as outlined in the following paragraph.

BLEED FUEL SYSTEM. On gravity flow fuel systems (Fig. L3-3), loosen fuel line fitting on injection pump and allow fuel to flow until free of air, then retighten fitting.

On models equipped with cartridge type filter (Fig. L3-4), unscrew bleed screw on filter housing and allow fuel to flow until free of air.

On models equipped with a fuel transfer pump (L-Fig. L3-4), loosen fuel supply line fitting on injection pump. Manually operate fuel pump primer lever until fuel flows from fitting, then retighten fitting.

If engine fails to start after bleeding air from filter and fuel line, loosen high pressure line at injector. Rotate crankshaft to operate injection pump until fuel is discharged at injector line fitting. Retighten injector line and start engine.

INJECTION PUMP TIMING

Injection pump timing is adjusted using shim gaskets (G-Fig. L3-3 and L3-4) between pump body and mounting surface of crankcase. To check pump timing, unscrew injector line (D) fitting from pump delivery union (1-Fig. L3-5). Unscrew delivery union and remove spring (3), washer (4) and delivery valve (5). Do not remove valve seat (7). Reinstall delivery union into pump body. Move throttle control lever to full speed position. Rotate crankshaft in normal direction of rotation so piston is on compression stroke. Note that fuel should flow from delivery union. Stop crankshaft rotation at moment fuel ceases to spill out of union. This is beginning of injection and timing dot (R-Fig. L3-6) on crankshaft pulley should be aligned with injection timing dot (I) on fan shroud.

If fuel stops flowing before timing dots are aligned (timing advanced), add shims under pump. If fuel stops flowing late (timing retarded), reduce thickness of shim gaskets.

After pump timing is properly adjusted, tighten pump mounting cap screws to a torque of 30 N·m (22 ft.-lbs.). Bleed air from system as previously outlined.

REPAIRS

TIGHTENING TORQUES

Refer to the following table for tightening torques.

Connecting rod.............34 N·m
(25 ft.-lbs.)
Cylinder head...............34 N·m
(25 ft.-lbs.)
Flywheel..................145 N·m
(110 ft.-lbs.)
Injection pump ....30 N·m
(22 ft.-lbs.)
Injector ....12 N·m
(9 ft.-lbs.)
Main bearing support:
  Flywheel end..............30 N·m
(22 ft.-lbs.)
  Gear train end............44 N·m
(26 ft.-lbs.)
Oil pan....................25 N·m
(18 ft.-lbs.)
Oil pump...................12 N·m
(9 ft.-lbs.)
Oil pump gear..............25 N·m
(18 ft.-lbs.)
Rocker arm cover...........20 N·m
(15 ft.-lbs.)
Timing gear cover..........44 N·m
(33 ft.-lbs.)

VALVE CLEARANCE ADJUSTMENT

To adjust valve clearance, first remove rocker arm cover and rotate crankshaft until piston is at TDC on compression stroke. Clearance between valve stem end and rocker arm should be 0.15 mm (0.006 inch) for both intake and exhaust with engine cold. Note that there are two adjusting screws (Fig. L3-7) in exhaust rocker arm on models equipped with compression release mechanism. Inner adjusting screw (V) is

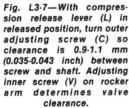

Fig. L3-7—With compression release lever (L) in released position, turn outer adjusting screw (C) so clearance is 0.9-1.1 mm (0.035-0.043 inch) between screw and shaft. Adjusting inner screw (V) on rocker arm determines valve clearance.

Lombardini

SMALL DIESEL

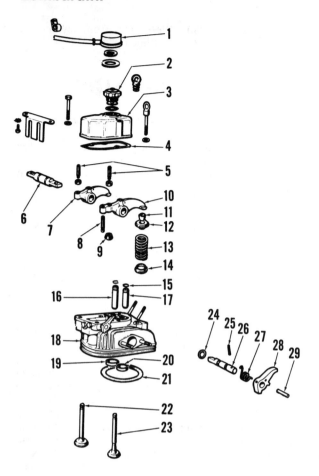

Fig. L3-8—Exploded view of cylinder head assembly.

1. Breather
2. Fill cap
3. Rocker cover
4. Gasket
5. Valve adjusting screws
6. Rocker arm shaft
7. Intake rocker arm
8. Compression release adjusting screw
9. Valve seal
10. Exhaust rocker arm
11. Valve keepers
12. Spring retainer
13. Spring
14. Spring seat
15. Locating rings
16. Intake valve guide
17. Exhaust valve guide
18. Cylinder head
19. Intake valve seat
20. Exhaust valve seat
21. Head gasket
22. Intake valve
23. Exhaust valve
24. "O" ring
25. Pin
26. Compression release shaft
27. Spring
28. Compression release lever
29. Pin

Fig. L3-9—Measure piston height in cylinder and refer to text to determine cylinder head gasket thickness.

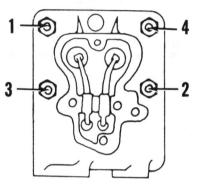

Fig. L3-10—Tighten cylinder head nuts in steps using a crossing pattern.

used to adjust valve clearance while outer screw (C) is used to adjust compression release gap.

COMPRESSION RELEASE

Some models are equipped with a manual compression release so the exhaust valve may be held open to aid starting. Compression release components (24 through 29–Fig. L3-8) are mounted in the cylinder head. Rotating shaft (26) will force the exhaust rocker arm (10) to slightly open the exhaust valve.

The compression release is adjusted by turning outer adjusting screw (C–Fig. L3-7) in exhaust valve rocker arm. Adjust compression release gap AFTER adjusting exhaust valve clearance. With Compression lever (L) in off position, clearance between adjusting screw and shaft should be 0.9-1.1 mm (0.035-0.043 inch).

Diameter of compression release shaft (26–Fig. L3-8) is 9.37-10.00 mm (0.369-0.393 inch) while lobe height is 8.45-8.50 mm (0.333-0.334 inch).

CYLINDER HEAD AND VALVE SYSTEM

Cylinder head should not be removed while hot as it may warp as a result. If cylinder head mounting surface is warped or pitted, up to 0.3 mm (0.012 inch) of material may be lapped from head to true surface.

Valve face angle is 45° for both valves. Renew valve if head margin is less than 0.5 mm (0.002 inch). Valve seat angle is 45° and recommmended seat width is 1.4-1.6 mm (0.055-0.065 inch). Renewable valve seat inserts are used on intake and exhaust. Before installing new inserts, cylinder head should first be heated to 160°-189°C (320°-355°F).

With valves installed, measure distance valve head is recessed from cylinder head surface. Intake valve should be recessed 0.70-0.90 mm (0.028-0.035 inch) and exhaust valve (except models with compression release) should be recessed 0.25-0.75 mm (0.010-0.030 inch). With compression release, exhaust valve recession should be 0.55-0.95 mm (0.022-0.037 inch). If recession is less than specified, regrind valve seat. If valve recession exceeds specified limit, renew valve seat insert.

Valve stem diameter is 6.98-7.00 mm (0.2748-0.2756 inch). Desired valve stem clearance in guide is 0.03-0.08 mm (0.001-0.003 inch) for both valves and maximum allowable clearance is 0.15 mm (0.006 inch). Valve guides are renewable and oversize guides are available. Outer diameter of oversize guide must be machined to provide a 0.05-0.06 mm (0.0020-0.0023 inch) interference fit in cylinder head bore. Cylinder head should be heated to 160°-180°C (320°-355°F) prior to installing new guides. A locating ring around top of guide determines distance guide is pressed into head. After installation, check guide inside diameter and ream as necessary to provide desired valve stem clearance.

Intake and exhaust valve springs are identical. Spring free length should be 42 mm (1.653 inches). Valve spring pressure should be 219-233 N (49-52 pounds) at 32 mm (1.260 inches).

The cylinder head gasket is available in varying thicknesses to adjust clearance between cylinder head surface and top of piston. Clearance must be 0.6-0.7 mm (0.024-0.027 inch) with piston at TDC. To determine required gasket thickness, measure from piston crown to gasket seating surface of cylinder as shown in Fig. L3-9. Subtract measurement (if piston is below sealing surface) or add measurement (if piston is above sealing surface) to 0.6-0.7 mm (0.024-0.027 inch) to obtain required gasket thickness. Gaskets are available

SERVICE MANUAL

Lombardini

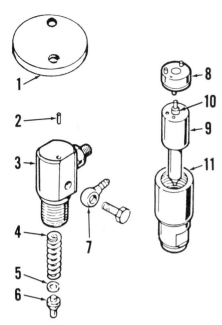

Fig. L3-11—Exploded view of injector.

1. Clamp plate
2. Dowel pin
3. Nozzle body
4. Spring
5. Shim
6. Spring seat
7. Return line fitting
8. Spacer
9. Nozzle tip
10. Valve
11. Nozzle holder nut

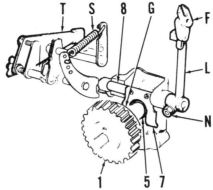

Fig. L3-12—View of governor mechanism. Refer to text for operation.

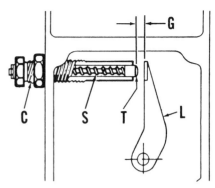

Fig. L3-13—View of torque control screw. Refer to text for adjustment.

in thicknesses of 0.5 mm (0.020 inch), 0.6 mm (0.023 inch), 0.7 mm (0.027 inch) and 0.8 mm (0.031 inch).

Tighten cylinder head nuts in steps using a crossing pattern as shown in Fig. L3-10. Final torque should be 34 N·m (25 ft.-lbs.).

INJECTOR

REMOVE AND REINSTALL. To remove injector, first clean dirt from injector, injection line, return line and cylinder head. Disconnect fuel return line and injection line and immediately cap or plug all openings. Unscrew retainer plate nuts and lift off retainer plate (1–Fig. L3-11) being careful not to lose dowel pin (2). Injector may now be carefully removed from cylinder head. Do not lose shims between injector and cylinder head.

Tighten injector retaining plate nuts to 12 N·m (9 ft.-lbs.). If accessible, measure protrusion of nozzle into combustion chamber. Nozzle tip should extend 2.5-3.0 mm (0.100-0.118 inch) above adjacent combustion chamber surface. Adjust position of nozzle by installing shims between injector and cylinder head. Shims are available in thicknesses of 0.5 mm (0.020 inch) and 1.0 mm (0.040 inch).

TESTING. A suitable test stand is required to check injector operation. Only clean, approved testing oil should be used to test injector.

WARNING: Fuel leaves the injection nozzle with sufficient force to penetrate the skin. When testing, keep yourself clear of nozzle spray.

Connect injector to tester and operate tester lever to purge air from nozzle and to make sure nozzle valve is not stuck. When operating properly, injector nozzle will emit a buzzing sound and cut off quickly with no leakage at the tip.

Opening pressure should be 19610-21575 kPa (2845-3130 psi) on Models 500, 520 and 530. On all other models, opening pressure should be 18630-19610 kPa (2700-2845 psi). If a new spring (4–Fig. L3-11) is being used, set opening pressure 1000 kPa (145 psi) higher than specified pressure settings to compensate for spring taking a set during initial operation. On all models, opening pressure is adjusted by varying thickness of shims (5).

To check for leakage past nozzle valve, operate tester lever slowly to maintain pressure at 2100 kPa (300 psi) below opening pressure. If a drop of fuel forms on nozzle tip within a 10 second period, nozzle valve must be overhauled or renewed. Slight wetness at the tip is permissible.

Operate tester lever briskly and check for an even and well atomized spray pattern. If pattern is wet, ragged or intermittent nozzle must be overhauled or renewed.

OVERHAUL. Clamp nozzle body (3–Fig. L3-11) in a vise with nozzle tip pointing upward. Remove nozzle holder nut (11). Remove nozzle valve (9) and spacer (8). Invert nozzle body and remove spring seat (6), shims (5) and spring (4).

Thoroughly clean all parts in a suitable solvent. Do not use steel wire brush or sharp metal tools to clean injector components. Clean inside orifice end of nozzle valve with wooden cleaning stick. The orifice spray holes may be cleaned by inserting a cleaning wire slightly smaller in diameter than the spray holes. Spray hole diameter is 0.20 mm (0.008 inch) on Models 500, 520 and 530 or 0.24 mm (0.009 inch) on all other models. Make certain nozzle needle (10) slides freely in bore of nozzle valve set (9). If needle sticks, reclean or renew nozzle assembly.

When reassembling injector, be sure all components are clean and wet with diesel fuel. Recheck injector operation as outlined in TESTING paragraph.

INJECTION PUMP

Refer to Fig. L3-5 for an exploded view of injection pump. To remove pump, disconnect fuel lines and immediately plug all openings to prevent entry of dirt. Remove retaining screws and lift out pump assembly. Be sure to retain pump mounting gaskets for use in reassembly. Pump timing is adjusted by changing gasket thickness.

It is recommended that injection pump be tested and serviced only by a shop qualified in diesel fuel injection repair.

When reinstalling pump, assemble correct thickness of mounting shim gaskets and be sure to engage control rack pin with governor arm. Tighten pump mounting screws to a torque of 25 N·m (18 ft.-lbs.). Loosen clamp nut (N–Fig. L3-12), then move throttle lever (T) to full speed position. Push governor lever (L) in until it stops thus moving injection pump control sleeve to maximum delivery. Tighten clamp nut (N).

Torque control screw (C–Fig. L3-13) serves as the full-load stop for governor linkage (L). Control screw is equipped with a spring-loaded tip which allows additional fuel delivery under high torque load. To adjust torque control screw, run engine at high idle with no load. Turn screw so there is a gap (G) between tip (T) and lever (L) of 2.3-2.7 mm (0.090-0.106 inch) on Model 500, 1.3-1.5 mm (0.050-0.060 inch) on Model 520 or 2.1-2.3 mm (0.083-0.090 inch) on all other models.

Illustrations Courtesy Lombardini U.S.A., Inc.

Lombardini — SMALL DIESEL

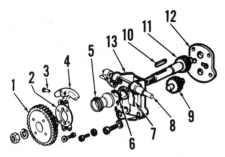

Fig. L3-14—Exploded view of governor and oil pump assembly.

1. Drive gear
2. Governor frame
3. Pins
4. Weights
5. Sleeve
6. Stop
7. Fork
8. Spindle
9. Gear
10. Key
11. Gear & shaft
12. Cover
13. Oil pump body

GOVERNOR

All models are equipped with a flyweight or ball type centrifugal governor which is attached to the back of oil pump drive gear as shown in Fig. L3-14. The oil pump drive gear (1) is driven by the crankshaft and rotates governor flyweights or balls. Movement of flyweights or balls causes sleeve (5) to move against fork (7) which rotates attached governor shaft (8). As governor shaft rotates, the governor lever (L-Fig. L3-12) forces arm (F) against a pin in the injection pump control sleeve thereby controlling fuel delivery to cylinder. Throttle lever (T) operates through governor spring (S) to control engine speed.

Governor components must move freely for proper governor operation. Governor spring (S-Fig. L3-12) free length should be 56.9-57.0 mm (2.240-2.244 inches). Spindle (8-Fig. L3-14) diameter should be 7.95-7.96 mm (0.3130-0.3134 inch). Desired clearance between spindle and bore in oil pump housing (13) is 0.06-0.10 mm (0.002-0.004 inch) with a maximum allowable clearance of 0.15 mm (0.006 inch).

OIL PUMP

Refer to Fig. L3-14 for an exploded view of oil pump. The oil pump is accessible for removing crankcase cover (3 – Fig. L3-15). Clearance between gears and pump body walls must not exceed 0.15 mm (0.006 inch). Renew oil pump if components are excessively worn or damaged. Tighten pump mounting screws evenly to 12 N·m (9 ft.-lbs.).

CAMSHAFT, CAM FOLLOWERS AND PUSH RODS

The camshaft rides directly in crankcase cover and crankcase bulkhead and is accessible after removing crankcase cover (3 – Fig. L3-15). Cam followers (7 and 8) pivot on stud (9) and transfer motion to push rods (26) which pass through tube (28) to rocker arms. In addition to valve actuating lobes, a lobe is ground on the camshaft to operate the fuel injection pump. On Models 6LD260/C, 6LD325/C and 6LD360/C, the camshaft also serves as the power takeoff.

Camshaft bearing journal diameters are 19.937-19.970 mm (0.7849-0.7862 inch) and 25.937-25.950 mm (1.0211-1.0216 inches). If wear exceeds 0.10 mm (0.004 inch), renew camshaft.

Cam follower pivot stud (9) diameter should be 9.4-9.6 mm (0.370-0.378 inch). Maximum allowable clearance between pivot stud and cam followers is 0.10 mm (0.004 inch).

Install camshaft so timing marks (M – Fig. L3-16) are aligned. If timing marks are absent from gears, proceed as follows: Position piston at top dead center (TDC) then install camshaft so intake cam follower is on opening side of cam lobe and exhaust cam follower is on closing side of cam lobe. If necessary, remesh gears so cam followers are at same height. Mark gears for future reference.

Depth of camshaft in crankcase must not be greater than 0.10 mm (0.004 inch) as measured from thrust face (TF – Fig. L3-16) to crankcase gasket surface (G). Camshaft end play should be 0.10-0.30 mm (0.004-0.012 inch) and is adjusted by varying thickness of crankcase cover gasket (4 – Fig. L3-15). Apply Loctite to crankcase cover (3) screws and tighten to 45 N·m (33 ft.-lbs.).

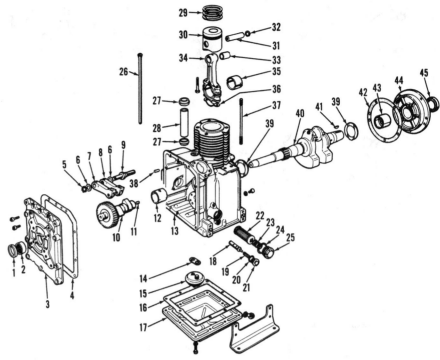

Fig. L3-15—Exploded view of engine.

1. Seal
2. Roller bearing
3. Crankcase cover
4. Gasket
5. Snap ring
6. Washer
7. Exhaust cam follower
8. Intake cam follower
9. Stud
10. Camshaft
11. Plug
12. Bushing
13. Engine block
14. Gasket
15. Oil pickup
16. Gasket
17. Oil pan
18. Oil pressure relief valve
19. Spring
20. Gasket
21. Plug
22. Oil filter
23. Spring
24. "O" ring
25. Plug
26. Push rods
27. Seal
28. Push rod tube
29. Piston rings
30. Piston
31. Piston pin
32. Snap ring
33. Bushing
34. Connecting rod
35. Rod bearing
36. Lockplate
37. Studs
38. Dowel pins
39. Thrust washers
40. Crankshaft
41. Key
42. Gasket
43. Bushing
44. Support
45. Seal

Fig. L3-16—View of camshaft and crankshaft gear timing marks (M). Measure depth of camshaft thrust face (TF) from crankcase gasket surface (G) as outlined in text.

SERVICE MANUAL

Lombardini

Fig. L3-17—Install piston so depression (D) is nearer flywheel side of engine. Some pistons may have an arrow on crown and arrow must point towards flywheel.

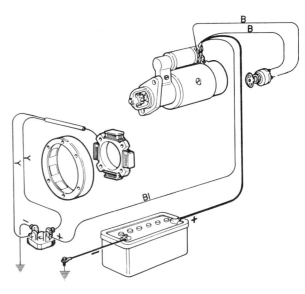

Fig. L3-18—Typical wiring schematic for engines equipped with alternator and electric starter.

The push rods are contained in tube (28) and must cross between cam followers and rocker arms. Push rod nearer cylinder connects intake cam follower and rocker arm while outer push rod connects exhaust cam follower and rocker arm.

PISTON AND ROD UNIT

Piston and connecting rod may be removed after removing cylinder head and oil pan.

When reinstalling piston and rod, note that depression (D – Fig. L3-17) in piston crown is closer to one side of piston. Install piston so depression side of piston is aligned with injector. Some pistons also have an arrow embossed in piston crown as shown in Fig. L3-17. Properly installed, arrow on piston crown will point towards flywheel.

The connecting rod and cap have machined serrations which must mate during assembly. Match marks on rod and cap must be on same side. Tighten connecting rod screws to 34 N·m (25 ft.-lbs.).

PISTON, PIN, RINGS AND CYLINDER

The piston on early models (before engine number 1098888) is equipped with two compression rings and one oil control ring. Late model engines are equipped with three compression rings and one oil control ring.

On Models 500, 520 and 530, ring end gap should be 0.25-0.40 mm (0.010-0.015 inch) for all compression rings and 0.20-0.35 mm (0.008-0.013 inch) for the oil ring. On all other models, ring end gap should be 0.25-0.45 mm (0.010-0.018 inch) for top compression ring, 0.30-0.45 mm (0.012-0.018 inch) for second and third compression rings and 0.25-0.40 mm (0.010-0.015 inch) for oil ring. Be sure rings are positioned squarely in cylinder when checking end gap.

To check piston ring groove wear, install rings onto piston and measure side clearance between top of ring and top of groove. Side clearance should be 0.11-0.15 mm (0.005-0.006 inch) for top groove, 0.06-0.10 mm (0.002-0.004) inch for second and third compression ring grooves and 0.05-0.10 mm (0.002-0.004 inch) for oil ring groove. Renew piston if side clearance is excessive.

Clearance between piston pin and connecting rod bushing should be 0.01-0.03 mm (0.0004-0.0012 inch). If clearance exceeds 0.07 mm (0.0027 inch), renew pin and bushing. Pin should be a thumb push fit in piston.

Standard piston diameter, measured at bottom of skirt perpendicular to pin bore, is 69.90-69.93 mm (2.7519-2.7531 inches) for Models 6LD260, 500 and 6LD260/C; 77.89-77.92 mm (3.0665-3.0677 inches) on Models 6LD325, 520 and 6LD325/C; 81.88-81.89 mm (3.2236-3.2240 inches) on Models 6LD360, 530, 6LD360/C and 6LD360/V; 85.91-85.93 mm (3.3823-3.3830 inches) on Model 6LD400. Renew piston if skirt wear exceeds 0.05 mm (0.002 inch). Piston and rings are available in standard size and oversizes of 0.5 mm (0.020 inch) and 1.0 mm (0.040 inch).

Cylinder standard inside diameter is 70.00-70.02 mm (2.7559-2.7567 inches) on Models 6LD260, 500 and 6LD260/C; 78.00-78.02 mm (3.0708-3.0716 inches) on Models 6LD325, 520 and 6LD325/C; 82.00-82.02 mm (3.2283-3.2291 inches) on Models 6LD360, 530, 6LD360/C and 6LD350/V; 86.00-86.02 mm (3.3858-3.3866 inches) on Model 6LD400. If cylinder taper or out-of-round exceeds 0.1 mm (0.004 inch), rebore cylinder to appropriate oversize.

When reistalling rings onto piston, stagger ring end gaps 180° apart. Be sure no end gaps are in line with piston pin bore. Lubricate piston, rings and cylinder with clean engine oil prior to reassembly.

CONNECTING ROD

The connecting rod small end is fitted with a renewable bushing. Clearance between piston pin and bushing should be 0.01-0.03 mm (0.0004-0.0012 inch). An insert type bearing is used in connecting rod big end. Crankshaft crankpin diameter should be 39.99-40.00 mm (1.5744-1.5748 inches) and desired clearance in bearing is 0.03-0.06 mm (0.0012-0.0024 inch). Maximum allowable clearance is 0.10 mm (0.004 inch). Bearings are available in undersizes of 0.25 mm (0.010 inch) and 0.50 mm (0.020 inch) as well as standard size.

CRANKSHAFT AND CRANKCASE

The crankshaft is supported by bushing (12 – Fig. L3-15) in the crankcase bulkhead, bushing (43) in support (44) on flywheel side and by a roller bearing (2) in the crankcase cover (3).

Desired bearing clearance for center and flywheel end main bearings is 0.03-0.06 mm (0.0012-0.0024 inch). Crankshaft journal diameter for center and flywheel end bearings is 39.99-40.00 mm (1.5744-1.5748 inches). Special bearing installing tool 7271.3595.047 is available from Lombardini to properly

Illustrations Courtesy Lombardini U.S.A., Inc.

Lombardini

install main bearings. Be sure oil hole in center bearing is aligned with hole in crankcase.

Crankshaft end thrust is taken by thrust washers (39 – Fig. L3-15). Thrust washer thickness should be 2.31-2.36 mm (0.090-0.092 inch). Desired crankshaft end play is 0.10-0.30 mm (0.004-0.012 inch). End play is adjusted by removing or adding gaskets (42) between support (44) and crankcase.

Inspect oil seals (1 and 45) and renew if necessary. Be sure to lubricate lip of seals before reassembling. Tighten flywheel side support plate cap screws to a torque of 30 N·m (22 ft.-lbs.) and crankcase cover cap screws to 45 N·m (33 ft.-lbs.).

SMALL DIESEL

ALTERNATOR

The internal alternator is mounted on the flywheel end of engine. The stator is secured to the engine crankcase while a ring of permanent magnets is carried by the flywheel. Note wiring schematic in Fig. L3-18.

LOMBARDINI

Model	No. Cyls.	Bore	Stroke	Displ.
7LD600/I, 710	1	90 mm (3.543 in.)	94 mm (3.700 in.)	598 cc (36.5 cu. in.)
7LD665/I, 720	1	95 mm (3.740 in.)	94 mm (3.700 in.)	666 cc (40.6 cu. in.)
7LD740/I	1	100 mm (3.937 in.)	94 mm (3.700 in.)	738 cc (45.0 cu. in.)

All models are four-stroke, air-cooled diesel engines. The crankcase and cylinder head are aluminum and cylinder barrel is cast iron. Crankshaft rotation is counterclockwise at pto end.

MAINTENANCE

LUBRICATION

Recommended engine oil is SAE 10W for temperatures below 0°C (32°F), SAE 20W for temperatures between 0°C (32°F) and 20°C (68°F), and SAE 40 for temperatures above 20°C (68°F). API classification for oil should be CD. Oil sump capacity is 2.2 liters (2.3 quarts). Manufacturer recommends renewing oil after every 300 hours of operation.

A renewable oil filter is located inside the engine block and can be renewed after removing cover plate (O–Fig. L4-1). Manufacturer recommends renewing filter after every 300 hours of operation.

All models are equipped with a pressurized oil system. Refer to Fig. L4-2 for a diagram of the oil circuit.

ENGINE SPEED ADJUSTMENT

Idle speed is adjusted by turning idle speed screw (I–Fig. L4-3). Idle speed should be 1000-1100 rpm. Maximum governed speed is adjusted by turning high speed screw (H). Adjust high idle (no-load) speed 180 rpm above specified rated full-load speed to compensate for governor droop during operation. Maximum full-load speed is normally 3000 rpm, but speed may vary depending upon application.

FUEL SYSTEM

FUEL FILTER. The fuel filter is located inside the fuel tank as shown in Fig. L4-4. Renew fuel filter after every 300 hours of operation or sooner if required.

BLEED FUEL SYSTEM. To bleed fuel system, loosen fuel line fitting on injection pump and allow fuel to flow until air-free, then retighten fitting. Loosen high pressure injection line at injector,

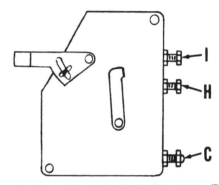

Fig. L4-3—View of speed adjusting screws: (I) Low idle speed; (H) High idle speed; (C) Torque control. Refer to text for adjustment.

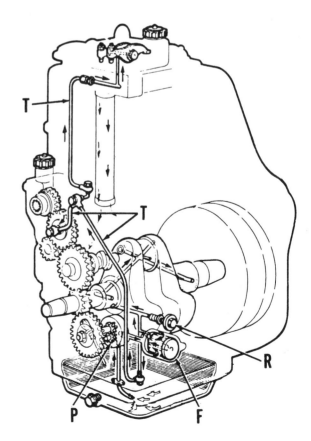

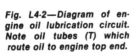

Fig. L4-2—Diagram of engine oil lubrication circuit. Note oil tubes (T) which route oil to engine top end.
F. Oil filter
P. Oil pump
R. Oil relief valve
T. Oil tubes

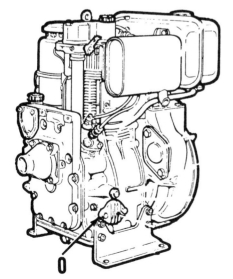

Fig. L4-1—Remove plate (O) for access to oil filter.

Lombardini SMALL DIESEL

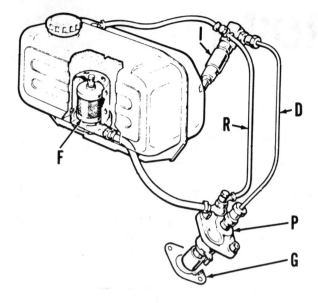

Fig. L4-4—Diagram of fuel circuit.
D. High pressure line
F. Fuel filter
G. Shim gasket
I. Injector
P. Injection pump
R. Return line

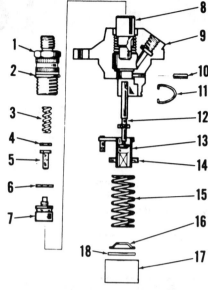

Fig. L4-5—View of fuel injection pump.

1. Delivery valve holder
2. "O" ring
3. Spring
4. Washer
5. Delivery valve
6. Gasket
7. Delivery valve seat
8. Barrel
9. Pump body
10. Pin
11. Clip
12. Plunger
13. Control sleeve
14. Spring seat
15. Spring
16. Spring retainer
17. Tappet
18. Spacer

then rotate engine crankshaft to operate injection pump until air-free fuel flows from injection line. Retighten injection line.

INJECTION PUMP TIMING

Injection pump timing is adjusted by changing thickness of shim gaskets (G–Fig. L4-4) between injection pump body and crankcase mounting surface. To check injection pump timing, disconnect high pressure delivery line from pump and remove delivery valve holder (1–Fig. L4-5). Remove spring (3), washer (4) and delivery valve (5). Do not remove valve seat (7). Reinstall delivery valve holder. Move throttle control to full speed position. Rotate crankshaft in normal direction so piston is on compression stroke. Note that fuel should spill out of delivery valve holder during this operation. Continue to rotate crankshaft until fuel stops flowing from delivery valve holder. Stop crankshaft rotation at moment fuel ceases to flow out. At this point, beginning of injection occurs and timing dot (R–Fig. L4-6) on fan plate should align with timing dot (I) on fan shroud.

Injection timing is 23°-25° BTDC on Models 710 and 720, 26°-28° BTDC on Model 7LD740 and 24°-26° BTDC on all other models. To advance injection timing, remove pump and reduce thickness of shim gaskets (G–Fig. L4-4). To retard injection timing, add pump shim gaskets. Reinstall delivery valve components after checking timing. Tighten delivery valve holder to a torque of 34 N·m (25 ft.-lbs.). Tighten injection pump retaining screws to 25 N·m (18 ft.-lbs.). Bleed air from fuel system as outlined in previous paragraph.

REPAIRS

TIGHTENING TORQUES

Refer to the following table for tightening torques.

Connecting rod.............49 N·m (36 ft.-lbs.)
Crankcase cover............25 N·m (18 ft.-lbs.)
Cylinder head..............60 N·m (44 ft.-lbs.)
Flywheel..................295 N·m (217 ft.-lbs.)
Idler gear.................20 N·m (15 ft.-lbs.)
Injection pump.............25 N·m (18 ft.-lbs.)
Injector retainer plate........10 N·m (7 ft.-lbs.)
Main bearing support........30 N·m (22 ft.-lbs.)
Oil pan....................25 N·m (18 ft.-lbs.)
Oil pump..................10 N·m (7 ft.-lbs.)
Oil pump gear..............49 N·m (36 ft.-lbs.)
Rocker arm cover...........20 N·m (15 ft.-lbs.)
Rocker arm shaft...........25 N·m (18 ft.-lbs.)

VALVE CLEARANCE ADJUSTMENT

To adjust valve clearance, first remove rocker arm cover. Rotate crankshaft until piston is at TDC on compression stroke. Clearance between valve stem end and rocker arm should be 0.10-0.15 mm (0.004-0.006 inch) for both intake and exhaust with engine cold.

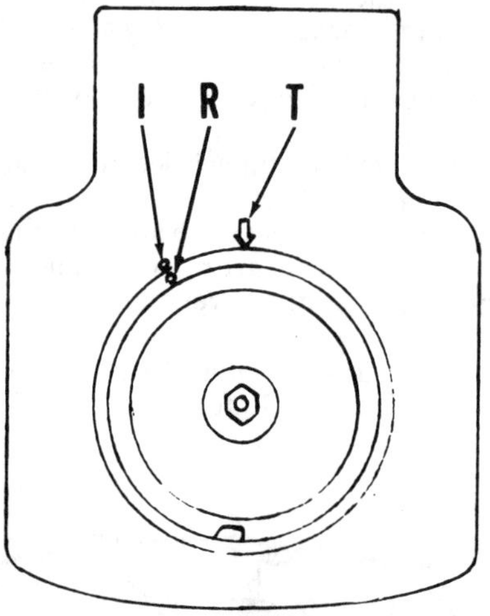

Fig. L4-6—Injection should occur when timing dot (R) on fan plate is aligned with injection timing dot (I) on fan shroud. Piston is at TDC when timing dot (R) and arrow (T) are aligned.

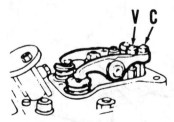

Fig. L4-7—Adjust exhaust valve clearance by turning adjustment screw (V). Adjust compression release by turning adjustment screw (C). Refer to text.

SERVICE MANUAL

Lombardini

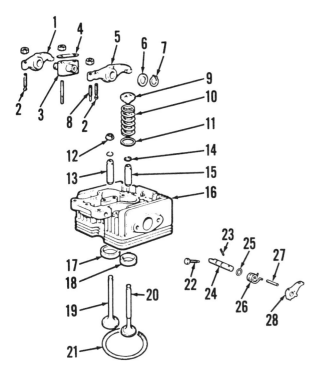

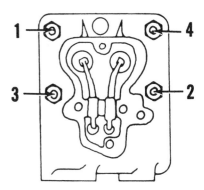

Fig. L4-8—Exploded view of cylinder head.

1. Intake rocker arm
2. Valve adjusting screws
3. Rocker stand
4. Plate
5. Exhaust rocker arm
6. Washer
7. Snap ring
8. Compression release adjusting screw
9. Spring retainer
10. Valve spring
11. Washer
12. Valve seal
13. Intake valve guide
14. Snap ring
15. Exhaust valve guide
16. Cylinder head
17. Intake valve seat
18. Exhaust valve seat
19. Intake valve
20. Exhaust valve
21. Head gasket
22. Locating screw
23. Pin
24. Compression release shaft
25. Washer
26. Spring
27. Pin
28. Compression release lever

Fig. L4-9—Tighten cylinder head nuts in steps using a crossing pattern as shown. Final torque should be 60 N·m (44 ft.-lbs.).

Note that there are two adjusting screws (Fig. L4-7) in exhaust rocker arm. Adjusting screw (V) nearer rocker arm shaft is used to adjust valve clearance while outer screw (C) adjusts compression release gap.

COMPRESSION RELEASE

A manual compression release is located in the cylinder head so the exhaust valve can be held open to aid starting. Rotating shaft (24–Fig. L4-8) forces the exhaust rocker arm to slightly open the exhaust valve.

The compression release is adjusted by turning outer adjusting screw (C–Fig. L4-7) in exhaust valve rocker arm. Adjust compression release gap AFTER adjusting exhaust valve clearance. With compression release lever in disengaged position, clearance between adjusting screw and shaft should be 0.9-1.1 mm (0.035-0.043 inch).

Diameter of compression release shaft (24–Fig. L4-8) is 9.37-10.00 mm (0.369-0.393 inch) while lobe height is 8.45-8.50 mm (0.333-0.334 inch).

CYLINDER HEAD AND VALVE SYSTEM

Cylinder head should not be removed when hot as head may deform. If cylinder head mounting surface is warped or pitted, up to 0.3 mm (0.012 inch) of material may be lapped from head to true surface.

Valve face angle is 45° for both valves. Renew valve if head margin is less than 0.4 mm (0.016 inch). Valve seat angle is 45° and recommended seat width is 1.4-1.6 mm (0.055-0.065 inch). Renewable valve seat inserts are used on intake and exhaust. Before installing new inserts, cylinder head should first be heated in an oven to 160°-180°C (320°-355°F).

With valves installed, measure distance valve head is recessed from cylinder head surface. Intake valve should be recessed 0.60-0.80 mm (0.024-0.031 inch) and exhaust valve (without compression release) should be 0.075-1.25 mm (0.030-0.049 inch). With compression release, exhaust valve recession should be 0.55-0.95 mm (0.022-0.037 inch). If recession is less than specified, regrind valve seat. If valve recession exceeds specified limit, renew valve seat insert.

Valve stem diameter is 7.98-8.00 mm (0.3142-0.3149 inch). Desired valve stem clearance in guide is 0.03-0.08 mm (0.001-0.003 inch) and maximum allowable clearance is 0.15 mm (0.006 inch). Valve guides are renewable and oversize guides are available. Outer diameter of oversize guide must be machined to provide a 0.05-0.06 mm (0.0020-0.0023 inch) interference fit in cylinder head bore. Cylinder head should be heated to 160°-180°C (320°-355°F) prior to installing new guides. A locating ring (14–Fig. L4-8) around top of guide determines distance guide is pressed into head. After installation, check guide inside diameter and ream as necessary to provide desired valve stem clearance.

Intake and exhaust valve springs are identical. Spring free length should be 52 mm (2.047 inches). Valve spring pressure should be 304-323 N (68-72 pounds) when compressed to a length of 25.8 mm (1.015 inches).

Desired clearance between rocker arms and shaft is 0.03-0.06 mm (0.001-0.002 inch). Maximum allowable clearance is 0.1 mm (0.004 inch). Tighten rocker arm shaft retaining screws to a torque of 25 N·m (18 ft.-lbs.).

Tighten cylinder head nuts gradually in steps using a crossing pattern as shown in Fig. L4-9. Final torque should be 60 N·m (44 ft.-lbs.).

INJECTOR

REMOVE AND REINSTALL. To remove injector, first clean dirt from injector, injection line, return line and cylinder head. Disconnect fuel return

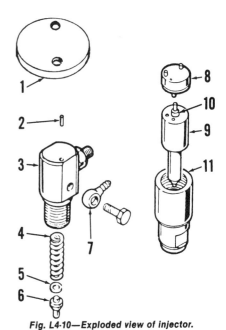

Fig. L4-10—Exploded view of injector.

1. Clamp plate
2. Dowel pin
3. Nozzle body
4. Spring
5. Shim
6. Pressure pin
7. Return line fitting
8. Spacer
9. Nozzle
10. Valve
11. Nozzle holder nut

211

Lombardini — SMALL DIESEL

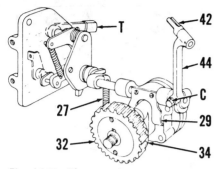

Fig. L4-11—Diagram of governor assembly. Refer to text for operation and adjustment.

- C. Clamp screw
- T. Throttle lever
- 27. Spring
- 29. Fork
- 32. Oil pump drive gear
- 34. Flyball assy.
- 42. Arm
- 44. Governor lever

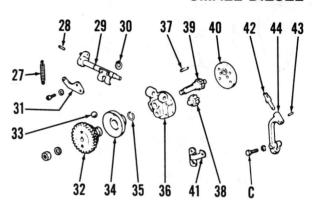

Fig. L4-12—Exploded view of governor and oil pump.
- 27. Governor spring
- 28. Pin
- 29. Shaft & fork
- 30. Snap ring
- 31. Spring plate
- 32. Oil pump drive gear
- 33. Governor balls (3)
- 34. Flyball housing
- 35. Snap ring
- 36. Oil pump body
- 37. Key
- 38. Driven gear
- 39. Drive gear & shaft
- 40. Oil pump plate
- 41. Bracket
- 42. Governor arm
- 43. Pin
- 44. Governor lever

line and injection line and immediately cap or plug all openings. Unscrew retainer plate (1 – Fig. L4-10) being careful not to lose dowel pin (2). Injector may now be carefully removed from cylinder head. Do not lose shims between injector and cylinder head.

Tighten injector retaining plate nuts to 10 N·m (7 ft.-lbs.). If accessible, measure protrusion of nozzle into combustion chamber. Nozzle tip should extend 3.5-4.0 m (0.138-0.157 inch) above adjacent combustion chamber surface. Adjust position of nozzle by installing 0.5 mm (0.020 inch) shims between injector and cylinder head.

TESTING. A suitable test stand is required to check injector operation. Only clean, approved testing oil should be used to test injector.

WARNING: Fuel leaves the injection nozzle with sufficient force to penetrate the skin. When testing, keep yourself clear of nozzle spray.

Connect injector to tester and operate test lever to purge air from nozzle and make sure nozzle valve is not stuck. When operating properly, injector nozzle will emit a buzzing sound and cut off quickly with no leakage at the tip.

Opening pressure should be 20595-21575 kPa (2985-3130 psi). Opening pressure is adjusted by varying number and thickness of shims (5 – Fig. L4-10). Operate tester lever slowly to maintain pressure at 2100 kPa (300 psi) below opening pressure and check for leakage past nozzle tip. If a drop appears within a 10 second period, nozzle valve must be overhauled or renewed. A slight wetness at tip is permissible. Operate tester lever briskly and check for even and well atomized spray pattern. If pattern is wet, ragged or intermittent, nozzle must be overhauled or renewed.

OVERHAUL. To disassemble nozzle, secure nozzle body (3 – Fig. L4-10) in a vise with nozzle tip pointing upward. Remove nozzle holder nut (11) and withdraw nozzle valve (9 and 10) and spacer (8). Invert nozzle body (3) and remove spring seat (6), shim (5) and spring (4).

Thoroughly clean all parts in a suitable solvent. Do not use steel wire brush or sharp metal tools to clean injector components. Clean inside the orifice end of nozzle valve with a wooden cleaning stick. The orifice spray holes may be cleaned by inserting a cleaning wire slightly smaller in diameter than the spray hole diameter of 0.28 mm (0.011 inch). Make certain nozzle needle (10) slides freely in bore of nozzle valve seat (9). If needle sticks, reclean or renew nozzle valve assembly.

When reassembling injector, be sure all parts are clean and wet with diesel fuel. Recheck injector operation outlined in TESTING paragraph.

INJECTION PUMP

Refer to Fig. L4-5 for an exploded view of injection pump. To remove pump, disconnect fuel lines and immediately plug all openings to prevent entry of dirt. Remove retaining screws and lift out pump assembly. Be sure to retain pump mounting shim gaskets for use in reassembly. Pump timing is adjusted by changing gasket thickness.

It is recommended that injection pump be tested and serviced only by a shop qualified in diesel fuel injection repair.

When reinstalling pump, assemble correct thickness of mounting shim gaskets and be sure to engage control rack pin with governor arm. Tighten pump mounting screws to a torque of 25 N·m (18 ft.-lbs.). If pump was overhauled or renewed, check and adjust injection timing as outlined in MAINTENANCE section.

To adjust control linkage, first move throttle lever (T – Fig. L4-11) to full speed position. Loosen clamp screw (C) and push governor lever (44) in until it stops thus moving injection pump control sleeve to maximum delivery. Retighten clamp screw (C).

GOVERNOR

All models are equipped with a flyball centrifugal type governor which is attached to the back of the oil pump drive gear (32 – Fig. L4-11 and L4-12). The oil pump drive gear (32) is driven by the crankshaft and rotates governor flyball assembly (34). Flyball housing (34) is interlocked with fork (29). As the flyballs move, the shaft attached to fork (29) is rotated thereby moving governor lever (44) and arm (42). Arm (42) mates with the pin on the injection pump control sleeve (13 – Fig. L4-5) to regulate fuel flow to cylinder. Throttle lever (T – Fig. L4-11) operates through governor spring (27) to control engine speed.

Torque control screw (C – Fig. L4-13 and L4-14) allows additional fuel usage under high torque load. The screw tip (T – Fig.

Fig. L4-13—Exploded view of throttle mechanism.
- C. Torque control
- H. High speed screw
- I. Idle speed screw
- S. Stop lever
- T. Throttle lever

SERVICE MANUAL

Lombardini

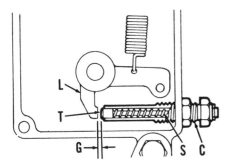

Fig. L4-14—Diagram of torque control screw. Refer to text for adjustment.

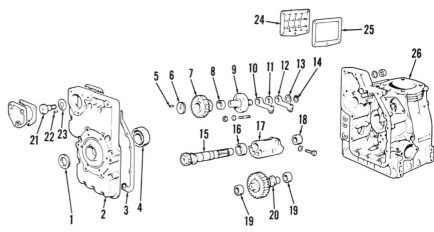

Fig. L4-15—Exploded view of camshaft, balancer and cam follower assemblies.

1. Seal	8. Bushing	15. Balancer shaft	21. Stud
2. Crankcase cover	9. Pivot shaft	16. Bearing	22. Pin
3. Gasket	10. Exhaust cam follower	17. Balancer	23. Washer
4. Bearing	11. Spacer	18. Bearing	24. Side cover
5. Pin	12. Intake cam follower	19. Bearing	25. Gasket
6. Washer	13. Washer	20. Camshaft	26. Crankcase
7. Idler gear	14. Snap ring		

L4-14) is backed by spring (S). To adjust torque control screw, run engine at high idle with no load and turn screw so gap (G) between tip (T) and lever (L) is 2.1 mm (0.083 inch) on Models 710 and 7LD600 or 2.3 mm (0.090 inch) on Models 720 and 7LD655.

Inspect governor components and renew any which are damaged or excessively worn. Mechanism must move freely for proper governor operation.

OIL PUMP

R&R AND OVERHAUL. The oil pump is mounted on the crankcase bulkhead and is accessible after removing crankcase cover. Refer to Fig. L4-12 for an exploded view of oil pump.

Clearance between gears and body walls must not exceed 0.15 mm (0.006 inch). Be sure pump shaft turns freely and axial end play does not exceed 0.15 mm (0.006 inch). Tighten pump mounting cap screws to a torque of 10 N·m (7 ft.-lbs.) and pump drive gear retaining nut to 49 N·m (37 ft.-lbs.).

CAMSHAFT, CAM FOLLOWERS AND PUSH RODS

R&R AND OVERHAUL. The camshaft rides in the crankcase bulkhead and crankcase cover (2–Fig. L4-15) and is accessible after removing cover. Cam followers (10 and 12) pivot on shaft (9) and transfer motion to push rods (56–Fig. L4-16) which pass through tube (57) to rocker arms. A lobe is ground on the camshaft to actuate the fuel injection pump.

On Models 710 and 720, intake and exhaust lobe height should be 29.36-29.56 mm (1.156-1.164 inches) and injection pump lobe height should be 42.6-42.8 mm (1.677-1.685 inches). On all other models, intake and exhaust lobe height should be 29.465-29.565 mm (1.160-1.164 inches) and injection pump lobe height should be 43.40-43.50 mm (1.708-1.712

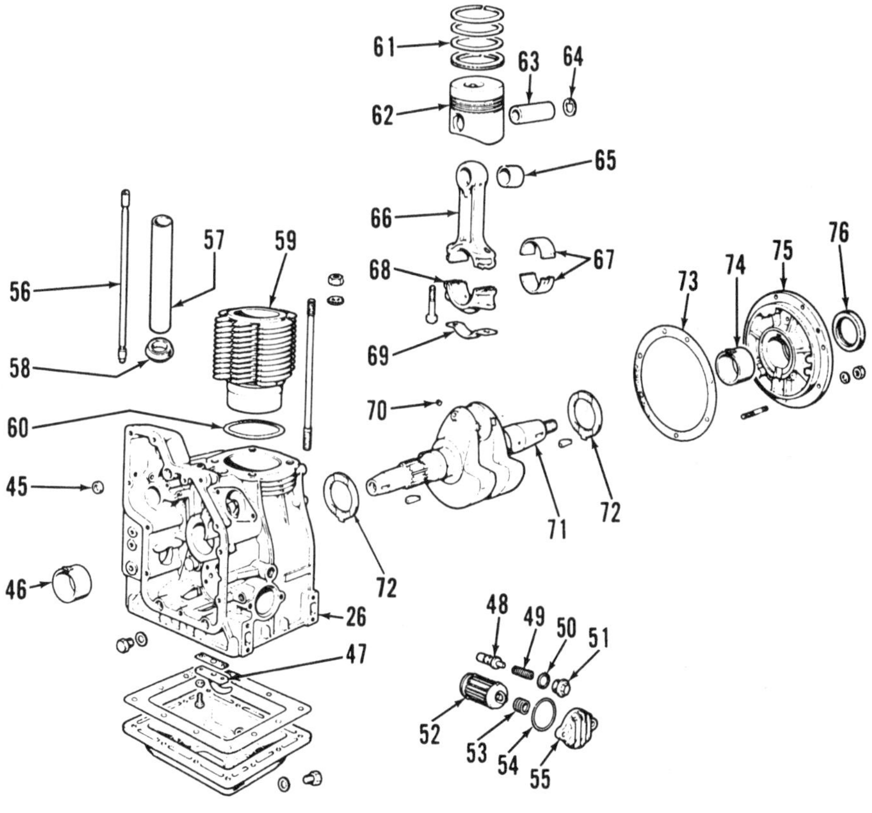

Fig. L4-16—Exploded view of crankshaft, piston and rod assemblies.

26. Crankcase	52. Oil filter	60. Gasket	68. Rod cap
45. Hollow dowel (3)	53. Spring	61. Piston rings	69. Lockplate
46. Main bearing	54. Gasket	62. Piston	70. Expansion plug
47. Oil pickup	55. Plate	63. Piston pin	71. Crankshaft
48. Oil pressure relief valve	56. Push rods	64. Snap rings	72. Thrust washers
49. Spring	57. Push rod tube	65. Bushing	73. Gasket
50. Washer	58. Grommets (2)	66. Connecting rod	74. Main bearing
51. Plug	59. Cylinder	67. Bearing	75. Bearing support
			76. Seal

inches). Bearing journal diameter should be 19.98-20.00 mm (0.7866-0.7874 inch) for both journals on all models.

Cam follower pivot shaft (9–Fig. L4-15) diameter should be 14.97-15.00 mm (0.589-0.590 inch). Maximum allowable clearance between shaft and cam followers is 0.10 mm (0.004 inch).

Illustrations Courtesy Lombardini U.S.A., Inc.

Lombardini

SMALL DIESEL

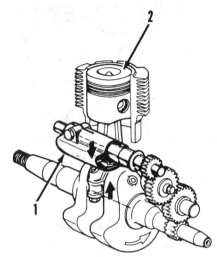

Fig. L4-17—When reinstalling balancer (1), note that balancer weight is down when piston (2) is at top dead center.

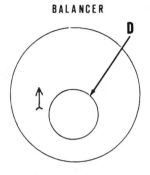

Fig. L4-18—Depression (D) in piston crown should be nearer injection pump and arrow should point towards balancer.

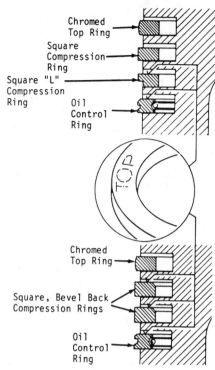

Fig. L4-19—Cross-sectional view of piston and rings showing correct installation.

Note that roller pin in cam follower is welded to one side of cam follower. Install cam followers so welds are on opposite sides of followers and not together. Install camshaft so timing marks on camshaft, crankshaft and idler gears are aligned. If timing marks are absent, proceed as follows: Position piston at top dead center, then install camshaft so intake cam follower is on opening side of cam lobe (exhaust lobe is adjacent to injection pump lobe). If necessary, remesh gears so cam followers are at same height. Mark gears for future reference.

The push rods are contained in a tube (57–Fig. L4-16) and must cross between cam followers and rocker arms. Push rod nearer cylinder connects intake cam follower and rocker arm while outer push rod connects exhaust cam follower and rocker arm. Push rod tube (57) is sealed by grommets (58) at both ends.

BALANCER

REMOVE AND REINSTALL. To remove balancer weight and shaft, remove crankcase cover (2–Fig. L4-15) and side cover (24). Loosen clamp screw and while holding weight, withdraw balancer shaft (15). Remove balancer weight (17) out side of engine.

Reverse disassembly procedure to install balancer. Note that with piston at top dead center (TDC), balancer weight is down and slightly inclined towards injection pump (Fig. L4-17).

PISTON AND ROD UNIT

Piston and connecting rod may be removed after removing cylinder head and oil pan.

When reinstalling piston and rod, note that depression (D–Fig. L4-18) in piston crown is offset to one side of piston. Install piston so depression side of piston is nearer injection pump side of engine. Some pistons also have an arrow embossed on top of piston. When properly installed, arrow on piston crown will point towards the balancer. Tighten connecting rod cap screws to 49 N·m (36 ft.-lbs.).

Rotate crankshaft to position piston at top dead center. Top of piston should be level with top of cylinder. To adjust, install shim gaskets (60–Fig. L4-16) between cylinder and crankcase as needed.

PISTON, PIN, RINGS AND CYLINDER

The piston is equipped with three compression rings and one oil control ring. On Models 710 and 720, ring end gap should be 0.35-0.55 mm (0.014-0.021 inch) for all compression rings and 0.25-0.40 mm (0.010-0.015 inch) for oil control ring. On all other models, ring end gap should be 0.40-0.65 mm (0.015-0.025 inch) for all compression rings and 0.30-0.60 mm (0.012-0.024 inch) for oil control ring.

To check piston ring groove wear, install new rings onto piston, then measure side clearance between top of ring and top of groove using a feeler gage. Renew piston if clearance exceeds 0.15 mm (0.006 inch) for top compression ring and 0.01 mm (0.004 inch) for all other rings.

Standard piston diameter, measured at bottom of skirt perpendicular to piston pin, is 89.83-89.86 mm (3.5366-3.5378 inches) on Models 710 and 7LD600, 94.83-94.85 mm (3.7335-3.7342 inches) on Models 720 and 7LD655 or 99.80-99.81 mm (3.9291-3.9295 inches) on Model 7LD740. If wear exceeds 0.05 mm (0.002 inch), renew piston. Oversize pistons and rings as well as standard size are available.

Cylinder bore standard inside diameter is 90.00-90.02 mm (3.5433-3.5441 inches) on Models 710 and 7LD600, 95.00-95.02 mm (3.7401-3.7409 inches) on Models 720 and 7LD655 or 100.00-100.02 mm (3.9370-3.9378 inches) on Model 7LD740. If cylinder bore wear exceeds 0.1 mm (0.004 inch), rebore cylinder to appropriate oversize.

Piston pin diameter is 27.995-28.005 mm (1.1022-1.1025 inches). Clearance between piston pin and connecting rod bushing should be 0.015-0.030 mm (0.0006-0.0012 inch) and maximum allowable clearance is 0.07 mm (0.003 inch).

When installing rings onto piston, stagger ring end gaps 180° apart. Be sure no end gaps are aligned with piston pin bore. Fig. L4-19 shows correct installation of rings on piston.

CONNECTING ROD

The connecting rod small end is fitted with a renewable bushing (65–Fig. L4-16). Clearance between piston pin and rod bushing should be 0.015-0.025

SERVICE MANUAL

Lombardini

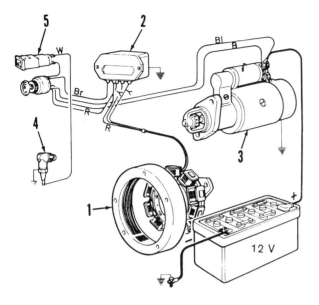

Fig. L4-20—Wiring schematic for models equipped with internal alternator and electric starter.
1. Alternator
2. Regulator
3. Starter motor
4. Oil pressure switch
5. Oil pressure light

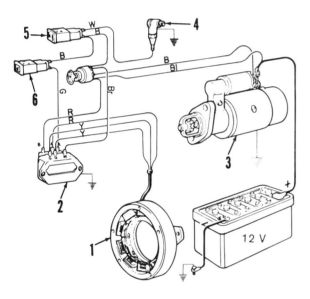

Fig. L4-21—Wiring schematic for models using an alternator warning light (6). Refer to Fig. L4-20 legend.

CRANKSHAFT AND CRANKCASE

The crankshaft is supported by sleeve bearings (46 and 74—Fig. L4-16) in the crankcase bulkhead and bearing support (75) and by bearing (4—Fig. L4-15) in the crankcase cover (2).

The crankshaft has drilled oil passages to distribute oil to the bearings. Expansion plugs, located adjacent to crankpin, may be removed to clean oil passages, however, be sure new plugs are installed securely.

On Models 710 and 720, desired bearing clearance for center and flywheel end main bearings is 0.03-0.07 mm (0.001-0.003 inch). On all other models, desired bearing clearance is 0.05-0.07 mm (0.002-0.003 inch). Standard diameter for center and flywheel end main bearing journals is 44.99-45.00 mm (1.7712-1.7716 inches) on all models.

Center and flywheel end main bearings are available in standard size and undersizes. When installing new bearings, be sure lubrication oil holes are aligned. Special bearing installing tool 7271.3595.047 is available from manufacturer.

Renew front oil seal (1—Fig. L4-15) and rear oil seal (76—Fig. L4-16). Be sure to lubricate seal lips prior to reassembly.

Thrust washer (72—Fig. L4-16) thickness should be 2.31-2.36 mm (0.091-0.093 inch). Crankshaft end play is adjusted by varying thickness of support gasket (73). Desired end play is 0.10-0.30 mm (0.004-0.012 inch). Tighten support retaining nuts to 30 N·m (22 ft.-lbs.) and crankcase cover cap screws to 25 N·m (18 ft.-lbs.). Tighten flywheel retaining nut to 295 N·m (217 ft.-lbs.).

ELECTRICAL SYSTEM

Some models are equipped with an alternator, regulator and starting motor. Refer to schematic drawing shown in Fig. L4-20 or L4-21. The alternator stator is attached to the crankcase cover while a ring of magnets is carried by the flywheel.

mm (0.0006-0.001 inch) with a maximum allowable clearance of 0.07 mm (0.003 inch).

An insert type bearing is used in connecting rod big end. Desired rod bearing clearance is 0.03-0.07 mm (0.001-0.003 inch). Maximum allowable clearance is 0.10 mm (0.004 inch). Bearing inserts are available in standard and undersizes. Crankshaft standard crankpin diameter is 49.99-50.00 mm (1.9681-1.9685 inches).

LOMBARDINI

Model	No. Cyls.	Bore	Stroke	Displ.
8LD600-2, 904	2	90 mm (3.543 in.)	94 mm (3.700 in.)	1196 cc (80 cu. in.)
8LD665-2, 914	2	95 mm (3.740 in.)	94 mm (3.700 in.)	1332 cc (81.3 cu. in.)
8LD665-2/L, L20	2	95 mm (3.740 in.)	94 mm (3.700 in.)	1332 cc (81.3 cu. in.)
8LD740-2	2	100 mm (3.937 in.)	94 mm (3.700 in.)	1476 cc (90 cu. in.)

Engines covered in this section are four-stroke, air-cooled deisel engines. The crankcase and cylinder head are aluminum and the cylinders are cast iron. Crankshaft rotation is counterclockwise at pto end. Number one cylinder is nearer flywheel.

MAINTENANCE

LUBRICATION

Recommended engine oil is SAE 10W for temperatures below 0°C (32°F), SAE 20W for temperatures between 0°C (32°F) and 20°C (68°F), and SAE 40 for temperatures above 20°C (68°F). API classification for oil should be CD. Oil sump capacity is 2.5 liters (2.6 quarts). Manufacturer recommends renewing oil after every 300 hours of operation.

A renewable oil filter is mounted on side of engine crankcase. Manufacturer recommends renewing filter after every 300 hours of operation.

All models are equipped with a pressurized oil system. Refer to Fig. L5-1 for a diagram of the oil circuit.

ENGINE SPEED ADJUSTMENT

Low idle speed is adjusted by turning idle speed screw (I – Fig. L5-2). Recommended idle speed is 900-950 rpm on Models 904, 914 and L20; 1000-1200 rpm on Models 8LD600-2, 8LD665-2 and 8LD740-2; or 1100-1400 rpm on Model 8LD665-2/L. Engine should be at normal operating temperature when adjusting engine speeds.

Maximum governed speed is adjusted by turning high speed screw (H). Set high idle (no-load) speed 150 rpm above specified governed full-load speed to compensate for governor droop during

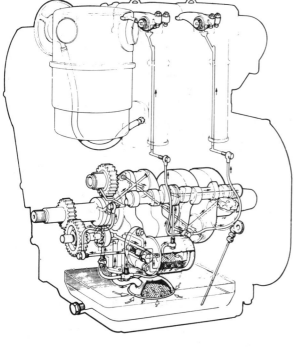

Fig. L5-1—Drawing of lubrication system.

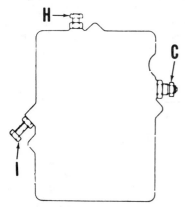

Fig. L5-2—Drawing of governor and throttle control linkage showing location of low idle speed adjusting screw (I), high idle speed adjusting screw (H) and torque control screw (C).

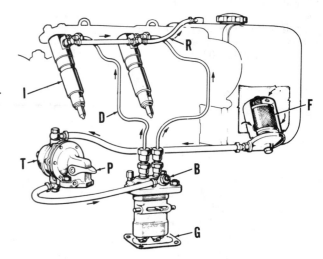

Fig. L5-3—Fuel circuit diagram.

- B. Pump bleed screw
- D. High pressure delivery line
- F. Fuel filter
- G. Shim gasket
- I. Injector
- P. Primer lever
- R. Fuel return line
- T. Fuel transfer pump

SERVICE MANUAL

Lombardini

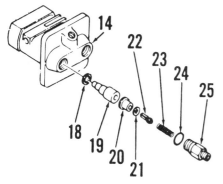

Fig. L5-4—Partial exploded view of fuel injection pump.

14. Pump body
18. Packing
19. Barrel
20. Delivery valve seat
21. Gasket
22. Delivery valve
23. Spring
24. "O" ring
25. Delivery valve holder

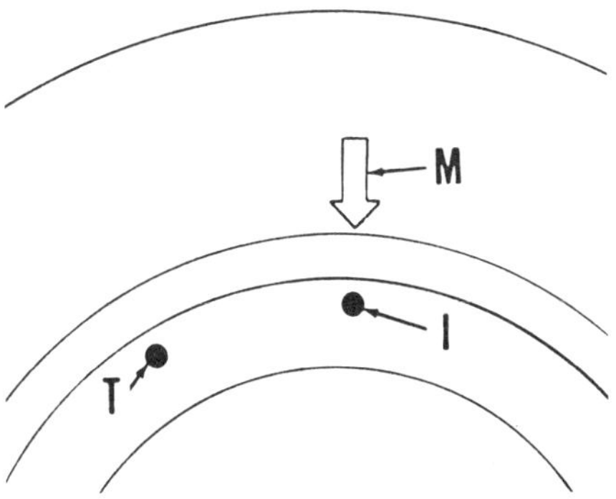

Fig. L5-6—Drawing of crankshaft pulley timing marks used on some models.
I. Injection
M. Reference mark
T. Top dead center

operation. Full-load speed is normally 3000 rpm on all models except 8LD655-2/L and L20, which are limited to 2200 rpm. Governed speed may vary depending upon application.

FUEL SYSTEM

FUEL FILTER. A renewable fuel filter (F – Fig. L5-3) is located in the fuel tank. Manufacturer recommends renewing filter after every 300 hours of operation. However, if engine indicates signs of fuel starvation (loss of power or surging), renew filter regardless of hours of operation. After renewing filter, air must be bled from system as outlined in the following paragraph.

BLEED FUEL SYSTEM. To bleed air from fuel system, first loosen bleed screw (B – Fig. L5-3) on fuel injection pump. Manually operate fuel transfer pump primer lever (P) until air-free fuel flows from bleed screw connection, then retighten bleed screw.

If engine fails to start after performing initial bleeding operation, loosen the high pressure injection lines (D) at the injectors. Rotate engine crankshaft to operate injection pump until fuel is discharged at injector line fittings.

Retighten injection lines, then start engine.

INJECTION PUMP TIMING

Injection pump timing is adjusted by changing thickness of shim gaskets (G – Fig. L5-3) between pump body and crankcase mounting surface. Injection should occur at 27°-28° BTDC on Models 904, 914 and L20 or 24°-26° BTDC on all other models.

To check injection pump timing, first disconnect injection line of number 1 cylinder from injection pump delivery valve holder. Unscrew delivery valve holder (25 – Fig. L5-4) and remove spring (23) and delivery valve (22). Do not remove delivery valve seat (20). Reinstall delivery valve holder. Move throttle control to full speed position. Operate fuel transfer pump primer lever while rotating crankshaft in normal direction so number 1 piston is on compression stroke. Note that fuel will flow out of delivery valve holder during this procedure. Stop rotating the crankshaft at the moment fuel ceases to flow which is beginning of injection. At this point, timing marks (I and M – Fig. L5-5 or L5-6) should be aligned.

To advance injection timing, remove shim gaskets (G – Fig. L5-3). Increase shim gasket thickness to retard beginning of injection. Tighten injection pump retaining screws to 25 N·m (18 ft.-lbs.). Reinstall delivery valve and tighten delivery valve holder to 34 N·m (25 ft.-lbs.). Bleed air from system as previously outlined.

FAN BELT TENSION

All models are equipped with a belt-driven cooling fan. Belt tension is adjusted by varying the number of shims (2 – Fig. L5-7) between fan pulley halves.

COOLING FAN

All models are equipped with an axial cooling fan to force air past the cylinders. The fan housing (4 – Fig. L5-7) is mounted on the crankcase. Alternator (7) is contained in the fan housing with the alternator rotor mounted on shaft (6).

Overhaul is evident after inspection of unit and referral to Fig. L5-7.

VALVE CLEARANCE ADJUSTMENT

To adjust valve clearance, first remove rocker arm covers and rotate crankshaft until number 1 piston is at TDC on compression stroke. Clearance between rocker arm and valve stem end should be 0.15 mm (0.006 inch) for both intake and exhaust valves with engine cold. Note that there are two adjusting screws (Fig. L5-8) in exhaust rocker arm. Adjusting screw (V) nearer rocker arm shaft is used to adjust valve clearance while outer screw (C) adjusts compression release.

After adjusting valve clearance on number 1 cylinder, rotate crankshaft

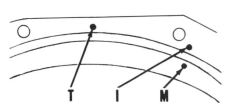

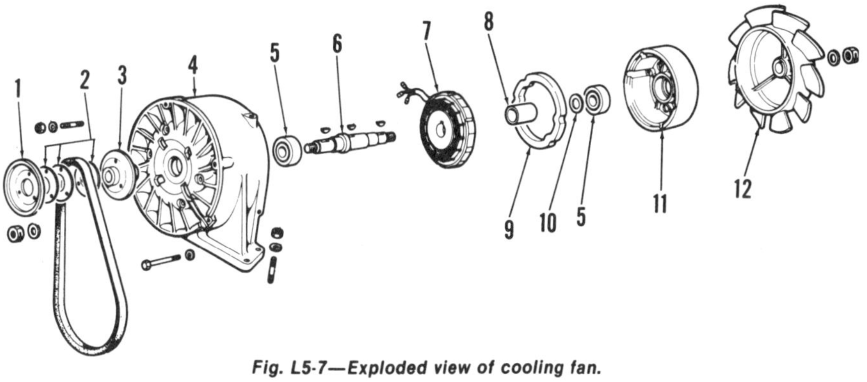

Fig. L5-5—Drawing of flywheel timing marks used on some models.
I. Injection
M. Reference mark
T. Top dead center

Fig. L5-7—Exploded view of cooling fan.

1. Pulley half
2. Shims
3. Pulley hub
4. Fan housing
5. Bearing
6. Shaft
7. Alternator
8. Spacer
9. Spacer
10. Washer
11. Alternator housing
12. Fan

Illustrations Courtesy Lombardini U.S.A., Inc.

Lombardini

SMALL DIESEL

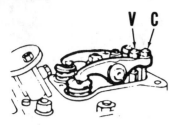

Fig. L5-8—Exhaust valve rocker arm is equipped with two adjusting screws. Inner screw (V) is used to adjust valve clearance and outer screw (C) is used to adjust compression release.

one complete revolution and adjust valves on number 2 cylinder.

COMPRESSION RELEASE

A manual compression release is located on each cylinder head so the exhaust valve can be held open to aid starting. Rotating shaft (28–Fig. L5-9) forces the exhaust rocker arm to slightly open the exhaust valve.

REPAIRS

TIGHTENING TORQUES

Refer to the following table for tightening torques.

Camshaft gear	195 N·m (145 ft.-lbs.)
Camshaft retainer	25 N·m (18 ft.-lbs.)
Connecting rod	49 N·m (36 ft.-lbs.)
Crankshaft pulley	440 N·m (325 ft.-lbs.)
Cylinder head	60 N·m (44 ft.lbs.)
Flywheel	295 N·m (220 ft.-lbs.)
Injection pump	25 N·m (18 ft.-lbs.)
Injector retainer	10 N·m (8 ft.-lbs.)
Main bearing support:	
Center	40 N·m (30 ft.-lbs.)
End	25 N·m (18 ft.-lbs.)
Oil pan	25 N·m (18 ft.-lbs.)
Oil pump	25 N·m (18 ft.-lbs.)
Oil pump gear	44 N·M (33 ft.-lbs.)
Rocker arm shaft	25 N·m (18 ft.-lbs.)
Timing gear cover	25 N·m (18 ft.-lbs.)

The compression release is adjusted by turning outer adjusting screw (C–Fig. L5-8) in exhaust valve rocker arm. Adjust compression release gap AFTER adjusting exhaust valve clearance. With compression release

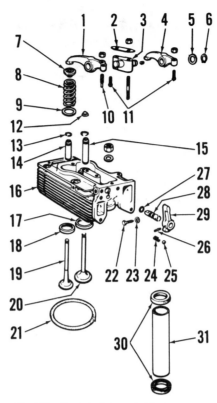

Fig. L5-9—Exploded view of cylinder head.

1. Exhaust rocker arm
2. Lockplate
3. Rocker arm stand
4. Intake rocker arm
5. Washer
6. Snap ring
7. Spring retainer
8. Valve spring
9. Spring seat
10. Compression release adjusting screw
11. Valve adjusting screws
12. Oil seal
13. Snap ring
14. Exhaust valve guide
15. Intake valve guide
16. Cylinder head
17. Intake valve seat
18. Exhaust valve seat
19. Exhaust valve
20. Intake valve
21. Head gasket
22. Locating screw
23. Washer
24. Spring
25. Detent ball
26. Pin
27. "O" ring
28. Compression release shaft
29. Compression release lever
30. Seals
31. Push rod tube

lever in disengaged position, clearance between adjusting screw and shaft should be 0.9-1.1 mm (0.035-0.043 inch).

CYLINDER HEAD AND VALVE SYSTEM

Cylinder head should not be removed when engine is hot as head may warp. If cylinder head mounting surface is warped or pitted, up to 0.3 mm (0.012 inch) of material may be lapped from head to true the surface.

Valve face angle is 45° for intake and exhaust. Renew valve if head margin is less than 0.5 mm (0.020 inch). Valve seat angle is also 45° and recommended seat width is 1.4-1.6 mm (0.055-0.065 inch). Renewable valve seat inserts are used on intake and exhaust. Before installing new inserts, cylinder head should first be heated to 160°-180°C (320°-355°F).

With valves installed, measure distance valve head is recessed from cylinder head surface. Intake valve should be recessed 0.60-0.80 mm (0.024-0.031

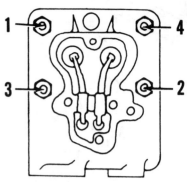

Fig. L5-10—Use a crossing pattern when tightening cylinder head nuts. Tighten nuts evenly in steps to a final torque of 60 N·m (44 ft.-lbs.).

inch) and exhaust valve (without compression release) should be recessed 0.75-1.25 mm (0.030-0.049 inch). With compression release, exhaust valve recession should be 0.55-0.95 mm (0.022-0.037 inch). If recession is less than specified, regrind valve seat. If recession exceeds specified limit, renew valve seat insert.

Valve stem diameter is 7.98-8.00 mm (0.314-0.315 inch). Desired valve stem clearance in guide is 0.03-0.08 mm (0.001-0.003 inch) and maximum allowable clearance is 0.15 mm (0.006 inch). Valve guides are renewable and oversize guides are available. Outer diameter of oversize guide must be machined to provide a 0.05-0.06 mm (0.0020-0.0023 inch) interference fit in cylinder head bore. Cylinder head should be heated to 160°-180°C (320°-355°F) prior to installing new guides. Use suitable installing tool when installing new guides to prevent damage to guide. After installation, check inside diameter of guide and ream, if necessary, to provide specified clearance for valve stem. A valve stem seal is used on intake valve.

Intake and exhaust valve springs are identical. Spring free length should be 52 mm (2.047 inches). Valve spring pressure should be 304-323 N (68-72 pounds) when compressed to a length of 25.8 mm (1.015 inches).

Desired clearance between rocker arms and pivot shaft (3–Fig. L5-9) is 0.03-0.06 mm (0.001-0.002 inch). Maximum allowable clearance is 0.1 mm (0.004 inch). Tighten rocker arm shaft retaining nuts to 25 N·m (18 ft.-lbs.).

When reinstalling cylinder head, be sure oil tubes to head are properly connected as shown in Fig. L5-1. Renew push rod tube seals if necessary. Before tightening cylinder head nuts, install exhaust and intake manifolds to correctly position heads. Tighten manifold nuts to 25 N·m (18 ft.-lbs.). Tighten cylinder head nuts gradually using a crossing pattern as shown in Fig. L5-10. Final torque should be 60 N·m (44 ft.-lbs.).

SERVICE MANUAL

Lombardini

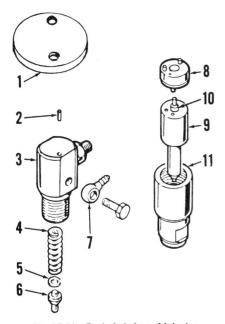

Fig. L5-11—Exploded view of injector.

1. Clamp plate
2. Dowel pin
3. Nozzle body
4. Spring
5. Shim
6. Spring seat
7. Return line fitting
8. Spacer
9. Nozzle
10. Valve
11. Nozzle holder nut

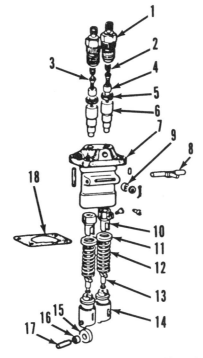

Fig. L5-12—Partially exploded view of fuel injection pump assembly.

1. Delivery valve holder
2. Spring
3. Delivery valve
4. Delivery valve seat
5. Seal
6. Barrel
7. Pump body
8. Control rack
9. Roller
10. Control sleeve & pinion
11. Spring seat
12. Spring
13. Plunger
14. Guide
15. Outer roller
16. Inner roller
17. Pin
18. Shim gasket

INJECTOR

REMOVE AND REINSTALL. To remove injector, first clean dirt from injector, injection line, return line and cylinder head. Disconnect fuel return line and injection line and immediately cap or plug all openings. Unscrew retainer plate (1–Fig. L5-11) being careful not to lose dowel pin (2). Injector may now be carefully removed from cylinder head. Do not lose shims between injector and cylinder head.

Tighten injector retainer plate nuts to 10 N·m (8 ft.-lbs.). If accessible, measure protrusion of nozzle into combustion chamber. Nozzle tip should extend 4.0-4.5 mm (0.157-0.177 inch) above adjacent combustion chamber surface. Adjust position of nozzle by installing 0.5 mm (0.020 inch) shims between injector and cylinder head.

TESTING. A suitable test stand is required to check injector operation. Only clean, approved testing oil should be used to test injector.

WARNING: Fuel leaves the injector nozzle with sufficient force to penetrate the skin. When testing, keep yourself clear of nozzle spray.

Connect the injector to tester and operate tester to purge air from nozzle and to make sure nozzle valve is not stuck. When operating properly, injector nozzle will emit a buzzing sound and cut off quickly with no leakage at the tip.

Opening pressure should be 20595-21575 kPa (2985-3130 psi) with a used nozzle valve spring (4–Fig. L5-11). If a new spring is used, opening pressure should be set at 21575-22555 kPa (3130-3270 psi) to compensate for the spring taking a set during operation. Opening pressure is adjusted by varying the number and thickness of shims (5).

Operate tester lever slowly to maintain pressure at 1470 kPa (215 psi) below opening pressure and check for leakage past nozzle valve. If a drop forms at nozzle tip within a 10 second period, nozzle valve must be overhauled or renewed. A slight wetness at tip is permissible.

Operate tester lever briskly and check for an even and well atomized spray pattern. If pattern is wet, ragged or intermittent, nozzle must be overhauled or renewed.

OVERHAUL. To disassemble injector, secure nozzle body (3–Fig. L5-11) in a vise with nozzle tip pointing upward. Remove nozzle holder nut (11) and withdraw nozzle valve (9 and 10) and spacer (8). Invert nozzle body and remove spring seat (6), shim (5) and spring (4).

Thoroughly clean all parts in a suitable solvent. Do not use a steel wire brush or sharp metal tools to clean injector components. Clean inside the orifice end of nozzle valve with a wooden cleaning stick. The orifice spray holes may be cleaned by inserting a cleaning wire slightly smaller in diameter than the spray hole diameter of 0.28 mm (0.011 inch). Make certain the nozzle needle (10) slides freely in bore of nozzle valve seat (9). If needle sticks, reclean or renew nozzle valve assembly.

When reassembling injector, be sure all parts are clean and wet with diesel fuel. Recheck injector operation as outlined in TESTING paragraph.

INJECTION PUMP

To remove injection pump, disconnect fuel lines and immediately plug all openings to prevent entry of dirt. Remove retaining screws and lift out pump assembly. Be sure to retain pump mounting shim gaskets for use in reassembly. Pump timing is adjusted by changing gasket thickness.

An exploded view of fuel injection pump is shown in Fig. L5-12. It is recommended that injection pump be tested and serviced only by a shop qualified in diesel fuel injection repair.

When reinstalling injection pump, assemble correct thickness of mounting shim gaskets and be sure to engage control rack pin with governor fork. Tighten pump mounting screws to a torque of 25 N·m (18 ft.-lbs.). If pump was overhauled or renewed, check and adjust injection timing as outlined in MAINTENANCE section.

PISTON AND ROD UNITS

REMOVE AND REINSTALL. Piston and connecting rod may be removed after removing cylinder head, oil pan and oil pickup.

When installing piston and rod, note that depression (D–Fig. L5-12A) in piston crown is closer to one side of piston. In-

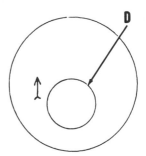

Fig. L5-12A—Arrow on top of piston crown should point towards intake side of engine and depression (D) should be aligned with injector tip.

Lombardini

SMALL DIESEL

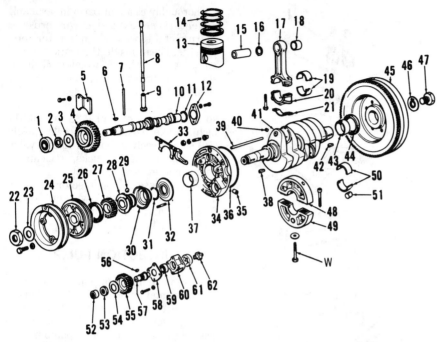

Fig. L5-13—Exploded view of crankshaft, camshaft and oil pump assemblies.

1. Bearing
2. Nut
3. Washer
4. Camshaft gear
5. Retainer
6. Key
7. Fuel pump push rod
8. Push rod
9. Tappet
10. Camshaft
11. "O" ring
12. Cover plate
13. Piston
14. Piston rings
15. Piston pin
16. Snap ring
17. Connecting rod
18. Bushing
19. Rod bearing
20. Rod cap
21. Lockplate
22. Nut
23. Washer
24. Rope pulley
25. Fan pulley
26. Seal
27. Gear
28. Governor hub
29. Balls
30. Governor cup
31. Snap ring
32. Seal
33. Governor arm & shaft
34. End bearing support
35. Bushing
36. "O" ring
37. Main bearing
38. Key
39. Governor rod
40. Plug
41. Crankshaft
42. Key
43. Main bearing
44. Seal
45. Flywheel
46. Lockwasher
47. Cap screw
48. Upper center bearing support
49. Lower center bearing support
50. Center main bearing
51. Round nut
52. Bearing
53. Nut
54. Washer
55. Gear
56. Key
57. Drive shaft
58. Bearing retainer
59. Bearing
60. Oil pump cover
61. Outer rotor
62. Inner rotor
W. Support screw

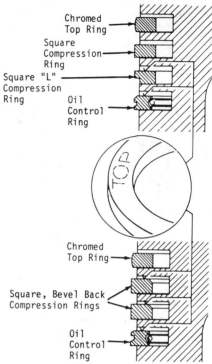

Fig. L5-16—Cross-sectional view of piston and rings showing correct installation of different types of rings.

stall piston so depression side of piston is nearer injector. Some pistons also have an arrow embossed on piston. Properly installed, arrow on piston crown will point toward injection pump. Match alignment marks on rod and cap and tighten rod screws to 49 N·m (36 ft.-lbs.).

Refer to CYLINDER section and measure piston height in cylinder.

PISTON, PIN AND RINGS

The piston is equipped with three compression rings and one oil control ring. On Models 904, 914 and L20, recommended ring end gap is 0.35-0.55 mm (0.014-0.021 inch) for all compression rings and 0.25-0.40 mm (0.010-0.016 inch) for oil control ring. On all other models, ring end gap should be 0.40-0.65 mm (0.016-0.025 inch) for all compression rings and 0.30-0.60 mm (0.012-0.023 inch) for oil control ring.

To check piston ring groove wear, install new rings onto piston, then measure side clearance between top of ring and top of groove using a feeler gage. Renew piston if clearance exceeds 0.15 mm (0.006 inch) for top compression ring and 0.10 mm (0.004 inch) for all other rings.

Standard piston diameter, measured at bottom of skirt perpendicular to piston pin, is 89.85-89.86 mm (3.5374-3.5378 inches) on Model 904; 89.83-89.86 mm (3.5366-3.5378 inches) on Model 8LD600-2; 94.85-94.86 mm (3.7342-3.7346 inches) on Model 914 and

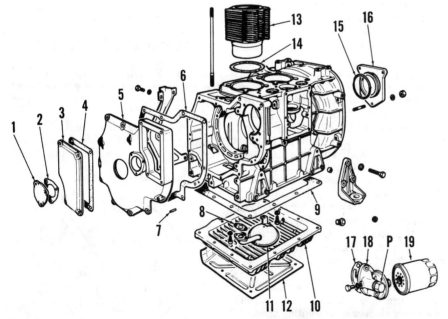

Fig. L5-14—Exploded view of crankcase assembly. Remove plug (P) for access to oil pressure relief valve.

1. Cover
2. Gasket
3. Cover
4. Gasket
5. Timing gear cover
6. Gasket
7. Pin
8. Gasket
9. Gasket
10. Oil pan
11. Oil pickup
12. Air shroud
13. Cylinder
14. Shim gasket
15. "O" ring
16. Cover
17. Gasket
18. Filter adapter
19. Oil filter

Illustrations Courtesy Lombardini U.S.A., Inc.

SERVICE MANUAL

Lombardini

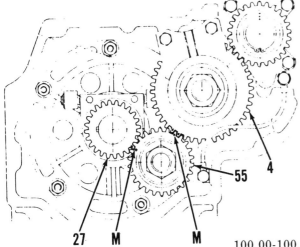

Fig. L5-17—View showing location of timing marks (M) on camshaft gear (4), crankshaft gear (27) and oil pump gear (55).

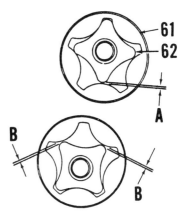

Fig. L5-18—Refer to text for clearances (A and B) between oil pump outer rotor (61) and inner rotor (62).

L20; 94.83-94.85 mm (3.7335-3.7346 inches) on Models 8LD655-2 and 8LD655-2/L and 99.80-99.81 mm (3.9291-3.9295 inches) on Model 8LD740-2. If wear exceeds 0.05 mm (0.002 inch), renew piston. Difference in piston weights must not exceed 6 grams (0.2 ounce).

Piston pin diameter is 27.995-28.005 mm (1.1022-1.1025 inches). Desired clearance between piston pin and connecting rod bushing is 0.015-0.035 mm (0.0006-0.0014 inch). Maximum allowable clearance is 0.07 mm (0.003 inch). Pin should be a thumb press fit piston.

When installing rings onto piston, stagger ring end gaps 180° apart. End gaps should not be in line with piston pin bore. Fig. L5-16 shows correct installation of rings on piston.

CONNECTING ROD

The connecting rod small end is fitted with a renewable bushing. Clearance between piston pin and rod bushing should be 0.015-0.035 mm (0.0006-0.0014 inch) with a maximum allowable clearance of 0.07 mm (0.003 inch).

A precision, insert type bearing is used in crankpin end of connecting rod. Desired rod bearing clearance is 0.03-0.07 mm (0.001-0.003 inch) with a maximum allowable clearance of 0.10 mm (0.004 inch). Bearings are available in standard and undersizes.

Difference in connecting rod weights must not exceed 10 grams (0.35 ounce).

CYLINDERS

All engines are equipped with removable cylinders. Standard cylinder diameter is 90.00-90.02 mm (3.5433-3.5441 inches) for Models 904 and 8LD600-2; 95.00-95.02 mm (3.7401-3.7409 inches) for Models 914, 8LD655-2, 8LD655-2/L and L20; and 100.00-100.02 mm (3.9370-3.9378 inches) for Model 8LD740-2. If cylinder wear exceeds 0.1 mm (0.004 inch), rebore cylinder to the next appropriate oversize.

Cylinder height is adjusted using shim gaskets (14 – Fig. L5-14) between bottom of cylinder and crankcase mounting surface. With piston at top dead center, top of piston must be even with top edge of cylinder. Shim gaskets are available in thicknesses of 0.1 and 0.3 mm (0.004 and 0.012 inch). The 0.8 mm (0.031 inch) thick cylinder head gasket ensures adequate clearance between piston crown and cylinder head.

TIMING GEARS

REMOVE AND REINSTALL. Remove belt guard and fan belt. Unscrew nut, then using a suitable puller, pull pulley off crankshaft. Remove timing gear cover.

Use a suitable puller to remove gears. Note that retainer (5 – Fig. L5-13) must be removed before pulling off camshaft gear (4). When reinstalling camshaft gear, place gear on shaft so retainer groove is out. Align timing marks (M – Fig. L5-17) on models so equipped, when installing gears. If timing marks are not present on timing gears, proceed as follows: The cylinder head, push rod tube and push rods for number 1 cylinder must be removed. Position number 1 piston at top dead center. If not previously removed, detach camshaft gear from camshaft. Rotate camshaft so number 1 cylinder intake valve cam follower is opening (rising) and exhaust valve cam follower is closing (going down), then stop rotation when cam followers are same height from top surface of crankcase. Without disturbing camshaft position, install camshaft gear. Mark crankshaft, oil pump and camshaft gears for future reference. Reinstall cylinder head.

Tighten camshaft gear nut to 195 N·m (145 ft.-lbs.) and oil pump gear nut to 44 N·m (33 ft.-lbs.). Renew crankshaft oil seal, located in timing gear cover, and lubricate seal lip with grease prior to reinstalling timing gear cover. Tighten timing gear cover cap screws to 25 N·m (18 ft.-lbs.).

OIL PUMP

R&R AND OVERHAUL. The oil pump is mounted on end main bearing support (34 – Fig. L5-13). To remove pump, remove timing gear cover and oil pump gear (55). Unscrew pump housing screws and disassemble pump.

Refer to Fig. L5-18 and measure clearance between inner and outer rotors (61 and 62). Clearance (A) should be 0.01-0.06 mm (0.0004-0.0023 inch) with a maximum allowable clearance of 0.10 mm (0.004 inch), and clearance (B) should be 0.02-0.10 mm (0.0008-0.004 inch) with a maximum allowable clearance of 0.20 mm (0.008 inch). Width of inner and outer rotors should be 14.95-14.97 mm (0.5886-0.5894 inch) and difference in rotor widths must not be greater than 0.02 mm (0.0008 inch). Outer rotor outer diameter is 40.54-40.57 mm (1.5960-1.5972 inches). Pump housing bore is 40.60-40.63 mm (1.5984-1.5996 inches). Clearance between outer rotor and pump housing bore should be 0.03-0.09 mm (0.001-0.003 inch) with a maximum allowable clearance of 0.13 mm (0.005 inch). With pump cover (60 – Fig. L5-13) and bearing retainer (58) installed and retaining screws torqued, inner rotor end play should be 0.03-0.11 mm (0.01-0.004 inch). Inspect bearing (52) in timing gear cover and renew if damaged.

To reassemble oil pump, reverse disassembly procedure. Install outer rotor (61) with rounded outer edge towards pump housing. Apply Loctite to outer surface of bearing (59) outer race. Tighten oil pump screws to 25 N·m (18 ft.-lbs.). Refer to TIMING GEARS sec-

Lombardini — SMALL DIESEL

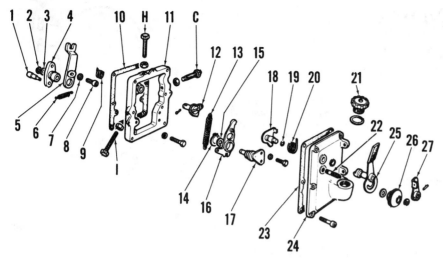

Fig. L5-19—Exploded view of governor and throttle control linkage.

1. Pin
2. Snap ring
3. Washer
4. Pivot flange
5. Governor arm
6. Start spring
7. Washer
8. Allen screw
9. Spring
10. Gasket
11. Plate
12. Pivot arm
13. Governor spring
14. Snap ring
15. Washer
16. Lever
17. Pivot
18. Arm
19. Washer
20. Spring
21. Oil fill cap
22. Stud
23. Gasket
24. Cover
25. Throttle lever
26. Knob
27. Stop lever
C. Torque idle speed screw
H. High idle speed screw
I. Low idle speed screw

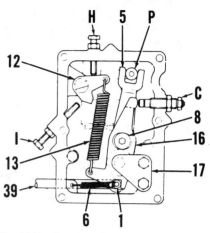

Fig. L5-20—Drawing of governor and throttle control linkage. Refer to Fig. L5-19 legend for parts identification.

tion and align oil pump gear timing marks. Tighten oil pump gear nut to 44 N·m (33 ft.-lbs.).

Oil pressure with engine at normal operating temperature and running at full speed should be 320-350 kPa (45-50 psi). To adjust oil pressure, remove plug (P–Fig. L5-14) from oil filter housing and add or remove shims to vary relief valve spring pressure.

CAMSHAFT AND CAM FOLLWERS

To remove camshaft, first remove cylinder head, push rod tubes and push rods. Remove timing gear cover and camshaft retainer (5–Fig. L5-13). Remove fuel injection pump and fuel transfer pump. Using suitable tools, raise cam followers away from camshaft. Secure cam followers so they will not fall into crankcase when camshaft is removed. Withdraw camshaft from crankcase.

Inspect front roller bearing (1–Fig. L5-13) and renew if necessary. Inspect camshaft lobes and bearing journals for wear or other damage. Center bearing journal diameter should be 40.94-40.96 mm (1.6118-1.6126 inches) and rear bearing journal should be 29.94-29.96 mm (1.1787-1.1795 inches) on Models 904, 914 and L20. On all other models, center bearing journal diameter should be 39.94-39.96 mm (1.5724-1.5732 inches) and rear bearing journal should be 27.987-28.000 (1.1018-1.1023 inches). Camshaft operates directly in unbushed bores in crankcase. If journal clearance in crankcase exceeds 0.10 mm (0.004 inch), camshaft and/or crankcase must be renewed.

Cam follower outer diameter is 13.97-13.98 mm (0.5500-00.5504 inch). Clearance in crankcase bore should be 0.02-0.05 mm (0.001-0.002 inch) with a maximum allowable clearance of 0.1 mm (0.004 inch).

Camshaft end play is controlled by retainer plate (5). Camshaft end play should be 0.2-0.4 mm (0.008-0.016 inch). Retainer plate thickness should be 5.7-5.8 mm (0.224-0.228 inch) and groove in camshaft gear (4) should be 6.0-6.1 mm (0.236-0.240 inch). Renew plate and/or gear if end play is excessive.

Tighten camshaft gear retaining nut to a torque of 195 N·m (145 ft.-lbs.) and retainer plate cap screws to 25 N·m (18 ft.-lbs.). Be sure timing marks are aligned as outlined in TIMING GEARS section.

GOVERNOR

All models are equipped with a flyball type governor mounted on the crankshaft. As the flyballs (29–Fig. L5-13) move in and out against cup (30), fork and lever assembly (33) forces push rod (39) against pin (1–Fig. L5-19 or L5-20). Pivot flange (4), lever (16) and governor arm (5) are forced to rotate thereby moving fuel injection pump control rack pin (P). Throttle lever (25) operates through pivot arm (12) and governor spring (13) to control engine speed.

The torque control screw (C) limits injection pump fuel delivery at full-load speed. On some models, the control screw tip is spring-loaded to provide additional fuel delivery under maximum load condtions. Spring-loaded torque control screw assembly is calibrated for specific engine application. If device is disassembled, do not change spring washer thickness. Torque control screw is available only as an assembly.

To adjust governor, proceed as follows: With engine stopped, remove cover (24–Fig. L5-19) and back out torque control screw (C) 5 or 6 turns. Check to be sure start spring (6–Fig. L5-19 and L5-20) has removed slack in governor mechanism. Loosen governor arm screw (8) and move governor arm (5) towards torque control screw until fuel injection pump control rack pin (P) is in maximum fuel position, then retighten screw. Reinstall control cover and run engine at high idle (no-load) speed. Turn torque control screw (C) in until engine speed just begins to decrease, then turn torque control screw out 2¼ turns on Models L20 and 8LD655-2/L or 1½ turns on all other models. Tighten torque screw locknut.

For access to governor flyball assembly, remove timing gear cover and crankshaft gear.

CRANKSHAFT AND BEARINGS

The crankshaft rides in sleeve bearings in the crankcase bulkhead and end bearing support (34–Fig. L5-13) and in insert bearings in center support halves 48 and 49).

To remove the crankshaft, remove flywheel, oil pan, cylinder head, pistons and rods and timing gear cover. Using suitable pullers, remove crankshaft gear and governor assembly from crankshaft. Remove end support (34). Remove crankshaft and center support assembly

SERVICE MANUAL

Lombardini

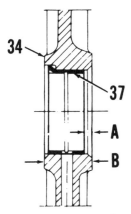

Fig. L5-21—Depth (A) of bearing (37) in end support (34) should be 5 mm (0.020 inch).

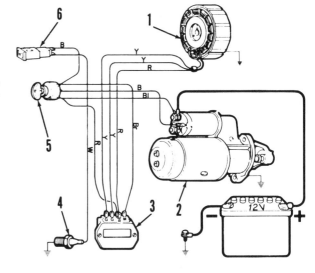

Fig. L5-22—Wiring schematic of models not equipped with an alternator warning light.
1. Alternator assy.
2. Starter motor
3. Regulator
4. Oil pressure sender
5. Switch
6. Oil pressure light

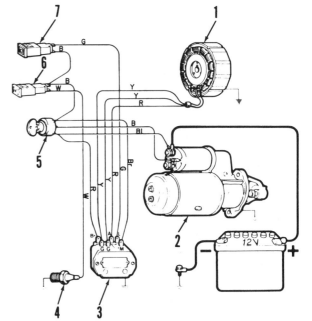

Fig. L5-23—Wiring schematic of models equipped with alternator light (7). Refer to Fig. L5-22 legend.

from crankcase. Unbolt and remove center support halves (48 and 49).

Standard diameter of center main bearing journal is 55.34-55.35 mm (2.1787-2.1791 inches). Standard diameter of outer main bearing journals is 54.94-54.95 mm (2.1630-2.1634 inches). Bearing clearance should be 0.05-0.09 mm (0.002-0.0035 inch) for center main bearing and 0.05-0.07 mm (0.002-0.003 inch) for outer main bearings. Standard and undersize bearings are available.

Standard crankpin journal diameter is 49.99-50.00 mm (1.9681-1.9685 inch). Recommended rod bearing clearance is 0.03-0.07 mm (0.001-0.003 inch) with a maximum allowable clearance of 0.10 mm (0.004 inch). Standard and undersize rod bearings are available.

Serrated parting surfaces of center support halves must be aligned during assembly. With support screws tightened to 25 N·m (18 ft.-lbs.), outside diameter of support should be 154.98-154.99 mm (6.1016-6.1019 inches). Maximum out-of-round is 0.01 mm (0.0004 inch).

Special tool 7271.3595.047 is available from manufacturer to install bearing (37–Fig. L5-13) in end support (34). When installing bearing, distance (A–Fig. L5-21) from bearing (37) to inside surface of support should be 5 mm (0.020 inch). Make certain lubricating oil holes are aligned.

Crankshaft end play should be 0.15-0.25 mm (0.006-0.010 inch) and is not adjustable. A worn end support (34–Fig. L5-13) or crankshaft will cause excessive end play.

Reassembly is reverse of disassembly. Be sure to lubricate lip of oil seals (26 and 44) before reinstalling. Tighten screws securing center support halves to 25 N·m (18 ft.-lbs.). Tighten center support assembly mounting screws to 40 N·m (30 ft.-lbs.) and end support cap screws to 25 N·m (18 ft.-lbs.).

ELECTRICAL SYSTEM

Refer to Fig. L5-22 or L5-23 for wiring schematic. Note that circuit in Fig. L5-23 includes an alternator warning light (7) and the voltage regulator is different than the regulator used in circuit shown in Fig. L5-22. The alternator assembly (1) is contained in the fan housing.

Lombardini

SMALL DIESEL

LOMBARDINI

Model	No. Cyls.	Bore	Stroke	Displ.
9LD560-2	2	90 mm (3.453 in.)	88 mm (3.464 in.)	1120 cc (68.3 cu. in.)
10LD360-2	2	82 mm (3.228 in.)	68 mm (2.677 in.)	718 cc (43.8 cu. in.)
10LD400-2	2	86 mm (3.386 in.)	68 mm (2.677 in.)	790 cc (48.2 cu. in.)

All engines covered in this section are four-stroke, air-cooled, direct injection diesel engines. The crankcase and cylinder heads are aluminum and the cylinders are cast iron. Crankshaft rotation is counterclockwise at pto end.

MAINTENANCE

LUBRICATION

Recommended engine oil is SAE 10W for temperatures below 0°C (32°F), SAE 20W for temperatures between 0°C (32°F) and 20°C (68°F), and SAE 40 for temperatures above 20°C (68°F). API classification for oil should be CD. Oil sump capacity is 2.8 liters (2.9 quarts) for Model 9LD560-2 and 2.5 liters (2.6 quarts) for all other models. Manufacturer recommends renewing oil after every 300 hours of operation.

A renewable oil filter is located in side of crankcase (Fig. L6-1). Manufacturer recommends renewing filter after every 300 hours of operation.

Fig. L6-2—Drawing showing location of low speed adjusting screw (I) and high speed adjusting screw (H).

ENGINE SPEED ADJUSTMENT

Engine should be at normal operating temperature when adjusting engine speed. Low idle speed is adjusted by turning idle speed screw (I – Fig. L16-2). Recommended low idle speed is 1200-1300 rpm.

Maximum governed speed is adjusted by turning high speed screw (H). To compensate for governor droop during operation, set high idle (no-load) speed 180 rpm above specified governed full-load speed on Model 9LD560-2 or 200 rpm above full-load speed on all other models. Specified full-load speed is normally 3000 rpm on Models 9LD560-2 and 10LD400-2. On all other models, full-load speed is normally 3600 rpm. Governed speed may vary depending upon application.

FUEL SYSTEM

FUEL FILTER. A renewable fuel filter (F – Fig. L6-3) is located in the fuel tank. Manufacturer recommends renewing filter after every 300 hours of opera-

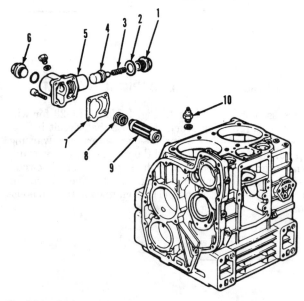

Fig. L6-1—Engine oil filter is located in side of crankcase. Oil pressure relief valve is located in filter housing.
1. Plug
2. Gasket
3. Spring
4. Relief valve poppet
5. Housing
6. Plug
7. Gasket
8. Spring
9. Oil filter
10. Oil pressure sender

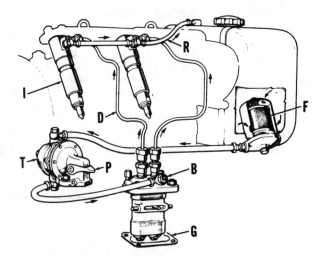

Fig. L6-3—Typical fuel circuit design.
B. Pump bleed screw
D. High pressure delivery line
F. Fuel filter
G. Shim gasket
I. Injector
P. Primer lever
R. Fuel return line
T. Fuel transfer pump

Illustrations Courtesy Lombardini U.S.A., Inc.

SERVICE MANUAL

Lombardini

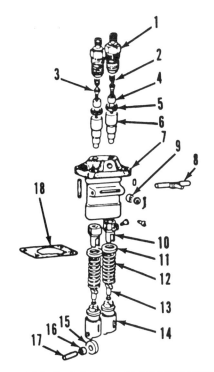

Fig. L6-4—Exploded view of fuel injection pump.

1. Delivery valve holder
2. Spring
3. Delivery valve
4. Delivery valve seat
5. Seal
6. Barrel
7. Pump body
8. Control rack
9. Roller
10. Control sleeve & pinion
11. Spring seat
12. Spring
13. Plunger
14. Guide
15. Outer roller
16. Inner roller
17. Pin
18. Shim gasket

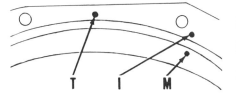

Fig. L6-5—Timing marks (I and M), located in flywheel end of engine, should be aligned when injection to number 1 cylinder occurs. Piston is at top dead center when marks (M and T) are aligned.

tion. However, if engine indicates signs of fuel starvation (loss of power or surging), renew filter regardless of hours of operation. After renewing filter, air must be bled from system as outlined in the following paragraph.

BLEED FUEL SYSTEM. To bleed air from fuel system, first loosen bleed screw (B–Fig. L6-3) on fuel injection pump. Manually operate fuel transfer pump primer lever (P) until air-free fuel flows from bleed screw connection, then retighten bleed screw.

If engine fails to start after performing initial bleeding operation, loosen the high pressure injection lines (D) at the injectors. Rotate crankshaft to operate injection pump until fuel is discharged at injection line fitting. Retighten injection lines, then start engine.

INJECTION PUMP TIMING

Injection pump timing is adjusted by changing the thickness of shim gaskets (G–Fig. L6-3) between pump body and crankcase mounting surface. Injection should occur at 24°-26° BTDC on all models.

To check injection pump timing, first disconnect injection line of number 1 cylinder from injection pump delivery valve holder. Unscrew delivery valve holder (1–Fig. L6-4) and remove spring (2) and delivery valve (3). Do not remove delivery valve seat (4). Reinstall delivery valve holder. Move throttle control to full speed position. Operate fuel transfer pump prime lever while rotating crankshaft in normal direction so number 1 piston is on compression stroke. Note that fuel should be flowing out of delivery valve holder opening during this procedure. Stop rotating crankshaft at the moment fuel ceases to flow from delivery valve. At this point, beginning of injection occurs and timing marks (I and M–Fig. L6-5) should be aligned.

To advance injection timing, reduce shim gasket (18–Fig. L6-4) thickness. Increase shim gasket thickness to retard beginning of injection. Tighten injection pump mounting screws to a torque of 25 N·m (18 ft.-lbs.). Reinstall delivery valve and tighten delivery valve holder to 34 N·m (25 ft.-lbs.). Bleed air from system as previously outlined.

REPAIRS

TIGHTENING TORQUES

Refer to the following table for tightening torques.
Camshaft gear..............60 N·m
(44 ft.-lbs.)
Connecting rod:
 9LD....................30 N·m
(22 ft.-lbs.)
 10LD...................34 N·m
(25 ft.-lbs.)
Cylinder head:
 9LD.....................6 N·m
(44 ft.-lbs.)
 10LD...................44 N·m
(33 ft.-lbs.)
Flywheel:
 9LD...................295 N·m
(215 ft.-lbs.)
 10LD..................176 N·m
(130 ft.-lbs.)
Injection pump...............25 N·m
(18 ft.-lbs.)
Injector retainer.............12 N·m
(9 ft.-lbs.)
Main bearing support:
 Center..................30 N·m
(22 ft.-lbs.)
 End.....................25 N·m
(18 ft.-lbs.)
Oil pan.....................25 N·m
(18 ft.-lbs.)
Oil pump:
 9LD....................25 N·m
(18 ft.-lbs.)
 10LD...................20 N·m
(15 ft.-lbs.)
Oil pump gear...............34 N·m
(25 ft.-lbs.)
Rocker arm shaft............25 N·m
(18 ft.-lbs.)
Timing gear cover...........25 N·m
(18 ft.-lbs.)

VALVE CLEARANCE ADJUSTMENT

To adjust valve clearance, first remove rocker arm cover and rotate crankshaft until number 1 piston is at TDC on compression stroke. Clearance between rocker arm and valve stem end should be 0.15-0.20 mm (0.006-0.008 inch) for both intake and exhaust valves with engine cold. After adjusting valve clearance on number 1 cylinder, rotate crankshaft one complete revolution and adjust valves on number 2 cylinder.

CYLINDER HEAD AND VALVE SYSTEM

Cylinder head should not be removed when engine is hot as head may warp. If cylinder head mounting surface is warped or pitted, up to 0.3 mm (0.012 inch) of material may be lapped from head to true the surface.

Valve face angle is 45° for intake and exhaust. Renew valve if head margin is less than 0.5 mm (0.020 inch). Valve seat angle is also 45° and recommended seat width is 1.4-1.6 mm (0.055-0.065 inch). Renewable valve seat inserts are used on intake and exhaust. Before installing new inserts, cylinder head should first be heated to 160°-180°C (320°-355°F).

With valves installed, measure distance valve head is recessed from cylinder head surface. Intake valve should be recessed 0.60-0.80 mm (0.024-0.031 inch) and exhaust valve (without compression release) should be recessed 0.80-1.25 mm (0.031-0.049 inch) on Model 9LD560-2 or 0.75-1.20

Illustrations Courtesy Lombardini U.S.A., Inc.

Lombardini — SMALL DIESEL

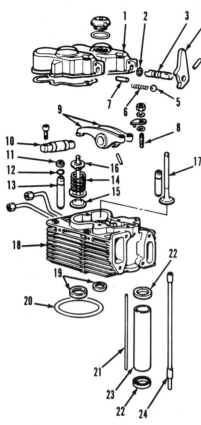

Fig. L6-6—Exploded view of cylinder head assembly.

1. Rocker arm cover
2. "O" ring
3. Compression release shaft
4. Compression release lever
5. Detent ball
6. Spring
7. Pin
8. Adjusting screw
9. Rocker arms
10. Rocker arm shaft
11. Oil seal (intake only)
12. Snap ring
13. Valve guide
14. Valve spring
15. Intake valve
16. Spring retainer
17. Exhaust valve
18. Cylinder head
19. Valve seat inserts
20. Cylinder head gasket
21. Push rod
22. Oil seals
23. Push rod tube
24. Oil supply tube

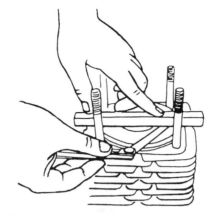

Fig. L6-7—Measure piston height in cylinder and refer to text to determine required cylinder head gasket thickness.

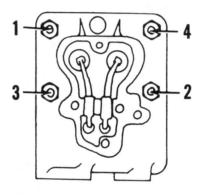

Fig. L6-8—Use a crossing pattern as shown when tightening cylinder head retaining nuts.

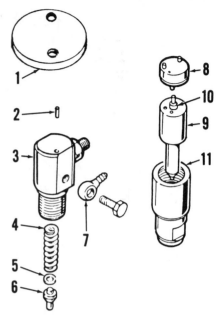

Fig. L6-9—Exploded view of injector assembly.

1. Clamp plate
2. Dowel pin
3. Body
4. Spring
5. Shim
6. Spring seat
7. Return line fitting
8. Spacer
9. Nozzle seat
10. Nozzle valve
11. Nozzle holder nut

mm (0.030-0.049 inch) on Model 9LD560-2 or 0.75-1.20 mm (0.030-0.047 inch) on all other models. With compression release, exhaust valve recession should be 0.55-0.95 mm (0.022-0.037 inch) on all models. If recession is less than specified, regrind valve seat. If recession exceeds specified limit, renew valve seat insert.

Valve stem diameter is 7.98-8.00 mm (0.314-0.315 inch). Desired operating clearance of valve stem in guide is 0.03-0.08 mm (0.001-0.003 inch) and maximum allowable clearance is 0.15 mm (0.006 inch). Valve guides are renewable and oversize guides are available. Outer diameter of oversize guide must be machined to provide a 0.05-0.06 mm (0.0020-0.0023 inch) interference fit in cylinder head bore. Cylinder head should be heated to 160°-180°C (320°-355°F) prior to installing new guides. Use suitable installing tool when installing new guides to prevent damage to guide. After installation, check inside diameter of guide and ream, if necessary, to provide specified clearance for valve stem. A valve stem seal is used on intake valves.

Intake and exhaust valve springs are identical. Spring free length should be 52 mm (2.047 inches). Valve spring pressure should be 304-323 N (68-72 pounds) when compressed to a length of 25.8 (1.015 inches).

Desired clearance between rocker arms and pivot shaft (10 – Fig. L6-6) is 0.03-0.06 mm (0.001-0.002 inch). Maximum allowable clearance is 0.1 mm (0.004 inch). Tighten rocker arm shaft retaining screws to 25 N·m (18 ft.-lbs.).

The cylinder head gasket is available in varying thicknesses to adjust clearance between cylinder head surface and top of piston. Clearance must be 0.75-0.90 mm (0.030-0.035 inch) on Model 9LD560-2 or 0.70-0.85 mm (0.028-0.033 inch) on all other models. To determine required gasket thickness, position piston at TDC and measure from piston crown to gasket seating surface of cylinder as shown in Fig. L6-7. Subtract the measurement (if piston is below sealing surface) or add the measurement (if piston is above sealing surface) to specified clearance dimension to calculate required gasket thickness.

When reinstalling cylinder head, attach intake and exhaust manifolds to correctly position cylinder heads before tightening cylinder head nuts. Then, tighten cylinder head nuts in gradual steps using a crossing pattern as shown in Fig. L6-8. Final torque should be 60 N·m (44 ft.-lbs.) on Model 9LD560-2 or 44 N·m (33 ft.-lbs.) on all other models.

INJECTOR

REMOVE AND REINSTALL. To remove injector, first clean dirt from injector, fuel lines and surrounding area. Disconnect fuel return line and high pressure line and plug all openings immediately. Remove retainer plate (1 – Fig. L6-9) being careful not to lose dowel pin (2). Withdraw injector from cylinder head. Retain shims between injector and head for use in installation.

Fig. L6-10—Arrow on piston crown should point towards intake manifold side of engine and depression (D) should be aligned with injector.

SERVICE MANUAL

Lombardini

If accessible, measure protrusion of nozzle tip into combustion chamber. Nozzle tip should extend 3.0-3.5 mm (0.118-0.138 inch) and may be adjusted by installing 0.5 mm (0.020 inch) shims between injector and cylinder head. Tighten injector clamp plate nuts to 12 N·m (9 ft.-lbs.).

TESTING. A suitable test stand is required to check injector operation. Only clean, approved testing oil should be used to test injectors.

WARNING: Fuel leaves the injector nozzle with sufficient force to penetrate the skin. When testing, keep yourself clear of spray.

Connect injector to tester and operate tester to purge air from nozzle and to make sure nozzle valve is not stuck. When operating properly, injector will emit a buzzing sound and cut off quickly with no leakage at the tip.

Opening pressure should be 20595-21575 kPa (2985-3130 psi) with a used nozzle valve spring (4–Fig. L6-9). If a new spring is used, opening pressure should be set at 21575-22555 kPa (3130-3270 psi) to compensate for the spring taking a set during operation. Opening pressure is adjusted by varying the number and thickness of shims (5).

Operate tester lever slowly to maintain pressure at 1470 kPa (215 psi) below opening pressure and check for leakage past nozzle valve. If a drop forms at nozzle tip within a 10 second period, nozzle valve must be overhauled or renewed. A slight wetness at the tip is permissible.

Operate tester lever briskly and check for an even and well atomized spray pattern. If pattern is wet, ragged or intermittent, nozzle must be overhauled or renewed.

OVERHAUL. To disassemble injector, secure nozzle body (3–Fig. L6-9) in a vise with nozzle tip pointing upward. Remove nozzle holder nut (11) and withdraw nozzle valve (9 and 10) and spacer (8). Invert nozzle body and remove spring seat (6), shim (5) and spring (4).

Thoroughly clean all parts in a suitable solvent. Do not use a steel wire brush or sharp metal tools to clean injector components. Clean inside the orifice end of nozzle valve with a wooden cleaning stick. The orifice spray holes may be cleaned by inserting a cleaning wire slightly smaller in diameter than the spray hole diameter which is 0.28 mm (0.011 inch) for Model 9LD560-2 or 0.25 mm (0.010 inch) for all other models. Make certain the nozzle needle (10) slides freely in bore of nozzle valve seat (9). If needle sticks, reclean or renew nozzle valve assembly.

When reassembling injector, be sure all parts are clean and wet with diesel fuel. Recheck injector operation as outlined in TESTING paragraph.

INJECTION PUMP

To remove injection pump, disconnect fuel lines and immediately plug all openings to prevent entry of dirt. Remove retaining screws and lift out pump assembly. Be sure to retain pump mounting shim gaskets (18–Fig. L6-4) for use in reassembly. Pump timing is adjusted by changing gasket thickness.

It is recommended that injection pump be tested and serviced only by a shop qualified in diesel fuel injection repair.

When reinstalling injection pump, assemble correct thickness of mounting shim gaskets and be sure to engage control rack pin with governor fork. Tighten pump mounting screws to a torque of 15 N·m (18 ft.-lbs.). If pump was overhauled or renewed, check and adjust injection timing as outlined in MAINTENANCE section.

PISTON AND ROD UNITS

The piston and connecting rod may be removed after removing cylinder head and oil pan. Identify piston and rod units as they are removed so units can be reinstalled in their original position.

When installing piston and rod, note that depression (D–Fig. L6-10) in piston crown is closer to one side of piston. Install piston so depression will be aligned with injector. Some pistons also have an arrow embossed on piston. Properly installed, arrow on piston crown will point towards intake manifold side of engine. Tighten connecting rod cap screws to a torque of 30 N·m (22 ft.-lbs.) on Model 9LD560-2 or 34 N·m (25 ft.-lbs.) on all other models.

Clearance between piston crown and cylinder head is adjusted by changing thickness of head gasket. Refer to CYLINDER HEAD section for adjustment procedure.

PISTON, PIN AND RINGS

The pistons are equipped with three compression rings and one oil control ring. On Model 9LD560-2, recommended ring end gap is 0.40-0.65 mm (0.016-0.025 inch) for all compression rings and 0.25-0.40 mm (0.010-0.016 inch) for oil control ring. On all other models, compression ring end gap should be 0.30-0.50 mm (0.012-0.020 inch) and oil control ring end gap should be 0.25-0.50 mm (0.010-0.020 inch).

To check piston ring groove wear, install new rings onto piston, then measure side clearance between top of rings and top of grooves using a feeler

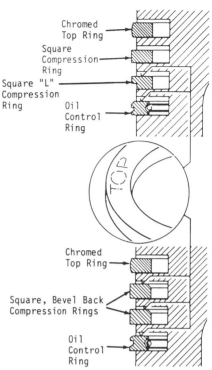

Fig. L6-11—Cross-sectional view of rings and piston showing correct installation of various types of rings.

gage. Renew piston if clearance exceeds 0.15 mm (0.006 inch) for top compression ring and 0.10 mm (0.004 inch) for all other rings.

Standard piston diameter, measured at bottom of skirt perpendicular to piston pin, is 89.86-89.88 mm (3.5378-3.5386 inches) for Model 9LD560-2, 81.90-81.92 mm (3.2244-3.2252 inches) for Model 10LD360-2 and 86.00-86.02 mm (3.3858-3.3866 inches) for Model 10LD400-2. If wear exceeds 0.05 mm (0.002 inch), renew piston. Difference in piston weights must not exceed 6 grams (0.2 ounce).

Piston pin diameter is 24.995-25.000 mm (0.9840-0.9842 inch) for Model 9LD560-2 and 19.995-20.000 mm (0.7872-0.7874 inch) for all other models. Clearance between piston pin and connecting rod bushing should be 0.020-0.035 mm (0.0008-0.0014 inch) for Model 9LD560-2 and 0.010-0.030 mm (0.0004-0.0012 inch) for all other models. Maximum allowable clearance is 0.07 mm (0.0027 inch) for all models. Pin should be a thumb press fit in piston.

When installing rings onto piston, stagger ring end gaps 180° apart. Be sure no end gaps are aligned with piston pin bore. Fig. L6-11 shows correct installation of rings on piston.

CONNECTING ROD

The connecting rod small end is fitted with a renewable bushing. Clearance

Lombardini — SMALL DIESEL

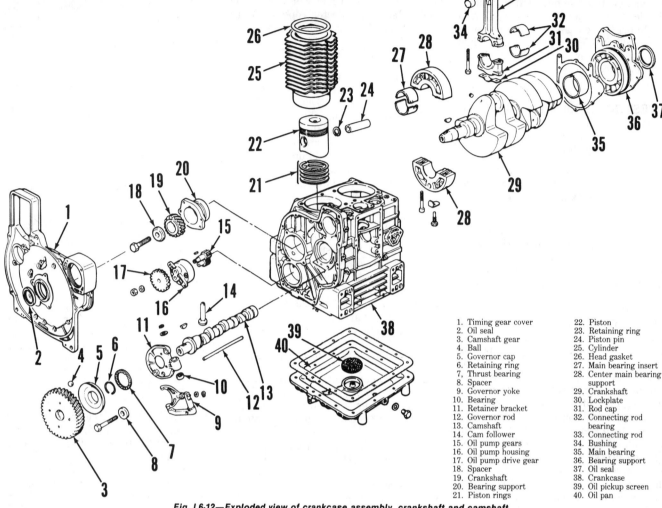

Fig. L6-12—Exploded view of crankcase assembly, crankshaft and camshaft.

1. Timing gear cover
2. Oil seal
3. Camshaft gear
4. Ball
5. Governor cap
6. Retaining ring
7. Thrust bearing
8. Spacer
9. Governor yoke
10. Bearing
11. Retainer bracket
12. Governor rod
13. Camshaft
14. Cam follower
15. Oil pump gears
16. Oil pump housing
17. Oil pump drive gear
18. Spacer
19. Crankshaft
20. Bearing support
21. Piston rings
22. Piston
23. Retaining ring
24. Piston pin
25. Cylinder
26. Head gasket
27. Main bearing insert
28. Center main bearing support
29. Crankshaft
30. Lockplate
31. Rod cap
32. Connecting rod bearing
33. Connecting rod
34. Bushing
35. Main bearing
36. Bearing support
37. Oil seal
38. Crankcase
39. Oil pickup screen
40. Oil pan

between piston pin and rod bushing should be 0.020-0.035 mm (0.0008-0.0014 inch) for Model 9LD560-2 and 0.010-0.030 mm (0.0004-0.0012 inch) for all other models. Maximum allowable clearance is 0.07 mm (0.0027 inch) for all models.

A precision, insert type bearing is used in crankpin end of connecting rod. Desired rod bearing clearance is 0.05-0.08 mm (0.002-0.003 inch) for Model 9LD560-2 and 0.03-0.06 mm (0.0012-0.0023 inch) for all other models. Bearings are available in standard size and undersizes.

Difference in connecting rod weights must not exceed 10 grams (0.35 ounce) on Model 9LD560-2 or 5 grams (0.18 ounce) on all other models.

CYLINDERS

All models are equipped with removable, cast iron, reborable cylinders. Standard cylinder diameter is 90.00-90.02 mm (3.5433-3.5441 inches) for Model 9LD560-2; 82.00-82.02 mm (3.2283-3.2291 inches) for Model 10LD360-2 and 86.00-86.02 mm (3.3858-3.3866 inches) for Model 10LD400-2. If cylinder wear exceeds 0.1 mm (0.004 inch), rebore cylinder to next appropriate oversize.

TIMING GEARS

REMOVE AND REINSTALL. Remove fan housing, crankshaft pulley, fan and alternator assembly. Unbolt and remove timing gear cover (1–Fig. L6-12). Remove gears from shafts.

When reinstalling gears, be sure timing mark on camshaft gear (3–Fig. L6-13) is aligned with timing marks on crankshaft gear (1). Oil pump gear (2) does not require timing. Tighten camshaft gear cap screw to a torque of 60 N·m (44 ft.-lbs.) and oil pump drive gear retaining nut to 34 N·m (25 ft.-lbs.).

OIL PUMP

The oil pump is mounted in the crankcase and is driven by the crankshaft gear. To remove oil pump, remove timing gear cover (1–Fig. L6-12) and oil pump drive gear (17).

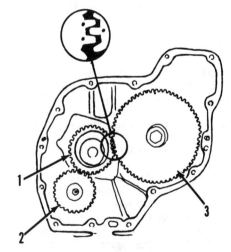

Fig. L6-13—View showing location of timing marks on camshaft gear (3) and crankshaft gear (1).

1. Crankshaft gear
2. Oil pump gear
3. Camshaft gear

SERVICE MANUAL

Lombardini

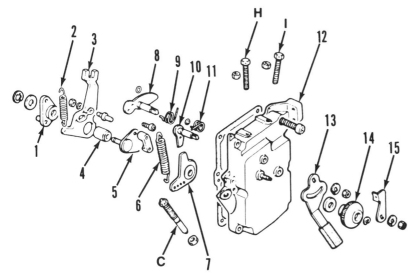

Fig. L6-14—Exploded view of governor and throttle control linkage.

1. Pivot flange
2. Start spring
3. Governor arm
4. Bushing
5. Pivot bracket
6. Governor spring
7. Governor lever
8. Arm
9. Spring
10. Arm
11. Spring
12. Cover
13. Throttle lever
14. Knob
15. Stop lever
C. Torque control screw
H. High idle speed screw
I. Low idle speed screw

Remove pump housing mounting cap screws and withdraw housing (16) and gears (15).

Inspect housing and gears for excessive wear and other damage. Renew pump if clearance between tips of gear teeth and pump body exceeds 0.15 mm (0.006 inch).

To reinstall oil pump, reverse the disassembly procedure. Tighten oil pump cap screws to a torque of 25 N·m (18 ft.-lbs.) on Model 9LD560-2 or 20 N·m (15 ft.-lbs.) on all other models. Tighten oil pump drive gear cap screw to 34 N·m (25 ft.-lbs.).

Oil pressure with engine at normal operating temperature and running at full speed should be 320-350 kPa (45-50 psi). Oil pressure regulating valve (4 – Fig. L6-1) is located in oil filter housing mounted on side of crankcase. Pressure regulator valve spring free length should be 37 mm (1.456 inches).

CAMSHAFT AND CAM FOLLOWERS

To remove camshaft, first remove cylinder head, push rod tubes and push rods. Remove timing gear cover, fuel injection pump and fuel transfer pump. Remove camshaft gear (3 – Fig. L6-12), governor assembly and retainer bracket (11). Using suitable tools, raise cam followers (14) away from camshaft (13). Secure cam followers so they will not fall into crankcase when camshaft is removed. Withdraw camshaft from crankcase.

Camshaft operates directly in unbushed bearing bores in crankcase. Camshaft center bearing journal diameter should be 39.94-39.96 mm (1.5724-1.5732 inches) and end journal diameter should be 27.987-28.000 mm (1.1018-1.1023 inches). Inspect cam lobes and cam follower surfaces for excessive wear or other damage and renew if necessary. Cam follower outer diameter should be 13.97-13.98 mm (0.5500-0.5504 inch). Desired clearance between cam followers and crankcase bores is 0.02-0.05 mm (0.0008-0.0020 inch) and maximum allowable clearance is 0.10 mm (0.004 inch).

Camshaft end play is controlled by retainer plate (11). Desired end play is 0.10-0.26 mm (0.004-0.010 inch). Tighten retainer plate cap screws to a torque of 25 N·m (18 ft.-lbs.). Tighten camshaft gear cap screw to 60 N·m (44 ft.-lbs.) Tighten camshaft gear cap screw to 60 N·m (44 ft.-lbs.). Be sure timing marks on camshaft gear and crankshaft gear are aligned as shown in Fig. L6-13.

GOVERNOR

All models are equipped with a flyball type governor mounted on the camshaft. As the flyball (4 – Fig. L6-12) moves in and out against cup (5), fork assembly (9) forces push rod (12) against pivot flange (1 – Fig. L6-14). Pivot flange (1) and governor control arm (3) are forced to rotate thereby moving injection pump control rack. Throttle lever (13) operates through pivot arms and governor spring (6) to control engine speed. Torque control screw (C) limits governor linkage movement thus limiting fuel delivery under full load. Start spring (2) returns the injection pump control rack to maximum fuel delivery position with engine stopped to aid in starting.

To adjust governor, remove cover (12) and back out torque srew (C) about 5 turns. With engine stopped, loosen screw securing governor arm (3) to pivot flange (1). Move governor arm to place injection pump control rack in maximum fuel position and make certain all slack is removed from push rod (12 – Fig. L6-12). Retighten governor arm screw. Install control cover, then run engine at full (no-load) speed. Turn torque control screw (C – Fig. L6-14) in until engine speed just begins to decrease, then turn control screw back out 1½ turns.

For access to governor flyball assembly, remove timing gear cover and camshaft gear.

CRANKSHAFT AND BEARINGS

To remove crankshaft, remove flywheel, blower fan (Model 9LD560-2), timing gear cover, oil pan, cylinder head and pistons and connecting rods. Remove crankshaft gear and outer bearing support (20 – Fig. L6-12). Remove bearing support (36). Remove center bearing support mounting cap screws, then withdraw crankshaft and center bearing assembly from crankcase. Unbolt and remove center bearing support halves (28) from crankshaft.

Standard diameter of center main bearing journal is 55.34-55.35 mm (2.1787-2.1791 inches). Standard diameters of outer main bearing journals are 54.93-54.95 mm (2.1626-2.1634 inches) and 71.98-72.00 mm (2.8338-2.8346 inches). Recommended bearing clearance is 0.05-0.09 mm (0.002-0.0035 inch) for center main bearing, 0.05-0.08 mm (0.002-0.003 inch) for smaller diameter outer main bearing and 0.07-0.11 mm (0.003-0.004 inch) for larger diameter outer main bearing. Standard and undersize bearings are available.

Standard crankpin diameter is 45.500-45.516 mm (1.7913-1.7919 inches) on Model 9LD560-2 and 39.984-40.00 mm (1.5742-1.5748 inches) on all other models. Recommended rod bearing clearance is 0.05-0.08 mm (0.002-0.003 inch) for Model 9LD560-2 and 0.03-0.06 mm (0.001-0.002 inch) for all other models. Standard and undersize rod bearings are available.

Serrated parting surfaces of center support halves (28) must be aligned during assembly. With support screws tightened to 25 N·m (18 ft.-lbs.), outside diameter of support should be 147.010-147.020 mm (5.7878-5.7882 inches) for Model 9LD560-2 and 128.000-128.018 mm (5.0394-5.0401 inches) for all other models. Maximum allowable out-of-round is 0.01 mm (0.0004 inch).

Lombardini

SMALL DIESEL

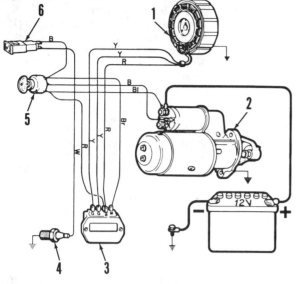

Fig. L6-15—Typical wiring schematic of models not equipped with an alternator light.

1. Alternator
2. Starter motor
3. Regulator
4. Oil pressure switch
5. Switch
6. Oil pressure light

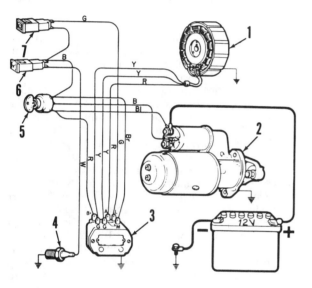

Fig. L6-16—Typical wiring schematic of models equipped with an alternator warning light (7). Refer to Fig. L6-15 legend.

Special tool 7271.3595.047 is available from manufacturer to install bearing (35) in end support (36). When installing bearing, make certain oil hole in bearing is aligned with oil hole in support.

Crankshaft end play should be 0.15-0.30 mm (0.006-0.012 inch) for all other models. End play is adjusted by changing thickness of outer bearing support (36) mounting gasket.

Reassembly is the reverse of disassembly. Be sure to inspect and renew oil seals (2 and 37) if necessary. Lubricate lip of oil seals before reinstalling. Tighten screws securing the center support halves to 25 N·m (18 ft.-lbs.). Tighten center support mounting cap screws to 30 N·m (22 ft.-lbs.) and end support cap screws to 25 N·m (18 ft.-lbs.). Tighten flywheel mounting cap screw to a torque of 295 N·m (215 ft.-lbs.) on Model 9LD560-2 or 176 N·m (130 ft.-lbs.) on all other models.

ELECTRICAL SYSTEM

Refer to Fig. L6-15 or L6-16 for wiring schematic. Note that circuit in Fig. L6-16 includes an alternator warning light (7) and the voltage regulator is different than the regulator used in circuit shown in Fig. L6-15. The alternator assembly (1) is contained in the fan housing.

Illustrations Courtesy Lombardini U.S.A., Inc.

SERVICE MANUAL

MITSUBISHI

Mitsubishi

MITSUBISHI ENGINE NORTH AMERICA, INC.
610 Supreme Drive
Bensenville, Illinois 60106

Model	No. Cyls.	Bore	Stroke	Displ.
K3A	3	65 mm (2.555 in.)	78 mm (3.065 in.)	776 cc (47.35 cu.in.)
K3C	3	70 mm (2.751 in.)	78 mm (3.065 in.)	900 cc (54.92 cu.in.)
K3D	3	73 mm (2.869 in.)	78 mm (3.065 in.)	979 cc (59.72 cu.in.)
K3H	3	78 mm (3.065 in.)	90 mm (3.537 in.)	1290 cc (78.69 cu.in.)
K3M	3	84 mm (3.301 in.)	90 mm (3.537 in.)	1496 cc (91.26 cu.in.)

All models are four-stroke, indirect injection, liquid cooled diesel engines. Number one cylinder is at timing gear end of engine. Metric fasteners are used throughout the engine.

MAINTENANCE

LUBRICATION

Engine oil and filter should be renewed after every 100 hours of operation. Recommended engine lubricant is API classification CC/CD motor oil. Ambient temperature should be used as a guide to weight of engine oil. Below −20° C (4° F) use SAE 5W or 5W-20 oil. In temperatures between −20° and 0° C (−4° to 32° F), use SAE 10W oil. In temperatures between 0° and 20° C (32° to 68° F), SAE 20W oil is recommended. In temperatures between 10° and 35° C (50° to 95° F), SAE 30 oil is recommended. In temperatures above 30° C (86° F), use SAE 40 or 20W-40 oil. SAE 10W-30 engine oil may be used in all temperatures between −20° and 35° C (−4° to 95° F).

FUEL SYSTEM

FUEL FILTER. A renewable paper element cartridge type filter is located in the fuel filter; refer to Fig. MB1. The filter element cannot be cleaned. Element should be renewed after every 500 hours of operation or sooner if engine is misfiring or loss of power is evident. Be sure "O" rings (4 and 6) are positioned correctly when installing new filter element. Do not attempt to start engine until air is bled from fuel system.

BLEED FUEL SYSTEM. Air must be bled from fuel system if fuel tank is allowed to run dry, if fuel lines, filter or other components within the fuel system have been disconnected or removed, or if engine has not been operated for a long period of time.

To bleed system, open fuel shut-off valve (2—Fig. MB1) and loosen vent screws on fuel filter base. When fuel flow from filter base is free of air, tighten the vent screws, left screw first. Loosen vent screw on injection pump and allow fuel to flow from pump until free of air. Tighten vent screw and start engine.

If engine will not start, loosen high pressure fuel line at each fuel injector. Continue to crank engine until fuel escapes from loosened connections. Tighten pressure line connections and start engine.

INJECTION PUMP TIMING

Injection pump timing is adjusted by varying the thickness of the shim gasket (17—Fig. MB2) between injection pump body and mounting surface on housing. Use of thicker shim gasket retards the timing; a thinner shim gasket will advance the timing. Beginning of injection should occur at 23 degrees BTDC on engine Model K3D and at 21 degrees BTDC on other models.

To check timing, shut off fuel supply and remove No. 1 (front) injector line. Remove No. 1 delivery valve holder (1—Fig. MB2), spring (3) and delivery valve assembly (5). Remove delivery valve

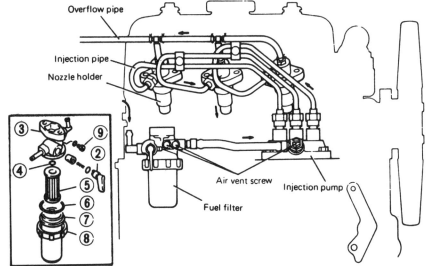

Fig. MB1—Illustration showing fuel filter assembly and location of bleed screws.
2. Shut-off valve
3. Filter body
4. "O" ring
5. Filter element
6. "O" ring
7. Filter cup
8. Retaining nut
9. Bleed screw

Mitsubishi

SMALL DIESEL

spool from valve seat, then reinstall seat and holder and tighten to a torque of 39-49 N·m (30-35 ft.-lbs.).

NOTE: Do not crank engine with delivery valve holder removed or loosened.

On the rear edge of crankshaft pulley, there are three closely spaced marks (3—Fig. MB3). The center mark indicates 21 degrees BTDC; the other two marks indicate 2 degrees on either side of the 21 degree mark (upper mark is 19 degrees BTDC, lower mark is 23 degrees BTDC). Turn crankshaft clockwise slowly until No. 1 piston is coming up on compression stroke. Stop turning crankshaft when the three timing marks on crankshaft pulley are approaching timing pointer (2). With fuel supply turned on, fuel should flow freely from delivery valve holder discharge port. Continue to turn crankshaft slowly until fuel just stops flowing from delivery valve holder. The moment that fuel stops flowing is beginning of fuel injection. At this point, the lower (23 degrees BTDC) mark should be aligned with timing pointer (2) on engine Model K3D, and the center (21 degrees BTDC) mark should be aligned with pointer on all other models.

If injection timing requires adjustment, remove the fuel injection pump as outlined in REPAIRS section and install either a thicker shim gasket to retard timing or a thinner shim gasket to advance timing as required. Changing shim gasket thickness by 0.1 mm (0.004 inch) will vary timing by approximately 1 degree. Reinstall pump and repeat timing check. When timing is correct, reinstall delivery valve spool and spring using a new gasket (4—Fig. MB2) and "O" ring (2). Tighten delivery valve holder to a torque of 39-49 N·m (30-35 ft.-lbs.). Bleed air from system as previously outlined.

GOVERNOR

All models are equipped with a flyweight type governor mounted on the front end of injection pump camshaft.

To adjust engine speeds, proceed as follows: Loosen locknuts on high speed stop screw (8—Fig. MB4), low speed stop screw (9) and damper spring adjusting screw (4). Back out damper spring adjusting screw until tension is relieved on damper spring. Operate engine until normal operating temperature is reached. Place throttle lever in wide open throttle position and adjust high speed stop screw to obtain 2940 engine rpm, then tighten locknut. Apply a small amount of thread locking compound to damper spring adjusting screw. Turn damper spring screw clockwise to increase engine speed to 2950 rpm, then tighten locknut and install adjusting screw cover (5). Move throttle lever to slow speed position and adjust low speed stop screw (9) to obtain 900 engine rpm. Tighten stop screw locknut, then recheck speed settings. Install a new seal wire on high speed adjusting screw.

Fig. MB2—Exploded view of multiple plunger injection pump used on all models.
1. Delivery valve holder
2. "O" ring
3. Spring
4. Gasket
5. Delivery valve assy.
6. Plunger
7. Control sleeve
8. Upper seat
9. Spring
10. Lower seat
11. Shim
12. Tappet
13. Adjusting plate
14. Plate
15. Tappet guide pin
16. Control rack
17. Shim gasket
18. Idle set plate
19. Smoke set plate
20. Air bleed screw
21. Fuel inlet
22. Injection pump body

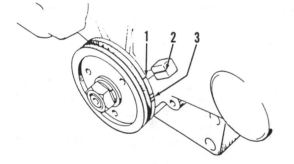

Fig. MB3—Timing marks are located on crankshaft front pulley. Center mark (3) of three closely spaced marks indicates 21 degrees BTDC. The outside marks are 2 degrees on either side of 21 degrees.
1. TDC mark
2. Timing pointer
3. Injection timing marks

Fig. MB4—View of governor linkage adjustment screws.
1. Damper spring
2. Tie rod
3. Locknut
4. Adjusting screw
5. Seal cap
6. Cover
7. Governor arm
8. High speed stop screw
9. Low speed stop screw

REPAIRS

TIGHTENING TORQUES

Refer to the following table for special tightening torques. Metric fasteners are used throughout the engine.

Connecting rod:
Models K3H,K3M . . . . 54-58 N·m
(40-43 ft.-lbs.)
All other models . . . . 31-34 N·m
(23-25 ft.-lbs.)

SERVICE MANUAL

Mitsubishi

Cylinder head:
(Refer to Fig. MB6
for bolt numbers.)
  Models K3H, K3M,
    bolts 1, 2 & 3 . . . . 98-108 N·m
    (72-80 ft.-lbs.)
  Models K3H, K3M,
    bolts 4-11 . . . . 146-157 N·m
    (108-116 ft.-lbs.)
  All other models,
    bolts 1, 2 & 3 . . . . 69-79 N·m
    (51-58 ft.-lbs.)
  All other models,
    bolts 4-11 . . . . 108-117 N·m
    (80-87 ft.-lbs.)
Flywheel . . . . 113-122 N·m
  (83-90 ft.-lbs.)
Injector nozzle holder . . . . . 15-20 N·m
  (11-14 ft.-lbs.)
Injection pump
  delivery valve holder . . . . . 39-49 N·m
  (30-35 ft.-lbs.)
Crankshaft pulley nut . . . 196-245 N·m
  (145-180 ft.-lbs.)
Main bearing cap bolts . . . . 49-54 N·m
  (36-54 ft.-lbs.)

WATER PUMP

To remove water pump, drain cooling system and remove radiator. Remove fan belt and cooling fan. Remove coolant inlet hose and bypass hose from pump. Unbolt and remove water pump assembly.

Water pumps are not serviceable and must be renewed as a unit. Reinstall pump by reversing removal procedure.

THERMOSTAT

The bypass type thermostat is located in the coolant outlet elbow. Thermostat should start to open at approximately 82° C (180° F) and be fully open at 95° C (203° F). When installing thermostat, be sure spring side is toward engine. Install water outlet with arrow mark upward.

VALVE ADJUSTMENT

Recommended clearance between valve stem and rocker arm is 0.25 mm (0.010 inch) for both intake and exhaust valves on all models. Valve clearance should be adjusted by turning screw on rocker arm with engine cold and each piston at top dead center of compression stroke. Firing order for all models is 1-3-2.

To check and adjust clearance, turn crankshaft until No. 1 TDC mark (Fig. MB3) on crankshaft pulley is aligned with timing mark on timing gear cover. Both rocker arms for No. 1 cylinder should be loose; if not, piston is on exhaust stroke and crankshaft must be turned one complete revolution. Use a feeler gage to check clearance of intake and exhaust valves for No. 1 cylinder. Adjust clearance if necessary, then rotate crankshaft 240 degrees to bring No. 3 piston to top dead center of compression stroke and adjust valve clearance for this cylinder if necessary. Rotate crankshaft an additional 240 degrees to bring No. 2 piston to top dead center and adjust valve clearance for this cylinder if necessary.

CYLINDER HEAD

To remove cylinder head, proceed as follows: Drain cooling system. Disconnect battery cables. If applicable, remove muffler, side panels, hood and air cleaner. Disconnect fuel return line and remove injector pipes; cap all openings in fuel system to prevent entry of dirt. Disconnect glow plug wiring, water temperature sending unit wire and, if fitted, tachometer drive cable. Remove cap screws securing alternator brace to cylinder head. Remove rocker cover, unbolt and remove rocker arm assembly and push rods. Loosen cylinder head bolts in sequence shown in Fig. MB5. Lift cylinder head from engine and remove cylinder head gasket.

Remove injector nozzles and glow plugs from cylinder head. Remove valve assemblies taking care to separate and identify parts for each cylinder. Thoroughly clean head and remove carbon deposits. Inspect for cracks, evidence of water leakage and other damage. Check cylinder head with straightedge as shown in Fig. MB7. If a 0.1 mm (0.004 inch) feeler gage can be inserted between cylinder head surface and straightedge, cylinder head should be resurfaced or renewed.

When reinstalling cylinder head, be sure all surfaces are clean and dry. Do not apply any sealant to cylinder head gasket. Tighten cylinder head bolts gradually in three steps following tightening sequence shown in Fig. MB6. On Models K3H and K3M, final tightening torque is 98-108 N·m (72-80 ft.-lbs.) for bolts on push rod side of head and 146-157 N·m (108-116 ft.-lbs.) for remaining bolts. On all other models, final tightening torque is 69-79 N·m (51-58 ft.-lbs.) for M10 bolts and 108-117 N·m (80-87 ft.-lbs.) for M12 bolts.

VALVE SYSTEM

Intake and exhaust valves seat directly in cylinder head. Valve face and seat

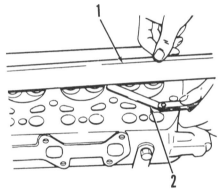

Fig. MB7—Use a straightedge (1) and feeler gage (2) to check cylinder head distortion. Repair or renew head if a 0.1 mm (0.004 inch) feeler gage strip can be inserted between straightedge and head surface.

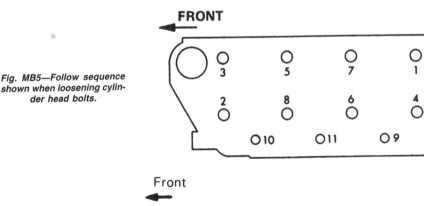

Fig. MB5—Follow sequence shown when loosening cylinder head bolts.

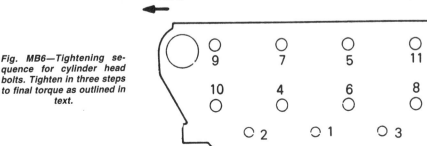

Fig. MB6—Tightening sequence for cylinder head bolts. Tighten in three steps to final torque as outlined in text.

Mitsubishi

angles are 45 degrees. Desired valve seat width is 1.3-1.8 mm (0.050-0.070 inch). After refacing valves, check valve margin (thickness at outer edge of face). Minimum allowable margin is 0.5 mm (0.020 inch); new margin dimension is 1.5 mm (0.060 inch) on engine Models K3H and K3M, and 1 mm (0.040 inch) on other models.

Standard valve stem diameter is 8 mm (0.315 inch) on engine Models K3H and K3M, and 6.6 mm (0.260 inch) on other models. Maximum allowable stem to guide clearance is 0.1 mm (0.004 inch) for intake valves and 0.15 mm (0.006 inch) for exhaust valves. Check valve stem with micrometer to determine whether stem wear, guide wear or both is cause of excessive clearance and renew valve and/or guide as necessary.

To remove worn guides, use suitable driver and remove guide from bottom side upward. Intake and exhaust valve guides are alike. Install new guide from top side of head so that upper end of guide is 12 mm (0.5 inch) above top surface of head; refer to Fig. MB10. Valve guides are presized, but installing guide in head may distort guide bore. If stem to guide clearance is insufficient, use proper size reamer to obtain desired fit. All valves are equipped with cup type stem seals which should be renewed whenever valves are serviced.

Valve springs are interchangeable between intake and exhaust valves. Renew springs which are distorted, heat discolored, or fail to meet test specifications. For engine Models K3H and K3M, valve spring free length is 45.9 mm (1.81 inches) and minimum free length is 44.5 mm (1.75 inches); springs should test 132-136 N (27.6-30.6 lbs.) at 41.8 mm (1.65 inches). For all other engine models, valve spring free length is 43 mm (1.69 inch) and minimum free length is 41.7 mm (1.64 inches). Springs for engine Models K3A and K3C should test 131-144 N (29.4-32.4 lbs.) at 36 mm (1.42 inches) and for engine Model K3D, springs should test 132-136 N (27.6-30.6 lbs.) at 41.8 mm (1.65 inches).

Intake and exhaust valve rocker arms are interchangeable. Standard inside diameter of rocker arm is 18.9 mm (0.744 inch). Maximum allowable rocker arm to shaft clearance is 0.2 mm (0.008 inch). Inspect valve stem contact surface of rocker arm for excessive wear and renew if necessary. When reassembling rocker arms and shaft, be sure the 3 mm (0.118 inch) hole in front end of shaft faces injection pump side of engine (see Fig. MB9). Also, be sure to install plate washer on front and rear shaft mounting brackets.

INJECTORS

REMOVE AND REINSTALL. Before removing an injector or loosening injector lines, thoroughly clean injector, lines and surrounding area using suitable solvent and compressed air.

To remove injector unit, first remove high pressure line leading from injection pump to injector. Disconnect leak-off line and remove nozzle holder mounting cap screws. Withdraw the injector assembly; use suitable puller to remove injector if stuck in cylinder head. Do not pry or hammer on nozzle holder.

NOTE: If nozzle seal washer did not come out with injector, thread a cap screw into washer and pull washer from bottom of injector bore.

When reinstalling injectors, be sure bore in cylinder head is clean. Insert injector with new seal washer. Tighten nozzle holder cap screws to a torque of 15-20 N·m (11-14 ft.-lbs.). Reconnect fuel lines and bleed air from system as previously outlined.

TESTING. A complete job of testing and adjusting the injector requires use of special equipment. Nozzle should be tested for opening pressure, seat leakage and spray pattern. Before conducting tests, operate tester lever until fuel flows from tester pipe, then attach nozzle.

WARNING: Fuel emerges from injector with sufficient force to penetrate the skin. When testing injector, keep yourself clear of nozzle spray.

SMALL DIESEL

With valve to tester gage closed, operate tester lever a few quick strokes to clear air from tester and to be sure nozzle valve is not stuck. Nozzle should open with a buzzing sound and cut off quickly at end of injection. If fuel drops collect at nozzle tip, nozzle is dirty or defective.

OPENING PRESSURE. With valve to tester gage open, operate tester lever slowly while observing gage pressure reading. Opening pressure should be 11770-12750 kPa (1710-1850 psi) on engine Models K3A, K3C and K3D, and 15690-16665 kPa (2275-2420 psi) on engine Models K3H and K3M.

Opening pressure is adjusted by changing thickness of shim (3—Fig. MB10). A change in shim thickness of 0.1 mm (0.004 inch) will change opening pressure approximately 965 kPa (140 psi).

SPRAY PATTERN. Operate tester lever quickly and observe spray. The spray should be finely atomized mist in a straight line from nozzle tip. If spray is at an angle, drops are observed, or is ragged or intermittent, nozzle must be overhauled or renewed.

OVERHAUL. Hard or sharp tools, emery cloth, wire brush (other than brass), or grinding compound must never be used. An approved nozzle cleaning kit is available from a number of specialized sources.

Clean all dirt and loose carbon from exterior of injector assembly. Secure nozzle in a soft-jawed vise or holding

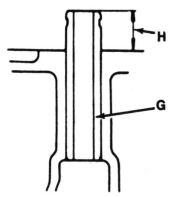

Fig. MB8—Install new valve guides so that top of guide is 12 mm (0.5 inch) above head surface.

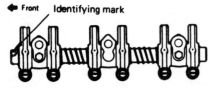

Fig. MB9—Assemble rocker arm shaft with identification mark on shaft toward front of engine and facing injection pump (right) side of engine.

Fig. MB10—Exploded view of fuel injector assembly used on all models.

1. Nozzle holder
2. Mounting flange
3. Shim
4. Spring
5. Spring seat
6. Spacer
7. Nozzle valve assy.
8. Retainer nut

SERVICE MANUAL

Mitsubishi

fixture, then remove nozzle retainer nut (8—Fig. MB10). Carefully remove parts and place in clean calibrating oil or diesel fuel as they are removed. Make certain that parts from each injector are kept together separately from other injector units. It may be necessary to soak nozzle in an approved carbon solvent and use a brass wire brush to remove carbon deposits from outside of nozzle. Do not handle nozzle valve unless hands are wet with calibrating oil or diesel fuel. Rinse parts in clean diesel fuel immediately after cleaning to neutralize solvent and prevent etching of polished surfaces.

Inspect all lapped surfaces for nicks or scratches and renew if necessary. Clean nozzle spray hole from inside using a pointed hardwood stick. Scrape carbon from pressure chamber using a brass scraper. Reclean all parts by rinsing in clean diesel fuel. Check nozzle valve fit by holding nozzle body in a vertical position and lifting needle valve up about one-third of its length, then releasing it. Needle must slide down to seat by its own weight. If needle movement is rough or sticky, reclean or renew nozzle valve assembly as necessary.

Reassemble injector while parts are immersed in diesel fuel to avoid contamination. Tighten nozzle retaining nut to a torque of 60-78 N·m (44-57 ft.-lbs.). Test injector after assembly.

GLOW PLUGS

Glow plugs are connected in parallel with each plug grounding through its mounting threads. The start switch is provided with a "HEAT" position which is used to energize glow plugs for starting. If indicator element fails to light when start switch is held in "HEAT" position, check for loose wiring connections or defective switch, indicator element or glow plugs. Specified resistance of glow plugs is 1.0 ohm (hot) and 0.16 ohm (cold). Glow plug tightening torque is 15-20 N·m (11-14 ft.-lbs.).

FUEL SHUT-OFF SOLENOID

An electrically operated fuel shut-off solenoid is used on all engines. When reinstalling shut-off solenoid, proceed as follows: Remove injection pump inspection plate from right side of cylinder block. Coat threads of solenoid with Loctite Thread Sealer 592, then position solenoid in cylinder block. Working through inspection cover opening, hold injection pump rack forward (fuel shut-off position) and thread solenoid in until plunger contacts rack. Then, back solenoid out about 1/8 turn which will provide a clearance of approximately 0.2 mm (0.008 inch) between plunger and pump rack. Tighten locknut to secure solenoid adjustment. Be sure arrow on rubber cap end is pointing up which will place water drain hole at bottom. Connect solenoid wires and install injection pump side cover.

INJECTION PUMP

To remove injection pump, first thoroughly clean pump and surrounding area using suitable solvent and compressed air. Shut off fuel supply, then disconnect fuel inlet hose. Disconnect and remove injector high pressure lines. Cap or plug all openings to prevent entry of dirt. Remove cover from side of pump housing and disconnect governor linkage tie rod (1—Fig. MB11) and stopper spring (2) from pump control rack. Remove pump mounting cap screws and lift pump from engine. Be sure to retain pump mounting shim gaskets which are used to set injection pump timing.

To reinstall pump, reverse removal procedure using same thickness of shim gaskets as were removed. Bleed air from system as previously outlined.

TIMING GEAR COVER

To remove timing gear cover, proceed as follows: If applicable, remove engine side covers, hood assembly, grille, and battery. Drain cooling system, then disconnect coolant hoses and remove radiator. Disconnect throttle control rod from governor control lever (3—Fig. MB11). Remove inspection plate from side of injection pump housing and disconnect stopper spring (2) and tie rod (1) from pump control rack. Unbolt and remove alternator, cooling fan and crankshaft pulley. Remove timing gear cover mounting cap screws and withdraw cover assembly.

Crankshaft front oil seal (2—Fig. MB12) may be renewed at this time. Be sure to lubricate seal lip prior to reinstalling cover. Inspect camshaft thrust plug (3) for excessive wear and renew if necessary. Inspect governor linkage (1) and pivot bearings. Renew if worn or damaged.

To reinstall timing gear cover, reverse removal procedure.

INJECTION PUMP CAMSHAFT

To remove injection pump camshaft, first remove injection pump, engine oil pump and timing gear cover as outlined in other paragraphs, then proceed as follows:

Remove governor flyweight assembly from camshaft drive gear. Working through one of the holes in camshaft

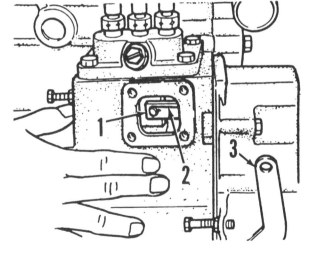

Fig. MB11—Remove inspection plate and disconnect governor linkage tie rod (1) and stopper spring (2) from injection pump control rack before removing timing gear cover.

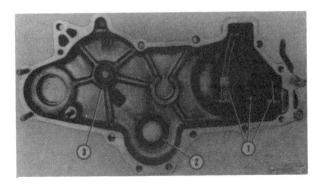

Fig. MB12—View of inner side of timing cover. Be sure to reinstall tie rod connecting governor linkage and injection pump when installing timing gear cover.
1. Governor linkage
2. Oil seal
3. Thrust plug

235

Mitsubishi SMALL DIESEL

gear, remove camshaft bearing retaining cap screw and washer that are located at top side of camshaft. Using a brass drift, lightly tap camshaft forward out of cylinder block.

Inspect camshaft for wear, scuffing or pitting and renew if necessary. Standard cam lobe height is 44 mm (1.732 inches) and minimum allowable height is 43 mm (1.693 inches). Inspect drive gear for broken teeth, excessive wear or other damage. Gear is a shrink fit on camshaft. To remove gear, first heat gear evenly to a temperature of 140°-150° C (285°-300° F), then press shaft out of gear. To install gear, heat gear and press onto shaft until gear bottoms against inner race of camshaft front bearing. Be sure timing mark on gear faces forward.

To reinstall camshaft, reverse removal procedure. Be sure oil pump drive is properly engaged in rear of camshaft. Check injection pump timing as previously outlined.

TIMING GEARS

With timing gear cover removed, inspect gears for wear or damage. Use a dial indicator to check gear backlash which should not exceed 0.3 mm (0.012 inch) between any two gears. If excessive backlash is indicated, check for wear of camshaft bushings and idler gear bushing before renewing gears.

The idler stub shaft and the idler gear with bushing are both renewable. The injection pump camshaft gear, engine camshaft gear and crankshaft gear are keyed and press fitted onto their shafts. To renew gears, it is recommended that camshafts and crankshaft be removed from engine. Use a press to remove and install gears. Heat new gears to 140°-150° C (285°-300° F) before installing gears onto shafts.

When reassembling, be sure to align timing marks as shown in Fig. MB13.

OIL PUMP AND RELIEF VALVE

A gerotor type oil pump is used on all models. Pump is mounted at rear of fuel injection pump and is driven by rear of injection pump camshaft. A spin-on cartridge type oil filter is mounted on rear of oil pump.

Oil pressure relief valve (items 8 through 11—Fig. MB14) is located in oil pump rear cover (7). Relief valve is set to open when oil pressure reaches 395 kPa (57 psi). To check oil pressure, remove oil pressure switch (2) and install a pressure gage. Minimum allowable oil pressure at 2400 engine rpm and oil at normal operating temperature is 195 kPa (28 psi).

Thoroughly clean oil pump and surrounding area before removing pump. With pump removed, separate rear cover (7) and rotor assembly (6) from pump body (3).

Inspect all parts for excessive wear or other damage. Using a feeler gage, check clearance between outer rotor and body (Fig. MB15). Standard clearance is 0.15-0.20 mm (0.006-0.008 inch) and maximum allowable clearance is 0.3 mm (0.012 inch). Clearance between inner and outer rotor (Fig. MB16) should be 0.05-0.12 mm (0.002-0.005 inch); maximum allowable clearance is 0.25 mm (0.01 inch). Standard end clearance between rotors and rear cover (Fig. MB17) is 0.03-0.07 mm (0.001-0.003 inch) and maximum allowable clearance is 0.20 mm (0.008 inch).

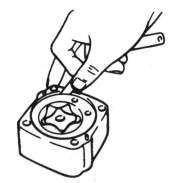

Fig. MB15—Measure clearance between outer rotor and pump body. Refer to text.

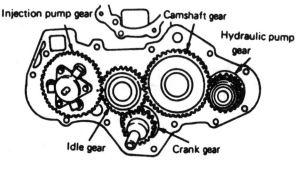

Fig. MB13—Timing gears are correctly timed if match marks on camshaft gear, injection pump gear and crankshaft gear are in register with match marks on idler gear.

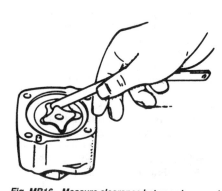

Fig. MB16—Measure clearance between inner and outer rotors.

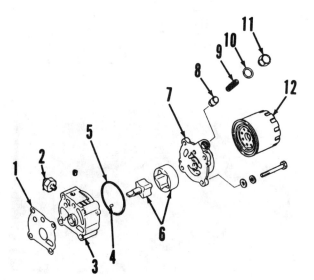

Fig. MB14—Exploded view of engine oil pump assembly.
1. Gasket
2. Pressure switch
3. Pump body
4. Dowel pin
5. "O" ring
6. Rotor assy.
7. Rear cover
8. Relief plunger
9. Spring
10. Gasket
11. Plug
12. Oil filter

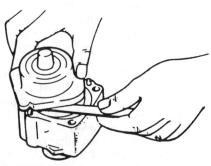

Fig. MB17—Measure clearance between rotor assembly and pump cover.

SERVICE MANUAL

GOVERNOR

The governor linkage is located inside timing gear housing and governor flyweights are mounted on injection pump camshaft gear. With timing gear cover removed as outlined in previous paragraph, inspect governor linkage for excessive wear, binding or other damage prior to disassembly. If governor parts appear normal and work smoothly, problem may be in injection pump instead of governor.

To disassemble governor components, unbolt and remove flyweight assembly (8—Fig. MB19) and sliding thrust plate (10) from camshaft gear. Remove pipe plug from side of timing gear housing. Remove governor spring (4), spring arm (15), shaft retainer pin (14) and governor lever set screw (12). Pull pivot shaft (13) out through pipe plug hole. Remove nut securing governor spring arm (5), then withdraw exterior control lever (1). Remove needle bearings (4—Fig. MB18) using a suitable puller.

Inspect parts and renew as necessary. To reassemble, reverse disassembly procedure.

CAMSHAFT AND BEARING

To remove camshaft, proceed as follows: Remove timing gear cover, cylinder head, push rods and cam followers. If so equipped, remove tachometer drive unit from left side of cylinder block. Carefully withdraw camshaft and gear assembly from front of cylinder block.

Inspect camshaft for excessive wear or other damage. Standard cam lobe height is 35.76 mm (1.408 inches); minimum wear height is 34.76 mm (1.368 inches). Maximum allowable clearance between camshaft journals and cylinder block bearing bores is 0.15 mm (0.006 inch). Front cam bearing bore has renewable bushing; if installing new bushing, be sure oil hole in bushing is aligned with hole in block as shown in Fig. MB20. Front edge of bushing should be slightly below flush with cylinder block.

Inspect cam followers for wear or flaking on cam contact surface and renew if necessary. Standard outside diameter of cam follower is 23 mm (0.905 inch). Maximum clearance of cam follower in cylinder block bore is 0.15 mm (0.006 inch).

A renewable thrust plug (3—Fig. MB12) in timing gear cover controls camshaft end play. Inspect plug for wear and renew as necessary.

Camshaft drive gear is press fit on camshaft and is located by a key. When installing gear on shaft, heat gear to 140°-150° C (285°-300° F), be sure keyway in gear is aligned with key and timing mark is toward front, then press gear onto shaft until bottomed against shoulder on shaft.

Reinstall camshaft and gear assembly by reversing removal procedure, making certain timing marks are aligned as shown in Fig. MB13.

PISTON AND ROD UNITS

Connecting rod and piston assemblies can be removed from above after removing cylinder head and oil pan. Remove carbon ridge from top of cylinders before removing piston and rod assemblies. During removal, identify piston and rod units with their respective cylinder numbers so parts can be reinstalled in their original positions. Be especially careful to retain rod caps with their respective rods.

Engine Models K3A, K3C and K3D use semi-floating piston pins which are

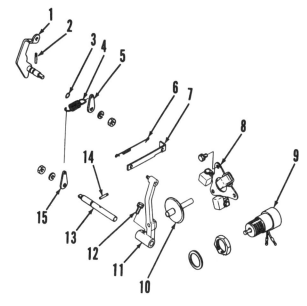

Fig. MB19—Exploded view of governor control linkage.
1. Control lever
2. Pin
3. "O" ring
4. Spring
5. Spring arm
6. Stopper spring
7. Tie rod
8. Flyweight assy.
9. Fuel shut-off solenoid
10. Sliding shaft
11. Governor lever
12. Set screw
13. Pivot shaft
14. Pin
15. Spring arm

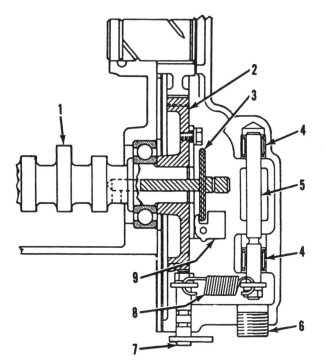

Fig. MB18—Cross-sectional view of governor control linkage.
1. Injection pump camshaft
2. Camshaft drive gear
3. Sliding shaft
4. Needle bearings
5. Pivot shaft
6. Pipe plug
7. Control lever (external)
8. Governor spring
9. Flyweight assy.

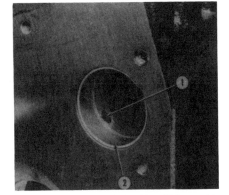

Fig. MB20—A renewable bushing (2) is used in front camshaft bearing bore. Be sure oil hole (1) in bushing is aligned with oil hole in block.

press fitted in connecting rods. Do not attempt to remove piston from rod if piston and rod unit is serviceable as assembled or if you do not have suitable special tools for this job. Special piston pin removal and installation tools are available from a number of sources and tools conforming to piston size and shape must be used to avoid damage to piston and connecting rod. To remove piston pin, place piston in special support block and used piloted push rod to press pin from assembly. On disassembly, carefully check connecting rod and piston pin bore for damage and discard if damage is noted. Use either a connecting rod heater to facilitate piston and rod assembly or use special pin inserter tools in a press. Sunnen press fit lubricant #300 is recommended for press installation. Be sure arrow mark on piston crown is aligned with connecting rod identification mark as shown in Fig. MB21 when installing piston on rod. Be sure pin is centered in connecting rod and piston rotates smoothly on pin.

Engine Models K3H and K3M are equipped with full floating piston pins. Pin is retained by a snap ring at each end of piston pin bore. To disassemble, remove snap rings and push pin from piston and rod unit. New snap rings must be used on reassembly. When assembling, heat piston to about 80° C (180° F) to allow easy insertion of piston pin. Lubricate parts with clean oil and assemble with arrow mark on piston crown aligned with identification mark on connecting rod as shown in Fig. MB21. Install new snap rings with sharp edge facing away from piston pin.

Refer to following paragraph for proper installation of piston rings in ring grooves on piston. Thoroughly lubricate piston, pin, rings and cylinder wall with clean engine oil. Be sure ring end gaps are equally spaced around piston and not aligned. Use a suitable ring compressor when installing piston and rod units in cylinders and be sure arrow mark on piston crown is toward front of engine. Use rubber protector boots on connecting rod bolts to avoid damage to crankshaft rod journals. Tighten connecting rod cap nuts to a torque of 31-34 N·m (23-25 ft.-lbs.) on engine Models K3A, K3C and K3D, and to a torque of 54-58 N·m (40-43 ft.-lbs.) on Models K3H and K3M.

PISTONS, PINS, RINGS AND CYLINDERS

Engine Models K3A, K3C and K3D are equipped with four-ring pistons (three compression rings and one oil control ring). Engine Models K3H and K3M are equipped with three-ring pistons (two compression rings and one oil ring). On all models, pistons operate in unsleeved bores. Pistons are available in oversizes as well as standard size.

Check each cylinder bore for wear, scoring, cracks or other damage. Measure bore at top and bottom to determine wear and taper of cylinder. If cylinder wear exceeds 0.2 mm (0.008 inch), rebore cylinder to oversize. Standard cylinder bore for each engine model is as follows:

Standard Cylinder Diameter—
  K3A . . . . 65 mm
                            (2.559 in.)
  K3C . . . . 70 mm
                            (2.756 in.)
  K3D . . . . 73 mm
                            (2.874 in.)
  K3H . . . . 78 mm
                            (3.071 in.)
  K3M . . . . 84 mm
                            (3.307 in.)

Inspect pistons for wear, scuffing or other damage. Measure piston skirt diameter at lower end of skirt and 90 degrees from pin bore. Subtract piston skirt diameter from cylinder bore measurement to determine operating clearance. Maximum allowable piston skirt to cylinder bore clearance is 0.3 mm (0.012 inch). Measure piston pin to pin bore clearance in piston; maximum allowable clearance is 0.08 mm (0.003 inch) on all models. Standard piston pin diameter is 19 mm (0.748 inch) on models with semi-floating pin and 23 mm (0.9055 inch) on models with full floating pin.

Measure side clearance of new piston ring in groove with feeler gage. On top half-keystone ring, hold ring face flush with piston groove lands and measure clearance between bottom of ring and piston groove. Ring side clearance specifications are as follows:

Ring Side Clearance—
Top Ring
  Standard . . . . 0.06-0.12 mm
                          (0.002-0.005 in.)
  Wear limit . . . . 0.3 mm
                             (0.012 in.)
Second Compression Ring—
  Standard . . . . 0.05-0.09 mm
                          (0.002-0.0035 in.)
  Wear limit . . . . 0.2 mm
                             (0.008 in.)
Third Compression Ring
(If Used)—
  Standard . . . . 0.04-0.08 mm
                          (0.0016-0.003 in.)
  Wear limit . . . . 0.2 mm
                             (0.008 in.)
Oil Ring—
  Standard . . . . 0.03-0.07 mm
                        (0.0012-0.003 in.)
  Wear limit . . . . 0.2 mm
                             (0.008 in.)

Renew piston if scuffed, cracked, or if piston skirt, piston pin and/or piston ring side clearances are excessive.

Check piston ring end gap in cylinder bore; desired end gap is 0.15-0.40 mm (0.006-0.016 inch) with maximum allowable end gap of 1.5 mm (0.060 inch).

When installing rings on piston, refer to Fig. MB22. Place oil ring expander in groove of piston, then install oil ring with end gap 180 degrees from point expander ends butt together. Place remaining rings on piston so that end gaps of all rings are spaced at equal distance around piston.

CONNECTING RODS AND BEARINGS

Connecting rod bearings are slip-in precision type, renewable from below

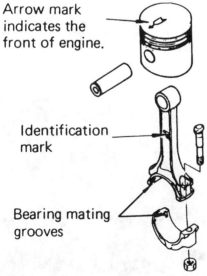

Fig. MB21—When assembling piston and connecting rod, be sure arrow mark on piston crown and identification mark on rod are facing in same direction.

Fig. MB22—Cross-sectional view of piston rings used on four-ring piston (A) and three-ring piston (B). Be sure rings are installed properly in grooves and that rings having identification mark on one side are installed with identification mark up.

SERVICE MANUAL

Mitsubishi

after removing oil pan. Desired bearing to crank journal clearance is 0.04-0.10 mm (0.001-0.004 inch) and maximum allowable clearance is 0.15 mm (0.006 inch). Before installing new bearings, carefully check crankshaft journal for scoring or excessive wear and proceed to renew crankshaft or have crankshaft reground if these conditions are noted.

If irregular bearing wear or wear on thrust faces of crank journal end of rod is noted, remove rod and piston unit as previously outlined, renew rod if defect is obvious or check rod with connecting rod alignment tools. Maximum alignment distortion allowable is 0.05 mm (0.002 inch).

Check side clearance of connecting rod on crankshaft journal. Desired rod side clearance is 0.10-0.35 mm (0.004-0.014 inch). If side clearance exceeds 0.50 mm (0.020 inch), renew rod and/or crankshaft as necessary.

On engine Models K3H and K3M with full floating piston pin, check pin to connecting rod pin bore clearance. Maximum allowable clearance is 0.10 mm (0.004 inch). On other models, pin is press fit in connecting rod.

When installing new connecting rod bearing, be sure that tangs (projections) on bearing shell engage slots in rod and cap and that tangs are to same side of assembly. Be sure that cap is being installed on matching rod. Tighten bearing cap bolts to a torque of 31-34 N·m (23-25 ft.-lbs) on engine Models K3A, K3C and K3D, or to a torque of 54-58 N·m (40-43 ft.-lbs.) on engine Models K3H and K3M.

CRANKSHAFT AND MAIN BEARINGS

Crankshaft is supported in four main bearings and crankshaft end play is controlled by flanges on No. 3 main bearing (see Fig. MB23). Desired crankshaft end play is 0.06-0.03 mm (0.003-0.012 inch); maximum allowable end play is 0.5 mm (0.020 inch). Renew crankshaft if end play is excessive with new main bearings. Main bearings are slip-in precision (insert) type. Desired main journal oil clearance is 0.04-0.10 mm (0.001-0.004 inch); maximum clearance is 0.15 mm (0.006 inch).

Main bearing caps are numbered and marked with an arrow to indicate front of engine. Front and rear caps are sealed at top and sides with sealant. Main bearing caps are not renewable separately from cylinder block and are not interchangeable. Removal of crankshaft requires removal of engine.

Inspect crankshaft bearing journals and seal contact surfaces for wear, scoring or bearing metal adhering to journals from lack of lubrication or engine overheating. Check crankshaft bearing journals with micrometer. Refer to following crankshaft dimensions:

Main Journal Standard
Diameter—
   Models K3A, K3C and K3D . . 52 mm
                            (2.047 in.)
   Models K3H and K3M . . . . . . 57 mm
                            (2.244 in.)

Crankpin Standard
Diameter—
   Models K3A, K3C and K3D . . 42 mm
                              (1.654 in.)
   Models K3H and K3M . . . . . . 48 mm
                              (1.890 in.)

Crankshaft wear is usually noted by scoring of journals, out-of-round condition or bearing taper. Crankshaft journals may be polished if wear or scoring is minimal. If connecting rod journal taper or out-of-round measurement exceeds 0.01 mm (0.0004 inch), the crankshaft should be reground to fit undersize bearings or be renewed. If main bearing journals are worn so that bearing clearance is excessive, regrind or renew crankshaft.

When installing crankshaft, be sure to place main bearing caps in proper numbered position with arrow mark pointing forward. With No. 3 main bearing cap bolts loosened, pry crank back and forth to align bearing thrust surfaces. Apply a light coat of Loctite Gasket Eliminator 515 to upper surface of front and rear bearing caps prior to installation. Apply a thin coat of Loctite 515 to bearing cap side seals, then install seals into grooves of front and rear bearing caps with rounded corner of seals outward as shown in Fig. MB24. Tighten main bearing cap bolts to a torque of 49-54 N·m (36-40 ft.-lbs.).

CRANKSHAFT REAR OIL SEAL

The crankshaft rear oil seal (18—Fig. MB23) is mounted in a seal retainer (17) bolted to rear of cylinder block and can be renewed after removing flywheel. Install new seal in retainer with lip fac-

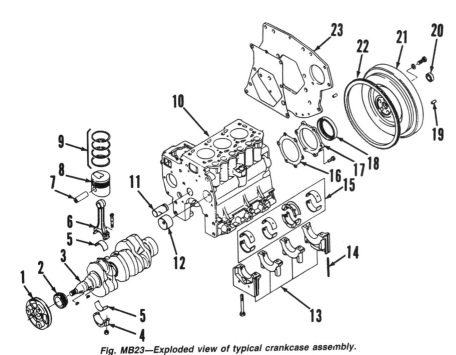

Fig. MB23—Exploded view of typical crankcase assembly.
1. Crankshaft pulley
2. Crankshaft gear
3. Crankshaft
4. Connecting rod cap
5. Bearing insert
6. Connecting rod
7. Piston pin
8. Piston
9. Piston rings
10. Cylinder block
11. Idler gear shaft
12. Camshaft front bearing
13. Main bearing caps
14. Cap side seals
15. Main bearing inserts
16. Gasket
17. Seal retainer
18. Rear oil seal
19. Dowel pin
20. Pilot bearing
21. Flywheel
22. Starter ring gear
23. Rear plate

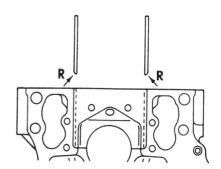

Fig. MB24—Be sure side seals for front and rear main bearing caps are installed with rounded corners (R) facing outward and toward top of cylinder block.

239

Mitsubishi

ing forward. Lubricate seal lip with clean oil and install assembly over crankshaft using new gasket (16).

ELECTRICAL SYSTEM

ALTERNATOR AND REGULATOR

A Mitsubishi alternator with a Tirrill externally mounted regulator is used on all models. The regulator consists of a voltage relay and pilot lamp relay. Regulator has no provision for adjustment. Specified alternator output should be 14 volts and 30 amperes minimum at 2500 rpm. Regulated voltage should be 14 to 15 volts.

To disassemble alternator, scribe mating marks on end frames for proper alignment at reassembly, then proceed as follows: Remove end frame mounting screws and separate stator (8—Fig. MB25) and end frame (11) assembly from drive end frame (4). Remove pulley nut (1), then withdraw rotor (7) from drive end frame.

Check rotor bearings and renew if necessary. Inspect rotor slip ring; minimum wear diameter is 32 mm (1.259 inches). Ohmmeter reading between slip rings should be about 7.9 ohms; higher reading would indicate open circuit and lower reading a short circuit. Ohmmeter reading between rotor shaft and either slip ring should indicate an open circuit; a reading of near zero would indicate grounded circuit.

Disconnect stator coil from rectifier and check windings as follows: Check coil windings for continuity by placing an ohmmeter probe to neutral lead wire "N" and the other probe to each of the three coil wires. A high reading would indicate an open circuit. Connect ohmmeter leads between each stator lead and the stator frame. A very low reading would indicate a grounded circuit. A short circuit within the stator windings cannot be readily determined with an ohmmeter because of low resistance of the windings. However, a stator with a short circuit will usually have burned coil wires in the area of the short.

To test diodes, remove rectifier assembly (10) from end frame. Using an ohmmeter, check positive side diode trio in both directions. Ohmmeter should indicate continuity in one direction and no continuity when probes are reversed. Negative side diode trio may be tested in a similar manner. If no continuity is noted in either direction, one or more of the diodes are defective and rectifier must be renewed.

Inspect brushes and springs for wear and damage. Standard brush length is 18 mm (0.71 inch) and minimum wear length is 8 mm (0.31 inch). Be sure brush contact surfaces are free of grease before unit is assembled.

To reassemble alternator, reverse disassembly procedure.

STARTING MOTOR

Mitsubishi starting motors are used on all models. Starter Model M002T50371 is used on engine Models K3A and K3C. Starter Model M002T50381 is used on engine Model K3D. Starter Model M002T56271 is used on engine Models K3H and K3M. Starter service specifications are as follows:

SMALL DIESEL

Models M002T50371-M00250381
No-load test—
  Volts 11.5
  Current draw (max.)......90 amps
  Rpm (min.) 3600
  Brush length-standard ......17 mm
    (0.67 in.)
  Wear limit ....11.5 mm
    (0.45 in.)
  Pinion gap..........0.5-2.0 mm
    (0.020-0.080 in.)
  Pinion shaft end play
    (max.)..................0.5 mm
    (0.020 in.)

Model M002T56271
No-load test—
  Volts 11.0
  Current draw (max.).....130 amps
  Rpm (min.) 3850

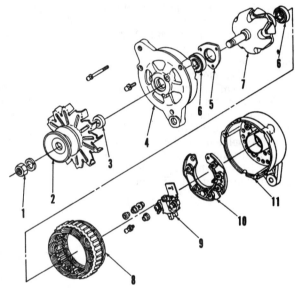

Fig. MB25—Exploded view of alternator assembly used on all models.
1. Nut
2. Pulley & fan assy.
3. Spacer
4. Drive end frame
5. Retainer plate
6. Bearings
7. Rotor
8. Stator
9. Brush holder
10. Rectifier
11. End frame

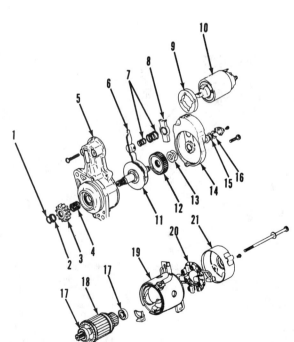

Fig. MB26—Exploded view of starter motor used on engine Model K3M. Other models are similar.
1. Retaining ring
2. Pinion stop
3. Pinion gear
4. Spring
5. Front housing
6. Shift lever
7. Springs
8. Retainer
9. Shims
10. Solenoid
11. Pinion shaft
12. Reduction gear
13. Shims
14. Center housing
15. Washer
16. Retainer
17. Bearings
18. Armature
19. Field coil & housing
20. Brush holder assy.
21. End housing

SERVICE MANUAL

Brush length-standard .... 17 mm
(0.67 in.)
Wear limit .... 11.5 mm
(0.45 in.)
Pinion gap............. 0.5-2.0 mm
(0.020-0.080 in.)
Pinion shaft end play
(max.).................... 1.0 mm
(0.040 in.)

Refer to Fig. MB26 for exploded view of typical starting motor. To disassemble, unbolt and remove solenoid (10). Retain shims (9) for use in reassembly. Remove through-bolts and separate end cover (21), brush holder (20), field coil and housing (19) and the armature (18). Remove retaining ring (1), stop (2), pinion gear (3) and spring (4). Remove cap screws securing front and center housings (5 and 14). Remove rear retainer (16) from pinion shaft (11), then withdraw shaft and reduction gear (12).

Check armature windings for short circuit and renew if necessary. There should not be any continuity between armature commutator and shaft. Check for continuity across both ends of field coils. If open circuit is noted, renew field coil assembly (19). Check armature bearings for roughness and renew as needed. Inspect pinion gear, reduction gear and pinion shaft for wear or damage.

When reassembling starter drive, install shims (13) between rear housing (14) and reduction gear (12) as necessary so that pinion shaft end play is less than shown for starter in specification chart. To check and adjust pinion gear end gap, first remove connector from terminal "M" on solenoid. Connect battery positive cable to solenoid "S" terminal and connect battery negative terminal to starter housing to shift pinion gear outward. Lightly push pinion gear inward to remove slack and measure gap between gear and stop washer (2). Add or remove solenoid mounting shims (9) to obtain desired pinion gap of 0.5-2.0 mm (0.020-0.080 inch).

MWM DIESEL

MWM DIESEL, INC.
3200 Pointe Parkway
Norcross, Georgia 30092

Model	No. Cyls.	Bore	Stroke	Displ.
D202-2	2	95 mm (3.74 in.)	105 mm (4.13 in.)	1490 cc (90.8 cu. in.)
D202-3	3	95 mm (3.74 in.)	105 mm (4.13 in.)	2230 cc (136.2 cu. in.)
D302-1	1	95 mm (3.74 in.)	105 mm (4.13 in.)	750 cc (45.5 cu. in.)
D302-2	2	95 mm (3.74 in.)	105 mm (4.13 in.)	1490 cc (90.8 cu. in.)
D302-3	3	95 mm (3.74 in.)	105 mm (4.13 in.)	2230 cc (136.2 cu. in.)

All MWM (Murphy) engines in this section are four-stroke direct injection diesels with vertical cylinders. D202 engines are liquid-cooled and D302 engines are air-cooled. Crankshaft rotation is counterclockwise viewed from flywheel end on all engines. Firing order is 1-3-2 on three-cylinder engines.

MAINTENANCE

LUBRICATION

Engine crankcase oil must meet API service classification CC/SF, CD/SF, CC or CD. Use of multigrade oil is approved but oil must meet API classification CD/SF or CD. If sulfur content in the fuel exceeds 0.5 percent, a CD or CD/SF oil must be used.

If ambient temperature is permanently below 0°C (32°F), SAE 10W-30 or 5W-20 oil is preferred but SAE 10W may also be used. If temperature varies between minus 10°C (14°F) and 10°C (50°F), SAE 20W-20 is preferred but 15W-40 may be used. If temperature is between 0°C (32°F) and 30°C (85°F), preferred oil is SAE 30 but 15W-40 may be used. If temperature remains above 30°C (85°F), SAE 40 oil is recommended.

Recommended oil change interval is every 250 hours of operation or every six months, whichever occurs first. The oil filter should also be renewed at this time. Manufacturer recommends using only genuine MWM Diesel replacement oil filter.

Standard crankcase capacity with filter change is approximately 4.5 liters (4.7 quarts) for D202-2, 7.75 liters (8.2 quarts) for D202-3, 3.1 liters (3.25 quarts) for D302-1, 5.5 liters (5.8 quarts) for D302-2 and 7.6 liters (8.1 quarts) for D302-3.

With engine warm and running at full speed, normal oil pressure is 400-500 kPa (68-72 psi). An oil pressure relief valve assembly is located in the oil filter mounting housing on D202-3 and D302-3 engines. On all other engines, relief valve assembly is mounted in a drilled passage in the crankcase next to the oil filter.

D302-3 engines are equipped with an engine oil cooler located in the cooling air housing. A cooler bypass valve is set to open at 250 kPa (36 psi) to divert oil from the cooler when oil is cold.

FUEL SYSTEM

FUEL FILTER AND BLEEDING. The fuel filter should be renewed after every 1000 hours of operation or sooner if loss of engine power is evident. If engine fails to operate properly after filter element has been renewed, note that there is a filter screen located in the inlet section of fuel feed pump which may become plugged. Screen can be removed and cleaned after removing top cover from fuel feed pump.

Whenever fuel filter is serviced or fuel lines are disconnected, air must be bled from the fuel system in the following manner. Loosen bleed screw on fuel filter mounting base. Operate priming lever on fuel feed pump (6 – Fig. M1-1) until air-free fuel is discharged at bleed screw, then retighten screw. If engine fails to start at this point, loosen high pressure fuel line fittings at the injec-

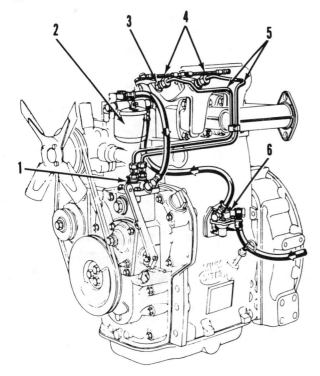

Fig. M1-1 — View of typical fuel system components. Some engines are equipped with two fuel filters.

1. Injection pump
2. Fuel filter
3. Injector
4. Fuel leak-off lines
5. High pressure injection pipes
6. Fuel feed pump

SERVICE MANUAL

MWM

tors. Position injection pump control in "run" position, then crank engine until fuel is discharged at loosened fittings. Retighten fittings and start engine.

INJECTION PUMP TIMING. To check pump timing, it is first necessary to establish an accurate top dead center for number 1 piston as follows: Remove belt guard and rocker arm cover. Rotate crankshaft until intake valve opens about 5-6 mm (¼ inch), then install special tool M2021 to hold intake valve in the open position. Fabricate a reference timing pointer and install over the front pulley. Rotate crankshaft counterclockwise until piston just contacts the intake valve and scribe a reference mark on front pulley in line with the timing pointer. Rotate crankshaft clockwise until piston contacts intake valve and scribe a second reference mark opposite timing pointer. Select a timing decal from degree decal set number M2010 which matches diameter of front pulley. Position the decal (3–Fig. M1-2) equally between the two reference marks on the pulley. Remove M2021 tool from intake valve.

Remove the plug, located in front of injection pump, from the timing gear cover. On pumps with Version I or II governors, push injection pump control rack into the pump so it will not be in the "start-fuel" position. On Version III governors, install special tool M2022 (2) in the plug hole until it bottoms to move control rack out of "start-fuel" position. Position throttle at half speed position.

NOTE: Timing cannot be checked with pump control rack in "start-fuel" position as injection timing is retarded approximately 12° in this position.

Disconnect number one injector line at pump, then remove delivery valve holder, spring and delivery valve from pump. Reinstall delivery valve holder with a drip tube (1) if available. Rotate crankshaft about ½ turn counterclockwise. Manually actuate primer lever on fuel feed pump and note that fuel should flow from delivery valve holder, then slowly turn crankshaft clockwise until fuel flow just stops. At this point, port closure occurs and start of injection begins. If timing is correct, the following reading on degree decal should be aligned with timing pointer: 24° BTDC for D202-2, 23° BTDC for D302-1 or 25° BTDC for all other engines.

Pump timing is adjusted by changing thickness of shims (Fig. M1-3) between injection pump and mounting surface of crankcase. Adding shims will retard timing and removing shims will advance timing. Each 0.1 mm (0.004 inch) change in shim thickness will change pump timing about 1°. Shims are available in 0.2 mm (0.008 inch) and 0.3 mm (0.012 inch) thicknesses.

REPAIRS

TIGHTENING TORQUES

Refer to the following table for special tightening torques. All fasteners are metric.

Camshaft nut* . . . . .200-210 N·m
(148-155 ft.-lbs.)
Connecting rod bolts
 M12-10.9. . . . . .75-80 N·m
(55-59 ft.-lbs.)
 M12-12.9. . . . . .95-100 N·m
(70-74 ft.-lbs.)
Crankshaft
 counterweight . . . . .140-150 N·m
(104-110 ft.-lbs.)
Crankshaft pulley . . . . . . .120-125 N·m
(89-92 ft.-lbs.)
Cylinder head
 D202 engine. . . . . .170-180 N·m
(126-132 ft.-lbs.)
 D302 engine. . . . . .50-55 N·m
(37-40 ft.-lbs.)
Flywheel bolts. . . . . .120-125 N·m
(89-92 ft.-lbs.)
Flywheel housing. . . . . .60-65 N·m
(44-48 ft.-lbs.)
Injector retainer. . . . . .10-15 N·m
(8-11 ft.-lbs.)
Main bearing caps. . . . . . .130-140 N·m
(96-103 ft.-lbs.)
Oil pan. . . . . .20-25 N·m
(15-18 ft.-lbs.)
Oil pump. . . . . .20-25 N·m
(15-18 ft.-lbs.)
Rocker arm bracket. . . . . . .35-40 N·m
(26-29 ft.-lbs.)

*Left-hand threads.

Fig. M1-2—When checking injection timing, install pump rack tool M2022 (2) on engines with Version III governors to move control rack out of automatic "start-fuel" position.
1. Drip tube
2. Pump rack tool M2022
3. Timing decal

COMPRESSION PRESSURE

Compression pressure may be checked to establish relative condition of engine before proceeding with engine disassembly. All injectors should be removed when checking compression pressure. With engine cold, normal compression pressure is 2500-3000 kPa (365-435 psi) at cranking speed. Minimum allowable compression pressure is 2200 kPa (319 psi) and pressure difference between cylinders must not exceed 300 kPa (43 psi). Some causes of low compression pressure are: valves leaking, piston rings worn, cylinder scored or clearance between piston and cylinder head too large.

VALVE ADJUSTMENT

Valve clearance is adjusted with engine cold. Remove rocker arm cover, then turn crankshaft until piston for cylinder being adjusted is at top dead center on compression stroke (both valves closed). Loosen locknut and turn rocker arm adjusting screw until 0.2 mm (0.008 inch) feeler gage can be inserted between rocker arm and valve stem end. Tighten adjusting screw locknut and recheck clearance. Repeat procedure for remaining cylinders. Manufacturer recommends checking valve clearance after every 500 hours of operation.

CYLINDER HEAD

An individual cylinder head is used for each cylinder. Be sure to identify cylinder heads before removal so they can be reinstalled in their original location. Keep shims and gaskets from each cylinder with respective cylinder head for use in reassembly.

Fig. M1-3—Shims located between injection pump and crankcase mounting surface are used to adjust injection timing.

243

MWM

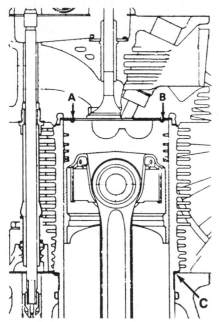

Fig. M1-4—To check piston compression clearance, place lead wires at position "A" and "B" on top of piston. Adjustment on D302 engines is made by changing shim thickness "C". Refer to text.

Clearance between top of piston and cylinder head should be checked as follows: With head removed, place two pieces of lead wire 2 mm (0.080 inch) in diameter at locations (A and B—Fig. M1-4) on piston. Install head with gasket and tighten nuts or bolts to specified torque. Rotate crankshaft by hand until piston has passed top dead center. Remove cylinder and measure thickness of wires. Average thickness of wires should be 0.9-1.2 mm (0.036-0.047 inch). If thickness is not within these specifications on D302 engines, add or remove shims (C) between bottom of cylinder barrel and crankcase. On D202 engines, check rod bearing clearance, piston pin clearance or connecting rod straightness if clearance exceeds specified value.

When reinstalling cylinder head, make certain contact surfaces are clean. On D202 engines, install two M1006 guide studs (or equivalent) diagonally from each other in two cylinder head bolt holes. Note the word "TOP" on head gasket and install accordingly. Position cylinder head over the guide studs and install two bolts (threads oiled) in holes opposite guide studs. Do not install push rods until after cylinder head bolts are tightened. Use a straightedge to align machined surfaces. Remove guide studs and install the other two bolts. Tighten evenly in steps using a criss-cross pattern to 170-180 N·m (126-132 ft.-lbs.). Install push rods and rocker arm assemblies and tighten mounting cap screws to 35-40 N·m (26-29 ft.-lbs.). Run engine for about one hour, then stop and allow to cool. With engine cold, loosen head bolts ¼ turn and retorque to specified setting. Readjust valve clearance.

On D302 engines, install cylinder head with new gasket. Lubricate bottom of stud nuts and threads of nuts and studs with Molykote "G-N" paste or equivalent. Install nuts finger tight, then use a straightedge to align machined surfaces of intake and exhaust port. Tighten nuts evenly in steps using a criss-cross pattern to 45-50 N·m (34-36 ft.-lbs.). Operate engine for about ½ hour, then stop and allow to cool. With engine cold, loosen all nuts ¼ turn and retighten in steps using the criss-cross pattern to 50-55 N·m (37-40 ft.-lbs.) Readjust valve clearance.

VALVE SYSTEM

Intake and exhaust face angle should be 45½° and valve seat angle should be 45°. Valve stem diameter for both valves is 8.952-8.970 mm (0.3525-0.3531 inch) with a wear limit of 8.949 mm (0.3523 inch). Valve guide inner diameter should be 9.013-9.028 mm (0.3548-0.3554 inch) with a wear limit of 9.060 mm (0.3567 inch). When renewing valve guides, cylinder head should first be heated to about 200°C (440°F). After installation, let cool and regrind valve seats.

Desired valve seat width is 2.10-2.45 mm (0.083-0.096 inch). With valves installed, the valve heads should be recessed 1.00-1.45 mm (0.040-0.057 inch) below combustion surface. Maximum allowable valve recession is 1.80 mm (0.071 inch). D302 engines are equipped with renewable valve seat inserts. The inserts are an interference fit in cylinder head counterbores and head should be heated to about 200°C (400°F) prior to installing new inserts.

Rocker arm bracket shaft diameter is 15.966-15.984 mm (0.6286-0.6293 inch) with a wear limit of 15.950 mm (0.6280 inch). Rocker arm bushing inside diameter should be 16.000-16.018 mm (0.6299-0.6306 inch) and maximum allowable diameter is 16.030 mm (0.6311 inch).

INJECTOR

REMOVE AND REINSTALL. To remove injector, disconnect fuel lines and immediately plug all openings to prevent entry of dirt. Remove mounting bracket nuts and withdraw injector and seal washer.

NOTE: Do not reuse nozzle seal washers. When reinstalling new washer, be sure metal side is towards the head.

SMALL DIESEL

To reinstall, reverse the removal procedure. Tighten mounting bracket nuts to 10-15 N·m (8-11 ft.-lbs.).

TESTING. A special nozzle tester is required to check injector operation. Injector should be checked for specified opening pressure, nozzle leakage and correct spray pattern.

WARNING: Fuel leaves the injector nozzle with sufficient force to penetrate the skin. When testing, keep yourself clear of nozzle spray.

Connect injector nozzle test pump, then operate tester lever several quick strokes to purge air from injector and to make sure nozzle valve is not stuck. Then operate tester lever slowly while observing tester gage. Nozzle opening pressure should be 18000-18500 kPa (2610-2680 psi) for standard output engines or 22000-22500 kPa (3190-3260 psi) for high output engines. Opening pressure may be adjusted by turning the adjusting screw located at the top of injector.

To check nozzle for leakage, operate tester lever slowly to maintain pressure at 2000 kPa (300 psi) below opening pressure and observe nozzle tip for leakage. If a drop falls from nozzle tip within a 10 second period, nozzle leaks and should be overhauled or renewed. During this test, be careful not to confuse fuel leaking from leak-off fitting at top of injector with leakage at nozzle tip.

To check spray pattern, operate the tester lever several quick strokes (about one stroke per second). All four nozzle spray holes must be open. Spray should be uniform and well atomized and nozzle should emit a "chatter" sound when operating properly.

INJECTION PUMP

The fuel injection pump pumping elements are actuated by lobes on the camshaft. The injection timing is adjusted by varying the thickness of shims located between the pump mounting flange and the crankcase. Refer to MAINTENANCE section for pump timing procedure.

When removing injection pump, be sure to plug all fuel line openings to prevent entry of dirt. It is recommended that injection pump be tested and overhauled only by a shop qualified in diesel fuel injection repair.

When reinstalling pump, use the original mounting shim pack. If shims were misplaced, required shim thickness can be determined as follows: Turn crankshaft until one lobe on camshaft for fuel pump is pointing downward. Measure distance from mounting sur-

SERVICE MANUAL

MWM

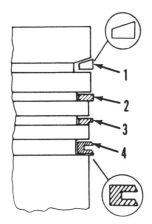

Fig. M1-5—Cross-sectional view of piston and piston rings. Observe "T" or "TOP" mark on rings for correct installation.

1. Keystone top ring
2. Rectangular second compression ring
3. Notched third compression ring
4. Oil control ring

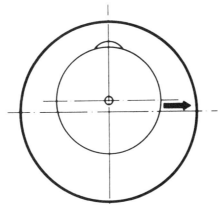

Fig. M1-7—Install piston so arrow on piston crown is pointing towards flywheel end of engine.

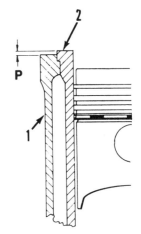

Fig. M1-8—On D202 engines, top of cylinder liner (2) should project (P) 0.05-0.10 mm (0.002-0.004 inch) above crankcase (1) surface when properly installed.

face of crankcase to base of camshaft lobe. Assemble shim thickness necessary to obtain mounting dimension of 82.7-82.9 mm (3.256-3.263 inches). Check injection pump timing as outlined in MAINTENANCE section.

PISTON, PISTON PIN AND RINGS

The piston is equipped with a keystone type top ring (Fig. M1-5), a rectangular second compression ring, a notched third compression ring and an oil control ring. When renewing pistons, note the difference between high output piston (HO–Fig. M1-6) and standard output piston (SO). If converting engine from one piston to the other, be sure to install correct fuel injection nozzles. High output piston should not be used in D302-1 engine.

Piston diameter should be 94.540-94.560 mm (3.7220-3.7228 inches) for all engines. Minimum allowable piston diameter is 94.50 mm (3.7205 inches). Ring groove width should be 3.060-3.080 mm (0.1205-0.1212 inch) for second and third compression rings with a wear limit of 3.120 mm (0.1228 inch). Maximum allowable ring side clearance for second and third rings is 0.20 mm (0.008 inch). Oil control ring groove width should be 5.050-5.070 mm (0.1988-0.1996 inch) with a wear limit of 5.110 mm (0.2012 inch). Oil control ring side clearance in groove should not exceed 0.15 mm (0.006 inch).

With ring positioned squarely in cylinder bore, measure ring end gap with a feeler gage. Desired end gap is 0.40-0.65 mm (0.016-0.026 inch) for all rings and maximum allowable end gap is 2.0 mm (0.079 inch).

The full floating piston pin is retained in the piston by two snap rings. Piston pin diameter is 31.994-32.000 mm (1.2596-1.2598 inches) with a wear limit of 31.990 mm (1.2595 inches). Inner diameter of pin bore in piston should be 32.002-32.006 mm (1.2599-1.2601 inches) with a wear limit of 32.015 mm (1.2604 inches).

When reassembling, be sure to note "T" or "TOP" marking on one side of ring and install with that side facing upward. Refer to Fig. M1-5. Be sure piston is reinstalled with arrow on piston crown pointing towards flywheel end of engine

(Fig. M1-7). Check piston height in cylinder as outlined in CYLINDER HEAD section.

CYLINDER

D202 engines are equipped with removable, wet type cylinder liners. D302 engines are equipped with removable, cast iron cylinder barrels.

Standard cylinder bore is 95.000-95.022 mm (3.7402-3.7410 inches) with a wear limit of 95.250 mm (3.750 inches). If excessively worn or scored, cylinders may be rebored for installation of 0.5 mm (0.020 inch) or 1.0 mm (0.040 inch) oversize piston and rings.

On D202 engines, the cylinder liner should project (P–Fig. M1-8) 0.05-0.10 mm (0.002-0.004 inch) above crankcase surface. If liner projection is not within specified limits, check for foreign material between liner flange and crankcase counterbore.

On D302 engines, clearance between piston and cylinder head should be checked as outlined in CYLINDER HEAD section whenever cylinder barrel is renewed. Shims located between bottom of cylinder and crankcase mounting surface are used to adjust piston compression clearance.

CONNECTING ROD

The connecting rod uses a bushing in the small end and a precision, insert type bearing in the big end.

Piston pin bushing inside diameter should be 32.030-32.080 mm (1.2610-1.2630 inches) with a wear limit of 32.150 mm (1.2657 inches). Piston pin outer diameter is 31.994-32.000 mm (1.2596-1.2598 inches). Desired

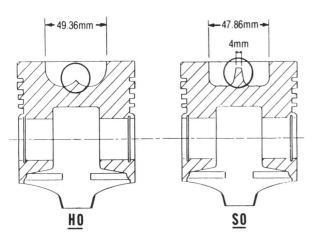

Fig. M1-6—Cross-sectional view of high output (HO) piston and standard output (SO) pistons showing visual differences.

MWM SMALL DIESEL

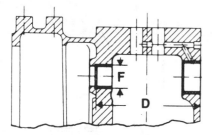

Fig. M1-9—Camshaft bore (F) at flywheel end may be rebored for installation of bushing. End of bore (D) should be 156-157 mm (6.14-6.18 inches) from front surface of crankcase on D302-1 engine. Refer to Fig. M1-10 for all other engines.

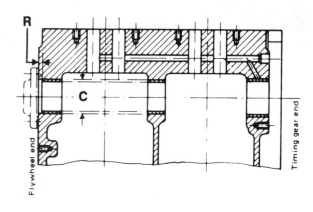

Fig. M1-10—Camshaft bores (C) for two and three-cylinder engines may be bore oversize for installation of bushings. Rear edge of bushings must be recessed (R). Refer to text for details.

clearance between piston pin and bushing is 0.03-0.07 mm (0.0012-0.0027 inch).

Standard diameter of crankpin is 57.951-57.970 mm (2.2815-2.2822 inches). Desired operating clearance for crankpin in rod bearing is 0.07-0.10 mm (0.0027-0.0039 inch). Connecting rod bearing must be renewed when the third layer of bearing material wears off which will be indicated by a shiny bronze contact pattern appearing on bearing surface.

The connecting rod bearing is positioned in big end of rod with locating pin. When assembling connecting rod and bearing, it is recommended that lower half of bearing insert be installed without bearing cap to ensure notch in bearing is indexed with pin. Then, reinstall bearing cap and tighten to 75-80 N·m (55-59 ft.-lbs.) for M12-10.9 bolts or 95-100 N·m (70-74 ft.-lbs.) for M12-12.9 bolts.

CAMSHAFT

The camshaft is supported at the front by a ball bearing located in the crankcase front cover, a renewable bushing in crankcase front bulkhead and directly in bores of crankcase at the rear.

To remove camshaft, first remove cylinder heads and push rods, crankshaft pulley and belt, cooling fan, injection pump, fuel transfer pump and timing gear cover assembly. Use suitable magnetic tools to hold the mushroom type cam followers up away from camshaft assembly. Identify cam followers as they are removed for use in reassembly.

Renew camshaft if bearing journals or cam lobes are scored. Camshaft journal diameter for ball bearing is 24.980-24.993 mm (0.9835-0.9840 inch). Journal diameter at front bushing location is 38.043-38.059 mm (1.4978-1.4984 inches). Bushing inside diameter should be 43.000-43.025 mm (1.6929-1.6939 inches) and maximum allowable inner diameter is 43.050 mm (1.6949 inches). On D302-1 engine, camshaft rear journal diameter is 29.940-29.960 mm (1.1787-1.1795 inches) with a wear limit of 29.920 mm (1.1780 inches). Crankcase bore for rear journal should be 30.000-30.025 mm (1.1811-1.1821 inches) with a wear limit of 30.050 mm (1.1831 inches). On all other engines, diameter of rear journals is 42.940-42.960 mm (1.6906-1.6913 inches) with a wear limit of 42.920 mm (1.6898 inches). Inner diameter of crankcase bores should be 43.000-43.025 mm (1.6929-1.6939 inches) and maximum allowable diameter is 43.050 mm (1.6949 inches).

Cam follower outer diameter should be 17.983-17.994 mm (0.7080-0.7084 inch) with a minimum allowable diameter of 17.975 (0.7077 inch). Cam follower bores in crankcase should be 18.000-18.015 mm (0.7087-0.7092 inch) with a wear limit of 18.020 mm (0.7094 inch).

If camshaft journals or crankcase bores are excessively worn, the journals can be ground 0.3 mm (0.012 inch) undersize and the crankcase bores at flywheel end can be rebored for installation of bushings. On D302-1 engine, machine crankcase bore (F–Fig. M1-9) inner diameter to 34.000-34.025 mm (1.3386-1.3395 inches). Note that end of bore (D) should be 156-157 mm (6.14-6.18 inches) from front surface of crankcase. On all other engines, machine camshaft bores (C-Fig. M1-10) to inner diameter of 47.000-47.025 mm (1.8504-1.8514 inches). Press rear bushing in until rear edge of bushing is recessed (R) 4 mm (0.157 inch) from rear surface of crankcase. On all engines, make certain oil hole in bushing is aligned with drilled oil passage in crankcase. Bushing inner diameter must be finish machined according to camshaft being installed (new dimensions or undersize). On D302-1 engine, inner diameter of rear bushing should be 30.000-30.021 mm (1.1811-1.1891 inches) for a new camshaft or 29.700-29.721 mm (1.1693-1.1701 inches) for an undersize camshaft. On all other engines, finished size of bushing bore should be 43.000-43.025 mm (1.6929-1.6939 inches) for a new camshaft or 42.700-42.725 mm (1.6811-1.6821 inches) for an undersize camshaft. Be sure to thoroughly clean crankcase after machining is completed.

Camshaft gear is a press fit on camshaft. Heat gear to approximately 220°C (425°F) prior to installation to obtain shrink fit. Note that gear has a scribe mark on hub which must be in line with a scribe mark on camshaft (Fig. M1-11) to provide correct valve timing.

To reinstall camshaft, reverse the removal procedure. Make certain "O" mark on crankshaft gear is meshed with "O-O" mark on camshaft gear.

BALANCE WEIGHT

D302-1 Engine

A counterbalance weight is used on D302-1 engines. Weight and gear may be removed after first removing timing gear cover.

Inspect ball bearing in timing gear cover for wear and renew if needed. Counterweight shaft bushing is a press fit in crankcase. Inside diameter of bushing when new is 32.005-32.030 mm (1.2600-1.2610 inches) and maximum

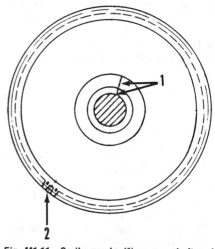

Fig. M1-11—Scribe marks (1) on camshaft and camshaft gear must be aligned when installing gear for correct timing. The "O-O" marks (2) must be aligned with "O" mark on crankshaft gear.

SERVICE MANUAL

MWM

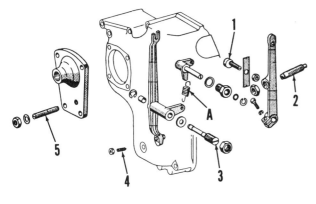

Fig. M1-12—Exploded view of Version I governor linkage and adjustment points.
A. Governor spring
1. High idle adjusting screw
2. Low idle adjusting screw
3. Start-fuel adjusting screw
4. Stop-limit adjusting screw
5. Throttle lock screw

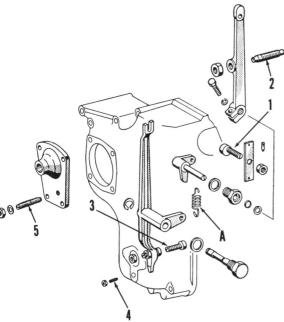

Fig. M1-13—Exploded view of Version II governor linkage and adjustment points. Refer to Fig. M1-12 for identification of components.

allowable bore diameter is 32.100 mm (1.2638 inches). Counterweight shaft outer diameter at bushing end is 31.950-31.975 mm (1.2579-1.2589 inches) with a wear limit of 31.900 mm (1.2559 inches).

Drive gear is a shrink fit on counterweight shaft. Shaft diameter at gear location should be 34.049-34.065 mm (1.3405-1.3411 inches). When renewing gear, first heat gear to about 220°C (425°F). Press gear on shaft making sure scribe mark on face of gear is aligned with "V" notch on counterweight.

Counterweight assembly must be properly timed with crankshaft gear when reinstalled. Make certain the "1" mark on crankshaft gear is aligned between the two chamfered teeth located on back side of counterweight gear. DO NOT use scribe mark on gear for gear timing.

GOVERNOR

The governor flyweight assembly is mounted on the oil pump drive gear. The governor linkage is mounted on the inside of the timing gear cover. For access to governor assembly, remove timing gear cover. Governor components should be checked for excessive wear or binding and renewed if needed.

Three different versions of governor linkage have been used. Refer to Figs. M1-12, M1-13 and M1-14. The main difference between the governors is the Version III contains two springs (A and B – Fig. M1-14) instead of one. The additional spring (B) provides an automatic "start-fuel" feature while a "start-fuel"

Fig. M1-14—Exploded view of Version III governor linkage and adjustment points. Refer to Fig. M1-12 for identification of components except for the following: Spring (B) provides automatic positioning of pump rack in "start-fuel" position. Screw (5) is used to adjust maximum fuel delivery.

button is used on the injection pump with Version I and II governors to provide extra fuel delivery during starting.

The other governor spring (A) used on all governors controls the engine operating speed range. Different springs are available for use at a specific speed range. Using a spring in an rpm range that it is not designed for will result in unsatisfactory governor performance. Consult with local engine distributor for listing of different springs and their rpm range. If engine speed range is changed, stamp the new maximum full load rpm on the engine name plate.

Adjustments

HIGH IDLE. The engine high idle (no-load) speed is adjusted using adjusting screw (1 – Fig. M1-12, M1-13 or M1-14) located on timing gear cover behind the throttle lever. An eccentric disc is located on inner end of this screw and only one turn is required to change setting from minimum to maximum. The high idle rpm should be adjusted to 3-5 percent above the maximum rated full load rpm. Specified full load speed is stamped on the engine name plate.

NOTE: If adjusting screw is turned too far in or out, internal governor lever may not contact the eccentric causing governor instability at high rpm. To correct this, proceed as follows:

Remove camshaft cover from timing gear cover and visually observe eccentric. Turn screw in or out as required until proper contact is made with governor lever. Reinstall camshaft cover, then adjust screw to obtain recommended high idle rpm. Install a seal wire to secure screw adjustment.

LOW IDLE. The engine low idle speed is adjusted using adjusting screw (2 – Fig. M1-12, M1-13 or M1-14) located on throttle lever. The adjusting screw also contains a spring-loaded pin which allows throttle lever to be moved past low idle position to shut off fuel delivery. Minimum low idle speed is 650 rpm.

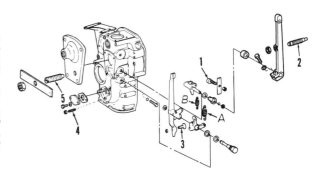

MWM

SMALL DIESEL

START-FUEL. The "start-fuel" adjusting screw (3–Fig. M1-12) in Version I governor is eccentric. One turn changes adjustment setting from minimum to maximum. If screw is turned too far in or out, governor lever may bind against front cover or fuel pump which could result in engine overspeed. Be sure pump control rack moves freely after adjustment.

CAUTION: DO NOT adjust "start-fuel" setting with engine running.

To adjust "start-fuel", remove crankshaft pulley. Remove lower cover plate, camshaft cover (Version III only) and top plug from timing gear cover. Move throttle to maximum speed position. Use a depth micrometer through plug opening in gear cover (Fig. M1-15) to measure distance from end of pump rack (3) to surface of front cover. The distance should be 11.0-11.4 mm (0.433-0.449 inch) on D302-1, 7.0-7.4 mm (0.276-0.291 inch) on D202-2 or 6.8-7.2 mm (0.268-0.283 inch) on all other engines. Turn adjusting screw (3–Fig. M1-12, M1-13 or M1-14) to obtain specified dimension. Tighten adjusting screw locknut to secure adjustment.

STOP-LIMIT. The stop-limit adjustment provides a mechanical stop to prevent injection pump control rack from bottoming in pump when control linkage is moved to shut-off position. The adjusting screw (4–Fig. M1-12, M1-13 or M1-14) is located on timing gear cover just below the lower cover plate.

CAUTION: Turning stop-limit adjusting screw in too far will interfere with governor weight operation and result in engine overspeed. DO NOT perform stop-limit adjustment with engine running.

To adjust stop-limit, place throttle lever in shut-off position. Back out adjusting screw several turns. Using a depth micrometer (5–Fig. M1-15) through plug opening at top of timing gear cover, measure distance from surface of gear cover to end of control rack (3) with rack bottomed in pump. Subtract 0.5 mm (0.020 inch) from this reading and adjust micrometer to this dimension. Reposition depth micrometer (with adjusted setting) through plug opening, then turn stop-limit adjusting screw in until control rack just starts to lift depth micrometer away from cover. Tighten adjusting screw locknut at this position.

THROTTLE LOCK (VERSION I and II). The throttle lock adjusting screw (5–Fig. M1-12 or M1-13) can be used as a low speed adjustment for a dual speed application. The adjusting

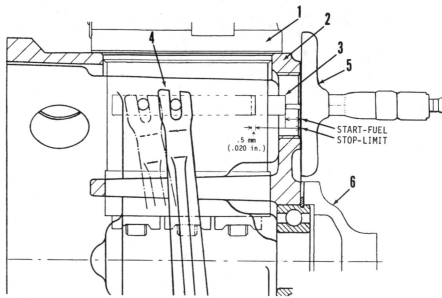

Fig. M1-15—Use a depth micrometer through plug hole at top of timing gear cover when adjusting "start-fuel" and "stop limit" settings. Refer to text.

1. Fuel injection pump
2. Timing gear cover
3. Pump control rack
4. Governor lever
5. Depth micrometer
6. Camshaft cover

screw, if turned all the way in, will also lock the internal governor lever against the high speed screw eccentric. The engine will then run at maximum governed speed with no way to slow it down. The screw is normally left "backed out" as with either situation just described, use of the throttle lever to control engine speed or to shut-off engine will be eliminated.

MAXIMUM FUEL DELIVERY (VERSION III). The maximum fuel delivery adjusting screw (5–Fig. M1-14) is used to set maximum full load horsepower with Version III governors. The screw, located on the camshaft cover, is not interchangeable with Version I and II governors. A dynamometer is required to properly adjust full load horsepower setting. DO NOT attempt adjustment without a dynamometer as engine damage may result from overfueling. Turning screw in will decrease maximum fuel output and turning screw out will increase fuel delivery.

CRANKSHAFT AND BEARINGS

To remove camshaft, first remove oil pan, cylinder heads, pistons and connecting rods, timing gear cover assembly, crankshaft gear (D302-1 engine), governor, oil pump, flywheel and flywheel housing. Mark position of crankshaft counterweights (D302-1 engine), then remove counterweights. Remove crankcase rear cover. On D302-1 engine, withdraw crankshaft through rear opening in crankcase. On all other engines, remove main bearing caps (noting location numbers on caps), then lift out crankshaft.

On all engines, standard diameter of crankshaft main journals is 64.951-64.970 mm (2.5571-2.5578 inches) and minimum allowable diameter is 64.940 mm (2.5567 inches). Crankpin standard diameter is 57.951-57.970 mm (2.2815-2.2823 inches) with a wear limit of 57.940 mm (2.2811 inches). Recommended main bearing clearance is 0.07-0.12 mm (0.0028-0.0047 inch) for D302-1 engine. For all other engines, main bearing clearance should be 0.08-0.11 mm (0.0031-0.0043 inch) and clearance for rear thrust bearing should be 0.09-0.12 mm (0.0035-0.0047 inch). Rod bearing clearance should be 0.07-0.10 mm (0.0028-0.0039 inch) for all engines. Main and rod bearings should be renewed if third layer of bearing material is worn off which is indicated by shiny bronze contact patterns on bearing surface. Main bearings are a press fit on D302-1 engine. Use tool M1016 (or equivalent) to remove and install bearings. Bearings are available in 0.20, 0.50, 0.75 and 1.0 mm (0.008, 0.020, 0.030 and 0.040 inch) undersizes.

On D302-1 engine, crankshaft end clearance is controlled by thrust washers at gear end of shaft. Washer thickness when new is 2.98-3.00 mm (0.117-0.118 inch) and minimum allowable thickness is 2.95 mm (0.116 inch). Recommended end clearance is 0.10-0.22 mm (0.004-0.009 inch). When installing thrust washers, be sure oil

SERVICE MANUAL

MWM

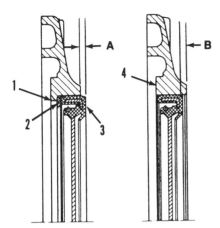

M1-16—Crankshaft rear oil seal may be installed offset (A or B) in crankcase rear cover to obtain contact with unworn portion of crankshaft flange. Refer to text.

1. Cover with flange
2. Shim
3. Oil seal
4. Cover without flange

grooves in rear washer face web of crankshaft gear and grooves in front washer face crankshaft gear.

On all engines except D302-1, crankshaft end clearance is controlled by the flanged rear main bearing. Desired end clearance is 0.08-0.23 mm (0.003-0.009 inch). Thrust bearing is available in 0.10, 0.25 and 0.50 mm (0.004, 0.010 and 0.020 inch) oversize widths as well as an optional 1.0 mm (0.040 inch) oversize width on a limited supply basis.

To reinstall crankshaft, reverse the removal procedure while observing the following points. Make certain bearings are properly positioned on locating pins. Tighten main bearing cap screws to 130-140 N·m (96-103 ft.-lbs.). On D302-1 engine, the crankshaft gear can be installed in only one position with all four bolt holes aligned. Tighten gear mounting bolts to 30-40 N·m (22-29 ft.-lbs.). On all other engines, crankshaft gear is positioned on crankshaft with a locating pin. The gear is a shrink fit and must be heated to 220°C (425°F) prior to installation. On all engines, backlash between crankshaft gear and camshaft gear should be 0.11-0.27 mm (0.005-0.010 inch). Be sure "O" mark on crankshaft gear is aligned with "O-O" marks on camshaft gear. On D302-1 engine, the "1" mark on crankshaft gear must also be aligned with the two chamfered teeth located on back side of counterweight gear. DO NOT use scribe mark on front of gear for timing purposes. Renew crankshaft front and rear oil seals as outlined in OIL SEALS section. Tighten flywheel mounting cap screws to 120-125 N·m (89-92 ft.-lbs.), flywheel housing cap screws to 60-65 N·m (44-48 ft.-lbs.) and crankcase rear cover screws to 20-25 N·m (15-18 ft.-lbs.).

OIL SEALS

Crankshaft front oil seal is a rope type seal located in the timing gear cover. Renewal is obvious after removing front pulley and hub. Apply grease to seal surface prior to reassembly.

Crankshaft rear oil seal is a lip type seal located in the crankcase rear cover. If a groove is worn into crankshaft flange at seal location, seal may be installed offset to provide a new sealing surface. If rear cover has an inner flange (1–Fig. M1-16), install a shim (part number 6.000.0.340.004.4) between flange and new seal (3) to offset seal outward (A). If rear cover does not have a flange (4), install seal offset 2 mm (0.080 inch) inward (B) to obtain new seal contact surface. Apply grease to seal lip and crankshaft flange prior to reassembly.

MWM DIESEL

Model	No. Cyls.	Bore	Stroke	Displ.
D226-2	2	105 mm (4.13 in.)	120 mm (4.72 in.)	2078 cc (127 cu. in.)
D327-2	2	100 mm (3.94 in.)	120 mm (4.72 in.)	1885 cc (115 cu. in.)

The MWM (Murphy) engines in this section are four-stroke, direct injection diesels with vertical cylinders. D226-2 engine is liquid-cooled and D327-2 engine is air-cooled. Crankshaft rotation is counterclockwise viewed from flywheel end on all engines.

MAINTENANCE

LUBRICATION

Engine crankcase oil must be good quality oil meeting API service classification CC/SF, DC/SF, CC or CD. Use of multigrade oil is approved but oil must meet API classification CD/SF or CD. If sulfur content in the fuel exceeds 0.5 percent a CD or CD/SF oil must be used.

If ambient temperature is permanently below 0°C (32°F), SAE 10W-30 or 5W-20 oil is preferred but SAE 10W may also be used. If temperature varies between minus 10°C (14°F) and 10°C (50°F), SAE 20W-20 oil preferred but 15W-40 may also be used. If temperature is between 0°C (32°F) and 30°C (85°F), SAE 30 oil is recommended but 15W-40 may also be used. If temperature remains above 30°C (85°F), SAE 40 oil is recommended.

NOTE: Use the same oil in the injection pump as required for engine crankcase.

Recommended oil change interval is every 250 hours of operation or every six months, whichever occurs first. The oil filter should also be renewed at this time. Manufacturer recommends using only genuine MWM Diesel replacement oil filter.

Standard crankcase capacity is 5.2 liters (5.5 quarts) for D226-2 engine and 4.7 liters (5.0 quarts) for D327-2 engine.

The Robert Bosch injection pump and governor assembly is lubricated with oil. The pump oil level should be checked at the oil level plug every 125 hours of operation. Due to normal fuel leakage past the pump plungers, oil level will increase during operation and must be drained off at the level check plug. On some continuous load applications, fuel leakage may be higher than normal and oil level check/drain interval may have to be lower than 125 hours. Manufacturer recommends renewing injection pump lubricating oil after every 750 hours of operation. Use same oil as recommended for engine crankcase.

FUEL SYSTEM

FUEL FILTER AND BLEEDING. The fuel filter should be renewed after every 1000 hours of operation or sooner if loss of engine power is evident. Note that there is also a filter screen located in the inlet section of the fuel feed pump which may become plugged. If engine fails to operate properly after renewing the fuel filter, remove and clean feed pump filter screen.

Whenever fuel filter is serviced or fuel lines are disconnected, air must be bled from the fuel system. Loosen bleed screw at fuel filter housing. Actuate fuel feed pump priming lever until air-free fuel flows from bleeder, then retighten bleed screw. Loosen bleed screw at top, rear of fuel injection pump housing. Operate feed pump priming lever until air is purged from pump, then retighten bleed screw. If engine fails to start at this point, loosen high pressure fuel line fittings at the injectors. Place pump controls in "run" position, then crank engine until fuel is discharged at loosened fittings. Retighten fittings and start engine.

ENGINE SPEED ADJUSTMENT. Before attempting engine speed adjustments, the governor identification number (stamped on tag located on governor housing) must be known. Two governors which may be used on these engines are as follows: EP/RSV325..1500 A2B 550DR governor regulates speed from a minimum of 650 rpm to a maximum of 3000 rpm with an average speed droop of 5-7 percent. The EP/RSV300...1000 A7B 505DR governor regulates speed from a minimum of 600 rpm to a maximum of 2000 rpm with an average speed droop of 2-3 percent.

NOTE: Changing the maximum speed setting may result in a change in speed droop, governor instability, power or smoke level. This will require additional adjustments which must be made with the use of a pump test bench.

To increase maximum speed setting, back out the throttle lever stop screw. This will also decrease droop. To decrease maximum speed setting, turn throttle lever stop screw in. This will also increase speed droop.

If a surge develops, if droop is excessive or if desired maximum rpm cannot be obtained, the main spring screw (Fig. M2-1) must be adjusted as follows:

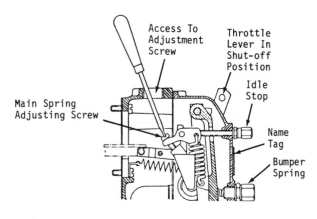

Fig. M2-1—View of governor control adjustment points.

SERVICE MANUAL

MWM

INJECTION TIMING

Engine	Pump Gear Equipped With Timing Device	Rpm Range	Timing Degrees Before TDC
D226	No	1500-1800	28
D226	No	1801-2300	30
D226	No	2301-2500	34
D226	No	3000	36
D226	No	3300	39
D226	Yes	650-3000	28
D327	No	1500-1800	28
D327	No	1801-2400	30
D327	No	3000 Constant	28
D327	No*	1500-1800	25½
D327	Yes	2401-2800	24
D327	Yes	650-3000	25
D327	Yes*	650-3000	25½

*Hand Start Only.

Fig. M2-2—Injection pump timing specification table.

Remove oil filler plug or breather from top of governor housing. Back out the idle stop screw and bumper spring screw. Place throttle lever in shut-off position. Turn main spring adjusting screw clockwise until it bottoms. On A2B governors, turn screw back out 2 turns. On A7B governors, turn screw back out 2 turns for 1000-1200 rpm setting, 3 turns for 1200-1600 rpm setting or 4 turns for 1600-2000 rpm setting.

After main spring screw is adjusted, start engine and adjust idle stop screw until engine speed is 50 rpm below desired idle speed. Adjust bumper spring screw until desired idle speed is obtained. Move throttle against maximum speed stop screw (be careful not to overspeed engine), then turn stop screw until desired maximum speed is obtained.

INJECTION PUMP TIMING. To check injection pump timing, it is first necessary to establish an accurate top dead center for number 1 piston as follows: Remove rocker arm cover. Rotate crankshaft clockwise until the intake valve opens about 5-6 mm (¼ inch), then install special tool M2021 on rocker arm bracket to hold intake valve in the opened position. Fabricate a reference timing pointer and attach to front cover over the crankshaft pulley. Rotate crankshaft counterclockwise until piston just contacts the intake valve and scribe a reference mark on front pulley in line with the timing pointer. Then, rotate crankshaft clockwise until piston just contacts intake valve and scribe a second reference mark on pulley in line with timing pointer. Select a timing decal from degree decal set number M2010 that matches diameter of pulley. Position the decal equally between the two reference marks on the pulley. Remove special tool M2021 from intake valve.

Remove number 1 injector fuel line from pump and install a drip tube in its place. Connect a nozzle tester pump to injection pump inlet using line kit number M2020. Move the pump "run/stop" control lever to the midrange position. Note that injection timing will be retarded about 9° when control lever is in "run" position. If control lever is in "stop" position, port closure (beginning of injection) will not occur. Rotate crankshaft counterclockwise until timing pointer is at approximately 40° BTDC. Pump the nozzle tester handle until air-free fuel flows from drip tube. Actuate tester pump to maintain a flow from drip tube and rotate crankshaft slowly clockwise until fuel flow just stops. At this point beginning of injection occurs and timing pointer should be aligned with specified degree mark on decal (within ½ degree) as listed in table shown in Fig. M2-2.

If necessary, pump timing may be adjusted as follows: Injection pump may be shifted in its slotted mounting holes for approximately 7° of adjustment. Moving top of pump towards the engine will advance the timing. The cam gear to pump drive gear mounting holes are also slotted to allow approximately 26° of pump timing adjustment. To adjust drive gear, engine front cover must first be removed. Move crankshaft to the specified static timing degrees position and secure "run/stop" lever in midrange position. Loosen cap screws securing camshaft gear to pump drive gear, then rotate injection pump shaft counterclockwise until drive gear cap screws bottom in the slots. While actuating nozzle test pump so fuel flows from drip tube, slowly rotate pump shaft clockwise until fuel flow just stops. At this point, tighten drive gear mounting screws to 55-60 N·m (41-44 ft.-lbs.) for M8-12.9 screws or 30-35 N·M (23-25 ft.-lbs.) for M8-10.9 screws. Recheck port closure timing and, if necessary, make final adjustments by rotating pump in slotted mounting holes.

Relieve pressure in nozzle tester hose to pump, then disconnect hose. Reconnect fuel inlet line and number 1 injector line to pump. If crankshaft front hub bolts or flywheel bolts were used to rotate crankshaft, be sure to check torque on these bolts and retighten to specifications if necessary.

REPAIRS

TIGHTENING TORQUES

Refer to the following appropriate table for special tightening torques. Torque values marked with asterisk* apply to engines with 9 digit engine serial numbers (small crank engines). All fasteners are metric.

Model D226-2
Camshaft gear
  Durlok M8-12.9 . . . . 55-60 N·m
   (41-44 ft.-lbs.)
  M8-10.9* . . . . 30-35 N·m
   (23-25 ft.-lbs.)
Connecting rod . . . . 95-100 N·m
   (70-74 ft.-lbs.)
Crankshaft
 counterweight* . . . . 140-150 N·m
   (104-110 ft.-lbs.)
Cranshaft hub
  M16-10.9 . . . . 285-295 N·m
   (211-217 ft.-lbs.)
  M16-12.9 . . . . 345-355 N·m
   (255-261 ft.-lbs.)
  M12-10.9* . . . . 120-125 N·m
   (89-92 ft.-lbs.)
  M12-12.9* . . . . 145-150 N·m
   (107-110 ft.-lbs.)
Cylinder head
  M14-12.9 . . . . 200-210 N·m
   (148-154 ft.-lbs.)
  M14-10.9* . . . . 170-175 N·m
   (126-129 ft.-lbs.)
Flywheel
  M16-10.9 . . . . 285-295 N·m
   (211-217 ft.-lbs.)
  M16-12.9 . . . . 345-355 N·m
   (255-261 ft.-lbs.)
  M12-10.9* . . . . 120-125 N·m
   (89-92 ft.-lbs.)
  M12-12.9* . . . . 145-150 N·m
   (107-110 ft.-lbs.)

MWM

Model D226-2 (Cont.)

Injector..................10-15 N·m
  (8-11 ft.-lbs.)
Injection pump gear.......60-70 N·m
  (45-71 ft.-lbs.)
Main bearing cap
  M14-10.9..............160-170 N·m
    (118-125 ft.-lbs.)
  M14-10.9*.............130-140 N·m
    (96-103 ft.-lbs.)
Oil Pan..................20-25 N·m
  (15-18 ft.-lbs.)
Oil pump.................30-40 N·m
  (23-29 ft.-lbs.)
Rocker arm bracket.......35-40 N·m
  (23-29 ft.-lbs.)

Model D327-2

Camshaft gear
  Durlock M8-12.9........60-70 N·m
    (45-51 ft.-lbs.)
  M8-10.9*...............30-35 N·m
    (23-25 ft.-lbs.)
Connecting rod
  M12-12.9...............95-100 N·m
    (70-73 ft.-lbs.)
  M12-10.9*..............75-80 N·m
    (56-59 ft.-lbs.)
Crankshaft
  counterweight*........140-150 N·m
    (104-110 ft.-lbs.)
Crankshaft hub
  M16-10.9..............285-295 N·m
    (211-217 ft.-lbs.)
  M12-10.9*.............120-125 N·m
    (89-92 ft.-lbs.)
  M12-12.9*.............150-155 N·m
    (111-114 ft.-lbs.)
Cylinder head...........50-55 N·m
  (37-40 ft.-lbs.)
Flywheel
  M16-10.9..............285-295 N·m
    (211-217 ft.-lbs.)
  M12-10.9*.............120-125 N·m
    (89-92 ft.-lbs.)
  M12-12.9*.............150-155 N·m
    (111-114 ft.-lbs.)
Injector................10-15 N·m
  (8-11 ft.-lbs.)
Injection pump gear......60-70 N·m
  (45-51 ft.-lbs.)
Main bearing caps
  M14-10.9..............160-170 N·m
    (118-125 ft.-lbs.)
  M14-10.9*.............130-140 N·m
    (96-103 ft.-lbs.)
Oil pan..................20-25 N·m
  (15-18 ft.-lbs.)
Oil pump.................30-40 N·m
  (23-29 ft.-lbs.)
Rocker arm bracket.......35-40 N·m
  (26-29 ft.-lbs.)

COMPRESSION PRESSURE

Compression pressure may be checked to establish relative condition of engine

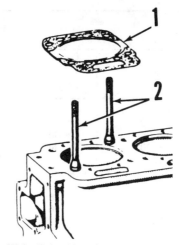

Fig. M2-3—Use guides studs (2) to align cylinder heads and gaskets (1) when reassembling.

before proceeding with engine disassembly. To check compression pressure, remove both injectors and install test adapter M3002 in one of the injector bores in cylinder head. Crank engine 3 to 5 revolutions. With engine cold, normal compression pressure at cranking speed is 2200-2700 kPa (320-390 psi) for D226-2 engine and 2500-2700 kPa (363-390 psi) for D327-2 engine. Compression pressure should not be less than 2000 kPa (290 psi) for D226-2 engine or 2200 kPa (320 psi) for D327-2 engine. On all engines, pressure difference between cylinders should not vary more than 300 kPa (43 psi). Some causes of low compression pressure are: valves leaking, piston rings worn, cylinder scored or clearance between piston and cylinder head too large.

VALVE ADJUSTMENT

Valve clearance is adjusted with engine cold. Remove rocker arm covers, then turn crankshaft until piston for cylinder being adjusted is at top dead center on compression stroke (both valves closed). Loosen locknut and turn rocker arm adjusting screw until a 0.20 mm (0.008 inch) feeler gage can be inserted between rocker arm and valve stem end. Tighten adjusting screw locknut and recheck clearance. Repeat procedure for the other cylinder.

CYLINDER HEAD

An individual cylinder head is used for each cylinder. Be sure to identify heads before removal so they can be reinstalled in their original location.

Whenever piston, cylinder, connecting rod, crankshaft or crankcase have been renewed, piston height in cylinder should be checked to ensure adequate clearance between top of piston and

SMALL DIESEL

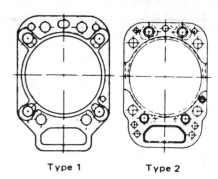

Fig. M2-4—Two different types of head gaskets are available. Type 1 is used on big crank engines and Type 2 is used on small crank engine. Refer to text.

cylinder head before reinstalling cylinder head. On D327-2 engine, install spacer tubes and tighten nuts on cylinder studs to hold cylinders firmly against crankcase surface. With piston at top dead center, use a dial gage to measure distance from top of piston to flat surface of cylinder barrel on D327-2 engine. On D226-2 engine with 9 digit engine number, piston should be 0.28-0.60 mm (0.0110-0.0236 inch) below surface of crankcase. On D226-2 engine with 10 digit engine number, piston should be 0.25-0.57 mm (0.010-0.022 inch) above surface of crankcase. On D327-2 engine, piston clearance should be 4.57-4.67 mm (0.180-0.184 inch). Change thickness of shims between bottom of cylinder and crankcase mounting surface to obtain specified piston height on D327-2 engine.

To reinstall cylinder head on D226-2 engine, proceed as follows: Install two alignment studs M1006 (or equivalent) diagonally from each other in two cylinder head bolt holes (Fig. M2-3). Place cylinder head gaskets over guide studs.

NOTE: Make certain correct gaskets are used. Type 1 gasket (Fig. M2-4) is 1.6 mm (0.063 inch) thick, is brown or white in color and is used on 10 digit engine number (big crank engines. Type 2 gasket is 0.9 mm (0.035 inch) thick, is grey in color and is used on 9 digit engine number (small crank) engines.

Assemble cylinder heads over alignment studs and install two bolts (with threads oiled) finger tight. Use a straightedge to align cylinder heads at intake and exhaust port machined surfaces. Remove alignment studs and install the other two bolts. Tighten all bolts evenly in steps using a crosswise pattern to 170-175 N·m (126-129 ft.-lbs.) for 9 digit identification number engines or 200-210 N·m (148-154 ft.-lbs.) for 10 digit identification number engines. Run

SERVICE MANUAL

engine for about one hour, then stop and allow to cool. With engine cold, loosen head bolts ¼ turn and retorque to specified setting. Readjust valve clearance.

To reinstall cylinder head on D327-2 engine, proceed as follows: Assemble cylinder heads with new gaskets onto cylinder studs. Lubricate bottom of stud nuts and threads of nuts and studs with Molykote "G-n" paste or equivalent. Use a straightedge to align cylinder heads at intake and exhaust port machined surfaces. Tighten stud nuts evenly in small increments using a crosswise tightening pattern to 45-50 N·m (34-36 ft.-lbs.). Operate engine for about ½ hour, then stop and allow to cool. Loosen stud nuts ¼ turn, then retighten in steps using the crosswise pattern to 50-55 N·m (37-40 ft.-lbs.). Recheck valve clearance.

VALVE SYSTEM

Intake and exhaust valve face angle should be 45½° and valve seat angle should be 45°. Desired valve seat width is 1.35-1.80 mm (0.053-0.070 inch) and maximum width is 2.5 mm (0.10 inch) for both valves. With valves installed, the valve heads should be recessed 1.03-1.42 mm (0.040-0.055 inch) below combustion surface of cylinder head. Valve seat inserts should be renewed if valve recession exceeds 1.80 mm (0.070 inch).

Valve seat inserts are interference fit in cylinder head. Standard cylinder head counterbore diameter is 43.030-43.046 mm (1.6941-1.6947 inches) for intake and 38.960-38.976 mm (1.5338-1.5344 inches) for exhaust. If worn or damaged, insert counterbores may be machined oversize 0.20 mm (0.0078 inch) for installation of oversize inserts. Cylinder head should be heated to about 200°C (400°F) prior to installing new inserts. After head cools, regrind valve seats.

Valve stem diameter for both valves is 8.952-8.970 mm (0.3525-0.3531 inch) with a wear limit of 8.949 mm (0.3523 inch). Valve guide inner diameter should be 9.013-9.028 mm (0.3549-0.3554 inch) with a wear limit of 9.060 mm (0.3567 inch). Valve guides are a press fit in cylinder head. If bore in cylinder head is worn or damaged, bore may be machined for installation of guides with 0.1 or 0.2 mm (0.004 or 0.008 inch) oversize outside diameter. After installation of new guides, valve seats must be reground.

Rocker arm bracket shaft diameter is 15.966-15.984 mm (0.6285-0.6293 inch) with a wear limit of 15.950 mm (0.6280 inch). Rocker arm bushing inner diameter should be 16.000-16.018 mm (0.6299-0.6306 inch) and maximum allowable diameter is 16.030 mm (0.6311 inch).

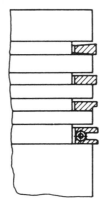

Fig. M2-5—Cross-sectional view showing correct installation of piston rings.

INJECTOR

REMOVE AND REINSTALL. To remove injector, disconnect fuel lines and immediately plug all openings to prevent entry of dirt. Remove mounting bracket nuts and withdraw injector and seal washer.

To reinstall, reverse the removal procedure using a new nozzle seal washer. Tighten mounting bracket nuts to 10-15 N·m (8-11 ft.-lbs.).

TESTING. A special nozzle test pump is required to check injector operation. Injection should be checked for specified opening pressure, nozzle leakage and correct spray pattern.

WARNING: Fuel leaves the injector nozzle with sufficient force to penetrate the skin. When testing, keep yourself clear of nozzle spray.

Connect injector to nozzle test pump, then operate tester lever several quick strokes to purge air from injector and to make sure nozzle valve is not stuck, then operate tester lever slowly while observing tester gage. Nozzle opening pressure should be 18000-18500 kPa (2610-2680 psi) for all engines. Opening pressure may be adjusted by changing thickness of spring adjusting shims. Changing shim thickness 0.1 mm (0.004 inch) will change pressure about 1000 kPa (145 psi).

To check nozzle for leakage, operate tester lever slowly to maintain pressure at 2000 kPa (300 psi) below opening pressure and observe nozzle tip for leakage. If a drop falls from nozzle tip within a 10 second period, nozzle valve is not seating and injector should be overhauled or renewed. During this test, be careful not to confuse fuel leaking from leak-off fitting at top of injector with leakage at nozzle tip.

To check spray pattern, operate the tester lever several quick strokes (about one stroke per second). All four nozzle spray holes must be open. Spray should be uniform and well atomized and nozzle should produce a "chatter" sound when operating properly.

INJECTION PUMP

A Robert Bosch fuel injection pump is used on all engines. The pump contains its own oil supply for lubrication of pump camshaft and governor components. Refer to MAINTENANCE section for lubrication details.

To remove injection pump, first disconnect fuel lines and immediately plug all openings. Disconnect speed control and engine stop linkage from pump. Remove crankshaft pulley and timing gear cover. Remove camshaft drive gear and injection pump drive gear or timing device (if so equipped) using suitable gear puller such as tool number M2024. Remove pump mounting nuts and remove injection pump.

It is recommended that injection pump be tested and overhauled only by a shop qualified in diesel fuel injection repair.

When reinstalling pump, be sure timing marks on pump drive gear and the drive gear on the camshaft are aligned. Tighten injection pump gear retaining nut to 60-70 N·m (45-51 ft.-lbs.). Adjust injection pump timing as previously outlined in MAINTENANCE section.

PISTON, PISTON PIN AND RINGS

The piston is equipped with a keystone type top ring, a rectangular second ring, notched third ring and an oil control ring at the bottom. Refer to Fig. M2-5. Piston standard diameter for D226-2 engine is 104.650-104.680 mm (4.1201-4.1212 inches) and minimum allowable diameter is 104.500 mm (4.1142 inches). Piston standard diameter for D327-2 engine is 99.450-99.470 mm (3.9154-3.9161 inches) and minimum allowable diameter is 99.400 mm (3.9134 inches).

On all engines, ring groove width should be 2.060-2.080 mm for second and third ring grooves with a wear limit of 2.120 mm (0.083 inch). Width of oil control ring groove should be 4.040-4.060 mm (0.1590-0.1598 inch) and maximum width is 4.100 mm (0.1614 inch). Top ring groove is keystone type and a special tool is required to check ring groove wear. Maximum allowable side clearance between rings and grooves is 0.20 mm (0.008 inch) for second and third rings and 0.15 mm (0.006 inch) for oil control ring.

MWM   SMALL DIESEL

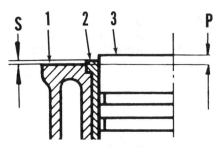

Fig. M2-5A—When cylinder liner is properly installed on D226-2 engine, top of liner must project above surface of crankcase. Refer to text for specified dimensions (S and P).

1. Crankcase
2. Cylinder liner
3. Piston

Piston ring end gap is measured with ring positioned squarely in cylinder bore. Desired end gap is 0.20-0.25 mm (0.008-0.010 inch) for all rings and maximum allowable end gap is 2.0 mm (0.079 inch).

The full floating piston pin is retained in the piston by two snap rings. Piston pin diameter is 34.994-35.000 mm (1.3777-1.3779 inches) for all engines with 10 digit engine numbers with a wear limit of 34.990 mm (1.3776 inches). Pin diameter for engines with 9 digit engine number is 31.994-32.000 mm (1.2596-1.2598 inches) with a minimum allowable diameter of 31.990 mm (1.2595 inches). Inner diameter of pin bore in piston should be 35.003-35.009 mm (1.3781-1.3783 inches) with a wear limit of 35.014 mm (1.3785 inches) for engines with 10 digit number. Piston pin bore for engines with 9 digit number should be 32.003-32.009 mm (1.2600-1.2602 inches) with a wear limit of 32.014 mm (1.2604 inches).

When reassembling, be sure third ring is installed notched side down. Be sure ring end gaps are staggered at 180° intervals around piston. Make certain arrow on piston crown is pointing towards flywheel end of engine when reinstalling piston. Check piston height in cylinder as outlined in CYLINDER HEAD section.

CYLINDER

Model D226-2 Engine

The D226-2 engine is equipped with renewable, dry type cylinder liners. Replacement liners are available semifinished which are a press fit in cylinder bores, or prefinished which are a slip fit in cylinder bores. The prefinished liner is available with a standard flange and a 0.50 mm (0.002 inch) oversize flange.

Cylinder liner standard diameter is 105.000-105.022 mm (4.1339-4.1347 inches) with a wear limit of 105.250 mm (4.1437 inches). If excessively worn or damaged, liners may be rebored for installation of 0.5 mm (0.020 inch) oversize pistons and rings.

Use special tool M1026 (or equivalent) to remove and install liners. Be sure to thoroughly clean cylinder bore and outside of liner before installing. When installing press fit liners, apply a light coat of oil to top of cylinder bore and bottom of liner before pressing into crankcase. After installation, top of liner flange (2 – Fig. M2-5A) must project a distance (S) above surface of crankcase of 0.02-0.07 mm (0.0008-0.0027 inch) for engines with 9 digit engine numbers or 0.05-0.10 mm (0.002-0.004 inch) for engines with 10 digit numbers. If projection is excessive, remove liner and check crankcase counterbore for dirt or other obstructions. If projection is less than specified, install liner with an oversize flange.

After pistons are installed, piston height (P) in sleeve should be measured to ensure assembly is correct. On engines with 9 digit number, top of piston should be 0.28-0.60 mm (0.0110-0.0236 inch) below surface of crankcase. On engines with 10 digit number, top of piston should be 0.25-0.57 mm (0.010-0.022 inch) above surface of crankcase.

Model D327-2 Engine

The D327-2 engine is equipped with removable, individual, cast iron cylinder barrels. Standard cylinder bore is 100.000-100.022 mm (3.9370-3.9378 inches) with a wear limit of 100.250 mm (3.9468 inches). If cylinders are excessively worn or damaged, they may be rebored for installation of 0.5 or 1.0 mm (0.020 or 0.040 inch) oversize pistons and rings.

After installing new cylinder barrels, check piston height in cylinder with piston at top dead center. Install two pipe spacers diagonally on cylinder studs and tighten two stud nuts to seat cylinder firmly against crankcase. The distance from the flat surface of cylinder to top of piston must be 4.57-4.67 mm (0.180-0.184 inch). Piston compression height is adjusted using shims between bottom of cylinder and mounting surface of crankcase.

CONNECTING ROD

The connecting rod uses a bushing in the small end. Bushing inside diameter should be 35.030-35.080 mm (1.3791-1.3811 inches) on engines with 10 digit engine number with a wear limit of 35.150 mm (1.3838 inches). Outside diameter of piston pin is 34.994-35.000 mm (1.3777-1.3779 inches) with a wear limit of 34.990 mm (1.3776 inches). On engines with a 9 digit engine number, bushing inside diameter should be 32.030-3.080 mm (1.2610-1.2630 inches) and maximum allowable inner diameter is 32.150 mm (1.2657 inches). Piston pin diameter is 31.994-32.000 mm (1.2596-1.2598 inches) with a wear limit of 32.990 mm (1.2595 inches). Desired operating clearance for all engines is 0.03-0.08 mm (0.0012-0.0031 inch).

A precision, insert type bearing is used in the connecting rod big end. Standard diameter of crankpin is 62.951-62.970 mm (2.4784-2.4791 inches) on engines with 10 digit number (big crank) or 57.951-57.970 mm (2.2815-2.4791 inches) on engines with 9 digit number (small crank). Desired operating clearance for crankpin to rod bearing is 0.07-0.10 mm (0.0027-0.0039 inch) for all engines. Connecting rod bearing must be renewed when the third layer of bearing material wears off which will be indicated by a shiny bronze contact pattern appearing on the bearing surface. Connecting rod side clearance between big end and crankshaft should be 0.30-0.50 mm (0.0118-0.0197 inch).

The connecting rod bearing is positioned in big end of rod with a locating pin. When assembling connecting rod and bearing, it is recommended that lower half of bearing insert be installed without bearing cap to ensure notch in bearing is indexed with pin. Then, install bearing cap and tighten screws to 95-100 N·m (70-74 ft.-lbs.) on engines with 10 digit engine number or 75-80 N·m (56-59 ft.-lbs.) on engines with 9 digit number.

TIMING GEARS AND FRONT COVER

To remove crankcase front cover, first remove engine cooling fan and crankshaft front pulley. Remove tachometer drive (if so equipped) from front cover. Remove front cover mounting screws and withdraw cover. Crankshaft oil seal is located in front cover and may be renewed at this time.

Remove gears as necessary. Use a suitable puller to remove injection pump gear or timing device. Crankshaft gear is a shrink fit. Gear must be heated to 270°C (518°F) prior to installation. Backlash between any two timing gears should be 0.12-0.28 mm (0.005-0.011 inch). Renew gears if backlash is excessive. Backlash between oil pump gear and crankshaft gear should be 0.2 mm (0.008 inch). Loosen oil pump mounting screws and move pump housing to obtain desired backlash.

SERVICE MANUAL

When reassembling gears, align timing marks as follows: The "0" mark (1 – Fig. M2-6) on injection pump timing device or drive gear must be aligned with "0-0" marks (2) on drive gear on camshaft hub. The "0" mark on the crankshaft gear (3 – Fig. M2-7) must be aligned with "0-0" marks on camshaft gear (2). The oil pump gear (4) is installed in any position.

CAMSHAFT

The camshaft is supported at the front by a renewable bushing in crankcase front bulkhead. Center and rear camshaft journals operate directly in unbushed bores in crankcase.

To remove camshaft, first remove cylinder heads and push rods, crankshaft pulley and timing gear cover. Unbolt and remove camshaft gears. Remove camshaft retaining plate. Turn engine upside down and move cam followers away from camshaft. Carefully withdraw camshaft from crankcase. Identify location of cam followers as they are removed for use in reassembly.

Inspect camshaft and cam followers for scoring and renew if necessary. Camshaft journal diameter is 42.940-42.960 mm (1.6906-1.6913 inches) with a wear limit of 42.920 mm (1.6898 inches). Camshaft bushing bore diameter should be 43.000-43.039 mm (1.6929-1.6944 inches) and maximum allowable diameter is 43.060 mm (1.6953 inches). Center and rear crankcase bore diameter should be 43.000-43.025 mm (1.6929-1.6939 inches) with a wear limit of 43.040 mm (1.6945 inches). Desired operating clearance for camshaft journals is 0.04-0.08 mm (0.0016-0.0031 inch).

If camshaft journals or crankcase bores are excessively worn, the journals may be ground 0.30 mm (0.012 inch) undersize the crankcase center and rear bores can be rebored for installation of bushings. Machine crankcase bores to an inner diameter of 47.000-47.025 mm (1.8504-1.8513 inches), then press in new bushings. The end of rear bushing should be recessed 4 mm (0.157 inch) from rear surface of crankcase. Be sure to align hole in bushing with oil passage hole in crankcase. Inner diameter of bushings must be finish machined to match camshaft being installed (new dimension or undersize). Bushing bore should be 43.000-43.025 mm (1.6929-1.6939 inches) for new dimension camshaft or 42.700-42.725 mm (1.6811-1.6821 inches) for undersize camshaft. After installing bushings a different cover (number 6.305.0.787.0.011.4) must be installed to seal rear bore.

Cam follower outer diameter should be 17.983-17.994 mm (0.7080-0.7084 inch) and minimum allowable diameter is 17.975 mm (0.7077 inch). Crankcase bore diameter is 18.000-18.015 mm (0.7087-0.7092 inch) with a wear limit of 18.020 mm (0.7095 inch).

Camshaft end play is controlled by the retainer bracket which engages the groove in end of camshaft. Bracket thickness when new is 6.850-7.050 mm (0.2697-0.2775 inch) and minimum allowable thickness is 6.750 mm (0.2658 inch). Width of groove in camshaft should be 7.100-7.190 mm (0.2795-0.2830 inch) with a wear limit of 7.220 mm (0.2843 inch). Desired end play is 0.10-0.29 mm (0.0039-0.0114 inch).

To reinstall camshaft, reverse the removal procedure. Align timing marks on gears as outlined in TIMING GEARS AND FRONT COVER section. Tighten camshaft gear retaining screws to specified torque.

CRANKSHAFT AND BEARINGS

When removing crankshaft, note that main bearing caps are numbered for correct reassembly. If main bearing inserts will be reused, keep inserts with their respective caps so they can be reinstalled in original positions.

On engines with 10 digit engine number (big crank), standard diameter of main journals is 69.951-69.970 mm (2.7540-2.7547 inches) and minimum allowable diameter is 69.940 mm (2.7535 inches). Crankpin standard diameter is 62.951-62.970 mm (2.4784-2.4791 inches) with a wear limit of 62.940 mm (2.4780 inches).

On engines with 9 digit engine number (small crank), standard diameter of main journals is 64.951-64.970 mm (2.5572-2.5578 inches) and minimum allowable diameter is 64.940 mm (2.5567 inches). Crankpin standard diameter is 57.951-57.970 mm (2.2815-2.2822 inches) with a wear limit of 57.940 mm (2.2811 inches).

On all engines, recommended main bearing clearance is 0.08-0.11 mm (0.0031-0.0043 inch) and rod bearing clearance is 0.07-0.10 mm (0.0027-0.0039 inch). Bearing inserts must be renewed if third layer of bearing material is worn off which is indicated by shiny bronze contact patterns on bearing surface. Bearings are available in undersizes of 0.25, 0.50, 0.75 and 1.0 mm (0.010, 0.020, 0.030 and 0.040 inch) as well as standard size.

Crankshaft end play is controlled by thrust rings on each side of rear main bearing on big crank engines with 10 digit engine number. Recommended end play is 0.04-0.25 mm (0.002-0.010 inch). Thrust rings are available in standard width as well as 0.10, 0.20, 0.30 and 0.50 mm (0.004, 0.008, 0.012 and 0.020 inch) oversize widths. Make certain thrust rings are installed with grooved side facing crankshaft web and flange surfaces.

On small crank engines with 9 digit engine number, crankshaft end play is controlled by flanges on each side of rear main bearing. Recommended end play is 0.08-0.23 mm (0.003-0.009 inch) on D226-2 engine and 0.12-0.21 mm (0.005-0.008 inch) on D327-2 engine. Bearings are available in standard thrust width as well as 0.10, 0.25 and 0.50 mm (0.004-0.010 and 0.020 inch) oversize widths. An optional 1.0 mm (0.040 inch) oversize width is also available on a limited supply basis.

When reinstalling crankshaft, make certain the bearings are properly positioned on locating pins and bearing caps are reinstalled in their original positions. Tighten main bearing cap screws to specified torque. Be sure "0" mark on crankshaft gear is aligned with "0-0" marks on camshaft gear. Renew

Fig. M2-6—When installing timing gears, "0" mark (1) on injection pump gear must be aligned with "0-0" marks (2) on drive gear on camshaft hub.

Fig. M2-7—View of timing gears. Injection pump drive gear is mounted to rear of camshaft gear (2).

1. Injection pump gear
2. Camshaft gear
3. Crankshaft gear
4. Oil pump gear

MWM — SMALL DIESEL

crankshaft front and rear oil seals as outlined in OIL SEALS section.

OIL SEALS

Crankshaft front oil seal is located in timing gear cover. To renew seal, first remove crankshaft pulley and hub. Then, pry old seal out of cover being careful not to damage cover bore. Clean hub surface and polish with fine emery cloth. If wear is evident on hub surface (groove exceeding 1 mm (0.039 inch), renew hub. Apply film of grease to outer diameter of seal and seal lip. Using a suitable installing tool, press seal into cover until it bottoms against inner flange (if so equipped) or until flush with outer suface of cover. Apply grease to hub surface, then reinstall on crankshaft. Tighten mounting screws to specified torque.

Crankshaft rear oil seal is located in crankcase rear cover. If a groove is worn into crankshaft flange at original seal location, new seal may be installed offset to provide a new sealing surface. If rear cover has a flange (1 – Fig. M2-8), install a shim (part number 6.206.0.340.004.4) between flange and new seal (3) to offset seal outward (A). If rear cover does not have a flange, install seal offset 2 mm (0.080 inch) inward (B) to obtain new seal contact area. Apply grease to seal lip and crankshaft flange prior to reassembly.

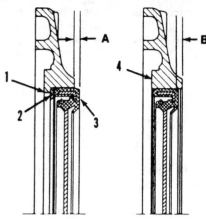

Fig. M2-8—Crankshaft rear oil seal may be installed offset (A or B) in crankcase rear cover to obtain contact with unworn portion of crankshaft flange if necessary.
1. Cover with flange
2. Shim
3. Oil seal
4. Cover without flange

SERVICE MANUAL

ONAN
1400 73rd Avenue N.E.
Minneapolis, Minnesota 55432

Model	No. Cyls.	Bore	Stroke	Displ.
DJA	1	3¼ in. (82.55 mm)	3⅝ in. (92.08 mm)	30 cu. in. (492 cc)
DJB	2	3¼ in. (82.55 mm)	3⅝ in. (92.08 mm)	60 cu. in. (983 cc)
DJBA	2	3¼ in. (82.55 mm)	3⅝ in. (92.08 mm)	60 cu. in. (983 cc)
DJC	4	3¼ in. (82.55 mm)	3⅝ in. (92.08 mm)	120 cu. in. (1966 cc)
DJE	2	3½ in. (88.90 mm)	3⅝ in. (92.08 mm)	(70 cu. in.) (1147 cc)
RDJC	4	3¼ in. (82.55 mm)	3⅝ in. (92.08 mm)	120 cu. in. (1966 cc)
RDJE	2	3½ in. (88.90 mm)	3⅝ in. (92.08 mm)	70 cu. in. (1147 cc)
RDJEA	2	3½ in. (88.90 mm)	3⅝ in. (92.08 mm)	70 cu. in. (1147 cc)
RDJF	4	3½ in. (88.90 mm)	3⅝ in. (92.08 mm)	140 cu. in. (2294 cc)

These ONAN engines make up a family of diesel-fueled power units having many features and specifications in common. Models with an "R" prefix, such as Model RDJC, are liquid-cooled while all other models are air-cooled. All models are equipped for electric starting only, with current production engines using a 12 volt automotive type starting motor. Some earlier models of DJA engine-generator sets are exciter cranked by switching 12 volt battery current through DC windings of direct coupled generator which serves as a starter.

Maintenance and repair procedures pertain to all engines in this group with special attention directed to particular differences among models.

OPERATION

PREHEATING AND STARTING. All engines are equipped with a glow plug in each cylinder to heat the precombustion chamber and a manifold heater to heat intake air.

CAUTION: Do not use ether as a starting aid. Heat from glow plugs or manifold heater could ignite ether vapors causing an explosion resulting in severe engine damage.

If engine is cold and air temperature is 55°F (13°C) or above, energize preheat circuit for approximately 20 seconds. If air temperature is below 55°F (13°C), preheat for one minute. While holding preheat switch on, engage starting switch until engine starts. Continue to preheat until engine runs smoothly. If engine fails to start after 30 seconds of cranking, preheat for one minute more and attempt to start again. Absence of exhaust smoke during cranking indicates no fuel delivery.

NOTE: Do not energize preheat circuit longer than one minute to prevent glow plug and heater element damage. Do not apply booster voltage in excess of 12 volts to starting system as preheat circuit will be damaged.

All engines are equipped with a low oil pressure safety switch and some models may also be equipped with a high temperature safety switch. If safety switches do not close and engine fails to start, check electrical circuit for continuity and check switch controls for proper adjustment.

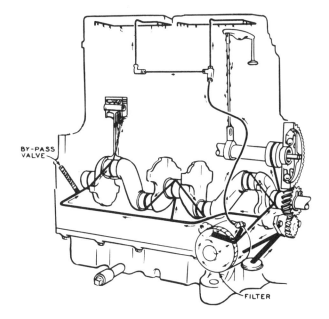

Fig. O200—View of full-pressure oil system used on all engines. Bypass valve is nonadjustable and normally opens at 25 psi (172 kPa). All models are equipped with a full-flow, spin-on type oil filter.

Onan

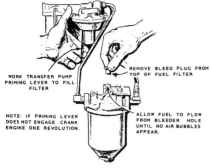

Fig. O201 – View showing bleeding air on early model fuel system. Refer to text.

SMALL DIESEL

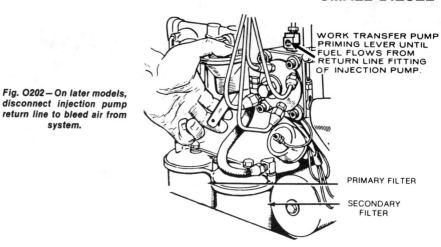

Fig. O202 – On later models, disconnect injection pump return line to bleed air from system.

MAINTENANCE

LUBRICATION

Recommended oil for crankcase use is API service classification CD/SF or CD/SE. Recommended oil change interval is after every 100 hours of operation with filter changed simultaneously. Check oil level daily.

Crankcase capacities (with filter change) are as follows:

DJA .... 3 qts.
(2.8 L)
DJB, DJBA, DJE .... 3.5 qts.
(3.3 L)
RDJE, RDJEA .... 3 qts.
(2.8 L)
DJC, RDJC, RDJF .... 6.5 qts.
(6.1 L)

If operating temperatures fall below 0°F (+18°C), SAE 5W-30 weight oil is recommended. From 0°F (+18°C) to 32°F (0°C), use SAE 10W-30 or 10W-40. Above 32°F (0°C), SAE 30 or SAE 40 weight oil may be used.

FUEL

A good quality number 2 diesel fuel is recommended for most conditions. Number 1 diesel fuel may be used for cold weather operation or for long periods of light duty operation. All fuel must have a minimum cetane number of 45.

NOTE: It is extremely important that fuel be kept clean and free of water. Due to precise tolerances of diesel injection components, dirt and water in the system can cause severe damage to injection pump and nozzles.

FUEL FILTERS. The fuel filter system consists of a sediment bowl, located at inlet of fuel transfer pump, and primary and secondary filters, located between transfer pump and injection pump. Water and sediment should be drained periodically from sediment bowl and both filters. Manufacturer recommends renewing primary filter after every 600 hours of operation and secondary filter after every 3000 hours of operation. However, fuel filter renewal interval is more dependent upon fuel quality and cleanliness than length of service or operating conditions. If engine indicates signs of fuel starvation (loss of power or surging), renew filters regardless of hours of operation. After renewing filters, air must be bled from system as outlined in following paragraph.

BLEED FUEL SYSTEM. To bleed air from fuel system, loosen bleed plug (Fig. O201) on top of secondary fuel filter (early models) or disconnect injection pump fuel return line (Fig. O202) on late models. Then, actuate primer lever on fuel transfer pump until air-free fuel flows. Return primer lever to normal operating position after bleeding operation is completed.

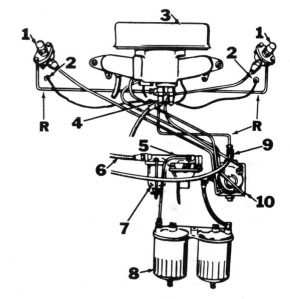

Fig. O203 – View of diesel fuel system typical of late models.
1. Injection nozzles
2. Glow plugs
3. Air cleaner
4. Manifold heater
5. Fuel transfer pump
6. Fuel supply line
7. Sediment bowl
8. Fuel filters
9. Fuel return line
10. Injection pump
R. Leak-off (return) lines

NOTE: If cam which drives transfer pump is at full lift, priming lever cannot be operated. Turn crankshaft one revolution to reposition cam.

INJECTION PUMP

Late Model DJE is equipped with either a Bryce or Kiki injection pump. Model DJA is equipped with an American Bosch Model PLB injection pump while all other models are equipped with an American Bosch Model PSU injection pump. Refer to the following pump sections for injection pump timing.

BOSCH MODEL PLB. On Model DJA, injection starts at 17° BTDC which corresponds to "PC" (port closing) mark stamped on flywheel. Adjustment is made by changing the thickness of a shim pack which is fitted between pump mounting pad on crankcase and pump body mounting flange.

SERVICE MANUAL

Onan

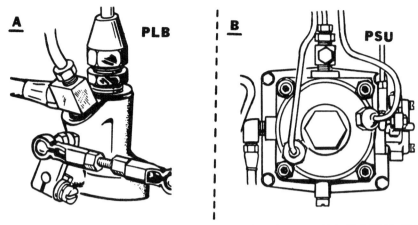

Fig. O204 — American Bosch injection pumps are used on all engines except late Model DJE which is equipped with a Bryce or Kiki injection pump. Model PLB (View A) is fitted to DJA engines and Model PSU is used on all other engines.

To adjust timing, shut off fuel supply. Disconnect fuel lines and remove injection pump. Cap all connections as they are loosened to prevent entry of dirt into system. Turn crankshaft until piston is on compression stroke and "PC" flywheel mark is aligned with pointer. Using a depth micrometer, measure the distance from pump tappet to pump mounting surface of crankcase as shown in Fig. O205. Subtract 1.670 inches (42.42 mm), standard Port Closing Dimension, from the measured depth to determine shim pack thickness required. Shim pack thickness should be 0.006-0.052 inch (0.15-1.32 mm). If the calculated thickness does not fall within these limits, recheck measurement and/or check for camshaft or tappet wear or improper assembly. Shims are available in an installation shim kit which contains all necessary thicknesses.

BOSCH MODEL PSU. Models DJB, DJBA, DJC, RDJC, RDJE, RDJEA, RDJF and early Model DJE are equipped with an American Bosch Model PSU injection pump.

Injection timing for PSU pumps calls for insertion of a timing button of proper thickness between pump plunger and tappet. Refer to Fig O207. All engines are factory-timed and port closing mark appears at a point on flywheel which may be at 17°, 19° or 21° BTDC depending on production series or specification letter of a particular engine. Coding of timing button and port closing dimension are stamped on pump flange as shown in Fig. O207.

Two methods for timing injection pumps are used. If pump is renewed, use procedure number one to time new pump to engine. Second method is required if timing dimensions are unavailable or if major parts such as crankcase, crankshaft or flywheel are renewed.

METHOD I. Shut off fuel supply, disconnect all lines, unbolt and remove pump. Protect open lines from dirt and damage. Use the following formula to calculate button thickness:

EXAMPLE

Port closing dimension
 (old pump)................1.109 in.
  (28.168 mm)
Button thickness
 (old pump).............. + 0.107 in.
  (2.718 mm)
   Total: 1.216 in.
  (30.886 mm)

Now, subtract:
Port closing dimension
 (new pump).............. − 1.094 in.
  (27.787 mm)
Button thickness required
 for new pump............0.122 in.
  (3.099 mm)

Refer to table, Fig. O208, and select button to correspond. In this example, Code 12 or M, part number 147-0190, will be correct. Install button under tappet as in Fig. O207.

NOTE: When removing tappet, be sure to hold pump drive securely to pump body to prevent pump from coming apart. Make certain metering sleeve does not drop off the plunger when plunger is removed.

Reinstall injection pump as follows: Rotate crankshaft until number 1 cylinder is on compression stroke and port clos-

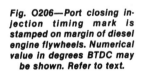

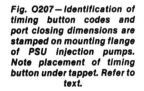

Fig. O205 — To adjust injection pump timing on Model DJA, use depth micrometer to measure distance from pump tappet to mounting surface of crankcase as shown. Refer to text.

Fig. O206 — Port closing injection timing mark is stamped on margin of diesel engine flywheels. Numerical value in degrees BTDC may be shown. Refer to text.

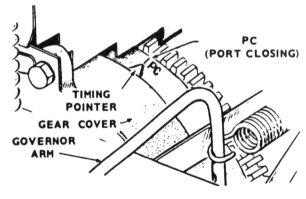

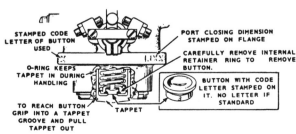

Fig. O207 — Identification of timing button codes and port closing dimensions are stamped on mounting flange of PSU injection pumps. Note placement of timing button under tappet. Refer to text.

259

GROUP 1				GROUP 2				GROUP 3			
CODE	PART NO.	SIZE		CODE	PART NO.	SIZE		CODE	PART NO.	SIZE	
		Inch	mm			Inch	mm			Inch	mm
16 or S	147-0186	.134	3.404	1 or A	147-0147	.119	3.023	6 or F	147-0152	.101	2.565
15 or R	147-0187	.131	3.357	2 or B	147-0148	.116	2.946	7 or H	147-0153	.098	2.489
14 or P	147-0188	.128	3.251	3 or C	147-0149	.113	2.870	8 or I	147-0154	.095	2.413
13 or N	147-0189	.125	3.175	4 or D	147-0150	.110	2.794	9 or K	147-0155	.092	2.337
12 or M	147-0190	.122	3.099	5 or E	147-0151	.107	2.718	10 or L	147-0156	.089	2.261
				11 or STD	147-0161	.104	2.642				

Fig. O208—Table of timing button codes and thicknesses for setting injection timing on PSU pumps. Refer to text.

ing mark on flywheel is aligned with timing pointer as shown in Fig. O206. Remove timing hole screw from mounting flange of pump and insert a ⅛-inch (3mm) brass wire into timing hole as shown in Fig. O209. Rotate pump face gear until gear locks as wire engages timing recess. Reinstall pump onto crankcase using correct thickness of mounting shims as stamped on the crankcase. This shim pack controls mesh of pump drive and driven gears. Tighten pump mounting cap screws to a torque of 15-16 ft.-lbs. (20-22 N·m). Remove brass wire and reinstall timing hole plug. Reconnect fuel lines and throttle linkage. Bleed air from fuel system.

METHOD II. Use this procedure to time pump when dimensions from old pump are not available. Begin by installing pump which has been fitted with standard timing button (may be unmarked or marked number 11). Refer to Fig. O210 and remove cap nut from delivery valve and delivery valve holder, then lift out delivery valve spring. Reinstall holder and cap nut. Note that early models do not have a delivery valve holder.

Rotate flywheel until PC mark on flywheel is about 15° from timing pointer in compression stroke of number 1 cylinder. Place fuel control at full speed position. Disconnect high pressure line from number 1 injector. Operate transfer pump priming lever (fuel should flow from disconnected injector line) while slowly turning flywheel clockwise. When fuel flow from line stops, this is port closing point and beginning of injection. At this point, port closing (PC) mark on flywheel should be aligned with timing pointer if timing button is correct.

If timing pointer is ahead of port closing mark, timing is early and a thinner button is needed. If timing pointer is between PC and TDC marks, timing is late and a thicker button must be installed. To select correct button size, carefully measure space between PC mark and timing pointer. Each 0.100 inch (2.54 mm) on flywheel circumference is equivalent to 0.003 inch (0.076 mm) button thickness. Refer to table in Fig. O208.

EXAMPLE. Standard button installed in pump for test is 0.104 inch (2.642 mm) thick. Flow of fuel stops 0.2 inch (5.08 mm) after PC mark is passed indicating late timing. Referring to table in Fig. O208, it will be noted that button of Code 4 or D is 0.110 inch (2.794 mm) thick or 0.006 inch (0.152 mm) thicker than standard button. Installing this button in place of test button should time injection pump correctly.

After pump is correctly timed, reinstall delivery valve spring. Tighten delivery valve holder to a torque of 65-70 ft.-lbs.. (90-95 N·m) and cap nut to 55-60 ft.-lbs. (75-80 N·m). Complete installation of pump as previously outlined in this section.

BRYCE AND KIKI INJECTION PUMPS. Later Model DJE may be equipped with a Bryce or Kiki injection pump. Bryce and Kiki injection pumps are similar with no difference in servicing. Refer to Fig. O211 for view of pump. Injection pump timing should occur at 14-22° BTDC. Shims between the injection pump and adapter are used to adjust injection pump timing. If injection pump, adapter or gasket is new, follow procedure in Method I to time injection pump. If pump, adapter or gasket is not renewed or pump timing is to be checked, follow timing procedure in Method II.

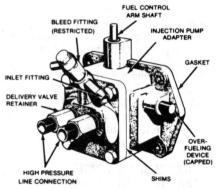

Fig. O211—View showing Bryce or Kiki fuel injection pump used on later Model DJE.

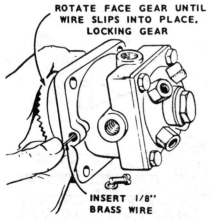

Fig. O209—Insert a brass wire into pump timing hole to lock PSU pump when timing and installing pump.

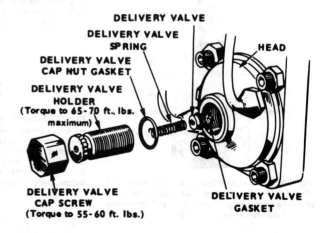

Fig. O210—Exploded view of delivery valve assembly of PSU injection pump. Note that only delivery valve spring is removed when using Method II to time two and four cylinder engines. Refer to text for procedure.

SERVICE MANUAL

Onan

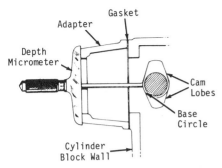

Fig. O212—Measure distance from adapter face to base circle of cam lobe to determine shim thickness for correct injection timing. Refer to text.

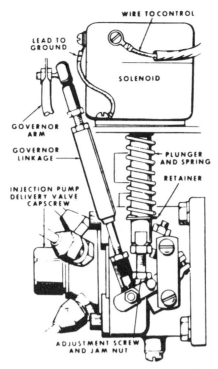

Fig. O214—Installed view of fuel solenoid. Refer to text.

METHOD I. To determine thickness of shims for correct injection pump timing, install gasket and adapter then secure adapter with cap screws tightened to 20-24 ft.-lbs. (27-33 N·m). Using a depth mircrometer, measure distance from adapter face to base circle of either camshaft lobe as shown in Fig. O212. Subtract this reading from the standard dimension of 3.2598 inches (82.8 mm) and the result is the required shim thickness between pump and adapter. Note that on early DJE models the standard dimension is 3.2540 inches)82.652 mm) and "3.2540" is printed adjacent to pump on engine. Shims are available in thicknesses of 0.002, 0.003, 0.006, 0.010, 0.014 and 0.018 inch (0.051, 0.076, 0.152, 0.254, 0.356 and 0.457 mm). Tighten pump mounting screws to 20-24 ft.-lbs. (27-33 N·m).

METHOD II. If pump, adapter or gasket have not been renewed, the following spill timing procedure may be used to check injection pump timing.

Disconnect high pressure fuel line of cylinder being timed from injection pump. Unscrew delivery valve holder (2–Fig. O213) and remove valve spring (13), volume reducer (3) and delivery valve (4), then reinstall delivery valve holder (2). Attach a spill pipe made of old injector line with a gland nut to delivery valve holder and aim spill pipe at a receptacle to catch discharged fuel. Move fuel control arm toward front of engine to full fuel position while simultaneously holding plunger in on fuel shut-off solenoid. Operate fuel transfer pump while simultaneously rotating flywheel slowly clockwise until fuel just stops flowing from spill pipe. Injection pump port is now closed and PC mark on flywheel and timing pointer should be aligned as shown in Fig. O206. If PC mark is not present on flywheel, measure around flywheel rim from TC mark. A distance of 0.1 inch (2.54 mm) equals one degree of crankshaft rotation. Injection timing is adjusted using shims (Fig. O211) installed between pump and adapter. Shims are available in thicknesses of 0.002, 0.003, 0.006, 0.010, 0.014 and 0.018 inch (0.051, 0.076, 0.152, 0.254, 0.356 and 0.457 mm). Tighten pump retaining screws to 20-24 ft.-lbs. (27-33 N·m). Reinstall delivery valve (4–Fig. O213), volume reducer (3) and valve spring (13) after timing adjustment. Tighten delivery valve holder (2) to 29-33 ft.-lbs. (40-45 N·m) on Bryce pump or 44-47 ft.-lbs. (60-64 N·m) on Kiki pump.

Injection pump timing for remaining cylinder may be checked using above procedure. Timing differential between cylinders should be 2.5 degrees or less.

GOVERNOR (FUEL) SOLENOID. This special-purpose solenoid is an optional item usually associated with two-speed governors furnished on some engines. It is referred to as a governor solenoid because it overrides governor control of injection pump throttle lever.

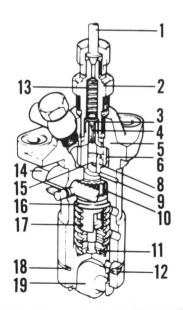

Fig. O213 – Cross-sectional view of Bryce or Kiki injection pump.

1. Delivery tube
2. Delivery valve holder
3. Volume reducer
4. Delivery valve
5. Pump housing
6. Suction chamber
7. Plunger barrel
8. Pin
9. Control sleeve
10.
11. Lower spring seat
12. Guide pin
13. Delivery valve spring
14. Control rack
15. Plunger
16. Upper spring seat
17. Plunger spring
18. Snap ring
19. Tappet

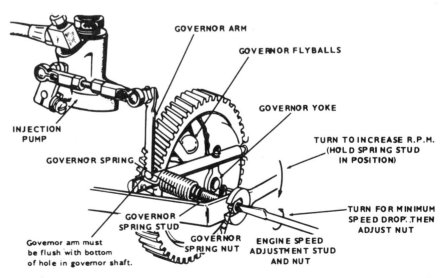

Fig. O215—View of early governor mechanism to show points of adjustment. Governor shown is typical of constant speed type used on early air-cooled diesel models.

Onan

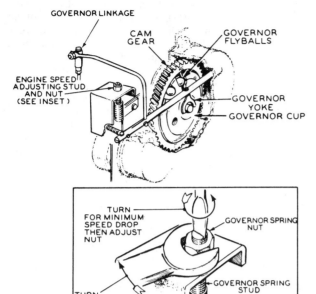

Fig. O216—View of governor linkage on early liquid-cooled models.

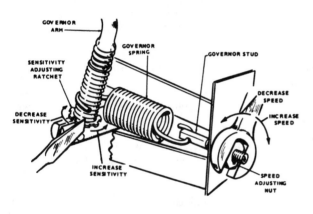

Fig. O217—Late governor mechanism showing adjustment on later air-cooled and liquid-cooled models. Refer to text.

Refer to Fig. O214 for typical mounting arrrangement.

When solenoid is energized by current flowing through its windings, plunger is retained within solenoid body and control of throttle is by governor action. When no current is flowing and solenoid de-energizes, plunger spring forces plunger outward against pump operating lever to hold lever in fuel shut-off position.

TESTING. Use a 12 volt input to check operation of solenoid plunger. Series-connected ammeter should indicate about 1 amp through holding coil when plunger is withdrawn into solenoid. If current exceeds 1 amp by a significant amount, then switch contacts for separate retractor winding did not open when plunger was drawn up. Resulting excess current will overheat solenoid. Check for plunger sticking in solenoid recess or for excessive tension on high speed governor spring. Plunger must fully retract to contact this switch, so a thorough check of possible causes of trouble is important. Renew solenoid assembly if defective.

ADJUSTMENT. With current "OFF", solenoid de-energized, loosen jam nut and turn adjustment screw (Fig. O214) as necessary to hold pump operating lever in fuel shut-off position. Tighten jam nut, then test for proper start-stop operation.

SMALL DIESEL

GOVERNOR

Flyball type mechanical governors are used on all engines. Governed speeds are 1800 rpm for 60 Cycle AC Generator, 1500 rpm for 50 Cycle AC Generator and 1750 rpm for DC Battery Charger. Recommended governed speed is stamped on generator nameplate. It should be possible to adjust governed speed to within 3 cycles (hertz) per second (90 engine rpm) and an adjustment of 2 cps (60 rpm) is usually attainable. For accurate speed adjustment, use of a reed type frequency meter on AC generator output is recommended by manufacturer over use of a mechanical tachometer.

To adjust governor linkage, proceed as follows: With engine not running, disconnect injection pump throttle link from governor arm and adjust the link, if necessary, until link can be reconnected with injection pump throttle arm in wide open position.

Two different types of governor linkage have been used. Refer to Fig. O215 for linkage on early air-cooled models or to Fig. O216 for early liquid-cooled models. Sensitivity is adjusted by turning adjustment stud in or out of governor spring. Engine speed is adjusted by turning governor spring nut while holding stud from turning. On late air-cooled and liquid-cooled models, governor spring nut controls engine speed while sensitivity is adjusted by turning the ratchet nut on governor arm as shown in Fig. O217.

Two- and four-cylinder engines (DJB, DJBA, DJC) may be furnished with a variable speed governor as shown in Fig. O218 instead of constant speed type. Variable speed is controlled by a ratchet locking throttle lever or by a solenoid control attached to auxiliary

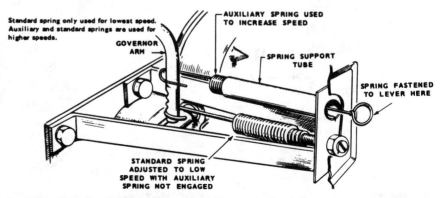

Fig. O218—View of variable speed governor auxiliary spring in relation to constant speed governor spring shown in preceding figures. Refer to text.

SERVICE MANUAL

Onan

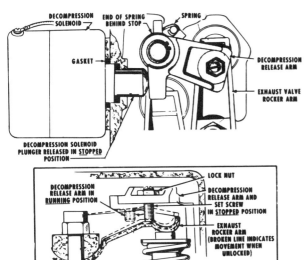

Fig. O219—Two-position view of compression release mechanism used on Model DJA engine. Upper view is downward, in line with cylinder bore. Lower view shows cross section with decompression release holding rocker arm down and exhaust valve open as discussed in text.

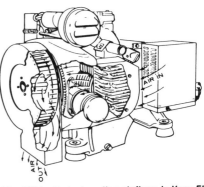

Fig. O221—Typical cooling air flows in Vacu-Flo system used for closed compartment installation.

governor spring (Fig. O218) to override standard constant speed spring. Basic speed and sensitivity adjustments remain the same with variable higher speeds set at desired levels by throttle control or solenoid position.

COMPRESSION RELEASE MECHANISM

Refer to Fig. O219 for top and cross section view of compression release mechanisim used on Model DJA (one cylinder) engine. Decompression is obtained by the release arm adjusting screw riding up the ramp of exhaust valve rocker arm, holding exhaust valve slightly open when release mechanism is effective. The compression release arm pivots on its axle pin and is spring actuated to both the RELEASE and RUNNING positions. The coil spring on release arm hub is tensioned to move arm into RUNNING position. A stronger spring located on solenoid plunger moves the arm to RELEASE position when solenoid is NOT energized. When current is routed to solenoid, the plunger is drawn away from release arm and arm is allowed to move to RUNNING position.

ADJUSTMENT. To adjust the compression release mechanism, first remove decompression solenoid and rocker arm cover. Turn crankshaft until exhaust valve is closed and push rod is loose. Loosen locknut and back out adjusting screw on compression release arm about two turns; then, check to be sure release arm pivots back and forth easily without binding. Arm should move counterclockwise against spring pressure and should snap back easily to RUNNING position when released. Pivot arm fully counterclockwise against its stop and turn adjusting screw until it just touches rocker arm (rocker arm must be in contact with valve stem). Turn screw ONE additional turn to open valve the specified amount and secure by tightening locknut.

CAUTION: If screw is tightened by more than one turn, valve head might strike piston. Manually move release arm to RUNNING position (clockwise) and check to be sure that some clearance exists between valve rocker arm and set screw.

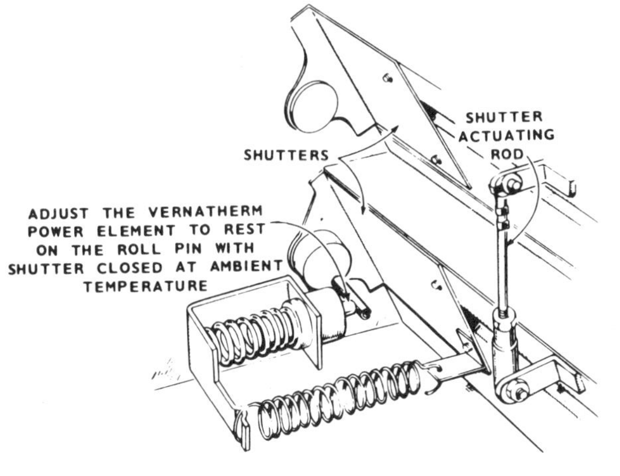

Fig. O220—View of thermostat controlled power shutter for pressure-cooled engines. Note adjustment. Refer to text for details.

On models with centrifugal switch, breaker point gap should be adjusted to 0.040 inch (1.0 mm).

AIR COOLING SYSTEM

Air cooling systems are used on Models DJA, DJB, DJBA, DJC and DJE to cool engine. Pressure cooling or Vacu-Flo cooling systems are used as outlined in the following sections.

PRESSURE COOLING. On models equipped with pressure cooling, free air is drawn by flywheel rotation into engine sheet metal housing through flywheel grille opening and is forced through cylinder cooling fins and out through a rear or side aperature.

Some engines may be equipped with a thermostat controlled shutter (Vernatherm) which allows engine compartment air to reach 120°F (50°C) before opening and becomes wide open at 140°F (60°C) for full ventilation of enclosure. Opening temperature of sensing element is not adjustable. To determine if this operating element is in working order, remove two screws which retain it to mounting bracket (note slotted holds for adjusting position) and test it by application of heat. Opening should begin at 120°F (50°C) and plunger should be fully extended at 140°F (60°C). Total movement should be at least 13/64 inch (5 mm). Reinstall so that plunger when fully withdrawn into element body just touches roll pin as in Fig. O220 with shutter completely closed at ambient (free air) temperature.

If shutter operation is unsatisfactory, check for a weak shutter return (closing) spring and examine nylon shutter bearings for dirt or damage. Clean and renew as necessary.

VACU-FLO COOLING. This system is designed for cooling industrial power plants which are installed in a closed compartment. Note in Fig. O221 that

263

Onan

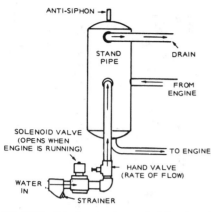

Fig. O222—Drawing showing coolant flow to and from stand pipe.

SMALL DIESEL

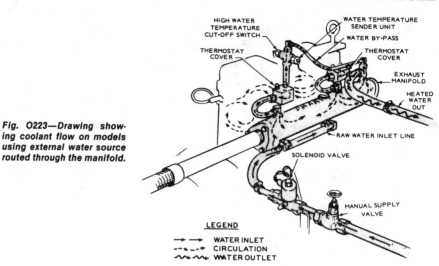

Fig. O223—Drawing showing coolant flow on models using external water source routed through the manifold.

flow of coolant air is drawn through engine shroud and cooling fins and forced out by flywheel blower through a vent or outside duct. Flow is in reverse direction from that of pressure-cooled engines.

IMPORTANT. If flywheel or flywheel blower is renewed, be sure that new part is correct for engine cooling system, whether pressure or Vacu-Flo.

Air volume requirement for proper cooling of these engines, expressed in cubic feet per minute is specified for each engine in factory-furnished operator's manual. Dependent upon engine size, this may range from 300 to 1600 cfm (8.5-45.3 m^3). Duct and vent sizes are detailed for each model and type of cooling system.

HIGH TEMPERATURE CUT-OFF. Some larger engine models are equipped with a high temperature safety switch for protection from overheating. Switch is normally closed, but opens to stop engine if compartment air temperature rises to 240°F (115°C) due to problem in cooling system caused by blockage or shutter failure to open. When engine compartment temperature drops to about 190°F (90°C), switch will automatically close and engine can be restarted.

LIQUID-COOLED SYSTEM

Liquid-cooled engines RDJC, RDJE, RDJEA and RDJF use either a radiator or external water source to cool engine. Refer to following sections for servicing.

RADIATOR SYSTEM. An automotive type radiator cooling system is used on some models. Note that cooling fan forces air through radiator and a belt driven water pump is mounted on front of engine. A bypass line is connected from the water pump to the water outlet on the cylinder head so water is circulated when the thermostat is contained inside the cylinder head water outlet.

Radiator capacity for all models is three gallons (11.3 liters). Do not use antileak type antifreeze on models equipped with coolant filters. Be sure adequate air circulation is possible around radiator, ducts and vents. Heated air should not be allowed to reenter cooling air stream.

EXTERNAL WATER SYSTEM. Some models may be equipped with a cooling system which routes water supplied from an external source through cooling passages in the engine block. No water pump is used as pressure is supplied at the water source. A solenoid valve shuts off water when the engine is not running while a supply valve controls rate of water flow. Three types of external water systems are used. In the direct flow system, water is routed directly to the engine block then discharged. In the standpipe system shown in Fig. O222, water circulates through the engine then returns to a standpipe to be mixed with fresh, cool water. The hottest water rises in the standpipe and is drained off. In the wet manifold system shown in Fig. O223, water is circulated around the exhaust manifold after passing through the engine block. Two 160°F (70°C) thermostats contained in housings attached to the cylinder heads must open before water enters the water jacket surrounding the exhaust manifold. A bypass line circulates water prior to thermostat opening. On all systems, the water discharge line must be sufficiently large so water flow is not impeded.

CRANKCASE BREATHER

The crankcase breather is located inside the rocker cover as shown in Fig. O224. Crankcase breather (41) vents crankcase fumes into the intake port. Inspect breather after every 200 hours of operation.

REPAIRS

TIGHTENING TORQUES

Recommended tightening torques for all models are as follows:

Connecting rod.........27-29 ft.-lbs.
(37-39 N·m)
Cylinder head..........44-46 ft.-lbs.
(60-62 N·m)
Exhaust manifold........13-15 ft.-lbs.
(18-20 N·m)
Flywheel...............65-70 ft.-lbs.
(88-95 N·m)
Fuel pump..............15-20 ft.-lbs.
(20-27 N·m)
Gearcase cover.........15-20 ft.-lbs.
(20-27 N·m)
Glow plug..............10-15 ft.-lbs.
(18-20 N·m)
Injection nozzle.......20-21 ft.-lbs.
(27-28 N·m)
Injection pump.........15-16 ft.-lbs.
(20-22 N·m)
Intake manifold........13-15 ft.-lbs.
(18-20 N·m)
Rear bearing plate.....40-45 ft.-lbs.
(54-61 N·m)

CYLINDER HEAD

To remove the cylinder head, it is first necessary on liquid-cooled models to drain coolant and disconnect interfering hoses. Remove the decompression

SERVICE MANUAL

Onan

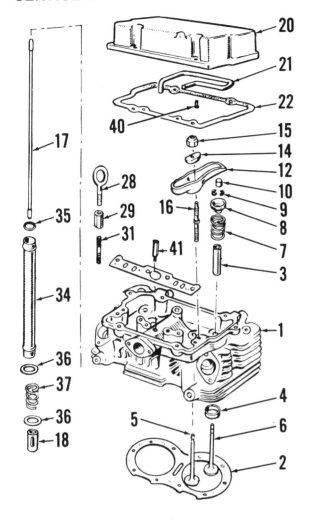

Fig. O224—Exploded view of typical cylinder head assembly. Model DJB is shown.

1. Cylinder head
2. Head gasket
3. Valve guide
4. Valve seat insert
5. Intake valve
6. Exhaust vlave
7. Valve spring
8. Spring retainer
9. Retainer locks
10. Valve stem cap
12. Rocker arm
14. Rocker ball pivot
15. Rocker locknut
16. Rocker arm stud
17. Tappet push rod
18. Tappet
21. Rocker oil line
22. Cover gasket
28. Lifting eyebolt
29. Extension nut
31. Cylinder stud
34. Push rod shield
35. "O" ring
36. Washer
37. Shield spring
40. Oil line screw
41. Crankcase breather

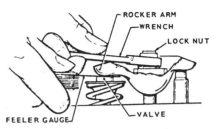

Fig. O226—Valve clearance is adjusted by turning rocker arm locknut while inserting proper size feeler gage between rocker arm and valve stem end. Refer to text.

(60-62 N·m) following same tightening sequence.

Adjust valve clearance and compression release linkage (Model DJA) as outlined in appropriate paragraphs. Complete installation by reversing removal procedure. Cylinder head should be retorqued and valves readjusted after the first 50 hours of engine operation.

VALVE CLEARANCE ADJUSTMENT

Valve clearance should be set with engine cold and piston 10°-45° past TDC on power stroke. Valve clearance is adjusted by turning rocker arm stud locknut (Fig. O226) while inserting a feeler gage of proper thickness between rocker arm and valve stem.

NOTE: A torque of 4-10 ft.-lbs. (5-13 N·m) should be required to turn self-locking rocker arm nut. If turning torque is not within this range, threads are damaged and nut and/or rocker arm stud should be renewed.

On two-cylinder engines, be sure to move each piston past TDC on its power stroke before adjusting valves for each cylinder. On four-cylinder models, adjust valves in regular engine firing order (1-2-4-3). Begin with number 1 cylinder with piston 10°-45° past TDC on power stroke. Rotate crankshaft 180° (½ turn) for each successive cylinder in firing order.

Valve clearance should be 0.011 inch (0.28 mm) for intake and 0.008 inch (0.20 mm) for exhaust on Model DJA. Model DJB, prior to Specification D, calls for both intake and exhaust to be 0.004 inch (0.10 mm). Beginning with Specification D, intake should be 0.009 inch (0.23 mm) and exhaust 0.007 inch (0.18 mm). Models DJBA and DJC should also be set at 0.009 inch (0.23 mm) intake and 0.007 inch (0.18 mm) for exhaust. Model RDJC valve clearance is 0.011 inch (0.28 mm) for intake and 0.016 inch (0.40 mm) for exhaust. On Models RDJE, RDJEA and RDJF, clearance for both intake and exhaust is 0.017 inch (0.43 mm).

solenoid (DJA only), rocker box cover, fuel injector and manifolds. Push rods and push rod tube shields will be loose when head is removed. Refer to Fig. O224. Both intake and exhaust valve are equipped with release type valve rotator caps. Lift off and identify the caps as head is removed, so they can be reinstalled on the same valves when unit is reassembled.

To reinstall cylinder head, proceed as follows: Assemble cylinder head with new gasket onto cylinder block. Before tightening cylinder head cap screws on four-cylinder models, install exhaust manifold and tighten manifold nuts evenly to a torque of 13-15 ft.-lbs. (18-20 N·m). This will align exhaust ports with manifold. On all models, tighten cylinder head bolts in appropriate sequence shown in Fig. O225 to first step torque of 25-30 ft.-lbs. (34-41 N·m). Then tighten cap screws in same sequence to final torque of 44-46 ft.-lbs. (60-62 N·m). Wait a few minutes to allow gasket to compress, then retorque to 44-46 ft.-lbs.

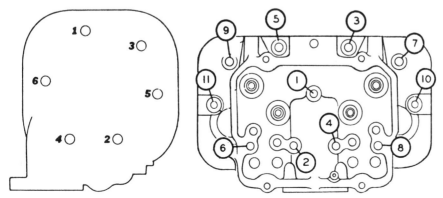

Fig. O225—Tightening sequence for cylinder head cap screws. Left view is for Model DJA and right view is for all other models. Refer to text.

265

Onan SMALL DIESEL

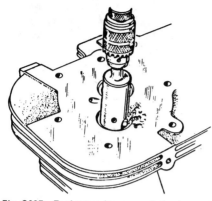

Fig. O227—Tool setup for removal of valve seat inserts. Adjust to cut 0.015 inch (0.4 mm) from edge of seat and take care not to bottom tool in counterbore.

Fig. O229—Release-type valve rotators are used on all models. Clearance between valve stem end and cap should be checked whenever valves are serviced. Refer to text.

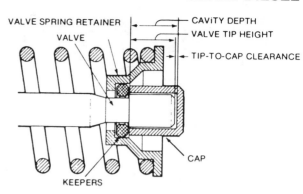

Manufacturer recommends readjusting valve clearance after every 500 hours of operation.

VALVE SYSTEM

All engines are equipped with renewable, insert type valve seats. Valve seat angle for intake and exhaust should be 45°. Intake valve face angle should be ground to 42° on all models and exhaust valve face angle should be 45°. Valve seat width should be 3/64 to 1/16 inch (1.2-1.6 mm). Manufacturer does not recommend handlapping valves to seats as sharp contact at seating surface could be destroyed. Apply Prussian Blue to valve face and rotate against seat to check seating surface.

Hardened valve seats are renewable. To remove old seat, use ONAN tool 420-0272 mounted in a drill press as shown in Fig. O227. Set tool to cut 0.015 inch (0.04 mm) from outer edge of seat. Seat material which remains may be easily peeled from counterbore. Use care not to damage counterbore.

Valve seat inserts are available in 0.002, 0.005, 0.010 and 0.025 inch (0.051, 0.127, 0.254 and 0.635 mm) oversizes as well as standard. Seat insert should have a 0.002-0.004 inch (0.05-0.10 mm) interference fit in its counterbore. Be sure counterbore is clean and free of burrs before installing insert. To make installation easier, heat cylinder head to 325°F (165°C) in an oven and chill valve seats with dry ice. Install immediately making certain insert is bottomed in counterbore.

After new seat has been installed and ground, temporarily install valve and check clearance between valve head and gasket surface of cylinder head. If clearance is less than 0.030 inch (0.76 mm), valve seat must be reground until recommended clearance is obtained.

Exhaust valve stem diameter for all models is 0.3405-0.3415 inch (8.649-8.674 mm). Intake valve stem diameter is as follows: Early Model DJA is 0.3405-0.3410 inch (8.649-8.661 mm) and late Model DJA is 0.3381-0.3420 inch (8.588-8.687 mm). Early Model DJB is 0.3405-0.3410 inch (8.649-8.661 mm) and late Model DJB has a tapered stem (Fig. O228) which is 0.3385-0.3397 inch (0.860-0.863 mm) measured 2.96 inches (75.2 mm) from end and 0.3401-0.3411 inch (8.639-8.664 mm) at 1.18 inches (30.0 mm) from end. All DJBA models are 0.3405-0.3410 inch (8.649-8.661 mm) and all DJE, RDJC, RDJE and RDJEA models are 0.3405-0.3415 inch (8.649-8.674 mm). Early RDJE models are 0.3405-0.3415 inch (8.649-8.674 mm) while late models use a tapered stem intake valve (Fig. O228) which measures 0.3385-0.3397 inch (8.597-8.628 mm) at 2.96 inches (75.2 mm) from retainer end and 0.3405-0.3410 inch (8.649-8.661 mm) at 1.18 inches (30.0 mm) from end. Recommended stem to guide clearance is 0.0015-0.003 inch (0.038-0.076 mm) for intake and 0.003-0.005 inch (0.076-0.127 mm) for exhaust on all models.

Valve guides are interchangeable for intake and exhaust valves. After installation, both guides must be reamed to provide recommended valve stem operating clearance. When properly installed, guides should extend 11/32 inch (8.7 mm) above machined spring seat of head.

Valve springs are identical for intake and exhaust valves. Springs should have a free length of 1⅞ inches and should test 45-49 pounds (200-218 N) when compressed to a length of 1.528 inches (38.81 mm).

All engines are equipped with release-type valve rotators on both intake and exhaust valves. The rotator cap (Fig. O229) releases keeper tension as the valve is pushed off its seat. This allows the valve to float and rotate during operation. For proper operation, clearance between end of valve and cap should be between 0.001-0.005 inch (0.025-0.127 mm). Too little clearance will prevent valve rotation and too much clearance may cause valve breakage. To check clearance, remove cap and measure depth of cavity in cap. Measure valve tip height from keeper and subtract this dimension from cap depth to determine clearance. If clearance is not within specifications, renew parts as necessary. When reassembling, place a drop of oil on end of valve stem before installing cap.

INJECTOR

WARNING: Fuel emerges from injector with sufficient force to penetrate the skin. When testing, keep your person clear of the nozzle spray.

All engines except later DJE models are equipped with American Bosch pintle type injector nozzles. Opening pressure is factory adjusted to 1900-1950 psi (13100-13450 kPa). However, pressure may decrease to approximately 1750 psi (12065 kPa) as nozzle spring takes a set. Later Model DJE is equipped with Diesel Kiki injection nozzles. Opening pressure is factory adjusted to an initial setting of 2130-2200 psi (14685-15170 kPa). Pressure may decrease to approximately 1950 psi (13445 kPa) as nozzle spring takes a set.

If engine does not start or does not run properly and the injector is suspected of being at fault, remove and reposition the unit where spray pattern can be observed or attach unit to a nozzle tester. Turn engine over with starter or actuate

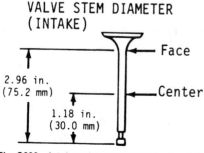

Fig. O228—Intake valve stem is tapered on late Models DJB, DJC and RDJF. Measure stem diameter at points indicated when checking condition of valve. Refer to text.

SERVICE MANUAL

Onan

Fig. O230—Comparison view of poor (left) and good (right) injector spray patterns. Refer to details.

tester handle. Spray pattern should be conical in shape with a solid core as shown in right-hand view, Fig. O230. If spray pattern is ragged or not symmetrical or if nozzle drips, install a new or rebuilt unit or disassemble, clean and adjust the injector unit. If spray pattern is satisfactory and tester unit is available, adjust opening pressure to pressure specified in preceding paragraph. Pressure is adjusted by turning pressure adjusting screw shown in Fig. O231. Screw may be turned after unscrewing plug and loosening cover. Tighten cover to 45-50 ft.-lbs. (61-68 N·m) after final screw adjustment and reinstall plug. Unscrew cover (C) and loosen locknut (L) on Kiki and early Bosch injectors to turn adjusting screw. After final screw adjustment, tighten locknut (L) and cover (C) to 45-50 ft.-lbs. (61-68 N·m). Do not disassemble a unit which can be returned to service without disassembly.

NOTE: Do not attempt to disassemble, clean or overhaul an injector unit unless the proper tools are available.

Fig. O231 shows a cross-sectional view of nozzle holder and nozzle valve. Be extremely careful when nozzle unit is handled not to strike or damage the protruding nozzle valve pintle. Any damage will deform the spray pattern and make the unit unfit for further use. A diesel nozzle tool set, ONAN part number 420-0208, is available for servicing nozzles. When nozzle is disassembled, be sure to keep all parts for each nozzle separated as parts must not be interchanged. Clean dirt and carbon from nozzle components using solvent and tools from cleaning kit.

Nozzle valve needle must slide freely in valve body. To check, first thoroughly rinse needle and seat in clean diesel fuel. With parts wet with diesel fuel, withdraw needle about one-third of the way out of the valve body. With valve held at 45° angle, needle should slide back to its seat by its own weight. If it does not, reclean or renew nozzle valve assembly.

When reassembling injector, nozzle body must be perfectly centered in cap nut. A centering sleeve (Fig. O232), part of tool kit 420-0208, is used to center nozzle body and initially tighten cap nut. Then, remove centering sleeve and tighten cap nut to a torque of 50-55 ft.-lbs. (70-75 N·m).

When reinstalling injector, be sure to renew nozzle heat shield and gasket assembly in cylinder head injector bore. Tighten injector retaining cap screws to a torque of 20-21 ft.-lbs. (27-28 N·m).

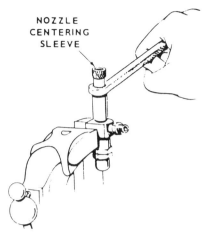

Fig. O232—Centering sleeve, part of ONAN tool set number 420-0208, is used to center nozzle body in cap nut when reassembling nozzle.

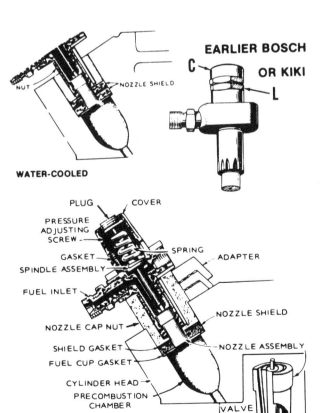

Fig. O231—View of injector assemblies. On late Bosch injector, a plug is unscrewed for access to pressure adjusting screw. On Kiki and early Bosch injectors, cover (C) must be removed for access to pressure adjusting screw.

INJECTION PUMP

Model DJA is equipped with an American Bosch Model PLB injection pump, later Model DJE is equipped with either a Bryce of Kiki injection pump while all other models are equipped with an American Bosch PSU injection pump.

Unless the shop is equipped with the necessary calibration stand and overhaul data, disassembly of pump should not be attempted.

When injection pump is removed, time and bleed the unit during reinstallation, and adjust governor linkage, as outlined in previous MAINTENANCE para-

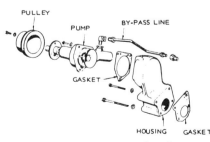

Fig. O233—Exploded view of water pump assembly.

Onan   SMALL DIESEL

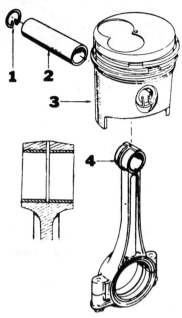

Fig. O234—Exploded view of piston and connecting rod assembly. Two bushings are used in upper end of rod to provide an oil groove at center of rod bore.

1. Retaining ring
2. Piston pin
3. Piston
4. Connecting rod

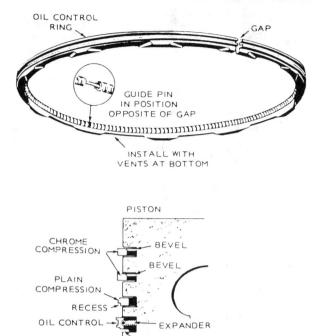

Fig. O235—Cross section of piston rings and piston showing ring arrangement. Ring end gaps should be staggered equally around piston but should not be in line with piston pin bore.

graphs. Tighten injection pump mounting screws alternately and evenly to a torque of 18-21 ft.-lbs. (24-28 N·m) on Model DJA and 15-16 ft.-lbs. (20-22 N·m) on all other models when unit is reinstalled.

WATER PUMP

Models RDJC, RDJE, RDJEA and RDJF equipped with a radiator cooling system use a water pump to circulate coolant. Refer to Fig. O233 for an exploded view of water pump assembly. Pump bearings are permanently sealed and packed with a lubricant. Bearings and impeller seal are renewable.

Manufacturer recommends using a mixture of ethylene glycol antifreeze and water as a coolant in both summer and winter. The antifreeze contains rust and corrosion inhibitors that protect the cooling system. Coolant should be changed every year.

CONNECTING ROD

Connecting rod and piston unit can be removed from above after removing cylinder head and engine oil base. Be sure to remove ring ridge from top of cylinder bore before attempting to remove piston. The forged connecting rod contains renewable, precision bearing inserts in large end. Bearings are available in undersizes of 0.010, 0.020 and 0.030 inch (0.254, 0.508 and 0.762 mm) as well as standard.

Standard crankpin diameter for all models is 2.0597-2.0605 inches (52.316-52.337 mm) with recommended diametral clearance of 0.001-0.0033 inch (0.025-0.084 mm).

Two bushings are used in piston pin end of rod. Bushings should be pressed in from each side of bore until flush with outer surfaces of rod. This will leave an oil groove space of approximately 1/16 inch (1.6 mm) in center of rod bore. Drill a 3/16 inch (5 mm) oil hole through bushings at top of rod before reaming bushings. Final sizing of bushing calls for a diametral clearance of 0.0002-0.0007 inch (0.005-0.018 mm). Standard diameter of piston pin is 0.9899-0.9901 inch (25.143-25.148 mm) with 0.002 inch (0.05 mm) oversize pin available.

When installing assembled piston and connecting rod, notch or "FRONT" mark on piston should face front of engine and witness marks on connecting rod and cap must be aligned. On engines after Specification P, valve relief cut-outs on head of piston should face in same direction as stamped reference numbers on connecting rod or raised witness marks of no number is imprinted on rod. Install piston and rod unit with reference numbers and valve relief cut-outs facing camshaft side of engine. On all models, tighten connecting rod cap screws to a torque of 27-29 ft.-lbs. (37-39 N·m).

PISTON, PIN, RINGS AND CYLINDER

Piston is equipped with three compression rings and one expander type oil control ring. Compression rings are marked "TOP" or otherwise identified for correct installation. Oil control ring should be installed with vents toward bottom of piston. Refer to Fig. O235. Rings should be postitioned with end gaps staggered equally around piston. Guide pin in expander should be opposite (180°) from end gap of oil control ring. No gap should be in line with piston pin bore.

Piston ring end gap should be measured with ring positioned squarely in cylinder bore. Recommended end gap for all rings is 0.010-0.020 inch (0.25-0.50 mm). Piston ring groove wear can be checked by installing a new ring in piston groove and measuring ring side clearance in groove. Recommended clearance is 0.003-0.005 inch (0.08-0.14 mm). Piston should be renewed if clearance exceeds 0.006 inch (0.15 mm).

Standard cylinder bore diameter is 3.4995-3.5005 inches (88.887-88.913 mm) for Models DJE, RDJE, RDJEA and RDJF. Standard cylinder bore diameter for all other models is 3.2495-3.2505 inches (82.537-82.563 mm). Some engines are factory equipped with 0.005 inch (0.13 mm) oversize piston and rings. These engines are identified by an "E" following the serial number. Pistons and rings are available in oversizes of 0.005, 0.010, 0.020, 0.030 and 0.040 inch (0.127, 0.254, 0.0508, 0.762 and 1.016 mm) as well as standard size. Recommended piston skirt clearance in cylinder is 0.0055-0.0075 inch (0.140-0.191 mm) when measured 90° from piston pin.

Piston pin should be a thumb push fit into piston at room temperature. If pin is excessively loose, renew pin and/or

SERVICE MANUAL

Onan

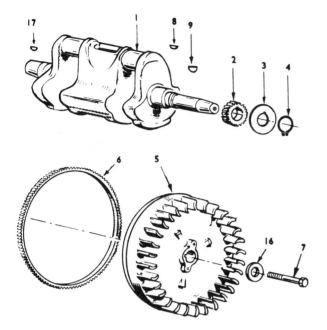

Fig. O236—Exploded view of crankshaft and flywheel typical of "J" series. Model DJB shown.

1. Crankshaft
2. Timing gear
3. Retainer washer
4. Lock ring
5. Flywheel
6. Ring gear
7. Flywheel cap screw
8. Timing gear key
9. Flywheel key
16. Flywheel washer
17. Drive gear key

piston. A 0.002 inch (0.05 mm) oversize piston pin is available. For easier pin installation, piston may be heated in hot water or an oven prior to assembly.

CRANKSHAFT AND BEARINGS

The crankshaft is supported in two lead-bronze sleeve bearings pressed into front of crankcase and rear bearing plate. Four-cylinder engines have an additional split center main bearing. Refer to Fig. O237.

To remove crankshaft, first remove oil pan, cylinder head, pistons and rods, flywheel and timing gear cover. Remove gear from front of crankshaft using a suitable puller. Remove rear bearing plate and center main bearing cap (four-cylinder engine). Withdraw crankshaft through rear opening in crankcase.

Crankshaft crankpin diameter is 2.0597-2.0605 inches (52.316-52.337 mm) and recommended connecting rod bearing clearance is 0.001-0.0033 inch (0.025-0.084 mm). Main journal diameter is 2.2437-2.2445 inches (56.99-57.01 mm) for Models DJA, DJB, DJBA, DJE, RDJE and RDJEA or 2.2427-2.2435 inches (56.965-56.985 mm) for Models DJC, RDJC and RDJF. Recommended main bearing clearance is as follows: All Models DJA, DJB and DJE and early Models RDJE and RDJEA should be 0.0014-0.0052 inch (0.036-0.132 mm). Early Models DJC, RDJC and RDJF should be 0.0024-0.0062 inch (0.061-0.157 mm). Late Model DJC should be 0.0024-0.0049 inch (0.061-0.124 mm). Late Model DJBA clearance is 0.0020-0.0033 inch (0.051-0.084 mm). Late Models RDJE and RDJEA should be 0.0030-0.0043 inch (0.076-0.109 mm). Late Model RDJF front and rear bearing clearance is 0.0030-0.0043 inch (0.076-0.109 mm) while center bearing is 0.0024-0.0052 inch (0.061-0.132 mm). Main and rod bearings are available in undersizes of 0.010, 0.020 and 0.030 inch (0.254, 0.508 and 0.762 mm) as well as standard size.

Crankshaft end play is controlled by rear bearing plate. To adjust end play, vary the thickness of gasket pack between rear bearing plate and crankcase until recommended end play of 0.010-0.015 inch (0.25-0.38 mm) is obtained. If more than 0.015 inch (0.38 mm) total gasket thickness is required, use a combination of gaskets and steel shims to avoid excessive gasket compression.

If crankpin journal must be resized, crankpin fillets must be shot peened to relieve metal stress. If facilities for shot

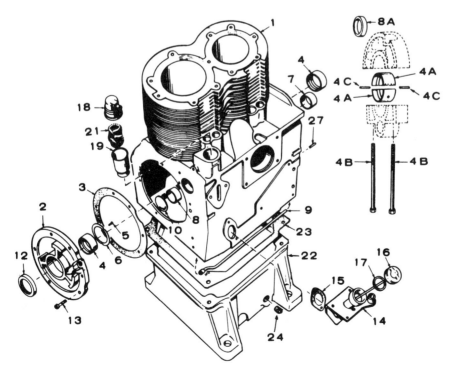

1. Block assy.
2. Bearing plate
3. Gasket
4. End main bearings
*4A. Center main bearing
*4B. Bolt, center main
*4C. Pin, center main
5. Thrust washer pin
6. Thrust washer
7. Cam bearing, front
8. Cam bearing, rear
*8A Cam bearing, center
9. Oil tube
10. Camshaft plug
12. Crank seal, rear
13. Bearing plate bolt
14. Oil fill tube
15. Gasket
16. Cap & sleeve
17. Cap gasket
18. Breather tube cap
19. Breather tube
21. Baffle
22. Oil base
23. Gasket
24. Drain plug
27. Gear cover dowel
*Four-cylinder models only.

Fig. O237—Exploded view of crankcase and oil base assembly typical of this engine series. Model DJBA is shown.

Onan — SMALL DIESEL

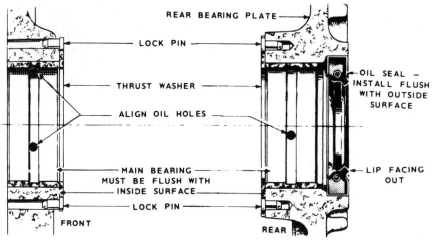

Fig. O238—Cross-sectional view of crankshaft main bearings to show proper installation. Note rear oil seal placement with open side of seal facing inward.

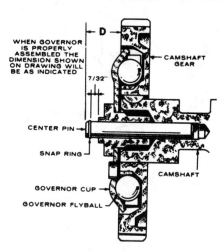

Fig. O242—Cross-sectional view of governor mechanism in camshaft gear. In order to limit governor cup movement of 7/32 inch (5.5 mm) as shown, dimension (D) must be 25/32 inch (19.8 mm) as described in text.

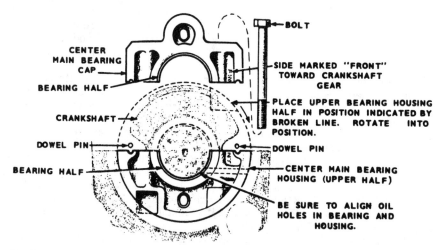

Fig. O239—Installation of center main bearing used on four-cylinder engines. Refer to procedure in text.

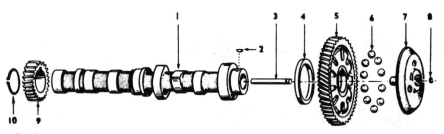

Fig. O240—Exploded view of camshaft and governor sensor assemblies. Model DJB is shown.
1. Camshaft
2. Cam gear key
3. Center pin
4. Thrust washer
5. Camshaft gear (incl. spacer).
6. Flyballs
7. Governor cap
8. Snap ring
9. Injection pump drive gear
10. Snap ring

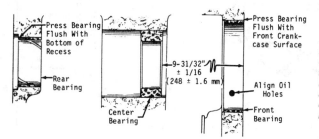

Fig. O241—Camshaft bushings should be pressed into crankcase as shown. Center bearing is used on four-cylinder models. Bushings are available in standard size only and do not require boring or reaming for proper fit.

peening are not available, renew crankshaft instead of regrinding.

Front oil seal, located in gear cover, and rear oil seal, located in, rear bearing plate, may be renewed at this time. Be sure open side of seal faces inward and outer side of seal is flush with outer surface of front cover or rear plate. Lubricate seal lips with grease prior to reassembly.

When renewing main bearing inserts, press front and rear bearings into place as shown in Fig. O238. Make certain oil holes in bearings are aligned with holes in housing. Install thrust washers and lock pins. Tighten rear bearing plate mounting cap screws to a torque of 40-45 ft.-lbs. (54-61 N·m).

On four-cylinder engines, center main bearing is assembled after crankshaft is installed. Refer to Fig. O239. Rotate upper half into place making sure side marked "FRONT" faces front of crankshaft. Position dowel pins into grooves of bearing cap. Install bearing cap with lower bearing half and tighten cap screws to a torque of 97-102 ft.-lbs. (131-138 N·m).

Heat crankshaft gear to approximately 350°F (175°C). Align keyway in gear with crankshaft key, then drive gear into place. Be sure timing marks on crankshaft gear and camshaft gear are aligned (Fig. O243). Complete installation by reversing the removal procedure.

CAMSHAFT AND GOVERNOR

The camshaft gear is a press fit on shaft. The assembled unit must be removed from flywheel end of housing after removing flywheel, timing gear cover, injection pump and fuel lift pump.

SERVICE MANUAL

Onan

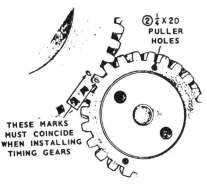

Fig. O243—Align crankshaft and camshaft timing marks as shown when installing camshaft.

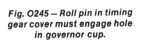

Fig. O245—Roll pin in timing gear cover must engage hole in governor cup.

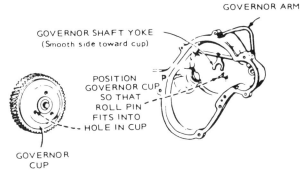

Remove rocker arms and push rods, then lay cylinder block on its side to prevent tappets from dropping. Tappets cannot be withdrawn until after the camshaft is out.

Camshaft bushings are precision type which may be renewed as shown in Fig. O241 after camshaft is removed. If carefully installed, bushings do not require reaming or honing for proper fit. Recommended diametral clearance is 0.0015-0.0030 inch (0.038-0.076 mm) and maximum allowable clearance before renewal is 0.0037 inch (0.094 mm).

Governor weight unit is mounted on camshaft gear and governor cup rides on a center pin pressed into camshaft as shown in Fig. O242. Check the distance center pin extends from front face of camshaft gear. Distance (D) must be 25/32 inch (19.84 mm) as shown to give the proper travel to governor cup.

Governor cup must be a free spinning fit on center pin. Renew the cup if race surface is grooved or rough, or if cup is excessively loose on center pin. Renew any flyballs if they have flat spots, grooves, or rust pitting. Center pin must be renewed if removed from camshaft for any reason.

When reinstalling the camshaft, align timing marks as shown in Fig. O243. Install crankshaft washer and snap ring, then check camshaft end play as shown in Fig. O244. Clearance (Y) should be 0.007-0.039 inch (0.18-0.99 mm) with camshaft pushed rearward and crankshaft pulled forward. If clearance is excessive, renew spacer washer (1) or crankshaft washer (3).

When renewing timing gear cover oil seal, drive old seal out inner side of cover. Using a suitable driver, install new seal until flush with outside surface of front cover. Be sure open side of seal faces inward. Lubricate seal lip prior to reassembly.

When reinstalling gear cover, be certain stop pin (Fig. O245) fits into one of the holes in governor cup. Tighten cap screws to a torque of 18-20 ft.-lbs. (24-27 N·m). Complete installation by reversing removal procedure. Bleed air from fuel system and check injection pump timing as outlined in MAINTENANCE section.

OIL PUMP

The engine is fully pressure lubricated and all oil is circulated through a full flow filter. Oil pressure should be 25 psi (172 kPa) or higher with engine at operating speed and temperature. A bypass valve is located in rear bearing plate. If oil pressure is too high or too low, inspect bypass valve plunger for sticking or wear.

The gear type oil pump is driven by crankshaft gear and is accessible after removing timing gear cover and oil base. Component parts of pump are not individually available. If pump is damaged or excessively worn, it must be renewed as a complete unit. Before installing pump, fill pump with oil to be sure it is primed. Tighten pump mounting cap screws to a torque of 15-20 ft.-lbs. (20-27 N·m) and oil base mounting screws to 45-50 ft.-lbs. (61-68 N·m).

FLYWHEEL

The flywheel is tapered fit on crankshaft. Use a suitable puller (such as ONAN tool 420-0100) to remove flywheel. When reinstalling, tighten retaining cap screw to a torque of 65-70 ft.-lbs. (90-95 N·m).

Replacement flywheel is supplied without timing markings. To accurately

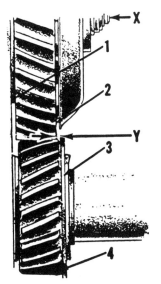

Fig. O244—Procedure to check camshaft gear end play. Refer to text.

X. Camshaft pushed in
Y. Measure end play here
1. Spacer washers
2. Camshaft gear
3. Crankshaft washer
4. Crankshaft gear

Fig. O246—To accurately mark port closure point on a replacement flywheel, piston drop must be measured as outlined in text.

MODEL	PISTON DROP DIMENSION	PORT CLOSURE (BTDC)
DJA	0.102 in. (2.591 mm)	17°
DJB, DJBA, DJC (Prior to Spec P)	0.155 in. (3.937 mm)	21°
DJB, DJBA, DJC (Spec P and later)	0.128 in. (3.251 mm)	19°
DJE, RDJE, RDJEA	0.128 in. (3.251 mm)	19°
RDJF (prior to Spec P)	0.155 in. (3.937 mm)	21°
RDJF (Spec P and later)	0.128 in. (3.251 mm)	19°

271

Onan

SMALL DIESEL

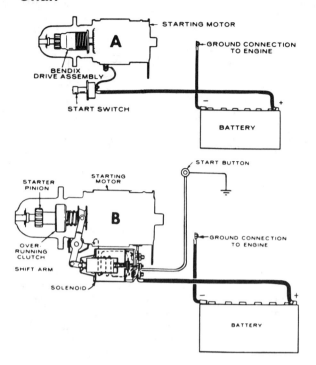

Fig. O247—Views of Bendix-drive starter and solenoid-shift starter in basic electric circuits. Refer to text.

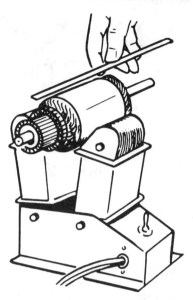

Fig. O248—Use of growler to check armature for short circuit. Follow procedure in text.

determine top dead center and port closing points and mark flywheel accordingly, number 1 piston travel must be measured with a depth gage. With cylinder head removed, rotate crankshaft until number 1 piston is at TDC of the compression stroke. Mark this point on the flywheel in line with timing pointer. Next, rotate flywheel counterclockwise until piston drops approximately 0.250 inch (0.6 mm). Then, rotate flywheel clockwise until piston is at its specified piston drop (port closure) dimension as listed in table shown in Fig. O246. Mark port closure (PC) point on flywheel.

Flywheel ring gear is renewable. To remove old ring gear, cut or grind part way through, then use a hammer and chisel to break gear. Heat new ring gear to about 400°F (200°C), then quickly install onto flywheel. Be sure gear bottoms squarely against flywheel shoulder.

NOTE: Do not heat ring gear with a torch as it may warp or heat treatment may be destroyed.

ELECTRIC STARTERS

Two styles of battery-driven electric starter motors are used. Bendix-drive starter shown at A–Fig. O247 is designed to engage teeth of flywheel ring gear when starter switch is depressed to close circuit causing starter motor to turn. Engagement of starter pinion with ring gear by means of spiral shaft screw within Bendix pinion is cushioned by action of its coiled drive spring so starting motor can absorb the sudden loading shock of engagement. Engine manufacturer recommends that complete Bendix drive unit be renewed in case of failure, however, if a decision is made to overhaul starter drive by obtaining parts from manufacturer of the starter (Prestolite), be sure that correct drive spring is used. Length of spring is critical to mesh and engagement of starter pinion to flywheel ring gear. There are no procedures for adjustment of this starter.

Service is generally limited to cleaning and careful lubrication, renewal of starting motor brushes (4 used), brush tension springs and starter motor and drive housing bearings. Refer to STARTER MOTOR TESTS for electrical check-out procedures for starter armature and field windings.

Solenoid-shift style starter (B–Fig. O247) uses a coil solenoid to shift starter pinion into mesh with flywheel ring gear and an overrunning (one-way) clutch to ease disengagement of pinion from flywheel as engine starts and runs.

NOTE: Starter clutch will burn out if held in contact with flywheel for overlong periods. Starter switch must be released quickly as engine starts.

All parts of stater are available for service. Solenoid unit and starter clutch are renewed as complete assemblies.

STARTER MOTOR TESTS

ARMATURE SHORT CIRCUIT. Place aramature in growler as shown in Fig. O248 and hold a hack saw blade or similar piece of thin steel stock above and parallel to core. Turn growler "ON". A short circuit is indicated by vibration of blade and attraction to core. If this condition appears, renew armature.

GROUNDED ARMATURE. Check each segment of commutator for grounding to shaft (or core) using ohmmeter setup as shown in Fig. O249. A low Rx1 scale) continuity reading indicates that armature is grounded and renewal is necessary.

It is good procedure to mount armature on a test bench or between lathe centers to check for runout of commutator or shaft. If shaft is worn badly, renewal is recommended. If commutator runout exceeds 0.004 inch (0.10 mm), reface by turning.

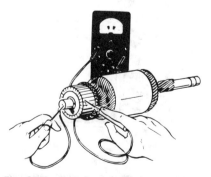

Fig. O249—Test for grounded armature commutator by placing ohmmeter test probes a shown. Probe shown touching shaft may also be held against core. Refer to text.

SERVICE MANUAL

Onan

Fig. O250—Use ohmmeter as shown to check field coil for suspected internal grounding.

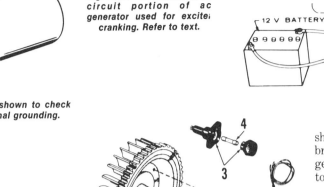

Fig. O252—Schematic of dc circuit portion of ac generator used for exciter cranking. Refer to text.

Ohmmeter used to check for breaks or opens in field coil windings. Be sure to check lead wires.

Fig. O251—Ohmmeter used to check for breaks or opens in field coil windings. Be sure to check lead wires.

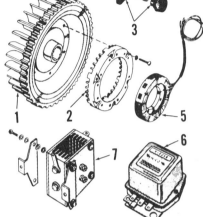

Fig. O253—Typical flywheel alternator shown in exploded view. In some models, regulator (6) and rectifier (7) are combined in a single unit.

1. Flywheel
2. Rotor
3. Fuse holder
4. Fuse (20A)
5. Stator & leads
6. Regulator, 2-step
7. Rectifier assy.

GROUNDED FIELD COILS. Refer to Fig. O250 and touch one ohmmeter probe to a clean, unpainted spot on frame and the other to connector as shown after unsoldering field coil shunt wire. A low range reading (Rx1 scale) indicates grounded coil winding. Be sure to check for possible grounding at connector lead which can be corrected, while grounded field coil cannot be repaired and calls for renewal.

OPEN IN FIELD COILS. Use procedure shown in Fig. O251 and check all four brush holders for continuity. If there is no continuity or if a high resistance reading appears, renewal is necessary.

EXCITER CRANKING

Exciter cranking, with cranking torque furnished by switching battery current through a separate series winding of generator field coils and the brushes using dc portion of generator armature is wired as shown in Fig. O252.

This starting procedure may be used on Model DJA. In cases where exciter cranking is inoperative, due to battery failure or other cause, unit may be started by use of a manual rope starter. In some cases, a recoil type "Readi-Pull" starter may be furnished for standby use.

In case exciter cranking system will not operate, isolate starter solenoid switch and battery from dc field windings and perform a routine continuity check of all components by use of a volt-ohmmeter. Refer to Fig. O252 for possible test points. If problem does not become apparent as caused by battery (low voltage), defective starter solenoid short or open circuit in lead wires or dc brushes, it will be necessary to check out generator in detail and may involve factory service.

FLYWHEEL ALTERNATOR

This battery charging system is simple and basically trouble-free. Flywheel-mounted permanent magnet rotor provides a rotating magnetic field to induce ac voltage in fixed stator coils. Current is then routed through a two-step mechanical regulator to a full-wave rectifier which converts this regulated alternating current to direct current for battery charging. Late models are equipped with a fuse between negative (−) side of rectifier and ground to protect rectifier from accidental reversal of battery polarity. Refer to schematic, Fig. O254. Maintenance services are limited to keeping components clean and ensuring that wire connections are secure.

TESTING. Check alternator output by connecting an ammeter in series between the positive (+), red terminal of rectifier and ignition switch. Refer to Fig. O254. At 1800 engine rpm), a discharged battery should cause about 8 amps to register on the ammeter. As battery charge builds up, current should decrease. Regulator will switch from high charge to low charge at about 14½ volts with low charge current of about 2 amps. Switch from low charge to high

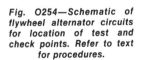

Fig. O254—Schematic of flywheel alternator circuits for location of test and check points. Refer to text for procedures.

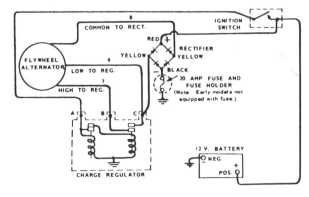

273

Onan — SMALL DIESEL

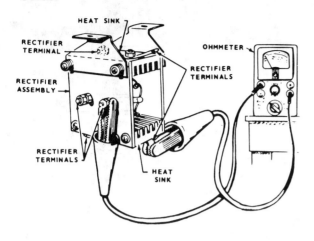

Fig. O255—Test each of four diodes in rectifier using volt-ohmmeter hook-up as shown. Refer to text for procedure.

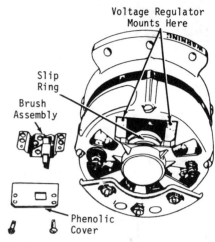

Fig. O256—View of optional belt driven, external alternator used on some models.

charge occurs at about 13 volts. If output is inadequate, test as follows:

Check rotor magnetism with a piece of steel. Attraction should be strong.

Check stator for grounds after disconnecting by grounding each of the three leads through a 12-volt test lamp. If grounding is indicated by lighted test lamp, renew stator assembly.

To check stator for shorts or open circuits, use an ohmmeter of proper scale connected across open leads to check for correct resistance values. Identify leads by reference to schematic.

From lead 7 to lead 8........0.25 ohms
From lead 8 to lead 9........0.95 ohms
From lead 9 to lead 7........1.10 ohms

Variance by over 25 percent from these values calls for renewal of stator.

RECTIFIER TESTS. Use an ohmmeter connected across a pair of terminals as shown in Fig. O255. All rectifier leads should be disconnected when testing. Check directional resistance through each of the four diodes by comparing resistance reading when test leads are reversed. One reading should be much higher than the other.

NOTE: Forward-backward ratio of a diode is on the order of several hundred ohms.

If a 12 volt test lamp is used instead of an ohmmeter, bulb should light, but dimly. Full bright or no light at all indicates that diode being tested is defective.

Voltage regulator may be checked for high charge rate by installing a jumper lead across regulator terminals (B and C – Fig. O254). With engine running, battery charge rate should be about 8 amps. If charge rate is low, then alternator or its wiring is defective.

If charge rate is correct (near 8 amps), defective regulator or its power circuit is indicated. To check, use a 12 volt test lamp to check input at regulator terminal (A). If lamp lights, showing adequate input, regulator is defective and should be renewed.

NOTE: Regulator, being mechanical, is sensitive to vibration. Be sure to mount it on bulkhead or firewall separate from engine for protection from shock and pulsating action.

Engine should not be run with battery disconnected, however, this alternator system will not be damaged if battery terminal should be accidentally separated from binding post.

BELT DRIVEN ALTERNATOR

An optional belt driven, battery charging, external alternator may be used on some models. Refer to Fig. O256. Normal alternator output voltage is 13.8 to 14.8 volts. If alternator output is not within specifications, perform regulator test as follows: With engine running, connect a jumper wire from positive output post on alternator to field post and recheck alternator output. If voltage reading is now within specifications or higher, regulator is defective. Voltage regulator and brush assembly may be renewed without disassembling alternator.

Before disassembling alternator, place a scribe mark across end frame and front cover to provide proper alignment when reassembling. Check stator and rotor windings for open or grounded circuits. Diodes should have continuity in one direction only. Renew faulty components or alternator assembly as required.

CLUTCH

When optional Rockford clutches are furnished with these engines, an adapter flange is fitted to engine output shaft for mounting clutch unit and a variety of housings are used dependent upon application and model of engine or clutch used. Refer to Fig. O257 for guidance in adjustment and proceed as follows:

Remove plate from top of housing and rotate engine manually until lock screw (1 – Fig. O257) is at top of ring (2) as shown. Loosen lock screw and turn adjusting ring clockwise (as facing through clutch toward engine) until toggles cannot be locked over center. Then, turn ring in reverse direction until toggles can just be locked over center by a very firm pull on operating lever. If a new clutch plate has been installed, slip under load to knock off "fuzz" and readjust. Lubricate according to instructions on unit plate.

Fig. O257—Procedure for adjustment of Rockford clutch. Refer to text.

SERVICE MANUAL

REDUCTION GEAR ASSEMBLIES

Typical reduction gear unit is shown in Fig. O258. Ratio of 1:4 is common in industrial applications. Lubrication calls for use of SAE 50 weight motor oil or SAE 90 gear oil. Refer to instructions printed on gear case for guidance. In most cases, a total of six plugs are fitted into case for lubricant fill or level check. Plug openings to be used are determined by positioning of gearbox in relation to horizontal or vertical. It is recommended that square plug heads be cut off those plugs not to be used to fill, check or drain so as to eliminate chance of error by overfill or underfill. All parts shown are available for renewal if needed in overhaul.

NOTE: In some installations, no shaft seal is fitted between engine crankcase and reduction gear housing. In these cases, with a common oil supply, engine oil lubricates gears and bearings of reduction gear unit and gear oil is not used. Be sure to check nameplate or operator's manual.

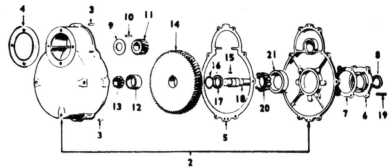

Fig. O258—Exploded view of typical reduction gear set. Refer to text for service details.

2. Housing & cover
3. Dowel pins (2)
4. Gasket (engine)
5. Cover gasket
6. Bearing retainer
7. Shims
8. Oil seal
9. Pinion washer
10. Pinion key
11. Pinion gear
12. Bearing cup
13. Bearing cone
14. Driven gear
15. Gear key
16. Snap ring
17. Bearing spacer
18. Shaft
19. Key
20. Bearing cone
21. Bearing cup

ONAN

Model	No. Cyls.	Bore	Stroke	Displ.
L317	3	89 mm (3.50 in.)	92 mm (3.62 in.)	1700cc (105 cu. in.)
L423	4	89 mm (3.50 in.)	92 mm (3.62 in.)	2300 cc (140 cu. in.)

Engines covered in this section are four-stroke liquid-cooled diesel engines. All models are equipped with a 12 volt electric starting motor. Crankshaft rotation is counterclockwise when viewed from the rear (flywheel end) of engine.

OPERATION

PREHEATING AND STARTING. When engine is cold and air temperature is below 21°C (70°F), engine glow plugs must be preheated to aid in starting.

CAUTION: Do not use ether as a starting aid. Heat from glow plugs could ignite ether vapor causing an explosion resulting in severe engine damage.

If engine is equipped with a preheat indicator light, engine is ready for starting when preheat light goes out. If not equipped with a preheat indicator light, energize glow plug circuit as indicated in the following table.

Ambient Temperature	Preheat Time
Above 21°C (70°F)	10-20 seconds
−18°C to 21°C (0°-70°F)	25-35 seconds
Below −18°C (0°F)	30-45 seconds

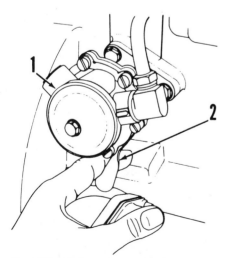

Fig. O300—Fuel transfer pump (1) is located at right rear side of engine. Transfer pump is equipped with a priming lever (2) which may be manually operated when bleeding air from fuel system.

If engines fails to start after 30 seconds of cranking, wait two minutes before repeating preheat and starting procedure.

MAINTENANCE

LUBRICATION

Recommended engine oil is API service classification CD/SF or CD/SE. The following SAE viscosity grades are recommended: SAE 10W for temperatures below 0°F (32°F), SAE 20W-20 for temperatures between −10°C (15°F) and 15°C (60°F) or SAE 30W or SAE 15W-40 if temperature range will be consistently above 0°C (32°F).

Recommended oil change interval is after every 100 hours of operation. Spin on type oil filter should be renewed at each oil change interval. Oil should be drained with engine at operating temperature. Crankcase capacity is approximately 5.7 L (6 quarts) with filter change. Do not overfill.

FUEL SYSTEM

A good quality Number 2 diesel fuel is recommended for most conditions. For cold weather operation, Number 1 fuel or a blend of Number 1 and Number 2 fuel may be used. All fuel must have a minimum Cetane number of 40 to ensure satisfactory engine performance.

NOTE: It is extremely important that fuel be kept clean and free of water. Due to precise tolerances of diesel injection components, dirt and water in the system can cause severe damage to injection pump and injection nozzles.

FUEL FILTER. The fuel filter is mounted on right side of engine. Fuel filter renewal interval is more dependent upon fuel quality and cleanliness than length of service or operating conditions. If engine indicates signs of fuel starvation (loss of power or surging), fuel filter should be renewed. Manufacturer also recommends renewing fuel filter yearly regardless of hours of operation or engine performance.

To renew filter, shut off fuel supply. Thoroughly clean filter and surrounding area. Release filter retaining clips and remove filter element. Install new filter element and secure with retaining clips. Bleed air from system as outlined in the following paragraph.

BLEED FUEL SYSTEM. To bleed air from fuel system, loosen fuel inlet line fitting at the injection pump. Actuate fuel transfer pump priming lever (2–Fig. O300) until air-free fuel flows from fitting, then retighten fitting.

NOTE: If camshaft lobe which drives transfer pump is at full lift position, priming lever cannot be operated. Rotate crankshaft one revolution to reposition camshaft lobe.

If engine fails to start after bleeding low pressure portion of fuel system, it will be necessary to bleed injector high pressure lines. Loosen high pressure line fitting at each injector. With stop control in "Run" position and throttle in high speed position, crank engine with starter motor until fuel is discharged from each injector line. Do not operate starter motor longer than 30 seconds at a time to prevent overheating. Tighten injector line fittings and start engine.

CAUTION: Do not use ether as a starting aid as glow plugs could ignite ether vapor causing an explosion resulting in severe engine damage.

ENGINE SPEED ADJUSTMENT. To check engine speed, first run engine until operating temperature is obtained. Move throttle control to low idle position and high idle position making sure injec-

SERVICE MANUAL

Onan

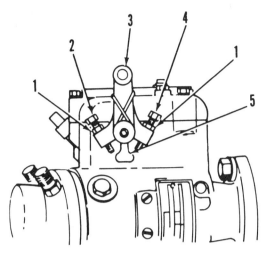

Fig. O301—View of injection pump control lever and speed adjustment points.
1. Locknut
2. High idle adjusting screw
3. Speed control lever
4. Low idle adjusting screw
5. Lever stop

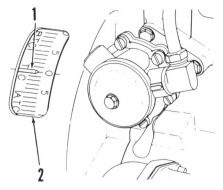

Fig. O303—Engine timing marks are located on flywheel. Remove timing window cover from flywheel housing to observe timing marks and pointer (1) through timing window (2).

tion pump control lever (3–Fig. O301) contacts low idle and high idle stop screws. Low idle speed should be within the range of 755 to 825 rpm depending upon engine application. High idle speed (no-load) should be 7 to 10 percent greater than speed stamped on pump name plate on variable speed applications or 3 to 4.5 percent greater on fixed rpm applications.

NOTE: Adjustments made by non-qualified personnel which alter exhaust emission control settings may violate Federal, state and/or local regulations. Manufacturer's warranty on injection pump and engine may also be voided by unauthorized adjustments to injection pump.

To adjust engine speed setttings, disconnect throttle control from pump control lever (3–Fig. O301). With engine running, push speed control lever forward until low idle adjusting screw (4) contacts lever stop (5). Turn screw in to increase speed or out to decrease speed. Hold adjusting screw and tighten locknut to secure adjustment. Move control lever rearward until high idle adjusting screw (2) contacts stop. Turn adjusting screw out to increase speed or in to decrease speed. After specified speed setting is obtained, hold adjusting screw and tighten locknute. Be sure to install a new seal wire on high speed adjusting screw.

SPEED DROOP ADJUSTMENT. Injection pumps used on fixed rpm (generator) applications are equipped with an external speed droop adjusting screw (1–Fig. O302). The droop screw, located at rear of injection pump housing, varies governor sensitivity by changing governor spring rate. Turning adjusting screw in makes governor regulation less sensitive and turning droop screw out increases governor sensitivity to changing engine loads. Droop screw adjustment will affect both full-load and no-load settings and may require readjustment of high speed stop screw as previously outlined. After each adjustment of screw, engine must be shut off briefly to allow governor spring to unload before making another adjustment. Be sure to install new seal wires after adjustment is completed.

PUMP STATIC TIMING. To check injection pump to engine static timing, first shut off fuel supply. Thoroughly clean outside of pump. Place a pan beneath fuel pump to catch fuel, then remove timing window cover from side of pump (Fig. O304). Remove timing window cover from side of flywheel housing (Fig. O303). Turn crankshaft in normal direction of rotation until number one piston is coming up on compression stroke. Continue to slowly turn crankshaft until timing pointer in flywheel housing is aligned with specified degree mark on flywheel. (Specified static timing is stamped on engine model plate.) At this point, pump timing marks (visible through timing window) on cam ring and governor weight retainer should be aligned as shown in Fig. O304.

If pump timing marks are not aligned, loosen pump mounting stud nuts and rotate injection pump until timing marks are aligned. Tighten pump stud nuts. Turn crankshaft back (clockwise viewed from rear) about ½ revolution, then turn crankshaft in normal direction until specified timing mark on flywheel is aligned with pointer again and recheck pump timing mark alignment. Reinstall pump timing window cover and bleed air from system if necessary.

COOLING SYSTEM

All engines are liquid-cooled using a belt driven water pump to circulate coolant through the engine block and radiator or heat exchanger. Either one or two thermostats may be used, depending upon engine cooling requirements, to regulate engine operating temperature. A high temperature cutoff switch senses coolant temperature

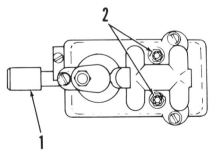

Fig. O302—Speed droop adjusting screw (1) is used to adjust governor sensitivity on fixed rpm engine applications. Electric fuel solenoid terminals are located at (2).

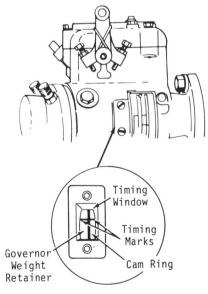

Fig. O304—Injection pump timing marks are located under timing window cover on side of pump housing.

Onan

SMALL DIESEL

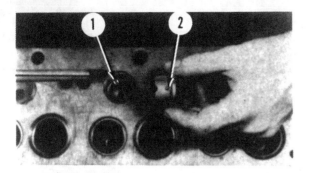

Fig. O305—When reinstalling hot plugs in cylinder head, align guide (2) on plug with slot (1) in cylinder head bore.

and shuts off engine if operating temperature exceeds a safe limit. Examine cooling system to determine cause if cutoff switch is activated.

Manufacturer recommends cleaning and flushing cooling system, including the engine block, every two years or 3000 hours of operation. A mixture of ethylene glycol antifreeze should be used as an engine coolant even when freezing temperatures are not expected. In addition to lowering the freezing point of water, antifreeze contains additives which prevent rust and corrosion of cooling system components. Antifreeze solution should be renewed yearly. Maintain coolant level approximately one inch below filler opening.

REPAIRS

TIGHTENING TORQUES

Refer to the following table for engine assembly torques. Fastener threads must be clean and lubricated with engine oil before torquing.

Connecting rod..............85 N·m
(63 ft.-lbs.)
Cylinder head ....123 N·m
(90 ft.-lbs.)
Flywheel...................68 N·m
(50 ft.-lbs.)
Front pulley...............133 N·m
(98 ft.-lbs.)
Glow Plug..................19 N·m
(14 ft.-lbs.)
Injection pump gear..........52 N·m
(38 ft.-lbs.)
Injector nozzle..............69 N·m
(51 ft.-lbs.)
Main bearing cap...........123 N·m
(90 ft.-lbs.)
Rocker arm cover ....18 N·m
(13 ft.-lbs.)
Rocker arm studs............52 N·m
(39 ft.-lbs.)

CYLINDER HEAD

To remove cylinder head, first drain cooling system and disconnect battery ground cable. Remove components such as muffler, air cleaner and shrouds as necessary to gain access to cylinder head. Remove fan and water pump assembly. Disconnect and remove high pressure fuel lines and fuel return lines from injectors. Disconnect wiring to glow plugs. Unbolt and remove intake and exhaust manifolds. Remove rocker arm cover. Disconnect and remove rocker arm oiling tube. Remove rocker arms and push rods keeping all parts in order so they can be reinstalled in their original positions. Remove cylinder head mounting cap screws and lift off cylinder head.

Thoroughly clean cylinder head and block, then inspect for cracks, warpage or other damage. If warped, a maximum of 0.25 mm (0.010 inch) of material may be removed from cylinder head and 0.05 mm (0.002 inch) may be removed from cylinder block to true mating surfaces. If warped beyond these limits, components must be renewed. Inspect hot plugs (Fig. O305) for damage (hairline cracks are normal and acceptable) or erosion and renew if necessary. When installing hot plug, be sure to align guide on hot plug with groove in cylinder head. Hot plug must not protrude more than 0.065 mm (0.026 inch) beyond cylinder head surface or be recessed more than 0.25 mm (0.001 inch) below surface of head. If hot plug is loose in its bore, remove plug and coat outer surface with Loctite 325 before reinstalling.

When reinstalling cylinder head, be sure cap screw holes in cylinder block are free of dirt, corrosion or an excessive amount of oil or water which would interfere with proper tightening of cap screws. Machined surfaces of cylinder head and block must be thoroughly clean. Do not use any sealer or gasket cement on head gasket or mounting surfaces. Be sure gasket is installed with side upward as marked on gasket. Apply light coat of engine oil to cap screw threads, then tighten gradually in steps following appropriate tightening sequence shown in Fig. O306 or O307. Final tightening torque is 123 N·m (90 ft.-lbs.) for all models. Reinstall push rods and rocker arms in their original positions. Lubricate threads of rocker arm locknuts. A minimum torque of 3.4 N·m (30 in.-lbs.) should be required to turn locknuts. If less torque is required, nut must be renewed. A new locknut should require a minimum of 6.2 N·m (55 ft.-lbs.) initial torque to turn.

NOTE: Do not use air wrench to remove or reinstall rocker arm locknuts as self-locking threads will be damaged.

Adjust valve clearance as outlined in the following paragraph. Install new in-

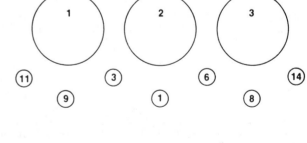

Fig. O306—Follow sequence shown when tightening cylinder head cap screws on Model L317.

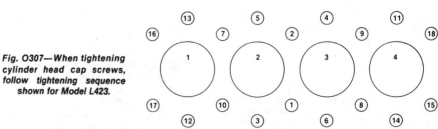

Fig. O307—When tightening cylinder head cap screws, follow tightening sequence shown for Model L423.

SERVICE MANUAL

Onan

Fig. O308—Injector nozzle seals (1) should be renewed whenever injectors are removed. Be sure seals are installed with concave side facing down.

Fig. O309—Exhaust manifold mounting cap screws should be tightened evenly following the sequence shown. Tightening procedure should be performed twice to ensure all cap screws are at specified torque.

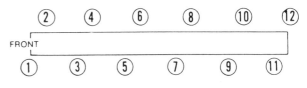

jection nozzle seals into nozzle bore with concave (inward curve) side facing down (Fig. O308).

NOTE: Incorrect installation of nozzle seals will cause nozzles to overheat and possibly stick. Do not attempt to reuse nozzle seals.

Tighten rocker cover mounting nuts to a torque of 18 N·m (13 ft.-lbs.). Tighten intake manifold cap screws to a torque of 23 N·m (17 ft.-lbs.). Tighten exhaust manifold cap screws evenly following the tightening sequence shown in Fig. O309. Final torque for cap screws with flatwashers is 28 N·m (21 ft.-lbs.) and for flange head cap screws is 35 N·m (26 ft.-lbs.). After completing initial tightening sequence, repeat tightening sequence to ensure all cap screws are at specified torque. Exhaust manifold cap screws should be retorqued after two hours of operation. Complete installation by reversing the removal procedure.

VALVE CLEARANCE ADJUSTMENT

Clearance between rocker arm and end of valve stem should be 0.20 mm (0.008 inch) for intake valves and 0.30 mm (0.012 inch) for exhaust valves with engine cold. Valve clearance is adjusted by turning rocker arm stud locknut while inserting a feeler gage of proper thickness between rocker arm and valve stem.

NOTE: A minimum torque of 3.4 N·m (30 in.-lbs.) should be required to start turning rocker arm locknut with threads oiled. Renew locknuts which do not meet minimum requirements. A new locknut with threads oiled should require a minimum initial torque of 6.2 N·m (55 in.-lbs.) to turn nut.

Adjust valves of each cylinder in regular engine firing order with piston at TDC on compression stroke. Firing order is 1-3-2 on Model L317 and 1-2-4-3 on Model L423.

On Model L317, turn crankshaft in normal direction of rotation until number 1 piston is at TDC on compression stroke (both rocker arms should be loose). Adjust intake and exhaust valve clearance as specified. Rotate crankshaft 240° (2/3 revolution) and adjust valves for number 3 cylinder. Turn crankshaft an additional 240° and adjust valves for number 2 cylinder.

On Model L423, turn crankshaft in normal direction of rotation until number 1 piston is at TDC on compression stroke (both rocker arms should be loose). Adjust intake and exhaust valves for number 1 cylinder as specified. Rotate crankshaft 180° (1/2 turn) and adjust valves for number 2 cylinder. Repeat procedure for cylinders 4 and 3.

VALVE SYSTEM

Valve face angles are 29° for intake and 44° for exhaust valves. Valve seat angles are 30° for intake and 45° for exhaust. Manufacturer recommends against handlapping which could destroy the interference angle seating surface. Width of valve seats should be 0.88-2.30 mm (1/32-3/32 inch). Depth of valve head below cylinder head surface should be 0.48-1.13 mm (0.019-0.044 inch) for intake valves and 0.67-1.33 mm (0.026-0.052 inch) for exhaust valves. If depth is less than specified, grind seat to obtain desired depth. If depth exceeds 2.13 mm (0.084 inch) on intake or 2.33 mm (0.092 inch) on exhaust, install new valve seat insert.

Hardened valve seat inserts are standard for exhaust valve seats. Seat inserts are also available through service parts for intake seats. Cylinder head must be machined to accept intake seat inserts. Oversize inserts are also available for both exhaust and intake. Refer to the chart in Fig. O310 for seat insert dimensions and corresponding counterbore dimensions required in cylinder head.

Valve stem diameter for all valves should be 7.93-7.95 mm (0.312-0.313 inch). Clearance between valve stem and guide should be 0.04-0.08 mm (0.002-0.003 inch), If clearance exceeds 0.16 mm (0.006 inch), valve guide and/or valve should be renewed. Valve guides may be renewed by pressing old guide out through bottom of cylinder head. Use appropriate valve guide driver to avoid damaging guides during pressing

VALVE SEAT INSERT	INSERT O.D.	COUNTERBORE O.D.	COUNTERBORE DEPTH
Exhaust - Standard	37.00-37.03mm (1.4567-1.4578in.)	36.93-36.95mm (1.4539-1.4547in.)	8.49-8.59mm (0.334-0.338in.)
Oversize 0.25mm	37.25-37.28mm (1.4665-1.4677in.)	37.18-37.20mm (1.4638-1.4645in.)	8.49-8.59mm (0.334-0.338in.)
0.50mm	37.50-37.53mm (1.4764-1.4775in.)	37.43-37.45mm (1.4736-1.4744in.)	8.49-8.59mm (0.334-0.338in.)
Intake - Standard	40.37-40.40mm (1.5894-1.5905in.)	40.30-40.32mm (1.5866-1.5874in.)	8.40-8.50mm (0.331-0.334in.)
Oversize 0.25mm	40.62-40.65mm (1.5992-1.6004in.)	40.55-40.57mm (1.5965-1.5972in.)	8.40-8.50mm (0.331-0.334in.)
0.50mm	40.87-40.90mm (1.6091-1.6102in.)	40.80-40.82mm (1.6063-1.6070in.)	8.40-8.50mm (0.331-0.334in.)

Fig. O310—Chart shows cylinder head counterbore machining specifications for installation of standard and oversize valve seat inserts.

Onan

Fig. O311—Exploded view of injector nozzle used on all models.

1. Upper body
2. Shim
3. Spring
4. Spring seat
5. Stop plate
6. Nozzle needle
7. Nozzle valve body
8. Nozzle holder

operation. Exhaust valve guides are 6 mm (0.236 inch) longer than intake guides and are not interchangeable. Guides must be pressed into cylinder head until top of guide protrudes a specified distance from top surface of head. On Spec. A and B engines, guide height should be 15.6-16.4 mm (0.614-0.645 inch). On Spec. C engines with valve rotators or spring wear plates, guide height should be 22.3-23.1 mm (0.878-0.909 inch). Inside diameter of replacement guides is semifinished and must be reamed or honed after installation to provide specified valve stem-to-guide operating clearance. Valve seats must be reground after installing new guides.

All valves are equipped with valve stem seals which should be renewed whenever valves are removed from cylinder head. Lubricate valve stem and place a plastic seal protector over end of valve stem when installing new seals. Do not remove valves after new seals are installed. Do not reuse old seals.

Some engines are equipped with positive valve rotators. Since it is difficult to determine if a rotator is good or bad, manufacturer recommends renewing rotators at time of major engine overhaul or if a build-up of carbon is evident on valve face and seat.

Valve springs are interchangeable for intake and exhaust valves. Valve spring free length should be approximately 51.0 mm (2.0 inches). Spring tension should test 287-307 N (65-69 pounds) when compressed to 42.5 mm (1.67 inches). Renew spring if test load is less than 258 N (58 pounds) at 42.5 mm (1.67 inches).

INJECTOR NOZZLE

WARNING: Fuel energes from injector with sufficient force to penetrate the skin. When testing injector, keep yourself clear of nozzle spray.

REMOVE AND REINSTALL. To remove injectors, first thoroughly clean injector, lines and surrounding area. Remove high pressure lines and fuel return lines from injectors. Cap all openings to prevent entry of dirt. Unscrew injector from cylinder head using a 27 mm deep socket. Remove and discard injector nozzle seals from injector bores.

When reinstalling injector, make certain new nozzle seal is installed with concave (inward curve) side facing down (Fig. O308).

NOTE: Incorrect installation of seal will cause nozzle valve to overheat and possibly stick. Do not reuse nozzle seals.

Tighten injectors to a torque of 69 N·m (51 ft.-lbs.). Reconnect fuel lines and bleed air from system as outlined in MAINTENANCE section.

TESTING. A complete job of testing and adjusting injectors requires use of special test equipment. Only clean, approved test oil (Viscor 1487) should be used in tester tank. Nozzle should be tested for opening pressure, leakage and spray pattern.

Before conducting tests, connect nozzle to tester and operate tester lever quickly to purge air from nozzle and make sure nozzle is not stuck.

OPENING PRESSURE. Open valve to tester gage and operate tester lever slowly while observing gage reading. Opening pressure should be 13000-13800 kPa (1890-2000 psi).

Opening pressure is adjusted by changing thickness of shim (2—Fig. O311). A change in shim thickness of 0.04 mm (0.0016 inch) will result in a pressure change of approximately 500 kPa (72 psi).

SEAT LEAKAGE. To check for leakage, operate tester lever to maintain gage pressure 1035 kPa (150 psi) below opening pressure. If a drop falls from nozzle tip within 10 seconds, nozzle valve must be cleaned or renewed.

SPRAY PATTERN. To check spray pattern, tester lever must be actuated to produce 4-6 nozzle opening cycles per

SMALL DIESEL

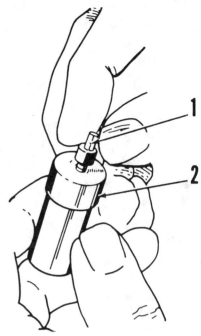

Fig. O312—Nozzle needle (1) must slide freely in nozzle body (2). Reclean or renew nozzle valve assembly if needle fails to slide down to its seat by its own weight with body held at 45° angle.

second. At slower speeds, fuel will not be properly atomized. Spray pattern should be uniform and well atomized emerging in a straight axis from nozzle tip. When opening properly, nozzle will operate with a high pitched chatter.

OVERHAUL. Hard or sharp tools, emery cloth, steel wire wheels or grinding compounds should never be used when servicing injectors. An approved nozzle cleaning kit is available from Onan and from a number of other specialized sources. Do not attempt to disassemble nozzle without proper test equipment to adjust opening pressure.

Clean outside of nozzle prior to disassembly. Secure nozzle body in a vise or holding fixture. Remove nozzle valve holder (8—Fig. O311) and separate components from nozzle body. Place all parts in clean calibrating oil or diesel fuel as they are removed.

NOTE: Nozzle valve components are mated parts and must be kept together. Never interchange components between nozzle assemblies.

Soak all parts in an approved carbon solvent, if necessary, to loosen hard carbon deposits. Rinse parts in clean diesel fuel or calibrating oil immediately after cleaning to neutralize the solvent and prevent etching of polished surfaces. Clean nozzle valve seat using tools from special cleaning kit. Perform the following test to nozzle valve after cleaning.

SERVICE MANUAL

Onan

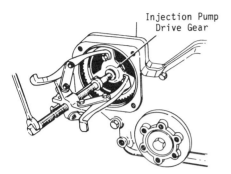

Fig. O313—Injection pump drive gear may be removed through opening in timing gear cover using a suitable gear puller.

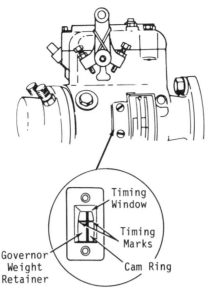

Fig. O314—Injection pump static timing is correct if pump timing marks are aligned when number 1 piston is on compression stroke and specified timing mark on flywheel is aligned with pointer.

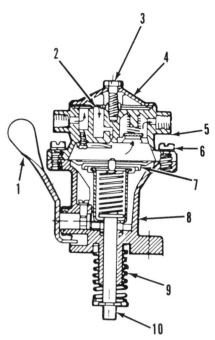

Fig. O315—Exploded view of fuel transfer pump used on all models.

1. Manual priming lever
2. Filter
3. Cap screw
4. Cap
5. Pump cover
6. Screw
7. Diaphragm
8. Pump body
9. Push rod spring
10. Push rod

With needle and seat thoroughly rinsed in clean diesel fuel, pull needle about one third of its length out of valve body (Fig. O312). When released, needle must slide down to bottom of seat by its own weight. If needle will not slide freely, reclean valve seat or renew nozzle valve needle and seat assembly.

When reassembling injector, do not handle polished surfaces unless hands are wet with diesel fuel. Thoroughly rinse all parts in clean diesel fuel and assemble while wet. Tighten nozzle body holder to a torque of 69 N·m (51 ft.-lbs.). Retest injector as previously outlined.

GLOW PLUGS

Glow plugs are parallel connected with each individual glow plug grounding through its mounting threads. To test glow plugs, first disconnect wires from all glow plug terminals. Using an ohmmeter, check for continuity of each glow plug. An "infininite" resistance reading indicates an open circuit and glow plug must be renewed. Glow plug tightening torque is 19 N·m (14 ft.-lbs.).

INJECTION PUMP

All models are equipped with a distributor type Roosa Master Model DB2 injection pump. The injection pump should be adjusted and repaired only by qualified injection pump service facility. Adjustments made to pump by non-qualified personnel may violate Federal, state and/or local exhaust emissions regulations and may also void manufacturer's warranty on pump and engine.

REMOVE AND REINSTALL. Before removing pump, first thoroughly clean pump and surrounding area. Shut off fuel supply. Remove timing window covers from flywheel housing and injection pump. Rotate crankshaft in normal direction until number 1 piston is on compression stroke and specified timing mark on flywheel is aligned with timing pointer. (Specified static timing is stamped on engine model plate.) At this point, pump timing marks (Fig. O314) should be aligned as shown. Disconnect throttle cable from pump control lever. Disconnect all fuel lines from pump and plug all openings. Disconnect electrical wiring from fuel shut-off solenoid. Remove cover plate from front of timing gear housing. Remove nut securing injection pump drive gear to pump shaft. Use a suitable puller to remove drive gear from pump shaft (Fig. O313). Radiator and shroud will need to be removed on most applications to provide working clearance for gear removal. Remove pump mounting stud nuts and withdraw pump from engine.

To reinstall pump, first position pump on mounting studs and tighten stud nuts finger tight. Rotate pump shaft to align timing marks on governor weight retainer and cam ring (Fig. O314). Make sure number 1 piston is on compression stroke and specified timing mark on flywheel is aligned with timing pointer. Install pump drive gear onto pump shaft and tighten retaining nut to a torque of 52 N·m (38 ft.-lbs.). Recheck alignment of pump timing marks and rotate pump to align marks if necessary. Tighten pump stud nuts to a torque of 23 N·m (17 ft.-lbs.). Complete installation by reversing the removal procedure. Bleed air from system as outlined in MAINTENANCE section.

FUEL TRANSFER PUMP

A diaphragm type fuel transfer pump, mounted on right side of engine is used to maintain positive fuel pressure to inlet of fuel injection pump. Transfer pump is driven by a cam lobe on engine camshaft. Transfer pump operating pressure should be 26-48 kPa (4-7 psi) at engine idle speed.

To remove pump, shut off fuel and disconnect fuel lines. Unbolt and remove pump from engine.

Prior to disassembly, scribe mating marks on pump cover and body to aid alignment during reassembly. Remove cap (4 – Fig. O315) and filter (2). Remove six screws retaining cover (5), then carefully remove cover. Compress push rod spring (9) and remove spring retaining ring. Withdraw spring, push rod (10) and diaphragm (7).

Clean and inspect all parts for wear or damage. A kit is available for renewal of gaskets, diaphragm and filter.

To reassemble, install diaphragm, push rod and spring. Compress spring and install retaining ring. Assemble cover to body aligning mating marks. Tighten retaining screws to a torque of 2-3 N·m (17-26 in.-lbs.). Position filter with spacer legs facing upward, then install cap and tighten cap screw to a torque of 2-3 N·m (17-26 in.-lbs.).

Reinstall pump and bleed air from system as outlined in MAINTENANCE section.

WATER PUMP

A centrifugal type water pump is used to circulate coolant through the cooling

Onan SMALL DIESEL

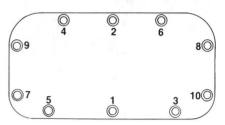

Fig. O317—Tighten oil cooler (models so equipped) mounting cap screws evenly using torque sequence shown. Retorque cap screws after two hours of operation.

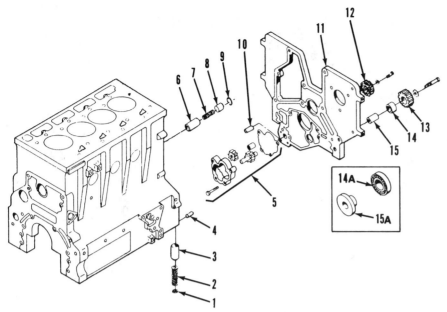

Fig. O316—Exploded view of engine oil pump and pump drive gears. Pressure relief valve (6 through 10) is used only on engines equipped with an oil cooler. Idler gear retainer (15A) and bearing (14A) are used on engines equipped with front pto.

1. Washer
2. Spring
3. Pressure regulator piston
4. Valve retaining pin
5. Oil pump assy.
6. Relief valve body
7. Spring
8. Relief valve piston
9. Snap ring
10. Bushing
11. Timing gear back plate
12. Pump drive gear
13. Idler gear
14. Bushing
15. Idler shaft

system. Two sizes of pumps along with two different impellers for each pump and a variety of different drive pulleys are used depending upon engine application and cooling requirements.

Service parts are not available for water pump repair. If seal is leaking or bearing feels rough or binds, water pump must be renewed as a unit.

OIL PUMP

The gear type oil pump is mounted to rear of timing gear housing back plate and is driven by the crankshaft through an idler gear. An oil pressure regulating valve, located in main oil gallery at front of cylinder block, maintains oil pressure at 345-380 kPa (50-55 psi) at high idle speed. On engines equipped with an oil cooler, a safety relief valve is also used to limit pump discharge pressure. Safety relief valve opens at 830 kPa (120 psi). Relief valve is located in front of cylinder block just above the oil pump.

Oil pump is accessible after removing engine oil pan. Note location of shorter cap screw when removing oil pickup from oil pump body. Remove oil pump drive gear (12 – Fig. O316), then separate pump gears from body.

Inspect all parts for excessive wear or other damage. Radial clearance between pump gears and body bore should not exceed 0.175 mm (0..07 inch). Maximum allowable end clearance of gears to body is 0.150 mm (0.006 inch). Maximum allowable drive shaft clearance in body is 0.07 mm (0.003 inch). Drive shaft clearance in bushing must not exceed 0.15 mm (0.006 inch). Idler shaft clearance in gear must not exceed 0.06 mm (0.002 inch). Renew pump assembly if wear limits are exceeded.

To reinstall pump, reverse the removal procedure. Tighten oil pump and pick up tube mounting cap screws evenly in two steps to a final torque of 23 N·m (17 ft.-lbs.). Be sure oil pump gears rotate freely after tightening mounting screws. Tighten oil pump drive gear and idler gear cap screws to a torque of 23 N·m (17 ft.-lbs.). Tighten oil pan mounting screws, starting at the center and working outward to each end, to a torque of 11 N·m (8 ft.-lbs.).

OIL COOLER

Some engines are equipped with an oil cooler mounted on left side of engine block. If a drop in oil pressure occurs due to oil overheating, the oil cooler could be restricted. Oil cooler should be cleaned or renewed. If lubrication system is contaminated with metal particles from an engine failure, oil cooler should be renewed. Do not attempt to clean metal particles from oil cooler.

Oil cooler may be pressure tested for leaks using air pressure not to exceed 138 kPa (20 psi). Renew cooler core if faulty.

When reinstalling oil cooler, tighten mounting cap screws gradually and evenly using tightening sequence shown in Fig. O317. Tighten to a final torque of 22 N·m (16 ft.-lbs.). Retorque cooler cap screws after two hours of operation.

TIMING GEARS AND COVER

To remove timing gear cover, first drain coolant and remove radiator. Remove fan belt, fan, alternator and mounting bracket and crankshaft pulley. Loosen oil pan mounting cap screws and lower front end of oil pan. Remove timing gear cover cap screws, tap cover to loosen gasket seal and withdraw cover from engine. Crankshaft front oil seal is located in timing gear cover and may be renewed at this time.

Inspect timing gears for excessive wear, chipped teeth or other damage. Maximum allowable backlash between any two gears is 0.30 mm (0.012 inch). Idler gear end play should not exceed 0.40 mm (0.016 inch). Idler gear bushing and stub shaft are renewable. Bushing inside diameter should be 40.030-40.054 mm (1.576-1.577 inches) and stub shaft outer diameter should be 39.986-40.000 mm (1.5742-1.5748 inches). Maximum allowable bushing bore to shaft clearance is 0.10 mm (0.004 inch). Idler gear bushing is furnished semifinished and bushing bore must be reamed to correct size after pressing into gear. Camshaft gear and crankshaft gear are press fitted to their respective shafts. Refer to appropriate sections for shaft removal and installation if gears require renewal.

When reassembling timing gears, make certain timing marks (Fig. O318) on idler gear are aligned with corresponding marks on crankshaft gear and camshaft gear. There are no timing marks on injection pump drive gear. Refer to INJECTION PUMP section for gear removal and installation and pump timing procedure.

Renew crankshaft front oil seal and wear sleeve as outlined in CRANKSHAFT OIL SEAL paragraph. Lubricate lip of oil seal with oil or grease before reinstalling cover. Tighten cover mounting cap screws to a torque of 23

SERVICE MANUAL

Onan

Fig. O318—Valve timing is correct when single dot on idler gear (3) is aligned with single dot on crankshaft gear (2) and two dots on idler gear are aligned with two dots on camshaft gear (4).
1. Oil pump idler gear
2. Crankshaft gear
3. Idler gear
4. Camshaft gear
5. Oil pump drive gear

screws to a torque of 23 N·m (17 ft.-lbs.). Align timing marks as shown in Fig. O318 and install idler gear. Tighten idler gear cap screw to a torque of 52 N·m (39 ft.-lbs.). Complete installation by reversing removal procedure.

PISTON AND ROD UNITS

Piston and connecting rod are removed from the top after removing oil pan, oil pickup tube and cylinder head. Remove carbon ridge and ring ridge (if present) from top of cylinder before attempting to remove piston. A piece of rubber tubing may be placed over ends of connecting rod bolts during removal and installation to avoid scratching cylinder wall or crankpin journal.

NOTE: Mark and number each piston and connecting rod assembly so it can be returned to its respective cylinder. The connecting rod and cap are marked with identifying numbers and must be kept together.

When reinstalling piston and rod assembly, thoroughly lubricate piston and cylinder with clean engine oil. Piston and rod units should be reinstalled in their original locations. Make certain connecting rod identification marks (1–Fig. O319) are facing camshaft side of engine and piston swirl chamber is towards intake side of engine. Tighten rod cap nuts evenly in steps to a final torque of 85 N·m (63 ft.-lbs.).

PISTON, PIN AND RINGS

All models are equipped with two compression rings and one oil control ring. Piston and rings are available in oversizes as well as standard size. Piston to cylinder clearance, measured at right angle to piston pin bore, should be 0.127-0.173 mm (0.0050-0.0068 inch). If clearance exceeds 0.250 mm (0.010 inch), cylinders should be bored oversize and pistons and rings renewed.

Full floating piston pins are retained in piston by snap rings at each end. Standard piston pin diameter is 31.995-32.00 mm (1.2596-1.2598 inches). Pin bore in piston should be 32.003-32.009 mm (1.2599-1.2601 inches). Rec-

N·m (17 ft.-lbs.). Tighten oil pan cap screws to a torque of 11 N·m (8 ft.-lbs.). Complete installation by reversing the removal procedure.

CAMSHAFT AND BEARINGS

REMOVE AND REINSTALL. To remove camshaft, first remove timing gear cover and cylinder head as previously outlined.

NOTE: Camshaft may be removed without removing cylinder head by first removing push rods, then use a pencil magnet to lift cam followers out of their bores in cylinder block.

With cylinder head removed, withdraw cam followers keeping all parts separated in order of removal so they can be reinstalled in their original locations. Remove fuel transfer pump. Rotate camshaft to remove retaining plate cap screws through openings in camshaft gear. Withdraw camshaft assembly from cylinder block being careful not to damage camshaft bushings. Remove idler gear from stub shaft for timing purposes.

Using a feeler gage, check end clearance between retainer plate and camshaft gear. Desired end play is 0.10-0.50 mm (0.004-0.020 inch). If end play exceeds 0.70 mm (0.027 inch), renew retainer plate and/or gear. Camshaft gear is keyed and press fitted to camshaft. Use a press to remove and install gear. Press gear onto shaft until it bottoms against shaft shoulder. Check cam lobes for scoring, excessive wear or other damage. Renew camshaft if height from base to tip of intake or exhaust cam lobes is less than 42.50 mm (1.673 inches). If camshaft is renewed, it is recommended that cam followers be renewed also. Camshaft bearing journal outside diameter should be 50.000-50.014 mm (1.9685-1.9691 inches). Recommended operating clearance in cam-

shaft bushings is 0.030-0.116 mm (0.0011-0.0046 inch) and maximum allowable clearance is 0.15 mm (0.006 inch). Maximum allowable journal runout, measured at center bearing journal, is 0.10 mm (0.004 inch).

To remove camshaft bushings, the crankshaft, timing gear housing back plate and flywheel housing must first be removed. Use a suitable cam bushing puller to remove bushings. All bushings except the rear one should be pulled rearward from cylinder block bores. Pull rear bushing forward from its bore. When installing new bushings, reverse the removal procedure starting at front bushing and pulling all bushings except rear one rearward into their bores. Pull rear bushing in from the rear towards the front.

NOTE: Front bushing is wider and has two oil holes in it. Be sure both holes are aligned with corresponding oil passages in cylinder. Make certain single oil hole in each of the other bushings is aligned with oil passage from main bearing.

Cam bushings should be checked for nicks or burrs after installation. Bushings are precision type and do not require reaming or honing. Apply sealing compound to rear expansion plug before installing.

When reinstalling camshaft, thoroughly lubricate cam bushings and camshaft. Tighten retaining plate cap

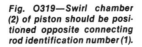

Fig. O319—Swirl chamber (2) of piston should be positioned opposite connecting rod identification number (1).

283

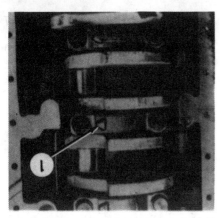

Fig. O320—Raised arrow (1) on main bearing caps should point towards camshaft side of engine.

ommended operating clearance of pin in piston is 0.003-0.014 mm (0.0001-0.0005 inch) and maximum allowable clearance is 0.03 mm (0.0012 inch).

Piston ring groove wear may be checked between ring and top of ring groove with a feeler gage. Renew piston if side clearance exceeds 0.25 mm (0.010 inch) for top and second compression ring groove. Maximum allowable clearance for oil control ring groove is 0.15 mm (0.006 inch).

Piston ring end gap is checked by inserting ring squarely into cylinder bore and measuring end gap with a feeler gage. Recommended end gap for both compression rings is 0.31-0.57 mm (0.012-0.022 inch). End gap for oil ring should be 0.22-0.58 mm (0.009-0.023 inch). Maximum allowable end gap for all rings is 1.50 mm (0.060 inch). Filing ring ends to increase end gap is not recommended.

When reinstalling rings, assemble oil ring end gap 180° from guide pin in ring expander. Install compression rings in correct grooves with sides marked "Up 2nd" and "Up Top" or with a dot towards top of piston. Be sure ring end gaps are staggered evenly around piston.

Heating piston in hot water prior to assembly will make piston pin installation easier. Be sure pin retaining rings are installed with sharp edge facing outward. Piston must be assembled with swirl chamber (2–Fig. O319) positioned opposite identification number (1) side of connecting rod.

CONNECTING ROD AND BEARINGS

Connecting rods are equipped with a renewable bushing in the small end and precision, insert type bearings in the large end.

Recommended operating clearance between crankpin and bearing is 0.030-0.091 mm (0.0012-0.0036 inch) and maximum allowable clearance is 0.15 mm (0.006 inch). Connecting rod side clearance should be 0.05-0.045 mm (0.002-0.018 inch). Renew connecting rod if side clearance exceeds 0.45 mm (0.018 inch).

When renewing piston pin bushing, be sure oil hole in bushing is aligned with oil passage in top end of connecting rod. Bushings must be reamed or honed to correct size after being pressed into rod. Recommended inside diameter of bushing is 32.01-32.02 mm (1.2602-1.2606 inches). Desired clearance between piston pin and bushing is 0.010-0.025 mm (0.0004-0.0010 inch) and maximum allowable clearance is 0.05 mm (0.002 inch). A new piston pin installed in a new bushing at 21°C (70°F) should be a thumb push fit.

CRANKSHAFT AND BEARINGS

To remove crankshaft, first remove oil pan, cylinder head, pistons and rods, timing gear cover, idler gear, camshaft, injection pump, timing gear back plate, flywheel and flywheel housing. Unbolt and remove main bearing caps, then carefully lift out crankshaft.

Inspect crankshaft journals for scoring, excessive wear or signs of overheating. Out-of-round or taper of journals should not exceed 0.005 mm (0.0002 inch). Crankshaft may be reground undersize. Bearings are available in 0.25, 0.50 and 0.75 mm undersizes.

Main bearing journal standard diameter is 75.99-76.01 mm (2.9917-2.9925 inches). Main bearing operating clearance should be 0.03-0.10 mm (0.0012-0.0039 inch) and maximum allowable clearance is 0.15 mm (0.006 inch).

Crankpin journal standard diameter is 55.99-56.01 mm (2.2043-2.2051 inches). Connecting rod bearing operating clearance should be 0.03-0.09 mm (0.0012-0.0036 inch) and maximum allowable clearance is 0.15 mm (0.006 inch).

Crankshaft end play is controlled by thrust flanges on rear main bearing. Recommended end play is 0.10-0.33 mm (0.004-0.013 inch) and maximum allowable end play is 0.38 mm (0.015 inch).

Inspect crankshaft gear for chipped teeth or excessive wear. Use a suitable gear puller to remove gear. When installing new gear, first heat gear to about 175°C (350°F). Align gear with key in crankshaft and make sure timing mark on gear faces forward. Drive or press gear onto crankshaft until it bottoms against shaft shoulder.

To reinstall crankshaft, proceed as follows: Note that main bearing insert halves are marked UPPER or LOWER and also marked for position in cylinder block. Assemble upper bearing halves in their designated positions in block. Be sure bearing tangs engage slots in block. Lubricate with clean engine oil, then carefully lower crankshaft into block. Assemble lower bearing halves in their respective caps. (Bearing caps are marked in numerical order and also front and rear caps are marked "F" and "R" respectively. Be sure bearing tangs engage slots in bearing caps. Install bearing caps. Install bearing caps in proper locations with raised arrow (1–Fig. O320) on caps pointing towards camshaft side of engine. Push crankshaft towards front of engine to align rear bearing thrust surfaces, then tighten bearing cap mounting cap screws alternately and in steps to a final torque of 123 N·m (90 ft.-lbs.). Be sure crankshaft turns freely after cap screws are torqued. Complete installation by reversing the removal procedure.

CRANKSHAFT OIL SEALS

FRONT SEAL. The front oil seal is located in timing gear cover. A wear sleeve is available with replacement seal to service a damaged or grooved crankshaft.

To renew seal, timing gear cover must be removed as previously outlined. Be sure new seal is pressed squarely into cover bore and is bottomed against cover shoulder. Outer diameter of a new seal is coated with a sealant which eliminates the need for using a sealing compound when installing seal. Be sure open side of seal faces inward and seal lip is lubricated with oil or grease prior to assembly.

To remove a previously installed wear sleeve from crankshaft, use a chisel to make two marks across wear sleeve about three-quarters of the way through sleeve. This will expand sleeve to allow easier removal. Be careful not to damage crankshaft surface.

Apply Loctite 271 to crankshaft and inside of new wear sleeve, then install sleeve onto crankshaft making certain outside chamber of sleeve faces forward. A special installing tool, Number 420-0418, is available from Onan to properly position and install wear sleeve. Clean excess Loctite from crankshaft and wear sleeve.

REAR SEAL. Crankshaft rear oil seal is located in seal retainer plate which is bolted to rear of block. A wear sleeve is pressed onto crankshaft flange to protect crankshaft from wear and to provide a smooth sealing surface for oil seal.

SERVICE MANUAL

Onan

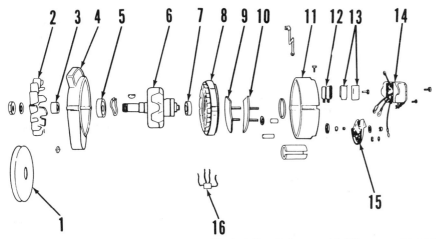

Fig. O321—Exploded view of alternator assembly. Isolation diode assembly (15) was used on early models and diode trio (16) is used on late models.

1. Pulley
2. Fan
3. Spacer
4. Front cover
5. Bearing
6. Rotor
7. Bearing
8. Stator
9. Negative diode rectifier
10. Positive diode rectifier
11. End frame
12. Brush assy.
13. Cover plates
14. Voltage regulator
15. Isolation diode assy.
16. Diode trio

To renew seal, first remove flywheel and flywheel housing. Remove seal retainer plate and drive out old seal. Press a new seal into retainer plate using a suitable seal installing tool until outer surface of seal is flush with machined surface of retainer plate. A special oil seal and wear sleeve installing tool (number 420-0417) is available from Onan. Be sure open side of seal faces inward. Lubricate seal lip with oil or grease prior to assembly onto crankshaft.

To renew wear sleeve, use a hammer and chisel to make chisel marks across wear sleeve about three-quarters of the way through. This will expand sleeve to allow removal from crankshaft. Be careful not to damage crankshaft flange.

NOTE: If wear sleeve special installing tool is not available, note position of wear sleeve on crankshaft flange so new sleeve can be installed in same position.

Apply Loctite 271 to crankshaft flange and inside of new sleeve. Position sleeve on crankshaft so outer chamfer side is facing rearward. Press sleeve onto crankshaft flange with special installing tool until tool bottoms against end of crankshaft. Wipe excess Loctite from crankshaft and sleeve.

Reinstall seal retainer plate making sure seal is centered on crankshaft flange. Tighten retainer plate mounting cap screws to a torque of 23 N·m (17 ft.-lbs.) and oil pan cap screws to 11 N·m (8 ft.-lbs.). Tighten flywheel housing cap screws to a torque of 84 N·m (62 ft.-lbs.) and flywheel mounting cap screws to 68 N·m (50 ft.-lbs.).

FLYWHEEL AND RING GEAR

The flywheel is indexed to crankshaft with a dowel pin and retained by six cap screws. Flywheel can be installed in one position only.

If engine is equipped with a clutch, inspect flywheel friction surface for grooves, cracks or other damage. A maximum of 0.40 mm (0.016 inch) of material may be removed to provide a smooth surface before renewal is required. Inspect clutch shaft pilot bearing in flywheel bore and renew if worn or damaged. Press against bearing outer race only when installing new bearing.

Check ring gear for worn or broken teeth and renew if necessary. To remove ring gear, use a grinder to split ring gear at the base of one of the teeth. Then, drive ring gear off the flywheel. Heat new ring gear to 185°-195°C (365°-385°F) prior to installing onto flywheel.

NOTE: Do not overheat ring gear as gear may warp or heat treatment may be destroyed.

Drive new gear down squarely against shoulder of flywheel. Gear must be driven on quickly before it cools and contracts to flywheel in the wrong position.

To reinstall flywheel, reverse the removal procedure. Tighten mounting cap screws using a star pattern sequence in two steps to a final torque of 68 N·m (50 ft.-lbs.).

ELECTRICAL SYSTEM

ALTERNATOR AND REGULATOR. Refer to Fig. O321 for an exploded view of alternator assembly. Normal alternator output voltage should be between 13.8-14.8 volts with engine operating at approximately 1000 rpm. If output voltage is too low, connect a jumper wire from positive output post to field post and recheck output voltage. If

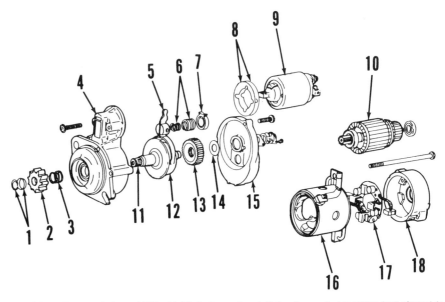

Fig. O322—Exploded view of Mitsubishi starter motor. A Delco-Remy starter motor (not shown) is used on some models.

1. Pinion stop ring
2. Pinion gear
3. Spring
4. Front housing
5. Lever
6. Springs
7. Seal
8. Fiber washers
9. Solenoid
10. Armature
11. Pinion shaft
12. Overrunning clutch
13. Reduction gear
14. Washer
15. Center housing
16. Frame & field coil assy.
17. Brush holder & brushes
18. End cover

285

voltage is unchanged, alternator is defective. If voltage is now within specifications or higher, regulator is defective. The solid state regulator is not adjustable.

Voltage regulator and brushes may be renewed without disassembling alternator. Renew brushes if less than 4.2 mm (0.165 inch) of brush extends below bottom of holder.

Before disassembling alternator, place scribe mark across end frame and front cover to provide proper alignment when reassembling. Check stator and rotor windings for open or grounded circuits. Diodes should have continuity in one direction only. Renew faulty components or alternator assembly as required.

STARTER. Refer to Fig. O322 for an exploded view of electric starter motor. When disassembling starter, be sure to retain all shims and washers for use in reassembly.

Check armature for continuity between commutator segments. No continuity indicates an open circuit and armature must be renewed. Test for continuity between armature shaft and each commutator segment. Continuity indicates a grounded circuit and armature must be renewed. Use a growler to check armature for short circuits.

To check field coil, test for continuity between brushes. No continuity indicates a defective field coil. There should be no continuity between field coil and starter frame.

Minimum brush length is 11.5 mm (0.435 inch). Check for continuity between positive side of brush holder and brush holder base. Continuity indicates a short circuit and holder assembly should be renewed.

Inspect reduction gears and overrunning clutch for wear and damage. Pinion gear should be free to turn in one direction and lock when turned in opposite direction.

NOTE: Do not clean overrunning clutch in solvent. Washing clutch will cause grease to leak out.

When reassembling, apply light coat of grease to all gears, armature bearings sleeve bearing and sliding portion of lever assembly. Install adjusting washers (14 – Fig. O322) between center housing (15) and reduction gear (13) to obtain recommended pinion shaft end play of 0.1-0.8 mm (0.004-0.031 inch). To adjust pinion gap, disconnect positive terminal "M" on solenoid. Connect positive terminal of battery to "S" terminal on solenoid and negative terminal to starter body. This will shift pinion gear forward into cranking position. Then, lightly push pinion rearward and measure the gap. Desired pinion gap is 0.3-2.0 mm (0.012-0.079 inch). If necessary, change number of fiber washers (8) used between solenoid and front housing. Adding washers will decrease pinion gap.

PERKINS

PERKINS ENGINES, INC.
32500 Van Born Road
Wayne, Michigan 48184

Model	No. Cyls.	Bore	Stroke	Displ.
3.152	3	91.44 mm (3.6 in.)	127 mm (5.0 in.)	2502 cc (152 cu. in.)
D3.152, AD3.152	3	91.44 mm (3.6 in.)	127 mm (5.0 in.)	2502 cc (152 cu. in.)
3.1522	3	91.44 mm (3.6 in.)	127 mm (5.0 in.)	2502 cc (152 cu. in.)
3.1524, T3.1542	3	91.44 mm (3.6 in.)	127 mm (5.0 in.)	2502 cc (152 cu. in.)

All engines in this section are liquid-cooled, four-stroke diesels with vertical, inline cylinders. The 3.152 engine uses indirect type fuel injection while all other engines are direct injection diesels. "Front" of engine refers to crankshaft pulley end of engine. Firing order is 1-2-3 with number 1 cylinder at "front" of engine. Crankshaft rotation is clockwise viewed from the "front". Model T3.1524 is equipped with a turbocharger. The engine identification number is stamped on the right side of engine cylinder block, forward of the fuel lift pump on all engines.

MAINTENANCE

LUBRICATION

Recommended crankcase oil is SAE 10W for temperatures between minus 18°C (0°F) to minus 1°C (30°F), SAE 20W for temperatures between minus 1°C (30°F) and 27°C (80°F) and SAE 30 for temperatures above 27°C (80°F). Use of multigrade oils is acceptable. Oils for naturally aspirated engines should meet MIL-L-46152 or MIL-L-2104C specifications. However, if engine is new or just overhauled, manufacturer does not recommend using MIL-L-2104C specification oil in naturally aspirated engines until after the first oil change. Lubricating oil for turbocharged engines must meet MIL-L-2104C specification.

Engine oil and filter element should be renewed after every 250 hours of operation. Crankcase capacity is approximately 7.4 liters (7.8 quarts) for 3.152 and D3.152 engines and 6.8 liters (7.2 quarts) for all other engines. Capacity can vary depending on engine application; fill to "FULL" mark on dipstick.

Oil pressure relief valve setting is 345-448 kPa (50-65 psi).

COOLING SYSTEM

Manufacturer recommends using a mixture of water and ethylene glycol antifreeze in cooling system. In addition to providing protection against freezing, the antifreeze solution contains additives to protect against corrosion and also raises the boiling point of the coolant. If an antifreeze solution is not used, an approved corrosion inhibitor should be added to the water.

FUEL SYSTEM

FUEL FILTER AND BLEEDING.
All engines are equipped with a renewable, paper element type fuel filter. Some engines are equipped with two filter elements, an intermediate element and a final filter element. A water trap and sediment bowl may also be used between the fuel tank and fuel lift pump. The water trap should be cleaned after every 250 hours of operation or more often if water or sediment is evident in the bowl. The fuel filter elements should be renewed after every 500 hours of operation or once a year, whichever comes first. However, if engine indicates signs of fuel starvation (surging or loss of power), filters should be renewed regardless of hours of operation.

Whenever fuel filters are renewed or fuel lines have been disconnected, air must be bled from the fuel system as follows: Loosen vent screw (1 – Fig. PR1) on side of injection pump governor cover and screw (2) on side of fuel pump body. Manually actuate primer lever on fuel lift pump until air-free fuel flows from vent screws. Retighten pump body screw first, then governor cover screw. If engine fails to start at this point, loosen high pressure fuel line fittings at the injectors. Place pump control in "run" position and throttle in full speed position, then crank engine until fuel is discharged from loosened fuel line fittings. Retighten fittings and start engine.

ENGINE SPEED ADJUSTMENT.
The engine maximum no-load speed setting can be established from the identification plate. A typical code is XW50E600/8-2470. The last four numbers indicate the specified no-load speed which is 2470 rpm in this case. If a replacement pump is being installed, be sure to stamp the setting code number from the old pump on the identification plate of the new pump.

Fig. PR1-1—Loosen vent screws (1 and 2) on side of injection pump to bleed air from fuel system.

Perkins

SMALL DIESEL

Fig. PR1-2—The "E" scribe line on injection pump rotor should align with square end of snap ring when number 1 piston is at specified static timing position. Refer to text.

The engine must be at normal operating temperature and an accurate tachometer used when adjusting engine speed setting. Adjust throttle lever stop screw until correct speed is obtained. After completing adjustment, install a seal wire on adjusting screw to secure adjustment.

INJECTION PUMP TIMING. The injection pump drive shaft has a milled slot in forward end which engages a dowel pin in pump drive gear. Thus, injection pump can be removed and reinstalled without regard to timing position.

To check the pump timing, shut off fuel and remove timing window cover from side of pump housing. A snap ring is provided inside pump for timing purposes (Fig. PR1-2). When scribed line "E" on pump rotor is aligned with square end of snap ring, beginning of injection to number 1 cylinder should occur. If correct setting of snap ring is in doubt, pump must be removed to adjust snap ring position. Connect an injector test pump to number 1 outlet fitting, then actuate test pump to obtain pressure of 9750 kPa (440 psi). Rotate injection pump shaft by hand until it locks up. Adjust snap ring position so square end is aligned with line marked "E" on pump rotor. Reinstall pump.

Remove valve cover and rotate crankshaft until number 1 piston is at TDC on compression stroke (exhaust valve on number 3 cylinder will be fully open.) Remove valve spring from number 1 intake valve and allow valve to rest on piston crown.

NOTE: Install a piece of wire or rubber band around valve stem to prevent valve from accidentally falling into cylinder.

Mount a dial indicator on end of the intake valve stem, then turn crankshaft and zero indicator at top dead center. Turn crankshaft in opposite direction of rotation approximately 1/8 turn. Then turn crankshaft in normal direction until piston is at specified BTDC position as listed in the following timing table. The correct timing figures can be found by reference to prefix letters and numbers of the setting code on the injection pump name plate.

Prefix Code	Static Timing (BTDC)	Piston Position (BTDC)
3.152 Engine:		
LW45, LW49	16°	3.15 mm (0.124 in.)
LW51, LW52	18°	3.94 mm (0.155 in.)
LW58, LW59	22°	5.84 mm (0.230 in.)
PW43, PW44	16°	3.15 mm (0.124 in.)
PW50	16°	3.94 mm (0.155 in.)
D3.152 Engine:		
MW47E, MW49, MW53E, MW57	24°	6.86 mm (0.270 in.)
RW50, RW52, RW54, SW46	24°	6.86 mm (0.270 in.)
SW46*, SW46L	29½°	10.43 mm (0.411 in.)
SW48*	26°	8.00 mm (0.315 in.)
SW52*, SW57, SW58	24°	6.86 mm (0.270 in.)
TW48E	20°	4.88 mm (0.192 in.)
TW50	22°	5.84 mm (0.230 in.)
WW, WW74E, WW48	16°	3.15 mm (0.124 in.)
WW49L	20°	4.88 mm (0.192 in.)
3.1522 Engine:		
XW50E	14°	2.37 mm (0.092 in.)
3.1524 Engine:		
WW47E, WW49L	20°	4.87 mm (0.192 in.)
ZW	24°	6.98 mm (0.275 in.)
T3.1524 Engine:		
YW	16°	3.15 mm (0.124 in.)

*Used in Lincoln electric welding sets.

With piston at specified BTDC position, scribed line "E" on pump rotor

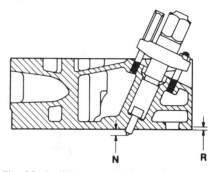

Fig. PR1-3—When remachining cylinder head surface on all engines except 3.152, a maximum of 0.30 mm (0.012 inch) of material (R) may be removed. Nozzle tip protrusion (N) must not exceed 4.67 mm (0.184 inch) on D3.152 engine or 6.32 mm (0.249 inch) on 3.1522, 3.1524 and T3.1524 engines.

should be in line with square end of snap ring. If timing mark is not aligned, loosen pump mounting nuts and move pump in slotted mounting holes until mark is aligned. If timing mark cannot be aligned by shifting pump on mounting studs, timing gear cover should be removed and gear timing checked.

REPAIRS

TIGHTENING TORQUES

Refer to the following table for special tightening torques.

Balance weight.............75 N·m
(55 ft.-lbs.)
Camshaft gear ....28 N·m
(21 ft.-lbs.)
Connecting rod nuts:
  Nonplated, self-locking......95 N·m
(70 ft.-lbs.)
  Cadmium plated...........61 N·m
(45 ft.-lbs.)
  Phosphated...............81 N·m
(60 ft.-lbs.)
Crankshaft pulley:
  With 4.8 mm (0.189 in.)
    thick washer............142 N·m
(105 ft.-lbs.)
  With 8.9 mm (0.350 in.)
    thick washer............325 N·m
(240 ft.-lbs.)
Cylinder head..........Refer to text.
Flywheel..................106 N·m
(78 ft.-lbs.)
Idler gear.................68 N·m
(50 ft.-lbs.)
Injector....................16 N·m
(12 ft.-lbs.)
Main bearing caps..........150 N·m
(110 ft.-lbs.)

VALVE ADJUSTMENT

Valve clearance is adjusted with engine cold. On 3.152 and D3.152

SERVICE MANUAL

Perkins

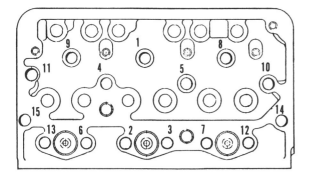

Fig. PR1-4—Cylinder head tightening sequence for 3.152 engine.

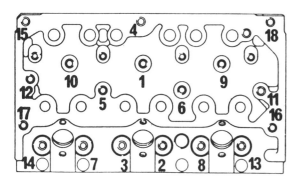

Fig. PR1-5—Cylinder head tightening sequence for all engines except 3.152 engine.

engines, clearance should be 0.30 mm (0.012 inch) for both intake and exhaust valves. On 3.1522, 3.1524 and T3.1524 engines, clearance should be 0.20 mm (0.008 inch) for exhaust valves and 0.32 mm (0.0125 inch) for intake valves.

To adjust valve clearance on engines equipped with a TDC mark on flywheel, turn crankshaft until TDC mark is in center of flywheel housing inspection hole and number 1 piston is on compression stroke (both valves closed). Check and adjust clearance as necessary on valves 1, 2, 3 and 5 (number 1 valve at front). Turn crankshaft one complete revolution until TDC mark is again visible in inspection hole, then adjust clearances for number 4 and 6 valves.

To adjust valves on engines not equipped with a TDC mark, first turn crankshaft until intake valve for number 1 cylinder is opening and the exhaust valve is closing. Check and adjust clearances on number 4 and 6 valves. Place temporary marks on crankshaft and on adjacent point on front cover, then turn crankshaft one complete revolution until marks are realigned. Check and adjust clearances on valves 1, 2, 3 and 5.

CYLINDER HEAD

To remove cylinder head, drain coolant and disconnect radiator hose from thermostat housing. Detach cold starting connections. Disconnect and remove fuel lines and injectors. Plug all fuel system openings to prevent entry of dirt. Disconnect oil pipe from cylinder head and remove intake pipe, exhaust pipe and turbocharger (if so equipped). Remove rocker arm cover, then remove rocker arm assembly and oil pipe. Remove cylinder head retaining nuts and cap screws in reverse order of tightening sequence shown in Fig. PR1-4 or PR1-5. Lift off cylinder head assembly.

Clean cylinder head and inspect for wear and damage. Check mounting surface for distortion. If head is warped more than 0.08 mm (0.003 inch) crosswise or 0.15 mm (0.006 inch) lengthwise, mounting surface should be remachined. On 3.152 engine, material may be removed as necessary as long as overall head thickness is not reduced below 75.69 mm (2.980 inches). On all other engines, a maximum of 0.30 mm (0.012 inch) may be removed providing nozzle protrusion (N–Fig. PR1-3) does not exceed 4.67 mm (0.184 inch) on D3-152 engine or 6.32 mm (0.249 inch) on 3.1522, 3.1524 and T3.1524 engines.

NOTE: Shim washers or extra nozzle seal washers must not be used to reduce nozzle projection.

After head surface has been machined, check valve head depths as outlined in VALVE SYSTEM section and correct to specifications if necessary.

To reinstall cylinder head, proceed as follows: Cylinder head and crankcase surfaces must be clean. Do not use any gasket sealing compounds on cylinder head gasket. Different methods of retaining the cylinder head have been used. On early 3.152 and D3.152 engines, studs and nuts were used to secure cylinder head. On these engines, refer to appropriate tightening sequence shown in Fig. PR1-4 or PR1-5 and tighten retaining nuts in steps to 81 N·m (60 ft.-lbs.). On late 3.152 engines, the studs were replaced by cap screws except for the six used at position 13, 6, 2, 3, 7 and 12 in Fig. PR1-4. On late D3.152 engines, all but two of the studs (13 and 14–Fig. PR1-5) were replaced by cap screws of three different lengths. The four shortest 82.5 mm (3¼ inches) screws are used at positions 7, 3, 2 and 8; the mid-length 92 mm (3⅝ inches) screws should be at positions 15, 4 and 18; and the longest 98.4 mm (3⅞ inches) are used at remaining positions. On all engines with both nuts and cap screws, tighten in steps following sequence shown in Fig. PR1-4 or PR1-5 to 95 N·m (70 ft.-lbs.). On 3.1522, 3.1524 and T3.1524 engines, studs are used at positions 13 and 14 (Fig. PR1-5) and also at positions 4, 15 and 18 (some early engines have cap screws at these three positions). Two different lengths of cap screws are used – the shorter screws are used at positions 2, 3, 7 and 8 and the longer screws are used at remaining positions. Nuts and cap screws should be tightened in steps following sequence shown in Fig. PR1-5 to a final torque of 95 N·m (70 ft.-lbs.).

Reinstall rocker arm shaft assembly. Before tightening retaining nuts on early engines, be sure slot in front end of rocker shaft (Fig. PR1-6) is positioned as shown (pointing towards sharp corner at

Fig. PR1-6—On early engines, slot (S) in end of rocker arm shaft should be positioned so it is pointing towards sharp corner at top of bracket (P). On late engines, slot is removed and shaft is correctly located by a dowel through one of the brackets.

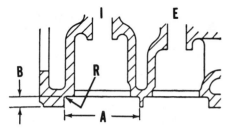

Fig. PR1-7—View of valve seat insert counterbore for engines so equipped. Refer to text for specified dimensions.

tip of bracket). This will ensure correct lubrication of shaft assembly. On late engines, the slot has been removed and shaft is correctly positioned by a dowel through one of the mounting brackets. Complete installation by reversing the removal procedure.

Operate engine until normal operating temperature is reached. Then, while engine is hot, remove rocker shaft assembly and injectors. Retighten nuts and cap screws in recommended sequence to specified torque. If a nut or cap screw does not turn before specified torque is reached, loosen 1/12 to 1/6 of a turn, then retighten to specified torque. After retightening all nuts and screws, recheck the first 10 positions (without loosening) to be certain they are still at correct torque. Reinstall removed components and readjust valve clearance.

VALVE SYSTEM

All 3.152 engines have 45° valve face and seat angles. Early D3.152 engines (prior to engine number U584638F) have 45° face and seat angles, except for some engines rated above 2250 rpm which have 35° face and seat angles. Late D3.152 engines and all 3.1522, 3.1524 and T3.1524 engines have 35° face and seat angles.

When reconditioning valves and seats, do not remove any more material than necessary, otherwise, valve recession (distance valve head is below surface of cylinder head) may exceed the specified limit. Specified valve recession is 1.68-3.50 mm (0.066-0.140 inch) for 3.152 engines. On D3.152 engines with 45° seats, valve recession should be 1.50-2.13 mm (0.059-0.084 inch) for intake and 1.47-2.13 mm (0.058-0.084 inch) for exhaust. On D3.152 engines with 35° seats, valve recession should be 1.32-1.83 mm (0.054-0.072 inch) for intake and 1.60-2.11 mm (0.063-0.083 inch) for exhaust. On all 3.1522, 3.1524 and T3.1524 engines, valve recession should be 1.32-1.75 mm (0.052-0.069 inch) for intake and 1.60-2.18 mm (0.063-0.086 inch) for exhaust.

Renewable valve seat inserts are used on intake and exhaust for 3.152 engines and on exhaust only on D3.152 engines with 45° seats. When renewing valve seat inserts, new valve guides should also be installed to ensure accurate machining of cylinder head counterbore. Using suitable cutters, refer to Fig. PR1-7 and machine cylinder head to the following dimensions:

Intake—3.152
Depth (B)............6.30-6.35 mm
(0.248-0.250 in.)
Diameter (A).......47.60-47.62 mm
(1.874-1.875 in.)
Radius (R)............1.02-1.30 mm
(0.040-0.050 in.)

Exhaust—3.152
Depth (B)............6.30-6.35 mm
(0.248-0.250 in.)
Diameter (A).......41.25-41.28 mm
(1.624-1.625 in.)
Radius (R)............1.02-1.30 mm
(0.040-0.050 in.)

Exhaust—D3.152
Depth (B)............7.87-7.92 mm
(0.310-0.312 in.)
Diameter (A).......42.62-42.64 mm
(1.678-1.679 in.)
Radius (R)—max..........0.38 mm
(0.015 in.)

Valve seat inserts should be pressed (not hammered) squarely into cylinder head counterbores using the valve guides as a pilot. Inserts must bottom firmly in counterbore. On D3.152 engines, machine a 30° flare to a depth of 3.3-3.5 mm (0.130-0.138 inch), measured at inner diameter of exhaust insert. Machine all seats as required to recess valve the recommended distance from cylinder head gasket surface.

All engines are equipped with renewable valve guides. Inside diameter of guides should be 7.98-8.01 mm (0.3141-0.3155 inch). Valve stem outer diameter is 7.90-7.92 mm (0.311-0.312 inch). Desired operating clearance of valves in guides is 0.05-0.11 mm (0.0021-0.0045 inch). Maximum allowable clearance is 0.152 mm (0.006 inch) for intake and 0.14 mm (0.0055 inch) for exhaust. Guides should have interference fit in cylinder head of 0.04-0.08 mm (0.0016-0.0031 inch).

When renewing guides on 3.152 and D3.152 engines, note that both ends of guides are chamfered, one 20° and the other 45°. Be sure to install guides with 45° chamfer end towards top of head. Press in until end of guide protrudes 14.73-15.09 mm (0.580-0.594 inch) above spring seating surface of head.

When renewing valve guides on 3.1522, 3.1524 and T3.1524 engines, note that intake guides are shorter than exhaust guides and are not interchangeable. The inside diameter of both guides is counterbored at one end and this end must be installed towards the valve seat. Press guides in from the top of head until top of intake guide protrudes 9.19-9.55 mm (0.362-0.376 inch) above spring seating surface of head and exhaust guide protrudes 14.73-15.09 mm (0.580-0.594 inch) above spring seating surface.

Inspect rocker arms and shaft for excessive wear and other damage. Be sure to identify parts in order of removal so they can be reinstalled in their original positions. Rocker shaft diameter is 15.81-15.84 mm (0.6223-0.6238 inch). Rocker arm bore diameter is 15.86-15.89 mm (0.6244-0.6256 inch). Desired operating clearance of rocker arms on shaft is 0.02-0.09 mm (0.0008-0.0035 inch) and maximum allowable clearance is 0.13 mm (0.005 inch). Rocker arm bushings are not renewable. When reassembling rocker arm assembly, be sure slot in end of shaft (early engines) will be towards front of engine. Refer to Fig. PR1-6 for proper position of slot. On late engines, shaft is correctly located by a dowel through one of the rocker shaft brackets and slot has been removed from end of shaft.

The valve tappets operate in bores in the cylinder head. Outside diameter of tappets when new is 15.81-15.84 mm (0.6225-0.6236 inch) and inside diameter of bores in cylinder head is 15.86-15.90 mm (0.6244-0.6260 inch). Desired operating clearance of tappets in cylinder head bores is 0.02-0.09 mm (0.0008-0.0035 inch).

On some engines, two valve springs (inner and outer) are used on the intake valves. One spring is used on exhaust valves. Exhaust valve spring and intake valve outer spring free length should be 45.29-45.80 mm (1.783-1.803 inches). Spring test load is 9.45-11.25 kg (20.75-24.75 pounds) when compressed to a length of 38.10 mm (1.5 inches). Inner valve spring (if so equipped) free length should be 34.67-35.69 mm (1.365-1.405 inches). Inner spring test load is 3.18-4.08 kg (7-9 pounds) when compressed to a length of 30.16 mm (1.1875 inches). Renew springs if they fail to meet specifications or if they are rusted or distorted.

COOLING SYSTEM

The engine coolant is circulated by a centrifugal type pump mounted on the front of timing gear cover. The pump is belt driven from the engine crankshaft. After renewing the drive belt, it is recommeneded that belt tension be rechecked after the first two hours of operation. Adjust tension so belt deflects about 10 mm (3/8 inch) when

SERVICE MANUAL

Perkins

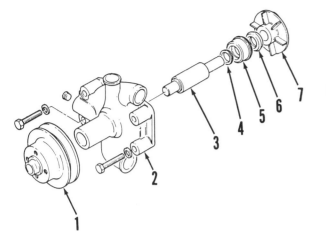

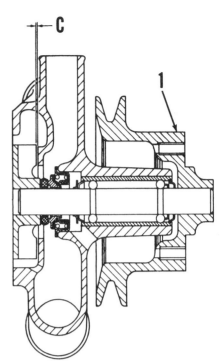

Fig. PR1-8—Exploded view of typical water pump.
1. Fan hub & pulley
2. Pump housing
3. Shaft & bearing assy.
4. Thrower
5. Seal
6. Ceramic seat
7. Impeller

depressed with moderate thumb pressure at a point midway between water pump and alternator pulleys.

A thermostat is mounted in the cylinder head water outlet. The bellows type thermostat should start to open at 77°-82°C (170°-180°F) and be fully open at 98°C (208°F).

To overhaul water pump, first remove pump pulley (1–Fig. PR1-8) using a suitable puller. Press pump shaft with bearings out through rear of pump housing. Press shaft out of impeller (7). Remove ceramic seat (6), seal (5) and thrower (4).

Pump shaft and bearing assembly is renewed as a unit. Press shaft (shorter end first) into rear of pump housing until front end of bearing is flush with front surface of pump housing. Assemble thrower onto shaft. Install seal with carbon face to rear and install ceramic seat with ceramic insert toward seal. Support front end of pump shaft, then press impeller onto shaft until clearance (C–Fig. PR1-9 or PR1-10) of 0.25-0.51 mm (0.010-0.020 inch) is obtained between impeller blades and pump body. Support rear end of shaft (not the impeller), then press fan pulley onto shaft until front end of pulley is flush with front end of shaft (Fig. PR1-9); or if fan mounting hub projects forward from pulley, press pulley on until fan mounting surface is 140.49 mm (5.531 inches) from rear surface of pump housing (D–Fig. PR1-10).

INJECTOR

When removing injectors, it is of utmost importance to plug all openings in fuel system to prevent entry of dirt. Be sure to remove copper washer from bottom of injector bore in cylinder head. Thoroughly clean cylinder head bore and injector nozzle before reinstalling injector. It is recommended that copper seal washer and rubber dust seal be renewed whenever injectors are removed. A thin metal heat shield is used around the injectors on 3.1522 engine. The head shield must be renewed whenever the injector is removed. Tighten injector mounting nuts evenly to 16 N·m (12 ft.-lbs.).

TESTING. A special nozzle test pump is required to test and adjust injectors. The injector should be checked for specified opening pressure, nozzle leakage and correct spray pattern.

WARNING: Fuel leaves the injector nozzle with sufficient force to penetrate the skin. When testing, keep yourself clear of nozzle spray.

Connect injector to nozzle test pump, then operate tester lever several quick strokes to purge air from injector and to make sure nozzle valve is not stuck. Then operate tester lever slowly while noting nozzle opening pressure reading on tester gage. The specified opening pressure can be determined by referring to the injector code letters stamped on the injector body and the following table.

Code letters	Opening Pressure
3.152 Engines	
DD, DE	12160-12665 kPa
	(1765-1840 psi)
GC	17225-18745 kPa
	(2500-2720 psi)
D3.152 Engines	
BV, CR, CS	17225-18745 kPa
	(2500-2720 psi)
DF	17225-17730 kPa
	(2500-2570 psi)
DN, FS*, GW, UB	17225-18745 kPa
	(2500-2720 psi)
EE	17730-19250 kPa
	(2570-2790 psi)
GM	17730-18745 kPa
	(2570-2720 psi)
3.1522 Engines	
GS	23810-25330 kPa
	(3455-3675 psi)
3.1524 Engines	
EE	24320-25840 kPa
	(3530-3750 psi)
T3.1524 Engines	
HN	24320-25840 kPa
	(3530-3750 psi)

*When servicing "FS" injectors, they should be converted to "DN" type by installing appropriate nozzle valve assembly.

Fig. PR1-9—When reassembling, press impeller onto shaft until clearance (C) is 0.25-0.51 mm (0.010-0.020 inch). Fan hub (1) should be pressed on until flush with end of shaft.

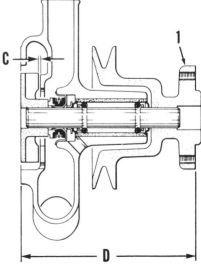

Fig. PR1-10—If equipped with fan hub (1) that extends forward from pulley, press hub on until dimension (D) is 140.49 mm (5.531 inches). Impeller clearance (C) should be 0.25-0.51 mm (0.010-0.020 inch.)

Perkins

Fig. PR1-11—When installing fuel injection pump, align scribe marks (T) on pump flange and rear of timing gear housing. Pump timing marks are located behind cover (W).

To check nozzle for leakage, operate tester lever slowly to maintain pressure at 2000 kPa (300 psi) below opening pressure and observe tip for leakage. If a drop falls from nozzle tip within a 10 second period, nozzle valve is not seating and injector should be overhauled or renewed. During this test, be careful not to confuse fuel leaking from leak-off fitting at top of injector with leakage at nozzle tip.

To check spray pattern, operate the tester lever several quick strokes (about one stroke per second). Spray should be uniform and well atomized and nozzle should produce a "chatter" sound when operating properly.

INJECTION PUMP

A C.A.V. type D.P.A. fuel injection pump is used on all engines. It is recommended that injection pump be tested and serviced only by a shop qualified in diesel fuel injection repair.

To remove injection pump, disconnect and remove fuel lines from pump and immediately plug all openings. Disconnect speed control and shut-off control linkage at the pump. Remove cover plate from front of timing gear cover. On early 3.152 and D3.152 engines, remove inspection plate from side of timing gear cover and wedge injection pump drive gear in position to prevent it from moving out of mesh with idler gear. All other engines have a lug in gear case which holds gear in position when retaining screws are removed.

CAUTION: Do not turn crankshaft with fuel pump removed as damage could result to timing gear case and gears.

Remove pump gear mounting cap screws being careful not to drop them into timing case. Remove the injection pump mounting stud nuts, then withdraw the pump from rear of timing gear case.

To reinstall injection pump, align slot

SMALL DIESEL

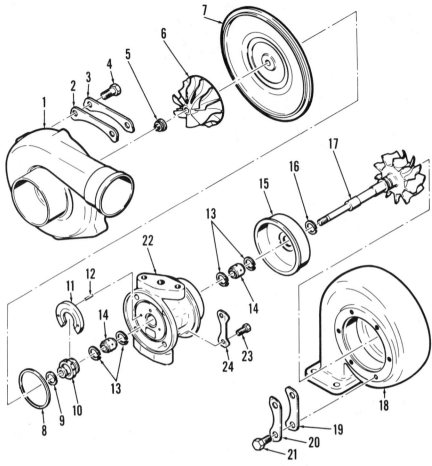

Fig. PR1-12—Exploded view of AiResearch T-31 turbocharger.

1. Compressor housing
2. Clamp
3. Lockplate
4. Cap screw
5. Nut
6. Compressor wheel
7. Backplate
8. Seal ring
9. Piston ring
10. Thrust collar
11. Thrust bearing
12. Spring pin
13. Snap ring
14. Bearing
15. Turbine wheel shroud
16. Piston ring
17. Turbine wheel & shaft
18. Turbine housing
19. Clamp
20. Lockplate
21. Cap screw
22. Center housing
23. Cap screw
24. Lockplate

in pump shaft with dowel in drive gear and position pump on rear of timing case. Install pump stud nuts finger tight. Secure drive gear to pump shaft with the three cap screws. Adjust pump position until scribe line on pump flange is aligned with scribe line on rear of timing gear housing (T–Fig. PR1-11), then tighten stud nuts securely. Complete installation and bleed air from system. If pump was renewed, check and adjust maximum no-load speed as outlined in MAINTENANCE section.

FUEL LIFT PUMP

The lift pump is a diaphragm type which is actuated by the engine camshaft. To check pump output, disconnect the fuel outlet line and install a 0-70 kPa (0-10 psi) pressure gage to pump outlet. Crank engine for 10 seconds and note gage pressure reading. If pressure is less than 31 kPa (4.5 psi), pump is defective. Also, note time required for pump pressure to drop to one half of reading obtained when engine was being cranked. If time is less than 30 seconds, pump is defective. Individual parts are available for pump repair.

TURBOCHARGER

Model T3.1524

An AiResearch T-31 turbocharger is used. To ensure satisfactory turbocharger operation, it is important that turbocharger receives adequate engine lubricating oil pressure, minimum 207 kPa (30 psi), and a clean supply of inlet air. Maximum boost pressure should be 41 kPa (6 psi) when engine is running at maximum speed and full load.

To disassemble turbocharger, first scribe match marks across compressor housing, center section and turbine housing for reassembly purposes. Remove compressor and turbine housings (1 and 18–Fig. PR1-12) being careful not to damage compressor or turbine wheel blades. Bent or cracked

SERVICE MANUAL

Perkins

blades cannot be repaired. Use a T-handle wrench to remove compressor wheel retaining nut to prevent possible bending of shaft. Remove compressor wheel (6), then withdraw turbine wheel and shaft assembly (17) from center section (22). Remove backplate (7) and thrust bearing (11). Remove outer retaining rings (13) and bearings (14) from center section. Remove and discard seal rings.

Inspect parts for signs of damage, corrosion and wear. Compressor and turbine must not show signs of rubbing, damage from foreign material, cracks or excessive wear. Clean parts in non-caustic carbon solvent. Renew all parts in which there is any doubt about serviceability. Always renew "O" ring, seal rings, bearing retaining rings, shaft bearings, compresson wheel locknut, turbine housing bolts and lock plates.

To reassemble, install bearings and retaining rings in center section and lubricate with clean oil. Position turbine shaft with seal ring into center section. Assemble thrust bearing (11) over thrust collar (10). Install seal ring on thrust collar, then position thrust collar over shaft so bearing is flat against housing and engages housing locating pins. Install seal ring in groove of housing, then install backplate with thrust spring over shaft, thrust collar and seal ring. Tighten backplate cap screws to 8.9 N·m (79 in.-lbs.) and secure with lock plates. Install compressor wheel. Oil threads of locknut and tighten to 2.26 N·m (20 in.-lbs.). Then, use a T-handle wrench to further tighten nut until shaft length increases by 0.14-0.16 mm (0.0055-0.0065 inch). If equipmment is not available to measure shaft stretch, this alternate tightening method may be used: After tightening nut to 2.26 N·m (20 in.-lbs.), continue to tighten nut through an angle of 110 degrees. Install compressor housing while aligning match marks made prior to disassembly. Tighten mounting screws to 14 N·m (130 in.-lbs.). Install turbine housing while aligning match marks on housing and center section. Tighten mounting screws to 14 N·m (130 in.-lbs.). Be sure rotating assembly turns freely and does not contact housings.

When reinstalling turbocharger, connect oil drain pipe, then pour clean engine oil into oil inlet port of center housing while turning shaft assembly by hand to lubricate bearings. Connect air inlet and outlet pipes and exhaust pipe. With oil supply pipe loose and fuel pump control in stop position, crank engine with starter motor until oil flows from oil supply pipe. Connect oil pipe to center section. Start engine and run at idle speed for about five minutes to allow lubricating oil to flow before increasing engine speed or load.

PISTON, PISTON PIN AND RINGS

The pistons can be removed from the top after removing cylinder head, oil pan, oil pump and connecting rod caps. Be sure to remove ring ridge or carbon deposits from top of cylinders before attempting to remove pistons. The pistons are marked 1, 2 or 3 on piston crown, number 1 being at front of engine. Be sure to keep pistons and connecting rods together as assemblies for proper reassembly.

Piston pins are retained in pistons by two snap rings. Heating piston to about 38°-49°C (100°-120°F) will ease removal and installation if pin is tight. Manufacturer recommends renewing snap rings during reassembly if they have been in service for a long period of time. Piston pin diameter is 31.744-31.750 mm (1.2498-1.2500 inches) and bore diameter in piston is 31.747-31.753 mm (1.2499-1.2501 inches). Pin is a transition fit in piston.

Piston ring end gap is measured with ring positioned squarely in cylinder bore. Refer to the following table for specified end gap values.

3.152 Engine with Chromed Liners
  All rings............0.28-0.61 mm
               (0.011-0.024 in.)
3.152 Engine with Cast Iron Liners and D3.152 Engines Rated Up to 2250 Rpm
  Top ring............0.36-0.76 mm
               (0.014-0.030 in.)
  2nd ring and
    slotted scraper......0.28-0.69 mm
              (0.011-0.027 in.)
D3.152 Engine Rated Above 225 Rpm
  Top and 4th rings......0.36-0.69 mm
               (0.014-0.027 in.)
  2nd and 3rd rings......0.28-0.61 mm
               (0.011-0.024 in.)
AD3.152 Engine (Up-rated)
  Top ring............0.36-0.69 mm
               (0.014-0.027 in.)
  2nd and 3rd rings......0.28-0.61 mm
               (0.011-0.024 in.)
  4th ring............0.25-0.74 mm
               (0.010-0.029 in.)
3.1522, 3.1524 and T3.1524 Engines
  Top ring............0.30-0.74 mm
               (0.012-0.023 in.)
  2nd and 3rd rings......0.20-0.64 mm
               (0.008-0.025 in.)
  Oil control ring........0.25-0.76 mm
               (0.010-0.030 in.)

Inspect piston for wear and other damage. Renew piston if scoring is evident on piston skirt. Clean and inspect piston ring grooves for wear. Refer to the following table for specified ring groove and ring side clearance values.

3.152 and D3.152
Ring groove width:
  Top and 2nd..........2.43-2.46 mm
               (0.0957-0.0968 in.)
  3rd................3.23-3.25 mm
               (0.1272-0.1279 in.)
  4th and 5th..........6.40-6.43 mm
               (0.2520-0.2531 in.)
Ring clearance in groove:
  Top.................0.05-0.10 mm
               (0.0020-0.0039 in.)
  Maximum.............0.18 mm
               (0.007 in.)
  2nd, 4th and 5th.......0.05-0.10 mm
               (0.0020-0.0039 in.)
AD3.152 Engine
Ring groove width:
  Top, 2nd and 3rd......2.43-2.46 mm
               (0.0957-0.0968 in.)
  4th................4.81-4.84 mm
               (0.1894-0.1905 in.)
Ring clearance in groove:
  Top.................0.05-0.10 mm
               (0.0020-0.0039 in.)
  Maximum.............0.18 mm
               (0.007 in.)
  2nd and 3rd..........0.05-0.10 mm
               (0.0020-0.0039 in.)
  4th.................0.05-0.11 mm
               (0.0020-0.0043 in.)

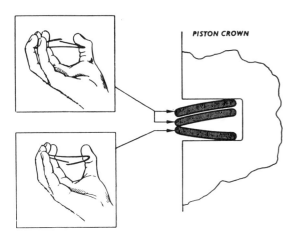

Fig. PR1-13—View showing correct installation of three-segment laminated compression ring used on some early engines.

Perkins   SMALL DIESEL

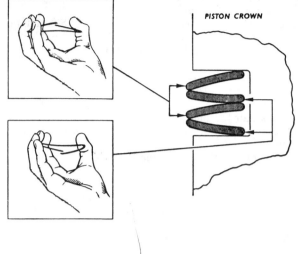

Fig. PR1-13A—View showing correct installation of four-segment, laminated compression ring used on some late production engines.

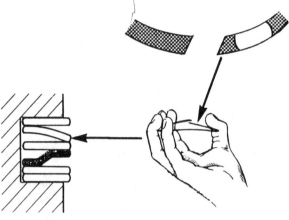

Fig. PR1-14—View showing correct installation of laminated oil scraper ring used on some engines.

3.1522, 3.1524 and
T3.1524 Engines
Ring groove width:
  Top and 2nd..........2.43-2.46 mm
                       (0.0957-0.968 in.)
  Oil control...........4.81-4.84 mm
                       (0.1894-0.1905 in.)
Ring clearance in groove:
  Top................0.05-0.10 mm
                       (0.0020-0.0039 in.)
  Maximum............0.18 mm
                       (0.007 in.)
  2nd and oil
  oil control.........0.05-0.10 mm
                       (0.0020-0.0039 in.)

Fit rings to pistons in the following positions (reading from top of piston):

3.152 Engine with Chrome Plated Liners.
  1. Plain compression.
  2. Taper face compression.*
  3. Laminated compression.†
  4. Slotted scraper.
  5. Slotted scraper.

3.152 Engine with Cast Iron Liners and D3-152 and AD3.152 Engines Rated up to 2250 rpm.
  1. Chrome plated compression.
  2. Plain compression.

  3. Laminated compression.†
  4. Spring-loaded, laminated scraper or internal expander scraper.
  5. Slotted scraper.

*One side of taper face compression ring is marked "T" or "TOP" and must be installed with this side facing upward.

†The laminated compression ring is composed of three segments on early engines or four segments on late engines. When assembling, be sure ring segments are positioned correctly as shown in appropriate Fig. PR1-13 or PR1-13A. The spring-loaded, laminated scraper ring must be assembled as shown in Fig. PR1-14. The internal expander type scraper ring (Fig. PR1-15) should be assembled as shown. Be sure the rail end gaps are equally spaced around the piston.

D3.152 Engines Rated Above 2250 rpm.
  1. Chrome plated compression.
  2. Plain compression.
  3. Internally stepped compression.
  4. Coil spring loaded scraper.

AD3.152 Engine.
  1. Chrome plated compression.
  2. Internally stepped compression.
  3. Internally stepped compression.
  4. Coil spring loaded scraper.

3.1522 Engine.
  1. Chrome compression.
  2. Internally stepped compression.
  3. Coil spring loaded scraper.

3.1524 and T3.1524 Engines.
  1. Tapered face, chrome compression.
  2. Tapered face, plain comression.
  3. Coil spring loaded scraper.

When installing coil spring loaded scraper ring install coil spring first, then install ring over spring with ring end gap positioned 180° opposite from coil spring ends. When installing compression rings with internal step (notch), the stepped edge of ring should be towards top of piston.

If reinstalling original pistons, note that pistons are stamped 1, 2 or 3 and should be assembled on respective connecting rods stamped 1, 2 or 3 and installed in respective cylinder (number 1 at front). Be sure the letter "F", word "Front" or arrow mark on piston crown is towards front of engine and numbered side of connecting rod and cap is towards injection pump side of engine. If identification marks have been removed from piston crown, assemble piston to connecting rod so offset cavity in piston crown is towards side of rod marked with identification number. Mark front side of piston crown for future reference.

If installing new pistons, the correct piston height in relation to top surface of cylinder block must be maintained. With piston at top dead center, piston crown must be 0.00-0.13 mm (0.000-0.005 inch) below top surface of cylinder block for 3.152 engine; between 0.102 mm (0.004 inch) above to 0.025 mm (0.001 inch) below top of block for D3.152 engine; 0.03-0.15 mm (0.001-0.006 inch) below top of block for AD3.152 engine; between 0.03 mm (0.001 inch) above to 0.17 mm (0.0067 inch) below cylinder block surface for 3.1522 engine; and 0.025-0.152 mm (0.001-0.006 inch) below top of cylinder block on 3.1524 and T3.1524 engines. Pistons for 3.152, D3.152 and AD3.152 engines have extra material in the piston crown area which must be machined off to provide the correct piston height. On 3.1522, 3.1524

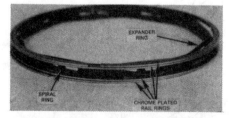

Fig. PR1-15—View of internal expander type oil scraper ring. Install internal expander first, then assemble two lower rails, spiral ring and two upper rails in that order.

and T3.1524 engines, three different height grades ("H" for high, "M" for medium and "L" for low) of pistons are used in production to obtain desired piston height. The grade letters are stamped on top of piston. Replacement pistons are stamped on top of piston. Replacement pistons are available in "H" and "L" grades. "H" grade pistons may be used to replace "M" or "L" grade pistons, but piston crown will have to be machined to obtain specified piston height.

When assembling new pistons to connecting rods, be sure offset cavity in top of piston is towards side of rod marked with cylinder identification numbers. Reinstall connecting rod and piston with numbered side of rod towards injection pump side of engine. Stamp top of pistons with appropriate cylinder number and front end identification for future reference.

CONNECTING ROD

The connecting rod uses a renewable bushing in small end. Use suitable installing tools when renewing bushings to prevent damage to bushing or rod. Be sure oil hole in bushing is aligned with oil hole in top of rod. Bushings must be reamed to final size after installation. Bushing inside diameter should be 31.76-31.79 mm (1.2505-1.2515 inches). Piston pin outer diameter is 31.744-31.750 mm (1.2498-1.2500 inches). Desired operating clearance of pin in bushing is 0.01-0.04 mm (0.0004-0.0016 inch).

A precision, insert type bearing is used in the connecting rod big end. Standard inside diameter of bearings is 57.19-57.21 mm (2.2515-2.2525 inches) and outside diameter of crankpins is 57.11-57.13 mm (2.2485-2.2492 inches). Desired bearing running clearance is 0.06-0.10 mm (0.002-0.004 inch). Connecting rod big end side clearance on crankpin should be 0.24-0.50 mm (0.010-0.020 inch).

When reinstalling connecting rods, be sure side stamped with cylinder identification number is toward injection pump side of engine. Manufacturer recommends renewing connecting rod cap nuts whenever they are removed. Three types of connecting rod nuts have been used on 3.152, D3.152 and AD3.152 engines. The original nuts were self-locking. These were later replaced by cadmium plated (bright finish) nuts which have been replaced in latest engines which phosphated (dull black finish) nuts. Model 3.1522, 3.1524 and T3.1524 engines used only cadmium plated and phosphated nuts. The nuts should be tightened to the following torque values: 95 N·m (70 ft.-lbs.) for self-locking nuts, 61 N·m (45 ft.-lbs.) for cadmium plated nuts and 81 N·m (60 ft.-lbs.) for phosphated nuts.

CYLINDER BLOCK AND LINERS

The cylinder block is equipped with renewable thin wall, dry type cylinder liners. Model 3.152 and D3.152 engines are equipped with cast iron or chrome plated liners. All other engines use cast iron liners. Cast iron and chrome plated liners are not interchangeable due to difference in liner flange dimensions.

Production liners are an interference fit in cylinder bores. Service replacement liners are a transitional fit in cylinder block and are prefinished so no machining is required after installation. If desired, unbored cast iron, production style liners (with interference fit in cylinder block) are available for service installation. These liners must be bored and honed to correct size after installation. Liners with oversize outside diameter are available for installation in cylinder block when the standard block bore must be enlarged to remove damage. These liners are available in chromed, prefinished, 0.13 mm (0.005 inch) oversize and cast iron, unbored 0.25 mm (0.010 inch) oversize. When an oversize liner is used, the oversize amount should be stamped on top of cylinder block between the liner and side of block for future reference.

When removing and installing cylinder liners, use suitable tools to prevent damage to cylinder block or liners. Special tool 21825052 and adapter 21825054 are available from manufacturer for this purpose. Be sure cylinder block bore and liner outside diameter are thoroughly clean before proceeding with installation.

Shims are available for installation under liner flange, if necessary, to obtain recommended liner installed height. Before installing liner, measure flange thickness and cylinder block counterbore depth to determine shim thickness required to adjust liner flange height. Shims are 0.13 mm (0.005 inch) thick. Depth of counterbore for chrome liner flange should be 1.17-1.24 mm (0.046-0.049 inch) and for cast iron liner flange, depth should be 3.76-3.86 mm (0.148-0.152 inch). On 3.152 and D3.152 engines with chromed liners, top of liner flange should be 0.02-0.18 mm (0.001-0.007 inch) below surface of cylinder block when properly installed. On all engines with cast iron liners, top of flange should be between 0.10 mm (0.004 inch) above to 0.10 mm (0.004 inch) below top surface of block.

Cylinder block bore should be 93.66-93.69 mm (3.6875-3.6885 inches). Production style liners should have a 0.03-0.08 mm (0.001-0.003 inch) interference fit in block. Service liners should have a 0.03 mm (0.001 inch) loose to 0.03 mm (0.001 inch) tight transition fit in block. Inside diameter of a finished production liner should be 91.48-91.50 mm (3.6015-3.6025 inches) and inside diameter of service liner should be 91.50-91.53 mm (3.6025-3.6035 inches). Renew liner if bore wear exceeds 0.178 mm (0.007 inch).

On some engines, restriction plugs are installed in some of the water passages in the top of cylinder block. If plugs are removed, be sure correct size of plug (Fig. PR1-16) is reinstalled in correct hole as shown. Plugs must be flush with top surface of cylinder block. Front plug, if renewed, will require machining prior to installation to fit flush with top surface of block.

TIMING GEARS AND FRONT COVER

To remove timing gear cover, drain coolant and disconnect hose from water pump. Remove fan belt, water pump, alternator and alternator brackets and crankshaft front pulley. Remove cover mounting cap screws and withdraw cover.

The crankshaft front oil seal (12 – Fig. PR1-17) is located in the front cover (11)

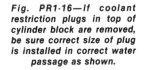

Fig. PR1-16—If coolant restriction plugs in top of cylinder block are removed, be sure correct size of plug is installed in correct water passage as shown.

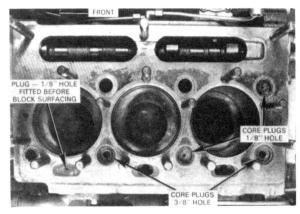

Perkins

SMALL DIESEL

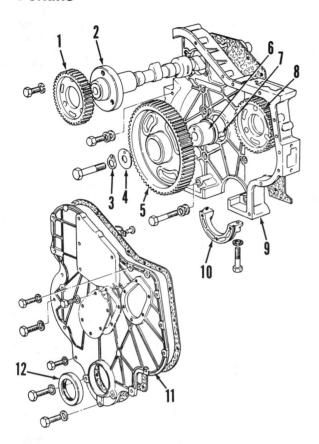

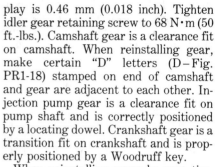

Fig. PR1-17—Exploded view of timing gears, housing and cover used on 3.1522 engine. Other engines are similar.

1. Camshaft gear
2. Camshaft
3. Locking washer
4. Retainer washer
5. Idler gear
6. Idler stub shaft
7. Locating dowel
8. Injection pump gear
9. Housing
10. Bottom cover
11. Front cover
12. Crankshaft oil seal

play is 0.46 mm (0.018 inch). Tighten idler gear retaining screw to 68 N·m (50 ft.-lbs.). Camshaft gear is a clearance fit on camshaft. When reinstalling gear, make certain "D" letters (D—Fig. PR1-18) stamped on end of camshaft and gear are adjacent to each other. Injection pump gear is a clearance fit on pump shaft and is correctly positioned by a locating dowel. Crankshaft gear is a transition fit on crankshaft and is properly positioned by a Woodruff key.

When reinstalling gears be sure timing marks (Fig. PR1-18) are aligned properly. When all marks are aligned, number 1 piston will be at TDC on the compression stroke.

CAMSHAFT

The camshaft is supported at the front by a renewable bushing in the crankcase on some engines. All other camshaft journals operate directly in unbushed bores in crankcase.

To remove camshaft, first remove timing gear cover as previously outlined. Rotate crankshaft until marked teeth on crankshaft gear, idler gear and camshaft gear are aligned. Remove fuel transfer pump. Remove rocker shaft assembly, then raise tappets and support in raised position. Carefully withdraw camshaft and gear assembly from cylinder block.

Inspect camshaft for scoring or wear and renew if necessary. Front journal diameter is 47.47-47.50 mm (1.869-1.870 inches), center journal diameter is 47.22-47.24 mm (1.859-1.860 inches) and rear journal diameter is 46.71-47.74 mm (1.839-1.840 inches). Renew camshaft if journals are worn or out-of-round in ex-

and may be renewed at this time. Early engines were equipped with a black nitrile seal and a crankshaft oil thrower. Later engines used a red silicone seal and the oil thrower was replaced by a spacer. The oil thrower should not be used with later style seal. Current production engines use a black Viton type seal with a protruding lip at the front. Refer to Fig. PR1-17A. The new seal should be pressed into cover until front face of seal is distance (X) below outer surface of cover. On early engines (except AD3.152), distance (X) should be 16 mm (0.625 inch). On early AD3.152, distance (X) should be 11.43 mm (0.450 inch). On all late engines with black Viton seal, installed distance should be 9.35 mm (0.375). Be sure spring-loaded lip faces inward on all engines.

Inspect all gears for excessive wear and other damage and renew as necessary. When renewing gears be sure there is a minimum backlash of 0.08 mm (0.003 inch). Idler gear bushing bore diameter should be 53.98-54.02 mm (2.1252-2.1268 inches) and idler stub shaft diameter is 53.92-53.94 mm (2.1229-2.1236 inches). Clearance of gear on stub shaft should be 0.03-0.09 mm (0.0012-0.0035 inch). Idler gear end play should be 0.13-0.38 mm (0.005-0.015 inch) and maximum allowable end

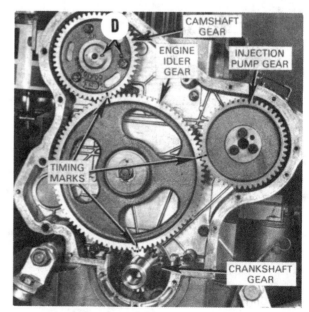

Fig. PR1-18—View of engine timing gear train. Timing marks must be aligned when engine is assembled. However, with gears installed, the marks will align only once every 18 crankshaft revolutions.

Fig. PR1-17A—Current production Viton type front crankshaft seal has a dust seal lip (1) protruding forward. Spring-loaded lip (2) must be facing inward. Refer to text for distance (X) that seal should be below outer surface of timing gear cover (3).

SERVICE MANUAL

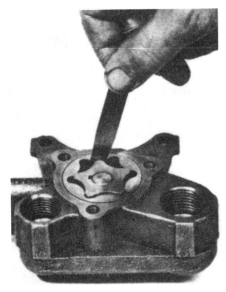

Fig. PR1-19—Use a feeler gage to check clearance between inner and outer oil pump rotors.

cess of 0.05 mm (0.002 inch). Inner diameter of front bushing (if used) should be 47.55-47.60 mm (1.872-1.874 inches) and running clearance should be 0.05-0.13 mm (0.002-0.005 inch). Cylinder block front bore diameter without bushing is 47.60-47.68 mm (1.874-1.877 inches) and running clearance should be 0.10-0.20 mm (0.004-0.008 inch). Crankcase bore for center journal is 47.35-47.42 mm (1.864-1.867 inches) and for rear journal is 46.84-46.91 mm (1.844-1.847 inches). Desired clearance for camshaft in either bore is 0.10-0.20 mm (0.004-0.008 inch).

If camshaft gear was removed, note that "D" marks (D–Fig. PR1-18) on end of camshaft and on gear must be adjacent to each other. To reinstall camshaft

Fig. PR1-20—Pump rotor end clearance may be checked with a straightedge and feeler gage as shown.

reverse the removal procedure. Be sure to align timing marks on gears. The camshaft end thrust is taken by a steel spring riveted to inside of timing gear cover.

OIL PUMP

The rotary type oil pump is mounted on front main bearing cap and is driven from the crankshaft gear thorough an idler gear. The oil pressure relief valve is located in the pump body. Normal oil pressure with engine at full speed and normal operating temperature is 207-414 kPa (30-60 psi). Relief valve opening pressure setting is 345-448 kPa (50-65 psi).

To remove oil pump, first drain oil and remove oil pan. Remove the small lower section of timing gear housing which extends below crankshaft. Remove idler gear. Disconnect pump discharge pipe. Remove mounting screws and withdraw pump assembly.

Remove pump end plate. Do not remove pump drive gear. Individual parts are not available for pump repair. If components fail to meet specified values, pump assembly must be renewed. Check rotor clearance with a feeler gage as shown in Fig. PR1-19. Check rotor end clearance with a straightedge and feeler gage as shown in Fig. PR1-20. Refer to the following table for desired clearance.

3.152, D3.152 and AD3.152 Engines

Pump Part Nos. 41314026,
  41314046, 41314078
  Inner rotor to
    outer rotor........0.01-0.06 mm
                    (0.0005-0.0025 in.)
  Inner rotor
    end clearance......0.04-0.08 mm
                    (0.0015-0.0030 in.)
  Outer rotor end
    clearance........0.01-0.06 mm
                    (0.0005-0.0025 in.)

Pump Part Nos. 41314043,
  41314079, 41314081
  Inner rotor to
    outer rotor........0.03-0.15 mm
                    (0.001-0.006 mm)
  Inner and outer
    rotor end
      clearance.......0.03-0.13 mm
                    (0.001-0.005 in.)

Pump Part No. 41314124
  Inner rotor to
    outer rotor........0.06-0.11 mm
                    (0.0025-0.0045 in.)
  Inner and outer
    rotor end
      clearance.......0.03-0.08 mm
                    (0.001-0.003 in.)

Fig. PR1-21—Note that main bearing caps are numbered and also marked with a serial number. When correctly installed, serial number on cap should be on same side as serial number stamped on bottom of cylinder block.

3.1522, 3.1524 and T3.1524 Engines

Inner rotor to
  outer rotor........0.06-0.11 mm
                  (0.0025-0.0045 in.)
Inner rotor end
  clearance.........0.04-0.09 mm
                  (0.0015-0.0035 in.)
Outer rotor
  end clearance......0.03-0.08 mm
                  (0.001-0.003 in.)

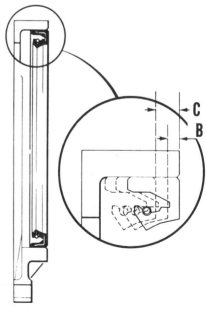

Fig. PR1-22—Early style crankshaft rear oil seal may be pressed further inward into rear housing to obtain a new seal contact area on crankshaft flange. Seal is installed flush with rear face of housing in production.
B. 3.2 mm (⅛ inch)
C. 6.4 mm (¼ inch)

297

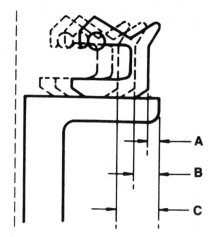

Fig. PR1-23—Late style rear oil seal has a dust seal lip protruding rearward. Spring-loaded lip must face inward. Seal may be installed in three different positions. Position "A" is production installation for late style seal.
A. 2.2 mm (3/32 inch)
B. 4.6 mm (3/16 inch)
C. 6.9 mm (9/32 inch)

Inspect oil pump idler gear and shaft for wear and other damage. Inside diameter of gear bushing when new is 16.67-16.69 mm (0.6563-0.6570 inch). Outer diameter of stub shaft should be 16.63-16.64 mm (0.6548-0.6553 inch). Desired running clearance of gear on shaft is 0.02-0.06 mm (0.0008-0.0024 inch). Gear end play should be 0.20-0.58 mm (0.008-0.023 inch).

CRANKSHAFT AND BEARINGS

The crankshaft operates in four renewable main bearings. The bearings are precision insert type. Crankshaft end thrust is taken by split thrust washers located on both sides of rear main bearing. If main bearings will be reused, keep inserts with their respective caps so they can be reinstalled in their original positions.

Crankshaft standard main journal diameter is 69.81-69.83 mm (2.7485-2.7492 inches) with a wear limit of 69.77 mm (2.7470 inches). Desired main bearing clearance is 0.04-0.10 mm (0.0016-0.0040 inch).

Crankpin standard diameter is 57.11-57.13 mm (2.2485-2.2492 inches) with a wear limit of 57.07 mm (2.2470 inches). Desired rod bearing running clearance is 0.06-0.10 mm (0.0024-0.0040 inch).

Main bearings and connecting rod bearings are available in 0.25, 0.51 and 0.76 mm (0.010, 0.020 and 0.030 inch) undersizes. The crankshaft is hardened by the Tufftride process and must be rehardened after regrinding. If facilities are not available for Tufftriding, crankshaft should be renewed.

Crankshaft end play should be 0.05-0.38 mm (0.002-0.015 inch). Standard thrust washer thickness when new is 3.07-3.12 mm (0.121-0.123 inch). Oversize thrust washers are available which have a thickness of 3.26-3.31 mm (0.1285-0.1305 inch). When installing thrust washers, be sure steel side of thrust washers is against main bearing housing.

To reinstall crankshaft and bearings, reverse the removal procedure. Note that main bearing caps are numbered and also marked with a serial number (Fig. PR1-21). When correctly installed, serial number on cap should be on same side as serial number stamped on bottom of cylinder block. Tighten bearing cap mounting screws to 150 N·m (110 ft.-lbs.).

REAR OIL SEAL

ROPE TYPE SEAL. A rope type crankshaft rear oil seal is used on some 3.152 and D3.152 engines. The rope seal fits into grooves of two-piece retainer which bolts around crankshaft flange. The rope type seal is precision cut to length and must be installed in retainer halves so 0.25-0.50 mm (0.010-0.020 inch) of the seal projects from each end of the retainer halves. Use a round bar to roll seal into grooves to compress the seal. Lubricate the exposed surface of seal with graphite grease. Apply sealing compound to gasket surfaces, then install retainer halves. Make sure retainer with oil shroud is towards top of engine. Tighten retainer clamp bolts to initial torque of 8 N·m (6 ft.-lbs.), then tighten seal retainer mounting screws to 16 N·m (12 ft.-lbs.). Finally, tighten clamp bolts to final torque of 16 N·m (12 ft.-lbs.).

LIP TYPE SEAL. A spring loaded, lip type crankshaft rear oil seal is used on most engines. An early style seal had

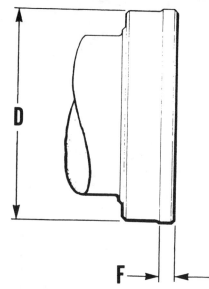

Fig. PR1-24—Oil seal contact area of crankshaft rear flange may be remachined to a minimum diameter (D) of 133.17 mm (5.243 inches). However, original diameter must be maintained for distance (F) of 4.8 mm (3/16 inch) from rear face of flange.

a flat face as shown in Fig. PR1-22 while later style seal has a dust seal lip protruding rearward as shown in Fig. PR1-23.

On original production engines, early style seal is installed flush with rear face of retainer and late style seal is installed deeper into housing as shown at A – Fig. PR1-23. When installing a new seal on a worn crankshaft, seal may be pressed further inward (position "B" or "C" – Fig. PR1-22 or PR1-23) to obtain an unworn seal contact point on crankshaft flange.

If all three seal positions have been used, the crankshaft flange may be remachined undersize to a minimum diameter (D – Fig. PR1-24) of 133.17 mm (5.243 inches). However, original diameter must be maintained for distance (F) of 4.8 mm (3/16 inch) from rear face of crankshaft flange for flywheel mounting. When crankshaft flange has been reground, new seal should be installed in original production location.

Be sure to lubricate crankshaft flange and oil seal lip with engine oil prior to reassembly. Using suitable tools, install seal assembly being careful to protect lip of seal.

PERKINS

Model	No. Cyls.	Bore	Stroke	Displ.
4.99	4	76.2 mm (3.0 in.)	88.9 mm (3.5 in.)	1620 cc (98.9 cu. in.)
4.107	4	79.4 mm (3.125 in.)	88.9 mm (3.5 in.)	1760 cc (107.4 cu. in.)
4.108	4	79.4 mm (3.125 in.)	88.9 mm (3.5 in.)	1760 cc (107.4 cu. in.)
4.154	4	88.9 mm (3.5 in.)	101.6 mm (4.0 in.)	2520 cc (154 cu. in.)

All engines in this section are liquid-cooled, four-stroke, indirect injection diesels with vertical, inline cylinders. Crankshaft rotation is clockwise viewed from front. Firing order is 1-3-4-2. The engine identification number is stamped on the cylinder block behind the injection pump.

MAINTENANCE

LUBRICATION

Recommended crankcase oil is SAE 10W for temperatures from minus 18°C (0°F) to minus 1°C (30°F), SAE 20W for temperatures between minus 1°C (30°F) and 27°C (80°F) or SAE 30 for temperatures above 27°C (80°F). Use of multiviscosity oils is also acceptable. Oil must meet MIL-L-46152 or MIL-L2104C specifications. However, if engine is new or just overhauled, manufacturer does not recommend using MIL-L-2104C specification oil until after the first oil change.

Engine oil and filter element should be renewed after every 150 hours of operation or every 3 months, whichever occurs first. Crankcase capacity varies according to application.

COOLING SYSTEM

Manufacturer recommends using a mixture of water and ethylene glycol antifreeze in the cooling system. In addition to providing protection against freezing, the antifreeze solution contains additives to protect against corrosion and also raises the boiling point of the coolant. If an antifreeze solution is not used, an approved corrosion inhibitor should be added to the water.

FUEL SYSTEM

FUEL FILTER AND BLEEDING.
The paper element type fuel filter should be renewed after every 450 hours of operation or once a year, whichever comes first. However, if engine indicates signs of fuel starvation, filter should be renewed regardless of hours of operation.

Whenever fuel filter has been renewed or fuel lines have been disconnected, air must be bled from the fuel system as follows: loosen the vent screws on top of fuel filter housing, the hydraulic head locking screw (B–Fig. PR2-1) on side of pump body and the vent screw (A) on top of governor housing. Manually actuate fuel lift pump priming lever until air-free fuel flows from each vent screw.

NOTE: If fuel lift pump priming lever cannot be moved, fuel pump is at maximum lift position on camshaft lobe. Turn crankshaft one revolution to reposition camshaft.

Retighten vent screws in the following order: Filter vent screw, head locking screw and governor vent screw. Then loosen fitting at injection pump fuel inlet (F). Operate hand primer until fuel flows from around the threads, then retighten fitting. Loosen high pressure fuel line fittings at the injectors. With stop control in "run" position and throttle at full speed position, crank engine with starter motor until fuel is discharged from the fuel pipes. Retighten fittings and start engine.

ENGINE ADJUSTMENT SPEED.
The engine should be at normal operating temperature when checking and adjusting engine speeds. Low idle speed is adjusted using screw (6–Fig. PR2-1). Speed should be adjusted in conjunction with setting the antistall device. Recommended low idle speed may vary depending on engine application.

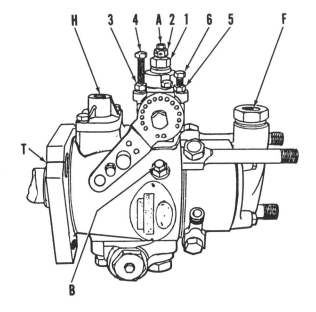

Fig. PR2-1—Illustration of typical C.A.V. injection pump equipped with hydraulic governor and antistall device.

A. Governor housing bleed screw
B. Pump body bleed screw
F. Fuel inlet fitting
H. Timing inspection cover & fuel return fitting
T. Timing scribe mark
1. Locknut
2. Antistall device adjusting screw
3. Locknut
4. High speed adjusting screw
5. Locknut
6. Idle speed adjusting screw

Perkins

SMALL DIESEL

Fig. PR2-2—Remove valve spring from either valve stem on number 1 cylinder and install a dial indicator on end of valve stem as shown to find piston TDC. Set injection pump timing as outlined in text.

Fig. PR2-3—Align proper timing line (A or C) on injection pump rotor with square end of snap ring when number 1 piston is at specified static timing position. Refer to text.

To adjust, loosen antistall screw (2) two complete turns. Adjust idle speed screw (6) until specified speed is obtained. Turn antistall screw in until there is a slight increase in speed, then turn screw back out ½ turn and tighten locknut (1). To check adjustment, accelerate to maximum speed, then return control lever to idle position and note time required for speed to return to idle. If return time exceeds three seconds, the device is screwed in too far. If engine should stall, the device has been screwed out too far. Readjust if necessary.

The maximum no-load speed adjusting screw (4) is sealed at the factory and should not be altered while engine is under warranty without factory authority. The maximum no-load speed setting will vary according to engine application. The specified speed setting can be determined from the injection pump setting code stamped on the pump name plate. The last four numbers of the code indicate the maximum no-load speed. A typical setting code is EH39/1200/0/4480 with 4480 being the specified maximum speed setting. Adjust maximum speed screw (4) as necessary to obtain recommended speed. Do not exceed specified speed. Install a seal wire after completing adjustment to secure adjustment.

INJECTION PUMP TIMING. If engine timing gears are correctly installed and timed and scribed line (T–Fig. PR2-1) on pump mounting flange is aligned with scribed line on rear of timing gear housing (pump with mechanical governor) or scribed line on flange of pump mounting adapter housing (pump with hydraulic governor), injection timing should be correct. Scribe lines may be aligned by loosening pump mounting nuts and shifting pump in slotted mounting holes.

To check pump internal timing marks, the fuel return adapter plate (pump with hydraulic governor) or inspection cover (pump with mechaninical governor) must be removed. The pump rotor is marked with a number of scribed lines, each one identified with a letter. A snap ring, with one end having a square edge, is also located inside the pump housing. When the appropriate scribed line is aligned with square edge of snap ring, beginning of injection for number 1 cylinder should occur. Refer to Fig. PR2-3.

NOTE: On early pumps, the snap ring had a line scribed on one end for timing purposes.

If correct setting of snap ring is in doubt, the pump must be removed to check and adjust snap ring position. Connect an injector test pump to number 1 outlet fitting of injection pump, then actuate test pump to obtain a pressure of 9750 kPa (440 psi). Rotate injection pump shaft by hand until it locks up. At this point, adjust snap ring position so square edge is aligned with scribed line (Fig. PR2-3) marked with letter "A" (pump with hydraulic governor) or letter "C" (pump with mechanical governor). Reinstall pump.

To check pump to engine timing, proceed as follows: Remove valve cover, then rotate crankshaft until number 1 piston is at TDC on compression stroke. At TDC position, number 4 cylinder exhaust valve should be open and timing pin in front cover should engage hole in rear of crankshaft pulley. Remove valve spring from one of the valves on number 1 cylinder and allow valve to rest on top of piston.

NOTE: Install a piece of wire or rubber band around the valve stem to prevent valve from accidentally falling into cylinder.

Position a dial indicator against end of valve stem and zero indicator gage at piston TDC (Fig. PR2-2). Turn crankshaft in opposite direction of normal rotation (clockwise viewed from front) until gage indicates specified BTDC position as listed in the following table. The correct timing figures can be found by reference to the prefix letters and numbers of the setting code on the injection pump name plate.

Prefix Code	Static Timing (BTDC)	Piston Position (BTDC)
4.99 Engine		
AH28, BH26	26°	5.74 mm (0.226 in.)
CH35	19°	3.05 mm (0.120 in.)
DH19	26°	5.74 mm (0.226 in.)
LH20, LH26	20°	3.40 mm (0.134 in.)
LH29	22°	4.06 mm (0.160 in.)
MH26	18°	2.75 mm (0.108 in.)
4.107 Engine		
CH35	19°	3.05 mm (0.120 in.)
LH23, LH28, LH29, LH31	20°	3.40 mm (0.134 in.)
LH31/900/2/2770, LH31/900/2/3130, LH31/900/9/1880	22°	4.06 mm (0.160 in.)
MH27	18°	2.75 mm (0.108 in.)
PH28, PH30	18°	2.75 mm (0.108 in.)
PH34	20°	3.40 mm (0.134 in.)
4.108 Engine		
EH34E, EH39	18°	2.75 mm (0.108 in.)
LH30	22°	4.06 mm (0.160 in.)
PH23E, PH25E	18°	2.75 mm (0.108 mm)
PH25E/500/9/2090	21°	3.73 mm (0.147 in.)
PH27, PH30	18°	2.75 mm (0.108 in.)
PH30/500/5/2450, PH30/500/6/1570, PH30/500/9/1990, PH30/500/9/2090	21°	3.73 mm (0.147 in.)
RH30E, SH33E	21°	3.73 mm (0.147 in.)
TH23E	20°	3.40 mm (0.134 in.)

SERVICE MANUAL

Perkins

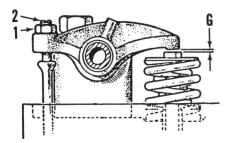

Fig. PR2-4—Loosen locknut (1) and turn adjusting screw (2) until recommended valve clearance (G) is obtained.

With piston at specified BTDC position, appropriate scribed line ("A" for pump with hydraulic governor or "C" for pump with mechanical governor) should be aligned with square edge of snap ring. If timing mark is not aligned, loosen pump mounting nuts and move pump in slotted mounting holes until mark is aligned. If timing mark cannot be aligned by shifting pump on mounting studs, timing gear front cover should be removed and gear timing checked.

REPAIRS

TIGHTENING TORQUES

Refer to the following table for special tightening torques. Fastener threads should be lightly oiled before assembly. Whenever self-locking nuts are removed, they must be renewed.

Models 4.99 and 4.107

Connecting rod	57 N·m (42 ft.-lbs.)
Crankshaft pulley	203 N·m (150 ft.-lbs.)
Cylinder head	57 N·m (42 ft.-lbs.)
Flywheel	81 N·m (60 ft.-lbs.)
Idler gear	49 N·m (36 ft.-lbs.)
Injector	16 N·m (12 ft.-lbs.)
Main bearing caps	57 N·m (42 ft.-lbs.)

Model 4.108

Connecting rod	57 N·m (42 ft.-lbs.)
Crankshaft pulley	Refer to text.
Cylinder head	81 N·m (60 ft.-lbs.)
Flywheel	81 N·m (60 ft.-lbs.)
Idler gear	49 N·m (36 ft.-lbs.)
Injector	16 N·m (12 ft.-lbs.)
Main bearing caps	115 N·m (85 ft.-lbs.)

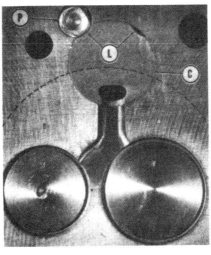

Fig. PR2-5—View of cylinder head showing installed combustion chamber cap (C). Broken line indicates approximate location of cylinder wall when head is installed.

C. Combustion chamber cap
L. Locating slots
P. Expansion plug

Model 4.154

Camshaft gear	68 N·m (50 ft.-lbs.)
Connecting rod:	
Plain nuts	61 N·m (45 ft.-lbs.)
Phosphated nuts	82 N·m (60 ft.-lbs.)
Crankshaft pulley	166 N·m (123 ft.-lbs.)
Cylinder head	115 N·m (85 ft.-lbs.)
Flywheel	108 N·m (80 ft.-lbs.)
Idler gear	28 N·m (21 ft.-lbs.)
Injector	16 N·m (12 ft.-lbs.)
Main bearing caps	115 N·m (85 ft.-lbs.)

VALVE ADJUSTMENT

Clearance between rocker arms and valve stems should be checked and adjusted with engine cold. Recommended clearance for both intake and exhaust valves is 0.30 mm (0.012 inch). Adjust valves in the following sequence by loosening locknut (1–Fig. PR2-4) and turning adjusting screw (2) until correct size feeler gage can be inserted at (G). Recheck clearance after tightening locknut.

Be sure fuel pump control is in "stop" position. Rotate crankshaft in normal direction until valves on number 4 cylinder are in "overlap" position (intake opening and exhaust closing), then adjust valves of number 1 cylinder. Similarly, with valves of number 2 cylinder in overlap, adjust valves of number 3 cylinder. With valves of number 1 cylinder in overlap, adjust valve clearance on number 4 cylinder. With valves of number 3 cylinder in overlap, adjust valves of number 2 cylinder.

CYLINDER HEAD

To remove cylinder head, first drain cooling system and disconnect hose at water outlet. Remove valve cover. Disconnect starting aid fuel line and electrical wire. Disconnect fuel lines and plug all openings. Remove injectors. Remove intake and exhaust pipes. Loosen cylinder head screws in opposite order of tightening sequence shown in Fig. PR2-6 or PR2-7. Remove rocker arms and shaft assembly and push rods, then lift off cylinder head.

Clean cylinder head and inspect for wear and other damage. Check cylinder head distortion using a straightedge and feeler gage. Maximum allowable distortion measured crosswise is 0.08 mm (0.003 inch) concave and 0.13 mm (0.005

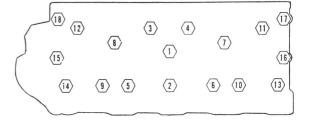

Fig. PR2-6—On 4.99, 4.107 and 4.108 engines, follow sequence shown when tightening cylinder head nuts. Refer to text for tightening procedure.

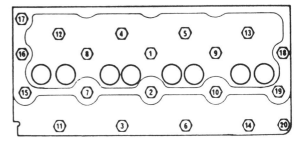

Fig. PR2-7—On 4.154 engines, follow sequence shown when tightening cylinder head cap screws. Refer to text.

Perkins

SMALL DIESEL

inch) convex. Maximum allowable distortion measured lengthwise is 0.15 mm (0.006 inch). Remachining of cylinder head surface is not permitted.

Precombustion chamber caps (C – Fig. PR2-5) are retained in cylinder head by expansion plug (P). To remove combustion chamber cap, insert a brass drift through injector bore and tap out cap and expansion plug. When installing cap, align opening with machined channel in cylinder head and secure the cap by installing a new expansion plug. Height of insert in relation to cylinder head surface must be 0.051 mm (0.002 inch) above to 0.051 mm (0.002 inch) below on 4.99, 4.107 and 4.108 engines or 0.02 mm (0.001 inch) above to 0.07 mm (0.003 inch) below on 4.154 engine.

Model 4.99 and 4.107 engines use either a copper and asbestos or a copper, steel and asbestos cylinder head gasket. These gaskets should be installed using a light coating of sealing compound on both sides. Both types of gaskets are marked "TOP FRONT" to indicate correct installation position.

Model 4.108 engine uses a "Klinger" type cylinder head gasket which is made of black composite material. Do not use gasket sealing compound with this type gasket. It must be installed dry. It is very important that gasket is positioned correctly on cylinder block with the words "TOP FRONT" properly located, otherwise the steel beading will be pinched between cylinder head and top of cylinder liners.

On 4.154 engines, cylinder head gasket must be installed dry. Do not use gasket sealing compound.

To reinstall cylinder head on 4.99, 4.107 and 4.108 engines, proceed as follows: Position cylinder head and gasket onto cylinder block. Lubricate threads of stud nuts with engine oil. Tighten nuts following sequence shown in Fig. PR2-6 progressively in three steps to a final torque of 57 N·m (42 ft.-lbs.) for 4.99 and 4.107 engines or 81 N·m (60 ft.-lbs.) for 4.108 engine. Final tightening sequence should be repeated to be sure all nuts are properly tightened. Install push rods and rocker arm assembly. Tighten rocker shaft oil feed line finger tight at this time. Tighten rocker bracket mounting nuts evenly to 20 N·m (15 ft.-lbs.), then tighten oil feed line fitting securely. Complete installation and run engine until it reaches operating temperature. Retighten cylinder head nuts in recommended sequence to specified torque. If a nut does not move before specified torque is reached, loosen nut about 1/3 turn, then retighten to specified torque. After retightening all nuts, recheck the first 10 nuts in sequence to ensure correct torque. Do not loosen nuts during this final check. After engine cools, readjust valve clearance as previously outlined. Manufacturer also recommends that cylinder head nuts be retorqued again between 25 to 50 hours of engine operation.

On 4.154 engine, install cylinder head, push rods and rocker shaft assembly. Be sure notch in shaft is positioned on front mounting stud and oil passage in front rocker support is aligned with oil passage in cylinder head. Install cylinder head mounting screws and tighten progressively in three steps following sequence shown in Fig. PR2-7. Final torque should be 115 N·m (85 ft.-lbs). Repeat final torque tightening step to ensure all cap screws are correctly tightened. Complete installation and run engine until normal operating temperature is reached. Retorque cylinder head cap screws (while hot) following recommended tightening sequence. Manufacturer recommends that cylinder head cap screws be retorqued again after 20 to 40 hours of engine operation.

VALVE SYSTEM

On early engines, valves are numbered and the cylinder head has corresponding numbers to ensure correct assembly. On later engines, the valves are not numbered. If valves are to be reused, they should be suitably identified so they can be reinstalled in original positions.

Valve face and seat angles are 45° for both intake and exhaust. The valves seat directly in the cylinder head making it important that a minimum amount of material be removed when machining valve seats. Valve recession (depth valve head is below surface of cylinder head) should be measured to determine condition of cylinder head and valves. Specified valve recession for 4.99, 4.107 and 4.108 engines is 0.711-0.991 mm (0.028-0.039 inch) for intake and 0.533-0.813 mm (0.021-0.032 inch) for exhaust. Maximum allowable recession is 1.22 mm (0.048 inch) for both valves. On 4.154 engine, specified valve recession is 0.74-1.04 mm (0.029-0.041 inch) for intake and 0.69-1.02 mm (0.027-0.040 inch) for exhaust. Maximum allowable recession is 1.52 mm (0.060 inch) for both valves.

Valve seat inserts are available for installation on 4.99, 4.107 and 4.108 engines to correct excessive valve recession. Before installing valve seat inserts, the valve guides must first be renewed to provide an accurate pilot bore for machining operations required to fit inserts. After inserts are installed, grind valve seats as required to obtain minimum limit of specified valve recession.

All engines are equipped with renewable valve guides. On 4.154 engine, inside diameter of guides should be 7.99-8.01 mm (0.3145-0.3155 inch). Valve stem diameter is 7.92-7.95 mm (0.312-0.313 inch) for intake and exhaust. Desired clearance of valve stem in guide is 0.04-0.09 mm (0.0015-0.0035 inch). Maximum allowable clearance is 0.13 mm (0.005 inch) for intake and 0.15 mm (0.006 inch) for exhaust. On 4.99, 4.107 and 4.108 engines, valve guide inner diameter should be 7.978-8.014 mm (0.3141-0.3155 inch). Valve stem diameter is 7.925-7.950 mm (0.312-0.313 inch) for intake valve and 7.912-7.937 mm (0.3115-0.3125 inch) for exhaust. Desired clearance of valve stem in guide is 0.028-0.089 mm (0.0011-0.0035 inch) for intake and 0.041-0.102 mm (0.0016-0.004 inch) for exhaust. Maximum allowable clearance is 0.13 mm (0.005 inch) for intake and 0.15 mm (0.006 inch) for exhaust.

Valve guides are an interference fit in cylinder head. When renewing guides, use suitable tools to remove and install guides to prevent damage to guides or cylinder head. Guides should be pressed in from the top until top of guide protrudes 20.32-20.70 mm (0.800-0.815 inch) above surface of cylinder head on 4.99, 4.107 and 4.108 engines. Guides should protrude 16.20-16.81 mm (0.638-0.662 inch) on 4.154 engine.

Some engines use inner and outer valve springs. Manufacturer advises that new valve springs be installed at time of a major engine overhaul. Inspect springs for distortion or other damage. Springs must meet the following values:

Models 4.99, 4.107 and 4.108

Outer spring
  Test length..............42.51 mm
                            (1.780 in.)
Load at test
  length.............24.13-26.67 kg
                        (53.2-58.8 lbs.)
Inner spring (if used)
  test length...............38.86 mm
                          (1.530 in.)
Load at test
  length................1.21-13.9 kg
                        (26.6-30.6 lbs.)

Model 4.154

Outer spring
  Test length..............40.38 mm
                            (1.59 in.)
Load at test
  length...............17.2-19.0 kg
                          (38-42 lbs.)
Inner spring
  Test length..............37.84 mm
                            (1.49 in.)
Load at test
  length............12.05-13.35 kg
                        (26.6-29.4 lbs.)

SERVICE MANUAL

Perkins

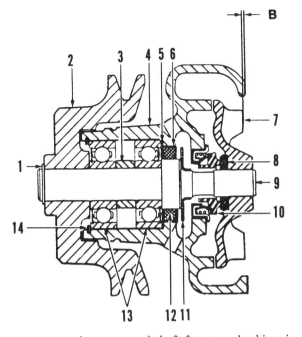

Fig. PR2-8—Cross section view of separate bearing type water pump used on some 4.99, 4.107 and 4.108 engines. Dimension (B) should be 0.00-0.15 mm (0.000-0.006 inch).

1. Snap ring
2. Pulley
3. Spacer
4. Housing
5. Retainer flange
6. Seal retainer
7. Impeller
8. Counterface
9. Shaft
10. Seal assy.
11. Thrower
12. Felt seal
13. Bearings
14. Snap ring

Inspect rocker arms and shaft for excessive wear and other damage. Be sure to identify parts in order of removal so they can be reinstalled in their original positions. On 4.99, 4.107 and 4.108 engines, rocker shaft diameter is 15.805-15.843 mm (0.6223-0.6237 inch) and rocker arm bushing inner diameter should be 15.862-15.894 mm (0.6223-0.6237 inch). Recommended operating clearance of rocker shaft in bushing is 0.019-0.089 mm (0.0008-0.0035 inch) with a wear limit of 0.13 mm (0.005 inch). On 4.154 engine, rocker shaft outer diameter is 15.83-15.86 mm (0.6234-0.6244 inch) and rocker arm bushing inner diameter is 15.87-15.89 mm (0.625-0.6258 inch). Recommended operating clearance of shaft in bushing is 0.01-0.06 mm (0.0006-0.0024 inch) with a wear limit of 0.13 mm (0.005 inch). Rocker arm bushings are renewable. When installing new bushings, be sure oil hole in bushing is aligned with oil hole in rocker arm. On late 4.154 engines, replacement bushing is supplied without an oil hole. The bushing must be drilled, after installation, with a 2 mm (5/64 inch) drill through the rocker arm oil hole.

COOLING SYSTEM

The engine coolant is circulated by a centrifugal type pump which is belt driven from the crankshaft pulley. Belt tension should be adjusted so belt deflects about 10 mm (3/8 inch) when depressed with moderate thumb pressure. After renewing drive belt, it is recommended that belt tension be rechecked after the first two hours of operation to allow for initial stretch of a new belt.

A thermostat is located in the cylinder head water outlet. The thermostat should start to open at 79.5°-83.5°C (175°-182°F) and be fully open at about 96°C (205°F).

Models 4.99, 4.107 and 4.108

Two different types of water pumps are used. In one type of pump, the impeller shaft is supported by two separate bearings (Fig. PR2-8), and in the other type of pump, the shaft and bearings are a combined assembly (Fig. PR2-9).

WATER PUMP (SEPARATE BEARING TYPE). To disassemble pump, first remove pulley (2–Fig PR2-8). Press shaft (9) rearward out of bearings (13), then press impeller (7) off shaft and remove seal assembly. Remove retaining ring (14) and press bearings forward out of pump housing.

To reassemble pump, position thrower (11) in pump body. Install seal retainer (6), felt seal (12) and retaining flange (5) with dish of flange towards the seal. Press bearings and spacer onto impeller shaft with shielded side of bearings facing outward towards front and rear ends of shaft. Lubricate inner surfaces of bearings with high melting point grease, then press the bearings and shaft assembly into the pump housing and install retaining ring (14). Press

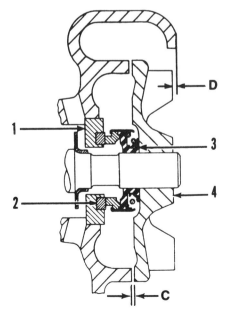

Fig. PR2-8A—View of seal assembly used on early style pumps. Some pumps did not use a counterface (2) in the insert (1). Refer to text for dimension (C) or (D).

1. Insert
2. Counterface
3. Seal assy.
4. Impeller

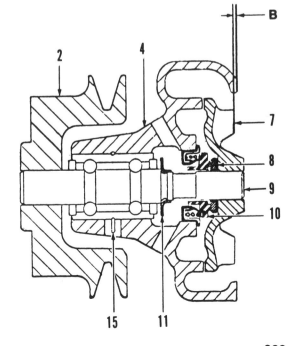

Fig. PR2-9—Cross section view of combined bearing and shaft style water pump assembly used on some 4.99, 4.107 and 4.108 engines. Dimension (B) should be 0.00-0.15 mm (0.000-0.006 inch).

2. Pulley
4. Housing
7. Impeller
8. Counterface
9. Shaft & bearing assy.
10. Seal assy.
11. Thrower
15. Clip

303

Perkins

SMALL DIESEL

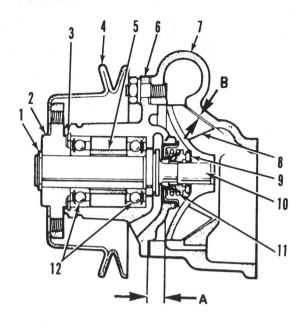

Fig. PR2-10—Cross section view of water pump with two-piece housing used on some 4.154 engines. Refer to text for dimensions (A) and (B).

1. Snap ring
2. Hub
3. Snap ring
4. Pulley
5. Spacer
6. Bearing housing
7. Impeller housing
8. Impeller
9. Counterface
10. Shaft
11. Seal assy.
12. Bearings

thrower onto the shaft. Install any pump mounting cap screws that cannot be installed after pulley is in position, then support end of shaft and press pulley on until it contacts front bearing. Reinstall seal assembly onto shaft. Refer to Fig. PR2-8 for assembled view of late style seal and PR2-8A for early style seal. Support front end of shaft, then press on impeller until rear face of impeller is 0.00-0.15 mm (0.000-0.006 inch) below rear face of body (B – Fig. PR2-8) on late style pump. On early style pump without a seal counterface (2 – Fig. PR2-8A), press impeller on until clearance (C) between front face of impeller and pump body is 0.12-0.25 mm (0.005-0.010 inch). On early style pump with a seal counterface (2), press impeller on until rear face of vanes is 0.74-0.84 mm (0.029-0.033 inch) below rear face of body (D).

WATER PUMP (COMBINED BEARING AND SHAFT TYPE). To disassemble pump, remove pulley (2 – Fig. PR2-9) using a suitable puller. Remove retainer clip (15) through opening in pump body. Press shaft and bearing assembly (9) rearward from pump body. Press shaft out of impeller (7) and remove seal assembly (10).

To reassemble pump, press bearings and shaft (with smaller diameter end rearward) into pump body until retaining clip (15) can be installed in bearing groove. Install any pump mounting cap screws that cannot be installed after pulley is in position, then support rear end of shaft and press pulley on until front of pulley is flush with end of shaft. Press thrower (11) onto shaft. Be sure carbon face of seal (10) and counterface (8) are clean, then assemble into pump housing. Support front end of shaft and

press impeller onto shaft until rear face of vanes is 0.00-0.15 mm (0.000-0.006 inch) below rear face of pump body (B).

Model 4.154

WATER PUMP. Two different styles of water pumps have been used on this engine. One type of pump used a two-piece housing (Fig. PR2-10), while the other type of pump uses a one-piece housing (Fig. PR2-10A).

To disassemble either pump, remove retaining ring (1), then use a suitable puller to remove hub (2). Remove impeller housing (7 – Fig. PR2-10) on two-piece housing pumps. On both pumps, press shaft (10) complete with seal and impeller out through rear of housing. Remove snap ring (3), then press bear-

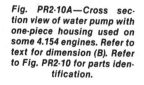

Fig. PR2-10A—Cross section view of water pump with one-piece housing used on some 4.154 engines. Refer to text for dimension (B). Refer to Fig. PR2-10 for parts identification.

ings (12) and spacer (5) out through front of housing. Press shaft out of impeller and remove seal assembly.

To reassemble, press bearings and spacer onto shaft with the shielded side of bearings facing outward towards front and rear ends of shaft. Pack inner surfaces of bearings into housing and install retaining ring (3). Support impeller end of shaft and press hub onto shaft. Install retaining ring (1). Install seal (11) and counterface (9) with ceramic face of counterface mating against carbon face of seal. Support front end of shaft, then press impeller onto shaft until rear face of impeller is 0.38-0.76 mm (0.015-0.030 inch) below rear surface of pump body (B – Fig. PR2-10A) on one-piece pump housings. On pumps with two-piece housings, press impeller onto shaft until clearance (A – Fig. PR2-10) between front of impeller and pump body is 9.90-10.49 mm (0.390-0.413 inch). Install impeller housing (7) and check running clearance (B) between impeller vanes and housing. Clearance should be 0.10-0.14 mm (0.004-0.045 inch). On either style pump, be sure pump shaft turns freely after assembly.

Reinstall pump assembly and tighten mounting cap screws to 20 N·m (15 ft.-lbs.).

INJECTOR

When removing injectors, it is of utmost importance to immediately plug all openings in fuel system to prevent entry of dirt. Be sure to remove copper seal washer from injector bore in cylinder head.

Before reinstalling injectors, be sure bore in cylinder head is clean. It is recommended that copper seal washers be renewed whenever injectors are

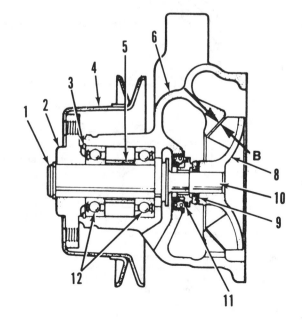

SERVICE MANUAL

removed. Tighten injector mounting nuts to 16 N·m (12 ft.-lbs.).

TESTING. A special nozzle test pump should be checked for specified opening pressure, nozzle leakage and correct spray pattern.

WARNING: Fuel leaves the injector nozzle with sufficient force to penetrate the skin. When testing, keep yourself clear of nozzle spray.

Connect injector to nozzle test pump, then operate tester lever several quick strokes to purge air from injector and to make sure nozzle valve is not stuck. Then operate tester lever slowly while noting nozzle opening pressure reading on tester gage. Specified opening pressure is 13680-15200 kPa (1980-2200 psi). Opening pressure is adjusted by removing cap nut, loosening locknut and turning adjusting screw until desired pressure is obtained.

To check nozzle valve for leakage, operate tester lever slowly to maintain pressure at 2000 kPa (300 psi) below opening pressure and observe nozzle tip for leakage. If a drop falls from nozzle tip within a 10 second period, nozzle valve is not seating and injector should be overhauled or renewed.

To check spray pattern, operate the tester lever several quick strokes (about one stroke per second). Spray should be uniform and well atomized and nozzle should produce a "chatter" sound when operating properly.

INJECTION PUMP

A D.P.A. type injection pump manufactured by C.V.A. is used on all engines. Some pumps are equipped with a hydraulic governor while other pumps use a mechanical flyweight type governor. It is recommended that the injection pump be tested and serviced only by a shop qualified in diesel fuel injection repair.

To remove injection pump, first clean pump and surrounding area. Disconnect and remove fuel lines from pump and immediately plug all openings. Disconnect speed control linkage and shut-off control. On mechanical governor pumps, remove inspection cover from timing gear cover and remove pump drive gear mounting cap screws. On all pumps, remove pump mounting nuts and withdraw pump assembly.

To reinstall pump, reverse the removal procedure. On hydraulic governor pumps, pump shaft has a master spline that mates with drive hub spline. On mechanincal governor pump, pump shaft has a dowel on the end which mates with slot in pump drive gear.

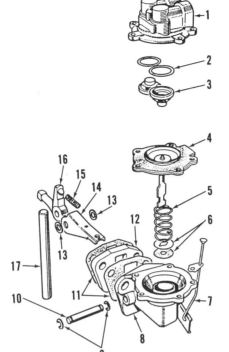

Fig. PR2-11—Exploded view of fuel lift pump. On early pumps, valves (3) were held in position by a clip and two screws (not shown).

1. Pump cover
2. "O" rings
3. Valves
4. Diaphragm
5. Spring
6. Spring retainer & gasket
7. Manual priming lever
8. Pump body
9. "C" clips
10. Pin
11. Gaskets
12. Spacer
13. Washers
14. Lever
15. Spring
16. Rocker
17. Push rod

Tighten drive gear mounting cap screws to 28 N·m (21 ft.-lbs.). Before tightening pump mounting nuts, align timing scribe marks on pump mounting flange (T–Fig. PR2-1). Bleed air from fuel system as outlined in MAINTENANCE section. If pump was renewed, check and adjust maximum no-load speed setting.

FUEL LIFT PUMP

The lift pump is a diaphragm type which is actuated by the engine camshaft. To check pump output, disconnect the fuel outlet line and install a 0-70 kPa (0-10 psi) pressure gage to pump outlet. Crank engine for 10 seconds and note gage pressure reading. If pressure is less than 25 kPa (3.75 psi), pump is defective. Also, note time required for pump pressure to drop to one half of reading obtained when engine was being cranked. If time is less than 30 seconds, pump is defective.

Service parts for pump repair are available separately. Refer to Fig. PR2-11 for an exploded view of fuel lift pump assembly. Inspect all parts for wear and damage and renew if necessary. Thoroughly clean both valve bores in cover (1) and install new valve gaskets

Perkins

(2). Insert a valve in the fuel inlet bore with its spring facing towards pump diaphragm. Install outlet valve in opposite direction. Press valves into cover using a piece of soft tubing having an ID of 9/16 inch and an OD of ¾ inch. Stake valves into cover in six places, between original stake marks, around each valve. Note that early pump valves are held in position by a retaining plate and two screws. Do not attempt to stake these valves into place. Install pump rocker arm and lever assembly into body in opposite order of removal being sure to properly locate lever washers (13) between outside of lever (14) and body (8). Install rubber seal and steel spring seat (6) and diaphragm return spring (5) into body. Install diaphragm (4) so flat of diaphragm pull rod aligns with lever (14). Push diaphragm down slightly and rotate it 90 degrees in either direction to engage lever. Install cover (1) on body (8), lining up scribe marks, and secure into position.

Install pump onto engine in opposite order of removal and reconnect fuel lines. Bleed fuel system as previously outlined.

PISTON, PISTON PIN AND RINGS

The pistons can be removed from the top after removing cylinder head, oil pan and connecting rod caps. Be sure to remove ring ridge or carbon deposits from top of cylinder before attempting to remove pistons. Before disassembling pistons from connecting rods, mark piston crown to indicate "front" in relation to "FRONT" marking cast on connecting rod. Also, number pistons to coincide with cylinder numbers so they can be reinstalled in original positions if reused. Keep pistons and connecting rods together as assemblies for proper reassembly.

Full floating piston pins are retained in pistons by two snap rings. Heating piston to 40°-50°C (100°-120°F) will ease removal and installation if pin is tight. Piston pin diameter is 22.225-22.23 mm (0.8750-0.8752 inch) for early 4.99 engines, 23.812-23.817 mm (0.09375-0.09377 inch) for late 4.99 and all 4.107 engines, 26.987-26.993 mm (1.0625-1.0627 inches) for 4.108 engines and 31.74-31.75 mm (1.2498-1.2499 inches) for 4.154 engines. Pin is a transition fit in piston bore. Inside diameter of piston bore should be 22.22-22.23 mm (0.8748-0.87525 inch) for early 4.99 engines, 23.81-23.82 mm (0.9374-0.9378 inch) for late 4.99 and all 4.107 engines, 26.989-26.994 mm (1.06255-1.06275 inches) for 4.108 engine with standard pistons, 26.992-26.996 mm (1.0627-1.0629 inches) for 4.108 engine

Perkins

with controlled expansion pistons and 31.74-31.75 mm (1.2495-1.250 inches) for 4.154 engines.

Piston ring end gap is measured with ring positioned squarely in cylinder bore. Refer to the following table for specified end gap values.

Models 4.99 and 4.107

Compression and oil
   control rings . . . . 0.30-0.43 mm
   (0.012-0.017 in.)

Model 4.108

Standard piston:
  Compression and oil
    control ring . . . . 0.23-0.43 mm
    (0.009-0.017 in.)
Controlled expansion
piston (three ring):
  Top ring . . . . 0.31-0.59 mm
    (0.012-0.023 in.)
  Second ring . . . . 0.23-0.51 mm
    (0.009-0.020 in.)
  Oil control ring . . . . . 0.31-0.59 mm
    (0.012-0.023 in.)
Controlled expansion
piston (four ring):
  Top ring . . . . 0.30-0.58 mm
    (0.012-0.023 in.)
  Second and third
    rings . . . . 0.23-0.50 mm
    (0.009-0.020 in.)
  Oil control ring . . . . 0.25-0.53 mm
    (0.010-0.021 in.)

Model 4.154

Top ring . . . . 0.36-0.69 mm
  (0.014-0.027 in.)
Second and third rings . . . 0.28-0.61 mm
  (0.011-0.024 in.)
Fourth ring . . . . 0.36-0.69 mm
  (0.014-0.027 in.)
Fifth ring . . . . 0.28-0.61 mm
  (0.011-0.024 in.)

Inspect piston for wear and other damage. Renew piston if scoring is evident on piston skirt. Clean and inspect piston ring grooves for wear. Refer to the following table for specified ring width and ring side clearance values.

Models 4.99 and 4.107

Ring groove width:
  Top . . . . 2.034-2.06 mm
    (0.0801-0.0811 in.)
  Second and third . . . . 1.638-1.664 mm
    (0.0645-0.0655 in.)
  Fourth and fifth . . . . . 4.826-4.851 mm
    (0.190-0.191 in.)
Ring side clearance in
groove:
  Compression rings . . . 0.051-0.102 mm
    (0.002-0.004 in.)
  Oil control rings . . . . . 0.064-0.114 mm
    (0.0025-0.0045 in.)

Model 4.108

Standard piston:
  Ring groove width—
    Top . . . . 2.045-2.070 mm
    (0.0805-0.0815 in.)
    Second and third . . . 1.638-1.664 mm
    (0.0645-0.0655 in.)
    Fourth . . . . 3.200-3.225 mm
    (0.126-0.127 in.)
    Fifth . . . . 4.826-4.851 mm
    (0.190-0.191 in.)
Ring side clearance
in groove—
    Top . . . . 0.061-0.112 mm
    (0.0024-0.0044 in.)
    Second and third . . . 0.051-0.102 mm
    (0.002-0.004 in.)
    Fifth . . . . 0.063-0.114 mm
    (0.0025-0.0045 in.)

Model 4.108

Controlled Expansion Piston
(Three Rings):
  Ring groove width—
    Top . . . . 2.035-2.086 mm
    (0.080-0.082 in.)
    Second . . . . 2.53-2.55 mm
    (0.099-0.1005 in.)
    Third . . . . 4.80-4.826 mm
    (0.1890-0.1900 in.)
Ring side clearance
in groove—
    Top . . . . 0.063-0.140 mm
    (0.0025-0.0055 in.)
    Second . . . . 0.039-0.089 mm
    (0.0015-0.0035 in.)
    Third . . . . 0.039-0.089 mm
    (0.0015-0.0035 in.)
Controlled Expansion Piston
(Four Rings):
  Ring groove width—
    Top . . . . 2.035-2.086 mm
    (0.0801-0.0821 in.)
    Second and third . . . 1.64-1.65 mm
    (0.064-0.065 in.)
    Fourth . . . . 4.79-4.81 mm
    (0.1887-0.1895 in.)
Ring side clearance
in groove—
    Top . . . . 0.05-0.13 mm
    (0.002-0.005 in.)
    Second and third . . . 0.04-0.09 mm
    (0.0015-0.0035 in.)
    Fourth . . . . 0.05-0.09 mm
    (0.0020-0.0035 in.)

Model 4.154

Ring groove width:
  Top, second and
    third . . . . 2.44-2.46 mm
    (0.096-0.097 in.)
  Fourth and fifth . . . . . . 4.81-4.84 mm
    (0.1895-0.1905 in.)
Ring side clearance in
groove:
  All rings . . . . 0.05-0.10 mm
    (0.002-0.004 in.)

SMALL DIESEL

Fig. PR2-12—Install laminated oil scraper ring segments as shown. Ends of segments should point as shown when squeezed between thumb and finger.

Piston rings vary according to engine type and application. Be sure correct rings are used. Assemble rings on pistons in the following positions (reading from top of piston):

Models 4.99 and 4.107
1. Plain, parallel faced compression
2. Internally stepped compression
3. Internally stepped compression
4. Spring-loaded, chrome scraper
5. Slotted scraper

Model 4.108 (standard piston)
1. Plain, parallel face compression
2. Internally stepped compression
3. Internally stepped compression
4. Laminated segment oil control*
5. Slotted scraper

Model 4.108 (controlled expansion piston—three rings)
1. Chrome, taper face
2. Internally stepped taper face
3. Spring-loaded scraper

Model 4.108 (controlled expansion piston—four rings)
1. Chrome, parallel face compression
2. Internally stepped compression
3. Internally stepped compression
4. Spring-loaded scraper

Model 4.154
1. Chrome, parallel face compression
2. Internally stepped compression
3. Internally stepped compression
4. Chrome, slotted scraper
5. Slotted scraper

*Model 4.108 rated at 3000 rpm and below uses a slotted scraper oil ring in fourth groove.

One side of taper face and internally stepped (notched) compression rings should be marked "TOP" or "BTM" (bottom) to indicated correct assembly position. Be sure to install accordingly. The slotted scraper rings can be installed either side up. The laminated oil scraper ring must be assembled as shown in Fig. PR2-12. To assemble spring-loaded scra-

SERVICE MANUAL

Perkins

Fig. PR2-13—To assemble spring-loaded scraper ring, install internal expander (1), two bottom rail rings (3), spiral ring (2) and two top rail rings (4) in that order.

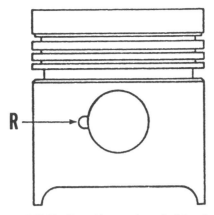

Fig. PR2-14—If marking on top of piston indicating front of piston has been removed by machining the crown, then the side of piston with small recess (R) in pin bore should face front of engine.

Fig. PR2-15—Vent holes (arrows) are provided in cylinder block between upper and lower cylinder sleeve sealing rings to allow coolant which may leak past upper seal ring to escape.

per rings, install internal expander (1–Fig. PR2-13) two bottom rail rings (3), spiral ring (2) and two top rail rings (4) in that order. Be sure rail ring gaps are staggered around piston.

If reinstalling original pistons, be sure they are reinstalled on same connecting rod and in same cylinder bore. On some engines, the piston pin is offset slightly in the piston. Make certain these pistons are installed with the word "FRONT" or arrow on top of piston facing front of engine. If these marks have been removed, position piston so the small recess (R–Fig. PR2-14) in pin bore is towards the front.

If installing new pistons, the specified piston height in relation to top surface of cylinder block must be maintained. Pistons are provided with a machining allowance on piston crown to enable necessary material to be removed to obtain specified piston height. To determine amount of material to be removed, install piston and connecting rod assemblies in their respective bores. With piston at TDC, measure distance from top of piston to top of block and compare reading to the following specifications. After machining is completed, mark piston crown to indicate cylinder number and front of piston.

Model	Piston Height Above Cylinder Block
4.99	0.22-0.30 mm (0.0085-0.012 in.)
4.107	0.22-0.30 mm (0.0085-0.012 in.)
4.108	0.05-0.15 mm (0.002-0.006 in.)
4.154 (before S.N. GA...000398C)	0.00-0.10 mm (0.000-0.004 in.)
4.154 (S.N. GA... 000398C and after)	0.10-0.20 mm (0.004-0.008 in.)

Premachined pistons are available for 4.99 engine in three grades to provide specified piston height in block. The grade letter is stamped on piston crown. Service grade F is used to replace production grades "B, D and F"; service grade "L" replaces production grades "H, J and L", and service grade "P" replaces production grades "N and P". Service grade "F" pistons may also be machined to provide the other two grade heights if desired.

CONNECTING ROD

The connecting rod uses a renewable bushing in the small end. When renewing bushing, use suitable tools to avoid damaging bushing or rod and make sure oil hole in bushing is aligned with oil hole in rod. Bushing must be reamed to final size after installation. Specified bushing inside diameter is 22.24-22.26 mm (0.8757-0.8762 inch) for early 4.99 engines, 23.83-23.84 mm (0.9382-0.9387 inch) for late 4.99 and all 4.107 engines, 27.004-27.005 mm (1.06315-1.0632 inch) for 4.108 engines and 31.76-31.79 mm (1.2505-1.2515 inches) for 4.154 engines. Recommended operating clearance of piston pin in bushing is 0.01-0.03 mm (0.0005-0.0012 inch) for 4.99 and 4.107 engines, 0.0114-0.0178 mm (0.0005-0.0007 inch) for 4.108 engines and 0.01-0.04 mm (0.0005-0.0016 inch) for 4.154 engines.

A precision, insert type bearing is used in the connecting rod big end. Inside diameter of bearing should be 50.838-50.863 mm (2.0015-2.0025 inches) for 4.99, 4.107 and 4.108 engines or 57.16-57.19 mm (2.2504-2.2512 inches) for 4.154 engines. Recommended rod bearing running clearance is 0.036-0.081 mm (0.0014-0.0032 inch) for 4.99, 4.107 and 4.108 engines or 0.04-0.08 mm (0.0016-0.0031 inch) for 4.154 engines.

Connecting rod side clearance on crankshaft should be 0.19-0.27 mm (0.0075-0.0105 inch) for early 4.99 engines, 0.16-0.27 mm (0.0065-0.0105 inch) for late 4.99 and all 4.107 and 4.108 engines and 0.24-0.33 mm (0.0095-0.0131 inch) for 4.154 engines.

When reinstalling connecting rods, be sure side marked "FRONT" is facing front of the engine. Make certain locating tabs of bearing inserts are correctly positioned in machined notches of rod and cap. Lubricate bearing surfaces prior to assembly. Manufacturer recommends renewing connecting rod nuts on 4.154 engine whenever they are removed. Tighten connecting rod cap screws or nuts to torque recommended in TIGTENING TORQUES section.

CYLINDER BLOCK AND LINERS

Models 4.99 and 4.107

The cylinder block is equipped with cast iron, wet type, push fit cylinder liners. The liners are sealed at the bottom by two "O" rings (one "O" ring on early 4.99 engines) located in machined recesses in cylinder block. If cylinder liners are going to be removed but reused, identify liners so they can be reinstalled in original cylinder bore and in same relative position.

Use a suitable tool to remove liners. When reinstalling liners, be sure sealing surfaces are smooth and free of any corrosion. Be sure to renew "O" ring seals. Coat liners and "O" rings with soapy water, then hand push liner into cylinder bore until bottomed. When correctly installed, top of liner flange should be between 0.076 mm (0.003 inch) above to 0.025 mm (0.001 inch) below surface of cylinder block.

The area of the cylinder bore between the sealing rings is vented to outside of cylinder block by small holes as shown in Fig. PR2-15. Vent holes allow any coolant which may leak past the upper

Perkins

SMALL DIESEL

Fig. PR2-16—View of timing gear train used on 4.99, 4.107 and 4.108 engines showing timing marks correctly aligned.

- D. Camshaft gear to hub alignment marks
- P. Injection pump gear to drive hub alignment marks
- T. Gear train timing marks
- 1. Crankshaft gear
- 2. Camshaft gear
- 3. Idler gear
- 4. Injection pump gear

Fig. PR2-17—Cap screw holes in timing idler gear hub are oversized to permit shifting hub position to adjust timing gear backlash on 4.99, 4.107 and 4.108 engines.

Fig. PR2-18—When installing camshaft gear on 4.99, 4.107 and 4.108 engines, make sure to use round mounting holes in gear and that "D" marks are adjacent to each other.

seal ring to escape rather than entering engine oil pan. Make sure vent holes are open when servicing the cylinder liners. Note that slight leakage of coolant in a newly overhauled engine is normal during initial period of operating, but should stop after seal rings seat.

Inside diameter of liners is prefinished to 76.20-76.225 mm (3.00-3.001 inches) for 4.99 engines and 79.374-79.400 mm (3.125-3.126 inches) for 4.107 engines. Renew liners if bore wear exceeds 0.15 mm (0.006 inch).

Models 4.108 and 4.154

These engines are equipped with renewable, cast iron, dry type, interference fit cylinder liners. The liners for 4.108 engines are thin wall type and must be renewed if excessively worn. The liners used in 4.154 engines may be rebored for installation of 0.76 mm (0.030 inch) oversize pistons.

When checking cylinder liner wear, each one should be measured at top, center and bottom at right angle and parallel to centerline of cylinder block thereby giving six readings for each cylinder. Standard inside diameter of liners is 79.375-79.40 mm (3.125-3.126 inches) for 4.108 engines and 88.92-88.95 mm (3.501-3.502 inches) for 4.154 engines. Maximum allowable bore wear for either engine is 0.15 mm (0.006 inch).

To remove cylinder liners, use a suitable press and shouldered plate to push old liners out from the bottom through top of cylinder block. Inspect cylinder block bore for damage. Bore diameter should be 82.525-82.550 mm (3.249-3.250 inches) for 4.108 engines and 96.84-96.86 mm (3.8125-3.8135 inches) for 4.154 engines. On 4.108 engines, cylinder block bore may be bored oversize to remove damage and a 0.25 mm (0.010 inch) oversize outside diameter liner installed. If an oversize liner is installed at the factory, the liner oversize is stamped on top of cylinder block between the liner and edge of block.

To install new liners, lightly lubricate outside of liner with engine oil. When properly installed, top of liner must protrude above surface of cylinder block 0.584-0.686 mm (0.023-0.027 inch) on 4.108 engines. A solid stop of the correct thickness should be positioned on top of cylinder block to limit press travel and establish correct liner protrusion. On 4.154 engines, liner flange seats against counterbore in top of cylinder block. When liner is correctly installed, top of liner collar should protrude 0.66-0.79 mm (0.026-0.031 inch) above top surface of cylinder block. On both engines, cylinder liner inner diameter must be bored and finish honed to final desired dimension after installation.

TIMING GEARS

Models 4.99, 4.107 and 4.108

The timing gear train consists of the crankshaft gear, camshaft gear, injection pump gear as shown in Fig. PR2-16. Due to the odd number of idler gear teeth, timing marks will align only once in 38 crankshaft revolutions, therefore, nonalignment of timing marks does not necessarily mean the gears are mistimed.

NOTE: Before removing timing gears, first remove the rocker arm cover and rocker shaft assembly to prevent pistons from contacting valves if crankshaft is turned while gears are removed.

Idler gear (3) and hub are retained by two cap screws. Idler gear bushing bore diameter when new is 39.687-39.728 mm (1.5625-1.5641 inches). Idler hub diameter is 39.654-39.668 mm (1.5612-1.5619 inches) and desired running clearance is 0.008-0.041 mm (0.0003-0.0016 inch). When reinstalling idler gear, be sure timing marks (T) are properly aligned. Oversize mounting holes (Fig. PR2-17) are provided in idler hub to obtain necessary backlash adjustment by shifting position of hub. Specified backlash between idler and other three gears is 0.038-0.064 mm (0.0015-0.0025

SERVICE MANUAL

Perkins

Fig. PR2-19—When installing injection pump drive gear on 4.99, 4.107 and 4.108 engines, be sure scribe marks (P) are aligned.

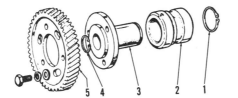

Fig. PR2-20—Exploded view of injection pump drive gear, hub and bushing assembly used on engines equipped with an injection pump with a hydraulic governor.

1. Snap ring
2. Bushing
3. Drive hub
4. Snap ring
5. Gear

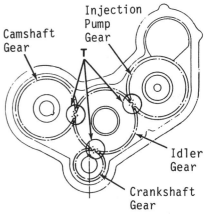

Fig. PR2-21—View of gear train with timing marks (T) correctly aligned for 4.154 engine.

inch). Tighten mounting cap screws to 49 N·m (36 ft.-lbs.). Check idler gear end play using a feeler gage. Recommended end play is 0.076-0.208 mm (0.003-0.008 inch) with a wear limit of 0.25 mm (0.010 inch).

Camshaft gear (2—Fig. PR2-16) is attached to camshaft flange with three cap screws. When reinstalling gear, make certain the stamped "D" markings on gear and hub are aligned and cap screws are installed in round holes (not the oblong holes) as shown in Fig. PR2-18. Tighten the mounting screws to 27 N·m (20 ft.-lbs.).

The injection pump drive gear (4—Fig. PR2-16) is retained to pump drive hub with three cap screws. Attaching holes in gear are slotted as shown in Fig. PR2-19. When reinstalling gear, be sure to align scribe lines (P) on hub and gear, then tighten retaining cap screws to 27 N·m (20 ft.-lbs.). Note that replacement pump drive gear is not marked and must be installed and marked as follows: Turn crankshaft until timing gear marks are aligned and number 1 piston is at TDC on compression stroke. Remove fuel pump gear, then turn crankshaft to position number 1 piston at static timing position as outlined in INJECTION PUMP TIMING section. Turn pump drive hub until scribe line on rotor marked "A" (hydraulic governor pump) is aligned with square end of pump timing snap ring. Install new injection pump gear making sure timing gears and pump shaft do not move. Recheck injection pump static timing and scribe timing line on new gear to coincide with mark on hub if timing is correct. Rotate crankshaft so number 1 piston is at TDC, then stamp two dots on the new gear to coincide with the single dot on idler gear (Fig. PR2-16).

To remove injection pump drive hub (3—Fig. PR2-20), first remove pump drive gear and injection pump. Remove retaining snap ring and withdraw drive hub. Inspect bushing (2) for wear. Inside diameter of bushing should be 33.34-33.78 mm (1.3125-1.3135 inches) and outside diameter of drive hub is 33.287-33.312 mm (1.3105-1.3115 inches). Desired running clearance of hub in bushing is 0.079-0.129 mm (0.0031-0.0051 inch). Bushing is a press fit in cylinder block bore. With drive hub reinstalled, check end play with a feeler gage between front face of bearing and rear face of drive hub. End play should be within the limits of 0.051-0.254 mm (0.0052-0.010 inch).

The crankshaft gear is an interference fit on crankshaft and is located by a key. Use a suitable puller to remove gear.

Model 4.154

The timing gear train consists of the crankshaft gear, camshaft gear, injection pump gear and an idler gear that connects the other three gears as shown in Fig. PR2-21. When timing marks (T) are aligned properly, number 1 piston will be at TDC on compression stroke. However, due to the odd number of teeth on the idler gear, the marks will not align every rotation of crankshaft when number 1 piston is at TDC on compression stroke. Nonalignment of timing marks does not necessarily mean the gears are mistimed.

NOTE: To reinstall timing gears, the rocker arm shaft must be removed so pistons will not contact valves when crankshaft is turned. Since cap screws securing rocker shaft are also securing cylinder head, engine coolant should be drained, injectors should be removed and cap screws should be loosened gradually to prevent distortion.

The idler gear is attached by two nuts and the idler hub is an interference fit in the timing case. Idler gear bushing inner diameter when new is 44.01-44.04 mm (1.7327-1.7337 inches) and idler hub outer diameter is 43.95-43.97 mm (1.7303-1.7313 inches). Desired running clearance of gear on hub is 0.03-0.09 mm (0.0014-0.0034 inch). Recommended idler gear end play is 0.15-0.30 mm (0.006-0.012 inch) with a wear limit of 0.38 mm (0.015 inch). Before reinstalling idler hub, be sure oil holes in cylinder block and hub are open. Install idler gear aligning timing marks (T) as shown in Fig. PR2-21. Renew self-locking retaining nuts and tighten to 28 N·m (21 ft.-lbs.).

The camshaft gear is an interference fit on the camshaft. Use a suitable puller to remove gear. When reinstalling gear, draw the gear onto the shaft using the retaining washer and cap screw. Tighten cap screw to 68 N·m (45 ft.-lbs.).

The injection pump drive gear is retained by three cap screws. On engines equipped wtih an injection pump with a hydraulic governor, gear is properly located by aligning scribe marks on the gear and drive hub. On mechanical governor pumps, gear is doweled to pump shaft. When installing a new pump gear with hydraulic governor pump, the gear is supplied without timing marks. To install and time new gear, turn crankshaft until number 1 piston is at TDC on compression storke. Remove injection pump gear, then turn crankshaft to position number 1 piston at static timing position as outlined in INJECTION PUMP TIMING section. Remove fuel return plate from injection pump and turn pump drive hub until scribe line on rotor marked "A" is aligned with square end of pump gear. Install new injection pump gear making sure timing gears and pump shaft do no move. Recheck injection pump static timing. If timing is correct, rotate crankshaft so number 1 piston is at TDC. Scribe a line on gear to coincide with line on drive hub and stamp two

309

Perkins SMALL DIESEL

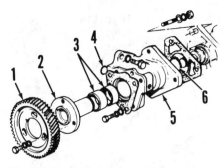

Fig. PR2-22—Exploded view of injection pump drive hub and adapter housing used on 4.154 engines equipped with an injection pump with a hydraulic governor.

1. Drive gear
2. Drive hub
3. Bushings
4. "O" ring
5. Adapter housing
6. Snap ring

dots on gear to coincide with the single dot on idler gear.

Backlash between idler gear and fuel pump gear on hydraulic governor engines is adjusted by loosening the fuel pump adapter housing mounting screws and shifting housing as necessary. Recommended backlash is 0.10-0.20 mm (0.004-0.008 inch). Gear backlash is not adjustable on mechanical governor engines.

To remove pump drive hub (hydraulic governor pumps), first remove pump drive gear and injection pump. Remove snap ring (6–Fig. PR2-22) from rear of drive hub (2), then withdraw hub forward from housing (5). Inspect bushings (3) located in adapter housing for wear and renew if necessary. Inside diameter of bushings when new is 31.75-31.79 mm (1.250-1.2516 inches) and outside diameter of hub is 31.70-31.72 mm (1.248-1.249 inches). Desired running clearance of hub in bushings is 0.02-0.09 mm (0.001-0.0035 inch). With drive hub reinstalled, check end play. Recommended end play is 0.13-0.28 mm (0.005-0.011 inch). Be sure "O" ring (4) is renewed if adapter housing (5) is removed.

The crankshaft gear is a light interference fit on crankshaft and is located by a key. Use a suitable puller to remove gear.

CAMSHAFT

To remove the camshaft, it will be necessary to mount the engine on a suitable engine stand so it can be turned upside down. The purpose of this is to allow the mushroom type cam followers to move away from the cam lobes to provide clearance for camshaft removal. If this cannot be done, then provision should be made to secure the cam followers in their uppermost position through use of suitable retaining clips.

Inspect all parts for scoring, wear and other damage. Renew camshaft if journal wear exceeds 0.05 mm (0.002 inch).

Renew components if they fail to meet the following specified values.

Models 4.99, 4.107 and 4.108

Camshaft journal diameter:
    No. 1 . . . . 45.491-45.517 mm
    (1.791-1.792 in.)
    No. 2 . . . . 45.237-45.263 mm
    (1.781-1.782 in.)
    No. 3 . . . . 45.034-45.060 mm
    (1.773-1.774 in.)
Cylinder block bore diameter:
    No. 1 . . . . 45.568-45.606 mm
    (1.794-1.7955 in.)
    No. 2 . . . . 45.314-45.390 mm
    (1.784-1.787 in.)
    No. 3 . . . . 45.110-45.161 mm
    (1.776-1.778 in.)
Journal clearance:
    No. 1 . . . . 0.051-0.114 mm
    (0.002-0.0045 in.)
    No. 2 . . . . 0.051-0.152 mm
    (0.002-0.006 in.)
    No. 3 . . . . 0.051-0.127 mm
    (0.002-0.005 in.)
Cam follower bore
    diameter . . . . 14.275-14.307 mm
    (0.562-0.563 in.)
Cam follower outer
    diameter . . . . 14.224-14.249 mm
    (0.560-0.561 in.)
Cam follower
    clearance in bore . . . . 0.025-0.082 mm
    (0.001-0.003 in.)
Thrust plate
    thickness . . . . 4.060-4.115 mm
    (0.160-0.162 in.)
Camshaft end play:
    Desired . . . . 0.076-0.228 mm
    (0.003-0.009 in.)
    Maximum . . . . 0.51 mm
    (0.020 in.)

Model 4.154

Camshaft journal diameter:
    No. 1 . . . . 51.91-51.94 mm
    (2.0437-2.0448 in.)
    No. 2 . . . . 51.66-51.69 mm
    (2.0339-2.0350 in.)
    No. 3 . . . . 51.41-51.44 mm
    (2.0241-2.0252 in.)
    No. 4 . . . . 51.16-51.19 mm
    (2.0142-2.0154 in.)
Cylinder block bore
  diameter:
    No. 1 . . . . 52.0-52.03 mm
    (2.0472-2.0484 in.)
    No. 2 . . . . 51.75-51.78 mm
    (2.0374-2.0386 in.)
    No. 3 . . . . 51.5-51.53 mm
    (2.0275-2.0287 in.)
    No. 4 . . . . 51.25-51.28 mm
    (2.0177-2.0189 in.)
Journal clearance:
    All journals . . . . 0.06-0.12 mm
    (0.0024-0.0047 in.)

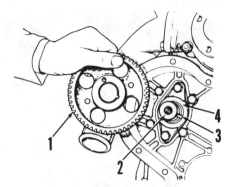

Fig. PR2-23—On 4.154 engines, camshaft gear (1) is keyed to camshaft and retained by a cap screw and washer.

1. Camshaft gear
2. Camshaft
3. Key
4. Thrust plate

Thrust plate thickness . . . 5.92-6.00 mm
    (0.233-0.236 in.)
Camshaft end play:
    Desired . . . . 0.03-0.18 mm
    (0.0012-0.0071 in.)
    Maximum . . . . 0.38 mm
    (0.015 in.)

Thoroughly lubricate cam followers and camshaft with engine oil prior to reassembly. Reinstall cam followers and camshaft in opposite order of removal. Be sure to align timing gear marks.

OIL PUMP

The rotary type oil pump fits into a machined bore in cylinder block and is driven through spiral gears from the camshaft. The oil pressure relief valve is contained within the oil pump end cover (Fig. PR2-30). Normal oil pressure with engine at full speed and normal operating temperature should be 207-414 kPa (30-60 psi).

To remove oil pump assembly, first drain oil and remove oil pan. Disconnect

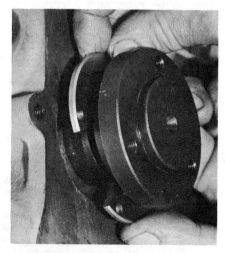

Fig. PR2-24—On 4.99, 4.107 and 4.108 engines, remove thrust washer halves as camshaft is withdrawn.

SERVICE MANUAL

Perkins

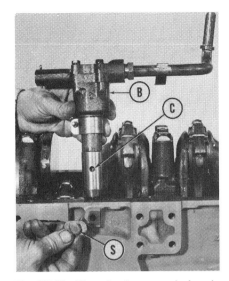

Fig. PR2-25—View showing removal of engine oil pump.

B. Pump body
C. Locating hole
S. Set screw

Fig. PR2-26—Oil pump rotor clearance may be measured with a feeler gage as shown. Refer to text for specified clearance.

Fig. PR2-27—Use a feeler gage to measure clearance between outer rotor and oil pump body. Refer to text.

Fig. PR2-28—End clearance of rotors may be checked using a straightedge and feeler gage as shown.

oil pressure pipe. Remove pump locating cap screw (S–Fig. PR2-25) from side of cylinder block, then withdraw pump assembly.

Oil pump spiral drive gear is a press fit on shaft and is not keyed or pinned. Gear can be removed using a suitable puller or press. Pump rotors, body and cover are available as a unit assembly only. Check rotor clearance as shown in Figs. PR2-26 and PR2-27. Measure end clearance between rotors and machined surface of pump body as shown in Figs. PR2-26 and PR2-27. Measure end clearance between rotors and machined surface of pump body as shown in Fig. PR2-28. Refer to the following table for specified pump clearances.

Fig. PR2-29—When reinstalling oil pump outer rotor, be sure chamfered edge (C) is towards bottom of housing bore.

Fig. PR2-30—Oil pump relief valve is contained within oil pump end cover.
1. Cotter pin
2. Retainer
3. Spring
4. Valve poppet
5. Pump end cover

Models 4.99, 4.107 and 4.108
Inner rotor to outer
 rotor..............0.013-0.063 mm
  (0.0005-0.0025 in.)
Outer rotor to pump
 body................0.28-0.33 mm
  (0.011-0.013 in.)
End clearance:
 Inner rotor.........0.038-0.076 mm
  (0.0015-0.0030 in.)
 Outer rotor.........0.013-0.063 mm
  (0.0005-0.0025 in.)

Model 4.154
Inner rotor to outer
 rotor ....0.05-0.15 mm
  (0.002-0.006 in.)
Outer rotor to pump
 body................0.13-0.25 mm
  (0.0052-0.010 in.)
End clearance
 (both rotors)..........0.02-0.13 mm
  (0.001-0.005 in.)

When reassembling pump, make certain outer rotor is installed with chamfered edge down as shown in Fig. PR2-29. Be sure pump rotates smoothly after assembly. Prime the pump with engine oil before reinstalling. Before starting engine, crank with starter motor with injection pump in "stop" position for about 20 seconds to allow pump and pipes to fill with oil. Start engine and run at low speed until normal oil pressure is obtained.

CRANKSHAFT AND BEARINGS

Models 4.99, 4.107 and 4.108

The crankshaft operates in three precision, insert type main bearings. End thrust is taken by thrust washers located on either side of rear main bearing.

NOTE: On late 4.99 and 4.107 engines, the annular oil groove in the cylinder block main bearing bores has been deleted. Late style main bearing inserts have an annular groove machined in the bearing (inner) face along the centerline of the oil feed hole. The later style bearing inserts may be used in both early and late style engines, but under no circumstances should early style bearings (without grooves) be installed in late style cylinder block (without grooves).

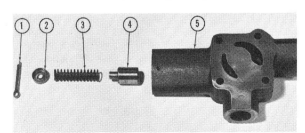

Perkins — SMALL DIESEL

Fig. PR2-31—On some engines, crankshaft pulley has a timing mark which must be aligned with timing mark on end of crankshaft.

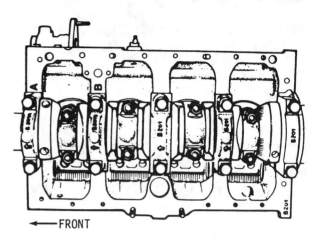

Fig. PR2-32—View of main bearing caps on 4.154 engine. All caps are identified by a serial number that corresponds with serial number stamped on rear bottom surface of cylinder block. An arrow is cast into each cap and should point to front of engine.

Crankshaft main journal standard diameter is 57.099-57.112 mm (2.2480-2.2488 inches) for front and center journals and 57.086-57.099 mm (2.2475-2.2480 inches) for rear journal. If journals are out-of-round in excess of 0.01 mm (0.0005 inch) or wear exceeds 0.03 mm (0.001 inch), crankshaft should be reground or renewed. Recommended main bearing running clearance is 0.051-0.089 mm (0.002-0.0035 inch) for front and center bearings and 0.063-0.102 mm (0.0025-0.004 inch) for rear bearing.

Crankpin standard diameter is 50.78-50.80 mm (1.9993-2.0001 inches). Crankpin wear limit is 0.03 mm (0.001 inch) and out-of-round limit is 0.01 mm (0.0005 inch). Recommended rod bearing running clearance is 0.036-0.081 mm (0.0014-0.0032 inch).

Main journals and crankpins may be reground to 0.25, 0.51 and 0.76 mm (0.010, 0.020 and 0.030 inch) undersize. However, the crankshafts used in most 4.108 and some 4.107 engines are Tufftrided and must be properly rehardened after grinding. If facilities are not available for Tufftriding, the 20-hour nitriding process may be substituted. If neither of these processes can be performed, crankshaft must be renewed. The Tufftrided crankshafts can be identified by their part number stamped on nose or number 1 web. The following part numbers are Tufftrided: 31315741, 31315827, 31315828, 31315829, 31315831, 31315836, 31315838, 31316111, 31316112, 31316113, 31316114, 31316122, 31316128 and 31316121. All other crankshafts used are induction hardened and do not require rehardening after regrinding.

When reinstalling crankshaft, note that main bearing caps are numbered and are not interchangeable. Tighten main bearing cap mounting cap screws to 115 N·m (85 ft.-lbs.). Be sure crankshaft turns freely. Check crankshaft end play using a feeler gage. Recommended end play is 0.051-0.381 mm (0.002-0.015 inch) and maximum allowable end play is 0.51 mm (0.020 inch). Thrust washers are available in 0.19 mm (0.0075 inch) oversize to adjust end play if necessary. Renew front and rear oil seals if necessary. When reinstalling crankshaft pulley on late engines, the pulley is marked with a punch mark (Fig. PR2-31) which must be aligned with scribed line on crankshaft nose. Tighten pulley retaining cap screw as follows: On 4.108 engines with 39.6 mm (1.56 inch) long screw and on all 4.99 and 4.107 engines, tightening torque is 203 N·m (150 ft.-lbs.). On 4.108 engines with cadmium plated (shiny finish) retaining screw, torque is 250 N·m (190 ft.-lbs.). On 4.108 engines using a phosphated (dull black finish) cap screw, torque is 310 N·m (230 ft.-lbs.). Tighten flywheel retaining screws to 81 N·m (60 ft.-lbs.).

Model 4.154

The crankshaft is supported in five precision, insert type main bearings. End thrust is taken by thrust washers located on both sides of center main bearing.

Crankshaft main journal standard diameter is 69.81-69.82 mm (2.7485-2.7488 inches). Recommended main bearing running clearance is 0.06-0.11 mm (0.0025-0.0045 inch). Crankpin standard diameter is 57.10-57.12 mm (2.2485-2.249 inches). Recommended rod bearing runnning clearance is 0.04-0.08 mm (0.0015-0.003 inch). If main journals or crankpins are out-of-round in excess of 0.01 mm (0.0005 inch) or wear exceeds 0.03 mm (0.001 inch), crankshaft should be reground or renewed. Main journals and crankpins may be reground 0.25, 0.51 and 0.76 mm (0.010, 0.020 or 0.030 inch) undersize.

When reinstalling crankshaft, note that main bearing caps are identified in relation to the cylinder block by a serial number stamped on each cap and also bottom face of cylinder block (Fig. PR2-32). An arrow is cast into each cap and must point towards front of engine for correct installation. Number 1 cap (beginning at front) is stamped with an "A", number 2 cap is stamped with a "B". Corresponding letters are stamped on cylinder block. The third cap carries the thrust washers and is easily identified. The number 4 cap has no identification marks, but it is similar to number 1 and 2 caps which are identified, so its position is predetermined. The number 5 cap is the rear main, which is drilled and tapped to hold the bottom half of oil seal retainer, and therefore, should not be mistaken for any of the other caps. Tighten bearing cap mounting screws to 115 N·m (85 ft.-lbs.). Renew front and rear oil seals as outlined in the following CRANKSHAFT OIL SEALS section. Be sure timing mark on crankshaft pulley is aligned with mark on end of crankshaft. Tighten pulley retaining screw to 167 N·m (123 ft.lbs.). Tighten flywheel retaining screws to 108 N·m (80 ft.-lbs.)

CRANKSHAFT OIL SEALS

FRONT SEAL. The front seal is located in the timing gear front cover. On early 4.99, 4.107 and 4.108 engines, a black nitrile oil seal and a crankshaft oil thrower were used. On late engines, a red silicon seal with front dust lip is used and a spacer is used in place of oil thrower. When using red silicone seal to replace early black nitrile seal, the oil thrower must also be removed and replaced with spacer. On all engines, press new seal into cover from the front making sure spring-loaded lip faces inward. On 4.99, 4.107 and 4.108 engines, press

SERVICE MANUAL

seal in until it bottoms against front cover flange. On 4.154 engines, press seal in until front face of seal is 6.35-7.62 mm (0.250-0.300 inch) below front surface of cover as shown at (X–Fig. PR2-33). Be sure to lubricate seal lip and crankshaft with oil before reinstalling.

REAR SEAL. The rear seal consists of rope type seal which fits into grooves of a two-piece retainer. The retainer halves are bolted around rear flange of crankshaft. The rope type seal is precision cut to length and must be installed in retainer halves so 0.25-0.50 mm (0.010-0.020 inch) of the seal projects from each end of the retainer halves. Use a round bar to roll and compress seal into retainer groove (Fig. PR2-34). Lubricate the exposed surface of seal with graphite grease before reinstalling. Apply sealing compound to gasket surfaces, then reinstall retainer halves and install all mounting cap screws finger tight. Tighten retainer clamping bolts to initial torque of 8 N·m (6 ft.-lbs.), then tighten seal retainer mounting screws to 16 N·m (12 ft.-lbs.). Finally, tighten clamping bolts to final torque of 16 N·m (12 ft.-lbs.).

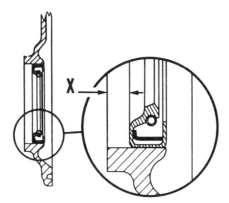

Fig. PR2-33—When renewing front oil seal on 4.154 engines, press seal into timing gear cover until distance (X) from front of cover to front of seal is 6.35-7.62 mm (0.250-0.300 inch).

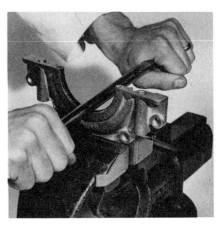

Fig. PR2-34—Use a round bar to compress rope type rear seal into retainer groove when renewing seal.

Peugeot

SMALL DIESEL

PEUGEOT

**PEUGEOT MOTORS OF AMERICA, INC.
INDUSTRIAL ENGINE DIVISION
One Peugeot Plaza
Lyndhurst, New Jersey 07071**

Model	No. Cyls.	Bore	Stroke	Displ.
XDP4.88	4	88 mm (3.465 in.)	80 mm (3.150 in.)	1946 cc (118.7 cu.in.)
XDP4.90	4	90 mm (3.543 in.)	83 mm (3.268 in.)	2112 cc (128.9 cu.in.)
XD2P4.94	4	94 mm (3.700 in.)	83 mm (3.268 in.)	2304 cc (140.6 cu.in.)
XD2PS	4	94 mm (3.700 in.)	83 mm (3.268 in.)	2304 cc (140.6 cu.in.)
XC2PT	4	94 mm (3.700 in.)	83 mm (3.268 in.)	2304 cc (140.6 cu.in.)

Peugeot engines covered in this section are four-stroke, liquid-cooled diesel engines. Models XD2PS and XD2PT are turbocharged. All engines have a cast iron cylinder block and aluminum cylinder head.

Cylinders are numbered 1 through 4 with number 1 cylinder nearest flywheel. Firing order is 1-3-4-2. Crankshaft rotation is counterclockwise at flywheel end.

MAINTENANCE

LUBRICATION

Recommended oil is API classification CD. Oil viscosity should be SAE 30 for ambient temperatures above 15°C (59°F), SAE 20 for ambient temperatures between minus 7°C (20°F) and plus 15°C (59°F), and SAE 10 for ambient temperatures below minus 7°C (20°F).

The engine is equipped with a pressurized oil system. Oil pressure at working speed should be 300 kPa (44 psi) with oil temperature at 80°C (176°F).

ENGINE SPEED ADJUSTMENT

Low idle speed should be 700-750 rpm on Models XDP 4.88 and XDP 4.90, 800-850 rpm on Model XD2P 4.94 and 750-800 rpm on Models XD2PS and XD2PT.

Low idle speed on models equipped with Bosch fuel injection pump is adjusted by turning screw (I-Fig. P1-1).

To adjust low idle speed on engines equipped with a Roto Diesel injection pump and variable speed control, be sure idle speed control is set for slow idle speed (not fast idle) and turn adjuster at pump end of idle control cable. To adjust idle speed on fixed speed engines with Roto Diesel injection pump, turn idle speed screw (I – Fig. P1-2). Adjust throttle stop screw so engine decelerates to idle speed without stalling. Note on fixed speed engines that an electrical device stops engine, however, in case electrical stop device malfunctions, idle lever stop (P) may be rotated so idle speed lever will completely close pump fuel circuit thereby stopping engine.

FUEL SYSTEM

FUEL FILTER. The fuel filter should be drained after every 50 hours of operation or more often to remove contaminants. Filter element should be renewed after every 200 hours of operation.

BLEED FUEL SYSTEM. To bleed fuel system on engines equipped with

Fig. P1-1—Low idle speed on Bosch fuel injection pump is adjusted by turning screw (I).

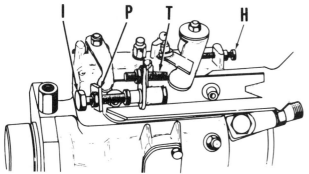

Fig. P1-2—Drawing showing location of idle speed screw (I) and high speed screw (H) on fixed speed engines with Roto Diesel injection pump. Refer to text for adjustment.

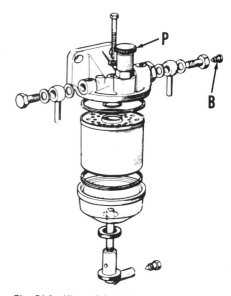

Fig. P1-3—View of fuel filter used on engines equipped wtih Roto Diesel fuel injection pump.

SERVICE MANUAL

Peugeot

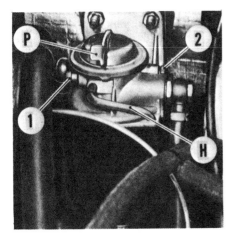

Fig. P1-4—View of fuel filter used on models equipped wtih Bosch fuel injection pump.

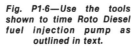

Fig. P1-6—Use the tools shown to time Roto Diesel fuel injection pump as outlined in text.
4. Dial gage
5. Gage holder
6. Timing pin
7. Anti-backlash tool
8. Holder screws
9. Gage mounting screw

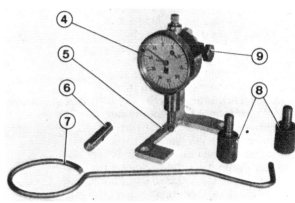

Roto Diesel injection pump, open fuel tank valve and fuel filter bleed screw (B – Fig. P1-3). Operate pump handle (P) until air-free fuel flows from bleed screw hole. Close filter bleed screw. Open fuel injection pump bleed screw and operate filter pump handle until air-free fuel exits pump bleed screw hole. Close pump bleed screw. Loosen fuel injection lines at injectors. With throttle open, rotate crankshaft until air-free fuel flows from injection lines, then reconnect lines.

To bleed fuel system on engines equipped with Bosch injection pump, open fuel tank valve and open bleed screw (1 – Fig. P1-4). Operate filter pump handle (P) until air-free fuel is discharged from hose (H), then close bleed screw (1). Open screw (2) and operate pump handle until resistance is felt, then close screw. Loosen injection pump inlet fitting and allow fuel to flow until air is expelled, then tighten fitting. Loosen fuel injection lines at injectors. With throttle open, rotate crankshaft until air-free fuel flows from injection lines, then reconnect lines.

INJECTION PUMP TIMING. Number 4 piston (number 4 cylinder is nearest timing gear end of engine) must be placed in firing position to check injection timing or install and time injection pump. To determine piston position, remove valve cover without dislodging push rod from cam follower, disconnect exhaust rocker arm of number 4 cylinder from push rod. Rotate rocker arm away from number 4 cylinder exhaust valve. Using a suitable spring compressor, remove exhaust valve spring. Allow exhaust valve to rest on piston, then mount a dial gage so gage contacts valve stem thereby reading piston travel. Be sure valve does not bind in valve guide. Rotate crankshaft so number 4 piston is on compression and at firing position listed in the following table:

Model	Degrees BTDC	Piston Position BTDC
XDP 4.88*	24°	4.29-4.39 mm (0.169-0.173 in.)
XDP 4.90	25°	4.86-4.96 mm (0.191-0.195 in.)
XD2P 4.94	24°	4.49-4.59 mm (0.177-0.180 in.)
XD2PS, XD2PT	...	0.78-0.82 mm (0.031-0.032 in.)

Fig. P1-5—View of Roto Diesel fuel injection pump with inspection plate (1) removed showing timing groove (2) and timing pin guide (3).

*Timing specifications in table are for Model XDP 4.88 engines which operate at speeds in excess of 3000 rpm. For Model XDP 4.88 engines operated at 3000 rpm or less, degrees BTDC are 15° and piston position BTDC is 1.67-1.77 mm (0.066-0.069 inch).

After number 4 piston has been placed in timing position, refer to appropriate following section. Stamp timing marks on engine for future reference.

ROTO DIESEL PUMP. With number 4 piston in firing position as previously outlined, remove pump inspection plate (1 – Fig. P1-5) and note timing groove (2) in pump drive sleeve. If pump is separated from engine, rotate drive gear so groove (2) is aligned with timing pin guide (3). If pump is mounted on engine, groove (2) should be near timing pin guide (3), if not, remove pump and rotate drive gear. Install timing pin (6 – Fig. P1-6) and dial gage as shown in Fig. P1-7. If not installed, mount pump on engine and tighten support plate screws to 20-24 N·m (15-18 lbs.-ft.). Pump retaining nuts must be just loose enough to rotate pump. Install anti-backlash tool (7 – Fig. P1-6) as shown in Fig. P1-7, then rotate pump until maximum dial gage reading is obtained and tighten pump retaining nuts to 15-20 N·m (11-15 ft.-lbs.). Timing marks should be stamped on pump support plate and pump flange for future reference. Remove timing tools and install inspection plate.

BOSCH PUMP. With number 4 piston in firing position as previously outlined,

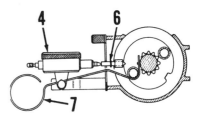

Fig. P1-7—Install timing tools as shown to time Roto Diesel fuel injection pump.

315

Peugeot — SMALL DIESEL

unscrew plug (2–Fig. P1-8) and install a dial gage to measure pump plunger travel. If pump is separated from engine, install pump noting drive gear master spline (1). Tighten pump retaining screws so pump is secure but can still be rotated. If not previously marked, stamp marks on block and flywheel or pulley to indicate number 4 piston timing position. Rotate crankshaft and zero dial gage at bottom dead center of pump plunger travel. Rotate crankshaft and return to number 4 piston timing position. Rotate pump so dial gage indicates 0.29-0.31 mm (0.0114-0.0122 inch) and tighten pump screws to 15-20 N·m (11-15 ft.-lbs.). Timing marks should be stamped on pump support plate and mounting flange for future reference. Remove timing tools.

Fig. P1-8—View of Bosch fuel injection pump. Note drive gear master spline (1).

COOLING SYSTEM

All engines are liquid-cooled and equipped with a water pump to circulate coolant. Due to aluminum construction of cylinder head, an ethylene glycol based antifreeze solution or anticorrosive additive should be used in coolant.

Thermostat opening temperature is 65°C (150°F).

REPAIRS

TIGHTENING TORQUES

Refer to the following table for special tightening torques. All fasteners are metric.

Camshaft retainer.........15-20 N·m
  (11-15 ft.-lbs.)
Connecting rod..........52-61 N·m
  (38-45 ft.-lbs.)
Cylinder head
  With flanged
    injectors.............64-68 N·m
      (47-50 ft.-lbs.)
  With threaded
    injectors:
    XDP4.88, XDP4.90.......59 N·m
      (44 ft.-lbs.)
    XD2P4.94, XD2PS,
      XD2PT................78 N·m
        (58 ft.-lbs.)
Exhaust manifold........20-24 N·m
  (15-18 ft.-lbs.)
Flywheel
  XDP4.88............64-68 N·m
    (47-50 ft.-lbs.)
  All other engines......74-78 N·m
    (55-58 ft.-lbs.)
Injection pump..........15-20 N·m
  (11-15 ft.-lbs.)
Injection pump gear
  Bosch.................5-7 N·m
    (45-60 in.-lbs.)
  Roto Diesel..........22-25 N·m
    (16-19 ft.-lbs.)

Injection pump support
  plate..................20-24 N·m
    (15-18 ft.-lbs.)
Injector
  Flange type...........15-20 N·m
    (11-15 ft.-lbs.)
  Threaded type..........88 N·m
    (65 ft.-lbs.)
Intake manifold..........20-24 N·m
  (15-18 ft.-lbs.)
Main bearing............98-117 N·m
  (72-86 ft.-lbs.)
Oil pan..................8-12 N·m
  (6-9 ft.-lbs.)
Oil pump cover............5-7 N·m
  (45-60 in.-lbs.)
Timing gear cover.........8-12 N·m
  (6-9 ft.-lbs.)
Water pump...............8-12 N·m
  (6-9 ft.-lbs.)
Water pump pulley nut......30-39 N·m
  (22-29 ft.-lbs.)

VALVE ADJUSTMENT

Valve clearance is adjusted with engine cold. During normal engine operation, valve clearance should be checked and adjusted after every 400 hours of operation. Refer to the following table for recommended clearances.

Model	Intake
XDP4.88, XDP4.90	*0.15-0.20 mm (0.006-0.008 in.)
XD2P4.94	0.30-0.35 mm (0.012-0.014 in.)
XD2PS, XD2PT	0.15 mm (0.006 in.)

Model	Exhaust
XDP4.88, XDP4.90	*0.25-0.30 mm (0.010-0.012 in.)
XD2P4.94	0.30-0.35 mm (0.012-0.014 in.)
XD2PS, XD2PT	0.25 mm (0.010 in.)

*After removing and reinstalling cylinder head on engines equipped with flanged injectors, valve clearance should be initially adjusted to 0.25-0.30 mm (0.010-0.012 inch) for intake and 0.35-0.40 mm (0.014-0.016 inch) for exhaust. After 20 to 50 hours of operation, readjust to specifications given in table.

ROCKER ARMS AND SHAFT

R&R AND OVERHAUL. Remove valve cover for access to rocker arm shaft assembly. Unscrew retaining nuts and lift off rocker shaft. Push rods may be removed by carefully separating push rod from cam follower so cam follower does not leave its bore.

Rocker arms and shaft are lubricated by engine oil routed through an external oil line to cylinder head, then through an oil passageway to oil sleeve (4–Fig. P1-9).

Minimum rocker shaft diameter is 18.96 mm (0.746 inch). Rocker arm bushing is renewable.

When assembling rocker shaft assembly, install oil sleeve (4) on shaft so chamfered hole in shaft is aligned with screw hole in sleeve. Place copper washer (6) on oil sleeve locating screw (5) then install and tighten screw to 5-7

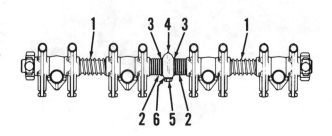

Fig. P1-9—View of rocker shaft assembly on Model XD2P4.94. Other models are similar.
1. Strong spring
2. Weak spring
3. Washer
4. Oil sleeve
5. Locating screw
6. Copper washer

SERVICE MANUAL

Peugeot

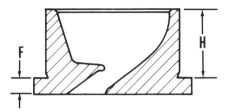

Fig. P1-10—Turbulence chamber flange thickness (F) is 3.975-4.025 mm (0.1565-0.1585 inch) while height (H) is 18.50-18.80 mm (0.728-0.740 inch).

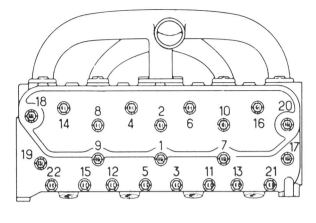

Fig. P1-11—Cylinder head bolt tightening sequence for XDP4.88 and XDP4.90 engines equipped with flanged injectors.

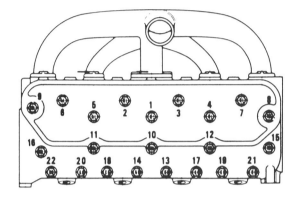

Fig. P1-12—Cylinder head bolt tightening sequence for all engines except XDP4.88 and XDP4.90 engines equipped with flanged injectors.

N·m (45-60 in.-lbs.). Note that washers (3) and weak springs (2) are adjacent to oil sleeve. On models with rocker shaft stands at end of shaft, insert a 0.1 mm (0.004 inch) thick shim between numbers 1 and 4 intake rocker arms and outer shaft stands. Insert push rods, then install rocker shaft assembly being sure rubber gasket is installed on oil sleeve nipple. Tighten rocker shaft retaining nuts to 49 N·m (36 ft.-lbs.), and on models so equipped, tighten outer rocker shaft stand screws to 20 N·m (15 ft.-lbs.). Remove shims from rocker shaft ends and be sure rocker arms rotate freely.

CYLINDER HEAD

R&R AND OVERHAUL. Drain coolant, then detach coolant hoses and water pump belt. If desired, remove water pump. Disconnect fuel injection lines, remove injectors and cap or plug fuel openings to prevent contamination. Disconnect glow plug wire. Remove intake and exhaust manifolds. Detach external oil line from head. Remove rocker shaft and push rods as previously outlined. Unscrew cylinder head screws and remove head.

Cylinder head surface should be machined if warped more than 0.2 mm (0.008 inch) or unevenness exceeds 0.1 mm (0.004 inch). Minimum cylinder head height is 89.35 mm (3.518 inches).

NOTE: Turbulence chambers must be removed from head prior to machining head surface.

Turbulence chamber standout above head surface should be 0.0-0.03 mm (0.0-0.001 inch). To remove turbulence chamber, insert a malleable punch through injector hole and drive out chamber. Flange thickness (F-Fig. P1-10) should be 3.975-4.025 mm (0.1565-0.1585 inch) while height of chamber (H) should be 18.50-18.80 mm (0.728-0.740 inch). Clearance between bottom of bore in head and chamber should be 0.1-0.6 mm (0.004-0.023 inch). Oversize turbulence chambers are available. Be sure retaining ball is installed in groove when installing turbulence chamber.

Valve seats and guides are renewable with oversizes available. Install valve guide so tip of guide is 27.95-28.05 mm (1.100-1.104 inches) from head mating surface. Ream guide to inner diameter of 8.520-8.542 mm (0.335-0.336 inch). Intake valve seat angle is 45° on Models XDP4.88 and XDP4.90 and 30° on all other models. Exhaust valve seat angle is 45° for all models.

Three thicknesses of cylinder head gasket are used. Head gasket with thickness of 1.40 mm (0.055 inch) is used on Models XDP4.88 and XDP4.90. To select gasket thickness on all other models, measure piston height above block surface with piston at TDC. If piston standout is lower than 0.84 mm (0.033 inch) on Model XD2P4.94 or lower than 0.79 mm (0.031 inch) on Models XD2PS or XD2PT, install 1.58 mm (0.062 inch) thick head gasket. If piston standout is higher than 0.84 mm (0.033 inch) on Model XD2P4.94 or higher than 0.79 mm (0.031 inch) on Models XD2PS or XD2PT, install 1.70 mm (0.067 inch) thick head gasket.

Cylinder head and gasket surfaces must be clean and dry. Do not apply any sealing compound to gasket. Refer to appropriate head bolt tightening sequence shown in Fig. P1-11 or P1-12 and tighten bolts as follows: On engines equipped with flanged injectors, tighten head bolts in two steps with a final torque of 64-68 N·m (47-50 ft.-lbs.). Then, loosen each bolt ¼ turn one at a time and retighten to specified torque. Adjust valve clearance as outlined in VALVE ADJUSTMENT paragraph. After the first 1 to 2 hours of operation and again between 20 to 50 hours of operation, cylinder head bolts should be loosened ¼ turn and retorqued (using appropriate tightening sequence). On engines equipped with threaded injectors, tighten head bolts in two steps (following sequence shown in Fig. P1-12) to torque specified in TIGHTENING TORQUES paragraph. Then, loosen and retorque bolts one at a time in proper sequence. Start engine and operate at 3000 rpm for 10 minutes. Stop engine and allow to cool for four hours, then loosen bolts one at a time and retighten to specified torque two times following proper sequence each time.

VALVE SYSTEM

Both valves ride in renewable valve guides and seat in renewable seat inserts. Refer to CYLINDER HEAD section for guide and seat installation.

Intake valve face and seat angles are 45° on Models XDP 4.88 and XDP 4.90 and 30° for all other models. Exhaust valve face and seat angles are 45° for all models. Intake and exhaust valve guide inside diameter is 8.520-8.542 mm (0.335-0.336 inch). Intake valve stem diameter is 8.473-8.495 mm (0.3336-

Peugeot   SMALL DIESEL

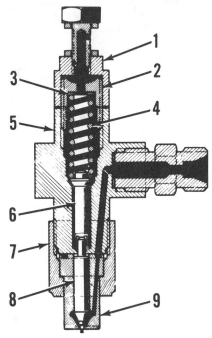

Fig. P1-13—Cross-sectional view of Roto Diesel injector.

1. Cap nut
2. Adjusting nut
3. Washer
4. Spring
5. Injector body
6. Push rod
7. Nozzle nut
8. Nozzle valve
9. Nozzle

0.3344 inch) and exhaust valve stem diameter is 8.453-8.475 mm (0.3328-0.3336 inch). Distance from top of valve to head surface should be 0.75-1.15 mm (0.030-0.045 inch).

Inner valve spring free length is 41 mm (1.614 inches) and outer valve spring free length is 44.6 mm (1.756 inches). Inner valve spring pressure is 181 newtons (41 pounds) at 21.5-22.5 mm (0.846-0.885 inch). Outer valve spring pressure is 451 newtons (101 pounds) at 25.5-26.5 mm (1.004-1.043 inches).

INJECTOR

WARNING: Fuel emerges from injector with sufficient force to penetrate the skin. When testing injector, keep yourself clear of nozzle spray.

Model XDP 4.88, XDP 4.90 and XD2P 4.94 are equipped with flange type Roto-Diesel injectors. Models XD2PS and XD2PT are equipped with screw type Bosch injectors. Refer to appropriate following section for service.

ROTO DIESEL INJECTORS. Prior to removing injector, thoroughly clean injector, lines and surrounding area using compressed air and a suitable solvent.

TESTING. A complete job of testing and adjusting injectors requires use of special test equipment. Only clean, approved testing oil should be used to test injectors. Injector nozzle should be tested for opening pressure, seat leakage and spray pattern.

When operating properly during test, injector nozzle will emit a buzzing sound and cut off quickly with no fluid leakage at seat.

Before conducting test, operate tester lever until test oil flows, then attach injector. Close valve to tester gage and pump tester lever a few quick strokes to be sure nozzle valve is not stuck, which would indicate that injector may be serviceable without disassembly.

Opening pressure is adjusted by turning adjusting nut (2 – Fig. P1-13). Refer to following table for opening pressure on Roto Diesel injection nozzles:

Injector Type	Opening Pressure
RDN OSDC 6577	11000-12000 kPa (1595-1740 psi)
RDN 12 SD 6236	13500-14500 kPa (1960-2100 psi)
RDN 12 SD 6517	12300-13300 kPa (1785-1930 psi)
RDN 45 D 6432	11500-12500 kPa (1670-1815 psi)

The nozzle tip should not leak when injector fluid is pressurized to 2000 kPa (300 psi) below opening pressure. Hold pressure for 10 seconds; if drops appear or nozzle tip becomes wet, valve is not seating and injector must be overhauled.

The injector spray pattern should be well atomized, conical and emerge from nozzle tip in a straight axis. If pattern is wet, ragged or intermittent, nozzle must be overhauled or renewed.

NOTE: Be sure injector tester is operating properly and connections are tight, especially if similar malfunctions are found in a series of injectors.

OVERHAUL. Hard or sharp tools, emery cloth, grinding compound or other than approved solvents or lapping compounds must never be used. An approved nozzle cleaning kit is available through a number of specialized sources.

Wipe all dirt and loose carbon from exterior of nozzle and holder. Refer to Fig. P1-13 for a cross-sectional view of injector.

Secure injector body (5) in a soft jawed vise or holding fixture and disassemble injector, being careful not to drop or damage nozzle valve (8). Place all parts in a clean calibrating oil or diesel fuel as they are removed using a compartmented pan and using extra care to keep parts from each injector together and separate from other units.

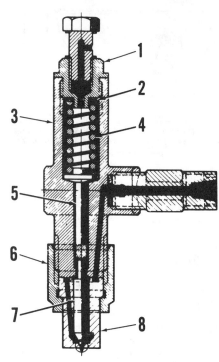

Fig. P1-14—Cross-sectional view of Bosch injector.

1. Cap nut
2. Shim
3. Body
4. Spring
5. Push rod
6. Nozzle nut
7. Nozzle valve
8. Nozzle

Clean exterior surfaces with a brass wire brush, soaking in an approved carbon solvent if necessary, to loosen hard carbon deposits. Rinse parts in clean diesel fuel or calibrating oil immediately after cleaning to neutralize the solvent and prevent etching of polished surfaces. Tighten nozzle nut (7) to 64 N·m (47 ft.-lbs.).

BOSCH INJECTORS. Prior to removing injector, thoroughly clean injector, lines and surrounding area using compressed air and a suitable solvent.

TESTING. The testing procedure for Bosch injectors is same as Roto Diesel injectors; refer to Roto Diesel section while noting the following.

Opening pressure for Bosch injectors should be 11000-12000 kPa (1595-1740 psi) for XDP4.90 and XD2P4.94 engines or 12500-13500 kPa (1815-1960 psi) on XD2PS and XD2PT engines. Opening pressure is adjusted by varying thickness of shim (2 – Fig. P1-14). A change in shim thickness of 0.10 mm (0.004 inch) will change opening pressure approximately 500 kPa (75 psi).

GLOW PLUGS

Each cylinder is equipped with a glow plug. Glow plugs are connected in parallel with each glow plug grounded through mounting threads. Before

SERVICE MANUAL

Peugeot

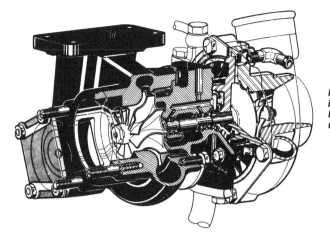

Fig. P1-15—Cutaway drawing of Garret AiResearch Model T03 turbocharger used on Models XD2PS and XD2PT.

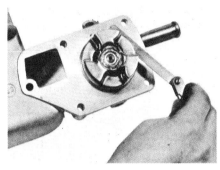

Fig. P1-17—Measure clearance between water pump flange and impeller. Maximum clearance is 1mm (0.040 inch).

suspecting a glow plug malfunction, determine that current is reaching glow plugs. To check individual glow plugs, remove electrical strap between glow plugs. With negative battery terminal grounded to engine, connect a test light between positive battery terminal and glow plug electrode. If test light turns on, then glow plug is good. If test light remains off, then glow plug is defective and must be renewed.

INJECTION PUMP

To remove injection pump, disconnect fuel lines and cap openings to prevent contamination. Disconnect control cables and remove rear pump support. On Models XD2PS and XD2PT, unscrew two Allen screws retaining injection pump and separate pump from engine. On all other models, remove screws securing pump support plate and separate pump with plate from engine.

The injection pump should be tested and overhauled by a shop qualified in diesel injection pump repair.

If removed, install gear on gear driven models or splined hub on chain driven models. Refer to INJECTION PUMP TIMING section and install pump. Refer to ENGINE SPEED ADJUSTMENT section if idle or high engine speed requires adjustment.

TURBOCHARGER

Models XD2PS and XD2PT

A Garret AiResearch Model T03 turbocharger is used on Models XD2PS and XD2PT. Refer to Fig. P1-15 for a cutaway drawing of turbocharger. Oil to lubricate and cool turbocharger bearings is provided by pressurized engine oil routed through an external oil line. Return oil is directed to a fitting in side of engine block.

The turbocharger is equipped with an integral wastegate which regulates turbocharger at a maximum pressure of 60 kPa (8.7 psi). The wastegate valve (V-Fig. P1-16) is seated in turbine housing by spring (S) with the valve stem attached to diaphragm (D). When pressure in compressor housing (C) reaches 60 kPa (8.7 psi), diaphragm (D) overcomes spring pressure and moves the valve off its seat. Exhaust gases bypass the turbocharger turbine thereby reducing turbine and compressor speed which results in reduced pressure in compressor housing. Spring force seats the wastegate valve (V) and the cycle repeats.

REMOVE AND REINSTALL. Removal of unit is apparent with inspection. Turbocharger must be serviced as a unit assembly as overhaul is accomplished at factory. When installing turbocharger, oil cavity of center housing should be filled with oil and oil line from engine should be primed with oil. This procedure is necessary to prevent dry starting of turbocharger and possible early failure of bearings.

WATER PUMP

R&R AND OVERHAUL. To remove water pump, drain coolant, disconnect hoses and remove drive belt. Unbolt and remove water pump. Using a suitable puller, pull impeller off shaft. Disassemble remainder of pump.

When assembling water pump, press impeller on shaft so there is a maximum clearance of 1 mm (0.040 inch) between impeller and pump flange (Fig. P1-17).

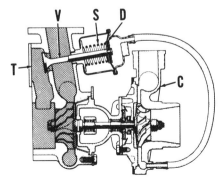

Fig. P1-16—Cross-sectional view of turbocharger and wastegate used on Models XD2PS and XD2PT.

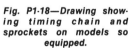

Fig. P1-18—Drawing showing timing chain and sprockets on models so equipped.
1. Rubbing block
2. Injection pump sprocket
3. Idler sprocket
4. Crankshaft sprocket
5. Tensioner
6. Tensioner shoe
7. Camshaft sprocket
C. Copper plated link
M. Timing marks
S. Adjusting screw

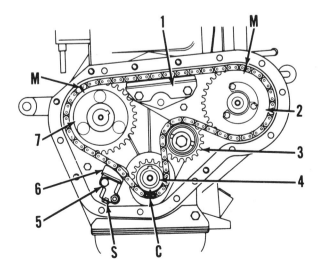

Peugeot

SMALL DIESEL

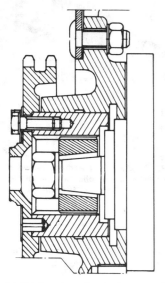

Fig. P1-19—Cross-sectional view of injection pump sprocket carrier on models equipped with a timing chain.

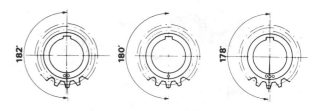

Fig. P1-20—View of crankshaft sprocket timing marks on factory installed sprockets.

TIMING CHAIN AND SPROCKETS

All Models So Equipped

CHAIN TENSION ADJUSTMENT. Chain slack is removed during engine operation by a hydraulically actuated chain tensioner (5–Fig. P1-18) which uses pressurized engine oil to force chain tensioner shoe (6) against the timing chain.

To adjust chain tensioner, loosen idler (3) retaining nut. Turn idler eccentric counterclockwise so gap between chain tensioner body (5) and shoe (6) is 0.5-1.0 mm (0.020-0.040 inch). Tighten idler retaining nut to 49 N·m (36 ft.-lbs.). Turn tensioner adjusting screw (S) so shoe (6) bounces against chain. Loosen rubbing block (1) nuts and place a straightedge on chain from camshaft to injection pump sprockets to back up chain. Press rubbing block (1) against chain and tighten nuts.

REMOVE AND REINSTALL. Timing chain and sprockets are accessible after removing chain cover. The injection pump must be removed if chain and sprockets are to be removed. Loosen idler sprocket retaining nut and rotate idler eccentric to loosen chain. Unscrew injection pump sprocket (2–Fig. P1-18), then remove sprocket while disengaging chain. If required, remove camshaft, crankshaft and idler sprockets.

Refer to Fig. P1-19 for a cross-sectional view of injection pump sprocket carrier. If sprocket or carrier assembly is renewed, check clearance between sprocket hub and timing chain cover with gasket installed. Clearance should be 0.06-0.94 mm (0.002-0.037 inch).

The camshaft sprocket must be heated to 250°C (482°F) prior to installation on camshaft end. There must be 0.05-0.15 mm (0.002-0.006 inch) clearance between camshaft sprocket hub and camshaft retainer.

Note in Fig. P1-20 the three timing mark configurations used on the crankshaft sprocket of new engines. Replacement crankshaft sprockets have only one timing mark.

To install timing chain, wrap chain around injection pump sprocket so marks (M–Fig. P1-18) on chain and sprocket are aligned. Engage chain with camshaft, crankshaft and idler sprockets so copper plated link (C) is opposite crankshaft sprocket timing mark and marks (M) on chain and camshaft sprocket are aligned. Tighten injection pump sprocket screws to 5-7 N·m (45-60 in.-lbs.) on Bosch pump or 22-25 N·m (16-19 ft.-lbs.) on Roto Diesel pump. Refer to CHAIN TENSION ADJUSTMENT section and adjust timing chain tension as outlined. Install injection pump and timing chain cover. Tighten timing cover screws and nuts to 8-12 N·m (6-9 ft.-lbs.).

TIMING GEARS

All Models So Equipped

Crankshaft, camshaft and idler gears are accessible after removing timing gear cover. Remove injection pump to service injection pump drive gear.

To install camshaft gear, heat gear to 250°C (482°F) prior to installation on camshaft end. There must be 0.05-0.15 mm (0.002-0.006 inch) clearance between camshaft retainer and gear hub.

End play of idler gear on idler shaft should be 0.05-0.35 mm (0.002-0.014 inch).

Timing gears are stamped with timing marks which must be aligned during assembly. Tighten timing cover screws and nuts to 8-12 N·m (6-9 ft.-lbs.). Install injection pump as previously outlined.

OIL PUMP

R&R AND OVERHAUL. To remove oil pump, remove oil pan, then unscrew flange screw (F–Fig. P1-21) and taper screw (R). Oil pump is serviced as a unit assembly.

To install pump, insert pump into cylinder block so taper hole in body is aligned with taper screw (R) and install screw with copper washer. Tighten screw to 20-24 N·m (15-18 ft.-lbs.). Install flange screw (F) without a gasket and lightly tighten. Measure gap between flange screw (F) and boss as shown in Fig. P1-21. Select and install a shim 0.05-0.10 mm (0.002-0.004 inch) thicker than measured gap. Tighten flange screw to 79-98 N·m (58-72 ft.-lbs.).

PISTON AND ROD UNITS

REMOVE AND REINSTALL. Piston and connecting rod may be removed after removing cylinder head and oil pump. Unscrew rod cap retaining nuts, remove rod cap and extract piston and rod.

Connecting rod marks and indentation in piston crown must be on same side. Install piston and rod so piston crown indentation is towards injection pump side of engine. Tighten rod nuts to 52-61 N·m (38-45 ft.-lbs.).

PISTON AND RINGS

Model XDP4.88 and XDP4.90 pistons are equipped with three compression rings and an oil control ring (Fig. P1-22). Model XD2P4.94, XD2PS and XD2PT pistons are equipped with two compression rings and an oil control ring (Fig. P1-23).

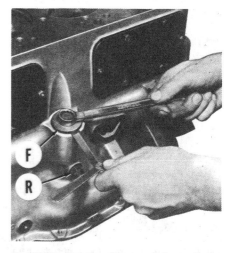

Fig. P1-21—Measure clearance between flange screw (F) and boss as outlined in text. Taper screw (R) holds oil pump in block.

SERVICE MANUAL

Pistons of Models XCP4.88 and XDP 4.90 are classed "A" or "B" according to piston size. Piston size should be matched with cylinder liner. Refer to CYLINDER LINER section. Piston class letter is stamped on piston crown. Class A piston diameter is 87.700-87.730 mm (3.4528-3.4539 inches) for Model XDP4.88 and 89.705-89.735 mm (3.5317-3.5329 inches) for Model XDP4.90. Class B piston diameter is 87.720-87.750 mm (3.4535-3.4547 inches for Model XDP4.88 and 89.725-89.755 mm (3.5325-3.5336 inches for Model XDP4.90. Piston diameter is measured perpendicular to piston pin, 74 mm (2-15/16 inches) froom bottom of piston shirt. Piston clearance should be 0.27-0.32 mm (0.0126 inch) for Model XDP4.88 or 0.295-0.315 mm (0.0116-0.0124 inch) for Model XDP4.90 using piston diameter previously listed.

Pistons of Models XD2P4.94, XD2PS and XD2PT are classed "A" through "H" according to piston size. Piston class letter is stamped on piston crown. Class A pistons are standard size while class C, E and G pistons are major oversize pistons. Class B, D, F and H pistons are slightly larger than standard or oversize pistons. For instance, a class C piston is 0.200 mm (0.0079 inch) oversize and a class D piston is 0.215 mm (0.0084 inch) larger than standard. Where possible, the slightly larger piston may be installed without boring to the next oversize. Measure piston diameter perpendicular to piston pin., 76.5 mm (3 inches) from bottom of piston skirt on Model XD2P4.94 or 73.4 mm (2⅞ inches) from bottom of piston skirt on Models XD2PS and XD2PT. Piston clearance should be 0.34-0.39 mm (0.0134-0.0154 inch) on Model XD2P4.94 or 0.37-0.42 mm (0.0145-0.0165 inch) on Models XD2PS and XD2PT.

Top piston ring end gap should be 0.30-0.45 mm (0.012-0.018 inch) on Model XDP4.88, 0.38-0.63 mm (0.015-0.025 inch) on Model XDP 4.90, 0.40-0.65 mm (0.016-0.026 inch) on Model XD2P4.94 and 0.25-0.45 mm (0.010-0.018 inch) on Models XD2PS and XD2PT. Piston ring end gap for middle compression rings should be 0.30-0.45 mm (0.012-0.018 inch) on Model XDP 4.88, 0.38-0.63 mm (0.015-0.025 inch) on Model XDP4.90, 0.35-0.60 mm (0.014-0.024 inch) on Model XD2P4.94 and 0.25-0.45 mm (0.010-0.018 inch) on Models XD2PS and XD2PT. Oil control ring end gap should be 0.15-0.30 mm (0.006-0.012 inch) on Models XD2P4.94, XD2PS and XD2PT. Oil control ring on Models XDP4.88 and XDP4.90 must not be altered.

When installing rings, top compression ring (1 – Fig. P1-22 or P1-23) is a barrel face ring and may be installed

Peugeot

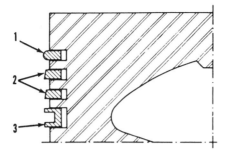

Fig. P1-22—Cross-sectional view showing proper installation of piston rings on XDP4.88 and XDP4.90 engines. Center compression rings (2) have one side marked "TOP" and must be installed accordingly.

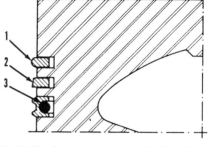

Fig. P1-23—Cross-sectional view of piston rings used on all engines except XDP4.88 and XDP4.90. Note "TOP" marking on middle compression ring (2) and oil control ring (3) when installing rings.

either side up. Intermediate compression rings (2) have one side marked "TOP" and must be installed accordingly. U–FLEX type oil control ring (3 – Fig. P1-22) may be installed either side up, but GOETZE expander type oil ring (3 – Fig. P1-23) must be installed with side marked "TOP" facing upward.

CYLINDER LINER

Models XDP4.88 and XDP4.90

Models XDP4.88 and XDP4.90 are equipped with wet type cylinder liners. Cylinder liner lower end is sealed by a rubber ring. Cylinder liner standout above block surface should be 0.025-0.085 mm (0.001-0.003 inch).

Cylinder liners are classed acccording to bore size to match with pistons. Refer to PISTON AND RINGS section. The liner is marked on the outer edge at bottom of bore with one or two notches. Class "A" cylinder liners have one notch and class "B" cylinder liners have two notches. Cylinder liner bore diameter of class A liners is 88.00-88.02 mm (3.4646-3.4653 inches) for Model XDP 4.88 or 90.00-90.02 mm (3.5433-3.5441 inches) for Model XDP4.90. Cylinder liner bore diameter of class B liners is 88.02-88.04 mm (3.4654-3.4661 inches) for Model XDP 4.88 or 90.02-90.04 mm (3.5441-3.5449 inches) for Model XDP 4.90.

PISTON PIN

A full floating piston pin is used in all models. Piston pin diameter is 30.000-30.006 mm (1.1811-1.1813 inches) for Models XD2PS and XD2PT or 28.000-28.006 mm (1.1024-1.1026 inches) for all other models.

CONNECTING ROD AND BEARINGS

Connecting rods are equipped with a renewable bushing in the small end and insert type bearings in the big end. Bushing inner diameter is 30.007-30.020 mm (1.1814-1.1819 inches) on Models XD2PS and XD2PT or 28.007-28.020 mm (1.1026-1.1031 inches) on all other models. Clearance between big end bearing and crankpin should be 0.040-0.092 mm (0.0016-0.0036 inch) on all models. Rod side play should be 0.010-0.025 mm (0.004-0.010 inch).

Connecting rods are graded according to weight and stamped with a number (1 through 5) or letter (A through H). Renew rod with one of same weight grade.

CAMSHAFT AND TAPPETS

R&R AND OVERHAUL. To remove camshaft, remove valve cover and push rods. Remove tappet cover and extract tappets. Remove oil pan and oil pump. Detach timing chain or gear cover, and on models so equipped, remove timing chain. Unscrew camshaft retainer plate screws and withdraw camshaft.

Camshaft journal diameter is 41.925-41.950 mm (1.6506-1.6516 inches). Camshaft bearing clearance should be 0.05-0.11 mm (0.002-0.004 inch). Standard cam follower diameter is 23.95-23.96 mm (0.9429-0.9433 inch). Oversize cam follower with a diameter of 24.15-24.16 mm (0.9508-0.9512 inch) is available. Cam follower should be 0.04-0.08 mm (0.0016-0.0031 inch).

To install camshaft, reverse removal procedure. Be sure camshaft journals and lobes are adequately lubricated. Refer to TIMING GEARS AND SPROCKETS or TIMING GEARS section to install camshaft sprocket or gear. Be sure camshaft journals and lobes are adequately lubricated. Align gear timing marks on models so equipped. Tighten camshaft retainer plate screws to 15-20 N·m (11-15 ft.-lbs.). On models equipped with a timing chain, refer to TIMING CHAIN AND SPROCKETS section and install chain. Refer to OIL PUMP section for pump installation.

321

Peugeot

SMALL DIESEL

CRANKSHAFT AND BEARINGS

R&R AND OVERHAUL. To remove crankshaft, remove flywheel and timing gear or chain cover. Remove timing chain on models so equipped. Remove pistons and rods. The crankshaft can be removed after detaching main bearing caps.

Standard main journal diameter is 54.994-55.021 mm (2.1651-2.1661 inches). Main bearing clearance should be 0.040-0.098 mm (0.0016-0.0038 inch) for all journals. Undersize main bearings are available.

Crankshaft end play is controlled by thrust washers attached to center main bearing. Crankshaft end play should be 0.08-0.29 mm (0.003-0.011 inch). Thrust washers are available in thicknesses of 2.30-2.33 mm (0.090-0.092 inch) and 2.50-2.53 mm (0.098-0.099 inch). Install thrust washers with smooth side towards main bearings.

Crankpin standard journal diameter is 54.994-55.021 mm (2.1651-2.1661 inches) on Models XD2PS and XD2PT, and 49.984-50.011 mm (1.9679-1.9689 inches) on all other models. Rod bearing undersizes are available.

To install crankshaft, reverse removal procedure. Tighten main bearing cap screws to 98-117 N·m (72-86 ft.-lbs.). Rolling torque necessary to turn crankshaft with pistons installed should not exceed 59 N·m (44 ft.-lbs.).

SLANZI

SLANZI OF NORTH AMERICA, INC.
P.O. Box 885
Minneapolis, MN 55440

Model	No. Cyls.	Bore	Stroke	Displ.
DVA515	1	86 mm (3.386 in.)	88 mm (3.465 in.)	511 cc (31.2 cu. in.)
DVA1030	2	86 mm (3.386 in.)	88 mm (3.465 in.)	1022 cc (62.4 cu. in.)

These engines are vertical cylinder, four-stroke, air-cooled direct injection diesels. Crankshaft rotation is counterclockwise at flywheel end.

MAINTENANCE

LUBRICATION

Recommended engine oil is SAE 20W-20 for temperatures below 0°C (32°F). When ambient temperatures will be above 0°C (32°F), use SAE 30 for winter operation or SAE 40 for summer operation. Only premium grade, heavy-duty, diesel engine oil should be used.

Under normal operating conditions, manufacturer recommends changing engine oil after every 200 hours of operation. When operating under severe duty conditions, such as high temperature and heavy load, oil should be changed after every 100 hours of operation. Oil should be drained while the engine is hot. Manufacturer recommends renewing the oil filter element after every 400 hours of operation. Crankcase capacity for DVA515 is 1.6 liters (1.75 quarts) without filter change and 1.75 liters (1.85 quarts) with filter change. Capacity for DVA1030 is 3.4 liters (3.6 quarts) without filter change and 3.55 liters (3.75 quarts) with filter change.

FUEL SYSTEM

FUEL FILTER AND BLEEDING. All engines are equipped with a renewable fuel filter. On DVA515 engines, filter is located inside the fuel tank behind the fuel outlet cover. On DVA1030 engines, fuel filter element is externally mounted on flywheel end of engine.

Manufacturer recommends renewing fuel filter element after every 200 hours of operation or sooner if engine indicates signs of fuel starvation (surging or loss of power). After renewing fuel filter, air must be bled from fuel system as follows:

To bleed air from fuel system, loosen bleeder screw (3–Fig. S1-1) at fuel injection pump. On DVA515 engine, allow gravity flow to force air out bleeder. Tighten bleed screw when fuel flow is free of air. On DVA1030, manually operate fuel transfer pump (2) until air-free fuel flows from bleeder valve, then retighten bleed screw.

If engine fails to start at this point, loosen high pressure fuel line (6) fitting at the injector. Set throttle at full speed position, then crank engine until fuel is discharged from loosened fuel line. Tighten fitting and start engine.

ENGINE SPEED ADJUSTMENT

The engine should be at normal operating temperature before adjusting engine speeds. Low and high idle (no-load) speeds are adjusted using throttle lever stop screws (1 and 2–Fig. S1-2) located on governor cover. Turning screw (1) clockwise will increase low idle speed and turning screw (2) counterclockwise will increase maximum

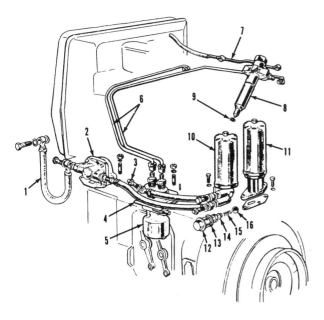

Fig. S1-1—View of fuel system, oil filter and oil pressure relief valve assembly for DVA1030. On DVA515, fuel filter is located in fuel tank and fuel transfer pump (2) is not used.

1. Fuel inlet line
2. Fuel transfer pump
3. Bleed screw
4. Shim gasket
5. Injection pump
6. High pressure injector lines
7. Fuel return line
8. Injector
9. Nozzle gasket
10. Fuel filter
11. Oil filter
12. Plug
13. Gasket
14. Oil pressure adjusting screw
15. Spring
16. Relief valve ball

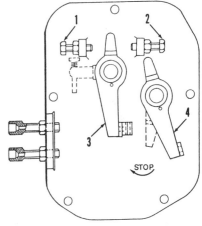

Fig. S1-2—Low and high speed adjusting screws are located on the governor cover.

1. Low speed screw
2. High speed screw
3. Throttle lever
4. Stop lever

Slanzi   SMALL DIESEL

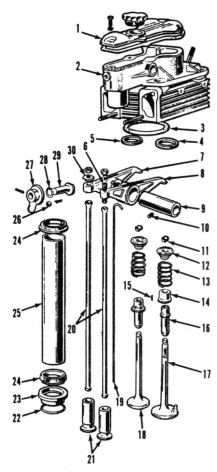

Fig. S1-3—Exploded view of cylinder head and valve system. Compression release mechanism is used on DVA515 engine.

1. Valve cover
2. Cylinder head
3. Gasket
4. Valve seat insert (intake)
5. Valve seat insert (exhaust)
6. Adjusting screw
7. Rocker arm (exhaust)
8. Rocker arm (intake)
9. Rocker arm shaft
10. Set screw
11. Valve keepers
12. Retainer
13. Valve spring
14. Cap nut (intake only)
15. Locating pin
16. Valve guide
17. Intake valve
18. Exhaust valve
19. Oil tube
20. Push rods
21. Cam followers
22. "O" ring
23. Bushing
24. Seals
25. Push rod tube
26. Detent ball
27. Compression release lever
28. "O" ring
29. Compression release shaft
30. Plate

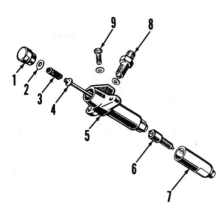

Fig. S1-4—Exploded view of fuel injector assembly.

1. Cap nut
2. Shim
3. Spring
4. Push rod
5. Injector body
6. Nozzle valve assy.
7. Nozzle holder nut
8. Fuel inlet fitting
9. Leak-off banjo bolt

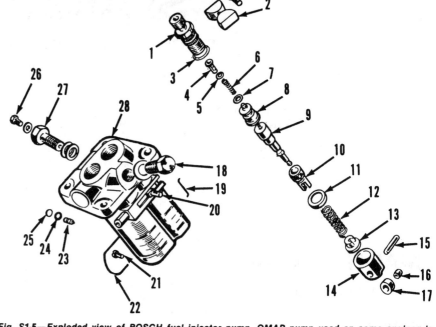

Fig. S1-5—Exploded view of BOSCH fuel injector pump. OMAP pump used on some engines is similar.

1. Delivery valve holder
2. Locks
3. Gasket
4. Spring guide
5. Shim
6. Spring
7. Gasket
8. Delivery valve
9. Pumping element
10. Sleeve
11. Retainer
12. Spring
13. Retainer
14. Tappet
15. Pin
16. Inner roller
17. Outer roller
18. Plug
19. Retainer
20. Control rack pin
21. Pin
22. Clip
23. Pin
24. Gasket
25. Plug
26. Bleed screw
27. Fuel inlet banjo bolt
28. Fuel pump body

governed speed. Adjust screws as needed to obtain specified governed speed as required for engine application. Do not exceeed maximum governed limit of 3200 rpm.

INJECTION PUMP TIMING

The injection pump is actuated by the engine camshaft. Injection timing is adjusted using shim gaskets (4 – Fig. S1-1) between pump body and mounting surface of crankcase. Engines may be equipped with BOSCH or OMAP injection pump. Injection timing is the same for either pump.

To check injection timing, proceed as follows: Rotate crankshaft until "PM" mark on air shroud (13 – Fig. S1-6) is aligned with "PMS" mark on flywheel cover (12) which indicates number 1 piston is at top dead center on compression stroke.

NOTE: It is possible to mount flywheel cover in four different positions, but only one can be used for timing purposes. If timing mark accuracy is in doubt, remove valve spring from a valve for number 1 cylinder. Allow valve to rest on top of piston, then use a dial indicator on valve stem end while turning crankshaft to double check TDC markings.

Disconnect injector line for number 1 cylinder from injection pump. Connect a fuel flow sight gage (special tool number 59.40.50) to injection pump delivery valve holder. If tool is not available, remove delivery valve holder (1 – Fig. S1-5), spring (6) and valve (8), then reinstall delivery valve holder. Remove governor cover and position pump control rack at half speed setting. Maintain this setting during timing check. Connect fuel supply to pump. Note that fuel should flow from delivery valve holder at this time if delivery valve is removed. Rotate crankshaft counterclockwise approximately ½ turn, then turn slowly clockwise (normal direction of rotation) until first drop of fuel is observed at sight gage tool, or until fuel flow just stops if delivery valve is removed. At this point, beginning of injection occurs and "A1" mark on flywheel cover should be aligned with "PM" mark on air shroud. If timing marks are not aligned, increase pump mounting gasket thickness to retard timing or reduce shim gasket thickness to advance timing.

REPAIRS

TIGHTENING TORQUES

Refer to the following table for special tightening torques. All fasteners are metric.

Camshaft . . . . 147 N·m
(108 ft.-lbs.)

SERVICE MANUAL

Slanzi

Connecting rod........... 34-44 N·m
(25-32 ft.-lbs.)
Cylinder head............ 58-69 N·m
(43-51 ft.-lbs.)
Flywheel................ 216-255 N·m
(160-195 ft.-lbs.)
Main bearing support –
DVA1030................. 49 N·m
(36 ft.-lbs.)
Nozzle holder nuts.......... 20 N·m
(15 ft.-lbs.)

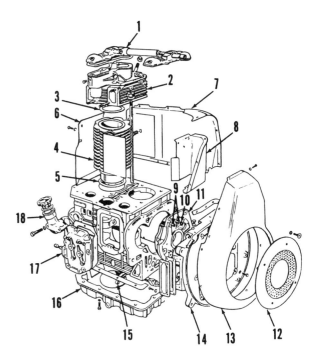

Fig. S1-6—Exploded view of DVA1030 crankcase assembly.

1. Valve cover
2. Cylinder head
3. Head gasket
4. Cylinder
5. Shim gasket
6. Air shield
7. Air shroud
8. Air shield
9. Oil tubes
10. Gasket
11. Main bearing support
12. Flywheel fan cover
13. Flywheel shroud
14. Crankcase end cover
15. Oil pickup
16. Oil pan
17. Governor cover
18. Oil filler

VALVE ADJUSTMENT

To adjust valve clearance, first remove valve cover and rotate crankshaft until piston is at TDC on compression stroke (both rocker arms should be free to move). With engine cold, valve clearance should be 0.15 mm (0.006 inch) for intake and exhaust. Turn adjusting screw (6 – Fig. S1-3) until appropriate size feeler gage can be inserted between rocker arm and valve stem end. Tighten adjusting screw locknut and recheck adjustment.

After adjusting valves on first cylinder of DVA1030 engine, turn crankshaft one complete revolution and adjust valves on second cylinder.

CYLINDER HEAD AND VALVE SYSTEM

The cylinder head is equipped with renewable valve guides and valve seat inserts for intake and exhaust. Valve stem diameter is 7.963-7.972 mm (0.3135-0.3138 inch) for intake and 7.961-7.970 mm (0.3134-0.3138 inch) for exhaust. Inside diameter of valve guides should be 7.975-8.011 mm (0.3140-0.3145 inch). Desired valve stem clearance in guide is 0.003-0.048 mm (0.0001-0.0019 inch) for intake and 0.005-0.050 mm (0.0002-0.0020 inch) for exhaust. Use a suitable installing tool when pressing new guides into cylinder head to prevent damage to guide. Press in from top of head until guide shoulder bottoms. Guides are presized and should not require reaming if carefully installed.

Valve seat inserts should have an interference fit in cylinder head counterbore of 0.125-0.170 mm (0.005-0.007 inch). Chilling insert in dry ice prior to installation will ease assembly. When properly installed, valve seat should be recessed 0.15-0.30 mm (0.006-0.012 inch) below cylinder head surface. With valves installed, measure distance valve heads project from cylinder head surface. Valve protrusion should be 1.10-1.50 mm (0.043-0.059 inch). If protrusion is excessive, regrind valve seat. If protrusion is less than specified, renew valve seat.

Injector nozzle tip should extend 3.3-3.7 mm (0.130-0.146 inch) beyond cylinder head surface. Adjust nozzle position, if necessary, using shim gaskets between nozzle valve and bottom of cylinder head nozzle bore.

When reassembling cylinder head, be sure pin (15 – Fig. S1-3) is in place to locate cap nut (14) on intake valve. Be sure lubricating oil tube (19) is properly connected. Renew push rod tube seals (24) if necessary. Tighten cylinder head stud nuts evenly using a crossing pattern to 58-69 N·m (43-51 ft.-lbs.). Adjust valve clearance as previously outlined.

INJECTOR

REMOVE AND REINSTALL. Engines may be equipped with BOSCH or OMAP injector. Service procedures are identical for either injector. To remove injector, first clean injector, fuel lines and surrounding area of cylinder head. Disconnect return line and high pressure line and plug all openings. Unscrew retaining nuts and withdraw injector from cylinder head. Do not lose shim gasket between injector and cylinder head.

To reinstall injector, reverse the removal procedure. Tighten retaining nuts to 20 N·m (15 ft.-lbs.).

TESTING. A suitable test stand is required to check injector operation. Only clean, approved testing oil should be used to test injector.

WARNING: Fuel leaves the injector nozzle with sufficient force to penetrate the skin. When testing, keep yourself clear of nozzle spray.

Connect injector to tester and operate tester lever a few quick strokes to purge air from injector and to make sure nozzle valve is not stuck. When operating properly, injector nozzle will emit a buzzing sound and produce a uniform spray pattern.

Slowly operate tester lever while observing pressure gage reading. Opening pressure should be 19615-20595 kPa (2845-2985 psi). Opening pressure is adjusted by varying the thickness of shim (2 – Fig. S1-4).

Operate tester lever slowly to maintain pressure at 2000 kPa (300 psi) below opening pressure and check for leakage past nozzle valve. If a drop forms at nozzle tip within a 10 second period, nozzle valve is not seating and injector must be overhauled or renewed.

OVERHAUL. To disassemble injector, secure nozzle body (5 – Fig. S1-4) in a vise with nozzle tip facing upward. Remove nozzle holder nut (7) and withdraw nozzle valve (6). Invert nozzle body and remove plug (1), shim (2), spring (3) and rod (4).

Thoroughly clean all parts in a suitable solvent. Do not use steel wire brush or sharp metal tools to clean injector components. Make certain the nozzle valve needle slides freely in nozzle valve body. If needle sticks, reclean or renew nozzle valve assembly.

When reassembling injector, be sure all parts are clean and wet with diesel fuel. Recheck injector operation as outlined in TESTING section.

INJECTION PUMP

Engines may be equipped with a BOSCH or OMAP fuel injection pump.

Illustrations Courtesy North American Diesel, Inc.

Slanzi

SMALL DIESEL

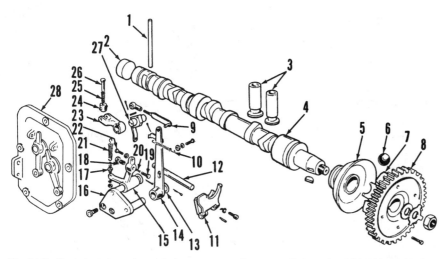

Fig. S1-7—Exploded view of camshaft, governor and governor linkage for DVA1030. DVA515 is similar.

1. Fuel pump push rod
2. Plug
3. Cam followers
4. Camshaft
5. Governor plate
6. Flyball
7. Collar
8. Camshaft gear
9. Injection pump link
10. Governor spring
11. Pivot plate
12. Push rod
13. Washer
14. Governor arm
15. Pivot arm
16. Bracket
17. Adjusting screw
18. Adjusting screw
19. Pin
20. Spring
21. Spring
22. Adjusting screw
23. Throttle arm
24. Connector
25. Spring
26. Pin
27. Lever
28. Governor cover

If connecting rod or piston was renewed, refer to CYLINDER section and measure piston height in cylinder.

PISTON, PIN AND RINGS

The piston is equipped with two compression rings and an oil control ring. Recommended piston ring end gap is 0.30-0.45 mm (0.012-0.018 inch) for compression rings and 0.25-0.40 mm (0.010-0.016 inch) for oil control ring. Standard piston diameter, measured at bottom of skirt perpendicular to piston pin, is 85.890-85.920 mm (3.3815-3.3826 inches).

Piston pin is retained in the piston by two snap rings. Piston pin diameter is 25.995-26.000 mm (1.0234-1.0236 inches) and piston pin bore is 25.991-26.000 mm (1.0233-1.0236 inches). Pin is a transition fit in piston bore. Initial fit when cold is 0.009 mm (0.0004 inch) interference to 0.005 mm (0.0002 inch) loose. Heating piston will ease removal and installation of pin.

CONNECTING ROD

The connecting rod small end is fitted with a renewable bushing. Bushing inside diameter should be 26.010-26.020 mm (1.0240-1.0244 inches). Recommended clearance between piston pin and bushing is 0.010-0.030 mm (0.0004-0.0012 inch).

Service procedures are similar for either pump.

To remove injection pump, disconnect fuel lines and immediately plug all openings to prevent entry of dirt. Remove retaining nuts and lift out injection pump. Be sure to note pump mounting shim gasket thickness for use in reassembly. Pump timing is adjusted by changing shim gasket thickness.

An exploded view of BOSCH pump is shown in Fig. S1-5. OMAP pump is similar. It is recommended that injection pump be tested and serviced only by a shop qualified in diesel fuel injection repair.

When reinstalling pump, assemble original thickness of mounting shim gasket and be sure to engage control rack pin (20) with governor linkage. If original mounting gasket thickness is unknown, required gasket thickness can be determined as follows: Rotate crankshaft until one lobe of injection pump camshaft is pointing downward. Measure distance from mounting surface of crankcase to base of camshaft lobe. Assemble gasket thickness necessary to obtain mounting dimension of 82.6-83.0 mm (3.252-3.268 inches) for BOSCH pump or 82.9-83.0 mm (3.264-3.268 inches) for OMAP pump. Check injection pump timing as outlined in MAINTENANCE section.

PISTON AND ROD UNIT

REMOVE AND REINSTALL. Piston and connecting rod may be removed after removing cylinder head, oil pan and oil pickup.

When reinstalling piston and rod, be sure depression in piston crown will be aligned with the injector nozzle tip. Make certain locating tabs of connecting rod bearing inserts engage notches in rod and cap and are located on the same side. Tighten connecting rod cap nuts to 34-44 N·m (25-32 ft.-lbs.).

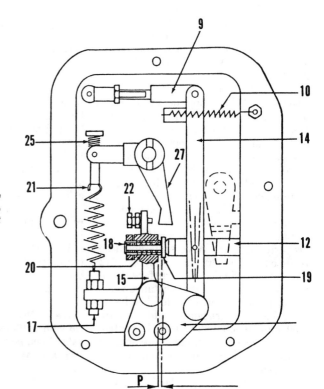

Fig. S1-8—Sectional view of governor linkage. Refer to Fig. S1-7 for component identification.

SERVICE MANUAL

Slanzi

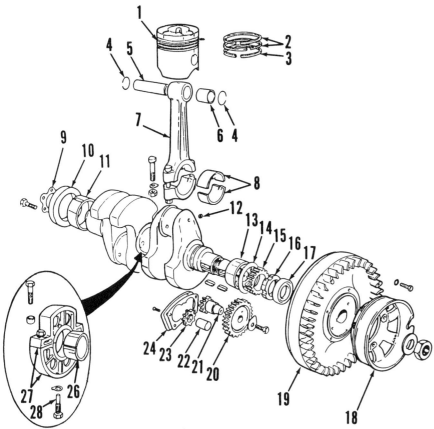

Fig. S1-9—Exploded view of crankshaft and oil pump assemblies. Center main bearing (26) and support (27) is used on DVA1030 engines.

1. Piston
2. Compression rings
3. Oil control ring
4. Retaining rings
5. Piston pin
6. Bushing
7. Connecting rod
8. Bearing
9. Washer
10. Oil seal
11. Main bearing
12. Plug
13. Main bearing
14. Thrust washer
15. Crankshaft gear
16. Nut
17. Oil seal
18. Starting pulley
19. Flywheel
20. Idler gear
21. Oil pump drive gear
22. Pin
23. Driven gear
24. Oil pump cover
26. Main bearing
27. Center main bearing support
28. Locating screw

A precision, insert type bearing is used in crankpin end of connecting rod. With rod bearing installed and rod cap nuts torqued to 34-44 N·m (25-32 ft.-lbs.), inside diameter of bearing should be 47.689-47.719 mm (1.8775-1.8786 inches). Recommended rod bearing clearance is 0.040-0.081 mm (0.0016-0.0032 inch). Bearings are available in standard size and undersizes.

CYLINDER

All engines are equipped with removable cylinders. Standard cylinder diameter is 86.000-86.020 mm (3.3858-3.3866 inches). Desired clearance between piston skirt and cylinder bore is 0.080-0.130 mm (0.003-0.005 inch). Cylinders that are damaged or excessively worn may be rebored for installation of 0.5 mm (0.020 inch) oversize piston and rings.

Piston height in cylinder is adjusted using shim gasket (5 – Fig. S1-6) between bottom of cylinder and crankcase mounting surface. With piston at top dead center, top of piston must be 0.0-0.2 mm (0.0-0.008 inch) below top edge of cylinder.

GOVERNOR

All engines are equipped with a flyball type governor mounted on the camshaft drive gear. See Fig. S1-7. As the flyball (6) move in and out against plate (5), the fork (11) and push rod (12) transfer this movement through the governor linkage thereby moving injection pump control rack. The throttle control lever operates through lever (23), pivot arm (15) and spring (21) to control engine speed.

Injection pump maximum fuel delivery under load is adjusted by screw (22 – Fig. S1-7 or S1-8). Turning screw one complete turn will change fuel delivery approximately 10 percent. Turning screw out (counterclockwise) will increase fuel flow.

Injection pump control rack is automatically placed in increased fuel delivery position for starting by spring (10).

Governor sensitivity (reaction to change in load) can be increased by turning adjusting screw (17) to increase tension on springs (21 and 25) and also by adjusting length of governor link (9).

The tendency for the governor to surge during speed changes is adjusted by turning set screw (18) to increase or decrease spring pin protrusion (P – Fig. S1-8). Increasing pin protrusion will reduce surging effect of governor. Dimension (P) must not exceed 0.5 mm (0.020 inch) measured without compressing spring.

For access to governor flyball assembly, remove flywheel shroud, flywheel, crankcase cover housing and camshaft gear.

CAMSHAFT AND CAM FOLLOWERS

To remove camshaft, first remove cylinder head, push rod tube and push rods, fuel injection pump and fuel transfer pump (DVA1030) and push rod (1 – Fig. S1-7). Remove flywheel shroud, flywheel and crankcase cover housing. Using suitable tools, raise cam followers (3) away from camshaft and hold in that position. Remove camshaft gear (8) and governor plate (5). Unbolt and remove retainer plate (11), then carefully withdraw camshaft. Cam followers may be removed from the bottom after removing the oil pan. Be sure to identify cam followers so they can be reinstalled in their original positions.

The camshaft is supported in unbushed bores in the crankcase. A camshaft with oversize bearing journals is available if crankcase bores become excessively worn for standard size camshaft. Standard journal diameter for DVA515 is 35.950-35.975 mm (1.4153-1.4163 inches) for first journal and 23.959-23.980 mm (0.9432-0.9440 inch) for second journal. Crankcase bore diameter is 36.000-35.025 mm (1.4173-1.4183 inches) and 24.000-24.021 mm (0.9448-0.9456 inch). Clearance should be 0.025-0.075 mm (0.001-0.003 inch) for first journal and 0.020-0.062 mm (0.0008-0.0024 inch) for second journal. For DVA1030, standard journal diameter is 42.459-42.475 mm (1.6716-1.6722 inches) for first journal, 35.950-35.975 mm (1.4153-1.4163 inches) for second journal and 23.967-23.978 mm (0.9436-0.9440 inch) for third journal. Crankcase bore inner diameter is 42.500-42.525 mm (1.6732-1.6742 inches) for first bore, 36.000-36.025 mm (1.4173-1.4183 inches) for second bore and 24.000-24.021 mm (0.9449-0.9457 inch) for third bore. Clearance should be 0.025-0.066 mm (0.001-0.0026 inch) for first journal, (0.025-0.075 mm (0.001-0.003 inch) for second journal and

Slanzi

SMALL DIESEL

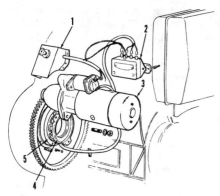

Fig. S1-10—On engines equipped with electric starting, an internally mounted alternator is located in the flywheel.

1. Key switch
2. Voltage regulator
3. Starter motor
4. Stator
5. Rotor

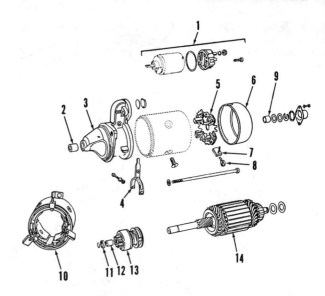

Fig. S1-11—Exploded view of starter motor used on engines equipped with electric starting.

1. Solenoid
2. Bushing
3. Drive cover
4. Lever
5. Brush frame
6. Cover
7. Brush
8. Spring
9. Bushing
10. Field coil & housing
11. Retainer
12. Bushing
13. Starter drive
14. Armature

0.022-0.054 mm (0.001-0.002 inch) for third journal.

Camshaft end play is controlled by retainer plate (11) and retaining groove in front of camshaft. Desired end play is 0.030-0.180 mm (0.0012-0.0071 inch).

To reinstall camshaft, reverse the removal procedure making certain the timing marks on camshaft gear, idler gear and crankshaft gear are aligned. Tighten camshaft gear retaining nut to 147 N·m (108 ft.-lbs.).

OIL PUMP

The oil pump gears (21 and 23–Fig. S1-9) are located in the crankcase cover housing. To remove pump gears, remove flywheel shroud and flywheel. Unbolt and remove cover housing. Remove idler gear (20), pump cover (24) and gears.

Inspect pump gears and housing for excessive wear or other damage. Renew as needed and reassemble by reversing the removal procedure.

The oil pressure relief valve is also located in the crankcase cover housing. Relief valve pressure setting may be adjusted with adjusting screw (14–Fig. S1-1).

CRANKSHAFT AND BEARINGS

The crankshaft rides in sleeve bearings (11 and 13–Fig. S1-9) located in crankcase housing and bearing support housing. DVA1030 engines are also equipped with a split center main bearing (26) mounted in bearing support (27).

To remove the crankshaft, remove flywheel, crankshaft pulley, oil pan, cylinder head, piston and connecting rod and crankcase cover housing. On DVA1030 engine, remove center main bearing support locating screw (28). On all engines, carefully withdraw crankshaft assembly out rear of crankcase.

On DVA515 engine, standard diameter of main journals is 44.946-44.962 mm (1.7695-1.7701 inches). Standard inner diameter of main bearings is 45.00-45.025 mm (1.7716-1.7726 inches). Desired operating clearance is 0.038-0.079 mm (0.0015-0.0031 inch). Standard and undersize main bearings are available. Replacement bearings must be bored to provide desired clearance. Standard crankpin diameter is 47.638-47.649 mm (1.8743-1.8759 inches) and desired rod bearing clearance is 0.040-0.081 mm (0.0016-0.0032 inch).

On DVA1030 engine, standard diameter of front main journal is 51.940-51.960 mm (2.0446-2.0454 inches), rear main journal diameter is 57.935-57.955 mm (2.2810-2.2818 inches) and center main journal diameter is 55.437-55.457 mm (2.1825-2.1833 inches). Desired main bearing clearance is 0.040-0.090 mm (0.0016-0.0035 inch) for front bearing, 0.045-0.095 mm (0.0018-0.0037 inch) for rear bearing and 0.043-0.093 mm (0.0017-0.0036 inch) for center bearing. Standard and undersize bearings are available. Front and rear replacement bearings must be bored to provide desired clearance. Crankpin standard diameter is 47.638-47.649 mm (1.8755-1.8759 inches) and desired rod bearing clearance is 0.040-0.081 mm (0.0016-0.0032 inch).

Crankshaft end play should be 0.050-0.150 mm (0.0020-0.0059 inch). End play is adjusted by changing thrust washer (14) thickness.

Reassembly is the reverse of disassembly. Be sure to renew crankshaft oil seals (10 and 17). Lubricate lip of seals with grease prior to reassembly. On DVA1030 engine, tighten center main bearing support cap screws to 49 N·m (36 ft.-lbs.). Make certain timing marks on camshaft gear, idler gear and crankshaft gear are aligned.

ELECTRICAL SYSTEM

On electric start engines, a flywheel mounted alternator is used to charge the battery. See Fig. S1-10. Electrical system is negative grounded on all engines. To avoid damage to charging system, do not disconnect battery cables or other electrical connections while engine is running. Refer to Fig. S1-11 for an exploded view of starter assembly.

SERVICE MANUAL

SLANZI

Model	No. Cyls.	Bore	Stroke	Displ.
DVA1500	2	92 mm (3.622 in.)	88 mm (3.465 in.)	1755 cc (107 cu. in.)

The DVA1500 engine is a four-stroke, air-cooled, direct injection diesel with vertical inline cylinders. Crankshaft rotation is counterclockwise at flywheel end.

MAINTENANCE

LUBRICATION

Recommended engine oil is SAE 20W-20 for temperatures below 0°C (32°F). When ambient temperature will be above 0°C (32°F), use SAE 30 for winter operation or SAE 40 for summer operation. Only premium grade, heavy-duty, diesel engine oil meeting API CC specifications should be used.

Under normal operating conditions, manufacturer recommends changing engine oil after ever 200 hours of operation. When operating under severe duty conditions, such as high temperature and heavy load, oil should be changed after every 100 hours of operation. Oil should be drained while the engine is hot. Manufacturer recommends renewing the oil filter element after every 400 hours of operation. Crankcase capacity is approximately 4 liters (4.3 quarts) without filter change and 4.6 liters (4.9 quarts) with filter change.

The engine oil pressure relief valve assembly is located in the oil filter adapter housing (9 – Fig. S2-1). Pressure setting is adjustable using screw (4).

FUEL SYSTEM

FUEL FILTER AND BLEEDING. Manufacturer recommends renewing the fuel filter element after every 200 hours of operation. However, if engine indicates signs of fuel starvation (surging or loss of power), filter should be renewed regardless of hours of operation. After renewing fuel filter, air must be bled from the fuel system.

NOTE: A plastic filter screen is located in fuel transfer pump inlet. If engine fails to run properly after main fuel filter element has been renewed, inspect for plugged pump filter screen.

To bleed air from fuel system, loosen bleed screw (11 – Fig. S2-2) on fuel filter mounting base. Operate fuel transfer pump manually until air-free fuel flows from bleeder, then retighten screw. Loosen bleed screw (4) at injection pump fuel inlet. Operate fuel transfer pump until fuel flows from bleeder, then retighten screw. Continue to actuate transfer pump until resistance is felt to pressurize system.

If engine fails to start at this point, loosen high pressure fuel line (3) fittings at the injectors. Set throttle control in full speed position. Actuate decompression lever, then crank engine until fuel is discharged from loosened fuel lines. Retighten fittings and start engine.

ENGINE SPEED ADJUSTMENT

Engine should be at normal operating temperature when adjusting engine speeds. Low and high idle (no-load) speeds are adjusted using the throttle lever stop screws located on top of governor cover. If necessary, adjust screws to obtain specified governed speed as required for engine application.

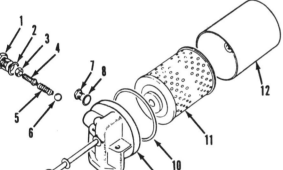

Fig. S2-1—Exploded view of engine oil filter assembly and oil pressure relief valve.
1. Plug
2. Washer
3. Nut
4. Adjusting screw
5. Spring
6. Ball
7. Plug
8. Washer
9. Adapter housing
10. Gasket
11. Oil filter element
12. Canister

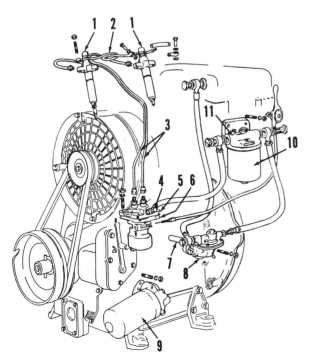

Fig. S2-2—View of fuel system components. Pump mounting gasket (6) is used to adjust injection timing.
1. Injector
2. Leak-off line
3. High pressure lines
4. Bleeder screw
5. Injection pump
6. Shim gasket
7. Push rod
8. Fuel transfer pump
9. Oil filter assy.
10. Fuel filter assy.
11. Bleeder screw

Illustrations Courtesy North American Diesel, Inc.

Slanzi

SMALL DIESEL

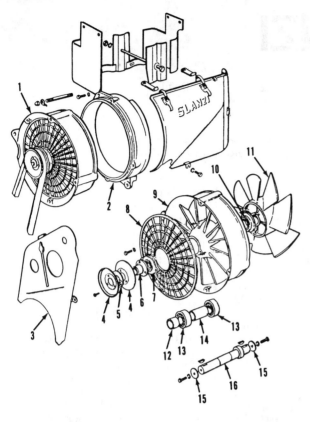

Fig. S2-3—Exploded view of axial flow engine cooling fan. Belt tension is adjusted by changing thickness of washer (5).

1. Fan assy.
2. Mounting housing
3. Belt guard
4. Pulley halves
5. Shim washer
6. Hub
7. Flange
8. Guard
9. Stator housing
10. Spacer
11. Fan
12. Spacer
13. Bearings
14. Spacer
15. Retaining washers
16. Fan shaft

stops if delivery valve is removed. At this point, beginning of injection occurs and "TDP" mark on flywheel should be aligned with reference mark made previously on flywheel housing. If timing marks are not aligned, remove injection pump and increase thickness of pump mounting gasket to retard timing or decrease thickness of gasket to advance timing.

REPAIRS

TIGHTENING TORQUES

Refer to the following table for special tightening torques. All fasteners are metric.

Connecting rod...........61-65 N·m
(45-48 ft.-lbs.)
Crankshaft
counterweights..........59-64 N·m
(44-47 ft.-lbs.)
Cylinder head..............83 N·m
(61 ft.-lbs.)
Flywheel..................236 N·m
(174 ft.-lbs.)
Injector...................20 N·m
(15 ft.-lbs.)
Main bearing center support...45 N·m
(33 ft.-lbs.)

COOLING FAN

An axial flow fan, belt driven by the crankshaft, provides the engine cooling air flow. Fan drive belt tension is ad-

Do not exceed maximum governed limit of 2750 rpm.

INJECTION PUMP TIMING

The injection pump is actuated by the injection pump camshaft. Injection timing is adjusted using shim gaskets (6–Fig. S2-2) between pump body and mounting surface of crankcase. Engine may be equipped with a BOSCH or OMAP injection pump. Injection timing is the same for either pump.

To check injection timing, top dead center of number 1 piston must first be determined. Rotate crankshaft until number 1 piston is on compression, then remove valve spring from a valve on number 1 cylinder to allow valve to rest on top of piston. Position a dial indicator against end of valve stem, then slowly turn crankshaft to locate piston TDC. Mark a reference mark on flywheel housing in alignment with arrow mark on flywheel. Reinstall valve spring.

Disconnect injector line for number 1 cylinder from injection pump. Connect a fuel flow sight gage (special tool 59.40.50) to number 1 discharge fitting on pump. If tool is not available, remove delivery valve holder (1–Fig. S2-6), spring (3) and delivery valve (5), then reinstall delivery valve holder. Remove governor cover and position pump control rack at half speed setting. Maintain this setting during timing check. Connect fuel supply to pump. Note that fuel should flow from delivery valve holder

at this time if delivery valve is removed. Rotate flywheel clockwise approximately ½ turn, then turn slowly counterclockwise (normal direction of rotation) until first drop of fuel is observed at sight gage tool, or until fuel flow just

Fig. S2-4—Exploded view of valve assembly and compression release mechanism.

1. Pin
2. Compression release lever
3. Seals
4. Shaft
5. Spring clip
6. Pin
7. Connector
8. Spacer
9. Screw
10. Rocker arm (intake)
11. Rocker arm (exhaust)
12. Adjusting screw
13. Valve keepers
14. Retainer
15. Valve spring
16. Cap nut (intake only)
17. Locating pin
18. Valve guide
19. Intake valve
20. Exhaust valve
21. Cam followers
22. Seal
23. Cup washer
24. Spring
25. Push rod tube
26. Rocker arm support
27. Washer
28. Snap ring
29. Push rods

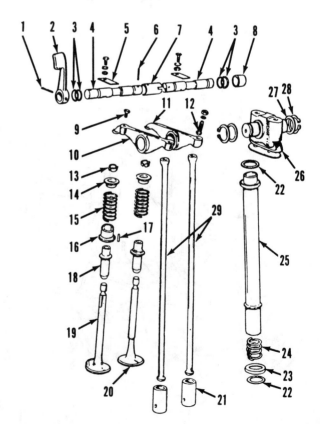

Illustrations Courtesy North American Diesel, Inc.

SERVICE MANUAL

justed by changing thickness of washer (5 – Fig. S2-3) between fan pulley halves (4).

To disassemble fan, remove belt guard, fan pulley and drive belt. Unbolt and remove fan assembly (1) from mounting housing (2). Remove pulley hub (6), flange (7) and guard (8). Remove fan retaining screw and washer and withdraw fan (11) and spacer (10). Tap shaft (16), bearings (13) and spacers out of stator housing (9).

To reassemble, reverse the disassembly procedure.

VALVE ADJUSTMENT

To adjust valve clearance, first remove valve covers and rotate crankshaft until number 1 piston is at TDC on compression stroke (both rocker arms should be free to move). With engine cold, valve clearance should be 0.20 mm (0.008 inch) for intake and 0.25 mm (0.010 inch) for exhaust. Turn rocker arm adjusting screws (12 – Fig. S2-4) until appropriate size feeler gage can be inserted between rocker arm and valve stem end. Recheck adjustment after retightening adjusting screw locknut.

After adjusting valves on number 1 cylinder, turn crankshaft one complete revolution and adjust valves on second cylinder.

CYLINDER HEAD AND VALVE SYSTEM

To remove cylinder heads, first remove blower fan assembly and shrouds. Remove intake and exhaust manifolds. Disconnect fuel lines from injectors, then remove injectors. Remove cylinder head stud nuts and lift off cylinder heads. Remove valve covers with decompression shaft and rocker arm support (26 – Fig. S2-4) with rocker arms for access to valves.

Valve stem diameter is 8.960-8.975 mm (0.3527-0.3533 inch) for intake and exhaust valves. Valve guide inner diameter should be 8.976-9.006 mm (0.3534-0.3545 inch). Desired clearance of valve stem in guide is 0.010-0.046 mm (0.0004-0.0018 inch). Valve guides are renewable. When renewing guides, use a suitable installing tool to avoid damaging guides. Press guides into cylinder head until they bottom against their shoulders. Guides are presized and should not require reaming if carefully installed.

Renewable valve seat inserts are used for intake and exhaust. Inserts should have an interference fit of 0.125-0.170 mm (0.005-0.007 inch) in cylinder head counterbore. Chilling inserts with dry ice prior to installation will ease assembly. When properly installed,

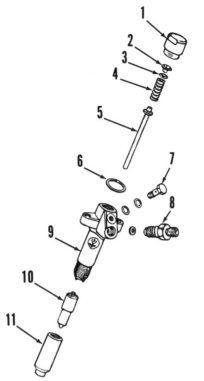

Fig. S2-5—Exploded view of fuel injector assembly.

1. Cap nut
2. Spring seat
3. Shim
4. Spring
5. Push rod
6. Gasket
7. Leak-off banjo bolt
8. Fuel inlet fitting
9. Injector body
10. Nozzle valve assy.
11. Nozzle holder nut

valve seat should be recessed 0.15-0.30 mm (0.006-0.012 inch) below surface of cylinder head. Position valves in cylinder and measure distance valve head extends from cylinder head surface. Both intake and exhaust valve should extend 0.90-1.25 mm (0.035-0.049 inch). If valve projection is excessive, regrind valve seat. If projection is less than specified, renew valve seat insert.

The injector nozzle tip should extend 3.0-3.5 mm (0.118-0.138 inch) beyond cylinder head surface. Adjust nozzle position, if necessary, using shim gaskets between injector nozzle and bottom of cylinder head nozzle bore.

When reassembling cylinder head, be sure locating pin (17) is in place to retain cap nut (16) on intake valve stem. Renew push rod tube oil seals (22) if necessary. Tighten cylinder head stud nuts evenly using a crossing pattern to 83 N·m (61 ft.-lbs.). Adjust valve clearance as previously outlined.

INJECTOR

REMOVE AND REINSTALL. Before removing injector, thoroughly clean injector, fuel lines and surrounding area.

Slanzi

Disconnect fuel return line and high pressure line and plug all openings. Unscrew retaining nuts and withdraw injector from cylinder head.

To reinstall injector, reverse the removal procedure. Tighten retaining nuts to 20 N·m (15 ft.-lbs.).

TESTING. A suitable test stand is required to check injector operation. Only clean, approved testing oil should be used to test injector.

WARNING: Fuel leaves the injector nozzle with sufficient force to penetrate the skin. When testing, keep yourself clear of nozzle spray.

Connect injector to tester and operate tester lever a few quick strokes to purge air from injector and to make sure nozzle valve is not stuck. When operating properly, nozzle will emit a buzzing sound and produce a uniform spray pattern.

To check opening pressure, slowly operate tester lever while observing pressure gage reading. Opening pressure should be 19615-20595 kPa (2845-2985 psi). Opening pressure is adjusted by varying the thickness of shim (3 – Fig. S2-5).

Operate tester lever slowly to maintain pressure at 2000 kPa (300 psi) below opening pressure and check for leakage past nozzle valve seat. If a drop forms at nozzle tip within a 10 second period, nozzle valve is not seating and injector must be overhauled or renewed.

OVERHAUL. To disassemble injector, secure nozzle body (9 – Fig. S2-5) in a vise or holding fixture. Remove nozzle holder nut (11) and withdraw nozzle valve (10). Remove cap (1) and remove spring (2), shim (3), spring (4) and rod (5).

Thoroughly clean all parts in a suitable solvent. Do not use steel wire brush or sharp metal tools to clean injector components. Make certain the nozzle valve needle slides freely in nozzle valve body. If needle sticks, reclean or renew nozzle valve assembly.

When reassembling injector, be sure all parts are clean and wet with diesel fuel. Recheck injector operation as outlined in TESTING section.

INJECTION PUMP

Engine may be equipped with a BOSCH or OMAP fuel injection pump. Service procedures are similar for either pump. An exploded view of OMAP pump is shown in Fig. S2-6. It is recommended that the injection pump be tested and serviced only by a shop qualified in diesel fuel injection repair.

Slanzi

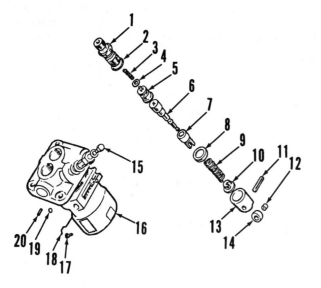

Fig. S2-6—Exploded view of OMAP fuel injection pump. BOSCH pump used on some engines is similar.

1. Delivery valve holder
2. Gasket
3. Spring
4. Gasket
5. Delivery valve
6. Pumping element
7. Bushing
8. Cup
9. Spring
10. Cup
11. Pin
12. Inner roller
13. Tappet
14. Outer roller
15. Bleed screw
16. Pump housing
17. Pin
18. Retainer
19. Plug
20. Pin

To remove injection pump, disconnect fuel lines and immediately plug all openings to prevent entry of dirt. Remove inspection cover from side of crankcase and disconnect governor rod from pump control rack pin. Remove pump mounting nuts and lift out pump assembly. Be sure to note pump mounting gasket thickness for use in reassembly.

When reinstalling injection pump, assemble original thickness of mounting shim gasket. Be sure to engage pump control rack pin with governor rod. If original mounting gasket thickness is unknown, required gasket thickness can be determined as follows: Rotate crankshaft until one lobe of injection pump camshaft is pointing downward. Measure distance from mounting surface of crankcase to base of camshaft lobe. Assemble gasket thickness necessary to obtain mounting dimension of 82.6-83.0 mm (3.252-3.268 inches) for BOSCH pump or 82.9-83.0 mm (3.264-3.268 inches) for OMAP pump. Check injection timing as outlined in MAINTENANCE section.

PISTON AND ROD UNIT

The piston and connecting rod may be removed after removing the cylinder head and oil pan. Number pistons and connecting rods as they are removed so they can be reinstalled in their original locations.

When reinstalling piston and rod, be sure piston is positioned so depression in piston crown will be aligned with the injector nozzle tip. Make certain locating tabs of connecting rod bearing inserts engage notches in rod and cap and are located on the same side. Tighten connecting rod cap nuts to 61-65 N·m (45-48 ft.-lbs.).

If connecting rod or piston was renewed, refer to CYLINDER section and measure piston height in cylinder.

PISTON, PIN AND RINGS

The piston is equipped with three compression rings and an oil control ring. Recommended piston ring end gap is 0.35-0.55 mm (0.014-0.022 inch) for compression rings and 0.25-0.40 mm (0.010-0.016 inch) for oil control ring. Standard piston diameter, measured at bottom of skirt perpendicular to piston pin, is 95.880-95.900 mm (3.7748-3.7756 inches). Piston and rings are available in standard size and 0.5 mm (0.020 inch) oversize.

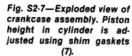

Fig. S2-7—Exploded view of crankcase assembly. Piston height in cylinder is adjusted using shim gaskets (7).

1. Valve cover
2. Cylinder stud
3. Cylinder head
4. Valve seat inserts
5. Head gasket
6. Cylinder
7. Shim gasket
8. Cover
9. Governor cover
10. Crankcase front cover
11. Cover
12. Crankcase
13. Oil pan
14. Main bearing support
15. Cover

SMALL DIESEL

Piston pin is retained in piston by two snap rings. Piston pin diameter is 27.995-28.000 mm (1.1021-1.1023 inches). Piston pin bore should be 27.991-28.000 mm (1.1019-1.1023 inches). Pin is a transition fit in piston bore. Initial pin fit when cold is 0.009 mm (0.0004 inch) interference to 0.005 mm (0.0002 inch) loose. Heating piston will ease removal and installation of pin.

CONNECTING ROD

The connecting rod small end is fitted with a renewable bushing. Bushing inside diameter should be 28.000-28.021 mm (1.1023-1.1031 inches). Recommended clearance between piston pin and bushing is 0.0-0.025 mm (0.0-0.001 inch).

A precision, insert type bearing is used in big end connecting rod. With bearing installed in rod and cap nuts torqued to 61-65 N·m (45-48 ft.-lbs.), inside diameter of bearing should be 57.175-57.201 mm (2.2510-2.2520 inches). Recommended clearance between crankpin and rod bearing is 0.051-0.090 mm (0.0020-0.0035 inch). Bearings are available in standard size and undersizes.

CYLINDER

Engine is equipped with removable cast iron cylinders. Standard cylinder bore is 96.000-96.020 mm (3.7795-3.7803 inches). Recommended clearance between piston skirt and cylinder is 0.100-0.140 mm (0.0039-0.0055 inch). Cylinders that are damaged or ex-

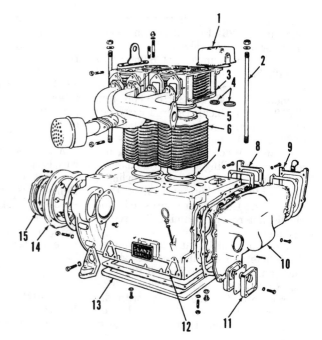

Illustrations Courtesy North American Diesel, Inc.

SERVICE MANUAL

Slanzi

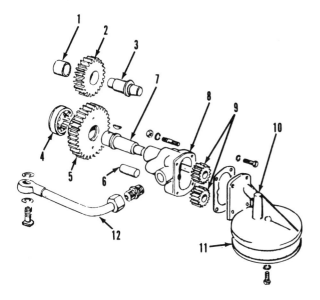

Fig. S2-8—Exploded view of engine oil pump assembly.
1. Bushing
2. Idler gear
3. Shaft
4. Bearing
5. Drive gear
6. Shaft
7. Pump drive shaft
8. Pump housing
9. Gears
10. Pump cover & oil pickup
11. Filter screen
12. Oil pressure tube

cessively worn may be rebored for installation of 0.5 mm (0.020 inch) oversize piston and rings.

Piston height in cylinder is adjusted using shim gasket (7–Fig. S2-7) between bottom of cylinder and crankcase mounting surface. With piston at top dead center, top of piston must be 0.0-0.2 mm (0.0-0.008 inch) below top edge of cylinder.

TIMING GEARS AND FRONT COVER

To gain access to timing gears, first remove belt guard, fan belt and engine cooling fan. Remove side cover (8–Fig. S2-7) and disconnect governor linkage from injection pump. Remove crankshaft front pulley, then unbolt and remove front cover (10) with governor linkage. Remove gears as needed. Be sure to align timing marks on gears when reassembling.

OIL PUMP

The gear type oil pump (Fig. S2-8) is mounted internally to front of crankcase. The pump is driven by the crankshaft gear through an idler gear (2).

To remove the oil pump, first drain the engine oil. Remove the oil pan and disconnect oil pressure tube (12). Remove timing gear cover as previously outlined. Remove pump drive gear (5), then unscrew pump mounting nuts and withdraw pump assembly.

Remove cap screws securing pump body (8) to oil pickup housing (10) and separate components. Inspect gears, shafts and pump body for damage or excessive wear and renew if necessary.

To reinstall oil pump, reverse the removal procedure.

GOVERNOR

The flyball type governor (Fig. S2-9) is located in the crankcase front cover. The governor is mounted on the front of camshaft gear. Camshaft rotation causes flyballs (4) to move in and out, depending upon engine speed, which moves governor cup (2) on bushing (1). The fork (9) transfers this motion through the governor linkage to the injection pump control rack to control injection pump fuel delivery.

To gain access to governor and governor linkage, remove governor side cover and crankcase front cover. Inspect all parts for wear and damage. Governor linkage must move freely to operate properly.

VALVE CAMSHAFT AND CAM FOLLOWERS

The camshaft is supported by unbushed bores in crankcase. A camshaft with oversize bearing journals is available if crankcase bores become excessively worn for standard size camshaft. Barrel type cam followers (7–Fig. S2-9) are used.

To remove camshaft, first remove cylinder head, push rods, push rod tubes and crankcase front cover as previously outlined. Lift out cam followers. Be sure to identify position of each cam follower as it is removed for correct reassembly.

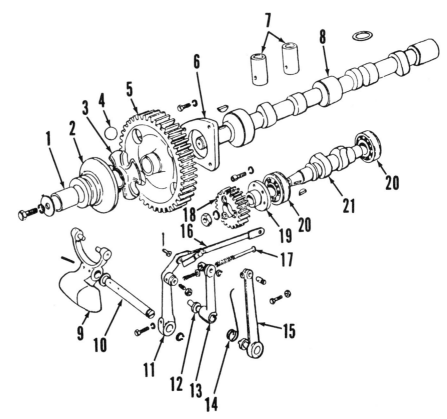

Fig. S2-9—Exploded view of valve camshaft (8), injection pump camshaft (21) and governor assembly.
1. Bushing
2. Governor cup
3. Collar
4. Flyball
5. Camshaft gear
6. Retainer plate
7. Cam followers
8. Valve camshaft
9. Governor fork
10. Pivot pin
11. Governor arm
12. Pin
13. Throttle arm
14. Spring
15. Throttle lever
16. Governor link
17. Pin
18. Gear
19. Hub
20. Bearings
21. Injection pump camshaft

Illustrations Courtesy North American Diesel, Inc.

Slanzi SMALL DIESEL

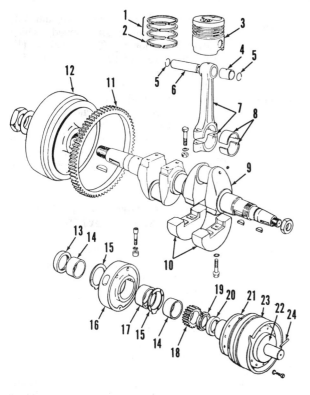

Fig. S2-10—Exploded view of piston and crankshaft assembly.

1. Compression rings
2. Oil control ring
3. Piston
4. Bushing
5. Retaining rings
6. Piston pin
7. Connecting rod
8. Rod bearing
9. Crankshaft
10. Counterweights
11. Ring gear
12. Flywheel
13. Oil seal
14. Main bearing
15. Thrust washers
16. Center bearing carrier
17. Main bearing
18. Crankshaft gear
19. Nut
20. Oil seal
21. Flange
22. Flange
23. Pulley
24. Pin

Remove governor retaining cap screw and remove governor assembly and camshaft gear. Remove camshaft retainer plate (6), then carefully withdraw camshaft.

Inspect bearing journals, cam lobes and cam followers for scoring, wear or other damage. Diameter of camshaft bearing journals should be 44.945-44.964 mm (1.7696-1.7702 inches) for front journal, 35.950-35.965 mm (1.4154-1.4159 inches) for center journal and 27.955-27.975 mm (1.1005-1.1013 inches) for rear journal. Diameter of crankcase bearing bores should be 45.000-45.025 mm (1.7716-1.7726 inches) for front bore, 36.000-36.025 mm (1.4173-1.4183 inches) for center bore and 28.000-28.021 mm (1.1023-1.1031 inches) for rear bore. Desired clearance between camshaft and support bores is 0.035-0.080 mm (0.0014-0.0031 inch) for front journal, 0.035-0.075 mm (0.0014-0.0030 inch) for second journal and 0.025-0.066 mm (0.0010-0.0026 inch) for rear journal.

Camshaft end play is controlled by retainer plate (6). Desired end play is 0.10-0.20 mm (0.004-0.008 inch).

To reinstall camshaft, reverse the removal procedure. Be sure to align timing marks on gears.

INJECTION PUMP CAMSHAFT

To remove the injection pump camshaft (21—Fig. S2-9), remove injection pump, fuel transfer pump and crankcase front cover. Remove drive gear (18) and hub (19), then remove camshaft (21) and bearings (20) from crankcase.

Inspect all parts and renew if necessary. To reinstall, reverse the removal procedure. Refer to INJECTION PUMP TIMING section to check and adjust injection timing.

CRANKSHAFT AND BEARINGS

The crankshaft rides in three sleeve type main bearings. Front bearing is located in crankcase bulkhead, rear bearing is mounted in crankcase rear cover. The center bearing is split and is contained in the two-piece center bearing support (16—Fig. S2-10).

To remove crankshaft, first remove the oil pan, flywheel and crankcase rear cover. Remove front pulley flange (22), then remove nut from front of crankshaft and withdraw front pulley (23) and flange (21) assembly. Unscrew retaining nut (19) and remove crankshaft gear (18). Remove connecting rod caps and push connecting rods and pistons up out of the way. Remove center bearing support retaining screws, then carefully withdraw crankshaft assembly from rear of crankcase. Separate center bearing support and bearing from crankshaft.

Standard diameter of main journals is 57.980-58.000 mm (2.2826-2.2834 inches) for front and rear journals and 62.970-62.983 mm (2.4793-2.4796 inches) for center journal. Standard main bearing inner diameter should be 58.045-58.060 mm (2.2852-2.2858 inches) for front and rear bearings and 63.031-63.055 mm (2.4815-2.4825 inches) for center bearing. Desired main bearing clearance is 0.045-0.080 mm (0.0018-0.0031 inch) for front and rear journals and 0.048-0.085 mm (0.019-0.0033 inch) for center journal. Standard and undersize main bearings are available. Replacement bearings must be bored to provide desired clearance.

Standard crankpin diameter is 57.111-57.124 mm (2.2484-2.2489 inches). Desired rod bearing operating clearance is 0.051-0.090 mm (0.002-0.0035 inch). Undersize bearings are available.

Crankshaft end play is controlled by thrust washers (15) located on either side of center main bearing support (30). Desired end play is 0.08-0.30 mm (0.003-0.012 inch). When installing thrust washers, be sure grooved side faces away from bearing support.

Inspect oil seals (13 and 20) and renew if necessary. Be sure to lubricate seal lips with grease prior to reassembly. To reinstall crankshaft, reverse the removal procedure. Tighten center bearing carrier screws to 45 N·m (33 ft.-lbs.), counterweight retaining screws to 59-64 N·m (44-47 ft.-lbs.) and flywheel nut to 236 N·m (174 ft.-lbs.).

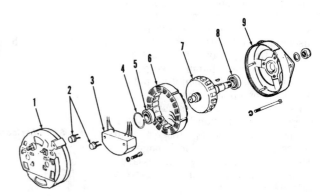

Fig. S2-11—Exploded view of alternator used on engines with electric starting.

1. End cover
2. Diodes
3. Regulator
4. Retainer
5. Bearing
6. Stator
7. Rotor
8. Bearing
9. Cover

Illustrations Courtesy North American Diesel, Inc.

SERVICE MANUAL

Slanzi

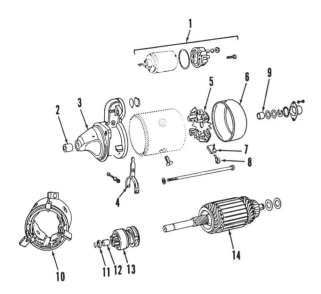

Fig. S2-12—Exploded view of electric starting motor.
1. Solenoid
2. Bushing
3. Drive end housing
4. Lever
5. Brush frame
6. Cover
7. Brush
8. Spring
9. Bushing
10. Field coil & housing
11. Retainer
12. Bushing
13. Starter drive
14. Armature

ELECTRICAL SYSTEM

The electrical system (if so equipped) is a 12-volt, negative ground system. The alternator is mounted to front of the engine and driven by the cooling fan drive belt. Belt tension is adjusted by changing thickness of washer (5–Fig. S2-3) between fan drive pulley halves (4). An integral, solid-state regulator (3–Fig. S2-11) controls alternator output.

A BOSCH or HARELLI starter motor is used on engines equipped with electric starting. See Fig. S2-12 for an exploded view of typical starter motor.

Slanzi

SLANZI

SMALL DIESEL

Model	No. Cyls.	Bore	Stroke	Displ.
DVA1550	3	86 mm (3.39 in.)	88 mm (3.46 in.)	1533 cc (93.5 cu. in.)
DVA1750	3	92 mm (3.62 in.)	88 mm (3.46 in.)	1755 cc (107 cu. in.)

These engines are four-stroke, air-cooled, direct injection diesel with vertical inline cylinders. Crankshaft rotation is counterclockwise at flywheel end. Number one cylinder is closest to crankshaft pulley end of engine. Firing order 1-3-2.

MAINTENANCE

LUBRICATION

Recommended engine oil is SAE 20W-20 for temperatures below 0°C (32°F). When ambient temperatures will be above 0°C (32°F), use SAE 30 for winter operation or SAE 40 for summer operation. Only premium grade, heavy-duty, diesel engine oil should be used.

Under normal operating conditions, manufacturer recommends changing engine oil after every 200 hours of operation. When operating under severe duty conditions, such as high temperature and heavy load, oil should be changed after every 100 hours. Oil should be drained while the engine is hot. Recommended oil filter renewal interval is every 400 hours of operation.

FUEL SYSTEM

FUEL FILTER AND BLEEDING. Manufacturer recommends renewing the fuel filter element after every 200 hours of operation. However, if engine indicates signs of fuel starvation (surging or loss of power), filter should be renewed regardless of hours of operation. After renewing fuel filter, air must be bled from the fuel system.

The fuel transfer pump also contains a plastic filter screen (1–Fig. S3-1) in the pump inlet. If engine fails to run properly after main fuel filter has been renewed, inspect for plugged pump filter screen as follows: Disconnect fuel lines from pump. Separate fuel inlet and outlet housing (4) from pump body to expose inlet valve (2) and filter screen. Remove and clean filter screen, then reinstall screen and reassemble pump.

To bleed air from fuel system, loosen bleeder screw on fuel filter mounting base. Operate fuel transfer pump until air-free fuel flows from bleeder, then retighten screw. If engine fails to start at this point, loosen high pressure fuel line fittings at the injectors. Set throttle control at full speed position. Crank engine with starter until fuel is discharged from loosened fuel lines. Retighten fittings and start engine.

ENGINE SPEED ADJUSTMENT

Engine should be at normal operating temperature when adjusting engine speed settings. Low and high idle (no-load) speeds are adjusted using throttle lever stop screws (1 and 3–Fig. S3-2) on governor cover. Turning screw (1) clockwise will decrease low speed setting and turning screw (3) clockwise will increase maximum governed speed. Adjust screws as required to obtain specified governed speeds depending on engine application. Do not exceed maximum governed limit of 3200 rpm.

Injection pump maximum fuel delivery is adjusted using stop lever screw (5). Turning screw clockwise will increase fuel delivery.

INJECTION PUMP TIMING

The injection pump is actuated by the injection pump camshaft (18–Fig. S3-9). Injection timing is adjusted by loosening timing gear (17) retaining screws and rotating camshaft within limits of retaining screw slots in timing gear.

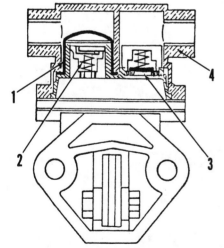

Fig. S3-1 — A plastic filter screen (1) is located in the inlet side of fuel transfer pump. If engine indicates signs of fuel starvation, check for plugged screen.

1. Filter screen
2. Inlet valve
3. Discharge valve
4. Valve housing

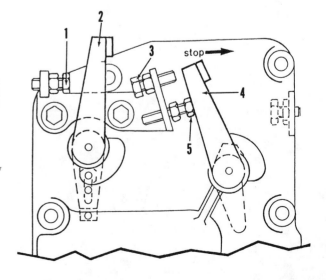

Fig. S3-2 — View of governor cover showing engine speed and maximum fuel delivery adjustment points.

1. Low idle adjusting screw
2. Throttle lever
3. Maximum speed adjusting screw
4. Stop lever
5. Fuel delivery adjusting screw

SERVICE MANUAL

Slanzi

To check injection timing, rotate crankshaft pulley clockwise until number 1 piston is on compression and second mark (top dead center) on pulley is aligned with timing pin on crankcase front cover. Disconnect injector line for number 1 cylinder from injection pump. Connect a fuel flow sight gage (special tool number 59.40.50) to number 1 discharge fitting on pump. If tool is not available, remove delivery valve holder (1 – Fig. S3-6), spring (6) and delivery valve (8), then reinstall delivery valve holder. Remove governor cover and position pump control rack at half speed setting. Maintain this setting during timing check. Connect fuel supply to pump. Note that fuel should flow from delivery valve holder at this time if delivery valve is removed. Rotate crankshaft pulley counterclockwise about ½ turn, then slowly turn clockwise (normal direction of rotation) until first drop of fuel is observed at sight gage tool, or until fuel flow just stops if delivery valve is removed. At this point, beginning of injection occurs and first mark on crankshaft pulley should be aligned with timing pin. If timing mark is not aligned, remove camshaft gear inspection cover from crankcase front cover. Loosen camshaft gear mounting screws, then rotate injection pump camshaft within limits of camshaft gear screw slots to obtain correct timing. Rotating camshaft clockwise will advance timing.

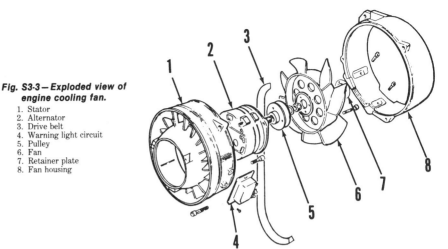

Fig. S3-3 – Exploded view of engine cooling fan.
1. Stator
2. Alternator
3. Drive belt
4. Warning light circuit
5. Pulley
6. Fan
7. Retainer plate
8. Fan housing

COOLING FAN

The engine cooling fan (6 – Fig. S3-3) is mounted on the alternator and fan drive pulley (5). Fan belt tension should be checked every 100 hours of operation. Tension is adjusted by changing thickness of shim (11 – Fig. S3-11) between crankshaft pulley halves on early production engines or by moving an idler pulley on late production engines.

REPAIRS

TIGHTENING TORQUES

Refer to the following table for special tightening torques. All fasteners are metric.

Connecting rod............31-37 N·m
(23-27 ft.-lbs.)
Crankshaft counterweight..58-64 N·m
(43-47 ft.-lbs.)
Cylinder head.............54-58 N·m
(40-43 ft.-lbs.)
Flywheel.................64-65 N·m
(47-48 ft.-lbs.)
Injector.................15-16 N·m
(11-12 ft.-lbs.)
Main bearing supports.....45-49 N·m
(33-36 ft.-lbs.)

VALVE ADJUSTMENT

To adjust valve clearance, first remove valve covers. Rotate crankshaft until number 1 piston is at TDC on compression stroke (both rocker arms should be free to move). With engine cold, valve clearance should be 0.15 mm (0.006 inch) for intake and exhaust. Turn rocker arm adjusting screw (22 – Fig. S3-4) until appropriate size feeler gage can be inserted between rocker arm and valve stem end. Tighten adjusting screw locknut and recheck adjustment.

Rotate crankshaft and adjust valves of each remaining cylinder in order of firing (1-3-2).

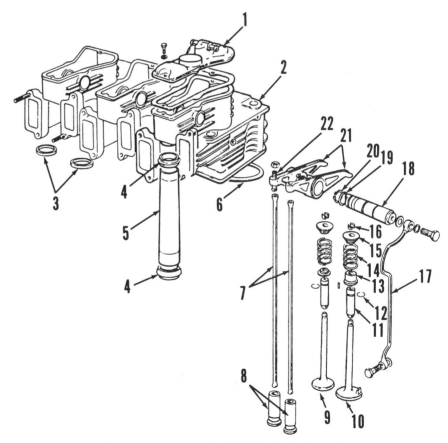

Fig. S3-4 – Exploded view of cylinder head and valve system.
1. Valve cover
2. Cylinder head
3. Valve seat inserts
4. Seals
5. Push rod tube
6. Head gasket
7. Push rods
8. Cam followers
9. Exhaust valve
10. Intake valve
11. Valve guide
12. Snap ring
13. Cap nut (intake only)
14. Valve spring
15. Retainer
16. Valve keepers
17. Oil tube
18. Rocker shaft
19. Gasket
20. Snap ring
21. Rocker arms
22. Adjusting screw

Illustrations Courtesy North American Diesel, Inc.

Slanzi SMALL DIESEL

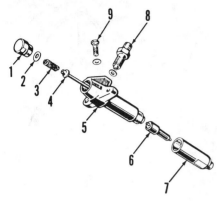

Fig. S3-5 — Exploded view of injector used on DVA1550 engine.

1. Cap nut
2. Shim
3. Spring
4. Push rod
5. Injector body
6. Nozzle valve assy.
7. Nozzle holder nut
8. Fuel inlet fitting
9. Leak-off banjo bolt

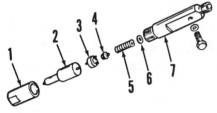

Fig. S3-5A — Exploded view of injector used on DVA1750 engine.

1. Nozzle holder nut
2. Nozzle valve
3. Spacer
4. Spring seat
5. Spring
6. Shim
7. Body

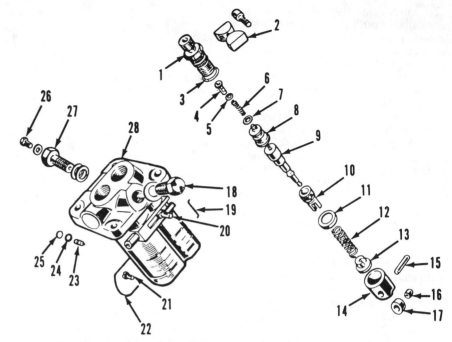

Fig. S3-6 — Exploded view of BOSCH fuel injection pump used on all engines.

1. Delivery valve holder
2. Lock
3. Gasket
4. Spring guide
5. Shim
6. Spring
7. Gasket
8. Delivery valve
9. Pumping element
10. Sleeve
11. Gasket
12. Spring
13. Sleeve
14. Tappet
15. Pin
16. Inner roller
17. Outer roller
18. Plug
19. Retainer
20. Control rack pin
21. Screw
22. Retainer
23. Eccentric
24. Gasket
25. Plug
26. Bleed screw
27. Fuel inlet banjo bolt
28. Pump housing

CYLINDER HEADS AND VALVE SYSTEM

An individual cylinder head is used for each cylinder. To remove cylinder heads, remove blower fan and air shields, crankcase breather and tubes and intake and exhaust manifolds with air cleaner and muffler. Disconnect and remove injector high pressure and return fuel lines. Remove lubricating oil tubes from cylinder heads. Remove cylinder head stud nuts and lift off head. Remove valve cover, rocker arm shaft (18 – Fig. S3-4) and rocker arms (21) for access to valves.

Valve stem diameter is 7.976-7.985 mm (0.3140-0.3144 inch) for intake and exhaust valves. Desired clearance of valve stem in guide is 0.015-0.039 mm (0.0006-0.0015 inch) for intake and 0.020-0.044 mm (0.0008-0.0017 inch) for exhaust. Valve guides are renewable. When pressing in new guides, use suitable installing tool to prevent damage to guide. A snap ring (12) at top of guide determines installed height of guides. After installation, make certain valve stem move freely in guide. Guides are presized and should not require reaming if carefully installed.

Valve seat inserts are used for intake and exhaust. Inserts should have an interference fit of 0.130-0.180 mm (0.005-0.007 inch) in cylinder head counterbore. Chilling inserts in dry ice prior to installation will ease assembly. When properly installed, valve seat should be recessed 2.35-2.50 mm (0.093-0.098 inch) below cylinder head surface. With valves positioned in cylinder head, measure the distance valve head projects from head surface. Both intake and exhaust valves should project 0.90-1.10 mm (0.035-0.043 inch). If projection is excessive, regrind valve seat. If projection is less than specified, renew valve seat.

Injector nozzle tip should extend 2.3-2.7 mm (0.091-0.106 inch) beyond cylinder head surface. Adjust nozzle position, if necessary, using shim gaskets between nozzle valve and bottom of cylinder head nozzle bore.

If pistons or cylinders were renewed, check and adjust piston height in cylinder as outlined in CYLINDERS section before reinstalling head. If necessary, renew push rod tube seals. Tighten cylinder head stud nuts evenly using a crossing pattern to 54-58 N·m (40-43 ft.-lbs.). Complete installation by reversing removal procedure. Adjust valve clearance as previously outlined.

INJECTOR

REMOVE AND REINSTALL. Prior to removing injectors, thoroughly clean injectors, fuel lines and surrounding area. Disconnect and remove return lines and high pressure fuel lines and plug all openings to prevent entry of dirt. Unscrew retaining nuts and withdraw injector from cylinder head.

Make sure nozzle bore in cylinder head is clean before reinstalling injector. Use same thickness of seal washer to maintain nozzle tip protrusion in cylinder head. Tighten retaining nuts to 15-16 N·m (11-12 ft.-lbs.).

TESTING. A suitable test stand is required to check injector operation. Only clean, approved testing oil should be used to test injectors.

WARNING: Fuel leaves the injector nozzle wih sufficient force to penetrate the skin. When testing, keep yourself clear of nozzle spray.

Connect injector to tester and operate tester lever a few quick strokes to purge air from injector and to make sure nozzle valve is not stuck. When operating properly, injector nozzle will emit a buzzing sound and produce a uniform spray pattern.

SERVICE MANUAL

Slanzi

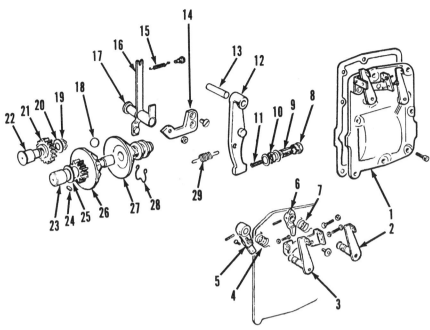

Fig. S3-7 – Exploded view of governor and governor linkage.

1. Governor cover
2. Stop lever
3. Throttle lever
4. Spring
5. Lever
6. Cam
7. Spring
8. Plug
9. Adjusting screw
10. Nut
11. Spring
12. Pivot arm
13. Pin
14. Bracket
15. Spring
16. Governor lever
17. Eccentric pin
18. Flyball
19. Snap ring
20. Thrust washer
21. Idler gear
22. Idler shaft
23. Governor shaft
24. Set screw
25. Thrust washer
26. Collar
27. Governor cup
28. Spring
29. Spring

all parts are clean and wet with diesel fuel. Recheck injector operation as outlined in TESTING section.

INJECTION PUMP

To remove injection pump, disconnect fuel lines and immediately plug all openings to prevent entry of dirt. Remove retaining nuts and lift out injection pump.

An exploded view of injection pump is shown in Fig. S3-6. It is recommended that injection pump be tested and serviced only by a shop qualified in diesel fuel injection repair.

To reinstall injection pump, reverse the removal procedure. Note that pump mounting gasket thickness will affect pump timing. To determine correct gasket thickness, rotate crankshaft until one lobe of injection pump camshaft is pointing downward. Measure the distance from mounting surface of crankcase to base of camshaft lobe. Assemble gasket thickness necessary to obtain mounting dimension of 82.6-83.0 mm (3.252-3.267 inches). Check injection pump timing as outlined in MAINTENANCE section.

GOVERNOR

The flyball type governor is located in the side of crankcase behind the governor cover assembly (1 – Fig. S3-7). The governor is driven by a gear on the injection pump camshaft via an idler gear (21). Governor rotation causes flyballs (18) to move in and out which moves the

To check opening pressure, slowly operate tester lever while observing pressure gage reading. Opening pressure should be 19615-20595 kPa (2845-2985 psi). Opening pressure is adjusted by varying the thickness of shim (2 – Fig. S3-5 or 6 – Fig. S3-5A).

Operate tester lever slowly to maintain pressure at 2000 kPa (330 psi) below opening pressure and check for leakage past nozzle valve. If a drop forms at nozzle tip within a 10 second period, nozzle valve is not seating and injector must be overhauled or renewed.

OVERHAUL. To disassemble DVA1550 injector, refer to Fig. S3-5 and proceed as follows: Secure nozzle body (5) in a vise, then remove nozzle holder (7) and nozzle valve (6). Remove end cap (1) and withdraw shim (2), spring (3) and push rod (4).

To disassemble DVA1750 injector, refer to Fig. S3-5A and proceed as follows: Secure injector body (7) in a vise or holding fixture, then remove nozzle holder nut (1), nozzle valve (2), spacer (3), spring seat (4), spring (5) and shim (6).

NOTE: When disassembling more than one injector, keep parts separated as they should not be interchanged.

Thoroughly clean all parts in a suitable solvent. Do not use steel wire brush or sharp metal tools to clean injector components. Make certain nozzle needle slides freely in nozzle valve seat. If needle sticks, reclean or renew nozzle valve assembly.

When reassembling injector, be sure

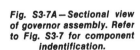

Fig. S3-7A – Sectional view of governor assembly. Refer to Fig. S3-7 for component indentification.

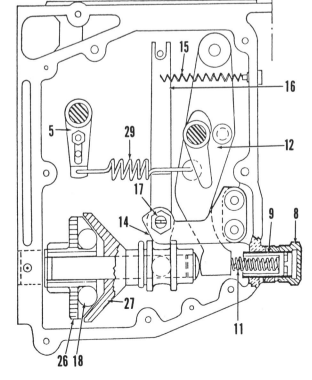

Slanzi

SMALL DIESEL

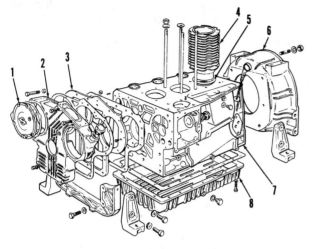

Fig. S3-8 — Exploded view of crankcase and front cover assembly.

1. Cover plate
2. Front cover
3. Plate
4. Cylinder
5. Shim gasket
6. Flywheel housing
7. Crankcase
8. Oil pan

pin, is 85.890-85.920 mm (3.3815-3.3827 inches) for DVA1550 and 91.885-91.905 mm (3.6175-3.6183 inches) for DVA1750. Recommended piston ring end gap for all engines is 0.40-0.65 mm (0.016-0.026 inch) for the compression rings and 0.25-0.45 mm (0.010-0.018 inch) for the oil control ring.

Piston pin is retained in piston by two snap rings. Pin diameter is 25.995-26.000 mm (1.0234-1.0236 inches) and piston pin bore is 25.991-26.000 mm (1.0233-1.0236 inches). Pin is a transition fit in piston bore. Initial fit when cold is 0.009 mm (0.0003 inch) interference to 0.005 mm (0.0002 inch) loose. Heating piston will ease removal and installation of pin.

governor cup (27). The cup movement is transferred through the governor linkage to actuate injection pump control rack.

Remove governor cover (1) to gain access to governor flyball assembly. Remove set screw (24) securing governor shaft, then remove governor linkage and flyball assembly.

Inspect all parts for wear or other damage and renew if necessary. Be sure all parts move freely. Reinstall by reversing the removal procedure.

To adjust governor linkage, proceed as follows: To reduce tendency of governor to surge during changes in speed or load, remove plug (8 — Fig. S3-7 or S3-7A) and turn adjusting screw (9) inward to increase spring (11) projection. Note that this adjustment will also alter engine speed adjustment. If governor still does not operate smoothly, loosen nut and rotate eccentric pin (17) to move axis of pin upward slightly.

To increase governor sensitivity in reacting to change in load (speed droop), loosen nut and shorten lever (5) to increase tension on spring (29).

Engine idle speed is adjusted by turning stop screw (1 — Fig. S3-2). Maximum governed speed is adjusted using screw (3). Maximum injection pump fuel delivery is adjusted with screw (5). Turning screw inward (clockwise) increases maximum fuel delivery.

Injection pump control rack is automatically set in increased fuel delivery position for starting by tension of spring (15 — Fig. S3-7).

PISTON AND ROD UNIT

Pistons and connecting rods may be removed after first removing cylinder heads, oil pan and oil pickup. Be sure to number piston and rod units as they are removed so units can be reinstalled in their original position.

When reinstalling piston and rod, be sure to position piston so depression in piston crown will be aligned with injector nozzle tip when cylinder head is installed. Make certain locating tabs of connecting rod bearing insert engage notches in rod and cap and are located on the same side. Tighten connecting rod cap nuts to 31-37 N·m (23-27 ft.-lbs.).

If connecting rod or piston was renewed, refer to CYLINDER section and measure piston height in cylinder.

PISTON, PIN AND RINGS

The piston is equipped with two compression rings and an oil control ring. Standard piston diameter, measured at bottom of skirt perpendicular to piston

CONNECTING ROD

The connecting rod small end is fitted with a renewable bushing. Bushing inside diameter should be 26.010-26.020 mm (1.0240-0.0244 inches). Recommended clearance between piston pin and bushing is 0.010-0.030 mm (0.0004-0.0012 inch).

A precision, insert type bearing is used in crankpin end of connecting rod. With rod bearing installed and rod cap nuts tightened to 31-37 N·m (23-27 ft.-lbs.), inside diameter of bearing should be 53.073-53.108 mm (2.0895-2.0909 inches). Recommended rod bearing clearance is 0.038-0.086 mm (0.0015-0.0034 inch). Bearings are available in standard size and undersizes.

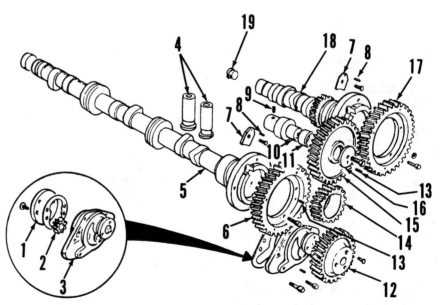

Fig. S3-9 — View of timing gears, valve camshaft, fuel injection camshaft and oil pump assembly.

1. Cover
2. Pump cover
3. Oil pump housing
4. Cam followers
5. Valve camshaft
6. Gear
7. Retainer plate
8. Locating pin
9. Set screw
10. Idler shaft
11. Bushing
12. Oil pump gear
13. Locating pin
14. Crankshaft gear
15. Idler gear
16. Retaining plate
17. Gear
18. Fuel injection camshaft
19. Plug

Illustrations Courtesy North American Diesel, Inc.

SERVICE MANUAL

Slanzi

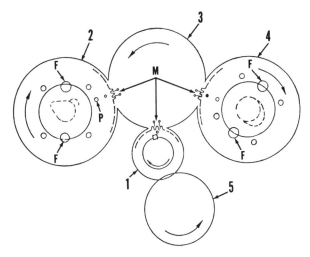

Fig. S3-10 – When reinstalling timing gears, timing marks (M) must be aligned as shown. Refer to text.
1. Crankshaft gear
2. Valve camshaft gear
3. Idler gear
4. Injection pump camshaft gear
5. Oil pump gear

pin hole (P) are positioned as shown. Install gears (2 and 4) aligning their timing marks with the respective marks on idler gear and insert locating pin into hole in camshaft gear and shaft flange. Secure gears with socket head retaining screws. Check injection pump timing as outlined in MAINTENANCE section.

VALVE CAMSHAFT AND CAM FOLLOWERS

The camshaft rides in a bored sleeve bearing at front of crankcase and in unbushed bores at center and rear of crankcase. The camshaft must be removed before the mushroom type cam followers can be removed.

To remove camshaft, first remove timing gear cover and camshaft drive gear (6 – Fig. S3-9) as outlined in TIMING GEAR section. Remove cylinder heads, push rods and push rod tubes. Remove fuel transfer pump and push rod. Raise cam followers off cam lobes and support with suitable tools. Remove camshaft retaining plate (7), then carefully withdraw camshaft. Cam followers (4) may now be removed through bottom of crankcase after removing oil pan. Be sure to identify cam followers as they

CYLINDER

Engines are equipped with removable, cast iron cylinders. Standard cylinder bore is 85.980-86.000 mm (3.3850-3.3858 inches) for DVA1550 and 92.000-92.020 mm (3.6220-3.6228 inches) for DVA1750. Desired piston skirt clearance in cylinder is 0.080-0.120 mm (0.003-0.005 inch) for DVA1550 and 0.095-0.135 mm (0.004-0.005 inch) for DVA1750. If cylinder bore is damaged or excessively worn, cylinder may be rebored for installation of 0.5 mm (0.020 inch) oversize piston and rings.

Piston height in cylinder is adjusted by varying shim gasket (5 – Fig. S3-8) thickness between bottom of cylinder and crankcase mounting surface. With piston at top dead center, top of piston must be even with to 0.2 mm (0.008 inch) below top edge of cylinder.

OIL PUMP

The engine oil pump (3 – Fig. S3-9) is mounted on the front of the crankcase and is driven by the crankshaft gear. The oil pressure relief valve assembly is located in drilled passage in the side of crankcase behind the fuel filter.

To gain access to pump, crankshaft front pulley and timing gear cover must first be removed. Unbolt and remove pump drive gear, then remove pump housing assembly.

Inspect pump gears and housing for excessive wear or damage and renew if necessary. To reinstall pump, reverse the removal procedure.

TIMING GEARS AND FRONT COVER

To gain access to timing gears, first remove crankshaft front pulley and cooling fan housing. Unbolt and remove crankcase front cover. Rotate crankshaft until number 1 piston is at TDC on compression stroke and timing marks (M – Fig. S3-10) on gears are aligned as shown. Then, remove gears (Fig. S3-9) as needed.

To reassemble, first install idler gear (3 – Fig. S3-10) aligning center timing marks with mark on crankshaft gear (1). Turn valve camshaft and injection pump camshaft so flange holes (F) and locating

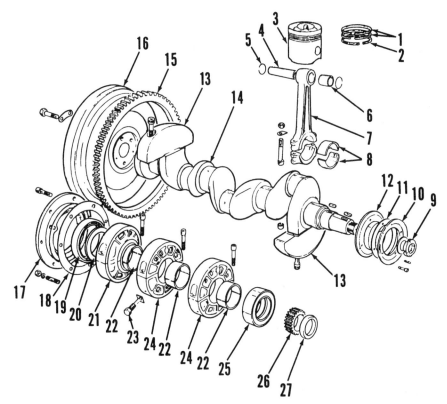

Fig. S3-11 – Exploded view of piston, connecting rod and crankshaft assembly.
1. Compression rings
2. Oil control ring
3. Piston
4. Piston pin
5. Retaining rings
6. Bushing
7. Connecting rod
8. Rod bearing
9. Nut
10. Pulley half
11. Shim
12. Pulley half
13. Counterweight
14. Crankshaft
15. Ring gear
16. Flywheel
17. Seal retainer
18. Gasket
19. Oil seal
20. Thrust washer
21. Rear bearing carrier
22. Main bearings
23. Locating screw
24. Center bearing carrier
25. Front main bearing
26. Crankshaft gear
27. Oil seal

Illustrations Courtesy North American Diesel, Inc.

341

Slanzi

SMALL DIESEL

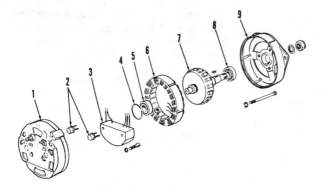

Fig. S3-12 — Exploded view of alternator assembly.
1. Stator housing
2. Diodes
3. Voltage regulator
4. Retaining ring
5. Bearing
6. Stator
7. Rotor
8. Bearing
9. End cover

are removed so they can be reinstalled in their original locations.

Inspect cam lobes and bearing journals for scoring, wear and other damage. Diameter of bearing journals (from front to rear) is 41.950-41.975 mm (1.6516-1.6526 inches) for first journal, 40.950-41.975 mm (1.6122-1.6132 inches) for second journal, 39.950-39.975 mm (1.5728-1.5738 inches) for third journal and 29.959-29.980 mm (1.1795-1.1803 inches) for rear journal. Inner diameter of camshaft support bores in the crankcase (from front to rear) is 42.020-42.045 mm (1.6543-1.6553 inches) for front bored sleeve bearing, 41.020-41.045 mm (1.6150-1.6160 inches) for second bore, 40.020-40.045 mm (1.5756-1.5766 inches) for third bore and 30.020-30.040 mm (1.1819-1.1827 inches) for rear bore. Desired camshaft running clearance in crankcase is 0.040-0.081 mm (0.0016-0.0032 inch) for rear journal and 0.045-0.095 mm (0.0018-0.0037 inch) for all other journals. Camshaft end play is set by retaining plate (7). Desired end play is 0.040-0.165 mm (0.0016-0.0065 inch).

Thoroughly lubricate cam followers, camshaft bearing journals and lobes during installation. Reinstall by reversing the removal procedure. Refer to TIMING GEARS section for proper installation and timing of camshaft gear.

INJECTION PUMP CAMSHAFT

To remove injection pump camshaft (18 – Fig. S3-9), first remove injection pump, timing gear cover and camshaft drive gear (17). Remove retaining plate (7), then withdraw camshaft.

Inspect camshaft bearing journals and cam lobes for scoring, wear or other damage. If cam lobes are worn or damaged, inspect injection pump tappet rollers also. Camshaft front journal diameter should be 49.950-49.975 mm (1.9665-1.9675 inches) and rear journal diameter should be 29.959-29.980 mm (1.1795-1.1803 inches). Crankcase bearing bore inner diameter should be 50.000-50.025 mm (1.9685-1.9695 inches) for front sleeve bearing and 30.000-30.021 mm (1.1811-1.1819 inches) for rear bore. Desired operating clearance in crankcase is 0.025-0.075 mm (0.001-0.003 inch) for front journal and 0.020-0.062 mm (0.0008-0.0024 inch) for rear journal. Camshaft end play is controlled by retaining plate (7). Desired end play is 0.040-0.165 mm (0.0016-0.0065 inch).

Reinstall camshaft by reversing the removal procedure. Install and time the drive gear as outlined in TIMING GEARS section. Check injection pump timing as outlined in MAINTENANCE section.

CRANKSHAFT AND BEARINGS

The crankshaft rides in four main bearings. The front bearing (25 – Fig. S3-11) is a one-piece assembly while the two center bearings and the rear bearing are split bearing inserts (22) contained in two-piece bearing supports (21 and 24).

To remove the crankshaft, first remove oil pan, front pulley and timing gear cover. Remove flywheel and rear seal retainer (17). Remove connecting rod caps and push rods and pistons into cylinders for clearance. Remove center bearing support locating screws (23), then carefully withdraw crankshaft out rear of crankcase.

Standard diameter of main journals is 59.960-59.980 mm (2.3606-2.3614 inches). Main bearing inner diameter should be 60.011-60.054 mm (2.3626-2.3643 inches). Desired main bearing clearance is 0.031-0.094 mm (0.0012-0.0037 inch). Undersize main bearings are available.

Crankpin standard diameter is 53.022-53.035 mm (2.0875-2.0880 inches). Desired rod bearing clearance is 0.038-0.086 mm (0.0015-0.0034 inch). Undersize bearings are available.

Crankshaft end play should be 0.070-0.259 mm (0.003-0.010 inch). End play is adjusted by changing thrust washer (20) thickness.

When reinstalling crankshaft, be sure to align timing gear timing marks as shown in Fig. S3-10. Renew oil seals (19 and 27), if necessary, and lubricate lip of seals with grease prior to reassembly. Tighten bearing support socket head screws to 45-49 N·m (33-36 ft.-lbs.). Tighten counterweight (13) cap screws, if removed, to 58-64 N·m (43-47 ft.-lbs.). Tighten flywheel cap screws to 64-65 N·m (47-48 ft.-lbs.). Complete installation by reversing the removal procedure.

ELECTRICAL SYSTEM

The electrical system is 12-volt negative ground. The alternator is mounted in front of the engine cooling fan and is belt driven by the crankshaft. Alternator and blower fan belt tension is adjusted by adding or removing shims (11 – Fig. S3-11) between crankshaft pulley halves on early production engines or by moving an idler pulley on late production engines. An integral solid-state regulator controls alternator output. See Fig. S3-12 for an exploded view of alternator.

A BOSCH or HARELLI starter motor is used on all engines. Refer to Fig. S3-13 for an exploded view of starter assembly.

Fig. S3-13 — Exploded view of electric starter motor.
1. Solenoid
2. Bushing
3. Drive end housing
4. Lever
5. Brush frame
6. Cover
7. Brush
8. Spring
9. Bushing
10. Field coil & housing
11. Retainer
12. Bushing
13. Starter drive
14. Armature

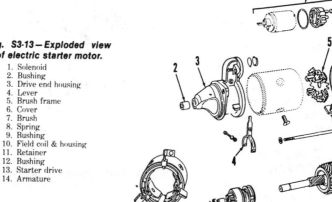

SLANZI

Model	No. Cyls.	Bore	Stroke	Displ.
DVA1800	2	100 mm (3.937 in.)	114 mm (4.488 in.)	1792 cc (109 cu. in.)
DVA2700	3	100 mm (3.937 in.)	114 mm (4.488 in.)	2686 cc (164 cu. in.)
DVA3600	4	100 mm (3.937 in.)	114 mm (4.488 in.)	3585 cc (219 cu. in.)
DVA3600TB	4	100 mm (3.937 in.)	114 mm (4.488 in.)	3585 cc (219 cu. in.)

All engines are four-stroke, air-cooled, direct injection diesels with vertical inline cylinders. Crankshaft rotation is counterclockwise at flywheel end. Number one cylinder is nearest crankshaft pulley end of engine. Firing order for DVA2700 is 1-3-2 and for DVA3600 and DVA3600TB is 1-3-4-2. DVA3600TB engine is equipped with a turbocharger.

MAINTENANCE

LUBRICATION

Recommended engine oil is SAE 20W-20 when ambient temperature is below 0°C (32°F). When temperature will be above 0°C (32°F), use SAE 30 for winter operation or SAE 40 for summer operation. Multigrade oil is acceptable for use in all temperatures. Manufacturer suggests using SAE 15W-40. Regardless of oil viscosity, use only premium grade, heavy-duty, diesel engine oil meeting API CC specifications.

Manufacturer recommends changing the engine oil after every 150 hours of operation. Oil should be drained while engine is hot. The engine oil filter should be changed after every 450 hours of operation. Crankcase capacity is approximately 6.7 liters (7.1 quarts) for DVA1800, 10 liters (10.5 quarts) for DVA2700 and 13.5 liters (14.2 quarts) for DVA3600 and DVA3600TB.

The engine oil pressure relief valve (1 – Fig. S4-1) and oil cooler bypass valve (4), if so equipped, are located in the oil filter adapter housing (6).

FUEL SYSTEM

FUEL FILTER AND BLEEDING.
All engines are equipped with a renewable fuel filter. Manufacturer recommends renewing filter element after every 200 hours of operation or more often if engine indicates signs of fuel starvation (surging or loss of power). After renewing fuel filter, air must be bled from the fuel system.

NOTE: A plastic filter screen is located in fuel transfer pump inlet. If engine fails to run properly after main fuel filter element has been renewed, remove and clean pump filter screen (3 – Fig. S4-3).

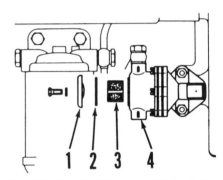

Fig. S4-3—A plastic filter screen (3) is located on inlet side of fuel transfer pump (4).

1. Cover
2. Gasket
3. Filter screen
4. Fuel transfer pump

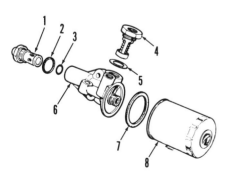

Fig. S4-1—Engine oil pressure relief valve (1) and oil cooler bypass valve (4), if equipped, are located in oil filter adapter housing.

1. Pressure relief valve
2. Gasket
3. "O" ring
4. Oil cooler bypass valve
5. Gasket
6. Filter adapter housing
7. Gasket
8. Oil filter

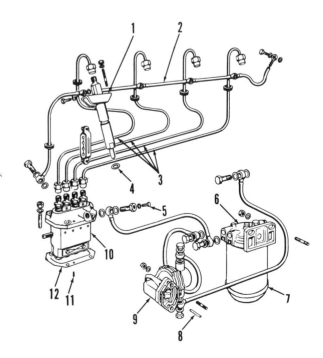

Fig. S4-2—Typical fuel system components used on DVA3600 engine. Other engines are similar.

1. Injector
2. Fuel return line
3. High pressure injector lines
4. Nozzle gasket
5. Bleed screw
6. Bleed screw
7. Fuel filter
8. Push rod
9. Fuel transfer pump
10. Injection pump
11. Pin
12. Shim gasket

Slanzi SMALL DIESEL

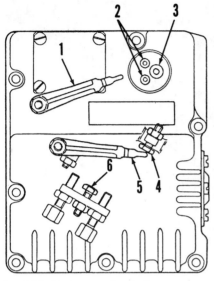

Fig. S4-4—Engine speed adjusting screws are located on governor cover.
1. Stop lever
2. Fuel limiting device screws
3. Extra fuel knob
4. Low speed screw
5. Throttle lever
6. High speed screw

Fig. S4-6—Exploded view of axial flow engine cooling fan.
1. Belt guard
2. Pulley
3. Alternator belt
4. Stator housing
5. Fan
6. Snap ring
7. Bearings
8. Spacer
9. Fan shaft
10. Idler & belt tightener
11. Drive belt

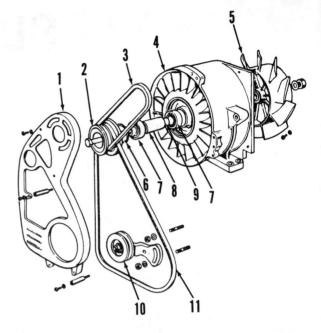

To bleed air from fuel system, loosen bleed screw (6–Fig. S4-2) on filter mounting bracket. Manually operate transfer pump until air-free fuel flows from bleeder, then retighten screw. Loosen bleed screw (5) on injection pump fuel inlet. Actuate transfer pump until fuel flows from bleeder, then retighten the screw. Continue to operate transfer pump until resistance is felt indicating fuel system is pressurized.

If engine fails to start at this point, loosen high pressure fuel lines (3) at injectors. Set throttle control in full speed position. Crank engine until fuel is discharged at loosened fuel line connections, then retighten fittings and start engine.

ENGINE SPEED ADJUSTMENT

Engine should be at normal operating temperature when checking and adjusting engine speed settings. Low and high idle (no-load) speeds are adjusted using throttle lever stop screws (4 and 6–Fig. S4-4) on outside of governor housing. Turning screw (4) clockwise will decrease low idle speed and turning screw (6) clockwise will increase maximum governed speed. Adjust screws as required to obtain governed speeds as specified for engine application.

Injection pump maximum fuel delivery is adjusted by loosening the two screws (2) and rotating fuel limiting device. Pulling knob (3) of fuel limiting device outward will allow pump control rack to move to maximum fuel delivery position to provide extra fuel delivery for starting.

INJECTION PUMP TIMING

The injection pump is actuated by the injection pump camshaft (2–Fig. S4-15). Injection timing is adjusted by loosening timing gear (1) retaining screws and rotating camshaft within limits of retaining screw slots in timing gear.

To check injection timing, first rotate crankshaft pulley clockwise until number 1 piston is on compression and second mark (top dead center) on pulley is aligned with timing pointer on front cover. Disconnect injector line for number 1 cylinder from injection pump. Connect a fuel flow sight gage (special tool number 59.40.50) to number one discharge fitting on pump. If tool is not available, remove delivery valve holder (1–Fig. S4-9), spring (3) and delivery valve (5), then reinstall delivery valve holder. Remove governor cover and position pump control rack at half speed setting. Maintain this setting during timing check. Connect fuel supply to pump. Note that fuel should flow from delivery valve holder at this time if delivery valve is removed. Rotate crankshaft pulley counterclockwise about ½ turn, then slowly turn clockwise (normal direction of rotation) until first drop of fuel is observed at sight gage tool, or until fuel flow just stops if delivery valve is removed. At this point, beginning of injection occurs and first mark on crankshaft pulley should be aligned with timing pointer. If timing mark is not aligned, remove camshaft gear inspection cover from crankcase front cover. Loosen camshaft gear mounting screws, then rotate injection pump camshaft within limits of camshaft gear screw slots to obtain correct timing. Rotating camshaft clockwise will advance timing and counterclockwise will retard timing.

DRIVE BELTS

The cooling fan drive belt and alternator drive belt tension should be checked after every 150 hours of operation. Under thumb pressure, cooling fan belt should deflect 13-15 mm (17/32-19/32 inch) as shown at (B–Fig. S4-5). Belt tension is adjusted by moving the idler pulley (3). The alternator drive belt should deflect 8-10 mm (5/16-3/8 inch) under finger pressure as shown at (A). To adjust, loosen nuts securing alternator (1) and reposition alternator to ob-

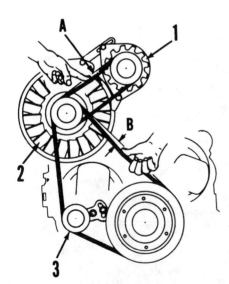

Fig. S4-5—Drive belt tension (A and B) should be checked after every 150 hours of operation. Refer to text for adjustment.
1. Alternator
2. Cooling fan
3. Idler

344

Illustrations Courtesy North American Diesel, Inc.

SERVICE MANUAL

tain recommended tension. After adjusting belt tension, check for interference between alternator and belt guard. If alternator contacts the belt guard, renew alternator drive belt.

REPAIRS

TIGHTENING TORQUES

Refer to the following table for special tightening torques. All fasteners are metric.

Connecting rod..........57-61 N·m
(42-45 ft.-lbs.)
Crankcase crossbeam......42-46 N·m
(31-34 ft.-lbs.)
Crankshaft counterweight..59-63 N·m
(44-47 ft.-lbs.)
Cylinder head.............76-80 N·m
(56-59 ft.-lbs.)
Flywheel................105-111 N·m
(78-82 ft.-lbs.)
Injector..................18-20 N·m
(13-15 ft.-lbs.)
Main bearing cap........172-182 N·m
(127-134 ft.-lbs.)

COOLING FAN

An axial flow fan, belt driven by the crankshaft, provides engine cooling air flow. To disassemble fan, first remove belt guard (1 – Fig. S4-6) and drive belts (3 and 11). Remove fan retaining nut and withdraw fan (5). Remove snap ring (6), then tap fan shaft (9) and bearings (7) out of stator housing (4).

Inspect all parts and renew if necessary. To reinstall, reverse the removal procedure. Refer to DRIVE BELTS section to adjust belt tension.

VALVE ADJUSTMENT

To adjust valve clearance, first remove valve covers and rotate crankshaft until number 1 piston is at TDC on compression stroke (both rocker arms should be free to move). With engine cold (room temperature), valve clearance should be 0.15 mm (0.006 inch) for intake and exhaust. Turn rocker arm adjusting screw (6 – Fig. S4-7) until appropriate size feeler gage can be inserted between rocker arm and valve stem end. Recheck adjustment after retightening screw locknut.

Rotate crankshaft to position next piston in firing order at TDC on compression stroke and adjust valves for that cylinder.

CYLINDER HEAD AND VALVE SYSTEM

To remove cylinder heads, first remove intake and exhaust manifolds

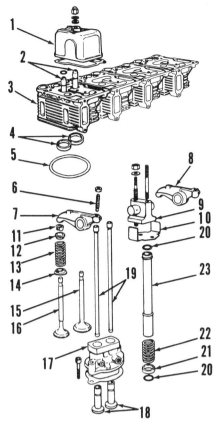

Fig. S4-7—Exploded view of cylinder heads and valve system.

1. Valve cover
2. Valve guides
3. Cylinder heads
4. Valve seat inserts
5. Head gasket
6. Adjusting screw
7. Rocker arm (exhaust)
8. Rocker arm (intake)
9. Rocker arm shaft
10. Rocker arm retainer
11. Valve keepers
12. Retainer
13. Valve spring
14. Spring seat
15. Intake valve
16. Exhaust valve
17. Cam follower guide
18. Cam followers
19. Push rods
20. Gasket
21. Retainer
22. Spring
23. Push rod tube

and cooling air shroud. Disconnect and remove fuel lines from injectors and immediately plug all openings. Remove the injectors. Remove cylinder head stud nuts and lift off cylinder heads. Remove valve cover and rocker arm support for access to valves.

Valve stem diameter is 8.961-8.970 mm (0.3528-0.3531 inch) for intake and exhaust. Valve guide inner diameter should be 9.005-9.020 mm (0.3545-0.3551 inch). Desired clearance of valve stem in guide is 0.035-0.059 mm (0.0014-0.0023 inch). Valve guides are renewable. When installing new guides, use a suitable installing tool to avoid damaging guides. After installation, check fit of valve stem in guide. Replacement guides are presized and should not require reaming if carefully installed.

Renewable valve seat inserts are used for intake and exhaust. Inserts should have an interference fit of 0.115-0.160 mm (0.0045-0.0063 inch) in cylinder head counterbore. Chilling inserts in dry ice prior to installation will ease

Slanzi

assembly. When properly installed, valve seat should be recessed 1.03-1.18 mm (0.040-0.046 inch) below cylinder head surface. Position valves in cylinder head and measure distance top of valve is recessed below cylinder head surface. Valve recession should be 0.9-1.1 mm (0.035-0.043 inch). If valve recession is excessive, renew valve seat. If recession is less than specified, regrind valve seat.

Rocker arm pivot pin diameter is 20.035-20.043 mm (0.7888-0.7891 inch). Rocker arm bearing inner diameter should be 20.030-20.060 mm (0.7886-0.7897 inch). Desired rocker arm operating clearance is 0.012-0.045 mm (0.0005-0.0018 inch).

The injector nozzle tip should extend 2.3-2.7 mm (0.090-0.106 inch) beyond cylinder head surface. Adjust nozzle position using shim gaskets between nozzle valve and bottom of cylinder head nozzle bore.

If pistons or cylinders were renewed, refer to CYLINDERS section and check and adjust piston height in cylinder before reinstalling cylinder heads. To reinstall heads, reverse the removal procedure. Be sure to renew push rod tube seals if necessary. Tighten cylinder head stud nuts evenly to 76-80 N·m (56-59 ft.-lbs.). Adjust valve clearance as previously outlined.

INJECTOR

REMOVE AND REINSTALL. Before removing injectors, thoroughly clean injector, fuel lines and surrounding area. Disconnect fuel return line and high pressure line at injector and plug all openings. Unscrew retaining nuts and withdraw injector from cylinder head.

Before reinstalling injector, make certain nozzle bore in cylinder head is clean. Reinstall injector and tighten stud nuts to 18-20 N·m (13-15 ft.-lbs.).

TESTING. A suitable test stand is required to check injector operation. Only clean, approved testing oil should be used to test injector.

WARNING: Fuel leaves the injector nozzle with sufficient force to penetrate the skin. When testing, keep yourself clear of nozzle spray.

Connect injector to tester and operate tester lever a few quick strokes to purge air from injector and to make sure nozzle valve is not stuck. When operating properly, injector nozzle will emit a buzzing sound and produce a uniform spray pattern.

To check opening pressure, slowly operate tester lever while observing the pressure gage reading. Opening pressure should be 20000-21000 kPa

Slanzi   SMALL DIESEL

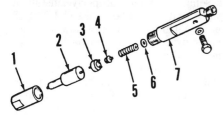

Fig. S4-8—Exploded view of injector assembly.
1. Nozzle holder nut
2. Nozzle valve
3. Spacer
4. Spring seat
5. Spring
6. Shim
7. Injector body

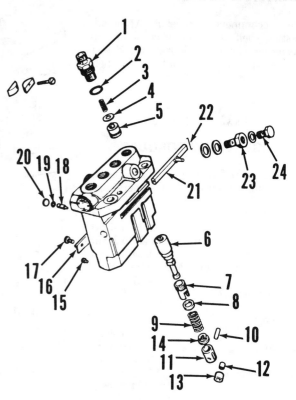

Fig. S4-9—Exploded view of injection pump assembly.
1. Delivery valve holder
2. Gasket
3. Spring
4. Gasket
5. Delivery valve
6. Pumping element
7. Sleeve
8. Retainer
9. Spring
10. Pin
11. Tappet
12. Inner roller
13. Outer roller
14. Retainer
15. Screw
16. Retainer plate
17. Screw
18. Eccentric
19. Gasket
20. Plug
21. Control rack
22. Retainer
23. Fuel inlet banjo bolt
24. Bleed screw

(2900-3045 psi). Opening pressure is adjusted by varying the thickness of shim (6–Fig. S4-8).

Operate tester lever slowly to maintain pressure at 2000 kPa (300 psi) below opening pressure and check for leakage past nozzle valve seat. If a drop forms at nozzle tip within a 10 second period, nozzle valve is not seating and injector must be overhauled or renewed.

OVERHAUL. To disassemble injector, secure nozzle body (7–Fig. S4-8) in a vise or holding fixture. Remove nozzle holder nut (1) and withdraw nozzle valve (2), spacer (3), spring seat (4), spring (5) and shim (6). If disassembling more than one injector, be sure to keep parts for each injector separated as they should not be interchanged.

Thoroughly clean all parts in a suitable solvent. Do not use steel wire brush or sharp metal tools to clean injector components. Make certain the nozzle valve needle slides freely in nozzle valve body. If needle sticks, reclean or renew nozzle valve assembly. Nozzle valve spray holes are 0.27 mm (0.010 inch) in diameter. Spray holes may be cleaned by inserting a cleaning wire that is slightly smaller in diameter than the spray hole.

When reassembling injector, be sure all parts are clean and wet with diesel fuel. Recheck injector operation as outlined in TESTING section.

INJECTION PUMP

To remove injection pump, first remove cooling air shroud. Disconnect and remove fuel lines and immediately plug all openings to prevent entry of dirt. Remove retaining nuts and lift out injection pump.

An exploded view of injection pump is shown in Fig. S4-9. It is recommended that injection pump be tested and serviced only by a shop qualified in diesel fuel injection repair.

When reinstalling pump, note that pump mounting gasket thickness will affect pump timing. To determine correct gasket thickness, rotate crankshaft until one lobe of injection pump camshaft is pointing downward. Measure distance from mounting surface of crankcase to base of camshaft lobe. Assemble gasket thickness necessary to obtain mounting dimension of 82.6-83.0 mm (3.252-3.268 inches). Check injection pump timing as outlined in MAINTENANCE SECTION.

PISTON AND ROD UNIT

The pistons and connecting rods may be removed after removing cylinder heads and oil pan. Number pistons and rods as they are removed so they can be reinstalled in their original locations.

To reinstall piston and rod, reverse the removal procedure. Make certain locating tabs of connecting rod bearing inserts engage notches in rod and cap and are located on the same side. Tighten rod cap nuts to 57-61 N·m (42-45 ft.-lbs.).

If connecting rod or piston was renewed, refer to CYLINDER section and measure piston height in cylinder.

PISTON, PIN AND RINGS

The piston is equipped with two compression rings and an oil control ring. Recommended piston ring end gap is 0.40-0.60 mm (0.016-0.023 inch) for top compression ring, 0.35-0.55 mm (0.014-0.021 inch) for second compression ring and 0.25-0.45 mm (0.010-0.018 inch) for oil control ring. Standard piston skirt diameter, measured at bottom of skirt perpendicular to piston pin, is 99.878-99.893 mm (3.9322-3.9328 inches).

Piston pin is retained in piston by two snap rings. Pin diameter is 31.994-32.000 mm (1.2596-1.25987 inches) and piston bore is 31.991-31.997 mm (1.2595-1.2597 inches). Pin is a transition fit in piston bore. Initial fit when cold is 0.009 mm (0.0003 inch) interference to 0.003 mm (0.0001 inch) loose. Heating piston will ease removal and installation of pin.

CONNECTING ROD

The connecting rod small end is fitted with a renewable bushing. The bushing inside diameter should be 32.010-32.020 mm (1.2602-1.2606 inches). Desired clearance between piston pin and bushing is 0.010-0.026 mm (0.0004-0.001 inch).

A precision, insert type bearing is used in big end of connecting rod. With bearing installed and rod cap nuts torqued to 57-61 N·m (42-45 ft.-lbs.), bearing inside diameter should be 59.993-60.018 mm (2.3619-2.3629 inches). Recommended rod bearing clearance is 0.050-0.088 mm (0.002-0.0035 inch). Bearings are available in standard size and undersizes.

CYLINDER

Removable cast iron cylinders are used on all engines. Standard cylinder bore is 100.000-100.020 mm (3.9370-3.9378 inches). Recommended clearance between piston skirt and cylinder is 0.107-0.142 mm (0.004-0.0055 inch). Cylinders that are damaged or excessively worn may be rebored and fit-

SERVICE MANUAL

Slanzi

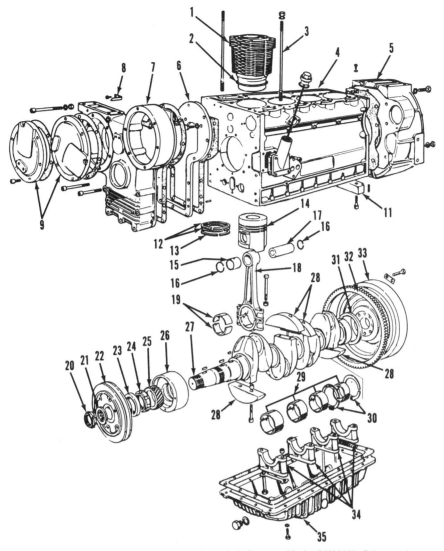

Fig. S4-10—Exploded view of crankcase and crankshaft assembly for DVA3600. Other engines are similar except for number of cylinders.

1. Cylinder
2. Shim gasket
3. Cylinder stud
4. Crankcase
5. Flywheel housing
6. Plate
7. Front cover
8. Timing pointer
9. Timing gear covers
11. Crossbeam
12. Compression rings
13. Oil control ring
14. Piston
15. Bushing
16. Retaining rings
17. Piston pin
18. Connecting rod
19. Rod bearing
20. Nut
21. Washer
22. Pulley
23. Oil seal
24. Nut
25. Gear
26. Front main bearing
27. Crankshaft
28. Counterweights
29. Main bearing inserts
30. Thrust washers
31. Oil seal
32. Ring gear
33. Flywheel
34. Main bearing caps
35. Oil pan

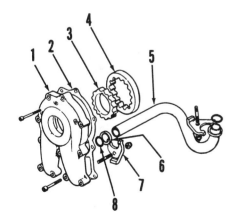

Fig. S4-11—Exploded view of geroter type engine oil pump.

1. Pump housing
2. Gasket
3. Inner rotor
4. Outer rotor
5. Pickup tube
6. Washer
7. Retainer
8. Gasket

ted with 0.5 mm (0.020 inch) oversize piston and rings.

Piston height in cylinder is adjusted by varying thickness of shim gasket (2–Fig. S4-10) between bottom of cylinder and crankcase mounting surface. With piston at top dead center, top of piston must be 0.0-0.2 mm (0.0-0.008 inch) below top edge of cylinder.

OIL PUMP

The gerotor type oil pump is mounted to front of crankcase and is enclosed by the crankcase front cover. The crankshaft extends through the oil pump and drives the pump inner rotor (3–Fig. S4-11).

To gain access to the oil pump, the crankcase front cover must first be removed as outlined in TIMING GEARS AND FRONT COVER section. Remove crankshaft gear, then unbolt and remove oil pump assembly.

Inspect pump components for scoring, excessive wear or other damage. Outer diameter of outer rotor (4) should be 101.511-101.549 mm (3.9965-3.9980 inches) and inner diameter of pump housing (1) should be 101.752-101.852 mm (4.0060-4.0099 inches). Desired clearance between outer rotor and pump housing is 0.203-0.341 mm (0.008-0.013 inch). Pump rotor end clearance

Illustrations Courtesy North American Diesel, Inc.

Slanzi SMALL DIESEL

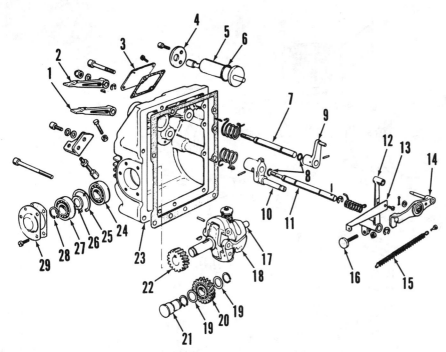

Fig. S4-12—Exploded view of governor assembly.

1. Throttle lever
2. Stop lever
3. Cover
4. Retainer plate
5. Fuel limiting device
6. "O" ring
7. Shaft
8. "O" rings
9. Arm
10. Throttle arm
11. Shaft
12. Governor arm
13. Lever
14. Counter lever
15. Governor spring
16. Adjusting screw
17. Push rod
18. Flyweight assy.
19. Thrust washers
20. Idler gear
21. Shaft
22. Drive gear
23. Governor housing
24. Inner bearing
25. Snap ring
26. Spacer
27. Outer bearing
28. Snap ring
29. Cover

should be 0.008-0.084 mm (0.0003-0.0033 inch) for DVA1800 and DVA2700 or 0.010-0.087 mm (0.0004-0.0034 inch) for DVA3600 and DVA3600TB.

GOVERNOR

All engines are equipped with a flyweight type governor. The governor housing is mounted on the side of crankcase and is driven by a gear on the injection pump camshaft through an idler gear (20 – Fig. S4-12).

To remove governor assembly, first remove cover (3) and disconnect governor spring (15) from governor lever (12). Remove housing mounting cap screws and withdraw governor assembly.

To disassemble, remove nuts securing stop lever (2) and throttle lever (1) and remove levers. Remove retaining rings, then withdraw inner stop lever (9) and shaft (7) and throttle lever (10) and shaft (11). Remove retaining plate (4) and withdraw fuel limiting device (5). To remove flyweight assembly (18), remove bearing cover (29) and snap ring (28). Install special removal tool 59.60.22 (available from manufacturer) or similar tool to governor housing as shown in Fig. S4-13. Turn the screw (1) clockwise to push the flyweight shaft out of the outer bearing and remove flyweight assembly. Remove inner bearing and gear (22 – Fig. S4-12) from flyweight shaft. Flyweight assembly must be renewed as a complete unit.

To reassemble, proceed as follows: Install snap ring (25) and outer bearing (27) in housing bore. Assemble gear, inner bearing and spacer (26) onto flyweight shaft. Remove push rod (17) from flyweight assembly, then position assembly in housing. Using special installing tool 59.69.16 (2 – Fig. S4-14) or similar tool, pull flyweight shaft into outer bearing until outer snap ring (28 – Fig. S4-12) can be installed. Insert push rod (17) making certain rounded end faces outward. Complete reassembly of governor by reversing the disassembly procedure.

Refer to ENGINE SPEED ADJUSTMENT section and adjust low and high idle speeds if necessary.

TIMING GEARS AND FRONT COVER

To service timing gears, first rotate crankshaft until number 1 piston is on compression stroke and second timing mark (TDC) on crankshaft pulley is aligned with timing pointer on front cover. Remove crankshaft pulley, drive belts and cooling fan assembly. Unbolt and remove crankcase front cover. Then, remove gears as necessary. Refer to Fig. S4-15.

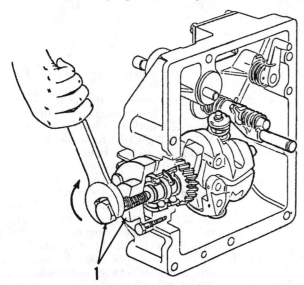

Fig. S4-13—Special tool (1) number 59.60.22 is available from manufacturer to remove governor flyweight.

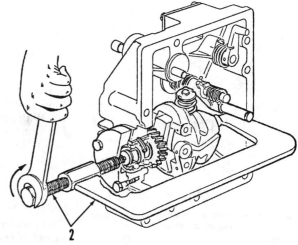

Fig. S4-14—Special tool (2) number 59.69.16 should be used to reinstall flyweight assembly.

Service Manual

Slanzi

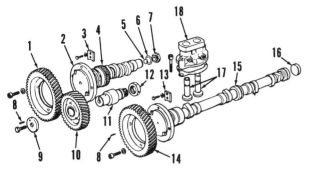

Fig. S4-15—Exploded view of timing gears, valve camshaft and fuel injection camshaft.

1. Gear
2. Injection pump camshaft
3. Retainer
4. Gear
5. Coupling
6. Gasket
7. Plug
8. Locating pin
9. Washer
10. Idler gear
11. Idler shaft
12. Nut
13. Retainer
14. Gear
15. Valve camshaft
16. Plug
17. Cam followers
18. Cam follower guide

To reassemble, reverse the disassembly procedure. Be sure timing marks on the gears are aligned and locating pin (8) is installed in valve camshaft gear (14). Check and adjust injection timing as outlined in INJECTION PUMP TIMING section.

VALVE CAMSHAFT AND CAM FOLLOWERS

The camshaft is supported by a bored sleeve bearing at front of crankcase and by unbushed bores in center and rear of crankcase. Mushroom type cam followers (17–Fig. S4-15) operate in cam follower housings (18).

To remove the camshaft, first remove cylinder head, push rods, push rod tubes and crankcase front cover as previously outlined. Disconnect oil lines from cam follower housings, then unbolt and remove housings with cam followers. Identify components as they are removed so they can be reinstalled in their original locations. Rotate crankshaft until timing marks on gears are aligned, then remove camshaft drive gear (14) and pin (8). Remove camshaft retaining plate (13), then carefully withdraw camshaft.

Inspect camshaft lobes, cam followers, bearing journals and crankcase bearing bores for scoring, wear or other damage. Outside diameter of cam followers should be 14.966-14.984 mm (0.5892-0.5899 inch) and inside diameter of cam follower guide bores should be 15.000-15.018 mm (0.5905-0.5912 inch). Desired operating clearance of cam followers in guide bores is 0.016-0.052 mm (0.0006-0.0020 inch). Camshaft front journal diameter is 47.930-47.955 mm (1.8870-1.8880 inches), center journal diameter is 45.930-45.955 mm (1.8083-1.8093 inches) and rear journal diameter is 44.930-44.955 mm (1.7689-1.7699 inches). Inner diameter of bearing bores in crankcase are 48.000-48.025 mm (1.8898-1.8907 inches) for front sleeve bearing, 46.000-46.025 mm (1.8110-1.8120 inches) for center bore and 45.000-45.025 mm (1.7716-1.7726 inches) for rear bore. Desired clearance between camshaft and crankcase bearing bores is 0.045-0.095 mm (0.0018-0.0037 inch).

To reinstall camshaft, reverse the removal procedure. Camshaft end play should be 0.020-0.098 mm (0.0008-0.0038 inch) and is controlled by camshaft retainer plate (13). Be sure to reinstall cam gear indexing pin (8) and align timing marks on crankshaft gear, idler gear (10) and camshaft gear.

INJECTION PUMP CAMSHAFT

To remove the injection pump camshaft (2–Fig. S4-15), first remove injection pump, fuel transfer pump and crankcase front cover. Rotate crankshaft until timing marks on crankshaft gear, idler gear (10) and camshaft gear (1) are aligned. Remove camshaft gear and camshaft retainer plate (3). Carefully withdraw camshaft from crankcase.

Inspect camshaft journals, cam lobes and injection pump cam followers for scoring, wear or other damage. Camshaft front journal diameter is 57.930-57.950 mm (2.2807-2.2815 inches) and rear journal diameter is 29.940-29.960 mm (1.1787-1.1795 inches). Crankcase front sleeve bearing inner diameter should be 58.000-58.030 mm (2.2835-2.2846 inches) and rear bearing bore should be 30.000-30.025 mm (1.1811-1.1821 inches). Desired operating clearance is 0.050-0.100 mm (0.0020-0.0039 inch) for front journal and 0.040-0.088 mm (0.0016-0.0034 inch) for rear journal.

To reinstall camshaft, reverse the removal procedure. Camshaft end play should be 0.020-0.098 mm (0.0008-0.0038 inch) and is controlled by retainer plate (3). Be sure to align timing gear marks. Check injection pump timing as outlined in MAINTENANCE SECTION.

CRANKSHAFT AND BEARINGS

To remove the crankshaft, first remove crankshaft pulley (22–Fig. S4-10), crankcase front cover (7), flywheel, oil pan, oil pickup tubes and crankcase crossbeam (11). Remove nut (24) and crankshaft gear (25) from front of crankshaft. Remove connecting rod caps and remove rods and pistons or push upward out of the way. Remove main bearing caps, then carefully withdraw crankshaft from crankcase.

Main journal standard diameter is 74.985-75.000 mm (2.9522-2.9527 inches) for front journal and 69.810-69.825 mm (2.7484-2.7490 inches) for all other journals. Front bearing inner diameter should be 75.045-75.085 mm (2.9545-2.9561 inches) and inner diameter of all other bearings (caps installed) should be 69.888-69.926 mm (2.7515-2.7530 inches). Desired main bearing clearance is 0.045-0.100 mm (0.0018-0.0039 inch) for front bearing and 0.063-0.116 mm (0.0025-0.0045 inch) for all other bearings. Standard and undersize bearings are available.

Crankpin standard diameter is 59.930-59.943 mm (2.3595-2.3600 inches) and bearing inner diameter is 59.993-60.018 mm (2.3619-2.3629 inches). Desired rod bearing operating clearance is 0.050-0.088 mm (0.0020-0.0035 inch). Standard and undersize bearings are available.

Crankshaft end play is controlled by

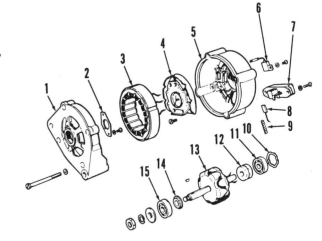

Fig. S4-16—Exploded view of alternator.

1. End frame
2. Bearing retainer
3. Stator
4. Rectifier
5. End cover
6. Condenser
7. Voltage regulator
8. Brush
9. Spring
10. Washer
11. Bearing
12. Commutator ring
13. Rotor
14. Spacer
15. Bearing

Illustrations Courtesy North American Diesel, Inc.

Slanzi — SMALL DIESEL

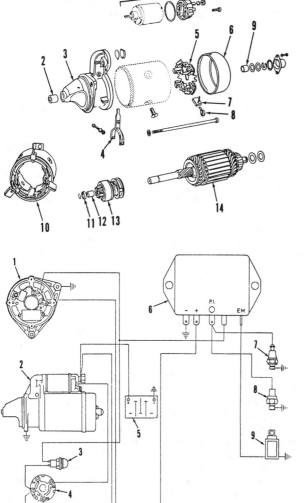

Fig. S4-17—Exploded view of electric starter motor.
1. Solenoid
2. Bushing
3. Drive end housing
4. Lever
5. Brush frame
6. Cover
7. Brush
8. Spring
9. Bushing
10. Field coil & housing
11. Retainer
12. Bushing
13. Starter drive
14. Armature

Fig. S4-18—Wiring diagram for engines equipped with safety shut-off system.
1. Alternator
2. Starter
3. Warning light
4. Key switch
5. Battery
6. Safety shut-off module
7. Oil pressure switch
8. Temperature switch
9. Shut-off solenoid

thrust washers (30) mounted on each side of rear main bearing. Washers are available in standard thickness and oversizes of 0.10 mm (0.004 inch) and 0.20 mm (0.008 inch). Desired crankshaft end play is 0.06-0.12 mm (0.0024-0.0047 inch). Be sure to install thrust washers with grooved side facing away from main bearing.

Inspect oil seals (23 and 31) and renew if necessary. Be sure to lubricate oil seal lips with grease prior to reassembly. Reinstall crankshaft by reversing the removal procedure. Tighten main bearing cap bolts to 172-182 N·m (127-134 ft.-lbs.) and crankshaft counterweight (if removed) cap screws to 59-63 N·m (44-47 ft.-lbs.). Tighten crankcase crossbeam (11) screws to 42-46 N·m (31-34 ft.-lbs.) and flywheel cap screws to 105-111 N·m (80-82 ft.-lbs.). Be sure to align timing gear marks when reinstalling crankshaft gear.

ELECTRICAL SYSTEM

The standard electrical system is a 12-volt negative ground system. The externally mounted alternator is controlled by a solid-state, nonadjustable voltage regulator. Refer to Fig. S4-16 for an exploded view of alternator assembly and Fig. S4-17 for an exploded view of starter motor.

Some engines may be equipped with a safety shut-off switch which stops engine if overheating or loss of oil pressure occurs. Refer to Fig. S4-18 for electrical system wiring schematic which shows safety shut-off components.

Illustrations Courtesy North American Diesel, Inc.

SERVICE MANUAL

VOLKSWAGEN

VOLKSWAGEN OF AMERICA, INC.
420 Barclay Blvd.
Lincolnshire, Illinois 60069

Model	No. Cyls.	Bore	Stroke	Displ.
068.2	4	76.5 mm (3.012 in.)	80.0 mm (3.150 in.)	1471 cc (90 cu. in.)
068.5	4	76.5 mm (3.012 in.)	86.4 mm (3.400 in.)	1588 cc (97 cu. in.)
068.A	4	76.5 mm (3.012 in.)	86.4 mm (3.400 in.)	1588 cc (97 cu. in.)

These engines are four-stroke, liquid-cooled type with vertical cylinders and overhead camshaft. The cylinder block is cast iron and the cylinder head is aluminum. The cylinders are numbered 1 through 4 with number 1 cylinder nearest crankshaft pulley. Firing order is 1-3-4-2. Crankshaft rotation is clockwise as viewed from crankshaft pulley end. Engine Model 068.A is turbocharged.

MAINTENANCE

LUBRICATION

Recommended motor oil is API classification CC or CD or VW Standard 505 00 for naturally aspirated engines and VW Standard 505 00 only for turbocharged engines.

Single weight SAE 40 oil should be used in ambient temperatures above 20° C (70° F), SAE 30 for temperatures 0°-20° C (32°-70° F), SAE 20W-20 for temperatures −10° to 0° C (15°-32° F) and SAE 10W for temperatures below −10° C (15° F).

Multi-grade SAE 10W-30 or 10W-40 oil may be used in ambient temperatures up to 10° C (50° F), SAE 15W-40 or 15W-50 oil in temperatures from −15° to 30° C (5°-85° F), and SAE 20W-40 or 20W-50 oil in temperatures of −10° to 30° C (15°-85° F).

Recommended engine oil change interval is every 150 hours of operation. Oil capacity with filter change is 4.5 liters (4.7 quarts) on engines equipped with cast aluminum oil pan, 4.0 liters (4.2 quarts) on late production engines equipped with stamped steel pan and 3.5 liters (3.7 quarts) on earlier models with stamped steel oil pan.

Engine oil pressure should be a minimum of 200 kPa (29 psi) with engine running at 2000 rpm and at operating temperature of 80° C (175° F). An oil cooler is mounted between oil filter base and filter on turbocharged engines.

FUEL SYSTEM

FUEL FILTER. Fuel filter drain plug in bottom of filter canister and vent screw on filter mounting bracket should be loosened periodically to drain water and sediment trapped in the filter.

Filter should be renewed after every 300 hours of operation. Clean filter base and filter before removing canister. Moisten filter "O" ring with diesel fuel before installing new filter. Bleed air from system as outlined in following paragraph.

BLEED FUEL SYSTEM. To bleed system, loosen bleed screw at top of filter bracket and allow fuel to flow until free of air. Tighten filter bleed screw and loosen inlet line fitting on fuel injection pump; allow fuel to flow until free of air and tighten inlet line fitting.

If engine fails to start at this point, loosen high pressure lines at fuel injectors, move throttle to run position and crank engine until fuel is discharged at loosened fittings. Tighten fittings at injectors to a torque of 25 N·m (18 ft.-lbs.) and start engine.

INJECTION PUMP TIMING

To check fuel injection pump timing, first check to be sure pump is mounted on engine front plate with timing mark (M—Fig. V1-1) scribed on pump housing aligned with mark (N) on front plate. If not, loosen pump mounting bolts and rotate pump assembly to bring marks into alignment and tighten mounting bolts to a torque of 25 N·m (18 ft.-lbs.). Turn crankshaft so that number 1 piston is on compression stroke and TDC mark on flywheel is aligned with boss in window of flywheel housing. The timing mark on injection pump drive gear (belt sprocket) should then be aligned with marks (M and N) on pump and mounting plate. If not, refer to TIMING BELT paragraph and correctly install timing belt.

NOTE: If flywheel housing is not installed, a special tool for setting engine crankshaft at TDC is available as shown in Fig. V1-2.

If validity of timing mark scribed on fuel injection pump flange is in doubt,

Fig. V1-1—View showing location of idle speed (I) and high speed (H) governor adjusting screws. When installing pump, align mark (M) on pump flange with notch (N) on injection pump mounting plate.

Volkswagen

timing can be checked using dial indicator to check injection pump plunger lift as follows:

Turn crankshaft to put number 1 piston on compression stroke and align TDC mark on flywheel (either by boss in timing window or by using special tool as shown in Fig. V1-2). Remove plug from rear of fuel injection pump and install adapter (VW tool No. 2066) and dial indicator as shown in Fig. V1-3. Preload indicator so dial reading is about 2 mm (0.080 inch), then turn crankshaft counterclockwise (in reverse of normal rotation) until dial indicator stops moving and set dial indicator gage to zero reading.

NOTE: Do not use crankshaft pulley bolt to reverse engine rotation on models with front pto.

Then, turn crankshaft in normal rotation until TDC mark on flywheel is aligned with boss or gage. At this point, the dial indicator reading should be 0.78-0.88 mm (0.031-0.034 inch) on naturally aspirated engine or 0.83-0.97 mm (0.033-0.038 inch) on turbocharged engine. If not, loosen pump mounting bolts and rear mounting plate bolt. Turn pump to bring dial indicator reading into specified range, then tighten pump retaining bolts to a torque of 25 N·m (18 ft.-lbs.). Scribe a new timing mark on pump mounting flange in line with the timing mark on mounting plate. Remove dial indicator and adapter and reinstall plug using new gasket. Tighten the plug to a torque of 15-25 N·m (11-18 ft.-lbs.).

GOVERNOR

The governor is an integral part of the fuel injection pump assembly. Low and high idle (no-load) speeds are adjusted using screws (I and H—Fig. V1-1) on injection pump. Low idle speed adjusting screw (I) is on engine side of pump. With engine at normal operating temperature, turn adjusting screws as required to obtain governed speeds specified on engine identification plate.

NOTE: Some engine identification plates may not list a low idle speed. On these engines, set low idle speed to 920-980 rpm.

COMPRESSION PRESSURE

Checking compression pressure requires a special gage adapter that is installed in place of a fuel injector. All injectors should be removed, then the gage and adapter are installed in each cylinder in turn. At cranking speed, desired compression pressure is 3400 kPa (495 psi); minimum allowable compression pressure is 2800 kPa (405 psi). Readings of all four cylinders should not vary more than 500 kPa (73 psi).

REPAIRS

TIGHTENING TORQUES

Refer to the following table for special tightening torques.

Camshaft bearing caps . . . . 20 N·m
  (15 ft.-lbs.)
Camshaft sprocket . . . . 45 N·m
  (33 ft.-lbs.)
Intermediate shaft
  sprocket . . . . 45 N·m
  (33 ft.-lbs.)
Connecting rod . . . . See Text
Crankshaft sprocket . . . . See Text
Cylinder head . . . . See Text
Intake & exhaust
  manifolds . . . . 25 N·m
  (18 ft.-lbs.)
Flywheel . . . . See Text

SMALL DIESEL

Turbocharger mounting
  bolts . . . . 45 N·m
  (33 ft.-lbs.)
Turbocharger exhaust pipe
  bolts . . . . 25 N·m
  (18 ft.-lbs.)
Turbocharger oil return
  connector . . . . 25 N·m
  (18 ft.-lbs.)
Injection pump mounting . . . . 25 N·m
  (18 ft.-lbs.)
Injection pump sprocket . . . . . 45 N·m
  (33 ft.-lbs.)
Fuel injector . . . . 70 N·m
  (52 ft.-lbs.)
Glow plug . . . . 30 N·m
  (22 ft.-lbs.)
Main bearing . . . . 65 N·m
  (48 ft.-lbs.)
Oil pan, sheet metal . . . . 20 N·m
  (15 ft.-lbs.)
Oil pan, cast aluminum . . . . . 10 N·m
  (8 ft.-lbs.)
Oil pump cover . . . . 10 N·m
  (8 ft.-lbs.)
Oil pump mounting screws . . . 20 N·m
  (15 ft.-lbs.)
Timing belt tensioner . . . . 45 N·m
  (33 ft.-lbs.)

WATER PUMP AND COOLING SYSTEM

Water pump belt tension is adjusted by varying the number of spacers between the halves of the water pump pulley. Belt tension is checked by applying thumb pressure midway between pulleys. Belt should deflect 5-10 mm (0.2-0.4 inch).

It is recommended that a mixture of clean, soft water and antifreeze with anti-corrosion additives be used in the cooling system.

THERMOSTAT

The thermostat is mounted in lower end of water pump cover. To remove thermostat, drain coolant by detaching the hose from thermostat housing and the lower hose from rear of water pump cover. Thermostat should start to open at 80° C (175° F) and be fully open at 91° C (195° F). Minimum stroke of thermostat should be 7 mm (0.28 inch).

VALVE ADJUSTMENT

Mechanical Tappets

Check valve clearance with engine warm. Measure clearance between camshaft lobe and tappet using feeler gage and with lobe pointing upward. Desired valve clearance is 0.20-0.30 mm (0.008-0.011 inch) for intake valves and 0.40-

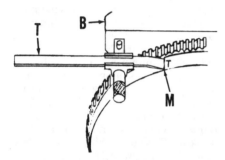

Fig. V1-2—A special tool (2068) is available for setting engine at TDC when flywheel housing is removed. Set vernier scale to 112.8 mm, attach tool to block (B) as shown and align flywheel TDC mark (M) with end of tool.

Fig. V1-3—View showing special tool adapter 2066 and dial indicator installed in rear end of fuel injection pump to check validity of timing mark (M—Fig. V1-1) scribed on fuel injection pump.

SERVICE MANUAL

Volkswagen

0.50 mm (0.016-0.019 inch) for exhaust valves.

After cylinder head overhaul, valve clearance may be adjusted on bench prior to installing cylinder head on engine. Set valve clearance to 0.15-0.25 mm (0.006-0.010 inch) for intake valves and 0.35-0.45 mm (0.014-0.018 inch) for exhaust valves.

Clearance is adjusted by varying the thickness of a disc (shim) located in a recess in the top of each tappet. Shims are available in thicknesses of 3.00 to 4.25 mm in steps of 0.05 mm. Shim thickness in millimeters is etched on the bottom side of the shim. Shims may be reused if no damage to cam contact surface is noted. Use thicker shim to decrease valve clearance and a thinner shim to increase clearance. Select a shim that will give as close to mid-range of desired adjustment as possible. Be sure that shims are installed with thickness marked side down.

A special tool is used to depress tappets for removing and installing shims. When depressing tappets to change shims, engine crankshaft should be turned 1/4 turn from TDC position so that valves will not strike pistons. If tool is not available, camshaft must be removed to change shim thickness.

Hydraulic Tappets

Engine Models 068.5 and 068.A, 1987 production and later, are equipped with nonadjustable hydraulic valve tappets. If valve noise is noted after engine has been running and oil temperature has reached 60° C (140° F), stop engine and remove valve cover. Turn engine so that cam lobe is up on suspected noisy tappet and using wood or plastic wedge, push down on tappet. If free travel exceeds 0.1 mm (0.004 inch), tappet should be renewed.

To renew hydraulic tappets, camshaft must be removed. After installing new tappets, engine should not be started for one-half hour as there is danger of valves striking the pistons.

TIMING BELT

A flat, internal tooth timing belt is used to drive the sprockets (gears) on camshaft and fuel injection pump. The intermediate (oil pump drive) shaft has a flat drive pulley riding against back side of belt. Belt must be installed with crankshaft, camshaft and injection pump drive shaft sprockets in time. A flat idler mounted on an eccentric shaft runs against the back side of timing belt to maintain proper tension. A timing belt tension indicator such as Volkswagen's tool VW210 should be used to adjust belt tension; if using tool VW210, adjust belt to obtain a scale value of 12 to 13. Refer to tool instruction sheet if using another indicator.

Later production engines are fitted with a two-piece fully enclosed belt guard. On these engines, the fuel injection pump mounting plate is redesigned to accommodate a thicker intermediate shaft pulley. Only the late style pulley is available for service and when installing this pulley on earlier engines, the clearance between the pulley and fuel injection pump mounting plate must be checked. If clearance is less than 0.5 mm (0.020 inch), the edge of the plate behind the pulley must be reworked to provide proper pulley clearance.

NOTE: As timing belt failure with engine running can cause serious engine damage, timing belt should be inspected periodically and the belt should be renewed if fraying, cracking, excessive wear or any damage from fuel or oil is noted.

To remove timing belt, first remove timing belt cover and valve cover. Turn crankshaft so that slot in rear of cam is horizontal (Fig. V1-4), TDC timing mark on flywheel is aligned with boss on flywheel housing timing window, and timing mark on fuel injection pump is aligned with mark on mounting plate and pump flange (see Fig. V1-1). Both camshaft lobes on number 1 cylinder should be pointing upward at this time. Loosen the bolt securing the belt tensioner and remove timing belt from pulleys.

NOTE: Do not turn engine crankshaft or camshaft with timing belt removed.

Prior to installing timing belt, be sure camshaft, crankshaft and injection pump drive pulleys are properly "timed" as described in removal procedure. Loosen camshaft sprocket bolt one-half turn and tap rear side of sprocket to loosen sprocket on camshaft taper. On late models with fully enclosed timing belt, use a hammer and punch through the 6 mm (1/4 inch) hole provided in the rear cover to tap rear of sprocket. Refer to Figs. V1-4 and V1-5 and insert a bar (use VW tool No. 2065 if available) in camshaft slot to hold camshaft slot in perfect alignment with cylinder head surface. Install a pin (VW tool No. 1064 if available) through injection pump pulley and mounting plate (see Fig. V1-6) to lock pulley in proper

Fig. V1-4—View of timing slot (S) in rear end of engine camshaft. Refer also to Fig. V1-5.

Fig. V1-6—Insert dowel through holes in injection pump sprocket and pump mounting plate to hold pump in timing position while installing timing belt. VW tool number for dowel is 2065.

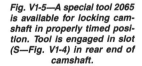

Fig. V1-5—A special tool 2065 is available for locking camshaft in properly timed position. Tool is engaged in slot (S—Fig. V1-4) in rear end of camshaft.

Volkswagen

position. Refer to Fig. V1-7 and place timing belt on sprockets, intermediate shaft pulley and idler pulley as shown in diagram. Remove locating pin from injection pump sprocket and tighten belt by turning tensioner hub clockwise; hold tensioner hub and tighten tensioner bolt to a torque of 45 N·m (33 ft.-lbs.). Tighten camshaft sprocket bolt to a torque of 45 N·m (33 ft.-lbs.) and remove bar from slot in rear end of camshaft. Turn crankshaft several turns, strike belt once between camshaft and injection pump sprockets with plastic mallet to remove any slack and recheck belt tension.

After belt is installed and properly tensioned, recheck timing marks on flywheel and injection pump sprocket, be sure slot in rear end of camshaft is in horizontal position and number 1 cylinder cam lobes are both pointing up. Then, reinstall valve cover and timing belt cover.

CAMSHAFT

The camshaft can be removed after removing engine timing belt as outlined in preceding paragraph. Remove sprocket from front end of camshaft to allow renewal of camshaft oil seal. On models with fully enclosed timing belt, insert a punch through the 6 mm (1/4 inch) hole in rear cover plate and tap the sprocket off front end of camshaft. Note position numbers on camshaft bearing caps; head and caps are align bored and caps cannot be changed from original position. Loosen all camshaft bearing cap bolts evenly and a small amount at a time until camshaft is free of valve spring pressure, then remove the caps and camshaft.

With camshaft removed, the tappets can be removed from cylinder head. Be sure to identify each tappet so they can be installed in their original positions. Be careful not to lose adjustment shim disks from tops of mechanical lift tappets. Inspect tappet barrels and cylinder head bores for excessive wear or scoring. Hydraulic tappets should be placed in clean compartmented container with cam contact surface down.

Inspect camshaft, cylinder head and bearing caps for excessive wear and damage. Place camshaft in head without tappets and install front and rear bearing caps; locate rear cap before tightening bolts by bumping cam to front and to rear, then check camshaft end play. Camshaft end play should not exceed 0.15 mm (0.006 inch). Runout of center camshaft journal should not exceed 0.01 mm (0.0004 inch). Check camshaft journal clearance with Plastigage; bearing clearance should not exceed 0.11 mm (0.0043 inch).

To install camshaft, first be sure engine crankshaft is turned to position number 1 piston at TDC on compression stroke. Lubricate tappets and replace in same bores from which they were removed. Lubricate camshaft oil seal and place on front end of camshaft with seal lip to rear. Lubricate camshaft and position camshaft in head with number 1 lobes up and with seal properly located on shaft and in head. Lubricate camshaft bearing caps and place caps in proper positions, taking care that oil seal is not moved out of correct placement. Tighten the bearing cap nuts evenly and

SMALL DIESEL

a small amount at a time until all caps are pulled evenly to head, then tighten cap bolts to a torque of 20 N·m (15 ft.-lbs.).

Reinstall camshaft sprocket and timing belt as outlined in TIMING BELT paragraph.

CYLINDER HEAD

To remove cylinder head, first drain coolant and disconnect coolant hose from cylinder head. Clean fuel injector lines and surrounding area, then remove fuel injector lines and cap all openings. Remove turbocharger if so equipped. Unbolt and remove exhaust and intake manifolds from head. Remove timing belt cover, loosen belt tensioner and remove timing belt. On late models with fully enclosed timing belt, insert a punch through the hole provided in timing belt rear cover and tap camshaft sprocket off camshaft tapered end. Remove valve cover, loosen cylinder head bolts in reverse order of sequence shown in Fig. V1-9, and remove cylinder head.

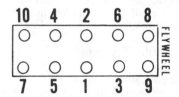

Fig. V1-9—Cylinder head bolt tightening sequence. When removing head, reverse sequence to loosen head bolts.

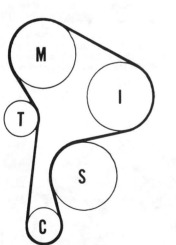

Fig. V1-7—Schematic diagram for timing belt installation.

C. Crankshaft sprocket
I. Injection pump sprocket
M. Camshaft sprocket
S. Intermediate shaft pulley
T. Belt tensioner

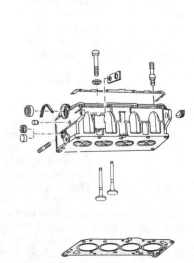

Fig. V1-8—Exploded view of cylinder head assembly. Thickness of shim (S) in top of mechanical valve tappet is varied for valve clearance adjustment; hydraulic tappets are not fitted with shims.

SERVICE MANUAL

NOTE: It is not necessary to remove camshaft; however, if camshaft is not removed, be careful not to damage valves protruding from bottom of head while handling head.

Clean cylinder head gasket surface and check head with straightedge and feeler gage; maximum allowable distortion of cylinder head surface is 0.1 mm (0.004 inch). If cylinder head distortion exceeds maximum amount, Volkswagen recommends that head be renewed. Cracks may appear between valve seat inserts and are usually not detrimental to reuse of cylinder head unless cracks appear deep and are over 0.5 mm (0.020 inch) wide. Refer to VALVE SYSTEM paragraphs for valve information.

Cylinder head gaskets are available in different thicknesses to obtain desired valve-to-piston clearance. Thickness of gasket is indicated by notches in edge of gasket as shown in Fig. V1-10. Use same gasket thickness as original gasket unless new pistons have been installed or installing a short block assembly.

If a short block or new pistons have been installed, measure height of each piston above cylinder block surface at top dead center position. Refer to following table for gasket selection using highest piston height measurement:

Engine Model	Piston Height	Gasket Notches
068.2	0.43-0.63 mm (0.017-0.025 in.)	2
	0.64-0.82 mm (0.025-0.032 in.)	3
	0.83-0.92 mm (0.032-0.036 in.)	4
	0.93-1.02 mm (0.036-0.040 in.)	5
068.5, 068.A*	0.67-0.82 mm (0.026-0.032 in.)	1
	0.83-0.92 mm (0.032-0.036 in.)	2
	0.93-1.02 mm (0.036-0.040 in.)	3
068.5, 068.A**	0.66-0.86 mm (0.026-0.034 in.)	1
	0.87-0.90 mm (0.034-0.035 in.)	2
	0.91-1.02 mm (0.036-0.040 in.)	3

*With mechanical valve tappets.
**With hydraulic valve tappets. Gasket has additional 16 mm diameter breather hole between number 1 and 2 cylinders.

Check to see that head contact surface of manifolds is not distorted and that exhaust manifold is not cracked. Resurface or renew manifolds if defects are noted.

To install cylinder head, proceed as follows: Turn engine crankshaft so that number 1 piston is at top of cylinder and TDC mark on flywheel is aligned with boss of timing window. Turn fuel injection pump sprocket so timing mark on sprocket is aligned with marks on pump flange and mounting plate. Install long guide studs in front and rear cylinder block bolt holes on manifold side of engine and place new head gasket on block. Turn camshaft so that number 1 cylinder lobes are both pointing up and set cylinder head down over the guide studs. Install head bolts in the eight open bolt holes and tighten finger tight. Remove the two guide studs and install the two remaining head bolts finger tight. After tightening head bolts as outlined in following paragraphs, install manifolds using new gaskets and tighten manifold retaining bolts to a torque of 25 N·m (18 ft.-lbs.).

NOTE: Two different types of head bolts have been used and a different tightening procedure must be used with each type bolt. Early engines used M11 hex socket head bolts; later models use M12 12-point socket head bolts. The later type M12 bolts cannot be reused. Refer to appropriate following paragraph:

M11 HEAD BOLTS. When installing early M11 hex socket head bolts, refer to tightening sequence in Fig. V1-9 and tighten bolts in three steps as follows:

Tightening Torque
Step 1 .... 50 N·m
(37 ft.-lbs.)
Step 2 .... 70 N·m
(52 ft.-lbs.)
Step 3 .... 90 N·m
(66 ft.-lbs.)

After engine is started, allow to run until oil temperature is above 50° C (120° F) and retighten head bolts to a torque of 90 N·m (66 ft.-lbs.) in sequence shown in Fig. V1-9. After engine has been operated for approximately 20 hours, loosen bolts (engine hot or cold) 30 degrees (1/12 turn) one at a time in

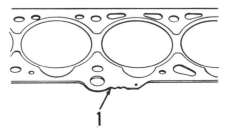

Fig. V1-10—Number of notches (1) in side of head gasket indicates gasket thickness; refer to text.

Volkswagen

sequence shown in Fig. V1-9 and retorque to 90 N·m (66 ft.-lbs.).

M12 HEAD BOLTS. When removing cylinder head with later style 12-point socket head bolts, discard the bolts and use new bolts to install head. Refer to Fig. V1-9 and tighten the bolts in three steps as follows:

Tightening Torque
Step 1 .... 40 N·m
(30 ft.-lbs.)
Step 2 .... 60 N·m
(44 ft.-lbs.)
Step 3 .... 1/4 turn additional

After engine has been started, allow to run until oil temperature reaches 50° C (120° F), then tighten head bolts in one movement an additional 1/4 turn in sequence shown in Fig. V1-9. After engine has been operated for approximately 20 hours, follow sequence shown in Fig. V1-9 and again tighten head bolts in one movement (engine hot or cold) an additional 1/4 turn.

VALVE SYSTEM

Remove cylinder head for access to valve system components. Remove camshaft and valve tappets from head as outlined in CAMSHAFT paragraph.

Use a valve spring compression tool for recessed valve springs (VW tool 2037 or correct diameter cylinder with slot in side for access to valve split keys) to compress valve springs. Remove valve split keys, spring retainer, inner and outer springs and valves. Early production valves had only one key groove, later production (May, 1981 and later) valves have triple key grooves. Keys and valve spring retainers are not interchangeable between early and late valves.

Check valve spring free height. For models with mechanical tappets, standard spring height is 40.2 mm (1.583 inches) for outer spring and 33.9 mm (1.335 inches) for inner spring. Spring height for models with hydraulic tappets is 37.0 mm (1.457 inches) for outer spring and 32.5 mm (1.280 inches) for inner spring. Renew any spring that is obviously distorted or if free height is not close to new dimension.

Pull valve stem seals from guides and remove valve spring seats. Using new valve, check valve guide wear by holding tappet end of valve flush with guide and measuring side play of valve head with dial indicator. Valve head must not move more than 1.3 mm (0.050 inch). If guide is excessively worn, use suitable driver with 8 mm pilot to drive guide out to top (camshaft) side of head. Drive new guide in from top side of head until flange on guide contacts head. If

Volkswagen

necessary, use 8 mm (0.315 inch) reamer to size guides after installation.

Valve face and seat angle is 45 degrees for both intake and exhaust. Valve seat width should be 2 mm (0.080 inch) for intake seats and 2.4 mm (0.095 inch) for exhaust seats. Maximum distance from cylinder head surface to valve head surface cannot exceed 1.5 mm (0.059 inch) after reworking seat.

Minimum valve margin (thickness of valve head at outer edge) is 0.5 mm (0.020 inch) for intake valves; exhaust valves should not be refaced. Valve stem diameter new is 7.97 mm (0.314 inch) for intake valves and 7.95 mm (0.313 inch) for exhaust valves.

On models with hydraulic tappets, measure distance from end of valve stem to valve gasket surface of cylinder head using a straightedge and depth gage. Minimum dimension is 35.8 mm (1.409 inches) for intake valves and 36.1 mm (1.421 inches) for exhaust valves. If dimension is less than specified distance, hydraulic lifters cannot operate properly.

Always use new valve seals when reassembling cylinder head. Reassemble by reversing disassembly procedure. On models with mechanical valve tappets, adjust valve tappet gap on bench before installing head.

INJECTORS

REMOVE AND REINSTALL. Before removing an injector or loosening injector lines, thoroughly clean injectors, lines and adjacent area using suitable solvent and compressed air. Disconnect high pressure and bypass lines from injectors and cap or plug all openings to prevent dirt entry. Unscrew injector from cylinder head. Remove and discard heat shields (9—Fig. V1-11).

Before installing injector, be sure injector bore and seating surface in head is clean and free of carbon. Install new heat shield (9) with concave side inward as shown in Fig. V1-12. Install injector and tighten to 70 N·m (51 ft.-lbs.). Install injector lines, leaving fittings loose at injectors. Open throttle and crank engine until fuel flows from loose fittings, then tighten fittings to a torque of 25 N·m (18 ft.-lbs.).

TESTING. A complete job of testing and adjusting injectors requires use of special test equipment. Only clean, approved testing oil should be used in tester tank. Injector nozzle should be tested for opening pressure and seat leakage. When operating properly, injector nozzle will emit a buzzing sound and cut off quickly with no fluid leakage at seat.

Before conducting test, operate tester lever until oil flows from open line, then attach injector. Close valve to tester gage and operate tester lever a few quick strokes to be sure nozzle valve is not stuck.

WARNING: Fuel emerges from injector with sufficient force to penetrate the skin. When testing injector, keep yourself clear of nozzle spray.

OPENING PRESSURE. Open valve to tester gage and operate tester lever slowly while observing gage reading. Opening pressure for new injectors is 13000-13800 kPa (1885-2000 psi) for naturally aspirated engines and 15500-16300 kPa (2250-2365 psi) for turbocharged engines. Minimum allowable opening pressure for used injectors is 12000 kPa (1740 psi) for naturally aspirated engines and 14000 kPa (2030 psi) for turbocharged engines.

Opening pressure is adjusted by varying thickness of the shim (2—Fig. V1-11). Shims are available in thickness of 1.00 to 1.95 mm in steps of 0.05 mm. Changing shim thickness by 0.05 mm will change opening pressure by about 500 kPa (70 psi). When adjusting opening pressure, set pressure to new injector specifications.

SMALL DIESEL

SEAT LEAKAGE. Injector nozzle tip should not leak at a pressure of 11000 kPa (1560 psi) when held at this pressure for a period of 10 seconds. If drop appears at nozzle tip, nozzle valve is not seating properly and injector must be renewed or overhauled as outlined in following paragraph.

OVERHAUL. Hard or sharp tools, emery cloth, grinding compound or other than approved solvents or lapping compounds must never be used. An approved nozzle cleaning kit is available through a number of specialized tool sources.

Wipe all dirt and loose carbon from outside of injector assembly. Refer to Fig. V1-11 for exploded view of injector and proceed as follows:

Secure injector body (1) in a soft jawed vise or holding fixture and remove nozzle nut (8). Remove parts and place in clean diesel fuel or calibrating fluid, taking care to keep parts from each injector separated from other injector units.

Clean exterior surfaces with a brass wire brush, soaking in an approved carbon solvent if necessary to loosen hard carbon deposits. Rinse parts in clean diesel fuel or calibrating oil immediately after cleaning to neutralize the solvent and prevent etching of polished surfaces. Make certain nozzle valve (6) slides freely of its own weight in nozzle body (7). If valve sticks, reclean or renew the nozzle and valve assembly.

Reassemble injector while holding in pan of clean diesel fuel or calibrating oil. Tighten nozzle retaining nut (8) and upper body (1) to a torque of 70 N·m (51 ft.-lbs.). Retest injector and reinstall if found usable.

INJECTION PUMP

Refer to Fig. V1-1 for view of injection pump installed on engine. The injection pump and integral governor assembly should be tested and overhauled by a qualified diesel injection system repair shop.

To remove injection pump, first thoroughly clean injection pump, injection lines and surrounding area using suitable solvent and compressed air. Disconnect fuel injection lines and fuel supply and return lines and immediately cap or plug all openings to prevent dirt entry. Disconnect control cables. Remove timing belt cover and turn crankshaft so that timing mark on fuel injection pump drive sprocket is aligned with timing marks on pump flange and mounting plate and TDC mark on flywheel is aligned with boss in timing window. Remove injection pump sprocket retaining bolt. Loosen belt tensioner and slide timing belt from injection

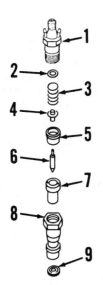

Fig. V1-11—Exploded view of diesel fuel injector.
1. Upper body
2. Shim
3. Spring
4. Thrust pin
5. Nozzle holder
6. Needle valve
7. Nozzle
8. Lower body
9. Heat shield

Fig. V1-12—Cross-sectional drawing of injector heat shield showing proper installation.

SERVICE MANUAL

Volkswagen

pump sprocket. Using suitable pullers, remove sprocket from pump shaft. Remove bolt from pump rear bracket (do not unbolt rear bracket from pump) and unbolt and remove pump from mounting plate.

Injection pumps supplied for turbocharged models have the breather tube for the BPE (boost pressure enrichment) housing closed. Remove the clamp holding the hose pinched closed or cut off end of closed rubber cap so that unit can breath.

When installing pump, align timing mark (M—Fig. V1-1) on pump flange with timing mark (N) on pump mounting plate, then tighten mounting bolts (including rear bracket bolt) to a torque of 25 N·m (18 ft.-lbs.). Complete reassembly by reversing removal procedure, leaving injection line fittings at injectors loose until system is bled. Refer to TIMING BELT paragraph for installing timing belt.

TURBOCHARGER

Model 068.A

REMOVE AND REINSTALL. Thoroughly clean turbocharger and surrounding area prior to disconnecting oil lines or removing turbocharger. Remove air inlet and discharge hose. Disconnect oil supply and return lines. Remove cap screws securing turbocharger to intake manifold, remove exhaust pipe stud nuts and remove turbocharger assembly.

The turbocharger is serviced as a complete assembly only. Refer to following paragraph for testing turbocharger and renew unit if found defective.

To reinstall turbocharger, reverse removal procedure. Before torquing any fasteners, install turbocharger mounting bolts and oil return pipe to engine bolts finger tight. Tighten intake manifold to turbocharger cap screws to a torque of 45 N·m (33 ft.-lbs.) and tighten exhaust pipe nuts to a torque of 25 N·m (18 ft.-lbs.). Tighten oil return to turbocharger bolts to a torque of 25 N·m (18 ft.-lbs.) and tighten oil return line fitting to a torque of 30 N·m (22 ft.-lbs.). Before connecting oil supply line, fill turbocharger oil inlet with clean motor oil to ensure start-up lubrication. With injection pump controls in "shut off" position, crank engine with starter until oil is discharged from oil supply line, then connect supply line to turbocharger. When first starting engine, allow engine to operate at idle speed for about one minute to allow oil to circulate to turbocharger.

TESTING TURBOCHARGER. The turbocharger boost pressure is limited by an internal wastegate to 64-72 kPa (9.3-10.4 psi). An air dump valve is located in the intake manifold to serve as a safety relief valve if turbocharger boost pressure becomes excessive. The air dump valve is set to open at 76-86 kPa (11.0-12.5 psi). Prior to testing turbocharger, it should be determined that engine settings such as injection pump timing, maximum no-load speed, engine valve clearance, fuel injectors, and engine compression pressures are to specification.

To check boost pressure, disconnect boost pressure line between intake manifold and injection pump. Connect a turbocharger boost pressure gage into boost line using a tee fitting, then reconnect boost line to open end of tee. Boost pressure must be checked with engine at full throttle and under full load.

If boost pressure is too high, wastegate is malfunctioning and turbocharger should be renewed. If boost pressure is too low, disconnect air dump valve hose at air intake and plug hose opening. Repeat boost pressure check. If specified boost pressure is now obtained, air dump valve is defective. If boost pressure is still low and all other engine and injection pump settings are satisfactory, turbocharger is defective and should be renewed.

INTERMEDIATE SHAFT

All Models

An intermediate shaft carried in two renewable sleeve bearings is used only to drive the engine oil pump on diesel engines. The intermediate shaft seal (5—Fig. V1-13) can be renewed after removing timing belt, intermediate shaft pulley (4) and seal retainer (6). Refer to TIMING BELT paragraph for belt removal procedure. The shaft (8) can be withdrawn from front of block after removing seal retainer. Install oil seal with lip to inside. Lubricate seal, place new "O" ring (7) on retainer and install retainer in cylinder block. Tighten oil seal retainer bolts to a torque of 25 N·m (18 ft.-lbs.). Tighten pulley bolt (1) to a torque of 45 N·m (33 ft.-lbs.).

OIL PUMP AND RELIEF VALVE

The engine oil pump is driven from gear on rear end of the intermediate shaft (8—Fig. V1-13) through a gear and shaft assembly mounted vertically in crankcase. Top bearing housing for oil pump drive gear is inserted from top side of block. Pump drive gear can be removed from above after removing "Y" clamp and bearing housing. Install new sealing "O" ring on bearing housing when reassembling. Slot in bottom of gear must engage tang on oil pump shaft before gear will fully engage.

The oil pump assembly can be unbolted and removed from bottom of block after removing oil pan. Early models have a one-piece pump cover and oil pickup (17—Fig. V1-13). Later models have a two piece pump cover and pickup. The oil pressure relief valve is located in the pump cover on all models.

To inspect pump, unbolt and remove cover from bottom of pump. Standard gear backlash is 0.05 mm (0.002 inch); maximum allowable backlash is 0.2 mm (0.008 inch). Check end play of gears using straightedge and feeler gage; maximum allowable end play is 0.15 mm (0.006 inch). Pump should be renewed if cover or pump body is scored or grooved.

Pack pump with grease or heavy oil on reassembly so pump will prime easily and tighten cover retaining bolts to a torque of 10 N·m (8 ft.-lbs.). Place pump in block with drive tang aligned with slot in drive gear, then tighten bolts

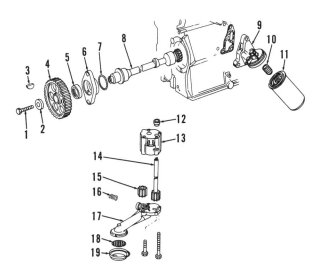

Fig. V1-13—Exploded view of intermediate shaft, oil filter and early oil pump assembly. Late pump has two-piece pump cover and pickup instead of one-piece unit (17) shown. Oil pump drive gear and bearing are not shown.

1. Sprocket bolt
2. Flat washer
3. Woodruff key
4. Sprocket
5. Oil seal
6. Seal retainer
7. "O" ring
8. Intermediate shaft
9. Filter mount
10. Pipe nipple
11. Oil filter
12. Bushing
13. Pump body
14. Drive shaft & gear
15. Driven gear
16. Relief spring
17. Pump cover & pickup
18. Screen
19. Cap

357

Volkswagen

retaining pump to crankcase to a torque of 20 N·m (15 ft.-lbs.).

PISTON AND ROD UNITS

Piston and rod units are removed from above after removing cylinder head and oil pan. Pistons, connecting rods and caps should be identified with cylinder number so they can be reinstalled in same position from which they were removed. On turbocharged models, the cylinder has additional oil jet lubrication and piston skirt is notched to clear the oil jet at bottom of stroke.

Be sure arrow mark on piston crown and forging marks (M—Fig. V1-15) on connecting rod and cap are to same side of assembly and install rod and piston unit with these marks forward.

NOTE: Two different connecting rod bolts have been used; refer to Fig. V1-14. If equipped with late type stretch bolt (B), the bolts must be discarded and new bolts installed on reassembly.

On early models with rigid style connecting rod bolts (A—Fig. V1-14), lubricate bolt and nut threads and tighten nuts to a torque of 45 N·m (33 ft.-lbs.).

On later models with stretch type connecting rod bolts (B), be sure to assemble using new bolts. Lubricate bolt and nut threads and tighten nuts to a torque of 30 N·m (22 ft.-lbs.). Then, tighten nuts an additional 1/2 turn (180 degrees).

NOTE: If assembling unit with new bolts only for purpose of checking bearing clearance with Plastigage, tighten to 30 N·m (22 ft.-lbs.) only and do not tighten the additional 1/2 turn. These bolts can be reused for final assembly.

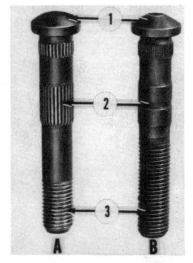

Fig. V1-14—View showing early rigid type connecting rod bolt (A) and late stretch type bolt (B). Bolts can be distinguished by differences in head (1), center section (2) and thread length (3). Refer to text for tightening procedure.

CYLINDER BORES

The pistons ride directly in unsleeved cylinder bores. Standard cylinder bore diameter is 76.51 mm (3.012 inches). If cylinder bore wear exceeds 0.08 mm (0.003 inch) at point of maximum wear, cylinder should be rebored and honed to next appropriate oversize. Pistons and rings are available in oversizes of 0.25, 0.50 and 1.0 mm (0.010, 0.020 and 0.040 inch).

PISTONS, PINS, AND RINGS

Pistons are equipped with two compression rings and one oil control ring and a full floating piston pin retained by snap ring at each end of pin in piston pin bore.

Check pistons for cracks, scoring or other damage. Measure piston skirt at right angle to piston pin and about 15 mm (0.59 inch) from bottom of skirt. Piston standard diameter new is 76.48 mm (3.011 inches). Skirt to cylinder bore clearance should be 0.03 mm (0.0012 inch). Pistons are available in oversizes with skirt diameters of 76.73 mm (3.021 inches), 76.98 mm (3.031 inches) and 77.48 mm (3.050 inches). Renew piston if skirt diameter is 0.04 mm (0.0016 inch) less than standard diameter. Cylinders must be rebored and honed to next oversize to install larger diameter oversize pistons.

NOTE: Although pistons for all engines are same diameter, compression height (center of piston pin hole to top of piston) is 44.9 mm (1.768 inches) for Model 068.2 and 41.7 mm (1.642 inches) for other models. Turbocharged models also require a different piston with oil jet relief notch in bottom of skirt.

Check ring side clearance in piston groove using new rings and a feeler gage. Wear limit is 0.2 mm (0.008 inch) for compression rings and 0.15 mm (0.006 inch) for oil ring on naturally aspirated models. For turbocharged pistons,

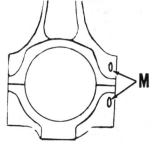

Fig. V1-15—Connecting rod and cap forging marks (M) are installed to same side and toward front of engine.

SMALL DIESEL

maximum allowable ring side clearance is 0.25 mm (0.010 inch) for compression rings and 0.15 mm (0.006 inch) for oil ring. Renew pistons if ring grooves are worn past maximum clearance limit.

Piston pin should be a thumb press fit in connecting rod and a slight interference fit in piston. To remove pin, remove snap rings at each end of piston pin bore. Using a piloted drift, tap pin out of piston and connecting rod. If pin is tight in piston, heat piston to about 60° C (140° F). When installing pin, be sure that side of rod with forging marks (M—Fig. V1-15) is facing same direction as arrow on piston crown. Always use new snap rings and do not compress snap rings any more than necessary to install in piston pin bore grooves.

Piston rings are available in standard diameter size as well as in 0.25, 0.50 and 1.0 mm (0.010, 0.020 and 0.040 inch) oversize to fit standard and oversize pistons. Piston ring end gap in unworn cylinder bore should be 0.3-0.5 mm (0.012-0.020 inch) for compression rings and 0.25-0.40 mm (0.010-0.016 inch) for oil control ring. Cylinders should be rebored for next oversize pistons and rings if piston ring end gap exceeds 1.0 mm (0.039 inch) with new piston rings.

Install new rings on pistons with side having identification mark up. Stagger ring end gaps equally 1/3 of the way around piston.

CONNECTING RODS AND BEARINGS

Connecting rod is fitted with a pressed in renewable piston pin bushing and a precision insert type bearing at crankpin end. The piston pin bushing must be reamed or honed to fit piston pin after installing bushing in rod.

Crankpin standard diameter is 45.958-45.978 mm (1.8094-1.8101 inches) on Model 068.2 and 47.758-47.778 mm (1.8802-1.8810 inches) on other models. Connecting rod bearing inserts are available in standard size and 0.25, 0.50 and 0.75 mm (0.010, 0.020 and 0.030 inch) undersizes.

Connecting rod bearing to crankpin oil clearance should be 0.28-0.88 mm (0.0011-0.0035 inch); maximum allowable oil clearance is 0.12 mm (0.0047 inch). Maximum allowable side clearance of rod on crankpin is 0.37 mm (0.0146 inch).

Refer to PISTON AND ROD UNITS paragraph for important information on connecting rod bolts.

CRANKSHAFT AND MAIN BEARINGS

To remove crankshaft, drain coolant and lubricating oil. Remove water

SERVICE MANUAL

Volkswagen

pump, alternator and starting motor. Remove fuel injection pump, cylinder head, oil pan, oil pump and the piston and rod units. Remove crankshaft V-belt pulley or optional front power take off drive. On turbocharged models, remove crankshaft vibration dampener. Remove flywheel, crankshaft sprocket and front and rear crankshaft seal carriers; refer to CRANKSHAFT SPROCKET paragraph. Be sure bearing caps are numbered so they can be reinstalled in same position. Unbolt and remove main bearing caps and lift crankshaft from block.

Refer to CONNECTING ROD AND BEARINGS paragraph for crankshaft crankpin information.

Standard crankshaft main journal diameter is 53.958-53.978 mm (2.1243-2.1251 inches) on all models. Crankshaft bearings are available in standard size and undersizes of 0.25, 0.50 and 0.75 mm (0.010, 0.020 and 0.030 inch).

Main bearing oil clearance should be 0.03-0.08 mm (0.0012-0.0031 inch) with maximum allowable clearance being 0.17 mm (0.0067 inch). Crankshaft end play, controlled by thrust bearings at center main journal, should be 0.07-0.17 mm (0.0028-0.0067 inch) with maximum allowable end play being 0.37 mm (0.0145 inch).

To reinstall crankshaft, reverse removal procedure. Before tightening center main bearing bolts, pry crank to front, then to rear, to center the thrust bearing. Tighten main bearing bolts to a torque of 65 N·m (48 ft.-lbs.). Be sure crankshaft turns freely in bearings before proceeding further.

CRANKSHAFT SPROCKET

EARLY ENGINES. On early engines, crankshaft timing sprocket was located on crankshaft by a Woodruff key and a M12 × 1.5 mm cap screw. On these engines, apply thread locking compound to cap screw and tighten to a torque of 80 N·m (60 ft.-lbs.). A Duralok M12 × 1.5 mm bolt is also used; do not apply thread locking compound to this bolt. Oil the Duralok M12 bolt before installing and tighten to a torque of 150 N·m (110 ft.-lbs.).

LATE ENGINES. On late engines, crankshaft timing sprocket is located on crankshaft by a notch in the front end of crankshaft and a lug (tab) on sprocket which engages the notch. Sprocket retaining cap screw is a hex head Duralok M14 × 1.5 mm bolt. Oil threads, install bolt and tighten to a torque of 200 N·m (148 ft.-lbs.). This bolt has been changed in most recent engine production to a 12-point head bolt, and this bolt can be used as a service replacement for the hex head M14 bolt. The 12-point bolt must always be renewed if removed. Tighten the 12-point bolt to a torque of 90 N·m (66 ft.-lbs.), then bolt an additional 180 degrees (1/2 turn). The additional 1/2 turn can be made in stages; however, it is important that an exact 1/2 turn is maintained.

FRONT PTO. A front pto drive with extended crankshaft is optional on some engines. On these models, timing sprocket is locked to crankshaft by a heavy flat washer, a thrust ring and two tapered lapping clamping rings. As there is no key or tab on the front pto timing sprocket, timing belt should be installed and camshaft and fuel injection pump timing set before tightening crankshaft bolt. Apply thread locking compound and tighten the bolt to a torque of 85 N·m (63 ft.-lbs.).

NOTE: If turning crankshaft using wrench on crankshaft bolt on these models, always turn bolt clockwise (to right). Turning the bolt in opposite direction may loosen bolt and cause timing sprocket to shift on crankshaft.

CRANKSHAFT OIL SEALS

Crankshaft seals are located in seal carriers attached to front and rear of cylinder block. Seals may be renewed after removing flywheel and crankshaft timing belt sprocket without removing seal carriers. If removing seal carriers, note that bolts are installed in each carrier through engine oil pan. Lubricate lips of seals with clean engine oil before installing over crankshaft. Tighten seal carrier retaining bolts to a torque of 10 N·m (8 ft.-lbs.). Tighten pan to oil carrier bolts to 20 N·m (15 ft.-lbs) if equipped with stamped steel pan, or to 10 N·m (8 ft.-lbs.) if equipped with aluminum oil pan.

ELECTRICAL SYSTEM

Engine may be equipped with a 12 volt or optional 24 volt electrical system. Follow standard procedure in testing alternator, voltage regulator and starter.

GLOW PLUGS

Each cylinder is equipped with a glow plug as a cold starting aid. Glow plugs are connected in parallel with each glow plug grounded through mounting threads. Before suspecting glow plug malfunction, check to see that current is reaching glow plugs when circuit is activated. To check individual glow plugs, remove connector strap between glow plugs. Connect a test light between positive battery terminal to each glow plug electrode. If test light turns on, glow plug should be good. If test light will not light, glow plug is defective.

If inner end of glow plug is burned off, a faulty fuel injector should be suspected. Glow plug may be damaged if installed with tightening torque more than the specified 30 N·m (22 ft.-lbs.).

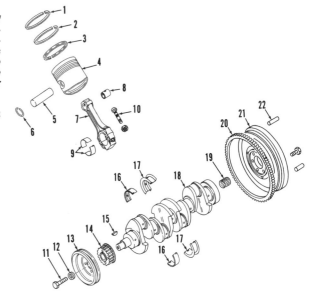

Fig. V1-16—Exploded view of crankshaft, piston and rod assemblies. Flanged type center main bearing (17) is shown; some engines are equipped with separate thrust bearings at center main position.

1. Top compression ring
2. Second compression ring
3. Oil control ring
4. Piston
5. Piston pin
6. Snap rings (2)
7. Connecting rod
8. Rod pin bushing
9. Crankpin bearings
10. Rod bolts (see Fig. V1-14)
11. Crankshaft bolt
12. Flat washer
13. Crankshaft pulley
14. Crankshaft sprocket
15. Woodruff key (early models)
16. Main bearing inserts
17. Center main bearing
18. Crankshaft
19. Pilot bearing
20. Flywheel ring gear
21. Flywheel
22. Clutch cover dowels

WESTERBEKE

J. H. WESTERBEKE CORP.
Avon Industrial Park
Avon, Massachusetts 02322

Model	No. Cyls.	Bore	Stroke	Displ.
W13 (4.4 KW)	2	70 mm (2.756 in.)	78 mm (3.071 in.)	600 cc (36.6 cu. in.)
W18 (6 KW)	3	65 mm (2.559 in.)	78 mm (3.071 in.)	776 cc (47.4 cu. in.)
W21 (8 KW)	3	73 mm (2.874 in.)	78 mm (3.071 in.)	979 cc (59.7 cu. in.)
W27 (11 KW)	4	73 mm (2.874 in.)	78 mm (3.071 in.)	1305 cc (79.6 cu. in.)
W33 (12.5 KW)	4	78 mm (3.071 in.)	78 mm (3.071 in.)	1490 cc (90.9 cu. in.)

All models listed are four-stroke cycle, water cooled, vertical cylinder, valve in head type indirect injection diesel engines. Direction of engine rotation is counterclockwise as viewed from flywheel end. Engines are for marine use and are available as either direct power source or marine diesel generators. Numbers in parenthesis following engine model number is size of marine diesel generator unit. Firing order is 1-3-2 on three-cylinder models and 1-3-4-2 on four cylinder engines.

FUEL FILTER. A primary fuel filter element (18—Fig. WB1-1) is located in the electric fuel pump (12) and a secondary filter element (28) is located between the electric pump (12) and fuel injection pump. Recommended replacement interval for both filter elements is after each 200 hours of use. Excessive sediment or water in fuel may require renewing filter elements earlier than specified use interval. Bleed system as outlined in following paragraph after servicing fuel filters.

BLEED FUEL SYSTEM. Be sure fuel supply valve is open and loosen the bleed screw (29—Fig. WB1-1) on secondary filter. Turn the ignition switch to "ON" position to actuate electric pump. When fuel free of air bubbles is flowing out bleed screw opening, tighten the bleed screw. If engine will not start, loosen the injector line (3 and 4) fittings at injectors (40) and crank engine until fuel flows at the loosened fittings. Tighten the fittings and start engine.

MAINTENANCE

LUBRICATION

Recommended engine lubricant is API classification CC or CD. SAE 10W-30 motor oil may be used in all ambient temperatures. If single grade oil is used, use SAE 10 in temperatures below 0° C (32° F), SAE 20 in temperatures from 0°-27° C (0°-80° F) and SAE 30 in temperatures above 27° C (80° F). Oil pan capacity is as follows:

Model	Capacity
W13	2.5 liters (2.65 qts.)
W18, W21	3.6 liters (3.80 qts.)
W27	4.0 liters (4.25 qts.)
W33	4.5 liters (4.75 qts.)

FUEL SYSTEM

Refer to Fig. WB1-1 for exploded view of 2-cylinder engine fuel system; other models are similar.

Fig. WB1-1—Exploded view of Model W13 fuel system. Other models are similar except for number of pumping elements in injection pump, injection lines and injectors.

1. Injection pump
2. Timing shims
3. Injector line
4. Injector line
5. Fuel return hose
6. Fuel return line
7. Banjo bolt
8. Washers
9. Banjo connector bolt
10. Aluminum washer
11. Washers
12. Electric fuel pump
14. Line to filter
15. Fuel supply line
18. Fuel filter element
19. Gasket
20. Filter cap
21. Pump ground wire
22. Magnet
23. Banjo bolt
24. Aluminum washer
25. Washer
26. Filter bowl retainer
27. Filter bowl
28. Filter element
29. Bleed screw
30. Filter base
31. "O" rings
32. Washers
33. Banjo bolt
34. Aluminum washer
35. Banjo fitting
36. Glow plug solenoid
37. Filter & pump bracket
38. Clamp
39. Seal washer
40. Fuel injector
41. Washers
42. Banjo bolt
43. Injector leak-off line

Illustrations Courtesy J.H. Westerbeke Corp.

SERVICE MANUAL

Westerbeke

INJECTION PUMP TIMING

Injection timing should be 23 degrees BTDC for all marine engines and 12.5 KW generator set. Timing for all generator sets except the 12.5 KW unit should be 19 degrees BTDC. To check injection timing, remove No. 1 cylinder fuel injection line and the No.1 cylinder injection pump delivery valve holder (H—Fig. WB1-2) and spring (S). Lift out delivery valve (V) and reinstall spring and delivery valve holder. Turn crankshaft in normal direction of rotation and stop when fuel stops flowing from open delivery valve holder. The correct timing mark on engine crankshaft pulley (see Fig. WB1-3) should be aligned with mark (M) on boss of timing cover. Repeat procedure several times to be sure you are obtaining accurate reading.

If injection timing is not correct, it will be necessary to remove the fuel injection pump and change the thickness of the shim (2—Fig. WB1-1) under the pump mounting flange. Changing shim thickness by 0.1 mm will result in a change in timing of about one degree. Shims are available in nine different thicknesses from 0.2 mm to 1.0 mm in steps of 0.1 mm. Decrease shim thickness to advance timing, or increase shim thickness to retard timing.

GOVERNOR

The flyweight type governor is mounted on the front face of the governor gear on Model W13 and on fuel injection pump drive gear on all other models. All governor linkage is mounted in the timing gear cover.

On all power units, the high speed no-load adjustment screw is adjusted and secured with a wire and lead seal at the factory; refer to Fig. WB1-4. If high idle speed is found to be incorrect cut seal wire, loosen locknut and turn screw in to decrease engine speed or out to increase speed to about 8 percent above rated speed on engine nameplate. With high idle speed correctly adjusted, tighten locknut and install new wire and lead seal.

On marine diesel generator units, the adjustment screw is not used and engine speed when generator is providing power is controlled by a solenoid connected to governor control arm. The solenoid bracket and throttle linkage must be adjusted so that solenoid operates freely and solenoid plunger is bottomed in solenoid when activated. Adjust engine speed under load to rated speed on unit nameplate.

Adjust low-idle stop screw or throttle linkage to obtain about 900 rpm on all models.

EXCESS FUEL DEVICE

An internal feature of the fuel injection pump is an excess fuel device that permits throttle operation without engine running to override the maximum fuel delivery stop. This provides an excess amount of fuel for aid in starting engine. Adjustment of this device should not be attempted outside authorized diesel system service station. If engine smokes excessively under load, this device may be at fault.

REPAIRS

TIGHTENING TORQUES

When making repairs, refer to the following special tightening torques:
Cylinder head bolts:
  M10 bolts, Models W18,
    W21, W27, W33 . . . . 69-78 N·m
      (51-58 ft.-lbs.)
  M12 bolts, Models W18,
    W21 . . . . 108-117 N·m
      (80-86 ft.-lbs.)
  M12 bolts, Models W13,
    W27 . . . . 118-127 N·m
      (87-94 ft.-lbs.)
  M14 bolts, Model W33* 147-157 N·m
      (109-115 ft.-lbs.)

*See Fig. WB1-9; M14 bolt numbers 3, 4, 9 and 10 are tightened to a torque of 108-117 N·m (80-86 ft.-lbs.).
Crankshaft pulley nut:
  Model W13 . . . . 147-196 N·m
      (109-144 ft.-lbs.)
  All other models . . . . . 196-245 N·m
      (144-180 ft.-lbs.)
Main bearing cap bolts,
  all . . . . 49-54 N·m
      (36-43 ft.-lbs.)
Connecting rod cap bolts:
  Model W33 . . . . 54-58 N·m
      (38-43 ft.-lbs.)
  All other models . . . . 32-34 N·m
      (23-25 ft.-lbs.)
Flywheel bolts:
  With separate
    washers . . . . 113-122 N·m
      (84-90 ft.-lbs.)
  Washer attached
    bolts . . . . 128-137 N·m
      (95-100 ft.-lbs.)
Oil pan drain plug . . . . 49-58 N·m
      (37-43 ft.-lbs.)
Oil filter . . . . 11-12 N·m
      (8-9 ft.-lbs.)
Delivery valve,
  injection pump . . . . 40-49 N·m
      (29-36 ft.-lbs.)
Injector mounting
  bolts . . . . 15-19 N·m
      (11-14 ft.-lbs.)
Nozzle retaining nut . . . . . . 59-78 N·m
      (44-57 ft.-lbs.)
Glow plug . . . . 15-19 N·m
      (11-14 ft.-lbs.)

WATER PUMP AND COOLING SYSTEM

Cooling is provided by a closed fresh water cooling system which is kept within operating temperature by a sea water heat exchanger. The fresh water system is a closed system having a pressure cap and coolant recovery tank. It is recommended that the fresh water

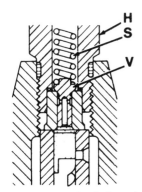

Fig. WB1-2—Cross-sectional view of injection pump delivery valve (V), holder (H) and spring (S). Delivery valve is temporarily removed to check injection timing.

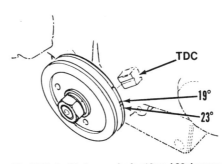

Fig. WB1-3—Timing marks for 19 and 23 degrees BTDC are located on crankshaft pulley. Alignment mark (M) is on timing cover.

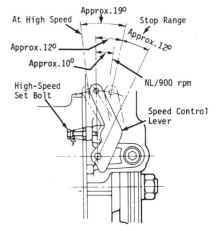

Fig. WB1-4—View showing governor speed control lever and high speed set bolt.

Illustrations Courtesy J.H. Westerbeke Corp.

Westerbeke

SMALL DIESEL

system be filled with a 50:50 mixture of water and permanent type antifreeze. Two water pumps, a fresh water circulating pump and a sea water lift pump, are used. A zinc rod installed in the heat exchanger is used to combat corrosion in the sea water system.

SEA WATER PUMP. Refer to Figs. WB1-5 and WB1-6 for exploded views of the two different types of sea water pumps used. On Model W13, the sea water pump (Fig. WB1-5) is mounted on front of timing cover. On all other models, the pump (Fig. WB1-6) is mounted on rear side of engine front plate. On most models, a pump drive gear unit (2—Figs. WB1-15 and WB1-16) is driven from engine camshaft timing gear. On some generator unit, the sea water pump is belt driven.

Disassembly of either type sea water pump is evident from reference to exploded view in Fig. WB1-5 or Fig. WB1-6 and examination of unit. The rubber impeller can be inspected and renewed if necessary after removing pump cover plate (1); it is not necessary to remove water pump from engine to service impeller. When installing impeller, apply grease to impeller cavity and impeller blades and be sure all impeller blades are curved backward in trailing position. Renew cover plate if worn.

When pump is serviced, be sure to check quickly for sea water flow after starting engine. If pump does not prime quickly, damage to impeller will result from overheating.

FRESH WATER PUMP. The belt driven fresh water circulating pump is serviced as a complete assembly only. If pump seal is leaking, bearings are loose or noisy, or if other damage is noted, renew the fresh water pump.

THERMOSTAT

Two different types of thermostats have been used. One type is a common valve type thermostat; the second type has a bypass valve disc at bottom of unit. It is important that the correct type thermostat be installed when renewing same. To check thermostat, place it in boiling water; the thermostat should open fully.

VALVE ADJUSTMENT

Valve rocker arm to valve stem clearance is 0.25 mm (0.010 inch) for intake and exhaust with engine cold on all models. To check valve clearance adjustment, remove valve cover and turn piston to TDC of compression stroke while checking clearance for that cylinder. Repeat for each cylinder of engine.

CYLINDER HEAD

To remove cylinder head, drain engine coolant, disconnect coolant hose and remove intake and exhaust manifolds. Remove fuel injection lines and immediately cap all openings. Remove fuel filter unit and fuel injector assemblies from cylinder head. Remove oil pressure tube to cylinder head on Model W13. Remove all other attachments and connections to cylinder head such as thermostat housing and glow plug wiring. Remove valve rocker arm cover, rocker arm assembly and push rods. Gradually loosen cylinder head bolts in reverse order of tightening sequence shown in Fig. WB1-7, WB1-8 or WB1-9. Remove cylinder head and gasket.

Carefully clean cylinder head and block surfaces and inspect head for distortion, cracks or other damage. Cylinder head must be renewed or resurfaced if distortion (warpage) exceeds 0.1 mm (0.004 inch). The precombustion chambers (swirl chambers) are pressed into cylinder head and should not be removed. Check valve heads for sinkage; valve head should not be more than 1 mm (0.040 inch) below cylinder head gasket surface.

To reinstall cylinder head, be sure all gasket surfaces are thoroughly clean. Do

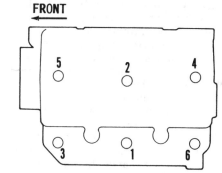

Fig. WB1-7—Cylinder head bolt tightening sequence for two-cylinder engine.

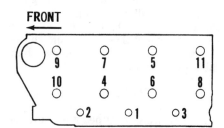

Fig. WB1-8—Cylinder head bolt tightening sequence for three-cylinder engine.

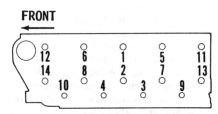

Fig. WB1-9—Cylinder head bolt tightening sequence for four-cylinder engine.

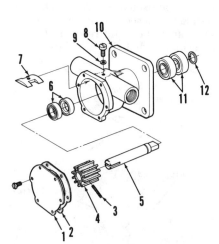

Fig. WB1-5—Exploded view of sea water pump used on Model W13. Refer to Fig. WB1-6 for other models.

1. Cover plate
2. Gasket
3. Impeller drive screw
4. Rubber impeller
5. Pump shaft
6. Water seal assy.
7. Pump cam
8. Cam retaining screw
9. Washer
10. Housing
11. Bearings
12. Snap ring

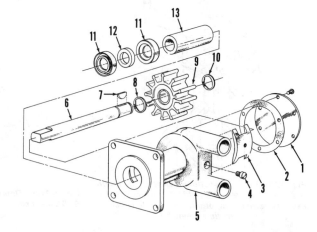

Fig. WB1-6—Exploded view of sea water pump used on most models except Model W13. Refer to Fig. WB1-5 for Model W13 pump.

1. Cover plate
2. Gasket
3. Pump cam
4. Cam retaining screw
5. Pump housing
6. Pump shaft
7. Woodruff key
8. Snap ring
9. Rubber impeller
10. Snap ring
11. Seals
12. Water slinger
13. Graphite bushing

Illustrations Courtesy J.H. Westerbeke Corp.

SERVICE MANUAL

Westerbeke

not use gasket sealer of any kind on the gasket or gasket surfaces of head and block. Refer to Fig. WB1-7, WB1-8 or WB1-9 and tighten cylinder head bolts in several steps in sequence shown until final tightening torque is reached. All M10 (10 mm) head bolts should be tightened to a torque of 69-78 N·m (51-58 ft.-lbs.). M12 bolts should be tightened to a torque of 108-117 N·m (80-86 ft.-lbs.) on Models W18 and W21, and to a torque of 118-127 N·m (87-94 ft.-lbs.) on Models W13 and W27. On Model W33, tighten the M14 bolts in sequence numbers 3, 4, 9 and 10 to a torque of 108-117 N·m (80-86 ft.-lbs.) and all other M14 bolts to a torque of 147-157 N·m (109-115 ft.-lbs.).

Complete reassembly by reversing disassembly procedure. After the first 50 hours of operation after cylinder head installation, the head bolts should be retorqued to specification. Retorque rocker arm shaft support bolts and check valve clearance.

VALVE SYSTEM

Refer to Fig. WB1-10 for exploded view of typical cylinder head and valve system. Both the intake and exhaust valves seat directly on machined seat in cylinder head. Valve face and seat angle is 45 degrees. Valve guides are renewable on all models. Stem diameter is 6.6 mm (0.260 inch) on Models W13, W18, W21 and W27, and 8.0 mm (0.315 inch) on Model W33. Valve margin (thickness of head at outer edge of face) is 1.0 mm (0.040 inch) on all models except Model W33, which has a valve margin of 1.5 mm (0.059 inch).

Valve seat width should be 1.3-1.8 mm (0.051-0.071 inch) with maximum allowable seat width being 2.5 mm (0.098 inch). If guides are to be renewed, seats should be cut or ground after installing guides. It may be necessary to ream valve guides after installation to obtain a desirable valve stem to guide fit. After refacing and reseating valves, check distance from valve head to cylinder head gasket surface. If more than 1.0 mm (0.040 inch) with new valve, cylinder head must be renewed.

When testing valve springs, refer to the following specifications:

Model W13
Spring free length, new . . . . . 43 mm
(1.693 in.)
Spring free length, min. . . . . 41.7 mm
(1.642 in.)
Spring installed height . . . . . . 36 mm
(1.417 in.)
Pressure @ installed
  height . . . . 131-143 N
(29.5-32.2 lbs.)

Models W18-W21-W27
Spring free length, new . . . . . 43 mm
(1.693 in.)
Spring free length, min. . . . . 41.7 mm
(1.642 in.)
Spring installed height . . . . . 37.1 mm
(1.461 in.)
Pressure @ installed
  height . . . . 110-122 N
(24.7-27.3 lbs.)

Model W33
Spring free length, new . . . 45.85 mm
(1.805 in.)
Spring free length, min. . . . 44.5 mm
(1.752 in.)
Spring installed height . . . . . 41.8 mm
(1.646 in.)
Pressure @ installed
  height . . . . 123-136 N
(27.6-30.6 lbs.)

On all models, valve spring pressure at installed height may be up to 15 percent less than specified above.

ROCKER ARM ASSEMBLY

The rocker arm assembly (see Fig. WB1-10) can be unbolted and removed from top of head after removing valve rocker arm cover. With rocker arm assembly removed, lift out the push rods. Keep location of all removed parts noted so they may be reinstalled in same location. Rocker arms (5), spacer springs (10) and rocker arm supports (6) can be removed from shaft after removing bolt (11) and the snap ring (4) at each end of shaft.

Check all parts, including push rod ends, rocker arm adjusting screws, and rocker arm valve contact pads for wear or damage. Intake and exhaust rocker arms are alike. Rocker arm inside diameter is 18.9 mm (0.744 inch). Shaft to rocker arm clearance should be 0.05 mm (0.002 inch); maximum allowable clearance is 0.2 mm (0.008 inch).

Reassemble using new parts as necessary by reversing disassembly procedure. Note that identification mark (M—Fig. WB1-10), a 3 mm (0.118 inch) hole in rocker arm shaft, is assembled to front and to rocker arm screw adjusting side. Bolt (11) in front support must engage notch (N) in rocker arm shaft (9). Loosen locknuts (7) and loosen rocker arm adjusting screws before installing push rods and rocker arm assembly. Adjust valve clearance cold to 0.25 mm (0.010 inch) before starting engine.

INJECTORS

REMOVE AND REINSTALL. To remove fuel injectors, first thoroughly

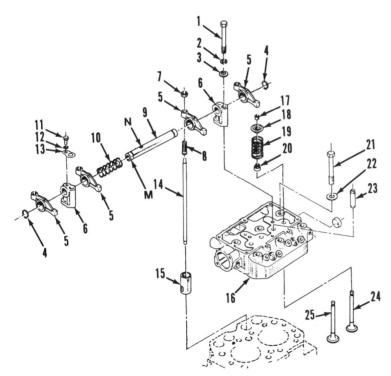

Fig. WB1-10—Exploded view of cylinder head and valve system parts for two-cylinder engine; other models are similar. Several different bolt arrangements are used for retaining valve rocker arm cover.

N. Locating notch
M. Shaft position mark
1. Rocker support bolt
2. Lockwasher
3. Flat washer
4. Snap ring
5. Rocker arms
6. Rocker shaft support
7. Adjusting screw nut
8. Adjusting screw
9. Rocker arm shaft
10. Spacer spring(s)
11. Shaft detent bolt
12. Lockwasher
13. Washer plate
14. Push rod
15. Valve lifter
16. Cylinder head
17. Split valve locks
18. Spring retainer
19. Valve spring
20. Valve stem seal
21. Head bolts
22. Head bolt washers
23. Valve guides
24. Exhaust valve
25. Intake valve

Illustrations Courtesy J.H. Westerbeke Corp.

Westerbeke

clean injectors, lines and surrounding area using suitable solvent and compressed air. Disconnect fuel return line from fittings on side of injectors and remove high pressure lines from pump to injectors. Immediately cap or plug all openings to prevent entry of dirt. Remove injectors from cylinder head by unscrewing the injector retaining cap screws. Remove gasket from bottom of injector bore in cylinder head.

Clean injector bores in cylinder head before reinstalling injectors. Install injectors using new gaskets. Install fuel injector lines to align injectors, then tighten injector retaining cap screws evenly to a torque of 15-19 N·m (11-14 ft.-lbs.).

TESTING. A complete job of testing and adjusting an injection nozzle requires use of special test equipment. Use only a clean approved testing oil in injector tester tank. Injection nozzle should be checked for opening pressure, spray pattern and seat leakage. Test injection nozzle as outlined in following paragraphs:

WARNING: Fuel emerges from injector with sufficient force to penetrate the skin. When testing injector, keep yourself clear of nozzle spray.

OPENING PRESSURE. Before conducting test, operate tester lever until fuel flows from tester line, then attach injection nozzle to line. Pump lever a few quick strokes to purge air and be sure nozzle valve is not stuck and spray hole is open.

Operate tester lever slowly and observe gage reading when nozzle opens. On all models, opening pressure should be 11770-12750 kPa (1707-1850 psi). If necessary to adjust opening pressure, refer to OVERHAUL paragraph for injector disassembly. Add shims to increase opening pressure or remove shim thickness to decrease pressure. Adjusting shims are available in eight different thicknesses of 0.1, 0.9, 1.0, 1.1, 1.2, 1.3, 1.35 and 1.4 mm, and are used as required to obtain specified opening pressure. Injector opening pressure varies approximately 980 kPa (142 psi) with each 0.1 mm change in shim thickness. Change shim thickness as necessary, reassemble nozzle and recheck opening pressure.

SPRAY PATTERN. The throttling type nozzle should emit a tight conical spray pattern with no branches, splits or dribbles. If incorrect spray pattern is observed, check for partially clogged or damaged spray hole or improperly seating or sticking nozzle valve.

SEAT LEAKAGE. Wipe nozzle tip dry, then operate tester lever as in spray pattern test. If a drop forms or undue wetness appears on nozzle tip after test, disassemble and clean or renew injector.

OVERHAUL. First, thoroughly clean outside of injector. Hold injector body (2—Fig. WB1-11) in soft jawed vise. Unscrew nozzle retaining nut (10) and disassemble nozzle. Be careful not to lose or mix adjusting shims (4). Place all parts in clean calibrating oil or diesel fuel and keep all parts of each injector separate from others.

Clean exterior surfaces with a brass wire brush. Soak parts in an approved carbon solvent if necessary to loosen hard carbon deposits. Rinse parts in clean diesel fuel or calibrating oil immediately after cleaning to neutralize the solvent.

Clean nozzle spray hole from inside using a hardwood scraper. Scrape carbon from nozzle chamber using a hooked brass scraper. Clean valve seat using a brass scraper, then polish seat using wood polishing stick and mutton tallow.

Back flush nozzle using a reverse flush adapter on injector tester. Reclean all parts by rinsing thoroughly in clean diesel fuel or calibrating oil and reassemble while parts are immersed in the fluid. Make sure to assemble with same shims as were removed. Tighten nozzle retaining nut to a torque of 59-78 N·m (44-57 ft.-lbs.) and retest injector.

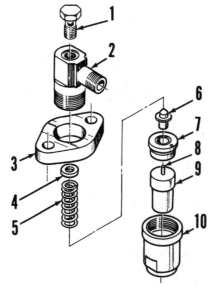

Fig. WB1-11—Exploded view of fuel injector assembly.

1. Return banjo bolt
2. Injector body
3. Clamp plate
4. Adjusting shims
5. Spring
6. Spring seat
7. Spacer
8. Nozzle valve
9. Nozzle
10. Cap nut

SMALL DIESEL

INJECTION PUMP

The fuel injection pump on Model W13 engine is mounted on top of the engine timing gear cover, and ball on pump rack engages slot in upper end of governor lever (20—Fig. WB1-12). On all other models, the fuel injection pump is mounted on the side of engine crankcase and the ball on pump rack is connected to governor lever by a tie rod (15—Fig. WB1-13) and spring (14).

To remove fuel injection pump, first thoroughly clean pump, injectors and lines, close fuel supply valve, then proceed as follows: Remove fuel injection lines, fuel supply and return lines and immediately cap all openings. Except on Model W13, remove cover (1—Fig. WB1-13) or stop lever and cover (1A) from side of cylinder block below fuel pump. Disconnect tie rod (15) and spring (14) from pump rack (16). On all models, unbolt and remove pump, taking care not to lose or damage any of the shims from under pump flange used to set injection timing. It is recommended that the pump not be disassembled outside of an authorized diesel system repair station.

Install pump in housing with same thickness of shims that were removed unless injection timing is to be changed. On Model W13, be sure ball on pump rack engages slot in upper end of governor control lever (20—Fig. WB1-12). Except on Model W13, reconnect tie rod (15—Fig. WB1-13) and spring (14) to pump rack (16) and install side cover (1) or stop lever cover (1A) with new gasket (2). Tighten the pump retaining bolts evenly. Complete reassembly by reversing removal procedure.

TIMING GEAR COVER

To remove timing gear cover, thoroughly clean engine and proceed as follows: On Model W13, unbolt and remove sea water cooling pump (1—Fig. WB1-12) and fuel injection pump. Except on Model W13, remove cover (1—Fig. WB1-13) or stop lever cover (1A) from side of cylinder block below fuel injection pump flange and disconnect spring (14) and tie rod (15) from pump rack (16). On all models, remove the fresh water pump and alternator. Unbolt and remove accessory pulley, if so equipped, from front of crankshaft pulley. Remove crankshaft pulley nut, washer and pulley. Disconnect any governor linkage from arm on timing cover. Unbolt and remove timing gear cover from front of crankcase on Model W13, or from engine front plate on all other models. Note that front plate is bolted to engine crankcase from inside the timing gear cover and is not removed with the gear cover. On Model

SERVICE MANUAL

W33, crankshaft front ball bearing is piloted in timing gear cover. On all models, be careful in removing cover from aligning dowel pins, the sea water pump ball bearing, and on Model W33, the crankshaft ball bearing.

With timing gear cover removed, install new crankshaft front oil seal in gear cover with lip of seal to inside. Inspect oil seal contact surface of crankshaft pulley (or on spacer sleeve on Model W13) and recondition or renew pulley or sleeve as necessary. Model W33 engines after serial #4844 have a renewable camshaft thrust plug in the timing gear cover.

To install front cover, be sure all governor linkage, the governor sliding plate and the sea water pump drive gear are in place. Install cover using new gasket and, on all models except Model W13, guide governor tie rod and spring into proper position. Install fuel injection pump on Model W13; on other models, reconnect governor spring and tie rod to fuel injection pump rack and install side cover (1—Fig. WB1-13) or stop lever cover (1A) with new gasket (2). Complete remainder of reassembly by reversing disassembly procedure. Tighten crankshaft pulley nut to a torque of 147-196 N·m (109-144 ft.-lbs.) on Model W13, and to a torque of 196-245 N·m (144-180 ft.-lbs.) on all other models.

GOVERNOR

Model W13

The governor unit is accessible for service after removing the engine timing gear cover. Refer to exploded view of governor unit in Fig. WB1-12. Remove sliding shaft (23) from front of governor gear (26), then unbolt and remove governor weight assembly (24) from gear. Remove snap ring (25) and pull gear from stub shaft (27). Remove the governor spring (14). Unscrew nut (12) and remove nut, washer (11) and lever (10) from control shaft (6) and pull shaft from timing cover. Remove spring lever (17) and governor lever (20) from shaft (18), then remove grooved pin (19) and shaft from cover.

Inspect all parts and renew as necessary. Reassemble by reversing disassembly procedure.

All Models Except W13

The governor unit is accessible for service after removing the engine timing gear cover. Refer to exploded view of governor unit in Fig. WB1-13 and to cross-sectional views of the governor unit in Fig. WB1-14. Remove the sliding shaft (24) from front end of fuel injection pump camshaft and unbolt and remove flyweight assembly (25) from pump drive gear (26). To remove linkage from inside of timing gear cover, proceed as follows: Remove governor spring (10). Remove nut (13), washer (12) and lever (11) and pull speed control lever from timing gear cover. Remove nut (17), washer (18) and lever (11), then remove governor lever cap screw (21). Remove groove pin (20) and shaft (19) from timing cover and governor lever (23). Remove tie rod (15) and spring (14) from governor lever.

Inspect all parts, including the needle bearings (28 and 29—Fig. 1-14) for governor lever shaft in timing cover, and renew as necessary. Reassemble by reversing disassembly procedure.

TIMING GEARS

Model W13

Timing gears are accessible for inspection after removing engine timing gear cover. The timing gears consist of crankshaft gear and camshaft gear. The governor gear is mounted on a stub shaft located in front end of engine crankcase. The sea water pump drive gear (2—Fig. WB1-15) is mounted in two ball bearings (1 and 3), one piloted in timing gear cover and the other in engine block.

To remove camshaft gear, it is necessary to remove the camshaft; refer to CAMSHAFT paragraph. To remove governor gear (26—Fig. WB1-12), remove slider (23), unbolt governor weight assembly (24), remove snap ring (25) from stub shaft (27) and pull gear from shaft. Stub shaft can be renewed if worn or scored. Crankshaft gear can be removed using suitable puller.

When reassembling be sure timing mark on camshaft gear is matched with timing mark on crankshaft gear.

All Models Except Model W13

The engine camshaft gear (6—Fig. WB1-16) is driven from crankshaft gear via an idler gear (7) mounted on a stub shaft in front of engine crankcase. The idler gear stub shaft can be pulled from the cylinder block. Install new shaft into front of crankcase with annular oil ring groove to inside. The stub shaft should protrude 23.5-24.5 mm (0.925-0.964 inch) on Model W33 and 26.0-27.0 mm (1.024-1.063 inches) on other models. The governor unit is mounted on the fuel injection pump drive gear (9) and this gear is driven from the idler gear;

Fig. WB1-12—Exploded view of timing cover, governor and governor linkage for Model W13. Refer to Fig. WB1-13 for other models.

1. Sea water pump
2. Gasket
3. Crankshaft oil seal
4. Timing gear cover
6. Throttle arm
7. Locknut
8. High speed screw
9. "O" ring
10. Arm
11. Washer
12. Nut
13. Pin
14. Governor spring
15. Nut
16. Washer
17. Arm
18. Governor shaft
19. Pin
20. Governor lever
21. Washer
22. Set screw
23. Sliding shaft
24. Governor weights
25. Snap ring
26. Governor gear
27. Stub shaft

Illustrations Courtesy J.H. Westerbeke Corp.

Westerbeke — SMALL DIESEL

refer to FUEL INJECTION PUMP CAMSHAFT paragraph. The sea water pump drive gear (2) is driven by the engine camshaft gear (6). Model W33 crankshaft ball bearing and the crankshaft timing gear on all models can be removed using suitable pullers.

When reassembling, refer to Fig. WB1-17 and align timing marks as follows: The "1" mark on idler gear is aligned with the "1" mark on crankshaft gear. (On Model W33, crankshaft mark "1" is not visible with front ball bearing installed. Align idler gear mark "1" with a mark line on side of crankshaft gear boss.) Align mark "2" on camshaft gear with mark "2" on idler gear. Align injection pump gear mark "3" with mark "3" on idler gear.

INJECTION PUMP CAMSHAFT

Model W13

On Model W13, cam lobes for operation of the fuel injection pump are located on the front end of engine camshaft (5—Fig. WB1-15). Refer to CAMSHAFT paragraph.

All Models Except W13

The fuel injection pump camshaft (15—Fig. WB1-16) can be removed after removing timing gear cover, engine oil pump and filter assembly, sliding plate and governor weight unit, and the fuel injection pump. Working through hole in injection pump drive gear (9), remove bearing retaining cap screw (10) and washers (11 and 12). Then, withdraw camshaft from front of engine.

With camshaft removed, check ball bearing (13) at front end of shaft, condition of drive gear, camshaft lobes and the drive slot for the engine oil pump at rear end of shaft. The rear bearing journal of camshaft rides directly in cylinder block bore. Standard cam lobe height is 44 mm (1.7323 inches); renew

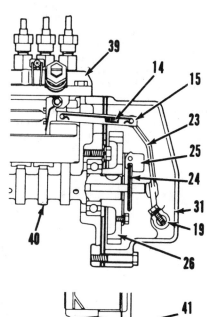

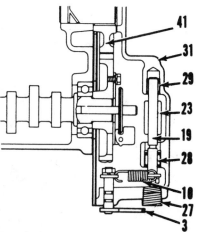

Fig. WB1-13—Exploded view of timing cover, governor and governor linkage for all models except Model W13. Refer to Fig. WB1-12 for Model W13.

1. Cover plate
1A. Stop lever & plate
2. Gasket
3. Throttle arm
4. Lock nut
5. High speed screw
6. Low idle screw
7. Locknut
8. Pin
9. "O" ring
10. Governor spring
11. Arms
12. Washer
13. Nut
14. Tie rod spring
15. Tie rod
16. Injection pump rack
17. Nut
18. Washer
19. Governor shaft
20. Pin
21. Set screw
23. Governor lever
24. Sliding shaft
25. Governor weights
26. Injection pump gear
27. Plug
28. Needle roller bearing
29. Needle roller bearing
30. Crankshaft oil seal
31. Timing gear cover
32. Gasket
33. Engine front plate
34. Thrust button (Model W33)
35. Gasket
36. Adapter plate
37. Gasket
38. Sea water pump

Fig. WB1-14—Cross-sectional views of governor and linkage for all models except Model W13.

3. Throttle arm
10. Governor spring
14. Tie rod spring
15. Tie rod
19. Governor shaft
23. Governor lever
24. Sliding shaft
25. Governor weights
26. Injection pump gear
27. Plug
28. Needle roller bearing
29. Needle roller bearing
31. Timing gear cover
39. Fuel injection pump
40. Injection pump camshaft
41. Idler gear

SERVICE MANUAL

Westerbeke

shaft if lobes are worn to 43 mm (1.693 inches). To renew ball bearing and/or drive gear, first press shaft from gear and remove locating key. Then, if necessary, press shaft from ball bearing.

Reinstall injection pump camshaft by reversing removal procedure.

CAMSHAFT AND BEARINGS

Camshaft can be removed after removing fuel injection pump, engine timing cover, cylinder head and cam followers (tappets). On Model W13, remove engine oil pump and drive coupling from rear end of camshaft. Except on Model W13, remove the tachometer drive shaft from camshaft side of cylinder block. Withdraw camshaft from front of cylinder block.

Check camshaft bearing journals, lobes, drive gear teeth and, on Model W13, engine oil pump drive slot for wear or damage. Cam lobe diameter is 35.76 mm (1.408 inches) for both intake and exhaust valve lobes. Minimum lobe diameter (wear limit) is 34.76 mm (1.369 inches).

To remove camshaft gear on Model W13, press camshaft forward out of gear. On all other models, press camshaft rearward out of camshaft gear. Be sure to retain the gear locating key for reassembly. When installing camshaft in gear, be sure side of gear with timing mark is facing forward. Insert key in shaft, align keyway in gear with key and press shaft through gear.

The Model W13 camshaft is supported by two unbushed bores in cylinder block. On all other models, the front camshaft bearing bore is fitted with a renewable bushing; rear camshaft journals ride directly in unbushed bore of cylinder block. Maximum camshaft bearing oil clearance is 0.15 mm (0.006 inch) for both renewable front bushing and unbushed bearing bores. If necessary to renew camshaft bushing, drive new bushing into block so that oil hole in bushing is aligned with oil passage in block and front edge of bushing is flush with cylinder block.

Reinstall camshaft by reversing removal procedure. Be sure timing marks are aligned as described in TIMING GEARS paragraph.

OIL PUMP AND RELIEF VALVE

The gerotor type oil pump is driven through a coupling at rear end of engine camshaft on Model W13, and slot in rear end of fuel injection pump camshaft on other models. Oil pump, relief valve and oil filter assembly can be unbolted and removed from outside of engine. Refer to Fig. WB1-18 for exploded view of unit for Model W13, and to Fig. WB1-19 for other models. Remove oil filter (5), then remove the four cap screws which hold pump to engine and pump components together.

Pressure relief valve in pump cover should open at 343 kPa (50 psi) on Model W13, 588 kPa (85 psi) on Models W18 and W21, and 392 kPa (57 psi) on Models W27 and W33. To check pressure, remove oil pressure sender and install a master test gage in opening.

Pump outer rotor to body clearance should be 0.15-0.2 mm (0.006-0.008 inch) with wear limit of 0.3 mm (0.012 inch). Clearance between lobes of inner and outer rotors should be 0.06-0.12 mm (0.0024-0.0047 inch) with wear limit of 0.25 mm (0.010 inch). Clearance between rotors and cover, measured by placing a straightedge across pump body, should be 0.03-0.07 mm (0.0012-0.0027 inch). Check for scoring in pump body and on cover plate.

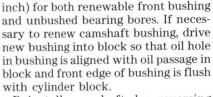

Fig. WB1-15—Exploded view of Model W13 engine camshaft and sea water pump drive gear unit. Fuel injection pump lobes are in front of camshaft gear. Refer to Fig. WB1-17 for all other models.
1. Ball bearing
2. Sea water pump gear
3. Ball bearing
4. Cam gear key
5. Camshaft
6. Camshaft gear

Fig. WB1-16—Exploded view of camshaft, injection pump camshaft, timing gears and sea water drive gear.
1. Ball bearing
2. Sea water pump gear
3. Ball bearing
4. Cam gear key
5. Camshaft
6. Camshaft gear
7. Idler gear
8. Bushing
9. Injection pump gear
10. Bearing retaining bolt
11. Lockwasher
12. Flat washer
13. Ball bearing
14. Pump gear key
15. Injection pump camshaft

Fig. WB1-17—View showing timing gear mark locations for all models except Model W13.

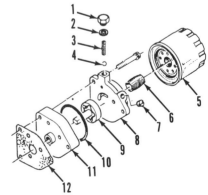

Fig. WB1-18—Exploded view of oil pump, relief valve and oil filter unit for Model W13. Refer to Fig. WB1-19 for other models.
1. Plug
2. Gasket
3. Spring
4. Relief valve ball
5. Oil filter
6. Adapter
7. Plug
8. Pump cover
9. Rotor set
10. "O" ring
11. Pump body
12. Gasket

Westerbeke

SMALL DIESEL

Reassemble using new parts as necessary. Pump inner and outer rotors are available as set only; all other parts are available individually. Lubricate pump parts with heavy oil to aid pump in priming when engine is started.

PISTON AND ROD UNITS

The piston and connecting rod units may be removed from above after removing cylinder head, oil pan and oil pickup screen. Before removing rod and piston unit, check side clearance of rod on crankpin. Rod side clearance should be 0.1-0.35 mm (0.004-0.014 inch); maximum allowable rod side clearance is 0.5 mm (0.020 inch).

The piston pin (3—Fig. WB1-20) is retained by being a press fit in connecting rod (5) except on Model W33 which has full floating piston pin (3A—Fig. WB1-21) retained by snap rings (4) at each end of pin bore in piston. Do not disassemble rod and piston units with press fit piston pin unless special removal and installation tools are available, and it is necessary to renew piston or connecting rod. When assembling piston and connecting rod on Model W33, heat piston to approximately 80° C (175° F) to facilitate installation of piston pin.

When installing piston and rod assembly, be sure piston ring end gaps are spaced equally at 90 degrees around piston. Lubricate piston rings, cylinder wall, rod bearing insert and crankpin with clean oil. Install piston and rod assembly with arrow on top of piston and identification marks on side of connecting rod towards front of engine. Tighten connecting rod cap bolts (6A—Fig. WB1-21) to a torque of 54-58 N·m (38-43 ft.-lbs.) on Model W33, and rod nuts (9—Fig. WB1-20) to a torque of 32-34 N·m (23-25 ft.-lbs.) on all other models.

PISTONS, PINS, RINGS AND CYLINDERS

Piston skirt diameter should be measured at right angle to piston pin and bottom of piston skirt. Maximum piston skirt to cylinder clearance is 0.3 mm (0.012 inch). If piston skirt to cylinder clearance is excessive, cylinders can be rebored and honed to next available oversize. Piston and rings are available in oversizes of 0.25, 0.50 and 0.75 mm (0.010, 0.020 and 0.030 inch).

Piston pin standard diameter is 23.0 mm (0.9055 inch) for Model W33 and 19.0 mm (0.748 inch) for all other models. Maximum allowable pin to pin bore in piston clearance is 0.08 mm (0.003 inch). Maximum clearance of piston pin in connecting rod is 0.10 mm (0.004 inch) for Model W33.

Pistons are fitted with three compression rings and one oil control ring. Piston ring end gap should be 0.15-0.40 mm (0.006-0.016 inch) for all rings on all models; maximum allowable ring end gap is 1.5 mm (0.059 inch). Side clearance of top compression ring in groove should be 0.06-0.11 mm (0.0024-0.0043 inch) with new ring; maximum allowable side clearance is 0.3 mm (0.012 inch). If equipped with taper ring, measure side clearance with outer face of ring held flush with piston.

Cylinder bores are not sleeved. Cylinders may be rebored and honed to fit next available oversize pistons and rings if cylinder bore wear measures more than 0.20 mm (0.008 inch). Standard cylinder bore diameter is 70.0 mm (2.756 inches) for Model W13, 65.0 mm (2.559 inches) for Model W18, 73 mm (2.874 inches) for Models W21 and W27, and 78 mm (3.071 inches) for Model W33.

If pistons and cylinder bores are not excessively worn or scored, new rings may be installed after removing ring ridge at top of cylinder and removing cylinder glaze with ball hone.

CONNECTING RODS AND BEARINGS

Connecting rods (5A—Fig. WB1-21) for Model W33 engines have full floating piston pin (3A) with pin being secured by a snap ring (4) at each end of pin bore in piston. Piston pin to connecting rod oil clearance should be a maximum of 0.1 mm (0.004 inch). When installing piston on rod, place piston in oil heated to a temperature of 80° C (175° F) for about five minutes. Identification marks on side of rod and arrow on top of piston must be in same direction. Install piston pin and retaining snap rings.

For all models except Model W33, piston pin (3—Fig. WB1-20) is a press fit in connecting rod (5). When installing piston on this type rod, align pin bores in piston and rod, then press pin into assembly so that it is centered in piston

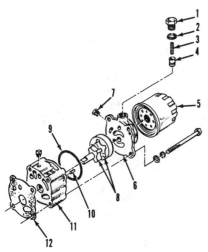

Fig. WB1-19—Exploded view of oil pump, relief valve and filter for all models except Model W13. Refer to Fig. WB1-18.

1. Plug
2. Gasket
3. Spring
4. Relief valve poppet
5. Oil filter
6. Pump cover
7. Plug
8. Rotor set
9. "O" ring
10. Dowel pin
11. Pump body
12. Gasket

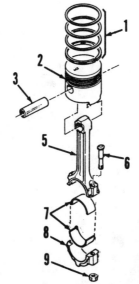

Fig. WB1-20—Exploded view of piston and rod unit typical of all models except Model W33. Piston pin (3) is press fit in connecting rod (5) on these engines.

1. Piston ring set
2. Piston
3. Piston pin
5. Connecting rod
6. Rod bolt
7. Bearing inserts
8. Connecting rod cap
9. Cap retaining nuts

Fig. WB1-21—Exploded view of piston and rod unit for Model W33. The full-floating piston pin (3A) is retained by snap rings (4) on W33 engine.

1A. Piston ring set
2A. Piston
3A. Piston pin
4. Snap rings
5A. Connecting rod
6A. Rod cap screws
7A. Bearing inserts
8A. Connecting rod cap

Illustrations Courtesy J.H. Westerbeke Corp.

SERVICE MANUAL

Westerbeke

with equal gaps between each side of rod and piston. Arrow on top of piston and identification marks on side of rod must be in same direction.

Connecting rod crankpin bearing (7—Fig. WB1-20 or 7A—Fig. WB1-21) is a precision slip-in insert type for all models. Crankpin standard diameter is 47.985-48.0 mm (1.8892-1.8897 inches) for Model W33 and 41.95-41.965 mm (1.6516-1.6522 inches) for all other models. Maximum rod bearing to crankpin oil clearance is 0.15 mm (0.006 inch) for all models. Crankpin bearings are available in undersizes of 0.25, 0.50 and 0.75 mm (0.010, 0.020 and 0.030 inch) as well as standard size.

CRANKSHAFT AND MAIN BEARINGS

Model W13

Crankshaft for Model W13 is supported in two main bearings. Front bearing (9—Fig. WB1-22) is pressed into main bearing retainer (8) and rear bearing (14) is pressed into rear end of crankcase (13). Crankshaft end play should be 0.06-0.3 mm (0.0024-0.012 inch) and is controlled by thrust washers (7) at each side of front main bearing retainer. Check end play before removing crankshaft.

To remove crankshaft, remove flywheel, timing gear cover and the rod and piston units. Remove crankshaft sleeve (4—Fig. WB1-22) and pull crankshaft timing gear (5) from front end of shaft (8) and remove thrust plate (6) and thrust washer (7). Unbolt front bearing retainer (2) and pull retainer from front end of cylinder block. Remove the rear thrust washer and keep front and rear washers identified because of wear pattern if they are to be reinstalled. Slide crankshaft forward out of rear main bearing and cylinder block. Pry oil seal (15) from rear end of block.

Check crankshaft main journal, crankpin journal, rear oil seal surface and front thrust surface for excessive wear, scoring or other damage. Main journal standard diameter is 58.945-58.960 mm (2.3207-2.3213 inches). Crankpin standard diameter is 41.950-41.965 mm (1.6516-1.6522 inches). Main and rod bearings are available in undersizes of 0.25, 0.50 and 0.75 mm (0.010, 0.020 and 0.030 inch) as well as standard size. Check wearing surface of thrust plate (6—Fig. WB1-22). Thrust washers (7) are available in standard thickness only.

If necessary to install new main bearings (9 and 14), old bearings must be removed and new bearings installed using a suitable bushing driver. Be sure oil hole in bearings are aligned with oil passages in cylinder block and front bearing retainer. Front bearing should not protrude from either front or rear of bearing retainer. Install new oil seal in rear end of crankcase with lip forward (to inside).

Lubricate rear oil seal, main bearings and crankshaft main journals, then slide crankshaft through crankcase into rear main bearing and oil seal. Take care not to damage rear main bearing or sealing lip on crankshaft oil seal. Using light grease, stick rear thrust washer (7) in recess at rear side of bearing retainer (8) with tab on washer engaging slot in retainer and oil groove side of washer out (to rear). Carefully install bearing retainer to avoid displacement of thrust washer and damage to front main bearing. Tighten retaining cap screws evenly and insert front thrust washer in recess of bearing retainer with tab engaging slot in retainer and oil groove side of washer forward. Install thrust plate with chamfered side to rear. Install crankshaft gear with stepped side toward thrust plate. Install sleeve with chamfered end to front.

Complete the reassembly of engine by reversing disassembly procedure. Crankshaft end play may be checked before installing timing gear cover by temporarily installing crankshaft pulley and retaining nut. Check end play with dial indicator and if within specification and crankshaft turns freely in main bearings, remove crankshaft pulley, then proceed to reassemble engine.

Model W33

Crankshaft is supported in a ball bearing (5—Fig. WB1-25) at front side of crankshaft timing gear and five main bearings with precision slip-in type inserts (13 and 14—Fig. WB1-24) and removable main bearing caps. The ball bearing is a sliding fit in timing gear cover and a press fit on front end of crankshaft. Except for procedure of removing ball bearing from front of crankshaft to allow removal of engine front plate, all service procedures are same as outlined for other models in following paragraph. When reassembling, it is suggested that ball bearing not be installed until timing gear marks are aligned.

Three and Four Cylinder Models

Refer to preceding paragraph for special information on Model W33. Crankshaft is supported in four main bearings in three-cylinder engines and five main bearings in four cylinder engines. Crankshaft end play should be 0.06-0.3 mm (0.0024-0.0118 inch) and is controlled by thrust flanges on center main bearing insert (14—Fig. WB1-24).

To remove crankshaft, first remove flywheel and flywheel housing, rear crankshaft oil seal retainer, engine timing gear cover, timing gears (it is not necessary to remove crankshaft gear), camshafts, engine front plate (1—Fig. WB1-24) and the rod and piston units. Unbolt and remove the main bearing caps, then lift crankshaft from crankcase.

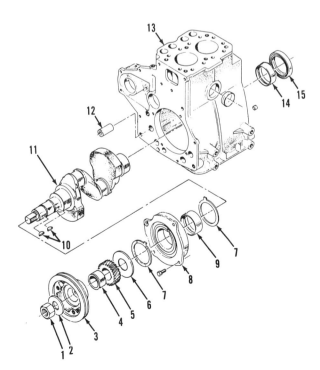

Fig. WB1-22—View showing Model W13 cylinder block and main bearings and crankshaft. Refer to Fig. WB1-24 for other models.

1. Crankshaft pulley nut
2. Washer
3. Crankshaft pulley
4. Spacer sleeve
5. Crankshaft gear
6. Thrust plate
7. Thrust washers
8. Main bearing support
9. Front main bearing
10. Pulley & gear keys
11. Crankshaft
12. Governor stub shaft
13. Cylinder block
14. Rear main bearing
15. Rear crankshaft seal

Westerbeke

SMALL DIESEL

Check crankshaft main bearing and rod bearing journals, thrust surfaces on journal with flanged bearing and rear oil seal contact area for excessive wear, scoring or other damage. Crankshaft standard main bearing journal diameter is 56.945-56.960 mm (2.2419-2.2425 inches) for Model W33 and 41.950-41.965 mm (1.6516-1.6522 inches) on other models. Standard crankpin bearing diameter is 47.985-48.000 mm (1.8892-1.8898 inches) on Model W33 and 41.950-41.965 mm (1.6516-1.6522 inches) on other models. Connecting rod and main bearings are available in undersizes of 0.25, 0.50 and 0.75 mm (0.010, 0.020 and 0.030 inch) as well as standard size.

To install crankshaft, proceed as follows: Check to see that block bearing surfaces are clean, then install upper main bearing halves with oil hole aligned with oil passage in cylinder block. Be sure all bearing journals and oil holes in crankshaft are clean, lubricate crankshaft and bearings and lower crankshaft into block. Apply sealant to front and rear main bearing cap mating surfaces in cylinder block. Insert lower bearing halves in main bearing caps, lubricate the bearings and install caps. Caps are marked with position number and embossed arrow mark. Tighten the main bearing cap bolts to a torque of 49-54 N·m (36-43 ft.-lbs.). Check to see that crankshaft turns freely in bearings and that end play is within limits of 0.06-0.30 mm (0.0024-0.0118 inch). It may be necessary to loosen cap with thrust bearing (14—Fig. WB1-24), pry crankshaft to front and rear to align thrust surfaces, then tighten cap bolts.

With crankshaft and main bearing caps installed, apply sealant to the side seals and install in front and rear bearing caps. Beveled ends of seals must be inserted first and with the bevels to outside of cap as shown in Fig. WB1-26. Complete the reassembly of engine by reversing disassembly procedure.

CRANKSHAFT REAR OIL SEAL

On Model W13, the crankshaft rear oil seal (15—Fig. WB1-22) is pressed into rear end of cylinder block. On all other models, the crankshaft rear oil seal (11—Fig. WB1-24) is carried in a seal retainer (10) bolted to rear face of cylinder block. When installing seal, lip must be toward inside of engine (forward). Install the seal retainer using new gasket (9) and tighten retaining cap screws evenly.

ELECTRICAL SYSTEM

An exploded view of typical electrical system is shown in Fig. WB1-27.

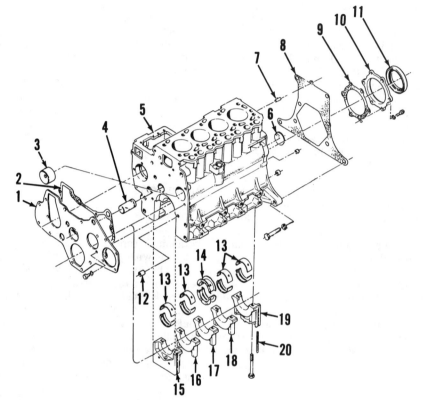

Fig. WB1-24—Exploded view of cylinder block and main bearings typical of all models except Model W13. A four-cylinder block is shown; three-cylinder block has four main bearings.

1. Engine front plate
2. Gasket
3. Camshaft front bearing
4. Idler gear stub shaft
5. Cylinder block
6. Camshaft rear plug
7. Dowel pin
8. Gasket
9. Gasket
10. Seal retainer
11. Crankshaft rear oil seal
12. Hollow dowel
13. Main bearings
14. Main bearing w/thrust flange
15. Front bearing cap
16. No. 2 bearing cap
17. Thrust bearing cap
18. No. 4 bearing cap
19. Rear main bearing cap
20. Front & rear cap side seals

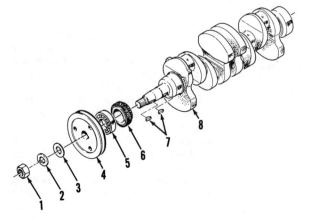

Fig. WB1-25—Model W33 crankshaft is shown; other models (except Model W13) are similar except they do not have ball bearing (5) support in timing gear cover.

1. Crankshaft pulley nut
2. Lockwasher
3. Flat washer
4. Crankshaft pulley
5. Ball bearing (W33 only)
6. Crankshaft gear
7. Pulley & gear keys
8. Crankshaft

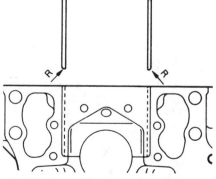

Fig. WB1-26—Install front and rear main bearing cap side seals with rounded ends (R) to outside as shown above.

Illustrations Courtesy J.H. Westerbeke Corp.

SERVICE MANUAL

Westerbeke

GLOW PLUGS

Glow plugs (1—Fig. WB1-27) are wired in parallel on all models. Glow plug circuit will continue to operate if one or more, but not all, glow plugs are defective. To check glow plugs, disconnect wires (2) to glow plug terminals and check individual glow plugs with an ohmmeter. Resistance should be 1 to 1.2 ohms. If no resistance is noted, renew glow plug.

ALTERNATOR AND REGULATOR

A 50 ampere alternator with internal voltage regulator is used on all models. Exploded view of alternator assembly is shown in Fig. WB1-28. Regulator has no provisions for adjustment. Alternator output cold should be 24 amperes at 1300 alternator rpm and 50 amperes at 2500 alternator rpm. Regulated voltage should be 14.1-14.7 volts at 20° C (68° F).

New brush length is 18 mm (0.709 inch). Renew brushes if worn to 8 mm (0.315 inch). Standard slip ring diameter is 33 mm (1.299 inches) and wear limit is 32.2 mm (1.268 inches).

To check stator assembly, use ohmmeter and check for continuity between each stator coil terminal. If there is no continuity, renew stator. Check for continuity between terminals and stator core. If continuity exists, renew the stator.

To check rotor, use ohmmeter to check for continuity between the two slip rings. If no continuity is found, renew the rotor. Check for continuity between slip ring and rotor core or shaft. If continuity exists, renew the rotor.

Using ohmmeter, check diodes in rectifier assembly. Ohmmeter should show infinite reading in one direction and continuity when ohmmeter leads are reversed. Renew rectifier assembly if continuity exists in both directions, or no continuity exists in either direction.

STARTING MOTOR

Refer to Fig. WB1-29 for exploded view of starter motor used on all models. Starter service specifications are as follows:

No-load test—
Volts 11.5
Current draw (max.) . . . . . . 90 amps
Rpm (min.) 3600
Brush length—Standard . . . . . . 17 mm
(0.669 in.)
Wear limit . . . . 11.5 mm
(0.453 in.)
Pinion gap . . . . 0.5-2.0 mm
(0.020-0.079 in.)

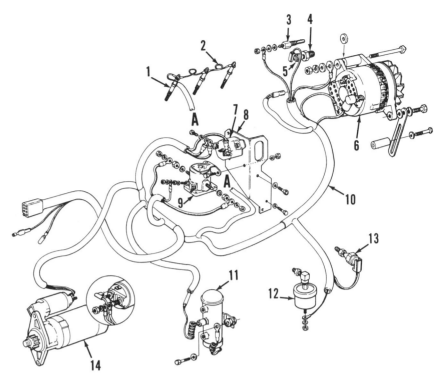

Fig. WB1-27—Drawing showing layout of typical engine electrical system.

1. Glow plugs
2. Glow plug leads
3. Temperature sender
4. Pipe bushing
5. Temperature alarm switch
6. Alternator
7. Circuit breaker
8. Bracket
9. Glow plug solenoid
10. Wiring harness
11. Electric fuel pump
12. Oil pressure sender
13. Oil pressure alarm switch
14. Starting motor

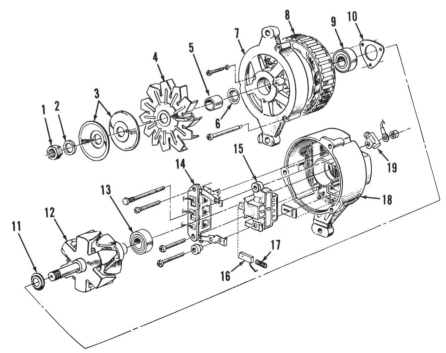

Fig. WB1-28—Exploded view of 50 ampere alternator used on all models.

1. Pulley nut
2. Lockwasher
3. Pulley
4. Fan
5. Spacer
6. Slinger
7. Front cover
8. Stator
9. Ball bearing
10. Bearing retainer
11. Slinger
12. Rotor
13. Ball bearing
14. Rectifier assy.
15. Regulator & brush holder
16. Brushes (2)
17. Brush springs (2)
18. Rear cover
19. Condenser

Illustrations Courtesy J.H. Westerbeke Corp.

Westerbeke
SMALL DIESEL

1. Retainer ring
2. Retainer
3. Pinion
4. Front bracket
5. Shims
6. Solenoid
7. Field lead
8. Pinion shaft
9. Spring
10. Retainer
11. Reduction gear
12. Washer
13. Center bracket
14. Washer
15. Clip
16. Cover
17. Ball bearing
18. Armature
19. Field coils & housing
20. Brush holder
21. Brush
22. Brush
23. Brush springs (4)
24. Brush
25. Ball bearing
26. Rear housing

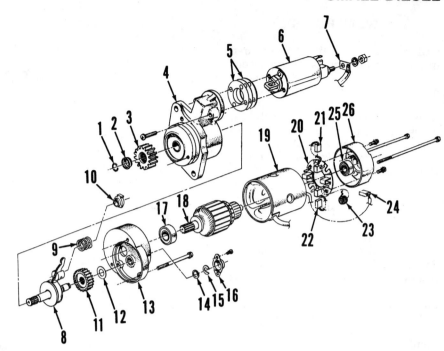

Fig. WB1-29—Exploded view of 12 volt starting motor used on all models.

Illustrations Courtesy J.H. Westerbeke Corp.

WESTERBEKE

Westerbeke

Model	No. Cyls.	Bore	Stroke	Displ.
DS, WPD3	1	76.2 mm (3.00 in.)	57.15 mm (2.25 in.)	261 cc (15.9 cu. in.)
W7, WPD4	1	76.2 mm (3.00 in.)	66.68 mm (2.625 in.)	304 cc (18.5 cu. in.)

All models included in this section are four-stroke, vertical cylinder, indirect injection diesel engines. Engines are water cooled with a fresh water closed system which is cooled by a sea water heat exchanger. Models DS and W7 are power units, Models WPD3 and WPD4 are marine diesel generator units. Both metric and inch size fasteners are used.

MAINTENANCE

LUBRICATION

Under normal operating conditions, recommended engine oil is to conform with minimum performance level of API CC or CD meeting MIL-L-2104B specifications. SAE 10W-30 multigrade oil can be used for all ambient operating temperatures. Oil and filter should be changed after each 150 hours of normal operation.

FUEL SYSTEM

Number 2 diesel fuel is recommended for most conditions. Number 1 diesel fuel should be used for light duty or cold weather operation. Refer to Fig. WB2-1 for drawing showing fuel system components.

FUEL FILTER. Fuel filter element (18—Fig. WB2-1) should be renewed after every 500 hours operation; however, filter may need to be renewed as early as 250 hours operation if fuel is not exceptionally clean. If engine shows signs of fuel starvation (surging or loss of power), renew fuel filter regardless of hours of operation. The fuel filter is mounted in the bottom of fuel tank on engines that have the tank attached to the engine.

BLEED FUEL SYSTEM. With a separately mounted fuel tank having a fuel lift pump (29—Fig. WB2-1), the pump priming lever must be operated to bleed fuel system. On engines fitted with a separately mounted fuel filter, loosen the two vent screws (15) on top of filter base (14). If engine is not fitted with a self-bleed fuel system, be sure fuel tank is full, move engine control lever to "RUN" position and vent fuel at pump through bleed screw (6) until flow of fuel is free of air. If engine will not start, loosen fuel injector line (5) fitting at injector and crank engine until fuel free of air flows from loosened joint, then tighten fitting and crank engine.

INJECTION PUMP TIMING

Injection timing is varied by changing shim (11—Fig. WB2-1) thickness between pump mounting flange and crankcase. Setting of STOP/RUN lever to check injection pump timing will vary with engine governor type and operating speed.

On models with fixed speed governor running at 3000 rpm or below, move STOP/RUN lever to RUN position. On models with fixed speed governor running above 3000 rpm and all models with variable speed governor controls, move STOP/RUN lever to about 10 degrees before vertical position and fix lever in this position. On all models with variable speed governor controls, move speed control lever to full speed position. On all models, do not operate the overload stop lever.

To check timing, proceed as follows: Close fuel supply valve or drain fuel tank and remove fuel injection line. Unscrew and remove delivery valve holder (1—Fig. WB2-2), spring (2) and "O" ring (3). Then, carefully remove delivery valve (4) without disturbing seat (6) or seal (5). Place the removed parts in clean diesel fuel. Place "O" ring (3) on delivery valve holder and reinstall holder in pump. Install a spill pipe (see Fig. WB2-3) on the delivery valve body

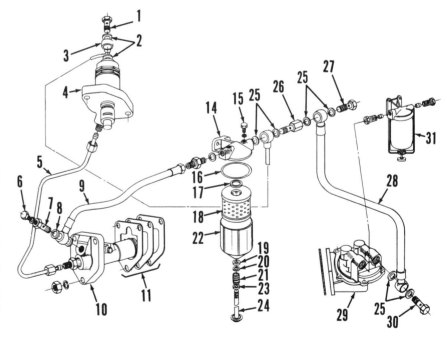

Fig. WB2-1—Drawing showing typical fuel system.

1. Banjo bolt
2. Copper washers
3. Return line
4. Fuel injector
5. Injector line
6. Bleed screw
7. Banjo bolt
8. Copper washers
9. Fuel supply line
10. Injection pump
11. Timing shims
14. Filter base
15. Bleed screws (2)
16. Bowl seal ring
17. Upper element seal
18. Filter element
19. Lower element seal
20. Washer
21. Spring
22. Filter bowl
23. Center bolt seal
24. Center bolt
25. Copper washers
26. Banjo bolt
27. Banjo bolt
28. Fuel pump feed line
29. Fuel feed pump
30. Banjo bolt
31. Sediment bowl & water trap

Illustrations Courtesy J.H. Westerbeke Corp.

Westerbeke
SMALL DIESEL

and turn crankshaft to position piston at 90 degrees BTDC.

Open fuel supply valve or pour some fuel in tank. Fuel should flow from spill pipe. If fuel lift pump is installed, it is necessary to operate primer lever to obtain fuel flow. Turn crankshaft slowly until fuel stops flowing from spill pipe and check timing mark through flywheel housing timing window (see Fig. WB2-4). Repeat check several times to be sure reading of timing is accurate. If timing is too slow, decrease shim thickness. Increase shim thickness if timing is too far advanced. Changing shim thickness by 0.1 mm (0.004 inch) will change timing about 1 degree. When timing is correct, remove spill pipe, reinstall delivery valve and spring using new ''O'' ring (3—Fig. WB2-2) and tighten delivery valve holder to a torque of 41 N·m (30 ft.-lbs.). Reinstall fuel injector line and bleed system before tightening line fitting at injector.

Refer to the following chart for injection pump timing specifications:

Fuel Injection Timing

Model	Speed Range, RPM	Degrees BTDC
DS, WPD3	Up to 2200	23
DS, WPD3	2201-2701	26
DS, WPD3	2701-3300	29
DS, WPD3	3301-3600	34
DS, WPD3	Variable speed	29
W7, WPD4	Up to 2200	26
W7, WPD4	2201-2700	28
W7, WPD4	2701-3300	32
W7, WPD4	3301-3600	33
W7, WPD4	Variable speed	28

GOVERNOR

The governor is a steel ball flyweight type located inside engine crankcase and is gear driven from engine camshaft gear. Governor controls may be fixed single speed or variable speed type, all with a Run/Stop control. An overload mechanism may be incorporated with governor controls; refer to following ''OVERLOAD STOP'' paragraph. Governor adjustments should be made only when engine is at normal operating temperature. Refer to the following paragraphs for adjustment of the different governor control units.

FIXED SPEED CONTROL. Refer to Fig. WB2-5; loosen locknut (2) and turn adjuster (1) to obtain desired engine speed. Engine no-load speed should be 4 percent above rated speed shown on engine nameplate. Tighten locknut when desired speed is obtained.

VARIABLE SPEED. Refer to Fig. WB2-6 and proceed as follows: Set speed control in idle position. Loosen locknut (2) and turn adjuster (3) to obtain idle speed of approximately 1000 rpm and tighten locknut.

Set speed control to full speed position, loosen locknut (7) and turn adjusting screw (6) to obtain no-load speed that is 8 percent higher than rated speed shown on nameplate. Tighten locknut and install new locking wire and lead seal. Recheck idle speed and readjust if necessary.

There should be a small amount of slack in control cable with speed control in idle position. If not, turn cable adjuster (5) in until control lever can be moved slightly before cable begins to move plunger (P).

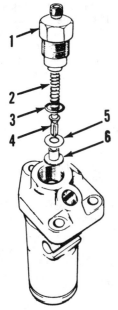

Fig. WB2-2—Delivery valve (4) is removed to check injection spill timing.

1. Delivery valve holder
2. Spring
3. ''O'' ring
4. Delivery valve
5. Seal
6. Delivery valve seat

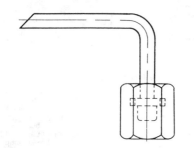

Fig. WB2-3—Spill pipe is attached to open delivery valve holder to check injection timing.

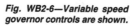

Fig. WB2-4—Flywheel timing marks can be viewed through timing window in flywheel housing. Desired timing mark must align with mark at side of window.

Fig. WB2-6—Variable speed governor controls are shown.
- S/R. STOP/RUN lever
1. Governor spring
2. Locknut
3. Idle speed adjuster
4. Throttle lever
5. Cable adjuster
6. High speed adjusting screw
7. Locknut

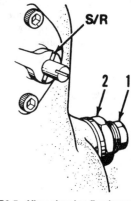

Fig. WB2-5—View showing fixed speed governor adjusting screw (1) and locknut (2). STOP/RUN lever is ''S/R''.

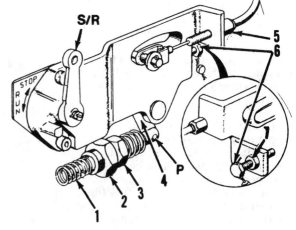

SERVICE MANUAL

OVERLOAD STOP

Some engines are equipped with an overload stop. Refer to Fig. WB2-16. The overload stop is factory adjusted and should not be disturbed unless wear or damage occurs to the mechanism or when installing a new fuel injection pump.

Overload stop can be adjusted only when engine is disassembled. Refer to OVERLOAD STOP paragraph in OVERHAUL section.

REPAIRS

TIGHTENING TORQUES

When making engine repairs, observe the following special tightening torque values:

Cylinder head bolts ....30 N·m
 (22 ft.-lbs.)
Rocker support nut ....32.5 N·m
 (24 ft.-lbs.)
Timing cover bolts ....12 N·m
 (9 ft.-lbs.)
Sump (oil pan) bolts ....12 N·m
 (9 ft.-lbs.)
Connecting rod bolts ....34 N·m
 (25 ft.-lbs.)
Camshaft gear bolt..........36 N·m
 (27 ft.-lbs.)
Camshaft extension
 shaft bolt ....19 N·m
 (14 ft.-lbs.)
Crankshaft gear bolt ....36 N·m
 (27 ft.-lbs.)
Crankshaft extension
 shaft bolt ....19 N·m
 (14 ft.-lbs.)
Flywheel nut...............210 N·m
 (155 ft.-lbs.)
Flywheel extension bolt .....36 N·m
 (27 ft.-lbs.)
Fuel injection pump
 mounting bolts............18 N·m
 (13 ft.-lbs.)
Fuel injection pump
 delivery valve ....20 N·m
 (15 ft.-lbs.)
Fuel injector stud nuts:
 Models DS, WPD3 ....14 N·m
 (10 ft.-lbs.)
 Models W7, WPD4 ....18 N·m
 (13 ft.-lbs.)
Air cell plug ....95 N·m
 (70 ft.-lbs.)

WATER PUMP AND COOLING SYSTEM

On Models W7 and WPD4, a fresh water pump circulates coolant through the engine, heat exchanger and surge tank. A sea water pump circulates sea water through the heat exchanger to cool the fresh water system and then is discharged out with the engine exhaust through an elbow attached to exhaust manifold. On Models DS and WPD3, a sea water pump circulates water directly through the engine.

Sea water pump impeller can be renewed after removing cover, gasket and impeller. Lubricate new impeller with water pump grease, insert impeller with twisting motion causing vanes to bend backward from pump rotation. Slide impeller onto shaft with flat of impeller inside diameter aligned with flat on shaft. Install gasket with new cover.

To completely overhaul pump, refer to the following paragraphs.

Models DS and W7 Sea Water Pump

Sea water pump, used to supply water to heat exchanger on Model W7 and for direct cooling of Model DS, is mounted on rear of transmission. To remove water pump, disconnect water lines from pump and remove the three cap screws (1—Fig. WB2-7) and cover plate (2). The pump body (6) with impeller (4) can now be removed from transmission. Inspect impeller and body and renew as necessary.

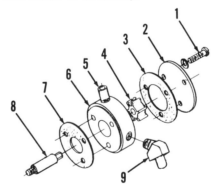

Fig. WB2-7—Exploded view of transmission mounted sea water pump. Shaft (8) is threaded into transmission shaft.

1. Cover bolts
2. Cover
3. Gasket
4. Rubber impeller
5. Water inlet
6. Pump body
7. Gasket
8. Pump shaft
9. Water outlet

Fig. WB2-8—Exploded view of timing gear cover mounted sea water pump. Shaft (11) is bolted to engine timing gear.

1. Cover bolts
2. Cover
3. Gasket
4. Rubber impeller
5. Water nipple
6. Pump body
7. Water outlet
8. Gasket
9. Seal
10. Adapter housing
11. Pump drive shaft

Check water pump drive shaft (8) for wear or damage. If necessary to renew shaft, use adjustable wrench to turn shaft clockwise (left-hand threads) out of transmission. Using suitable hooked puller, remove both shaft seals (not shown) from transmission. Install first (transmission oil) seal with spring loaded lip to inside and second (sea water) seal with spring loaded lip to outside (water pump). Lubricate shaft and seals and thread shaft counterclockwise into transmission.

To install pump, coat inside of body impeller cavity and impeller with water pump grease. Twist impeller into body with impeller vanes turned back from pump shaft rotation. Slide body and impeller onto shaft using new gasket and aligning flat in impeller with flat on shaft. Install cover with new gasket.

Model WPD4 Sea Water Pump

Model WPD4 sea water pump used to supply sea water to heat exchanger is mounted on engine timing gear cover and pump shaft is bolted to front of engine camshaft gear. To remove water pump, disconnect water lines from pump and remove the three cap screws (1—Fig. WB2-8) and cover plate (2). The pump body (6) with impeller (4) can now be removed from pump mounting adapter (10).

Inspect impeller and body and renew as necessary. Check pump drive shaft (11) for wear or damage. If necessary to renew shaft, it will be necessary to remove engine timing gear cover. Remove heat exchanger mounting plate from front of engine with heat exchanger and expansion tank attached, then remove timing gear cover. Unbolt and remove water pump extension shaft from front of camshaft gear. With timing cover removed, shaft seal can be renewed in cover. Install extension shaft and tighten bolts to a torque of 19 N·m (14 ft.-lbs.).

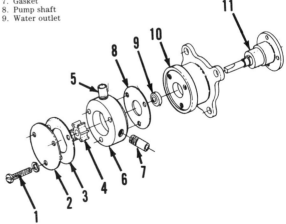

Westerbeke

To install pump, coat inside of body impeller cavity and impeller with water pump grease. Twist impeller into body with impeller vanes turned back from pump shaft rotation. Slide body and impeller onto shaft using new gasket (8) and aligning flat in impeller with flat on shaft. Install cover with new gasket (3); gasket must match pump cam lobe pattern in pump body.

Direct Sea Water Cooling Pump

For exploded view of pump used to directly cool engine with sea water, refer to Fig. WB2-9. Pump adapter (11) is mounted on front of timing gear cover and pump shaft (12) is bolted to camshaft gear.

To remove pump, disconnect water hose from pump body (6). Remove pump cover (2) and impeller (4). Remove nuts from pump body mounting studs and slide pump body forward off of pump shaft. Remove seal (5) from pump body. It is not necessary to remove cam (C). Slide water slinger (7) and seal washer (8) off pump shaft, then remove shims (9) and plate (10). If shaft (12) is excessively worn or damaged, or if timing cover oil seal is leaking, it will be necessary to remove engine timing cover. The shaft can then be unbolted and removed from camshaft gear and the oil seal can be renewed in timing gear cover.

Reinstall pump by reversing removal procedure. Note that sufficient shims (9) must be used to keep shaft from contacting cover (2). Lubricate impeller with water pump grease before installing in pump body.

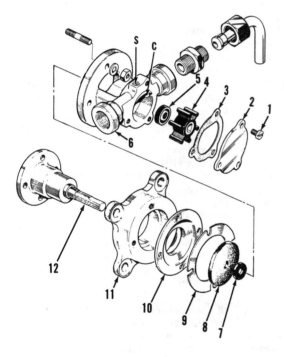

Fig. WB2-9—Pump used for direct sea water cooling. Shaft (12) is bolted to engine camshaft gear.

C. Impeller cam
S. Cam retaining screw
1. Cover screws
2. Cover
3. Gasket
4. Rubber impeller
5. Pump seal
6. Pump housing
7. Water slinger
8. Seal washer
9. Shims
10. Shim plate
11. Adapter housing
12. Pump shaft

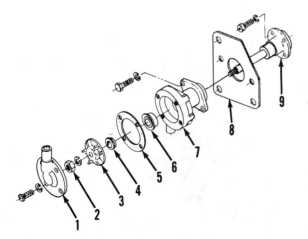

Fig. WB2-10—Exploded view of fresh water cooling system pump. Pump shaft (9) is bolted to engine crankshaft gear.

1. Cover
2. Impeller nut
3. Cast impeller
4. Ceramic ring & rubber washer
5. Gasket
6. Seal assy.
7. Pump body
8. Mounting plate
9. Pump shaft

SMALL DIESEL

Model W7 and WPD4 Fresh Water Pump

To remove pump, refer to Fig. WB2-10 and proceed as follows: Remove water hose and front cover (1) from pump. Lift decompression lever and turn crankshaft until hole in flywheel outer diameter is visible through timing hole in housing. Insert a steel rod in hole of flywheel to keep crankshaft from turning. Unscrew nut (2) from front of pump shaft (9) and remove rotor (3). Remove the two cap screws holding pump body (7) to timing gear cover and remove body from drive shaft.

With pump body removed, renew seals as follows: Press seal (6) out of housing. Lightly coat outside of new seal with Permatex #2 sealer, then carefully press seal into pump body taking care not to damage sealing surface. Pry ceramic ring (4) from impeller and remove the rubber washer. Insert new ceramic ring in new rubber washer, lightly oil outside of washer and press the ring and washer into impeller recess by hand.

Check pump drive shaft (9) for wear or damage. If necessary to renew shaft, it will be necessary to remove engine timing gear cover except for early shaft with drive tang and slot. Unbolt and remove pump extension shaft from front of crankshaft gear. With timing cover removed, shaft seal can be renewed in cover. Install extension shaft and tighten bolts to a torque of 19 N·m (14 ft.-lbs.).

To install pump, slide pump body over pump shaft taking care not to damage new seal. Assemble remainder of parts by reversing removal procedure.

SALT WATER ANODE

A zinc anode is threaded into the salt water chamber of heat exchanger of Models W7 and WPD4 to reduce corrosion. The anode should be checked and renewed if necessary at least once each season of operation. The zinc is worn out when only 25 percent of original material remains.

VALVE ADJUSTMENT

Valve rocker arm to valve tip clearance should be 0.1 mm (0.004 inch) for all models and adjustment may be checked after removing rocker arm cover. To adjust valves, turn engine to TDC of compression stroke and check gap between rocker arm and tip of valve stem. It may be necessary to bend a 0.1 mm (0.004 inch) gage strip in a right angle to reach valve gap. Loosen rocker arm adjusting screw locknut and turn adjusting screw (3—Fig. WB2-11) to obtain correct clearance. Tighten locknut and

SERVICE MANUAL

Westerbeke

recheck clearance, then install rocker arm cover with new gasket.

CYLINDER HEAD

Exploded view of cylinder and cylinder head is shown in Fig. WB2-11. To remove cylinder head, proceed as follows: Remove pump to injector line(s), injector fuel return line and fuel injector(s). Immediately cap or plug all fuel openings. Remove the manifolds and rocker arm cover. Remove cylinder head oiling line and cold weather starting aid priming plunger (12). Gradually loosen cylinder head stud nuts in diagonal sequence, then remove nuts and washers. Remove rocker arm support assembly, gasket and the push rods. Lift cylinder head from engine.

If engine has been disassembled, piston to cylinder head (bumping) clearance should be checked before final installation of cylinder head. Turn crankshaft so that piston is about 6 to 7 mm (1/4 inch) before top dead center. Place three pieces of soft lead wire or solder at equal spacing on top of piston so they are not aligned with valve heads or combustion chamber. Install cylinder head with gasket, then turn crankshaft so that piston moves past top dead center. Remove the cylinder head and measure thickness of flattened wires. The average of the three thicknesses is piston to head (bumping) clearance. Clearance should be 0.56-0.66 mm (0.022-0.026 inch). Remove cylinder and add shims (26) to increase clearance or remove shims to decrease clearance.

When installing cylinder head, use new cylinder head gasket and rocker arm support gasket. Be sure gasket surfaces are cleaned thoroughly. Lightly coat top of head gasket with "Golden Hermatite" to prevent gasket from seizing to the aluminum head, then install cylinder head. Place rocker arm support on retaining stud with new gasket and be sure support is aligned on dowel pin (7). Install the self-locking nut on rocker arm support stud. Install the remaining stud nuts and washers and turn down finger tight. Tighten the self-locking nut down just enough to be sure rocker arm support is level and is in contact with cylinder head. Then, starting with rocker arm support stud, tighten each of the five cylinder head nuts down 1/4 turn at a time in a diagonal pattern to final torque value. Engine rocker support nut should be tightened to a torque of 32.5 N·m (24 ft.-lbs.) and the remaining nuts to a torque of 30 N·m (22 ft.-lbs.). Complete reassembly by reversing removal procedure.

After 20 running hours from cylinder head installation, cylinder head nuts should be retorqued. Loosen each nut 1/4 turn, then tighten to specified torque in diagonal sequence starting with rocker arm support stud.

VALVE SYSTEM

Valves seat in renewable inserts in the cylinder head. Valve face and seat angle is 45 degrees on all models. Seat width should be 1.295-1.727 mm (0.051-0.068 inch). Distance from head of valve to cylinder head gasket surface (valve head clearance) should be 0.99-1.45 mm (0.039-0.057 inch). Maximum allowable valve head clearance is 1.93 mm (0.076 inch). Intake and exhaust valves, valve springs and guides are alike.

Cylinder head is equipped with renewable valve guides. Heat cylinder head in boiling water for at least two minutes prior to removing old guides. Support top side head on wood blocks and using suitable driver, remove old guides toward top of head. To install new guides, be sure valve guide bores in head are clean. Heat head in boiling water, then press new guides in from top of head until shoulder on guide contacts cylinder head. New guides are prefinished and have a special coating, and must not be reamed after installation. Resurface valve seats after new guides are installed.

DECOMPRESSOR LEVER

To adjust decompressor lever (23—Fig. WB2-11), remove valve cover and turn crankshaft so that exhaust valve is fully closed. With decompressor lever in vertical position, pin on decompressor shaft should just lift exhaust rocker arm sufficiently to take up valve clearance. If adjustment is necessary, hold decompressor lever in vertical position and with screwdriver inserted in slot in outer end of shaft (21), turn shaft to obtain proper adjustment. Friction bushing (24) will allow shaft to be turned in lever.

INJECTOR

The pintle type injector nozzle emits a fine spray pattern which is directed to the hole in the air cell located directly across the cylinder head combustion chamber. A properly performing fuel injector is very important in these engines.

REMOVE AND REINSTALL. First, remove air cleaner assembly, engine cowling and if so equipped, engine mounted fuel tank. Thoroughly clean fuel injector lines and surrounding area. Disconnect and remove the injection pump to injector line and fuel return line from injector. Remove the two injector retaining nuts and pull injector from cylinder head. Remove the fuel injector sealing washer.

Be sure injector bore in cylinder head is clean and insert new sealing washer in bore. Install injector and retaining nuts. Install pump to injector line and tighten fittings finger tight plus 1/3 turn with wrench. Tighten injector retaining nuts to a torque of 18 N·m (13 ft.-lbs.). Complete reassembly by reversing dis-

Fig. WB2-11—Exploded view of cylinder and cylinder head. Shims (26) are used to adjust piston to cylinder head clearance.

1. Rocker cover
2. Exhaust rocker arm
3. Adjusting screw
4. Rocker arm bushing
5. Rocker arm support
6. Gasket
7. Support dowel pin
8. Intake rocker arm
10. Chain
11. Lifting eye
12. Priming plunger
13. Valve guide & spring seat
14. Plunger "O" ring
15. Cylinder head
16. Lever stop pin
17. Gasket
18. Water port plug
19. Head gasket
20. "O" ring
21. Compression release shaft
22. Roll pin
23. Compression release lever
24. Friction bushing
25. Cylinder
26. Cylinder shims
27. Water port plug
28. Gasket
29. Split valve locks
30. Valve spring retainer
31. Valve spring, inner
32. Valve spring, outer
33. Valve
34. Push rod

Illustrations Courtesy J.H. Westerbeke Corp.

Westerbeke SMALL DIESEL

assembly procedure. Loosen injector line fitting at injector and bleed air from line, then tighten fitting.

TESTING. A complete job of testing and adjusting the injector requires use of special test equipment and approved calibrating oil. Nozzle should be tested for opening pressure, seat leakage and spray pattern. When tested, nozzle should open with high pitch buzzing sound (chatter) and cut off quickly at end of injection with minimum seat leakage.

WARNING: Fuel emerges from injector with sufficient force to penetrate the skin. When testing injector, keep yourself clear of nozzle spray.

Before conducting test, operate tester lever until fuel flows from open line, then attach injector. Close valve to tester gage and pump tester lever a few quick strokes to purge air from injector and to be sure nozzle valve is not stuck.

OPENING PRESSURE. Open valve to tester gage, then operate tester lever slowly while observing gage reading. Opening pressure should be 16200-18275 kPa (2350-2650 psi). Adjust opening pressure by removing cap nut (1—Fig. WB2-12) and turning adjusting screw (3).

SPRAY PATTERN. Spray pattern should be a straight cone from nozzle tip, well atomized and free of split or drops. If pattern is not correct, injector must be disassembled and the nozzle cleaned or renewed. A poor spray pattern will result in fuel spray missing the hole in the air cell, causing loss of power and excessive exhaust smoke.

SEAT LEAKAGE. Operate tester lever slowly to obtain pressure 1000 kPa (150 psi) below observed opening pressure and hold gage at this reading for 10 seconds. If drop of fuel forms at nozzle sufficient to cause continuous film on finger tip, clean or renew nozzle.

BACK LEAKAGE. Operate tester lever slowly to bring gage reading to 15175 kPa (2200 psi), then record time interval until gage reading drops to 10135 kPa (1470 psi). This time should be more than 6 seconds. If back leakage time is less than 6 seconds, nozzle should be renewed.

OVERHAUL. Wipe all dirt and loose carbon from exterior of injector assembly and secure injector body (7—Fig. WB2-12) in fixture or soft jawed vise. Remove cap nut (1), adjusting screw (3), sealing washers and spacer (2). Lift out spring pad (4), spring (5) and spring pressure rod (6). Unscrew nozzle nut (9) and remove nozzle body and valve (8). It may be necessary to push nozzle out of nut using copper or brass hollow drift. Do not drive out by striking nozzle end face.

Clean exterior surfaces with a brass wire brush. Soak parts in an approved carbon solvent, if necessary, to loosen hard carbon deposits. Rinse parts in clean diesel fuel or calibrating oil immediately after cleaning to neutralize solvent and prevent etching of polished surfaces.

Scrape carbon from pressure chamber (3—Fig. WB2-13) using a hooked scraper (4). Use suitable size wire or twist drill (1) to clean hole (2) to pressure chamber. Clean nozzle spray hole using a spray hole cleaner (5) or hardwood stick.

Back flush nozzle using reverse flush adapter on tester. Reclean all parts by rinsing thoroughly in clean diesel fuel or calibrating oil and assemble while immersed in clean fluid. Set opening pressure before installing cap nut (1—Fig. WB2-12).

FUEL AIR CELL

The fuel emitted from the injector nozzle is directed into an air cell directly across the cylinder head combustion chamber from the injector nozzle. The air cell hole can be cleaned of carbon with a soft wire by working through injector opening in cylinder head with head installed, or from combustion chamber in head after cylinder head has been removed.

INJECTION PUMP

The fuel injection pump is mounted on the right side of crankcase near timing gear cover and is operated from a lobe on engine camshaft. To remove injection pump, first thoroughly clean outside of engine and proceed as follows:

Close fuel supply valve or drain fuel tank. Disconnect and remove fuel lines and immediately cap or plug all openings. Remove rocker arm oil supply line. Turn crankshaft to position piston at TDC on exhaust stroke. Remove injection pump mounting stud nuts. Move STOP/RUN lever to about 10 degrees before vertical position to align fuel pump rack ball with cutaway in crankcase and remove pump and timing shims. Be careful not to lose or damage any of the shims. Disassembly of pump is not recommended outside of an authorized diesel fuel system repair facility. A new or reconditioned injection pump should be installed if pump is faulty.

To reinstall pump, be sure piston is at TDC on exhaust stroke. Place STOP/RUN lever at about 10 degrees before vertical position to align governor lever with cutaway in engine crankcase. Place timing shims on pump mounting studs. Install fuel injection pump, making sure ball on pump rack engages governor lever fork. Install the retaining nuts and tighten evenly to a torque of 18 N·m (13 ft.-lbs.). Complete remainder of reassembly by reversing disassembly procedure. Bleed fuel system as outlined in MAINTENANCE section.

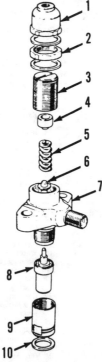

Fig. WB2-12—Exploded view of fuel injector.
1. Cap nut
2. Spacer
3. Adjusting screw
4. Spring seat
5. Spring
6. Push rod
7. Injector body
8. Nozzle & valve
9. Nozzle nut
10. Seal washer

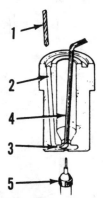

Fig. WB2-13—For cleaning nozzle, refer to text and cross-sectional view above.
1. Twist drill
2. Fuel passage
3. Fuel chamber
4. Brass scraper
5. Orifice cleaning tool

SERVICE MANUAL

If installing new pump, follow procedure in preceding paragraph, then refer to INJECTION PUMP TIMING paragraph in MAINTENANCE section and to OVERLOAD STOP paragraph in REPAIRS section.

TIMING GEAR COVER

Drain oil from engine and remove any mounting plates or accessories from cover. Remove the eight socket head cap screws holding cover to crankcase, then carefully remove cover from crankcase and aligning dowel pins. Remove all traces of gasket from cover and crankcase and install new camshaft and/or crankshaft extension shaft oil seals in cover.

To reinstall timing gear cover, reverse removal procedure using a new timing cover gasket. Align cover and gasket on dowel pin. Tighten cover cap screws to a torque of 12 N·m (9 ft.-lbs.).

TIMING GEARS

Timing gears are accessible after removing the timing cover. The camshaft gear and crankshaft gear are press fit and also are retained by a flat washer and cap screw. The oil pump drive gear and governor drive gear are retained on respective shafts by a hex nut.

The crankshaft gear (4—Fig. WB2-20) can be removed after removing crankshaft extension (1), gear retaining bolt (2) and washer (3). Then reinstall bolt without washer into crankshaft to protect crankshaft threads and thread puller bolts into crankshaft gear. When installing gear on crankshaft, first heat gear in hot oil. Be sure keyway is aligned and install gear on crankshaft using hollow driver. The single marked tooth on crankshaft gear must mesh between the two marked teeth on camshaft gear as shown in Fig. WB2-14. Install gear retaining washer and bolt and tighten bolt to a torque of 36 N·m (27 ft.-lbs.). Install extension shaft and tighten bolts to a torque of 19 N·m (14 ft.-lbs.).

To remove camshaft gear (3—Fig. WB2-18) with crankshaft gear installed, turn engine so that cut-out part of crankshaft gear shoulder is aligned with camshaft gear. Remove camshaft extension shaft, gear retaining bolt (1) and washer (2). Install bolt back in end of camshaft to protect camshaft threads and using suitable gear puller, remove camshaft gear. Take care not to lose key from camshaft. To install gear, install suitable threaded gear pusher tool and press gear onto camshaft. The single marked tooth of crankshaft gear must mesh between the two marked teeth of camshaft gear as shown in Fig. WB2-14. Install gear retaining washer and bolt and tighten bolt to a torque of 36 N·m (27 ft.-lbs.). Install extension shaft and tighten bolts to a torque of 19 N·m (14 ft.-lbs.).

To remove oil pump gear or governor gear, remove retaining nut and pull gear from shaft. Be careful not to lose key from shaft. Install key, gear and washer on shaft and tighten the self-locking nut.

OIL PUMP AND RELIEF VALVE

The gerotor type oil pump (7—Fig. WB2-15) is mounted in engine crankcase

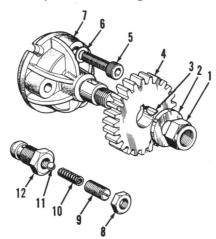

Fig. WB2-15—Exploded view showing oil pump (7) and pressure relief valve (8 through 12).

1. Hex nut
2. Washer
3. Woodruff key
4. Pump drive gear
5. Pump retaining screws
6. Lockwashers
7. Oil pump assy.
8. Locknut
9. Adjusting screw
10. Spring
11. Valve
12. Valve body

Fig. WB2-16—Cutaway view of crankcase with timing cover removed showing governor components. Refer to Fig. WB2-17 for part identification.

X. Shaft end play dimension
Y. Measured dimension
Z. Numerical value dimension

and oil pump gear (4) is driven by engine camshaft gear. The oil pressure relief valve assembly (8 through 12) is located in crankcase at left side of engine oil pump. To remove the engine oil pump, it is first necessary to remove the camshaft gear. Remove oil pump gear retaining nut (1) and washer and pull gear from pump shaft if pump is to be disassembled for inspection. Unbolt and remove the pump from engine crankcase. Install pump by reversing removal procedure.

The oil pump is serviced as a complete assembly only. Pump can be disassembled for inspection. Rotor end clearance should be 0.025-0.064 mm (0.001-0.0025 inch) with wear limit of 0.127 mm (0.005 inch). Clearance of inner rotor lobe to outer rotor lobe should be 0.051-0.127 mm (0.002-0.005 inch) and must not exceed 0.203 mm (0.008 inch). Shaft clearance in bearing bore should be 0.038-0.076 mm (0.0015-0.003 inch) and should not exceed 0.127 mm (0.005 inch). Shaft diameter new is 15.032-15.044 mm (0.5918-0.5923 inch). Renew pump if any component is excessively worn or damaged.

GOVERNOR AND LINKAGE

Refer to Fig. WB2-16 for view of governor installed in engine and to Fig. WB2-17 for exploded view. The governor unit can be serviced after removing timing gear cover from engine crankcase. If governor flyweight unit is to be disassembled for inspection, remove the governor gear retaining nut (41) and pull gear (40) from governor shaft. Be careful not to lose the key (38) from shaft. Unbolt and remove the governor flyweight assembly and housing (39) from cylinder block.

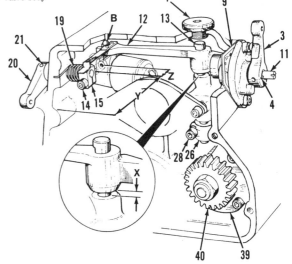

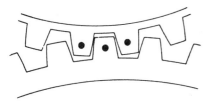

Fig. WB2-14—Timing marks on camshaft gear and crankshaft gear must be aligned as shown.

Illustrations Courtesy J.H. Westerbeke Corp.

Westerbeke

Carefully inspect governor unit for wear or damage to bearings (36), flyweight balls (34), drive plate (35) and sliding cone (33). If any part is worn or damaged beyond further use, the complete governor assembly must be renewed. Inspect governor shaft bushing (9—Fig. WB2-19) in engine crankcase and renew bushing if worn or damaged. Install governor flyweight unit by reversing removal procedure.

With timing gear cover removed, inspect governor linkage. To service governor linkage, remove camshaft gear and governor assembly. Unbolt and remove governor controls (see Fig. WB2-6) and the STOP/RUN control assembly (S/R). Loosen locknut (2) and unscrew governor spring unit (1). Remove plug (1—Fig. WB2-17) or, if so equipped, the crankcase breather assembly. Loosen clamp bolt (28) and remove arm (26) from bottom of lever shaft (13). Pull shaft and lever (12 or 12A) up and place a wood block in crankcase under lower end of shaft. Using a sleeve that will fit over upper end of shaft, drive lever (12 or 12A) downward until free of shaft. It may be necessary to use a thicker block than would originally fit under shaft to completely free the lever on shaft. Drive pin (21) from overload lever (20), if so equipped, and remove overload shaft (16) and cam (15) as an assembly without loosening cam bolt (14).

Clean and inspect all linkage parts and renew as necessary. To reassemble, proceed as follows: Install overload cam (15—Fig. WB2-17), shaft (16) and spring (19) in crankcase using new "O" ring (18). Position lever (12 or 12A) in crankcase and tap shaft (13) down through lever until it comes to stop. Install governor arm (26) and washer (27) on bottom of shaft and tighten bolt (28) enough to be sure arm will not fall off. Install governor spring unit (22 through 25). Install STOP/RUN mechanism (3 through 11) using new "O" ring (7) and gasket (10). To complete assembly and adjustment of linkage, the governor unit and fuel injection pump must be installed. Loosen bolt (28) in arm (26) and turn governor spring adjuster (22) in until arm is against governor, pushing the sliding cone (33) and drive plate (35) together. Operate overload lever (20), if so equipped, and push lever (12 or 12A) and fuel injection pump rack to maximum fuel position. With arm (26) and lever (12 or 12A) in this position, tighten clamp bolt (28). Note that arm (26) should be positioned on shaft (13) to allow an up/down movement (dimension "X"—Fig. WB2-16) in shaft of 0.25 mm (0.010 inch). To adjust overload stop, refer to OVERLOAD STOP paragraph. Complete reassembly by reversing disassembly procedure.

OVERLOAD STOP

The overload stop cam (15—Fig. WB2-16) will need to be adjusted if cam is loosened or removed from shaft, or if installing a new fuel injection pump.

To check overload stop adjustment, move STOP/RUN lever to STOP position and measure distance (dimension "Y") between end of fuel pump rack and straightedge placed across gasket face of crankcase. Move STOP/RUN lever to RUN position and move governor lever and pump rack so that distance between end of pump rack and gasket face of crankcase is equal to dimension "Y" plus 12.7 mm (0.500 inch) (dimension "Z"). The step on the overload cam (15) should then just touch the step in the governor lever (12A) spring. If not, loosen clamp bolt (14) and rotate cam to proper position, then tighten clamp bolt.

CYLINDER, PISTON AND ROD UNIT

The cylinder, piston and rod unit can be removed after removing cylinder head and oil sump (pan). Unbolt connecting rod cap and remove cylinder with piston and rod unit. Remove piston and rod from bottom end of cylinder. Shims (26—Fig. WB2-11) are installed between cylinder and crankcase to adjust piston to cylinder head clearance.

Standard cylinder bore is 76.20-76.23 mm (3.000-3.001 inches). If cylinder wear is 0.25 mm (0.010 inch) or more, cylinder must be rebored to next oversize of 0.50, 0.75 or 1.0 mm (0.020, 0.030 or 0.040 inch). Renew cylinder if wear exceeds rebore limit for largest oversize.

Piston pin location in piston is not offset and the flat-top piston may be installed either way on connecting rod. To install piston and rod unit, be sure the piston ring end gaps are positioned equally around the piston and pin retaining snap rings are securely installed. Lubricate piston, rings and cylinder bore with clean engine oil. Using ring compressor, install piston through bottom end of cylinder with numbers on connecting rod and cap toward push rod side of cylinder. Place rod bearing inserts in connecting rod and cap. Place cylinder shims on the cylinder studs, lubricate crankpin and install the cylinder, rod and piston unit. Install rod cap with assembly number on cap aligned with number on rod and tighten rod bolts to a torque of 34 N·m (25 ft.-lbs.).

If new piston is being installed, piston to cylinder head (bumping) clearance should be checked before final installation of cylinder head. Turn crankshaft so that piston is about 6 to 7 mm (1/4-inch) before top dead center. Place three

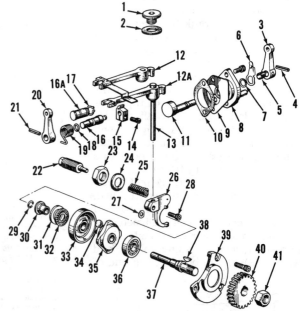

Fig. WB2-17—Exploded view of governor, governor controls and linkage. Refer to Fig. WB2-6 for optional variable speed controls and to Fig. WB2-16 for installed view of governor components. Governor lever (12A) and plug (16A) are used on engines not equipped with overload stop (items 14 through 21). Some engines may be equipped with crankcase breather instead of plug (1).

1. Plug
2. Gasket
3. STOP/RUN lever
4. Lever pin
5. Bushing
6. Lever detent
7. "O" ring
8. Mounting plate
9. Instruction plate
10. Gasket
11. STOP/RUN cam
12. Governor lever w/spring
12A. Governor lever, optional
13. Governor lever shaft
14. Clamp bolt
15. Overload cam
16. Overload cam shaft
16A. Plug
17. Snap ring
18. "O" ring
19. Lever return spring
20. Overload lever
21. Roll pin
22. Idle speed adjuster
23. Locknut
24. Gasket washer
25. Governor spring
26. Governor arm
27. "D" washer
28. Clamp bolt
29. Snap ring
30. Bushing
31. Thrust washer
32. Thrust bearing
33. Sliding cone
34. Weight balls
35. Drive plate
36. Ball bearing
37. Shaft
38. Woodruff key
39. Governor housing
40. Drive gear
41. Nut

SERVICE MANUAL
Westerbeke

pieces of soft lead wire or solder at equal spacing on top of piston so they are not aligned with valve heads or combustion chamber. Install cylinder head with gasket, then turn crankshaft so that piston moves past top dead center. Remove the cylinder head and measure thickness of flattened wires. The average of the three thicknesses is piston to head (bumping) clearance. Clearance should be 0.56-0.66 mm (0.022-0.026 inch). Remove cylinder and add shims (26—Fig. WB2-11) to increase clearance or remove shims to decrease clearance.

PISTON, PIN, AND RINGS

Piston is fitted with two compression rings and one oil control ring. With piston and rod unit removed as outlined in preceding paragraph, remove piston rings and the piston pin retaining snap rings (23—Fig. WB2-20). Check piston skirt for cracks, excessive wear or scoring. Check side clearance of new ring in top ring groove; side clearance must not be more than 0.25 mm (0.010 inch). Piston pin should not show any visible wear and should fit piston pin bore with a slight interference fit. Pistons and rings are available in oversizes of 0.50, 0.75 and 1.0 mm (0.020, 0.030 and 0.040 inch) as well as standard. Cylinder must be rebored and honed to install oversize piston and rings.

If new piston rings are being installed without reboring the cylinder, cylinder should be deglazed with medium grade carborundum cloth using hand motion to produce a matte surface or by use of a rotary brush cylinder hone. Wash cylinder in kerosene after deglazing to remove all traces of abrasive.

Piston ring end gap, new, should 0.30-0.43 mm (0.012-0.017 inch). Maximum allowable ring end gap is 1.14 mm (0.045 inch). Install compression rings with step or bevel on inside of ring up, and with step on outside of ring down. Oil control rings and compression rings without step or bevel should be installed with identification mark up; rings without step, bevel or mark may be installed either side up.

Place piston on rod and insert pin through bores in piston and rod. Be sure piston pin snap rings are securely installed in the grooves of piston pin bore. Sharp edge of snap ring should face away (out) from piston pin.

CONNECTING RODS AND BEARINGS

Connecting rod (19—Fig. WB2-20) is fitted with a pressed-in renewable bushing (21) at piston pin end and a slip-in precision fit bearing insert (18) at crankpin end. Piston pin should be a hand push sliding fit in bushing. If necessary to renew piston pin bushing, install new bushing with oil hole aligned with hole in connecting rod. After installing bushing, ream or hone to an inside diameter of 22.233-22.243 mm (0.8753-0.8757 inch). Crankpin bearing oil clearance should be 0.025-0.090 mm (0.001-0.0035 inch). Crankpin bearings are available in undersizes of 0.25 and 0.50 mm (0.010 and 0.020 inch) as well as standard size.

CAMSHAFT AND BEARINGS

To remove camshaft, first remove rocker arm support and push rods and fuel injection pump. Remove water pump and timing gear cover. Remove camshaft gear retaining bolt (1—Fig. WB2-18) and washer (2). To remove camshaft gear (3), turn crankshaft so that cut-out portion of drive shoulder on crankshaft gear (4—Fig. WB2-20) is aligned with camshaft gear. Hold cam followers up with magnet tools or invert engine so that followers fall away from camshaft. Working through holes in camshaft gear, remove cam thrust plate retaining screws (5—Fig. WB2-18) and pull cam and gear assembly from crankcase. Press camshaft from gear if necessary to renew gear, thrust plate or shaft.

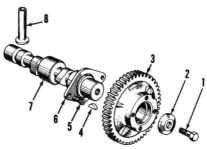

Fig. WB2-18—Exploded view of camshaft assembly. Extension shaft may be bolted to camshaft gear.

1. Camshaft gear bolt
2. Washer
3. Camshaft gear
4. Woodruff key
5. Retaining plate bolts
6. Retaining plate
7. Camshaft
8. Cam follower (tappet)

Fig. WB2-19—Exploded view of crankcase, oil sump and timing gear cover. On models with extension shaft(s), seal will be installed in timing cover in place of plug (P). Plate (3) is used on engines not equipped with fuel lift pump.

1. Crankcase
2. Gasket
3. Plate
4. Oil cap & dipstick
5. Filler neck
6. Governor shaft bushing
7. Camshaft bore plug
8. Camshaft bushing
9. Governor bushing
10. Camshaft bushing
11. Camshaft bushing
12. Dowel pin
13. Timing gear cover
14. Gasket
15. Oil drain plug
16. Oil sump
17. Gasket

With camshaft removed, remove cam followers (tappets) (8) and check for excessive wear or damage. Inspect camshaft bearings (8, 10 and 11—Fig. WB2-19) and renew if necessary. Camshaft bearings are available in standard size only and are a press fit in crankcase. Be sure when installing new bearings that oil holes in bearings are aligned with oil passages in cylinder block.

To install gear on camshaft, heat gear in hot oil, then press gear and thrust plate onto shaft. Reinstall camshaft and gear assembly making sure timing marks on camshaft and crankshaft gears are mated as shown in Fig. WB2-14 and tighten thrust plate bolts. Install gear retaining bolt and washer and tighten bolt to a torque of 36 N·m (27 ft.-lbs.). Camshaft end play should be 0.08-0.25 mm (0.003-0.010 inch). Install camshaft extension and tighten retaining cap screws to a torque of 19 N·m (14 ft.-lbs.).

CRANKSHAFT AND MAIN BEARINGS

The crankshaft is supported in two main bearings (7 and 9—Fig. WB2-20) and end play is controlled by a thrust washer (5) at each end of the crankshaft.

Crankshaft can be removed after removing crankshaft timing gear (4), the cylinder, rod and piston unit, and the flywheel. Unbolt and remove the rear main bearing housing. Remove crankshaft from rear end of cylinder block and remove the two thrust washers from crankshaft.

Crankshaft main and rod journal standard diameters are both 41.262-41.275 mm (1.6245-1.6250 inches). Inspect the crankshaft end thrust shoulders and check bearing journals with micrometer for wear. Main bearings are available in undersizes of 0.25 and 0.50 mm (0.010 and 0.020 inch) as well as standard size. Crankshaft main bearing oil clearance should be 0.02-0.077 mm (0.0008-0.0029

Westerbeke — SMALL DIESEL

inch) and crankpin bearing oil clearance should be 0.025-0.09 mm (0.0010-0.0035 inch).

To remove crankshaft front or rear main bearing, the crankcase or rear main bearing retainer should be heated in boiling water, then press bearing from crankcase or housing. To install new bearing, heat crankcase or bearing housing in boiling water and chill bearing in freezer. Press new bearing into crankcase or bearing housing so that oil hole in bearing is aligned with oil feed passage and with split side of bearing up. New bearings are prefinished and fit should be correct as installed.

To install crankshaft, first be sure all bearing journal surfaces and oil holes are clean. Place front crankshaft thrust washer on locating dowel pin (10) inside front end of crankcase and with tab on thrust washer engaging slot in crankcase. Bearing material side of thrust washer must face crankshaft shoulder. Lubricate crankshaft journals and main bearings and insert front end of crankshaft through thrust washer and front main bearing. Take care not to damage the bearing surfaces. Refer to CRANKSHAFT REAR OIL SEAL paragraph and install rear main bearing retainer. Be sure crankshaft turns smoothly and check crankshaft end play. End play should be 0.13-0.43 mm (0.005-0.017 inch) and should not exceed 0.51 mm (0.020 inch). Complete the reassembly of engine by reversing disassembly procedure.

CRANKSHAFT REAR OIL SEAL

The crankshaft rear oil seal (13—Fig. WB2-20) is mounted in the rear main bearing housing (12). Remove flywheel and rear bearing housing to renew seal. Be careful not to lose or damage rear crankshaft thrust washer. Check condition of oil seal surface of crankshaft. Remove oil seal (13) from housing and press new seal with lip to inside in squarely and flush with outside of housing.

Lubricate seal liberally with Lithium No. 2 grease or soak seal in engine oil for 24 hours before installation. Place rear crankshaft thrust washer on locating dowel pins in bearing housing with bearing side away from housing; tab on thrust washer must engage slot in bearing housing. Using new gasket (11), install gasket and bearing housing over rear end of crankshaft taking care not to damage seal on keyway edges on crankshaft. Tighten bearing housing retaining nuts in diagonal pattern. Check crankshaft end play which should be 0.13-0.43 mm (0.005-0.017 inch) and should not exceed 0.51 mm (0.020 inch). Complete the reassembly by reversing disassembly procedure.

FLYWHEEL

The flywheel (16—Fig. WB2-20) is mounted on rear end of crankshaft on a taper fit and retained by a hex nut and lock-tab washer (17). To remove flywheel retaining nut, crankshaft must be held from turning. Insert a steel rod through timing hole in bell housing into a hole located in outer diameter of flywheel. Bend tab back on washer and remove flywheel retaining nut. Remove steel rod and using gear puller bolts threaded into puller holes at center of flywheel, remove flywheel from crankshaft. Remove flywheel key (15) from crankshaft.

To install flywheel, insert key in slot of crankshaft and push flywheel onto shaft as far as possible. Use wood block or steel rod as in flywheel removal to keep crankshaft from turning. Place a new lock-tab washer on crankshaft, install nut and tighten to a torque of 210 N·m (155 ft.-lbs.).

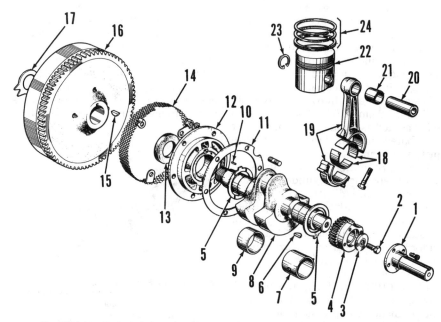

Fig. WB2-20—Exploded view showing crankshaft, flywheel and piston and rod unit.

1. Extension shaft
2. Crankshaft gear bolt
3. Washer
4. Crankshaft gear
5. Thrust washers
6. Woodruff key
7. Front main bearing
8. Crankshaft
9. Rear main bearing
10. Thrust washer dowels
11. Gasket
12. Rear main bearing housing
13. Crankshaft oil seal
14. Air inlet screen
15. Woodruff key
16. Flywheel
17. Tab washer
18. Bearing insert
19. Connecting rod
20. Piston pin
21. Bushing
22. Piston
23. Snap ring
24. Piston rings

SERVICE MANUAL   Westerbeke

WESTERBEKE

Model	No. Cyls.	Bore	Stroke	Displ.
10TWO	2	65 mm (2.559 in.)	68 mm (2.677 in.)	451 cc (27.5 cu. in.)

The Model 10TWO is a four-stroke cycle, water cooled, vertical cylinder, valve in head type indirect injection diesel engines. Direction of engine rotation is counterclockwise as viewed from flywheel end. Engine is for marine use and is available as either direct power source or 3KW marine diesel generator.

MAINTENANCE

LUBRICATION

Recommended engine lubricant is API classification CC or CD. SAE 10W-30 motor oil may be used in all ambient temperatures. If single grade oil is used, use SAE 10 in temperatures below 0° C (32° F), SAE 20 in temperatures from 0°-27° C (0°-80° F) and SAE 30 in temperatures above 27° C (80° F). Oil pan capacity including filter is 3.1 liters (3-1/4 quarts). Oil and filter should be changed after each 100 hours of operation.

FUEL SYSTEM

FUEL FILTER. A primary fuel filter element (4—Fig. WB3-1) is located in the electric fuel pump (2) and a secondary filter (13 through 18) is located between the electric pump and fuel injection pump (25). Recommended replacement interval for both filter elements is after each 200 hours of use or yearly, whichever comes first. Excessive sediment or water in fuel may require renewing filter elements earlier than specified use interval. Bleed system as outlined in following paragraph after servicing fuel filters.

BLEED FUEL SYSTEM. To bleed fuel system, be sure fuel is flowing to electric pump (2—Fig. WB3-1). Loosen secondary filter bleed screw (7) and actuate electric pump until fuel is flowing out bleed screw opening. Tighten bleed screw and actuate electric pump until fuel is flowing back to tank through return line (30).

It may be possible that air is trapped in the fuel injection lines (26) if engine will not start. Loosen the injection line fittings at the injectors, place throttle in full speed position and crank engine until pump control linkage in "Run" position, turn crankshaft slowly in normal direction of rotation and stop when fuel stops flowing from open delivery valve holder. The timing mark on engine crankshaft pulley should be aligned with timing mark on boss of timing gear cover.

If injection timing is not correct, it will be necessary to remove the fuel injectil fuel spurts from the loosened fittings. Tighten fittings and start engine.

INJECTION PUMP TIMING

Injection timing mark and TDC mark are on crankshaft pulley (Fig. WB3-2). To check injection timing, remove No. 1 fuel injection line and the No.1 injection pump delivery valve holder. Lift out delivery valve and reinstall spring and delivery valve holder. With injection tion pump and change the thickness of the shim (24—Fig. WB3-1) under the pump mounting flange. Changing shim thickness by 0.1 mm will result in a change in timing of about one degree. Shims are available in nine different thicknesses from 0.2 mm to 1.0 mm in steps of 0.1 mm. Decrease shim thickness to advance timing, or increase shim thickness to retard timing.

GOVERNOR

The flyweight type governor is mounted on the front face of the governor gear. All governor linkage is mounted inside the timing gear cover.

The high speed no-load adjustment screw is located in the timing gear cover and governor throttle lever contacts the screw at maximum speed position. If high idle speed is found to be incor-

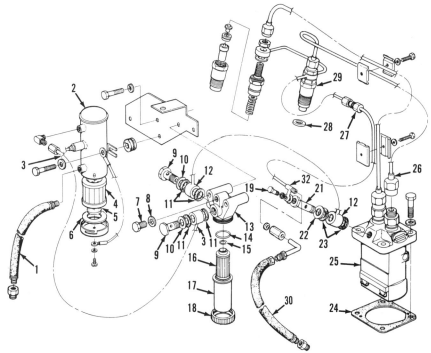

Fig. WB3-1—Exploded view of diesel fuel system.

1. Fuel supply line
2. Electric fuel pump
3. Secondary filter line
4. Primary filter element
5. Magnet
6. Cover gasket
7. Filter bleed screw
8. Sealing washer
9. Banjo bolt
10. Aluminum washer
11. Banjo washers
12. Return line
13. Filter base
14. Bowl seal ring
15. Element seal ring
16. Secondary filter element
17. Filter bowl
18. Bowl retainer
19. Banjo bolt
21. Banjo bolt
22. Aluminum washer
23. Banjo washers
24. Injection pump shim
25. Injection pump
26. Injector line
27. Fuel return line
28. Sealing washer
29. Fuel injectors
30. Fuel return hose

Illustrations Courtesy J.H. Westerbeke Corp.

Westerbeke

SMALL DIESEL

rect, cut seal wire, loosen locknut and turn screw in to decrease engine speed or out to increase speed to about 8 percent above rated speed shown on engine nameplate. With high idle speed correctly adjusted, tighten locknut and install new wire and lead seal.

Engine idle speed should be approximately 900 rpm. If necessary to adjust idle speed, loosen nuts (4—Fig. WB3-3) on idle speed stop (5) and turn nuts as necessary to obtain correct idle speed, then tighten nuts. It may be necessary to loosen throttle cable clamp (1) and move cable housing to obtain proper governor throttle arm (6) travel.

On marine diesel generator unit, the high idle speed adjustment screw is not used. Engine speed while generator is providing power is controlled by a solenoid connected to governor control arm. The solenoid throttle linkage must be adjusted so that solenoid operates freely and solenoid plunger is bottomed in solenoid when activated. Adjust length of solenoid to governor throttle arm linkage as necessary to obtain rated speed under load. Rated speed will be found on engine nameplate.

EXCESS FUEL DEVICE

An internal feature of the fuel injection pump is an excess fuel device that permits throttle operation to override the maximum fuel delivery stop when engine is not running. This provides an excess amount of fuel for aid in starting engine. Adjustment of this device should not be attempted outside authorized diesel system service station. If engine smokes excessively under load, this device may be at fault.

Glow plug .... 15-19 N·m
(11-14 ft.-lbs.)

WATER PUMP AND COOLING SYSTEM

Cooling is provided by a closed fresh water cooling system which is kept within operating temperature by a sea water heat exchanger. The fresh water system is a closed system having a pressure cap and coolant recovery tank. It is recommended that the fresh water system be filled with a 50:50 mixture of

water and permanent type antifreeze. Two water pumps, a fresh water circulating pump driven by belt from engine crankshaft and a sea water lift pump driven by belt from fresh water pump pulley, are used. A zinc rod installed in the heat exchanger is used to combat corrosion in the sea water system.

SEA WATER PUMP. Refer to Fig. WB3-4 for exploded view of the sea water pump. The rubber impeller (17) can be inspected and renewed if necessary

Fig. WB3-2—View of timing marks on timing gear cover and crankshaft pulley. Both TDC and injection timing marks are on pulley.

Fig. WB3-3—Exploded view of idle stop assembly.
1. Cable housing clamp
2. Cable housing
3. Idle stop bracket
4. Adjusting nuts
5. Idle stop
6. Throttle lever
7. Cable end

REPAIRS

TIGHTENING TORQUES

When making repairs, refer to the following special tightening torques:
Cylinder head bolts ...... 34-39 N·m
(25-29 ft.-lbs.)
Crankshaft pulley nut ... 146-195 N·m
(108-144 ft.-lbs.)
Connecting rod cap
  bolts .... 31-34 N·m
(23-25 ft.-lbs.)
Flywheel bolts .... 63-69 N·m
(46-51 ft.-lbs.)
Oil filter .... 11-12 N·m
(8-9 ft.-lbs.)
Delivery valve, injection
  pump .... 40-49 N·m
(29-36 ft.-lbs.)
Nozzle retaining nut ..... 59-78 N·m
(43-58 ft.-lbs.)

Fig. WB3-4—Exploded view of sea water pump assembly.
1. Set screw
2. Pulley
3. Bearing cover
4. Snap ring
5. Snap ring
6. Bearings
7. Bearing cover
8. Pump shaft
9. Pump body
10. Impeller cam
11. Washer
12. Cam screw
13. Seal
14. Dowel pin
15. "O" ring
16. Wear plate
17. Impeller
18. Impeller screw
19. Gasket
20. Cover plate

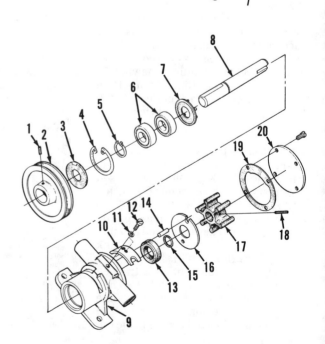

SERVICE MANUAL

Westerbeke

after removing pump cover plate (20). It is not necessary to remove water pump from engine to service impeller. When installing impeller, apply grease to impeller cavity and impeller blades and be sure all impeller blades are curved backward in trailing position. Renew cover plate if worn.

To disassemble water pump, remove cover (20), impeller (17), cam (10) and wear plate (16). Loosen set screw (1) and remove drive pulley (2). Pry out bearing cover (3) and remove large snap ring (4). Support pump housing and press shaft (8) and bearings (6) out drive pulley end of housing. Remove seal (13) from pump housing.

Lubricate seal lip with petroleum jelly and install seal with lip facing the impeller bore. Face of seal should be flush with impeller bore inner surface. Position "O" ring (15) and plastic spacer (7) on pump shaft, then install shaft and bearings into pump housing, pressing against outer race of bearings. Install snap ring (4) and outer cover (3). Apply small amount of sealant to mounting surface of cam (10) and retaining screw (12). Install cam, wear plate (16), impeller (17) and cover plate (20).

When pump is serviced, be sure to check quickly for sea water flow after starting engine. If pump does not prime quickly, damage to impeller will result from overheating.

FRESH WATER PUMP. The fresh water circulating pump is serviced as a complete assembly only. If pump seal is leaking, bearings are loose or noisy, or if other damage is noted, renew pump.

THERMOSTAT

To remove thermostat, refer to Fig. WB3-6 and loosen sea water pump mounting bolts and drive belt (18). Drain cooling system, remove hoses (2 and 3) from thermostat cover (4), then unbolt and remove thermostat cover. If gasket seal between sea water pump bracket (14) and cylinder head is broken, remove pump bracket and renew gasket (15). Two different types of thermostats have been used; one with a bypass valve plate at bottom of thermostat and the other a plain thermostat. It is important if renewing thermostat that a duplicate of the original is used.

To check thermostat, place in a pan of water, then heat water to boiling point. The thermostat should open fully.

VALVE ADJUSTMENT

Valve rocker arm to valve stem clearance is 0.25 mm (0.010 inch) for intake and exhaust with engine cold. To check valve clearance adjustment, remove valve cover and turn piston to TDC of compression stroke while checking clearance for that cylinder. Repeat for each cylinder of engine.

CYLINDER HEAD

Refer to Fig. WB3-7 for exploded view of cylinder head, rocker arm assembly and push rods.

To remove cylinder head, drain engine coolant, disconnect coolant hose and remove air intake connector and exhaust manifold. Remove fuel injection lines and immediately cap all openings. Remove fuel filter unit and fuel injector assemblies from cylinder head. Remove oil pressure tube to cylinder head. Remove all other attachments and connections, such as thermostat housing and sea water pump bracket, and glow plug wiring from cylinder head. Remove valve rocker arm cover, rocker arm assembly and push rods. Gradually loosen cylinder head bolts in reverse order of tightening sequence shown in Fig. WB3-8. Remove cylinder head and gasket.

Carefully clean cylinder head and block surfaces and inspect head for distortion, cracks or other damage. Cylinder head must be renewed or resurfaced

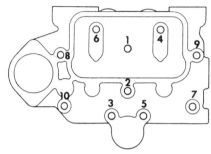

Fig. WB3-8—Cylinder head bolt tightening sequence.

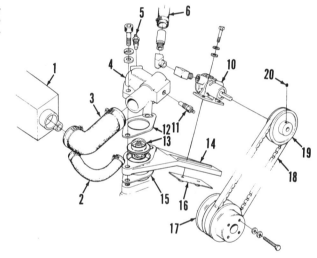

Fig. WB3-6—Exploded view of thermostat unit and sea water pump mounting arrangement.
1. Heat exchanger
2. Bypass hose
3. Outlet hose
4. Thermostat housing
5. Temperature sender
6. Hose to heat exchanger
10. Sea water pump
11. Temperature alarm sender
12. Gasket
13. Thermostat
14. Sea water pump bracket
15. Gasket
16. Welded nut plate
17. Fresh water pump pulley
18. Sea water pump belt
19. Sea water pump pulley
20. Set screw

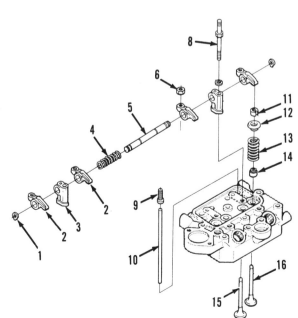

Fig. WB3-7—Exploded view of cylinder head assembly.
1. Snap ring
2. Rocker arms
3. Rocker shaft support
4. Spacer spring
5. Rocker arm shaft
6. Locknut
8. Rocker support bolt
9. Adjusting screw
10. Push rod
11. Split valve locks
12. Spring retainer
13. Valve spring
14. Valve stem seal
15. Intake valve
16. Exhaust valve

Westerbeke

SMALL DIESEL

if distortion (warpage) exceeds 0.05 mm (0.002 inch). The precombustion chambers (swirl chambers) are pressed into cylinder head and should not be removed. Check valve heads for sinkage; valve head should not be more than 1 mm (0.040 inch) below cylinder head gasket surface.

To reinstall cylinder head, be sure all gasket surfaces are thoroughly clean. Do not use gasket sealer of any kind on the gasket or gasket surfaces of head and block. Refer to Fig. WB3-8 and tighten cylinder head bolts in several steps in sequence shown until final tightening torque of 34-39 N·m (25-29 ft.-lbs.) is reached. Complete reassembly by reversing disassembly procedure.

VALVE SYSTEM

Both the intake and exhaust valves seat directly on machined seats in cylinder head. Valve face and seat angle is 45 degrees. Valve guides are renewable on all models. Valve stem diameter is 6.6 mm (0.260 inch). Valve margin (thickness of head at outer edge of face) is 1.0 mm (0.040 inch). Valve seat width should be 1.5-1.8 mm (0.059-0.071 inch). If guides are to be renewed, seats should be cut or ground after installing new guides. It may be necessary to ream valve guides after installation to obtain desired valve stem to guide fit. After refacing and reseating valves, check distance from valve head to cylinder head gasket surface using straightedge and feeler gage. If more than 1.0 mm (0.039 inch) with new valve, cylinder head must be renewed.

When testing valve springs, refer to the following specifications:

Spring free length, new . . . . . . 44 mm
(1.732 in.)
Spring installed height . . . . . . 37.6 mm
(1.480 in.)
Pressure @ installed height . 93-103 N
(21-23 lbs.)

ROCKER ARM ASSEMBLY

The rocker arm assembly can be unbolted and removed from top of head after removing valve rocker arm cover. Refer to Fig. WB3-7 for exploded view of assembly. With rocker arm assembly removed, lift out the push rods (10). Keep location of all removed parts noted so they may be reinstalled in same location. Rocker arms (2), spacer spring (4) and rocker arm supports (3) can be removed from shaft (5) after removing snap rings (1) at each end of shaft.

Check all parts, including push rod ends, rocker arm adjusting screws, and rocker arm valve contact pads for wear or damage. Intake and exhaust rocker arms are alike. Shaft to rocker arm clearance should be 0.04 mm (0.0015 inch).

Reassemble using new parts as necessary by reversing disassembly procedure. Loosen locknuts (6) and back rocker arm adjusting screws (9) out before installing push rods and rocker arm assembly. Adjust valve clearance cold to 0.25 mm (0.010 inch) before starting engine.

INJECTORS

REMOVE AND REINSTALL. To remove fuel injectors, first thoroughly clean injectors, lines and surrounding area using suitable solvent and compressed air. Remove high pressure lines (3—Fig. WB3-9) from pump to injectors. Remove hex nut (5) and the fuel return line (2) from injector. Remove injector from cylinder head by unscrewing the injector.

NOTE: Be sure that wrench is placed on injector nozzle nut (13) and not on injector body (7).

Remove sealing washer (1) from bottom of injector bore in cylinder head if not removed with injector.

Clean injector bores in cylinder head before reinstalling injectors. Install injectors using new copper washers. Tighten injector to a torque of 58-78 N·m (43-58 ft.-lbs.). Install fuel return lines and fuel pump to injector high pressure lines.

TESTING. A complete job of testing and adjusting an injection nozzle requires use of special test equipment. Use only a clean approved testing oil in injector tester tank. Injection nozzle

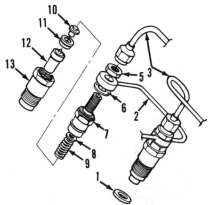

Fig. WB3-9—Exploded view of fuel injector assembly showing injection and return lines and complete injector.

1. Copper washer
2. Fuel return line
3. Injection lines
4. Hex nut
5. Hex nut
6. Sealing washer
7. Injector body
8. Adjusting shims
9. Spring
10. Spring seat/push rod
11. Spacer
12. Nozzle & valve
13. Nozzle nut

should be checked for opening pressure, spray pattern, seat leakage and back leakage.

WARNING: Fuel emerges from injector with sufficient force to penetrate the skin. When testing injector, keep yourself clear of nozzle spray.

OPENING PRESSURE. Before conducting test, operate tester lever until fuel flows from tester line, then attach injection nozzle to line. Pump lever a few quick strokes to purge air and be sure nozzle valve is not stuck and spray hole is open.

Operate tester lever slowly and observe gage reading when nozzle opens. On all models, opening pressure should be 14710-16670 kPa (2135-2415 psi). If necessary to adjust opening pressure, remove nozzle retaining nut (13—Fig. WB3-9), nozzle and needle (12), spacer (11), spring seat (10), spring (9) and shims (8). Add shims to increase opening pressure or remove shim thickness to decrease pressure. Adjusting shims are available in 16 different thicknesses from 1.00 to 1.95 mm in steps of 0.5 mm and are used as required to obtain specified opening pressure. Injector opening pressure varies approximately 980 kPa (142 psi) with each 0.1 mm change in shim thickness. Change shim thickness as necessary, reassemble nozzle and recheck opening pressure.

SPRAY PATTERN. The throttling type nozzle should emit a tight conical spray pattern with no branches, splits or dribbles. If incorrect spray pattern is observed, check for partially clogged or damaged spray hole or improperly seating or sticking nozzle valve.

SEAT LEAKAGE. Wipe nozzle tip dry, then operate tester lever slowly to obtain pressure 1000 kPa (150 psi) below observed opening pressure and hold pressure at this reading for 10 seconds. If a drop forms or undue wetness appears on nozzle tip during test, disassemble and clean or renew injector.

OVERHAUL. First, thoroughly clean outside of injector. Hold injector body (7—Fig. WB3-9) in soft jawed vise. Unscrew nozzle retaining nut (13) and disassemble nozzle. Place all parts in clean calibrating oil or diesel fuel and keep all parts of each injector separate from others.

Clean exterior surfaces with a brass wire brush. Soak parts in an approved carbon solvent if necessary to loosen hard carbon deposits. Rinse parts in clean diesel fuel or calibrating oil immediately after cleaning to neutralize the solvent.

SERVICE MANUAL

Westerbeke

Clean nozzle spray hole from inside using a hardwood scraper. Scrape carbon from nozzle chamber using a hooked brass scraper. Clean valve seat using a brass scraper, then polish seat using wood polishing stick and mutton tallow.

Back flush nozzle using a reverse flush adapter on injector tester. Reclean all parts by rinsing thoroughly in clean diesel fuel or calibrating oil and reassemble while parts are immersed in the fluid. Make sure to assemble with same shims as were removed. Tighten nozzle retaining nut to a torque of 59-78 N·m (43-58 ft.-lbs.) and retest injector.

INJECTION PUMP

The fuel injection pump is mounted on top of the engine timing gear cover, and pin (P—Fig. WB3-11) on pump rack engages slot in upper end of governor lever (13).

To remove fuel injection pump, first thoroughly clean pump, injectors and lines. Close fuel supply valve and remove fuel injection lines, fuel supply and return lines. Immediately cap all openings. Turn engine crankshaft so that piston is at TDC on exhaust stroke. Unbolt and remove pump, taking care not to lose or damage any of the shims (S) from under pump flange used to set injection timing. It is recommended that the pump not be disassembled outside of an authorized diesel system repair station.

Install pump in housing with same thickness of shims that were removed unless injection timing is to be changed. Turn engine crankshaft so that piston is at TDC of exhaust stroke. Be sure pin on pump rack engages slot in upper end of governor control lever. Check to see that pump flange is seated on dowel pin (4—Fig. WB3-10) in timing gear cover. Tighten the pump retaining bolts evenly. Complete reassembly by reversing removal procedure.

TIMING GEAR COVER

To remove timing gear cover, thoroughly clean engine, drain cooling system and proceed as follows: Remove the fresh water pump and alternator. Remove fuel injection pump. Remove crankshaft pulley nut, washer and pulley. Disconnect throttle linkage from arm (1—Fig. WB3-11) on timing cover and breather hose from nipple (6—Fig. WB3-10) at alternator side of engine. Unbolt and remove timing gear cover from front of crankcase. Be careful in removing cover from aligning dowel pins.

With timing cover removed, install new crankshaft front oil seal (2) in timing cover with lip of seal to inside. Inspect oil seal contact surface of crankshaft pulley spacer sleeve. Renew spacer sleeve if seal surface is grooved or scored.

To install front cover, be sure governor linkage is in place. Install cover using new gasket. Install fuel injection pump. Complete remainder of reassembly by reversing disassembly procedure. Tighten crankshaft pulley nut to a torque of 146-195 N·m (108-144 ft.-lbs.).

GOVERNOR

The governor unit is accessible for service after removing the engine tim-

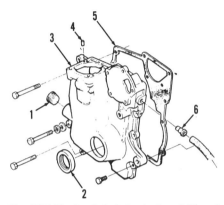

Fig. WB3-10—Crankshaft front oil seal (2) and governor linkage (see Fig. WB3-12) are mounted in engine timing cover (3).

1. Governor shaft plug
2. Crankshaft oil seal
3. Timing cover
4. Injection pump dowel
5. Gasket
6. Crankcase breather nipple

ing gear cover. Refer to drawing of governor in Fig. WB3-11 and to exploded view of governor unit in Fig. WB3-12. Remove sliding shaft (14—Fig. WB3-12) from weight assembly (15), then unbolt and remove governor weight assembly from gear (18). Remove snap ring (16) and washer (17) and pull gear from stub shaft (19). Remove the governor spring (7). Unscrew nut (6) and remove lever (4) from throttle control shaft (1). Drive out groove pin (2) and pull lever and shaft from timing cover. Remove governor lever (10), grooved pin (12) and governor lever (13) from shaft (11). Remove plug (22—Fig. WB3-11) and slide shaft from timing cover.

Inspect all parts and renew as necessary. Reassemble by reversing disassembly procedure.

INJECTION PUMP CAMSHAFT

Cam lobes for operation of the fuel injection pump are located on the front end of engine camshaft; refer to CAMSHAFT paragraph.

TIMING GEARS

Timing gears are accessible for inspection after removing engine timing gear cover. The timing gears consist of crankshaft gear (26—Fig. WB3-11) and camshaft gear (21). The governor gear (18) is mounted on a stub shaft located in front end of engine crankcase and gear is driven from camshaft gear. To remove

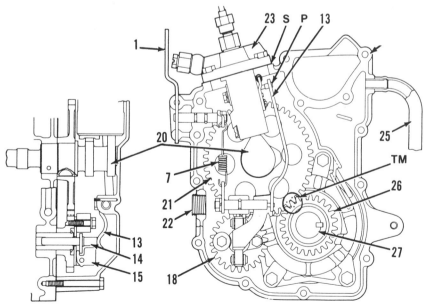

Fig. WB3-11—Cross section and installation views of governor and related parts. Refer also to Fig. WB3-12 for exploded view of governor.

P. Pump rack pin
S. Pump timing shims
TM. Timing gear marks
1. Throttle control lever
7. Governor spring
13. Governor lever
14. Sliding shaft
15. Governor flyweights
18. Governor gear
20. Injection pump cam lobes
21. Camshaft gear
22. Governor shaft plug
23. Injection pump
25. Crankcase breather hose
26. Crankshaft gear
27. Crankshaft

Westerbeke

camshaft gear, refer to CAMSHAFT paragraph. To remove governor gear, refer to GOVERNOR paragraph. Crankshaft gear can be removed using suitable puller.

When reassembling be sure timing marks (TM) on camshaft gear and crankshaft gear as aligned as shown in Fig. WB3-11.

OIL PUMP AND RELIEF VALVE

The gerotor type oil pump is driven through a coupling at rear end of engine camshaft. Refer to Fig. WB3-13 for exploded view showing oil filter (1), external oil line (4) to cylinder head and the pump assembly (7 through 19).

To remove pump cover (17), remove only the three long cap screws (19) extending through cover and pump housing to engine crankcase. Remove short cap screw (18) from housing flange to remove complete pump assembly.

Pressure relief valve (11) in pump housing (13) should open at 345 kPa (50 psi). To check pressure, remove oil pressure sender switch and install a master test gage in opening.

Pump outer rotor to body clearance should be 0.15-0.2 mm (0.006-0.008 inch) with wear limit of 0.3 mm (0.012 inch).

SMALL DIESEL

Clearance between lobes of inner and outer rotors should be 0.06-0.12 mm (0.0024-0.0047 inch) with wear limit of 0.25 mm (0.010 inch). Clearance between rotors and cover, measured by placing a straightedge across pump body, should be 0.03-0.07 mm (0.0012-0.0027 inch) with wear limit of 0.25 mm (0.010 inch). Check for scoring in pump body and on cover plate.

Reassemble using new parts as necessary. Pump inner and outer rotors are available as set only; all other parts are available individually. Lubricate pump parts with heavy oil or light grease to aid pump in priming when engine is started.

PISTON AND ROD UNITS

The piston and connecting rod units (Fig. WB3-14) may be removed from above after removing cylinder head, oil pan and oil pickup screen. Before removing rod and piston unit, check side clearance of rod on crankpin. Rod side clearance should be 0.1-0.35 mm (0.004-0.014 inch); maximum allowable rod side clearance is 0.5 mm (0.020 inch).

The piston pin (3) is retained by being a press fit in connecting rod (4). Do not disassemble rod and piston units unless necessary to renew piston or connecting rod. Disassembly and reassembly requires use of special piston pin

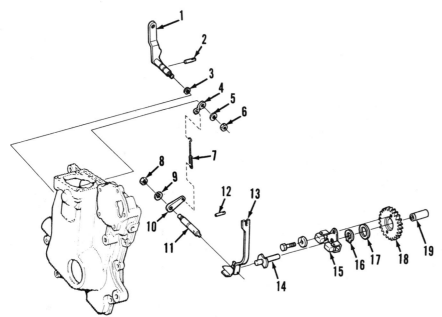

Fig. WB3-12—Exploded view of governor and governor linkage. Refer also to Fig. WB3-11.
1. Throttle control lever
2. Groove pin
3. "O" ring
4. Throttle lever
5. Lock washer
6. Hex nut
7. Governor spring
8. Hex nut
9. Lock washer
10. Governor spring lever
11. Governor lever shaft
12. Groove pin
13. Governor lever
14. Sliding shaft
15. Governor flyweight assy.
16. Snap ring
17. Washer
18. Governor gear
19. Stub shaft

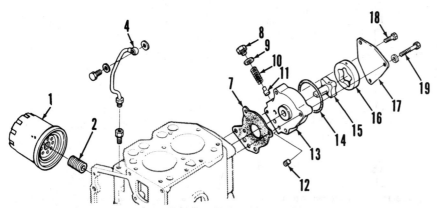

Fig. WB3-13—Exploded view of oil pump, cylinder head oil line and oil filter.
1. Filter cartridge
2. Filter adapter nipple
4. Cylinder head oil line
7. Gasket
8. Plug
9. Gasket
10. Spring
11. Pressure relief valve
12. Plug
13. Pump housing
14. "O" ring
15. Pump inner rotor
16. Pump outer rotor
17. Pump cover
18. Short cap screw
19. Long cap screws (3)

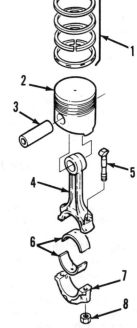

Fig. WB3-14—Exploded view of piston and connecting rod unit.
1. Piston ring set
2. Piston
3. Piston pin
4. Connecting rod
5. Rod bolts
6. Crankpin bearing insert
7. Rod cap
8. Rod nuts

SERVICE MANUAL

Westerbeke

insertion tools. If assembling piston and rod, be sure that arrow mark on top of piston and identification mark on side of connecting rod face same direction. Also, be sure that piston pin is centered in connecting rod.

When installing piston and rod assembly, be sure piston ring end gaps are spaced equally at 90 degrees around piston. Lubricate piston rings, cylinder wall, rod bearing insert and crankpin with clean oil. Install piston and rod assembly with arrow on top of piston and identification marks on side of connecting rod towards front of engine. Tighten connecting rod cap nuts to a torque of 31-34 N·m (23-25 ft.-lbs.).

PISTONS, PINS, RINGS AND CYLINDERS

Piston skirt diameter should be measured at right angle to piston pin and bottom of piston skirt. Maximum piston skirt to cylinder clearance is 0.3 mm (0.012 inch). If piston skirt to cylinder clearance is excessive, cylinders can be rebored and honed to next available oversize. Piston and rings are available in oversizes of 0.25, 0.50 and 0.75 mm (0.010, 0.020 and 0.030 inch).

Maximum allowable operating clearance between piston pin and bore in piston is 0.08 mm (0.003 inch). Piston pin is a press fit in connecting rod. Be sure that pin is centered in rod when assembling piston to rod. Make sure that arrow on top of piston and identification marks on side of connecting rod face in the same direction.

Pistons are fitted with three compression rings and one oil control ring. Piston ring end gap should be 0.20-0.50 mm (0.008-0.019 inch) for all rings on all models; maximum allowable ring end gap is 1.5 mm (0.059 inch). Side clearance of top compression ring in groove should be 0.06-0.11 mm (0.0024-0.0043 inch) with new ring; wear limit is 0.3 mm (0.012 inch). Side clearance of second and third compression rings and oil control ring should be 0.03-0.08 mm (0.001-0.003 inch) with wear limit of 0.2 mm (0.008 inch).

Cylinder bores are not sleeved. Cylinders may be rebored and honed to fit next available oversize pistons and rings if cylinder bore wear measures more than 0.20 mm (0.008 inch). Standard cylinder bore diameter is 65.0 mm (2.559 inches).

If pistons and cylinder bores are not excessively worn or scored, new rings may be installed after removing ring ridge at top of cylinder and removing cylinder glaze with ball hone.

CONNECTING RODS AND BEARINGS

Piston pin is a press fit in connecting rod. When installing piston on rod, align pin bores in piston and rod press pin into assembly so that it is centered in piston with equal gaps between each side of rod and piston. Arrow on top of piston and identification marks on side of rod must be in same direction.

Connecting rod crankpin bearing is a precision slip-in insert type for all models. Crankpin standard diameter is 42 mm (1.6535 inches). Bearing to crankpin oil clearance should be 0.035-0.050 mm (0.0014-0.0019 inch); wear limit is 0.15 mm (0.006 inch). Crankpin bearings are available in undersizes of 0.25, 0.50 and 0.75 mm (0.010, 0.020 and 0.030 inch) as well as standard size.

CAMSHAFT AND BEARINGS

Camshaft and gear assembly (see Fig. WB3-15) can be removed from cylinder block after removing fuel injection pump, engine timing cover, rocker arm assembly, push rods and cam followers (tappets). Remove engine oil pump and drive coupling from rear end of camshaft.

Check camshaft bearing journals, lobes, drive gear teeth and engine oil pump drive slot for wear or damage. Cam lobe diameter is 25.59 mm (1.007 inches) for both intake and exhaust valve lobes. Minimum lobe diameter (wear limit) is 25.49 mm (1.003 inches). Cam lobe diameter for injection pump lobes is 44.0 mm (1.732 inches); wear limit is 43.9 mm (1.728 inches).

To remove camshaft gear, press camshaft forward out of gear. Be sure to retain the gear locating key for reassembly. When installing camshaft in gear, be sure side of gear with timing mark is facing forward. Insert key in shaft, align keyway in gear with key and press shaft through gear.

The camshaft is supported by two unbushed bores in cylinder block. Maximum camshaft bearing oil clearance is 0.15 mm (0.006 inch).

Reinstall camshaft by reversing removal procedure. Be sure timing marks on camshaft gear and crankshaft gear are aligned as shown in Fig. WB3-11.

CRANKSHAFT AND MAIN BEARINGS

Crankshaft is supported in two main bearings. Front bearing (1—Fig. WB3-16) is pressed into main bearing retainer (2) and rear bearing (7) is pressed into rear end of crankcase. Crankshaft end play should be 0.1-0.3 mm (0.004-0.012 inch) and is controlled by thrust washer (7—Fig. WB3-17) at front side of front main bearing retainer (8) and by shoulder of crankshaft (9) against rear side of bearing retainer. Check end play with dial indicator before removing crankshaft.

To remove crankshaft, remove cylinder head, flywheel, timing gear cover, camshaft and the rod and piston units. Unbolt front bearing retainer and remove crankshaft, bearing retainer and crankshaft gear assembly from front end of block. Pull crankshaft gear (6) from crankshaft, remove key (10) and slide thrust washer and front main bearing retainer from shaft. Pry oil seal (8—Fig. WB3-16) from rear end of block.

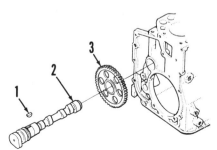

Fig. WB3-15—Camshaft gear (3) is removed from rear end of camshaft (2).

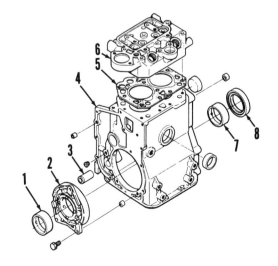

Fig. WB3-16—Crankshaft main bearings are pressed into front bearing support and in rear end of crankcase.
1. Front main bearing
2. Bearing support
3. Governor gear stub shaft
4. Crankcase
5. Head gasket
6. Cylinder head
7. Rear main bearing
8. Crankshaft rear oil seal

Illustrations Courtesy J.H. Westerbeke Corp.

Westerbeke — SMALL DIESEL

Check crankshaft main journal, crankpin journal, rear oil seal surface and front thrust surface for excessive wear, scoring or other damage. Main journal standard diameter is 54 mm (2.126 inches). Crankpin standard diameter is 42 mm (1.6535 inches). Main and rod bearings are available in undersizes of 0.25, 0.50 and 0.75 mm (0.010, 0.020 and 0.030 inch) as well as standard size.

If necessary to install new main bearings, old bearings must be removed and new bearings installed using a suitable bushing driver. Press bearings out toward inside of crankcase; press in same direction to install new bearings. Coat outside of bearings with engine oil before installing. Be sure oil holes in bearings are aligned with oil passages in cylinder block and front bearing retainer. Front bearing should be installed with front edge 2.5 mm (0.098 inch) below front thrust surface of bearing retainer. Install new oil seal in rear end of crankcase with lip forward (to inside). Lubricate front bearing and place bearing retainer on front end of crankshaft, then install thrust washer (chamfered inside diameter to rear), key and crankshaft gear with long shoulder and timing mark to front.

Lubricate rear oil seal, main bearings and crankshaft main journals and slide crankshaft through crankcase into rear main bearing and oil seal. Take care not to damage rear main bearing or sealing lip on crankshaft oil seal. Align front bearing retainer with dowel in cylinder block and tighten retaining cap screws evenly. Install sleeve (5—Fig. WB3-17) with chamfered end to front.

Complete the reassembly of engine by reversing disassembly procedure. Crankshaft end play may be checked before installing timing gear cover by temporarily installing crankshaft pulley with washers (2 and 4) and retaining nut. Check end play with a dial indicator.

CRANKSHAFT REAR OIL SEAL

The crankshaft rear oil seal (8—Fig. WB3-16) is pressed into rear end of cylinder block. If care is taken, seal can be renewed after removing flywheel. When installing seal, lip must be toward inside of engine (forward).

ELECTRICAL SYSTEM

GLOW PLUGS

Glow plugs are wired in parallel. Glow plug circuit will continue to operate if one glow plug is defective. To check glow plugs, disconnect wire to glow plug terminals and check individual glow plugs with an ohmmeter. Resistance should be 1 to 1.2 ohms. If no resistance is noted, renew glow plug.

ALTERNATOR AND REGULATOR

A 35 ampere alternator with internal voltage regulator is used on all models. Regulator has no provisions for adjustment. Alternator output cold should be 7 amperes at 1300 alternator rpm and 30 amperes at 2500 alternator rpm. Regulated voltage should be 14.1-14.7 volts at 20° C (68° F).

New brush length is 18 mm (0.709 inch). Renew brushes if worn to 8 mm (0.315 inch) or less. Standard slip ring diameter is 33 mm (1.2992 inches) and wear limit is 32.2 mm (1.2677 inches).

To check stator assembly, use ohmmeter and check for continuity between each stator coil terminal. If there is no continuity, renew stator. Check for continuity between terminals and stator core. If continuity exists, renew the stator.

To check rotor, use ohmmeter to check for continuity between the two slip rings. If no continuity is found, renew the rotor. Check for continuity between slip ring and rotor core or shaft. If continuity exists, renew the rotor.

Using ohmmeter, check diodes in rectifier assembly. Ohmmeter should show infinite reading when ohmmeter leads are connected in one direction and continuity when ohmmeter leads are reversed. Renew rectifier assembly if continuity exists in both directions, or if no continuity exists in either direction.

STARTING MOTOR

Starting motor service specifications are as follows:
No-load test—
  Volts 11.5
  Current draw (max.) . . . . . . 90 amps
  Rpm (min.) 3600
Brush length—Standard . . . . . . 17 mm
                            (0.669 in.)
  Wear limit . . . . 11.5 mm
                            (0.453 in.)
Pinion gap . . . . 0.5-2.0 mm
                         (0.020-0.079 in.)

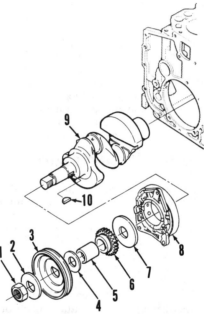

Fig. WB3-17—Crankshaft end play is controlled by thickness of bearing support (6), thrust plate (7) and shoulder on crankshaft front main bearing. Crankshaft front oil seal rides on spacer sleeve (5).

1. Hex nut
2. "D" washer
3. Crankshaft pulley
4. Flat washer
5. Spacer sleeve
6. Crankshaft gear
7. Thrust plate
8. Front bearing support
9. Crankshaft
10. Crankshaft gear key

SERVICE MANUAL

WISCONSIN

Model	No. Cyls.	Bore	Stroke	Displ.
TRD1-375	1	80 mm (3.14 in.)	75 mm (2.95 in.)	377 cc (23 cu. in.)
TRD1-380	1	80 mm (3.14 in.)	75 mm (2.95 in.)	377 cc (23 cu. in.)
TRD1-480	1	90 mm (3.54 in.)	75 mm (2.95 in.)	477 cc (29.1 cu. in.)
TRD1-540	1	90 mm (3.54 in.)	85 mm (3.34 in.)	540 cc (32.9 cu. in.)
TRD1-605	1	95 mm (3.74 in.)	85 mm (3.34 in.)	602 cc (36.7 cu. in.)

All models are four-stroke cycle, vertical cylinder, air-cooled direct injection diesel engines. Cooling fan is integral with flywheel. Engine rotation is counterclockwise when viewed from PTO end of engine. Model TRD1-375 has rope starter; all other models are electric starting.

MAINTENANCE

AIR CLEANER

All models are equipped with an oil bath air cleaner. Air filter element should be removed and cleaned in suitable solvent after each 50 hours of operation, or daily when operating in dusty or dirty conditions. Discard oil from air cleaner cup, wash cup in solvent and refill to proper level with clean engine oil.

LUBRICATION

Recommended engine lubricant is HD Series 3 or API classification CC/CD motor oil. If using single grade engine oil, use SAE 10W in ambient temperatures below 0° C (30° F), SAE 20W/20 in temperatures 0°-15° C (30°-59° F), SAE 30 in temperatures 15°-30° C (59°-86° F) and SAE 40 in temperatures above 30° C (86° F). Multi-grade oils may be used in all temperature ranges except above 30° C (86° F).

Crankcase lubricant capacity is 2 liters (2.11 quarts.). Engine oil and filter should be changed after every 100 hours of operation. Oil filter cartridge (1—Fig. W3-1) is located in side of cylinder block.

FUEL SYSTEM

FUEL FILTER. The fuel filter (2—Fig. W3-2 or Fig. W3-3) should be renewed after each 200 hours of operation. Filter cartridge (2—Fig. W3-3) for Model TRD1-375 is retained in bottom of fuel tank by filter cover (4) and cap screw (6). Filter (2—Fig. W3-2) for other models screws onto threaded nipple in bottom of fuel tank. It is necessary to drain fuel tank to service filter.

BLEED FUEL SYSTEM. All models have an automatic bleed valve located on fuel injection pump. Manual bleeding is not required. However, if injector line has been removed, it may be necessary to loosen injector line fitting at injector and crank engine until fuel flows from loosened fitting to purge air from line. Tighten fitting and start engine.

INJECTION PUMP TIMING

The injection pump is timed by shim gaskets placed between injection pump flange and cylinder block. If pump is removed, always reinstall with original number and thickness of shim gaskets.

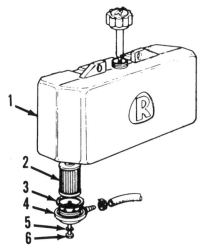

Fig. W3-3—Fuel filter (2) is mounted inside bottom of fuel tank on Model TRD1-375. Sealing ring (3) is provided with new fuel filter cartridge.
1. Fuel tank
2. Filter cartridge
3. Sealing ring
4. Filter cover
5. Sealing washer
6. Filter bolt

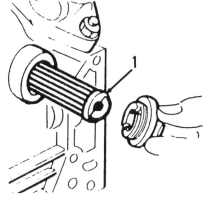

Fig. W3-1—Oil filter cartridge (1) is located in side of engine crankcase.

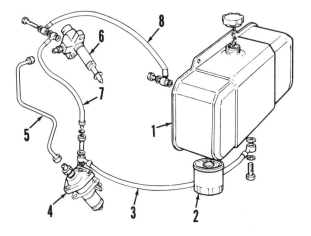

Fig. W3-2—Schematic drawing of fuel system for all models except TRD1-375.
1. Fuel tank
2. Fuel filter
3. Fuel supply line
4. Fuel injection pump
5. Fuel injection line
6. Fuel injector
7. Auto bleed-off line
8. Return line

Illustrations Courtesy Teledyne Total Power

Wisconsin

Gaskets are available in thicknesses of 0.1, 0.2 and 0.3 mm (0.004, 0.008 and 0.012 inch). Timing specifications are not stated by manufacturer.

GOVERNOR

Engine stop control and governor throttle levers are located in governor control cover as shown in Fig. W3-4. Exploded view of governor controls and linkage are shown in Fig. W3-13. Stop lever adjusting screws (B and C—Fig. W3-4) are located in flange on engine crankcase. Adjusting screws (12—Fig. W3-12) for throttle lever (13) are accessible after removing cover (17). Maximum rated speed for engine is shown on engine nameplate.

REPAIRS

When assembling engine, refer to the following special tightening torques:

Clutch plate nuts . . . . 245 N·m
(180 ft.-lbs.)
Connecting rod:
TRD1-375, 380 . . . . 34 N·m
(25 ft.-lbs.)
TRD1-480, 540, 605 . . . . 39 N·m
(29 ft.-lbs.)
Cylinder head nuts . . . . 49 N·m
(36 ft.-lbs.)
Flywheel nut . . . . 275 N·m
(202 ft.-lbs.)
Gear case cover bolts . . . . 22 N·m
(16 ft.-lbs.)
Injector nuts . . . . 22 N·m
(16 ft.-lbs.)
Oil sump cover bolts . . . . 13 N·m
(113 in.-lbs.)

VALVE ADJUSTMENT

Valves can be adjusted after removing rocker arm cover. Adjust valve clearance with engine cold to 0.15 mm (0.006 inch) for intake and exhaust. Turn crankshaft so that piston is at TDC on compression stroke (both push rods free to turn) and adjust both valves.

CYLINDER HEAD

To remove cylinder head, first remove air cleaner, exhaust muffler, cowling, rocker arm cover and fuel injector. Remove cylinder head nuts and washers and lift cylinder head assembly from studs. Remove the push rod tube and push rods.

With cylinder head removed, clean cylinder and head gasket surfaces and carefully inspect cylinder head. Install fuel injector in head with sealing washer(s) and check protrusion of nozzle tip from cylinder head surface. Tip should extend 2.25-2.75 mm (0.088-0.108 inch) from head. Add or remove 1.0 mm (0.040 inch) thick sealing washers to bring measurement within specifications.

To install cylinder head, reverse removal procedure, using new head gasket (10—Fig. W3-5) and push rod tube "O" rings (20—Fig. W3-6). End of push rod tube (21) with longer distance between "O" ring and end of tube is installed toward cylinder head. Using thin coating of grease, stick head gasket to bottom of cylinder head. Lubricate push rod tube "O" rings. Install head guiding the push rod tube into place. Alternately tighten cylinder head nuts to a final torque of 49 N·m (36 ft.-lbs.). Install fuel injector with number of washers determined as outlined in preceding paragraph. Adjust valve clearance before installing rocker arm cover.

VALVE SYSTEM

Exploded view of valve system is shown in Fig. W3-6. Valve face and seat angle is 45 degrees for both intake and exhaust valves. Valve seats are renewable and oversize seats are available so that valve seat counterbore can be reconditioned and maintain interference fit of seat to head. With valves installed, top of valve heads should be recessed 0.9-1.1 mm (0.035-0.043 inch) below surface of cylinder head. Maximum allowable valve head recess is 1.8 mm (0.070 inch); renew valves and/or seats if valve head recess is excessive.

Desired valve stem to guide clearance is 0.02-0.04 mm (0.0008-0.0015 inch) for intake valve and 0.04-0.065 mm (0.0015-0.0025 inch) on exhaust valve. Maximum allowable valve stem to guide clearance is 0.08 mm (0.003 inch) for intake valve and 0.1 mm (0.004 inch) for exhaust valves. Valve guides are renewable. To remove valve guide, press guide out toward top of head. To install new guide, position locating ring (8—Fig. W3-6) in groove on valve guide (6 or 7), then press guide downward into head until locating ring contacts head. Valve

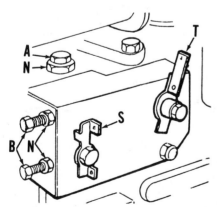

Fig. W3-4—View of engine stop control lever (S) and throttle lever (T). Moving stop lever to right stops engine. Moving throttle lever to right places governor in slow idle position; moving throttle lever to left increases engine speed.

A. Lever locating screw
B. Stop lever adjusting screws
N. Jam nuts
S. Stop lever
T. Throttle lever

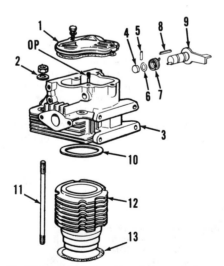

Fig. W3-5—Cylinder head is sealed to cylinder by head gasket ring (10). Combustion clearance is obtained by varying number of shims (13) located between cylinder and crankcase. Rocker arms are pressure lubricated through oil pipe (OP).

1. Rocker arm cover
2. Gasket
3. Cylinder head
4. Plug
5. Dowel pin
6. "O" ring
7. Spring
8. Taper pin
9. Compression release lever
10. Head gasket
11. Cylinder studs
12. Cylinder
13. Spacer shims

SMALL DIESEL

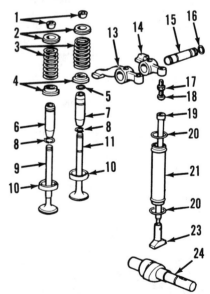

Fig. W3-6—Exploded view of valve system. Both intake and exhaust push rods are enclosed in a single tube (21).

1. Split valve locks
2. Valve spring retainers
3. Valve springs
4. Valve spring seats
5. Exhaust valve stem "O" ring
6. Intake valve guide
7. Exhaust valve guide
8. Guide ring
9. Intake valve
10. Valve seats
11. Exhaust valve
13. Intake rocker arm
14. Exhaust rocker arm
15. Rocker arm shaft
16. "O" rings
17. Jam nut
18. Adjusting screw
19. Push rod
20. "O" ring
21. Push rod tube
22. "O" ring
23. Cam followers
24. Camshaft

SERVICE MANUAL

guides are available with 0.1 mm (0.004 inch) oversize outside diameter for use if standard guide does not fit tightly in head. When installing oversize guides, ream guide holes in cylinder head for proper interference fit. Grind valve seats after installing guides.

Early models were equipped with a baffle plate fitted on intake valve guide. Baffle is located towards breather plug side of head by a dowel pin in the head. Early and late valves are not interchangeable, but early and late cylinder head assemblies may be interchanged.

Valve springs are alike for all models. Valve spring free length should be 35.6 mm (1.4 inches). At closed height of 25 mm (0.98 inch), spring pressure should be 14.6 kg (32 pounds). At valve open height of 17.5 mm (0.63 in.), spring pressure should be 25 kg (55 pounds). Limit of variation from spring specifications is 10 percent.

To remove rocker arms (13 and 14—Fig. W3-6) and shaft (15), remove shaft retaining pin and remove shaft from head. Inspect rocker arms, rocker arm shaft and push rods for excessive wear or other damage. Rocker arm to shaft clearance should be 0.03-0.06 mm (0.0012-0.0023 inch) with wear limit of 0.15 mm (0.006 inch). Install shaft with new "O" rings (16) and secure with pin.

INJECTORS

REMOVE AND REINSTALL. Prior to removing injector, thoroughly clean injector line, injection pump and injector using suitable solvent and compressed air. Remove injector fuel return line and unscrew fittings at each end of line and remove line from pump and injector. Remove the injector clamp stud nut and clamp plate, then remove injector from cylinder head. If injector sealing washer(s) was/were not removed with injector, carefully remove washer(s) from injector bore. Be careful not to lose any of the sealing washers as they are used to adjust injector tip position from combustion chamber surface of cylinder head.

When installing injector, be sure injector bore in cylinder head is clean. Using light grease, stick sealing washer(s) to nozzle tip.

NOTE: It is important that the same number and thickness of sealing washers be installed as were present on removal of injector to maintain correct piston to injector nozzle clearance.

Install the clamp plate and nut and tighten the nut to a torque of 22 N·m (16 ft.-lbs.). Install injection line and tighten fittings, leaving fitting at injector loose until system is bled. Connect fuel return line to injector. With controls in "Run" position, crank engine until fuel is discharged at loosened injector line fitting, then tighten fitting.

TESTING. A complete job of testing and adjusting the injector requires use of special test equipment. Only clean, approved testing oil should be used in tester tank. Injector should be tested for opening pressure, spray pattern and seat leakage.

WARNING: Fuel emerges from injector with sufficient force to penetrate the skin. When testing injector, keep yourself clear of nozzle spray.

Before conducting test, operate tester lever until fuel flows from tester line, then attach injector. Close tester gage valve and operate tester lever a few quick strokes to purge air from injector and to be sure nozzle valve is not stuck.

OPENING PRESSURE. Open valve to tester gage and operate tester lever slowly while observing gage reading. Nozzle should open at a pressure of 22065-23045 kPa (3200-3340 psi).

Opening pressure is adjusted by adding or removing shims (2—Fig. W3-7). Shims are available in thicknesses of 0.1 mm (0.004 inch) and 1.0 mm (0.040 inch).

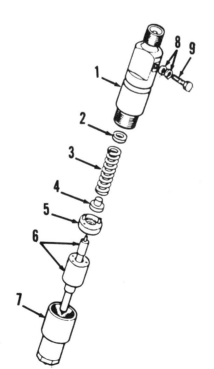

Fig. W3-7—Exploded view of typical fuel injector assembly.
1. Injector body
2. Shim washers
3. Spring
4. Spring seat
5. Spacer
6. Nozzle body & valve
7. Cap nut
8. Seal washers
9. Banjo fitting bolt

Wisconsin

SPRAY PATTERN. Spray pattern should be equal from all four nozzle tip holes and should be a fine mist. Spray angle and size of spray cones from each of the four holes should be equal. Place a sheet of blotting paper 30 cm (about 12 inches) below injector and operate tester one stroke. The spray pattern should be a perfect circle.

SEAT LEAKAGE. No oil should drip from injector nozzle after an injection. Operate tester lever two or three strokes, then wipe nozzle tip dry. Slowly bring tester gage pressure to 1960 kPa (285 psi) below opening pressure and hold this pressure for 5 seconds. Drop of oil should not form nor should oil drip from nozzle tip. Renew or overhaul injector if drop of oil is noted.

OVERHAUL. Wipe all loose carbon from outside of injector. Hold nozzle body (1—Fig. W3-7) in soft jawed vise or injector fixture and unscrew nozzle retaining nut (7). Remove nozzle body and nozzle valve (6), stop plate (5), spring seat (4), spring (3) and adjusting shims (2). Be careful not to lose any of the shims and place all parts in clean diesel fuel or calibrating oil as they are removed.

Clean exterior surfaces with a brass wire brush. Soak parts in approved carbon solvent if necessary to loosen hard carbon deposits. Rinse parts in clean diesel fuel or calibrating oil immediately after cleaning to neutralize solvent. Clean nozzle tip opening from inside with a wood cleaning stick. Clean the four 0.25 mm (0.010 inch) spray holes in nozzle tip by using a 0.2 mm (0.008 inch) cleaning wire in wire holder.

Check to see that needle does not bind in nozzle body. Needle should slide smoothly from its own weight. Rotate needle to several different positions when checking for binding.

Reassemble injector under clean diesel fuel or calibrating oil. If opening pressure was correct, use shim pack removed on disassembly. Add or remove shims if necessary to change opening pressure. Tighten nozzle retaining nut to a torque of 49 N·m (36 ft.-lbs.). Retest injector and reinstall in cylinder head if injector meets test specifications.

INJECTION PUMP

To remove injection pump, first thoroughly clean pump, injector and return lines and surrounding area using suitable solvent and compressed air. Drain fuel tank and disconnect fuel supply and return lines from injection pump. Remove line from pump to fuel injector. Cap or plug all openings immediately to prevent entry of dirt. Un-

Illustrations Courtesy Teledyne Total Power

Wisconsin

bolt and remove fuel injection pump from engine crankcase. Take care not to lose or damage gasket shims located between pump flange and crankcase.

It is recommended that the fuel injection pump be disassembled only in a qualified diesel injection system repair shop.

To reinstall pump, reverse removal procedure using original shim gaskets or new shim gaskets equal in thickness to shim gasket present when pump was removed. Shim gaskets are available in thicknesses of 0.1, 0.2 and 0.3 mm (0.004, 0.008 and 0.012 inch). Move stop lever to halfway position to aid fitting injection pump rack ball (B—Fig. W3-12) into governor fork (F). Tighten injection pump nuts to a torque of 22 N·m (16 ft.-lbs.). Fill fuel tank and allow air to escape from fuel supply and return line. Injection line fitting at injector should be left loose until line is bled of air by cranking engine.

TIMING GEAR COVER

To remove timing cover (gear case cover) (19—Fig. W3-8), first remove driven mechanism from crankshaft and remove pto drive key from shaft. Some engines may have taper end for pto connector. Drain engine lubricant, remove fuel injection pump, cylinder head and push rods, then unbolt and remove cover from crankcase. Be careful not to lose or damage the shim washers (1—Fig. W3-9) located between camshaft gear (2) and timing cover.

With cover removed, clean old gasket material from cover and face of crankcase. Install new crankshaft oil seal (20—Fig. W3-8) in cover with lip of seal to inside of engine.

To install cover, reverse removal procedure using new gasket. Be sure to place shim washers (1—Fig. W3-9) found on disassembly on camshaft as they con-

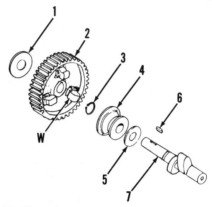

Fig. W3-9—Drawing of camshaft, gear and governor flyweight assembly.

1. Shim washers
2. Gear & weight assy.
3. Snap ring
4. Governor plate
5. Washer
6. Cam gear key
7. Engine camshaft

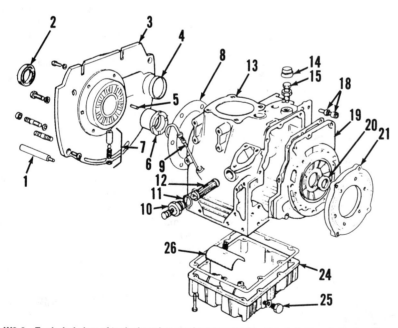

Fig. W3-8—Exploded view of typical engine crankcase assembly. Crankshaft end play is adjusted by varying shim gaskets (8) between crankcase and bearing support (3).

1. Cooling shroud spacer
2. Crankshaft oil seal
3. Bearing support
4. Plug (w/o electric start)
5. Main bearing dowel pin
6. Main bearing
7. Oil pump relief valve
8. Shim gaskets
9. Oil dipstick
10. Oil filter cap
11. Sealing ring
12. Oil filter cartridge
13. Crankcase
14. Cap
15. Governor lever screw
18. Stop lever adjusting screws (2)
19. Timing gear cover
20. Crankshaft oil seal
21. Pto adapter plate
24. Oil sump
25. Drain plug
26. Oil pickup screen

SMALL DIESEL

trol camshaft end play. If necessary to check camshaft end play, refer to CAMSHAFT paragraph before installing cover. Tighten cover retaining cap screws to a torque of 13 N·m (113 in.-lbs.). Adjust valve clearance after installing cylinder head. On models with tapered crankshaft end, tighten pto end bolt to a torque of 245 N·m (181 ft.-lbs.).

INJECTION PUMP CAMSHAFT

To remove injection pump camshaft and integral pump drive gear (4—Fig. W3-10), remove timing cover (gear end cover) as outlined in preceding paragraph, then withdraw injection gear and cam unit from engine oil pump cover (3). Cam lobe diameter should measure 35.95-36.05 mm (1.415-1.419 inches) with wear limit of 35.85 mm (1.411 inches). Injection gear shaft to housing oil clearance should be 0.07-0.09 mm (0.0027-0.0035 inch) with wear limit of 0.14 mm (0.0055 inch).

Reinstall injection pump camshaft being sure that timing marks on injection pump gear and crankshaft gear are aligned as shown in Fig. W3-11. It may be necessary to turn oil pump shaft so

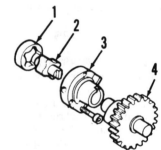

Fig. W3-10—Oil pump inner rotor (2) is driven by slot in inner end of injection pump gear and cam (4).

1. Oil pump outer rotor
2. Oil pump inner rotor
3. Oil pump cover
4. Drive gear & injection pump camshaft

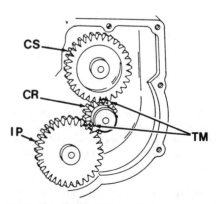

Fig. W3-11—Drawing showing timing gear alignment marks (TM). Single marked tooth of camshaft gear (CS) meshes between two marked teeth on crankshaft gear (CR). Single marked tooth of crankshaft gear meshes between two marked teeth on injection pump gear (IP).

SERVICE MANUAL

Wisconsin

that drive tang on oil pump shaft will engage slot in injection pump gear shaft and allow timing marks to be aligned.

TIMING GEARS

Timing gears (Fig. W3-11) are accessible for inspection after removing timing gear end cover. Fuel injection pump gear and cam can be withdrawn from oil pump cover. Crankshaft timing gear is integral part of crankshaft. Refer to CAMSHAFT paragraph for removal and installation of camshaft timing gear. Be sure all gear timing marks are aligned as shown in Fig. W3-11.

OIL PUMP AND RELIEF VALVE

The gerotor type oil pump (1, 2 and 3—Fig. W3-10) is located on engine crankcase and is driven from slot in inner end of injection pump camshaft (4). To remove oil pump, remove timing gear cover and the fuel injection pump camshaft and gear. Check pump inner rotor end play. Unbolt and remove oil pump cover (3) and rotors (1 and 2) from cylinder block.

Oil pump relief valve (7—Fig. W3-8) is located in the main bearing support at flywheel end of engine. To gain access to valve, remove flywheel and main bearing support plate.

Oil pump outer rotor to pump cover clearance should be 0.139-0.189 mm (0.005-0.007 inch) with wear limit of 0.339 mm (0.013 inch). Rotor end clearance should be 0.02-0.08 mm (0.0008-0.0031 inch) with wear limit of 0.13 mm (0.005 inch). Pump inner and outer rotor are serviced as a matched pair. Pump cover is available separately.

To install pump, coat parts with heavy oil or light grease such as Lubriplate and proceed as follows: Insert outer rotor in crankcase with chamfer to crankcase. Install internal rotor and pump cover, aligning cover oil passages with crankcase oil passages. Tighten pump cover bolts to a torque of 6 N·m (52 in.-lbs.). Check pump inner rotor end play. Complete reassembly by reversing removal procedures.

GOVERNOR

The governor flyweights (W—Fig. W3-9) are mounted on the inner side of the engine camshaft gear (2). Refer to CAMSHAFT paragraph for removal of camshaft and governor unit. Clearance of weight pins in camshaft gear should be 0.040-0.071 mm (0.001-0.002 inch) with wear limit of 0.12 mm (0.004 inch).

Refer to Fig. W3-12 for exploded view of governor linkage. Governor lever and pivot assembly (4) can be unbolted and removed from crankcase after unhooking governor spring (6). To remove internal stop lever (28) and throttle lever (7), remove external levers (18 and 22) and cover (17), then remove remainder of external parts. Remove snap rings (9 and 25) from lever shafts and remove speed control lever (7) and internal stop lever (28) from inside of crankcase gear housing. Renew "O" rings (8 and 29) and any damaged or worn parts.

Two different governor springs (6) are used. Spring with 36 coils should measure 58-60 mm (2.28-2.36 inches) free length and require a pull of 1 kg (2.2 pounds) to stretch spring to a length of 79-81 mm (3.11-3.19 inches). Free length of spring with 42 coils should be 60-63 mm (2.36-2.48 inches) and require a pull of 1 kg (2.2 pounds) to stretch spring to a length of 84-88 mm (3.30-3.46 inches).

CYLINDER, PISTON AND ROD UNITS

The cylinder, piston and connecting rod are removed from crankcase after removing cylinder head, oil sump and connecting rod cap. Remove piston and rod from bottom end of cylinder. Take care not to lose or damage shims (13—Fig. W3-5) located between cylinder flange and crankcase.

Piston is fitted with either two or three compression rings and one coil spring expander type oil control ring. See Fig. W3-13 or W3-14. Top compression ring may be installed either side up if side of ring does not have identification mark indicating top side of ring. Install second and, if so equipped, third compression rings with bevel on inside of ring up. Oil control ring may be installed either side up. Install spring expander, then oil ring with end gaps of expander and ring positioned 180 de-

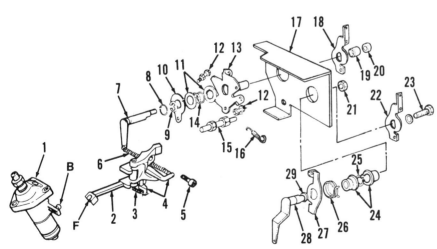

Fig. W3-12—Exploded view of governor linkage and controls.

B. Injection pump rack ball
F. Governor lever fork
1. Injection pump
2. Governor lever
3. Spring
4. Governor lever & pivot assy.
5. Pivot cap screws (2)
6. Governor spring
7. Speed control lever
8. "O" ring
9. Snap ring
10. Safety plate
11. Washers
12. Throttle lever adjusting screws
13. Throttle lever
14. Spring washers
15. Cover mounting stud
16. Spring
17. External linkage cover
18. External throttle lever
19. Nut
20. Spacer
21. Cover mounting nut
22. Stop control lever
23. Lever retaining bolt
24. Spacers
25. Snap ring
26. Stop lever spring
27. Stop plate
28. Internal stop lever
29. "O" ring

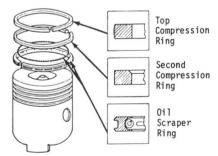

Fig. W3-13—Drawing showing correct installation of piston rings on piston fitted with three rings.

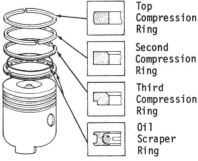

Fig. W3-14—Drawing showing correct installation of piston rings on piston fitted with four rings.

Illustrations Courtesy Teledyne Total Power

Wisconsin

grees apart. Equally space ring end gaps of rings around piston. Start ring end gap spacing by turning top ring so that it is 15 degrees from piston pin axis.

To reassemble, lubricate rod bearing, crankpin, piston, rings and cylinder with clean engine oil. Combustion chamber in piston crown must be toward injector as indicated in Fig. W3-15. Use suitable ring compressor to compress rings and push cylinder down over top end of piston. Place shims removed on disassembly on lower end of cylinder and install with combustion chamber in top of piston toward fuel injector. Install rod cap with reference marks (RM—Fig. W-16) on rod and cap aligned. Tighten rod bolts to a torque of 34 N·m (25 ft.-lbs.) on Models TRD1-375 and 380, and to a torque of 39 N·m (29 ft.-lbs.) on Models TRD1-480, 540 and 605.

Turn crankshaft so that piston is at TDC and using straightedge and feeler gage, measure clearance from top of piston to top end of cylinder. Clamp sleeve down to compress shims while taking measurement. Piston top to cylinder top clearance should be 0.25-0.35 mm (0.010-0.014 inch) on all models. Add or remove shims between cylinder and crankcase as necessary to bring clearance within specifications. Shims are available in thicknesses of 0.1 and 0.2 mm (0.004 and 0.008 inch).

PISTON, PIN AND RINGS

Piston skirt to cylinder bore clearance should be 0.07-0.10 mm (0.0027-0.004 inch) on Models TRD1-375 and 380, 0.06-0.09 mm (0.0023-0.0035 inch) on Model TRD1-480 and 0.065-0.095 mm (0.0025-0.0037 inch) on Models TRD1-540 and 605. Pistons and rings are available in oversizes of 0.5 mm (0.020 inch) and 1.0 mm (0.040 inch) as well as standard size.

Piston ring end gap should be 0.3-0.5 mm (0.012-0.020 inch) for compression rings and 0.25-0.40 mm (0.010-0.016 inch) for oil control ring.

Piston pin (4—Fig. W3-16) is retained by a snap ring (2) at each end of pin bore in piston. Piston pin bore to pin clearance should be 0.008-0.017 mm (0.0003-0.0007 inch) for Models TRD1-375, 380 and Type 9 Model 480; for all other Model 480 engines, clearance should be 0.002-0.008 mm (0.00008-0.0009 inch). Clearance for Models TRD1-540 and 605 should be 0.001-0.007 mm (0.00004-0.0003 inch).

CYLINDER

Standard cylinder bore diameter is 80.000-80.015 mm (3.1496-3.1502 inches) for Models TRD1-375 and 380, 90.000-90.015 mm (3.5433-3.5439 inches) for Models TRD1-480 and 540, and 95.000-95.015 mm (3.7402-3.7407 inches) for Model TRD1-605. Cylinder may be rebored and honed to 0.5 mm (0.020 inch) or 1.0 mm (0.040 inch) oversize if worn or scored.

CONNECTING ROD AND BEARINGS

The full floating piston pin (4—Fig. W3-16) rides directly in unbushed connecting rod bore. Crankpin end is fitted with a precision slip-in type insert (6). Connecting rod for Models TRD1-540 Type 0 and TRD1-605 Type 0 has a 4.5 mm (0.180 inch) lubrication hole.

Piston pin to connecting rod clearance should be 0.008-0.017 mm (0.0003-0.0007 inch) for Models TRD1-375 and 380 and Model TRD1-480 Type 9. All Model TRD1-480 except Type 9 should have pin to rod clearance of 0.023-0.038 mm (0.0009-0.0015 inch). For Models TRD1-540 and 605, pin to rod clearance should be 0.001-0.01 mm (0.00004-0.0004 inch).

Connecting rod to crankpin bearing oil clearance should be 0.015-0.07 mm (0.0006-0.0027 inch) for all models. Connecting rod crankpin bearing insert is available in undersizes of 0.25, 0.50 and 0.75 mm (0.010, 0.020 and 0.030 inch) as well as standard size.

CAMSHAFT

To remove camshaft, first drain engine lubricant, remove the oil filter, fuel injection pump, cylinder head and push rods and the timing gear cover (gear end cover). Be careful not to lose any of the shims (1—Fig. W3-9) from front end of camshaft. Invert engine so that cam followers (tappets) fall away from camshaft. Remove camshaft and governor assembly from cylinder block. Press camshaft from camshaft gear, being careful not to lose key (6) from shaft. Remove snap ring (3), governor plate (4) and washer (5) from camshaft. Refer to GOVERNOR paragraph for information on governor weights (W) and pins. To remove cam followers (23—Fig. W3-6), remove oil sump from bottom of cylinder block, then remove followers.

Camshaft lobe standard diameter is 34.25-34.35 mm (1.348-1.352 inches) and wear limit is 34 mm (1.338 inches). Camshaft end play is controlled by use of 0.3 and 0.5 mm (0.012 and 0.020 inch) thick shim washers placed on camshaft between front face of cam gear and timing gear cover. Clearance may be measured by placing straightedge across crankcase gasket surface and using feeler gage to measure distance between straightedge and shim washers. Clearance should be 0.1-0.2 mm (0.004-0.008 inch). Camshaft journal to crankcase and timing cover bores should be 0.040-0.071 mm (0.0015-0.0028 inch) with wear limit of 0.12 mm (0.004 inch).

Cam follower stem diameter is 8.98-8.99 mm (0.3535-0.3539 inch). Follower stem to bore clearance should be 0.01-0.035 mm (0.0004-0.0014 inch) with wear limit of 0.1 mm (0.004 inch).

To reinstall camshaft and followers, lubricate all parts thoroughly and reverse removal procedure.

FLYWHEEL

Flywheel is mounted on crankshaft taper and alignment is secured by a key. To remove flywheel, remove cooling fan shroud, flywheel nut (1—Fig. W3-17) and washer (2). On rope crank models, remove starter pulley (3). Remove air intake screen (5) and use suitable pullers and impact to remove flywheel from

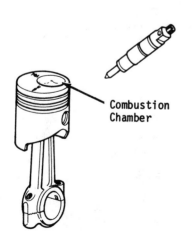

Fig. W3-15—Piston must be installed with combustion chamber toward fuel injector.

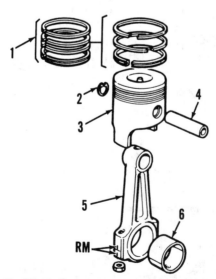

Fig. W3-16—Exploded view of piston, rings and connecting rod.
1. Piston ring set
2. Snap ring (2)
3. Piston
4. Piston pin
5. Connecting rod & cap
6. Bearing inserts
7. Rod cap nuts

SERVICE MANUAL

Wisconsin

crankshaft. On electric start models, check flywheel ring gear and renew if teeth are worn. Remove old ring gear with drift and hammer. Heat new gear until it will drop over shoulder on flywheel and allow to cool in place. Alternator rotor is bolted to inside face of flywheel/fan assembly.

To reinstall flywheel, be sure key is in place in crankshaft and place flywheel on taper with keyway in flywheel hub aligned with key. Install washer and retaining nut and tighten nut to a torque of 275 N·m (202 ft.-lbs.).

CRANKSHAFT AND MAIN BEARINGS

The crankshaft is supported in two sleeve type bearings (11—Fig. W3-17 and 6—Fig. W3-18), one in crankcase wall at pto end of crankshaft and the other in the bearing support at flywheel end of crankshaft. Crankshaft end play is controlled by flanges on the inboard side of each main bearing and by a shim gasket (8—Fig. W3-18) between the flywheel end bearing support plate (3) and crankcase.

To remove crankshaft, proceed as follows: Remove timing cover from pto end of engine. Remove oil sump, cylinder head and cylinder, piston and rod unit. Remove flywheel cooling shroud, flywheel and electric starting motor. Check crankshaft end play with dial indicator before proceeding farther. Crankshaft end play should be 0.1-0.2 mm (0.004-0.008 inch). Remove flywheel key, then unbolt and remove support plate from flywheel end of block. Check thickness of shim gasket between support plate and crankcase. Carefully withdraw crankshaft from cylinder block.

Standard crankshaft main journal diameter for all models is 41.984-42.000 mm (1.6529-1.6535 inches). Crankpin standard diameter is 39.994-40.010 mm (1.5746-1.5752 inches) for Models TRD1-375, 380 and 480, and 44.994-45.010 mm (1.7714-1.7720 inches) for Models TRD1-540 and 605. Crankshaft main journal to bearing oil clearance should be 0.030-0.086 mm (0.0012-0.0034 inch) and rod bearing to crankpin clearance should be 0.015-0.070 mm (0.0006-0.0027 inch). Check crankshaft and bearings for scoring, excessive wear or other damage. It is recommended that the oil passage plug (8—Fig. W3-17) be removed, oil passages cleaned and a new plug be installed whenever crankshaft is removed for engine overhaul.

Crankshaft main bearings are available in undersizes of 0.25, 0.50, 0.75 and 1.0 mm (0.010, 0.020, 0.030 and 0.040 inch) as well as standard size. Also, main bearings are available with outside diameter 1.0 mm (0.040 inch) oversize and inside diameter 1.0 mm (0.040 inch) undersize. Shim gasket for use between bearing support plate and crankcase are available in thicknesses of 0.1, 0.2 and 0.3 mm (0.004, 0.008 and 0.012 inch).

Bearings can be removed using suitable bearing driver and press. To install new main bearings, heat crankcase and/or bearing support to a temperature of 70°-80° C (158°-176° F) and push bearing into place. A notch (N—Fig. W3-17) in the bearing (11) flange must align with pin in crankcase or bearing support. To install bearing with 1.0 mm (0.040 inch) oversize diameter, the crankcase and bearing support must be assembled and the bearing bores align bored 1.0 mm (0.040 inch) oversize; standard bearing bore is 50.01-50.03 mm (1.9689-1.9697 inches).

If bearings are renewed, install crankshaft and flywheel end bearing support and check end play. Vary shim gasket thickness to obtain proper end play.

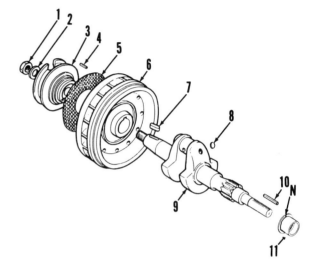

Fig. W3-17—Exploded view of crankshaft and flywheel components. Flywheel end bearing support (3—Fig. W3-18), bearing (6) and related parts are not included in this view.

N. Bearing flange notch
1. Flywheel nut
2. Washer
3. Rope start pulley
4. Start pulley dowel
5. Air intake screen
6. Flywheel
7. Flywheel key
8. Oil passage plug
9. Crankshaft
10. Pto end key
11. Main bearing

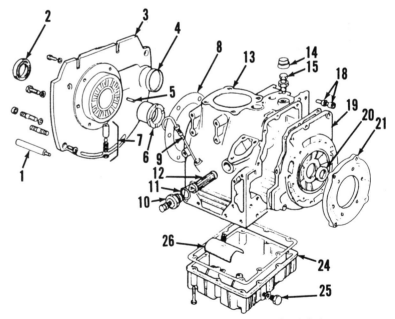

Fig. W3-18—Exploded view of typical engine crankcase assembly. Crankshaft end play is adjusted by varying shim gaskets (8) between crankcase and bearing support (3).

1. Cooling shroud spacer
2. Crankshaft oil seal
3. Bearing support
4. Plug (w/o electric start)
5. Main bearing dowel pin
6. Main bearing
7. Oil pump relief valve
8. Shim gaskets
9. Oil dipstick
10. Oil filter cap
11. Sealing ring
12. Oil filter cartridge
13. Crankcase
14. Cap
15. Governor lever screw
18. Stop lever adjusting screws (2)
19. Timing gear cover
20. Crankshaft oil seal
21. Pto adapter plate
24. Oil sump
25. Drain plug
26. Oil pickup screen

Illustrations Courtesy Teledyne Total Power

Wisconsin

SMALL DIESEL

CRANKSHAFT OIL SEALS

The pto end crankshaft seal (20—Fig. W3-18) is mounted in timing gear cover (19) and the flywheel end seal (2) is located in main bearing support (3). The seals should be removed whenever timing cover and/or bearing support are removed. Install seal with lip to inside. Lubricate seal and take care that seal is not damaged on crankshaft keyway when reinstalling cover or support.

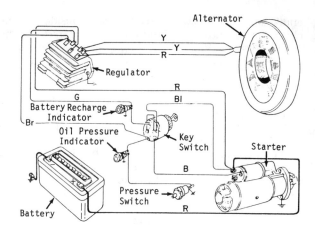

Fig. W3-19—Schematic diagram of electrical system for electric start models.
B. Black
Bl. Blue
Br. Brown
G. Green
R. Red
Y. Yellow

ELECTRICAL SYSTEM

All models except Model TRD1-375 are equipped with a 12 volt electrical starting and battery charging system. Refer to Fig. W3-19 for schematic diagram of electrical system and to Fig. W3-20 for exploded view. The flywheel alternator (5 and 6) can be serviced after removing engine flywheel.

With battery discharged (below 13 volts), charging current should be about 15 amperes at 3000 engine rpm. With battery at full charge, regulator limit is 14.5 volts and charging current should drop to 2 amperes. If test results are not within limits, renew voltage regulator. If renewing regulator does not correct problem, replace alternator assembly. Stator and rotor are serviced as a pair.

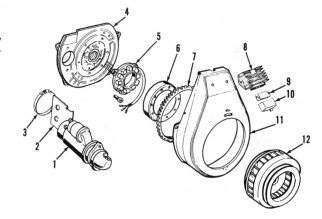

Fig. W3-20—Exploded view showing electric start components.
1. Starting motor
2. Motor bracket
3. Motor clamp
4. Main bearing support
5. Alternator stator
6. Alternator rotor
7. Starter ring gear
8. Voltage regulator
9. Wiring connector
10. Connector cover
11. Flywheel air shroud
12. Flywheel & fan assy.

Illustrations Courtesy Teledyne Total Power

SERVICE MANUAL

WISCONSIN

Model	No. Cyls.	Bore	Stroke	Displ.
TRD2-755	2	80 mm (3.15 in.)	75 mm (2.95 in.)	754 cc (46.01 cu. in.)
TRD2-850	2	85 mm (3.35 in.)	75 mm (2.95 in.)	851 cc (51.93 cu. in.)
TRD2-955	2	90 mm (3.54 in.)	75 mm (2.95 in.)	954 cc (58.21 cu. in.)
TRD2-1130	2	92 mm (3.62 in.)	85 mm (3.35 in.)	1130 cc (68.95 cu. in.)
TRD2-1205	2	95 mm (3.74 in.)	85 mm (3.35 in.)	1205 cc (73.53 cu. in.)

All models are four-stroke cycle, vertical cylinder, air-cooled direct injection type diesel engines. Engine rotation is counter-clockwise when facing pto shaft.

MAINTENANCE

AIR CLEANER

All models are equipped with an oil bath air cleaner. Air filter element should be removed and cleaned in suitable solvent after each 50 hours of operation, or daily when operating in dusty or dirty conditions. Discard oil from air cleaner cup, wash cup in solvent and refill to proper level with clean engine oil.

COOLING SYSTEM

All models are cooled by a centrifugal fan that is an integral part of engine flywheel. A flywheel fan shroud and cowling directs the cool air flow around cylinders. It is important that the flywheel screen and the cylinder fins be kept clean to avoid engine overheating and resulting engine damage. Manufacturer recommends daily check of cooling system.

LUBRICATION

Recommended engine lubricant is HD Series 3 motor oil or API classification CC/CD. If using single grade engine oil, use SAE 10W in ambient temperatures below 0° C (30° F), SAE 20W in temperatures 0°-15° C (30°-59° F), SAE 30 in temperatures 15°-30° C (59°-86° F) and SAE 40 in temperatures above 30° C (86° F). Multigrade oils may be used in all temperature ranges except above 30° C (86° F).

Crankcase lubricant capacity is 3 liters (3.17 quarts). Engine oil and filter should be changed after every 100 hours of operation.

FUEL SYSTEM

Fuel system for Models TRD2-755, 850 and 955 is shown in Fig. W4-1; fuel system for Models TRD2-1130 and 1205 is shown in Fig. W4-2.

FUEL FEED PUMP. All models are equipped with a diaphragm type fuel feed pump (4—Fig. W4-1 or W4-2) actuated by a cam on engine camshaft through a push rod. The pump is equipped with a hand lever for manual bleeding of fuel system.

FUEL FILTER. The fuel filter should be renewed after each 200 hours of operation. Filter (5—Fig. W4-1) on Models TRD2-755, 850 and 955 screws onto threaded nipple in bottom of fuel tank. Models TRD2-1130 and 1205 have a fil-

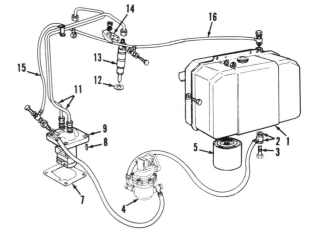

Fig. W4-1—Fuel system for Models TRD2-755, 850 and 955. Automatic bleed-off line (15) for filter (5) and injection pump (9) is connected to injector fuel return line (16).

1. Fuel tank
2. Sealing washers
3. Banjo fitting bolts
4. Fuel feed pump
5. Fuel filter
7. Shim gasket
8. Pump locating dowel
9. Fuel injection pump
11. Injection lines
12. Seal washer
13. Fuel injector
14. Injector clamp
15. Auto bleed-off line
16. Injector return line

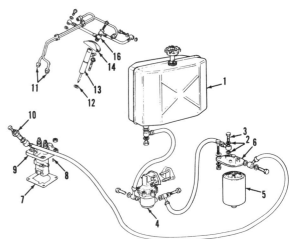

Fig. W4-2—Fuel system for Models TRD2-1130 and 1205. It is necessary to manually bleed the fuel filter and pump on these models. Bleed screws are located in top of filter base (6) and pump supply line banjo bolt (10).

1. Fuel tank
2. Sealing washers
3. Banjo fitting bolts
4. Fuel feed pump
5. Fuel filter
6. Filter base
7. Shim gasket
8. Pump locating dowel
9. Fuel injection pump
10. Bleed-off fitting
11. Fuel injection lines
12. Seal washer
13. Fuel injectors
14. Injector clamps
16. Injector return line

Illustrations Courtesy Teledyne Total Power

Wisconsin

399

Wisconsin

SMALL DIESEL

ter base (6—Fig. W4-2) and filter (5) mounted on engine crankcase.

BLEED FUEL SYSTEM. Models TRD2-755, 850 and 955 have automatic bleed of fuel injection pump through line (15—Fig. W4-1). Manual bleeding is not required.

To bleed air from Models TRD2-1130 and 1205, loosen bleed screws on fuel filter base (6—Fig. W4-2) and on injection pump inlet (10). Operate fuel transfer pump (4) manual lever until air is no longer present in fuel flow, then tighten the bleed screws.

On all models, it may be necessary to loosen injector line fittings at injectors and crank engine until fuel flows from loosened fittings. Tighten fittings and start engine.

INJECTION PUMP TIMING

Injection pump timing is set at factory by shim gaskets (7—Fig. W4-1 or W4-2) between the pump mounting flange and engine crankcase. To maintain timing, always save the shim gaskets when removing fuel injection pump and install new gaskets of same thickness when reinstalling pump. Specifications for timing fuel injection pump are not available. Gaskets are available in thicknesses of 0.1, 0.2 and 0.3 mm (0.004, 0.008 and 0.012 inch).

GOVERNOR

Engine stop control and governor throttle levers are located adjacent to fuel injection pump as shown in Fig. W4-3. Exploded view of governor controls and linkage is shown in Fig. W4-10. Throttle lever adjusting screws (H and L—Fig. W4-3) are located in flanged plate attached to crankcase. Engine rated speed is shown on engine nameplate.

REPAIRS

TIGHTENING TORQUES

When assembling engine, refer to the following special tightening torques:

Connecting rod . . . . 37-39 N·m
    (27-29 ft.-lbs.)
Crankcase bolts. . . . . 13 N·m
    (113 in.-lbs.)
Cylinder head nuts. . . . . 49 N·m
    (36 ft.-lbs.)
Flywheel nut. . . . . 275 N·m
    (202 ft.-lbs.)
Gear case cover bolts . . . . 13 N·m
    (113 in.-lbs.)
Injection pump nuts. . . . . 22 N·m
    (16 ft.-lbs.)
Injector nuts . . . . 22 N·m
    (16 ft.-lbs.)
Oil pump cover bolts . . . . 6 N·m
    (52 in.-lbs.)
Oil sump cover bolts . . . . 13 N·m
    (113 in.-lbs.)
Pto end bolt . . . . 245 N·m
    (180 ft.-lbs.)

VALVE ADJUSTMENT

Valves can be adjusted after removing rocker arm cover. Clearance between valve stem end and rocker arm with engine cold should be 0.15 mm (0.006 inch) for both intake and exhaust. Turn crankshaft so that number 1 piston is at TDC on compression stroke (both push rods free to turn) and check clearance for both valves using a feeler gage. If necessary, turn rocker arm adjusting screw (14—Fig. W4-4) to obtain recommended clearance. Turn crankshaft so that number 2 piston is at TDC on compression stroke and adjust valves on that cylinder.

CYLINDER HEAD

The cylinder heads are retained by long stud bolts threaded into crankcase at lower end of cylinders. To remove cylinder heads, first remove air cleaner, exhaust muffler, cowling, rocker arm covers, rocker arm assemblies and push rods and the fuel injectors. Remove external oiling line from crankcase to cylinder heads. Remove cylinder head nuts and washers and lift cylinder head assemblies from studs. Do not remove cylinders. Remove the push rod tubes.

With cylinder heads removed, clean cylinder and head gasket surfaces and carefully inspect cylinder head. Install

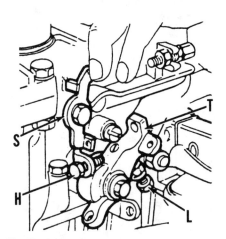

Fig. W4-3—View of stop lever (S) and throttle lever (T).
  H. High speed adjusting screw
  L. Low idle adjusting screw
  S. Stop lever
  T. Throttle lever

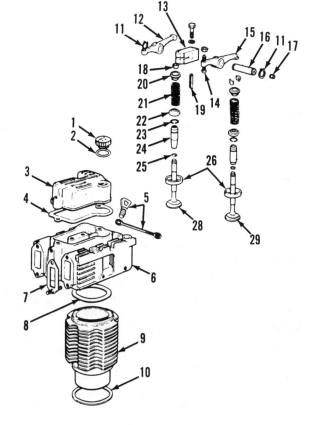

Fig. W4-4—Exploded view of cylinder head valve mechanism and cylinder. Piston to cylinder head clearance is adjusted by varying shim gasket (10) thickness. Cold weather starting is improved by removing plug (5) and adding a small quantity of engine oil to air intake to increase compression.
1. Breather cap
2. Sealing washer
3. Rocker arm cover
4. Cover gasket
5. Intake plug & tether
6. Cylinder head
7. Manifold gasket
8. Head gasket
9. Cylinder
10. Shim gasket
11. Snap rings
12. Intake rocker arm
13. Rocker shaft support
14. Adjusting screw
15. Exhaust rocker arm
16. Rocker arm shaft
17. Plug
18. Split valve locks
19. Support dowel pin
20. Valve spring retainers
21. Valve springs
22. Valve spring seats
23. Guide locating rings
24. Valve guides
25. Valve stem seals
26. Valve seats
28. Intake valve
29. Exhaust valve

SERVICE MANUAL

Wisconsin

fuel injector in head with sealing washer(s) and check protrusion of nozzle tip from cylinder head surface. Tip should extend 3.75-4.25 mm (0.147-0.167 inch) from head for type "B" (see Fig. W4-6) injectors or 2.25-2.75 mm (0.088-0.108 inch) for type "C" injectors (see Fig. W4-7). Add or remove sealing washers to bring measurement within specifications. Refer to VALVE SYSTEM paragraph for information on cylinder head assembly components.

To install cylinder head, reverse removal procedure using new head gasket (8—Fig. W4-4) and push rod tube "O" rings. Lubricate push rod tube "O" rings and install cylinder head guiding the push rod tube into place. Install cylinder head washers and nuts. Tighten nuts finger tight and align cylinder head manifold surfaces using a straightedge (B) as shown in Fig. W4-5. Tighten cylinder head nuts following sequence shown in Fig. W4-5 in 8 N·m (6 ft.-lbs.) increments to a final torque of 49 N·m (36 ft.-lbs.). Install fuel injector with number of washers determined as outlined in preceding paragraph. Install push rods and rocker arm assemblies, being sure that rocker arm supports (13—Fig. W4-4) fit down over aligning dowel pins (19). Adjust valve clearance before installing rocker arm cover.

VALVE SYSTEM

Valve face and seat angle is 45 degrees for both intake and exhaust valves. Valve seats are renewable and oversize seats are available so that valve seat counterbore in cylinder head can be reconditioned and maintain interference fit of seat to head. With valves installed, top of valve heads should be recessed 0.9-1.1 mm (0.035-0.043 inch) below surface of cylinder head. Renew valve and/or valve seat if valve head recession exceeds 1.8 mm (0.070 inch).

Desired valve stem to guide clearance for Models TRD2-755, 850 and 955 is 0.03-0.05 mm (0.0012-0.0020 inch) for intake valves and 0.045-0.065 mm (0.0017-0.0025 inch) for exhaust valves. Clearance wear limit is 0.1 mm (0.004 inch) for both intake and exhaust. For Models TRD2-1130 and 1205, desired valve stem to guide clearance is 0.02-0.04 mm (0.0008-0.0015 inch) for intake valve and 0.04-0.065 mm (0.0015-0.0025 inch) for exhaust valve. Maximum allowable valve stem to guide clearance is 0.08 mm (0.0031 inch) for intake valve and 0.1 mm (0.004 inch) for exhaust valves. Valve guides are renewable.

To remove valve guide, press guide out toward top of head. To install new guide, position locating ring (23—Fig. W4-4) in groove on valve guide, then press guide downward into head until locating ring contacts head. Valve guides are available with 0.10 mm (0.004 inch) oversize outside diameter. Use oversize outside diameter guides if standard guide does not fit tightly in head. Grind valve seats after installing guides to obtain specified valve head recess.

Intake and exhaust valve springs are alike. Limit of variation from following valve spring specifications is 10 percent.

Models TRD2-755, 850 & 955
Valve spring free length .... 35.6 mm
(1.40 in.)
Spring pressure
@ 25 mm (0.98 in) .... 14.6 kg
(32.2 lbs.)
Spring pressure
@ 17.5 mm (0.69 in.) .... 25 kg
(55.1 lbs.)

Models TRD2-1130 & 1205
Valve spring free length .... 50.1 mm
(1.97 in.)
Spring pressure
@ 34.8 mm (1.37 in.) .... 16.3 kg
(35.9 lbs.)
Spring pressure
@ 24.8 mm (0.98 in.) .... 27 kg
(59.5 lbs.)

Inspect rocker arms, rocker arm shaft and push rods for excessive wear or other damage. Rocker arms are retained on shaft by a snap ring at each end of shaft. Rocker arm to shaft clearance should be 0.03-0.055 mm (0.0012-0.0022 inch) with wear limit of 0.15 mm (0.006 inch).

INJECTORS

NOTE: Injectors are positioned in cylinder head by one or more sealing washers at bottom of injectors to adjust protrusion of injector tip from combustion surface of cylinder head.

REMOVE AND REINSTALL. Prior to removing injectors, thoroughly clean injector lines, injection pump and injectors using suitable solvent and compressed air. Remove injector fuel return line and unscrew injection line fittings at each end of lines and remove lines from pump and injectors. Immediately plug all openings to prevent entry of dirt in fuel system. Remove the injector clamp plates, then remove injectors from cylinder heads. If injector sealing washer(s) was/were not removed with injector, carefully remove washer(s) from injector bore.

When installing injectors, be sure injector bores in cylinder heads are clean. Using light grease, stick sealing washers on nozzle tip of injectors and insert injectors in cylinder heads. Be sure to use same number of washers that were removed with each injector. Install the clamp plates and stud nuts and tighten the nuts evenly to a torque of 22 N·m (16 ft.-lbs.). Install injection lines, leaving fittings at injectors loose until system is bled. Connect fuel return line to injectors.

TESTING. A complete job of testing and adjusting the injector requires use of special test equipment. Only clean, approved testing oil should be used in tester tank. Injector should be tested for opening pressure, spray pattern and seat leakage.

WARNING: Fuel emerges from injector with sufficient force to penetrate the skin. When testing injector, keep yourself clear of nozzle spray.

Before conducting test, operate tester lever until fuel flows from tester line, then attach injector. Close tester gage valve and operate tester lever a few quick strokes to purge air from injector and to be sure nozzle valve is not stuck.

OPENING PRESSURE. Open valve to tester gage and operate tester lever slowly while observing gage reading. Pressure at which nozzle opens varies with make and type of nozzle used; refer to the following information:

Fig. W4-5—View showing alignment bar (B) bolted to cylinder head manifold surfaces and cylinder head nut tightening sequence.

Illustrations Courtesy Teledyne Total Power

Wisconsin SMALL DIESEL

Type "B" nozzle (Fig. W4-6) should open at a pressure of 21575-22555 kPa (3128-3270 psi) and type "C" nozzle (Fig. W4-7) should open at 22065-23045 kPa (3200-3340 psi). Opening pressure is adjusted by adding or removing shims (2—Fig. W4-6 or W4-7).

SPRAY PATTERN. Spray pattern should be equal from all four nozzle tip holes and should be a fine mist. Spray angle and size of spray cones from each of the four holes should be equal. Place a sheet of blotting paper 30 cm (about 12 inches) below injector and operate tester one stroke. The spray pattern should be a perfect circle.

SEAT LEAKAGE. No oil should drip from injector nozzle tip after an injection. Operate tester lever two or three strokes, then wipe nozzle tip dry. Slowly bring tester gage pressure to 1960 kPa (285 psi) below opening pressure and hold this pressure for 5 seconds. Drop of oil should not form nor should oil drip from nozzle tip. Renew or overhaul injector if drop of oil is noted.

OVERHAUL. Wipe all loose carbon from outside of injector. Hold nozzle body in soft jawed vise or injector fixture and unscrew nozzle retaining nut (7—Fig. W4-6 or W4-7). Remove nozzle valve assembly (6), stop plate (5), spring seat (4), spring (3) and adjusting shims (2) from nozzle body (1). Be careful not to lose any of the shims and place all parts in clean diesel fuel or calibrating oil as they are removed. If more than one injector is being disassembled, do not intermix parts.

Clean exterior surfaces with a brass wire brush. Soak parts in approved carbon solvent if necessary to loosen hard carbon deposits. Rinse parts in clean diesel fuel or calibrating oil immediately after cleaning to neutralize solvent. Clean nozzle tip opening from inside with a wood cleaning stick. Clean the four 0.29 mm (0.0114 inch) spray holes in type "B" "Altecna" nozzle tip by using a 0.20 mm (0.008 inch) and then a 0.25 mm (0.010 inch) cleaning wire in wire holder. On other type "B" "Bosch" and "Omap" nozzles with 0.28 mm (0.011 inch) spray holes and all type "C" nozzles with 0.25 mm (0.010 inch) spray holes, clean the spray holes using a 0.20 mm (0.008 inch) cleaning wire only.

Check to see that needle does not bind in nozzle (6). Needle should slide smoothly from its own weight. Rotate needle to several different positions when checking for binding.

Reassemble injector under clean diesel fuel or calibrating oil. If opening pressure was correct, use shim pack removed on disassembly. Add or remove shims if necessary to change opening pressure. Tighten nozzle retaining nut to a torque of 49-63 N·m (36-47 ft.-lbs.) on type "B" injectors and 49 N·m (36 ft.-lbs.) on type "C" injectors. Retest injector and reinstall in cylinder head if injector meets test specifications.

INJECTION PUMP

To remove injection pump, first thoroughly clean pump, injector lines, fuel injectors and surrounding area using suitable solvent and compressed air. Close off fuel supply or drain fuel tank. Disconnect fuel supply and return line from injection pump. Remove lines from pump to fuel injectors. Cap or plug all openings immediately to prevent entry of dirt. Remove cover plate and stop lever assembly from side of crankcase below fuel injection pump. Remove the cotter pin (14—Fig. W4-8) and unhook tie rod (13) from injection pump rack (R). Unbolt and remove fuel injection pump (P) from engine crankcase. Take care not to lose or damage gasket shims located between pump flange and crankcase.

It is recommended that the fuel injection pump be disassembled only in a qualified diesel injection system repair shop.

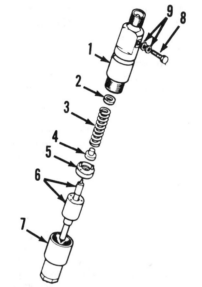

Fig. W4-7—Exploded view of type "C" fuel injector assembly. Fuel injection line attaches to top of injector body (1) and return line attaches to side of body with banjo bolt (8).

1. Injector body
2. Shims
3. Spring
4. Spring seat
5. Spacer
6. Nozzle and valve
7. Nozzle cap
8. Banjo bolt
9. Sealing washers

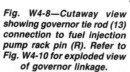

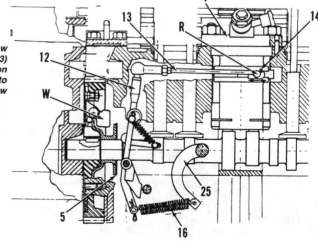

Fig. W4-8—Cutaway view showing governor tie rod (13) connection to fuel injection pump rack pin (R). Refer to Fig. W4-10 for exploded view of governor linkage.

P. Injection pump
W. Governor flyweights (3)
5. Governor plate
12. Governor linkage arm
13. Tie rod
14. Cotter pin
16. Governor spring
25. Throttle arm

Fig. W4-6—Exploded view of type "B" fuel injector assembly. Fuel injection line attaches to side of injector body (1) and return line attaches to top of injector with banjo bolt (8).

1. Injector body
2. Shims
3. Spring
4. Spring seat
5. Spacer
6. Nozzle & valve
7. Nozzle cap
8. Banjo bolt
9. Sealing washers

SERVICE MANUAL

To reinstall pump, reverse removal procedure and use new shim gaskets of same thickness as present when pump was removed. Shim gaskets are available in thicknesses of 0.1, 0.2 and 0.3 mm (0.004, 0.008 and 0.012 inch). Attach tie rod to injection pump rack pin and install new cotter pin. Tighten injection pump nuts to a torque of 22 N·m (16 ft.-lbs.). Refer to BLEED FUEL SYSTEM paragraph in MAINTENANCE section.

FUEL FEED PUMP

A diaphragm type fuel feed pump is located on injection pump side of cylinder block and is actuated by a push rod (push pin) riding on engine camshaft. To remove the pump, close fuel supply or drain fuel tank, clean pump and surrounding area and disconnect fuel lines from pump. Unbolt and remove pump and push rod from engine crankcase. Fuel pump is available as a complete assembly only. Renew pump if diaphragm is leaking or if it will not develop pressure.

Pump push rod length should be 34.0-34.2 mm (1.3386-1.33465 inches); renew pin if worn. With push rod in retracted position, it should protrude 1.05-1.45 mm (0.041-0.057 inch) from gasket face. Shim gaskets for fuel pump are available in thicknesses of 0.20 and 1.0 mm to bring protrusion of push rod within specifications.

INJECTION PUMP CAMSHAFT

Injection pump is actuated by two lobes at the center of the engine camshaft. Refer to CAMSHAFT paragraph for service information.

TIMING GEAR COVER

To remove timing cover (7—Fig. W4-9), first remove flywheel fan shroud (1), flywheel (6) and flywheel key. Drain engine lubricant, then unbolt and remove cover from crankcase. Be careful not to lose shim washers (1—Fig. W4-11) located between camshaft gear and timing cover.

With cover removed, clean old gasket material from cover and face of crankcase. Install new crankshaft oil seal in cover with lip of seal to inside of engine.

To install cover, reverse removal procedure using new gasket. If necessary to check camshaft end play, refer to CAMSHAFT paragraph before installing cover. Tighten cover retaining cap screws to a torque of 13 N·m (113 in.-lbs.). Place flywheel key in crankshaft and install flywheel with keyway aligned with crankshaft key. Tighten flywheel nut (2—Fig. W4-9) to a torque of 275 N·m (202 ft.-lbs.).

TIMING GEARS

Timing gears are accessible for inspection after removing timing gear end cover. Refer to CAMSHAFT paragraph for removal and installation of camshaft timing gear and to CRANKSHAFT AND BEARINGS paragraph for information on crankshaft gear. Be sure single marked tooth of crankshaft gear is meshed between two marked teeth of camshaft gear.

OIL PUMP AND RELIEF VALVE

The gerotor type oil pump (10, 11, 12 and 13—Fig. W4-15) is located in the engine crankcase and is driven from slot (S—Fig. W4-11) end of engine camshaft. To remove oil pump, remove cover at pto end of engine and withdraw inner and outer rotor.

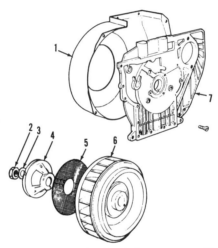

Fig. W4-9—View showing timing gear cover (7), flywheel (6) and flywheel air shroud (1). Crankshaft oil seal is mounted in timing gear cover.
1. Flywheel air shroud
2. Flywheel nut
3. Washer
4. Rope start pulley
5. Air intake screen
6. Flywheel & fan assy.
7. Timing gear cover

Oil pump rotor to crankcase bore clearance should be 0.09-0.14 mm (0.004-0.006 inch) with wear limit of 0.29 mm (0.011 inch). Check clearance with feeler gage inserted between outer rotor and crankcase bore. Rotor end clearance should be 0.01-0.05 mm (0.0004-0.0020 inch) with wear limit of 0.10 mm (0.004 inch). Check clearance using feeler gage with straightedge across crankcase gasket surface. Pump inner and outer rotor are serviced as a matched pair. Pump cover is available separately.

To install pump, coat parts with heavy oil or light grease such as Lubriplate and proceed as follows: Insert outer rotor in crankcase. Install internal rotor with drive tab aligned with slot in camshaft end. Install cover and tighten pump cover bolts to a torque of 6 N·m (52 in.-lbs.).

Oil pump relief valve in engines built prior to March 1983 was located in crankcase upper half. On all later models, the valve (20—Fig. W4-15) is located under the engine oil filter (21). To remove relief valve, unscrew oil filter and used puller bolt threaded into valve to pull valve from crankcase.

CAMSHAFT AND BEARINGS

Refer to Fig. W4-11 for exploded view of camshaft and governor flyweight assembly. Camshaft end play is controlled by shim washers (1) between cam gear (2) and timing gear cover.

To remove camshaft, drain engine lubricant, remove fuel injection pump, fuel feed pump and plunger, flywheel, timing cover (gear end cover), engine oil pump, rocker arm covers, rocker arms and push rods. Take care not to lose shim washers (1) from face of cam gear. Invert engine so that cam followers (tappets) fall away from camshaft. Unbolt and remove lower half of crankcase

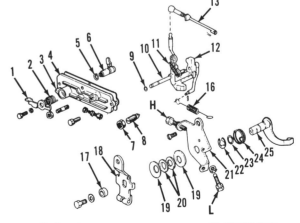

Fig. W4-10—Exploded view of governor linkage. Refer to Fig. W4-8 for cutaway view showing linkage installed. Adjusting screw (8) is mounted in cylinder block lower half; all other parts are located in upper half of cylinder block.
H. High speed adjusting screw
L. Low idle speed screw
1. Stop lever
2. Stop lever spring
3. Linkage cover plate
4. Gasket
5. "O" ring
6. Internal stop lever
7. Jam nut
8. Adjusting screw
9. "O" ring
10. Linkage shaft
11. Spring
12. Governor arm assy.
13. Tie rod
16. Governor spring
17. Spacer
22. Snap ring
23. "O" ring
24. Internal lever spring
25. Internal throttle lever
18. Throttle lever
19. Washers
20. Washers
21. Throttle stop plate

Wisconsin

Wisconsin — SMALL DIESEL

from upper half. Remove camshaft and governor assembly from crankcase. Press camshaft from camshaft gear. Remove cam followers (12), identifying them as they are removed so they may be installed in original position.

Camshaft intake and exhaust lobe diameter should measure 29.95-30.00 mm (1.1791-1.1811 inches) with wear limit of 29.70 mm (1.1693 inches) for Models TRD2-755, 850 and 955. For Models TRD2-1130 and 1205, valve lobes should measure 30.52-30.57 mm (1.2016-1.2036 inches) with wear limit of 30.25 mm (1.1909 inches). Injection pump lobes should measure 28.39-28.43 mm (1.1177-1.1193 inches) on all models.

Camshaft end play is controlled by use of 0.3 and 0.5 mm (0.012 and 0.020 inch) thick shim washers placed on camshaft between front face of cam gear and timing gear cover. To measure end play, place straightedge across crankcase gasket surface and use a feeler gage to measure distance between straightedge and shim washers. Clearance between straightedge and washers should be 0.1-0.2 mm (0.004-0.008 inch).

Clearance between camshaft journals and crankcase and timing cover bores should be 0.017-0.047 mm (0.0007-0.0018 inch) with wear limit of 0.1 mm (0.004 inch).

Tappet stem diameter is 11.98-11.99 mm (0.4717-0.4720 inch). Tappet stem to bore clearance should be 0.01-0.038 mm (0.0004-0.0015 inch) with wear limit of 0.1 mm (0.004 inch).

To reinstall camshaft and followers, lubricate all parts thoroughly and reverse removal procedure.

GOVERNOR

The governor weights (3—Fig. W4-11) are mounted on the engine camshaft gear (2). Refer to CAMSHAFT paragraph for removal of governor weights and weight plate (4). Refer to Fig. W4-10 for exploded view of governor linkage. Clearance of weight pins in camshaft gear should be 0.015-0.048 mm (0.0006-0.0019 inch) and wear limit is 0.1 mm (0.004 inch).

Governor linkage can be removed from upper half of crankcase after separating upper and lower crankcase halves. Governor tie rod (13—Fig. W4-10) length from center of ball socket to center of injection pump rack pin hole should be measured and if renewed, length of new tie rod should be adjusted to same length. Renew any damaged or worn parts. Reassemble using new "O" rings (5, 9 and 23) on stop lever arm (6), governor linkage shaft (10) and throttle arm (25).

CYLINDER, PISTON AND ROD UNITS

The cylinder, piston and connecting rod are removed from block after removing cylinder heads, oil sump cover and connecting rod caps. Remove piston and rod from bottom end of cylinder. Take care not to lose or damage shims (10—Fig. W4-4) located between cylinder flange and crankcase.

Pistons for Models TRD2-755, 850 and 955 are fitted with two compression rings and one coil spring expander type oil control ring. Models TRD2-1130 and 1205 have three compression rings and one coil expander type oil ring. Top compression ring may be installed either side up if side of ring does not have an identification mark indicating top side of ring. See Fig. W4-13. Install second and, if so equipped, third compression rings with bevel on inside of ring up. Oil control ring may be installed either side up. Install spring expander, then oil ring with end gaps of expander and ring positioned 180 degrees apart. Equally space ring end gaps of rings around piston. Start ring end gap spacing by turning top ring so that it is 15 degrees from piston pin axis.

To reassemble, lubricate rod bearing, crankpin, piston, rings and cylinder with clean engine oil. Use suitable ring compressor to compress rings and push cylinder down over top end of piston. Place shims removed on disassembly on lower end of cylinder and install in cylinder block. On Models TRD2-755, 850 and 955, side of rod with oil groove at piston pin end must be toward fuel injection pump side of engine. Install rod cap with reference numbers on rod and cap aligned. Tighten rod bolts to a torque of 37-39 N·m (27-29 ft.-lbs.).

Turn crankshaft so that piston is at TDC and using straightedge and feeler

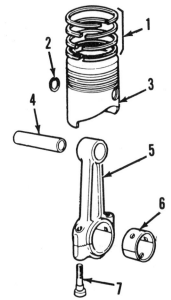

Fig. W4-12—Exploded view of piston and rod unit for Models TRD2-1130 and 1205. Other models have only three piston rings.

1. Piston ring set
2. Snap rings
3. Piston
4. Piston pin
5. Connecting rod & cap
6. Crankpin bearing
7. Rod bolts

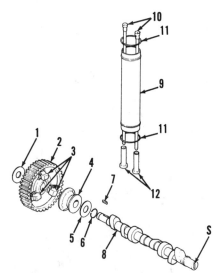

Fig. W4-11—Exploded view of camshaft, followers, push rod tube and push rods. Camshaft gear (2) and governor weights (3) are serviced as a complete assembly only.

S. Oil pump drive slot
1. Shim washers
2. Camshaft gear
3. Governor weights
4. Governor plate
5. Thrust washer
6. Snap ring
7. Gear key
8. Camshaft
9. Push rod tube
10. Push rods
11. "O" rings
12. Cam followers

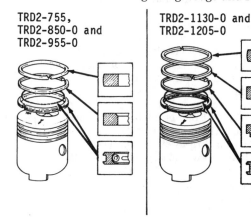

Fig. W4-13—Cross-sectional view of piston rings showing correct installation.

Illustrations Courtesy Teledyne Total Power

SERVICE MANUAL

Wisconsin

gage, measure clearance from top of piston to top end of cylinder. Hold sleeve down to compress shims while taking measurement. Piston top to cylinder top clearance should be 0.25-0.35 mm (0.010-0.014 inch) on all models. Add or remove shims between cylinder and crankcase as necessary to bring clearance within specifications. Shims are available in thicknesses of 0.1 and 0.2 mm (0.004 and 0.008 inch).

PISTONS, PINS, AND RINGS

Piston skirt to cylinder bore clearance should be 0.07-0.10 mm (0.0027-0.004 inch) on all models. Pistons and rings are available in oversizes of 0.5 mm (0.020 inch) and 1.0 mm (0.040 inch) as well as standard size for all models. On all models except Model TRD2-1205, pistons and rings are also available in 1.5 mm (0.060 inch) oversize.

Piston ring end gap should be 0.3-0.5 mm (0.012-0.020 inch) with wear limit of 0.8 mm (0.031 inch) for compression rings and 0.25-0.40 mm (0.010-0.016 inch) with wear limit of 0.7 mm (0.027 inch) for oil control ring.

Piston pin is retained by a snap ring (2—Fig. W4-12) at each end of pin bore in piston. It is very important that snap rings be installed with sharp edge to outside (away from pin). Piston pin bore to pin clearance should be 0.002-0.008 mm (0.00008-0.0003 inch) with wear limit of 0.05 mm (0.002 inch) for Models TRD2-755, 850 and 955. Clearance for Models TRD2-1130 and 1205 should be 0.001-0.01 mm (0.00004-0.0004 inch) with wear limit of 0.06 mm (0.0023 inch).

CYLINDER

Standard cylinder bore diameter is 85.000-85.015 mm (3.3465-3.3470 inches) for Models TRD2-755 and 850, 90.000-90.015 mm (3.5433-3.5439 inches) for Model TRD2-955, 92.000-92.015 mm (3.6220-3.6226 inches) for Model TRD2-1130 and 95.000-95.015 mm (3.7402-3.7407 inches) for Model TRD2-1205. Cylinder may be rebored and honed to 0.5 mm (0.020 inch) or 1.0 mm (0.040 inch) oversize on all models if worn or scored. On all models except Model TRD2-1205, cylinder may also be rebored and honed to 1.5 mm (0.060 inch) oversize.

CONNECTING RODS AND BEARINGS

The full floating piston pin rides directly in unbushed connecting rod bore. Crankpin end is fitted with a precision slip-in type insert.

Piston pin to connecting rod clearance should be 0.023-0.038 mm (0.0009-0.0015 inch) with wear limit of 0.07 mm (0.003 inch) for Models TRD2-755, 850 and 955. Models TRD2-1130 and 1205 should have pin to rod clearance of 0.001-0.007 mm (0.00004-0.0003 inch) with wear limit of 0.05 mm (0.002 inch).

Connecting rod to crankpin bearing oil clearance should be 0.020-0.072 mm (0.0008-0.0028 inch) for all models with wear limit of 0.17 mm (0.0067 inch). Connecting rod crankpin bearing insert is available in undersizes of 0.25, 0.50 and 0.75 mm (0.010, 0.020 and 0.030 inch) as well as standard size.

When assembling rod to piston on Models TRD2-755, 850 and 955, lubrication groove on connecting rod piston pin end must face injection pump side of engine. Be sure pin retaining snap rings (2—Fig. W4-12) are installed with sharp edge outward (away from piston pin).

FLYWHEEL

Flywheel is mounted on crankshaft taper and alignment is secured by a key. To remove flywheel, remove cooling fan shroud (1—Fig. W4-9) and flywheel nut (2). Use suitable pullers to remove flywheel from crankshaft. On electric start models, check flywheel ring gear and renew if teeth are worn. Alternator rotor is bolted to inside face of flywheel/fan assembly.

To reinstall flywheel, be sure key is in place in crankshaft and place flywheel on taper with keyway in flywheel hub aligned with key. Install washer and retaining nut and tighten nut to a torque of 275 N·m (202 ft.-lbs.).

CRANKSHAFT AND MAIN BEARINGS

The crankshaft is supported in three precision insert type bearings (3—Fig. W4-14). Bearings are located in webs of upper and lower crankcase halves (7 and 8—Fig. W4-15). Crankshaft end play is controlled by thrust surfaces of center main journal and center web of cylinder block halves.

To remove crankshaft, proceed as follows: Remove electric starting motor, flywheel and timing cover from flywheel end of engine. Remove oil sump cover, cylinder heads and the cylinder, piston and connecting rod units. Check crankshaft end play with dial indicator before proceeding farther. Crankshaft end play should be 0.1-0.2 mm (0.004-0.008 inch). Remove crankcase bolts and lift upper half from lower half. Carefully withdraw crankshaft from cylinder block and remove main bearings from upper and lower crankcase halves. Check bearings and crankshaft journals for excessive wear or scoring.

Standard crankshaft main journal diameter for all models is 45.005-45.015 mm (1.7718-1.7722 inches). Crankpin standard diameter is 44.994-45.010 mm (1.7714-1.7720 inches). Crankshaft main journal to bearing oil clearance should be 0.010-0.060 mm (0.0004-0.0023 inch) with wear limit of 0.15 mm (0.006 inch). Rod bearing to crankpin clearance should be 0.020-0.072 mm (0.0008-0.0028 inch). Check crankshaft and bearings for scoring, excessive wear or other damage. It is recommended that the two

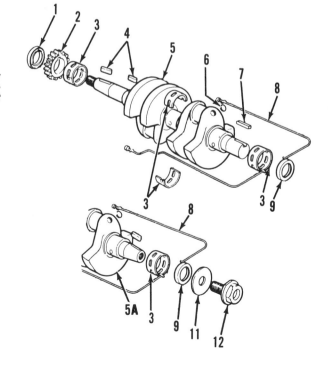

Fig. W4-14—Exploded view of crankshaft and main bearings. Pto end of crankshaft may be straight or tapered.
1. Oil seal
2. Crankshaft timing gear
3. Main bearings
4. Keys
5. Crankshaft (straight end)
5A. Crankshaft (tapered end)
6. Oil passage plugs
7. Straight pto key
7A. Tapered pto key
8. Cylinder block gasket
9. Crankshaft oil seal
11. Tapered end washer
12. Tapered end cap screw

Illustrations Courtesy Teledyne Total Power

Wisconsin

SMALL DIESEL

oil passage plugs (6—Fig. W4-14) be removed, oil passages cleaned and new plugs be installed whenever crankshaft is removed for engine overhaul.

Crankshaft main bearings are available in undersizes of 0.25, 0.50, 0.75 mm and 1.00 mm (0.010, 0.020, 0.030 and 0.040 inch) as well as standard size. Also, main bearings are available with outside diameter 1.0 mm (0.040 inch) oversize with inside diameter 1.0 mm (0.040 inch) undersize. Cylinder block halves must be assembled and the main journal bores be align bored to accommodate the oversize outside diameter main bearing inserts.

If crankshaft timing gear is worn or damaged, press shaft from gear. To install new gear, heat gear dry or in oil bath to a temperature of 70°-80° C (158°-176° F) and install with keyway aligned with key in crankshaft and with timing mark side of gear toward flywheel end of shaft.

Install crankshaft using new crankshaft oil seals and rubber sealing parts. Stick the eight "O" rings (18—Fig. W4-15) in grooves next to crankshaft main bearings. Install the rubber side seals (8—Fig. W4-14) in grooves in upper half of cylinder block using rubber adhesive. Place bearing inserts in cylinder block halves with aligning tabs and oil holes properly positioned. Lubricate crankshaft main journals and bearings and carefully lower crankshaft into upper half of cylinder block with timing marks on crankshaft gear and camshaft gear in mesh. Lubricate pto end crankshaft oil seal (9) and place seal in position on crankshaft and in upper cylinder block half. Install lower half of crankcase. Install the retaining bolts and washers and tighten bolts to a torque of 13 N·m (113 in.-lbs.).

Complete reassembly of engine by reversing disassembly procedure.

CRANKSHAFT OIL SEALS

The pto end crankshaft seal (9—Fig. W4-14) is mounted in crankcase and the flywheel end seal (1) is located in timing gear cover. The seals should be removed whenever timing cover is removed or the lower and upper crankcase halves are separated. Flywheel end seal can be renewed after removing flywheel and pto end seal can be renewed after removing drive coupling and coupling flange (9—Fig. W4-15). Carefully remove old seals to avoid damage to crankshaft, crankcase or timing gear cover. Lubricate seal lips and install seals with lip to inside. Take care that seals are not damaged on crankshaft keyways when installing.

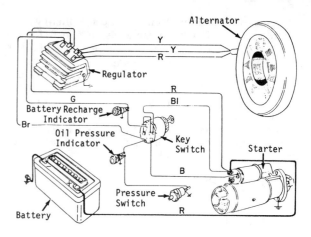

Fig. W4-16—Schematic diagram of electrical system for models with electric starting.

B. Black
Bl. Blue
Br. Brown
G. Green
R. Red
Y. Yellow

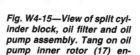

Fig. W4-15—View of split cylinder block, oil filter and oil pump assembly. Tang on oil pump inner rotor (17) engages slot (S—Fig. W4-11) in end of camshaft.

1. Timing cover gasket
2. Oil drain plug
4. Timing cover dowel
5. Gear cover plate
6. Gasket
7. Upper cylinder block half
8. Lower cylinder block half
9. Coupling flange
10. Oil pump cover
11. "O" ring
12. Oil pump inner rotor
13. Oil pump outer rotor
14. Oil pick-up screen
15. Oil sump cover
16. Gasket
17. Cylinder block dowel pins
18. "O" rings (8)
19. Pipe nipple
20. Relief valve assy.
21. Oil filter

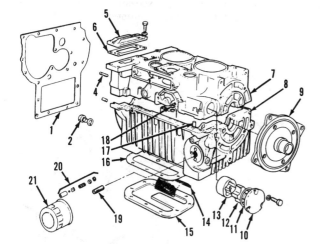

ELECTRICAL SYSTEM

All models are equipped with a 12 volt electrical starting and battery charging system. Refer to Fig. W4-16 for schematic diagram of electrical system. The flywheel alternator can be serviced after removing engine flywheel.

With battery discharged (below 13 volts), charging current should be about 15 amperes at 3000 engine rpm. With battery at full charge, regulator limit is 14.5 volts and charging current should drop to 2 amperes. If test results are not within limits, renew voltage regulator. If renewing regulator does not correct problem, replace alternator assembly. Stator and rotor are serviced as a pair.

SERVICE MANUAL

WISCONSIN

TELEDYNE TOTAL POWER
P. O. Box 181160
Memphis, Tennessee 38181-1160

Model	No. Cyls.	Bore	Stroke	Displ.
WD1-340	1	75 mm (2.953 in.)	78 mm (3.701 in.)	345 cc (21 cu. in.)
WD1-350	1	75 mm (2.953 in.)	78 mm (3.071 in.)	345 cc (21 cu. in.)
WD1-430	1	84 mm (3.307 in.)	78 mm (3.071 in.)	432 cc (26.3 cu. in.)
WD1-430	1	84 mm (3.307 in.)	78 mm (3.071 in.)	432 cc (26.3 cu. in.)
WD1-660	1	95 mm (3.740 in.)	95 mm (3.740 in.)	673 cc (41 cu. in.)
WD1-670	1	95 mm (3.740 in.)	95 mm (3.740 in.)	673 cc (41 cu. in.)
WD1-750	1	100 mm (3.937 in.)	95 mm (3.740 in.)	746 cc (45.5 cu. in.)

Engines covered in this section are air-cooled, single-cylinder, four-stroke diesel engines. Crankshaft rotation is counterclockwise at pto end. Metric fasteners are used throughout the engine.

MAINTENANCE

LUBRICATION

Recommended engine oil is SAE 10W or 5W-20 for temperatures below minus 18°C (0°F), SAE 10W or 10W-30 for temperatures between minus 18°C (0°F) and 0°C (32°F), SAE 20 or 15W-40 for temperatures between 0°C (32°F) and 20°C (68°F) and SAE 40 or 15W-40 for temperatures above 20°C (68°F). API oil classification should be CD. Oil sump capacity is 0.95 liters (1 quart) for Models WD1-340, WD1-350 and WD1-430 or 2.8 liters (3 quarts) for all other engines. Manufacturer recommends renewing oil after every 100 hours of operation. Oil should be drained while engine is hot.

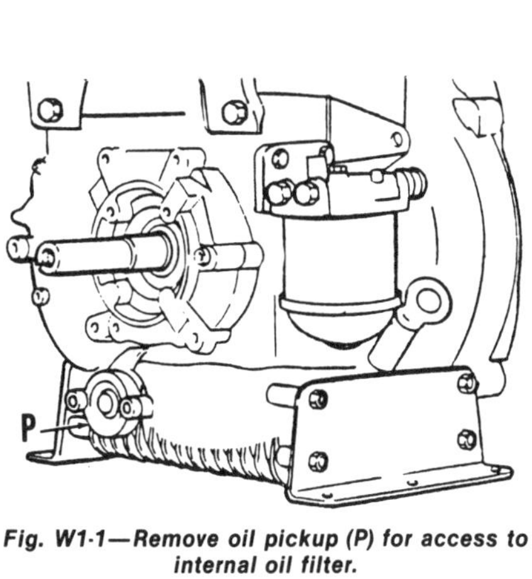

A reusable oil filter element is mounted internally on the oil pickup tube (P – Fig. W1-1). manufacturer recommends cleaning filter element after every 200 hours of operation. Renew element if damaged or if it cannot be cleaned. An externally mounted filter element is also used on WD1-660, WD1-670 and WD1-750 engines. Filter element should be renewed after every 200 hours of operation.

ENGINE SPEED ADJUSTMENT

Idle speed is adjusted by turning idle speed screw (I – Fig. W1-2) and high speed is adjusted by turning high speed screw (H). Idle speed should be 1500-1800 rpm while maximum governed speed under load may be 2200 or 3000 rpm depending on engine application.

FUEL SYSTEM

FUEL FILTERS. A fuel filter is located in the fuel tank on all models and some engines may be equipped with an external fuel filter. Renew fuel filters after every 200 hours of operation or sooner if required.

BLEED FUEL SYSTEM. On gravity-feed fuel systems, a fuel return line and check valve are connected to the fuel injection pump as well as the fuel supply line. Fuel supply line should bleed air automatically up to fuel injection pump on gravity flow systems. To bleed high pressure injection line, loosen fitting for high pressure line at injector, then rotate engine crankshaft to operate fuel injection pump until air-free fuel flows from injection line. Retighten injection line.

On engines equipped with a fuel transfer pump, loosen fuel supply line fitting on injection pump, then rotate engine crankshaft to operate fuel transfer pump until air-free fuel flows from fuel line. Retighten fuel supply line and loosen fitting for high pressure injection line at injector. Rotate engine crankshaft to operate fuel injection pump until air-free fuel flows from injection line. Retighten injection line.

Fig. W1-1—Remove oil pickup (P) for access to internal oil filter.

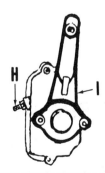

Fig. W1-2—View showing location of high speed adjusting screw (H). Idle speed screw is located behind speed control lever at (I).

Wisconsin   SMALL DIESEL

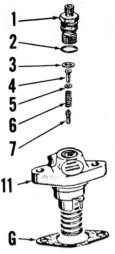

Fig. W1-3—Partial exploded view of fuel injection pump. Injection timing is adjusted using shim gaskets (G).

1. Delivery valve holder
2. "O" ring
3. Gasket
4. Spring guide
5. Shim
6. Spring
7. Delivery valve
11. Fuel injection pump

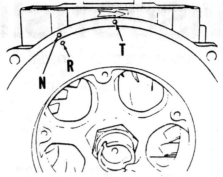

Fig. W1-4—Fuel injection should occur when flywheel reference mark (R) is aligned with crankcase mark (N). The piston is at top dead center when flywheel mark (R) is aligned with crankcase mark (T).

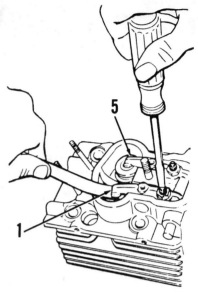

Fig. W1-5—Refer to text for adjustment of exhaust (1) and intake (5) valve clearance.

INJECTION PUMP TIMING

Injection pump timing is adjusted using shim gaskets (G–Fig. W1-3) between pump body and mounting surface of crankcase. Refer to INJECTION PUMP section for pump removal and installation.

To check injection pump timing, unscrew high pressure injection line from injection pump delivery valve holder (1). Unscrew delivery valve holder (1) and remove spring (6), shim (5), spring guide (4) and delivery valve (7), then screw delivery valve holder (1) back into pump. Move throttle control to full speed position. Rotate engine in normal direction (counterclockwise at pto) so piston is on compression stroke. Fuel will flow out of delivery valve holder – it may be necessary to connect a gravity-flow fuel tank to injection pump if engine is not so equipped. Stop engine rotation at moment fuel ceases to flow. Timing mark (R–Fig. W1-4) should be within 3 mm (1/8 inch) of injection timing mark (N) on crankcase. To advance injection timing, remove shim gaskets (G–Fig. W1-3); install shim gaskets to retard injection timing. Reinstall removed pump parts after checking timing.

REPAIRS

TIGHTENING TORQUES

Refer to the following table for special tightening torques. All fasteners are metric.

WD1-340, WD1-350 and WD1-430
Connecting rod .... 34 N·m
  (25 ft.-lbs.)
Crankcase .... 30 N·m
  (22 ft.-lbs.)
Cylinder head .... 44 N·m
  (33 ft.-lbs.)
Fan .... 16 N·m
  (12 ft.-lbs.)
Flywheel .... 176 N·m
  (130 ft.-lbs.)
Gear cover .... 25 N·m
  (18 ft.-lbs.)
Injection pump .... 25 N·m
  (18 ft.-lbs.)
Injection retainer plate ...... 10 N·m
  (7 ft.-lbs.)

WD1-660, WD1-670 and WD1-750
Connecting rod .... 49 N·m
  (36 ft.-lbs.)
Crankcase
  M8 .... 30 N·m
    (22 ft.-lbs.)
  M12 .... 69 N·m
    (51 ft.-lbs.)
Cylinder head .... 69 N·m
  (51 ft.-lbs.)
Fan .... 16 N·m
  (12 ft.-lbs.)
Flywheel .... 588 N·m
  (434 ft.-lbs.)
Gear cover .... 25 N·m
  (18 ft.-lbs.)
Injection pump .... 30 N·m
  (22 ft.-lbs.)
Injection retainer
  plate .... 10 N·m
    (7 ft.-lbs.)

VALVE ADJUSTMENT

Manufacturer recommends checking valve clearance after every 200 hours of operation. Valve clearance may be adjusted after removing rocker arm cover and turning crankshaft to position piston at TDC on compression stroke. Turn rocker arm adjusting screw (Fig. W1-5) until appropriate size feeler gage can be inserted between rocker arm and valve stem end. Recommended clearance with engine cold is 0.010 mm (0.004 inch) for intake and 0.15 mm (0.006 inch) for exhaust on WD1-340, WD1-350 and WD1-430, or 0.20 mm (0.008 inch) for both valves on WD1-660, WD1-670 and WD1-750 engines.

CYLINDER HEAD AND VALVE SYSTEM

R&R AND OVERHAUL. To remove cylinder head, disconnect high pressure and return fuel injection lines, then immediately cap fuel lines. Remove muffler, air cleaner, exhaust and intake manifolds. Remove rocker arm cover, unscrew cylinder head retaining nuts and remove cylinder head.

Valve face angle for both valves is 45½ degrees and valve seat angle is 45 degrees. Valve seats are renewable and must be installed with head heated to 200°C (392°F). There should be an interference fit of 0.079-0.120 mm (0.0031-0.0047 inch) between seat insert and cylinder head counterbore on WD1-340, WD1-350 and WD1-430 engines, or 0.127-0.159 mm (0.0050-0.0063 inch) on WD1-660, WD1-670 and WD1-750 engines. Renew valve if head margin is less than 0.6 mm (0.024 inch). With valves installed, top of valve heads should be recessed 0.60-0.95 mm (0.024-0.037 inch) below surface of cylinder head.

On WD1-340, WD1-350 and WD1-430 engines, valve stem diameter is 6.945-6.960 mm (0.2734-0.2740 inch). Valve guide inner diameter is

SERVICE MANUAL

Wisconsin

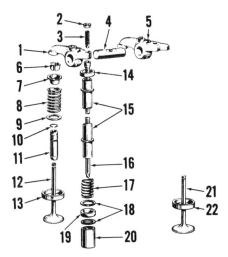

Fig. W1-6—Exploded view of valve train. On Models WD1-660, WD1-670 and WD1-750, an inner spring is located inside valve spring (8) and snap ring (10) is not used.

1. Exhaust rocker arm
2. Locknut
3. Adjuster
4. Rocker arm shaft
5. Intake rocker arm
6. Keys
7. Spring retainer
8. Valve spring
9. Washer
10. Snap ring
11. Valve guide
12. Exhaust valve
13. Exhaust valve seat
14. Seal
15. Push rod tube
16. Push rod
17. Spring
18. Washer
19. Seal
20. Cam follower
21. Intake valve
22. Intake valve seat

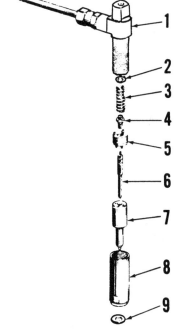

Fig. W1-7—Exploded view of injector.

1. Body
2. Shim
3. Spring
4. Push piece
5. Spacer
6. Valve
7. Nozzle
8. Nozzle holder
9. Gasket

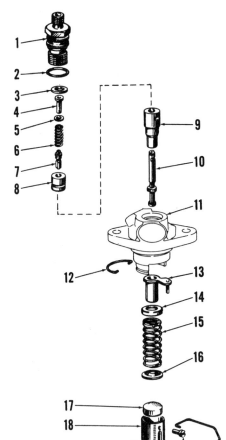

Fig. W1-8—Exploded view of fuel injection pump.

1. Delivery valve holder
2. "O" ring
3. Gasket
4. Spring guide
5. Shim
6. Spring
7. Delivery valve
8. Delivery valve seat
9. Barrel
10. Plunger
11. Body
12. Clip
13. Control sleeve
14. Spring seat
15. Spring
16. Spring retainer
17. Spacer
18. Tappet
19. Pin
20. Roller
21. Pin
22. Clip

7.000-7.015 mm (0.2756-0.2762 inch). Valve stem-to-guide clearance should be 0.040-0.070 mm (0.0016-0.0028 inch) with a wear limit of 0.2 mm (0.008 inch).

Valve guides are renewable. Heat cylinder head to 200°C (392°F) when removing or installing valve guides. To remove valve guide, press guide out top of cylinder head. When installing valve guide, note that guide may have a locating flange or a snap ring (10–Fig. W1-6). Press guide into head from rocker arm side of head so guide flange bottoms or snap ring seats in groove.

Reassemble by reversing disassembly procedure. Install new "O" rings on compression release shaft. Install new upper seals on push rod tubes. Tighten cylinder head nuts to 69 N·m (51 ft.-lbs.).

INJECTOR

REMOVE AND REINSTALL. To remove injector, first clean dirt from injector, injection line, return line and cylinder head. Disconnect return line and high pressure injection line and immediately cap or plug all openings. Unscrew retainer plate and remove injector and copper washer.

Reverse removal procedure to reinstall injector. Install a new copper washer (9–Fig. W1-7). Tighten injector retaining plate nuts to 10 N·m (7 ft.-lbs.).

TESTING. A suitable test stand is required to check injector operation. Only clean, approved testing oil should be used to test injector.

WARNING: Fuel leaves the injection nozzle with sufficient force to penetrate the skin. When testing, keep yourself clear of nozzle spray.

Attach injector to tester and operate pump lever a few quick strokes to purge air from injector and to make sure nozzle valve is not stuck. When operating properly during test, injector nozzle will emit a buzzing sound and cut off quickly with no leakage at seat. Spray pattern should be uniform and well atomized.

Opening pressure on WD1-340, WD1-350 and WD1-430 should be 20100-21085 kPa (2915-3055 psi). Opening pressure on WD1-660, WD1-670 and WD1-750 should be 24025-25000 kPa (3485-3625 psi). Opening pressure is adjusted by varying the thickness of shims (2–Fig. W1-7).

Operate tester to maintain a pressure of 17600 kPa (2550 psi). If a drop forms at nozzle tip within ten seconds, nozzle valve is not seating and injector must be overhauled or renewed.

OVERHAUL. Clamp injector body (1–Fig. W1-7) in a vise with nozzle pointing upward, unscrew nozzle holder nut (8) then remove injector components shown in Fig. W1-7. Thoroughly clean all parts in a suitable solvent. Clean inside orifice of nozzle tip with a wooden cleaning stick. When reassembling injector, make certain all components are clean and wet with clean diesel fuel oil. Tighten nozzle holder nut (8) to 49 N·m (36 ft.-lbs.).

Wisconsin

SMALL DIESEL

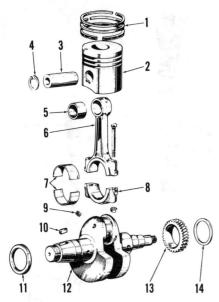

Fig. W1-9—Exploded view of crankshaft assembly.

1. Piston rings
2. Piston
3. Piston pin
4. Snap ring
5. Bushing
6. Connecting rod
7. Bearing
8. Rod cap
9. Plug
10. Key
11. Rubber ring
12. Crankshaft
13. Gear
14. Thrust washer

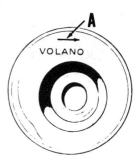

Fig. W1-10—Arrow (A) on piston crown must point towards flywheel.

reach inside side cover opening and place a magnet against guide pin (21). Detach clip (22) and use magnet to remove guide pin (21). Slide tappet (18) out of crankcase.

Refer to Fig. W1-8 for an exploded view of fuel injection pump. The injection pump should be tested and overhauled by a shop qualified in diesel fuel injection pump repair.

Reverse removal procedure to reinstall pump. Tighten pump retaining screws to 25 N·m (18 ft.-lbs.) on Models WD1-340, WD1-350 and WD1-430 or to 30 N·m (22 ft.-lbs.) on all other models. If pump is renewed or overhauled, or original shim gaskets are not used, refer to INJECTION PUMP TIMING section and adjust pump timing.

INJECTION PUMP

R&R AND OVERHAUL. To remove fuel injection pump, disconnect fuel lines and immediately cap all openings. Refer to GOVERNOR section and detach side control cover. Unscrew pump retaining screws and remove fuel injection pump. Do not lose shim gaskets (G–Fig. W1-3). Pump components (17 through 22–Fig. W1-8) will remain in crankcase. If tappet assembly must be removed,

PISTON, PIN, RINGS AND CYLINDER

R&R AND OVERHAUL. Remove cylinder head as previously outlined. If piston and cylinder do not require service but only removal, proceed as follows: Rotate crankshaft so piston is at top dead center. Lift cylinder up until piston pin is exposed, remove piston pin retainer, withdraw piston pin and remove cylinder and piston as a unit. If piston, rings or cylinder requires service, lift cylinder off crankcase, remove piston pin retainer, withdraw piston pin and remove piston.

Check cylinder bore for excessive wear or other damage. If cylinder taper or out-of-round exceeds 0.15 mm (0.006 inch), cylinder should be rebored and fitted with 0.5 mm (0.020 inch) or 1.0 mm (0.040 inch) oversize piston and rings.

Piston diameter is measured 12 mm (½ inch) from bottom of piston skirt on WD1-340, WD1-350 and WD1-430 or 18 mm (¾ inch) from bottom of skirt on all other engines. Standard pistons and cylinders are coded and matched according to size. Pistons are stamped with an A or B and cylinders are color coded with a red or yellow or a blue or green paint dot on outside of cylinder barrel. Refer to Table 1 for piston and corresponding cylinder sizes. Maximum allowable clearance between piston and cylinder is 0.2 mm (0.008 inch).

Check piston ring end gap with ring positioned squarely in cylinder bore. On WD1-340, WD1-350 and WD1-430 engines, recommended end gap is 0.25-0.40 mm (0.010-0.016 inch) for top ring, 0.30-0.45 mm (0.012-0.018 inch) for center ring and 0.25-0.40 mm (0.010-0.016 inch) for oil control ring. On WD-660, WD1-670 and WD-750 engines, ring end gap should be 0.35-0.55 mm (0.014-0.022 inch) for both compression rings and 0.25-0.40 mm (0.010-0.016 inch) for oil control ring.

Check side clearance of rings in their respective grooves with a feeler gage. On WD1-340, WD1-350 and WD1-430 engines, recommended side clearance is 0.050-0.082 mm (0.002-0.003 inch) for center ring and 0.040-0.072 mm (0.0016-0.0028 inch) for oil control ring. Maximum allowable side clearance is 0.15 mm (0.006 inch) for center ring and 0.13 mm (0.005 inch) for oil control ring. On WD1-660, WD1-670 and WD1-750 engines, groove side clearance should be 0.072-0.100 mm (0.0028-0.0040 inch) for center ring and 0.040-0.072 mm (0.0015-0.0028 inch) for oil control ring. Maximum allowable clearance is 0.25 mm (0.010 inch) for all rings.

Piston pin clearance in rod bushing should be 0.02-0.04 mm (0.0008-0.0016 inch) with a wear limit of 0.10 mm (0.004

TABLE 1—Standard piston and cylinder matching codes and dimensions.

Model	Code	Piston O.D.	Cylinder I.D.
WD1-340, WD1-350	A Red or Yel B Blue or Grn	74.930 - 74.940 mm (2.9500 - 2.9504 in.) 74.940 - 74.950 mm (2.9504 - 2.9508 in.)	75.000 - 75.010 mm (2.9527 - 2.9531 in.) 75.010 - 75.020 mm (2.9531 - 2.9535 in.)
WD1-430	A Red or Yel B Blue or Grn	83.925 - 83.935 mm (3.3041 - 3.3045 in.) 83.935 - 83.945 mm (3.3045 - 3.3049 in.)	84.000 - 84.011 mm (3.3071 - 3.3075 in.) 84.011 - 84.022 mm (3.3075 - 3.3079 in.)
WD1-660, WD1-670	A Red or Yel B Blue or Grn	94.920 - 94.930 mm (3.7370 - 3.7374 in.) 94.930 - 94.940 mm (3.7374 - 3.7378 in.)	95.000 - 95.011 mm (3.7402 - 3.7406 in.) 95.011 - 95.022 mm (3.7406 - 3.7410 in.)
WD1-750	A Red or Yel B Blue or Grn	99.890 - 99.900 mm (3.9326 - 3.9330 in.) 99.900 - 99.910 mm (3.9330 - 3.9335 in.)	100.000 - 100.011 mm (3.9370 - 3.9374 in.) 100.011 - 100.022 mm (3.9374 - 3.9379 in.)

SERVICE MANUAL

Wisconsin

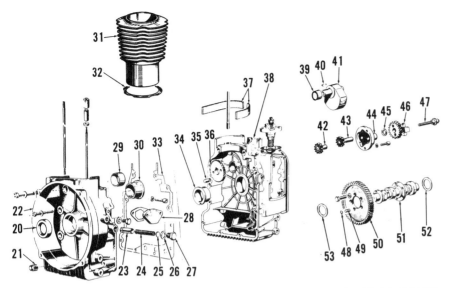

Fig. W1-11—Typical exploded view of crankcase. Balancer (41) is not used on Models WD1-340, WD1-350 and WD1-430.

20. Oil seal
21. Oil drain plug
22. Crankcase half
23. Oil pressure relief valve
24. Spring
25. Shim
26. Washer
27. Plug
28. Cover
29. Camshaft bushing
30. Main bearing
31. Cylinder
32. Shim gasket
33. Gasket
34. Main bearing
35. Dowel
36. Balancer gear
37. Felt
38. Crankcase half
39. Bushing
40. Key
41. Balancer
42. Driven gear
43. Drive gear & shaft
44. Oil pump body
45. Thrust washer
46. Balancer assy.
47. Push rod
48. Snap ring
49. Pin
50. Camshaft gear
51. Camshaft
52. Shims
53. Thrust washer

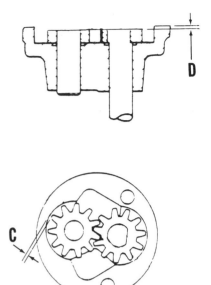

Fig. W1-13—Maximum oil pump gear side clearance (C) is 0.15 mm (0.006 inch) and maximum gear depth (D) is 0.10 mm (0.004 inch).

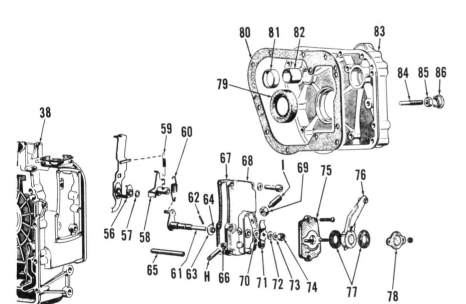

Fig. W1-12—Exploded view of governor linkage.

38. Crankcase half
56. Lever
57. Spacer
58. Pivot
59. Spring
60. Governor spring
61. Control shaft
62. Pin
63. Washer
64. "O" ring
65. Pivot shaft
66. Locknut
67. Gasket
68. Side cover
69. Locknut
70. Washer
71. Lever
72. Washer
73. Lockwasher
74. Nut
75. Cover
76. Control lever
77. Friction discs
78. Retainer
79. Seal
80. Gasket
81. Balancer bushing
82. Camshaft bushing
83. Gear cover
84. Max. fuel delivery screw
85. Nut
86. Plug

inch). Piston pin clearance in piston bore should be 0.002-0.011 mm (0.0001-0.0004 inch) with a wear limit of 0.06 mm (0.0024 inch).

When reinstalling piston rings, top compression ring and bottom oil control ring can be installed with either side up. On WD1-350, WD1-360 and WD1-430, center ring must be installed with side marked "TOP" facing upward. On WD1-660, WD1-670 and WD1-750, center ring must be installed with notched side facing downward. Position ring end gaps 120° apart.

Reverse disassembly procedure for reassembly. Arrow (A–Fig. W1-10) on piston crown must point towards flywheel end of engine. With piston at top dead center, clearance between piston crown and top edge of cylinder should be 0.8-0.9 mm (0.032-0.035 inch). Install cylinder shim gaskets (32–Fig. W1-11) as required to obtain desired clearance.

GOVERNOR

REMOVE AND REINSTALL. Refer to Fig. W1-12 for an exploded view of governor linkage. To remove or inspect governor linkage, unbolt side cover (68), rotate cover 90° clockwise to disengage governor spring (60) and remove side cover assembly. Remove gear cover (83) if easier access to governor is desired. Pull shaft (65) from crankcase to remove lever (56) and pivot (58), being careful not to drop spacer (57) into crankcase. Do not lose spacer (57), which must be installed in original position. Refer to OIL PUMP section if flyweight assembly (46–Fig. W1-11) must be removed.

Two different governor springs (60–Fig. W1-12) are used. Maximum governed speed is limited to 2200 rpm or 3000 rpm, depending on governor spring used. Long end of pivot spring (59) is connected to pivot (58).

Maximum fuel delivery screw (84) is located in gear cover (83) and should be adjusted to provide acceptable power without excessive smoke. With engine warm and under no load, accelerate engine quickly. If smoke is excessive, turn screw (84) clockwise. If additional fuel is needed, turn screw counterclockwise.

Illustrations Courtesy Teledyne Total Power

Wisconsin

SMALL DIESEL

Carefully turn screw in 1/8 turn or less increments.

OIL PUMP

R&R AND OVERHAUL. Refer to GOVERNOR section and remove governor and control linkage. Remove gear cover if not previously detached. Unbolt and remove oil pump and governor assembly. Press gear and governor assembly (46–Fig. W1-11) off oil pump gear shaft (43). Inspect oil pump gears and housing and renew if damaged or worn excessively. Maximum gear backlash is 0.30 mm (0.012 inch). Maximum gear depth (D–Fig. W1-13) is 0.10 mm (0.004 inch). Maximum radial clearance (C) between side of gear and housing is 0.15 mm (0.006 inch). Reverse disassembly procedure to assemble oil pump. Be sure thrust washer (45–Fig. W1-11) is on oil pump shaft before pressing governor assembly onto shaft.

The oil pressure relief valve (23–Fig. W1-11) is located in side of crankcase. Oil pressure is adjusted by removing or adding shims (25). Maximum oil pressure should be 380 kPa (55 psi) for WD1-340, WD1-350 and WD1-430 engines, or 275 kPa (40 psi) for WD1-660, WD1-670 and WD1-750 engines.

CONNECTING ROD

R&R AND OVERHAUL. To remove connecting rod, drain engine oil and detach flywheel. Remove piston and cylinder as previously outlined. Unscrew crankcase screws (eleven screws on Models WD1-340, WD1-350 and WD1-430; twelve screws on all other models) and position crankcase with gear cover (83–Fig. W1-12) down. Lift off flywheel side crankcase half (22–Fig. W1-11) and remove connecting rod.

Inspect connecting rod and crankpin. If crankshaft must be removed, refer to CRANKSHAFT AND CRANKCASE section.

Crankpin diameter should be 48.221-48.237 mm (1.8985-1.8991 inches) on Models WD1-340, WD1-350 and WD1-430, or 59.981-60.000 mm (2.3615-2.3622 inches) on all other models. Rod bearing clearance on Models WD1-340, WD1-350 and WD1-430 should be 0.030-0.072 mm (0.0012-0.0028 inch) with a maximum clearance of 0.10 mm (0.004 inch). Rod bearing clearance on Models WD1-660, WD1-670 and WD1-750 should be 0.040-0.10 mm (0.0016-0.0040 inch) with a maximum clearance of 0.12 mm (0.005 inch). Rod bearings are available in standard and 0.25 and 0.50 mm (0.010 and 0.020 inch) undersizes.

The small end rod bushing is renewable; refer to PISTON, PIN, RINGS AND CYLINDER section.

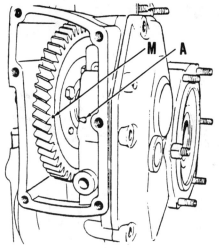

Fig. W1-14—View of camshaft gear timing mark (M) and crankcase mark (A) on Models WD1-340, WD1-350 and WD1-430.

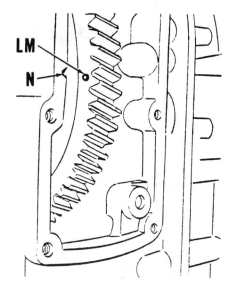

Fig. W1-15—View of large camshaft gear timing mark (LM) and crankcase notch (N) on Models WD1-660, WD1-670 and WD1-750.

Reverse removal procedure to install connecting rod. Tighten rod nuts to 34 N·m (25 ft.-lbs.) on Models WD1-340, WD1-350 and WD1-430, or 49 N·m (36 ft.-lbs.) on all other models. Be sure thrust washer (53–Fig. W1-11) is on camshaft end. Tighten eleven crankcase screws on Models WD1-340, WD1-350 and WD1-430 to 30 N·m (22 ft.-lbs.). On Models WD1-660, WD1-670 and WD1-750, tighten ten M8 crankcase screws to 30 N·m (22 ft.-lbs.) and two M12 crankcase screws to 69 N·m (51 ft.-lbs.).

CRANKSHAFT AND CRANKCASE

R&R AND OVERHAUL. To remove crankshaft, drain engine oil and detach flywheel. Unbolt side cover (68–Fig. W1-12), rotate cover 90° clockwise to disengage governor spring (60) and remove side cover assembly. Remove fuel injection pump, piston and cylinder as previously outlined. Remove cover (28–Fig. W1-11) or transfer fuel pump, if so equipped. Unscrew crankcase screws (eleven screws on Models WD1-340, WD1-350 and WD1-430; twelve screws on all other models) and position crankcase with gear cover (83–Fig. W1-12) down. Lift off flywheel side crankcase half (22–Fig. W1-11). Withdraw crankshaft and rod assembly, then remove rod if desired.

Crankshaft main bearing journal diameter should be 45.984-46.000 mm (1.8104-1.8110 inches) on Models WD1-340, WD1-350 and WD1-430, or 55.981-56.000 mm (2.2040-2.2047 inches) on all other models. Main bearing clearance should be 0.054-0.095 mm (0.002-0.004 inch) on Models WD1-340, WD1-350 and WD1-430. Main bearing clearance on Models WD1-660, WD1-670 and WD1-750 should be 0.04-0.08 mm (0.0016-0.0031 inch). Maximum main bearing clearance for all models is 0.12 mm (0.0047 inch). Main bearings (30 and 34–Fig. W1-11) are available in standard and 0.25 and 0.50 mm (0.010 and 0.020 inch) undersizes. If main bearings must be renewed, remove any components in crankcase which may be damaged by heat. Heat crankcase to 150°C (300°F), then using a suitable press, remove and install main bearings.

If camshaft bushing (29) must be renewed, heat crankcase half (22) to 150°C (300°F) prior to removal and installation.

If removed, install connecting rod on crankshaft and tighten rod nuts to 34 N·m (25 ft.-lbs.) on Models WD1-340, WD1-350 and WD1-430, or 49 N·m (36 ft.-lbs.) on all other models.

To install crankshaft on Models WD1-340, WD1-350 and WD1-430, rotate camshaft so timing mark (M–Fig. W1-14) on camshaft gear is aligned with arrow (A) on crankcase. Insert crankshaft into top crankcase half (38–Fig. W1-11) so crankpin is at top dead center. Recheck camshaft gear timing mark. With crankpin at top dead center, timing marks (M and A–Fig. W1-14) should be aligned.

To install crankshaft on Models WD1-660, WD1-670 and WD1-750, rotate camshaft so large timing mark (LM–Fig. W1-15) is aligned with notch (N) in crankcase. Insert crankshaft into pto crankcase half (38–Fig. W1-11) so red

SERVICE MANUAL

Wisconsin

paint mark (P–Fig. W1-16) is aligned with small camshaft gear timing mark (SM). Rotate crankshaft until red paint mark (P–Fig. W1-17) on crankshaft gear is adjacent to balancer gear. Red paint mark (P) must align with balancer gear mark (M). If marks (P and M) do not align, hold camshaft gear in place, withdraw crankshaft, rotate balancer so marks (P and M) will be aligned then reinsert crankshaft into crankcase. When properly mated, balancer gear mark (M) and red paint mark (P) will align on each crankshaft rotation, and red paint mark (P–Fig. W1-16) will align with small camshaft timing mark (SM) when large camshaft timing mark (LM–Fig. W1-15) is aligned with crankcase notch (N).

On all models, install rubber ring (11–Fig. W1-9) and oil seal (20–Fig. W1-11) with seal lip inward. Be sure thrust washer (53) is on camshaft end. Install crankcase half (22) but do not apply sealant to gasket or crankcase. Tighten eleven crankcase screws on Models WD1-340, WD1-350 and WD1-430 to 30 N·m (22 ft.-lbs.). On Models WD1-660, WD1-670 and WD1-750, tighten ten M8 crankcase screws to 30 N·m (22 ft.-lbs.) and two M12 crankcase screws to 69 N·m (51 ft.-lbs.)

Crankshaft end play should be 0.10-0.35 mm (0.004-0.014 inch) on Models WD1-340, WD1-350 and WD1-430, or 0.20-0.40 mm (0.008-0.016 inch) on all other models. If end play is excessive, remove crankshaft and install thicker thrust washer (14–Fig. W1-9).

Complete remainder of assembly by reversing disassembly procedure.

CAMSHAFT

R&R AND OVERHAUL. To remove camshaft, refer to previous section and disassemble crankcase but do not remove crankshaft. Remove gear cover (83–Fig. W1-12) and governor linkage. Set aside camshaft shims (52–Fig. W1-11) for future use. Detach gear (50) from camshaft (51) and withdraw camshaft from crankcase. Remove gear through side cover opening.

The camshaft is supported by bushings in the crankcase and gear cover. Maximum allowable camshaft journal clearance is 0.15 mm (0.006 inch). If the crankcase bushing (29–Fig. W1-11) requires renewal, refer to following section, as crankcase must be disassembled and heated for bushing installation. To renew bushing (81–Fig. W1-12), heat gear cover (83) to 150°C (300°F) before removal or installation. Be sure crankshaft seal (79) is not damaged by heat. Camshaft bushings are available in standard size and 0.25 and 0.50 mm (0.010 and 0.020 inch) undersizes.

Intake and exhaust cam lobe heights should be 32.775-33.025 mm (1.2904-

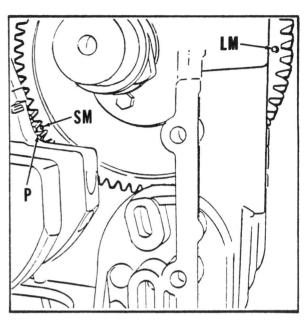

Fig. W1-16—View of large (LM) and small (SM) camshaft timing marks and red paint mark (P) on crankshaft gear of Models WD1-660, WD1-670 and WD1-750.

1.3002 inches) on Models WD1-340, WD1-350 and WD1-430. Intake cam lobe height on Models WD1-660, WD1-670 and WD1-750 should be 39.966-40.216 mm (1.5735-1.5833 inches) and exhaust cam lobe height should be 39.880-40.130 mm (1.5701-1.5799 inches). Fuel injection cam lobe height should be 34.925-35.075 mm (1.3750-1.3809 inches) on Models WD1-340, WD1-350 and WD1-430 or 40.857-41.107 mm (1.6085-1.6184 inches) on all other models.

To install camshaft, insert gear (50–Fig. W1-11) into crankcase through side cover opening and mate gear with crankshaft gear. On Models WD1-340, WD1-350 and WD1-430 with crankpin at top dead center, camshaft timing gear mark (M–Fig. W1-14) must be aligned with arrow (A) on crankcase. On Models WD1-660, WD1-670 and WD1-750, red paint mark (P–Fig. W1-16) on crankshaft gear must align with small camshaft gear timing mark (SM) while large camshaft timing mark (LM–Fig. W1-15) must align with crankcase notch (N). While holding camshaft gear in mesh with crankshaft gear, insert camshaft into crankcase and through camshaft gear. Rotate camshaft so locating pin (49–Fig. W1-11) in gear enters notch in camshaft flange. Install camshaft gear screws. Place original shims (52) on camshaft and install governor linkage and gear cover. Tighten gear cover screws to 25 N·m (18 ft.-lbs.). Complete remainder of assembly by reversing disassembly procedure, but do not install side cover.

Check camshaft end play, which should be 0.10-0.30 mm (0.004-0.012 inch) on Models WD1-340, WD1-350 and WD1-430 or 0.20-0.40 mm (0.008-0.016 inch) on all other models. Install or remove shims (52) as needed to obtain desired end play. Attach governor spring (60–Fig. W1-12) and install side cover.

BALANCER

Models WD1-660, WD1-670 And WD1-750

REMOVE AND REINSTALL. Models WD1-660, WD1-670 and WD1-750

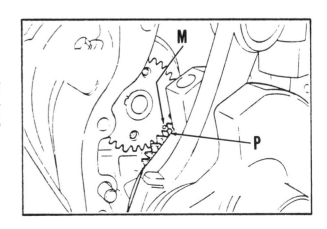

Fig. W1-17—Balancer gear mark (M) and red paint mark (P) on crankshaft gear should be aligned on Models WD1-660, WD1-670 and WD1-750.

Wisconsin SMALL DIESEL

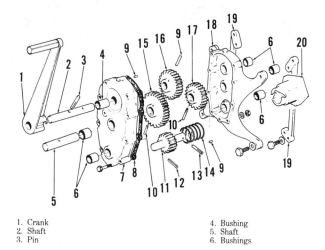

Fig. W1-18—Exploded view of crank type manual starter used on some models. Adapter (19) is used on Models WD1-660, WD1-670 and WD1-750.

1. Crank
2. Shaft
3. Pin
4. Bushing
5. Shaft
6. Bushings
7. Housing half
8. Gasket
9. Dowel
10. Pin
11. Output gear & shaft
12. Pin
13. Pin
14. Spring
15. Gear (29 teeth)
16. Gear (23 teeth)
17. Gear (21 teeth)
18. Housing half
19. Adapter
20. Crank jaw

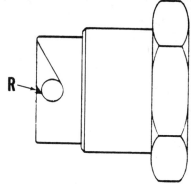

Fig. W1-19—File away area (R) on crank jaw to dimensions shown for easier disengagement.

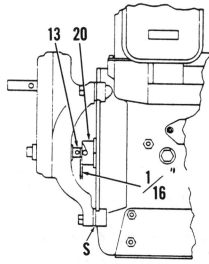

Fig. W1-20—Output shaft pin (13) must be at least 1.5 mm (1/16 inch) from crank jaw (20) when disengaged. Install shims (S) between starter housing and engine to obtain clearance.

are equipped with a balancer (41–Fig. W1-11) which is driven by gear (36) which meshes with the crankshaft gear.

To remove balancer, remove crankshaft as previously outlined then remove gear cover being careful not to lose camshaft shims (52). Use a suitable puller to remove gear (36) from balancer shaft.

Inspect balancer, gear and bushings (39–Fig. W1-11 and 82–Fig. W1-12). Renew if damaged or excessively worn. Crankcase or gear cover must be heated to 150°C (300°F) when removing or installing bushings. Remove any components which may be damaged by heat.

To install balancer, insert balancer into crankcase then use a suitable press to install gear (36–Fig. W1-11) onto balancer shaft with timing gear mark (M–Fig. W1-17) out. Install gear cover being sure to install camshaft shims (52–Fig. W1-11). Refer to CRANKSHAFT AND CRANKCASE section for crankshaft installation and reassembly.

MANUAL STARTER

R&R AND OVERHAUL. Refer to Fig. W1-18 for an exploded view of crank type manual starter used on some models. Unscrew housing mounting screws to remove starter from engine. Unscrew housing screws and separate housing halves (7 and 18) for access to starter components.

Inspect starter components for damage and excessive wear. Heat housing halves to 150°C (300°F) before removing and installing bushings. Manufacturer recommends renewing both bushings for a particular shaft.

When assembling starter, adequately grease all internal components. Be sure starter operates properly after assembly without binding or sticking. Manufacturer recommends filing down area (R–Fig. W1-19) of crank jaw so starter pin disengages easier. With starter mounted on engine, starter pin (13–Fig. W1-18) must be at least 1.5 mm (1/16 inch) from crank jaw as shown in Fig. W1-20. Install shims (S) between starter and engine to obtain desired clearance.

WISCONSIN

Model	No. Cyls.	Bore	Stroke	Displ.
WD2-860	2	84 mm (3.307 in.)	78 mm (3.071 in.)	864 cc (52.7 cu. in.)
WD2-1000	2	86 mm (3.386 in.)	86 mm (3.386 in.)	1000 cc (61.0 cu. in.)

Engines covered in this section are air-cooled, two-cylinder, four-stroke diesel engines. Crankshaft rotation is counterclockwise at pto end. Number 1 cylinder is nearer pto end of engine.

Metric fasteners are used throughout engine.

MAINTENANCE

LUBRICATION

Recommended engine oil is SAE 10W or 5W-20 when temperatures are below minus 18°C (0°F), SAE 10W or 10W-30 for temperatures between minus 18°C (0°F) and 0°C (32°F), SAE 20 or 15W-40 for temperatures between 0°C (32°F) and 20°C (68°F) and SAE 40 or 15W-40 for temperatures above 20°C (68°F). API oil classification should be CD. Manufacturer recommends renewing oil after every 100 hours of operation. Oil should be drained while engine is hot.

An external oil filter (F–Fig. W2-1) is located on side of engine. Manufacturer recommends renewing filter after every other oil change. A reusable filter element is located on the oil pickup tube (P–Fig. W2-2). Filter should be removed and cleaned after every 200 hours of operation. Renew filter if damaged or if unable to get filter completely clean.

Oil pressure with engine warm and running at full speed should be 275 kPa (40 psi). An oil pressure gage may be connected after unscrewing test port plug (T–Fig. W2-1). Refer to OIL PUMP section to adjust pressure relief valve setting.

ENGINE SPEED ADJUSTMENT

Idle speed is adjusted by turning idle speed screw (I–Fig. W2-3) and high speed is adjusted by turning high speed screw (H). Idle speed should be 1500-1800 rpm while maximum governed speed under load is 3000 rpm.

FUEL SYSTEM

FUEL FILTERS. A fuel filter is located in the fuel tank and an external fuel filter is mounted on side of engine. Renew fuel filters after every 200 hours of operation or sooner if required.

BLEED FUEL SYSTEM. On gravity-feed fuel systems, a fuel return line and check valve are connected to the fuel injection pump as well as the fuel supply line. Fuel supply line should bleed air automatically up to fuel injection pump on gravity flow systems. To bleed high pressure injection lines, loosen fittings for high pressure lines at injectors then rotate engine crankshaft to operate fuel injection pump until air-free fuel flows from injection lines. Retighten injection lines.

On engines equipped with a fuel transfer pump, loosen fuel supply line fitting on injection pump then rotate engine crankshaft to operate fuel transfer pump until air-free fuel flows from fuel line. Retighten fuel supply line and loosen fittings for high pressure injection lines at injectors. Rotate engine crankshaft to operate fuel injection pump until air-free fuel flows from injection lines. Retighten injection lines.

INJECTION PUMP TIMING

Injection pump timing is adjusted using shim gaskets (29–Fig. W2-4) between pump body and mounting surface on crankcase. Refer to INJECTION PUMP section for pump removal and installation.

To check injection pump timing, unscrew high pressure injection line of

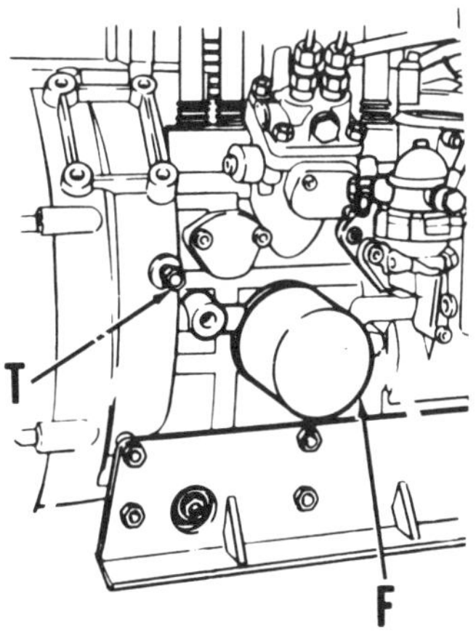

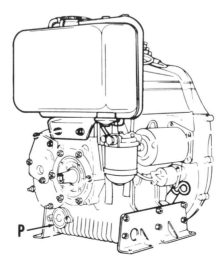

Fig. W2-2—Remove oil pickup (P) for access to internal oil filter.

Fig. W2-1—View showing location of external oil filter (F). Unscrew plug (T) and connect a pressure gage to measure oil pressure.

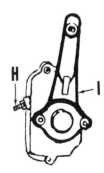

Fig. W2-3—View showing location of high speed adjusting screw (H). Idle speed screw is located behind speed control lever at (I).

Wisconsin   SMALL DIESEL

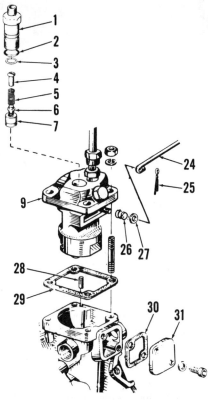

Fig. W2-4—Partial exploded view of fuel injection pump.

1. Delivery valve holder
2. "O" ring
3. Gasket
4. Spring guide
5. Spring
6. Delivery valve
7. Delivery valve seat
9. Fuel injection pump
24. Control rod
25. Cotter pin
26. Spring
27. Washer
28. Pin
29. Shim gasket
30. Gasket
31. Cover

so number 1 piston is on compression stroke – it may be necessary to connect a gravity-flow fuel tank to fuel injection pump if not so equipped. Note fuel will flow out of delivery valve holder. Stop engine rotation at moment fuel ceases to flow. Timing mark (R – Fig. W2-5) should be within 3 mm (1/8 inch) of injection timing mark (N) on crankcase. To advance injection timing, remove shim gaskets (29 – Fig. W2-4); install shim gaskets to retard injection timing. Reinstall removed pump parts after checking timing.

REPAIRS

TIGHTENING TORQUES

Refer to the following table for special tightening torques. All fasteners are metric.

Connecting rod .... 34 N·m
                                (25 ft.-lbs.)
Crankcase .... See text
Cylinder head .... 44 N·m
                                (33 ft.-lbs.)
Fan .... 34 N·m
                                (25 ft.-lbs.)
Flywheel .... 176 N·m
                                (130 ft.-lbs.)
Gear cover .... 25 N·m
                                (18 ft.-lbs.)
Injection pump .... 25 N·m
                                (18 ft.-lbs.)
Injector retainer plate .... 10 N·m
                                (7 ft.-lbs.)
Main bearing center support ... See text

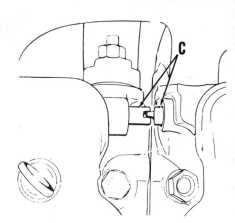

Fig. W2-7—Before removing cylinder head, rotate compression release shafts (C) so slot and tang in shaft ends are vertical.

VALVE ADJUSTMENT

Valve clearance may be adjusted after removing rocker arm cover and turning crankshaft to position piston for cylinder being adjusted at TDC on compression stroke. Turn rocker arm adjusting screw (Fig. W2-6) until appropriate size feeler gage can be inserted between rocker arm and valve stem end. Recommended clearance with engine cold is 0.15 mm (0.006 inch) for exhaust and 0.10 mm (0.004 inch) for intake.

CYLINDER HEAD AND VALVE SYSTEM

R&R AND OVERHAUL. To remove cylinder heads, disconnect high pressure and return fuel injection lines. Immediately cap fuel lines. Remove muffler, air cleaner, exhaust and intake manifolds. Turn compression release lever so connecting tang and slot on compression release shafts are vertical as shown in Fig. W2-7. Remove rocker arm covers, unscrew cylinder head retaining nuts and remove cylinder heads. Remove and discard rubber seals (R – Fig. W2-8) on cylinder head studs and upper push rod tube seals.

number 1 cylinder from injection pump delivery valve holder (1). Unscrew delivery valve holder (1) and remove spring (5), spring guide (4) and delivery valve (6), then screw delivery valve holder (1) back into pump. Move throttle control to full speed position. Rotate engine in normal direction (counterclockwise at pto)

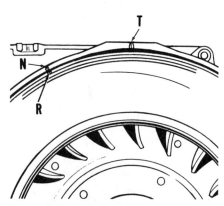

Fig. W2-5—Number 1 cylinder fuel injection should occur when flywheel reference mark (R) is aligned with crankcase mark (N). Number 1 piston is at top dead center when flywheel reference mark (R) is aligned with crankcase mark (T).

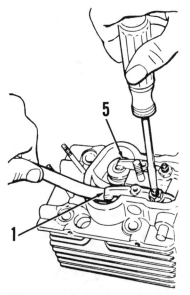

Fig. W2-6—Valve clearance should be 0.15 mm (0.006 inch) at exhaust rocker arm (1) and 0.10 mm (0.004 inch) at intake rocker arm (5).

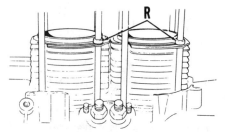

Fig. W2-8—View showing location of rubber seals (R) on cylinder head studs.

SERVICE MANUAL

Wisconsin

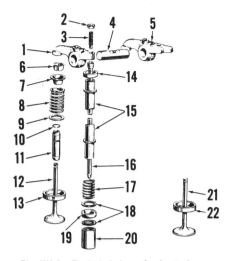

Fig. W2-9—Exploded view of valve train.

1. Exhaust rocker arm
2. Locknut
3. Adjuster
4. Rocker arm shaft
5. Intake rocker arm
6. Keys
7. Spring retainer
8. Valve spring
9. Washer
10. Snap ring
11. Valve guide
12. Exhaust valve
13. Exhaust valve seat
14. Seal
15. Push rod tube
16. Push rod
17. Spring
18. Washer
19. Seal
20. Cam follow
21. Intake valve
22. Intake valve seat

Valve face angle for both valves is 45½° and valve seat angle is 45°. Valve seats are renewable and must be installed with head heated to 200°C (392°F). There should be an interference fit of 0.079-0.120 mm (0.0031-0.0047 inch) between outside diameter of seat and bore in cylinder head. With valves installed, top of valve heads should be recessed 0.60-0.95 mm (0.024-0.037 inch) below surface of cylinder head.

Valve stem diameter is 6.945-6.960 mm (0.2734-0.2740 inch) for both valves and valve guide inside diameter is 7.000-7.015 mm (0.2756-0.2762 inch). Valve stem clearance should be 0.040-0.070 mm (0.0016-0.0028 inch) with a maximum allowable valve stem clearance of 0.2 mm (0.008 inch). Valve guides are renewable. To remove valve guide, press guide out top of cylinder head. To install valve guide, position locating ring (10—Fig. W2-9) on guide, then press guide into head from top side of head with ring end up. Press guide into head until locating ring is seated in head groove.

Reassemble by reversing disassembly procedure. Install new "O" rings on compression release shafts and install shafts so slot or tang on shaft end is towards inner side of cylinder head as shown in Fig. W2-7. Install new upper rubber seal on push rod tubes. Install new rubber seals on cylinder head studs shown in Fig. W2-8. Before tightening cylinder head nuts, install exhaust and intake manifolds to correctly position heads, then tighten cylinder head nuts to 44 N·m (33 ft.-lbs.).

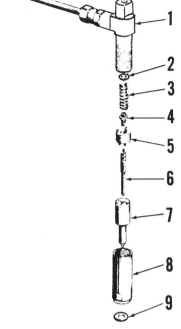

Fig. W2-10—Exploded view of injector.

1. Body
2. Shim
3. Spring
4. Push piece
5. Spacer
6. Valve
7. Nozzle
8. Nozzle holder
9. Gasket

INJECTOR

REMOVE AND REINSTALL. To remove injector, first clean dirt from injector, injection line, return line and cylinder head. Disconnect return line and high pressure injection line and immediately cap or plug all openings. Unscrew retainer plate and remove injector and copper washer.

Reverse removal procedure to reinstall injector. Install a new copper washer (9—Fig. W2-10). Tighten injector retaining plate nuts to 10 N·m (7 ft.-lbs.).

TESTING. A suitable test stand is required to check injector operation. Only clean, approved testing oil should be used to test injector.

WARNING: Fuel leaves the injection nozzle with sufficient force to penetrate the skin. When testing, keep yourself clear of nozzle spray.

Attach injector to tester and operate pump lever a few quick strokes to purge air from injector and to make sure nozzle valve is not stuck. When operating properly during test, injector nozzle will emit a buzzing sound and cut off quickly with no leakage at seat. Spray pattern should be uniform and well atomized.

Opening pressure should be 19600-20600 kPa (2845-2985 psi). Opening pressure is adjusted by varying the thickness of shims (2—Fig. W2-10).

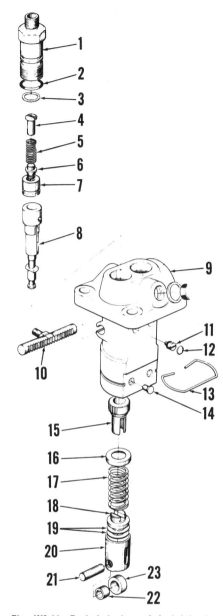

Fig. W2-11—Exploded view of fuel injection pump.

1. Delivery valve holder
2. "O" ring
3. Gasket
4. Spring guide
5. Spring
6. Delivery valve
7. Delivery valve seat
8. Plunger
9. Body
10. Control rack
11. Pin
12. Plug
13. Clip
14. Pin
15. Control sleeve
16. Spring seat
17. Spring
18. Spring seat
19. Shims
20. Tappet
21. Pin
22. Inner roller
23. Outer roller

Operate tester slowly to maintain pressure of 16200 kPa (2350 psi) and check for nozzle valve leakage. If a drop forms at nozzle tip within 10 seconds, nozzle valve is not seating and injector must be overhauled or renewed.

OVERHAUL. Clamp injector body (1—Fig. W2-10) in a vise with nozzle pointing upward. Unscrew nozzle holder nut (8), then remove injector components as shown in Fig. W2-10. Thor-

Illustrations Courtesy Teledyne Total Power

Wisconsin

SMALL DIESEL

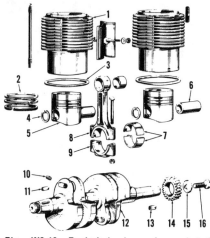

Fig. W2-12—Exploded view of crankshaft assembly.

1. Cylinder
2. Piston rings
3. Gasket
4. Snap ring
5. Piston
6. Piston pin
7. Rod bearing
8. Connecting rod
9. Rod cap
10. Plug
11. Key
12. Crankshaft
13. Key
14. Gear
15. Slotted washer
16. Cap screw

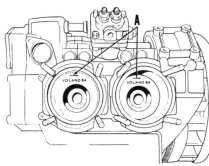

Fig. W2-13—Arrow (A) on piston crown must point towards flywheel.

Table 1—Color codes and corresponding dimensions for standard pistons and cylinders

Model	Color Code	Piston O.D.	Cylinder I.D.
WD2-860	Red or Yellow	83.925-83.935mm (3.3041-3.3045in.)	84.000-84.011mm (3.3071-3.3075in.)
WD2-860	Blue or Green	83.935-83.945mm (3.045-3.3049in.)	84.011-84.022mm (3.3075-3.3079in.)
WD2-1000	Red or Yellow	85.920-85.930mm (3.3827-3.3831in.)	86.000-86.011mm (3.3858-3.3862in.)
WD2-1000	Blue or Green	85.930-85.940mm (3.3831-3.3835in.)	86.011-86.022mm (3.3863-3.3867in.)

oughly clean all parts in a suitable solvent. Clean inside orifice of nozzle tip with a wooden cleaning stick. When reassembling injector, make certain all components are clean and wet with clean diesel fuel oil. Tighten nozzle holder nut (8) to 49 N·m (36 ft.-lbs.).

INJECTION PUMP

R&R AND OVERHAUL. To remove fuel injection pump, disconnect fuel lines and remove cover (31—Fig. W2-4) adjacent to pump. Place a clean rag underneath control rod (24) so cotter pin (25) cannot fall into crankcase. Then, remove cotter pin (25) and detach control rod end from pump control rack pin. Unscrew pump retaining nuts and remove fuel injection pump. Do not lose shim gaskets (29).

Refer to Fig. W2-11 for an exploded view of fuel injection pump. The injection pump should be tested and overhauled by a shop qualified in diesel fuel injection pump repair.

Reverse removal procedure for reinstallation. Tighten pump retaining nuts to 25 N·m (18 ft.-lbs.). If pump is renewed or overhauled, or original shim gaskets are not used, refer to INJECTION PUMP TIMING section and adjust pump timing.

PISTON, PIN, RINGS AND CYLINDER

R&R AND OVERHAUL. Remove cylinder head as previously outlined. If piston and cylinder do not require service but only removal, proceed as follows: Rotate crankshaft so piston to be removed is at top dead center. Lift cylinder up until piston pin is exposed, remove piston pin retainer, withdraw piston pin and remove cylinder and piston as a unit. If piston, rings or cylinder requires service, lift cylinder off crankcase, remove piston pin retainer, withdraw piston pin and remove piston.

Standard size pistons and cylinders are color-coded red, yellow, blue or green. A paint dot is located on inside of piston and on outside of cylinder. Piston diameter is measured 12 mm (1/2 inch) from bottom of piston skirt on Model WD2-860 or 15 mm (19/32 inch) from bottom of piston skirt on Model WD2-1000. Refer to Table 1 for piston and corresponding cylinder dimensions.

Check cylinder bore for excessive wear or other damage. Cylinders may be rebored for installation of 0.5 mm (0.020 inch) or 1.0 mm (0.040 inch) oversize pistons and rings. Clearance between piston skirt and cylinder bore should be 0.066-0.086 mm (0.0026-0.0034 inch) for WD2-860 and 0.071-0.091 mm (0.0028-0.0036 inch) for WD2-1000.

The top compression ring is a keystone type and second compression ring is rectangular. Piston ring end gap should be 0.30-0.45 mm (0.012-0.018 inch). Side clearance for second compression ring in piston groove should be 0.050-0.082 mm (0.002-0.003 inch) for WD2-860 and 0.070-0.102 mm (0.003-

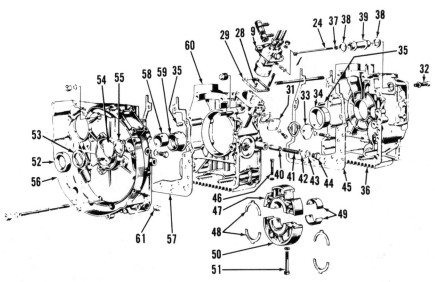

Fig. W2-14—Exploded view of crankcase assembly.

- 9. Fuel injection pump
- 24. Control rod
- 28. Pin
- 29. Shim gasket
- 31. Cover
- 32. Socket head screw
- 33. Cover
- 34. Main bearing
- 35. Pin
- 36. Pto crankcase section
- 37. Nut
- 38. Seal
- 39. Tube
- 40. Oil pressure relief valve
- 41. Spring
- 42. Shim
- 43. Washer
- 44. Plug
- 45. Gasket
- 46. Screw
- 47. Upper main bearing support half
- 48. Thrust washers
- 49. Main bearing
- 50. Lower main bearing support half
- 51. Screw
- 52. Inner seal ring
- 53. Oil seal
- 54. Cover
- 55. Gasket
- 56. Flywheel crankcase section
- 57. Gasket
- 58. Camshaft bushing
- 59. Main bearing
- 60. Center crankcase section
- 61. Oil drain plug

SERVICE MANUAL

Wisconsin

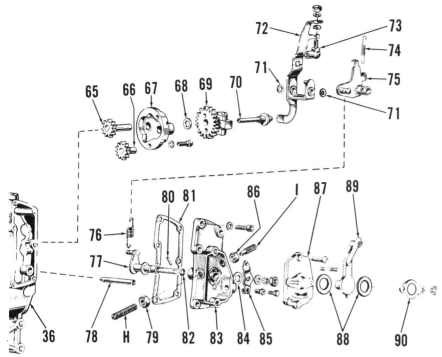

Fig. W2-15—Exploded view of oil pump, governor and linkage.

H. High speed screw	69. Governor assy.	76. Spring
I. Idle speed screw	70. Push rod	77. Lever & shaft
36. Pto crankcase section	71. Shims	78. Shaft
65. Drive gear	72. Governor arm	79. Locknut
66. Driven gear	73. Control rod end	80. Pin
67. Oil pump body	74. Spring	81. Gasket
68. Thrust washer	75. Pivot	82. "O" ring
		83. Side cover
		84. Washer
		85. Lever
		86. Locknut
		87. Cover
		88. Washers
		89. Speed control lever
		90. Plate

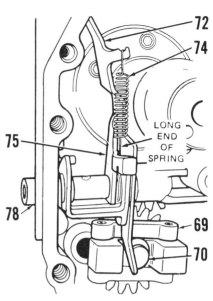

Fig. W2-16—View of governor linkage. Long end of spring (74) attaches to pivot (75).

0.004 inch) for WD2-1000. Oil control ring side clearance should be 0.040-0.072 mm (0.0016-0.0028 inch) for all engines. Maximum allowable side clearance for either ring on all engines is 0.25 mm (0.010 inch).

When reinstalling piston rings, center ring must be installed with side marked "TOP" facing upward. Top ring and oil control ring can be installed with either side up. Stagger ring end gaps 120° apart.

Reverse disassembly procedure for reassembly. Arrow (A–Fig. W2-13) on piston crown must point towards flywheel end of engine. With piston at top dead center, clearance between piston crown and top edge of cylinder should be 0.8-0.9 mm (0.032-0.035 inch). Install cylinder shim gaskets (3–Fig. W2-12) required to obtain desired clearance.

GOVERNOR

REMOVE AND REINSTALL. Refer to Fig. W2-15 for an exploded view of governor and control linkage. Remove cover (83) and gear cover (98–Fig. W2-18) for access to governor and linkage. Do not lose camshaft shims (95), which may remain on gear cover. Mark shims (71–Fig. W2-15) so they may be returned to original position.

Governor gear (69) is pressed on oil pump drive gear shaft (65), so oil pump and governor unit must be removed to press governor gear off shaft. Governor gear and flyweights (69) are available only as a unit assembly. Install thrust washer (68) on shaft before pressing governor gear assembly onto shaft.

Distance (D–Fig. W2-17) between speed control rod ends should be 129 mm (5 inches). Remove cover (31–Fig. W2-14) and place a rag underneath control rod so cotter pin cannot fall into crankcase. Detach cotter pin securing rod end and withdraw control rod (24). Adjust length of control rod by turning rod end (73–Fig. W2-17).

Reassemble by reversing removal procedure. Install shims (71–Fig. W2-15) in their original location. Install spring (74) so long end is connected to intermediate lever (75) as shown in Fig. W2-16. Tighten gear cover nuts to 25 N·m (18 ft.-lbs.).

Maximum fuel delivery screw (T-Fig. W2-18) is located in gear cover (98) and should be adjusted to provide acceptable power without excessive smoke. With engine warm and under no load, accelerate engine quickly. If smoke is excessive, turn screw (T) clockwise. If additional fuel is needed, turn screw counterclockwise. Carefully turn screw in 1/8 turn or less increments.

OIL PUMP

R&R AND OVERHAUL. Refer to GOVERNOR section and remove governor and oil pump unit. Press governor off oil pump gear shaft (65–Fig. W2-15). Inspect oil pump gears and housing and renew if damaged or worn excessively. Oil pump housing and gears are available individually. Reverse disassembly procedure to assemble oil pump. Be sure thrust washer (68) is on oil pump shaft before pressing governor assembly onto shaft.

The oil pressure relief valve (40–Fig. W2-14) is located adjacent to oil filter. Oil pressure is adjusted by removing or installing shims (42). Refer to LUBRICATION section for oil pressure testing.

CAMSHAFT

R&R AND OVERHAUL. To remove camshaft, refer to previous sections and remove cylinder heads and fuel injection pump. Remove push rods and push rod tubes. Extract and mark cam followers so they can be returned to original bores. Refer to GOVERNOR section and remove governor linkage. Remove fuel transfer pump on models so equipped. Set aside shims (95–Fig. W2-18) for future use. Detach gear (94) from camshaft (91), then rotate camshaft so

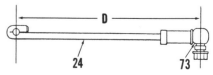

Fig. W2-17—Length (D) between control rod ends should be 129 mm (5 inches).

Wisconsin

SMALL DIESEL

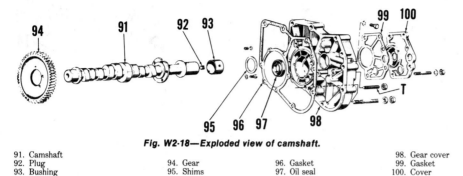

Fig. W2-18—Exploded view of camshaft.

91. Camshaft	94. Gear	98. Gear cover
92. Plug	95. Shims	99. Gasket
93. Bushing	96. Gasket	100. Cover
	97. Oil seal	

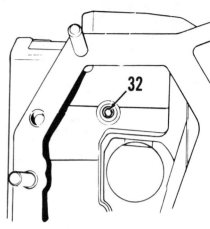

Fig. W2-21—View showing location of socket head screw (32) in pto crankcase section.

flat (F—Fig. W2-19) on camshaft flange coincides with crankcase projection (P). Carefully withdraw camshaft.

The camshaft is supported by bushings in the gear cover and crankcase. To inspect bushing (58—Fig. W2-14) in the crankcase, remove flywheel and cover (54). If camshaft bushing (58) must be renewed, disassemble crankcase. Maximum allowable camshaft journal clearance is 0.2 mm (0.008 inch). Camshaft bushings are available in undersizes of 0.25 and 0.50 mm (0.010 and 0.020 inch).

To install camshaft, rotate crankshaft so number 1 piston is at top dead center. Insert camshaft gear (94—Fig. W2-18) through control linkage opening in crankcase so dowel pin (101—Fig. W2-20) side of gear is towards end of crankcase. Insert camshaft through camshaft gear into crankcase while noting flat (F—Fig. W2-19) on camshaft flange which must coincide with crankcase projection (P). Engage camshaft gear with crankshaft gear so camshaft gear timing mark (M—Fig. W2-20) is aligned with crankcase timing arrow (A). Hold camshaft gear then rotate camshaft so camshaft flange notch engages camshaft gear pin (101). Install camshaft screws and recheck timing marks. Place original shims (95—Fig. W2-18) on camshaft and install governor linkage and gear cover. Tighten gear cover nuts to 25 N·m (18 ft.-lbs.). Check camshaft end play, which should be 0.15-0.25 mm (0.006-0.010 inch). Install or remove shims (95) as needed to obtain desired end play. Complete reassembly by reversing removal procedure.

CRANKSHAFT AND CRANKCASE

R&R AND OVERHAUL. Refer to Fig. W2-14 for an exploded view of crankcase, which is constructed in three sections: flywheel end, center and pto end.

To remove crankshaft, remove cylinders, pistons and camshaft as previously outlined. Remove electric starter, flywheel and alternator. Remove covers (33) at crankcase joints—there is only one cover if engine is equipped with a fuel transfer pump. Unscrew eleven retaining nuts from crankcase flywheel section and separate flywheel section from remainder of crankcase. Remove socket head screw (32—Fig. W2-14 or W2-21) adjacent to camshaft bore. Separate center crankcase section with crankshaft from pto crankcase section. Remove control rod tube (39) and seals (38). Detach connecting rods. Unscrew two socket head screws (51) in bottom of center crankcase which secure main bearing support in crankcase. Press crankshaft and main bearing support out of center crankcase. Unscrew main bearing support screws (46) and separate support halves (47 and 50) from crankshaft.

Crankcase main journal diameter should be 52.702-52.733 mm (2.0749-2.0761 inches) for center journal and 46.044-46.069 mm (1.8128-1.8137 inches) for end journals. Main bearing clearance should be 0.043-0.079 mm (0.0017-0.0031 inch) for center main bearing and 0.046-0.084 mm (0.0018-0.0033 inch) for end main bearings. Maximum allowable main bearing clearance is 0.2 mm (0.008 inch). Main bearings are available in undersizes of 0.25 and 0.50 mm (0.010 and 0.020 inch).

Crankpin diameter should be 48.267-48.294 mm (1.9003-1.9013 inches) and rod bearing clearance should be 0.030-

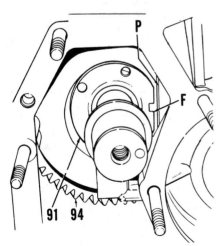

Fig. W2-19—Turn camshaft (91) so flat (F) on flange wall will pass crankcase projection (P).

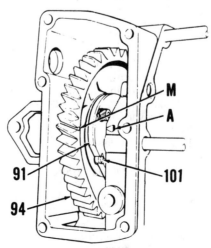

Fig. W2-20—Mate camshaft (91) with gear (94) as outlined in text.

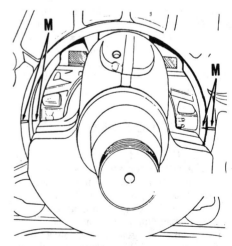

Fig. W2-22—Align crankshaft support and crankcase marks (M) as shown.

SERVICE MANUAL — Wisconsin

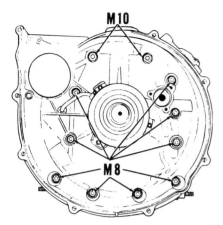

Fig. W2-23—View showing location of M8 and M10 nuts in flywheel crankcase section.

0.071 mm (0.0012-0.0028 inch) with a maximum allowable clearance of 0.2 mm (0.008 inch).

To reassemble crankshaft and crankcase, proceed as follows: Install main bearings (34 and 59) so flange is to inside of crankcase and oil hole is aligned in flywheel crankcase section bearing (59). Place thrust washer halves (48) on both sides of main bearing support halves (47 and 50) with grooved side out. Install main bearing support halves (47 and 50) on crankshaft and tighten screws (46) to 39 N·m (29 ft.-lbs.). With bearing support held towards one end of crankshaft, measure clearance between thrust rings (48) and side of crankshaft to determine crankshaft end play. Clearance should be 0.1-0.3 mm (0.004-0.012 inch). Thrust washer halves are available in standard and 0.1 mm (0.004 inch) oversize thicknesses.

Heat center crankcase to 150°C (300°F). With crankshaft pointing in correct direction, insert crankshaft into crankcase so mating line of bearing support is matched with crankcase alignment marks (M – Fig. W2-22) and immediately start support retaining screws but do not tighten.

Install connecting rods and tighten rod nuts to 34 N·m (25 ft.-lbs.). With control rod tube (39 – Fig. W2-14) and seals (38) in place, mate pto crankcase section (36) with center crankcase section (60). Install socket head screw (32 – Fig. W2-14 and W2-21) and tighten to 34 N·m (25 ft.-lbs.) only if center crankcase has cooled to ambient temperature. Mate flywheel crankcase section (56 – Fig. W2-14) with center crankcase. Tighten M8 nuts to 25 N·m (18 ft.-lbs.) and M10 nuts to 34 N·m (25 ft.-lbs.) (Fig. W2-23). If not previously tightened, tighten pto crankcase socket head screw (32 – Fig. W2-14 and W2-21) to 34 N·m (25 ft.-lbs.). Tighten main bearing support retaining screws (51 – Fig. W2-14) to 40 N·m (30 ft.-lbs.). Complete remainder of reassembly by reversing disassembly procedure.

Yanmar

SERVICE MANUAL

YANMAR

YANMAR DIESEL ENGINE (U.S.A.), INC.
1424 N. Hundley Street
Anaheim, California 92806

Model	No. Cyls.	Bore	Stroke	Displ.
1GM10	1	75 mm (2.95 in.)	72 mm (2.83 in.)	318 cc (19.4 cu. in.)
2GM20	2	75 mm (2.95 in.)	72 mm (2.83 in.)	636 cc (38.81 cu. in.)
3GM30	3	75 mm (2.95 in.)	72 mm (2.83 in.)	954 cc (58.21 cu. in.)
3HM35	3	80 mm (3.15 in.)	85 mm (3.35 in.)	1282 cc (78.23 cu. in.)

All models are four-stroke, indirect injection, liquid cooled diesel marine engines with gearbox driven from spring loaded flex plate attached to engine flywheel. Engines are available with either fresh water or sea water cooling. Models for fresh water cooling utilize sea water circulated through a heat exchanger to cool the fresh water being recirculated through engine and heat exchanger. Models for sea water cooling utilize expendable zinc elements in the engine water jacket to neutralize salt corrosion in the engine.

Engine model nameplate (see Fig. Y1-1) is attached to side of valve rocker arm cover on Model 1GM10 and to front face of engine timing gear cover on other models. The first number "1" in engine model number (M) indicates number of cylinders. The next two letters "GM" indicate engine type (marine engine in this example). The next two numbers "10" indicate engine series type. If there is no letter following engine series number, engine is equipped with direct sea water cooling. If letter "F" follows engine series number, engine is equipped with fresh water cooling. If letter "C" follows engine series number, engine is sail-drive. Engine serial number (S) is also stamped at top of block below fuel injector side of head on Model 1GM10, and on left side of cylinder block at crankshaft level on other models. Number one cylinder on 2 and 3-cylinder models is at flywheel end of engine.

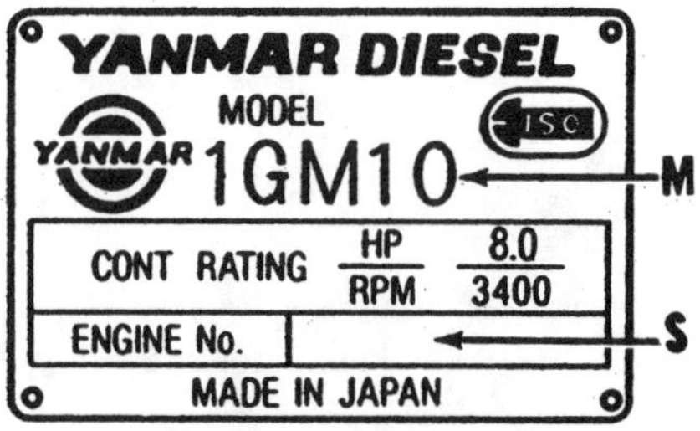

MAINTENANCE

LUBRICATION

Crankcase lubricant capacity is 1.3 liters (2.75 pints) for Model 1GM10; 2 liters (4.25 pints) for Model 2GM20; 2.6 liters (5.5 pints) for Model 3GM30; and 5.4 liters (11.4 pints) for Model 3HM35. Capacities are for engines mounted at 8 degree mounting angle.

Oil and filter should be changed after 20 hours of operation for new or reconditioned engines, and a second change after 30 hours of use from that change. Recommended oil change interval after first two changes is after every 100 hours of use.

Recommended lubricant is API classification CC or CD engine oil. In operating temperatures below 10° C (50° F), use of 10W, 20/20W, 20W or 20W/40 oils are recommended. From 10°-20° C (50°-68° F), use of SAE 20, 20W, 20/20W, or 20W-40 oil is recommended. From 20°-35° C (68°-95° F), SAE 30 or SAE 40 oil is recommended. Above 35° C (95° F), SAE 50 oil is recommended.

FUEL SYSTEM

FUEL FILTER. A renewable element cartridge (7—Fig. Y1-3) is located in the fuel filter. The filter element cannot be cleaned. Element should be renewed each 250 hours of operation, or sooner if misfiring or loss of power is evident. Thoroughly clean outside of filter before unscrewing retaining ring (9) and removing filter cup (8). Clean cup thoroughly and install new filter cartridge (7) with new sealing "O" ring (6). Open fuel shut-off valve and allow filter cup to fill with fuel before tightening retaining ring. Bleed system as outlined in following paragraph.

BLEED FUEL SYSTEM. Air must be bled from fuel system if fuel tank is allowed to run dry, if fuel lines, filter or other components within the fuel sys-

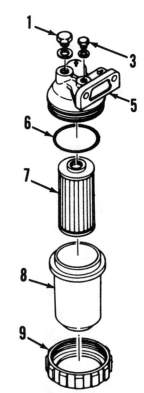

Fig. Y1-3—Exploded view of fuel filter assembly used on all models.
1. Plug
3. Bleed screw
5. Filter body
6. "O" ring
7. Filter element
8. Filter bowl
9. Retaining ring

Fig. Y1-1—Engine identification plate is attached to valve rocker arm cover on Model 1GM10 and to front face of timing gear cover on other models.

422

Illustrations Courtesy Yanmar Diesel Engine, Inc.

SERVICE MANUAL

Yanmar

tem have been disconnected or if engine has not been operated for a long period of time.

To bleed system, open fuel shut-off valve and loosen air bleed bolt (3—Fig. Y1-3). If fuel tank is mounted above filter, allow fuel to flow from filter air bleed until free of air. If tank is below filter, engage decompression levers and operate engine starting motor to actuate fuel feed pump until fuel is flowing from filter air bleed. Tighten filter air bleed bolt and loosen air bleed screw (6—Fig. Y1-4) on fuel injection pump inlet line (5) banjo bolt and operate engine starting motor until fuel flows from pump air bleed screw. Tighten injection pump bleed screw, disengage decompression and start engine.

If engine will not start, loosen high pressure fuel line (8) at each injector (9) and continue cranking engine until fuel escapes from loosened connections. Tighten pressure line connections and start engine.

INJECTION PUMP TIMING

Injection pump timing is adjusted by varying the thickness of shims (19—Fig. Y1-5) between injection pump body (10) and top of timing gear housing. Also, the timing of individual pump elements can be adjusted by varying the thickness of shims (25) located in pump plunger guide (26); however, this should not be attempted without use of fuel injection pump test stand. Beginning of injection should occur at 15 degrees BTDC for Models 1GM10 and 2GM20, at 18 degrees BTDC for Model 3GM30 and 21 degrees BTDC for Model 3HM35.

To check timing, clean pump and lines thoroughly using suitable solvent and compressed air and remove No. 1 (next to flywheel) injection line (8—Fig. Y1-4). Be sure system is bled of all air as described in preceding paragraph. Install a timing pipe on No. 1 delivery valve as shown in Fig. Y1-6. Turn crankshaft several revolutions to be sure timing pipe is filled, then turn to a position so that No. 1 piston is coming up on compression stroke. Move governor lever to midposition and slowly turn engine to bring timing marks (see Figs. Y1-7 and Y1-8) into alignment. Fuel should start to drip from timing pipe as marks are aligned. If fuel starts before marks are aligned, thickness of shims (S—Fig. Y1-6) must be increased; decrease shim thickness if fuel starts to flow after timing mark passes position of alignment. Changing shim thickness by 0.1 mm (0.004 inch) will change timing approximately one degree.

If timing is correct, reinstall No. 1 injection line and bleed line before tightening fitting at injector. If pump is re-

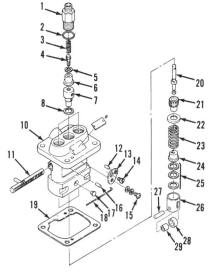

Fig. Y1-5—Exploded view of three-cylinder fuel injection pump. Components of one and two-cylinder pumps are similar. Pump should not be disassembled except by personnel having proper pump test equipment in a clean test room.

1. Delivery valve holder
2. "O" ring
3. Delivery valve spring
4. Delivery valve
5. Sealing washer
6. Delivery valve seat
7. Pump plunger body
8. Sealing washer
9. Pump body
10. Pump body
11. Pump control rack
12. Pin
13. Fuel injection adjusting plate
14. Eccentric screw
15. Set screw
16. Plunger barrel pin
17. Plunger guide pin
18. Pin retainer
19. Shims
20. Pump plunger
21. Fuel control pinion
22. Spring retainer
23. Plunger spring
24. Spring retainer
25. Adjusting shims
26. Plunger guide
27. Roller pin
28. Outer roller
29. Inner roller

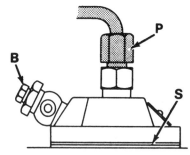

Fig. Y1-6—Install timing pipe in place of No. 1 injector line to check timing. Timing is adjusted by varying thickness of shims (S). Pump bleed screw (B) is located in inlet line banjo bolt.

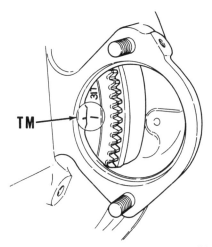

Fig. Y1-7—On all models, timing marks are visible after removing engine starting motor.

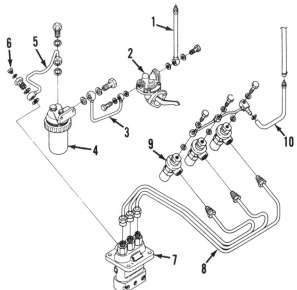

Fig. Y1-4—View showing components of three-cylinder fuel system. Arrangement of components for one and two-cylinder models is similar.

1. Fuel supply line
2. Fuel lift pump
3. Filter inlet line
4. Filter assy.
5. Filter to pump line
6. Injection pump bleed screw
7. Fuel injection pump
8. Injector lines
9. Fuel injectors
10. Fuel return line

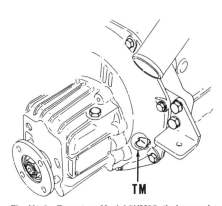

Fig. Y1-8—Except on Model 3HM35, timing marks are also located at rear side of flywheel cover plate.

Illustrations Courtesy Yanmar Diesel Engine, Inc.

Yanmar

moved to change shim thickness, bleed the system as previously outlined.

GOVERNOR

All models are equipped with a flyweight type governor mounted on front end of engine crankshaft. Refer to Fig. Y1-9 for drawing showing governor components for single cylinder engine and to Fig. Y1-10 for two and three-cylinder models. No-load maximum speed should be 3600-3650 rpm on Model 3HM35 and 3825-3875 rpm on all other models. Idle speed should be 825-875 rpm on all models.

If governed engine speed requires adjustment, loosen locknuts on high speed adjusting screw (4—Fig. Y1-9 or Fig. Y1-10) and idle adjusting screw (3). Operate engine until normal operating temperature is reached. Place throttle lever in wide open position and adjust high speed screw to obtain 3600-3650 rpm on Model 3HM35 and 3825-3875 rpm on other models, then tighten high speed screw locknut. Move throttle to slow speed position and adjust idle speed stop screw to obtain 825-875 rpm, then tighten idle speed screw locknut.

INJECTION LIMITER

An injection limiter (smoke stop) (10—Fig. Y1-9 or Y1-10) is located in engine timing cover. Refer to Fig. Y1-11 for exploded view of unit. If engine does not accelerate smoothly, remove the cap nut (1—Fig. Y1-11) and loosen the locknut (3). Back the injection limiter shaft (7) out slightly. If turned out too much, the engine will smoke under acceleration and when under load. If governor unit and fuel injection pump have been removed, adjust injection limiter as follows:

Move governor lever to idle speed position and remove oil filler cap on Model 1GM10, or side cover (7—Fig. Y1-10) on other models from timing gear housing. Remove cap nut (1—Fig. Y1-11), loosen locknut (3) and back injection control shaft (7) out so that spindle (6) in shaft does not contact governor lever (11—Fig. Y1-9 or Fig. Y1-10). Move governor lever to contact spindle, then turn injection control shaft in slowly until center mark (M—Fig. Y1-12) on rack is aligned with reference face (F) of pump body. Tighten locknut and reinstall sealing washer and cap nut. Adjustment points will be visible through oil filler cap opening or adjustment cover opening. The injection control shaft locknut and the no-load maximum speed screw locknut are secured with a wire and lead seal; be sure to reseal when adjustment is complete.

REPAIRS

TIGHTENING TORQUES

Refer to the following table for special tightening torques. Metric fasteners are used throughout the engine.

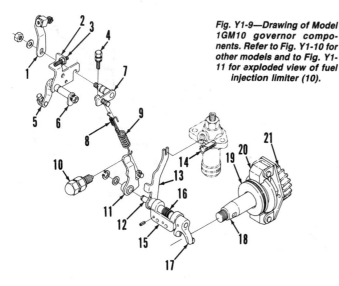

Fig. Y1-9—Drawing of Model 1GM10 governor components. Refer to Fig. Y1-10 for other models and to Fig. Y1-11 for exploded view of fuel injection limiter (10).

1. Speed control arm
2. Slow idle bracket
3. Slow idle speed screw
4. High idle speed screw
5. Engine stop lever
6. Engine stop cam
7. Speed control lever
8. Secondary governor spring
9. Primary governor spring
10. Fuel injection limiter
11. Control lever
12. Governor lever shaft
13. Rack control lever
14. Injection pump rack
15. Governor shaft bracket
16. Starting spring
17. Governor fork
18. Engine crankshaft
19. Governor sleeve
20. Flyweight assy.
21. Crankshaft timing gear

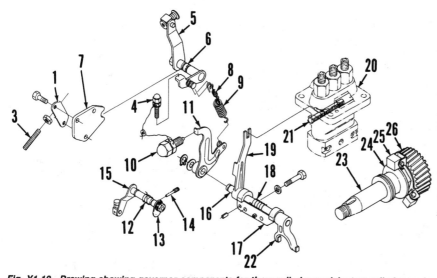

Fig. Y1-10—Drawing showing governor components for three-cylinder models; two-cylinder engine governor is similar. Refer to Fig. Y1-11 for exploded view of fuel injection limiter (10).

1. Idle adjuster bracket
3. Slow idle speed screw
4. High idle speed screw
5. Speed control arm
6. Speed control lever
7. Timing cover plate
8. Secondary governor spring
9. Primary governor spring
10. Fuel injection limiter
11. Control lever
12. Engine stop spring
13. Engine stop cam
14. Locking screw
15. Engine stop lever
16. Governor lever shaft
17. Governor shaft bracket
18. Starting spring
19. Rack control lever
20. Fuel injection pump
21. Injection pump rack
22. Governor fork
23. Engine crankshaft
24. Governor sleeve
25. Flyweight assy.
26. Crankshaft timing gear

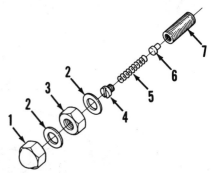

Fig. Y1-11—Exploded view of fuel injection limiter (smoke stop).

1. Cap nut
2. Sealing washers
3. Locknut
4. Spring retainer
5. Spring
6. Plunger
7. Injection limiter shaft

SERVICE MANUAL

Yanmar

Connecting rod—
  Model 3HM35 . . . . 44 N·m
    (32 ft.-lbs.)
  All other models . . . . 24 N·m
    (18 ft.-lbs.)
Cylinder head—
  Model 1GM10 . . . . 73 N·m
    (54 ft.-lbs.)
  Models 2GM20-3GM30,
    M12 bolts . . . . 103 N·m
      (76 ft.-lbs.)
  Models 2GM20-3GM30,
    M8 bolts . . . . 24 N·m
      (18 ft.-lbs.)
  Model 3HM35, M12 bolts . . 127 N·m
    (94 ft.-lbs.)
  Model 3HM35, M8 bolts . . . . 29 N·m
    (21 ft.-lbs.)
Flywheel . . . . 64-68 N·m
  (47-50 ft.-lbs.)
Crankshaft pulley . . . . 98 N·m
  (72 ft.-lbs.)
Crankshaft gear nut . . . . . . 79-98 N·m
  (58-72 ft.-lbs.)
Camshaft bearing
  set screw . . . . 20 N·m
    (15 ft.-lbs.)
Camshaft gear nut . . . . 69-78 N·m
  (51-58 ft.-lbs.)
Oil pump mounting bolts . . . . . 9 N·m
  (7 ft.-lbs.)
Rear main bearing bolts . . . . . 24 N·m
  (18 ft.-lbs.)
Intermediate main
  bearing bolts—
    Model 3HM35 . . . . 44-49 N·m
      (33-36 ft.-lbs.)
    All other models . . . . . . 30-34 N·m
      (22-25 ft.-lbs.)
Intermediate main
  set bolt—
    Model 3HM35 . . . . 69-73 N·m
      (51-54 ft.-lbs.)
    All other 2 & 3 cyl.
      models . . . . 44-49 N·m
        (33-36 ft.-lbs.)
Injector nozzle holder nuts . . . 19 N·m
  (14 ft.-lbs.)

Injection pump delivery
  valve holder . . . . 39-44 N·m
    (29-32 ft.-lbs.)
Injection pump retaining
  bolts . . . . 24 N·m
    (18 ft.-lbs.)

WATER PUMP AND COOLING SYSTEM

FRESH WATER COOLING. Models for fresh water cooling have two water pumps; one to circulate fresh water within the engine cooling system and another to pump sea water through a heat exchanger (see Fig. Y1-13) to cool the fresh water. The heat exchanger is also utilized as a water cooled exhaust manifold.

Refer to Fig. Y1-14 for exploded view of typical fresh water recirculating pump that is mounted on water inlet housing (8) attached to cylinder head. Component parts of this pump except the housing (3) are serviced, although it is recommended that complete water pump be renewed if unit is excessively worn or damaged.

To remove fresh water pump, drain sea water from heat exchanger and drain fresh water from heat exchanger and cylinder block. Loosen pump drive belt and remove alternator. Unbolt and remove water pump assembly.

To disassemble pump, unbolt and remove belt pulley from pulley hub (1—Fig Y1-14). Support pump housing and press pump shaft and bearing assembly (2) out of impeller (6) and pump housing (3). Remove shaft seal (5) from housing.

Inspect all parts for wear, corrosion or other damage and renew if necessary. To reassemble, press shaft and bearing assembly into pump housing until face of bearing is flush with front end of housing; apply press force against outer race of bearing only. Apply small amount of sealant to outer diameter of new shaft seal (5) and install seal in housing. Support front end of pump shaft, then press impeller onto shaft. Clearance between impeller fins and pump body should be 0.3-1.1 mm (0.012-0.043 inch). Distance between rear of impeller and straightedge placed across pump body should be 0.5 mm (0.020 inch).

The sea water heat exchanger pump is shown in Fig. Y1-15. Internal parts are

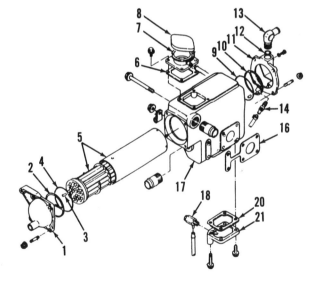

Fig. Y1-13—Exploded view of sea water-fresh water cooling system heat exchanger for Model 2GM20. Unit also functions as water cooled exhaust manifold.

1. Cover
2. Gasket
3. Dowel pin
4. "O" ring
5. Cooler pipe assy.
6. Gasket
7. Filler neck
8. Coolant cap
9. "O" ring
10. Gasket
11. Cover
12. "O" ring
13. Elbow
14. Drain cock
16. Exhaust manifold gasket
17. Exchanger housing
18. Drain valve
20. Gasket
21. Cover

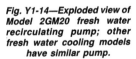

Fig. Y1-12—Adjust fuel injection limiter so that mark (M) on injection pump rack (R) is flush with machined reference face (F) of pump housing.

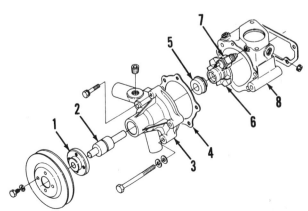

Fig. Y1-14—Exploded view of Model 2GM20 fresh water recirculating pump; other fresh water cooling models have similar pump.

1. Pulley flange
2. Shaft & bearing assy.
3. Pump body
4. Gasket
5. Seal assy.
6. Impeller
7. Coolant temperature switch
8. Pump housing

Yanmar

available for overhaul of this pump; the pump body (2) is not available separately. Refer to following paragraph for service information on this pump.

SEA WATER COOLING. A belt driven water pump with rubber impeller similar to that shown in Fig. Y1-15 is used on 2 and 3-cylinder models for direct sea water cooling. Water pump for Model 1GM10 is similar except that pump is internally driven by tang on rear end of pump shaft which engages slot in front end of engine oil pump. All components except the pump body (2) are available for service.

To disassemble pump, unbolt and remove cover plate (8—Fig. Y1-15) and pull impeller (6) off pump shaft. Remove belt pulley (11), if used, and spacer (12) from pump shaft. Remove snap ring (13). Support pump housing and press shaft (18) and bearings (14) out front of housing. Remove seal (20) and seal washer (19) from housing. Remove cam (5) from housing if necessary.

When reassembling pump, coat inside of impeller cavity and impeller with grease that will not be injurious to the rubber impeller. Impeller vanes should be folded back in direction opposite to pump rotation when impeller is installed in housing.

When servicing direct sea water cooling system, check the expendable zinc anti-corrosion elements (3—Fig. Y1-16) which should be renewed after each 500 hours, or when 50 percent corroded away. Dimensions of new element is 20 mm diameter by 20 mm long (0.787 × 0.787 inch) for Model 1GM10 and 20 mm diameter by 30 mm long (0.787 × 1.181 inch) for all other models. If elements appear in good condition, remove oxidized surface with wire brush and reinstall in engine.

THERMOSTAT

FRESH WATER COOLING. The thermostat (3—Fig. Y1-17) for fresh water cooling is located in the water pump mounting housing (4). The thermostat has an integral bypass valve and it is important that original replacement type thermostat be installed. The thermostat should start to open at 71° C (160° F) and be fully open at 85° C (185° F).

SEA WATER COOLING. A wax pellet type thermostat (3—Fig. Y1-18) is installed under the water outlet coupling (1). Two thermostats are used in the Model 3HM35 engine. Thermostat should start to open at 40°-44° C (104°-111° F) and be fully open at 50°-54° C (122°-129° F). When installing thermostat, be sure arrow on water outlet coupling (1) is pointing upward.

VALVE ADJUSTMENT

Valve clearance should be adjusted with engine cold and with piston of each cylinder at top dead center of compression stroke when adjusting valves of that cylinder. Recommended clearance is 0.2 mm (0.008 inch) for both the intake and exhaust valve.

SMALL DIESEL

CYLINDER HEAD

To remove cylinder head, drain coolant from engine and proceed as follows: Thoroughly clean fuel injectors, lines and pump using suitable solvent and

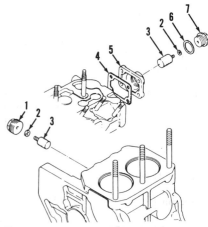

Fig. Y1-16—View showing expendable zinc elements (3) to deter corrosion in direct sea water cooling system of Model 2GM20; other sea water cooling models have zinc elements in similar locations.

1. Plug
2. Gasket
3. Zinc elements
4. Gasket
5. Cylinder head cover
6. Gasket
7. Plug

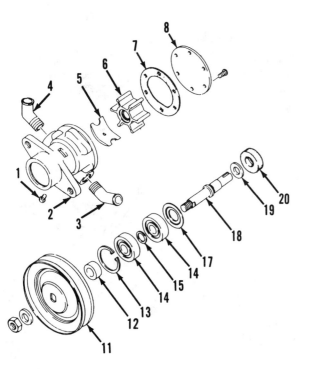

Fig. Y1-15—Exploded view of Model 2GM20 sea water pump for heat exchanger cooling of fresh water system; direct sea water cooling pumps are similar.

1. Cam screw
2. Pump body
3. Outlet elbow
4. Inlet elbow
5. Pump cam
6. Rubber impeller
7. Gasket
8. Cover plate
11. Drive pulley
12. Spacer
13. Snap ring
14. Bearings
15. Spacer
17. Bearing cover
18. Pump shaft
19. Water seal ring
20. Seal

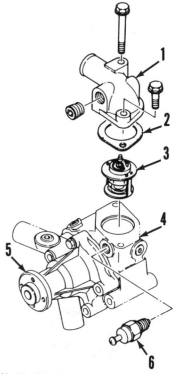

Fig. Y1-17—A spring loaded thermostat (3) with bypass valve is used in fresh water cooling models.

1. Water outlet
2. Gasket
3. Thermostat
4. Water pump mounting housing
5. Water pump assy.
6. Temperature sender unit

Illustrations Courtesy Yanmar Diesel Engine, Inc.

SERVICE MANUAL

Yanmar

compressed air. Remove high pressure lines and fuel return line from injectors and cap all openings. Remove air cleaner and intake from cylinder head. Disconnect coolant hose and remove water cooled manifold or fresh water heat exchanger assembly from cylinder head. Remove rocker arm cover, rocker arm shaft assembly and push rods. Upper ends of valve stems on all models are fitted with lash caps; be careful not to lose the caps. On Model 1GM10, rocker arms are retained on stand by bosses in rocker arm cover as shown in Fig. Y1-22. Refer to tightening sequence diagram (Fig. Y1-19, Y1-20 or Y1-21) and loosen cylinder head bolts in reverse of tightening sequence. Lift cylinder head from engine.

Thoroughly clean mating surfaces of head and cylinder block. Inspect surfaces for cracks and check surfaces for distortion using a straightedge and feeler gage. Maximum allowable distortion of cylinder head surface is 0.07 mm (0.003 inch).

Before reinstalling head, coat both sides of head gasket with "Three Bond 50" or equivalent (silicone non-solvent type liquid sealant). Place gasket on cylinder block with flat side upward (side marked "TOP" up) and coat bolt threads with oil.

To reinstall cylinder head, reverse removal procedure and tighten cylinder head bolts in several steps in sequence shown in Fig. Y1-19, Y1-20 or Y1-21 until proper head bolt torque is reached; refer to TIGHTENING TORQUES specification listing. Loosen locknuts and back rocker arm adjusting screws out before installing rocker arm support or shaft assembly and push rods. Adjust valves for proper clearance before installing rocker arm cover. Before tightening injector lines, bleed fuel system.

VALVE SYSTEM

Intake and exhaust valves seat directly in cylinder head. Valve face and seat angles are 45 degrees. Valve head is recessed 1.25 mm (0.049 inch) on Model 3HM35 and 0.95 mm (0.037 inch) on other models. Maximum allowable valve recess is 1.55 mm (0.061 inch) on Model 3HM35 and 1.25 mm (0.049 inch) on other models.

Nominal valve stem diameter for both intake and exhaust valves is 7.0 mm (0.2756 inch) on all models. Valve should be renewed if stem diameter measures 6.9 mm (0.2717) or less. Valve stem ends are fitted with lash caps (8—Fig. Y1-23); be careful when disassembling engine not to lose any of the caps.

Valve guides are renewable in all models. Desired valve stem to guide clearance is 0.045-0.07 mm (0.0018-0.0028 inch); maximum allowable clearance is 0.15 mm (0.006 inch). Renew guides and/or valves if clearance is excessive. Intake and exhaust valve guides are alike on Model 1GM10; exhaust guides on other models have a counterbore in combustion chamber end of guide. To remove old guides, use suitable piloted driver and drive guides out from bottom side of head. Drive new guides into top side of cylinder head until guide protrudes a distance of 7 mm (0.276 inch) above valve spring seating surface of head. On all models except Model 1GM10, top end of guides have an annular groove; these guides will be installed at correct position if guide is driven into head until lower edge of groove is flush with head surface. Ream new guides to obtain desired valve stem to guide clearance.

Valve spring free length is 38.5 mm (1.516 inches). Renew springs if distorted, discolored from heat or if free length is 37 mm (1.457 inches) or less. Spring pressure at installed height should be 14.43 kg (31.8 pounds) at 30.2 mm (1.189 inches) for Model 3HM35 and 16.16 kg (35.6 pounds) at 29.2 mm (1.150 inches) for all other models. Renew spring if load at specified height is 12.2 kg (26.9

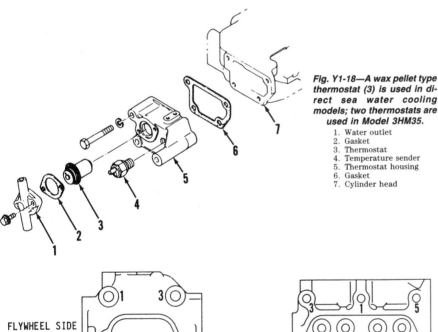

Fig. Y1-18—A wax pellet type thermostat (3) is used in direct sea water cooling models; two thermostats are used in Model 3HM35.

1. Water outlet
2. Gasket
3. Thermostat
4. Temperature sender
5. Thermostat housing
6. Gasket
7. Cylinder head

Fig. Y1-19—Tightening sequence for Model 1GM10 cylinder head bolts. When removing head, loosen bolts in reverse of sequence shown.

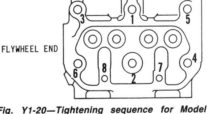

Fig. Y1-20—Tightening sequence for Model 2GM20 cylinder head bolts. Loosen bolts in reverse of order shown when removing cylinder head.

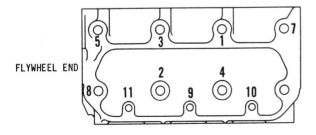

Fig. Y1-21—Tighten cylinder head bolts in order shown for Models 3GM30 and 3HM35. Reverse the order shown to loosen bolts when removing cylinder head.

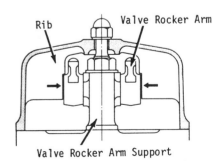

Fig. Y1-22—Rocker arms of Model 1GM10 are retained on support shaft by ribs inside rocker arm cover.

Yanmar

SMALL DIESEL

pounds) or less on Model 3HM35 or 13.7 kg (30.2 pounds) or less on other models.

Rocker arm bushings are not available separately from rocker arm assembly. On Model 1GM10, rocker arm shaft and support are integral; shaft is separate from supports on all other models. Intake and exhaust rocker arms are alike on one and two-cylinder models, and are different on three-cylinder engines. Rocker arm shaft diameter is 12 mm (0.472 inch) on Model 1GM10 and 14 mm (0.551 inch) on other models. Desired rocker arm to shaft clearance is 0.016-0.052 mm (0.0006-0.002 inch). Renew rocker arms and/or rocker arm shaft if clearance exceeds 0.15 mm (0.006 inch).

Be sure that the rocker arm support engages the dowel pin (7—Fig. Y1-23 or Fig. Y1-24) that holds support in alignment on cylinder head before tightening support retaining nuts. Bosses on the inside of the rocker arm cover on the single-cylinder engine hold the rocker arms on the shaft (see Fig. Y1-22). Tighten rocker arm support stud nuts to a torque of 36 N·m (27 ft.-lbs.).

DECOMPRESSION MECHANISM

A decompression lever (1—Fig. Y1-25) and cam (2) are fitted in the valve rocker arm cover at each exhaust valve. Moving the lever to decompression position will hold the exhaust valve open a slight amount allowing engine to be cranked easily. Adjustment of the decompression device is not provided. A check ball (4) and spring holds the lever in released position.

INJECTORS

REMOVE AND REINSTALL. Before loosening injector lines, carefully clean injectors, lines and surrounding area using suitable solvent and compressed air.

To remove an injector, first remove high pressure line leading from injection pump to injector. Disconnect fuel return (bleed) line from injector by removing banjo fitting bolt. Remove the two nuts (1—Fig. Y1-26) from injector retaining studs and pull injector (3) from cylinder head. With injector removed, the precombustion chamber and seals (4 through 8) can be removed from cylinder head.

Clean exterior surface of injector assembly and bore in cylinder head, being careful to keep all surfaces free of dust, lint and other small particles until injector is reinstalled. Clean all carbon from precombustion cups and install with new seals, then install fuel injector. Tighten nozzle retaining nuts down evenly to a torque of 20 N·m (15 ft.-lbs.). Reinstall high pressure lines and fuel return line and bleed the fuel system before tightening line connections.

TESTING. A complete job of testing and adjusting the injector requires use of special equipment. Nozzle should be tested for opening pressure, seat leakage and spray pattern. Before conducting tests, operate tester lever until fuel flows from tester pipe, then attach nozzle to pipe.

WARNING: Fuel emerges from injector with sufficient force to penetrate the skin. When testing injector, keep yourself clear of nozzle spray.

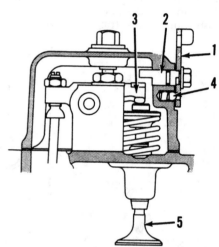

Fig. Y1-25—Cross-sectional view showing decompression mechanism. Cam (2) holds exhaust valve rocker arm (3) so that valve (5) cannot completely close when lever is moved to decompression position. Check ball (4) holds lever in released position.

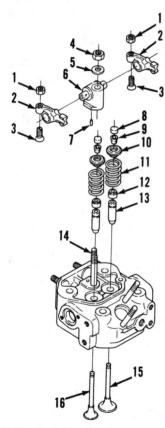

Fig. Y1-23—Exploded view of cylinder head, valves and rocker arm assembly on Model 1GM10 engine. Be especially careful not to lose valve lash caps (8) when removing rocker arm assembly.

1. Locknut
2. Rocker arm
3. Adjusting screw
4. Rocker support nut
5. Washer
6. Rocker arm support
7. Dowel pin
8. Valve lash caps
9. Split valve keys
10. Valve spring retainer
11. Valve springs
12. Valve stem seals
13. Valve guides
14. Rocker support stud
15. Intake valve
16. Exhaust valve

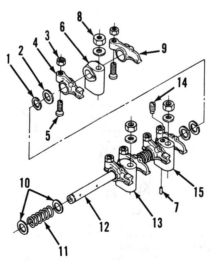

Fig. Y1-24—Exploded view of three-cylinder rocker arm assembly. Rocker arm shaft is secured by set screw (14) in rear (No. 1 cylinder) rocker arm support. Model 2GM20 rocker shaft assembly is similar except for number of rocker arms used and that intake and exhaust rocker arms are alike on this model.

1. Snap ring
2. Thrust washer
3. Locknut
4. Exhaust rocker arm
5. Adjusting screw
6. Rocker arm support
7. Dowel pin
8. Rocker support nut
9. Intake rocker arm
10. Washers
11. Spring
12. Rocker arm shaft
13. Rocker shaft support
14. Shaft retaining screw
15. Rear shaft support

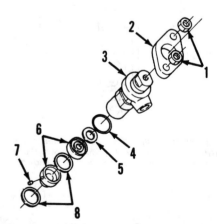

Fig. Y1-26—After fuel injector (3) is removed, precombustion chamber components (4 through 8) can be removed. Side of injector plate (2) with notches must be installed down (toward injector).

1. Injector retaining nuts
2. Injector retaining plate
3. Fuel injector
4. "O" ring
5. Insulator washer
6. Precombustion chamber
7. Locating pin
8. Sealing rings

SERVICE MANUAL

Close valve to tester gage and operate tester pump lever a few quick strokes to purge air from injector and to be sure that nozzle valve is not stuck or plugged.

OPENING PRESSURE. Open valve to tester gage and operate tester lever slowly while observing gage reading. Opening pressure should be 15190-16157 kPa (2205-2345 psi) for Model 3HM35 injectors and 16160-17150 kPa (2347-2489 psi) for all other injectors.

Opening pressure is adjusted by adding or removing shims (3–Fig. Y1-27). Each 0.1 mm (0.004 inch) change in shim pack thickness will change opening pressure approximately 689-979 kPa (142 psi).

SPRAY PATTERN. The spray should be well atomized and the pattern should be conical without splits, dribble or intermittent action. Otherwise, nozzle must be reconditioned or renewed.

SEAT LEAKAGE. Wipe tip of nozzle with clean blotting paper, then slowly bring gage pressure to a reading about 1965 kPa (285 psi) lower than observed opening pressure and hold this gage pressure for about 10 seconds. If any oil appears on nozzle tip, overhaul and retest injector.

OVERHAUL. Hard or sharp tools, emery cloth, grinding compound or other than approved solvents or lapping compounds must never be used. Wipe all dirt and loose carbon from exterior of nozzle and holder assembly. Refer to Fig. Y1-27 for exploded view of injector and proceed as follows:

Hold nozzle body (7) in a soft jawed vise or holding fixture and remove the spring holder (1) and nozzle cap nut (10). Place all parts in clean calibrating oil or diesel fuel as they are removed, taking care to keep parts from each injector separate from other injector units which may be disassembled at this time.

Clean exterior surfaces with a brass wire brush and soak parts in approved carbon solvent if necessary to loosen hard carbon deposits. Rinse parts in clean diesel fuel or calibrating oil immediately after cleaning to neutralize the solvent and prevent etching of polished surfaces.

Clean nozzle spray hole from inside using a pointed hardwood stick. Scrape carbon from pressure chamber using a hooked brass scraper. Clean valve seat using brass scraper, then polish seat using wood polishing stick and mutton tallow. Check nozzle needle (8) and nozzle body (9) for wear, scoring or other damage. With nozzle assembly wet with fuel oil and held in vertical position, nozzle needle should drop smoothly onto nozzle body seat by its own weight. If needle sticks, reclean or renew nozzle assembly.

Reclean all parts by rinsing thoroughly in clean diesel fuel or calibrating oil and assemble while parts are immersed in the fluid. Make sure adjusting shim pack is intact. Tighten nozzle retaining nut (10) to a torque of 98 N·m (72 ft.-lbs.) and tighten nozzle spring nut (1) to a torque of 69-78 N·m (51-58 ft.-lbs.). Do not overtighten as this may cause distortion and damage to nozzle. If nozzle is leaking, disassemble and clean again. Retest assembled injector as previously outlined.

INJECTION PUMP

To remove fuel injection pump from top of timing gear cover, first clean pump and fuel lines using suitable solvent and compressed air. Remove the pump to injector high pressure lines and disconnect fuel supply and return lines from pump. Remove the oil filler cap on Model 1GM10 or side cover from timing gear cover on other models. Unbolt injection pump from timing cover and, looking through filler cap or side cover opening, move governor lever to clear pump rack. Remove pump taking care not to lose or damage timing shims located between pump body and timing cover. A notch is provided in timing cover for clearance of pump rack.

With pump removed, hold pump assembly on side and move control rack to top of stroke. Rack should then slide down smoothly by its own weight. If not, rack is binding and should be washed with clean diesel fuel and rechecked. If binding cannot be eliminated, pump should be serviced at qualified diesel injection repair station.

To reinstall pump, take care to install pump with same thickness of timing shims as when removed. Look through oil filler cap or timing gear side cover opening to engage pin on injection pump rack with notch in end of governor lever. Install and tighten pump retaining nuts and complete installation by reversing removal procedures. Bleed fuel system before tightening fuel lines.

TIMING GEAR COVER

To remove timing gear cover, first remove fuel injection pump, then proceed as follows: Remove alternator. Drain cooling water from cylinder block and remove water pump(s). Unscrew crankshaft pulley nut and remove pulley. Unbolt and remove timing gear cover from engine, taking note that ball bearing on front of crankshaft is carried in bore of timing cover. Bearing should stay in timing cover as cover is removed from crankshaft.

With timing cover removed, check ball bearing and renew if loose or rough. Install new crankshaft seal with spring loaded lip to inside of cover. Inspect seal contact area of crankshaft to be sure it is not worn or scored. Reinstall cover by reversing removal procedure.

INJECTION PUMP CAMSHAFT

The fuel injection pump camshaft (2–Fig. Y1-28) is mounted on the front end of the engine camshaft (9), and can be removed after removing fuel injection pump and timing cover. On models with hand crank option, remove set screw (6) from front end of camshaft and remove the hand crank pin (7). Unscrew hex nut (1) and remove injection pump cam from engine camshaft.

Fuel cam standard height is 45 mm (1.7717 inches) and wear limit is 44.90 mm (1.7677 inches).

Reverse removal procedure to install new injection pump cam. Tighten hex nut to a torque of 69-78 N·m (51-58 ft.-lbs.), then install timing gear cover and remaining parts.

TIMING GEARS

Timing gears are accessible after removing timing gear cover. Backlash between crankshaft gear and camshaft gear and between crankshaft gear and oil pump gear should be 0.05-0.13 mm (0.002-0.005 inch); maximum allowable clearance is 0.3 mm (0.012 inch). Cam-

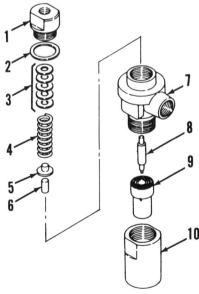

Fig. Y1-27—Exploded view of fuel injector assembly. Valve (8) and nozzle body (9) are matched set and not available separately.

1. Nozzle spring nut
2. Sealing washer
3. Adjustment shims
4. Nozzle spring
5. Spring seat
6. Inner spindle
7. Nozzle holder
8. Nozzle valve
9. Nozzle body
10. Nozzle cap nut

Illustrations Courtesy Yanmar Diesel Engine, Inc.

Yanmar
SMALL DIESEL

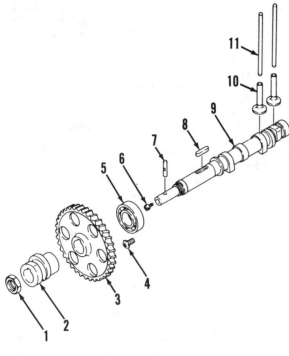

Fig. Y1-28—Fuel injection pump camshaft (2) is mounted on front end of engine camshaft (9). Parts for two-cylinder engine are shown; components for one and three-cylinder engine are arranged in similar manner.
1. Hex nut
2. Injection pump cam
3. Camshaft gear
4. Bearing retaining screw
5. Ball bearing
6. Crank pin screw
7. Hand crank pin
8. Gear & cam key
9. Camshaft
10. Cam followers
11. Push rods

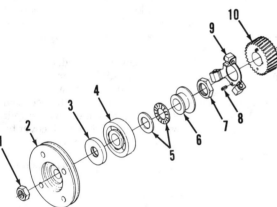

Fig Y1-29—View showing parts mounted on front end of engine crankshaft. Flyweight assembly for Model 1GM10 has only two weights. Pin (8) locates flyweight plate to crankshaft gear (10).
1. Pulley nut
2. Crankshaft pulley
3. Crankshaft oil seal
4. Ball bearing
5. Governor thrust bearing
6. Governor sleeve
7. Governor retaining nut
8. Dowel pin
9. Flyweight assy.
10. Crankshaft gear

On other models, the valve assembly is located in the oil filter mounting adaptor (Fig. Y1-32). Minimum oil pressure should be 49 kPa (7 psi) at 850 engine rpm on all models. Oil pressure should be 295-390 kPa (43-57 psi) at 3400 rpm on Model 3HM35 and at 3600 rpm on other models.

A gerotor type pump is used on all models. Exploded view of Model 1GM10 pump is shown in Fig. Y1-35 and explod-

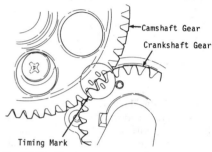

Fig. Y1-30—Single mark on crankshaft gear tooth should mesh between two marked teeth on camshaft gear.

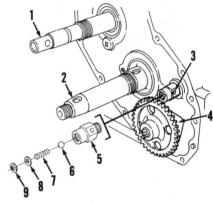

Fig. Y1-31—Oil pressure regulating valve (3) is located in front of cylinder block under timing gear cover on Model 1GM10.
1. Camshaft
2. Crankshaft
3. Oil pressure valve
4. Oil pump assy.
5. Valve body
6. Ball
7. Spring
8. Spring retainer
9. Snap ring

shaft gear (3—Fig. Y1-28) can be removed after removing the injection pump camshaft. Crankshaft gear (10—Fig. Y1-29) can be removed from front of crankshaft after removing governor thrust bearing (5), sleeve (6), retaining nut (7) and governor flyweight assembly (9). Oil pump drive gear and pump are removed as a unit.

Refer to Fig. Y1-30 for crankshaft gear to camshaft gear timing marks when reinstalling gears. Timing of oil pump gear to crankshaft gear is not required.

OIL PUMP AND RELIEF VALVE

Oil pump and drive gear are removed as a unit after removing timing gear cover. On Model 1GM10, pump retaining bolts are removed through openings in pump gear; pump retaining bolts on other models are located outside drive gear circumference.

The Model 1GM10 oil pump relief valve assembly (3—Fig. Y1-31) is located above pump drive gear in engine block.

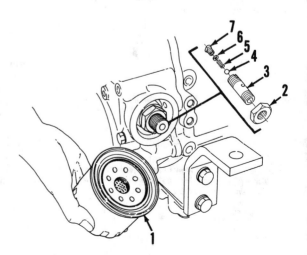

Fig. Y1-32—On all models except Model 1GM10, oil pressure regulating valve is located in oil filter (1) mounting fitting.
1. Oil filter
2. Nut
3. Valve body
4. Ball
5. Spring
6. Shim
7. Spring retainer

Illustrations Courtesy Yanmar Diesel Engine, Inc.

SERVICE MANUAL

Yanmar

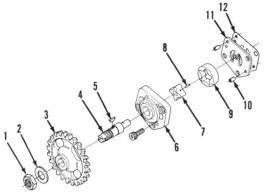

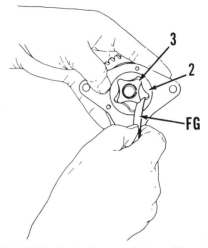

Fig. Y1-35—Exploded view of Model 1GM10 oil pump assembly.
1. Hex nut
2. Washer
3. Drive gear
4. Oil pump shaft
5. Woodruff key
6. Pump body
7. Inner rotor
8. Drive pin
9. Outer rotor
10. Cover plate pins
11. Cover plate
12. Mounting gasket

Fig. Y1-38—Check clearance between lobes of outer rotor (2) and inner rotor (3) using feeler gage (FG).

ed view of pump used in all other models is shown in Fig. Y1-36. Clearance between outer rotor and housing, and between inner and outer rotors should be 0.05-0.1 mm (0.002-0.004 inch); wear limit is 0.13 mm (0.005 inch). Rotor end clearance should be 0.03-0.08 mm (0.0012-0.003 inch); wear limit is 0.13 mm (0.005 inch). Refer to Figs. Y1-37, Y1-38 and Y1-39 for rotor clearance measurements. Shaft to housing clearance should be 0.015-0.05 mm (0.0006-0.002 inch). Renew oil pump if parts are scored or worn beyond limits.

weight assembly. A dowel pin (18) locates flyweight bracket (21) on engine crankshaft gear (22).

GOVERNOR

All models are equipped with a flyweight type governor mounted on engine crankshaft at front side of crankshaft timing gear. Refer to Fig. Y1-40 for exploded view of governor mechanism for Model 2GM20, governor components for other models are similar. Refer to MAINTENANCE section for governor adjustment and other views of governor mechanism.

With timing gear cover removed, remove thrust washer (15—Fig. Y1-40), needle thrust bearing (16) and governor sleeve (17) from engine crankshaft. Remove the flyweight assembly retaining nut from crankshaft, then remove fly-

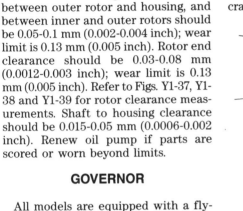

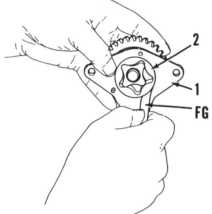

Fig. Y1-37—Check clearance between outer rotor (2) and pump body (1) using feeler gage (FG).

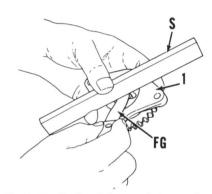

Fig. Y1-39—Check end clearance between rotors and face of pump body (1) using straightedge (S) and feeler gage (FG).

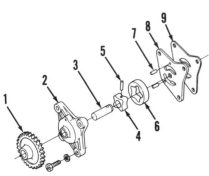

Fig. Y1-36—Exploded view of engine oil pump used on all models except Model 1GM10.
1. Drive gear
2. Pump body
3. Pump shaft
4. Inner rotor
5. Drive pin
6. Outer rotor
7. Cover plate pins
8. Cover plate
9. Mounting gasket

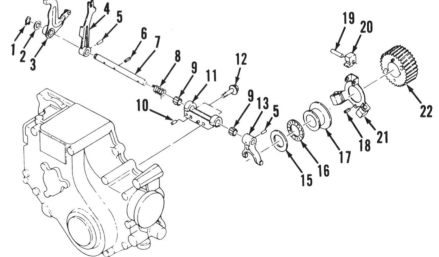

Fig. Y1-40—Exploded view of Model 2GM20 governor mechanism. Other models are similar except that Model 1GM10 has only two flyweights (20).
1. Snap ring
2. Washer
3. Control lever
4. Rack control lever
5. Taper pin
6. Spring pin
7. Governor lever shaft
8. Starting spring
9. Needle bearings
10. Dowel pin
11. Governor shaft bracket
12. Bracket bolts
13. Governor fork
15. Thrust washer
16. Thrust bearing
17. Governor sleeve
18. Dowel pin
19. Flyweight pin
20. Governor flyweights
21. Flyweight plate
22. Crankshaft gear

Illustrations Courtesy Yanmar Diesel Engine, Inc.

Yanmar

Sleeve (17) should slide smoothly on engine crankshaft. Check all parts for wear, scoring or other damage and renew as necessary. Reassemble by reversing disassembly procedure and adjust governor as outlined in MAINTENANCE section.

PISTON AND ROD UNITS

Connecting rod and piston assemblies are removable from above after removing cylinder head, oil pan and connecting rod caps.

When reinstalling, be sure ring end gaps are evenly spaced around piston. Piston may be installed in either direction on rod. When installing assembly, place alignment marks (see Figs. Y1-41 or Y1-42) on rod and cap together and toward camshaft side (exhaust side) of engine.

PISTONS, PINS, RINGS AND CYLINDERS

Three-ring pistons (two compression and one oil control ring) are used on all models. Piston pin is full floating and retained by a snap ring at each end of pin. Pistons and rings are available in standard size only. On all models, pistons operate in unsleeved cylinder bores. New pistons include piston rings; piston pin and retaining rings are sold separately.

Inspect pistons for cracks, wear, scuffing or other visible damage. Measure piston skirt diameter at right angle to piston pin and 9 mm (0.354 inch) from bottom edge of skirt. Check piston ring side clearance in groove using a feeler gage and new piston ring. Refer to the following piston assembly specifications:

Piston skirt diameter—
   Model 3HM35 . . . . 79.9-79.93 mm
      (3.146-3.147 in.)
   Wear limit . . . . 79.84 mm
      (3.143 in.)
   All other models . . . 74.91-74.94 mm
      (2.949-2.950 in.)
   Wear limit . . . . 74.85 mm
      (2.947 in.)
Ring side clearance—
   Top ring groove . . . . . 0.065-0.10 mm
      (0.0026-0.004 in.)
   Second ring groove . . 0.035-0.07 mm
      (0.0014-0.003 in.)
   Oil ring groove . . . . . . 0.02-0.055 mm
      (0.0008-0.002 in.)
Cylinder diameter—
   Model 3HM35 . . . . . . 80.0-80.03 mm
      (3.1496-3.1508 in.)
   Wear limit . . . . 80.10 mm
      (3.1535 in.)
   All other models . . . . 75.0-75.03 mm
      (2.9528-2.9540 in.)
   Wear limit . . . . 75.10 mm
      (2.9567 in.)

Cylinder out-of-round—
   All models, wear limit . . . . 0.02 mm
      (0.0008 in.)

Maximum allowable ring side clearance in groove is 0.2 mm (0.008 inch) for top and second compression rings and 0.15 mm (0.006 inch) for oil ring. Renew piston if side clearance is excessive.

Piston ring end gap in cylinder should be 0.25-0.45 mm (0.010-0.018 inch) on Model 3HM35 top compression and oil control ring, and 0.2-0.4 mm (0.008-0.016 inch) for all other rings, including the second compression ring on Model 3HM35. Maximum allowable piston ring end gap is 1.75 mm (0.069 inch) for top compression ring and oil control ring on Model 3HM35, and 1.5 mm (0.060 inch) for all other rings.

Piston pin diameter, new, is 22.99-23 mm (0.9051-0.9055 inch) for Model 3HM35, and 19.99-20 mm (0.7870-0.7874 inch) for all other models. Renew piston pin if wear equals or exceeds 0.02 mm (0.0008 inch). Desired fit of pin in piston is −0.005 mm (−0.0002 inch) interference fit to a clearance of 0.017 mm (0.0007 inch). Always install new piston pin retaining snap rings if rings have been removed.

To install new piston on connecting rod, heat piston in oil to a temperature of approximately 80° C (180° F) to facilitate installation of pin in piston. Install pin retaining snap rings in grooves at outer ends of pin bore in piston with any bow in snap ring ends to outside of piston.

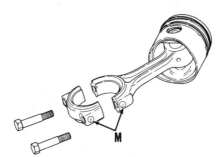

Fig. Y1-41—Alignment marks (M) are stamped in side of connecting rod on all models except Model 3HM35.

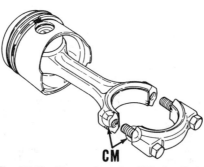

Fig. Y1-42—Alignment marks (CM) for Model 3HM35 are cast into connecting rod.

SMALL DIESEL

When installing rings on piston, refer to Fig. Y1-43 and observe the following: Install oil control ring expander in bottom groove, hold expander ends butted together, then install oil control ring with end gap positioned 180 degrees away from expander ends. Install compression rings with identification mark on side of rings up, and with end gaps of all rings spaced evenly around piston; end gap of top compression ring should be above piston pin.

CONNECTING RODS AND BEARINGS

Connecting rod is fitted with slip-in precision bearing inserts at crankpin end and a bushing at piston pin end. Side clearance of connecting rod on crankpin should be 0.2-0.4 mm (0.008-0.016 inch). Check connecting rod on rod alignment fixture; maximum distortion allowable is 0.08 mm (0.003 inch) per 100 mm (4 inches) length.

Desired crankpin bearing clearance is 0.036-0.092 mm (0.0014-0.0036 inch) for Model 3HM35, and 0.028-0.086 mm (0.0011-0.0034 inch) on all other models. Wear limit is 0.13 mm (0.005 inch) clearance on all models. Bearing inserts are available in standard and 0.25 mm (0.010 inch) undersize.

Desired piston pin to bushing clearance is zero to 0.02 mm (0.0008 inch) (a thumb press fit). Maximum allowable pin to rod bushing clearance is 0.1 mm (0.004 inch). Renew bushing or connecting rod if clearance is excessive.

CAMSHAFT AND BEARINGS

The camshaft on one and two-cylinder models is supported at front end by a ball bearing (5—Fig. Y1-44). Rear end of camshaft rotates directly in unbushed bore in cylinder block. Three cylinder models have two intermediate journals which ride directly in cast iron cylinder block in addition to front ball bearing and rear journal.

To remove camshaft, first remove the cylinder head, fuel lift pump and timing gear cover. If equipped with optional hand starter, unscrew set screw (6) and

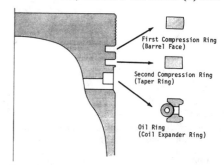

Fig. Y1-43—Cross-sectional view showing proper installation of piston rings in grooves.

SERVICE MANUAL

remove crank pin (7). Unscrew hex nut (1) and remove injection pump camshaft (2), camshaft timing gear (3) and key (8). Remove the screw (4) retaining front camshaft ball bearing in cylinder block. Use magnetic tools to raise cam followers (10) away from camshaft, or tip engine over so lifters fall away from camshaft, then remove camshaft from front of cylinder block. If necessary to remove cam followers, engine oil pan must be removed.

Measure camshaft lobes and journals and check against following values:
Rear journal diameter—
  Model 1GM10 . . . . 20 mm
    (0.7874 in.)
  All other models . . . . 30 mm
    (1.1811 in.)
Center journal diameter—
  3-cylinder . . . . 41.5 mm
    (1.6339 in.)
Desired bearing clearance—
  All models . . . . 0.05-0.1 mm
    (0.002-0.004 in.)
  Maximum allowable
    clearance . . . . 0.15 mm
      (0.006 in.)
Intake & exhaust cam lobe
  diameter—
  Model 1GM10 . . . . 29 mm
    (1.1417 in.)
  Wear limit . . . . 28.7 mm
    (1.1292 in.)
  All other models . . . . 35 mm
    (1.378 in.)
  Wear limit . . . . 34.7 mm
    (1.3661 in.)
Fuel lift pump lobe
  diameter—
  Model 1GM10 . . . . 22 mm
    (0.8661 in.)
  Models 2GM20 & 3GM30 . . . . 33 mm
    (1.2992 in.)
  Model 3HM35 . . . . 33.5 mm
    (1.3189 in.)

When supported at ends in "V" blocks or lathe centers, maximum runout measured at intermediate journal on 3-cylinder camshaft should be less than 0.004 mm (0.0016 inch) as measured with dial indicator.

Fuel injection pump cam lobe diameter should be 45 mm (1.7717 inch) on all models. Renew injection pump cam if lobe diameter measures 44.9 mm (1.7677 inch) or less.

Inspect cam followers and renew if contact face is cracked, pitted or if wear is not circular indicating follower is not turning normally. Cam follower stem diameter is 10 mm (0.3937 inch) for all models; renew if worn to diameter of 9.95 mm (0.3917 inch) or less. Desired stem to block bore clearance is 0.025-0.06 mm (0.001-0.0024 inch) for Model 1GM10, and 0.01-0.04 mm (0.0004-0.0016 inch) for all other models. Maximum allowable cam follower stem to block bore diameter is 0.10 mm (0.004 inch).

Lubricate cam followers, insert in cylinder block bores with block inverted or with magnetic holders to keep them away from camshaft. Lubricate cam journals and lobes, install in cylinder block, then complete reassembly of engine by reversing disassembly procedure. Align double timing marks on camshaft gear with single mark on crankshaft gear. Gear may be installed on key in camshaft, then the unit moved into place with marks aligned. Retaining screw for ball bearing can be installed through hole in web of camshaft gear. Install fuel injection pump cam with end surface marked "O" toward front.

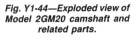

Fig. Y1-44—Exploded view of Model 2GM20 camshaft and related parts.
1. Hex nut
2. Injection pump cam
3. Camshaft gear
4. Bearing retaining screw
5. Ball bearing
6. Hand crank pin screw
7. Hand crank pin
8. Key
9. Camshaft
10. Cam followers
11. Push rods

Yanmar

CRANKSHAFT AND MAIN BEARINGS

Crankshaft for Model 1GM10 is supported in two main bearings; the front bearing is a circular bushing (4—Fig. Y1-45) pressed into front bore of cylinder block and the rear bearing is a circular bushing (10) pressed into rear bearing housing (12). Full circle thrust washers (3 and 5) located at front and rear of front main bearing control crankshaft end play. When reassembling, place thrust washer with square tang to rear and thrust washer with rounded tang and hole to front of front main bearing.

Crankshaft for Model 3GM30 and Model 3HM35 is supported in the two front and rear bushings as Model 1GM10,

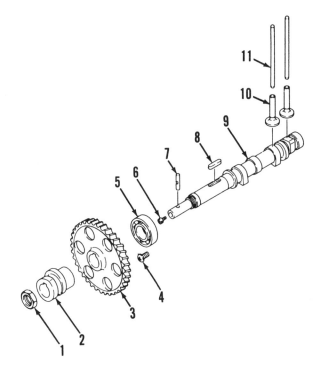

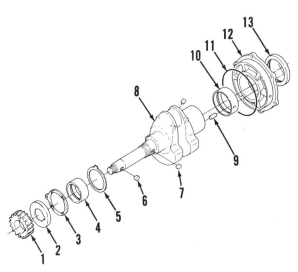

Fig. Y1-45—Exploded view of Model 1GM10 crankshaft and related parts. Front main bearing (4) is pressed into bore of cylinder block and rear bearing (10) is pressed into bore of rear bearing support. Note that front thrust washer (3) has round tang with hole and rear thrust washer (5) has square tang.
1. Crankshaft gear
2. Thrust plate
3. Front thrust washer
4. Front main bearing
5. Rear thrust washer
6. Crankshaft pulley key
7. Crankshaft gear key
8. Crankshaft
9. Flywheel dowel pin
10. Rear main bearing
11. "O" ring
12. Rear main bearing support
13. Crankshaft rear oil seal

Illustrations Courtesy Yanmar Diesel Engine, Inc.

Yanmar

and also has split shell type intermediate main bearings (3 and 6—Fig. Y1-46) retained in bearing blocks (2 and 5). Flanges on the rear intermediate bearing control crankshaft end play. Set bolts (1) threaded into the lower halves of the bearing blocks hold the intermediate bearings in position. See Fig. Y1-47.

Crankshaft for Model 2GM20 has one intermediate split shell bearing similar to the rear intermediate bearing (6—Fig. Y1-46) for the 3-cylinder models.

Prior to removing crankshaft from engine, check crankshaft end play. Desired crankshaft end play is 0.06-0.19 mm (0.0024-0.0075 inch) for Model 1GM10, and 0.09-0.19 mm (0.0035-0.0075 inch) on 2 and 3-cylinder models. Maximum allowable end play is 0.3 mm (0.012 inch) on all models.

To remove crankshaft, first remove connecting rod and piston units, crankshaft pulley, timing gear cover and ball bearing, governor assembly and crankshaft gear. On 2 and 3-cylinder models, remove the intermediate bearing set bolts (1—Fig. Y1-47). Remove gearbox and flywheel, then unbolt and remove rear bearing housing and pull crankshaft from rear end of cylinder block. Remove the bolts (B—Fig. Y1-46) from intermediate main bearing housings on 2 and 3-cylinder models.

Clean and inspect the crankshaft, cylinder block and bearings and check parts against the following specifications:

Standard Crankshaft Journal Diameters:

Front & Intermediate Main Journals—
 Model 3HM35 .... 46.95-46.964 mm
  (1.8484-1.849 in.)
 All other models .. 43.95-43.964 mm
  (1.7303-1.7309 in.)
Rear Main Journal—
 Model 3HM35 .... 64.95-64.964 mm
  (2.5571-2.5576 in.)
 All other models .. 59.95-59.964 mm
  (2.3602-2.3608 in.)
Crankpin Journal—
 Model 3HM35 .... 43.95-43.964 mm
  (1.7303-1.7309 in.)
 All other models .. 39.95-39.964 mm
  (1.5728-1.5734 in.)

Crankshaft Bearing Oil Clearance:

Front & Intermediate—
 Model 3HM35 ..... 0.036-0.095 mm
  (0.0014-0.0037 in.)
 All other models ... 0.036-0.092 mm
  (0.0014-0.0036 in.)
Rear main journal—
 Model 3HM35 ..... 0.036-0.099 mm
  (0.0014-0.0039 in.)
 All other models ... 0.036-0.095 mm
  (0.0014-0.0037 in.)

Crankpin bearing—
 Model 3HM35 ..... 0.036-0.092 mm
  (0.0014-0.0036 in.)
 All other models ... 0.028-0.086 mm
  (0.0011-0.0034 in.)

Maximum allowable oil clearance is 0.15 mm (0.006 inch) on all main bearings and 0.13 mm (0.005 inch) for crankpin connecting rod bearings. Crankshaft should be reground or renewed if journal wear exceeds 0.05 mm (0.002 inch), or if journal taper or out-of-round exceeds 0.01 mm (0.0004 inch). Bearings are available in standard size and 0.25 mm (0.010 inch) undersize.

To renew front and rear main bearings, carefully press out old bearings and press in new bearings using piloted fixture. Any deviation from pressing the bearing straight into bearing bore will result in damage to the bearing resulting in early failure.

To reinstall crankshaft, reverse removal procedure. On 2 and 3-cylinder models, assemble intermediate main bearing housings to crankshaft prior to crankshaft installation. Be sure on 3-cylinder models that bearing with thrust flanges is installed at rear intermediate position. On all 2 and 3-cylinder models, install intermediate bearing blocks with "F" mark (Fig. Y1-48) toward flywheel and with arrow marks on upper and lower bearing block halves aligned. Install rear bearing support with oil hole in support aligned with oil hole in cylinder block. Tighten rear bearing support bolts, then tighten intermediate bearing set bolts on 2 and 3-cylinder models. Complete reassembly of engine by reversing disassembly procedure.

CRANKSHAFT REAR OIL SEAL

The crankshaft rear oil seal (13—Fig. Y1-45) is located in the rear main bearing housing (12) and can be renewed after removing flywheel. Care must be exercised in removal of old seal so as not to damage crankshaft oil seal surface or seal bore of rear main bearing housing. Install new seal with sealing lip inward.

SMALL DIESEL

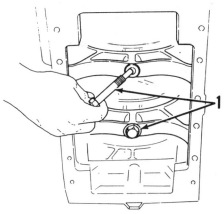

Fig. Y1-47—The intermediate bearing blocks are retained in position by set bolts (1). Tighten set bolts after installing rear main bearing support.

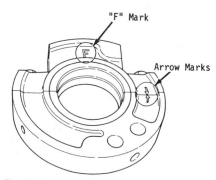

Fig. Y1-48—Assemble main bearing blocks on crankshaft so that arrow marks are aligned and "F" mark is facing flywheel end of crankshaft.

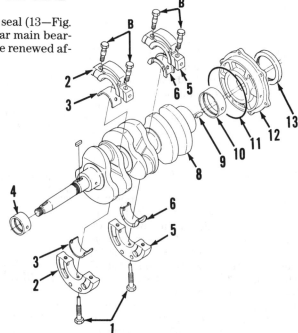

Fig. Y1-46—Intermediate bearings for 3-cylinder models are supported in bearing blocks (2 and 5). Model 2GM20 uses intermediate bearing support similar to (5) above.

1. Set bolts
2. Bearing block
3. Main bearing insert
4. Front main bearing
5. Bearing block
6. Main/thrust bearing
8. Crankshaft
9. Flywheel dowel pin
10. Rear main bearing
11. "O" ring
12. Main bearing support
13. Crankshaft rear oil seal

SERVICE MANUAL

ELECTRICAL SYSTEM

ALTERNATOR AND REGULATOR

Either a Hitachi LR155-20 55 ampere or Hitachi LR135-105 35 ampere alternator is used along with a Hitachi TR1Z-63 voltage regulator mounted internally. Regulator has no provisions for adjustment. Specified alternator output should be 53 amperes (minimum) at 5000 rpm for Model LR155-20 and 30 amperes (minimum) at 5000 rpm for Model LR135-105. Regulated voltage should be 14.2-14.8 volts.

New brush length is 16 mm (0.630 inch). A wear limit line is etched on the brushes; renew brushes if worn to or over the wear limit line. Standard slip ring diameter is 31.6 mm (1.244 inches) and wear limit is 30.6 mm (1.205 inches).

To check stator assembly, use ohmmeter and check for continuity between each stator coil terminal. If there is no continuity, renew stator. Check for continuity between terminals and stator core. If continuity exists, renew the stator.

To check rotor, use ohmmeter to check for continuity between the two slip rings. If no continuity is found, renew the rotor. Check for continuity between slip ring and rotor core or shaft. If continuity exists, renew the rotor.

Using ohmmeter, check diodes in rectifier assembly. Ohmmeter should show infinite reading in one direction and continuity when ohmmeter leads are reversed. Renew rectifier assembly if continuity exists in both directions, or no continuity exists in either direction.

STARTING MOTOR

A Model S12-77A starting motor is used on engine Model 3HM35 and a Model S114-303 starter is used on all other models. Starter service specifications are as follows:

Model S114-303
No-load test—
  Volts 12
  Current draw (max.) ......60 amps
  Rpm (min.) 7000
Brush length—
  Standard ....16 mm
                           (0.630 in.)
  Wear limit ....12 mm
                           (0.472 in.)
Pinion gap.............0.3-2.5 mm
                      (0.012-0.098 in.)

Model S12-77A
No-load test—
  Volts 12
  Current draw (max.) ......90 amps
  Rpm (min.) 4000
Brush length—
  Standard ....22 mm
                           (0.866 in.)
  Wear limit ....14 mm
                           (0.551 in.)
Pinion gap.............0.2-1.5 mm
                      (0.008-0.059 in.)

Yanmar

SMALL DIESEL

Model	No. Cyls.	Bore	Stroke	Displ.
2T73A	2	73 mm (2.87 in.)	75 mm (2.95 in.)	627 cc (38.26 cu. in.)
2TR13A	2	75 mm (2.95 in.)	75 mm (2.95 in.)	662 cc (40.39 cu. in.)
2T84A	2	84 mm (3.31 in.)	90 mm (3.54 in.)	997 cc (60.84 cu. in.)
2TR20A-X	2	90 mm (3.54 in.)	90 mm (3.54 in.)	1145 cc (69.87 cu. in.)
3T84A	3	84 mm (3.31 in.)	90 mm (3.54 in.)	1496 cc (91.29 cu. in.)

All models in this section are liquid cooled, indirect injection four-stroke diesel engines. Engine rotation is clockwise as viewed from front (timing gear end) of engine.

MAINTENANCE

LUBRICATION

Oil and filter should be changed after 20 hours of operation for new or reconditioned engines, and a second change after 30 hours of use from that change. Recommended oil change interval after first two changes is after every 100 hours of use.

Recommended engine lubricant is API classification CC or CD engine oil. In operating temperatures below 10° C (50° F), use of 10W, 20/20W, 20W or 20W/40 oils are recommended. From 10°-20° C (50°-68° F), use of SAE 20, 20W, 20/20W, or 20W-40 oil is recommended. From 20°-35° C (68°-95° F), SAE 30 or SAE 40 oil is recommended. Above 35° C (95° F), SAE 50 oil is recommended.

Crankcase oil sump capacity is as follows:

Model
2T73A . . . . 2.2 liters
(2.3 qts.)
2TR13A . . . . 2.4 liters
(2.5 qts.)
2T84A . . . . 4.0 liters
(4.2 qts.)
2TR20A-X . . . . 4.0 liters
(4.2 qts.)
3T84A . . . . 6.4 liters
(6.8 qts.)

FUEL SYSTEM

FUEL FILTER. The fuel system includes a fuel filter that should be checked daily, cleaned at least every 100 hours of operation and renewed whenever plugged with contaminants or at least every 300 hours of operation. The fuel filter incorporates a shut-off valve and two air vent (bleed) plugs (35—Fig. Y2-1 or Y2-2).

BLEED FUEL SYSTEM. The fuel system should be bled if fuel tank is allowed to run dry, if fuel filter, fuel lines or other components within the system are disconnected, or if engine has not been operated for a long period of time. If engine fails to start or if it starts, then stops, the cause could be air in the fuel system which can be removed by bleeding the system. Two bleed screws (35—Fig. Y2-1 or Y2-2) are located at top of fuel filter housing (34) and another bleed screw (36) is located on fitting of fuel pump inlet line. Loosen bleed screws at top of filter, wait until fuel

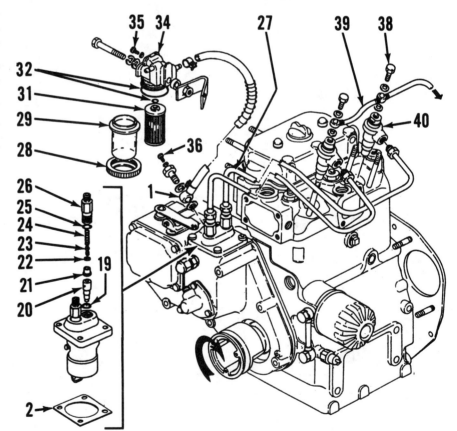

Fig. Y2-1—Partially exploded view of fuel injection system used on Models 2T73A and 2TR13A.

1. Inlet banjo fitting
2. Timing shims
19. Gasket
20. Barrel
21. Delivery valve seat
22. Gasket
23. Delivery valve
24. Valve spring
25. "O" ring
26. Delivery valve holder
27. Injector line
28. Retainer
29. Bowl
31. Filter element
32. Gasket
34. Filter base
35. Bleed screw
38. Banjo fitting
39. Bleed-off line
40. Injector assy.

Illustrations Courtesy Yanmar Diesel Engine, Inc.

SERVICE MANUAL

Yanmar

flows freely, then tighten screws. Loosen bleed screw on injection pump inlet line and allow fuel to flow until free of air, then tighten the screw.

Move speed control to "Run" position and attempt to start engine. If engine does not start, loosen high pressure fuel line at each injector and crank engine until fuel escapes from loosened connections. Tighten the fuel line compression nuts to a torque of 27 N·m (20 ft.-lbs.) and start engine.

INJECTION PUMP TIMING

Beginning of injection timing should occur as follows:

Engine Model	Timing Degrees BTDC
2T73A, 2TR13A	23-25
2T84A, 2TR20A-X	20-24
3T84A	25

Refer to Fig. Y2-1 for Models 2T73A and 2TR13A, and to Fig. Y2-2 for other models. Proceed as follows to check and adjust injection timing:

Close fuel shut-off valve and remove high pressure line (27) to No. 1 (rear) injector. Unscrew delivery valve holder (26) from injection pump, remove delivery valve (23) and spring (24), then reinstall delivery valve holder. Attach timing pipe to No. 1 delivery valve holder, pull out decompression knob, then turn engine slowly in clockwise direction until No. 1 (rear) piston is coming up on compression stroke. Turn fuel on and continue to turn engine until fuel just stops flowing from timing pipe. Point at which fuel just stops flowing is beginning of injection and should occur as mark on crankshaft pulley aligns with notch as shown in Fig. Y2-3. If fuel does not stop flowing when timing marks are aligned, piston may be on exhaust stroke. If crankshaft pulley mark is past notch at beginning of injection, remove some shims (2—Fig. Y2-1 or Y2-2) from between injection pump body and timing gear cover. If mark has not yet

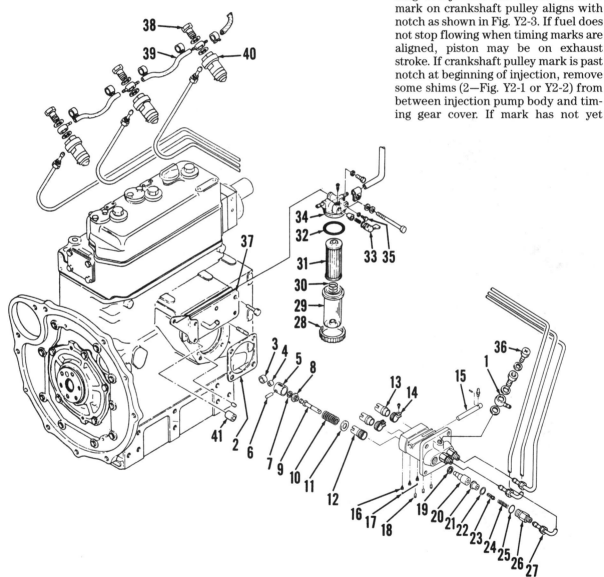

Fig. Y2-2—Exploded view of fuel injection system used on engine Model 3T84A. System is similar for two-cylinder engine Models 2T84A and 2TR20A-X.

1. Inlet banjo fitting
2. Timing shims
3. Roller
4. Bushing
5. Guide
6. Pin
7. Shim
8. Retainer
9. Plunger
10. Spring
11. Spring retainer
12. Fuel control pinion & sleeve
13. Sleeve
14. Pinion
15. Pump control rack
16. Guide stopper
17. Stopper pin
18. Plunger barrel pin
19. Gasket
20. Barrel
21. Delivery valve seat
22. Gasket
23. Delivery valve
24. Valve spring
25. "O" ring
26. Delivery valve holder
27. Injector line
28. Retainer
29. Bowl
30. Spring
31. Filter element
32. Gasket
33. Shut-off valve
34. Filter base
35. Bleed screw
36. Bleed screw
37. Cover
38. Banjo fitting
39. Bleed-off line
40. Injector assembly
41. Speed control shaft bushing

Illustrations Courtesy Yanmar Diesel Engine, Inc.

Yanmar

SMALL DIESEL

reached notch at beginning of injection, add shims (2) between pump and timing gear cover. Changing shim thickness 0.1 mm (0.004 inch) will change injection timing approximately one degree. After timing is correct, remove timing pipe and delivery valve holder. Install delivery valve, spring and delivery valve holder using new "O" ring (25) and copper ring (22). Tighten delivery valve holder to a torque of 39-44 N·m (29-32 ft.-lbs.).

GOVERNOR

Models 2T73A and 2TR13A

To adjust governor, remove cover (C—Fig. Y2-5) from timing gear cover, and cap nut (9) from spring body (3). Move governor lever (18) lightly in direction of spring (1) to be sure rack contacts pin (7), then release governor lever. Check location of punch mark (B) on injection pump control rack which should be aligned with machined edge of pump housing as shown in inset. If mark (B) is not at correct position, loosen locknut (5) and turn spring body (3) as necessary to align mark (B) with housing. Lock setting by tightening locknut (5) and install cap nut (9), then install cover (C).

High idle speed is adjusted at stop screw (6—Fig. Y2-6) and locked by locknut (8).

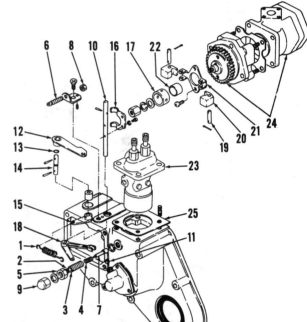

Fig. Y2-6—Exploded view of governor and timing cover for Models 2T73A and 2TR13A. Shaft (10) rides on steel ball (11) at bottom of shaft bore. High speed stop screw (6) is sealed.

1. Governor spring
2. Washer
3. Spring body
4. Spring
5. Locknut
6. High speed stop screw
7. Control pin
8. Locknut
9. Cap nut
10. Governor lever shaft
11. Steel ball
12. External speed control lever
13. "O" ring
14. External lever shaft
15. Speed control lever
16. Follower lever
17. Governor sleeve
18. Governor lever
19. Pins
20. Governor weights
21. Bracket
22. Sleeve guide
23. Fuel injection pump
24. Hydraulic pump & drive
25. Injection timing shims

All Models except 2T73A and 2TR13A

Several adjustments must be accomplished while assembling governor controls to permit correct governor action.

Torque springs, shaft (7—Fig. Y2-7), spring retainers and nuts (4) must be adjusted for proper operation. Setting (A) should be approximately 14-15 mm (0.551-0.590 inch) for two-cylinder

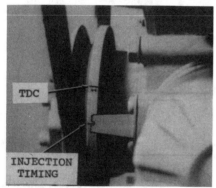

Fig. Y2-3—View of mark on crankshaft pulley indicating injection timing aligned with notch on timing pointer. "TDC" mark for No. 1 cylinder is also shown.

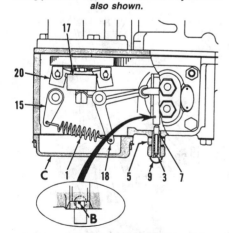

Fig. Y2-5—Cross-sectional view showing governor assembly typical of that used on engine Models 2T73A and 2TR13A. Adjustment of spring body (3) is outlined in text.

B. Rack alignment mark
C. Governor cover
1. Governor spring
3. Spring body
5. Locknut
7. Control pin
9. Cap nut
15. Internal speed control lever
17. Governor sleeve
18. Governor lever
20. Governor weights

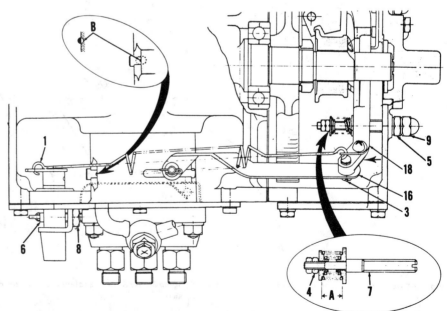

Fig. Y2-7—Cross-sectional view of governor assembly for Model 3T84A. Governors for Models 2T84A and 2TR20A-X are similar. Refer to text for adjustment procedures.

A. Spring setting distance
B. Rack alignment mark
1. Governor spring
3. Adjusting nut
4. Adjusting nuts
5. Locknut
6. High speed stop screw
7. Torque spring shaft
8. Locknut
9. Cap nut
16. Governor link
18. Governor lever

SERVICE MANUAL

Yanmar

models and 11.8-12.2 mm (0.465-0.480 inch) for three-cylinder models. The factory setting distance (A) should not be disturbed.

Assemble and connect governor and injection pump parts as shown in Fig. Y2-7. Disconnect rear end of spring (1) and loosen adjustment nut (3). Turn torque spring shaft (7) into timing gear cover (toward camshaft gear) far enough so that governor lever (18) will not contact torque spring retainer when pushed in that direction. Move link (16) toward the rear as far as possible, then tighten and lock adjustment nut (3). Move link (16) forward until punch mark (B) on pump control rack is centered under machined face of pump housing as shown, then turn torque shaft (7) out until torque spring retainer just contacts governor lever (18). Lock torque shaft with nut (5) and cap nut (9).

Complete adjustment by attaching spring (1). Check injection pump timing as previously outlined.

High idle speed is adjusted at stop screw (6—Fig. Y2-8) and locked by nut (8). Slow idle speed is adjusted by turning nuts (25) to relocate pin (26) on control rod.

REPAIRS

TIGHTENING TORQUES

Refer to the following table for special tightening torques. Metric fasteners are used throughout the engine.

Connecting rod—
 Models 2T73A &
  2TR13A .... 45-49 N·m
            (33-36 ft.-lbs.)
 All other models .... 64 N·m
            (47 ft.-lbs.)
Rocker arm supports—
 Models 2T73A &
  2TR13A .... 79-98 N·m
            (58-72 ft.-lbs.)
 All other models .... 65 N·m
            (48 ft.-lbs.)
Cylinder head—
 Models 2T73A &
  2TR13A .... 152-157 N·m
            (112-116 ft.-lbs.)
 All other models ...... 172-181 N·m
            (127-134 ft.-lbs.)
Flywheel—
 Models 2T84A &
  2TR20A-X .... 372-411 N·m
            (275-304 ft.-lbs.)
 All other models .... 64-69 N·m
            (47-51 ft.-lbs.)
Crankshaft pulley—
 Models 2T73A &
  2TR13A .... 118-147 N·m
            (87-108 ft.-lbs.)
 All other models .... 64 N·m
            (47 ft.-lbs.)
Timing gear cover—
 Models 2T73A &
  2TR13A .... 23-29 N·m
            (17-21 ft.-lbs.)
 All other models .... 25-26 N·m
            (18-19 ft.-lbs.)
Camshaft bearing set screw .. 19 N·m
            (14 ft.-lbs.)
Rear main bearing bolts...... 37 N·m
            (28 ft.-lbs.)
Intermediate main bearing
  bolts—
   Model 3T84A .... 64-69 N·m
            (47-51 ft.-lbs.)
Intermediate main set bolt—
   Model 3T84A .... 78 N·m
            (58 ft.-lbs.)
Injector nozzle holder nuts ... 20 N·m
            (15 ft.-lbs.)
Injection pump delivery—
   valve holder .... 39-44 N·m
            (29-32 ft.-lbs.)

WATER PUMP

To remove water pump, drain cooling system, remove fan belt, disconnect coolant hose from pump and unbolt radiator shroud. Unbolt water pump (9—Fig. Y2-9) from housing (5), then withdraw pump and radiator shroud.

If pump seals leak or bearing is rough or loose, renew pump as a complete assembly.

VALVE ADJUSTMENT

Recommended valve rocker arm to stem clearance is 0.2 mm (0.008 inch) with engine cold for both intake and exhaust valves for Models 2T73A and 2TR13A, and 0.15 mm (0.006 inch) for both intake and exhaust valves on other models. Valve adjustment should be checked if rocker arms have been removed or after each 300 hours of operation.

The rocker arm cover for Models 2T84A and 2TR20A-X incorporates the inlet air passage. On all models, disconnect decompression linkage and disconnect or remove interfering hardware, then unbolt and remove the rocker arm cover. Check and adjust valve clearance when piston is at TDC on compression stroke. TDC mark for No. 1 (rear) cylinder is marked on crankshaft pulley as shown in Fig. Y2-3. On two-cylinder models, TDC for No. 2 (front) cylinder is 180 degrees (1/2 turn of crankshaft) from No. 1 cylinder. On three-cylinder models, TDC for No. 3 (front) cylinder occurs 240 degrees (2/3 revolution) after No. 1 (rear) TDC, and No. 2 (center) reaches TDC at 240 degrees after the front. Usually top dead center is marked for all cylinders and is identified by "2T" or "3T"; however, the crankshaft pulley of some models may have only

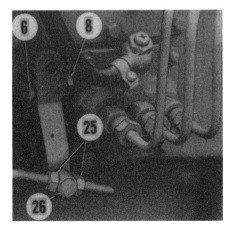

Fig. Y2-8—View showing high idle speed stop screw (6) and locknut (8), and slow idle speed adjusting nuts (25) for Model 3T84A. Adjustment points for Models 2T84A and 2TR20A-X are similar.

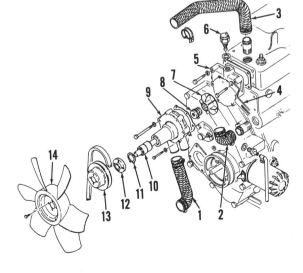

Fig. Y2-9—Exploded view of water pump used on most models.
1. Radiator to pump hose
2. Pump to block hose
3. Block to radiator hose
4. Alternator bracket
5. Mounting housing
6. Temperature sender
7. Pump impeller
8. Water pump seal
9. Water pump housing
10. Water pump shaft & bearing
11. Bearing retaining ring
12. Pulley hub
13. Pulley
14. Fan assy.

Illustrations Courtesy Yanmar Diesel Engine, Inc.

Yanmar

one TDC mark which is for No. 1 (rear) cylinder.

Reinstall rocker arm cover by reversing removal procedure. Adjust decompression lever clearance on models so equipped as follows: Remove the small covers from rocker arm cover above each exhaust valve, then turn crankshaft until No. 1 (rear) cylinder is at TDC on compression stroke. Move decompression control to decompression position. Loosen locknut (A—Fig. Y2-10) and turn screw (B) until it just contacts rocker arm, then turn one full turn additional. This adjustment will hold valve open approximately 0.8 mm (0.031 inch) when decompression device is engaged. Turn crankshaft to next cylinder TDC (180 degrees or 1/2 turn for 2-cylinder models, 240 degrees or 2/3 turn for front cylinder of 3-cylinder models) and adjust decompression lever for that cylinder. On 3-cylinder models, turn crankshaft 240 degrees (2/3-turn) from second position and adjust decompression lever for center (No. 2) cylinder.

CYLINDER HEAD

To remove cylinder head, first drain coolant from cylinder block. Thoroughly clean injector lines, injectors and fuel injection pump using suitable solvent and compressed air. Remove fuel injection pressure and return lines from injectors and pump. Remove fuel injector assemblies and precombustion chambers if head is to be overhauled. Remove rocker arm cover and back rocker arm adjusting screws out, then remove rocker arm support assemblies and push rods. Remove alternator and water pump. Loosen cylinder head nuts in reverse order shown in Fig. Y2-11 or Fig. Y2-12, then unbolt and remove cylinder head. Note that one cylinder head bolt (2—Fig. Y2-11) may be located in air intake passage.

Thoroughly clean cylinder head and block surfaces. Inspect head for cracks and check for warpage using straightedge and feeler gage. Head should be flat within 0.03 mm (0.001 inch); maximum distortion allowable is 0.1 mm (0.004 inch).

When reinstalling head, coat both sides of head gasket with "Three Bond No. 50" or "Dow Corning RTV Silicon Gasket." Position head gasket with rolled edge of gasket around cylinder bores upward. The larger cooling passages in gasket will be at rear (flywheel) end of engine. Lubricate bolt threads and install cylinder head retaining nuts. Tighten the nuts in several steps in sequence shown in Fig. Y2-11 or Fig. Y2-12. Head bolt final torque is 152-157 N·m (112-116 ft.-lbs.) on Models 2T73A and 2TR13A, and 172-181 N·m (127-134 ft.-lbs.) on all other models. Install push rods and rocker arm support and rocker arm assemblies. Tighten rocker arm support nuts to a torque of 79-98 N·m (58-72 ft.-lbs.) on Models 2T73A and 2TR13A, and 65 N·m (48 ft.-lbs.) on all other models. Adjust valve clearance and install rocker arm cover. Where applicable, check decompression controls and adjust if necessary. Complete remainder of installation by reversing removal procedure.

VALVE SYSTEM

Valve seat and face angle is 45 degrees for both intake and exhaust valves and valves seat directly in cylinder head. Recommended valve seat width is 1.77 mm (0.070 inch) for Models 2T73A and 2TR13A, and 2.12 mm (0.083 inch) for all other models. On 3-cylinder models, valve head is recessed 1.25 mm (0.049 inch) below surface of cylinder head; wear limit is 1.85 mm (0.073 inch). On 2-cylinder models, maximum valve head recess allowable is 0.5 mm (0.020 inch). Renew valve if margin is not even or is less than 1/2-thickness of new valve.

Valve guides are renewable by driving out worn guide and driving in new guide. New guides should be installed so that groove in upper end of guide is just above upper surface of cylinder head. Inside diameter of new guide must usually be reamed with suitable reamer to provide proper valve stem to guide fit. Intake and exhaust guides are not interchangeable; exhaust port end of exhaust guide is counterbored.

Install new valve stem seals whenever valves are serviced. Check valve stems and guides against the following values:

Models 2T73A and 2TR13A
Valve stem diameter—
    Standard............6.95-6.97 mm
        (0.2736-0.2744 in.)
    Wear limit ....6.9 mm
        (0.2717 in.)
Intake valve guide ID—
    Standard............7.0-7.015 mm
        (0.2756-0.2762 in.)
    Wear limit ....7.08 mm
        (0.2787 in.)
Exhaust valve guide ID—
    Standard............7.02-7.05 mm
        (0.2764-0.2776 in.)
    Wear limit ....7.08 mm
        (0.2787 in.)
Intake stem to guide clearance—
    Standard...........0.04-0.065 mm
        (0.0016-0.0026 in.)
    Wear limit ....0.15 mm
        (0.006 in.)
Exhaust stem to guide clearance—
    Standard...........0.045-0.07 mm
        (0.0018-0.0028 in.)
    Wear limit...............0.015 mm
        (0.006 in.)

Models 2T84A, 2TR20A-X and 3T84A
Valve stem diameter—
    Standard............7.96-7.97 mm
        (0.3134-0.3138 in.)

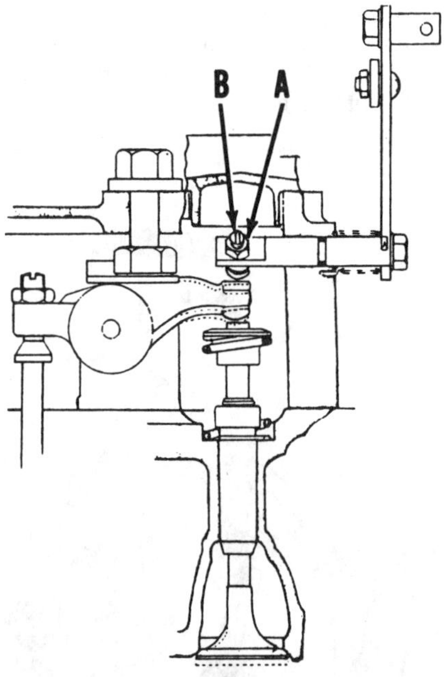

Fig. Y2-10—Cross-sectional view showing decompression lever adjusting screw (B) and locknut (A). Adjusting screw is accessible after removing cap (C) located above each exhaust rocker arm.

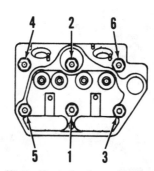

Fig. Y2-11—View showing cylinder head bolt tightening sequence for two-cylinder models. Loosen in reverse order when removing cylinder head.

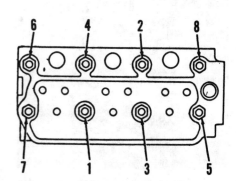

Fig. Y2-12—Cylinder head tightening sequence for three-cylinder models. When removing cylinder head, loosen cylinder head nuts in reverse order.

SERVICE MANUAL

Yanmar

Wear limit .... 7.9 mm
(0.311 in.)
Valve guide ID, all .... 8.01-8.025 mm
(0.3154-0.3159 in.)
Wear limit .... 8.08 mm
(0.3181 in.)
Valve stem to guide
  clearance, all .... 0.045-0.07 mm
(0.0018-0.0028 in.)
Wear limit .... 0.15 mm
(0.006 in.)

ROCKER ARMS AND PUSH RODS

Rocker arms and push rods can be removed after removing rocker arm cover. Rocker arms for each cylinder pivot on shaft of support stand which is retained to cylinder head by one stud nut and a locating dowel pin.

Rocker shaft diameter should be 13.98-14 mm (0.5504-0.5512 inch) with wear limit of 13.90 mm (0.547 inch) on engine Models 2T73A and 2TR13A. For all other models, rocker shaft diameter should be 16.98-17 mm (0.6685-0.6693 inch) with wear limit of 16.90 mm (0.665 inch).

Rocker arm bushing inside diameter should be 14.024-14.034 mm (0.5521-0.5525 inch) with wear limit of 14.10 mm (0.555 inch) for engine Models 2T73A and 2TR13A. For other models, rocker bushing inside diameter should be 17.016-17.034 mm (0.6699-0.6706 inch) with wear limit of 17.15 mm (0.675 inch). Clearance between bushing and shaft should not exceed 0.15 mm (0.006 inch) on any model engine.

Different rocker arms are used for intake and exhaust valves; the exhaust rocker arm has a decompression lever contact pad at valve end of rocker arm. Also, the No. 3 (front) cylinder rocker arms are different than for cylinder Nos. 1 and 2. Refer to Fig. Y2-13. Rocker arms should not be exchanged between rocker arm supports on any engine.

Check push rods by rolling on flat plate and renew any showing any curvature or if ends are galled, worn or chipped. Push rod length is 165.2 mm (6.504 inches) for Models 2T73A and 2TR13A, and 197 mm (7.776 inches) for other models.

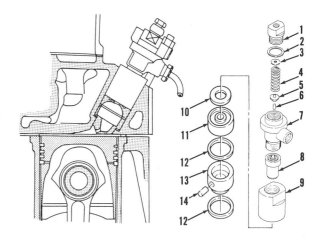

Fig. Y2-14—Cross-sectional and exploded views of fuel injector and precombustion chamber assemblies. Shims (3) are varied to adjust nozzle opening pressure. Precombustion chamber parts (10 to 14) can be removed from cylinder head after removing injector assembly.

1. Spring cap
2. Seal ring
3. Adjusting shims
4. Nozzle valve spring
5. Spring seat
6. Inner spindle
7. Injector body
8. Nozzle & needle
9. Nozzle cap
10. Heat insulator
11. Precombustion chamber half
12. Seal rings
13. Precombustion chamber half
14. Locating pin

On all models, back rocker arm adjusting screws out before installing push rods, rocker arm supports and rocker arms. Tighten rocker arm support nuts to a torque of 79-98 N·m (58-72 ft.-lbs.) on Models 2T73A and 2TR13A, and to a torque of 65 N·m (48 ft.-lbs.) on all other models. Adjust valves as previously outlined and install rocker arm cover.

INJECTORS

REMOVE AND REINSTALL. Before removing an injector or loosening injector lines, thoroughly clean injectors, lines and surrounding area using suitable solvent and compressed air. If engine is misfiring and faulty injector is suspected, loosen each injector line in turn with engine running at low idle speed. A faulty injector is suspect if loosening line does not affect engine operation.

To remove injector, first remove high pressure line and the fuel return (bleed-back) line from injector. Unbolt and remove injector hold-down clamp, then pull injector from cylinder head. Cap all openings to prevent dirt from entering fuel injector. The precombustion chamber (items 10 to 14—Fig. Y2-14) can be removed from injector cylinder head bore at this time.

Before installing injector, thoroughly clean bore in cylinder head and install precombustion chamber using all new seals. Insert injector and hold-down clamp, then tighten injector retaining nuts evenly to a torque of 20 N·m (15 ft.-lbs.). Bleed fuel injector line before tightening retaining fittings to a torque of 27 N·m (20 ft.-lbs.).

TESTING. A complete job of testing, cleaning and adjusting the fuel injector requires use of special test equipment. Use only clean approved testing oil in tester tank. Injector should be tested for opening pressure, seat leakage and spray pattern. Before connecting injector to test stand, operate tester lever until oil flows, then attach injector to tester line.

WARNING: Fuel emerges from injector with sufficient force to penetrate the skin. When testing injector, keep yourself clear of nozzle spray.

Close valve to tester gage and operate tester lever a few quick strokes to purge air from injector and to be sure nozzle valve is not stuck or plugged.

OPENING PRESSURE. Open the valve to tester gage and operate tester lever slowly while observing gage reading. Opening pressure should be 15200-16185 kPa (2205-2345 psi). Opening pressure is adjusted by changing thickness of shim stack (3—Fig. Y2-14). A change of shim thickness of 0.10 mm (0.004 inch) will change opening pressure approximately 980 kPa (142 psi).

SPRAY PATTERN. The spray pattern should be conical, well atomized, and emerging in a straight axis from nozzle tip. If spray is drippy, ragged or to one side, nozzle must be cleaned or renewed.

SEAT LEAKAGE. Wipe nozzle tip dry with clean blotting paper, then operate tester lever to bring gage pressure to approximately 1960 kPa (285 psi) below

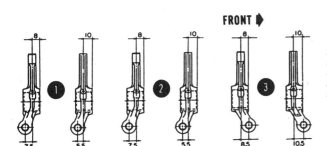

Fig. Y2-13—View showing rocker arms for three-cylinder engines. Note that rocker arms for front (No. 3) cylinder are different than for other cylinders.

Illustrations Courtesy Yanmar Diesel Engine, Inc.

Yanmar

opening pressure and hold this tester pressure for 10 seconds. If drop of oil appears at nozzle tip, or oil drips from nozzle, clean or renew nozzle.

OVERHAUL. Hard or sharp tools, emery cloth, grinding compounds or other than approved tools, solvents and lapping compounds, must never be used. Wipe all dirt and loose carbon from injector, then refer to Fig. Y2-14 and proceed as follows:

Secure injector body in holding fixture or soft jawed vise and loosen nuts (1 and 9). Unscrew adjusting spring nut (1) and remove parts (2 to 6) from top of injector. Unscrew nozzle nut (9) and remove nozzle and needle assembly (8). Place all parts in clean diesel fuel or calibrating oil as they are removed, and take care not to mix parts with those from other injector assemblies.

Clean exterior surfaces with brass wire brush to loosen carbon deposits; soak parts in approved carbon solvent if necessary. Rinse parts in clean diesel fuel or calibrating oil after cleaning to neutralize carbon solvent.

Clean nozzle spray hole from inside using pointed hardwood stick. Scrape carbon from nozzle pressure chamber using hooked scraper. Clean valve seat using brass scraper, then polish seat using wood polishing stick and mutton tallow.

Reclean all parts by rinsing thoroughly in clean diesel fuel or calibrating oil and assemble injector while immersed in fuel or oil. Install nozzle and needle assembly and retaining nut (9). Be sure shim pack is intact, then install spindle (6), spring seat (5), spring (4) and shim stack (3). Install cap nut (1) with new sealing ring (2). Tighten nozzle retaining nut (9) to a torque of 90-100 N·m (66-73 ft.-lbs.), then tighten spring retaining nut (1) to a torque of 70-80 N·m (52-59 ft.-lbs.). Retest assembled injector. Do not over-tighten nuts in attempt to stop any leakage; disassemble and clean injector if leak is noted.

THERMOSTART

The thermostart unit (see Fig. Y2-15) is installed in intake manifold and warms intake air by burning small quantities of diesel fuel in the intake manifold. Operation of the thermostart can be checked by viewing into intake manifold. With control engaged, inner coil of unit should glow bright red after about four seconds. After about ten seconds, burning fuel should start to drip from unit. If coils do not get hot, check for current at thermostart terminal. If current exists with control on, check for continuity between thermostart terminal and engine ground. Open circuit indicates burned out heating element.

The thermostart unit should not leak when air pressure of 138 kPa (20 psi) is applied to fuel inlet line.

INJECTION PUMP

Models 2T73A and 2TR13A

To remove injection pump, first clean pump, lines and surrounding area using suitable solvent and compressed air. Turn fuel off at filter, then disconnect all lines from pump. Cap or plug all openings immediately to prevent entrance of dirt. Unbolt and remove pump from engine timing gear cover. Be careful not to lose shims (2—Fig. Y2-1) located between injection pump and housing; these shims are used to set injection timing.

Before reinstalling pump, hold pump on side and lift control rack pin up to top of stroke; rack should slide to bottom of stroke by its own weight when rack pin is released. Have pump serviced at approved diesel repair station if pump control rack is binding.

When reinstalling pump, reverse removal procedure using same thickness of shims (2—Fig. Y2-1) as were removed. Check pump timing and vary shim thickness as required to provide correct timing. Bleed fuel system before tightening fuel injector line fittings to a torque of 27 N·m (20 ft.-lbs.).

Models 2T84A, 2TR20A-X and 3T84A

To remove injection pump, first clean pump, lines and surrounding area using suitable solvent and compressed air. Turn fuel off at filter, then disconnect all lines from pump. Cap or plug all openings immediately to prevent entrance of dirt. Detach speed control rod from governor lever. Remove cover (37—Fig. Y2-2) from pump chamber and cover from governor opening in timing gear cover. Disconnect governor spring (1—Fig. Y2-7) from internal speed control lever and remove clip from pin on pump control rack. Detach governor link (16) from rack pin, then unbolt and remove pump from engine. Be careful not to lose shims (2—Fig. Y2-2) located between injection pump and housing; these shims are used to set injection timing.

Before reinstalling pump, hold pump on side and lift rack pin up to top of stroke; rack should slide to bottom of stroke by its own weight when rack pin is released. Have pump serviced at approved diesel repair station if pump rack is binding.

When reinstalling pump, reverse removal procedure using same thickness of shims (2—Fig. Y2-2) as were removed. Check pump timing and vary shim thickness as required to provide correct timing. Bleed fuel system before tightening fuel injector line fittings to a torque of 27 N·m (20 ft.-lbs.).

TIMING GEAR COVER

Models 2T73A and 2TR13A

To remove timing gear cover, first disconnect tachometer cable and remove fan and fan belt if so equipped. Remove crankshaft pulley and breather/governor control cover. Disconnect governor control linkage and remove fuel injection pump. Unbolt and remove timing gear cover. Install new crankshaft seal in cover with lip facing inward.

When reinstalling cover, be sure all gasket material is removed from cover and engine block and coat new gasket with sealer before installation. Be sure governor sleeve (17—Fig. Y2-16) is in place against governor weights (20) and be careful not to damage governor control linkage when installing cover. Tighten cover retaining bolts to a torque of 23-29 N·m (17-21 ft.-lbs.). Lubricate crankshaft pulley oil seal contact surface, then install pulley and tighten retaining bolt to a torque of 118-147 N·m (87-108 ft.-lbs.). Reinstall fuel injection pump and complete reassembly by reversing cover removal procedure. Bleed fuel system, then start engine and check governor adjustments and readjust if necessary.

Models 2T84A, 2TR20A-X and 3T84A

To remove timing gear cover, first remove fan belt and fan and disconnect

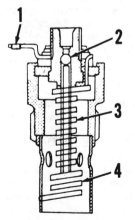

Fig. Y2-15—Cross-sectional drawing of thermostart unit.
1. Electrical terminal
2. Valve ball
3. Heater coil
4. Igniter coil

SERVICE MANUAL

Yanmar

tachometer cable if so equipped. Remove crankshaft pulley. Remove injection pump cover plate, then disconnect governor spring (1—Fig. Y2-17), remove clip pin and detach governor link (16) from pump control rack. Be careful not to deform governor spring when detaching spring from control lever. Unbolt and remove timing gear cover. Install new crankshaft oil seal in timing cover with lip facing inward.

Remove all gasket material from cover and engine crankcase. Coat new gasket with sealer before installation. Be sure that sleeve (17—Fig. Y2-17) is engaged with governor weights (20) and take care not to damage governor linkage when installing cover. Tighten timing cover retaining bolts to a torque of 24-26 N·m (18-19 ft.-lbs.). Lubricate seal surface of crankshaft pulley and install pulley. Tighten crankshaft pulley bolt to a torque of 64 N·m (47 ft.-lbs.). Bleed fuel injection system, start engine and check governor adjustments; readjust governed speeds if necessary.

INJECTION PUMP CAMSHAFT

Models 2T73A and 2TR13A

Fuel injection pump cam lobes (38—Fig. Y2-18) are mounted on front end of engine camshaft, and can be renewed after removing engine timing gear cover. Remove hex nut from front end of camshaft, then remove cam lobes (38) and spacer (39). Cam lobe height (large diameter) should be 45 mm (1.772 inches).

Models 2T84A, 2TR20A-X and 3T84A

Fuel injection pump cam lobes on these models are integral with engine camshaft; refer to CAMSHAFT paragraph for information.

TIMING GEARS

Models 2T73A and 2TR13A

Timing gears are accessible after removing timing gear cover. Backlash between camshaft gear and crankshaft gear should be 0.08-0.16 mm (0.003-0.006 inch) and maximum allowable backlash is 0.3 mm (0.012 inch).

Crankshaft gear can be pulled using threaded holes in gear and suitable puller. Align keyway with key and timing marks, then use threaded hole in front end of crankshaft to push gear onto shaft. Refer to Fig. Y2-21 for timing marks.

If camshaft gear is to be removed, first remove engine camshaft. Remove the hex nut, washer, fuel injection pump cam lobes (38—Fig. Y2-18) and spacer (39), then press camshaft from gear.

Models 2T84A, 2TR20A-X and 3T84A

Timing gears are accessible after removing engine timing gear cover. Backlash between camshaft gear and crankshaft gear and between crankshaft gear and oil pump drive gear should be 0.08-0.16 mm (0.003-0.006 inch) and maximum allowable clearance of 0.3 mm (0.012 inch).

The crankshaft gear can be removed using threaded holes in gear and suitable puller. To install gear, align keyway in gear with key on crankshaft and timing mark on gear with camshaft gear timing mark. Use threaded end of crankshaft and nut to push gear onto crankshaft. Refer to Fig. Y2-21 for timing marks.

If camshaft gear is to be removed, remove hex nut and governor flyweight assembly, then remove camshaft from engine and press shaft from gear.

OIL PUMP AND RELIEF VALVE

The gerotor type oil pump (15—Fig. Y2-18) is attached to the rear face of engine block and is driven from rear end of camshaft on engine Models 2T73A and 2TR13A. The pump can be removed after removing engine flywheel.

On all other models, the oil pump (15—Fig. Y2-19 or Fig. Y2-20) is attached to front face of cylinder block and is

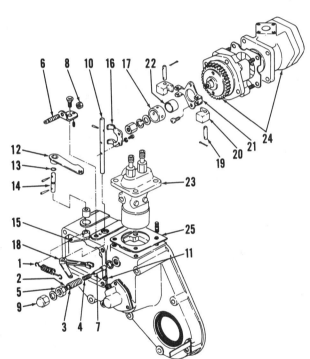

Fig. Y2-16—Exploded view of timing gear cover and related parts for Models 2T73A and 2TR13A.

1. Governor spring
2. Washer
3. Spring body
4. Spring
5. Locknut
6. High speed stop screw
7. Damper spring
8. Locknut
9. Cap nut
10. Pivot shaft
11. Steel ball
12. External speed control lever
13. "O" ring
14. Shaft
15. Internal speed control lever
16. Follower lever
17. Governor sleeve
18. Governor lever
19. Pins
20. Flyweights
21. Bracket
22. Guide sleeve
23. Injection pump
24. Hydraulic pump & drive gear

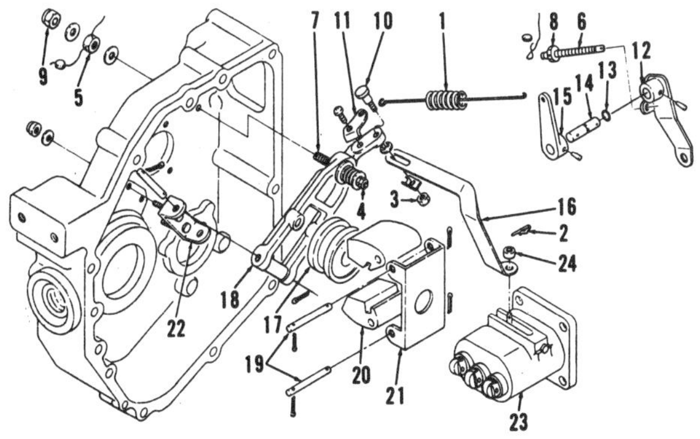

Fig. Y2-17—Exploded view of timing gear cover and related parts for Model 3T84A. Timing cover assembly for Models 2T84A and 2TR20A-X is similar.

8. Locknut
9. Cap nut
10. Attaching screw
11. Spring tab
12. External speed control lever
13. "O" ring
14. Shaft
15. Internal speed control lever
16. Governor link
17. Governor sleeve
18. Governor lever
19. Pins
20. Flyweights
21. Bracket
22. Pivot support
23. Injection pump
24. Spacer

1. Governor spring
2. Clip
3. Adjusting nut
4. Adjusting nuts
5. Locknut
6. High speed stop screw
7. Torque spring & shaft assy.

Illustrations Courtesy Yanmar Diesel Engine, Inc.

Yanmar SMALL DIESEL

gear driven from crankshaft gear. The oil pump can be removed after removing timing gear cover.

Pump rotor end clearance in housing can be measured by placing straightedge across pump body and using a feeler gage to check clearance between rotors and straightedge. Clearance should be 0.06-0.10 mm (0.0024-0.0039 inch) and should not exceed 0.3 mm (0.005 inch). Clearance between inner and outer rotors and between outer rotor and housing should be 0.05-0.10 mm (0.0024-0.004 inch) and should not exceed 0.15 mm (0.006 inch).

On all models, oil pressure relief valve (32—Fig. Y2-18, Fig. Y2-19 or Fig. Y2-20) is located in housing for oil filter. Desired oil pressure is 147-245 kPa (22-35 psi) for engine Models 2T73A and 2TR13A, and 245-343 kPa (35-50 psi) for all other models. Adding or removing one shim (35—Fig. Y2-20) will change oil pressure approximately 18 kPa (2.6 psi).

GOVERNOR

The governor for all models is a variable speed flyweight type. On engine Models 2T73A and 2TR13A, the governor weights (22—Fig. Y2-18) are located at front side of the hydraulic pump drive gear (40). On all other models, the governor weights (22—Fig. Y2-19 or Fig. Y2-20) are located at front of the camshaft timing gear (23).

To overhaul governor flyweight assembly, it is necessary to first remove engine timing gear cover. Refer to MAINTENANCE section for governor adjustments.

PISTON AND ROD UNITS

Connecting rod and piston units can be removed from above after removing cylinder head and engine oil pan.

When installing piston and rod, be sure to space piston ring end gaps at equal intervals around piston. Coat pistons and rings, sleeve and ring compressor with clean engine oil. Identification mark on connecting rod should be toward rear (flywheel) end of engine.

CYLINDER SLEEVES

Cylinder block is fitted with wet type removable cylinder sleeves (liners). Sleeves can be pulled from top of cylinder block after removing rod and piston units. Check cylinder sleeves for scoring or excessive wear. Cylinder sleeve inside diameter and wear limits are as follows:

Standard inside diameter—
Model 2T73A .... 73.0-73.03 mm
(2.874-2.875 in.)
Wear limit .... 73.17 mm
(2.881 in.)
Model 2TR13A .... 75.0-75.03 mm
(2.953-2.954 in.)
Wear limit .... 75.17 mm
(2.959 in.)
Models 2T84A, 3T84A 84.0-84.035 mm
(3.307-3.308 in.)
Wear limit .... 84.20 mm
(3.315 in.)
Model 2TR20A-X ..... 90.0-90.035 mm
(3.543-3.545 in.)
Wear limit .... 90.2 mm
(3.551 in.)

If cylinder sleeves are removed, be sure to thoroughly clean "O" ring grooves and counterbore at top of block. Install new "O" rings in grooves and paint outside of sleeve with waterproof paint and insert sleeve in block before paint dries. Check liner standout above cylinder block. Standout should be 0.05-0.13 mm (0.002-0.005 inch) for Models 2T73A and 2TR13A; 0.07-0.10 mm (0.003-0.004 inch) for Model 2TR20A-X; and 0.07-0.15 mm (0.003-0.006 inch) for Models 2T84A and 3T84A. If sleeve standout exceeds specified dimension, remove sleeve and check to be sure sleeve and block are clean and free of rust, scale or burrs.

PISTONS, PINS AND RINGS

Pistons for all models are fitted with three compression and one oil control

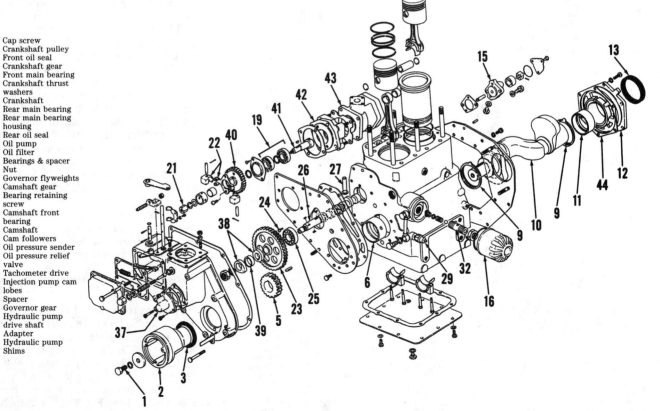

1. Cap screw
2. Crankshaft pulley
3. Front oil seal
5. Crankshaft gear
6. Front main bearing
9. Crankshaft thrust washers
10. Crankshaft
11. Rear main bearing
12. Rear main bearing housing
13. Rear oil seal
15. Oil pump
16. Oil filter
19. Bearings & spacer
21. Nut
22. Governor flyweights
23. Camshaft gear
24. Bearing retaining screw
25. Camshaft front bearing
26. Camshaft
27. Cam followers
29. Oil pressure sender
32. Oil pressure relief valve
37. Tachometer drive
38. Injection pump cam lobes
39. Spacer
40. Governor gear
41. Hydraulic pump drive shaft
42. Adapter
43. Hydraulic pump
44. Shims

Fig. Y2-18—Exploded view of cylinder block and related parts for Models 2T73A and 2TR13A.

SERVICE MANUAL
Yanmar

ring. Piston pin is floating fit in both piston and connecting rod and is retained by snap rings at each end of piston pin bore. Fit of pin in piston bore should be from −0.004 mm (−0.0002 inch) tight to +0.008 mm (0.0003 inch) loose, and fit of pin in connecting rod bushing should be a clearance of 0.025-0.049 mm (0.001-0.002 inch).

Inspect piston skirt for scoring or cracks and check skirt diameter at right angle to piston pin against the following dimensions:

Standard Piston Skirt Diameter—
Model 2T73A .... 72.89-79.92 mm
                     (2.870-2.871 in.)
  Wear limit .... 72.8 mm
                     (2.866 in.)
Model 2TR13A .... 74.89-74.92 mm
                     (2.948-2.949 in.)
  Wear limit .... 74.8 mm
                     (2.945 in.)
Models 2T84A,
3T84A .... 83.89-93.92 mm
                     (3.303-3.304 in.)
  Wear limit .... 83.8 mm
                     (3.299 in.)
Model 2TR20A-X ..... 89.86-89.89 mm
                     (3.538-3.539 in.)
  Wear limit .... 89.8 mm
                     (3.535 in.)

Measure both piston skirt and cylinder bore diameters. Piston skirt to cylinder bore clearance should be 0.108-0.173 mm (0.004-0.007 inch) for Models 2T73A, 2TR13A and 2TR20A-X. For models 2T84A and 3T84A, desired piston skirt to cylinder bore clearance is 0.077-0.142 mm (0.003-0.0056 inch). Maximum allowable piston skirt to cylinder bore clearance is 0.3 mm (0.012 inch) on all models.

Piston ring end gap for all rings on Models 2T73A and 2TR13A should be 0.2-0.4 mm (0.008-0.016 inch). On all other models, ring end gap should be 0.3-0.5 mm (0.012-0.020 inch). Maximum allowable ring end gap is 1.5 mm (0.059 inch) for all models. Specified ring side clearance in ring groove is 0.055-0.095 mm (0.002-0.004 inch) for the top ring and 0.020-0.055 mm (0.001-0.002 inch) for all other rings. Ring side clearance wear limit is 0.2 mm (0.008 inch) for compression rings and 0.15 mm (0.006 inch) for oil control ring.

Install all rings with marked side toward top of piston. When installing piston pin, heat piston to 80°-100° C (176°-210° F) for easy insertion of piston pin. Be sure pin retaining snap rings are in proper position in piston pin bore grooves.

CONNECTING RODS AND BEARINGS

Connecting rods are fitted with bushings pressed in piston pin end and slip-in precision type bearings at crankpin end. Check connecting rod piston pin bushing for proper pin to bushing clearance of 0.025-0.049 mm (0.001-0.002 inch) and renew bushing or connecting rod if clearance exceeds 0.11 mm (0.004 inch). Also, check to see that crankpin bearing insert has not been loose or turning in rod.

Crankpin bearing insert is available in 0.25 mm (0.010 inch) undersize as well as standard size. Standard crankpin diameter is 46.95-46.96 mm (1.8484-1.849 inches) for Models 2T73A and 2TR13A, and 53.95-53.96 mm (2.124-2.1245 inches) for all other models. Desired oil clearance is 0.036-0.095 mm (0.0014-0.0037 inch) and maximum allowable oil clearance is 0.15 mm (0.006 inch) for all models.

CAMSHAFT AND BEARINGS

Models 2T73A and 2TR13A

The camshaft is supported in a ball bearing behind camshaft timing gear and in two unbushed bores in cylinder block. Engine oil pump located at rear of cylinder block is driven by rear end of engine camshaft.

To remove camshaft, first remove timing gear cover, rocker arm assemblies and push rods. Use magnetic tools to lift cam followers away from camshaft, or turn engine upside down so that lifters fall away from cam. Remove camshaft

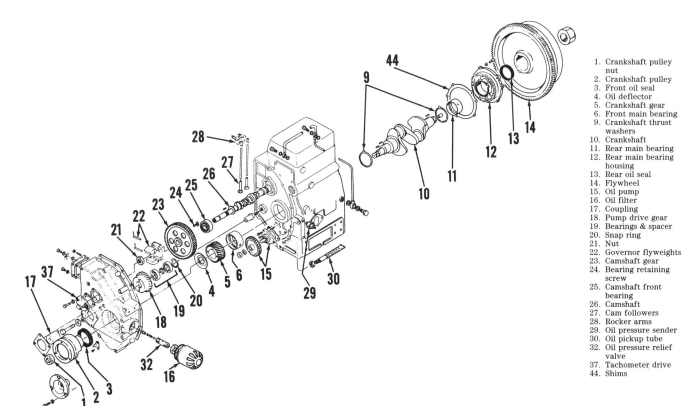

1. Crankshaft pulley nut
2. Crankshaft pulley
3. Front oil seal
4. Oil deflector
5. Crankshaft gear
6. Front main bearing
9. Crankshaft thrust washers
10. Crankshaft
11. Rear main bearing
12. Rear main bearing housing
13. Rear oil seal
14. Flywheel
15. Oil pump
16. Oil filter
17. Coupling
18. Pump drive gear
19. Bearings & spacer
20. Snap ring
21. Nut
22. Governor flyweights
23. Camshaft gear
24. Bearing retaining screw
25. Camshaft front bearing
26. Camshaft
27. Cam followers
28. Rocker arms
29. Oil pressure sender
30. Oil pickup tube
32. Oil pressure relief valve
37. Tachometer drive
44. Shims

Fig. Y2-19—Exploded view of cylinder block and related parts for Models 2T84A and 2TR20A-X.

Illustrations Courtesy Yanmar Diesel Engine, Inc.

Yanmar

SMALL DIESEL

bearing retaining screw (S—Fig. Y2-22), then pull camshaft and gear forward out of cylinder block. Camshaft can be pressed from gear and ball bearing after removing hex nut, fuel injection pump cam lobes and spacer from front of camshaft. Check camshaft against the following specifications:

Rear Camshaft Journal—
 Standard
  diameter ...... 29.939-29.960 mm
   (1.1787-1.1795 in.)
  Wear limit .... 29.90 mm
   (1.1772 in.)
 Oil clearance,
  desired .... 0.04-0.081 mm
   (0.0016-0.0032 in.)
  Wear limit .... 0.15 mm
   (0.006 in.)

Center Camshaft Journal—
 Standard
  diameter ...... 41.925-41.950 mm
   (1.6506-1.6516 in.)
 Oil clearance,
  desired .... 0.05-0.010 mm
   (0.002-0.004 in.)
  Wear limit .... 0.15 mm
   (0.006 in.)

Intake & exhaust lobe
 height, new .... 35.0 mm
   (1.3780 in.)

Models 2T84A, 2TR20A-X and 3T84A

The camshaft is supported in a ball bearing located behind the timing gear and in two unbushed bearing bores in cylinder block. The engine governor flyweight assembly is mounted on the camshaft at front side of the camshaft timing gear.

To remove camshaft, first remove engine timing gear cover, rocker arm cover, rocker arm assemblies and push rods. Use magnetic tools to hold cam followers away from camshaft, or invert engine so that followers fall away from camshaft. Remove bearing retaining screw (S—Fig. Y2-22) and pull camshaft and gear forward out of cylinder block. Camshaft can be pressed from gear and ball bearing after removing hex nut (left hand threads) and governor weight assembly from front end of shaft. Check camshaft against the following dimensions:

Rear Camshaft Journal—
 Standard
  diameter ...... 33.925-33.950 mm
   (1.3356-1.3366 in.)
  Wear limit .... 33.90 mm
   (1.3346 in.)
 Oil clearance,
  desired .... 0.05-0.10 mm
   (0.002-0.004 in.)
  Wear limit .... 0.15 mm
   (0.006 in.)

Center Camshaft Journal—
 Standard
  diameter ...... 45.925-45.950 mm
   (1.8081-1.8091 in.)
  Wear limit .... 45.88 mm
   (1.806 in.)
 Oil clearance,
  desired .... 0.05-0.010 mm
   (0.002-0.004 in.)
  Wear limit .... 0.15 mm
   (0.006 in.)

Intake & exhaust lobe
 height, new .... 38.63 mm
   (1.521 in.)
 Wear limit .... 38.13 mm
   (1.501 in.)

Fuel injection pump lobe
 height, new .... 34.9 mm
   (1.374 in.)
 Wear limit .... 34.8 mm
   (1.370 in.)

CRANKSHAFT AND MAIN BEARINGS

Two Cylinder Models

The crankshaft of two cylinder models is supported in two main bearings (6 and 11—Fig. Y2-18 or Fig. Y2-19) and end thrust is taken by thrust washers (9) located at each end of crankshaft.

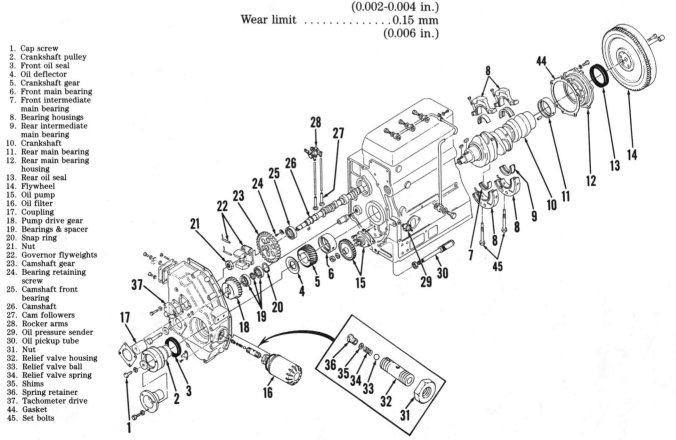

1. Cap screw
2. Crankshaft pulley
3. Front oil seal
4. Oil deflector
5. Crankshaft gear
6. Front main bearing
7. Front intermediate main bearing
8. Bearing housings
9. Rear intermediate main bearing
10. Crankshaft
11. Rear main bearing
12. Rear main bearing housing
13. Rear oil seal
14. Flywheel
15. Oil pump
16. Oil filter
17. Coupling
18. Pump drive gear
19. Bearings & spacer
20. Snap ring
21. Nut
22. Governor flyweights
23. Camshaft gear
24. Bearing retaining screw
25. Camshaft front bearing
26. Camshaft
27. Cam followers
28. Rocker arms
29. Oil pressure sender
30. Oil pickup tube
31. Nut
32. Relief valve housing
33. Relief valve ball
34. Relief valve spring
35. Shims
36. Spring retainer
37. Tachometer drive
44. Gasket
45. Set bolts

Fig. Y2-20—Exploded view of cylinder block and related parts for Model 3T84A. Exploded view of oil pressure relief valve is shown in inset. Rear intermediate bearing (9) has flanges to control crankshaft end play.

SERVICE MANUAL

Yanmar

To remove crankshaft, proceed as follows: Remove cylinder head and oil pan and remove the rod and piston units. Remove timing gear cover and pull crankshaft gear (5) from front end of shaft. Remove the oil pick-up tube (30—Fig. Y2-19) on Models 2T84A and 2TR20A-X. Check and record crankshaft end play for reference on reassembly. Remove flywheel and turn engine up on front end so that crankshaft is resting against front thrust washer. Unbolt and remove rear main bearing housing (12—Fig. Y2-18 or Fig. Y2-19). Lift crankshaft upward out of rear end of cylinder block, taking care not to lose or damage front thrust washer.

Front and rear main journal standard diameter is 64.95-64.964 mm (2.5571-2.5576 inches) for Models 2T73A and 2TR13A, and 69.950-69.964 mm (2.7539-2.7545 inches) for Models 2T84A and 2TR20A-X. Check journals for scoring, taper, out-of-round condition and excessive wear. Suggested wear limit is 64.91 mm (2.5555 inches) for Models 2T73A and 2TR13A, and 69.90 mm (2.752 inches) for Models 2T84A and 2TR20A-X. Desired oil clearance is 0.036-0.099 mm (0.0014-0.0039 inch) with maximum allowable clearance being 0.2 mm (0.008 inch).

The front main bearing (6—Fig. Y2-18 or Fig. Y2-19) is pressed into cylinder block and rear main bearing (11) is pressed into rear main bearing housing (12). When installing new bearings, be sure oil holes in bearings are aligned with oil passages in cylinder block and rear main bearing support. Rear seal (13) is installed with lip toward front (inside) of block.

The thrust washers (9) which limit crankshaft end play should be 2.9-3.0 mm (0.114-0.118 inch) thick with wear limit of 2.75 mm (0.108 inch). Crankshaft end play should be 0.1-0.2 mm (0.004-0.008 inch) with maximum allowable end play being 0.4 mm (0.016 inch). End play is controlled by varying thickness of shims (44) on reassembly.

To reassemble, proceed as follows: Be sure all bearings and seals are clean and lubricated with engine oil. Stand cylinder block on front end and place front thrust washer in block with bearing face toward crankshaft shoulder and with tab engaging notch in block. Lower crankshaft into block until seated against thrust washer. Use grease to stick thrust washer into rear bearing housing with bearing face toward crankshaft shoulder and with tab engaging notch in bearing housing.

If thrust washers and crankshaft are being reinstalled, refer to end play measurement taken prior to disassembly and use correct thickness of shims (44) for specified end play. Install rear bearing housing, tighten retaining bolts to a torque of 37 N·m (27 ft.-lbs.), then check to be sure crankshaft turns freely. Check crankshaft end play and if correct, complete reassembly by reversing disassembly procedure.

Three-Cylinder Model 3T84A

The crankshaft is supported in four main bearings and end play is controlled by a flanged main bearing (9—Fig. Y2-20) at third (from front) position. The front main bearing is pressed into crankcase bore and the rear bearing is pressed into rear bearing support (12). The two center bearings are carried in split supports (8) retained in position by set bolts (45).

To remove crankshaft, proceed as follows: Remove cylinder head and oil pan and remove the rod and piston units. Remove timing gear cover and pull crankshaft gear (5) from front end of shaft. Remove the oil pick-up tube (30). Remove flywheel and turn engine up on front end so that crankshaft is resting against front thrust washer. Remove the center main bearing support set bolts (46). Unbolt and remove rear main bearing housing (12); jack screw (threaded) holes are provided in support for easy removal. Lift crankshaft upward out of rear end of cylinder block. Remove bolts clamping bearing support (8) halves together and remove the supports and center main bearings (7 and 9).

Inspect crankshaft bearing journals and thrust surfaces for No. 3 main bearing flanges for scoring or excessive wear. Rear journal diameter should be 89.95-89.96 mm (3.5413-3.5417 inches) with suggested wear limit of 89.92 mm (3.5402 inches). The three front main bearing journals should measure 69.95-69.96 mm (2.7539-2.7543 inches) with suggested wear limit of 69.92 mm (2.7528 inches).

Main bearing oil clearance should be 0.036-0.099 mm (0.0014-0.0039 inch) for the three front journals and 0.066-0.132 mm (0.0026-0.0052 inch) for rear journal. Maximum allowable oil clearance is 0.15 mm (0.006 inch) for the three front journals and 0.2 mm (0.008 inch) for rear main journal.

Crankshaft end play can be determined by measuring end play of the assembled No. 3 main bearing and bearing support on the No. 3 main journal. End play should be 0.09-0.19 mm (0.0035-0.0075 inch).

To install crankshaft and main bearings, proceed as follows: Be sure all gasket material is removed from rear of cylinder block and rear main bearing support. Clean all parts thoroughly and lubricate bearings, crankshaft and seals with clean engine oil. Install the flanged main bearings (9) in bearing support with machined thrust flange surfaces. Assemble this bearing support on No. 3 crankshaft main journal with side of support marked "F" to flywheel end of crankshaft and tighten bolts to a torque of 64-69 N·m (47-51 ft.-lbs.). Assemble split bearing without flanges in other bearing support and install on No. 2 crankshaft journal with side of support marked "F" toward flywheel end of crankshaft and tighten bolts to a torque of 64-69 N·m (47-51 ft.-lbs.). Set cylinder block on supports with front end down and lower crankshaft assembly into block. Install, but do not tighten the two center main bearing set bolts (45). Install rear main bearing support (12) with new gasket (44) and with oil supply hole in cylinder block and bearing support aligned. Tighten the rear main bearing support bolts to a torque of 37 N·m (37 ft.-lbs.) and check to be sure crankshaft turns freely. Turn cylinder block to horizontal position and tighten No. 3 (thrust bearing) support set bolt to a torque of 78 N·m (58 ft.-lbs.), then tighten No. 2 bearing support set bolt to same torque. Again check to be sure crankshaft turns freely and crankshaft end play is within limits. Complete remainder of assembly by reversing disassembly procedure.

Fig. Y2-21—View of timing marks on camshaft and crankshaft gears; timing marks are typical of all models.

Fig. Y2-22—To remove camshaft assembly on all models, first remove ball bearing retaining screw (S).

Yanmar

SMALL DIESEL

CRANKSHAFT REAR OIL SEAL

The crankshaft rear oil seal (13—Fig. Y2-18, Y2-19 or Y2-20) is located in the rear main bearing support (12) and can be renewed after removing flywheel. Carefully remove seal to avoid damage to crankshaft and bore in bearing support, then install new seal with lip forward (to inside).

CRANKSHAFT FRONT OIL SEAL

The crankshaft front oil seal (3—Fig. Y2-18, Y2-19 or Y2-20) is located in the engine timing gear cover and can be renewed after removing crankshaft pulley (2). Carefully remove seal to avoid damage to timing cover and install new seal with lip to rear (to inside). Inspect seal surface of crankshaft pulley and renew or repair pulley if seal surface is scored or grooved.

ELECTRICAL SYSTEM

ALTERNATOR AND REGULATOR

Models 2T73A and 2TR13A

All models are equipped with a Kokusan alternator and current limiter. Refer to Fig. Y2-23 for exploded view of alternator. Current limiter (1) is a sealed unit and must be renewed if defective.

Models 2T84A, 2TR20A-X and 3T84A

All models are equipped with a Hitachi alternator and voltage regulator. Refer to Fig. Y2-24 for exploded view of alternator. Brushes are marked with a line to indicate wear limit; renew brushes if worn to mark. Voltage regulator is nonadjustable and must be renewed as a unit if defective.

STARTING MOTOR

Several different starting motors have been used. Be sure to check data plate on unit before attempting to service starter or acquire parts. Standard brush length is 16 mm (0.630 inch) for all models except Model 3T84A. Model 3T84A standard brush length is 20 mm (0.787 inch). Refer to Figs. Y2-25, Y2-26 or Y2-27 for exploded views.

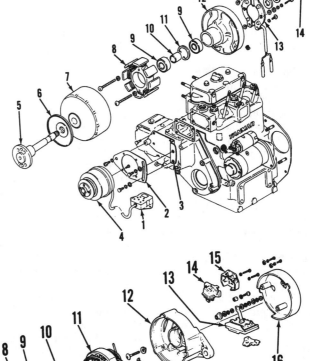

Fig. Y2-23—Exploded view of typical Kokusan alternator used on Models 2T73A and 2TR13A.

1. Current limiter
2. Adjustment bracket
3. Cover bracket
4. Alternator
5. Shaft
6. Pulley
7. Magnet wheel
8. Stationary windings
9. Bearings
10. Spacer
11. Snap ring
12. Housing
13. Diode assy.
14. Nut

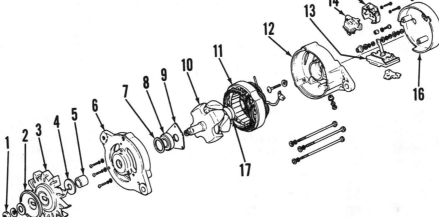

Fig. Y2-24—Exploded view of typical alternator used on Models 2T84A, 2TR20A-X and 3T84A.

1. Nut
2. Pulley
3. Fan
4. Washer
5. Spacer
6. Front cover
7. Retainer packing
8. Ball bearing
9. Retainer
10. Rotor
11. Stator
12. Housing
13. Diode assy.
14. Brushes & holder
15. Cover
16. Rear cover
17. Ball bearing

Illustrations Courtesy Yanmar Diesel Engine, Inc.

SERVICE MANUAL

Yanmar

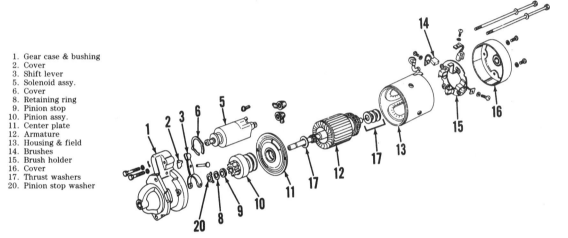

1. Gear case & bushing
2. Cover
3. Shift lever
4. Spring
5. Solenoid plunger
6. Cover shim
7. Solenoid coil
8. Retaining ring
9. Pinion stop
10. Pinion assy.
11. Center plate
12. Armature
13. Housing & field
14. Brushes
15. Brush holder
16. Cover
17. Thrust washers
18. "E" ring
19. Cap

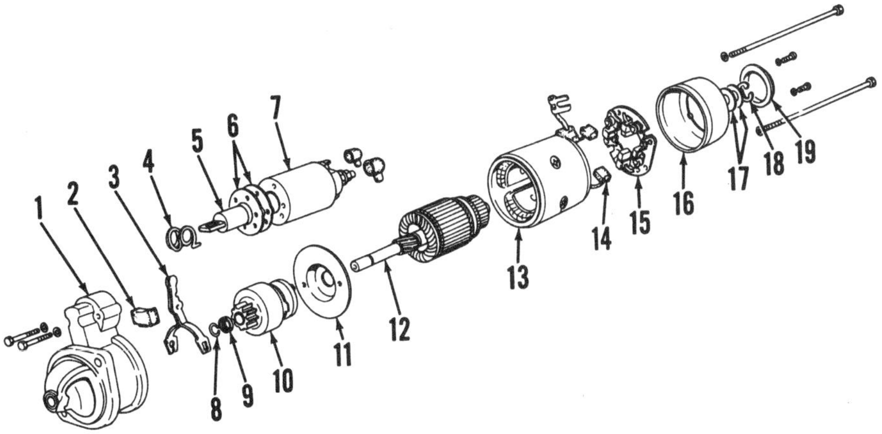

Fig. Y2-25—Exploded view of typical Hitachi starting motor used on most two-cylinder engines. Optional starter for Model 2TR20A-X is shown in Fig. Y2-26.

1. Gear case & bushing
2. Cover
3. Shift lever
5. Solenoid assy.
6. Cover
8. Retaining ring
9. Pinion stop
10. Pinion assy.
11. Center plate
12. Armature
13. Housing & field
14. Brushes
15. Brush holder
16. Cover
17. Thrust washers
20. Pinion stop washer

Fig. Y2-26—Exploded view of typical starter that is standard equipment for Model 3T84A and optional equipment for Model 2TR20A-X. Optional starter for Model 3T84A is shown in Fig. Y2-27.

1. Gear case & bushing
2. Cover
3. Shift lever
5. Solenoid assy.
6. Cover
8. Retaining ring
9. Pinion stop
10. Pinion assy.
11. Center plate
12. Armature
13. Housing & field
14. Brushes
15. Brush holder
16. Cover
17. Thrust washers
21. Cover
22. Thrust washer
23. Bushing
24. Reduction gear
25. Thrust washers
26. Needle bearing
27. Seal
28. "O" ring
29. Ball bearing
30. Ball bearing

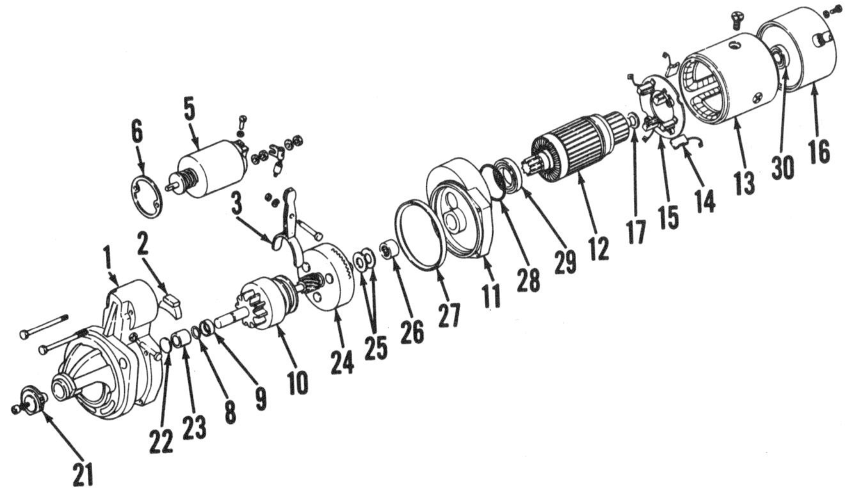

Fig. Y2-27—Exploded view of optionally available starter for Model 3T84A.

Illustrations Courtesy Yanmar Diesel Engine, Inc.

Yanmar

SMALL DIESEL

YANMAR

Model	No. Cyls.	Bore	Stroke	Displ.
3TN66E	3	66 mm (2.60 in.)	64.2 mm (2.53 in.)	658 cc (40.15 cu.in.)
3TNA72E	3	72 mm (2.83 in.)	72 mm (2.83 in.)	879 cc (53.64 cu.in.)
3TN75E	3	75 mm (2.95 in.)	75 mm (2.95 in.)	994 cc (60.66 cu.in.)
3TN82E	3	82 mm (3.23 in.)	86 mm (3.39 in.)	1362 cc (83.11 cu.in.)
3TN84E	3	84 mm (3.31 in.)	86 mm (3.39 in.)	1429 cc (87.20 cu.in.)
4TN82E	4	82 mm (3.23 in.)	86 mm (3.39 in.)	1816 cc (110.8 cu.n.)

All Series TN models are four-stroke, vertical cylinder, in-line, water cooled diesel engines for industrial use. Models 3TN66E and 3TNA72E are indirect fuel injection with pintle type nozzles; all other models are direct fuel injection with spray type nozzles. Models 3TN82E and 4TN82E (designated as 3TN82TE and 4TN82TE) are available with turbocharger. Camshaft is located in cylinder block and valves in head are operated through push rods and rocker arms. Engine rotation is clockwise as viewed from front (crankshaft pulley) end of engine. Engine serial number is stamped on cylinder block below the starting motor (exhaust manifold side of engine). Firing order is 1-3-2 on 3 cylinder models and 1-3-4-2 on four cylinder models. Number 1 cylinder is at flywheel end of engine.

MAINTENANCE

LUBRICATION

Recommended engine lubricant is SAE 10W-30 or SAE 30 engine oil with API classification CC or CD. Recommended oil change interval is after every 150 hours of use. Oil filter is a spin-on paper element cartridge. On new and reconditioned engines, lubricating oil and filter should be changed after first 50 hours of operation.

FUEL SYSTEM

FUEL FILTER. A fuel filter with renewable paper element (5—Fig. Y3-1) is used between the fuel feed pump and fuel injection pump. Filter element should be renewed after every 500 hours of use, or sooner if loss of power is evident. The filter element should be rated at 10-15 microns for indirect injection models and 8-9 microns for direct injection models.

BLEED FUEL SYSTEM. Air is automatically vented from filters on direct injection models. On indirect injection models, loosen bleed screw plug. Operate manual priming lever on left-hand side of fuel feed pump and allow fuel to flow until free of air, then tighten bleed plug. Loosen fuel return line check valve at injection pump and continue using feed pump primer lever until fuel flowing from loosened check valve is free of air. If engine will not start, loosen the fuel injector line fittings at the injectors and crank engine until fuel flows from all lines. Tighten the injector line fittings and start engine.

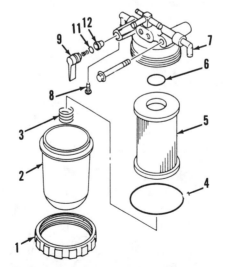

Fig. Y3-1—Exploded view of typical diesel fuel filter. On direct injection models, air is automatically vented; indirect injection models have air bleed screw (not shown).

1. Retaining ring
2. Filter bowl
3. Filter spring
4. "O" ring
5. Filter element
6. "O" ring
7. Filter base
8. Valve retaining screw
9. Fuel shut-off lever
11. "O" ring
12. Shut-off valve

FUEL FEED PUMP

A fuel feed pump is mounted on the injection pump camshaft housing and is actuated by cam of injection pump camshaft. Fuel feed pump is equipped with a hand lever for manual priming of fuel system. In some instances, it may be necessary to turn engine to move fuel pump cam away from pump before hand lever will operate pump. Although the pump can be disassembled for inspection, it is serviced as a complete assembly only. Renew pump if leaking or if it will not pump fuel.

INJECTION PUMP TIMING

Timing of fuel injection pump is accomplished by installing pump with timing gear marks aligned, and by loosening pump mounting bolts and rotating pump in slotted holes of mounting flange. Timing specifications are as follows:

Model & Type	Timing Degrees BTDC
3TN66E-S	13-15
3TN66E-G2	17-19
3TNA72E-S	15-17
3TNA72E-G2	17-19
3TN75E-S, -G1	15-17
3TN75E-G2	25-27
3TN82E-S, -G1	15-17
3TN82E-G2	25-27
4TN82E-S, -G1	15-17
4TN82E-G2	25-27
4TN82TE-S	14-16
4TN82TE-G1	13-15

To check injection pump timing, thoroughly clean injector, injector line and pump and remove fuel injection line to number 1 cylinder (cylinder at flywheel end of engine). Remove delivery valve holder lock, unscrew the delivery valve holder, remove delivery valve spring and reinstall holder. Connect a

SERVICE MANUAL

Yanmar

suitable spill pipe to delivery valve holder. Move governor control to "Run" position. While actuating fuel feed pump manually, turn crankshaft slowly in normal direction of rotation so that number 1 piston is starting up on compression stroke and observe fuel flow from open fuel delivery valve. When fuel just stops flowing, stop turning crankshaft and check injection timing mark (1—Fig. Y3-5) on engine flywheel. If timing mark is not aligned with matching mark (M) on flywheel housing, repeat check several times to be sure you are observing fuel flow correctly.

To adjust injection pump timing, loosen pump mounting bolts and rotate top of pump (or pump cam housing on Models 3TN66E and 3TNA72E) in toward engine to retard timing or away from engine to advance timing. Tighten pump mounting bolts and recheck timing. When injection pump timing is within specifications, install fuel delivery valve spring and tighten delivery valve holder to a torque of 35-39 N·m (26-28 ft.-lbs.). Install fuel injection line; bleed air from line before tightening fitting at injector.

GOVERNOR

The governor is located within the fuel injection pump cam housing on all models. Engine governed no-load speed depends upon engine application; refer to equipment manual for engine speed specifications. Refer to Fig. Y3-6 for maximum governed speed and idle speed adjustment screw location. Minimum low idle speed should be 900 rpm on Models 3TN66E and 3TNA72E; on all other models, minimum low idle speed should be 800 rpm.

REPAIRS

TIGHTENING TORQUES

Refer to the following special tightening torques:

Cylinder head bolts—
  Model 3TN66E . . . . 33-36 N·m
    (24-26 ft.-lbs.)
  Model 3TNA72E . . . . 59-63 N·m
    (44-47 ft.-lbs.)
  Model 3TN75E . . . . 67-70 N·m
    (50-52 ft.-lbs.)
  Models 3TN82E,
    4TN82E, 4TN82TE . . . . . 74-83 N·m
    (55-61 ft.-lbs.)
Connecting rod bolts—
  Models 3TN66E,
    3TNA72E . . . . 23-27 N·m
    (17-20 ft.-lbs.)
  Model 3TN75E . . . . 38-41 N·m
    (28-30 ft.-lbs.)
  Models 3TN82E,
    4TN82E, 4TN82TE . . . . 45-49 N·m
    (33-36 ft.-lbs.)
Valve rocker arm shaft . . . . 23-28 N·m
    (17-21 ft.-lbs.)
Flywheel bolts—
  Models 3TN66E,
    3TNA72E . . . . 81-86 N·m
    (60-63 ft.-lbs.)
  All other models . . . . 84-88 N·m
    (62-65 ft.-lbs.)
Main bearing bolts—
  Model 3TN66E . . . . 52-55 N·m
    (39-41 ft.-lbs.)
  Model 3TNA72E . . . . 74-83 N·m
    (55-65 ft.-lbs.)
  Model 3TN75E . . . . 77-80 N·m
    (57-59 ft.-lbs.)
  Models 3TN82E,
    4TN82E, 4TN82TE . . . . 94-102 N·m
    (69-76 ft.-lbs.)
Crankshaft pulley bolt—
  Models 3TN66E,
    3TNA72E . . . . 84-93 N·m
    (62-68 ft.-lbs.)
  All other models . . . . . . 108-127 N·m
    (80-94 ft.-lbs.)
Timing idler stub
  shaft bolts . . . . 54-63 N·m
    (40-47 ft.-lbs.)
Water pump mounting
  bolts . . . . 18-22 N·m
    (13-16 ft.-lbs.)
Injector nozzle nut,
  all models . . . . 40-44 N·m
    (29-32 ft.-lbs.)
Injector, indirect
  injection . . . . 49-53 N·m
    (37-39 ft.-lbs.)
Injector retainer, direct
  injection . . . . 4-5 N·m
    (35-43 inch-lbs.)
Glow plug, indirect injection
  models . . . . 15-19 N·m
    (11-14 ft.-lbs.)

WATER PUMP AND COOLING SYSTEM

To remove water pump, first drain cooling system. Unbolt and remove fan blade assembly and disconnect coolant hose. Loosen alternator mounting and brace to remove the fan belt. Unbolt and remove water pump (5—Fig. Y3-7) from cylinder head and block (Models 3TN66E, 3TN72E and 3TN75E), or remove water pump from cylinder block on Models 3TN82E and 4TN82E. The water pump is serviced as a complete assembly only; renew pump if seal is leaking or bearing is noisy or otherwise defective. Be sure mounting surfaces are clean and install pump using new gasket and, on Models 3TN66E, 3TN72E and 3TN75E, install a new sealing "O" ring on block connector pipe. Tighten retaining bolts to a torque of 18-22 N·m (13-16 ft.-lbs.). Fan belt should be ad-

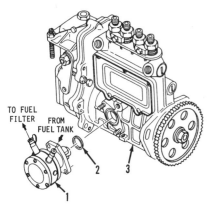

Fig. Y3-3—Direct injection models have fuel feed pump (1) mounted on the fuel injection pump (3). On indirect injection models, fuel feed pump (not shown) is mounted on injection pump camshaft housing.

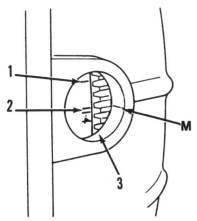

Fig. Y3-5—Timing marks are visible on flywheel after removing cover from hole in flywheel housing. When checking timing, watch for cylinders 1-4 TDC mark (2) and appropriate BTDC mark (1) on flywheel. Specified BTDC mark should be aligned with mark (M) at front side of hole.

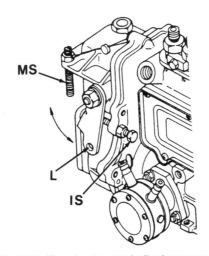

Fig. Y3-6—View showing speed adjusting screws on models with direct injection system. Models with indirect injection are similar.

Illustrations Courtesy Yanmar Diesel Engine, Inc.

Yanmar

justed to deflect 10-15 mm (0.4-0.6 inch) with hard finger pressure.

Radiator cap working pressure is 88 kPa (12.8 psi). Renew cap if defect is noted or if test reveals inaccurate working relief pressure. Cooling system should be pressurized to 88 kPa (12.8 psi) with engine warm to check for coolant leaks.

THERMOSTAT

The thermostat is mounted in the water pump housing on Models 3TN66E, 3TN72E and 3TN75E, and in the thermostat housing attached to front end of cylinder head on Models 3TN82E and 4TN82E. To remove thermostat, drain cooling system and unbolt water outlet (1—Fig. Y3-7 or Y3-8) from top of water pump or thermostat housing. Thermostat should start to open at 71° C (160° F) and be fully open at 85° C (185° F). Thermostat valve lift height fully open should be 4.5 mm (0.18 inch) on Model 3TN66E and 8 mm (0.31 inch) on all other models. When installing thermostat, be sure all gasket surfaces are clean and use new gasket and sealing ring.

VALVE ADJUSTMENT

Valve clearance should be adjusted to 0.2 mm (0.008 inch) on all models with engine cold. Adjustment is made by loosening locknut and turning adjusting screw at push rod end of rocker arm. To adjust valves, turn crankshaft in normal direction of rotation so that number 1 piston is at top dead center on compression stroke. Adjust intake and exhaust valves on number 1 cylinder. Continue to turn crankshaft until next piston in firing order is at top dead center on compression stroke and adjust valves for that cylinder in same manner. Repeat the procedure for each cylinder in order.

ROCKER ARMS AND PUSH RODS

Rocker arm shaft assembly (Fig. Y3-10) and push rods can be removed after removing rocker arm cover. Rocker arms for all cylinders pivot on full length shaft which is held by support brackets attached to cylinder head by cap screws. Intake and exhaust rocker arms and push rods are alike.

Rocker shaft diameter should be 9.97-9.99 mm (0.3925-0.3933 inch) with wear limit of 9.955 mm (0.392 inch) on engine Model 3TN66E and 11.966-11.984 mm (0.4711-0.4718 inch) with a wear limit of 11.955 mm (0.4706 inch) on Model 3TNA72E. For all other models, rocker shaft diameter should be 15.966-15.984 mm (0.6286-0.6293 inch) with wear limit of 15.955 mm (0.628 inch).

Rocker arm bushing inside diameter should be 10.00-10.02 mm (0.3937-0.3945 inch) with wear limit of 10.09 mm (0.397 inch) for engine Model 3TN66E and 12.00-12.02 mm (0.4724-0.4732 inch) with a wear limit of 12.09 mm (0.476 inch) on Model 3TNA72E. For all other models, rocker bushing inside diameter should be 16.00-16.02 mm (0.630-0.631 inch) with wear limit of 16.09 mm (0.633 inch). Clearance between bushing and

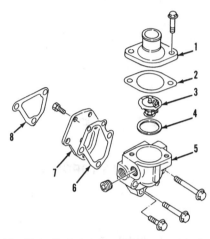

Fig. Y3-8—Exploded view of thermostat housing assembly for Models 3TN82E and 4TN82E.
1. Water outlet
2. Gasket
3. Thermostat
4. Sealing ring
5. Thermostat housing
6. Gasket
7. Cover plate
8. Gasket

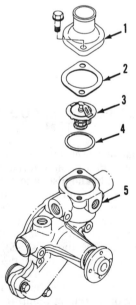

Fig. Y3-7—Water pump housing also serves as thermostat housing on Models 3TN66E, 3TN72E and 3TN75E. Water pump for Models 3TN82E and 4TN82E is separate unit from thermostat housing. Water pump is serviced as a complete assembly on all models.
1. Water outlet
2. Gasket
3. Thermostat
4. Sealing ring
5. Water pump

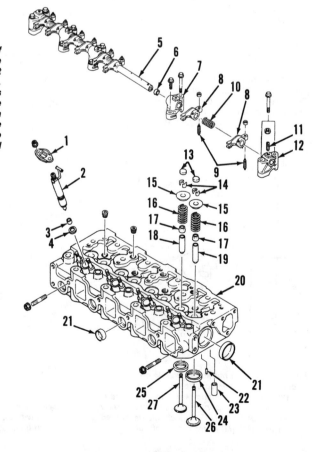

Fig. Y3-10—Exploded view of 4-cylinder head and valve mechanism. All other direct injection models are similar. Indirect injection models have a throttling pintle type injector and precombustion chamber in head below the injector. Standard models do not have valve seats (24 and 25).
1. Injector clamp
2. Fuel injector
3. Nozzle heat shield
4. Injector seat
5. Rocker shaft
6. Plug
7. Rocker shaft support
8. Rocker arm
9. Adjustment screw
10. Rocker spacer spring
11. Set screw
12. Front rocker support
13. Valve lash caps
14. Valve split locks
15. Valve spring retainers
16. Valve springs
17. Valve stem seals
18. Exhaust valve guide
19. Intake valve guide
20. Cylinder head
21. Cup plugs
22. Dowel pin
23. Plug
24. Intake valve seat
25. Exhaust valve seat
26. Intake valve
27. Exhaust valve

shaft should not exceed 0.135 mm (0.0053 inch) on any model engine.

Check push rods by rolling on flat plate and renew any showing any curvature or if ends are galled, worn or chipped. Push rod length is 114-115 mm (4.488-4.528 inches) on Model 3TN66E, 141-142 mm (5.551-5.590 inches) on Model 3TNA72E, 146.65-147.35 mm (5.774-5.801 inches) on Model 3TN75E, and 178.25-178.75 mm (7.018-7.037 inches) on all other models.

On all models, loosen rocker arm adjustment screw locknuts and back screws (9) out before installing push rods, rocker arm supports and rocker arms. Tighten rocker arm support bolts to a torque of 23-28 N·m (17-21 ft.-lbs.) on all models. Adjust valves as previously outlined and install rocker arm cover.

CYLINDER HEAD

To remove cylinder head, drain engine coolant and proceed as follows: Remove muffler, air cleaner and instrument panel. On turbocharged models, disconnect piping and remove turbocharger. Remove intake and exhaust manifolds, rocker arm cover, rocker arm assembly and the push rods. Remove fuel injector pressure and return lines and remove fuel injectors. Disconnect coolant hose from water outlet housing, or unbolt and remove water outlet housing from cylinder head. Remove cylinder head bolts and lift cylinder head from engine.

Before reinstalling cylinder head, thoroughly clean head and block gasket surfaces and inspect head for cracks, distortion or other defect. If alignment dowels (roll pins) were removed with head, pull pins from head and install in cylinder block. Check cylinder head gasket surface with straightedge and feeler gage. Head must be reconditioned or renewed if distortion (warp or twist) exceeds 0.15 mm (0.006 inch).

Place new gasket on block with trade mark or engine model number up and locate gasket on the two roll pins. Place head on block so that the two roll pins enter matching holes in head. Lubricate bolt threads and install the bolts, tightening in at least two steps in the sequence shown in Fig. Y3-11 or Fig. Y3-12 until proper torque is reached. Specified cylinder head bolt tightening torques are as follows:

Model	Cylinder Head Torque
3TN66E	33-36 N·m (24-26 ft.-lbs.)
3TNA72E	59-63 N·m (44-47 ft.-lbs.)
3TN75E	67-70 N·m (50-52 ft.-lbs.)
All other models	74-83 N·m (55-61 ft.-lbs.)

Complete reassembly of engine by reversing removal procedure. Yanmar recommends retightening cylinder head bolts after first 50 hours of use, then after each two years service.

VALVE SYSTEM

Valves seat directly in cylinder head on standard models; models with power pack (type -G2) have renewable valve seats. Valve seat and face angle is 30 degrees for intake valves and 45 degrees for exhaust valves. Recondition exhaust seats using 45, 60 and 30 degree stones or cutters and recondition intake seats using 30, 60 and 15 degree stones or cutters. Valve seat width specifications are as follows:

Intake Valve Seat
Model 3TN66E
  Desired width .... 1.15 mm
                         (0.045 in.)
  Maximum width .... 1.65 mm
                         (0.065 in.)
Model 3TNA72E
  Desired width .... 1.44 mm
                         (0.057 in.)
  Maximum width .... 1.98 mm
                         (0.078 in.)
Model 3TN75E
  Desired width .... 1.36-1.53 mm
                       (0.054-0.060 in.)
  Maximum width .... 1.98 mm
                         (0.078 in.)
All other models
  Desired width .... 1.07-1.24 mm
                       (0.042-0.049 in.)
  Maximum width .... 1.74 mm
                         (0.069 in.)

Exhaust Valve Seat
Model 3TN66E
  Desired width .... 1.41 mm
                         (0.056 in.)
  Maximum width .... 1.91 mm
                         (0.075 in.)
Model 3TNA72E
  Desired width .... 1.77 mm
                         (0.070 in.)
  Maximum width .... 2.27 mm
                         (0.089 in.)
Model 3TN75E
  Desired width .... 1.66-1.87 mm
                     (0.065-0.074 in.)
  Maximum width .... 2.27 mm
                         (0.089 in.)
All other models
  Desired width .... 1.24-1.45 mm
                     (0.049-0.057 in.)
  Maximum width .... 1.94 mm
                         (0.076 in.)

After reconditioning seats, check distance (D—Fig. Y3-13) valve head is below cylinder head surface. Standard valve head recess distance (sink) of Model 3TN66E is 0.4 mm (0.016 inch) for intake and 0.85 mm (0.034 inch) for exhaust valve. Model 3TNA72E intake valve recess should be 0.5 mm (0.020 inch) and exhaust valve recess should be 0.85 mm (0.034 inch). For all other models, both intake and exhaust valve recess should be 0.3-0.5 mm (0.012-0.020 inch). If valve head recess is 1.0 mm (0.039 inch) or more, install new valve and/or valve seat.

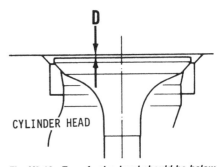

Fig. Y3-13—Top of valve head should be below cylinder head surface. Refer to text for specifications.

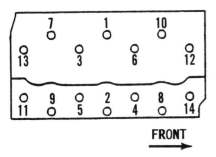

Fig. Y3-11—Cylinder head bolt tightening sequence for three cylinder engines.

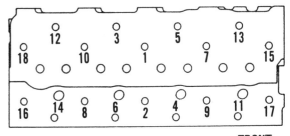

Fig. Y3-12—Cylinder head bolt tightening sequence for four cylinder engines.

Yanmar

If new valve seats are installed, recondition seating surface for full valve contact and proper seat width and check to be sure correct valve recess is maintained. Heads not equipped with renewable valve seats must be renewed or machined for installation of valve seats if seats cannot be ground.

Refer to the following valve stem specifications:

Model 3TN66E
Intake valve stem
  OD . . . . 5.460-5.475 mm
                    (0.215-0.216 in.)
  Wear limit . . . . 5.40 mm
                         (0.214 in.)
Exhaust valve stem
  OD . . . . 5.445-5.460 mm
                    (0.214-0.215 in.)
  Wear limit . . . . 5.40 mm

Model 3TNA72E
Intake valve stem
  OD . . . . 6.945-6.960 mm
                    (0.273-0.274 in.)
  Wear limit . . . . 6.90 mm
                         (0.272 in.)
Exhaust valve stem
  OD . . . . 6.945-6.960 mm
                    (0.273-0.274 in.)
  Wear limit . . . . 6.90 mm
                         (0.272 in.)

Model 3TN75E
Intake valve stem
  OD . . . . 6.945-6.960 mm
                    (0.273-0.274 in.)
  Wear limit . . . . 6.90 mm
                         (0.272 in.)
Exhaust valve stem
  OD . . . . 6.940-6.955 mm
                    (0.273-0.274 in.)
  Wear limit . . . . 6.90 mm
                         (0.272 in.)

All Other Models
Intake valve stem
  OD . . . . 7.960-7.975 mm
                    (0.313-0.314 in.)
  Wear limit . . . . 7.90 mm
                         (0.311 in.)
Exhaust valve stem
  OD . . . . 7.955-7.970 mm
                    (0.313-0.314 in.)
  Wear limit . . . . 7.90 mm
                         (0.311 in.)

Check valve guides against the following specifications and renew any guides with excessive wear. When installing new guides, drive old guide out toward top of head using suitable piloted driver. Install new guide with piloted driver so that top of guide (see Fig. Y3-14) is at guide height specified in following table:

Model 3TN66E
Guide height . . . . 7 mm
                         (0.276 in.)

Guide ID . . . . 5.50-5.15 mm
                    (0.2165-0.217 in.)
Wear limit . . . . 5.58 mm
                       (0.220 in.)

Model 3TNA72E
Guide height . . . . 9 mm
                         (0.354 in.)
Guide ID . . . . 7.005-7.02 mm
                  (0.2758-0.2764 in.)
Wear limit . . . . 7.08 mm
                       (0.279 in.)

Model 3TN75E
Guide height . . . . 12 mm
                         (0.472 in.)
Guide ID . . . . 7.005-7.02 mm
                  (0.2758-0.2764 in.)
Wear limit . . . . 7.08 mm
                       (0.279 in.)

All Other Models
Guide height . . . . 15 mm
                         (0.590 in.)
Guide ID . . . . 8.01-8.03 mm
                    (0.3154-0.316 in.)
Wear limit . . . . 8.10 mm
                       (0.319 in.)

Exhaust valve guides are identified by a groove around the guide. After installing guide, use suitable reamer to obtain proper fit of valve stem to guide.

Intake and exhaust valve springs are alike on all models. Spring free length, new, is 28 mm (1.102 inches) for Model 3TN66E, 37.4 mm (1.472 inches) on Model 3TNA72E, 42 mm (1.654 inches) on Model 3TN75E and 44.4 mm (1.748 inches) on all other models. Be sure to install springs with closer spaced coils next to cylinder head.

TURBOCHARGER

Models 3TN82TE and 4TN82TE are equipped with a turbocharger. As turbocharger internal parts are easily damaged, it is recommended that service work on the turbocharger be done by an authorized repair station.

The turbocharger is lubricated by a pressure line from the engine oil gallery and oil is then returned to engine crankcase. If seal and bearing wear are such that oil is leaking into the air intake system, evidenced by heavy exhaust smoke, the turbocharger must be repaired or renewed.

Engines with malfunctioning fuel injectors or loss of engine oil past the oil control ring will cause heavy carbon deposits on the turbocharger blower wheel. Also, poor maintenance of engine lubrication system may result in turbocharger bearing and seal wear. If engine is under major repair, the turbocharger should be inspected and renewed or repaired as necessary.

REMOVE AND REINSTALL. Thoroughly clean turbocharger and surrounding area prior to removal. Remove air inlet and discharge hose. Disconnect oil supply line and oil return line and plug openings. Unbolt and remove turbocharger from engine.

To reinstall turbocharger, reverse removal procedure. Before reconnecting oil return line, move speed control linkage to shut-off position and crank engine with starter until oil flows from turbocharger oil return port. Connect oil return line, start engine and operate at low speed for about one minute to allow oil to start circulating through turbocharger.

INJECTORS

REMOVE AND REINSTALL. Prior to removing injectors, thoroughly clean injectors, injector lines, fuel injection pump and surrounding area using suitable solvent and compressed air. Remove the fuel return line from top of injectors. Remove the fuel injection lines from injection pump and injectors. On Models 3TN66E and 3TNA72E, unscrew the fuel injectors from cylinder head. On all other models, remove the fuel injector retaining clamps and pull injectors from head. Immediately cap or plug all fuel openings. Remove seals from bottom of injector bores in cylinder head. On Models 3TN66E and 3TNA72E, remove precombustion chamber and packing from cylinder head injector bores.

When installing fuel injectors, first be sure injector bores in cylinder head are clean and all old seals are removed. Insert new seals and, on Models 3TN66E and 3TNA72E, install precombustion chambers in injector bores and install the injectors. On Models 3TN66E and 3TNA72E, tighten the nozzle to a torque of 49-53 N·m (37-39 ft.-lbs.). On all other models, tighten the two clamp nuts evenly to a torque of 4-5 N·m (35-43 in.-lbs.). Bleed air from injector lines before tightening line fittings at injector end.

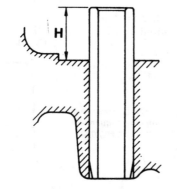

Fig. Y3-14—Valve guides must be installed to specified height (H) as outlined in text.

SERVICE MANUAL

Yanmar

TESTING. A complete job of testing, cleaning and adjusting the fuel injector requires use of special test equipment. Use only clean approved testing oil in tester tank. Injector should be tested for opening pressure, seat leakage and spray pattern. Before connecting injector to test stand, operate tester lever until oil flows, then attach injector to tester line.

WARNING: Fuel emerges from injector with sufficient force to penetrate the skin. When testing injector, keep yourself clear of nozzle spray.

Close valve to tester gage and operate tester lever a few quick strokes to purge air from injector and to be sure nozzle valve is not stuck.

OPENING PRESSURE. Open the valve to tester gage and operate tester lever slowly while observing gage reading. Opening pressure should be 19130-20100 kPa (2773-2915 psi) for Models 3TN66E and 3TNA72E and 11280-12260 kPa (1635-1778 psi) for all other models. Opening pressure is adjusted by changing thickness of shim stack (7—Fig. Y3-15 or Fig. Y3-16). Adding shim thickness will increase pressure.

SPRAY PATTERN. On Models 3TN66E and 3TNA72E, the spray pattern should be conical, well atomized, and emerging in a straight axis from nozzle tip. If spray is drippy, ragged or to one side, nozzle must be cleaned or renewed.

On all other models, spray pattern should be equal from all four nozzle tip holes and should be a fine mist. Spray angle (A—Fig. Y3-17) is 150 degrees and each spray cone should be 5-10 degrees (B). Place a sheet of blotting paper 30 cm (about 12 inches) below injector and operate tester one stroke. The spray pattern should be a perfect circle.

SEAT LEAKAGE. Wipe nozzle tip dry with clean blotting paper, then slowly actuate tester lever to obtain gage pressure approximately 1960 kPa (285 psi) below opening pressure and hold this tester pressure for 10 seconds. If drop of oil appears at nozzle tip, or oil drips from nozzle, clean or renew nozzle.

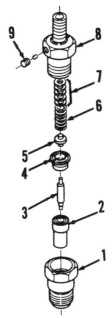

Fig. Y3-16—Exploded view of throttling pintle type fuel injector used on indirect injection models.
1. Nozzle nut
2. Nozzle body
3. Nozzle valve
4. Nozzle body spacer
5. Spring seat
6. Spring
7. Adjusting shims
8. Injector body
9. Fuel leak-off fitting

OVERHAUL. Hard or sharp tools, emery cloth, grinding compounds or other than approved tools, solvents and lapping compounds, must never be used. Wipe all dirt and loose carbon from injector, then refer to Figs. Y3-15 and Y3-16 and proceed as follows:

Secure injector nozzle in holding fixture or soft jawed vise and loosen nut (1). Unscrew nut and remove internal parts (2 to 7) from injector. Be careful not to lose pins (P) or shims (7). Place all parts in clean diesel fuel or calibrating oil as they are removed, and take care not to mix parts with those from other injector assemblies.

Clean exterior surfaces with brass wire brush to loosen carbon deposits; soak parts in approved carbon solvent if necessary. Rinse parts in clean diesel fuel or calibrating oil after cleaning to neutralize carbon solvent.

Clean nozzle spray hole from inside using pointed hardwood stick. Scrape carbon from nozzle pressure chamber using hooked scraper. Clean valve seat using brass scraper, then polish seat using wood polishing stick and mutton tallow. On direct injection nozzles, use a 0.2 mm (0.008 inch) diameter cleaning wire in a pin vise to clean the four spray holes. Inspect nozzle valve (3) and body (2) for scratches or wear. Be sure that nozzle valve slides smoothly in nozzle body bore; nozzle valve should slide down into nozzle body by its own weight. If valve sticks in nozzle body, reclean or renew nozzle assembly.

Reclean all parts by rinsing thoroughly in clean diesel fuel or calibrating oil and assemble injector while immersed in fuel or oil. Be sure shim stack is intact, then install shims (7), spring (6), spring seat (5), stop plate (4—Fig. Y3-15 or nozzle valve seat (4—Fig. Y3-16), nozzle valve (3) and nozzle body (2) and the nozzle nut. Tighten nozzle retaining nut (9) to a torque of 40-44 N·m (29-32 ft.-lbs.). Retest assembled injector. Do not over-tighten nozzle nut in attempt to stop any leakage; disassemble and clean injector if leak is noted.

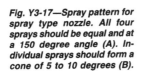

Fig. Y3-15—Exploded view of spray type fuel injector used on direct injection models.
P. Dowel pins
1. Nozzle nut
2. Nozzle body
3. Nozzle valve
4. Stop plate
5. Spring seat
6. Spring
7. Adjusting shims
8. Injector body
9. Fuel leak-off pipe

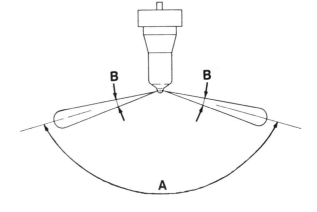

Fig. Y3-17—Spray pattern for spray type nozzle. All four sprays should be equal and at a 150 degree angle (A). Individual sprays should form a cone of 5 to 10 degrees (B).

Illustrations Courtesy Yanmar Diesel Engine, Inc.

Yanmar

INJECTION PUMP

Models 3TN66E and 3TNA72E

Thoroughly clean pump, fuel injectors, lines and surrounding area using suitable solvent and compressed air. Disconnect fuel return line from injectors and remove the high pressure fuel injection lines. Remove the cover from side of fuel injection pump camshaft housing and pull out the governor link snap pin. Unbolt and remove pump from top of camshaft housing.

To install fuel injection pump, reverse removal procedure. If injection pump camshaft housing has not been removed, timing will not be changed. Check oil level in pump cam housing and fill to center of camshaft with clean diesel engine motor oil. Pump timing is controlled by turning cam housing in slotted mounting holes. Leave fuel injection line fittings at injectors loose until air has been bled from lines.

All Models Except 3TN66E and 3TNA72E

To remove injection pump used on direct injection models, it will be necessary to remove timing gear cover, then unbolt and remove pump drive gear before pump can be removed from engine front plate. Thoroughly clean pump, injectors and lines using suitable solvent and compressed air. Disconnect throttle linkage from governor arm on pump. Disconnect injector fuel return line at pump and remove the pump to injector high pressure lines. Remove oil line from engine oil gallery to pump camshaft housing. If not present, scribe a timing mark on fuel injection pump mounting flange and engine front plate. Unbolt and remove pump from front plate.

To install pump, proceed as follows: Be sure pump mounting surfaces are clean and install pump on engine front plate using a new "O" ring. Align timing marks scribed on pump flange and engine front plate and tighten pump mounting bolts. Turn engine and pump so that drive gear can be installed with timing marks aligned as shown in Fig. Y3-19. Tighten pump gear retaining nut to a torque of 59-68 N·m (44-50 ft.-lbs.). Complete reassembly by reversing pump removal procedure, leaving fuel injection line fittings loose at injectors until air is bled from lines. Check pump camshaft housing for adequate lubricant; fill to oil line opening with clean diesel engine motor oil. If pump timing marks were not scribed on pump flange and front plate before pump was unbolted, or if a new pump is being installed, check fuel injection pump timing.

OIL PAN

Removal of oil pan is conventional. Use silicone gasket compound to seal pan to bottom of engine; a gasket is not supplied.

CRANKSHAFT FRONT OIL SEAL

The crankshaft front oil seal (3—Fig. Y3-20) is located in the timing gear cover and can be renewed after removing crankshaft pulley. Pry seal from timing cover, taking care not to damage the cover. Seal lip rides on hub of crankshaft pulley; inspect pulley hub for excessive wear at point of seal contact. Install new seal with lip to inside using suitable driver. Lubricate seal lip and crankshaft pulley hub, install pulley on crankshaft with hole in pulley hub aligned with dowel pin in crankshaft gear. Tighten pulley retaining bolt to a torque of 84-93 N·m (62-68 ft.-lbs.) on Models 3TN66E and 3TNA72E, and to a torque of 108-127 N·m (80-94 ft.-lbs.) on all other models.

TIMING GEAR COVER

To remove timing gear cover (4—Fig. Y3-20), drain cooling system and proceed as follows: Remove alternator, fan and water pump assembly and the crankshaft pulley. Remove tachometer drive mechanism from in front of camshaft. Remove hydraulic pump, if so equipped, from timing cover. Remove timing cover bolts from front end of oil pan, then unbolt and remove cover from front of engine.

Thoroughly clean timing cover and mounting gasket surfaces. Install new crankshaft oil seal (3—Fig. Y3-20) with lip to inside and lubricate seal. Run a bead of silicone gasket material on front cover and pan gasket surfaces; gaskets are not serviced. Install timing cover on engine, aligning holes in cover with the two hollow alignment dowel pins in front of engine. Clean crankshaft pulley hub and apply lubricant on seal contact surface. Install the crankshaft pulley and tighten retaining bolt to a torque of 84-93 N·m (62-68 ft.-lbs.) on Models 3TN66E and 3TNA72E, and to a torque of 108-127 N·m (80-94 ft.-lbs.) on all other models. To complete reassembly of engine, reverse removal procedure.

TIMING GEARS

The timing gears can be inspected after removing timing gear cover. Desired backlash between mating timing gears

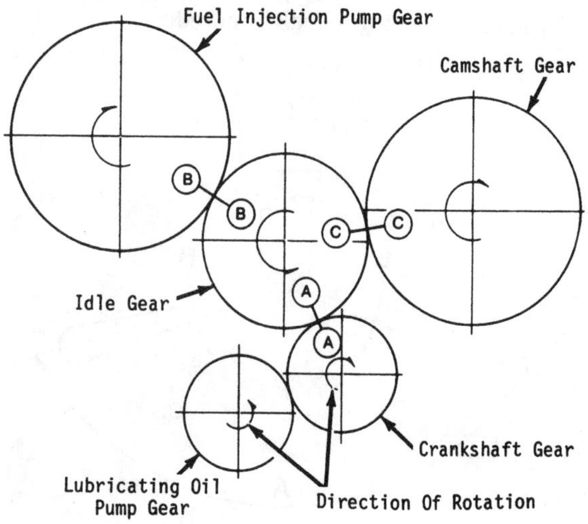

Fig. Y3-19—Schematic diagram of timing gear timing marks. Timing of oil pump gear is not required.

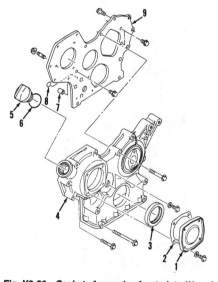

Fig. Y3-20—Gaskets for engine front plate (9) and timing gear cover (4) are not provided; use silicone gasket maker. Cover (1) is used when not equipped with hydraulic pump (not shown).

1. Cover
2. Gasket
3. Crankshaft seal
4. Timing gear cover
5. Oil filler cap
6. "O" ring
7. Hollow dowel pins
8. "O" ring
9. Engine front plate

SERVICE MANUAL

Yanmar

is 0.04-0.12 mm (0.0016-0.0047 inch); maximum allowable backlash is 0.2 mm (0.008 inch). Check gear teeth for excessive wear or tooth damage and renew as necessary. If necessary to renew gears, remove valve cover, rocker arm assembly and the push rods to relieve valve spring pressure on camshaft.

To remove idler gear and shaft on Models 3TN66E and 3TNA72E, refer to Fig. Y3-21 and remove snap ring (1), washer (2) and gear (3), then unbolt and remove stub shaft (5) from engine. Idler stub shaft diameter should be 19.959-19.980 mm (0.786-0.787 inches) and wear limit is 19.93 mm (0.785 inch). Desired bushing (4) to idler shaft oil clearance is 0.020-0.062 mm (0.0008-0.0024 inch) with maximum allowable clearance of 0.15 mm (0.006 inch). Renew bushing (4) if worn and idler gear (3) is otherwise serviceable. When installing stub shaft, tighten retaining bolts to a torque of 54-63 N·m (40-47 ft.-lbs.).

To remove direct injection engine timing gear idler, remove the two bolts (1—Fig. Y3-22) from idler shaft and remove the shaft (5) and idler gear (4). Idler shaft standard diameter is 45.950-45.975 mm (1.809-1.810 inches); renew if worn to 45.93 mm (1.808 inches). Bushing standard inside diameter is 46-46.025 mm (1.811-1.812 inches); wear limit is 46.08 mm (1.84 inches). Bushing (3) is renewable separately from idler gear (4) if gear is otherwise serviceable. Desired oil clearance for idler gear to shaft is 0.025-0.075 mm (0.001-0.003 inch). Renew shaft and/or bushing if clearance is 0.15 mm (0.006 inch) or more.

To remove crankshaft gear, use suitable pullers to remove gear from crankshaft. To remove camshaft gear, first remove camshaft and press shaft from gear. Fuel injection pump gear on direct injection models can be pulled from pump camshaft after removing retaining nut. On indirect injection models, gear is press-fit to pump camshaft.

When installing timing gears, be sure all timing marks are aligned as shown in Fig. Y3-19. Heat crankshaft timing gear to 180°-200° C (356°-392° F).

OIL PUMP AND RELIEF VALVE

The gerotor type oil pump (Fig. Y3-23) is mounted on front of engine front plate and gear on pump shaft is driven from crankshaft gear. The oil pump and drive gear are available as a complete assembly only. Renew pump assembly if excessive wear or other damage is noted.

On all models except turbocharged Model 4TN82TE, the system pressure relief valve (7—Fig. Y3-24) is located in the oil filter mounting adapter and regulates oil pressure to 295-392 kPa (43-56 psi). A bypass valve within the filter (1) opens when pressure difference between inlet and outlet side of filter is 78.4-117.7 kPa (11.4-17 psi).

On turbocharged Model 4TN82TE, the oil pump is fitted with a safety relief valve (V—Fig. Y3-23) to limit excessive system pressure. The system relief valve (12 to 16—Fig. Y3-25) is located in the

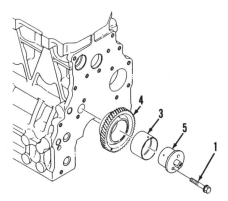

Fig. Y3-22—Idler gear shaft (5) for direct injection models is retained to front of block by two cap screws (1).

1. Cap screws
3. Idler gear bushing
4. Idler gear
5. Idler gear shaft

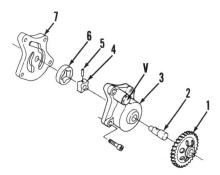

Fig. Y3-23—Engine oil pump is mounted inside timing gear cover to front face of engine front plate. Pump for turbocharged Model 4TN82TE has integral safety relief valve (V).

V. Safety relief valve
1. Drive gear
2. Pump shaft
3. Oil pump housing
4. Inner rotor
5. Pin
6. Outer rotor
7. End cover

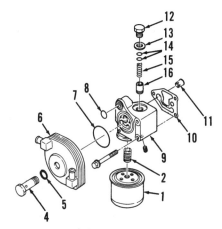

Fig. Y3-25—Oil filter base (9) for Model 4TN82TE has system relief valve (12 to 16) and oil cooler bypass valve (11). Engine oil cooler (6) is also mounted on filter base. Oil filter (1) has integral bypass valve.

1. Oil filter
2. Adapter pipe
4. Hollow bolt
5. "O" ring
6. Oil cooler
7. "O" ring
8. "O" ring
9. Filter base
10. Gasket
11. Cooler bypass valve
12. Retainer plug
13. Sealing washer
14. Adjusting shims
15. Spring
16. Relief valve poppet

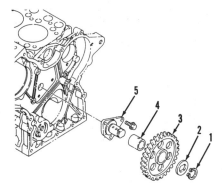

Fig. Y3-21—Idler gear (3) for indirect injection models is mounted on stub shaft (5).

1. Snap ring
2. Washer
3. Idler gear
4. Idler gear bushing
5. Stub shaft

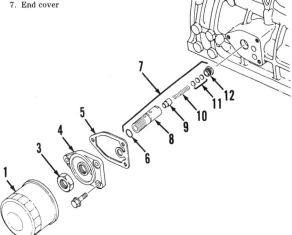

Fig. Y3-24—Exploded view of oil filter mounting used on most models. Adapter (8) has system relief valve inside. If disassembled, be sure valve retainer (12) is securely staked to adapter body (8). Oil filter (1) has integral by-pass valve.

1. Oil filter
3. Hex nut
4. Filter base
5. Gasket
6. "O" ring
7. Adapter & relief valve assy.
8. Adapter body
9. Relief valve poppet
10. Spring
11. Adjusting shims
12. Valve retainer

Illustrations Courtesy Yanmar Diesel Engine, Inc.

Yanmar

SMALL DIESEL

oil filter mounting bracket (9) and regulates normal system pressure to 294-392 kPa (43-57 psi). A bypass valve (11) in the oil filter bracket opens when pressure variation between oil cooler inlet and outlet reaches 118-147 kPa (17-21 psi). The bypass valve within the oil filter (1) opens if pressure difference between filter inlet and outlet is 78.4-117.7 kPa (11.4-17 psi).

To remove pump, first remove timing cover, then unbolt and remove pump from front of engine. The pump rear plate can then be removed from pump body to inspect pump. Clearance between outer rotor and pump body should be 0.10-0.17 mm (0.004-0.007 inch) with wear limit of 0.25 mm (0.010 inch). Clearance between lobes of inner and outer rotors should be 0.05-0.1 mm (0.002-0.004 inch) with maximum allowable clearance being 0.15 mm (0.006 inch). Rotor end clearance in pump body should be 0.03-0.09 mm (0.0012-0.0035 inch) and maximum end clearance is 0.13 mm (0.005 inch). Check for looseness of drive shaft in pump body and renew pump if loose or wobbly.

GOVERNOR

The flyweight type centrifugal governor is mounted in the rear end of the injection pump camshaft housing and disassembly should not be attempted outside of authorized diesel repair station.

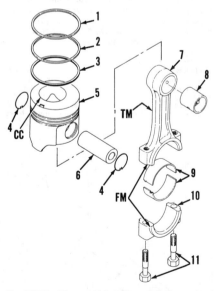

Fig. Y3-27—Disassembled view of piston and rod unit typical of all models in TN engine series. Refer to text for alignment of piston and rod with cylinder block.

CC. Combustion chamber
FM. Forging marks
TM. Trademark
1. Top compression ring
2. Second ring
3. Oil control ring
4. Snap rings
5. Piston
6. Piston pin
7. Connecting rod
8. Rod pin bushing
9. Bearing inserts
10. Rod cap
11. Cap screws

PISTON AND ROD UNITS

Piston and rod units are removed from above after removing oil pan, oil pump pickup tube and sleeve, and the cylinder head. On Model 4TN82TE turbocharged engine, remove the piston cooling oil nozzles. Keep rod bearing cap with matching connecting rod.

To install piston and rod units, be sure combustion chamber (CC—Fig. Y3-27) on top of piston is offset toward fuel injection pump side of engine, the trademark "YANMAR" (TM) on side of rod is toward rear of engine and forging marks (FM) on connecting rod and cap are aligned. Lubricate cylinder wall, crankpin, piston and rings and rod bearing. Place ring end gaps at equal spacing at 120 degrees around piston. Install piston and rod unit using suitable ring compressor. Lubricate connecting rod bolt threads and tighten bolts to a torque of 23-27 N·m (17-20 ft.-lbs.) on Models 3TN66E and 3TNA72E, 38-41 N·m (44-47 ft.-lbs.) on Model 3TN75E, and to a torque of 45-49 N·m (33-36 ft.-lbs.) on all other models.

PISTONS, PINS, AND RINGS

Pistons are fitted with two compression rings and one oil control ring with a coil type expander under the oil ring. Piston pin is full floating and is retained by a snap ring at each end of pin bore in piston. If necessary to remove piston from connecting rod, remove snap rings and push pin from piston and rod.

Inspect piston skirt for scoring, cracks and excessive wear. Measure piston skirt at right angle to pin bore and at a distance of 5 mm (0.2 inch) from bottom edge of skirt on Model 3TN66E, 8 mm (0.3 inch) on Model 3TNA72E, 12.5 mm (0.5 inch) on Model 3TN75E, and 24 mm (1 inch) on all other models. Piston skirt diameter specifications are as follows:

Model 3TN66E
Standard diameter . 65.927-65.957 mm
(2.596-2.597 in.)
Wear limit . . . . 65.85 mm
(2.593 in.)

Model 3TNA72E
Standard diameter . 71.922-71.952 mm
(2.832-2.833 in.)
Wear limit . . . . 71.81 mm
(2.827 in.)

Model 3TN75E
Standard diameter . 74.913-74.943 mm
(2.949-2.951 in.)
Wear limit . . . . 74.81 mm
(2.945 in.)

All Other Models
Standard diameter . 81.898-81.928 mm
(3.224-3.226 in.)
Wear limit . . . . 81.8 mm
(3.221 in.)

Piston pin to piston pin bore clearance should be zero to 0.018 mm (0-0.0007 inch) on Models 3TN66E, 3TNA72E and 3TN75E, and zero to 0.022 mm (0-0.0009 inch) on all other models. Maximum allowable piston pin clearance in piston is 0.045 mm (0.0018 inch) on all models. Be sure piston and connecting rod are positioned as shown in Fig. Y3-27 before installing piston pin. Be sure pin retaining snap rings are securely installed at each end of pin bore in piston.

Piston ring end gap for compression rings should be 0.15-0.30 mm (0.006-0.012 inch) for Model 3TN66E, 0.10-0.25 mm (0.004-0.010 inch) for Model 3TNA72E, 0.20-0.40 mm (0.008-0.016 inch) for Model 3TN75E, and 0.20-0.30 mm (0.008-0.012 inch) for all other models. Oil ring end gap should be 0.15-0.35 mm (0.006-0.014 inch) for Models 3TN66E and 3TNA72E, 0.20-0.40 mm (0.008-0.016 inch) for Model 3TN75E and 0.25-0.45 mm (0.010-0.018 inch) for all other models. Maximum allowable end gap is 1.5 mm (0.059 inch) for all rings.

Top ring side clearance in piston groove should be 0.065-0.10 mm (0.0026-0.0039 inch) for Model 3TN66E and 0.075-0.11 mm (0.003-0.0043 inch) for all other models. Second ring side clearance should be 0.03-0.065 mm (0.0012-0.0026 inch) for Models 3TN66E and 3TNA72E, and 0.045-0.080 mm (0.0018-0.0031 inch) for all other models. Maximum allowable side clearance for compression rings is 0.20 mm (0.008 inch) for Models 3TN66E and 3TNA72E, 0.25 mm (0.010 inch) for Model 3TN75E, and 0.30 mm (0.012 inch) for all other models. Oil control ring side clearance should be 0.02-0.055 mm (0.0008-0.0022 inch) for Models 3TN66E and 3TNA72E, and 0.025-0.060 mm (0.010-0.0024 inch) for all other models. Maximum allowable side clearance for oil control ring is 0.20 mm (0.008 inch) for all models. Renew pistons if ring side clearance is excessive.

When installing piston rings, place oil ring expander in groove, then install oil ring on top of expander with ring end gap at opposite side of piston from where expander ends meet. Install compression rings with side having identification mark up and with end gaps at 120 degree spacing from each other and oil ring end gap.

CYLINDERS

The cylinder blocks for series TN engines are sleeveless and the pistons and rings ride directly in the cylinder block bores. Pistons and rings are available in 0.25 mm (0.010 inch) oversize. Cylinders may be rebored and honed to this oversize if worn beyond limit for use of standard size pistons and rings. Check cyl-

SERVICE MANUAL

Yanmar

inder bores in block for cracks, scoring or excessive wear. Cylinder bore specifications are as follows:

Model TN66E
Standard cylinder
  bore . . . . 66.00-66.03 mm
    (2.599-2.600 in.)
  Wear limit . . . . 66.2 mm
    (2.606 in.)

Model 3TNA72E
Standard cylinder
  bore . . . . 72.00-72.03 mm
    (2.835-2.836 in.)
  Wear limit . . . . 72.2 mm
    (2.843 in.)

Model 3TN75E
Standard cylinder
  bore . . . . 75.00-75.03 mm
    (2.953-2.954 in.)
  Wear limit . . . . 75.2 mm
    (2.961 in.)

All Other Models
Standard cylinder
  bore . . . . 82.00-82.03 mm
    (3.228-3.230 in.)
  Wear limit . . . . 82.2 mm
    (2.236 in.)

Cylinder out-of-round condition should be less than 0.01 mm (0.0004 inch); maximum allowable out-of-round is 0.02 mm (0.001 inch). Rebore and hone cylinders or renew block if out-of-round condition or cylinder wear is excessive.

CONNECTING RODS AND BEARINGS

The connecting rod is fitted with a renewable bushing (8—Fig. Y3-27) at piston pin end and a slip-in precision fit bearing (9) at crankpin end. Rod twist or out of parallel condition should be less than 0.05 mm (0.002 inch) per 100 mm (4 inches) for Models 3TN66E and 3TNA72E, and less than 0.03 mm (0.0012 inch) per 100 mm (4 inches) for all other models. Maximum allowable rod distortion is 0.08 mm (0.003 inch) per 100 mm (4 inches).

Crankpin standard diameter is 35.97-35.98 mm (1.416-1.417 inches) for Model 3TN66E, 39.97-39.98 mm (1.5736-1.574 inches) for Model 3TNA72E, 42.952-42.962 mm (1.691-1.6914 inches) for Model 3TN75E, and 47.952-47.962 mm (1.8879-1.8883 inches) for all other models. Crankpin bearing oil clearance should be 0.02-0.07 mm (0.0008-0.0028 inch) for Models 3TN66E and 3TNA72E and 0.038-0.090 mm (0.0015-0.0035 inch) for all other models. Crankpin bearing oil clearance wear limit is 0.15 mm (0.006 inch) for all models.

Rod side clearance on crankpin journal for all models should be 0.2-0.4 mm (0.008-0.016 inch). Maximum allowable side clearance is 0.55 mm (0.022 inch). Renew connecting rod and/or crankshaft if clearance is excessive.

Piston pin standard diameter is 20 mm (0.7874 inch) for Model 3TN66E, 21 mm (0.827 inch) for Model 3TNA72E, 23 mm (0.906 inch) for Model 3TN75E, and 26 mm (1.024 inches) on all other models. Piston pin bushing to pin oil clearance should be 0.025-0.05 mm (0.001-0.002 inch).

CAMSHAFT AND BEARINGS

Camshaft, gear and thrust plate assembly can be removed from front of block after removing timing cover, valve cover, rocker arm assembly and push rods. Use magnetic tools to lift cam followers (6—Fig. Y3-28) from camshaft or invert engine so that followers fall away from cam. Remove camshaft thrust plate retaining bolts through web holes in camshaft gear and pull camshaft from cylinder block. If necessary to renew camshaft, thrust plate or gear, carefully press shaft from plate and gear.

Camshaft end play in block is controlled by the thrust plate (2). End play should be 0.05-0.15 mm (0.002-0.006 inch) for Models 3TN66E and 3TNA72E, and 0.05-0.20 mm (0.002-0.008 inch) on all other models. Renew thrust plate if end play exceeds 0.4 mm (0.016 inch) on all models; also check thrust surface wear on timing gear and camshaft.

Intake and exhaust cam lobe height should measure 29.97-30.03 mm (1.180-1.182 inches) on Model 3TN66E, 33.95-34.05 mm (1.337-1.341 inches) on Model 3TNA72E, and 38.635-38.765 mm (1.521-1.526 inches) on all other models. Wear limit is 29.75 mm (1.171 inches) on Model 3TN66E, 33.75 mm (1.329 inches) on

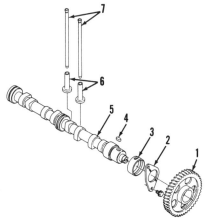

Fig. Y3-28—Timing gear (1) is press fit on camshaft (5). Block is fitted with renewable bushing (3) at front bearing position only.
1. Timing gear
2. Thrust plate
3. Camshaft bearing
4. Locating key
5. Camshaft
6. Cam followers
7. Push rods

Model 3TNA72E, and 38.4 mm (1.512 inches) on all other models.

The camshaft front journal rides in a renewable bearing (3—Fig. Y3-28) at front end of block, and center and rear journals ride directly in unbushed bores of block. Renew front bearing if oil clearance is 0.20 mm (0.008 inch) or more. Camshaft journal to unbushed block bore maximum oil clearance is also 0.20 mm (0.008 inch).

CAM FOLLOWERS

The mushroom type cam followers (valve tappets) (6—Fig. Y3-28) can be removed from bottom of block after removing camshaft and oil pan. Cam followers should be renewed if installing a new camshaft, or if bottom of follower shows cracks, pitting, excessive wear or if barrel of follower is excessively worn.

Follower outside standard diameter is 17.95-17.968 mm (0.7067-0.7074 inch) for Model 3TN66E, 20.927-20.96 mm (0.824-0.825 inch) for Model 3TNA72E, and 11.975-11.990 mm (0.472-0.473 inch) on all other models. Clearance in block should be 0.032-0.068 mm (0.0013-0.0027 inch) for Model 3TN66E, 0.04-0.094 mm (0.0016-0.0037 inch) for Model 3TNA72E, and 0.010-0.043 mm (0.0004-0.0017 inch) for all other models.

CRANKSHAFT AND MAIN BEARINGS

The crankshaft is supported in four main bearings in three cylinder engines, and in five main bearings (8—Fig. Y3-29) in four cylinder models. Crankshaft end play is controlled by split thrust washers (6 and 7) at each side of rear main bearing cap.

To remove crankshaft, remove piston and rod units, flywheel, flywheel housing, rear bearing seal retainer, timing cover, timing gears and engine front plate. Then, be sure that position number is marked on each main bearing cap and unbolt and remove main bearing caps and crankshaft. Check all main bearing journals, thrust surfaces at rear journal, and the crankpin journals for scoring or excessive wear.

Refer to CONNECTING RODS AND BEARINGS paragraph for information on crankpin journal specifications; main bearing journal specifications are as follows: Main journal diameter is 39.97-39.98 mm (1.5736-1.574 inches) for Model 3TN66E, 43.97-43.98 mm (1.731-1.732 inches) for Model 3TNA72E, 46.952-46.962 mm (1.8485-1.8489 inches) for Model 3TNA72E, and 49.952-49.962 mm (1.969-1.970 inches) for all other models. Specified main bearing oil clearance is 0.02-0.072 mm (0.0008-0.0028 inch) for

Illustrations Courtesy Yanmar Diesel Engine, Inc.

Yanmar

Models 3TN66E and 3TNA72E, and 0.038-0.093 mm (0.0015-0.0037 inch) on all other models. Maximum allowable main bearing oil clearance is 0.15 mm (0.006 inch) for all models.

Crankshaft end play should be 0.095-0.266 mm (0.004-0.010 inch) on Model 3TN66E, and 0.09-0.27 mm (0.004-0.011 inch) on all other models. Maximum allowable end play is 0.33 mm (0.013 inch) for all models.

When installing crankshaft, be sure that all bearing surfaces are clean and free of nicks or burrs. Install bearing shells with oil holes in cylinder block and bearing shells without oil holes in the main bearing caps. Bearing sets for all journals are alike. Lubricate bearings and crankshaft main journals and set crankshaft in bearings. Lubricate upper thrust washer halves (7) (those without locating projection) and insert thrust washers between block and crankshaft with bearing side of thrust washers towards crankshaft shoulders. Using light grease, stick lower thrust washer halves (6) in rear main bearing cap with tabs on washers aligned with notches in cap and with bearing side out. Lubricate bearing and install cap. Lubricate remaining bearing caps and install in same positions as they were removed. Note that side of caps with embossed letters "FW" are placed toward flywheel.

Tighten the main bearing cap bolts to a torque of 52-55 N·m (39-41 ft.-lbs.) on Model 3TN66E, 74-83 N·m (55-65 ft.-lbs.) on Model 3TNA72E, 77-80 N·m (57-59 ft.-lbs.) on Model 3TN75E, and 94-102 N·m (69-76 ft.-lbs.) on all other models. Check to be sure that crankshaft turns freely in bearings before completing reassembly of engine.

CRANKSHAFT REAR OIL SEAL

The crankshaft rear oil seal (1—Fig. Y3-30) can be renewed after removing flywheel and the rear seal retainer (2). Install new oil seal with lip to inside, then lubricate seal and install retainer with new gasket. Be sure retainer is aligned on locating dowel pins (4) before tightening bolts.

ELECTRICAL SYSTEM

GLOW PLUGS

Models with indirect injection utilize glow plugs as an aid in engine starting. The glow plugs are connected in parallel so that if one plug is burned out, the others will continue to operate. The glow plug indicator light will be on while the glow plugs are heating. A timer in the circuit will turn the indicator light off after about 15 seconds; at that time, turn switch to "OFF" position, then back to "START" to start engine.

The glow plugs can be checked with an ohmmeter; renew plug if there is no continuity. Glow plug resistance should be 1.35-1.65 ohms.

SMALL DIESEL

INTAKE AIR HEATER

Models with direct injection are equipped with an electric heater (1—Fig. Y3-31) mounted on the end of the intake manifold (3) to warm air entering the engine as a starting aid. To activate the heater, turn key switch to "ON" position and the heater pilot light will come on. After 15 seconds, a timer in the heater circuit will turn the pilot light off; at that time, turn key switch to "OFF" position, then to "START" to start engine. Heater requires 33.3 amperes current at 12 volts. Renew heater if no continuity is noted using ohmmeter attached to heater terminals.

ALTERNATOR AND REGULATOR

A Kokusan Model LR120-15C 12 volt, 20 ampere alternator with internal Model TRIZ-63 regulator is standard equipment; refer to Fig. Y3-32. A Kokusan 15 ampere, 12 volt "magneto-generator" is available as low cost optional equipment.

A wear limit line is etched on the alternator brushes. Standard brush length is 16 mm (0.630 inch) and wear limit is 9 mm (0.354 inch).

To check stator assembly, use ohmmeter and check for continuity between each stator coil terminal. If there is no continuity, renew stator. Check for continuity between terminals and stator core. If continuity exists, renew the stator.

To check rotor, use ohmmeter to check for continuity between the two slip rings. If no continuity is found, renew the rotor. Check for continuity between slip ring and rotor core or shaft. If continuity exists, renew the rotor.

Using ohmmeter, check diodes in rectifier assembly. Ohmmeter should show infinite reading in one direction and continuity when ohmmeter leads are reversed. Renew rectifier assembly if

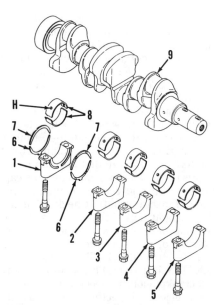

Fig. Y3-29—View showing main bearing thrust washer location on four-cylinder engine. Thrust washers are also located at rear main bearing on three-cylinder models which have four main bearings. Bearing halves with oil holes (H) must be installed in block.

H. Oil holes
1. Rear bearing cap
2. No. 2 bearing cap
3. No. 3 bearing cap
4. No. 4 bearing cap
5. Front bearing cap
6. Bottom thrust washers
7. Top thrust washers
8. Main bearing inserts
9. Crankshaft

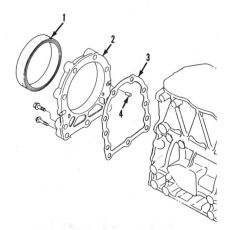

Fig. Y3-30—Rear crankshaft seal (1) is pressed into seal retainer (2) bolted to rear of block.

1. Crankshaft seal
2. Seal retainer
3. Gasket
4. Dowel pins

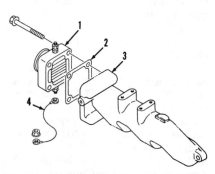

Fig. Y3-31—Optional air heater (1) starting aid is bolted to end of intake manifold (3).

1. Air heater
2. Gasket
3. Intake manifold
4. Ground lead

SERVICE MANUAL

Yanmar

continuity exists in both directions, or if no continuity exists in either direction.

STARTING MOTOR

Several different starters are used depending upon engine size and options such as higher compression ratios. Starter motor specifications are as follows:

Model 3TN66E-G2:
Starter Model S114-443
No-load volts 11.5
No-load amps (max.) 60
No-load rpm (min.) 7000

Models 3TN66E-S, 3TNA72E-G2, -S:
Starter Model 128000-1151
No-load volts 11.5
No-load amps (max.) 90
No-load rpm (min.) 3000
Brush height (min.) .... 7.7 mm
(0.303 in.)

Models 3TN75E-G2, -S:
Starter Model S114-349A
No-load volts 12
No-load amps (max.) 120
No-load rpm (min.) 5000
Pinion gap .... 0.3-1.5 mm
(0.012-0.059 in.)

Brush height (min.) .... 11 mm
(0.433 in.)

Model 3TN75E-G1:
Starter Model S114-146
No-load volts 12
No-load amps (max.) 70
No-load rpm (min.) 6000
Pinion gap .... 0.3-2.5 mm
(0.012-0.098 in.)
Brush height (min.) .... 11.5 mm
(0.453 in.)

Models 3TN82E-G2, -S:
Starter Model S114-257G
No-load volts 12
No-load amps (max.) 100
No-load rpm (min.) 4300
Pinion gap .... 0.3-0.5 mm
(0.012-0.020 mm.)
Brush height (min.) .... 12 mm
(0.472 in.)

Models 3TN82E-G1, 4TN82E-G1, 4TN82TE-G1:
Starter Model S12-77A
No-load volts 12
No-load amps (max.) 90
No-load rpm (min) 4000
Pinion gap .... 0.2-1.5 mm
(0.008-0.059 in.)
Brush height (min.) .... 8 mm
(0.315 in.)

Models 4TN82E-G2, -S, 4TN82TE-S:
Starter model S13-94
No-load volts 11
No-load amps (max.) 140
No-load rpm (min.) 3900
Pinion gap .... 0.3-1.5 mm
(0.012-0.059 in.)
Brush height (min.) .... 9 mm
(0.354 in.)

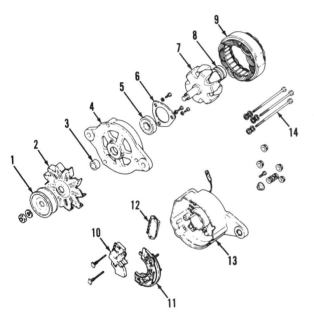

Fig. Y3-32—Exploded view of 20 ampere alternator used as standard equipment on all models.

1. Pulley
2. Cooling fan
3. Spacer
4. Front cover
5. Ball bearing
6. Bearing retainer
7. Rotor
8. Ball bearing
9. Stator
10. Brush holder assy.
11. Diode assy.
12. Voltage regulator
13. Rear cover
14. Through-bolts

Illustrations Courtesy Yanmar Diesel Engine, Inc.

YANMAR

SMALL DIESEL

Model	No. Cyls.	Bore	Stroke	Displ.
L40E	1	68 mm (2.68 in.)	55 mm (2.16 in.)	199 cc (12.14 cu.in.)
L60E	1	75 mm (2.95 in.)	62 mm (2.44 in.)	273 cc (16.66 cu.in.)
L75E	1	80 mm (3.15 in.)	70 mm (2.76 in.)	351 cc (21.42 cu.in.)
L90E	1	84 mm (3.31 in.)	70 mm (2.76 in.)	387 cc (23.61 cu.in.)

All models are single cylinder, four-stroke, direct injection air cooled diesel engines. Rotation of pto shaft is counterclockwise as viewed from outer end of shaft. If crankshaft is used as pto shaft (D-specification engine), engine rotation is clockwise as viewed from flywheel end; if camshaft is used as pto shaft (S-specification engine), engine rotation is counterclockwise as viewed from flywheel end.

Engine serial number is stamped on lower edge of engine crankcase cover. Always specify engine model, type and serial number when ordering parts.

MAINTENANCE

LUBRICATION

Standard oil sump capacity is 0.75 liters (1.6 pints) for Model L40E, 1.1 liters (2.3 pints) for Model L60E and 1.65 liters (3.5 pints) for Models L75E and L90E. Engines with large capacity oil pans are available; Model L40BE-D oil pan capacity is 2.8 liters (3 quarts) and Model L60BE-D oil pan capacity is 3.0 liters (3.2 quarts).

Recommended motor oil is API classification CC or CD. Multigrade SAE 5W-30 oil is recommended in operating temperatures of −30° to +10° C (−22° to 50° F), SAE 10W-30 in temperatures of −20° to +30° C (−18° to +86° F) and SAE 20W-40 in temperature range of −10° to +40° C (14° to 104° F).

Recommended oil change interval is each 100 hours of operation. Engines with standard size oil pan have a 60 mesh oil strainer (4—Fig. Y4-1) in the engine crankcase and the strainer should be removed and cleaned or renewed after each 500 hours of operation. Models with large capacity oil pan have a bypass type spin-on cartridge oil filter that should be renewed with each oil change.

FUEL SYSTEM

A Yanmar manufactured Bosch type fuel injection pump is mounted on engine crankcase and is operated by a lobe on engine camshaft. A Yanmar spray type fuel injector nozzle is used.

FUEL FILTER. On models with engine mounted fuel tank, fuel filter (14—Fig. Y4-2) is mounted in the tank. To remove filter, remove fuel shut-off valve (19) and plate (16) and remove filter element from bottom of tank. On models

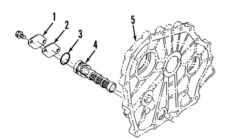

Fig. Y4-1—View showing oil filter screen in most models. Some engines have screen in different position. Refer to Fig. Y4-13.

1. Cover plate
2. Gasket
3. "O" ring
4. Filter screen
5. Crankcase cover

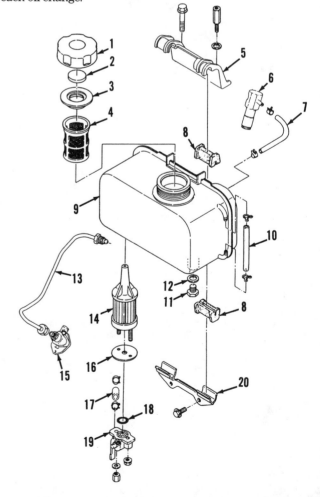

Fig. Y4-2—Exploded view of diesel fuel system with engine mounted fuel tank.

1. Filler cap
2. Seal
3. Gasket
4. Filter screen
5. Mounting clamp
6. Fuel injector
7. Fuel return line
8. Vibration dampers (4)
9. Fuel tank
10. Fuel gage pipe
11. Drain plug
12. Gasket
13. Fuel injector line
14. Fuel filter
15. Fuel injection pump
16. Plate
17. Fuel line to pump
18. "O" ring
19. Fuel shut-off valve
20. Mounting clamp

SERVICE MANUAL

Yanmar

with remote mounted fuel tank, an external filter is used. Fuel filter should be renewed after every 1000 hours of operation, or sooner if loss of engine power is evident.

BLEED FUEL SYSTEM. On models with gravity fuel feed to injection pump, loosen high pressure line (13—Fig. Y4-2) at injector (6), actuate compression release and crank engine until fuel flows from loosened fitting, then tighten fitting and start engine.

On models with electric fuel lift pump, loosen fuel supply line at fuel injection pump and run fuel lift pump until fuel is flowing from loosened connection. Tighten fuel supply line connection and loosen high pressure line fitting at injector. Crank engine until fuel is flowing from loosened fitting, then tighten fitting and start engine.

INJECTION PUMP TIMING

Refer to Fig. Y4-3 and Fig. Y4-4 for view showing flywheel timing marks. The marks are to right or left of TD mark depending upon whether engine has crankshaft (D-spec.) or camshaft (S-spec.) pto. Early models have a "TD" mark and an injection timing mark with no degrees indicated. Injection pump plunger was changed at engine serial number 63768 for Model L40E, 60183 for Model L60E, 01718 for Model L75E and 03818 for Model L90E, requiring different injection timing.

On early models with a "TD" mark and a single BTDC mark, injection should occur at the BTDC mark on flywheel. On models prior to serial numbers listed in preceding paragraph and with indicated BTDC degree marks, injection should occur at 13-15 degrees BTDC on Models L40E and L60E, and 12-14 degrees BTDC on Models L75E and L90E. On all late models (after serial number listed in preceding paragraph) with new type injection pump plunger, injection should occur at 19-21 degrees BTDC.

To check injection pump timing, remove fuel injection line and install a spill pipe on the injection pump. Set decompression lever in noncompression position. Move throttle control to run position and crank engine until fuel flows from spill pipe. Then turn flywheel slowly in normal direction of rotation so that piston is on compression stroke and observe when fuel just begins to flow from spill pipe. Injection timing mark on flywheel should be in line with "V" shaped mark on cylinder fin as shown in Fig. Y4-3 when fuel begins to flow. To be sure observation is correct, this step should be repeated three or four times.

Injection timing is varied by changing the thickness of shims between the fuel injection pump and engine crankcase. Shims are available in thicknesses of 0.2 mm (0.008 inch) and 0.3 mm (0.012 inch). Each 0.1 mm (0.004 inch) change of shim thickness will vary injection timing by 1 degree.

GOVERNOR

Engine rated speed is 3000 or 3600 rpm depending upon application. Maximum no-load governed speed is 3700-3800 rpm on models rated at 3600 rpm continuous output. Note that if checking engine speed at camshaft pto, engine speed is 2 times observed speed. Several different types of speed control devices have been used. Spring (S—Fig. Y4-5) connecting throttle control to governor arm should be connected to throttle arm (L) and governor arm (A) for the different control types as indicated in the chart shown in Fig. Y4-5A.

FUEL LIMITER

A spring loaded plunger (P—Fig. Y4-6) mounted on engine crankcase contacts the governor control arm (A) and limits governor arm travel under overload conditions. Adjustment of the fuel limiter is set at factory and adjustment

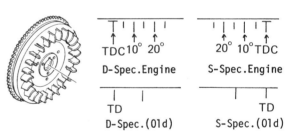

Fig. Y4-4—"TDC" and degrees BTDC marks will be found on flywheel of late model engines. Early engines have only a "TD" mark and one plain timing indicator mark. "D" specification engines turn clockwise; "S" specification engines turn counterclockwise.

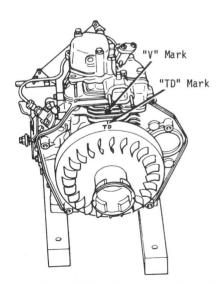

Fig. Y4-3—View showing location of top dead center "TD" timing mark on flywheel and V-shaped mark on engine cylinder. Refer also to Fig. Y4-4.

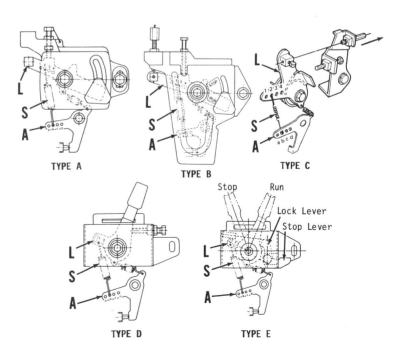

Fig. Y4-5—View showing different engine speed controls used. It is important that the spring (S) is hooked into the proper holes in speed control lever (L) and governor arm (A). Refer to chart shown in Fig. Y4-5A.

Illustrations Courtesy Yanmar Diesel Engine, Inc.

Yanmar

is sealed with wire and lead seal. To check fuel limiter adjustment, remove access cover from pump base to reveal access hole (H—Fig. Y4-7) and place throttle control lever in run position. The indicator (I) on pump control arm (15) should be aligned with mark (M) on pump base plate opening, and the governor arm (A—Fig. Y4-6) should just be contacting tip of fuel limiter plunger (P). To adjust fuel limiter, it will be necessary to cut the seal wire, loosen limiter locknut (N) and turn limiter assembly to obtain the above described conditions. After adjusting fuel limiter, install lock wire and new lead seal.

REPAIRS

TIGHTENING TORQUES

When servicing engine, observe the following tightening torques for special fasteners:

Cylinder head nuts—
  Model L40 . . . . 28-31 N·m
                            (20-23 ft.-lbs.)
  Model L60 . . . . 41-45 N·m
                            (30-33 ft.-lbs.)
  Models L75, L90 . . . . 57-61 N·m
                            (42-45 ft.-lbs.)
Cylinder head studs—
  Model L40 . . . . 13-15 N·m
                            (10-11 ft.-lbs.)
  Model L60 . . . . 15-16 N·m
                            (11-12 ft.-lbs.)
  Models L75, L90 . . . . 25-28 N·m
                            (18-21 ft.-lbs.)
Connecting rod—
  Models L40, L60 . . . . 18-19 N·m
                            (13-14 ft.-lbs.)
  Models L75, L90 . . . . 37-41 N·m
                            (27-30 ft.-lbs.)
Rocker arm support—
  Models L40, L60 . . . . 20-21 N·m
                            (15-16 ft.-lbs.)
  Models L75, L90 . . . . 42-46 N·m
                            (31-34 ft.-lbs.)

Flywheel nut—
  Models L40, L60 . . . . 98-108 N·m
                            (72-79 ft.-lbs.)
  Models L75,
    L90 (M16) . . . . 147-157 N·m
                          (109-115 ft.-lbs.)
  Models L75, L90
    (M18) . . . . 206-225 N·m
                          (152-166 ft.-lbs.)
Crankcase cover bolts—
  Model L40 . . . . 8-11 N·m
                            (6-8 ft.-lbs.)
  All other models . . . . 17-22 N·m
                            (12-16 ft.-lbs.)
Crankcase cover
  stiffener bolts . . . . 20-22 N·m
                            (15-16 ft.-lbs.)
Fuel injector clamp nuts . . . . 8-10 N·m
                            (6-7 ft.-lbs.)
Fuel injector clamp studs . . . 7-10 N·m
                            (5-7 ft.-lbs.)
Nozzle retaining nut . . . . . . 40-44 N·m
                            (29-32 ft.-lbs.)
Delivery valve holder . . . . . 25-34 N·m
                            (18-25 ft.-lbs.)
Injection pump studs . . . . 7-10 N·m
                            (5-7 ft.-lbs.)
Injection pump nuts . . . . 8-11 N·m
                            (6-8 ft.-lbs.)

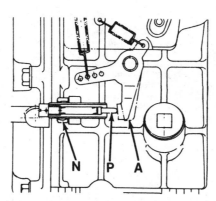

Fig. Y4-6—Fuel limiter plunger (P) should lightly contact governor arm (A), and indicator (I—Fig. Y4-7) on pump control lever (15) should be aligned with mark (M) in access hole (H) when engine is not running and throttle is in run position.

SMALL DIESEL

VALVE COVER AND DECOMPRESSION DEVICE

A compression relief (decompression) device is fitted in the valve cover. Refer to Fig. Y4-8. Moving decompression lever (1) to down position will cause lever shaft to contact exhaust rocker arm

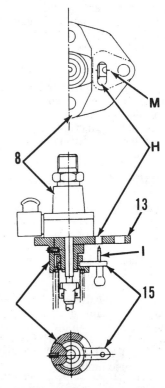

Fig. Y4-7—Indicator (I) should be aligned with mark (M) in access hole (H) and fuel limiter plunger (P—Fig. Y4-6) should just contact governor arm (A) when engine is not running and throttle is run position.

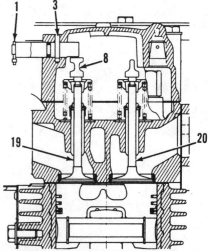

Fig. Y4-8—Cross-sectional view showing valves and compression release mechanism.
1. Decompression lever
3. Retaining pin
8. Exhaust rocker arm
19. Exhaust valve
20. Intake valve

ENGINE MODEL	GOVERNED RPM	GOVERNOR SPRING INSTALLATION HOLES		
		Types A, B, & C	Type D	Type E
L40, L60	3600	2-B	1-B	1-B
L40, L60	3000	2-C	1-C	2-C
L75, L90	3600	3-B	1-B	1-C
L75, L90	3000	3-C	1-C	2-C

Fig. Y4-5A—Governor spring installation chart.

SERVICE MANUAL

Yanmar

(8). Exhaust valve (19) is held partially open, thereby relieving compression making engine easier to crank. When using hand recoil starter, turn engine slowly until compression is felt through rope, then push lever down and crank engine. Lever automatically returns to "compression" (upright) position when engine is cranked.

VALVE ADJUSTMENT

To adjust valves, remove rocker arm cover and turn flywheel to position piston at top dead center on compression stroke (both valves should be loose). Adjust valve to rocker arm clearance with engine cold to 0.10-0.15 mm (0.004-0.006 inch) for both intake and exhaust.

ROCKER ARMS AND PUSH RODS

Rocker arms are mounted on a pedestal (10—Fig. Y4-9) retained by one bolt (6) and a locating dowel pin (17). Rocker arm pedestal, rocker arms (8 and 11) and push rods (7) can be removed after removing valve cover (4). Three different design push rod and rocker arm adjusting screw combinations have been used. Be sure when renewing rocker arm, adjusting screw (9) and/or push rod to obtain correct replacement parts.

Check rocker arm support shaft for excessive wear or scoring. Standard shaft diameter is 11.989-12.0 mm (0.4724-0.4729 inch) on Model L40 and L60, with wear limit of 11.90 mm (0.4685 inch). On Models L75 and L90, standard shaft diameter is 14.989-15.0 mm (0.5901-0.5906 inch) with wear limit of 14.90 mm (0.5866 inch).

Rocker arm standard inside diameter is 12.016-12.034 mm (0.4731-0.4738 inch) with wear limit of 12.10 mm (0.4764 inch) for Models L40 and L60, and 15.016-15.034 mm (0.5912-0.5919 inch) with wear limit of 15.10 (0.5945 inch) for Models L75 and L90.

Check push rods for obvious wear at either end and for distortion (bending) and renew if necessary.

When installing push rods and rocker arm assembly, loosen rocker arm adjusting screw locknuts and back screws out. Lubricate all parts and be sure rocker arm support is located properly on dowel pin, then tighten rocker arm support bolt to a torque of 20-21 N·m (15-16 ft.-lbs.) on Models L40 and L60, and to a torque of 42-46 N·m (31-34 ft.-lbs.) on Models L75 and L90. Adjust valve clearance as outlined in previous paragraph and complete reassembly by reversing disassembly procedure.

CYLINDER HEAD

Prior to removing cylinder head, thoroughly clean head, fuel injector and lines. Remove fuel injector line and injector. Immediately plug or cap openings in line, injection pump and injector. Remove rocker arm cover, rocker arm and support assembly and push rods. Unbolt and remove cylinder head.

To install cylinder head, make sure both the head and block surfaces are clean. If cylinder head studs have been removed, apply thread locking compound to block end of studs, install studs in block and tighten to a torque of 13-15 N·m (10-11 ft.-lbs.) for Model L40, 15-16 N·m (11-12 ft.-lbs.) on Model L60 and 25-28 N·m (18-21 ft.-lbs.) on Models L75 and L90. Place cylinder head gasket ring and push rod opening "O" ring on block, position cylinder head on studs and install retaining washers and nuts. Use corner-to-corner crossing sequence to tighten head nuts. First, tighten nuts evenly to a torque of 15 N·m (10 ft.-lbs.) on Model L40, 21 N·m (15 ft.-lbs.) on Model L60 and 29 N·m (20 ft.-lbs.) on Models L75 and L90. Then, tighten evenly to final torque of 28-31 N·m (20-23 ft.-lbs.) on Model L40, 41-45 N·m (30-33 ft.-lbs.) on Model L60 and 57-61 N·m (42-45 ft.-lbs.) on Models L75 and L90. Complete reassembly by reversing disassembly procedure.

VALVE SYSTEM

Valve face and seat angle is 45 degrees for both valves. Recommended valve seat width is 1.5-3.0 mm (0.060-0.118 inch). Check distance between top of valve and gasket surface of cylinder head using feeler gage and straightedge or a depth gage; recommended distance is 0.3-0.7 mm (0.012-0.028 inch). Renew valve and/or valve seat if valve head is more than 1.1 mm (0.043 inch) below head surface.

Valve guide inside diameter new is 5.5-5.515 mm (0.2165-0.2171 inch) for Model L40, 6.0-6.015 mm (0.2362-0.2368 inch) for Model L60 and 7.0-7.015 mm (0.2756-0.2762 inch) for Models L75 and L90. Valve guides are not available separately from cylinder head; renew head if valve guide inside diameter is worn to 5.58 mm (0.2197 inch) on Model L40, 6.08 mm (0.2394 inch) on Model L60, or 7.08 mm (0.2787 inch) on Models L75 and L90.

Check valves for excessive stem and tip wear, cracks in valve head and valve stem distortion and renew if necessary. Renew valve when distance between valve head and cylinder head surface is excessive with used valve, but within limits with new valve. Valve stem standard diameter is 5.450-5.456 mm

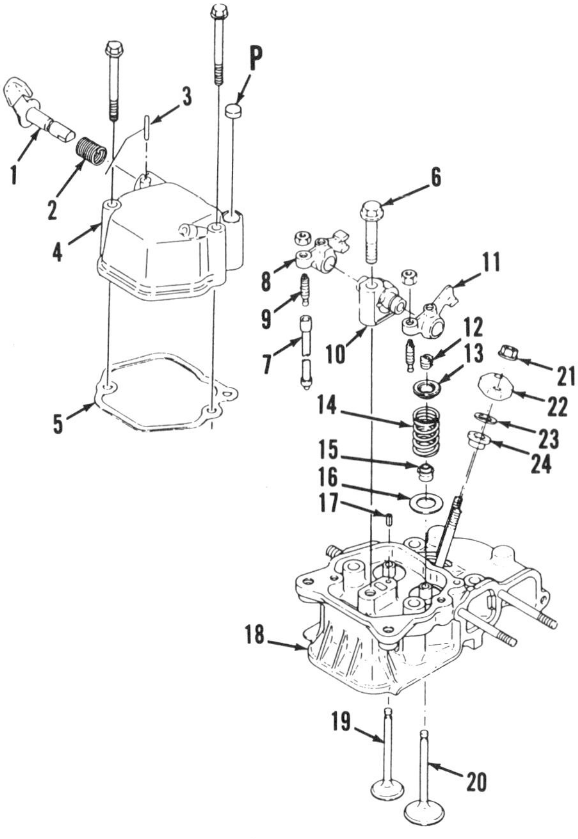

Fig. Y4-9—Exploded view of cylinder head and valve mechanism. The rocker arm cover on some models has a rubber plug (P) which can be removed for adding engine oil as cold weather starting aid.

1. Decompression lever
2. Return spring
3. Retaining pin
4. Rocker arm cover
5. Gasket
6. Rocker pedestal bolt
7. Push rod
8. Exhaust rocker arm
9. Adjusting screw
10. Rocker arm pedestal
11. Intake rocker arm
12. Split valve locks
13. Valve spring retainer
14. Valve spring
15. Valve stem seal
16. Spring seat washer
17. Pedestal dowel pin
18. Cylinder head
19. Exhaust valve
20. Intake valve
21. Injector clamp nuts
22. Injector clamp
23. Injector spacer
24. Injector gasket

Yanmar

SMALL DIESEL

(0.2146-0.2148 inch) for Model L40, 5.945-5.960 mm (0.2341-0.2346 inch) on Model L60 and 6.945-6.960 mm (0.2734-0.2740 inch) on Models L75 and L90.

Check valve springs for obvious defects such as distortion. If spring appears serviceable, check against the following specifications:

Valve spring standard free length—
- Model L40 . . . . 28 mm (1.102 in.)
- Model L60 . . . . 33 mm (1.299 in.)
- Models L75/L90 . . . . 37.7 mm (1.484 in.)

Valve spring minimum free length—
- Model L40 . . . . 26.5 mm (1.043 in.)
- Model L60 . . . . 31.5 mm (1.240 in.)
- Models L75/L90 . . . . 36.2 mm (1.425 in.)

When checking valve spring on spring tester, partially compress spring and observe tester gage reading. For each additional 1 mm (0.039 inch) of compression, tester gage should indicate 1.27 kg (2.8 pounds) for Model L40 spring, 1.41 kg (3.1 pounds) for Model L60 spring and 2.74 kg (6 pounds) for Model L75 and L90 spring.

Always install new valve stem seals when servicing valves. Lubricate valve stem with engine oil, then use properly fitting plastic seal driver to install seals so that distance from top of seals to valve spring seat is 9.5 mm (0.375 inch) on Model L40, 8.5 mm (0.335 inch) on Model L60 and 13.5 mm (0.530 inch) on Model L75 or L90.

INJECTOR

REMOVE AND REINSTALL. Prior to removing injector, thoroughly clean injector line, injection pump and injector using suitable solvent and compressed air. Remove injector fuel return line and high pressure line. Cap or plug all openings to prevent entry of dirt. Remove the injector clamp nuts and clamp plate, then remove injector from cylinder head. If injector spacer (23—Fig. Y4-9) and gasket (24) were not removed with injector, carefully remove these parts by screwing an M8 or M9 bolt into nozzle gasket, then pull on bolt to remove gasket and spacer.

When installing injector, be sure injector bore in cylinder head is clean and insert nozzle gasket and spacer in bore. Install the injector with locating pin aligned with slot in cylinder head. Install the clamp plate and nuts and tighten the nuts evenly to a torque of 40-44 N·m (29-32 ft.-lbs.). Install injection high pressure line, leaving fitting at injector loose until system is bled. Connect fuel return line to injector.

TESTING. A complete job of testing and adjusting the injector requires use of special test equipment. Only clean, approved testing oil should be used in tester tank. Injector should be tested for opening pressure, spray pattern and seat leakage.

Before conducting test, operate tester lever until fuel flows from tester line, then attach injector. Close tester gage valve and operate tester lever a few quick strokes to purge air from injector and to be sure nozzle valve is not stuck.

WARNING: Fuel emerges from injector with sufficient force to penetrate the skin. When testing injector, keep yourself clear of nozzle spray.

OPENING PRESSURE. Open valve to tester gage and operate tester lever slowly while observing gage reading. Pressure at which nozzle opens should be 19125-20105 kPa (2775-2915 psi). Opening pressure is adjusted by adding or removing shims (3—Fig. Y4-10). Standard thickness of the shim stack is 0.60-0.65 mm (0.0236-0.0256 inch). Changing shim stack thickness by 0.1 mm (0.004 inch) will change injector opening pressure by about 1960 kPa (285 psi).

SPRAY PATTERN. Spray pattern should be equal from all four nozzle tip holes and should be a fine mist. Spray angle (A—Fig. Y4-11) is 150 degrees and each spray cone should be a 5-10 degree cone (B). Place a sheet of blotting paper 30 cm (about 12 inches) below injector and operate tester one stroke. The spray pattern should be a perfect circle.

SEAT LEAKAGE. No oil should drip from injector nozzle after an injection. Operate tester lever two or three strokes, then wipe nozzle tip dry. Slowly bring tester gage pressure to 1960 kPa (285 psi) below opening pressure and hold this pressure for 5 seconds. Drop of oil should not form nor should oil drip from nozzle tip. Renew or overhaul injector if drop of oil is noted.

OVERHAUL. Wipe all loose carbon from outside of injector. Hold nozzle body in soft jawed vise or injector fixture and unscrew nozzle retaining nut (9—Fig. Y4-10). Remove nozzle body (8), stop plate (6), spring seat (5), spring (4) and adjusting shims (3). Be careful not to lose any of the shims and place all parts in clean diesel fuel or calibrating oil as they are removed.

Clean exterior surfaces with a brass wire brush. Soak parts in approved carbon solvent if necessary to loosen hard carbon deposits. Rinse parts in clean diesel fuel or calibrating oil immediately after cleaning to neutralize solvent. Clean nozzle tip opening from inside with a wood cleaning stick. Clean the four spray holes in nozzle tip by using a 0.2 mm (0.008 inch) cleaning wire in wire holder.

Check to see that needle does not bind in nozzle body (8). Needle should slide smoothly from its own weight. Rotate needle to several different positions when checking for binding. If needle sticks, reclean or renew nozzle assembly.

Reassemble injector under clean diesel fuel or calibrating oil. If opening pressure was correct, use shim pack removed on disassembly. Add or remove shims if necessary to change opening pressure. Tighten nozzle retaining nut to a torque of 40-44 N·m (29-32 ft.-lbs.). Retest injector and reinstall in cylinder head if injector meets test specifications.

INJECTION PUMP

To remove fuel injection pump, first thoroughly clean pump, injection line and fuel injector area using suitable solvent and compressed air. Close fuel supply valve and disconnect line from filter to pump. Remove the fuel injection

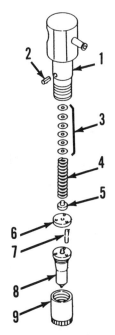

Fig. Y4-10—Exploded view of spray type fuel injector. Pin (2) locates injector in cylinder head.
1. Injector body
2. Dowel pin
3. Adjusting shims
4. Spring
5. Spring seat
6. Stop plate
7. Dowel pin
8. Nozzle & valve
9. Nozzle nut

SERVICE MANUAL

Yanmar

line. Cap or plug all openings to prevent entry of dirt. Unscrew nut at lower side of pump body and remove access cover and gasket. Then, unbolt and remove fuel injection pump and base plate. Be careful not to damage or lose any injection pump timing shims. Note governor fork connection with ball on lower side of pump control lever. Lift out injection pump tappet.

When reinstalling pump, be sure ball on pump control lever is engaged in fork on upper end of governor arm. Install the two retaining nuts at each side of pump body. Check to see that the pointer on pump control arm (15—Fig. Y4-7) is aligned with mark (M) on window of pump base plate when throttle control is in run position. If not, remove seal wire from fuel control limiter (P—Fig. Y4-6), loosen locknut (N) and turn limiter to bring pointer in line with mark. Refer also to FUEL LIMITER paragraph. Then, tighten locknut and install new seal wire. Bleed fuel system, then tighten fuel injection line.

CRANKCASE COVER

Prior to removing crankcase cover (20—Fig. Y4-14), first remove valve rocker arm assembly and push rods and remove the fuel injection pump and pump tappet to relieve pressure on camshaft. Remove any drive coupling or pulley from pto shaft (camshaft or crankshaft). Unbolt and carefully remove crankcase cover, along with oil pump and governor unit, from crankcase. The cover sealing gasket is aluminum and may be reused if not damaged or distorted.

Refer to OIL PUMP AND GOVERNOR paragraph if necessary to service oil pump and/or governor flyweight unit. If necessary to renew sleeve type main bearing (8—Fig. Y4-14), use appropriate piloted press tools and press old bearing from cover. Press new bearing into place making sure oil hole in bearing is aligned with oil passage in cover. Oil groove inside bearing must be toward top edge of cover. Install new pto shaft oil seal (1) with seal lip to inside. On crankshaft pto models, drive seal into cover until it is 4 mm (0.157 inch) below flush with cover. On camshaft pto models, drive seal in flush with cover.

If renewing the aluminum gasket (7), be sure to specify engine model and serial number as there are several different gaskets used. Before installing cover, refer to Fig. Y4-12 and be sure timing marks on balancer shaft gear and camshaft gear are aligned with crankshaft gear timing mark. (Gear driving balancer shaft is at inner end of crankshaft and camshaft drive gear is at cover end of crankshaft.) Oil pump gear is not timed to camshaft gear. Guide cover over balancer shaft bearing and pto shaft and be sure cover is in place on the two dowel pins before tightening cover cap screws. Tighten cap screws in a crisscross pattern until all are tightened to a torque of 8-11 N·m (6-8 ft.-lbs.) on Model L40 or 17-22 N·m (12-16 ft.-lbs.) on all other models.

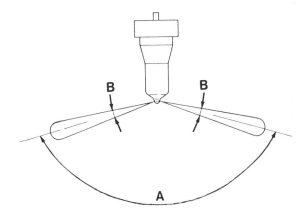

Fig. Y4-11—Angle between spray cones should be 150 degrees (A) and spray cone angle should be 5 to 10 degrees (B) and uniform from each of the four nozzle openings.

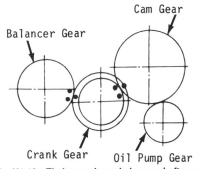

Fig. Y4-12—Timing marks on balancer shaft gear, crankshaft gear and camshaft gear should be aligned as shown.

OIL PUMP AND GOVERNOR

The gerotor type oil pump and governor assembly are mounted in the crankcase cover as shown in Fig. Y4-13. The oil pump gears (3) and governor (7) are serviced as complete assemblies only. The lubrication system does not have a relief valve, but a system oil pressure switch will provide a warning light and buzzer alarm if pressure falls below normal.

Oil pump cover (1) and pump outer rotor can be removed from outside of crankcase cover to inspect pump. However, it is necessary to remove crankcase cover (20), governor assembly (7) and oil pump drive pin (4) in order to remove inner rotor and shaft.

Clearance between outer rotor and housing should be 0.12-0.16 mm (0.0047-0.0063 inch). Clearance between inner and outer rotor lobes should be less than 0.14 mm (0.0055 inch). Rotor end clearance should be 0.02-0.08 mm (0.0008-

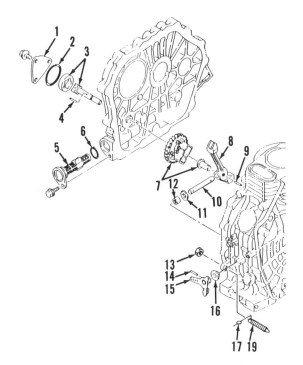

Fig. Y4-13—Exploded view showing oil pump and governor parts.

1. Oil pump cover
2. "O" ring
3. Oil pump rotors
4. Oil pump drive pin
5. Oil filter screen
6. "O" ring
7. Governor assy.
8. Governor arm
9. Dowel pin
10. Governor arm shaft
11. Flat washer
12. Needle bearing
13. Locknut
14. Dowel pin
15. Governor arm
16. Thrust bushing
17. Seal wire
19. Fuel limiter

Illustrations Courtesy Yanmar Diesel Engine, Inc.

Yanmar

0.003 inch). Outer rotor outside diameter should be 28.96-28.98 mm (1.1402-1.1409 inches) with wear limit of 28.9 mm (1.1378 inches). Rotor width should be 7.98-8.0 mm (0.314-0.315 inch) with wear limit of 7.9 mm (0.311 inch). Renew oil pump assembly if excessively worn or scored.

Oil pump housing (crankcase cover) inside diameter should be 29.1-29.121 mm (1.1457-1.1465 inches) and depth should be 8.02-8.05 mm (0.3157-0.3169 inch). Renew crankcase cover if pump housing is excessively worn or scored.

Inspect oil pump drive gear and governor weight assembly for damage or excessive wear and renew the assembly if defect is noted. Reassemble oil pump and governor units by reversing disassembly procedure.

PISTON AND ROD UNIT

The piston and connecting rod can be removed from above after removing cylinder head and crankcase cover, camshaft and balancer shaft. With piston at lower end of stroke, remove connecting rod nuts and bearing cap. Be careful not to lose washers located under the rod nuts on aluminum rods used in Models L40 and L60. Carefully turn crankshaft to move piston to top of stroke and remove piston and rod from top of block.

When assembling piston and connecting rod, align piston identification mark (ID) and match marks (MM) on rod as shown in Fig. Y4-15. When installing piston and rod unit, place piston ID mark toward crankcase cover side of engine. Lubricate rod bearing, piston rings, cylinder and crankpin. Be careful not to scratch or nick aluminum alloy rod used in Models L40 and L60. Turn crankshaft crankpin to bottom of crankcase and be sure crankshaft is properly positioned in crankcase bearing. Install piston and rod unit with bearing inserts. Install rod cap washers on Models L40 and L60. Tighten rod nuts to a torque of 18-19 N·m (13-14 ft.-lbs.) on Models L40 and L60, and to torque of 37-41 N·m (27-30 ft.-lbs.) on Models L75 and L90. Complete reassembly of engine by reversing disassembly procedure.

CYLINDER

The cast iron cylinder liner is integrally cast into the aluminum cylinder block. Cylinder standard inside diameter is 68.00-68.03 mm (2.6772-2.6783 inches) for Model L40, 75.00-75.03 mm (2.9528-2.9539 inches) for Model L60, 80.00-80.03 mm (3.1496-3.1508 inches) for Model L75 and 84.00-84.03 mm (3.3071-3.3083 inches) for Model L90. Check cylinder bore for scoring and/or excessive wear. Rebore to 0.25 or 0.50 mm (0.010 or 0.020 inch) oversize or renew cylinder block if cylinder bore is scored or worn more than 0.13 mm (0.005 inch) on Model L40 or 0.15 mm (0.006 inch) on any other model.

PISTON, PIN AND RINGS

Prior to Model L40 engine serial number 00126 or Model L60 engine serial number 29990, piston and connecting rod are available only as an assembly. Separate parts are available for all other engines. Engines with camshaft pto (counterclockwise engine rotation) require different piston than engines with crankshaft pto (clockwise engine rotation) due to offset of piston pin in piston. On Models L40 and L60, piston pin diameter and connecting rod width at pin end must be taken into account when obtaining new piston. Piston ring sets are included with service piston, and are also available separately from piston.

Inspect piston skirt for cracks, scoring or excessive wear. Measure piston skirt diameter at right angle to piston pin and at about 12 mm (1/2 inch) from bottom of piston. Check piston ring grooves for excessive wear or other damage and check fit of piston pin in piston pin bore.

New standard piston skirt diameter is 67.965 mm (2.6758 inches) for Model L40, 74.965 mm (2.9514 inches) for Model L60, 79.955 mm (3.1478 inches) for Model L75 and 83.955 mm (3.3053 inches) for Model L90. Renew piston if skirt diameter is worn more than 0.115 mm (0.0045 inch). Piston skirt to cylinder sleeve clearance should be 0.04-0.06 mm (0.0016-0.0024 inch) on Models L40 and L60, and 0.05-0.07 mm (0.002-0.0028 inch) on Models L75 and L90.

Piston pin to pin bore in piston clearance should be 0.015 mm (0.0006 inch) tight to 0.004 mm (0.0002 inch) loose on Model L40 and 0.017 mm (0.0007 inch)

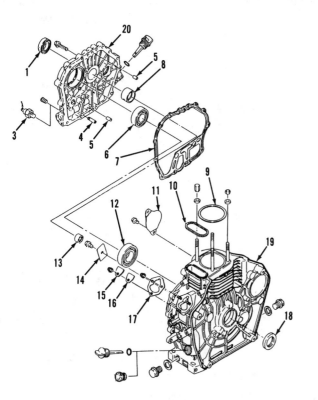

Fig. Y4-14—Exploded view showing crankcase cover, crankshaft bearings and camshaft bearings.
1. Oil seal
3. Oil pressure switch
4. Oil pipe
5. Dowel pins
6. Camshaft ball bearing
7. Gasket
8. Crankshaft main bearing
9. Cylinder head gasket
10. "O" ring
11. Cover
12. Crankshaft main bearing
13. Camshaft needle bearing
14. Main bearing retainer plate
15. Cover
16. Gasket
17. Injection pump timing shim
18. Crankshaft oil seal
19. Cylinder block

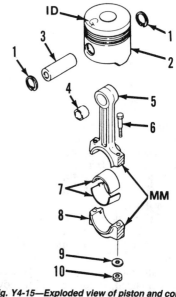

Fig. Y4-15—Exploded view of piston and connecting rod assembly. Piston pin bushing (4) is used only on Models L75 and L90. Washer (9) is used only on Models L40 and L60. On all models, be sure that piston identification mark (ID) is aligned with rod match marks (MM) as shown.

1. Snap rings
2. Piston
3. Piston pin
4. Pin bushing
5. Connecting rod
6. Rod cap bolt
7. Crankpin bearing
8. Rod cap
9. Washer
10. Nut

SERVICE MANUAL

tight to 0.005 mm (0.0002 inch) loose on all other models. Piston pin diameter on early Model L40 was 17 mm (0.6693 inch) and early Model L60 was 19 mm (0.7480 inch). On all other models, refer to following piston pin diameter specifications:

Piston pin standard
diameter:
    Model L40 . . . . . . 18.992-19.000 mm
                          (0.7477-0.7480 in.)
    Model L60 . . . . . . 20.991-21.000 mm
                          (0.8264-0.8268 in.)
    Models L75/L90 . . 22.991-23.000 mm
                          (0.9052-0.9055 in.)

Piston ring desired side clearance in groove on all models is 0.065-0.095 mm (0.0026-0.0037 inch) for top ring, 0.03-0.065 mm (0.0012-0.0026 inch) for second ring and 0.02-0.055 mm (0.0008-0.0022 inch) for oil ring. Renew piston if ring side clearance of new piston ring is 0.15 mm (0.006 inch) for any ring groove. Piston ring end gap should be 0.1-0.25 mm (0.004-0.010 inch) for top and second rings, and 0.15-0.35 mm (0.006-0.014 inch) for oil ring. Piston ring end gap should not exceed 1.0 mm (0.039 inch) on any ring.

When assembling piston to connecting rod, refer to Fig. Y4-15 and be sure piston and rod are facing in correct directions. It may be necessary to heat piston to 70°-80° C (158°-176° F) to facilitate installation of piston pin. Be sure retaining snap rings are securely installed in snap ring grooves at each end of piston pin. Install piston rings with side having identification mark up. Stagger the ring end gaps at 120 degrees around the piston.

CONNECTING ROD AND BEARING

Prior to Model L40 engine serial number 00126 or Model L60 engine serial number 29990, piston and connecting rod are available only as an assembly. Piston and rod are available separately for all other engines. On early Models L40 and L60, it is necessary to check piston pin diameter and rod thickness (width) at pin end before obtaining replacement part.

Connecting rod for Model L40 or Model L60 is aluminum alloy, and piston pin rides directly in connecting rod pin bore. On Model L75 and Model L90, connecting rod is forged steel and pin end is fitted with renewable piston pin bushing. Refer to the following connecting rod specifications for all except early Models L40 and L60:

Rod pin bore ID—
    Model L40 . . . . . . 19.012-19.024 mm
                          (0.7485-0.7490 in.)
    Model L60 . . . . . . 21.014-21.028 mm
                          (0.8273-0.8279 in.)
    Models L75/L90 . . 23.025-23.038 mm
                          (0.9065-0.9070 in.)
Rod crankpin
bearing ID—
    Model L40 . . . . . . 30.007-30.020 mm
                          (1.1814-1.1819 in.)
    Model L60 . . . . . . 36.007-36.020 mm
                          (1.4176-1.4181 in.)
    Models L75/L90 . . 40.014-40.028 mm
                          (1.5754-1.5759 in.)

Renew piston pin bushing (Models L75/L90) or connecting rod if pin bore is worn to 19.1 mm (0.752 inch) on Model L40, 21.1 mm (0.831 inch) on Model L60 or 23.1 mm (0.909 inch) on Models L75 and L90.

Standard crankpin diameter is 29.965-29.982 mm (1.1797-1.1804 inches) on Model L40, 35.965-35.982 mm (1.4160-1.4165 inches) on Model L60 and 39.965-39.982 mm (1.5734-1.5741 inches) on Models L75 and L90. Connecting rod bearing to crankpin oil clearance should be 0.025-0.055 mm (0.001-0.0022 inch) on Models L40 and L60, and 0.32-0.63 mm (0.0013-0.0025 inch) on Models L75 and L90. Rod bearings are available in undersizes of 0.25 mm (0.010 inch) and 0.50 mm (0.020 inch) as well as standard size.

BALANCE SHAFT AND BEARINGS

The engine balance shaft (3—Fig. Y4-16) is supported in two ball bearings (2) and is driven by a gear (5) at flywheel end of shaft that engages a balancer drive gear (8) on engine crankshaft. Balance shaft can be removed after removing crankcase cover (1). Renew the ball bearings if loose or rough. Gear on balance shaft is removable but is not available as separate part. When installing balance shaft, the balance shaft gear must be timed with crankshaft gear as shown in Fig. Y4-12.

Fig. Y4-16—Exploded view of engine balancer shaft, crankshaft and camshaft. Pto drive crankshaft and pto drive camshaft are both shown; actual engine will be fitted with one or the other. Camshaft does not extend through crankcase cover (1) on engines equipped with crankshaft pto.

1. Crankcase cover
2. Bearings
3. Balancer shaft
4. Key
5. Balancer gear
6. Camshaft drive gear
7. Crankshaft
8. Balancer drive gear
9. Flywheel
10. Camshaft
11. Drive gear

Yanmar

CAMSHAFT AND BEARINGS

The camshaft (10—Fig. Y4-16) is supported in a needle roller bearing at flywheel end of engine and by a ball bearing at crankcase cover end. Camshaft end thrust is taken by a flange on crankcase (needle roller bearing) end of shaft. The camshaft extends through the crankcase cover (1) and is used as the engine pto shaft on some models. On models with crankshaft pto, the crankcase cover does not have an opening for the camshaft.

To remove camshaft, first remove valve rocker arm support, push rods, fuel injection pump and crankcase cover. Turn engine upside down so that valve tappets and fuel injection pump tappet will fall away from camshaft, then withdraw camshaft from crankcase. Lift tappets from cylinder block, taking care to keep exhaust tappet and intake tappet identified for reassembly.

The camshaft gear (11) is removable from camshaft, but is not available as separate part. Check camshaft journal at crankcase (needle bearing) end for excessive wear. Journal standard diameter is 14.989-15.0 mm (0.5901-0.5906 inch) on all models. Renew camshaft if journal is worn to 14.92 mm (0.5874 inch).

Inspect needle bearing in engine crankcase and if necessary to renew bearing, use suitable extractor to pull old bearing from bore. When installing new bearing, install with lettered side of bearing cage toward bearing driver and drive bearing into bore so that outer edge of bearing is 1.5-2.0 mm (0.060-0.080 inch) below thrust face of cylinder block.

Ball bearing at crankcase cover end is a 0.03 mm (0.0012 inch) interference fit to 0.02 mm (0.0008 inch) loose fit on camshaft journal. Journal diameter for ball bearing is 24.980-24.993 mm (0.9839-0.9840 inch) on Model L40,

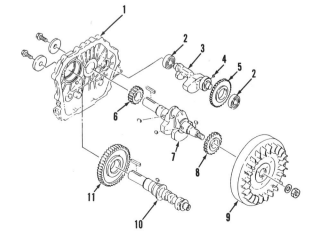

Yanmar

29.980-29.993 mm (1.1807-1.1808 inches) on Model L60 and 34.980-34.993 mm (1.3776-1.3777 inches) on Models L75 and L90. Renew camshaft if bearing is a loose fit on journal and journal measures 0.08 mm (0.003 inch) less than minimum standard diameter. Renew ball bearing if loose or rough. Camshaft end play on all models should be 0.043-0.277 mm (0.0017-0.0109 inch) with maximum allowable end play being 0.45 mm (0.0177 inch).

On models with camshaft pto, install new oil seal in crankcase cover with lip of seal to inside. Lubricate tappets and, if reinstalling used tappets, be sure to place exhaust and intake tappet in same bore as when removed. Lubricate needle bearing and camshaft lobes and install camshaft and gear assembly, being sure timing marks on camshaft gear and crankshaft gear are aligned as shown in Fig. Y4-12. Reverse removal procedure to complete reassembly of engine.

CRANKSHAFT AND MAIN BEARINGS

The crankshaft on all models is supported in a ball bearing (12—Fig. Y4-14) at flywheel end and a plain sleeve bushing (8) at crankcase cover end. Crankshaft can be removed after removing cylinder head, flywheel, crankcase cover, camshaft, balance shaft and the rod and piston unit. Remove ball bearing retaining plate (see Fig. Y4-17) and pull crankshaft from crankcase. Ball bearing should be a 0.002-0.025 mm (0.00008-0.0010 inch) interference (press) fit on crankshaft of Model L40, and 0.007-0.03 mm (0.0003-0.0012 inch) interference fit on crankshaft of all other models.

Crankshaft is fitted with a gear (5—Fig. Y4-16) at flywheel end to drive the engine balancer shaft and a gear (6) at crankcase cover end to drive engine camshaft. Although the gears can be removed from the crankshaft, they are not available separately.

Standard diameter of crankshaft main journal at crankcase cover end is 30.002-30.015 mm (1.181-1.182 inches) on Model L40, 35.002-35.018 mm (1.378-1.3787 inches) on Model L60 and 40.002-40.018 mm (1.5749-1.5755 inches) on Models L75 and L90. Desired bearing oil clearance is 0.03-0.08 mm (0.0012-0.0031 inch); maximum allowable oil clearance is 0.17 mm (0.0067 inch). Renew or regrind crankshaft if journal is worn to 29.91 mm (1.1776 inches) on Model L40, 34.91 mm (1.3744 inches) on Model L60, or 39.91 mm (1.5713 inches) on Model L75 or L90. Main bearing (8—Fig. Y4-14) can be renewed by pressing old bearing out of crankcase cover and installing new bearing with oil hole aligned with oil hole in cover and oil groove side of bearing up. The crankcase cover end bearing is available in undersizes of 0.25 mm (0.010 inch) and 0.50 mm (0.020 inch) as well as standard size.

Install new crankshaft oil seal (18—Fig. Y4-14)) flush with crankcase and with seal lip to inside. Refer to CAMSHAFT paragraph and renew camshaft needle bearing if necessary. Lubricate crankshaft and oil seal lip, then insert crankshaft and ball bearing into crankcase and install ball bearing retaining plate (14). If equipped with crankshaft pto, install new oil seal (1) in crankcase cover with seal lip facing inward. Install crankcase cover and complete reassembly of engine by reversing disassembly procedure.

ELECTRICAL SYSTEM

AIR HEATER

An optional air heater for warming intake air is available on models with charging alternator and battery. The

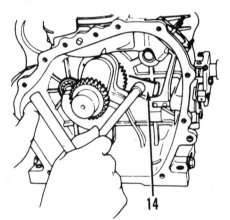

Fig. Y4-17—Removing crankshaft bearing retainer (14) from inside of crankcase (flywheel end of crankshaft).

SMALL DIESEL

heater is mounted on the air intake, and is used as a starting aid. The air heater is operated from the key switch on instrument panel. When heater switch is turned on, a timer (optional) activates a pilot light for a period of 15 seconds. After 15 seconds, turn key switch to "OFF" position, then turn to "START"; heater should warm the intake air on attempt to start engine. The heater requires 33.3 amperes current at 12 volts. Renew heater if there is no continuity between heater wire terminal and ground.

ALTERNATOR AND REGULATOR

A 6 volt alternator for lighting only and 12 volt alternators for lighting only or for battery charging to use with starting motor are available. All alternators available are located behind engine flywheel as shown in Fig. Y4-18. Refer to the following specifications:

Alternator Model GP9585—
  Wiring cover color . . . . Black
  Regulator model RS5112
  Rectifier model
    (half-wave) ZR2117
  DC Voltage at 3600 rpm 13
  DC Amperes at 3600 rpm . . . 0.7-1.3
  Minimum rpm for
    charging . . . . 1250 rpm
  AC output with regulator
    disconnected:
  3200 rpm . . . . 30.3 volts
  3750 rpm . . . . 35.7 volts
Alternator Model GP9589*—
  Wiring cover color . . . . Gray
  Regulator model RS5112
  Rectifier model
    (full wave) ZR2120
  DC Voltage at 3600 rpm 13
  DC Amperes at 3600 rpm . . . 1.6-1.8

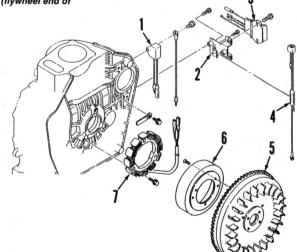

Fig. Y4-18—Alternator is mounted behind flywheel.
1. Rectifier
2. Bracket
3. Regulator
4. Fuse
5. Flywheel
6. Rotor
7. Stator

470

Illustrations Courtesy Yanmar Diesel Engine, Inc.

SERVICE MANUAL

Yanmar

Minimum rpm for
 charging............1000 rpm
AC output with regulator
 disconnected:
 3200 rpm ....46.0 volts
 3750 rpm ....53.8 volts
Alternator Model GP9587*—
 Wiring tube color..........Black
 Regulator
  model........RS5112 or RS2190
 DC Voltage at 3600 rpm.......13
 DC Amperes at 3600 rpm...2.6-3.3
 Minimum rpm for charging...1000
 AC output with regulator
  disconnected:
  3200 rpm ....39.0 volts
  3750 rpm ....45.4 volts
Alternator Model GP9595—
 Wiring tube color..........Yellow
 Regulator model..........RS5112
 DC Voltage at 3600 rpm........12
 DC Amperes at 3600
  rpm 14.8-16.0
 Minimum rpm for charging...1000
 AC output with regulator
  disconnected:
  3200 rpm ....35.5 volts
  3750 rpm ....41.5 volts
Alternator Model GP9591
 (Lighting only)—
 Wiring cover color..........Yellow
 Output at
  3600 rpm ....6 volts/15 watts

*Alternator Models GP9589 and GP9587 may be used without regulator or rectifier for lighting purposes only.

STARTING MOTOR

Conventional direct drive starters are used on all models; type S114-413 is used on models with camshaft pto and type S114-414 is used on models with crankshaft pto. Refer to Fig. Y4-19 for exploded view of typical starter. Refer to the following starting motor specifications:

Models S114-413, S114-414
 No-load test—
 Volts 11.5
 Current draw (max.)......60 amps

Fig. Y4-17—Exploded view of typical optional electric starter motor assembly.
1. Drive housing
2. Dust cover
3. Snap ring
4. Stop washer
5. Pinion assy.
6. Shift lever
7. Shift lever spring
8. Gaskets
9. Solenoid switch
10. Armature
11. Field coil
12. Brush
13. Brush holder
14. Bushings
15. Rear cover
16. Thrust washers
17. Snap ring
18. Dust cover

Rpm (min.)................7000
Brush length-Standard.....14 mm
  (0.551 in.)
Wear limit...............11 mm
  (0.433 in.)
Pinion gap............0.3-2.5 mm
  (0.012-0.098 in.)
Model S114-478
 No-load test—
 Volts 12
 Current draw (max.).....105 amps
 Rpm (min.)................4000
 Brush length-Standard.....16 mm
  (0.330 in.)
 Wear limit...............11 mm
  (0.433 in.)
 Pinion gap............0.3-1.5 mm
  (0.012-0.059 in.)

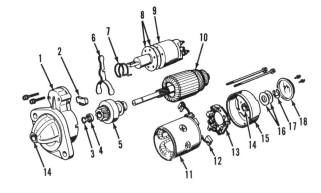

Illustrations Courtesy Yanmar Diesel Engine, Inc.

YANMAR

Model	No. Cyls.	Bore	Stroke	Displ.
TF50	1	74 mm (2.913 in.)	72 mm (2.835 in.)	309 cc (18.86 cu.in.)
TF60	1	75 mm (2.953 in.)	80 mm (3.150 in.)	353 cc (21.54 cu.in.)
TF70	1	78 mm (3.071 in.)	80 mm (3.150 in.)	382 cc (23.31 cu.in.)
TF80	1	80 mm (3.150 in.)	87 mm (3.425 in.)	437 cc (26.67 cu.in.)
TF90	1	85 mm (3.346 in.)	87 mm (3.425 in.)	493 cc (30.08 cu.in.)
TF110	1	88 mm (3.465 in.)	96 mm (3.780 in.)	583 cc (35.57 cu.in.)
TF120	1	92 mm (3.622 in.)	96 mm (3.780 in.)	638 cc (41.68 cu.in.)
TF140	1	96 mm (3.780 in.)	105 mm (4.134 in.)	760 cc (46.38 cu.in.)
TF160	1	102 mm (4.016 in.)	105 mm (4.134 in.)	857 cc (52.29 cu.in.)

All models are four-stroke, single horizontal cylinder, horizontal crankshaft, liquid cooled direct injection diesel engines. Cooling system is either closed radiator system or open water hopper above cylinder; models with water hopper will have a "-H" suffix on model number. Models TF80 through TF160 have two balance shafts mounted in engine crankcase. Engine rotation on all models is counterclockwise as viewed from flywheel end. Pto is at flywheel end only. Electric starting and alternator are optional.

MAINTENANCE

LUBRICATION

Recommended engine lubricant is API classification CC or CD motor oil. Initial oil change on new or reconditioned engine should be after 50 hours operation. Subsequent oil changes should be made after each 100 hours operation. Lubricating oil strainer (4—Fig. Y5-1) can be removed after unscrewing oil drain plug (1) and should be cleaned at each oil change. Take care not to lose the spring (3). An oil pressure indicator (OP—Fig. Y5-18) is located on crankcase at fuel injection pump side of engine. Oil pan capacities are as follows:

Model TF50 .... 1.2 liters
(2.5 pts.)
Models TF60/TF70 .... 1.8 liters
(3.8 pts.)
Models TF80/TF90 .... 2.2 liters
(4.6 pts.)
Models TF110/TF120 ...... 2.8 liters
(5.9 pts.)
Models TF140/TF160 ...... 3.0 liters
(6.3 pts.)

FUEL SYSTEM

FUEL FILTER. Refer to Fig. Y5-2 for exploded view of fuel filter assembly used on all models. Float (5) rises to indicate presence of water in sediment bowl (6). Close fuel shut-off valve (2) and remove retainer (7) and sediment bowl to drain water and sediment from bowl. Filter should be cleaned after every 300 hours of operation and the filter element (3) should be renewed after each 600 hours. When reinstalling filter and sediment bowl, allow bowl to fill completely with fuel before tightening bowl retainer.

BLEED FUEL SYSTEM. Disconnect fuel supply line (L—Fig. Y5-3) from fuel injection pump (FP) and open fuel valve (2—Fig. Y5-2) in filter base. When fuel is running freely with no air, reconnect supply line and loosen fuel injection line fitting (see Fig. Y5-3) at fuel injector (IN). Engage compression release lever (CR), place throttle control in run posi-

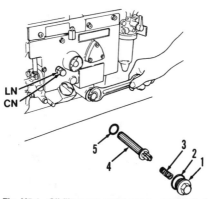

Fig. Y5-1—Oil filter screen (4) can be removed after removing crankcase oil drain plug (1). Cap nut (CN) covers end of governor speed limiter screw.
CN. Cap nut
LN. Lock nut
1. Oil drain plug
2. "O" ring
3. Spring
4. Filter screen
5. "O" ring

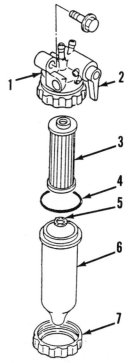

Fig. Y5-2—Exploded view of fuel filter assembly.
1. Filter body
2. Fuel shut-off valve
3. Fuel filter
4. "O" ring
5. Water float
6. Filter bowl
7. Bowl retainer

SERVICE MANUAL

Yanmar

tion and crank engine until fuel flows from loosened fitting. Tighten fitting and start engine.

INJECTION PUMP TIMING

Injection timing is adjusted by varying the thickness of shims between the fuel injection pump and engine crankcase. Injection timing should be as follows:

Model TF50 . . . . 12.5° BTDC
Models TF80, TF90 . . . . . . 18° BTDC
All other models . . . . 17° BTDC

To check injection pump timing, disconnect fuel injection line from fuel injector (see Fig. Y5-3). Place throttle control in run position and move decompressor lever (CR) to hold exhaust valve open. Turn flywheel several revolutions in normal direction of rotation until all air is bled from open injection line. Then turn flywheel slowly and observe point at which fuel just begins to flow from open line and check timing marks on flywheel (F—Fig. Y5-4) at timing indicator mark (M). If timing is not correct, remove fuel injection pump and add shim thickness to retard timing or remove shims to advance timing. A change of shim thickness by 0.1 mm (0.004 inch) will change timing by about one degree.

GOVERNOR

The flyweight type governor unit is mounted on engine crankshaft gear. Internal governor linkage is mounted on inside of crankcase (gearcase) cover. Governed speed at rated load is 2400 rpm. Speed limiter is set at factory; if necessary to readjust, remove cap nut (CN—Fig. Y5-1). Loosen locknut (LN) and turn speed limiter screw to obtain correct speed under load. Tighten locknut, recheck engine speed and install cap nut.

REPAIRS

TIGHTENING TORQUES

Refer to the following special tightening torques when reassembling engine:

Cylinder head—
  Models TF140, TF160 . . 186-206 N·m
    (138-151 ft.-lbs.)
  Models TF110, TF120 . . 129-138 N·m
    (95-101 ft.-lbs.)
  All other models . . . . 93-103 N·m
    (69-75 ft.-lbs.)
Connecting rod bolts—
  Models TF140, TF160 . . . . 56-61 N·m
    (42-45 ft.-lbs.)
  Models TF80, TF90,
    TF110, TF120 . . . . 52-56 N·m
    (38-41 ft.-lbs.)
  Models TF60, TF70 . . . . . . 37-41 N·m
    (28-30 ft.-lbs.)
  Model TF50 . . . . 22-23 N·m
    (16-17 ft.-lbs.)
Main bearing retainer—
  All models . . . . 23-28 N·m
    (17-20 ft.-lbs.)
Balance weight bearing
retainer—
  Models TF140, TF160 . . . . 82-91 N·m
    (60-67 ft.-lbs.)
  All other models . . . . 45-54 N·m
    (33-39 ft.-lbs.)
Flywheel nut—
  Models TF110, TF120,
    TF140, TF160 . . . . 295-340 N·m
    (217-253 ft.-lbs.)

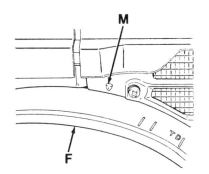

Fig. Y5-4—View showing flywheel timing marks.

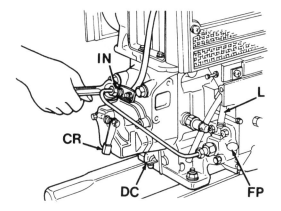

Fig. Y5-3—Loosen injection line nut at injector (IN) as final step of bleeding air from system. Refer to text for bleeding procedure.
L. Fuel supply line
CR. Compression release lever
DC. Cooling water drain cock
FP. Fuel injection pump
IN. Injector

Models TF60, TF70,
  TF80, TF90 . . . . 245-284 N·m
    (181-209 ft.-lbs.)
Model TF50 . . . . 167-196 N·m
    (123-144 ft.-lbs.)
Fuel cam nut,
  all models . . . . 89-107 N·m
    (66-79 ft.-lbs.)
Fuel injection pump
  mounting nut . . . . 23-28 N·m
    (17-20 ft.-lbs.)
Fuel injector retaining
nut—
  Models TF140, TF160 . . . . 23-28 N·m
    (17-20 ft.-lbs.)
  All other models . . . . 20 N·m
    (15 ft.-lbs.)
Injection pump delivery
valve holder—
  Models TF140, TF160 . . . . 45-49 N·m
    (33-36 ft.-lbs.)
  All other models . . . . 40-44 N·m
    (29-32 ft.-lbs.)
Fuel injector line
fittings—
  Model TF50 . . . . 26-29 N·m
    (19-21 ft.-lbs.)
  All other models . . . . 27-32 N·m
    (20-23 ft.-lbs.)

COOLING SYSTEM

The cooling system consists of radiator and cooling fan unit or a water hopper attached to top of cylinder block. The coolant drain cock (DC—Fig. Y5-3) is located on lower side of cylinder head. On models with a radiator, the cooling fan is driven by a V-belt from a pulley on engine crankshaft. To adjust the V-belt, loosen bracket clamp bolt and move idler bracket (B—Fig. Y5-5) so that a belt deflection of 15-20 mm (0.60-0.80

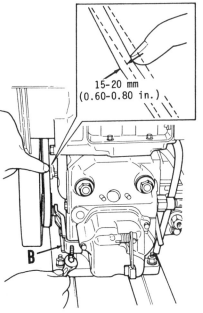

Fig. Y5-5—Move tension pulley bracket (B) to adjust radiator fan belt tension.

Yanmar

SMALL DIESEL

inch) is obtained from a finger pressure of 3-5 kg (7-11 pounds).

To remove water hopper, first drain coolant and remove hopper cover and core. Then remove bolts from inside hopper (Fig. Y5-6) and remove hopper from engine. The hopper is sealed to the block with silicone gasket material.

To remove radiator and cooling fan unit, first drain coolant and remove radiator cap, radiator cover and the screen at each side of unit. Remove the two lower bolts from cooling fan housing (1—Fig. Y-5-7) and unbolt radiator (2) from cylinder block. The radiator is sealed to the block with silicone gasket material.

FUEL TANK AND FILTER

The fuel tank and filter can be removed as a unit. First, close fuel supply valve and disconnect fuel supply line at fuel injection pump. Disconnect fuel return line from fuel injector. Unbolt fuel filter from engine crankcase cover. Remove the panel or lighting unit from end of fuel tank. Loosen fuel tank clamp bolts, remove self-locking nut and remove fuel tank. Reverse removal procedure to reinstall tank.

VALVE COVER AND DECOMPRESSION DEVICE

A compression relief (decompression) device is fitted in the valve cover (1—Fig. Y5-8). Moving compression relief lever (2) to straight out position will prevent exhaust valve from completely closing, thereby making engine easier to turn with crank. Decompression device is not adjustable.

VALVE ADJUSTMENT

Intake and exhaust valve clearance with engine cold should be 0.15 mm (0.006 inch) on Model TF50 and 0.2 mm (0.008 inch) on all other models. Valve clearance should be checked and adjusted, if necessary, after each 300 hours of operation. To check clearance, rotate flywheel to position piston at top dead center on compression stroke. Remove rocker arm cover and insert feeler gage (G—Fig. Y5-9) between rocker arm and valve stem. To adjust clearance, loosen locknut and turn rocker arm adjusting screw (S) as necessary.

ROCKER ARMS AND PUSH RODS

Rocker arms are mounted on a pedestal retained by one bolt and a locating dowel pin. Rocker arm pedestal, rocker arms and push rods can be removed after removing valve cover. Be sure to identify all parts as they are removed so that they can be installed in original positions if reused.

Desired rocker arm to shaft clearance is 0.016-0.052 mm (0.0006-0.002 inch) with maximum allowable clearance of 0.15 mm (0.006 inch). Check rocker arm support shaft for excessive wear or scoring. Standard shaft diameter is 13.982-14.000 mm (0.5505-0.5512 inch) on Models TF50, TF60, TF70, TF80 and TF90 with a wear limit of 13.90 mm (0.547 inch). On Models TF110, TF120, TF140 and TF160, standard shaft diameter is 15.982-16.000 mm (0.6292-0.6299 inch) with wear limit of 15.90 mm (0.626 inch).

Rocker arm standard inside diameter is 14.016-14.034 mm (0.5518-0.5525 inch) with wear limit of 14.10 mm (0.555 inch) for Models TF50, TF60, TF70, TF80 and TF90, and 16.016-16.034 mm (0.6306-0.6313 inch) with wear limit of 16.10 (0.634 inch) for Models TF70 and TF90. The exhaust rocker arm has a machined flat for the compression release mechanism.

Check push rods for obvious wear at either end and for distortion (bending) and renew if necessary. When installing push rods and rocker arm assembly, loosen rocker arm adjusting screw locknuts and back screws out. Lubricate all parts and be sure rocker arm support is located properly on dowel pin, then tighten rocker arm support bolt. Adjust valve clearance as outlined in previous paragraph and complete reassembly by reversing disassembly procedure.

CYLINDER HEAD

To remove cylinder head, drain engine coolant and then proceed as follows: Unbolt and remove the exhaust muffler assembly and the air cleaner assembly from head. Thoroughly clean fuel injector, injection line and fuel injection pump, then remove the fuel injection line. Remove the fuel injector and packing. Remove the valve cover, rocker arm assembly and push rods. Unbolt and remove cylinder head.

Before reinstalling cylinder head, thoroughly clean head and check for cracks, distortion or other damage. Maximum allowable distortion in cylinder head is 0.05 mm (0.002 inch). Install head using new gasket. Refer to Fig. Y5-10 and tighten head bolts evenly in several steps to final torque of 187-205 N·m (138-151 ft.-lbs.) on Models TF140 and TF160, to 94-103 N·m (69-75 ft.-lbs.) on Model TF50, and to 129-138 N·m (95-101 ft.-lbs.) on all other models.

Fig. Y5-6—Bolts attaching cooling water hopper to cylinder are accessible after removing hopper cover and inner core.

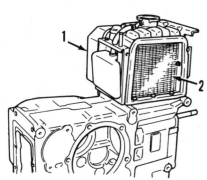

Fig. Y5-7—Removing cooling fan (1) and radiator assembly (2).

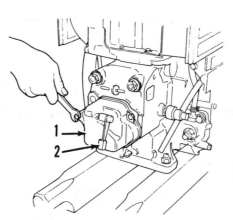

Fig. Y5-8—Valve cover (1) is fitted with a compression relief device (2).

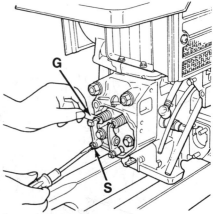

Fig. Y5-9—Use a feeler gage (G) to check clearance between rocker arm and valve stem. Turn screw (S) to adjust clearance.

SERVICE MANUAL

VALVE SYSTEM

Valve seat and face angle is 30 degrees for intake and 45 degrees for exhaust. Check valve margin (thickness of valve at outer edge of face); wear limit is 0.7 mm (0.028 inch) for intake valve and 0.5 mm (0.020 inch) for exhaust valve. Check valve stem diameter; wear limit is 8.85 mm (0.349 inch) for Models TF140 and TF160, 7.85 mm (0.309 inch) for Models TF110 and TF120, and 6.9 mm (0.272 inch) for all other models.

Check distance between top of valve and gasket surface of cylinder head using a feeler gage and straightedge or depth gage. Refer to the following valve recession specifications and renew valves and/or cylinder head if valves are recessed beyond specified wear limit.

Models TF50, TF60, TF70
Intake valve recession—
  Standard . . . . 0.35 mm
    (0.014 in.)
  Wear limit . . . . 0.60 mm
    (0.024 in.)
Exhaust valve recession—
  Standard . . . . 0.75 mm
    (0.030 in.)
  Wear limit . . . . 1.0 mm
    (0.040 in.)

Models TF80, TF90
Intake valve recession—
  Standard . . . . 0.30 mm
    (0.012 in.)
  Wear limit . . . . 0.60 mm
    (0.024 in.)
Exhaust valve recession—
  Standard . . . . 0.75 mm
    (0.030 in.)
  Wear limit . . . . 1.0 mm
    (0.040 in.)

Models TF110, TF120
Intake valve recession—
  Standard . . . . 0.70 mm
    (0.028 in.)
  Wear limit . . . . 1.0 mm
    (0.040 in.)
Exhaust valve recession—
  Standard . . . . 0.80 mm
    (0.031 in.)
  Wear limit . . . . 1.1 mm
    (0.043 in.)

Models TF140, TF160
Intake valve recession—
  Standard . . . . 0.35 mm
    (0.014 in.)
  Wear limit . . . . 0.70 mm
    (0.028 in.)
Exhaust valve recession—
  Standard . . . . 1.0 mm
    (0.040 in.)
  Wear limit . . . . 1.2 mm
    (0.047 in.)

The valve guides in Model TF50 are integral with the head, but guides are renewable in all other models. Desired valve stem to guide clearance is 0.03-0.06 mm (0.0012-0.0024 inch) for Model TF50 intake valve. For Model TF50 exhaust guide, and for both intake and exhaust guides of all other models, desired valve stem to guide clearance is 0.045-0.075 mm (0.0018-0.003 inch). Maximum allowable stem to guide clearance is 0.15 mm (0.006 inch) for both intake and exhaust guides of all models.

When renewing valve guides, drive old guide out toward outer (valve spring) side of head. Drive new guide in from outer side of head until height of guide above head is 12 mm (0.472 inch) on Models TF60 and TF70, 11 mm (0.433 inch) on Models TF80 and TF90, and 15 mm (0.591 inch) on all other models. Ream guide using suitable reamer to obtain desired valve stem to guide clearance.

Refer to the following valve spring specifications which apply to both the intake and exhaust valve spring:
Spring free length—
Models TF50, TF60, TF70,
  TF80, TF90
    New . . . . 36.12 mm
      (1.422 in.)
    Wear limit . . . . 34.5 mm
      (1.358 in.)
Models TF110, TF120
    New . . . . 42.00 mm
      (1.654 in.)
    Wear limit . . . . 40.5 mm
      (1.594 in.)
Models TF140, TF160
    New . . . . 43.50 mm
      (1.713 in.)
    Wear limit . . . . 42.0 mm
      (1.654 in.)

FUEL INJECTOR

REMOVE AND REINSTALL. Prior to removing injector, thoroughly clean injector line, injection pump and injector using suitable solvent and compressed air. Remove injector fuel return line (2—Fig. Y5-11), unscrew fittings at each end of injection line (3) and remove line from pump and injector. Remove the injector retaining nuts on Models TF140 and TF160; on other models remove clamp nuts and clamp plate (1), then remove injector from cylinder head. Remove injector heat shield (9—Fig. Y5-12) and seal washer (10) form bore in cylinder head if they were not removed with injector.

When installing injector, be sure injector bore in cylinder head is clean. Install seal washer, heat shield and injector in head. On Models TF140 and TF160, install and tighten nozzle retaining nuts to a torque of 23-28 N·m (17-20 ft.-lbs.). On all other models, install the clamp plate and nuts and tighten the nuts evenly to a torque of 20 N·m (15 ft.-lbs.). Install injection high pressure line, leaving fitting at injector loose until air is bled from system. Connect fuel return line to injector.

TESTING. A complete job of testing and adjusting the injector requires use of special test equipment. Only clean,

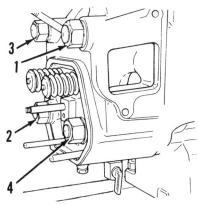

Fig. Y5-10—Head bolt tightening sequence for all models.

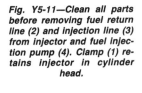

Fig. Y5-11—Clean all parts before removing fuel return line (2) and injection line (3) from injector and fuel injection pump (4). Clamp (1) retains injector in cylinder head.

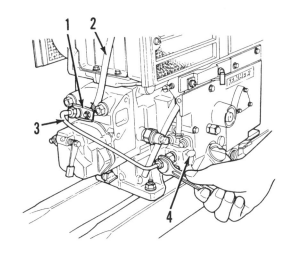

Yanmar

approved testing oil should be used in tester tank. Injector should be tested for opening pressure, spray pattern and seat leakage.

Before conducting test, operate tester lever until fuel flows from tester line, then attach injector. Close tester gage valve and operate tester lever a few quick strokes to purge air from injector and to be sure nozzle valve is not stuck.

WARNING: Fuel emerges from injector with sufficient force to penetrate the skin. When testing injector, keep yourself clear of nozzle spray.

OPENING PRESSURE. Open valve to tester gage and operate tester lever slowly while observing gage reading. Pressure at which nozzle opens should be 19614-20595 kPa (2845-2985 psi). Opening pressure is adjusted by turning adjusting screw (3—Fig. Y5-13) on Models TF140 and TF160, or by adding or removing shims (2—Fig. Y5-12) on other models. Changing shim stack thickness by 0.1 mm (0.004 inch) will change injector opening pressure by about 1960 kPa (285 psi).

SPRAY PATTERN. Spray pattern should be equal from all four nozzle tip holes and should be a fine mist. Spray angle between cones is 150 degrees and each spray cone should be a 5-10 degree cone. Place a sheet of blotting paper 30 cm (about 12 inches) below injector and operate tester one stroke. The spray pattern should be a perfect circle.

SEAT LEAKAGE. No oil should drip from injector nozzle after an injection. Operate tester lever two or three strokes, then wipe nozzle tip dry. Slowly bring tester gage pressure to 1960 kPa (285 psi) below opening pressure and hold this pressure for 5 seconds. Drop of oil should not form nor should oil drip from nozzle tip. Renew or overhaul injector if drop of oil is noted.

OVERHAUL. Wipe all loose carbon from outside of injector. Hold nozzle body in soft jawed vise or injector fixture and unscrew nozzle retaining nut (8—Fig. Y5-12 or 12—Fig. Y5-13). Remove nozzle body, nozzle valve, stop plate, spring seat, spring and adjusting shims if used. Be careful not to lose any of the shims (2—Fig. Y5-12) on models so equipped and place all parts in clean diesel fuel or calibrating oil as they are removed.

Clean exterior surfaces with a brass wire brush. Soak parts in approved carbon solvent if necessary to loosen hard carbon deposits. Rinse parts in clean diesel fuel or calibrating oil immediately after cleaning to neutralize solvent. Clean nozzle tip opening from inside with a wood cleaning stick. Clean the four spray holes in nozzle tip by using proper sized cleaning wire in wire holder.

Check to see that needle does not bind in nozzle. Needle should slide smoothly in nozzle body from its own weight. Rotate needle to several different positions when checking for binding. If nozzle needle sticks, reclean or renew nozzle assembly.

Reassemble injector under clean diesel fuel or calibrating oil. If opening pressure was correct, use shim pack (except Models TF140 and TF160) removed on disassembly; add or remove shims if necessary to change opening pressure. Tighten nozzle retaining nut to a torque of 69-73 N·m (51-54 ft.-lbs.). Retest injector and reinstall in cylinder head if injector meets test specifications.

SMALL DIESEL

INJECTION PUMP

To remove fuel injection pump, first thoroughly clean pump, injection line, fuel injector and surrounding area. Close fuel supply valve, disconnect fuel supply line from pump (P—Fig. Y5-14) and remove the fuel injection line. The pump can then be unbolted and removed from crankcase cover. Move the throttle control lever while removing pump so that pump rack clears governor lever and opening in crankcase cover. Be careful not to lose or damage timing adjustment shims.

Use same number and thickness of timing shims as removed when reinstalling fuel injection pump unless change of shim thickness is necessary to correct injection timing. Turn crankshaft so that pump cam lobe is away from injection pump. Move throttle control lever as necessary to prevent interference of fuel injection pump rack with governor lever. Tighten fuel injection pump retaining bolts to a torque of 23-28 N·m (17-20 ft.-lbs.). Complete balance of reassembly by reversing disassembly procedure. Refer to FUEL SYSTEM paragraphs in MAINTENANCE section and bleed fuel system.

CRANKCASE COVER

To remove crankcase (gearcase) cover (3—Fig. Y5-15), first remove oil drain plug (6), drain lubricant and remove oil filter screen. Close fuel supply valve and unbolt fuel filter base (4) from cover, leaving fuel lines attached. Remove fuel injection line and fuel injection pump. Unbolt and remove crankcase cover.

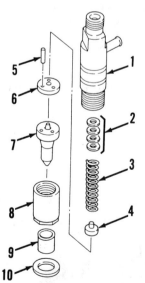

Fig. Y5-12—Exploded view of fuel injector assembly used on all models except Models TF140 and TF160. Opening pressure is adjusted by means of shims (2).

1. Injector body
2. Adjusting shims
3. Spring
4. Spring seat
5. Dowel pin
6. Stop plate
7. Nozzle & valve
8. Nozzle nut
9. Injector heat shield
10. Injector seal washer

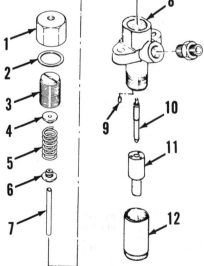

Fig. Y5-13—Exploded view of injector assembly used on Models TF140 and TF160. Opening pressure is adjusted by turning adjusting screw (3).

1. Cap nut
2. Seal washer
3. Adjusting screw
4. Spring seat
5. Spring
6. Spring seat
7. Push rod
8. Injector body
9. Dowel pin
10. Nozzle valve
11. Nozzle
12. Nozzle nut

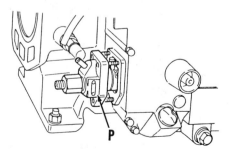

Fig. Y5-14—Removing fuel injection pump (P).

SERVICE MANUAL

With cover removed, inspect governor linkage and manual starting shaft. Renew manual starting shaft oil seal if leaking. If there is excessive wear on starting shaft and timing cover shaft bearing bore, both parts will need to be renewed. Starting shaft diameter is 24.972-24.993 mm (0.983-0.984 inch) with wear limit of 24.95 mm (0.9823 inch). Bore diameter in crankcase cover is 25.03-25.06 mm (0.9854-0.9866 inch) with wear limit of 25.2 mm (0.992 inch).

To install cover, first remove the oil pump cover (5) and the pump inner and outer rotors. Turn crankshaft so that governor flyweights are horizontal and make sure that governor sleeve engages the weights. Turn the starting shaft (2) in cover so that pin (1) is in position shown in Fig. Y5-15. Install cover and tighten retaining bolts evenly in a crisscross pattern. Install oil pump, fuel injection pump and fuel filter. Bleed air from fuel system as outlined in FUEL SYSTEM paragraphs in MAINTENANCE section.

INJECTION PUMP CAM

The injection pump cam (2—Fig. Y5-17) is mounted on the end of engine camshaft, and can be removed after removing crankcase (gearcase) cover assembly. Unscrew the hex nut (1) from end of camshaft and remove the injection pump cam.

Install injection pump cam on camshaft with pump lobe to outside. Tighten the retaining nut to a torque of 89-107 N·m (65-79 ft.-lbs.).

OIL PUMP AND RELIEF VALVE

The gerotor type oil pump and the oil pressure relief valve are mounted in the crankcase (gearcase) cover. Oil pump shaft is driven by a slot in end of engine crankshaft. An oil pressure indicator (OP—Fig. Y5-18) is threaded into an oil passage in the cylinder block. Indicator signal is blue for normal oil pressure or red for insufficient oil pressure. Remove indicator and install master gage fitting to check actual oil pressure. Oil pressure should be 245 kPa (35.5 psi) at normal operating temperature and engine rated speed.

Pump can be inspected after removing pump cover plate (5—Fig. Y5-15) and sealing "O" ring from outside of crankcase cover. Wear limit between outer rotor (Fig. Y5-19) and pump body (engine crankcase cover) is 0.20 mm (0.008 inch) on Models TF50, TF60, TF70, TF80 and TF90, and is 0.22 mm (0.0086 inch) on all other models. Wear limit between lobes of inner and outer rotors (Fig. Y5-20) is 0.20 mm (0.008 inch) on all models. End clearance of rotors in housing (Fig. Y5-21) is 0.09 mm (0.0035 inch) on Model TF50, 0.15 mm (0.006 inch) on Models TF140 and TF160, and is 0.12 mm (0.0047 inch) on all other models.

Clean and inspect pressure relief valve if low pressure is indicated and oil pump components are within specification. Renew pump components if wear is excessive. It will be necessary to renew the crankcase cover if oil pump cavity is deeply scored, or if rotor to cover clearances are excessive with new pump components.

GOVERNOR

The governor flyweight assembly (FW—Fig. Y5-22) is bolted to the engine balancer drive gear on Models TF80 through TF160, or to a spacer on engine crankshaft on Models TF50, 60 and 70. Governor linkage is mounted in engine crankcase cover. It is necessary to remove crankcase cover to inspect or service the governor. Renew parts as necessary.

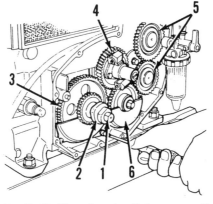

Fig. Y5-17—View of engine timing gears with crankcase (gearcase) cover removed. Balancer shafts and gears (5) are not used on some models.
1. Injection cam nut
2. Injection pump cam lobe
3. Camshaft gear
4. Crankshaft balancer drive gear
5. Balancer gears
6. Oil splash shield

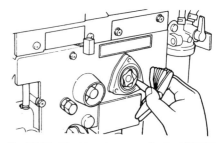

Fig. Y5-19—Checking clearance of outer oil pump rotor to crankcase (gearcase) cover.

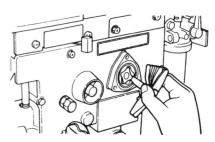

Fig. Y5-20—Checking inner to outer lobe clearance of engine oil pump.

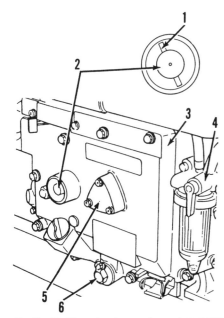

Fig. Y5-15—View showing crankcase (gearcase) cover. Pin (1) in starting shaft (2) must be turned as shown when installing crankcase cover.
1. Pin
2. Manual starter shaft
3. Crankcase cover
4. Fuel filter base
5. Engine oil pump
6. Oil drain plug

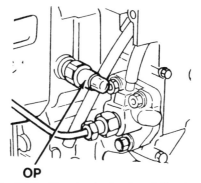

Fig. Y5-18—View showing location of oil pressure indicator (OP).

Fig. Y5-21—Checking rotor end clearance with straightedge and feeler gage.

Illustrations Courtesy Yanmar Diesel Engine, Inc.

Yanmar

PISTON AND ROD UNIT

The piston and connecting rod can be removed after removing cylinder head and cylinder block rear cover. Turn crankshaft so that piston is at bottom of cylinder and remove connecting rod cap. Then turn crankshaft so that piston is at top of cylinder and use a rod or hammer handle to push piston out of cylinder.

When installing piston and rod, be sure that triangle mark (T—Fig. Y5-23) on piston crown and matching mark (M) on connecting rod face upward. Lubricate cylinder, piston and rings, connecting rod bearing and crankpin. Set piston ring gaps so they are equally spaced around piston, then install piston and rod. Tighten connecting rod bolts to a torque of 22-23 N·m (16-17 ft.-lbs.) on Model TF50, 37-41 N·m (27-30 ft.-lbs.) on Models TF60 and TF70, 52-56 N·m (38-41 ft.-lbs.) on Models TF80, TF90, TF110 and TF120, and to a torque of 56-61 N·m (41-45 ft.-lbs.) on Models TF140 and TF160.

Check to see that engine turns over smoothly, then complete reassembly by reversing disassembly procedure.

CYLINDER SLEEVE

The wet type cylinder sleeve can be pulled from cylinder block using a suitable sleeve puller after removing connecting rod and piston unit. Sleeve should be renewed if scored, or if excessively worn. Refer to the following specifications:

Model TF50
Sleeve standard ID......74-74.03 mm
 (2.9134-2.9146 in.)
 Wear limit..............74.18 mm
 (2.9205 in.)
Model TF60
Sleeve standard ID.....75-75.03 mm
 (2.9528-2.9539 in.)
 Wear limit..............75.22 mm
 (2.9614 in.)
Model TF70
Sleeve standard ID.....78-78.03 mm
 (3.0709-3.0720 in.)
 Wear limit..............78.23 mm
 (3.080 in.)
Model TF80
Sleeve standard ID.....80-80.03 mm
 (3.1496-3.1508 in.)
 Wear limit..............80.18 mm
 (3.1567 in.)
Model TF90
Sleeve standard ID....85-85.035 mm
 (3.3465-3.3478 in.)
 Wear limit..............85.19 mm
 (3.354 in.)
Model TF110
Sleeve standard ID....88-88.035 mm
 (3.4646-3.4659 in.)
 Wear limit..............88.20 mm
 (3.4724 in.)
Model TF120
Sleeve standard ID....92-92.035 mm
 (3.6220-3.6234 in.)
 Wear limit..............92.21 mm
 (3.6303 in.)
Model TF140
Sleeve standard ID....96-96.035 mm
 (3.7795-3.7809 in.)
 Wear limit..............96.22 mm
 (3.7882 in.)
Model TF160
Sleeve standard ID..102-102.035 mm
 (4.0157-4.0171 in.)
 Wear limit............102.23 mm
 (4.0248 in.)

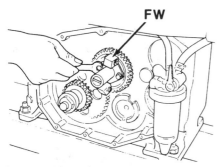

Fig. Y5-22—Removing governor flyweight assembly (FW) from balancer drive gear.

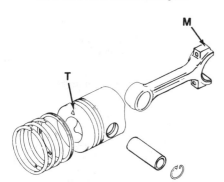

Fig. Y5-23—Triangle mark (T) on piston crown and matching marks (M) on rod and cap must be aligned to same side of piston and rod unit. Marks must be up when installed in engine.

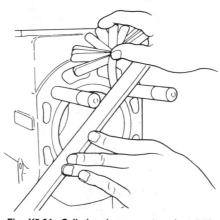

Fig. Y5-24—Cylinder sleeve must project 0.02-0.08 mm (0.0008-0.0031 inch) from cylinder block and can be checked with straightedge and feeler gage as shown.

SMALL DIESEL

Before installing sleeve, be sure that cylinder block surfaces are completely cleaned of rust and other deposits. Insert sleeve in block without the sealing "O" rings and check sleeve protrusion above cylinder block as shown in Fig. Y5-24. Sleeve should slide into block easily and should be 0.02-0.08 mm (0.0008-0.0031 inch) above block. If protrusion is excessive, check cylinder block and sleeve for rust or other deposits on stepped shoulder.

Install new sealing rings on outer diameter of sleeve, taking care that rings are not twisted in the grooves. Lubricate "O" rings and install sleeve in block.

PISTON, PIN AND RINGS

Piston for Model TF50 is fitted with two compression rings and one oil control ring. All other models have three compression rings and one oil control ring. Piston pin is retained by a snap ring at each end of pin bore in piston. Remove snap rings and push pin from piston and rod to disassemble rod and piston unit.

Check piston skirt for cracks, scoring or excessive wear. Refer to the following piston specifications:

Model TF50
Piston to cylinder
 clearance.........0.079-0.139 mm
 (0.0031-0.0055 in.)
 Wear limit..............0.30 mm
 (0.0118 in.)
Model TF60
Piston to cylinder
 clearance.........0.094-0.154 mm
 (0.0037-0.0061 in.)
 Wear limit..............0.30 mm
 (0.0118 in.)
Model TF70
Piston to cylinder
 clearance.........0.100-0.160 mm
 (0.0039-0.0063 in.)
 Wear limit..............0.31 mm
 (0.0122 in.)
Model TF80
Piston to cylinder
 clearance.........0.103-0.163 mm
 (0.0041-0.0064 in.)
 Wear limit..............0.32 mm
 (0.0126 in.)
Model TF90
Piston to cylinder
 clearance..........0.113-0.178 mm
 (0.0044-0.0070 in.)
 Wear limit..............0.34 mm
 (0.0134 in.)
Model TF110
Piston to cylinder
 clearance.........0.121-0.186 mm
 (0.0048-0.0073 in.)
 Wear limit..............0.35 mm
 (0.0138 in.)

Illustrations Courtesy Yanmar Diesel Engine, Inc.

SERVICE MANUAL

Yanmar

Model TF120
Piston to cylinder
 clearance .... 0.129-0.194 mm
 (0.0051-0.0076 in.)
 Wear limit .... 0.37 mm
 (0.0146 in.)

Model TF140
Piston to cylinder
 clearance .... 0.112-0.177 mm
 (0.0044-0.0070 in.)
 Wear limit .... 0.34 mm
 (0.0134 in.)

Model TF160
Piston to cylinder
 clearance .... 0.120-0.185 mm
 (0.0047-0.0073 in.)
 Wear limit .... 0.35 mm
 (0.0138 in.)

Piston pin should be a thumb press fit in piston when piston is at temperature of 50°-60° C (122°-140° F). Renew piston and pin if pin to piston clearance is 0.045 mm (0.0018 inch) or more, or if pin or pin bore in piston is scored or shows signs of overheating.

Check piston ring to groove clearance using new piston rings and feeler gage as shown in Fig. Y5-25. Renew piston if side clearance exceeds 0.2 mm (0.008 inch) for compression rings and 0.15 mm (0.006 inch) for oil control ring. Desired piston ring end gap for Model TF50 is 0.25-0.45 mm (0.010-0.018 inch) for compression rings and 0.10-0.30 mm (0.004-0.012 inch) for oil control ring. Ring end gap for all rings on Models TF60 and TF70 is 0.20-0.40 mm (0.008-0.016 inch). Ring end gap for top compression ring on Model TF80 is 0.20-0.40 mm (0.008-0.016 inch) and end gap for second and third compression rings and oil ring should be 0.30-0.50 mm (0.012-0.020 inch). Ring end gap for Model TF120 should be 0.30-0.50 mm (0.012-0.020 inch) for compression rings and 0.25-0.45 mm (0.010-0.018 inch) for oil control ring. Desired ring end gap for all rings on all other models is 0.30-0.50 mm (0.012-0.020 inch). Maximum allowable piston ring end gap is 1.5 mm (0.059 inch) for all rings on all models.

Install piston rings with identification mark on side of ring toward top of piston. Position ring end gaps at equal spacing around piston.

CONNECTING RODS AND BEARINGS

Piston pin bushing in connecting rod should be renewed if bushing is scored or excessively worn and connecting rod is otherwise serviceable. Maximum allowable piston pin to rod bushing clearance is 0.10 mm (0.0039 inch) for Model TF50 and 0.11 mm (0.0043 inch) for all other models.

Maximum allowable connecting rod distortion (twist or out of parallel) is 0.08 mm (0.003 inch) per 100 mm (4 inches) length.

Desired rod bearing to crankpin clearance is 0.018-0.077 mm (0.0007-0.0030 inch) on Model TF50 and 0.022-0.092 mm (0.0009-0.0036 inch) for Models TF140 and TF160. For all other models, desired crankpin bearing oil clearance is 0.028-0.086 mm (0.0011-0.0034 inch). Maximum allowable crankpin bearing clearance is 0.10 mm (0.004 inch) for all models.

ENGINE BALANCER SHAFTS

All Models Except TF50, TF60 and TF70

The balancer shafts are accessible after removing crankcase (gearcase) cover and removing the oil splash shield (6—Fig. Y5-17). Remove the idler gear (1—Fig. Y5-26) and the balance shaft bearing retainer (6), then pull balancer gears and shafts (7) from crankcase.

When installing balancer shafts, place the shaft with one timing mark on gear in top position and the other shaft in the lower position. The tooth with single timing mark on upper shaft gear should mesh between the two marked teeth on lower gear as shown in Fig. Y5-26. Install bearing retainer (6) and tighten retainer bolt to a torque of 82-91 N·m (60-67 ft.-lbs.) on Models TF140 and TF160, and to a torque of 45-54 N·m (33-39 ft.-lbs.) on all other models. Install idler gear (1) with single timing mark aligned with two marked teeth on crankshaft balancer drive gear (5) and double timing marks aligned with single mark on lower balancer gear as shown in Fig. Y5-26. Install oil splash shield and crankcase cover.

CAMSHAFT, TAPPETS AND BEARINGS

The camshaft rides directly in unbushed bore at flywheel side of the cylinder block and in a ball bearing at crankcase cover side of block. Fuel injection pump cam (2—Fig. Y5-26) and engine hand crank starting gear (4) are mounted on crankcase cover end of camshaft. Camshaft can be removed after removing crankcase cover, oil splash shield, rocker arm assembly and the push rods. Remove nut, injection pump cam and starter gear. Turn camshaft at least one revolution to push cam followers (tappets) forward. Take care when removing camshaft from cylinder block as exhaust cam lobe may strike rear end of cylinder liner.

With camshaft removed, it is possible to remove cam followers out through ball bearing bore of cylinder block. In some instances, it may be necessary to remove crankcase rear cover, and on models so equipped, remove balancer shafts in order to remove cam followers. Be careful to keep intake and exhaust cam followers identified so they may be reinstalled in same position. Renew cam followers if renewing camshaft. If wear on cam lobe face of cam follower is excessive, or face is scored or damaged, renew cam follower and consider renewing camshaft.

Cam follower barrel standard diameter is 12.984-12.996 mm (0.5112-0.5116 inch) on Models TF140 and TF160, and 9.980-9.995 mm (0.3929-0.0.3935 inch) on all other models. Cam follower wear

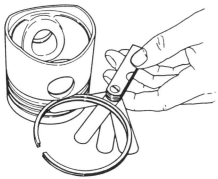

Fig. Y5-25—Measure ring side clearance in piston ring grooves using new rings and a feeler gage.

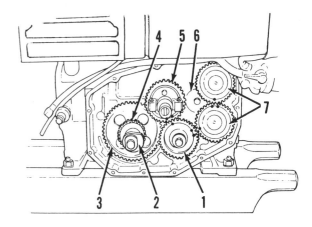

Fig. Y5-26—View showing timing mark alignment of idler gear (1), crankshaft balancer drive gear (5) and balancer shaft gears (7). Balancer shafts can be removed after unbolting retainer (6).

Illustrations Courtesy Yanmar Diesel Engine, Inc.

Yanmar

limit is 12.95 mm (0.510 inch) on Models TF140 and TF160, and 9.95 mm (0.392 inch) on all other models.

Check camshaft gear for damaged or worn teeth and renew gear if necessary. Check cam lobes and flywheel end bearing journal for scoring or excessive wear. Renew ball bearing at crankcase cover end of shaft if bearing is loose or rough.

Flywheel end bearing journal standard diameter is 19.939-19.960 mm (0.7850-0.7858 inch) on Model TF50, 24.939-24.960 mm (0.9819-0.9827 inch) on Models TF140 and TF160, and 21.939-21.960 mm (0.8637-0.8646 inch) on all other models. Desired journal to cylinder block bearing bore clearance on all models is 0.040-0.082 mm (0.0016-0.0032 inch) with maximum allowable clearance of 0.15 mm (0.006 inch).

Intake and exhaust cam lobe height is 34.3 mm (1.3504 inches) on Model TF50, 37.25 mm (1.4665 inches) on Models TF60 and TF70, 39.50 mm (1.5551 inches) on Models TF80, TF90, TF110 and TF120, and 44.8 mm (1.7638 inches) on Models TF140 and TF160. Renew cam if lobes are rough or if worn to 0.30 mm (0.012 inch) less than standard height.

Press new bearing onto camshaft and install cam gear with timing mark side out. Install starter gear and fuel injection pump cam and loosely install retaining nut. Lubricate cam followers and install them in cylinder block. Lubricate camshaft and ball bearing and carefully install camshaft in block; the exhaust lobe should be turned so that it will not strike cylinder sleeve. Turn camshaft so that single marked tooth of cam gear is between the two marked teeth of crankshaft gear. Tighten camshaft nut to a torque of 89-107 N·m (66-79 ft.-lbs.).

FLYWHEEL

Remove flywheel nut with impact wrench and socket, or with special hex wrench as shown in Fig. Y5-27. The flywheel should be removed using a puller seated against end of crankshaft and with puller bolts threaded into flywheel. Striking end of crankshaft to remove flywheel is not recommended.

On models with radiator cooling, the fan drive pulley is bolted to inside of flywheel. The starter ring gear on models with electric starting is mounted on inner face of flywheel. When installing flywheel, tighten retaining nut to a torque of 167-196 N·m (133-144 ft.-lbs.) on Model TF50; to a torque of 245-284 N·m (180-210 ft.-lbs.) on Models TF60, TF70, TF80 and TF90; and to a torque of 295-340 N·m (217-253 ft.-lbs.) on Models TF140 and TF160.

CRANKSHAFT AND MAIN BEARINGS

The crankshaft is supported in a ball bearing (2—Fig. Y5-28) at flywheel end and in a roller bearing (4) at crankcase cover end. Roller bearing should be renewed if crankshaft end play exceeds 0.45 mm (0.018 inch) on Model TF50, 0.21 mm (0.008 inch) on Models TF60 and TF70, 1.09 mm (0.043 inch) on Models TF80 and TF90, 1.24 mm (0.049 inch) on Models TF110 and TF120, and 1.61 mm (0.063 inch) on Models TF140 and TF160.

Crankshaft can be removed after removing crankcase cover, flywheel, cylinder head, connecting rod and piston unit, governor unit and spacer or balancer shaft drive gear and idler gear. Unbolt bearing housing (1) from flywheel side of block and withdraw crankshaft and main bearing retainer from block. Remove bolts from bearing retainer plate at inside of bearing housing and pull housing and crankshaft seal unit from crankshaft main bearing.

Renew crankshaft timing gear (5) if teeth are worn or damaged. Check main bearings for roughness or excessive wear and renew if such defects are noted. Check connecting rod journal for scoring or excessive wear and regrind journal for undersize bearing or renew crankshaft as necessary.

Crankpin standard diameter specifications are as follows:

Model TF50
Crankpin OD . . . . . . 35.965-35.982 mm
  (1.4159-1.4166 in.)
Wear limit . . . . 35.89 mm
  (1.413 in.)

Models TF60, TF70
Crankpin OD . . . . . . 42.956-42.972 mm
  (1.6912-1.6918 in.)
Wear limit . . . . 42.88 mm
  (1.688 in.)

Models TF80, TF90
Crankpin OD . . . . . . 44.956-44.972 mm
  (1.7699-1.7706 in.)
Wear limit . . . . 44.88 mm
  (1.767 in.)

Models TF110, TF120
Crankpin OD . . . . . . 47.952-47.973 mm
  (1.8879-1.8887 in.)
Wear limit . . . . 47.88 mm
  (1.885 in.)

Models TF140, TF160
Crankpin OD . . . . . . 53.953-53.978 mm
  (2.1241-2.1251 in.)
Wear limit . . . . 53.953 mm
  (2.1241 in.)

Desired connecting rod bearing to crankpin oil clearance is 0.018-0.077 mm (0.0007-0.0030 inch) for Model TF50, 0.022-0.092 mm (0.0009-0.0036 inch) on Models TF140 and TF160, and 0.028-0.086 mm (0.0011-0.0034 inch) on all other models. Maximum allowable crankpin bearing clearance is 0.1 mm (0.004 inch) on all models.

If removed, install bearing retaining plate and ball bearing on flywheel end of crankshaft. Install new oil seal in bearing housing with lip to inside, lubricate seal and bearing and install housing on crankshaft. Tighten bearing retainer plate bolts securely. Position roller bearing race against shoulder in crankcase with open end of race to inside. Install roller bearing and the timing gear on crankshaft. Lubricate ball bearing, place new gasket on bearing housing, then install the crankshaft, bearings and timing gear assembly in cylinder block, making sure that single marked tooth of camshaft gear is between the two marked teeth on crankshaft gear. Tighten bearing housing bolts to a torque of 23-28 N·m (17-20 ft.-lbs.).

Install spacer or balancer shaft drive gear and the governor flyweight assembly on crankshaft. On models so equipped, install balancer shaft idler gear with timing marks aligned with crankshaft balancer drive gear and lower balancer shaft gear as shown in Fig. Y5-26. Complete reassembly of engine

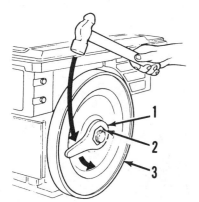

Fig. Y5-27—Removing engine flywheel nut (N) with special wrench (W) and hammer. An air impact wrench could also be used.

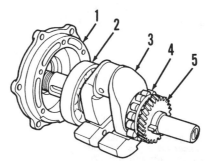

Fig. Y5-28—After removing crankshaft with flywheel end bearing support (1), remove retainer (not shown) and pull crankshaft with bearing (2) from support.

1. Bearing support
2. Ball bearing
3. Crankshaft
4. Roller bearing
5. Crankshaft gear

SERVICE MANUAL

Yanmar

by reversing disassembly procedure. Bleed the fuel system as previously outlined.

CRANKSHAFT OIL SEAL

The crankshaft oil seal at flywheel side of engine can be renewed after removing engine flywheel. A seal is not used at crankcase cover end of crankshaft as shaft does not extend through cover. Check to be sure crankshaft end play is not excessive. Carefully pull seal from main bearing housing and inspect sealing surface on crankshaft. Lubricate seal lip and install new seal with lip to inside.

ELECTRICAL SYSTEM

ALTERNATOR

A belt driven alternator is available on some models. Minimum charging current should be 6 amperes and 14 to 15 volts at 7000 alternator rpm.

STARTING MOTOR

The optional starting motor no-load speed should be 4000 rpm at 12 volts and maximum no-load amperage should be 105 amps. Brush length new is 12 mm (0.472 inch). Commutator standard diameter is 30 mm (1.181 inches) with wear limit of 29 mm (1.142 inches). Pinion gear projection with magnetic switch engaged should be 31.5-34.5 mm (1.240-1.358 inches).

Yanmar

SMALL DIESEL

YANMAR

Model	No. Cyls.	Bore	Stroke	Displ.
4JHE	4	78 mm (3.07 in.)	86 mm (3.39 in.)	1644 cc (100.3 cu.in.)
4JH-TE	4	78 mm (3.07 in.)	86 mm (3.39 in.)	1644 cc (100.3 cu.in.)
4JH-HTE	4	78 mm (3.07 in.)	86 mm (3.39 in.)	1644 cc (100.3 cu.in.)
4JH-DTE	4	78 mm (3.07 in.)	86 mm (3.39 in.)	1644 cc (100.3 cu.in.)

All Series JH engines are four-stroke, vertical cylinder, in-line, water cooled diesel engines for marine use and are equipped with marine gear box. Engine lubricant and fresh water coolant circulated in the engine are cooled by sea water heat exchangers. Model 4JHE is naturally aspirated; Model 4JH2-TE is turbocharged; and Models 4JH2-HTE and 4JH2-DTE are turbocharged and intercooled (intake air is cooled by sea water). Camshaft is located in cylinder block and valves in head are operated through push rods and rocker arms. Engine rotation is clockwise as viewed from front (crankshaft pulley) end of engine. Engine serial number is stamped on cylinder block below the starting motor (exhaust manifold side of engine).

MAINTENANCE

LUBRICATION

Engine crankcase capacity is 6.5 liters (6.86 quarts). Recommended lubricant is API classification CC or CD, SAE 10W-30 motor oil in ambient temperatures of −25°C to 30° C (−13°F to 74° F), or 20W-40 motor oil in temperatures from −10°C to 40° C (14° to 104° F). If diesel fuel has high sulfur content, API classification CD oil only should be used.

FUEL SYSTEM

Refer to Fig. Y6-1 for schematic view of early fuel line installation and to Fig. Y6-2 for late models with automatic air bleed for fuel filter.

FUEL FILTER. The spin-on cartridge type fuel filter is located in the fuel line between the fuel feed pump (mounted on side of fuel injection pump) and the fuel injection pump inlet. Filter element should be renewed after each 300 hours operation, or more often if dirt or water is appearing to block filter.

BLEED FUEL SYSTEM. To bleed air from fuel system of early production models, loosen bleed screw on top of filter housing, open fuel shut-off valve and operate hand primer lever on fuel lift pump until fuel flowing from open bleed screw is free of air and tighten screw. Loosen fuel return line from injection pump and continue to operate primer lever until fuel flows from open line. Tighten or reconnect line and crank engine. On late models, operate primer lever until you can hear fuel flowing to tank through return line. If engine will not start, loosen the fuel injection line fittings at injectors and crank engine until fuel is flowing from all openings. Tighten the fittings and start engine.

INJECTION PUMP TIMING

Timing of fuel injection pump is accomplished by installing pump with timing gear marks aligned, and by loosening pump mounting bolts and rotating pump in slotted holes of mounting flange. Injection should occur at 8-10 degrees BTDC on Model 4JH-E serial number 00574 and prior. On Model 4JH-E serial number 00575 and up, and for all other models, injection timing is 11-13 degrees BTDC.

To check injection pump timing, thoroughly clean injector, injector line and pump and remove fuel injection line to number 1 cylinder (cylinder at timing gear end of engine). Unscrew the delivery valve holder, remove delivery valve spring and reinstall holder. Place speed control linkage in "Run" position. While manually actuating fuel feed pump, turn crankshaft slowly until number 1 piston is starting up on compression stroke and observe open fuel delivery valve (fuel should be flowing from delivery valve). When fuel stops flowing, stop turning crankshaft and check injection timing mark (I—Fig. Y6-3) on engine flywheel. If timing mark is not aligned with match mark (M) on flywheel housing, repeat check several

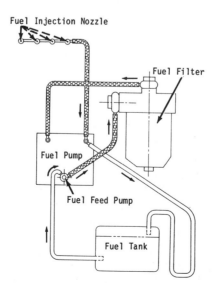

Fig. Y6-1—Schematic diagram of early fuel system. Open air bleed screw on top of filter and actuate primer lever on fuel feed pump to bleed air from system.

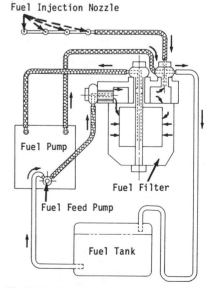

Fig. Y6-2—Schematic diagram of fuel system on later model engines. Air is bled from system automatically with operation of hand primer lever without opening air bleed screw.

SERVICE MANUAL

Yanmar

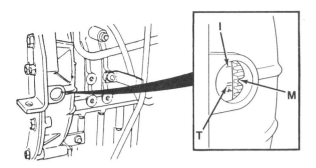

Fig. Y6-3—Timing marks on flywheel are visible through opening in flywheel housing. Inset view shows cylinders 1-4 TDC mark (T) aligned with match mark (M) in flywheel opening. The other mark (I) is injection timing mark.

times to be sure you are observing fuel flow correctly.

On later production engines with automatic timing advance, high speed advance timing can be checked with diesel timing light. Refer to the following table for timing advance specifications from static timing. (Add advanced angle to static timing.)

Model	Ser. No. Range	Advanced Angle
4JHE	00101-00574	7 degrees
4JHE	00575-01000	5.5 degrees
4JHE	01001 and up	4 degrees
4JH-TE	Below 11001	3.5 degrees
4JH-TE	11001 and up	2.5 degrees
4JH-HTE	All	2.5 degrees
4JH-DTE	All	2.5 degrees

To adjust injection timing, loosen pump mounting bolts and move top of pump in toward engine to retard timing or out to advance timing. Tighten pump mounting bolts and recheck timing. When injection pump timing is within specifications, install fuel delivery valve spring and tighten delivery valve holder to a torque of 35-39 N·m (26-28 ft.-lbs.). Install fuel injection line; bleed air from line before tightening fitting at injector.

GOVERNOR

The governor is located within the fuel injection pump camshaft housing on all models. Refer to Fig. Y6-4 for maximum governed speed and idle speed adjustment screw location. Minimum low idle speed should be 650 rpm; maximum no-load governed speed is 3900 rpm.

Adjustment of the high and low speed screws are adjusted at the factory and the setting of the high speed screw is sealed with a wire and lead seal. If necessary to readjust maximum no-load speed, be sure to install wire and new lead seal.

REPAIRS

TIGHTENING TORQUES

Refer to the following special tightening torques when servicing Series 4JH engines:

Cylinder head bolts .... 74-83 N·m
(55-61 ft.-lbs.)
Connecting rod bolts ...... 45-49 N·m
(33-36 ft.-lbs.)
Main bearing bolts .... 94-102 N·m
(84-90 ft.-lbs.)
Flywheel bolts .... 69-78 N·m
(51-57 ft.-lbs.)
Crankshaft pulley bolt ... 113-122 N·m
(84-90 ft.-lbs.)
Rocker arm
  support bolts.......... 24-27 N·m
(18-20 ft.-lbs.)
Engine oil pickup tube ...... 25 N·m
(19 ft.-lbs.)
Fresh water pump .... 7-10 N·m
(5-8 ft.-lbs.)
Fuel injection pump
  gear nut .... 59-68 N·m
(44-50 ft.-lbs.)
Pump timing advance
  plug .... 8-11 N·m
(2-8 ft.-lbs.)
Injector clamp,
  single-bolt .... 20-29 N·m
(15-21 ft.-lbs.)
Injector clamp, dual-bolt .... 4-5 N·m
(3-4 ft.-lbs.)

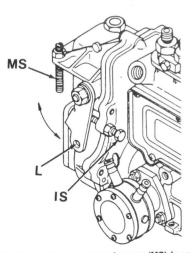

Fig. Y6-4—Maximum speed screw (MS) is set at factory and adjustment secured with safety wire and lead seal. Idle speed screw (IS) adjustment is not sealed. Governor control lever is (L).

WATER PUMP AND COOLING SYSTEM

Sea water is pumped to the air cooler, lube oil cooler and then to the heat exchanger where it cools the fresh water being circulated in the engine. Then, it is discharged by engine exhaust via the mixing elbow. Fresh water is circulated from the fresh water tank to cylinder block, turbocharger and cylinder head.

NOTE: Air cooler (intercooler), marine gearbox lube cooler and water cooled turbocharger are on Models 4JH-HTE and 4JH-DTE only.

SEA WATER PUMP. The sea water pump (1—Fig. Y6-5) is mounted at the back side of the engine front plate (3) and is driven from the engine camshaft gear. Front pump bearing (21) rides in bearing support (7) in the timing gear cover and rear bearing (18) rides in the pump housing. The rubber impeller (9) and wear plate (10) can be renewed after removing pump cover (8) and sealing "O" ring (13). When installing impeller, coat inside of pump housing with grease and position vanes of impeller as shown in Fig. Y6-6.

To remove pump, remove the sea water hoses, then unbolt and remove pump from rear face of engine front plate. Remove pump cover and take out the "O" ring, impeller and wear plate. Remove snap ring (11) from pump shaft. Working through slotted hole in drive gear (20—Fig. Y6-5), remove snap ring (19) from pump housing. Lightly tap impeller end of pump shaft (17) to remove the shaft, bearings and drive gear as an assembly. Tap shaft seal (12) out impeller end of pump housing.

Inspect all parts for wear, corrosion or other damage and renew as necessary. Refer to the following specifications:

Impeller width,
  standard .... 31.6-31.8 mm
(1.244-1.252 in.)
  Wear limit .... 31.3 mm
(1.232 in.)
Wear plate thickness .... 2 mm
(0.079 in.)
  Plate wear limit .... 1.8 mm
(0.071 in.)

Rotor end clearance at assembly should be 0-0.3 mm (0-0.012 inch); maximum allowable end clearance is 0.8 mm (0.031 inch). To reassemble pump, reverse disassembly procedure. Apply grease to inside of rotor housing, shaft splines, wear plate and to inside of cover plate. Be sure that impeller vanes are positioned as shown in Fig. Y6-6.

Yanmar

SMALL DIESEL

FRESH WATER PUMP. To remove fresh water pump (see Fig. Y6-8), drain the engine coolant. Loosen alternator mounting bolts and remove water pump belt. Unbolt and remove the pump assembly from front of cylinder head. The pump is serviced as an assembly only and must be renewed if leaking, if shaft bearing (9) is worn, rough or noisy, or if examination reveals internal damage to impeller (12). Refer to THERMOSTAT paragraph for information on cooling system thermostat located in top of water pump casting. Install pump with new gasket (16) and "O" ring (15). Tighten pump retaining bolts to a torque of 7-10 N·m (5-8 ft.-lbs.).

THERMOSTAT

The thermostat (3—Fig. Y6-8) is mounted in the top of the fresh water cooling pump. To remove thermostat, drain cooling system and unbolt thermostat cover (1) from top of water pump. Thermostat should start to open at 75°-78° C (167°-174° F) and be fully open at 90° C (194° F). Valve lift height with thermostat fully open should be 8 mm (0.315 inch). When installing thermostat, be sure all gasket surfaces are clean and use new gasket (2) and sealing ring (4). Renew thermostat after each year or 2000 hours of operation, whichever comes first.

PRESSURE CAP

The closed system pressure relief cap (5—Fig. Y6-9) is used as the filler cap for the fresh water supply in heat exchanger and engine. Coolant overflow is directed to a overflow tank and retrieved when the engine cools off via the overflow tube (12). Remove the cap only when fresh water system has cooled down. Cap should be tested with pressure system tester and should hold 74-103 kPa (10.7-14.9 psi) for a period of six seconds or more.

HEAT EXCHANGER

Refer to Fig. Y6-9 for exploded view of heat exchanger for cooling of engine coolant by sea water. The complete heat exchanger and exhaust manifold unit can be tested by plugging the water hose connectors, submersing the unit in water and applying compressed air at 195 kPa (28.5 psi) to the fresh water overflow outlet. There should be no air bubbles from submerged tank. Also, the heat exchanger may be tested by installing a pressure cap tester in place of the filler cap; pump pressure up to 147 kPa (21.3 psi). Pressure should hold if there are no leaks.

Heat exchanger cooling tube (3) can be removed from heat exchanger/exhaust manifold after removing the two end caps (1 and 10). Inspect cooling tube for corrosion or loose tubes and renew or repair as necessary. Clean mating surfaces thoroughly and install new "O" rings (2 and 9) when reassembling heat exchanger.

VALVE ADJUSTMENT

Valve clearance should be adjusted to 0.2 mm (0.008 inch) on all models with

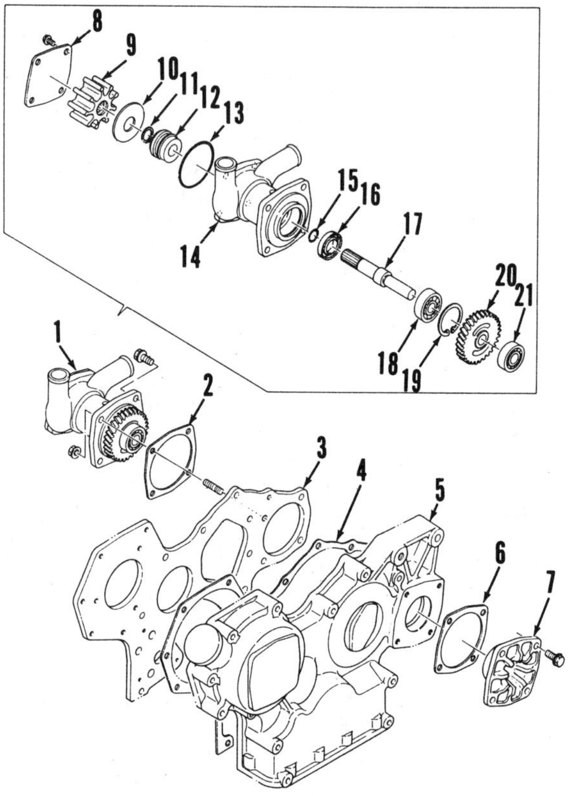

Fig. Y6-5—Sea water cooling pump (1) is mounted on back side of engine front plate (3) and is driven from camshaft gear.

1. Sea water pump
2. Gasket
3. Engine front plate
4. Gasket
5. Timing gear cover
6. Gasket
7. Bearing support
8. Cover plate
9. Rubber impeller
10. Wear plate
11. Snap ring
12. Seal assy.
13. "O" ring
14. Pump body
15. "O" ring
16. Oil seal
17. Pump shaft
18. Rear bearing
19. Snap ring
20. Drive gear
21. Front bearing

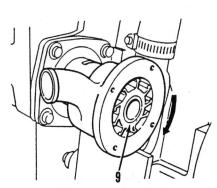

Fig. Y6-6—Salt water pump impeller (9) should be installed with blades curved as shown.

Fig. Y6-8—Fresh water circulating pump is mounted on front of cylinder head. A spacer (not shown) may be used between pulley (7) and flange (8) on some engines.

1. Water outlet
2. Gasket
3. Thermostat
4. Seal ring
5. Temperature sender
6. Temperature switch
7. Water pump pulley
8. Flange
9. Shaft & bearing assy.
10. Pump housing
11. Seal assy.
12. Impeller
13. Pump plate
14. Pipe flange
15. "O" ring
16. Gasket

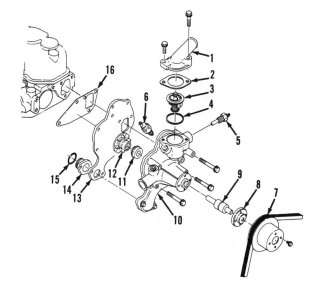

484

Illustrations Courtesy Yanmar Diesel Engine, Inc.

SERVICE MANUAL

Yanmar

engine cold. Adjustment is made by loosening locknut and turning adjusting screw (9—Fig. Y6-10) at push rod end of rocker arm. To adjust valves, turn crankshaft in normal direction of rotation so that number 1 piston is at top dead center on compression stroke, then adjust intake and exhaust valves on number 1 cylinder. Continue to turn crankshaft until next piston in firing order (1-3-4-2) is at top dead center on compression stroke and adjust valves for that cylinder in same manner. Repeat the procedure for each cylinder in order.

ROCKER ARMS AND PUSH RODS

Rocker arm shaft assembly and push rods can be removed after removing rocker arm cover. Be careful not to lose valve lash caps (13—Fig. Y6-10). Rocker arms (8) for all cylinders pivot on full length shaft (5) which is held by support brackets attached to cylinder head by cap screws.

Intake and exhaust rocker arms and push rods are alike. Check push rods by rolling on flat plate and renew any showing any curvature or if ends are galled, worn or chipped. Push rod diameter is 8 mm (0.315 inch) and length is 178.25-178.75 mm (7.018-7.037 inches).

Rocker shaft diameter should be 15.966-15.984 mm (0.6286-0.6293 inch) with wear limit of 15.955 mm (0.628 inch). Rocker arm bushing inside diameter should be 16.00-16.02 mm (0.630-0.631 inch) with wear limit of 16.09 mm (0.633 inch). Clearance between bushing and shaft should be 0.016-0.052 mm (0.0006-0.002 inch) and should not exceed 0.135 mm (0.0053 inch).

Loosen rocker arm adjustment screw locknuts and back screws out before installing push rods, rocker arm supports and rocker arms. Be sure all valve lash caps (13—Fig. Y6-10) are in place. Tighten rocker arm support bolts to a torque of 23-28 N·m (17-21 ft.-lbs.). Adjust valves as previously outlined and install rocker arm cover.

CYLINDER HEAD

To remove cylinder head, drain both fresh water and sea water from engine and proceed as follows: Remove muffler, air cleaner and instrument panel. On turbocharged models, disconnect piping and remove turbocharger. Remove intake manifold and exhaust manifold (heat exchanger), rocker arm cover, rocker arm assembly and the push rods. Remove fuel injector pressure and return lines and remove fuel injectors. Disconnect coolant hose from water outlet housing on fresh water pump, or unbolt and remove water pump from cylinder head. Remove cylinder head bolts and lift cylinder head from engine.

Before reinstalling cylinder head, thoroughly clean head and block gasket surfaces and inspect head for cracks, distortion or other defect. If alignment dowels (roll pins) were removed with head, pull pins from head and install in cylinder block. Check cylinder head gasket surface with straightedge and feeler gage. Head must be reconditioned or renewed if distortion (warp or twist) exceeds 0.15 mm (0.006 inch).

Place new gasket on block with trade mark or engine model number up and locate gasket on the two roll pins. Place head on block so that the two roll pins enter matching holes in head. Lubricate bolt threads and install the bolts, tightening in at least two steps in the sequence shown in Fig. Y6-11 until proper torque is reached. Specified cylinder head bolt final tightening torque is 74-83 N·m (55-61 ft.-lbs.).

VALVE SYSTEM

Valves seat in cylinder head on renewable valve seats (24 and 25—Fig. Y6-10).

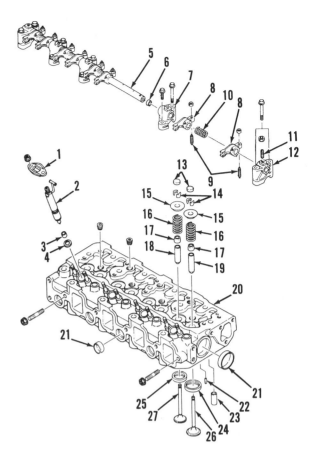

Fig. Y6-10—Exploded view of cylinder head assembly. Late type fuel injector clamp (1) with two retaining bolts is shown; early clamp had only one retaining bolt below the injector.

1. Injector clamp
2. Fuel injector
3. Nozzle heat shield
4. Injector seat
5. Rocker arm shaft
6. Rocker shaft plug
7. Rocker shaft support
8. Rocker arm
9. Adjustment screw
10. Rocker spacer spring
11. Set screw
12. Front rocker support
13. Valve lash caps
14. Valve split locks
15. Valve spring retainers
16. Valve springs
17. Valve stem seals
18. Exhaust valve guide
19. Intake valve guide
20. Cylinder head
21. Cup plugs
22. Dowel pin
23. Plug
24. Intake valve seat
25. Exhaust valve seat
26. Intake valve
27. Exhaust valve

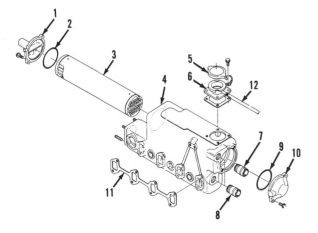

Fig. Y6-9—Exploded view of sea water/fresh water heat exchanger.

1. Sea water inlet/outlet
2. "O" ring
3. Cooling pipes
4. Housing
5. Radiator cap
6. Filler neck
7. Fresh water inlet nipple
8. Fresh water outlet nipple
9. "O" ring
10. End cover
11. Exhaust manifold gasket

Illustrations Courtesy Yanmar Diesel Engine, Inc.

Yanmar

SMALL DIESEL

Valve seat and face angle is 30 degrees for intake valves and 45 degrees for exhaust valves. Recondition exhaust seats using 45, 60 and 30 degree stones or cutters and recondition intake seats using 30, 60 and 15 degree stones or cutters. Valve seat width should be 1.28 mm (0.050 inch) on intake valves and 1.77 mm (0.070 inch) for exhaust valves. Maximum allowable seat width is 1.78 mm (0.070 inch) for intake valve seats and 2.27 mm (0.089 inch) for exhaust valve seats.

After reconditioning seats, check distance (D—Fig. Y6-12) that valve head is below cylinder head surface; standard valve head recess distance (sink) is 0.4-0.6 mm (0.016-0.024 inch) for both intake and exhaust valves. If valve recess distance exceeds 1.5 mm (0.060 inch), renew valve and/or valve seat.

If new valve seats are installed, grind valve seating surface to obtain full valve contact and proper seat width and check to be sure correct valve recess is maintained.

Standard valve stem diameter is 7.960-7.975 mm (0.3134-0.314 inch) for intake valves and 7.955-7.970 mm (0.3132-0.3138 inch) for exhaust valves. Renew valve if stem is worn to 0.13 mm (0.005 inch) below standard diameter.

Check valve guides and renew any guides with excessive wear. Valve guide standard inside diameter is 8.015-8.030 mm (0.3156-0.3161 inch); wear limit is 0.2 mm (0.008 inch) above standard diameter. When installing new guides, drive old guide out toward top of head using suitable piloted driver. Install new guide with piloted driver so that top of guide is 15 mm (0.590 inch) above top of cylinder head. Exhaust valve guides are identified by a groove around top end of the guide. After installing guide, use suitable reamer to obtain proper fit of valve stem to guide.

Intake and exhaust valve springs are alike on all models. Spring free length, new, is 44.4 mm (1.748 inches); renew any springs with free length of 43 mm (1.693 inches) or less. When spring is compressed to installed height of 40 mm (1.575 inches), spring pressure should be 12 kg (26.46 pounds); renew spring if pressure is 10 kg (22 pounds) or less. Be sure to install springs with closer spaced coils next to cylinder head.

TURBOCHARGER

Model 4JH-TE is equipped with an air cooled turbocharger and Models 4JH-HTE and 4JH-DTE are equipped with water cooled turbocharger. As turbocharger internal parts are easily damaged, it is recommended that service work on the turbocharger be done by an authorized repair station.

The turbocharger is lubricated by a pressure line from the engine oil gallery and oil is then returned to engine crankcase. If seal and bearing wear are such that oil is leaking into the air intake system, evidenced by heavy exhaust smoke, the turbocharger must be repaired or renewed.

Engines with malfunctioning fuel injectors or loss of engine oil past the oil control ring will cause heavy carbon deposits on the turbocharger blower wheel. Also, poor maintenance of engine lubrication system may result in turbocharger bearing and seal wear. If engine is under major repair, the turbocharger should be inspected and renewed or repaired as necessary.

REMOVE AND REINSTALL. Thoroughly clean turbocharger and surrounding area prior to removal. Drain fresh water cooling system on water cooled turbocharger and disconnect water lines. Remove air inlet and discharge hose. Disconnect oil supply line and oil return line and plug openings. Unbolt and remove turbocharger from engine.

To reinstall turbocharger, reverse removal procedure. Before reconnecting oil supply hose, fill turbocharger oil inlet with clean motor oil to ensure initial start-up lubrication. Before reconnecting oil return line to turbocharger, place speed control linkage in shut-off position and crank engine until oil flows from turbocharger oil return port. Then, connect oil return line and start engine. Operate engine at low speed for about one minute to allow oil to start circulating through turbocharger.

INJECTORS

REMOVE AND REINSTALL. Prior to removing injectors, thoroughly clean injectors, injector lines, fuel injection pump and surrounding area using suitable solvent and compressed air. Remove the fuel return line from top of injectors. Remove the fuel injection lines from injection pump and injectors. Remove the fuel injector retaining clamps (1—Fig. Y6-10) and pull injectors from head. Immediately cap or plug all fuel openings. Remove seals from bottom of injector bores in cylinder head.

When installing fuel injectors, first be sure that injector bores in cylinder head are clean and that old seal washers (4) are removed. Insert new seal washers in injector bores and install the injectors. Tighten the single bolt clamp nuts to a torque of 20-29 N·m (15-21 ft.-lbs.). On models with two injector clamp bolts, tighten the nuts evenly to a torque of 4-5 N·m (3-4 ft-lbs.). Bleed air from injector lines before tightening line fittings at injector end.

TESTING. A complete job of testing, cleaning and adjusting the fuel injector requires use of special test equipment. Use only clean approved testing oil in tester tank. Injector should be tested for opening pressure, seat leakage and spray pattern. Before connecting injector to test stand, operate tester lever until oil flows, then attach injector to tester line.

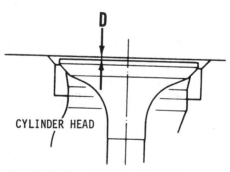

Fig. Y6-12—Top of valve head should be 0.4-0.6 mm (0.016-0.024 inch) below cylinder head surface as shown by dimension (D).

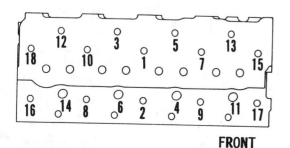

Fig. Y6-11—Cylinder head bolt tightening sequence for all models.

WARNING: Fuel emerges from injector with sufficient force to penetrate the skin. When testing injector, keep yourself clear of nozzle spray.

Close valve to tester gage and operate tester lever a few quick strokes to purge air from injector and to be sure nozzle valve is not stuck or plugged.

OPENING PRESSURE. Open the valve to tester gage and operate tester lever slowly while observing gage reading. Opening pressure should be 19130-

SERVICE MANUAL

Yanmar

20100 kPa (2773-2915 psi). Opening pressure is adjusted by changing thickness of shim stack (3—Fig. Y6-13). Adding shim thickness will increase pressure.

SPRAY PATTERN. Spray pattern should be equal from all four nozzle tip holes and should be a fine mist. Spray angle (A—Fig. Y6-14) is 150 degrees and each spray cone should be a 5-10 degree cone (B). Place a sheet of blotting paper 30 cm (about 12 inches) below injector and operate tester one stroke. The spray pattern should be a perfect circle.

SEAT LEAKAGE. Wipe nozzle tip dry with clean blotting paper, then slowly actuate tester lever to obtain gage pressure approximately 1960 kPa (285 psi) below opening pressure and hold this tester pressure for 10 seconds. If drop of oil appears at nozzle tip, or oil drips from nozzle, clean or renew nozzle.

OVERHAUL. Hard or sharp tools, emery cloth, grinding compounds or other than approved tools, solvents and lapping compounds, must never be used. Wipe all dirt and loose carbon from injector, then refer to Fig. Y6-13 and proceed as follows:

Secure injector nozzle in holding fixture or soft jawed vise and loosen nozzle nut (10). Unscrew nut and remove nozzle (9), valve (8), stop plate (6), spring seat (5), spring (4) and shims (3) from injector. Be careful not to lose pins (7). Keep shims (3) together. Place all parts in clean diesel fuel or calibrating oil as they are removed, and take care not to mix parts with those from other injector assemblies.

Clean exterior surfaces with brass wire brush to loosen carbon deposits; soak parts in approved carbon solvent if necessary. Rinse parts in clean diesel fuel or calibrating oil after cleaning to neutralize carbon solvent.

Clean nozzle spray hole chamber from inside using pointed hardwood stick. Scrape carbon from nozzle pressure chamber using hooked scraper. Clean valve seat using brass scraper, then polish seat using wood polishing stick and mutton tallow. Use a 0.2 mm (0.008 inch) diameter cleaning wire in a pin vise to clean the four spray holes.

Reclean all parts by rinsing thoroughly in clean diesel fuel or calibrating oil and assemble injector while immersed in fuel or oil. Be sure shim stack is intact, then install shims (3), spring (4), spring seat (5), stop plate (6), nozzle valve (8), nozzle body (9) and the nozzle nut (10), aligning nozzle body on dowel pins (7). Tighten nozzle retaining nut (10) to a torque of 40-44 N·m (29-32 ft.-lbs.). Retest assembled injector. Do not overtighten nozzle nut in an attempt to stop any leakage. Disassemble and clean injector if leak is noted.

FUEL FEED PUMP

A fuel feed pump is mounted on the fuel injection pump camshaft housing. Refer to Fig. Y6-16. A hand lever (8) is incorporated in the pump for manual priming of fuel system. In some instances, it may be necessary to turn crankshaft to move fuel pump cam away from pump before hand lever will operate pump.

Thoroughly clean the fuel feed pump and injection pump prior to removing feed pump. To disassemble, first scribe matching marks across cover (13), check valve housing (11) and pump body (1). Remove retaining screws (14) and separate cover and check valve housing from pump body. Remove diaphragm assembly (10) from pump body.

Inspect all parts for wear or damage and renew as necessary. To reassemble, reverse the disassembly procedure using new seal rings and gaskets. Be sure that match marks on cover, check valve housing and pump body made prior to disassembly are aligned. Tighten cover retaining screws evenly.

INJECTION PUMP

To remove injection pump, it will be necessary to remove timing gear cover, then unbolt and remove pump drive gear (7—Fig. Y6-17) on early models or timing advance mechanism (2) on later models fitted with automatic timing advance. Threaded holes are provided in drive gear or advance mechanism for installation of gear puller bolts. Thoroughly clean pump, injectors and lines using suitable solvent and compressed air. Disconnect throttle linkage from governor arm on pump. Disconnect fuel supply

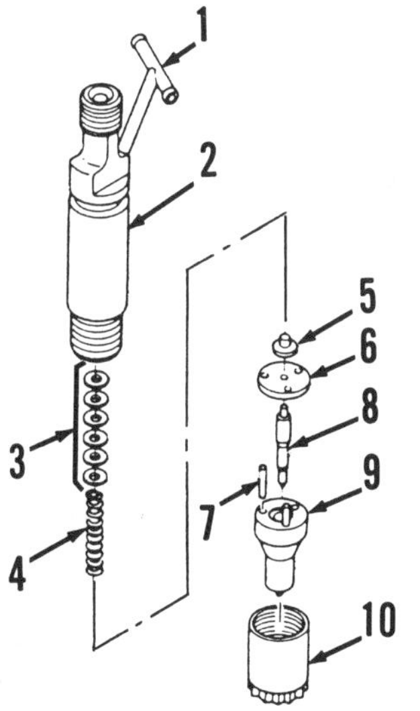

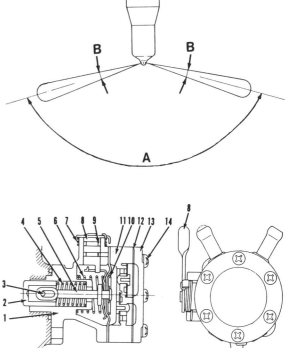

Fig. Y6-14—Angle between the nozzle spray cones should be 150 degrees (A). All four sprays should be uniform and have a cone angle of 5 to 10 degrees (B).

Fig. Y6-16—Fuel feed pump is mounted on side of fuel injection pump and is operated by cam on pump camshaft. Hand lever (8) is used to prime fuel system.

1. Pump body
2. Cam follower
3. Pin
4. Cam follower spring
5. Diaphragm rod
6. Diaphragm spring
7. Lever return spring
8. Hand lever
9. "O" ring
10. Diaphragm assy.
11. Check valve plate assy.
12. Gasket
13. Cover
14. Cover screws

Fig. Y6-13—Exploded view of spray type fuel injector typical of all model engines.

1. Fuel return pipe
2. Injector body
3. Shim stack
4. Spring
5. Spring seat
6. Stop plate
7. Dowel pins
8. Nozzle valve
9. Nozzle body
10. Nozzle retaining nut

Illustrations Courtesy Yanmar Diesel Engine, Inc.

Yanmar — SMALL DIESEL

and return lines at pump and remove the pump to injector high pressure lines. Remove engine lubricating oil line from pump camshaft housing. If mark is not present, scribe an alignment mark on fuel injection pump mounting flange and engine front plate. Unbolt and remove pump from front plate.

To install pump, proceed as follows: Be sure pump mounting surfaces are clean and install pump on engine front plate using a new "O" ring. Align timing marks scribed on pump flange and engine front plate and tighten pump mounting bolts. Turn crankshaft and injection pump camshaft so that pump drive gear is installed with timing marks aligned as shown in Fig. Y6-21. Tighten pump gear retaining nut to a torque of 59-68 N·m (44-50 ft.-lbs.). On models with timing advance, pack grease in area surrounding timing gear nut before installing cap plug (6—Fig. Y6-17). Complete pump installation by reversing removal procedure, leaving fuel injection line fittings loose at injectors until air is bled from lines. Check pump camshaft housing for adequate lubricant; fill to oil line opening with clean diesel engine motor oil. If pump timing marks were not scribed on pump flange and front plate before pump was unbolted, or if a new pump is being installed, check fuel injection pump timing.

TIMING ADVANCE MECHANISM

Some engines are equipped with a timing advance mechanism (2—Fig. Y6-17) that uses centrifugal force to automatically advance injection timing according to engine rpm. Timing advance mechanism may be removed from injection pump camshaft after removing timing gear cover.

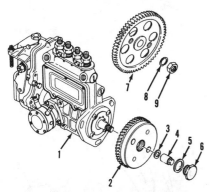

Fig. Y6-17—The fuel injection pump may be equipped with a solid drive gear (7) or a gear (2) with timing advance mechanism. Refer to Fig. Y6-18 for exploded view of the timing advance and gear unit.

1. Fuel injection pump
2. Gear & advance unit
3. Washer
4. Barrel nut
5. Seal
6. Cap plug
7. Drive gear
8. Flat washer
9. Nut

To disassemble timing unit, remove snap ring (1—Fig. Y6-18) and separate timer gear (3), weights (6) and weight holder (7). Be sure to retain shims (4) for use in reassembly.

Inspect parts for wear or damage and renew as necessary. If timer weights (6), springs (5) or gear (3) are renewed, timing advance angle must be reset as necessary using adjusting shims (4). End play of timer gear (3) on weight holder is adjusted by means of shims (2) and should be 0.02-0.10 mm (0.0008-0.0039 inch).

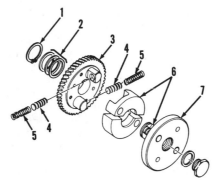

Fig. Y6-18—Exploded view of the fuel injection pump drive gear and timing advance mechanism used on some engines.

1. Snap ring
2. Spacer washers
3. Gear
4. Adjusting shims
5. Timing springs
6. Timing weights
7. Weight holder & hub

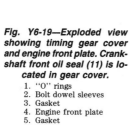

Fig. Y6-19—Exploded view showing timing gear cover and engine front plate. Crankshaft front oil seal (11) is located in gear cover.

1. "O" rings
2. Bolt dowel sleeves
3. Gasket
4. Engine front plate
5. Gasket
6. Oil filler cap
7. "O" ring
8. Timing gear cover
9. Tach drive opening cover
10. "O" ring
11. Crankshaft oil seal
12. Gasket
13. Cover plate

Install timer mechanism on injection pump camshaft, aligning timing marks as shown in Fig. Y6-21. Tighten timer retaining nut (4—Fig. Y6-17) to a torque of 59-68 N·m (44-50 ft.-lbs.). Pack area around the nut with grease, then install hex plug (6) with new "O" ring (5). Use a diesel timing light to check timing advance angle as outlined in INJECTION PUMP TIMING paragraph.

GOVERNOR

The flyweight type centrifugal governor is mounted in the rear end of the injection pump camshaft housing. It is recommended that disassembly should not be attempted outside of qualified diesel repair station. Refer to MAINTENANCE section for adjustment of low and high idle (no-load) speeds.

TIMING GEAR COVER

To remove timing gear cover (8—Fig. Y6-19), drain cooling system and proceed as follows: Remove alternator, fresh water pump assembly, sea water pump front bearing support and the crankshaft pulley. Remove hydraulic pump, if so equipped, from front of timing cover. Remove timing cover bolts from front end of oil pan, then unbolt and remove cover from front of engine.

Thoroughly clean timing cover and mounting gasket surfaces. Install new

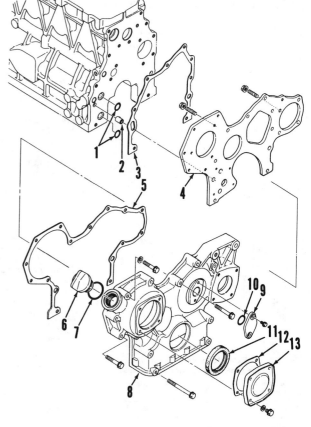

SERVICE MANUAL

Yanmar

crankshaft oil seal (11) with lip to inside and lubricate seal. Install timing cover on engine using new gasket and aligning holes in cover with the two hollow alignment dowel pins in front of engine. Lower ends of timing cover gasket may have to be trimmed even with oil pan surface. Clean crankshaft pulley hub and apply lubricant on seal contact surface. Install the crankshaft pulley and tighten retaining bolt to a torque of 113-122 N·m (84-90 ft.-lbs.). To complete reassembly of engine, reverse disassembly procedure.

TIMING GEARS

The timing gears can be inspected after removing timing gear cover. Desired backlash between mating timing gears is 0.04-0.12 mm (0.0016-0.0047 inch) and maximum allowable backlash is 0.2 mm (0.008 inch). Check gear teeth for excessive wear or tooth damage and renew as necessary. If necessary to renew gears, remove valve cover, rocker arm assembly and the push rods to remove valve spring pressure on camshaft.

To remove timing gear idler (1—Fig. Y6-20), remove the two bolts from idler shaft (3) and remove the shaft and idler gear. Idler shaft standard diameter is 45.950-45.975 mm (1.809-1.810 inches); renew if worn to 45.88 mm (1.8062 inches). Bushing (2) standard inside diameter is 46-46.025 mm (1.811-1.812 inches). Bushing may be renewed separately from idler gear if gear is otherwise serviceable. Desired oil clearance for idler gear bushing to shaft is 0.025-0.075 mm (0.001-0.003 inch). Renew shaft and/or bushing if clearance is 0.15 mm (0.006 inch) or more.

To remove crankshaft gear, use suitable pullers to remove gear from crankshaft. To remove camshaft gear, first remove camshaft from engine and press shaft from gear. Fuel injection pump gear can be pulled from shaft after removing retaining nut. On later production models with timing advance mechanism, remove cap plug (6—Fig. Y6-17), then unscrew retaining nut (4).

When installing timing gears, be sure all timing marks are aligned as shown in Fig. Y6-21. Heat crankshaft timing gear to 180°-200° C (356°-392° F).

OIL PAN

Remove the brackets that connect oil pan to flywheel housing. Removal of oil pan is otherwise conventional.

To install pan, proceed as follows: If new flywheel housing, engine front plate and/or timing cover gaskets were installed, trim ends of gaskets off flush with block surface. Apply sealer to joints of engine block to flywheel housing, front plate and timing cover. Install pan using a new gasket. Install the oil pan to flywheel housing brackets and tighten bracket bolts after oil pan retaining bolts are tight.

OIL PUMP AND RELIEF VALVE

The gerotor type oil pump (Fig. Y6-22) is mounted on front of engine timing cover plate, and gear on pump shaft is driven from crankshaft gear. On some models, the oil pump is fitted with a safety relief valve (V). The oil pump and drive gear is available as a complete assembly only. Renew pump assembly if excessive wear or other damage is noted.

To remove pump, first remove timing gear cover, then unbolt and remove pump from front of engine. The pump rear plate (7—Fig. Y6-22) can then be removed from pump body to inspect pump. Clearance between outer rotor (6) and pump body (3) should be 0.10-0.17 mm (0.004-0.007 inch) with wear limit of 0.25 mm (0.010 inch). Clearance between lobes of inner and outer rotors (4 and 6) should be 0.05-0.10 mm (0.002-0.004 inch) with maximum allowable clearance being 0.15 mm (0.006 inch). Rotor end clearance in pump body should be 0.03-0.09 mm (0.0012-0.0035 inch) and maximum end clearance is 0.13 mm (0.005 inch).

Check for looseness of drive shaft (2) in pump body and renew pump if shaft is loose or wobbly. Push oil pressure valve (V) piston in from rear side of pump and renew pump if piston does not return promptly.

The oil pressure regulating valve (4—Fig. Y6-23) is located in the oil filter

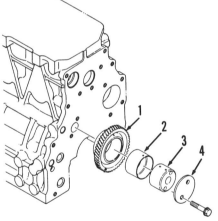

Fig. Y6-20—View showing mounting of idler gear (1) and shaft on front of crankcase. Shaft (3) and plate (4) may be integral part on some engines.

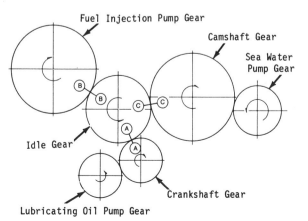

Fig. Y6-21—Schematic drawing of timing gears showing timing mark locations on crankshaft gear, idler gear, fuel injection pump drive gear and camshaft gear. No timing is required for oil pump and sea water pump drive gears.

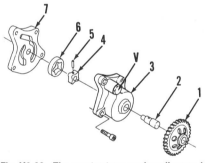

Fig. Y6-22—The gerotor type engine oil pump is mounted on front of engine front plate. A safety relief valve (V) is used in some pumps.

V. Safety relief valve
1. Drive gear
2. Pump shaft
3. Pump housing
4. Inner rotor
5. Drive pin
6. Outer rotor
7. End cover

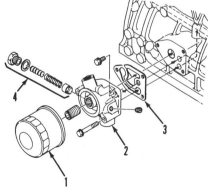

Fig. Y6-23—Engine oil pressure regulating valve (4) is located on oil filter bracket.

1. Oil filter
2. Oil filter bracket
3. Gasket
4. Oil pressure regulating valve

Illustrations Courtesy Yanmar Diesel Engine, Inc.

Yanmar SMALL DIESEL

mounting bracket (2) and regulates normal system oil pressure to 345-440 kPa (50-64 psi). A bypass valve within the oil filter (1) opens if pressure difference between filter inlet and outlet reaches 80-117 kPa (11.5-17 psi).

PISTON AND ROD UNITS

Piston and rod units are removed from above after removing oil pan, oil pump pickup tube and sleeve, piston cooling nozzles and the cylinder head. Keep rod bearing cap with matching connecting rod.

Note that combustion chamber (CC—Fig. Y6-24) in top of piston is offset to one side. When installing piston and rod units, be sure that offset is toward fuel injection side of engine, the trademark "YANMAR" on side of rod should be toward flywheel end of engine and match marks (FM) on rod and cap should be toward injection pump side of engine.

Lubricate cylinder wall, crankpin, piston and rings and rod bearing. Place ring end gaps at equal spacing at 120 degrees around piston. Install piston and rod unit using suitable ring compressor. Lubricate connecting rod bolt threads and tighten bolts to a torque of 45-49 N·m (33-36 ft.-lbs.). Install piston cooling nozzles and oil pickup tube. Complete reassembly by reversing disassembly procedures.

PISTONS, PINS, AND RINGS

Pistons are fitted with two compression rings and one oil control ring with a coil type expander under the oil ring. Pistons for turbocharged engines have valve recess in top; pistons for naturally aspirated engines, except for combustion chamber (CC—Fig. Y6-24) have flat top. Piston pin (6) is full floating and is retained by a snap ring (4) at each end of pin bore in piston. If necessary to remove piston from connecting rod, remove snap rings and push pin from piston and rod.

Inspect piston skirt for scoring, cracks and excessive wear. Measure piston skirt at right angle to pin bore and at a distance of 22 mm (0.866 inch) from bottom of skirt. Piston skirt diameter should be 77.91-77.94 mm (3.0673-3.0685 inches); renew piston if skirt is worn to diameter of 77.81 mm (3.063 inches) or less.

Piston pin to piston pin bore clearance should be zero to 0.022 mm (0-0.0009 inch). Maximum allowable piston pin clearance in piston is 0.045 mm (0.0018 inch). Be sure offset combustion chamber in top of piston and connecting rod with trade mark side are positioned as shown in Fig. Y6-24 before installing piston pin. Be sure pin retaining snap rings are securely installed at each end of pin bore in piston.

Piston ring end gap should be 0.25-0.40 mm (0.010-0.016 inch) for compression rings and 0.2-0.4 mm (0.008-0.016 inch) for oil control ring. Maximum allowable ring end gap is 1.5 mm (0.059 inch) for all rings.

Ring side clearance in piston grooves can be checked by inserting new ring in piston ring groove and measuring clearance between ring and land using a feeler gage. Top ring side clearance in piston groove should be 0.07-0.10 mm (0.0028-0.0039 inch). Side clearance for second compression ring should be 0.035-0.065 mm (0.0013-0.0025 inch). Oil control ring side clearance should be 0.030-0.060 mm (0.0011-0.0029 inch). Maximum allowable side clearance for all rings is 0.20 mm (0.008 inch). Renew piston if ring side clearance exceeds wear limit.

Top compression ring (Fig. Y6-25) is barrel face with top and bottom sides chrome plated. Second compression ring is tapered face. When installing piston rings, place oil ring expander in groove, then install oil ring on top of expander with ring end gap at opposite side of piston from where expander ends meet. Install compression rings with side having identification mark up and with end gaps at 120 degree spacing from each other and oil ring end gap.

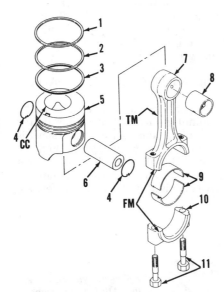

Fig. Y6-24—Exploded view of piston and connecting rod unit. Piston for naturally aspirated engine is shown. Pistons used with turbocharged engines have valve reliefs in piston crown.

CC. Combustion chamber
FM. Forging marks
TM. "YANMAR" trademark
1. Top compression ring
2. Second compression ring
3. Oil control ring
4. Snap rings
5. Piston
6. Piston pin
7. Connecting rod
8. Piston pin bushing
9. Crankpin bearings
10. Rod cap
11. Rod bolts

CYLINDER SLEEVES

The cylinder blocks of series JH engines are fitted with hand push fit dry cylinder sleeves (liners). Sleeves, pistons and rings are available for standard bore size only. Standard sleeve inside diameter is 78.00-78.03 mm (3.071-3.072 inches). Renew sleeve if worn to 78.12 mm (3.076 inches) or larger at maximum wear point.

Sleeves are removed from top of block after removing piston and connecting rod units. As they are installed with a hand push fit, removal should be relatively easy. Before installing new sleeves in cylinder block, be sure bores are cleaned of all rust and carbon deposits. Be sure counterbore at top of block is clean. Do not hone cylinder block bores as this may increase bore diameter and cause sleeve failure. Cylinder block bore (with sleeve removed) standard diameter is 82.00-82.03 mm (3.2283-3.2295 inches). Renew cylinder block if bore diameter exceeds 82.06 mm (3.2307 inches) or if out-of-round more than 0.02 mm (0.0008 inch).

Coat cylinder bore and outside of sleeve with oil and push sleeve into block by hand pressure only. Do not drive sleeve into place as this may distort the thin-wall sleeve.

After installation, gasket contact surface at top of sleeve should protrude a distance (D—Fig. Y6-26) of 0.03-0.09 mm (0.0012-0.0035 inch) from top of block. Rust or carbon not cleaned thoroughly from top of cylinder bore will cause excessive sleeve protrusion.

CONNECTING RODS AND BEARINGS

The connecting rod is fitted with a renewable bushing at piston pin end and a slip-in precision fit bearing at crankpin end. Crankpin standard diameter is 47.952-47.962 mm (1.8879-1.8883 inches). Crankpin bearing oil clearance should be 0.038-0.090 mm (0.0015-0.0035

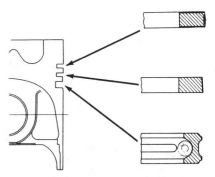

Fig. Y6-25—Cross-sectional view of piston and rings. Top compression ring is barrel face and chrome plated. Second compression ring is tapered face.

Illustrations Courtesy Yanmar Diesel Engine, Inc.

SERVICE MANUAL

Yanmar

inch) with wear limit of 0.16 mm (0.0063 inch). Piston pin standard diameter is 25.987-26.0 mm (1.0231-1.0236 inches). Piston pin bushing to pin oil clearance should be 0.025-0.051 mm (0.001-0.002 inch) with wear limit of 0.11 mm (0.0043 inch). When renewing piston pin bushing, be sure that oil hole in bushing is aligned with hole in connecting rod. New piston pin bushing must be honed or reamed to provide desired pin to bushing oil clearance.

Rod twist or out of parallel condition should be less than 0.05 mm (0.002 inch) per 100 mm (4 inches) length. Maximum allowable rod distortion is 0.07 mm (0.0028 inch) per 100 mm (4 inches). Rod side clearance on crankpin journal should be 0.20-0.40 mm (0.008-0.016 inch). Maximum allowable side clearance is 0.55 mm (0.022 inch); renew connecting rod and/or crankshaft if side clearance is excessive.

CAMSHAFT AND BEARINGS

Camshaft, gear and thrust plate assembly (see Fig. Y6-27) can be removed from front of block after removing timing gear cover, valve cover, rocker arm assembly and push rods. Use magnetic tools to lift and hold cam followers (7) away from camshaft or turn engine over so that followers move away from cam. Remove the two thrust plate retaining bolts through web holes in camshaft gear (1) and pull camshaft from block. If necessary to renew camshaft, thrust plate or gear, carefully press shaft from plate and gear.

Camshafts have different valve timing for different engines. Also, diameter (D—Fig. Y6-27) of gear mounting nose and inside diameter of gear have been changed; be sure if renewing cam or drive gear that correct parts are obtained.

Camshaft end play in block is controlled by the thrust plate (3). End play should be 0.05-0.25 mm (0.002-0.010 inch). Renew thrust plate if end play exceeds 0.40 mm (0.016 inch) on all models. Also check thrust surface wear on timing gear and camshaft and renew as necessary.

Intake cam lobe height on all models and exhaust cam lobe height of Model 4JHE should measure 38.66-38.74 mm (1.522-1.525 inches) with wear limit of 38.4 mm (1.512 inches). Exhaust cam lobe height should be 38.86-38.94 mm (1.530-1.533 inches) for all models except Model 4JHE with wear limit of 38.6 mm (1.519 inches).

The camshaft front journal rides in a renewable bearing (4) at front end of block and center and rear journals ride directly in unbushed bores of block. Oil clearance for camshaft journal at timing gear end should be 0.04-0.13 mm (0.0015-0.005 inch), clearance for intermediate journal should be 0.065-0.115 mm (0.0025-0.0045 inch) and clearance for journal at flywheel end should be 0.05-0.10 mm (0.002-0.004 inch). Oil clearance wear limit is 0.20 mm (0.008 inch) for all journals.

CAM FOLLOWERS

The mushroom type cam followers (valve tappets) (7—Fig. Y6-27) can be removed from bottom of block after removing camshaft and oil pan. Be sure to identify cam followers as they are removed so they can be installed in their original positions if reused.

Cam followers should be renewed if installing a new camshaft, or if bottom of follower shows cracks, pitting, excessive wear or if barrel of follower is excessively worn. Follower outside standard diameter is 11.975-11.990 mm (0.472-0.473 inch). Clearance in block should be 0.010-0.043 mm (0.0004-0.0017 inch).

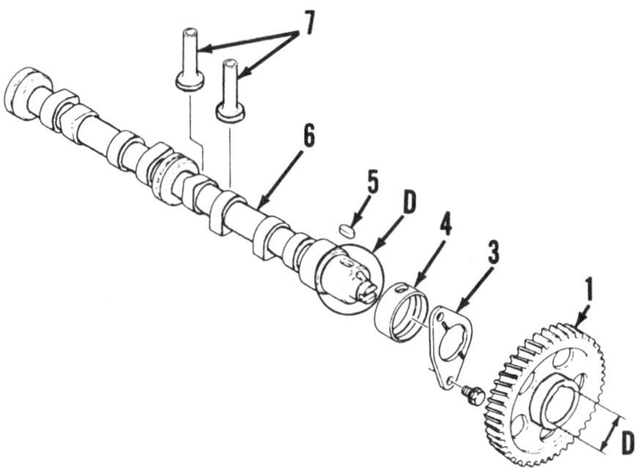

CRANKSHAFT AND MAIN BEARINGS

The crankshaft is supported in five main bearings (4—Fig. Y6-28). Crankshaft end play is controlled by split thrust washers (2 and 3) at each side of rear main bearing cap (1).

To remove crankshaft, remove piston and rod units, flywheel, flywheel housing, rear bearing seal retainer, timing gear cover, timing gears and engine front plate. Then be sure that position number is marked on each main bearing cap; if not, stamp numbers on caps. Unbolt and remove main bearing caps and lift crankshaft from cylinder block.

Check all main bearing journals, thrust surfaces at rear journal, and the crankpin journals for scoring or excessive wear. Crankpin standard diameter is 47.952-47.962 mm (1.8879-1.8883 inch). Crankpin bearing oil clearance should be 0.038-0.090 mm (0.0015-0.0035 inch) with wear limit of 0.16 mm (0.0063 inch).

Main journal standard diameter is 49.952-49.962 mm (1.9666-1.9670

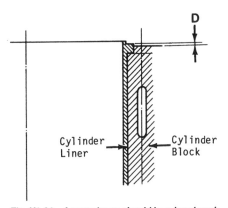

Fig. Y6-26—A new sleeve should be a hand push fit in a clean, lubricated cylinder bore. Gasket contact surface of sleeve should protrude a distance (D) of 0.03-0.09 mm (0.0012-0.0035 inch) above cylinder block.

Fig. Y6-27—Measure dimension (D) of camshaft and timing gear prior to ordering new parts. Camshaft retainer plate bolts are accessible through holes in gear web.

1. Camshaft gear
3. Retainer plate
4. Camshaft bearing
5. Camshaft gear key
6. Camshaft
7. Cam followers

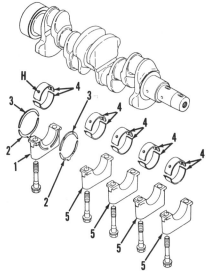

Fig. Y6-28—All five main bearing sets (4) are alike; bearing shells with oil holes (H) are installed in cylinder block.

H. Oil holes
1. Rear main bearing cap
2. Lower thrust washers
3. Upper thrust washers
4. Main bearing sets
5. Bearing caps

Yanmar

inches). Main journal bearing oil clearance should be 0.038-0.093 mm (0.0015-0.0037 inch). Maximum allowable main bearing oil clearance is 0.15 mm (0.006 inch). Crankshaft end play should be 0.09-0.271 mm (0.004-0.011 inch). Maximum allowable end play is 0.30 mm (0.012 inch) for all models.

NOTE: While crankshafts for all engines have same dimensions, a crankshaft for Models 4JHE and 4JH-TE should never be used in a Model 4JH-HTE or 4JH-DTE engine as a more durable crankshaft design is required for these models.

When installing crankshaft, be sure all bearing surfaces are clean and free of nicks or burrs. Install bearing shells with oil holes (H) in cylinder block and bearing shells without oil holes in the main bearing caps. Bearing sets for all journals are alike. Lubricate bearings and crankshaft main journals and set crankshaft in bearings. Lubricate upper thrust washer halves (3) (those without locating tab) and insert thrust washers between block and crankshaft with bearing side of thrust washers toward crankshaft shoulders. Using light grease, stick lower thrust washer halves (2) in rear main bearing cap (1) with tabs positioned in notches in cap, lubricate bearing and install cap. Lubricate remaining bearing shells and install caps in same positions as they were removed. Note that side of caps with embossed letters "FW" are placed toward flywheel.

Tighten the main bearing cap bolts to a torque of 94-102 N·m (69-76 ft.-lbs.). Loosen bolts on rear main bearing cap (1) and pry crankshaft back and forth to center thrust bearing surfaces, then retorque bolts. Check to be sure that crankshaft turns freely in bearings and crankshaft end play is correct before completing reassembly of engine.

CRANKSHAFT FRONT OIL SEAL

The crankshaft front oil seal is located in the timing gear cover and can be renewed after removing crankshaft pulley. Pry seal from timing cover, taking care not to damage the cover. Seal lip rides on hub of crankshaft pulley; inspect pulley hub for excessive wear at point of seal contact. Install new seal with lip to inside using suitable driver. Lubricate seal lip and crankshaft pulley hub, then install pulley on crankshaft with hole in pulley hub aligned with dowel pin in crankshaft gear. Tighten pulley retaining bolt to a torque of 112-122 N·m (83-90 ft.-lbs.).

CRANKSHAFT REAR OIL SEAL

The crankshaft rear oil seal (1—Fig. Y6-29) is mounted in the flywheel housing (3) and can be renewed after removing flywheel. Carefully pry seal from housing to avoid damage to crankshaft and housing. Lubricate end of crank and seal, then using suitable driver, install new oil seal with lip to inside.

If flywheel housing is removed, install new oil seal with lip to inside, then lubricate seal and install flywheel housing with new gasket. Be sure housing is aligned on locating dowel pins (4) before tightening bolts.

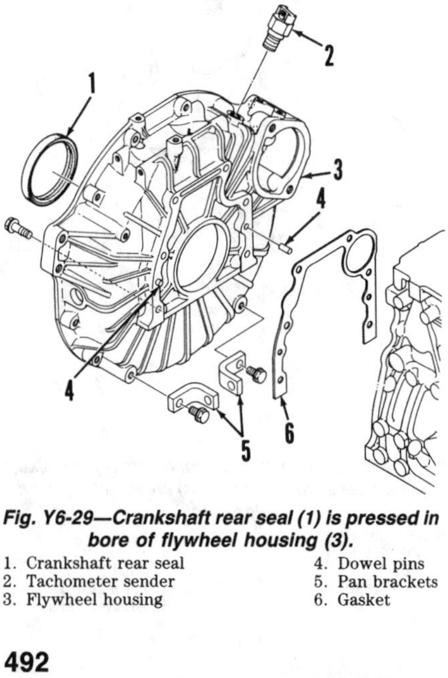

SMALL DIESEL ELECTRICAL SYSTEM

INTAKE AIR HEATER

An intake air heater is optionally available for cold weather starting. The 400 watt heater (3) is attached to the end of the intake manifold (4) as shown in Fig. Y6-30. The heater requires 33.3 amperes at 12 volts. Renew heater if no continuity can be found with ohmmeter between heater terminal and ground.

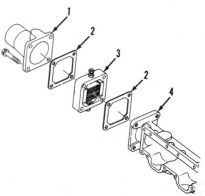

Fig. Y6-30—The optionally available intake air heater (3) is mounted on end of intake manifold (4).

1. Intake connector
2. Gaskets
3. Air heater
4. Intake manifold

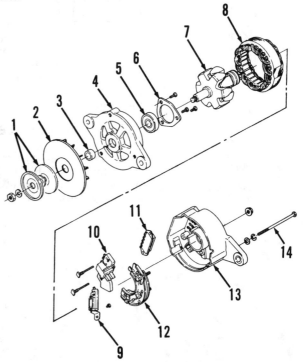

Fig. Y6-29—Crankshaft rear seal (1) is pressed in bore of flywheel housing (3).

1. Crankshaft rear seal
2. Tachometer sender
3. Flywheel housing
4. Dowel pins
5. Pan brackets
6. Gasket

Fig. Y6-31—Exploded view of alternator and internal regulator assembly typical of unit used on all engines.

1. Drive pulley
2. Fan
3. Spacer
4. Front cover
5. Ball bearing
6. Bearing retainer
7. Rotor
8. Stator coil
9. Capacitor
10. Brush holder
11. Voltage regulator
12. Rectifier diode assy.
13. Rear cover & bearing
14. Through-bolts

Illustrations Courtesy Yanmar Diesel Engine, Inc.

SERVICE MANUAL

ALTERNATOR AND REGULATOR

Standard alternator is a Hitachi Model LR155-20 55 ampere alternator with internal Model TRIZ-63 voltage regulator. A Hitachi Model LR180-03 80 ampere alternator with internal Model TRIZ-63 regulator is optionally available. Regulated voltage should be 14.2-14.8 volts at 5000 alternator rpm for both alternators. Refer to Fig. Y6-31 for exploded view of typical alternator unit.

STARTING MOTOR

A Model S12-77A starter motor is used for all engine applications. Starter motor specifications are as follows:

No-load volts 12.0
No-load amps (max.) 90
No-load rpm (min.) 4000
Brush height, standard . . . . 22 mm
(0.866 in.)
Wear limit . . . . 14 mm
(0.551 in.)
Commutator diameter,
 standard . . . . 43 mm
(1.693 in.)
Wear limit . . . . 40 mm
(1.575 in.)
Pinion gap . . . . 0.2-1.5 mm
(0.008-0.059 in.)

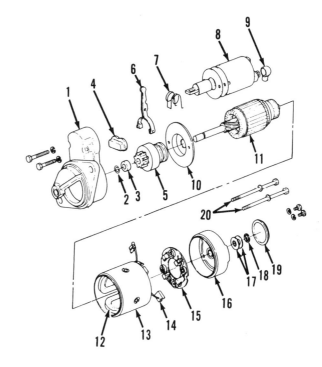

Fig. Y6-32—Exploded view of starter motor used on all models.

1. Drive case
2. Snap ring
3. Pinion stop
4. Dust cover
5. Drive pinion assy.
6. Shift lever
7. Shift lever spring
8. Magnetic switch
9. Terminal cover
10. Center plate
11. Armature
12. Field coil
13. Yoke
14. Brush
15. Brush holder
16. Rear cover
17. Thrust washers
18. Snap ring
19. Dust cover
20. Through-bolts

Illustrations Courtesy Yanmar Diesel Engine, Inc.

NOTES

NOTES

NOTES